POWER SYSTEMS AND PROTECTION

DR. N.K. DATTA

POWER SYSTEMS AND PROTECTION

By

Dr. N.K. Datta

B.Sc. E. Eng. (Glasgow University)
Ph.D. (Engg.) Jadavpur University, F.I.E., C. Eng. (India)
Professor Electrical Engineering Dept.
B.P. Poddar Institute of Management & Technology, Kolkata
Former Principal, M.I.T. Bishnupur, West Bengal

S.K. KATARIA & SONS®

Educational Publishers

4885/109, Prakash Mahal, Dr. Subhash Bhargav Lane,
Opposite Delhi Medical Association, Daryaganj, New Delhi-110002 (INDIA)
Phone : +91-11-43551243; Mobile: +91-9871775858, 7678425209
e-mail: katariabook@gmail.com; katariabooks@yahoo.com
Website: www.skkatariaandsons.com

Published by:

S.K. Kataria & Sons®

4885/109, Prakash Mahal, Dr. Subhash Bhargav Lane,
Opposite Delhi Medical Association, Daryaganj, New Delhi-110002 (INDIA)
Phone: +91-11-43551243; Mobile: +91-9871775858, 7678425209
e-mail: katariabook@gmail.com; katariabooks@yahoo.com
Website: www.skkatariaandsons.com

First Edition: 2010, Second Edition: 2012, Reprint: 2024

Price: ₹ 795/-

Laser Typesetting by :
Quick Media, Delhi

Printed at:
New A.S. Offset Press

Preface

This book has been prepared since 2006, while taking the classes on Power System, to meet the syllabus of Power System, I, II & III for B. Tech courses of all the Universities in India and abroad, including that of West Bengal University of Technology. Infact since writing of my first books Electrical Circuit Theory and Electrical Engineering Technology and their ready acceptance by the engineering faculty and the students of various Universities inspired me to write this book Power System and Protection. This book has been written with the objective of assisting students to acquire the basic concepts and tools of analyses, to equip them, in finding solutions to power system and protection engineering problems in practice with the help of simple, easily understood student friendly language and good number of worked examples on all the basic principles covered in this book.

This book provides a comprehensive coverage of the curricula and will perhaps serve as a very useful textbook for electrical engineering students at the undergraduate level. This book has been conceived and then prepared to provide a thorough understanding of the basic principles and techniques of power system analysis. Beginning with basic concepts of generation of Power and an exhaustive coverage of transmission line parameters, symmetrical and unsymmetrical and a symmetrical fault analysis, power flow studies, system stability, EHV, HVDC, compensation with FACTS, System Control, Insulation Coordination etc. as shown in detail in the contents. In the Appendix, Indian Electricity Act, 2003 is given with necessary abstract from the original publication by Government of India. WBUT has included this Act in the syllabus of Power System II. In the Appendix also given are the Question banks with answer for competitive examination and also for the semester examination.

The author acknowledges with regards that while writing this book many books and articles on the subject written by many authors in India and in U.K. have been consulted, for which sincere most gratefulness and gratitude is expressed. The author is also thankful to his colleagues particularly Asst. Profs. Chandrani Das, Sr. Lecturer, Madhumita Mandal and Sr. Lab Asst. Mr. Ranjit Pal and also few talented students of B.P. Poddar Institute of Management & Technology for some of their suggestions. Author is also extremely grateful to Prof. (Dr.) Saminar Chowdhury Ex Head, Power System, E. Eng. Dept. Jadavpur University and also Dr. Sankar Sen Ex-Power Minister, Govt. of W. Bengal and Ex-VC of Jadavpur University for their guidance while preparing this book. Thanks are also due to Dr. Partha Datta, Sr. Research Fellow, Cranfield University, U.K. for all his help and encouragement in completing this book. Last but not the least, I am extremely thankful to Sri Sanjiv Kataria, M.D. of S.K. Kataria & Sons and his staff for preparing and publishing this book.

Dr. N.K. Datta

Dedication

Dedicated to my best friend, Philospher and guide Sri Krishna Thakur and also to my parents & my wife, Manjusri

Contents

CHAPTER

1

Basics of Power System Growth

1.1 BACKGROUND OF POWER SYSTEM

INSIDE THIS CHAPTER

- Background at Power System
- Structure of Power Systems
- Conventional/Unconventional Sources of Electric Energy
- Unconventional energy sources
- Different types of energy sources and efficiency in their use
- Main parts of power system
- Growth of power system in India
- Load forecasting
- Power system engineering studies
- Use of computers and micro-processes
- Salient features of various types of power station & their comparison

Electrical powers/energy is an ingradient for the industrial and all-round development of any country. In fact, it is a converted form of energy because it can be generated centrally in builk and then transmitted economically over long distances—sometime hundreds of kilometers. Moreover, it can be adopted easily and efficiently to domestic and industrial application, particularly for lighting purposes, heating purposes like electric furnaces, and mechanical work like drives. The per capita consumption of electrical energy is a reliable indicator of a country's state of development.

However, to efficiently use electrical power an efficient power system is required. In a power system consisting of generators, switchgear, transformers, transmission and distribution circuits is invitable and sooner or later in such a large network some failure, (sometime have serious failure) will occur somewhere in the system. The probability of the failure or occurrence of abnormal condition is more on the power lines, simply because of their greater length, and exposure to the atmosphere and other abnormal conditions. Infact some of the causes of failures in power systems are :

1. Breakdown may occur at normal voltage on account of (*a*) the deterioration of insulation and (*b*) the damage due to unprediatable causes such as the perching of birds, accidental short circuiting by snakes, kite strings, tree branches etc.

2. Breakdown may occur because of abnormal voltages, though the insulation is otherwise healthy to withstand normal voltage. This may happen because of either (*a*) switching surges or surges caused by lightning.

However latest' present day practice is to provide a high insulation level of the order of 3 to 5 times the nominal value of the voltage. But still the pollution on an insulator string which is commonly caused by deposited soots or carbon dust in industrial areas and by salt deposited by wind-borne, sea-spray in coastal areas cause the insulation strength to decrease. This will initially lower the insulation resistance, and cause a small leakage current to be diverted, thus accelerating the deterioration. Even if the installation is enclosed, such as sheathed and armoured cables as well as metal-clad switchgear, deterioration of the insulation occurs because of ageing. Void formation in the insulating compound of underground cables due to the unequal expansions and contraction caused by the rise and fall of temperatures is another cause of insulation failure.

Moreover, the line and apparatus insulation may be subjected to transient over voltages because of the switching operations. The voltage which rises at a rapid rate, may achieve a peak value which approaches three times phase to line voltage. It is for this purpose that a higher insulation level is provided initially. If the insulation levels have been correctly chosen and they have not been impaired described in (1) above, the system will withstand these routine over-voltages. But if the insulation has developed some form of weakeness, it is at the time of switching that failure may be expected. Lightning produces a very high voltage surge in the power system of the order of million of volts and thus it is not feasible to provide an insulation which can withstand this abnormality in power system. In fact these surges travel with the velocity of light in the power system circuits, the limiting factors are the surge impedance and the line resistance.

The most serious result of a major uncleared fault is fire which may not only destroy the equipment of its origin but may spread in the system and cause total failure. The most common type of fault which is also the most dangerous one is the short circuit which may have any or some of the following consequences :

1. A large reduction of the line voltage over a major point of the power system circuit. This will lead to the breakdown of the electrical supply to the common and may produce wastage in production.

2. Damage to the circuit elements of the power system by the electrical arc which almost always accompanies a short circuit.

3. Due to abnormal mechanical forces being set up due to many short circuits, damage occurs to other electrical equipments in the system.

4. Disturbances to the stability of the electrical system and this may even lead to a complete shutdown of the power system.

5. A marked reduction in the voltage which may sometimes be so great that relays having pressure coils tend to fail.

6. Causes considerable reduction in the voltage on healthy feeders connected to the system having fault. This may cause either an abnormally high current being drawn by the motors, or the operation of the non-voltage coils of the meters. In the later case considerable loss of industrial production may result as the motors will have to be restarted.

It is of interest and also of immenese use for power system engineers to note that faults on overhead lines account for nearly half of the total number of faults. Below the frequency of fault occurrence in different lines of a power system can be given with fare amount of accuracy :

Equipment//Circuit Elements		Percentage of Total Faults Occurring to a Power System
Overhead (O.H.) lines	—	50%
Cables	—	10%
Transformers	—	10%
Switgears	—	15%
CTs & PTs	—	8%
Control equipments like relays, (Electromechanical & electronic), Computerised control system	—	3%
Miscelaeneous	—	4%

It is useful to have idea of the frequency of the incidence of faults on the different circuit elements in a power system as this information is of help in considering the problems of design and application of protection.

However, being faults on overhead lines around fifty percent of total, a designer or practical engineers will have to analyse the nature of fault on the overhead lines and arrange/design efficient protection systems to reduce the number of faults in the system. In fact, on a three-phase system the breakdown of insulation between one of the phases or all the phases to Earth is known as line to ground fault on a single phase length fault. In our country where the Neutral is grounded, that is, neutral is not kept isolated, these earth faults do produce fairly large magnitudes of fault current. So in designing a system of protection greater attention is given to the reliable operation of relays under line to ground fault conditions besides ensuring equipments like transformers, switchgears, CTs, PTs. Isolators etc. in overhead and underground cable system are kept healthy.

However, in some cases consumers are charged on the basis of KVA demand instead of kW to penalise loads of low power factor.

Two other factors used frequently are :

(*a*) $$\text{Plant Power Factor} = \frac{\text{Actual energy produced}}{\text{Maximum possible energy that could have been produced (based on installed plant capacity)}}$$

$$= \frac{\text{Average Demand}}{\text{Installed Capacity}}$$

(*b*) $$\text{Plant use Factor} = \frac{\text{Actual energy produced (kWh)}}{\text{Plant capacity (kW)} \times \text{Time (in Hours) the plant has been in operation}}$$

1.1.1 What is Meant by Tariff

The cost of electric power is normally given by the expression $(a + b \times kW + c \times kWh)$ per annum, where a is a fixed charge for the utility, independent of the power output, b depends on the maximum demand on the system and hence on the interest and depreciation on the installed power station; and c depends on the units produced and therefore on the full charge. Tariff structures may be such as to influence the load curve and to improve the load factor.

1.1.2 What is Load Forecasting

Power plant planning and construction require a gestation period of four to eight years or even longer for the present day super power station, energy and load demand forecasting plays a crucial role in power system studies. This necessitates long range forecasting. While sophisticated probilistic methods exist, the simple extrapolation technique is quite adequate for long range forecasting. Since weather has a much more influence on residential than the industrial components, it may be better to prepare forecast in constituent parts to obtain total, both powers and energy, forecasts are made. Multifactors involved render forecasting an involved process requiring experience and high analytical ability.

In fact, yearly forecasts are based on previous year's loading for the period under conideration updated by factors such as general load increases, major loads and weather trends.

However, in short-term load forecasting, hour-by-hour predictions are made for the particular day under consideration. A minor forecast error on low side might necessitate the use of efficient, oil-fired turbine generators or "peaking units" which are quite costly. On the other hand a high side forecast error would keep excessive generation in hot reserve. Accuracy is of the order of 01% is desirable. In fact a temperature difference of 2°C can vary the total load by 01%. This indicates the importance of reliable weather forecast to a good load forecast. The short term forecast problem is not a simple one as often random factors such as unexpected storm, strikes the sudden telecast of a good TV programme can upset the predictions. Regression analysis is often used for obtaining a short term load forecast which is very important and is required before solving unit commitment and economic load despatch. In India energy demand and installed generating capacity are both increasing exponentially, so is population growth which is a truly formidable combination. Power demand (65% of the total power required in India is for industrial consumption. Generation of electricity in India will have to be around 850 billion kWh in 2010-2011 compared to less than 200 billion kWh in 1986-87) has been roughly doubling energy ten years as in many other countries. In 1992, the total installed generation capacity in India was 70,000 MW. In fact as per the present indications, considering all the factors by 2020 it would go beyond 200,000 MW, which means, this would require corresponding development in energy resources like, coal; and of course nuclear reactors and Hydel and Solar power.

1.2 STRUCTURE OF POWER SYSTEMS

Generating stations, transmissions lines and the distribution systems are the main components of an electric power system. Generating stations and a distribution system are connected through transmission lines. For economical and technological reasons individual power systems are organised in the form of electrically connected areas or regional grids, also called power pools. Each area or regional grid operates technically and economically independently, but these are eventually interconnected to form a national grid, (which may even form an international grid, as in, Europe and USA). so that each area is contractually tied to other areas in respect to certain generation and sheduling features. India is in the process of almost completing a National Grid.

It must be stated here that interconnection has the economic advantage of reducing the reserve generation capacity in each area. In case of sudden increase in load or loss of generation in one area it is immediately possible to borrow power from adjoining interconnected areas. Interconnection causes larger current to flow on transmission lines under faulty conditions with a consequent increase in capacity of Circuit Breakers.

Also, the synchronous machines of all interconnected areas must operate stably and in a synchronous manner. It permits the construction of larger and more economical generating units and the transmission of large chunk of power from the generating plants to major load centres. It provides capacity savings by seasonal exchange of power between areas having opposing winter and summer requirements. It permits capacity savings from time zones and random diversity. It facilitates transmission of off-peak power. It, in fact also gives the flexibility to meet unexpected emergency loads.

1.2.1 Siting of Power Station

Power station siting will depend upon many factors-technical, economical and environmental. As it is considerably cheaper to transport bulk energy over Extra High Voltage (EHV) transmission lines than to transport equivalent quantities of coal over rail roads the recent trends in India and as well as abroads is to build super (very large) thermal power stations near coal mines with good Rail and Road transportation facilities. Bulk power can be transmitted to fairly long distances over transmission lines of 400 kV and above. However, the country's coal resources are located mainly in the eastern belt and some coal fired stations continue to be sited in distant western and southern regions. Hence the choice of site for coal-fired thermal station is more flexible. The following two alternatives are possible.

(*a*) Power stations may be built close to the load centres and coal is transported to them from the mines by rail road and in some cases by sea.

(*b*) Power station may be built close to coal mines (these powers stations are sometime called pit-head stations) and electric energy is transported to the load centres.

As for hydel-power stations or Hydropower stations the siting for these hydropower stations is determined by natural water power sources. However some hydropower stations are being operated by reserved water on the hill top. These hydropower stations may be far away from the load centres, hence the powers generated by hydropower stations are transported to load centres.

Since last few decades the importance of nuclear power stations is being realised by the Electricity Boards of India and the Government due to the expected shortage of coal and oil after several years, and also the danger of increased environmental pollution due to coal/oil fired power stations. Hence several Nuclear power stations have been constructed and generating power. Siting of nuclear stations are not constrained by the problems of fuel transport, fuel locations and the problem of air pollution. In fact, a greater flexibility exists in their siting, so that these stations are located to load centres while avoiding high density pollution areas to reduce the risks, however remote, of radioactivity leakage. Nuclear power station designers and operators seem to learn from the disaster of Chevernol Nuclear Power station accident in Russia. If properly designed, installed and controlled, Nuclear power stations are the answer to the coal/oil fired powers stations to meet the energy-demand. In fact in Europe, USA, Russia, Japan nuclear power stations are playing a major role in generating electric power. In India, at present about 60% of electric power used is generated in thermal plants using coal and 10% from nuclear energy based Thermal Plants, and the remaining 23% comes from hydro and as little as 7% from Solar and wind power generated electric power. These are all approximate figures as on 2008. This shows as on date coal is the fuel for most of the power stations.

1.2.2 Type of Generations

Electric power is generated at a voltage of 6.6 kV, 11 kV, 25 kV ranges which then is stepped up to the transmission levels in the range of 66 to 400 kV or higher. As the transmission capability of a line is proportional to the square of its voltage research is continuously being carried out to raise transmission voltages. Some of the countries are already employing 765 kV. The voltages are expected to rise to 1200 kV in the near future. This increase in voltage reduces the transmission losses. At present in India 400 kV lines are already in operation in about all the Electricity Boards. However, for long distances (over 600 km), it is economical to transmit bulk power by DC transmission. It also avoids some of the technical problems, associated with very long distance transmission. The DC voltages used are 400 kV and above, and the line is connected to the AC systems at the two ends through a transformer and converting/inverting equipment like silicon controlled rectifiers-SCR are employed for this purpose. Several DC transmission lines have been in use in Europe and the USA. In India also this process like HVDC has been commissioned and several others are under planning and construction stage.

The first step down in voltages is at the distribution substation. In practice, two distribution voltage levels are exmployed : (1) The primary on feeder voltage at 33 kV. (2) The secondary or consumer voltage level is set at 11 kV or 415 V 3-phase and 230 V 1-phase.

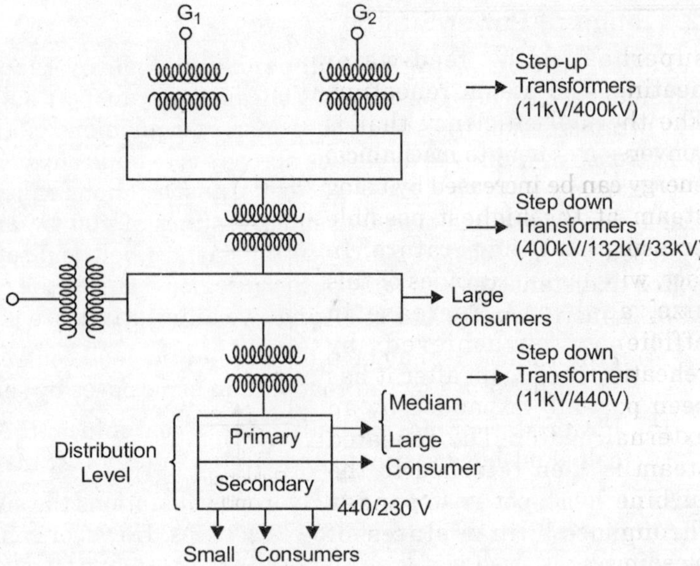

Fig. 1.1

The distribution system, fed from the distribution transformers stations, supplies power to the domestic or industrial and commercial consumers. Thus the power system operates at various voltage levels as schematically shown the structures of a power system, in the Fig. 1.1.

However, as the distribution system design, planning and operation are subjects of great importance, discussions will be made in separate chapters in the later part of the book.

1.3 CONVENTIONAL/UNCONVENTIONAL SOURCES OF ELECTRIC ENERGY

Thermal Generation of Power due to coal, oil, nuclear, gas sources and Hydro-generation are the main Conventional sources of Electric Energy. Due to expected shortage of fossil fuels in near future, and hence to conserve the fossil fuels the scientists and engineers across the world have been forced to search for unconvential sources of electric energy. Some of the unconventional sources being explored across the world are solar, wind and tidal sources. The conventional and some of the unconventional sources and techniques of energy generation are being briefly discussed below with a stress on future trends, particularly with reference to the Indian Energy scenario.

1.3.1 Thermal Power Stations (Coal Fired)

The coal has enormous chemical energy stored in it. This chemical energy stored in coal is transformed into electric energy in thermal power plants by releasing the heat by combustion of coal. This heat released produces steam in a boiler at high pressure and temperature, which when passed through a steam turbine gives off some of its internal energy as mechanical energy by the turbine. The axial-flow type of turbine is normally used with several cylinders on the same shaft. The steam turbine acts as a prime mover and drives the alternator that is electric generator whose shaft is coupled with the shaft of turbine. A simple schematic diagram of coal fired thermal plant is shown in Fig. 1.2. The efficiency of the overall conversion process is poor and its maximum value in about 40% because of the high heat losses in the combustion gases and the large quantity of heat rejected to the condenser which has to be given off in cooling. The steam power station operates on the principle of Rankin, modified to include superheating, feed-water heating, and steam reheating. The thermal efficiency that is conversion of heat to mechanical energy can be increased by using steam at the highest possible pressure and temperature. In fact, with steam turbines of this size, additional increase in efficiency is achieved by reheating the steam after it has been partially expanded by an external heater. The reheated steam is then returned to the turbine where it is expanded through the final states of bleeding.

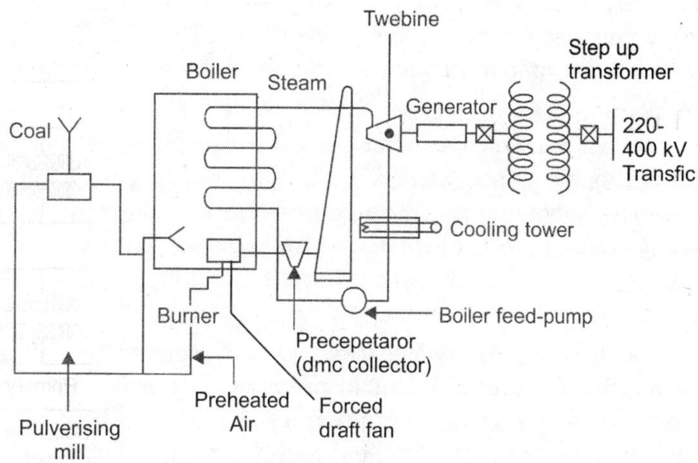

Fig. 1.2

However the present trend is to go in for longer sizes of units to take advantage of the principle of economy. Larger units can be installed at much lower cost per kilowatt. These are also cheaper to operate due to the higher efficiency of larger units. In fact the record says there is generally a saving of as high as 20% in capital cost per kilowatt by going up from 100 MW to 250 MW unit size and an additional saving in full cost becomes about 8% per KWh. Since larger units consume less fuel per kWh, they obviomly produce less air, thermal and waste pollution and this is a significant achievement in environmental pollution.

However, the only trouble in the case of a large unit is the tremendous shock to the power system when outage of such a large capacity unit occurs. This shock can be tolerated so long as this unit size does not exceed 10% of the on line capacity of large grid. The first 500 MW super thermal unit has been commissioned in India at Trombay. However today's maximum generation unit size is (around is 1200 MW) limited by the permissible current densities used in rotor and stator windings. In fact at present efforts are on to develop super conducting machines where the winding temperatures will be nearing absolute zero. Extreme high current and flux densities obtained in such machines could perhaps increase unit sizes to several GWs which would result in better generating economy.

However in a coal-fired steam plant air and thermal pollution is always present. The air polluting agents consisting of particulates and gases such as NOX, oxides of sulphur, etc. are emitted via the exhaust gases and thermal pollution is due to the rejected heat transferred from the condenser to cooling water. Cooling towers are used in situation where the stream/lake cannot withstand the thermal burden without excessive temperature rise. The problem of air pollution can be minimised through scrubbers and precipitators and by resorting to minimum emission despatch and passing clean Air Act like that, in 1970 passed in USA. In India, the problem with coal is its high ash content (up to 40%). To solve this, fluidised bed combustion technology is being developed and perfected. However, the fluidised bed boiler is undergoing extensive development and is being preferred due to its lower pollutant level and better efficiency. Direct ignition of pulverised coal is being introduced but initial oil firing support is needed.

What is Cogeneration

However considering the tremendous amount of waste heat generated in thermal power generation, it is advisable to save fuel by the simultaneous generation of electricity and steam (or hot water) for industrial use or space heating. These systems, now called cogeneration, have long been common, here in India, and abroad. In fact, currently there is renewed interest in these because of the overall increase in energy efficiencies which are claimed to be as high as 65%.

In fact cogeneration of steam and power is highly energy efficient and is particularly suitable for chemicals, paper, textiles, food, fertiliser and petroleum refining industries. These industries, thus can solve energy shortage problem in a big way. Moreover, they will not have to depend on the grid power which is not so reliable. They can of course sell the extra power to the government for use in deficient areas. They may also sell power to the neighbouring industries a concept called wheeling power.

In fact there are two possible ways of cogeneration of heat and electricity. (*a*) Topping cycle, in which case fuel is burnt to produce electrical and mechanical power and the waste heat from the power generation provides the process heat and (*b*) Bottoming cycle in which case fuel first produces process heat and the waste heat from the processes is then used to produce power. It is an established fact that coal-fired plants share environmental problems with some other types of fossil-fuel plants, like "acid rain" and the "greenhouse" effect.

1.3.2 Hydel (Hydro-Electric) Power Plant

Powers generation process by utilising the potential energy of water is perhaps the oldest and cheapest method of electricity generation. In fact this energy is obtained almost free of running cost and is completely pollution free. However, it involves high capital cost initially (which is however almost same if not less than that for thermal power station of the same rating) because of the heavy civil engineering construction works. Moreover it also requires a long gestation period of about five to eight years as compared to four to six years for steam plants. In fact hydroelectric stations are designed mostly as multipurpose projects such as rivers flood control, storage of irrigation and drinking waters, and navigation. In Fig. 1.3 a simple

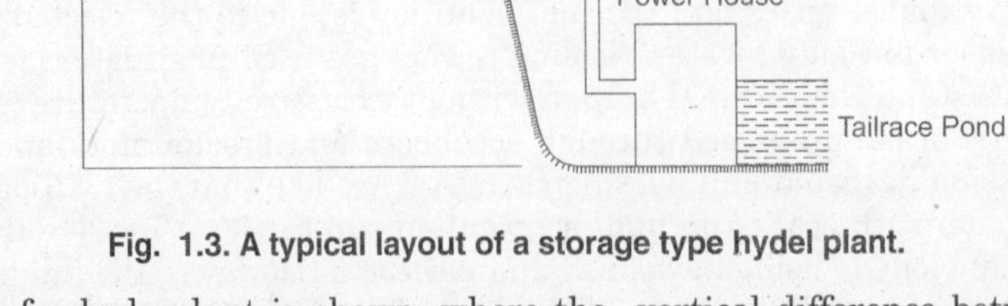

Fig. 1.3. A typical layout of a storage type hydel plant.

block diagram of a hydroplant is shown, where the vertical difference between the upper reservoir and tail race is called the head. There are different types of Hydel/Hydro plants such as (1) Run-of-Rivers (use of water with stream as it comes), (2) Poadase (Medium Head), and (3) Water-Reservoir type (of high-head) hydel plants employed for, bulk power generation. In fact cascaded plants are also constructed so that stream, on the same waters stream where the discharge of one plant becomes the inflow of a downstream plant.

Another remarkable advantage of Hydro-electric powers plant is that these plants are capable of starting quickly-about in five minutes. The rate of taking up load on the machines is of the order of 20 MW/min. Further, no losses are incurred at standstill. Thus, hydro-electric plants are ideal for meeting peak loads. In West Bengal, India, a Reservoir-type Hydro-electro plant on Ayodhya Hill, at Purulia of 900 MW capacity has been commissioned in March, 2008 to meet the demand of peak time. In fact the time from start up to the actual connection to the grid can be as short as 2 mins, may be sometime little less or more than that. Here it should be stated that the power available from a hydroplant is $P = 981 \;\rho WH\; W$, where ρ is the density 1000 kg/m^2, W is the water discharge in m^3/s through turbine, H = Head in metre (m). However, there are some problems present in Hydroelectric power plant, like (1) Silting, (2) Seepage, (3) Ecological damage to the surrounding region, (4) The base load must be used for peak sharing and energy saving in coordination with thermal plants.

India has tremendous potential of around 10,000 MW hydel plants in Himalayan region. Himachal J&K, Uttarakhand, UP, North-Eastern states like Assam, Sikkim which should be fully exploited to generate cheap and clean power for villages, particularly, located far away from the grid power. However, in areas where sufficient hydro-electric generation is not available, peak load may be handled by means of pumped storage that is the resevoir type hydel project. The pumped storage system presently is of great importance. The pumped storage system consists of upper and lower reservoirs and turbine-generator sets, which can also give about six hours of full load generation. Such a plant acts as a conventional hydro plant during the peak load period when production costs are the higher. The turbines are driven by water from the upper reservoir in the usual manner. During the light load period water in the lower reservoir is pumped back into the upper one so as to be ready for use in the next cycle of the peak load period. The generation in this period change to synchronous motor action and drive the turbines which now work as pumps. The electric power is supplied to the sets from the general power network as adjoining thermal plant. In fact the overall efficiency of the sets is normally as high as 60-70%. The pumped storage scheme in fact, analogous to the charging and discharging of a battery. It has added advantage that the synchronous machines can be used as synchronous condensors for VAR compensation of the power network if required. In a way, from the point of the thermal sector of the system the pumped storage scheme "sharves the peak" and fills the through of daily load-demand curve.

1.3.3 Nuclear Power Generating Station

Fossil fuels, particularly coal reserves are fast nearing the end in the not so distant future. The immediate practical alternative source of large scale electric energy generation is nuclear energy. The developed countries like UK, USA, France, Germany, Japan etc. have already, in a big way switched over to the use of nuclear energy for power generation.

In India however at present, that is, in the year 2009, this source accounts for only as low as 5% of the total power generation with power stations at Tarapur (Maharashtra), Kota (Rajasthan), Kalpakkam (Tamil Nadu). In 1992 it was planned that by year 2000 the nuclear power generation capacity would be raised to 10,000 MW, but for various political (national and international) reasons it could not be achieved. However, Indian Government presently reached an agreement with International Atomic Energy Commission and the Nuclear Club headed by USA for non-stop supply of nuclear fuels required for the present and future nuclear power plants to meet the growing demand of electric power for very fast growth of Indian economy.

The fuel for nuclear power plants is Uranium-235. A very simple description of Nuclear power generation is that when Uranium-235 is bombarded fission reaction takes place releasing nutrons and heat energy. These neutrons then participate in the chain reaction of fissioning some atoms of ^{235}U. In order that the freshly released neutrons be able to fission the Uranium atoms, their speeds must be reduced to critical value. Therefore, for the reaction to be sustained nuclear fuel rods must be embedded in neutron speed reducing agents (like graphite, heavy water etc.) called moderators. For reaction control, control rods made of neutron-absorbing material (made of Boron-steel) are used which when inserted into the reactor vessel control the amount of neutron flux thereby controlling the rate of action. But then rate can be controlled only within a narrow range. The schematic diagram of a nuclear power plant is shown in Fig. 1.4. The heat released by the nuclear reaction is transported to a heat exchange via primary coolant (like CO_2, water etc.). Steam in then generated in the heat exchanger, which is used in a conventional manner to generate electric energy by means of a steam turbine. In fact, various types of reactors are being used in practice for power plant purposes viz. Advanced Gas Reactor (AGR), Boiling Water Reactor (BWR), and Heavy Water Moderated Reactor (HWMR) etc.

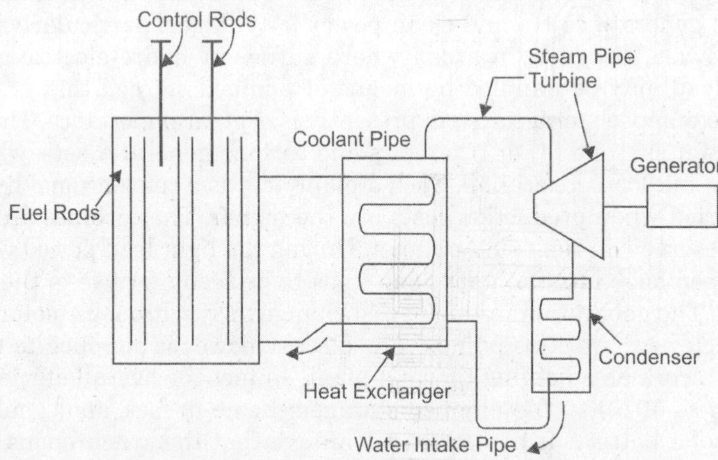

Fig. 1.4. Schematic view of Nuclear Power Plant

In fact CAND reactor Natural Uraniam (in Oxide form), pressurised heavy water moderated is adopted in India. a schematic diagram of CANDU Reactor is shown in Fig. 1.5.

Merits and demertis of Nuclear Power plants as compared to conventional thermal power plants are mentioned below :

Merits

1. Nuclear Power plants are totally free of any kind of air-pollution.

2. It requires little fuel in term of values and weight, and therefore poses no transportation problem and may be sited, independently of nuclear fuel supplies, close to load centres. However, safety consideration require that they be normally located away from populated areas.

Demerits

1. Nuclear reactors produce radioactive fuel wastes, the disposal of which posses serious, dangerous environmental hazards.

2. Because of relatively high capital cost as against running cost, the nuclear plant should operate continuously as a base load station. However, whereas possible, it is preferable to support such a station with a pumped storage scheme mentioned earlier.

3. The rate of nuclear reaction, can be lowered only by a small margin. So that the load on a nuclear power plant can only be permitted to be marginally reduced below its full load value. Nuclear power stations must, therefore, be reliably connected to a power network, as tripping of the lines connecting the station can be quite serious and may require shutting down of the reactor with all its consequences.

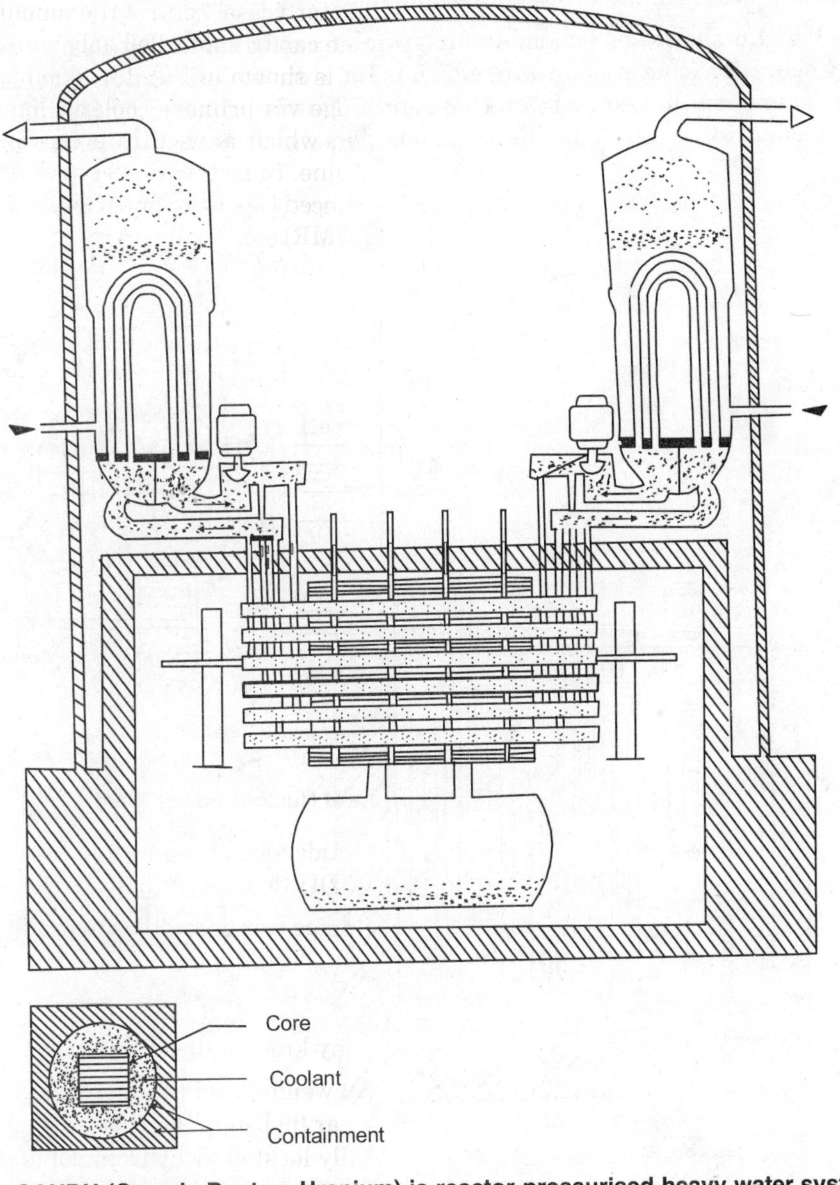

Fig. 1.5. CANDU (Canada Deutem Uranium) is reactor-pressurised heavy water system

1.3.4 Nuclear Fuel Availability

Nuclear power stations world-wide depend on supply of Uranium U^{235}. But the world Uranium resources are quite limited, and at the present rate may not last much beyond 50 to 70 years. However, there is a redeeming feature. In fact during the fusion of ^{235}U, some of the neutrons are absorbed by the more abundant Uranium isotope ^{238}U (enriched Uranium contains only about 3% of ^{235}U while most of it is ^{238}U) connecting it to plutonium (that is ^{239}U), which in itself is a fissionable material and can be extracted from the reactor fuel waste by a fuel reprocessing plant. Platinum would then be used in the next generation reactors (like FBRs first breeder reactors) thereby considerably extending the life of nuclear fuels. The FBR technology is being intensit developed as it will extend to availability of nuclear fuels at predicted rates of energy consumption to several countries.

The Fig. 1.6 shows the schematic diagram of an FBRs, on which it is essential that for breeding operation, conversion ratio (which is the ratio of fissil material generated and fossile material consumed) has to be more than unity. This is achieved by fast moving neutrons so that no moderator is needed. The neutrons do slow down a little through collisions with

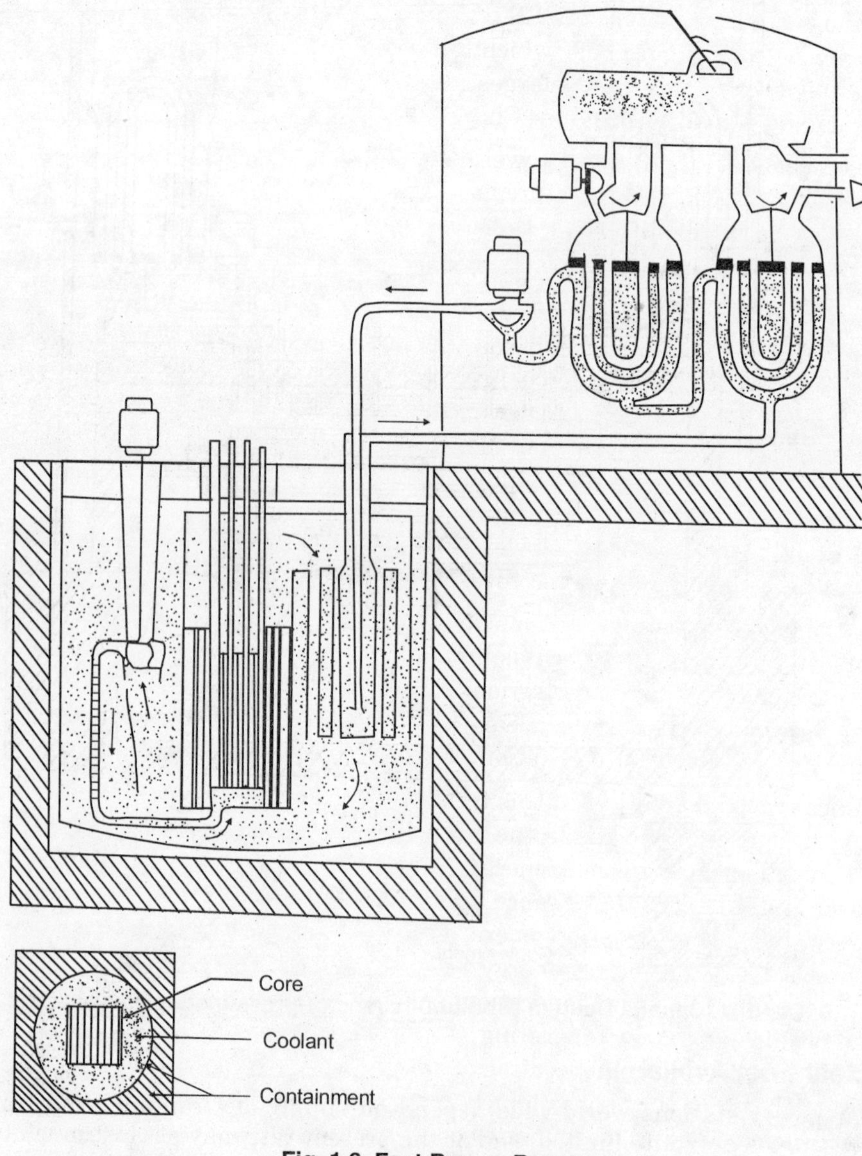

Fig. 1.6. Fast Breeze Reactor

structural and fuel element. The energy density/kg of fuel is very high and so the core is small. It is therefore necessary that the coolant should possess good thermal properties and hence liquid sodium is used. In fact the for an FBR consists of 20% plutonium plus 8% uranium oxide. The coolant liquid sodium, covers the reactor at 650°C at atmosphere pressure. The heat so transported is load to a secondary sodium circuit which transform it to a heat exchanges to generate steam at 540°C.

However, wih a breeder reactor the release of platinum an extremely toxic material, would make the environmental considerations most stringent. In fact FBR technology is expected to reduce the cast of electric energy so as to compare favourably with that of conventional thermal plants.

In fact, an important advantage of FBR technology is that it can also use thorium (as fertile material) which gets converted to ^{235}U which is fissionable. This holds great promise for India as we in India have one of the world's largest deposit of Thorium about 500,000 tons in form of sand in Kerala and along the Gopalpur Chatrapur Coast of Orissa.

It is important to note that typical power densities (that is MW/m^3) in fission reactor cores are as : gas cooled –0.53, high temperature gas cooled –7.75, Heavy Water 18.0, Boiling water-29-0, pressurised water –54.75 and Fast Breeder Reactor –760.00.

What is Fusion

Energy is produced in the fusion process by the combination of two light nuclei to form a single heavier one under sustained conditions of extremely high temperature which is generally in million degrees of centigrade. Fusion is futuristic. In fact generation of electricity via fusion would and could solve the long-term energy needs of the world with minimum environmental pollution hazards precipitated by the fossil fuelled generation. However, considering radioactive wastes, the impact of fusion reactors would be much less than the fission reactors. In case of succession in fusion technology in near future on a break through in the pollution-free solar energy, FBRs would become obsolete. However, today there is an intense need today to develop FBR technology as an insurance against future to develop these two technologies.

In fact in the past few decades, serious doubts have been raised about the safety claims of nuclear plants. In fact there have been as many as 150 near disaster nuclear accidents from the three-mile accident in USA to the serious Chernobyl accident in the former USSR. There is serious fear that this may put the nuclear energy development in reverse gear. The accident in 2011 at Fukuswave Nuclac Power Station raised sections question on safety factor indesign.

The technical world feared that if this type of serious accident goes on and not controlled there could be serious energy crisis in the world's developed and developing countries which have pitched their hopes on nuclear energy to meet their ever increasing energy needs. In fact, it has been in the record that France and Canada are possibly the two countries with a fairly clear record of nuclear energy generation. UK also comes next to these two countries on the safety record in their nuclear energy generating units. Hence, India which has taken a big step towards increasing power generation to meet the fast increasing energy needs setting good number of nuclear power generating stations throughout the country with carefully thought design, construction and operating strategies as it is committed to go in a big way for nuclear energy generation and hopes to achieve a capacity of 20,000 MW by 2015. However, as per Indian nuclear scientists, heavy water-based plants are most safe. But most conservative strategies must be adopted in design construction and operation of nuclear power plants, with no compromise on safety to life.

1.3.5 MHD (Magneto Hydrodynamic) Generation

In MHD technology electrical energy is directly generated by the hot gases produced by the combustion of fuel without the need for mechanical moving parts. In the process efficiency is much higher than the process of thermal generation of electrical energy due to two stages of energy conversion. (1) Heat released by the fuel is converted to rotate and mechanical energy by means of a thermo cycle, then (2) the mechanical energy is used to rotate the electric generator.

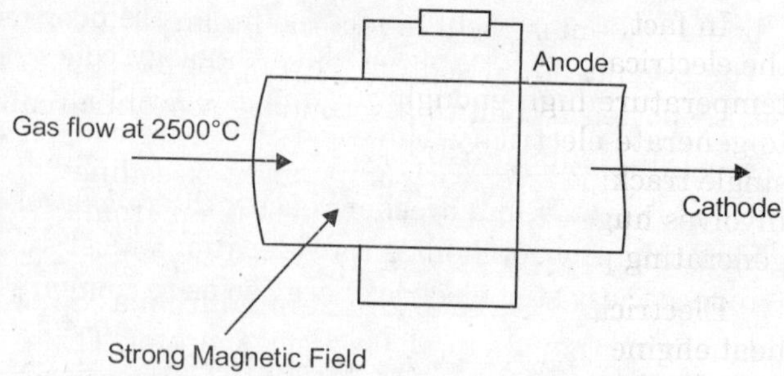

Fig. 1.7. The principle of MHD power generation.

In a MHD generator electrically conducting gas at a very high temperature is passed in a strong magnetic field thereby generating electricity. High temperature is needed to ionize the gas, so that it has good electrical conductivity. The conducting gas is obtained by burning a fuel and injecting on seeding material such as potassium carbonate in the products of combustion. The principle of MHD power generation is shown in Fig. 1.7. About 50% efficiency can be achieved if the MHD generator is operated in random with a conventional steam plant.

However, the technological feasibility of MHD generation has been although established its economic feasibility is yet to be practically demonstrated successfully and economically. India, in fact, has started a research and development project in collaboration with Soviet Union to install a pilot MHD plant based on coal as fuel, and generating 2 MW power. However after this operation is successfully developed the next step will be with natural gas as fuel.

Gas Turbine

Gas turbines are operated by using gas as fuel. However, compressed air is fed to the combustion chamber where continuous combustion of fuel oil is maintained. The hot gases are produced in this process and the gas thus produced are made to run a gas turbine. In fact, in normal running a gas turbine is less economical compared to a conventional steam power plant, but it is used as a prime mover for its ability to start and take up load quickly. Thus it is capable of meeting sudden peaks of system load. And fuel economy can be improved by using hot exhaust gases of the gas turbine in a conventional steam generation cycle.

1.4 UNCONVENTIONAL ENERGY SOURCES

Due to the limited availability of coal, natural gas, there is considerable international effort into the development of alternative/unconventional/new/renewable/clean sources of energy. In fact most of the new sources are nothing but the manifestation of solar energy, *e.g.*, wind, sea, waves, ocean thermal energy conversion-OTEC, etc. Below briefly discussion is made on the possibilities and potentialities of various methods of using solar energy and other non-conventional sources. In fact each topic itself requires detail discussion which is not in the scope of this book.

1.4.1 Solar Energy

The reasearch on the subject shows that the average incident solar energy received in earth's surface is about 600 W/m^2, but the actual value varies quites considerably. It has the advantage of being free of cost, non-exhaustable and completely pollution-free. On the other hand, it has several drawbacks also like-energy density per unit area in very atmospheric conditions greatly reduce the energy received. Therefore, in harnessing solar energy for electricity generation, challenging technological problems, exist. The most important being that of the collection and concentration of solar energy and its conversion to the electrical form through efficient and comparatively economical means.

In fact, at present, two technologies are being developed for conversion of solar energy to the electrical form. In case of one technology, collectors with contractors are employed to achieve temperature high enough (around 700°C+) to, operate a heat engine at reasonable efficiency to generate electricity. However, there are considerable engineering difficulties in building a single tracking bowl with a diameter exceeding 30 m to generate perhaps 200 kW. The scheme involves huge capital outlay which is far from being competitive with conventional electricity generating power stations.

Electricity may be also generated from a "Solar Pond" by using a special 'low temperature' heat engine coupled to an electric generator. This type of solar generation is not yet considered in India, but in Israel at Ein Borek, a solar pond produces a steady 150 kW from 0.74 hectare at a busbar with a cost of around Rs. 4/kWh. In fact solar power potential is unlimited as long as our sun is there.

Photovoltaic generation (Directly Converted Electricity). This direct conversion to Electricity from Solar Energy converts solar energy to the electrical form by means of silicon wafer photoelectric cells which are known as "Solar Cells". The main problem in this system is the cost and maintenance of Solar Cells. However with the likelihood of a breakthrough in the large scale production of cheap solar cells with amorphous silicon. This technology may compete with conventional methods of electricity generation, particularly as conventional fuels are expected to become searce in coming years. However, solar energy could at the most support upto 10% of the total energy demand.

It should be noted that in all solar thermal schemes, storage is necessary because of the fluctuating nature of Sun's energy. Of course, this is equally true with many other conventional sources as well as sources like wind. Hence fluctuating sources with fluctuating loads complicate still further the electricity supply.

Biofuels

The Biofuels are produced from biomass. The material of plants and animals is called biomass which is transformed by chemical and biological processes to produce intermediate biofuels by chemical and biological processes to produce intermediate biofuels such as methane gas, ethanol liquid or chemical solid. Biomass is burnt to provide heat for cooking, comfort heat, crop drying, factory processes and producing steam from electricity production and transport. In India approx out put for biomass is about 2000 mW and that for agricultural waste is about 6000 mW.

Wave Energy

The energy content of sea waves is very high. In India, with several hundreds of kilometers of coast line, a vast source of energy is available. The powers in the wave is proportional to the square of the amplitude and to the period of the motion. Therefore the long period (around 10 sec), large amplitude (2 metre) waves are of considerable intest for power generation, with energy fluxes commonly averaging between 50 and 70 kW/m width of oncoming wave. Though the engineering problem associated with wave power are formidable, the amount of energy that can be hernessed is large and development work is in progress.

1.4.2 Ocean Thermal Energy Conversion (OTEC)

The ocean is the world's largest solar collector. Important difference of 20°C between warm, solar absorbing surface water and cooler 'bottom' water can occur. This can provide a continually replenished store of thermal energy which is in principle available of some of this thermal energy into work and there into electricity.

Energy Storage

In Power System this is a most important task. There is a lot of problem in storing electricity in large quantities. Energy which can be converted into electricity can be stored in a number of ways. Storage of any nature is however very costly and its economics must be properly planned. The options are pumped storage, secondary batteries, compressed air, heat, superconducting coils etc. Methane gas, ethanol liquid on charcoal solid, biomass is burnt to provide heat for cooking, comfot heat (space heat), crop drying, factory processes and raising steam for electricity production and transport.

1.5 DIFFERENT TYPES OF ENERGY SOURCES AND EFFICIENCY IN THEIR USE

In the world major sources of energy continue to be still coal, gas, petroleum, but slowly but steadily nuclear power stations are coming up for producing electrical energy which gradually is becoming one of the major source of energy. However, even in today's modern world, in certain parts of the world major source of energy still continue to be human and animal muscle. Wind power, solid fuel, such as wood and animal dung, and water wheel are some other sources of primary energy, but they, are declining in significance due to increase of our sources like nuclear, coal, gas, petroleum, hydroelectric and improvement in distribution of this energy to the remote part of the world.

In past few decades the non-conventional sources like solar, wind, wave, etc. are also contributing to the requirement of world's energy.

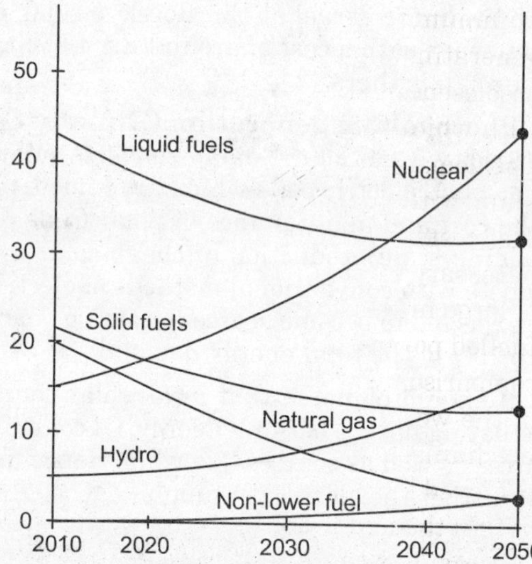

Fig. 1.8. Appex share of primary fuels in meeting world's energy need 2010-2050

From the Fig. 1.8, we can see the possible changes in percentage share of primary fuels in meeting the world's energy requirement over a period of next forty years, that is upto 2050. From the survey made in the U.K., in 2000, solid fuels contributed about 20 percent of the told share of primary fuels, the liquid fuels about 70 percent, natural gas about 20 percent, nuclear 15, Hydro 5 percent. But by 2050, Nuclear fuel impacted to become 40 percent and liquid fuels, 25 percent solid fuels 15 percent, Hydro 10 percent, Non-conventional and 10 percent. These forecasts are subject to 10–15% variations due to many variable factors.

Energy may be needed as heat, as light, as motive power or for direct use in electrical processes. It may be needed in large amounts or small, continuously or intermittantly, in locations readily accessible to its primary sources or to means of transportation or in remote in inaceesible areas. Therefore what is most economic in one set of circumstances may be quite inappropriate in others. Also proper attention has to be given to the form in which energy is to be made available. Electric power is the backbone of industrial world today. In fact, comfort, convenience and safety of large population all over the world depends upon electric power. It is now a proven fact that average wage level and standard of living in any country of the world, are dependent to a very large extent on the amount power used per capita in its industry. History, tradition, economic circumstances such as availability of natural resources all govern the development and use of electrical power in a country.

A growing proportion of energy is being met all over the world by electricity. This trend will further increase in faster rate because of increasing availability of clean electricity. This applies especially to developing countries because their industrial progress will be based on modern technologies which mostly use electricity. In other words the advance of the world economy depends on the continuous availability of electricity in right quantity and right quality at a reasonable tariff. Therefore planning of electricity supply is most important regarding its generation and distribution and maintenance of the whole system of power commonly called power systems.

For securing maximum benefits electricity supplies are planned in relation to the generation process taken entirely, planning will usually cover requirements upto to years ahead and it will examine the alternative strategies which may be adopted with the object of achieving minimum total system costs. In general, this may mean a study of proportion of the types of generating plants (oil or coal fred, nuclear, hydro, solar etc.) in the system on the effects of obsolesence, retirements of staffs, and transmission distances. Much will depend on the accuracy with which requirements can be forecast and accuracy may diminish in the future projection. The demand of electricity changes continuously and lowest system costs and obtained by running plant with cheapest running cost first duly allowing for transmission costs. But changes in fuel costs will affect total system costs of existing as well as future plant. It is accordingly necessary to consider the effects of possible rises in fuel costs. However, the generating costs of large nuclear power stations now compare very favourably with these of contemporary fossil-fuelled power stations in India and abroad, while their running costs are significantly lower in comparison with those of conventional power stations. In fact in most of the developed countries of the world like UK, USA, Russia, despite their refineries and networks of pipeline, highly mechanised coal mines and well developed transport systems. Competitiveness has been achieved since larger nuclear station operated at or near load can be economically absorbed within the larger and integrated electricity generating systems which much advanced countries possess. For countries with system nuclear power has a more difficult target to hit but even no some developing countries like India may find the installation of nuclear power stations in nuclear sizes more economical than developing their oil and coal industries.

In fact considering the gradual reduction in the oil and coal sources, all the countries of the world will have to start thinking seriously of developing nuclear power stations for generating electricity without which in the present world all the services every need will be stopped. Hence future of the world civilisation largely depends on the application of nuclear energy process to generate electricity.

1.6 MAIN PARTS OF POWER SYSTEM

Main parts of power system are (1) Generating stations, (2) Transmission system, (3) Distribution systems (4) Substations for transmission distribution of electricity consisting of Switchgear Transformer and the protection system to protect the equipment and the power system network.

For an efficient power system prediction of load energy requirements is most necessary and some of the methods of prediction are : (*a*) Load Survey (*b*) Methods of extrapolation, (*c*) Mathematical methods, (*d*) Mathematical methods using economic parameters.

In load survey method loads are grouped under different categories such as Industrial, Commercial, Domestic, Services like Hospital transport etc. Detailed information category-wise and area wise is collected to estimate the demand and its growth rate duration and time of occurrence of load and the energy requirement with peak demand etc. This enables a forecast of electrical load and energy. The method may cover a period of 5 to 10 years for planning of equipment and erection while the period may be longer- upto about 20 years for overall planning for equipment, land, human resource etc. with costs.

(*b*) Methods of extrapolation are used to estimate the yearly increase in energy consumption. These methods, in fact are basically statistical. For example in past few years since 2000, due to rapid growth of industries, increase in numbers of high income group of people and overall improvement of the life style, it has been found that depending upon annual electricity consumption per capita typical yearly increases in energy are as under :

100 to 500 kWh	5 to 10%
501 to 1500 kWh	10% to 20%
1500 to 2000 kWh	10% to 25%
2001 to 3000 kWh	8% to 12%
2001 to above	5 to 10%

This is first an approximate figure, which are likely to go up in next 10 years, and to be corrected every year.

(*c*) Mathematical methods use exponential or linear curve putting techniques for extrapolation of the curves showing energy consumption for past several years statistical data for a sufficiently long period must therefore be available for such methods.

(*d*) Methods using economic parameters are actually based on the assumption on that per capita consumption of electrical enough in any country depends on economic factors, some of these being specific gross investment industrial production, specific GNP, steel consumption, housing, transport etc. Indices for these parameters are found and suitably used for extrapolation by linear or exponential approximation. Let us now see the growth of power systems and the problems below.

1.7 GROWTH OF POWER SYSTEMS IN INDIA

We Indians are perhaps lucky that our country is fairly rich in natural resources like coal and lignite, thorium while some oil resources have been discovered and expected to get a reasonably good reserve of oil hence intense exploration is being undertaken in various regions of the country including coastal areas and sea near the coast. Moreover, India has immense water power resources also of which only around 25% have so far been utilised, *i.e.*, only 25,000 MW has so far been commissioned upto the end of the five-year plan. In fact, as per a recent report of CWPC-Central Water Power Commission, the total potential of hydropower is about 100,000 MW at 80% load factor. As regards nuclear power, India is deficient in Uranium, but has rich deposits of Thorium which can be utilised at a future date in fast breeder reactors. However, since independence, the country has made tremendous progress in the development of electric energy and today it has the largest system among the developing countries.

In 1947, the installed capacity was as low as 1900 MW in the early stages of the growth of power system, the major portion (about 90%) of generation was through small thermal stations, although due to economical reasons hydro-electric generation development received attention in areas like Punjab, UP, Kerala, Tamil Nadu.

However, in the beginning of the First Fire Year Plan (1951-56), the total installed capacity was around 2300 MW (560 MW hydro, 1004 MW thermal, 149 MW through oil stations and 587 MW through non-utilities). For transmitting this power to the load centres, transmission lines of upto 110 kV voltage level were constructed. Then the emphasis during the Second Five Year Plan (1956-61) was on the development of basic and heavy industries and thin there was a need to step up power generation. The total installed capacity which was around 3420 MW at the end of the First Five Year Plan because 5700 MW at the end of the Second Five Year Plan.

Then the introduction of 230 kV transmission line came up in Tamil Nadu and Punjab. During this plan, total about 1009 circuit Kilometres were energised. In 1955-56 the total installed is capacity was increased to 10,170 MW. During the Third Five Year Plan (1961-66) transmission growth took place very rapidly, with a nine-fold expansion in voltage level below 66 kV. Emphasis was on normal electrification. A signifacnt development in this phase was the emergence of an inter-state grid system. The country was divided into five region, each with a regional electricity board to promote integrated operation of the constituent power systems. Fig. 1.9 shows Five Region of the country, NORTH, EAST, SOUTH, WEST and NORTH-EAST with approximate projected installed capacity in the year, 2000-01.

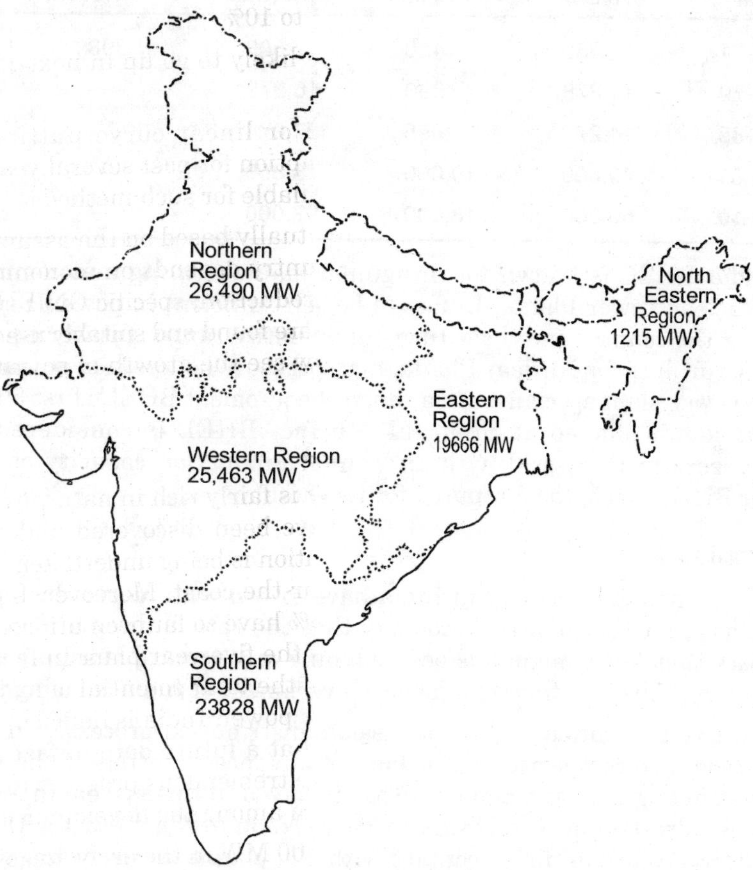

Fig. 1.9. Energy map of India showing five electricity region. Regional (Approx.) projected installed capacity for the year 2000.

It is noticed that during the Fourth Five Year Plan. India started generating nuclear power with Tarapur Nuclear Plant, where 2 × 210 MW units were commissioned in April-May, 1969. This station uses two boiling water reactors of American design. In August, 1972, the first unit of 220 MW of the Rajasthan Atomic Power Project of Kota was added to the nuclear generating capability. The total generating capacity at Kota is 430 MW with nuclear Reactors of Canadian design which use natural uranium as fuel and heavy water as a moderator and coolant. The third nuclear power station of 2 × 235 MW has been commissioned at Kalpakam, Tamil Nadu. This is the first nuclear station to be completely desiged, engineered and constructed by Indian scientists and engineers. A reactor research centre has been set up near the Madras Atomic Power Station to carry out study in fast breeden reactor technology. The fourth Nuclear Power Plant has been set up at Narora in U.P. It has two units of 235 MW each. The fifth will be in Kaiga in Karnataka and sixth in Gujarat near Surat.

Besides these nuclear plant there is possibility of ten more nuclear power plant expected to be started in the states like West Bengal, Bihar, Jharkhand, U.P. in next four to five years. The Table 1.1 gives an idea of the growth of generating capacity so far and approximate future projection for 2000-10.

Table 1.1. Growth of Installed Capacity in India (in MW)

Year	Hydro	Nuclear	Thermal	Diesel	SO_4	Total
1970-71	6383	420	7,503	398	—	14,704
1978-79	11,378	890	16,372	—	—	28,640
1984-85	14,271	1,095	27,074	—	—	42,240
2000-01	49,000	10,000	78,859	—	248	1,37,859
2009-10	60,000	18,000	92,000	—	750	00

In India BHEL is known for designing, manufacturing and commissioning of Thermal Hydro, Nuclear Power Plants. In fact to be self-sufficient in power, BHEL has plants spread out all over the country and these turn out an entire range of power equipment like turbosets, hydrosets, turbines for Nuclear Plants, high pressure boilers, power transformers switchgears etc. Each power plant specialises in a range of equipment BHEL's first 500 MW turbo-generator has been commissioned at Singrauli. In fact BHEL is considered among the top five manufacturers in the world with an annual production capacity of more than 5000 MW. Moreover BHEL has become reputed for its Quality.

1.7.1 Problems

The demands of electricity in India have grown with tremendous pace in past few years and it is good for the country's economy that the rate of growth of demand is increasing in such a pace that the demand has been running much ahead of supply in almost all the states of the country. This is a great challenge, Government of India is facing today, in 2011.

Electricity generation and transmission, distribution processes in India as on today, are very inefficient in comparison with those of advance countries of the world. The writer as a power system engineer and being directly involved in this system in almost all the electricity authorities from Jammu & Kashmir to Tamil Nadu to North-East with great apathy found in his last thirty years of direct contact with the engineers (from top to the bottom) of these electricity authorities, is bound to say that besides the want of enough installed capacity, the administration and control of the operation of transmission & distribution is so poor (although there are many knowledgeable engineers everywhere) that as a whole the supply of electricity is further detoriorating. Hence, to meet the increasing demand of eletricity in the country, both installed capacity and administration (technical & commercial) has to be improved to the standard of the advanced countries like Japan, UK, USA etc. Below a brief comparison with these advanced countries is given.

In India, generating capacity is utilised on an average for about 3,500 hours out of 8,760 hours in a year but in Japan, it is used 5,500 hours. If the utilisation factor could be increased, it should be possible to avoid power cuts. the transmission loss as in 2000 on a national basis was about 20% consisting of both technical losses in transmission lines and transformers and also non-technical losses caused by energy thefts and metres not being read or not being provided by the electricity authorities. It should be possible to achieve considerable saving by reducing this loss to 15% by the end of eight and/or Ninth Five Year Plans by using more efficient ways and means (as practiced is advanced countries) and also by adopting sound and almost flaw-less administrative and commercial practices.

In fact, every attempt should be made to improve system load factors by flattening the load curve by giving proper tariff incentives and taking other administrative measures. However, as for Central Electricity Authority's (CEA) fourteenth annual report, the all India load factor likely to present by 2010-11 is 65%. In the early sixties about 5 lacs of villages have been electrified and 88 lacs of pumpsets have been energised in India. By the year 2010 these figures are expected to double. The growing demand of fast-growing industrial and domestic and entertainment requirement, will need, by 2010, one can imagine the huge requirement of electricity in India. With present set up of generation of electricity it is difficult to meet this demand by 20% unless India starts installing few more nuclear power stations along with more Hydro and Coal/gas fired Power Stations.

Assuming a very modest average annual energy growth of 5% (aim though is 10%) India's eletrical energy requirement in the year 2010 will be enormously high. A difficult and challenging task of planning, engineering and constructing new power stations is immenent to meet this situation.

The Government has so far built several super thermal power stations such as Singrauli (UP) Farakka (W. Bengal), Korba (MP), Ramagundam (AP), Nevveli (TN)—all in coal mining areas each with capacity in the range of 2000 MW. Immediately, perhaps by 2015, few more super thermal plants are being planned by central government in India. By then, despite some objections from some group in Indian policies regarding nuclear agreement for getting nuclear fuels from advanced countries, few more nuclear power plant will be installed. For coal based Thermal Power plants, intensive work has been taken on Boiler furnaces to burn coal with high ash content. In fact NTPC (National Thermal Power Cooperation) has been in charge of the large scale, generation projects in India. Although power generated by these technologies are a bit expensive but it is still worth having Nuclear Power Plant and more coal/gas based Power plants. Another important factor is that in India technically administratively and financially more efficiently powerful private organisation like TATAS are notyet given responsibility to build Nuclear and based power plants. This policy of Indian Government has to be revised to solve this problem. Although many Industries particularly power-intensive Industries have installed their own captivepower plants, which mostly use diesel or gas which is not economical from national point of view, hence why not use these private Industries knowledge to build super thermal power houses—coal/gas based or even nuclear fuel-based. For this Government has to enforce now-laws considering national security and large power demand in coming decades).

As far as economical considerations are concerned, Hydro-electric power will continue to remain cheaper than other types for the next decade. In fact, it is regrettable that India has so far developed only around about 18% of its estimated total hydro-electric potential of 89000 MW. However, the utilisation of this perenial source of energy would involve massive investment in Dasus, Channels, and generation-transmission system besides load aequisition. Narmada project was built to Gujarat with many obstacles from so called Narmada-Bachao Movement. Today Gujarat and the country is enjoying the benefit of Narmada project. India today needs few more Narmada Projects.

The Central Electricity Authority, the Planning Commission and Ministry of Energy are Coordinating to work out a perspective plan to develop all hydro-electric sources by the year 2020 to be exceuted by the NHPC (National Hydro Power Corporation).

Considering "clean energy", in India, concerted efforts have started to develop SOLAR ENERGY and other non-conventional sources of energy, so that the growing demand can be met and depleting fossil fuel resources may be conserved. To meet the energy requirement, it is expected that the coal production to be increased to more than about 500 million ton in 2010. But considering the expected reduction in coal reserve in India in the world by 2050, immediate attention to be given to non-conventional energy and nuclear power sources. Nuclear energy today is perhaps the answer to the power scarcity problem faced by the world and India particularly.

As stated because of limited coal reserves and above all its poorquality. India today has no choice but to keep going on with its nuclear energy plans despite all the objection from few groups and some unfoundly fear of nuclear radiation hazard. However, according to the AEC-Atomic Energy Commission, India's nuclear power generation will increase to 20000 MW by 2020. Everything seems to be set for a take off in nuclear power production using India's own thorium reserves in breeder reactors.

Brief discussions have been carried out above on the problem and porspect of present and future power generation. Below, another important factor the problem and prospects of Transmission &Distribution is discussed briefly.

Problems & Prospects of T & D

Earlier there was a problem of developing a national grid. However, now a number of 400 kV lines are operating successfully and thus it helped in developing a national grid in the country. This helped to transmit power from one region to the other as and when required. However for better performance (as done in several advanced countries) there is a need in future to go for even higher voltages –750 kV/1000 kV. In fact by the year 2015 the remaining work of 400 kV lines and 220 kV lines would be in operation. Also lines may be series and shunt compensated to carry huge blocks of power with greater stability. Moreover, there is a need for constructing HV DC (High Voltage DC) links in the country since DC lines can carry considerably more power at the some voltage and require power conductors. A 400 kV Singrauli-Vindhyachal line of 500 MW capacity first HV DC back-to-back scheme has been commissioned by NTPC followed by first point-to-point bulk EHV DC transmission of 1500 MW at 500 kV over a distance of 915 km from Rihind to Delhi.

Infact due to very fast increase of demand of quality power throughout the country, the whole electrical-power energy scenario is changing so fast that perhaps by 2020, if Government planning is implemented properly future of electrical power availability will be pretty bright.

However, considering the urgency of achieving power supplies position by the year 2020, huge amount of funds will be required to be invested from now on word, otherwise India will continue to remain in a power shortage state as because of globalisation of Indian economy the growth of Industry cannot be slowed down. Moreover India is on the verge of reaching a world-power status within next ten years so. To do so India has to soone or later become a power surplus country. Besides this number of young talented engineers and managers are increasing in a phenomenon rate and to meet their job-demand industries have to expand in the same rate if not more. Hence this is the most serious problem India is facing today due to continuous shortfall in electrical energy generation. If our policy-makers of the planning commission does not give top most priority in taking all the necessary actions immediately to put the country in a power surplus position the country may face a serious turmoil-politically and also economically within next five to ten years. However, with the ever increasing complexity in Indian society and the growth of power network and their economic and integrated operation and perhaps considering above stated problems Central Government is planning to establish central load dispatch centres with computer control and also a monitoring cell to see the proper utilisation of the central funds released to various zones in the country.

1.8 LOAD FORECASTING

1.8.1 Load Management and Energy Conservation

Load management perhaps is one of the most important duties of the power system engineers. As power plant planning and construction requires a gestation period of four to eight years or sometime even longer for the present day super power stations, energy and load demand forecasting and load management plays a crucial role in power system studies and efficient operation.

This obviously needs long range forecasting while some sophisticated probablistic methods exist, the simple extrapolation technique is quite adequate for long range forecasting and this method is hence generally used. As in every country, weather has a much more influence on residential than the industrial demand, it may be better to prepare forecast in constituent parts to obtain total. Both power and energy forecasts are made. Multifactors involved randam forecasting an involved process requiring experience and high analytical ability of operation engineers. In fact, yearly forecasts are based on previous year's loading for the period under consideration updated by factors such as (1) general load increases, (2) major loads-running and expected during the period considered and (3) weather trends-rural and sudden changes as expected (from past record) during the period under consideration.

In case of short-term load forecasting, hour-by-hour predictions are made for the particular day under consideration. A minor forecoast error on low side might require the use of inefficient, oil fired turbine generators on "peaking units" which are quite costly. On the other hand, a high side forecast error would keep excessive generation in hot reserve. Hence in the short time forecast an accuracy of the order of 01% is most desirable. A temperature difference of 2°C can vary the total load by 01%. This shows the importance of reliable weather forecast to enable the power system engineers to forecast a good load forecast.

In fact the short term forecast problem is not a simple one as often random factors such as unexpected storms, thunder strikes, showers etc. the sudden telecast of a good and popular TV programme can upset the production for load requirement for the evening of any particular day. However regression analysis is often used for obtaining a short-term load forecast which is very important and is required before solving unit commitment and economic.

As stated above, in India, energy demand and installed generating capacity are both increasing exponentially (although demand is increasing at the same time—so is population growth). It has been noticed that power demand has been roughly doubling every ten years and so is in many other countries in the world. As per the present rate of growth of demand while by 2000 the demand was around 1,00,000 MW (in 1997 it was around 70,000 MW) but the demand by 2015 it may be around 2,00,000 MW and by 2020 be around 4,00,000 MW. This means the country require a corresponding development in sources of coal and nuclear energy.

1.8.2 Load Management

As stated above regarding various load management scheme for load-forecasting, it is possible, by these schemes to shift demand away from peak hours. In fact a direct method would be the control of the load either through modified tariff structures that encourage the individual customers to readjust their own electricity use schedules or direct electrical control of appliances in the form of remote timer controlled on/off switch with least inconvenience to the customer. Systems for load management are varied. Here detail discussion is not in the scope of this book. For detail reader can refer to IEEE paper. February 1979 on the subject by Morgen M.G. However, in Europe Ripple control has been tried for Load Management. Remote kWh meter reading by carrier systems is being also tried. In fact most of the potential for load control lies to the domestic sector. In UK, as found in many hotels, in many houses automatic switching off of the supply to the main switch is arranged the moment the occupants close the door from outside with noone inside while keeping the Fridge on. In such case control circuit is accordingly designed. In USA, Japan also similar control system for load circuit is being implemented where the consumers are agreeable. In India also Electricity authorities should give serious thought in such load control methods to avoide unnecessary wastage of electricity.

1.9 POWER SYSTEM ENGINEERING STUDIES & USE OF COMPUTER & MICROPROCESSOR

The power system engineering of the last two decades particularly from 1990 has to face a vareity of challanging tasks. In fact as the demands are increasing, and so is increasing the complexities of the system. Hence a power system engineer can meet these problems and challanges of power system complexities by keeping abreast of the recent scientific advances and the latest techniques. He has to have good experience on trouble-shooting and maintenance and power system network's complexities besides knowledge of load control and load management and the characteristics of the loads. A power system engineers has to have knowledge on generation of electricity and its transmission and distribution.

Then in planning side he has to make decision on how much electricity to be generated-where, when and by using what fuel. He has to be involved in construction tasks of great magnitude both in generation and transmission. Moreover he has to solve the problem of planning and coordinated operation of a vast and complex power network, so as to achieve a high degree of economy and reliability. Moreover he has to additionally face the perennial problem of power shortages and to evolve strategies for energy conservation and load management.

1.9.1 Energy Conservation

Here a brief discussion is to be done on energy conservation. Energy conservation is perhaps the cheapest new source of energy. There are various conservation measures such as cogeneration and use energy-efficient motors to avoid wasteful electricity uses. We can in fact, achieve considerable electric power savings by reducing unnecessary high lighting levels, oversized motors etc. All the consumers (Domestic, Industrial, entertainment) must be taught how consumption levels can be reduced without any essential lowering of comfort and application. In this rate restructuring can have incentives in this regard. There is no consciousness on energy accontability yet and no sense of urgency as in developed countries. Our electricity supply authorities can think of the present tariff structure, where the discount is given on the higher consumption by reducing the tariff is some cases. Considering the importance of energy conservation the higher the consumption higher should be the tariff in industries and in domestic consumption of eletricity.

In energy conservation process Energy Storage (of unused or excess electrical energy) plays an important role.

1.9.2 Energy Storage

Although there is a lot of problem in storing electricity in large quantities, but there are numbers of way by which electricity can be stored and connected into electricity when required. However, storage of any kind is very costly and its economics must be worked out property. Various options available are : Pumped Storage, compressed air, heat, hydrogen gas, secondary Batteries. For example, in West Bengal Purnlia Pumpa Storage plant has been commissioned. And this stored energy is used to meet the peak houe demand. The most widely used storage battery is the lead acid battery.

A fuel cell converts chemical energy of a fuel into electricity directly, with no intermediate computation cycle. In the fuel cell, hydrogen is supplied to the negative electrode and oxygen (or air) to the positive. Hydrogen and oxygen are combined to give water and electricity. The porous electrodes allow hydrogen ions to pass. But due to their high cost (amount Rs. 1,00,000 per KW) the fuel cells are not in wide use.

Maintenance

Proper maintenance and particularly the preventive maintenance play a very important role not only in efficient operation of power system. But also present unnecessary loss of eletricity energy and thus helping the conservation of energy. Therefore, maintenance must be on regular shedule to prevent system failure. Planning of maintenance shedule such that it includes proper training of maintenance manpower equipped with all the latest digital electronic test equipment and microprocessor so that before the failure occurs management gets the signal for remedial actions.

1.10 USE OF COMPUTERS MICROPROCESSOR

The very first method for solving various power system problem were AC and DC network analysis developed as early as in 1930s. AC analysis were used for load flow and stability studies whereas DC were preferred for short-circuit studies.

However, in 1940s analogue Computers were developed in 1940s and were used in conjunction with AC network analysis to solve various problem for off-line studies. In 1950s many analogue devices were developed to control the on-line function such as generation control, frequency and tie-line control. The 1950s also saw the advent of digital computers which were first used to solve a load flow problem in 1956. Power system studies by computers, in fact gave greater flexibity, accuracy, speed and of course economy. But till 1970s, there was a widespread use of computers in system analysis. Then with the entry of microprocessors in the arena, then besides main frame computers, mini, micro and personal computers are all increasingly being used to carry out various power system studies and solve power system problems for off-line and on-line application. Off-line applications include research, routine evaluation of system performance and data assimilation and retrieval. It is mainly used for planning and analysing some new aspects of the system. On-line and real time applications include data-logging and the monitoring of the system state.

A large central computer is used in central load despatch (CLD) centres for economic and secure control of large integrated systems.

Microprocessors and computers installed in generating station control various local processes such as starting up of a generator from the cold state etc. Table 1.2 gives the time scale of various hierarchical control problems to be solved by computers/microprocessors.

Table 1.2

Time Scale		Control Problems
Milli seconds	—	Relaying and system voltage control and excitation control
25 – 5 minutes	—	Automatic Generation Control (AGC)
10 min – few hours	—	Economic Despatch (ED)
-do-	—	Security Analysis
Few hours – 1 weak	—	Unit Commitment (UC)
1. month – 6 months	—	Maintenance Sheduling
1 years – 10 years	—	System planning (modified/extension)

1.11 SALIENT FEATURES OF VARIOUS TYPES OF POWER STATIONS COMPARISON

In Table 1.3 a comparative study is made on various types of generation processes :

Table 1.3. Salient Features of Various Types of Power Stations—A Comparison

Particulars	Hydro-electric	Thermal	Nuclear	Diesel
1	2	3	4	5
1. Site	Ample quantity of water at sufficient head and possibility of constructing a dam to store water in the catchment area are decisive factors. Transportation facilities should also be available.	Generally located near the load centre but other factors such as transportation of fuel, enough water for cooling the condensate, cost of land are also kept in view. Present thinking is to install thermal stations near coal mines.	Located near the load centre. Easy transportation of nuclear fuel, availability of cooling water and burial of radioactive waste determine the site for a nuclear power station.	Can be installed anywhere.
2. Cost (*a*) Initial cost (approximate) per kW, Rs.	1,500 to 4,000	1,000 to 1,500	2,500	1,500 to 2,500
(*b*) Fixed cost per annum (interest on capital, depreciation, taxes, insurance, as a percentage of initial cost)	10%	13%	15%	15%
(*c*) Variable cost made up of (*i*) Cost of fuel	Practically nil as no fuel is required	High	Low	Very high
(*ii*) Transpor-tation of fuel	Nil	Very high especially when the power stations are away from coal mines and railway sidings.	Since quantity fuel required is very small transportation cost is low.	High than that for nuclear stations but lower than that for thermal power stations.
(*iii*) Maintenance cost	Comparatively low	Higher as compared to hydro and diesel but comparable to nuclear stations.	On the high side as skilled and well trained staff is required to handle the equipment.	Comparatively low
(*iv*) Transmission liability	Very high	Low except in the case of mine-head stations.	Low	Nil

Particulars	Hydro-electric	Thermal	Nuclear	Diesel
1	2	3	4	5
(*d*) Running cost per kWh of energy generated (approximate figures) as in 2008	2 paise	6 paise	4 paise	40 paise
3. Time required for completion	About 10 years	3—4 years	10 years	
4. Simplicity and cleanliness	Simple and clean	Causes air pollution. Disposal of ash is another problem	Clean source of power generation which does not contribute to air pollution. Handling of equipment is very complicated. Nuclear wastes disposal is, however, a major problem.	Simpler and cleaner than thermal plants.
5. Field of application	Can be used to supply peak load or base load	Generally used to supply base load	Used to supply base load as the reactors cannot be easily controlled to respond quickly to load changes.	Used as a standby to supply part of load in a power system when required. Also used to supply peak load demand in conjunction with hydro, thermal or nuclear stations which supply base load.
6. Reliability	Simple, robust and most reliable	Less reliable	Reliable	Less reliable
7. Space	Need a lot of space for civil engineering construction work such as dams, etc. the building has to be much larger than that required for other types of plants.	Need much more space than Diesel electric stations but much less when compared with hydro-stations. A huge space is required for storage of fuel (*i.e.*, coal).	Need less space as compared to hydro and thermal stations.	Normally of small sizes and need less space.

QUESTIONS

1. Write a short note on (1) Thermal Power Station, (2) Nuclear Power Stations, (3) Hydel Power Station.
2. State what are the factors to be considred for selection of the site for (1) Thermal Power Station and (2) Nuclear Power Sation.
3. State what are the function of economist and superheater in a Thermal Power Station.
4. What are the types of fuels used in a (1) Thermal Power Plant, (2) Nuclear Power Plant.
5. State how the coal is utilised in Thermal Power Plant.
6. Write short note on
   1. Advantages and disadvantages of Thermal Power Plants.
   2. Advantage and diadvantages of Nuclear Power Station.
   3. Advantages and disadvantages of :
      (*a*) Hydel Power Plants
      (*b*) Solar Energy
7. State what are the salient features of various types of power stations.
8. Give a comparison of the performances of Nuclear, Thermal and Hydel Power Stations.
9. Write a short note on load Management and energy conservation.
10. State what are the major problems generally faced by Indian power Industry.
11. Write a note on the growth of power system in India.
12. State what are the different types of energy sources and efficiency in their use.

CHAPTER

2

Power System's Electrical Basics

2.1 POWER SYSTEM ELECTRICAL FUNDAMENTAL

INSIDE THIS CHAPTER

- Power system electrical fundamental
- Single phase transmission
- 3-phase transmission
- Complex power
- Load characteristics
- The P.U. (per unit) system

So long we discussed on basic ideas and Generation of Power of Power System. Below a brief discussion will be made on some very basic electrical engineering fundamentals of Power System Transmission lines which transmit the power generated to the load centres which may hundreds and thousands of kilometers away.

These transmission lines have basic three elements of electrical engineering. Resistance inductor and capacitor. It is known to the readers by now that the resistor consists ohmic or dissipative energy whereas the inductor and capacitor store in the positive half cycle and give away in the negative half cycle supply in the magnetic field and electric field energies respectively. The ohmic form of energy is dissipated into heat wherever a current flows in a resistive median as I^2Rt, where I is the current flowing for a period of seconds through a resistor of R ohm.

In case of an inductor the energy stored is 1/2 LI^2, where the current I flows through a coil of L Henry and in a magnetic field. And in case of a capacitor, the energy stored between the two "plates" of a capacitor in the form of electric field in 1/2 CV^2, where C is the capacitance and V is the voltage across the plates" of the cpacitor.

Let us consider first the power transmission using single phase (1-ph.) circuits assuming that in all the analysis the source is a perfect sinusoidal with fundamental frequency component only.

2.2 SINGLE PHASE TRANSMISSION

Considering an inductive circuit and the instantaneous voltage of $v = V_m \sin wt$.

Two the current will be $i = I_m \sin (wt - \phi)$ where ϕ is the angle by which the current lags the voltage as shown is Fig. 2.1.

Then the instantaneous power,

$$\begin{aligned} p &= vi = V_m \sin wt, I_m \sin (wt - \phi) \\ &= V_m - I_m \sin wt \sin (wt - \phi) \\ &= (V_m I_m/2) [\cos \phi - \cos (2\, wt - \phi)] \quad \text{...(2.1)} \end{aligned}$$

Now, the value p is positive when both v and i are either positive or negative and represents the rate at which the energy is being consumed by the load. In fact, in this case, the current flows in the direction of voltage drop. Then the power is negative when the current flows in the direction of voltage rise which means the energy is being transferred from the load into the network to which it is being connected. Now if the circuit is purely reactive, the voltage and current will be 90° out of phase and so the power (p) will have equal positive and negative half cycles and the average value of will be zero. From equation (2.1) the power (p) pulsates around the average power at double the supply frequency. Then, equation (2.2) can be written as :

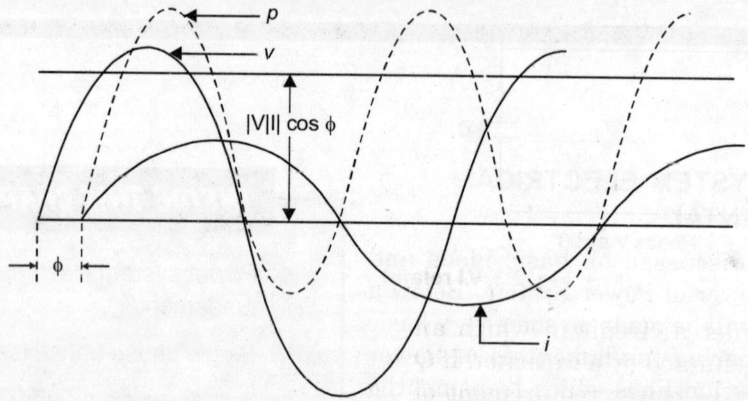

Fig. 2.1. Voltage current power in 1-ph circuit power at double the supply frequency.

$$p = VI \cos \phi \, (1 - \cos 2 \omega t) = VI \sin \phi \cdot \sin 2 \omega t \qquad ...(2.3)$$

In the Fig. 2.2 the instantaneous power has been decomposed into two components as :

1. The component p has been marked 1 pulsate around the power $VI \cos \phi$ but never goes negative as the factor $(1 - \cos 2 \omega t)$ can at the best become zero but the neves will go negative. Hence we define this average powers as the real powers p which physically means the useful powers being transmitted.

2. The component marked (2) contains the term $\sin \phi$ which is negative for capacitive circuit and is positive for inductive circuit. This component pulsates and has zero at, its average value. This component is known as reactive power as it travels back and forth on the line without doing any useful work.

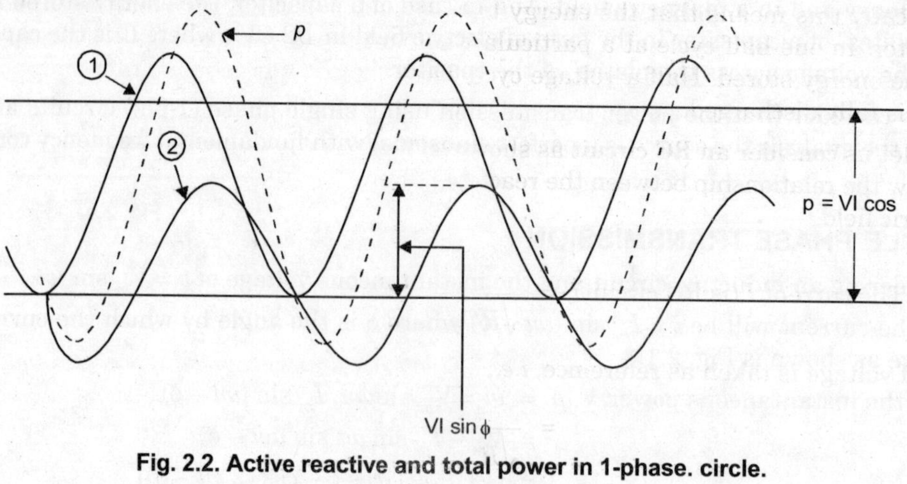

Fig. 2.2. Active reactive and total power in 1-phase. circle.

Equation (2· 3) is rewritten as $p = P(1 - \cos 2wt) - Q \sin 2wt$ ...(2.4)

Both P and Q have the same dimension of watts but to emphasis is the fact that Q represents a nonactive power, it is measured in term of voltamperes reactive, *i.e.* VAR.

Because of the interesting properly of sin ϕ which is $-ve$ for capacitive and is $+ve$ for inductive circuits. The term Q requires more attention. This, in fact, seem a capacitor is a generator of positive reactive VAR, a concept which is equally adopted by power system engineers. Hence, it is better to consider a capacitor supplying a lagging current rather than taking a leading current, as can be seen in Fig. 2.3.

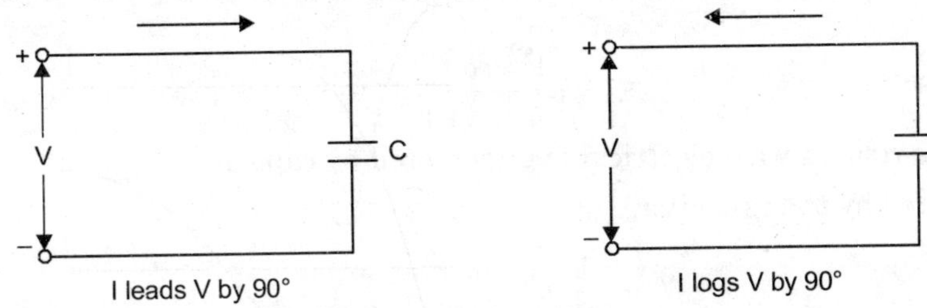

Fig. 2.3. V.I relations is a capacitor.

Let us consider a circuit in which an inductive load is shunted by a capacitor. If Q is the total reactive power requirement of the load and Q' is the reactive power that the capacitor can generate, the net reactive power to be transmitted over the line will be $(Q - Q')$. This is the very basic/fundamental concept of synchronous phase modifiers for controlling the voltage of the system. The phase modifier is basically a synchronous reactive working as a capacitor when over excited and as an inductor when under excited.

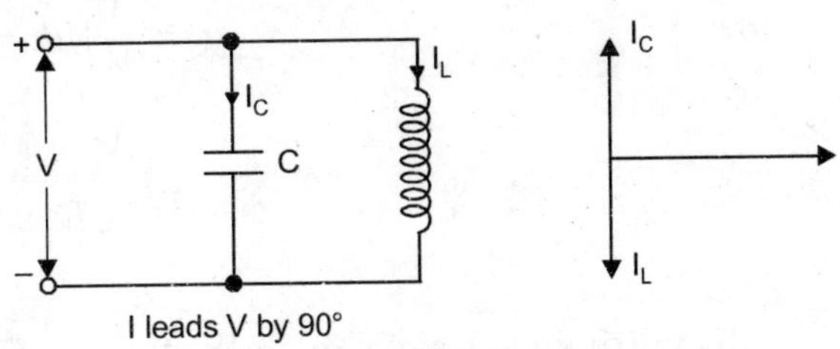

Fig. 2.4. Power flow a L-C cirnit.

However, it is interesting to consider the case when a capacitor and an inductor of the same reactive power requirement are connected is parallel as in Fig. 2.4. The current I_L and I_C are equal in magnitude and therefore, the power requirement is same. The line power will, therefore, be zero.

Physically this means that the energy travels back and forth between the capacitor and the inductor. In one half cycle at a particular moment the capacitor is fully charged and the coil has no energy stored. Half a voltage cycle later the coil stores maximum energy and the capacitor is fully discharged.

Now let us consider an RC circuit as shown in Fig. 2.5 to show the relationship between the reactive power and electric field.

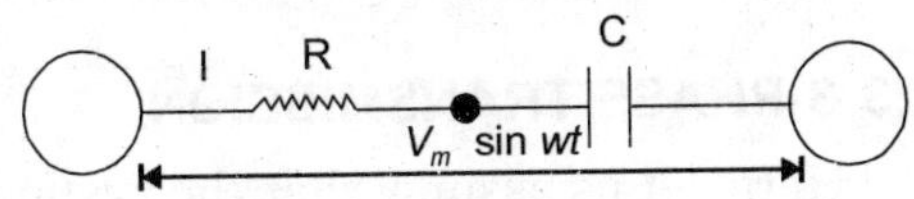

Fig. 2.5.

$$\therefore \quad \text{The current } I \text{ (is RC circuit)} = \frac{V}{\sqrt{R^2 + (1/wc)^2}} = \frac{V_w C}{\sqrt{R^2 w^2 c^2 + 1}} \quad ...(2.5)$$

and if voltage is taken as referemce, *i.e.*, $v = V_m \sin wt$, the current $i = I_m \sin(wt + \phi)$

$$\therefore \quad i = \frac{V_m WC}{\sqrt{R^2 w^2 c^2 + 1}} \cdot \sin(wt + \phi) \quad ...(2.6)$$

where $$\sin\phi = \frac{1/wc}{\sqrt{I^2R^2+(1/wc)^2}} = \frac{1}{\sqrt{R^2w^2c^2+1}} \quad ...(2.7)$$

Now, reactive power $$Q = VI\sin\phi \quad ...(2.8)$$

Substituting I and $\sin\phi$ in (2.8)

we get $$Q = V\cdot\frac{Vwc}{\sqrt{R^2w^2c^2+1}}\cdot\frac{1}{\sqrt{R^2w^2c^2+1}} = \frac{V^2wc}{R^2w^2c^2+1} \quad ...(2.9)$$

$\therefore$ Reactive power $$= \frac{V^2wc}{R^2w^2c^2+1}$$

Now this can be related with electrical energy stored by capacitor.

The energy stored by the capacitor,

$$W = \frac{1}{2}cv^2 \quad ...(2.10)$$

Now, $$v = \frac{1}{c}\int idt = \frac{1}{c}\cdot\frac{V_m wc}{\sqrt{R^2w^2c^2+1}}\cdot\frac{\cos(wt+\phi)}{w}$$

$$= \frac{V_m\cos(wt+\phi)}{\sqrt{R^2w^2c^2+1}} \quad ...(2.11)$$

$\therefore$ by putting values of v in equation (2.10) $$W = \frac{1}{2}c\cdot\frac{V_m^2\cos^2(wt+\phi)}{(R^2w^2c^2+1)} = \frac{V^2\cos^2(wt+\phi)}{R^2w^2c^2+1} \quad ...(2.12)$$

Now, $$\frac{dw}{dt} = \frac{V^2}{R^2w^2c^2+1}\cdot 2\cos(wt+\phi)\cdot\sin(wt+\phi)\,w\cdot c$$

$$= \frac{V^2wc}{R^2w^2c^2+1}\cdot\sin 2(wt+\phi) = Q\cdot\sin 2(wt+\phi) \quad ...(2.13)$$

Hence, from this it is clear that the rate of change of electric field energy is a harmonically varying quantity with a frequency double the supply frequency and has a peak value equal to Q.

In an R-L circuit the magnetic field energy and reactive power in a coil are similarly related.

2.3 3-PHASE TRANSMISSION

Here, let us assume that the system is balanced, that mean the three-phase voltages and currents are balanced. These quantities can be expressed mathematically as follows.

$$V_a = V_m\sin wt,\ V_b = V_m\sin(wt-120°),\ V_c = V_m\sin(wt+120°)$$

$$V_{ia} = I_m\sin(wt-\phi),\ i_b = I_m\sin(wt-\phi-120°),\ i_c = I_m\sin(wt-\phi+120°)$$

The total power transmitted equals the sum of the individual power is each phase.

$$P = V_a i_a + V_b i_b + V_c i_c$$

$$= V_m\sin wt\, I_m\sin(wt-\phi) + V_m\sin(wt-120°)\, I_m\sin(wt-120°-\phi)$$

$$+ V_m I_m \sin(wt + 120°) \sin(wt + 120° - \phi)$$

$$= VI[2 \sin wt \sin(wt - \phi) + 2 \sin(wt - 120°) \sin(wt - 120° - \phi) + 2 \sin(wt + 120°) \sin(wt + 120° - \phi)]$$

$$= VI[\cos\phi - \cos(2wt - \phi) + \cos\phi - \cos(2wt - 240° - \phi) + \cos\phi - \cos(2wt + 240° - \phi)] \quad ...(2.15)$$

$$= 3\, VI \cos\phi$$

From this we see that the total instantaneous 3-phase power is constant and is equal to three times the real power for phase, *i.e.*, $p = 3p$, where p is the power per phase. However, in case of single phase transmission we noted that the instantaneous power expression contained both the real and reactive power expression but in case of 3-phase it is found that instantaneous power is constant. This does not mean that the reactive power is of no importance in a 3 phase system.

In case of a 3 phase system the sum of three currents at any instant is zero, but this does not mean that the current in each phase is zero. Similarly even though the sum of reactive power instantaneously in 3-phase system is zero, but in each phase it does exist and is equal to $VI \sin\phi$, and, hence, for 3-phase, reactive power is equal to $Q_{3\text{-ph.}} = 3\, VI \sin\phi = 3Q$, where Q is the reactive power in each phase. However it is to be noted that the term $Q_{3\text{-ph}}$ makes a little physical sense as could be the concept of 3-phase currents, $I_{3\text{-ph}} = 3I$. Nevertheless the reactive power in a 3-phase system is expressed as Q_{3p}. This is done to maintain symmetry between the active and reactive power.

2.4 COMPLEX POWER

Let us consider a single phase network and say

$$V = |V|\, e^{ix} \text{ and } I = |I|\, e^{j\beta} \quad ...(2.16)$$

where α and β are the angles that V and I subtrend with respect to some reference axis. We calculate the real and reactive power by finding the product of V with conjugate of I, *i.e.*,

$$S = VI^* = |V|\, e^{j\alpha} \cdot |I|\, e^{-j\beta} = |V|\, |I|\, e^{j(\alpha - \beta)}$$

$$= |V|\, |I| \cos(\alpha - \beta) + j|V|\, |I| \sin(\alpha - \beta) \quad ...(2.17)$$

Here the angle $(\alpha - \beta)$ is the phase difference between the phasor V and I and is normally denoted by ϕ.

$$\therefore \quad S = |V|\, |I| \cos\phi + j\, |V|\, |I| \sin\phi \quad ...(2.18)$$

$$= P + jQ$$

The quantity S is called the complex power. The magnitude of $S = \sqrt{P^2 + Q^2}$ is termed as the apperent power and its with are volt-amperes and the larger unit are KVA or MVA. The practical significance of apperent power is as a rating unit of generators and transformers, as the apperent power rating is a direct indicator of heating of reactive which determines the rating of the machines. It is to be noted that Q is positive when $(\alpha - \beta)$ is positive, *i.e.*, when V leads I, *i.e.*, the load is inductive and Q is *–ve* when V lags I, *i.e.*, the load is capacitive. This agrees with the normal convention adopted in power system, *i.e.*, taking Q due to an inductive load as *+ve* and reactive power it is necessary to find out VI^* rather $V^* I$, which would reverse the sign for Q as,

$$V^* \cdot I = |V| \cdot e^{j\alpha} \cdot |I|\, e^{j\beta} = |V|\, Ie^{-\alpha(\alpha - \beta)}$$

$$= |V|\, |I| \cos(\alpha - \beta) - j\, |V|\, |I| \sin(\alpha - \beta)$$

$$= |V|\,|I| \cos\phi - j\,|V|\,|I| \sin\phi$$
$$= P - jQ \qquad ...(2.19)$$

2.5 LOAD CHARACTERISTICS

In practice, in an electrical power system, it is different to practice the load variation accurately because, the load devices may vary from a few watt night lamps to multi-megawatt induction motors, heating and ligating equipments, various types of electronic devices. Load requirements and this characteristics also different with different type of loads. For example, the heating load maintains constant resistance with voltage change and hence the power varies with (voltage)^2 whereas the lighting load is independent of frequency and power courrent varies as $V^{1.6}$ instead of V^2 in case of purely resistance lead. In case of an impedance, *I.e.*, a lumped load,

$$P = \frac{V^2}{R^2 + (Lw)^2} R = \frac{V^2}{R^2 + (2\pi fL)^2} \times R$$

and
$$Q = \frac{V^2}{R^2 + (wL)^2} \times Lw = \frac{V^2}{R^2 + (2\pi fL)^2} \times 2\pi fL \qquad ...(2.20)$$

From these expression of equation (2.20) above it is clean that both, P and Q increase as the square of voltage V. Also with increasing frequency, the active power P decrease whereas Q increase.

Hence above equation (2.20) can be expressed in the form

$$P = P[f, |V|] \text{ and } Q = Q[f, |V|] \qquad ...(2.21)$$

In case of composite loads, which, in fact, form a major part of the system load as also function of voltage and frequency and can in general be written as in equation (2.21). For this type of load however, no direct relationship can be available as for independence loads. For a particular composite load however, an empirical relation between the load, and the voltage and frequency can be obtained. In fact, normally we are concerned with incremental changes in P and Q as a function of incremental changes is P and Q as a function of incremental changes in P $|V|$ and f.

Now, from equation (2.21), we can write for incremental change ΔP in P as

$$\Delta P \simeq \frac{\delta P}{\delta |V|} + \frac{\delta P}{\delta f} \cdot \Delta f \qquad ...(2.22)$$

and
$$\Delta Q \simeq \frac{\delta Q}{\delta |V|} + \frac{\delta Q}{\delta f} \cdot \Delta f$$

Then, the four partial derivatives can be obtained empirically. But, it is remembered that whereas an impedance load P decreases with increasing frequency a composite will increase because a composite load mostly consists of induction motors (amount 60% to 70% of total load), which always will experience increased load, as frequency or speed increases. But the need for ensuring a high degree of service reliability in the operation of modern electric systems is considerate be sent important. To assume this, the supply should not only be reliable but should be of good quantity, which means, the voltage and frequency should vary within certain limits as specified by the B.I.S., otherwise operation of the system at subnormal frequency and lower voltage will result serious problem particularly induction motors including functional horse-powers motors used for various application in domestic and industrial areas. For example, for refrigeration reduced frequency and high consumption, as the motor draws largest current at reduced power factor and reduced voltage. The system operation at subnormal frequency and voltage loads to the loss of revenue to the suppliers the to accompanying reduction in load demand. However, perhaps the most serious effect of subnormal frequency and voltage is on the operation of the thermal power station anxiliairies. The output of the the auxiliaries goes down, as a result of which the generation is also decreased. This may result an complete shut-down of the plant if remedial Pearson, like say load shedding , is not resorted to load-shedding is, in fact, done with the help of under-frequency relays which automatically discount blocks of loads or sectionalise the transmission system depending upon the system requirements.

2.6 THE P.U. (PER UNIT) SYSTEM

P.U. system is found quite comensurate for analysis purpose in a complex power system. In fact, in a long inter-connected power system with various voltage levels and various capacity equipments it has been found very useful and convenient to work with P.U. (per unit) system of quantities for analysis purposes rather than in also the values of quantities. Sometimes, per cent values are used instead of P.U. system, but it is really always very such convenient to work with or use P.U. values.

The P.U. value of any quantity is defined as :

$$\frac{\text{The actual value of the quantity (many unit)}}{\text{The base or reference value (in the same unit)}}$$

It is known that in electrical engineering the three basic quantities are voltage, current and impedance. If we choose any two of them as the base as reference quantity the third one automatically will have a base or reference value depending upon the other two, that is if V and I over the base voltage and current in a system, the base impedance of the system is fixed and is given by, $Z = \frac{V}{I}$. And the rating of the equipments in a power system are given in terms of operating voltage and the capacity in KVA. Hence, it is found convenient and very much useful to select voltage and KVA as the base quantities.

Now, let V_b be the base voltage and KVA_b be the base kilo voltage powers, then

$$V_{P.U.} = \frac{V_{actual}}{V_b}$$

and the base current, $$I_b = \frac{KVA_b \times 100}{V_b}$$

$$\therefore \quad \text{P.U. current} = \frac{\text{Actual current}}{\text{Base current}} = \frac{\text{Actual current}}{KVA_b \times 100} \times V_b$$

$$\text{Base impedance} = \frac{\text{Base voltage}}{\text{Base current}} = \frac{V_b^2}{KVA_b \times 1000}$$

$$\text{P.U. impedance} = \frac{\text{Actual impedance}}{\text{Base impedance}} = \frac{z}{(V_b^2/KVA_b \times 1000)} = \frac{Z \cdot \text{KVA}_b \times 1000}{V_b^2}$$

$$= \frac{Z \cdot \text{MVA}_b}{(kV_b)^2}$$

From this expression, it means that the P.U. impedance is directly proportion to the base kVA and inversely proportion to the square of the base voltage. Normally P.U. impedance of various equipment corresponding to its own rating voltage and kVA are given and since we choose one common base kVA and voltage for the whole system, therefore, it is desired to find out the P.U., impedance of the various equipment corresponding to the common base voltgc and kVA.

Now, if the individual quantities are $Z_{\text{P.U. old}}$, kVA_{old} and V_{old} and the common base quantities are $Z_{\text{P.U.}}$ now, kVA_{now} and V_{now}, then making use of the relation above, we get,

$$Z_{\text{P.U. now}} = Z_{\text{P.U. old}} \frac{\text{kVA}_{\text{now}}}{\text{kVA}_{\text{old}}} \cdot \left(\frac{V_{\text{old}}}{V_{\text{now}}}\right)^2 \quad ...(2.23)$$

In power system analysis this is a very important relation. The P.U. impedance of an equipment corresponding to its own rating is given by, $Z_{\text{P.U.}} = \dfrac{IZ}{V}$ where Z is the also the value of the impedance of the equipment. It is observed that the P.U. representation of the impedance of an equipment is more meaningful than its value *e.g.*, saying that the impedance of machine is 15 ohm does not give any idea regarding the size of the machine. For a large machine 15 ohm appears to be quite large, whereas for small machines 15 own is very small whereas for equipment of the same general type the P.U. volt drops and losses are in the same order regardless of size.

In fact, with P.U. systems there is less chance of making mistake in phase and line voltage, single phase or three phase quantities. Also the P.U. impedance of the transformer is same whether referred to primary or secondary side of the transformer which actually is not the case when considering absolute value of these impedances. This can be explained better way below :

Say, the impedance of the transformer referred to primary side be Z_p and that on secondary side be Z_s, then,

$$Z_p = Z_s \left(\frac{V_p}{V_s}\right)^2$$

where V_p and V_s are the primary and secondary voltages of the transformer.

Now,

$$Z_{p\cdot(\text{P.U.})} = \frac{Z_p I_p}{V_p} = Z_s \left(\frac{V_p}{V_s}\right)^2 \cdot \frac{I_p}{V_p}$$

$$= Z_s \cdot \frac{V_p \cdot I_p}{V_s^2} = Z_s \cdot \frac{V_s I_s}{V_s^2} = \frac{Z_s \cdot I_s}{V_s} = Z_{s.\text{P.U.}}$$

From this, hence, it is clear that the P.U., impedance of the transformer referred to primary side $Z_{p.\text{ P.U.}}$ is equal to the P.U. impedance of the transformer referred to the secondary side. $Z_{s\text{ . P.U.}}$. This actually is great advantage of P.U. system of calculation.

In fact, the base values in a system are selected in such a way that the P.U. voltages and currents in system are approximately unity. In fact, sometimes the base kVA is chosen equal to the sum of the ratings of the various equipments on the system of equal to the capacity of the largest unit. The different voltage levels in a power systems are due to the pressure of transformers. Therefore the procedure for selecting base voltage is as follow :

A voltage corresponding to any part of the system could be taken as a base and the base voltages in other parts of the circuit separated from the original part by transformer is related through the turns of rate of the transformers. This is very important. For example, say, if the base voltage on primary side is V_{pp}, then on the secondary side of the transformer the base voltage will be $V_{ss} = V_{pp}(N_s/N_p)$, where N_s and N_p are the turns of the transformer on secondary and primary respectively. The example below will perhaps help to follow the procedure for selecting the base quantities in various parts of the system and their effect on the P.U. values of the impedance of the various equipments.

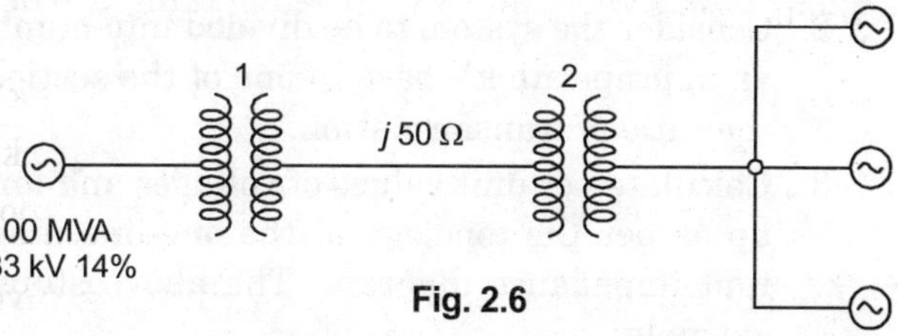

Fig. 2.6

Example 2.1. A 100 MVA, 33 kV, 3-ph generating unit has a subtransmit reactance of 14%. The generator is connected to the motors through a transmission line and transformers as shown in Fig. 2.6. The motors have rated inputs of 30 MVA, 20 MVA, 50 MVA at 30 kV and 20% subtransmit reactance. The 3 phase transformers are rated at 110 MVA, 32 kV/110 kV, $\Delta - y$ with leakage reactance 7.5%. The line has a reactance of 50 ohm selecting the generator circle, determine the base quantities in the generator circle, determine the base quantities in other parts of the system and evaluate the corresponding P.U. values.

Solution : Let us assume the base values as 100 MVA and 33 kV is the generator circuit, the P.U. reactance of generator will be 15%. The base value of voltage in the line will be

$$V_b = 33 \times \frac{110}{32} = 113.43 \text{ kV}$$

Now in the motor circuit $V_b \times \frac{V_p}{V_s} = 113.43 \times \frac{32}{110} = 33 \text{ kV}$

The reactance of the transform given is 7.5% corresponding 110 MVA, 32 kV. Therefore corresponding to 100 MVA and 33 kV, the P.U. reactance will be by using equation; 2.23 and refer to equt cct of Fig. 2.7 of Fig. 2.6.

$$= 0.075 \times \frac{100}{110} \times \left(\frac{32}{33}\right)^2 = 0.0683 \text{ P.U.}$$

The P.U. impedance of line

$$= \frac{50 \times 100}{(113.42)^2} = 0.3886 \text{ P.U.}$$

The P.U. reactance of motor (1)

$$= 0.2 \times \frac{100}{30} \times \left(\frac{30}{33}\right)^2 = 0.551 \text{ P.U.}$$

The P.U. reactance of motor (2)

$$= 0.2 \times \frac{100}{20} \times \left(\frac{30}{33}\right)^2 = 0.826$$

The P.U. reactance of motor (3)

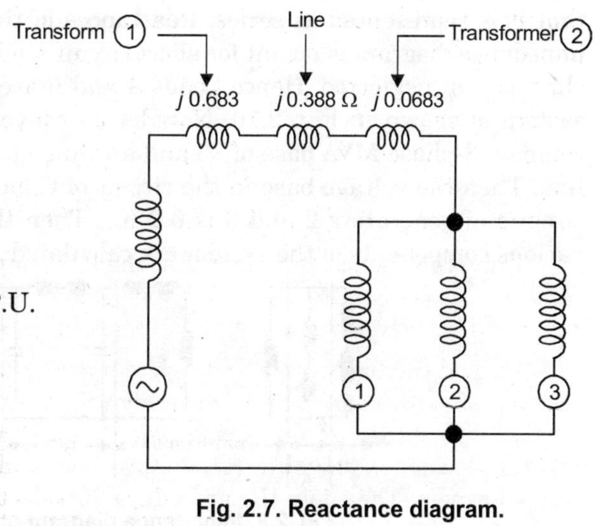

Fig. 2.7. Reactance diagram.

$$= 0.2 \times \frac{100}{50} \times \left(\frac{30}{33}\right)^2 = 0.330 \text{ P.U.}$$

2.6.1 P.U. (Per Unit) Diagram of a Power System

To study the power system, we can directly draw the impedence diagram by following the following steps, from a one line diagram of power system :

1. Choose an approximate common MVA (20 KVA) base for the system.
2. Consider the system to be divided into number of sections by the transformers. Choose an appropriate kV base in one of the section. Calculate KV bases of other sections in the ratio of transformation.
3. Calculate per unit values of voltages and impedances in each section and connect then up as per the topology of the one-line diagram. The result in the single-phase per unit impedance diagram. The above steps are better understood by the following example.

Example 2.2. Obtain the *pu* impedance diagram of the power system shown in the line diagram of Fig. 2.8.

Generator No. 1. 30 MVA, 10.5 KV,

$X'' = 1.6\Omega$.

Generator No. 2. 15 MVA, 6.6 KV

$X'' = 1.2\Omega$.

Generator No. 3. 25 MVa, 6.6 KV,

$X'' = 0.56\Omega$.

Transformer T_1 15 MVa, 33/11 KV, X = 15.2Ω/phase (3-phase) on *HT* side.

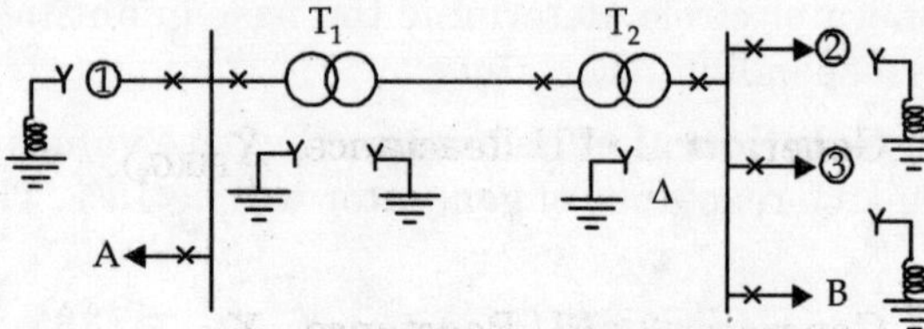

Fig. 2.8. One-line diagram of a simple power system.

Transformer T_2 15 MVa, 33/6.2 KV, X = 16Ω/phase (3-phase) on 1 + *T* side.

Transmission line, 20.5Ω/ph; Load A = 40 MW, 11 KV, 0.9 pf (kg), Load B = 40 MW, 66 KV .85 p.f. (kg).

Solution : The per phase impedance diagram of the power system of Fig. 2.8 is drawn in Fig. 2.9. We make following assumption (1) Line capacitance and resistances are neglected so that it is represented as series. Reactances of the system (2) We shall also assume that the impedance diagram is meant for short circuit studies. And current drawn by static loads under short circuit neglected. Hence Loads *A* and *B* are ignored. Then the reactance diagram of the system is shown in Fig. 2.10. Now let us convert all reactances to P.U. *X* form. Choose or common 3-phase MVA base of 30 and a voltage base of 33 KV line-to-line on the transmission line. Then the voltage base in the circuit of Generator 1 is 11 KV line-to-line and that in the circuits of generators 2 and 3 is 6.2 KV. Then the P.U. *X* (PU Reactance) written as X_{PU} of various components of the system are calculated as below.

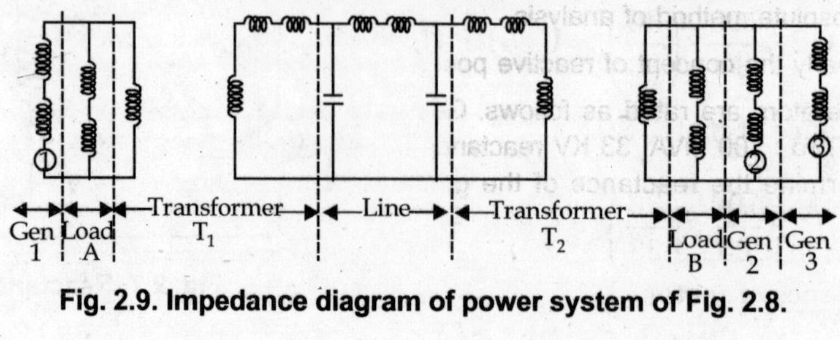

Fig. 2.9. Impedance diagram of power system of Fig. 2.8.

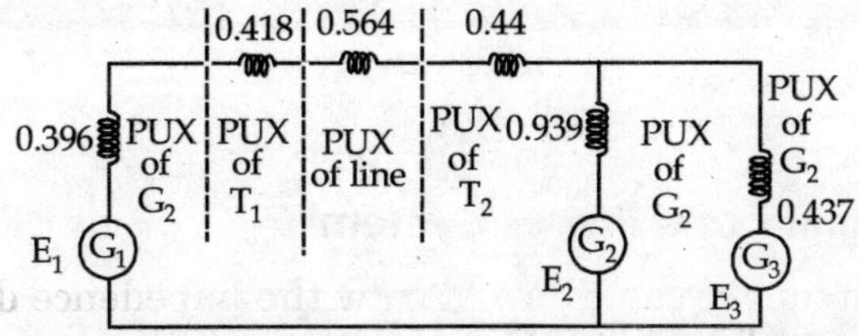

Fig. 2.10. Reactance diagram of the power system of Fig. 2.9 with loads neglected.

Transmission line PU Reactance, $X_{PU\,(line)} = \dfrac{X_{line} \times 30}{(33)^2} = \dfrac{20.5 \times 30}{(33)^2} = 0.564$

Transformer T_1, PU Reactance, $X_{PU(T_1)} = \dfrac{X_{T_1} \times 30}{(33)^2} = \dfrac{15.2 \times 30}{(33)^2} = 0.418$

Transformer T_2, PU Reactance, $X_{PU(T_2)} = \dfrac{X_{T_2} \times 30}{(33)^2} = \dfrac{15.2 \times 30}{(33)^2} = 0.44$

Generator 1 : PU Reactance, $X_{PU(G_1)} = \dfrac{X_{G_1} \times 30}{(33)^2} = \dfrac{1.6 \times 30}{(33)^2} = 0.396$

Generator 2 : PU Reactance, $X_{PU(G_2)} = \dfrac{X_{G_2} \times 30}{(33)^2} = \dfrac{1.2 \times 30}{(33)^2} = 0.936$

Generator 3 : PU Reactance, $X_{PU(G_3)} = \dfrac{X_{G_3} \times 30}{(33)^2} = \dfrac{0.56 \times 30}{(33)^2} = 0.43$

These PU Reactance Values are now put into the Reactance diagram of Fig. 2.10.

QUESTIONS

1. Show that the rate of change of electricfield energy is a haemonically varying quantity with a frequency double the supply frequency and has a peak value equal to Q the total reactive power.
2. Show that in a three phase transmission the total instantaneous 3-phase power is constant and is equal to three times the real power per phase, $p = 3\,p$, where p is the power per phase.
3. Prove that the complex power $s = P + jQ$.
4. Give a practical figure for the various category of loads in a power system.
5. Explain clearly whate is meant by good supply and discuss the effect of bad supply on the performance of the power system.
6. Explain the P.U. system of analyising power system problem. State the advantages of this method over the absolute method of analysis.
7. Explain clearly the concept of reactive power in single-phase and three-phase circuits.
8. Three generators are rated as follows. Generator one : 200 MVA, 33 KVA, Reactance : 10%, Generator Two : 100 MVA, 33 KV reactance 8%, Generator Three : 120, MVA, 32 KV, reactance 12%. Determine the reactance of the generator corresponding to base values of 250 MVA, 33 KV.

CHAPTER

3

Transmission Line Parameters

INSIDE THIS CHAPTER

- Introduction of R&C
- Inductance
- Flux linkages due to flux upto an external point
- Inductance of a single phase two-wire transmission line
- Types of conductors used in transmission line
- Flux linkages of conductor in a group
- Inductance of 3-phase transmission lines unsymmetrically spaced
- Transposition of transmission lines
- Composite conductors for transmission line & Inductance
- Geometric mean distance (GMD)
- Bundled conductors, Transposition of Lines
- Inductance of double circuit in 3-phase line
- Resistance
- Skin effect & proximity effect & Ferrauli effect
- Solved examples on inductance in transmission lines
- Capacitance of transmission line
- Potential difference between two conductors of a group of parallel conductors
- Capacitance of a 3-phase line with equal spacing
- Capacitance of a 3-phase line with unsymmetrical spacing
- Effect of earth on cpacitance of transmission line
- Capacitance of three phase line with more than one circuit
- Bundled conductors & capacitances in a 3-ph line
- Few solved examples

3.0 RESISTANCE & INDUCTANCE & CAPACITANCE

3.1 INTRODUCTION

An electric transmission line has four parameters which affect its ability to fulfill its function as part of power system are Resistance, Inductance, Capacitance and conductance and these parameters are symbolised as R, L, C and G respectively. Out of these four parameters R and G are least important in the sense that they do not affect much the total equivalent impedance of the line and hence the transmission capacity or performance. However they are very much important when transmission efficiency and economy are to be evaluated as they completely determine the real transmission line losses. If R is the effective resistance of the conductor and I is the current flowing through the conductor,

then, $$R = \frac{\text{Power Loss in Conductor (of transmission line)}}{I^2} \text{ohm}$$

However, the effective resistance of the conductor is equal to the d.c. resistance of the conductor only if the current is uniformly distributed throughout the section of the conductor. The difference in the d.c. resistance and effective resistance to frequencies less than 50 Hz is less than 01% for copper conductors of section less than 350,000 circular nuts. The loss on the overhead lines is due to (1) ohmic loss in the power conductors, (2) corona loss and (3) leakage at the insulators which support the lines at the towers. This leakage loss is different form the leakage in cables as in leakage on overhead lines is limited only to the insulators. This could be represented as conductors uniformly distributed along the line. Since the corona loss and the leakage over the insulators is negligibly small under normal operating conditions, the conductors between the conductor of an overhead line is assumed to be zero.

3.2 INDUCTANCE BASIC CONCEPT

Inductance is by far the most dominant of the parameters from a power system engineering point of view. In fact it is the inductance and the inductive reactance which limits the transmission capacity of a transmission line.

What is an Inductor/Inductance

An inductor is a device which stores energy in a magnetic field. By definition, the inductance L of an inductor is the ratio of its total magnetic flux linkages to current I. The inductance, that means,

$$L = \frac{N\phi_m}{I} = \frac{\lambda}{I}$$

This definition is satisfactory for a medium for which the permeability is constant. However, the permeability of ferrous media is not constant and for such cases the inductance is defined as the ratio of the infinitesimal change in flux linkage to the infinitesimal change in current producing it *i.e.*, $L = d\lambda/dI$. The unit of inductance is the Henry.

Mutual inductance between two circuits is defined as the flux linkages of one circuit due to the current in the second circuit per ampere of current in the second circuit. If the current I_2 produces λ_2 flux linkage with circuit 1, the mutual inductance is $M_{12} = \frac{\lambda_{12}}{I_2}$ Henries The phasor voltage drop in circuit (1) by the flux linkages of circuit (2) is

$$V_1 = j\,w\,M_{12}\,I_2 = j\,w\,\lambda_{12} \text{ volts}$$

The concept of mutual inductance is required while considering the coupling between parallel lines and the influence of power lines on telephone lines. In fact the mutual inductance between two circuits is defined as the flux linkages of one circuit due to the current in the second circuit per ampere current in the second circuit.

3.2.1 Inductance of a Current Carrying Conductor

In Power System, transmission lines are composed of parallel conductors which for all practical purposes can be considered as infinitely long. It is assumed that the return path for the current in the conductor (shown in cross-section in Fig. 3.1), to be so far away that magnetic field of the conductor is not affected, as the line of flux are concentric with the conductor. At first let us develop expressions for flux linkages of a long isolated current-carrying cylindrical conductor with return path lying at infinity. This system forms a single-turn circuit, flux linking which is in the form of circular lines concentric to the conductor. The total flux can be divided into two paths, that which is internal to the conductor and the flux external to the conductor. Such a division is helpful as the internal flux progressively links a smaller amount of current as we proceed inwards towards the centre of the conductor, while the external flux always links the total current inside the conductor in Fig. 3.1 the cross-sectional view of a long cylindrical conductor carrying current I, is shown.

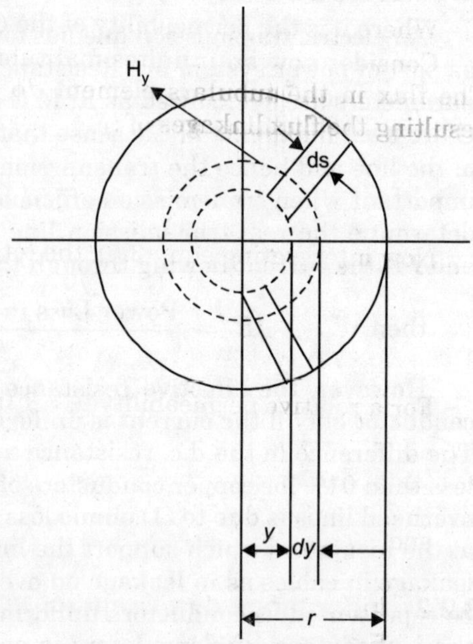

Fig. 3.1. Flux linkage due to internal flux in a current carrying conductor.

Flux Linkages due to Internal Flux

The mmf round a concentric closed circular path of radius y internal to the conductor as shown in the Fig. 3.1 is

$$= \int H_y \cdot ds \cdot I_y \text{ (Ampere s Law)} \quad \text{...(3.1)}$$

where H_y = magnetic field intensity (AT/m)

I_y = current enclosed (A)

By symmetry, H_y is constant and is in the director of ds all along the circular path.

Now using the same symbols as above, $2\pi y\, H_y = I_y = I$

or
$$H_y = \frac{I}{2\pi y} \text{ AT/m}$$

Therefore, from eqn. (3.1), we have

$$2\,\pi y\, H_y = I_y \quad \text{...(3.2)}$$

Now assuming uniform current density,

$$I_y = \left(\frac{\pi y^2}{\pi r^2}\right) I = \left(\frac{y^2}{r^2}\right) I \quad \text{...(3.3)}$$

Then form eqns. (3.2) and (3.3), we obtain

$$H_y = \frac{yI}{2\pi r^2} \cdot AT/m \quad \text{...(3.4)}$$

The flux density B_y, y metres from the centre of the conductor, is

$$B_y = \mu H_y = \frac{\mu y I}{2\pi r^2} \text{ Wb/m}^2 \quad ...(3.4)$$

Where μ is the permeability of the conductor.

Consider now an infinitesimal tubular element of thickness dy and length of one metre. The flux in the tubulars element $d\phi = B_y dy$ webers links the fractional turn $(I_y/I = y^2/r^2)$ resulting the flux linkages of :

$$d\lambda = \left(\frac{y^2}{r^2}\right) \cdot d\phi = \left(\frac{y^2}{r^2}\right) \frac{\mu y I}{2\pi r^2} \cdot dy \text{ Wb-T/m} \quad ...(3.5)$$

Now intergrating eqn. (3.5), the total internal flux linkages as :

$$\lambda_{int} = \int_0^r \frac{\mu I}{2\pi r^4} \cdot y^3 \cdot dy = \frac{\mu I}{8\pi} \text{ Wb} - \text{T/m} \quad ...(3.6)$$

For a relative permeability $\mu_r = 1$ (non-magnetic conductor) $\mu = 4\pi \times 10^{-7}$ H/m, therefore

$$\lambda_{int} = \frac{I}{2} \times 10^{-7} \text{ Wb} - \text{T/m} \quad ...(3.7)$$

and

$$L_{int} = \frac{1}{2} \times 10^{-7} \text{ H/m} \quad ...(3.8)$$

3.2.2 Inductance of a Conductor Due to External Flux

To determine inductance let us limit calculation only to those flux linkages of an isolated conductor which lie between two point P_1 and P_2 at distances D_1 and D_2 respectively from the centre of the conductor, as can be seen in Fig. 3.2. Since the flux paths are concentric circles around the conductor all the flux between P_1 and P_2 lies within the concentric cylindrical surfaces passing through P_1 and P_2. The flux $d\phi$ contained in the tubular element of thickness dy is, $d\phi = \frac{\mu I}{2\pi y} \cdot dy \cdot$ Wb/m length of conductor. The flux $d\phi$ being external to the conductor links all the current in the conductor which together with the return conductor at infinity form a single return, such that its flux linkages are given by

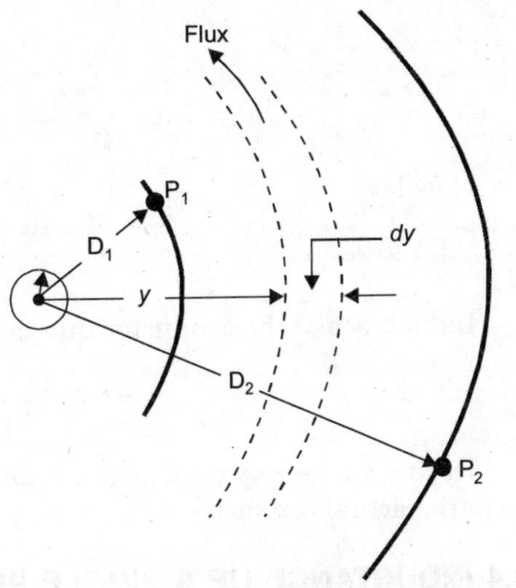

Fig. 3.2. Flux linkages due to flux external points P_1, P_2.

$$d\lambda = I \times d\phi = \frac{\mu I}{2\pi y} \cdot dy$$

Therefore, the total flux linkages of the conductor due to flux between point P_1 and P_2 is

$$\lambda_{12} = \int_{D_1}^{D_2} \frac{\mu I}{2\pi y} \cdot dy = \frac{\mu}{2\pi} I \cdot ln \cdot \frac{D_2}{D_1} \cdot \text{ Wb} - \text{T/m}$$

where In stands for natural logarithm.

Since, $\mu_r = 1, \mu = 4\pi \times 10^{-7}$

$$\lambda_{12} = 2 \times 10^{-7} \cdot I, ln \cdot \frac{D_2}{D_1} \cdot \text{ Wb/m} \quad ...(3.9)$$

Then the inductance of the conductor contributed by the flux included between point P_1 and P_2 in then

$$L_{12} = 2\times10^{-7}\, ln \frac{D_2}{D_1} \text{ H/m} \quad ...(3.10)$$

or

$$L_{12} = 0.461 \log \frac{D_2}{D} \cdot \text{mH/km} \quad ...(3.11)$$

3.3 FLUX LINKAGES DUE TO FLUX UPTO AN EXTERNAL POINT

To calculate the flux linkages due to flux up to an external point, let the external point be at distance D from the centre of the conductor. Flux linkages of the conductor due to external flux (from the surface of the conductor upto the external point) is obtained from eqn. (3.9) by substituting $D_1 = r$

and

$$D_2 = D_1, i.e., \lambda_{ext} = 2\times10^{-7} I \cdot l_n (D/r) \quad ...(3.12)$$

Total flux linkages of the conductor due internal and external flux are :

$$\lambda = \lambda_{int} + \lambda_{ext}$$

$$= \frac{I}{2}\times10^{-7} + 2\times10^{-7}\cdot I\cdot ln\cdot\frac{D}{r}$$

$$= 2\times10^{-7} I\left(\frac{1}{4} + ln\cdot\frac{D}{r}\right)$$

$$= 2\times10^{-7} I\cdot ln\cdot\frac{D}{re^{1/4}}$$

Now let

$$r^1 = r\cdot e^{-1/4} = 0.7788\, r$$

$\therefore$

$$\lambda_1 = 2\times10^{-7} I\cdot ln\cdot\frac{D}{r}\cdot \text{Wb} - \text{T/m} \quad ...(3.13)$$

Inductance of the conductor due to flux upto external point is therefore :

$$L_1 = 2\times10^{-7}\cdot ln\cdot\frac{D}{r_1}\cdot \text{H/m} \quad ...(3.14)$$

Here r can be regarded as the radius of a fictitious conductor with no internal inductance as in the actual conductor.

3.4 INDUCTANCE OF A SINGLE PHASE TWO-WIRE TRANSMISSION LINE

The Fig. 3.3 shows a single-phase two-wire transmission line. One conductor is the return circuit of the other. A line of flux set up by the current in conductor 1 at a distance equal to greater than $D + r_2$ from the centre of conductor 1 links with zero net current, as conductors 1 and 2 carry equal currents in opposite directions, and so no voltage is induced in the circuit for points so located.

If the conductor 1 and 2 carrying current I_1 and I_2 as shown in the Fig. 3.3, in a single phase line, as stated above, then

$$I_1 + I_2 = 0, \therefore I_2 = -I_1$$

Here it is very important to note that the effect of earth's presence on magnetic field geometry is insignificant. This is so because the relative permeability of earth is about the same as that of air and its electrical conductivity is relatively small. Now let us consider the flux linkages of the circuit caused by current in conductor 1 only.

Regarding flux linkages the following three points to be noted, with reference to Fig. 3.3.

(1) External flux from r_1 to $(D - r_2)$ links all the current I_1 in conductor 1. (2) External flux from $(D - r_2)$ links a current where magnitude progressively reduces for I_1 to zero along this distance, because of the effect of negative current flowing in conductor 2. (3) Flux beyond $(D + r_2)$ links a net current of zero.

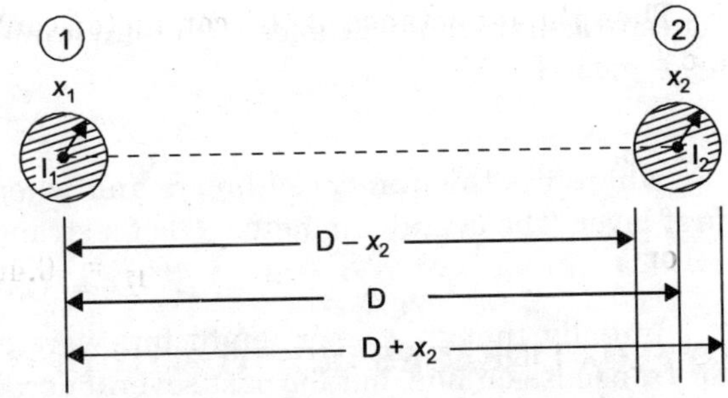

Fig. 3.3. Singl-Face two-wire line and the magnetic field due to in conductor 1 only

To calculate the total inductance due to current in conductor 1, a simplifying assumption is to be made. If D is such greater than r_1 and r_2, which is normally the case for overhead lines; it can be assumed that the flux for $(D - r_2)$ to the centre of conductor 2 links all the current I_1 and the flux from the centre of conductor 2 to $(D - r_2)$ links zero current.

Now, based on these assumptions, the flux linkages of the circuit caused by current in conductor 1 as per eqn. (3.13), are,

$$\lambda_1 = 2 \times 10^{-2} I_1 ln \frac{D}{r_1} \quad ...(3.15)$$

Then the inductance of the conductor due to current is conductor 1 only is

$$L_1 = 2 \times 10^{-7} \cdot ln \frac{D}{r_1^1} \quad ...(3.16)$$

Similarly the inductance of the circle due to current in conductor 2 is,

$$L_2 = 2 \times 10^{-7} \cdot ln \frac{D}{r_2} \quad ...(3.17)$$

Now, using the superposition theorem (of circuit theory of Electrical Engineering), the flux linkages and likewise the inductance of the circuit caused by current in each conductor considered separately may be added to obtain the total inductance of the total circuit.

Hence, $$L = L_1 + L_2 = 4 \times 10^{-7} \ ln \sqrt{r_1 r_2} \text{ H/m} \quad ...(3.18)$$

Now, if $$r_1 = r_2 = r_1 \text{ then } L = 4 \times 10^{-7} \ ln\ D/r \text{ mH/km} \quad ...(3.19)$$

or $$L = 0.921 \log D'/r \text{ mH/km} \quad ...(3.20)$$

3.5 TYPES OF CONDUCTORS USED IN TRANSMISSION LINE

Before going further we should be briefly acquainted with the types of conductors used in transmission lines.

Above we have been discussing transmission lines consisting of single solid cylindrical conductor for forward and return paths. But to provide the necessary flexibility for stringing, conductors used in practice are always stranded except for very small cross-sectional areas. Stranded conductors are composed of strands of wires, electrically in parallel, with alternate layers spiralled in opposite direction to prevent unwinding. The total number of strands (N) in concentrically stranded cables with total annular space filled with strands of uniform diameter (d) is given by

$$N = 3x^2 - 3x + 1 \quad \text{...(3.20)}$$

where x is the number of layers and where in the single central strand is counted as the first layer. The overall diameter (D) of a stranded conductor is

$$D = (2x - 1)\, d \quad \text{...(3.21)}$$

Initially, though, copper conductors were used for transmission line, but for past several decades, Aluminium is the most commonly employed conductor material. It has the advantage of being cheaper and lighter than copper though with less conductivity and tensile strength. Low current density and low conductivity of Aluminium result in larger coverall conductor diameter which results in reduced electrical stress at conductor surface for a given voltage so that the line is corona free. (The corona phenomenon will be discussed later in this book). The low tensile strength of Aluminium conductors is made up by providing centrol strands of high tansile strength steel. Such a conductor is known ACSR – Aluminium Conductor Steel Reinforced, and is commonly used in overhead transmission lines. Fig. 3.4 shows the cross-sectional view of an ACSR conductor with 24 strands of Aluminium and 7 strands of steel in the centre. In EHV (Extra High Voltage) transmission lines, expanded ACSR conductors are used. These are provided with paper or Hessian between various layers of strands so as to increase the overall conductor diameter to reduce electrical stress at conductor surface and prevent corona effect. In fact the most effective way of constructing corona-free EHV lines is to provide several conductors per phase in suitable geometrical configuration. These are turned as bundled conductors which now for EHV lines are a common practice.

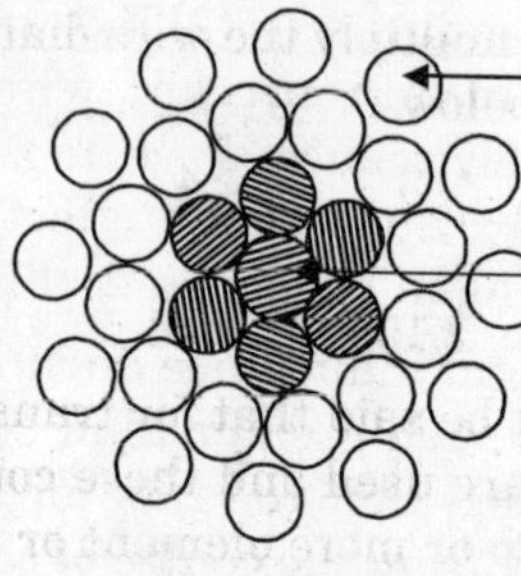

Fig. 3.4. Cross-sectional ACSR of 7 steel strand 24 al strands.

As stated ACSR conductors which combine the lightness with electrical conductivity and rustlessness of aluminium with the high tensile strength of steel are now used as overhead conductors on about every kind of system, low voltage distribution to the most important long distance transmission lines of the world. The reason for this can be summarised as follows :

1. The superior mechanical strength of ACSR can be utilised by using span of larger lengths which results in smaller number of supports for a particular length of transmission.

2. Steel-cored aluminium conductors are normally cheaper than copper conductors of equal resistance and this economy is obtained without sacrifice of efficiency of reliability or of length of useful life.

3. A reduction in the number of supports involves a corresponding reduction in the total cast of insulators, foundations erection and incidentally the cost of maintenance replacements and stores are similarly reduced.

4. The increase in span length is beneficial also. It is well known that the vast majority of shut down in the operation of an overhead line arise at points of supports, due to faulty insulators, flash-overs by birds and so on. As such a reduction in the number of points of supports will correspondingly reduce the risk of outages.

5. These conductors are corrosion resistant and are useful under unfavourable conditions of industrial atmosphere and several condition of exposure such as may occur.

6. Corona losses are reduced because of the larger diameter of ACSR conductors than copper conductor.

In fact the conductivity of a steel-cored aluminium conductor is taken as that of the aluminium portion alone and though steel core may add slightly to the current carrying capacity this is usually neglected. The specific resistance of hard drawn aluminium is approximately 1.6 times that of normal hard drawn copper and the sectional area of an aluminium conductor must therefore be 1.6 time that of the equivalent copper. In order to obtain the overall diameter of stranded conductor, multiply the wire diameter (diameter of one strand) D by the appropriate constant in the table below :

No of wires of: equal diameter	3	4	7	12	19	37	61	91
Dia-constant:	2.155	2.41	3	4.155	5	7	9	11

In the end it must be said that for transmission lines operating at high voltages normally stranded conductors are used and these conductors are known as composite conductors also, as they compose of two or more element or strands electrically in parallel.

3.6 FLUX LINKAGES OF CONDUCTORS IN A GROUP

Before the inductance of composite conductor lines is determined at first a study must be made on Flux Linkages of one conductor in a group.

As can be seen in Fig. 3.5, let us consider a group of n-parallel round conductors carrying phasor current $I_1, I_2, I_3 \ldots I_n$ whose sum equals to zero. Distances of these conductors from a remote point P are indicated as $D_1, D_2, D_3 \ldots D_n$. Now let us obtain an expression for the total flux linkages of the ith conductor of the group considering flux upto to point P only which is very far from the group.

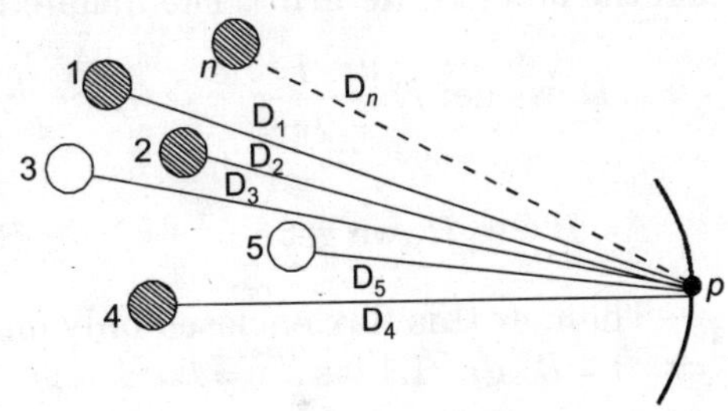

Fig. 3.5. Group of n-parallel conductors carrying conductors.

The flux linkages of conductor (1) due to its own current I_1 self linkages, are given by

$$\lambda_1 = 2 \times 10^{-7} \cdot I_2 \cdot ln \frac{D_i}{R_1'} \text{ Wb-T/m} \qquad \ldots(3.22)$$

$$(\text{As } \lambda_{1-P} = \lambda_1) = \frac{\mu_0 I_1}{8\pi} + \frac{\mu_0 I_1}{2\pi} ln \frac{D_1}{R_1} \quad \text{where } R_1 \text{ is the radius of conductor 1}$$

$$\lambda_{1P} = \lambda_1 = \mu_0 I_1/8\pi + \mu_0 I_1/2\pi \, ln \frac{D_1}{R_1} \quad \text{where } R_1 \text{ is the raidus of conductor 1}$$

$$= 2 \times 10^{-7} I_1 \, ln \frac{D_1}{R_1}$$

Proof is from fundamental theories of flux linkages – Internal flux linkages we know H = magnetic field intensity at any distance (say r which is less than R—the radius of conductor for internal flux linkage due to current I through it).

$$H = \frac{I_r}{2\pi R^2}$$

Now, $B = \mu H = b_r \mu_0 H = \frac{\mu_0 I}{2\pi R^2} \times r$ (as $\mu_r = 1$) for conductor

$\therefore$ Total internal flux linkages, $\lambda = \int_0^R d\lambda$

$\therefore$ Flux linkage for internal flux $\lambda = \int_0^R \frac{\mu_0 I}{2\pi R^4} \cdot r^3 \left(\text{from } d\lambda = \frac{\mu_0 I}{2\pi R^2} r\, de \cdot \frac{r^2}{R^2} = \frac{\mu_0 I r^3}{2\pi R^4} \right)$

$= \frac{\mu_0 I}{2\pi R^4} \int_0^R r^3 \cdot de$ (Note : This can be checked in detail with any Eletrical Engineering Text)

$= \mu_0 I/8\pi$ (This proves that flux linkage due to the internal flux is independent of conductor size)

External flux linkages are due to flux line outsid the conductor. However, there will be no flux line that encloses both the conductors. This is because r, the distance, $> D$ (D the diameter of conductor), the total current enclosed is zero (for single phase line : one conductor is 'go' and the other is 'Return'). The magnetic field intensity H, due to one conductor at any distance $\leq r < D$, we get $H = \frac{I}{2\pi r}$.

As, $B = \mu_0 H$, we get $B = \frac{\mu_0 I}{2\pi r}$ ($\mu_r = 1$ for air)

Then, as this flux enclosed only one conductor as in case of internal flux, $d\lambda = d\phi \cdot 1 = B \cdot dr \cdot 1 = B \cdot dr \cdot 1.1$ (as $d\phi = B \cdot dr \cdot 1$)

$$\therefore \quad d\lambda = \frac{\mu_0 I}{2\pi r} \cdot dr$$

Therefore, the total external flux linkages due to current flow is one conductor,

$$\lambda = \int_R^{D-R} d\lambda$$

$$= \int_R^{D-R} \frac{\mu_0 I}{2\pi r} \cdot dr = \frac{\mu_0 I}{2\pi} ln \frac{D-R}{R}$$

$$= \frac{\mu_0 I}{2\pi} \cdot ln \frac{D}{R} \left(\begin{array}{l} \text{as, } R \ll D_1 \\ \therefore D - R \simeq D_1 \end{array} \right)$$

$\therefore$ Total flux linkages = Internal flux linkages + External flux linkage.

$$= \frac{\mu_0 I}{8\pi} + \frac{\mu_0 I}{2\pi} \cdot ln \frac{D}{R}$$

$$= 2 \times 10^{-7} ln \frac{D}{R}$$

Hence it is shown above briefly has we got, λ_{11} = flux linkages of conductor 1 due to the current I_1 including the internal flux linkages.

$$\lambda_{11} = 2\times10^{-7}\, I_1 ln\frac{D_1}{R_1'}$$

Then the flux linkages λ_{12} to conductor 1 due to current in the conductor 2 are,

$$\lambda_{12} = 2\times10^{-7}\cdot I_2 ln\frac{D_2}{D_{12}}$$

It is to be also noted that flux due to conductor 2 that lies between conductors 2 and 1 will not link conductor 1 and therefore the distance involved are $D_{2P}\,(= D_2)$ and D_{12}.

Hence, the flux linkages of conductor 1 due to all the conductor in the group is

$$\lambda_{1P} = \lambda = 2\times10^{-7}\left[I_1\, ln\frac{D_{1P}(=D_1)}{R_1'} + I_2\, ln\frac{D_{2P}(=D_2)}{D_{12}} + ... + I_n\, ln\frac{D_{nP}(=D_n)}{D_{1n}}\right]$$

This equation can be be reorganised as

$$\lambda_{1B} = \lambda_1 = 2\times10^{7}\left[I_1\, ln\frac{1}{R_1'} + I_2\, ln\frac{2}{D_2} + ... + I_n\, ln\frac{1}{D_n}\right] \quad ...(3.23)$$

$$+ 2\times10^{-7}\left[I_1\, ln\, D_1 + I_2\, ln\, D_2 + ... + I_n\, ln\, D\right]$$

Now, $I_1 + I_2 + ... + I_n = 0$, we get $I_n = (-I_1 + I + ... + I_{n-1})$
Substituting this in the eqn. 3.23, we get,

$$\lambda_{1P} = \lambda_1 = 2\times10^{-7}\left[I_1\, ln\frac{1}{R_1'} + I_2\, ln\frac{1}{D_{12}} + ... + I_n\, ln\frac{1}{D_{1n}}\right] \quad ...(3.24a)$$

$$+\left[2\times10^{-7}\, I_1\, ln\, D_{1P} + I_2/n\, D_{2P} + ... + I_{n-1}\, ln\, D_{(n-1)}\, P\right]$$

$$-\left[I_1\, ln\, D_{nP} - I_2 D_{nP} + ... - I_{n-1}\, lnD_{nP}\right]$$

$$= 2\times10^{-7}\left[I_1\, ln\frac{1}{R_1'} + I_2/ln\frac{1}{D_{12}} + ... + Inln\frac{1}{D_{1n}}\right] \quad ...(3.24b)$$

$$+ 2\times10^{-7}\left[I_1\, ln\frac{D_{1P}}{D_{nP}} + I_2\, ln\frac{D_{2P}}{D_{nP}} + ... + I_{n-1}\, ln\frac{D_{(n-1)P}}{D_{nP}}\right]$$

But, as P is a point considered to be very far from the group of the conductors the ratios, $ln^{D_1 P}/D_{nP} \simeq ln^{D_2 P}/D_{nP} \simeq ... \simeq ln\, D_{1n-DP}/D_{nP} = 0$. The net flux linkage λ_{1P} therefore, become

$$\lambda_{1P} = \lambda_1 = 2\times10^{-7}\left[I_n\, ln\frac{1}{R_1'} + I_2\, ln\frac{1}{D_{12}} + ... + ln\frac{1}{D_{1n}}\right] \text{ Wb – Turn/metre} \quad ...(3.25)$$

3.7 INDUCTANCE OF 3-PHASE TRANSMISSION LINES UNSYMMETRICALLY SPACED

Above we have discussed single phase lines. The basic equation developed in this case can be easily applied to the calculation of the induction of 3-phase transmission lines. Fig. 3.6 shows the conductors of a three phase line with, unsymmetrical spacing, where three conductors *a*, *b* and *c* are carrying currents I_a, I_b and I_c respectively. The three conductors are placed unsymmetrically, that means D_{ab} (between *a* and *b*), D_{bc} (between *b* and *c* and) D_{ca} (between *c* and *a*) are not equal to each other.

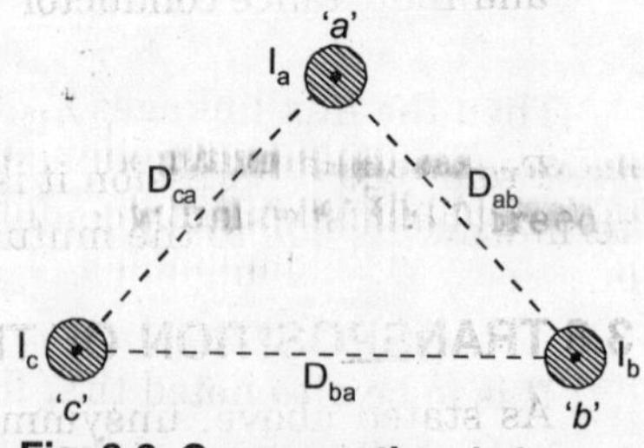

Fig. 3.6. Cross-sectional view of a 3-phase unsymmetrically spaced transmission line.

That is, $D_{ab} \neq D_{bc} \neq D_{ca}$. Also it is assumed that there is no neutral conductor, so that, $I_a + I_b + I_c = 0$. Unsymmetrical spacing causes the flux linkages and therefore the inductance of each phase also are to be different, which we shall calculate as shown below :

Now the flux linkage of conductor *'a'* due to I_a, I_b and I_c can be determined by using the eqn. (3.25).

Flux linkage of $\quad 'a' = \lambda_a = 2\times 10^{-7}\left[I_a\, ln\frac{1}{R^1} + I_b\, ln\frac{1}{D_{ab}} + I_c \cdot ln\frac{1}{D_{ca}}\right]$ ...(3.26)

Similarly : Flux linkage of

$$'b' = \lambda_b = 2\times 10^{-7}\left[I_a\, ln\frac{1}{D_{ab}} + I_b\cdot ln\frac{1}{R_a'} + I_c\cdot\frac{1}{D_{bc}}\right] \quad ...(3.27)$$

Flux linkage of $\quad 'c' = \lambda_c = 2\times 10^{-7}\left[I_a\, ln\frac{1}{D_{ca}} + I_b/_n\frac{1}{D_{bc}} + I_c\, ln\frac{1}{R_a'}\right]$ ...(3.28)

Now taking I_a as reference, $I_b = k^2 I_a$ and $I_c = k\, I_a$ when $k = (-0.5 + j\,0.866)$ substituting these values of I_b and I_c in the expression of λ above we get,

$$\lambda = 2\times 10^{-7}\left[\begin{array}{l} I_a\, ln\dfrac{1}{R_a'} + I_a(-0.5 - j\,0.866)\, ln\dfrac{1}{D_{ab}} + \\ \qquad\qquad I_a(-0.5 + j\,0.866)\, ln\dfrac{1}{D_{ca}}\end{array}\right]$$

$$= 2\times 10^{-7}\cdot I_a\left[ln\frac{1}{R_a'} - ln\frac{1}{\sqrt{bc}} - j\frac{\sqrt{3}}{2}\, ln\frac{b}{c}\right]$$

$\therefore$ Inductance of conductor $\quad 'a' = L_a = \dfrac{\lambda_a}{I_a} = 2\times 10^{-7}\left[ln\dfrac{1}{R'} - ln\dfrac{1}{\sqrt{bc}} - j\dfrac{\sqrt{3}}{2}\, ln\dfrac{b}{c}\right]$

where, $b = D_{ca}$, $c = D_{ab}$

Similarly,

Inductance of conductor $\quad 'b' = L_b = \dfrac{\lambda_b}{I_b} = 2\times 10^{-7}\left[ln\dfrac{1}{R_a'} - ln\dfrac{1}{\sqrt{ca}} - j\dfrac{\sqrt{3}}{2}\cdot ln\dfrac{c}{a}\right]$

where, $a = D_{bc}$, $c = D_{ab}$

and Inductance conductor 'c' $= L_c = \frac{\lambda_c}{I_c} = 2 \times 10^{-7} \left[ln\frac{1}{R_a'} - ln\frac{1}{\sqrt{ab}} - j\frac{\sqrt{3}}{2} ln\frac{a}{b} \right]$

where, $a = D_{bc}$, $b = D_{ca}$

From these expression it is noticed that L_a, L_b, L_c are unequal and they contain imaginary term which is due to the mutual inductance.

3.8 TRANSPOSITION OF TRANSMISSION LINES

As stated above, unsymmetrical spacing of lines causes flux linkage and therefore the inductance of each phase will be different, which results unbalanced receiving end voltages, even though sending-end voltage and line currents are balanced.

Also voltages will be induced in adjacent communication lines even when line current are balanced. This problem is tackled by power system engineers by exchanging position of the conductors at regulars intervals along the line such that each conductor occupies the original position of every other conductor over an equal distance. Such an exchange of conductor position is known as transposition of lines and is shown in Fig. 3.7. This arrangement causes each conductor to have the same average inductance over the transposition cycle. Over the length of one transposition cycle, the total flux linkages, that means net voltage induced in a nearby electrical circles communication/telephone line is zero. This is because over the length of the one complete transposition cycle of power line, the net voltage induced in the telephone line is zero, as stated as it is the sum of three induced voltages which are displaced by 120° in time phase. However, under unbalanced conditions where power currents flow in the earth or in overhead ground wires (zero sequence currents) voltages will be induced in nearby electrical circuit or communication lines and interference will take planer.

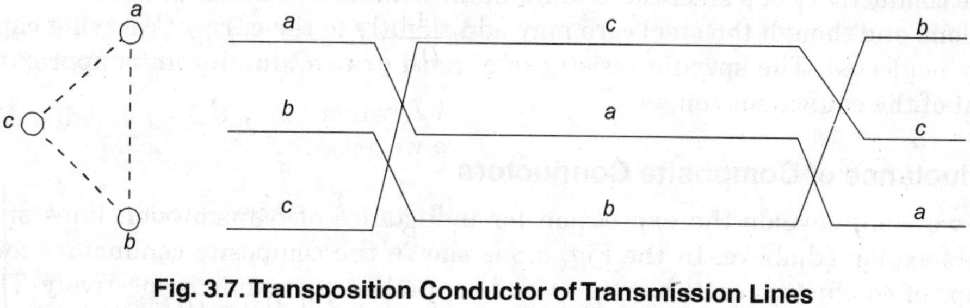

Fig. 3.7. Transposition Conductor of Transmission Lines

In fact modern power lines are normally not transposed. The transposition, however, may be affected at the intermediate switching station.

3.9 COMPOSITE CONDUCTORS FOR TRANSMISSION

In power system for transmission lines operating at high voltages stranded conductors are normally used. These conductors compose of two or more elements and strands and electrically in parallel. As they are composed of two or more segment or element these conductors are known composite conductors.

In fact the conductors used for transmission lines are stranded copper conductors, ACSR (Aluminium Conductors Steel Reinforced) conductors and copper-weld copper conductors. By using different proportions of steel and aluminium strands different tensile and current carrying capacity conductors can be obtained. Then by the use of filler such as paper, between the outer aluminium strands and the inner steel strands, a conductor of large diameter can be obtained for use in high voltages. This type of conductor is known as expanded ACSR.

However, sometimes hollow conductors are used to increase the effective diameters of the conductor so as to reduce corona (which will be discussed later in this book) loss and hence radio interference level. A typical hollow copper conductor (Anaconda) consists of a twisted copper *'I'* beam as a core about which strands of copper wire are wound. The *'I'* copper beam is twisted in a direction opposite to that of the inner layer strands. ACSR – Aluminium Conductor Steel Reinforced in fact, combine the lightness, electrical conductivity and rustlessness of aluminium with the high tensile strength of steel are now employed as overhead conductors on very kind of system, low voltage distribution to the most important long distance transmission lines of the world. The reasons for this be noted as follows.

1. The superior mechanical strength of ACSR can be utilised by using span of larger length which result smaller number of support for a particular length of transmission.

2. Corona losses are reduced because of the larger diameter of the conductor.

3. These conductors are corrosion resistant and are useful under unfavourable conditions of Industrial atmosphere and severe condition of exposure such as may occurs on the saline weather on the sea front.

4. These conductors are normally pretty cheaper than copper conductors of equal resistance and this economy is obtained without sacrificing efficiency reliability of length of useful life.

5. Superior mechanical strength of ACSR can be utilised by using span of larger lengths which result in smaller number of supports for a particular length of transmission line.

Reduction in number of supports obviously involves a corresponding reduction in the total cost of insulators foundation erection and incidentally the costs of maintenance, replacements and stores are also similarly reduced in quantity and cost.

6. The conductivity of a steel-cored aluminium conductor is taken as that of the aluminium portion alone and though the steel core may add slightly to the current carrying capacity, this is usually neglected. The specific resistance of hard drawn aluminium is approximately 1.6 times that of the equivalent copper.

3.9.1 Inductance of Composite Conductors

Now we shall develop the expression for inductance of transmission lines of composite conductors explained above. In the Fig. 3.8 is shown the composite conductors consisting of two groups of conductors each having m and n number of strands respectively. The current here is assumed to be equally divided amongst the strands. One group of conductors act as "go" conductor for the 1-phase line and the other as the "return". The current per strand then is I/m amperes in one group and $-I/n$ ampere in the other.

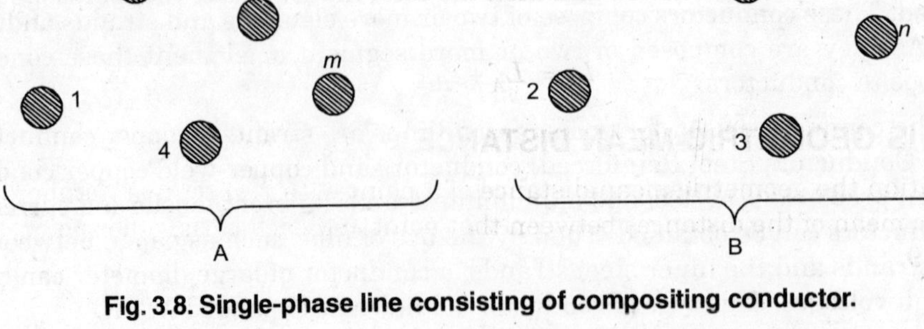

Fig. 3.8. Single-phase line consisting of compositing conductor.

Now by using the equation (3.25), the flux linkage of strand 1 in conductor A is given by

$$\lambda_1 = 2\times10^{-7}\frac{I}{m}\left[ln\frac{1}{R^1}+ln\frac{1}{D_{12}}+...+ln\frac{1}{D_{1n}}\right]-2\times10^{-7}\frac{I}{n}\left[ln\frac{1}{D_{11}^1}+ln\frac{1}{D_{12}^1}+...+ln\frac{1}{D_{1n}^1}\right]$$

$$= 2\times10^{-7}\,ln\frac{\sqrt[n]{D_{11}^1\,D_{12}^1\ldots D_{1n}^1}}{\sqrt[m]{R'\,D_{12}\,D_{13}\ldots D_{1n}}}\ \text{Wb-turns/metre}$$

$$\therefore\ L_1 = \frac{\lambda_1}{Im} = 2m\times10^{-7}\,ln\frac{\sqrt[n]{D_{11}^1\,D_{12}^1\ldots D_{1n}^1}}{\sqrt[m]{R_1^1\,D_{12}\,D_{13}\ldots D_{1m}}} \qquad ...(3.29)$$

Similarly the inductance of segment 2 is conductor A,

$$L_2 = \frac{\lambda_2}{Im} = 2m\times10^{-7}\,ln\frac{\sqrt[n]{D_{11}^1\,D_{12}^1\ldots D_{1n}^1}}{\sqrt[m]{R^1\,D_{21}\,D_{2n}\ldots D_{2m}}} \qquad ...(3.30)$$

Similarly for $L_3, L_4 \ldots L_m$

Now, the average inductance of m strands in conductor A,

$$L_A = \frac{L_{av}}{m} = \frac{L_1+L_2+...+L_m}{m^2} \qquad ...(3.31)$$

Then substituting the values of $L_1, L_2, ..., L_m$ in eqn. (3.31)

$$\text{Hence, } L_A = 2\times10^{-7}\,ln\frac{\sqrt[mn]{(D'_{11},D'_{12}\ldots D_{1n})(D'_2,D'_{22}\ldots D'_{2n})\ldots(D'_{1m}\,D'_{2m}\ldots D'_{mn})}}{\sqrt[m^2]{(R'\,D_{12}\,D_{13}\ldots D_{1m})(R'\,D_{21}\,D_{23}\ldots D_{2m})\ldots(R'\,D_{m1}\,D_{m2}\ldots D_{mm})}} \qquad ...(3.32)$$

Here the mnth root of the product of the mn distances between m strands of conductor A and n strands of conductor B is called geometric mean distance (GMD) and is denoted as D_n and the m^2th root of m^2 distances, that is, the distances of the various strands from one of the strands and the radius of the same strand, the distances of such m grouping constitute m^2 term in the denominator, is called the geometric mean radius (GMR) or self GMD and is denoted as D_s. The expression for inductance of conductor A consisting of m strands from equation (3.32) becomes,

$$L_A = 2\times10^{-7}\,I_n\frac{D_m}{D_s}\ \text{henry/metre} \qquad ...(3.32)$$

Similarly, the inductor of conductor B can also be similarly obtained and then the total inductance of the composite obtained and then the total inductance of the composite conductors is

$$L = L_A + L_B \qquad ...(3.33)$$

3.10 WHAT IS GEOMETRIC MEAN DISTANCE

By definition the geometric mean distance of a point with respect to a number of points is the geometric mean of the distances between that point and each of the other point, as referred to the Fig. 3.9.

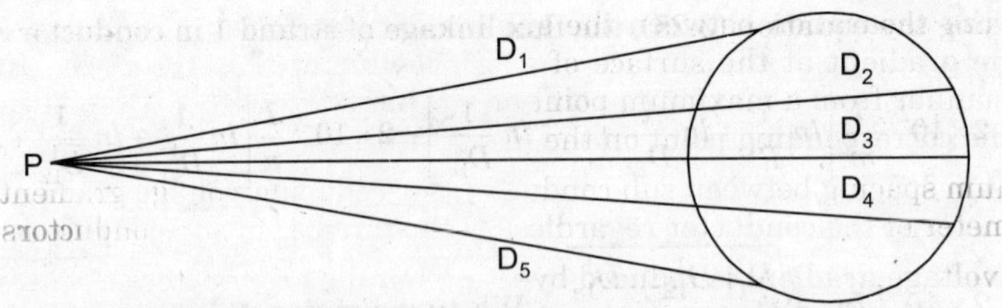

Fig. 3.9. Distances from an external point to five point on a circle.

In fact Geometric Mean Distance is a mathematical concept used for the calculation of Inductance.

With reference to the Fig. 3.9, we can write the geometric mean distance of point P with respect to five points on the circle is

$$GMD_P = 5\sqrt{D_1 D_2 D_3 D_4 D_5} \quad ...(3.34)$$

However, in case the number of points on the circle are increased to infinity, it can be seen intuitively that the geometric mean distance between the point P and the infinite points on the circle will be the geometric mean of all the distances and will correspond to the distance between the point P and centre of the circle.

The concept of GMD is applicable to areas also. The GMD between two circular areas will be distance between the centres of the two areas and such. The GMD method does not apply strictly to non-homogeneous conductors such as ACSR or when the current is not uniformly distributed over the section of the conductor. An approximate value of inductance for ACSR conductors can be calculated by assuming negligible current in the steel strands.

3.11 BUNDLED CONDUCTORS

A bundle conductor is a conductor made up of two or more sub-conductors and is used as one phase conductors. From various important technical factors for voltages in excess of 230 KV, it is in fact not possible to use a round single conductor. Instead of going in for a hollow conductors it is preferable to use more than one conductor per phase which is known as bunding of conductors.

It is found that the increase in transmission capacity justifies economically the use of two conductors bundles on 220 KV lines. The advantages in using bundled conductors are given as below :

(1) Reduced voltage gradient, (2) Reduced Reactance, (3) Reduced corona loss, (4) Reduced radio interference, (5) Reduced surge impedance.

Then the reactance of the bundle conductors is reduced because the self GMD of the conductors is increased and as we know reactance = $K \ln \frac{GMD}{GMR}$ and as GMR is increased the reactance is reduced.

In fact, theoretically, there is an optimum sub-conductor spacing for bundle conductors that will give minimum gradient on the surface for a sub-conductor and hence highest disruptive voltage.

For a two conductor bundle, the equation for maximum gradient at the surface of a sub-conductor is :

$$g = \frac{V(1 + (2r/s))}{2r \ln (d/\sqrt{rs})} \quad ...(3.34a)$$

where "S" is the separation between the sub-conductors. Because of the sub-conductors on each other, the gradient at the surface of a sub-conductor is not uniform. In fact it varies cosinusoidal manner from a maximum point on the outside surface on the line of centres, to a minimum at the corresponding point on the inside surface.

The optimum spacing between sub-conductors for reducing voltage gradient is eight to ten times the diameter of the conductor regardless of the number of sub-conductors per phase.

Since the voltage gradient is reduced by using bundled conductors the radio interference is also reduced. We also know that surge impedance of a line is given by $\sqrt{L/C}$ where L is the inductance and C is the Capacitance per unit length of the line. Since by bundling the self GMD is increased, the inductance is reduced and capacitance increased, as a result the surge impedance is reduced.

This in turn means that the maximum power that can be transmitted is increased. Hence, for large power transmission at higher voltages bundled conductors should be used. For calculating the reactance of the bundled conductor the procedure is same as for composite conductors, as done above. The basic difference between a composite conductor and bundled conductor is that the sub-conductor of a bundle conductor are separated from each other by a distance of almost 30 cm or more and the wires of a composite conductor touch each other.

The bundle conductors generally comprises two, three or four conductors arranged in configuration shown in Fig. 3.10. The current in fact, will not divide equally among the conductors. As said above, the more the number of conductors in a bundle the more is the self GMD reducing considerably the line inductance and hence with advantage of increased transmission capacity of the line. Hence it is economical to transmit large chunks of power over long distances by employing EHV lines. But the line voltages that can be used are severely limited by the corona phenomenon. As we know that the corona is the result of ionisation of the atmosphere when a certain field intensity (about 3000 kV/m at NTP) is reached corona discharge cause interference in nearby electrical circuits including communication circuit and is associated with power loss which reaches severity in bad weather condition. In fact critical line voltage for formation of corona can be raised considerably by the use of bundled conductor. However this increase in critical corona voltage is dependent on number of conductors in the group, the clearance between them and the distance between the groups forming separate phase. Lastly it is to be noted that the optimum spacing of conductor can be of the order of 8 – 10 times the conductor's diameter, irrespective of the number of conductors in the bundle.

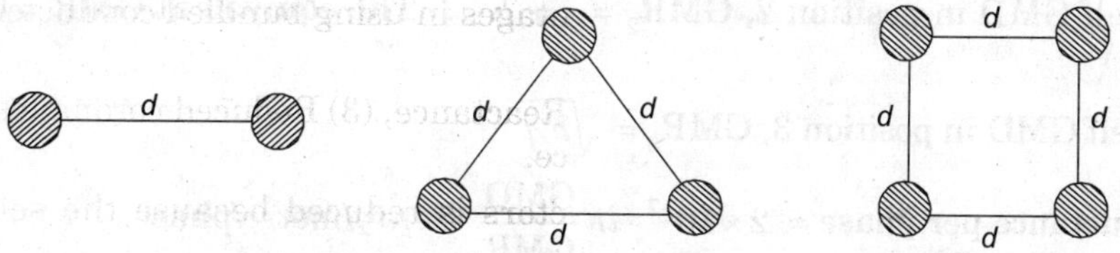

Fig. 3.10. Configuration of bundled conductors (d = distance between conductors).

3.12 INDUCTANCE OF DOUBLE CIRCUIT IN 3-PHASE LINE

In the Fig. 3.11, the layout of a transposed double circuit of a 3-phase transmission line is shown. The double circuit in this figure consists of three conductors in each circuit. The three conductors correspond to three phases a, b, c and a', b', c'. Here conductors, b, b' and c, c' from other phases. This means there are two strands conductors, per phase. Now, since the conductors are not symmetrically placed to calculate the inductance of the line, the conductors should be transposed. The three necessary position have been indicated in the Fig. 3.11.

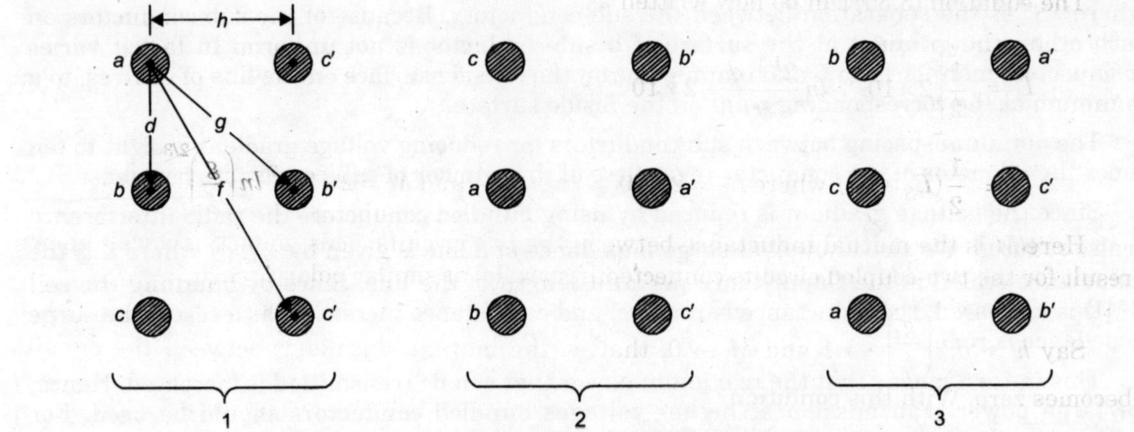

Fig. 3.11. Transposition double circuit line.

Then, GMD of the conductors in phase *'a'* with conductors in other two phases in position (1) shown in the Fig. 3.11.

$\text{GMD}_1 = (d \cdot 2d \cdot h \cdot g)^{1/4} = 2^{1/4} \cdot d^{1/2} (g \cdot h)^{1/4}$

Then, the GMD in the second position (2), for a,

$\text{GMD}_2 = (d \cdot d \cdot g \cdot g \cdot)^{1/4}$ (distances between conductor, $a \cdot b \cdot, a - a', a - c'$ and $a - b'$ and d, d, g, g respectively)

$= d^{1/2} \cdot g^{1/2}$

Then, the GMD in the 3rd position in the Fig. 3.11, for *'a'*, similarly is

$$\text{GMD}_3 = (d \cdot 2d \cdot h \cdot g)^{1/4} = 2^{1/4} \cdot d^{1/2} \cdot g^{1/2} \cdot h^{1/4}$$

Then, the equivalent GMD of the system is given as

$$\text{GMD} = [\text{GMD}_1 \cdot \text{GMD}_2 \cdot \text{GMD}_3]^{1/3} = 2^{1/6} \cdot d^{1/2} \cdot g^{1/3} \cdot h^{1/6}$$

GMR_1 = Self GMD of phase aa' conductors in position 1.

$$\text{GMR}_1 = \text{Self GMD}_1 = \sqrt{r_1' f}$$

GMR_2 = Self GMD in position 2, $\text{GMR}_2 = \sqrt{r'h}$, ∴ The equivalent GMR $= (r^3 \cdot f^2 \cdot h)^{1/6} = r^{1/2} \cdot f^{1/3} \cdot h^{1/6}$

GMR_3 = Self GMD in position 3, $\text{GMR}_3 = \sqrt{r' f}$

Hence inductance per phase $= 2 \times 10^{-7} \cdot ln \dfrac{\text{GMD}}{\text{GMR}}$ Henry/metre/phase.

$$= 2 \times 10^{-7} \cdot ln \frac{2^{1/6} d^{1/2} g^{1/3} k^{1/6}}{r'^{1/2} f'^{2/3} h'^{6}}$$

$$= 2 \times 10^{-7}\, ln \left[2^{1/6} \left(\frac{d}{r'}\right) \right]^{1/2} \left(\frac{g}{f}\right)^{1/3} \text{ H/metre/phase} \qquad ...(3.35)$$

Note : Here, the conductors of two phases are placed diagonally opposite rather than in the same horizontal plane, in all the three position. By doing this the self GMD (GMR) of the conductors is increased where the GMD is reduced and thereby the inductance per phase is lowered.

The equation (3.35) can be now written as,

$$L = \frac{1}{2}\left[2 \times 10^{-7} \cdot ln\frac{2^{1/3} \cdot d}{r^1} + 2 \times 10^{-7}\, ln\left(\frac{g}{f}\right)^{2/3}\right] \quad ...(3.36)$$

$$= \frac{1}{2}(L_S + M), \text{ where } L_S = 2 \times 10^{-7} \cdot ln\frac{2^{1/2} \cdot d}{r^1} \text{ and } M = 2 \times 10^{-7}\, ln\left(\frac{g}{f}\right)^{2/3}$$

Here M is the mutual inductance between the two circuits. This, in fact, is a well known result for the two-coupled circuits connected in parallel at similar polarity each.

Say $h \gg d$, $\left(\frac{g}{f}\right) \to 1$ and $M \to 0$, that is, the mutual impedance between the circuits becomes zero. With this condition

$$L = 1 \times 10^{-7} \cdot ln\frac{3\sqrt{2}\, d}{r^1} \quad ...(3.37)$$

It must be clearly stated that the GMD method applied above though, to a particular configuration of a double circuit is valid for any configuration as large n the circuits are eletrically parallel. While the GMD is valid for fully transpored lines, it is commonly applied for untransposed transmission lines as well giving quite accurate results. Also note that the conductors of two phases are placed diagonally opposite rather than in the same horizontal plane in all the three position shown above. And by doing so the self GMD (GMR) of the conductors is increased whereas the GMD is reduced, and as such the inductance per phase is lowered.

3.13 RESISTANCE

Above we discussed calculation of inductance (in power systems transmission lines) which has tremendous effect in the power system, but comparatively although the contribution of line resistance to series line impedance can be neglected in most cases, in fact, it is the main source of line power loss. Thus while considering transmission line economy, the presence of line resistance must be considered.

Then, the effective resistance is given by

$$R = \frac{\text{Average power loss in conductor in watts}}{I^2} \text{ Ohm} \quad ...(3.38)$$

Here I is the rms current in the conductor in amperes. Ohmic or DC resistance is given by the formula,

$$R_0 = l \cdot \frac{l}{A} \text{ ohms where } l = \text{Resistivity of conductors, ohm} \cdot m \quad ...(3.39)$$

l = length, m

A = Cross-sectional area, m^2

It is to be noted here that the effective resistance is given by eqn. (3.38) which is equal to the DC resitance of the conductor given by the eq. (3.39) but only if the current distribution is uniform throughout the conductor. However in case of some small changes in temperature, the resistance increases with temperature in accordance with the relationship

$$R_t = R_0\,(1 + \alpha\, t) \quad ...(3.40)$$

where R_0 = Resistance at 0°C temperature

α = Temperature coefficient of the conductor at 0°C

R_t = Resistance at t °C temperature

However, the Eqn. (3.40) can be used to find the resistance R_{t_2} at temperature t_2 °C, if resistance R_{t_1} at temperature t_1 °C is known

$$R_{t_2}/R_{t_1} = \frac{1/\alpha_0 + t_2}{1/\alpha_0 + t_1} \quad ...(3.41)$$

3.14 SKIN EFFECT

In a conductor when direct current (DC) flows, the current is uniformly distributed across the section of the conductor when Alternate Current (AC) flows, in the conductor the current is obviously not uniform, with the outer segments of the conductor carrying more current than segment closer to the centre. This means that the current density is higher at the surface (outer segments) of the conductor compared to the current density at its centre. This effect becomes more pronounced as frequency of the supply is increased and the size of the conductor. This phenomenon is known as SKIN EFFECT The inner filaments carrying current give rise to flux which links the inner segments/filaments only, whereas the flux due to current carrying outer filament enclose both the inner as well as the outer film. In fact, this non-uniformity of flux linkage is the main cause of 'Skin Effect'.

To explain this phenomenon further, let us imagine a solid round conductor to be composed of segment/filament of equal cross-sectional area. The flux linking the filament progressively decreases as well more towards the outer filaments for the simple reason that the flux inside a filament cannot/does not link it. The inductive reactance of the imaginary filament therefore decreases outwards with the result that the outer filaments connect more AC (alternating current) than the inner filament which being parallel. With increase of frequency the non-uniformity of inductive reactance of the filament becomes more pronounced so also the non-uniformity of current distribution. However, for large solid conductors the skin effect is quite significant even at 50 Hz. The analytical study of Skin Effect, infact requires the use of Bessel's function.

3.14.1 Proximity Effect

The alternating magnetic flux in a conductor caused by the current flowing in a neighbouring conductor gives rise to circulating current which cause an apparent increase in the resistance of the conductor. This phenomenon is called PROXIMITY EFFECT.

In a two-wire system more lines of flux link elements further apart than the elements nearest to each other. Therefore the inductance of the element further apart is more as compared to the element near each other and the current density is less in the element further apart than the current density in the elements near each other. The effective resistance, is therefore, increase due to non-uniform distribution of current.

The proximity effect is pronounced in case of cables (under ground cable particularly) where the distance between the conductors is small. Whereas for overhead lines with usual spacing the proximity effect is negligibly small.

However, from the above discussion it can be stated that both Skin Effect and proximity effects depend upon conductor size, frequency, distance between conductors and permeability of conductor material.

3.15 SOLVED EXAMPLE ON INDUCTANCE IN TRANSMISSION LINES

After studying the articles above thoroughly following examples can be solved.

Example 3.1. *A conductor is composed of seven identical copper strands each having a rdius "r", as shown in Fig. 3.12. Find the self GMD of the conductor.*

Solution : Referring to the Fig. 3.12 and the articles discussed above on self GMD. The self GMD of the seven conductor is the 49th root of the 49 distance.

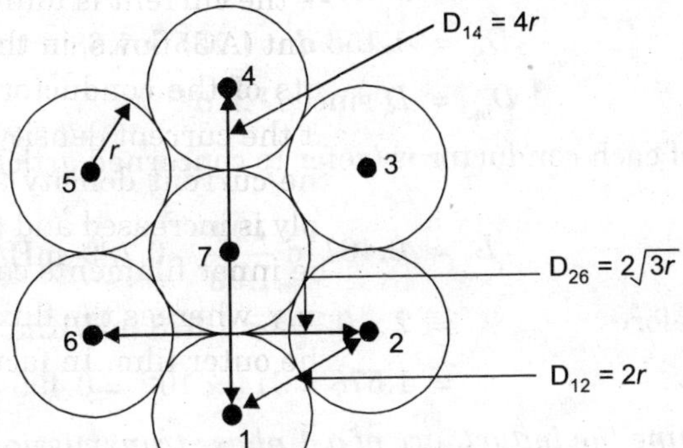

Fig. 3.12. Cross-section of a seven strand conductor.

Hence, $D_s = [(r^1)^7 (D_{12}^2 D_{26}^2 D_{14} D_{17})^6 (2r)^6]^{1/49}$

Now substituting the various distances values (as shown in Fig. 3.12, where r is the radius of each strand) we get,

$D_S = [(0.7788\, r)^7 (2^2 r^2 \times 3 \times 2^2 r^2 \times 2^2 r \times 2r \times 2r)^6]^{1/49}$

or
$$D_S = \frac{2r[3(0.7788]^{1/7}}{6^{1/49}} = 2.177\, r.$$

Example 3.2. *The outside diameter of single layer of aluminium strands of an ACSR conductor shown in Fig. 3.13 is 5.04 cm. The diameter of each strand is 1.68 cm. Determine the reactance at 1 metre spacing 50 Hz neglecting the effect of the centrol strand of steel and advance reason for the same.*

Solution : Again refer to the article discussing the subject, we have got the Fig. 3.13.

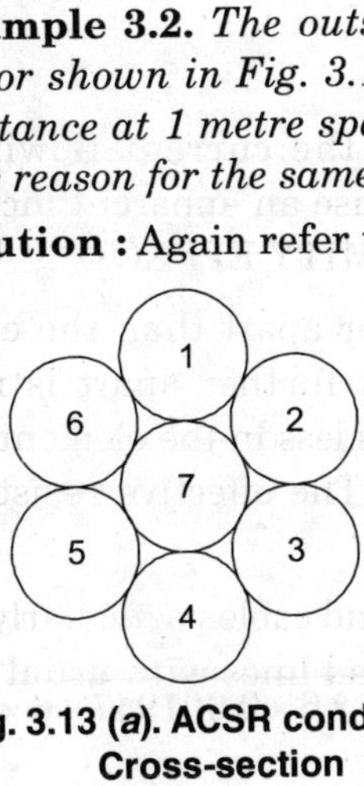

Fig. 3.13 (*a*). ACSR conductor s Cross-section

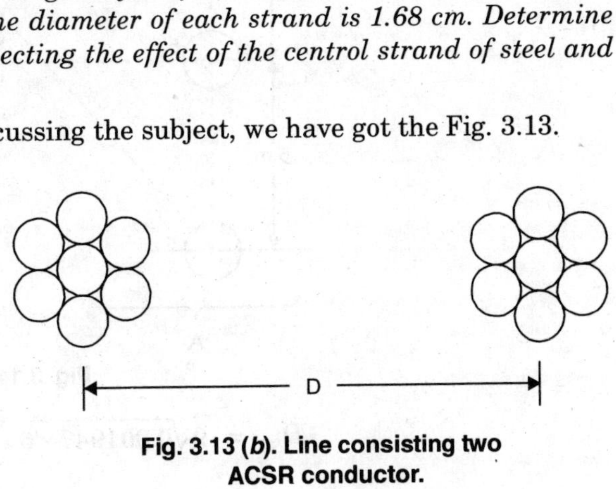

Fig. 3.13 (*b*). Line consisting two ACSR conductor.

It is known that the conductivity of steel being much poorer than that of aluminium and the internal inductance of steel strands being μ-times that of aluminium strands, the current conducted by the central strands of steel can be assumed to be zero.

Hence diameter of steel strand = 5.04 – 2 × 768 = 1.68 cm. Thus all strands are of the same diameters, say d. For the arrangement of strands see Fig. 3.13 (*a*).

$$D_{12} = D_{16} = d, D_{13} = D_{15} = \sqrt{3d},\ D_{14} = 2d$$

$$D_S = \left[\left[\left(\frac{d'}{2}\right)d^2\left(\sqrt{3d}\right)^2 (2d)\right]^6\right]^{1/36}$$

Now substituting $d' = 0.7788\ d$ and simplifying

$$D_S = 1.155\ d = 1.155 \times 1.68 = 1.93 \text{ cm.}$$

$$D_m = D \text{ since } D \gg d$$

Now the inductance of each conductor is (refer to concerned article on subject above)

$$L = 0.461 \ln \frac{100}{1.93} = 0.789 \text{ mH/Km.}$$

Loop inductance therefore $= 2 \times 0.798 = 1.578$ mH/Km.

Loop reactance $= 1.578 \times 314 \times 10^{-3} = 0.495$ ohm/km.

Example 3.3. *Determine the inductance of a 1-phase transmission line consisting of three conductors of 2.5 mm radii in the "Go" conductor and two conductor of 5 mm radii the return conductor. In the Fig. 3.14 the configuration is shown.*

Solution : At first let us find the self GMD of conductor in groop A = 0.001947 m. The self of GMD of conductor in Group A (refer to the Article on Subject).

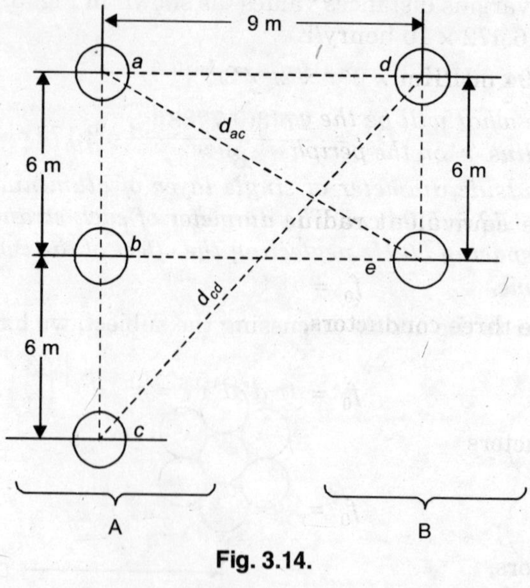

Fig. 3.14.

$$D_{SA} = \sqrt[9]{0.001947 \times 6 \times 12 \times 0.001947 \times 6 \times 6 \times 0.001947 \times 6 \times 12}$$

$$= 0.4809 \text{ metres}$$

Then, $$D_{SB} = \sqrt{5 \times 10^{-3} \times 0.7788 \times 6} = 0.1528 \text{ m.}$$

$$D_{ac} = \sqrt{9^2 + 6^2} = 10.81 \text{ m}, D_{cd} = \sqrt{12^2 + 9^2} = 15 \text{ m}$$

$$D_{mA} = 6\sqrt{9 \times 10.81 \times 10.81 \times 9 \times 15 \times 10.81} = 10.74 \text{ metres}$$

∴ Inductance $$L_A = 2 \times 10^{-7}\, ln\, \frac{10.74}{0.4809} = 0.62 \text{ mH/km.}$$

and $$L_B = 2 \times 10^{-7}\, ln\, \frac{10.74}{0.1528} = 0.8 \text{ mH/km.}$$

∴ Total inductance per km.

$$L_{Total} = L_A + L_B = 1.42 \text{ mH/Km.}$$

Example 3.4. *Determine the inductance of a 3-phase line operating at 50 Hz and conductors arranged as shown in Fig. 3.15. The conductor diameter is 0.8 cm.*

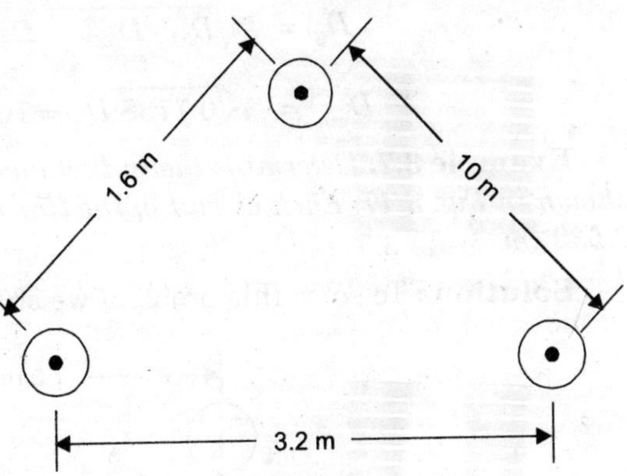

Fig. 3.15

Solution : The self GMDk of the conductor (GMR),

$$\text{GMR} = \frac{0.7788 \times 0.8}{2 \times 100} = 0.003115$$

The mutual GMD of the conductor

$$= 3\sqrt{1.6 \times 3.2 \times 1.6} = 2.015 \text{ m.}$$

$$\therefore \text{ Inductance per km} = 2 \times 10^{-4}\, ln \frac{2.015}{0.00315}$$

$$= 2 \times 6.472 \times 10 \text{ henry/Km}$$

$$= 1.294 \text{ mH/Km.}$$

Example 3.5. *State what will be the equipment radius of a bundle conductor having its part of conductors of radius 'r' on the periphery of a circle of dia 'l', if the number of conductors is 2, 3, 4, 6?*

Solution : Let the equivalent radius or geometric mean radius bc f_0, then for two conductors, we get.

$$f_0 = (rd)^{1/2} = r^{1/2} \cdot d^{1/2}$$

Now when there are three conductors,

$$f_0 = (r\, d'\, d^1)^{1/3} = r^{1/3}\, d^{2/3} \left(\frac{3}{4}\right)^{1/3}$$

Then for four conductors

$$f_0 = \left(r \cdot \frac{d}{\sqrt{2}} \cdot \frac{d}{\sqrt{2}} \cdot d\right)^{1/4} = r^{1/4} \cdot d^{3/4} \cdot \left(\frac{1}{4}\right)^{1/4}$$

And for six conductors,

$$f_0 = r^{1/6} \left(\frac{d}{2}\right)^{5/6} \cdot 6^{1/6} = \left[6r \cdot \left(\frac{d}{2}\right)^5\right]^{1/6}$$

Example 3.6. *A conductor consists of seven identical strands each having radius of r. Determine the factor by which r should be multiplied to find the self GMD of the conductor.*

Solution : From the Fig. 3.16 we can note as below :

$$D_{11} = r_1 \, D_{12} = D_{16} = 2r = D_{17}$$

$$D_{14} = 4r, D_{13} = D_{15} = \sqrt{D_{14}^2 \cdot D_{45}^2}$$

Now as the conductors through 6 tieat the circumference of a circle,

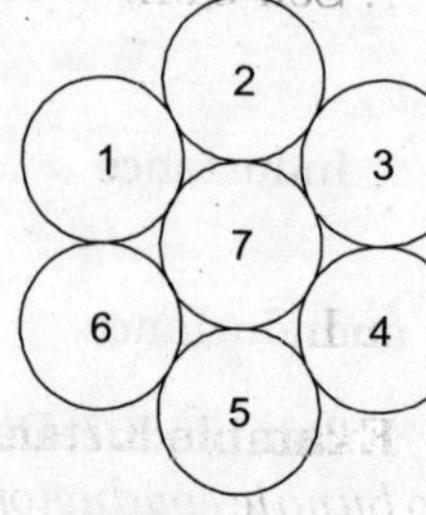

Fig. 3.16

$$\therefore \quad D_{13} = D_{15} = \sqrt{16r^2 - 4r^2} = \sqrt{12r^2} = 2\sqrt{3}r$$

$$\therefore \quad D_{51} = 7\sqrt{r \cdot 2r \cdot 2\sqrt{3}r \cdot 4r \cdot 2\sqrt{3}r \cdot 2r \cdot 2r}$$

$$\therefore \quad 7\sqrt{r^7 .128 \times 3} = 7\sqrt{384 \cdot r}$$

$$D_{57} = 7\sqrt{2r \cdot r \cdot 2r \cdot 2r \cdot 2r \cdot 2r \cdot 2r} = 7\sqrt{64\,r}$$

$$\therefore \quad D_5 = 7\sqrt{D_{51} \cdot D_{52} \ldots\ldots D_{57}} = 49\sqrt{384^6 \times 64} = 49\sqrt{2.04 \times 10^{17}} \cdot 26$$

$$D_{seq} = 7\sqrt{0.7788}\, D_S = 7\sqrt{0.7788} \cdot 49\sqrt{2.05 \times 10^{17}} \cdot r = \mathbf{2.176r.}$$

Example 3.7. *Determine the inductance per km of a transposed double circuit 3-phase line shown in Fig. 3.17. Each circuit of the line remain on its own side. The dia of the conductor is 2.532 cm.*

Solution : To solve this problem we shall refer to Fig. 3.17. Let us first find.

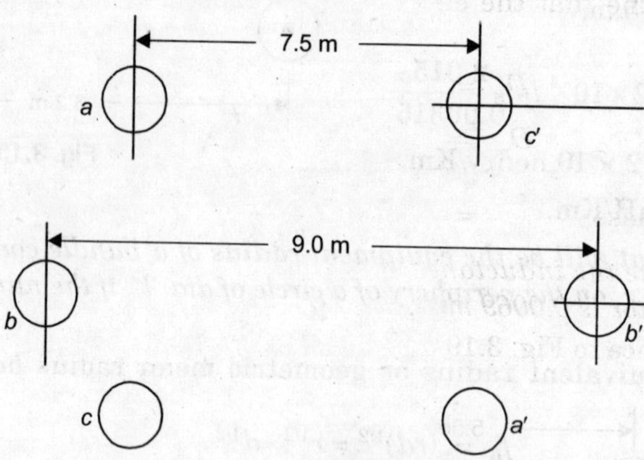

Fig. 3.17

Self GMD of each conductor $= 1.266 \times 0.7788$ cm.

$= 0.00986$ cm.

$$\therefore \quad D_{bc} = D_{ab} = \sqrt{4^2 + 0.75^2} = 4.0697 \text{ m}$$

$$D_{ab}' = \sqrt{4^2 + 8.25^2} = 9.1685 \text{ m}$$

$$D_{aa}' = \sqrt{8^2 + 7.5^2} = 10.965 \text{ m.}$$

$$D_{m_1} = 4\sqrt{4.0697 \times 8 \times 4.0697 \times 9.1685 \times 9.1685} = 6.1084$$

$$D_{m_2} = 4\sqrt{4.0697 \times 4.0697 \times 9.1685 \times 9.1685} = 6.1084$$

$$D_{m_3} = D_{m_1} = 6.878.$$

$$\therefore \quad D_m = 3\sqrt{D_{m_1} \cdot D_{m_2} \cdot D_{m_3}} = 6.61 \text{ m.}$$

$\therefore$ Self GMD of each phase $D_{S_1} = \sqrt{0.00986 \times 10.965} = 0.3288 = D_{S_3}$

$$D_{S_2} = \sqrt{0.00986 \times 9} = 0.29789$$

$$\therefore \quad D_S = 3\sqrt{D_{S_1} \cdot D_{S_2} \cdot D_{S_3}} = 0.318 \text{ m.}$$

$\therefore$ Inductance $= 2 \times 10^{-4}\, I_n \dfrac{6.61}{0.318}$ H/km/phase = 0.606 mH/km/phase

Example 3.7. *Determine the inductance per km per phase of a single circuit 460 kV line using two bundle conductors per phase as shown in Fig. 3.18. The dia of each conductor is 5.0 cm.*

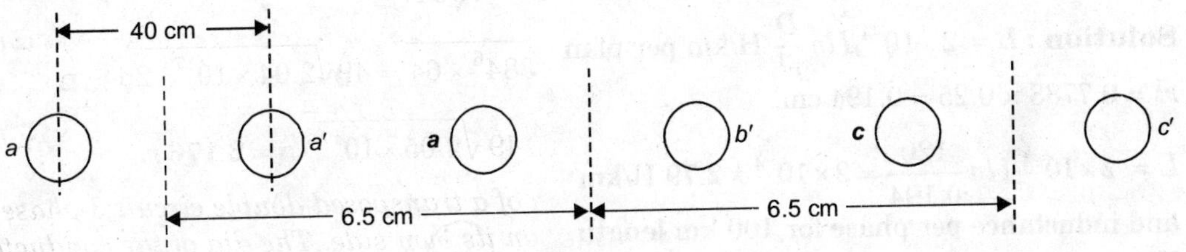

Fig. 3.18

Solution : Let us assume that the effect of transposition to be negligibly small in this arrangement :

$$\therefore \quad D_S = \sqrt{0.025 \times 0.4 \times 0.7788} = 0.08825$$

$$D_m = 3\sqrt{6.5 \times 13.0 \times 6.5} = 8.19$$

$\therefore$ Inductance per Km/phase $= 2 \times 10^{-4}\, I_n \dfrac{8.19}{0.8825} = 0.908$ mH/km/phase

Example 3.9. *Determine the inductance of the double circuit line shown in the Fig. 3.19. The self GMD of the conductor is 0.0069 metre.*

Solution : With reference to Fig. 3.19 and the article on the above subject, we find :

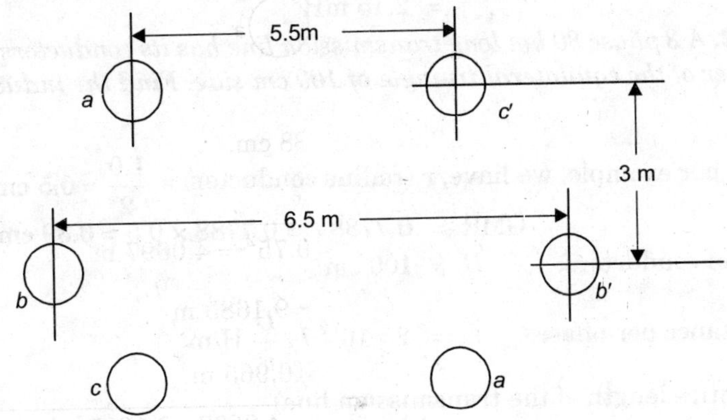

Fig. 3.19

$$D_{ab} = D_{bc} = \sqrt{3^2 + 0.5^2} = 3.04 \text{ m.}$$

$$D_{ac} = 6 \text{ m.}$$

$$D_{ab}' = \sqrt{3^2 + 6^2} = 6.708 \text{ m}$$

$$D_{aa}' = \sqrt{6^2 + 5.5^2} = 8.14 \text{ m}$$

$$D_{m_1} = 4\sqrt{3.04 \times 6 \times 5.5 \times 6.708} = 5.09 \text{ m} = D_{m_3}$$

$$D_{m_2} = 4\sqrt{3.04 \times 3.04 \times 6.708 \times 6.708}$$

$$= 4.515 \text{ m}$$

$$\therefore \quad D_m = 4.89 \text{ m}$$

Example 3.10. *A three phase transmission line of 100 km long has its conductor of 0.5 cm dia, spaced at the cones of an equilateral triangle of 200 cm side as shown in Fig. 3.20. Find the inductance per phase of the system.*

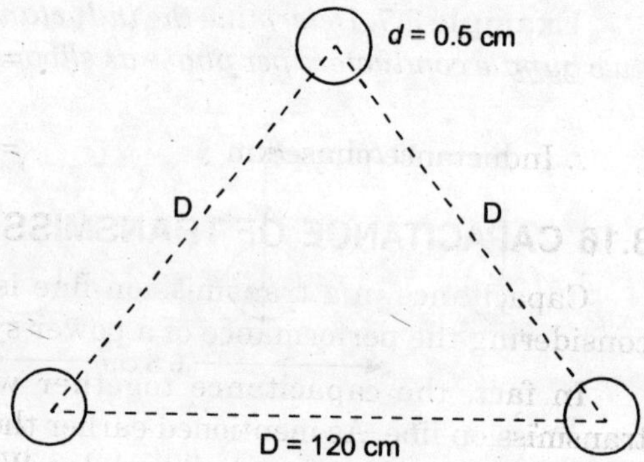

Fig. 3.20

Solution : $L = 2 \times 10^{-4}\, I\, ln \dfrac{D}{r^1}$ H/km per plan

$r^1 = 0.7788 \times 0.25 = 0.194$ cm.

$$L = 2 \times 10^{-4}\, I\, ln \frac{120}{0.194} = 2 \times 10^{-4} \times 2.79 \text{ H/km}$$

and inductance per phase for 100 km length.

Example 3.11. *A single phase line of 230 V has conductor spacing of 135 cm. The radius of conductor is 0.8 cm. Calculate the loop inductance in mH of the line per km.*

Solution : As per example, spacing between the conductors,

$$D = 135 \text{ cm}$$

Radius of conductors, $r = 0.8$ cm

Loop inductance of line $L = \left(1 + 4\, I\, ln \dfrac{D}{r}\right) \times 10^{-7}$ H/m

$$= \left(1 + 4\, I\, ln \frac{135}{0.8}\right) \times 10^{-7} \times 1000$$

$$= 2.15 \text{ mH}$$

Example 3.12. *A 3 phase 80 km long transmission line has its conductors of 1.0 cm diameter spaced at the corner of the equilateral triangle of 100 cm side. Find the inductance per phase of the system.*

Solution : As per example, we have, r (radius conductor) $= \dfrac{1.0}{2} = 0.5$ cm

$\therefore$ GMR $= 0.7788\, r = 0.7788 \times 0.5 = 0.39 \text{ cm} = r^1$

Then spacing of conductors, $D = 100$ cm

Now, as inductance per phase, $L = 2 \times 10^{-7}\, I_n \dfrac{D}{r'}$ H/m

$\therefore$ L (for the entire length of the transmission line)

$$= 2\times10^{-7}\ln\left(\frac{100}{0.39}\right)\times 80\times1000$$

$$= 0.8875 \text{ H. (as length of line = 80 km)}$$

Example 3.13. *Calculate the inductance of each conductor of 3-phase, 3-wire system when the conductors are arranged in a horizontal plane with spacing such that D_{31} = 4 m, $D_{12} = D_{20}$ = 2 m. The conductors are transposed and have a diameter of 3.00 m.*

Solution : The Fig. 3.21 shown the arrangement of conductors of 3-phase line. The conductor radius, r = 3/2 = 1.5 cm.

Equivalent equilateral spacing, $D_{eq} = (D_{12} \times D_{23} \times D_{31})^{1/9} = (2 \times 2 \times 4)^{1/9} = 2.52$ m.

∴ Inductance/phase/metre, $L = [0.5 + 2 \ln (D_{eq}/r)] \times 10^{-7}$ H

$$= \left(0.5+2\ln\frac{252}{1.5}\right)\times10^{-7} \text{ H} = 10.75\times10^{-9} \text{ H}$$

∴ Inductance/phase/km $= 10.75 \times 10^{-7} \times 1000 =$ **1.11 mH.**

3.16 CAPACITANCE OF TRANSMISSION LINE

Capacitance in a transmission line is so important that it has to be accounted for while considering the performance of a power system.

In fact, the capacitance together with conductance form the shunt admittance of a transmission line. As mentioned earlier the conductance is the result of leakage over the surface of insulators and is of negligible order. When an alternating voltage is applied to the line, the line capacitance draws a leading sinusoidal current called the charging current which is drawn even when the line is open circuited at the far and. The line capacitance being proportional to its length, the charging current is for lines less than 100 km long. But as said for larger lines it becomes an important factor.

In fact the flow of current through a conductor gives rise to a magnetic field and charging of conductor result in an electric field. A charge if brought in the vicinity of this electric field experiences a force. The intensity of this field at any point is defined as the force per unit charge and is termed as electric field intensity designated as E. The units of this field are newton per coloumb or volts per metre. The direction of electric field intensity is the same as the direction of the force experienced by the unit charge. As we are dealing with the transmission line conductors we can start discussion with this electric field due to infinite line of charge.

3.16.1 Electric Field of Infinity Long Straight Conductor

Let us imagine an infinitely long straight conductor removed reduced from other conductors including Earth, and carrying a uniform charge q coloumb/metre length.

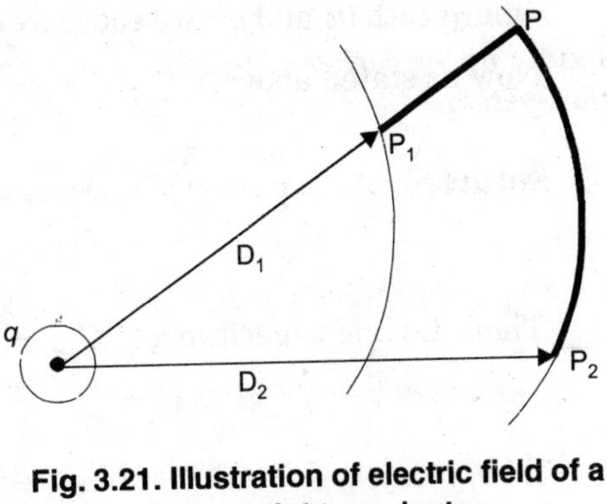

Fig. 3.21. Illustration of electric field of a long straight conductor.

We also know, from basic principles of electrostatic (can be found in detail in Electrical Engineering Technology, book meant for 1st year and 2nd year B.Tech course), by symmetry, the equipotential surfaces will be concentric cylinders when the line of electrostatic stress will be radial. Then, the electric field intensity at a distance say '*y*' from the axis of the conductor is $q/2\,\pi\,\varepsilon\,y$, where ε is the permittivity (In SI unit the permittivity of free space is $\varepsilon_0 = 0.85 \times 10^{-12}$ F/m, and relative permittivity for air $= \varepsilon_r = \varepsilon/\varepsilon_0 = 1$) of the medium.

Now, as can be seen in the Fig. 3.21, there are two points, say P_1 and P_2 located at distances, say, D_1 and D_2 respectively from the axis of the conductors. Then the potential difference V_{12}, between P_1 and P_2 is given by as :

$$V_{12} = \int_{D_1}^{D_2} \varepsilon d\, d_y = \int_{D_1}^{D_2} \frac{q}{2\,\pi\,\varepsilon\,y} \cdot d_y \text{ V} \qquad \text{...(3.42a)}$$

[As the potential difference is independent of the path, we can choose the path of integration since the path. PP_2 lies along an $P_1\,PP_2$ shown in sold line in Fig. 3.21 along a equi-potential medium. V_{12} is simply obtained by integrating along P_1P).

∴ By integrating the eqn. (3.42)

We get,
$$V_{12} = \frac{q}{2\pi\,\varepsilon}\, ln\, \frac{D_2}{D_1} \text{ Volts} \qquad \text{...(3.42b)}$$

where q is the instantaneous charge of the conductor in coloumbs/metre length.

3.16.2 Capacitance of A Two-Conductor Line

Now let us consider a two-conductor/wire line as shown in Fig. 3.22 which is being excited from a 1-phase source. Then the line develops equal and opposite sinusoidal charges on the two conductors which can be represented by q_a and q_b. This means $q_a = -q_b$.

Now, V_{ab} (due to q_a) $= \dfrac{q_a}{2\pi\varepsilon}\, ln \dfrac{D}{r_a}$ (refer to eqn. 3.42 *b*)

And, similarly V_{ab} (due to q_b) $= \dfrac{q_b}{2\pi\varepsilon}\, ln\, \dfrac{r_b}{D}$

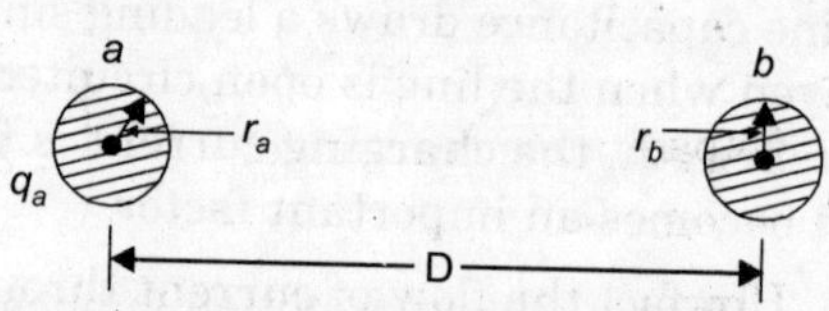

Fig. 3.22. Cross-sectional view of a two-conductor line.

These are in phasor notation hence we get,

$$V_{ab} = \frac{1}{2\,\pi\,\varepsilon}\left(q_a \cdot ln\frac{D}{r_a} + q_b\, ln\,\frac{r_b}{D}\right) \text{ volts}$$

where both q_a and q_b are complex quantity.

Now as stated above, $q_a = -q_b$ (for a two-wire line)

we get,
$$V_{ab} = \frac{q_a}{2\pi\varepsilon}\, ln\left(\frac{D^2}{r_a\,r_b}\right) = \frac{q_a}{\pi\,\varepsilon} \cdot ln\,\frac{D}{\sqrt{r_a\,r_b}} \qquad \text{...(3.43)}$$

Then the line capacitance, $C_{ab} = \dfrac{q_a}{V_{ab}} = \dfrac{\pi\,\varepsilon}{ln\,(D/\sqrt{r_a r_b})}$ (by using eqn. 3.43 for V_{ab})

Now, if say, $r_a = r_b = r$ then, $C_{ab} = \dfrac{\pi\,\varepsilon}{ln\,(D/r)}$ farads/metre ...(3.44)

Here, in this expression C_{ab} denotes the capacitance between conductors a and b, that means between the lines. And $\varepsilon = \varepsilon_0\ \varepsilon_r$. Now if we assume $\varepsilon_r = 1$ then $\varepsilon = \varepsilon_0 = \frac{1}{4\pi \times 9 \times 10^9}$ farads/metre. The substituting this value of ε in eqn. (3.44), we get

$$C_{ab} = \frac{0.0121}{\log(D/r)} \cdot \mu\,\text{F/km} \qquad \text{...(3.45)}$$

Now to determine the capacitance of each line to neutral let us see Fig. 3.23 (a) and (b) where the line-to-line capacitance can be equivalently considered as two equal capacitances in series.

Then the voltage across the lines divides equally between the capacitances so that natural point 'n' is at the ground potential so the capacitance of each line to neutral is given by :

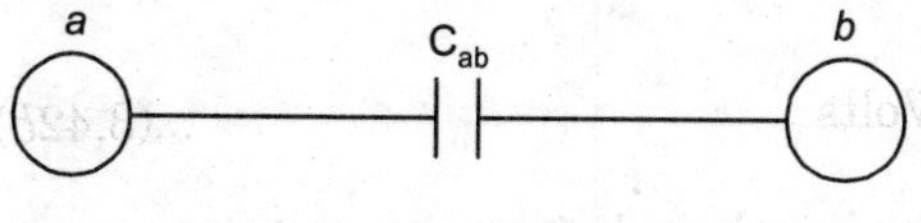

Fig. 3.23 (*a*). Line-to-line capacitance.

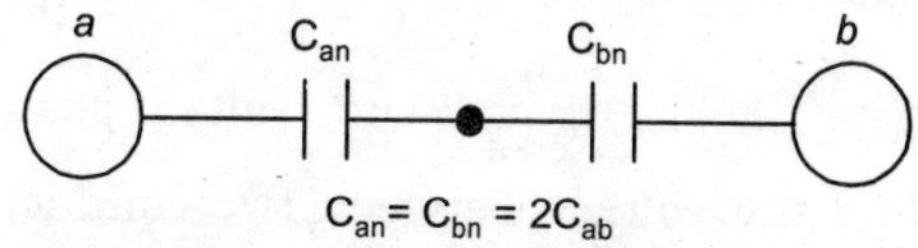

Fig. 3.23 (*b*). Line-to-line capacitance.

$$C_n = C_{an} = C_{bn} = 2C_{ab} = 2 \times \frac{0.0121}{\log(D/r)}$$

$$= \frac{0.0242}{\log(D/r)}\ \mu\text{F/km} \qquad \text{...(3.46a)}$$

Capacitance reactance, $\left(\therefore X_{cn} = \frac{1}{2\pi C_n}\right)$ ...(3.46b)

However, the assumption applicable in deriving the above equation are as below :

1. The charge on the surface of each conductor is being assumed to be uniformly distributed, but this is in practice, strictly not currect. However, if non-uniformity of charge distribution is considered, then,

$$C_n = \frac{0.0242}{\log\left(\frac{D}{2r} + \left(\frac{D^2}{4r^2} - 1\right)\right)}\ \mu\text{F/Km} \qquad \text{...(3.47)}$$

Here, if $D/2\,r \gg 1$ then we can see that the eqn. (3.47) reduces to that of eqn. (3.46a) and the error caused by the assumption of uniform charge distribution, is negligibly small.

2. The cross-section of both the conductors is assumed to be circular while in practice stranded conductors are used although, these strands are put together to in a circular form, Hence the use of the radius of the circumscribing circle for stranded conductor causes insignificant error.

Hence, it can be concluded that the expression of eqn. (3.46a) can be considered almost correct.

Another point must be noted here that the concept of self GMD as applicable for Inductance Calculation of transmission line is not applicable for the capacitance calculation. This is because the equation for inductance contains a constant term corresponding to the internal flux linkages whereas since charges reside on the surface of the conductor, similar term is absent in the capacitance expression. Because of this, the radius in the expression for capacitance is the actual outside radius of the conductor, whereas for inductance calculation the radius is the self GMD of the conductor.

3.17 POTENTIAL DIFFERENCE BETWEEN GROUP OF PARALLEL CONDUCTORS

When a number of conductros are arranged in such a manner that they are parallel to each other, then the voltage between any two of them can be found by applying the equation (3.42) (developed above for V_{12}), $V_{12} = \frac{q}{2\pi\varepsilon} \ln \frac{D_2}{D_1}$ volts repeatedly to find the voltage between two conductors in question due to the charge on each conductor in the group independently. Then the voltage drop between the two conductors is the sum of the voltage drops due to each charged conductors. This is the basic principle on which is based the method during an expression for such capacitance.

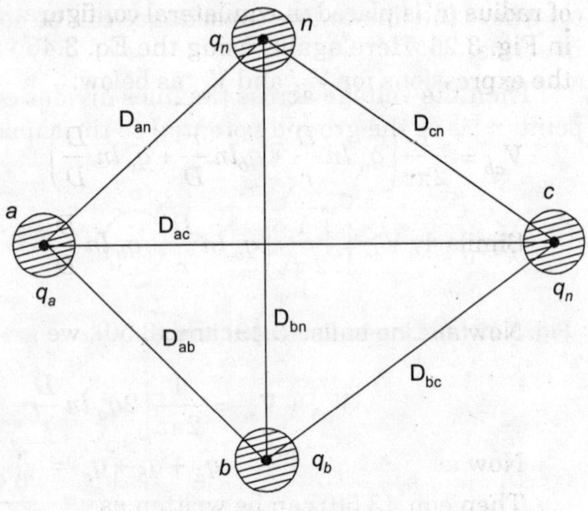

Fig. 3.24. A group of parallel conductors.

In the Fig. 3.24 is shown a group of parallel charged conductors. Let us assume that there are no other charged surfaces in the vicinity the sum of charges on conductors is zero. Then it is also assumed that the conductors are far removed from the ground (to have negligible effect) and are sufficiently removed from each other (which means the conductor radii are much smaller than the distances between the conductors). Practically also the spacing commonly used in overhead power transmission lines always meet these conditions. Moreover these assumptions imply that the charge on each conductor remains uniformly distributed around its periphery and length.

Then the potential difference between any two conductors of the group can be obtained by adding the contribution of the individual charged conductors by repeated application of eqn. (3.42).

Hence the potential difference between conductors a and b (that is voltage drop from a to b), in phasor notation is :

$$V_{ab} = \frac{1}{2\pi\varepsilon}\left(q_a \ln\frac{D_{ab}}{r_a} + q_b \ln\frac{r_b}{D_{ba}} + q_c \ln\frac{D_{cb}}{D_{ca}} + \ldots + q_n \ln\frac{D_{nb}}{D_{na}}\right) \quad \ldots(3.46)$$

In this equation each term is the potential drop from a to b caused by charge on one of the conductors of the group. Same way expressions on similar lines could be written for voltage drop between any two conductor of the group. However, if the charges vary sinusoidally so the voltages also change (this is the property of AC transmission line), the equation (3.46) still applies with changes per metre length, and voltages are regarded as phasors.

Now if the voltages between conductors and other conductors in the group are known the group of simultaneous equation expressing the voltage drops in terms of changes may be sloved for the changes, as the distances are known.

3.18 CAPACITANCE OF A 3-PHASE LINE WITH EQUAL SPACING

A 3-phase line composed of three identical conductors of radius 'r' is placed in equilateral configuration as shown in Fig. 3.25. Here again, using the Eq. 3.46 we can write the expressions for V_{ab} and V_{ac} as below:

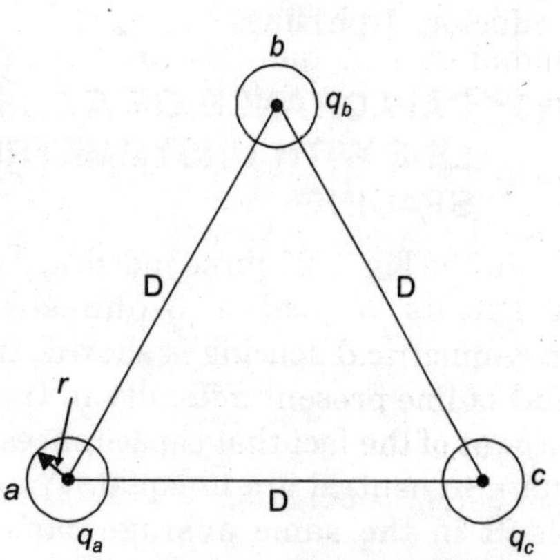

Fig. 3.25. Cross-section of 3-phase line with equilateral spacing.

$$V_{ab} = \frac{1}{2\pi\varepsilon}\left(q_a \ln\frac{D}{r} + q_b \ln\frac{r}{D} + q_c \ln\frac{D}{D}\right) \quad ...(3.49)$$

Similarly $$V_{ac} = \frac{1}{2\pi\varepsilon}\left(q_a \ln\frac{D}{r} + q_b \ln\frac{D}{D} + q_c \ln\frac{r}{D}\right) \quad ...(3.50)$$

Now adding eqns. (3.49) and (3.50), we get

$$V_{ab} + V_{ac} = \frac{1}{2\pi\varepsilon}\left[2q_a \ln\frac{D}{r} + (q_b + q_c)\ln\frac{r}{D}\right] \quad ...(3.51)$$

Now as $q_a + q_b + q_c = 0$, we get $q_a = -(q_b + q_c)$ or $(q_b + q_c) = -q_a$

Then eqn. (3.50) can be written as

$$V_{ab} + V_{ac} = \frac{1}{2\pi\varepsilon}\left[2q_c \ln\frac{D}{r} - q_a \ln\frac{r}{D}\right] = \frac{3q_a}{2\pi\varepsilon}\ln\frac{D}{r} \text{ Volts} \quad ...(3.52)$$

Now with balanced 3-phase voltages being applied to the line, t follows from the phasor diagram of Fig. 3.26, that

$$V_{ab} + V_{ae} = 3V_{an} \quad ...(3.53)$$

Now substituing the value of $V_{ab} + V_{ae}$ from eqn. (3.53) in eqn. (3.52), we have,

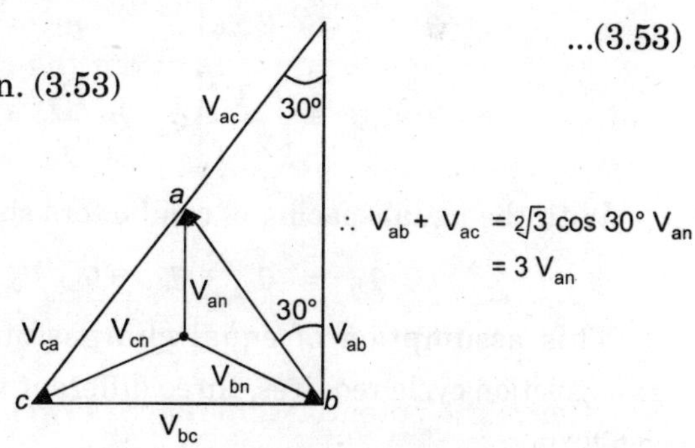

Fig. 3.26. Phasor diagram of balanced 3-phase.

$$3 V_{an} = \frac{3q_a}{2\pi\varepsilon} \cdot \ln\frac{D}{r} \text{ volts}$$

$$\therefore \quad V_{an} = \frac{q_a}{2\pi\varepsilon} \ln \cdot \frac{D}{r} \quad ...(3.54)$$

Hence $$C_{an} = \frac{q_a}{V_{an}} = \frac{q_a}{(q_a/2\pi\varepsilon)\ln\frac{D}{r}} = \frac{2\pi\varepsilon}{\ln\frac{D}{r}}$$

For air medium $\varepsilon_r = 1$

$$\therefore \quad C_n = \frac{0.0242}{\log(D/r)} \ \mu\text{F/Km} \quad (3.55)$$

Hence line changing current of phase a, I_a, can be found as $= \frac{V_{an}}{X_{cn}}$

As $$X_{cn} = \frac{1}{C_n \cdot 2\pi f} = \frac{1}{c_n \cdot \omega}$$

$\therefore \qquad I_a \text{ (line changing)} = j\,\omega\, C_n\, V_{an}$

Note that while calculating C_n assumption are same as in case of calculating C_{ab} for two-conductor, (1-ph) line.

3.19 CAPACITANCE OF A 3-PHASE LINE WITH UNSYMMETRICAL SPACING

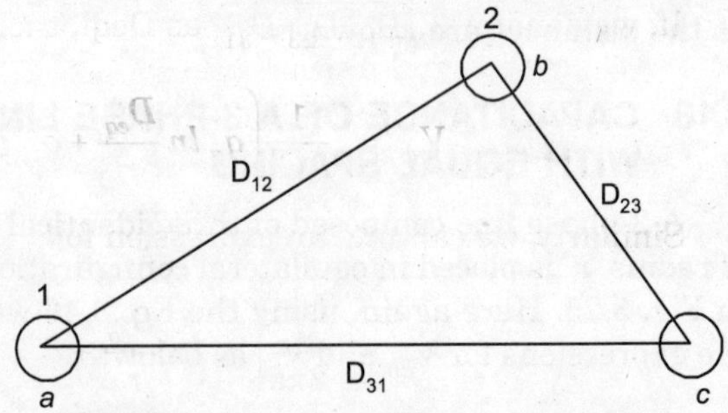

Fig. 3.27. Cross-section of a 3-phase line assymetrically spaced (fully transposed).

In the Fig. 3.27 three indentical conductors of radius '*r*', of a 3-phase line with unsymmetrical spacing is shown. In fact, this kind of line present difficulty in treatment on account of the fact that capacitances from each phase to neutral are unequal. Transpositions result in the same average capacitance to neutral of various phases over the entire length of the line.

Hence, we may calculate the capacitance of each phase of an untransposed line as the average value of the phase of the same line correctly transposed with negligible error on account of slight dissymmetry is actual practice.

Now as the conductors are rotated cyclically in the three sections of the transposition cycle, corresponding three expressions can be for V_{ab} can be written in three position as $V_{ab_1}, V_{ab_2}, V_{ab_3}$ respectively, so we have,

$$V_{ab_1} = \frac{1}{2\pi\varepsilon}\left[q_a\, ln\frac{D_{12}}{r} + q_b\, ln\frac{r}{D_{12}} + q_c\, ln\frac{D_{23}}{D_{31}}\right]$$

$$V_{ab_2} = \frac{1}{2\,\pi\,\varepsilon}\left[q_a\, ln\frac{D_{23}}{r} + q_b\cdot ln\frac{r}{D_{23}} + q_c\, ln\frac{D_{31}}{D_{12}}\right]$$

$$V_{ab_3} = \frac{1}{2\,\pi\,\varepsilon}\left[q_a\cdot ln\frac{D_{31}}{r} + q_b\, ln\frac{r}{D_{31}} + q_c\, ln\frac{D_{12}}{D_{23}}\right]$$

With the usual spacing of conductors sufficient accuracy is obtained by assuming,

$$q_{a_1} = q_{a_2} = q_{a_3} = q_a;\ q_{b_1} = q_{b_2} = q_{b_3} = q_b;\ q_{c_1} = q_{c_2} = q_{c_3} = q_c.$$

This assumption of equal charges/unit length of a line in the three sections of the transposition cycle requires, three different values of V_{ab} as V_{ab_1}, V_{ab_2} and V_{ab_3} in three positions as above.

Hence, the solution can be considerably simplified by taking V_{ab} as the average of three voltages.

That is, $$V_{ab\,(\text{average})} = \frac{1}{3}(V_{ab_1} + V_{ab_2} + V_{ab_3}) \qquad \text{...(3.56)}$$

Hence, by putting the values of $V_{ab_1}, V_{ab_2}, V_{ab_3}$ in this equation we get

$$V_{ab} = \frac{1}{6\pi\varepsilon}\left(q_a\, ln\frac{D_{12}\,D_{23}\,D_{31}}{r^3} + q_b\, ln\frac{r^3}{D_{12}D_{23}D_3} + q_c\, ln\frac{D_{12}\,D_{23}\,D_{31}}{D_{12}D_{23}D_{31}}\right)$$

$$= \frac{1}{2\pi\,\varepsilon}\left(q_a\, ln \frac{3\sqrt{D_{12} D_{23} D_{31}}}{r} + q_b\, ln \frac{r}{3\sqrt{D_{12}\, D_{23}\, D_{31}}}\right)$$

If, we denote, $3\sqrt{D_{12}\, D_{23}\, D_{31}}$ as Deqi. then

$$V_{ab} = \frac{1}{2\pi\varepsilon}\left(q_a\, ln \frac{D_{eq}}{r} + q_b ln \frac{r}{D_{eq}}\right) \quad ...(3.57)$$

Similarly, we can find an expression for V_{ac} as :

$$V_{ac} = \frac{1}{2\pi\varepsilon}\left(q_a\, ln \frac{D_{eq}}{r} + q_c\, ln \frac{r}{D_{eq}}\right) \quad ...(3.58)$$

Now from Fig. 3.24, we know $V_{ab} + V_{ac} = 3\, V_{an}$ and also we know, $q_b + q_c = -\, q_a$. Therefore, $V_{an} = \frac{1}{3}(V_{ab} + V_{ac})$, and putting the expression of $V_{ab} + V_{ac}$, we get,

$$V_{an} = \frac{1}{6\pi\varepsilon}\left(2q_a\, ln \frac{D_{eq}}{r} - q_c\, ln \frac{r}{D_{eq}}\right) = \frac{1}{2\,\pi\varepsilon} q_a\, ln \frac{D_{eq}}{r} \text{ volts}$$

And hence

$$C_n = \frac{q_a}{V_{an}} = \frac{2\,\pi\varepsilon}{ln\,(D_{eq}/r)} \text{ F/m} \quad ...(3.59)$$

or for Air medium, $\Sigma_r = 1$, we get

$$C_n = \frac{0.0242}{\log\,(D_{eq}/r)}\ \mu\text{F/Km} \quad ...(3.59\ (a))$$

Now it can be noted here that for equilateral spacing $D_{eq} = D_1$ the expression in Eq. (3.59 (*a*)) for C_n gives the exact result developed earlier.

Hence, the line charging current, in '*a*', $I_q = j\,\omega\, C_n\, V_{an}$ A/Km ...(3.60)

3.20 EFFECT OF EARTH ON CAPACITANCE OF TRANSMISSION LINE

While calculating the capacitance of transmission lines above, we have in fact ignored the presence of earth effect. It is known that the presence of earth alters the electric field of the line as a result its capacitance is also affected. In fact earth may be assumed to be a perfect conductor in the form of horizontal plane of infinite extent. Therefore the electric field of charged conductors is faced to conform to a presence of thin equipotential surface.

However, the effect of earth on capacitance can be conveniently taken into account by the method of images, which was originally suggested by Lord Kelvin of Glasgow University, U.K. Briefly it can be said in this method of image we equivalently replace the presence of ground below a charged conductor, simply by a fictitious conductor equal and opposite charges and located as fave below the surface of ground as the obverhead conductor above it. Such a fictitious conductors the "mirror image" of the overhead conductor. This method of creating the same electric field as in the presence of earth is known as the method of image, which as said, was originally suggesed by Lord Kelvin.

Now to calculate the effect of Earth on conductor of transmission line, we shall consider both (1) 1-phase line and (2) 3-phase line.

3.20.1 Capacitance of a 1-Phase Line (With Earth Effect)

Let us consider a 1-phase line shown in the Fig. 3.28, where two conductors 'a' and 'b', with radius 'r' for each conductor placed at height 'h' above Earth, that is ground. In this case, hence, for finding V_{ab} we shall take into account charges q_a and q_b due to conductors a and b, above the ground, and charges $-q_a$, $-q_b$ for the images of conductors a and b. That is a' and b' respectively below the ground at the same height h, as if they are above the ground.

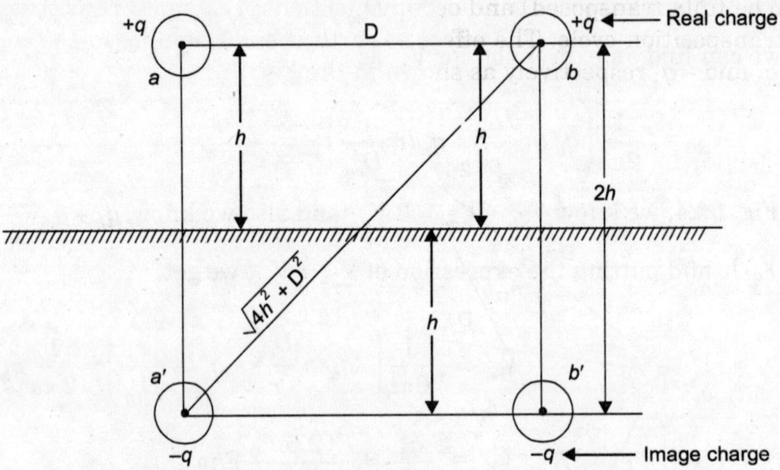

Fig. 3.28. 1-ph transmission line with images.

Now by following the same method of calculation in case of capacitance of a two-wire line, we can write;

$$V_{ab} = \frac{1}{2\pi\varepsilon}\left[q_a \ln\frac{D}{r} + q_b \ln\frac{r}{D}\right] + \frac{1}{2\pi\varepsilon}\left[(-q_a)\ln\frac{(4h^2+D^2)^{1/2}}{2h} - q_b \ln\frac{2h}{(4h^2+D^2)^{1/2}}\right]$$

$$= \frac{1}{2\pi\varepsilon}\left[q_a \ln\frac{2hD}{r\sqrt{D^2+4h^2}} + q_b \ln\frac{r\sqrt{D^2+4h^2}}{2hD}\right]$$

But as $q_a + q_b = 0$, $\therefore\ q_a = -q_b$ by simplifying the expression we get,

$$V_{ab} = \frac{q}{\pi\varepsilon}\ln\frac{2hD}{r(4h^2+D^2)^{1/2}} \quad ...(3.60)$$

$$\therefore \quad C_{ab} = \frac{\pi\varepsilon}{\ln\dfrac{D}{r\left(1+(D/2h)^2\right)^{1/2}}} \text{ F/m line to line} \quad ...(3.61)$$

and $$C_n = 2\times C_{ab} = \frac{2\pi\varepsilon}{\ln\dfrac{D}{r(1+D^2/4h^2)^{1/2}}} \text{ F/m to neutral} \quad ...(3.62)$$

3.20.2 Capacitance of 3-Phase Line (With Earth Effect)

For calculation of capacitance of a 3-phase line with earth effect, the method of images can similarly be applied as done is case of 1-phase line.

In the Fig. 3.29, the method of images is shown for calculation of capacitance of 3-phase line with Earth effect.

In this case, the conductors *a*, *b* anc *c* carry the charges q_a, q_b and q_c, (moreover the line is considered to be fully transposed) and occupy positions, 1, 2 and 3 respectively, in the first section of the transposition cycle. The effect of earth is simulated by image conductors with charges $-q_a$, $-q_b$ and $-q_c$ respectively as shown in the Fig. 3.29.

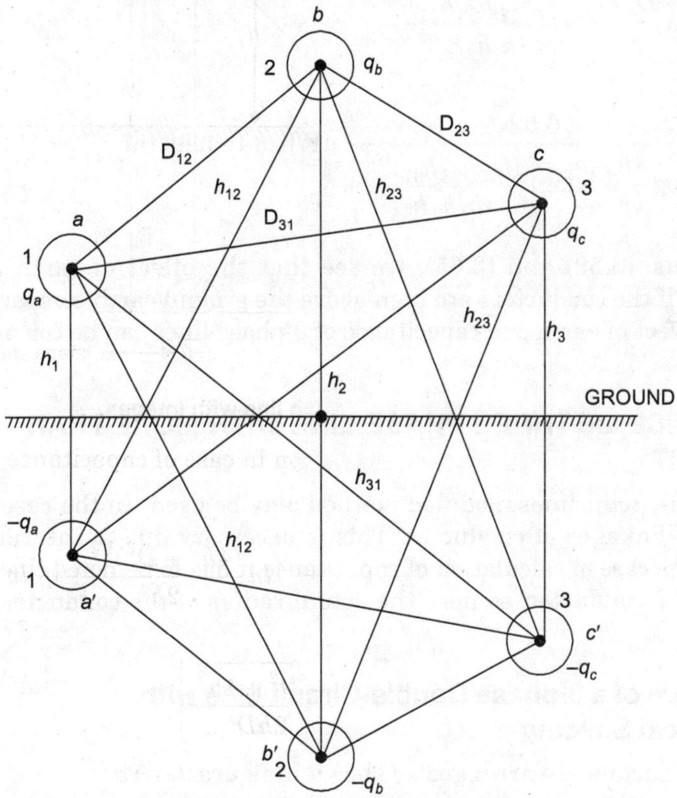

Fig. 3.29. 3-phase line with images.

The equations for the three sections of the transposition cycle can be written for the voltage V_{ab} as determined by the three charged conductors and their images. Hence, with conductor *a* in position 1, *b* in position 2 and *c* in position 3.

$$\therefore \quad V_{ab} = \frac{1}{2\pi\varepsilon}\left[q_a\left(ln\frac{D_{12}}{r} - ln\frac{h_{12}}{h_1}\right) + q_b\left(ln\frac{r}{D_{12}} - ln\frac{h_2}{h_{12}}\right) + q_c\left(ln\frac{D_{23}}{D_{31}} - ln\frac{h_{23}}{h_{31}}\right)\right] \quad ...(3.63)$$

Similar equation for V_{ab} can be written for the second and third sections of the transposition cycle. Now, if the fairly accurate assumption of constant charge per unit length of the conductor throughout the transmission cycle is mode, the average value of V_{ab} for the three sections of the cycle is given as :

$$V_{ab} = \frac{1}{2\pi\varepsilon}\left[q_a\left(ln\frac{D_{eq}}{r} - ln\frac{(h_{12}h_{23}h_{31})^{1/3}}{(h_1h_2h_3)^{1/3}}\right) + q_b\left(ln\frac{r}{D_{eq}} - ln\frac{(h_1h_2h_3)^{1/3}}{(h_{12}h_{23}h_{31})^{1/3}}\right)\right] \quad ...(3.64)$$

where $D_{eq} = (D_{12}D_{23}D_{31})^{1/3}$

The equation for the average value of the phasor V_{ac} is found in the similar manner then using $V_{ab} + V_{ac} = 3V_{an}$ and $q_a + q_b + q_c = 0$, (the conditions used in earlier calculation in the article 3.18 above), we finally obtain the following expression for the capacitance to the neutral;

$$C_n = \frac{2\pi\varepsilon}{ln\frac{D_{eq}}{r} - ln\left(\frac{(h_{12}h_{23}h_{31})^{1/3}}{(h_1h_2h_3)^{1/3}}\right)} \text{ F/m to neutral} \quad ...(3.65)$$

or $$C_n = \frac{0.0242}{\log\frac{D_{eq}}{r} - \log\left(\frac{(h_{12}h_{23}h_{31})^{1/3}}{(h_1h_2h_3)}\right)} \;\mu\text{F/Km to neutral} \quad ...(3.66)$$

Comparing eqns. (3.59) and (3.65), we see that the effect of earth is to increase the capacitance of line. If the conductors are high above the ground/earth compared to the distances among them, the effect of earth one capacitance of 3-phase lines can be considered as negligibly small.

3.21 CAPACITANCE OF THREE PHASE LINE WITH MORE THAN ONE CIRCUIT

In case of multi-circuit lines modified method may be used. In the case of calculation, we found self GMD of linkages of conductor. This is necessary due to internal flux linkages of conductor. But in the case of calculation of capacitance it has to be noted that all charges reside on the surface of the conductor, so here the actual radius of the conductor of the circuit will be used

3.21.1 Capacitance of a 3-phase Double-Circuit Line with Symmetrical Spacing

In this case conductors are arranged as shown in figure 3. 29 *a*

Then, $$V_{ab} = \frac{1}{2\pi\varepsilon}\left[q_a\left(ln\frac{D}{r} + ln\frac{\sqrt{3}D}{2D}\right) + q_b\left(ln\frac{r}{D} + ln\frac{2D}{\sqrt{3}D}\right) + q_c\left(ln\frac{D}{\sqrt{3}D} + ln\frac{\sqrt{3}D}{D}\right)\right]$$

$$= \frac{1}{2\pi\varepsilon}\left[q_a ln\left(\frac{\sqrt{3}D}{2r}\right) + q_b ln\left(\frac{2r}{\sqrt{3}D}\right) + q_c ln\left(\frac{\sqrt{3}D^2}{\sqrt{3}D^2}\right)\right]$$

$$= \frac{1}{2\pi\varepsilon}\left[q_a ln\frac{\sqrt{3}D}{2r} + q_b ln\frac{2r}{\sqrt{3}D}\right]$$

Similarly $$V_{ac} = \frac{1}{2\pi\varepsilon}\left[q_a ln\frac{\sqrt{3}D}{2r} + q_c ln\frac{2r}{\sqrt{3}D}\right]$$

We know, $V_{ab} + V_{ac} = 3V_{an}$, and $q_a + q_b + q_c = 0$

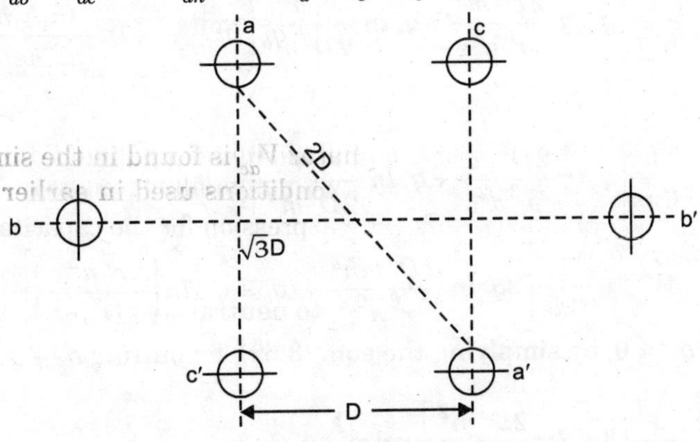

Fig. 3.29 (*a*)

$$\therefore \text{ Simplyfying, we get } \quad 3V_{an} = \frac{3q_a}{2\pi\varepsilon} ln\,\sqrt{3}\,D \text{ or } V_{an} = \frac{q_a}{2\pi\varepsilon} ln\frac{\sqrt{3}\cdot D}{2r} \quad ...(3.67)$$

$$\therefore \text{ Now, } \quad C_{an} = \frac{q_a}{V_{an}}$$

$$\therefore \text{ From eqn. (3.67), } C_{an} = \frac{q_a}{V_{an}} = \frac{2\pi\varepsilon}{ln\dfrac{\sqrt{3}\,D}{2r}} \text{ F/m} = \frac{0.0555}{ln\,(\sqrt{3}\,D/2r)}\ \mu\text{F/Km} \quad ...(3.67a)$$

And capacitance per phase, $C_n = 2C_{an}$

3.21.2 Capacitance of a 3-phase Double Circuit Line with Unsymmetrical Spacing & Fully Transposed

Conductor are arranged as in fugure 3.29 *b, c, d* corresponding to different conduction positions of transposition cycle.

∴ We can with the expression of V_{ab_1}, V_{ab_2} and V_{ab_3} as below :

$$V_{ab_1} = \frac{1}{2\pi\varepsilon}\left[q_a\left(ln\frac{D}{r} + ln\frac{m}{n}\right) + q_b\left(ln\frac{r}{D} + ln\frac{h}{m}\right) + q_c\left(ln\frac{D}{2D} + ln\frac{m}{n}\right)\right] \text{ volts}$$

$$V_{ab_2} = \frac{1}{2\pi\varepsilon}\left[q_a\left(ln\frac{D}{r} + ln\frac{m}{n}\right) + q_b\left(ln\frac{r}{D} + ln\frac{n}{m}\right) + q_c\left(ln\frac{2D}{D} + ln\frac{h}{m}\right)\right] \text{ volts}$$

$$V_{ab_3} = \frac{1}{2\pi\varepsilon}\left[q_a\left(ln\frac{2D}{r} + ln\frac{h}{n}\right) + q_b\left(ln\frac{r}{2D} + ln\frac{n}{h}\right) + q_c\left(ln\frac{D}{D} + ln\frac{m}{m}\right)\right] \text{ volt}$$

$$\therefore \quad V_{ab} = V_{\text{average}} = \frac{1}{3}\left(V_{ab_1} + V_{ab_2} + V_{ab_3}\right) \quad ...(3.68)$$

By putting the expressions of V_{ab_1}, V_{ab_2} & V_{ab_3} in eqn. (3.68) and simplifying, we get,

$$V_{ab} = \frac{1}{6\pi\varepsilon}\left[q_a ln \frac{2D^3 m^2}{r^3 n^2} + q_b ln \frac{r^3 n^2}{2D^3 m^2}\right] \text{volts}$$

Now, similarly,

$$V_{ac} = \frac{1}{6\pi\varepsilon}\left[q_a ln \frac{2D^3 m^2}{r^3 n^2} + q_c ln \frac{r^3 n^2}{2D^3 m^2}\right] \text{volts}$$

$$\therefore \quad V_{ab} + V_{bc} = 3V_{an} = \frac{1}{6\pi\varepsilon}\left[2q_a ln \frac{2D^3 mh^2}{r^3 n^2} + (q_b + q_c) ln \left(\frac{r^3 n^2}{2D^3 m^2}\right)\right] \quad ...(3.69)$$

Now as $q_a + q_b + q_c = 0$, by simplying the eqn. (3.69) by putting $q_b + q_c = -q_a$ we get,

$$3V_{an} = \frac{1}{6\pi\varepsilon}\left[3q_a ln \frac{2D^3 m^2}{r^3 n^2}\right] \text{volts}$$

$$\therefore \quad C_{an} = \frac{q_a}{V_{an}} = \frac{6\pi\varepsilon}{ln \frac{2D^3 m^2}{r^3 n^2}} = \frac{0.0555}{ln\left[2^{1/3}\frac{D}{r}\left(\frac{m}{n}\right)^{2/3}\right]} \mu\text{F/Km} \quad ...(3.70)$$

And capacitance per phase C_n will be $2\,C_{an}$.

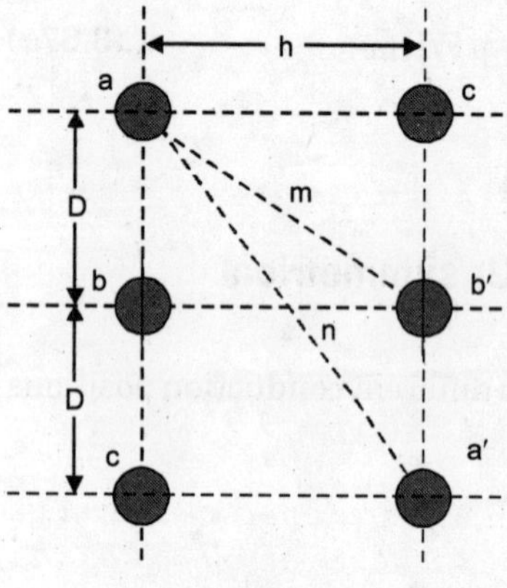

Fig. 3.29*b*

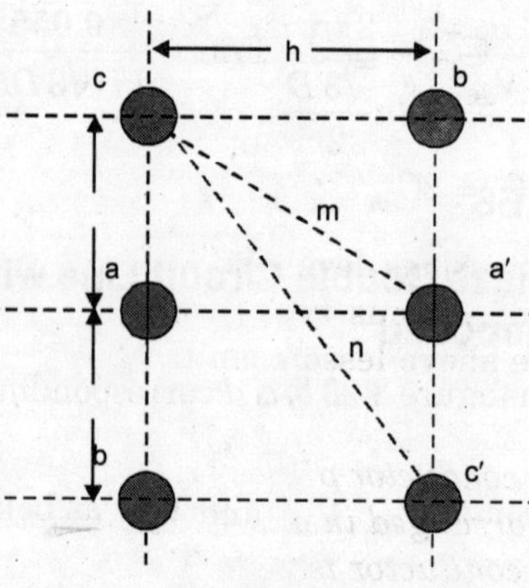

Fig. 3.29*b*

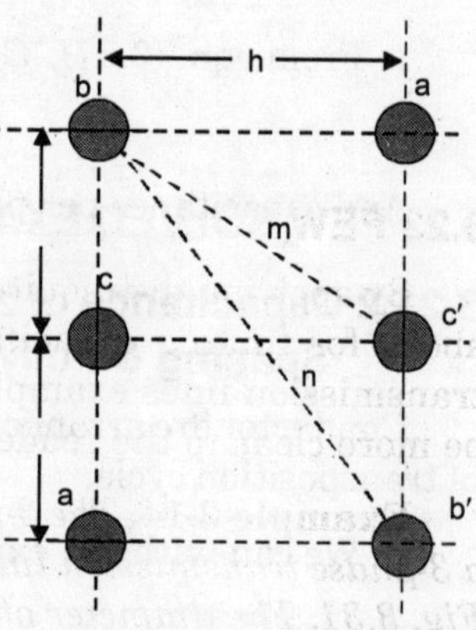

Fig. 3.29*b*

From comparison of above expressions for various constants for inductance and capacitance of transmission lines in above articles of the chapter, we find that the two are similar except, that in inductance expressions we use the fictitious conductor radius r' (= 0.77888 r), but in expressions for capacitance actual conductor radius r is used. This fact suggest that the method of GMD would be applicable in the calculations for capacitance as well provided it is modified by using the outer conductor radius for finding D_S (the self geometric mean distance).

3.22 BUNDLED CONDUCTORS & CAPACITANCES IN A 3-PHASE LINE

In the Fig. 3.30 a bundled conductor line arrangement. The conductors of any one bundle are in parallel, and it is assumed that charge per bundle divides equally among the conductors of the bundle as $D_{12} \gg d$. Also, $D_{12} - d \approx D_{12} + d \approx D_{12}$ for the same reasons. The result obtained with these assumptions are fairly accurate for usual spacings. Hence, if the charge of a is 'q_a', the conductors a and a' have a charge of $q_a/2$ each. Similarly the charge is equally divided for phase b and c.

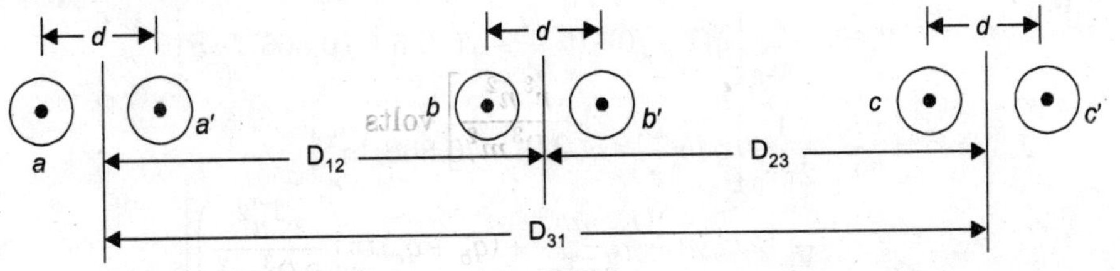

Fig. 3.30. Cross-section view of a bundle conductor in 3-ph transformer line

Now, we can write an equation for the voltage from conductor a to conductor b, we get

$$V_{ab} = \frac{1}{2\pi\varepsilon}\left[0.5q_a\left(ln\frac{D_{12}}{r} + ln\frac{D_{12}}{d}\right) + 0.5q_b\left(ln\frac{r}{D_{12}} + ln\frac{d}{D_{12}}\right)\right] \quad ...(3.68)$$

or
$$V_{ab} = \frac{1}{2\pi\varepsilon}\left(q_a\,ln\frac{D_{12}}{\sqrt{rd}} + q_b\,ln\frac{\sqrt{rd}}{D_{12}} + q_c\,ln\frac{D_{23}}{D_{31}}\right) \quad ...(3.69)$$

Now let us consider the line to be transposed and proceeding in the usual manner than the final result will after simplifying the eqn. (3.69), and applying conditions stated above

$$C_n = \frac{V_{ab}}{q_a}\frac{0.0242}{\log(Deq/\sqrt{rd})}\cdot \mu\text{F/Km to neutral}$$

3.23 FEW SOLVED EXAMPLES

By applying the required expressions developed above for finding capacitance of various type of transmission lines examples the above lesson can be more clear to the reader.

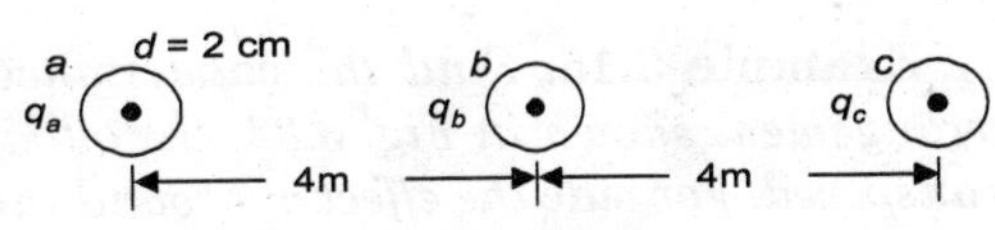

Fig. 3.31

Example 3.14. *The 3-phase conductor p 185 of a 3-phase transmission line are arranged in a horizontal plane and 4 metres apart as shown in Fig. 3.31. The diameter of each conductor is 2 cm. Calculate the capacitance developed on the conductors a, b and c.*

Solution : We can write the expression for V_a (as developed above)

Since the spacing between, conductors in very large compared with the radius of the conductor and assuming that there are no other charged surfaces and neglecting the effect of ground, $q_a + q_b + q_c = 0$. Then

$$V_a = \frac{1}{2\pi\varepsilon_0}\left[q_a ln\frac{1}{r} + q_b\,ln\frac{1}{D} + q_c\,ln\frac{1}{2D}\right]$$

We also know that in such line arrangement, $q_a + q_b + q_c = 0$. And being phasor quantity, $q_a = q\,(1 - j_0)$, $q_b = q\,(-0.5 - j\,0.866)$, $q_c = q\,(-5 + j\,0.8666)$. Now by simplification we get,

Then $$V_a = \frac{1}{2\pi\varepsilon_0}\left[q_a\, ln\frac{D}{R} - q_c\, ln\, 2\right] = \frac{1}{2\pi\varepsilon_0}\left[q_a\, ln\frac{D}{R} - q(-0.5 + j0.866) Inq\right]$$

$$= \frac{1}{2\pi\varepsilon_0}\left[q(1 + j0)\cdot ln\frac{D}{r} - q(-0.5 + j0.866)\, In\, q\right]$$

$$= \frac{1}{2\pi\varepsilon_0}\left[q\, ln\frac{D}{r} - q(-0.5 + j0.866)\ln^2\right]$$

$\therefore$ $$C_a = \frac{V_a}{q} = \frac{2\pi\varepsilon_0}{ln\frac{D}{r} + 0.3465 - j0.6} \times 10^3 \text{ F/km}$$

And putting values of D, r $$C_a = C_n \frac{2\pi \times (10^{-9}/36\pi)}{ln\frac{D}{r} + (0.5 + j0.866)\, Inq}$$

$\therefore$ We get $$r\,C_a = \frac{2\pi \times \frac{10^{-9}}{36\pi}}{5.99 + 0.3465 - j0.6} \times 10^3 \text{ F/Km} = 0.00865 + j0.000823\,\mu\text{F/Km}$$

Then, $$C_c = \frac{2\pi\varepsilon_0}{ln\frac{D}{r} - (-0.5 - j0.866)\, Inq} \text{ F/m}$$

Putting the values, $C_c = 0.00868 - j0.000823\ \mu$ F/Km

and $$C_b = \frac{2\pi\varepsilon_0}{ln\frac{D}{r}} = 0.009273\,\mu\text{F/Km}$$

Example 3.15. *Find the capacitance per Km per phase to neutral of a 3-phase line arrangement shown in Fig. 3.32. the outside dia of ACSR conductors is 2.60 cm. The line is transposed. For side the effect of ground into account.*

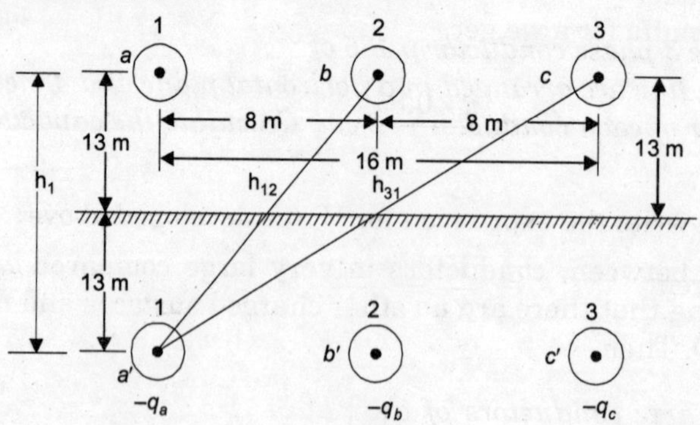

Fig. 3.32

Solution : The expression the capacitance per phase to neutral can be written as (as deduced above) :

$$C_n = \frac{2\pi\varepsilon}{ln\frac{D_{eq}}{r} - ln\,3\sqrt{\frac{h_{12}, h_{23}, h_{31}}{h_1 h_2 h_3}}}$$ (refer to Article 3.19.2)

where $h_{12} = \sqrt{64 + 676} = \sqrt{740} = 27.2\text{ m} = h_{23} = \sqrt{8^2 + 26^2}$ (refer to Fig. 3.32)

$h_{31} = \sqrt{676 + 256} = \sqrt{932} = 30\text{ m} = \sqrt{26^2 + 10^2}$ (refer to Fig. 3.32)

$\therefore\quad h_1 = h_2 = h_3 = 26\text{ m}$

and $D_{eq} = \sqrt[3]{8 \times 8 \times 16} = 10.01\text{ m}$

$$\therefore\quad C_n = \frac{2\pi\varepsilon}{ln\frac{10.01 \times 100}{1.3} - ln\left(\frac{3\sqrt{27.2 \times 27.2 \times 30.6}}{36}\right)} = \frac{2\pi\varepsilon}{ln\,779 - In\,1.16}$$

$$= \frac{2\pi\varepsilon}{6.65 - 0.148} = \frac{2\pi\varepsilon}{6.50}\text{ F/Km}$$

or $$C_n = \frac{2\pi \times 10^{-9} \times 10^3}{36\pi \times 6.50} = \frac{10^{-6}}{18 \times 6.50} = 8.5 \times 10^{-9}\text{ F/Km}$$

Example 3.16. *A 1-phase line of 250 V, 50 Hz, has conductor spacing of 1.5 metre. The diameter of each conductor 1.5 cm. Calculate (1) capacitance of the line in μF for a length of 50 Km, and (2) charging current.*

Solution : We know capacitance of the line, $C = \frac{\pi\varepsilon_0}{lnd/r}$ Farad per km.

(1) Here, as per exmple, l = 50 Km, = 50 × 10³ m, d = 1.5 m = 150 m, r = 1.5/2 = 0.75

∴ By using the formula for a we get,

$$C = \frac{\pi \times 0.8854 \times 10^{-12} \times 50 \times 10^3}{ln\,150/0.75} = 0.263\ \mu\text{F for 50 k}$$

Now charging current $= \frac{V}{X_c} = \frac{V}{1/\omega c} = \omega c V = 2\pi \times 50 \times 0.263 \times 10^{-6} \times \mathbf{250}$

$= 20.66\text{ mA.}$

Example 3.17. *Three conductors of a 3-phase overhead line arranged in horizontal plane, 6 metres apart. The diameter of each conductor is 1.24 cm. Find the capacitance per 100 km of the line in μF.*

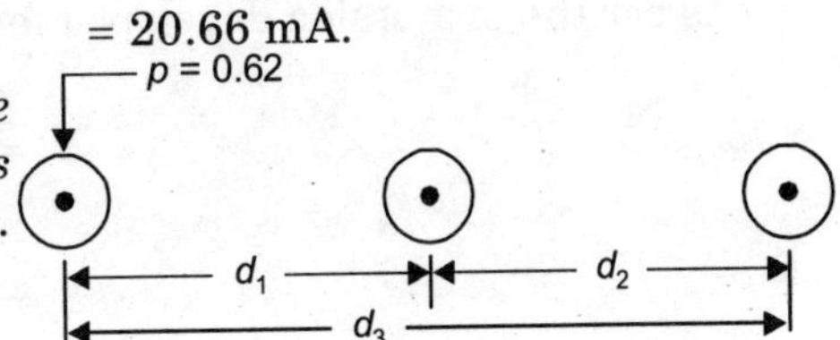

Fig. 3.33

Solution : In this case, $d_1 = d_2$ 6 m with ref. to Fig. 3.33 the distances d_1, d_2, d_3 are,

$d_1 = d_2 = 6$ m and $d_3 = 2\,d_1 = 2 \times 6 = 12$ m

$\therefore \quad d = 3\sqrt{d_1 d_2 d_3} = 3\sqrt{6 \times 6 \times 12} = 7.41$

l = length = 100 km.

The radius of the conductor $r = \frac{1.24}{2} = 0.62$ cm.

Hence capacitance of the line/phase,

$$C = \frac{2\pi\varepsilon_0}{ln\frac{d}{r}} \times l \text{ Farad} = \frac{2\pi \times 8.854 \times 10^{-12} \times 100 \times 10^3}{ln\,7.41/0.62} = \mathbf{0.784}\ \mu\text{F}$$

Example 3.18. *A single-circuit, three-phase, 50 Hz transmission line consists of three conductors arranged as shown in Fig. 3.34. If the conductors have diameter equal to 0.8 cm. Find the inductive reactance of 25 Km long line per kilometer per phase. Also calculate the capacitance and capacitive reactance of the transmission line.*

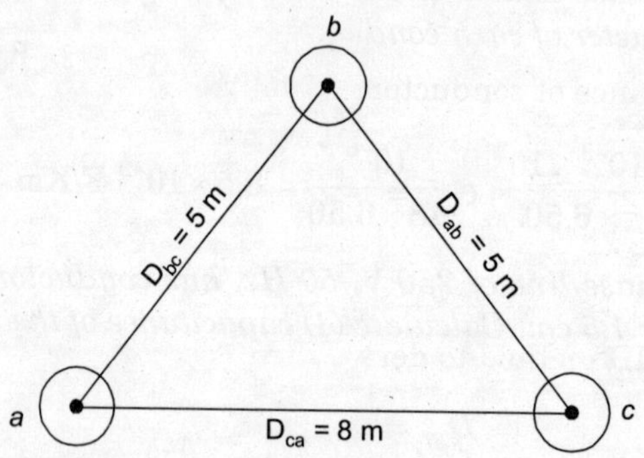

Fig. 3.34.

Solution : Let us finite find $D_{eq} = D_e$

$$D_e = D_{eq} = (D_{ab} \times D_{bc} \times D_{ca})^{1/3} = (5 \times 5 \times 8)^{1/3} = 5.848 \text{ m}$$

And radius of conductor, $r = \frac{1}{2} \times 0.8 \times 10^{-2}$ m

Hence, $$D_e r = \frac{5.848}{1/2 \times 0.8 \times 10^{-2}} = \frac{5.848 \times 2 \times 10^{+2}}{0.8} = 1462$$

Hence, $ln\,(D_e r) = ln\,(1462) = \mathbf{7.29}$

$\therefore$ As per the expression developed for,

$$L = 2\left(\frac{1}{4} + ln\frac{D}{r}\right) \times 10^{-7} \text{ H/m}$$

$$\therefore \quad L = 2\left(\frac{1}{4} + 7.29\right) \times 10^{-7} \times 10^3 = 1.51 \text{ mH/Km.}$$

$\therefore$ We find Inductive Reactance, $X_L = \omega L = 314 \times 1.51 \times 10^{-3} = \mathbf{0.473\ \Omega}$

(*b*) Now to determine X_C let us first find capacitance C.

We know from expression for C, as below :

$$C = \frac{\pi \varepsilon_0}{ln(D/r)} \text{ F/Km}$$

As, ε_0 8.854 × 10^{-12} F/m

$$\therefore \qquad C = \frac{\pi \varepsilon_0}{ln(D/r)} \text{ F/m}$$

$\therefore$ Capacitance in Farad for 25 km $= C \times 25 \times 10^3 = \dfrac{\pi \times 8.854 \times 10^{-12}}{ln \times 7.29} \times 25 \times 10^3$

$= 0.35\,\mu\text{F}$ for 25 Km

$\therefore$ Capacitive Reactance, $X_C = \dfrac{1}{C_\omega} = \dfrac{1}{0.350 \times 10^{-6} \times 314} = \mathbf{9.1\,K.ohm.}$

Example 3.19. *A 3-phase double circuit line is shown in Fig. 3.34. Find the capacitance per phase to neutral. Diameter of each conductor = 2.5 cm.*

Soltion : The capacitance of conductors to neutral is :

$$C_{an} = \frac{2\pi\varepsilon}{ln\left[2^{1/3}\left(\frac{D}{r}\right)\left(\frac{m}{n}\right)^{1/3}\right]} \text{ F/m}$$

Now we apply the GMD method to derive the expression for capacitance

$$D_{eq} = 3\sqrt{D_{ab} \times D_{bc} \times D_{ca}}$$

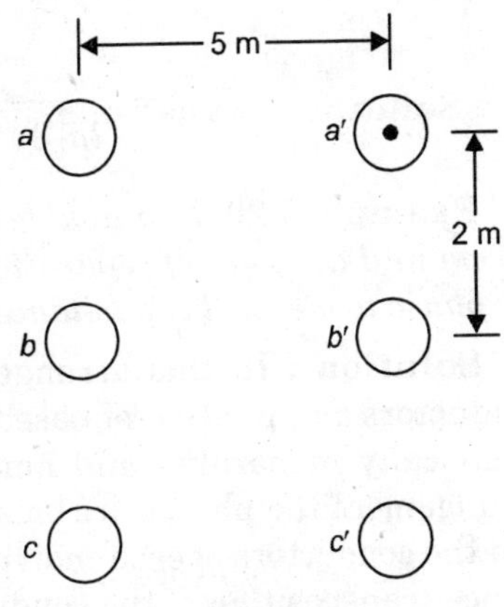

Fig. 3.34 (a)

where D_{ab} = neutral GMD between phases a and b in position 1.

D_{bc} = neutral GMD between phases b and c in position 1.

D_{ca} = neutral GMD between phases c and a in position 1.

Now, $D_{ab} = 4\sqrt{D_m \cdot D_n} = \sqrt{D_m}$,

$D_{bc} = \sqrt{D_m}$ and $D_{ca} = \sqrt{2Dh}$

$\therefore \quad D_{eq} = 3\sqrt{D_{ab} \cdot D_{bc} \cdot D_{ca}} = 2^{1/6} \cdot D^{1/2} \cdot m^{1/3} \cdot h^{1/6}$

Now, to find self GMD in position for 1 for the entire phase of conductor a and a'.

$$D_{S_1} = \sqrt[4]{r \times n \times r \times n} = \sqrt{rn}$$

Similarly, $D_{S_2} = \sqrt[4]{r \times h \times r \times h} = \sqrt{rh}$

and $\qquad D_{S_3} = \sqrt[4]{r \times n \times r \times n} = \sqrt{rn}$

$$\therefore \quad D_S = \sqrt[3]{D_{S_1} \cdot D_{S_2} \cdot D_{S_3}} = \sqrt[3]{(r^{1/2} \cdot n^{1/2}) \times (r^{1/2} \cdot h^{1/2}) \times (r^{1/2} \cdot n^{1/2})}$$

$$= r^{1/2} \cdot h^{1/6} \cdot n^{1/3}$$

$$\therefore \quad \frac{D_{eq}}{D_S} = \frac{2^{1/6} \cdot D^{1/2} \cdot m^{1/3} \cdot h^{1/6}}{r^{1/2} \cdot h^{1/6} \cdot n^{1/3}} = \frac{2^{1/6} \cdot D^{1/2} \cdot m^{1/3}}{r^{1/2} \cdot n^{1/3}} = \left[2^{1/3} \cdot \left(\frac{D}{r}\right)\left(\frac{m}{n}\right)^{2/3}\right]^{1/2}$$

Now, as per the expression

$$C_n = \frac{2\pi\varepsilon}{ln\left(\frac{D_{eq}}{D_S}\right)} = \frac{2\pi\varepsilon}{\frac{1}{2} ln\left[\sqrt[3]{2}\left(\frac{D}{r}\right)\left(\frac{m}{n}\right)^{2/3}\right]}$$

$$= 2 \times \frac{2\pi\varepsilon}{ln\left[\sqrt[3]{2}\left(\frac{D}{r}\right)\left(\frac{m}{n}\right)^{2/3}\right]} \text{ F/m}$$

Now, substituting the values of

$$m = \sqrt{5^2 + 2^2} = \sqrt{29}, \text{ and } n = \sqrt{5^2 + 4^2} = \sqrt{39}, D = 200$$

in the C_n

We get,

$$C_n = \frac{4\pi\varepsilon}{ln\left[\sqrt[3]{2}\left(\frac{200}{1.25}\right)\left(\frac{39}{29}\right)^{2/3}\right]}$$

$$= \frac{4\pi\varepsilon}{ln\left[3\sqrt{2}\,(160)(0.7440)^{1/3}\right]} = 0.02\ \mu\text{F/m}$$

Example 3.20. *In a double circuit 3-phase line in Fig. 3.35 has conductors of diameter of 2.5 cm and distance of separation (D) is 2 m in the hexagonal spacing arrangement, calculate the phase to mentral capacitance in micro farads per 100 Km. of the line.*

Solution : In this arrangement (Fig. 3.35) of six conductors, phase a is composed of conductors aa', phase b of based b', phase c of c and c'. The two conductors of each phase are electrically in parallel and hence the same charge. However, because of the symmetrical arrangement the phases are balanced and the conductors of each individual phase are balanced and the conductors of each individual phase are also balanced if the effect of ground is neglected. Hence transposition of the conductors is not necessary to balance the phases.

In the discussions made above on the subject in this chapter on capacitance, we have seen that,

$$C_n = 2C_{an} = \frac{2 \times 2\pi\varepsilon}{\ln\left(\frac{\sqrt{3}D}{2r}\right)} = \frac{2\pi\varepsilon}{\ln\left(\frac{\sqrt{3}D}{2r}\right)^{1/2}} \cdot \text{ F/m}$$

and, capacitance of one phase to neutral is expressed as $C_n = \dfrac{2\pi\varepsilon}{ln\left(\dfrac{D_m}{D_S}\right)}$. where D_m is the mentral GMD between the phases, and D_S in the self GMD between the conductors belonging to the same phase.

Than it may be said that, $D_m = \sqrt[4]{D \times \sqrt{3}D \times \sqrt{3}D \times D} = \sqrt[4]{3}D$ and $D_S = \sqrt{2Dr}$

$$\therefore \quad \frac{D_m}{D_S} = \frac{\sqrt[4]{3} \cdot D}{\sqrt{2Dr}} = \frac{\sqrt[4]{3}\sqrt{D}}{\sqrt{2}r} = \left(\frac{\sqrt{3} \cdot D}{2r}\right)^{1/2}$$

$$\therefore \quad C_n = \frac{2\pi\varepsilon}{ln\left(\dfrac{\sqrt{3}D}{2r}\right)^{1/2}} \text{ F/m}$$

Now, putting the values of $D = 200$, and $r = 2.5$, we get phase to neutral capacitance,

$$C_n = \frac{2\pi\varepsilon}{ln\left(\dfrac{\sqrt{3} \cdot 200}{2.5}\right)^{1/2}} = \frac{4\pi\varepsilon}{ln\dfrac{\sqrt{3} \times 200}{2.5}} = \frac{4\pi \times (10^9/36\pi)}{ln\left(\dfrac{\sqrt{3} \times 200}{2.5}\right)} = \frac{1}{9 \times 10^9 \cdot ln\left(\dfrac{\sqrt{3} \times 200}{2.5}\right)} \text{ F/m}$$

Hence, C_n for 100 Km length

$$= \frac{1 \times 100 \times 1000 \times 10^6}{9 \times 10^9 \cdot ln\dfrac{\sqrt{3} \times 200}{2.5}} = 2.26\ \mu\text{F}$$

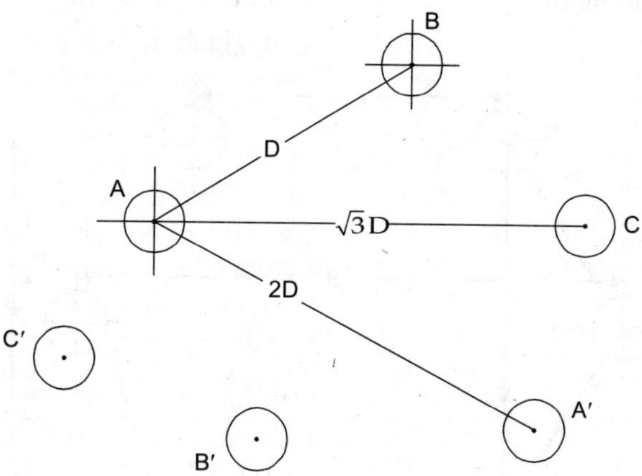

Fig. 3.35

Example 3.21. *Determine the capacitance and the charging current per km when the transmission line of 3-phade system with conductors of diameter 8 cm, and operating at 132 KV, 50 HZ as shown in Fig. 3.36.*

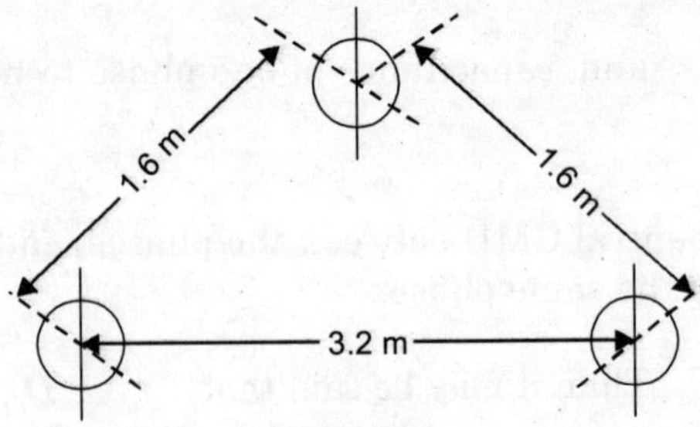

Fig. 3.36

Solution : The diameter of the conductor = 8 cm

$\therefore$ radius of the conductor = 4 cm

And the mutual GMD of conductors, D_m = 2.015 m

$$D_m = \sqrt[3]{1.6 \times 3.2 \times 1.6} = \mathbf{2.015} \text{ m}$$

$\therefore$ Capacitance per phase per metre

$$= \frac{2\pi\varepsilon_0}{ln\frac{D_m}{r}} = \frac{2\pi\varepsilon_0}{ln\frac{2.015}{0.4} \times 10^2}$$

$$= \frac{10^{-9}}{18 \times ln\frac{201.5}{0.4}}$$

$\therefore$ The carrying current $\quad = 8.928$ pF/metre

$$= \frac{132 \times 100}{\sqrt{3}} \times 8.928 \times 10^{-9} \times 314 = 8.928 \times 10^{-12} \times 10^3 \text{ F/km}$$

$= 0.2136$ amp/km $\qquad = 8.928 \times 10^{-9}$ F/km

Example 3.22. *Determine the capacitance and the charging current for km when the transmission line as shown in the Fig. 3.37. The diameter of the conductor is 2.5 cm and the line operates at 220 KV.*

Solution : At first we have to find $\quad D_m$ = Mutual GMD

Self GMD of each conductor $\quad = 1.25 \times 0.7788$ m

$= 0.00986$ m

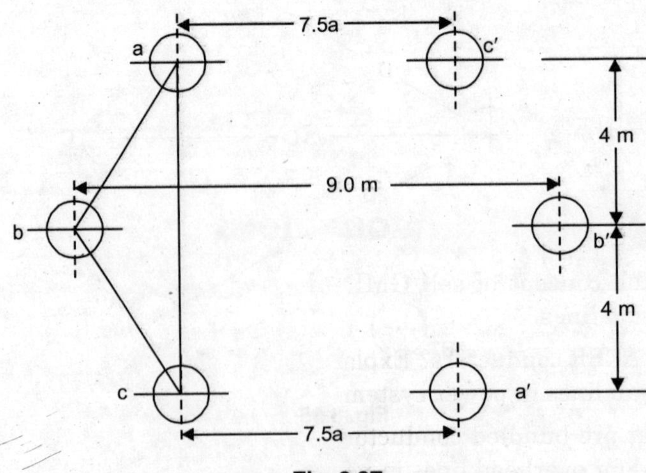

Fig. 3.37

$$D_{bc} = D_{ab} = \sqrt{4^2 + 0.75^2} = 4.0697 \text{ m}$$

$$D_{ab}' = \sqrt{4^2 + 8.25^2} = 9.1685 \text{ m}, \; D_{aa}' = \sqrt{8^2 = 7.5^2} = 10.965 \text{ m}$$

$$D_{m_1} = \sqrt[4]{4.0697 \times 7 \times 7.5 \times 9.168} = 6.878$$

$$D_{m_2} = \sqrt[4]{4.0697 \times 4.0697 \times 9.1685 \times 9.1685} = 6.1084$$

$$D_{m_3} = D_{m_1} = 6.878, \qquad \therefore D_m = \sqrt[3]{D_{m_1} \; D_{m_2} \; D_{m_3}} = \mathbf{6.61} \text{ m}$$

$$\therefore \text{Capacitance per km} = \frac{10^{-6}}{18 \, ln \dfrac{6.61}{0.3582}} = 0.01905 \; \mu\text{F/km}$$

$$\therefore \text{Charging current per km} = \frac{220 \times 1000}{\sqrt{3}} \times 314 \times 0.019 \times 10^{-6} = \mathbf{0.74} \text{ A/km}$$

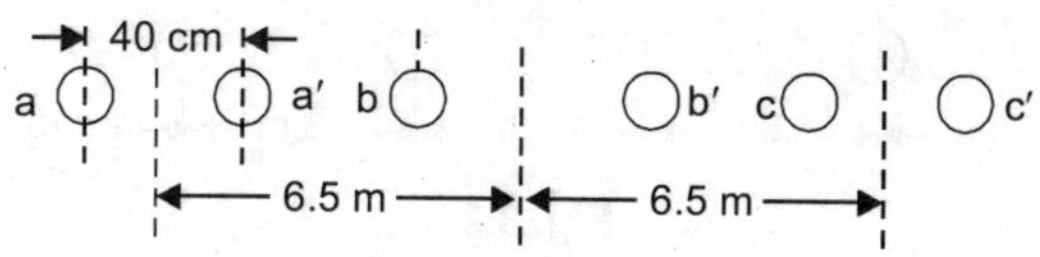

Fig. 3.38

Example 3.23. *Determine the capacitance and charging current per km of a single-phase 220 KV line using two bundle conductors per phase with each conductor having diameter of 4.5 cm. The Fig. 3.38 show the arrangement.*

Solution : Assuming the effect of transposition negligibly

small, $D_m = \sqrt[3]{6.5 \times 13.0 \times 6.5} = 8.19$ m

$$D_s = \sqrt{2.25 \times 10^{-2} \times 0.4} = 0.095 \text{ m}$$

$$\therefore \text{Capacitance per km} = \frac{10^{-6}}{18 \, ln \dfrac{8.12}{0.095}} = 0.0125 \; \mu\text{F}$$

$$\therefore \text{The charging current} = \frac{220 \times 1000}{\sqrt{3}} \times 314 \times 0.0125 \times 10^{-6} = \mathbf{0.5} \text{ amp}$$

QUESTIONS

1. Explain the concept of self GMD and mutual GMD for evaluating inductance of transmission lines.
2. What are ACSR conductors? Explain the advantages of ACSR conductors when used for overhead lines in power system.
3. State what are bundled conductors? Discuss the advantages of bundled conductors when used for overhead lines in power system.

4. Show that the inductance per unit length of an overhead line due to internal flux linkages is constant and is independent of size of conductor.

5. Determine the self GMD of the following types of conductors in terms of radius "r" of an individual strand, as shown in the Fig. 3.1. **(Ans.** (*a*) 1.773 r, (*b*) 1.7 r)

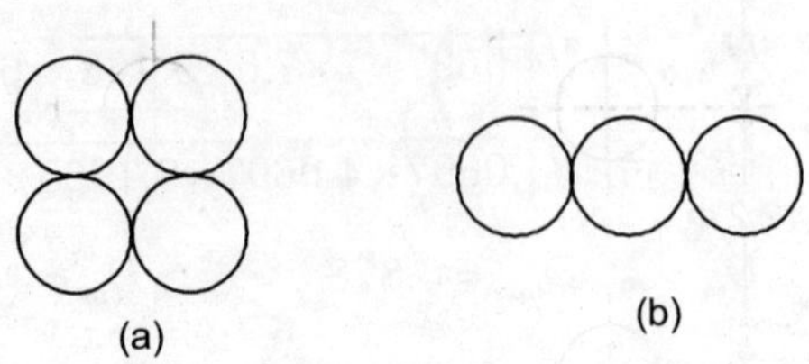

Fig. 3.1

6. Determine the inductance per km/phase of a double circuit, 3-phase transmission line. The radius of each conductor is 12 own, and the double circuit is shown in Fig. P.3.2. **(Ans.** 0.51 mH/km)

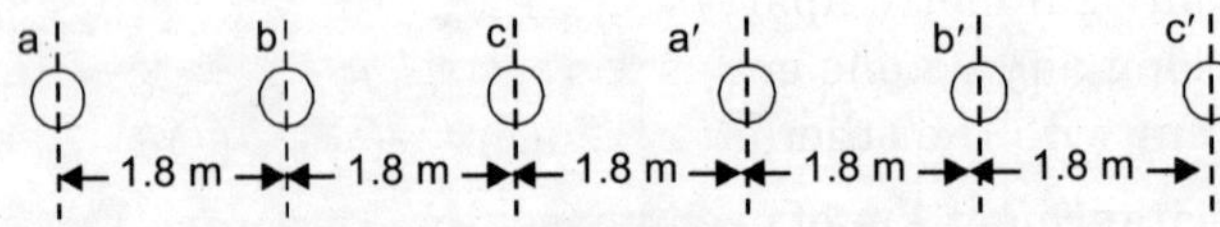

Fig. 3.2

7. Derive expressions for inductance of a 3-phase line with conductors untransposed. State what is the significance of imaginary term in the expression for inductance. Derive the expression for inductance for a completely transposed line.

8. (*a*) Derive an expression for the flux linkages of one conductor in a group of n conductors carrying currents whose sum in zero.

(*b*) Derive an expression for inductance of composite conductors of a single phase line consisting of "m". Strands in one conductor and "n" strands in the other conductor.

9. Determine the inductance per km of a 3-phase transmission line having conductors per phase and arranged as in Fig. 3.3. **(Ans.** 1.1 mH/km/phase)

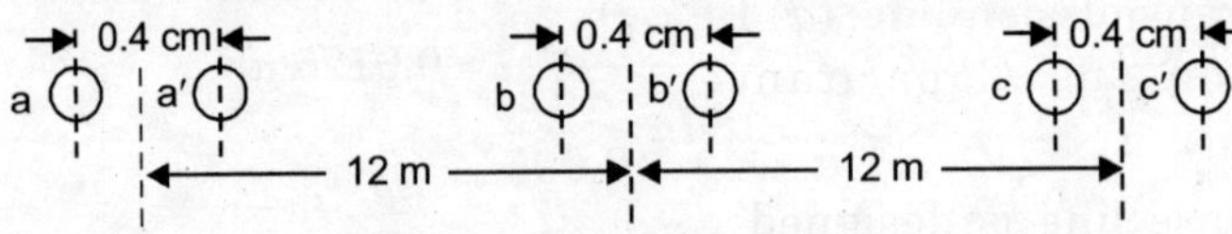

Fig. 3.3

10. State whether a constant term is obtained in the expression for capacitance as in case of inductance. Explain with reasons.

11. State what is the method of images. Deduce an expression for the capacitance per unit length of a 3-phase line completely transposed. Also state what is the effect of reaction the capacitance of the transmission line.

12. A phase double circuit line is hown in Fig. 3.5. The diameter of each conductor is 2 cm. Find the capacitance and charging current per km length of the line : The line is assumed to be transposed and operating at the voltage 220 KV.

(Ans. (*a*) 0.0202 μF/km, (*b*) 0.804 amp/km)

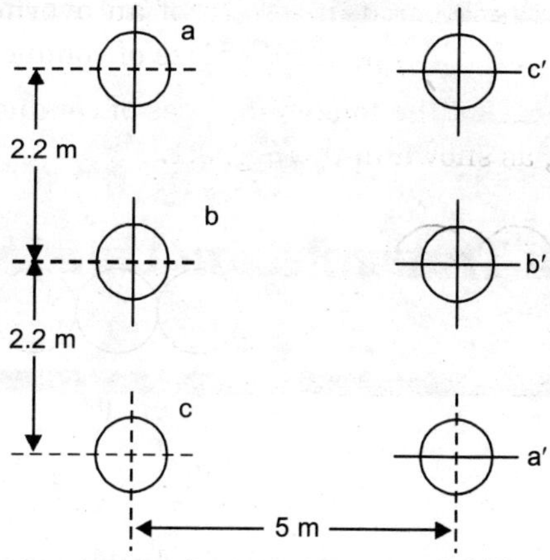

Fig. 3.6

13. A single-phase overhead line consists of a pair of parallel conductors 12 mm in diameter spaced uniformly 2.5 metre apart in air. Find the capacitance per km length. If the line is 30 km long and its one end is connected to 50 KV, 50 HZ system. What is the charging current with the other end open-circuit. (**Ans.** (*a*) 0.00461 μ/F, (*b*) 2.2 amps)

14. Find the capacitance per km of each phase (line-to-neutral) for the line arrangement as shown in Fig. 3.7, with conductor diameter of 2.5 mm. Assume that the supply currents to be balanced and equally divided between circuit. (**Ans.** 0.521 mH)

15. Derive expression of the capacitance per phase of a 3-phase line with (*a*) equilateral and (*b*) unsymmetrical spacings of the conductors. In case of (*c*) assume transposition.

16. Write short notes on :

(*a*) Proximity Effect, (*b*) Skin Effect, (*c*) Advantages of bundle Conductors for overhead transmission line, (*d*) Effect of Earth on the Capacitance of Conductors.

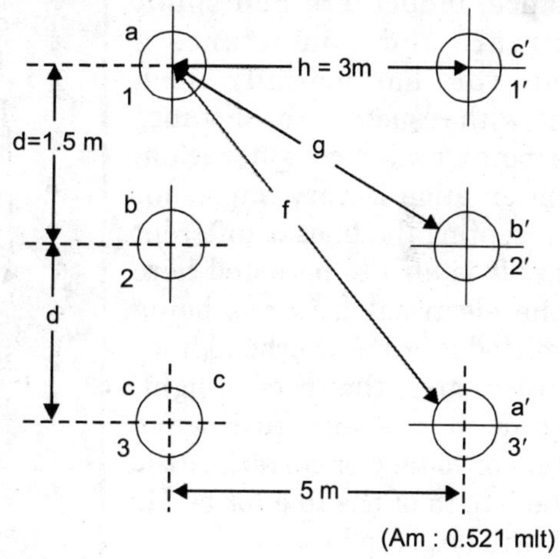

Fig. 3.7

17. A three phase line is designed with equilateral spacing of 2.85 metres. It is decided to build the line with horizontal spacing in the same so that $D_{13} = 2D_{12} = 2D_{23}$. What should be the spacing between adjoint conductor to keep the inductance same as in previous case. Conductors transposed :

(**Hints :** Find $= D_{eq} = (D_{12} \cdot D_{23} \cdot D_{13})^{1/3} = 2.44$. Then find D from, $\sqrt[3]{2D} = 2.88$ as per example.

$\therefore D = (2.88)/(2.88/\sqrt[3]{2}) = 2.3$ m. Inductance to be same as with an equilateral spacing of conductors are $D_{12} = 2.3$. And, $D_{12} = 2.3 = D_{23}$ and $D_{13} = 2D = 4.6$ m Am)

CHAPTER

4

Performance of Transmission Line in Power System

INSIDE THIS CHAPTER

- Rrepresentation of transmission lines
- Definition of performance factors in transmission lines
- Definition of constants A, B, C, D and values of a, b, c, d for short line
- Mixed conditions for a line
- Medium length lines
- Long transmission lines
- Equivalent circuit representation of long line
- Surge impedance and surge impedance loading of transmission line
- Ferranti effect
- Electrical length of a line
- Electrical short lossless lines components
- Input impedance of a transmission line
- Matching condition in a transmission line
- Losses in an open circuited line
- Effect of transformers (and on other impedances) on the performance of a transmission line
- Equivalent 'T' and 'π' Network of a long line
- Constant voltage transmission
- Circle Diagrams

4.1 REPRESENTATION OF TRANSMISSION LINES

A transmission line is a set of conductors run from one place to another place, supported on transmission towers. Such lines have four distributed parameter as discussed in Chapter 3, series resistance, inductance and shunt capacitance and conductance a currents vary harmonically along the line with respect to the distance of the point under consideration. This observation is very important in representing the lines of different lengths. It is also to be noted here that the electrical power is being transmitted over the overhead lines at approximately the speed of light. In order to get one full wave variation of voltage or current or the line the length of the line for 50 Hz supply will be given by

$$f \cdot \lambda = v \qquad \text{...(4.1)}$$

where f is the frequency of supply, λ is the wavelength. By wavelength we mean, the length of the line in this case, and v is the velocity of the wave, *i.e.*, the velocity of light.

Hence substituting for $f = 50$ Hz

$v = 3 \times 10^8$ m/sec.

We get

$$\lambda = \frac{v}{f} = \frac{3\times10^8}{50} = 6\times10^6 \text{ metres} = 6000 \text{ km} \qquad \text{...(4.2)}$$

This means that if the length of the line is 6000 Km, the voltage or current wave at the two ends of the line will be as shown in Fig. 4.1. For line length less than about 160 km the voltage or current variation on the line is not much and it can be said that for line length of about 160 Km the parameters could be assumed to be lumped and not distributed. Such lines are known as electrically short transmission lines. In power system these electrically short transmission lines are again categories as short transmission lines, and medium transmission lines where the effect of shunt capacitance is neglected. And lines above 80 km and below 160 km length are termed as medium length lines.

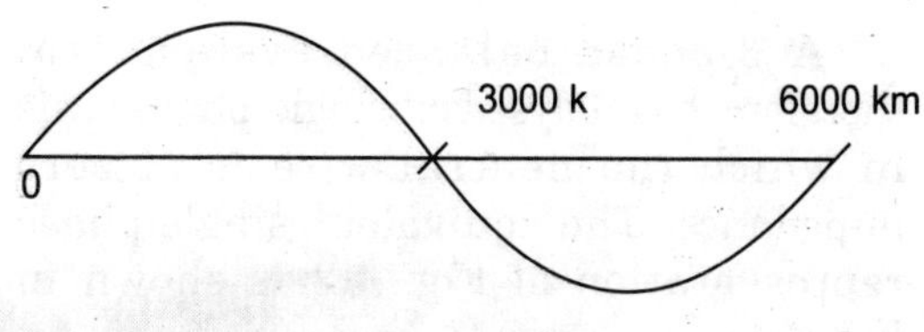

Fig. 4.1. Voltage or current wave at the two ends of the line.

In fact for medium length lines the shunt capacitance can be assumed to be lumped at the middle of the line, or half of the shunt capacitance can be assumed to be lumped at the middle of the line or half of the shunt capacitance may be considered to be lumped at each end of the line. The two representations of medium length are termed as nominal $-T$ and nominal π respectively.

For line lengths more than 160 Km the parameter are distributed and rigorous calculation are required to be made except in certain cases where lines up to 250 Km can be analysed using nominal $-\pi$ representation.

A typical 3-phase system is shown in Figs. 4.2 and 4.3. A 3-phase star load is connected to the generator through a 3-ph have transmission system as shown in Fig. 4.2. In Fig. 4.3 is shown a single-line diagram of a 3-phase system.

A 3-phase system is normally a balanced system irrespective of the fact that the conductors are not transposed, as the untransposed conductors introduces slight disymmetry which can be ignored for all practical purposes.

It is known that the sum of all the currents in a balanced polyphase network is zero and therefore, the current through the wire connected between the star point of the load and neutral of the system is zero. This mean that the star point of the load and neutral of the system are at the same potential.

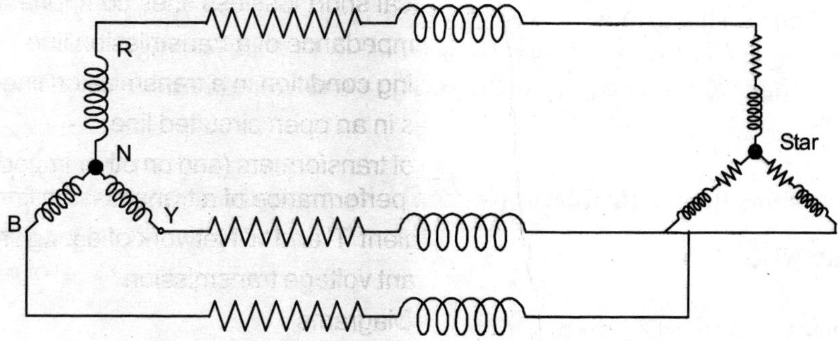

Fig. 4.2. A 3-phase system.

Fig. 4.3. Single-line diagram of a 3-phase system.

A 3-phase balanced system can, therefore, be analysed on single-phase basis in which the neutral wire is of zero impedance. The equivalent single-phase representation of Fig. 4.2 is shown in Fig. 4.4.

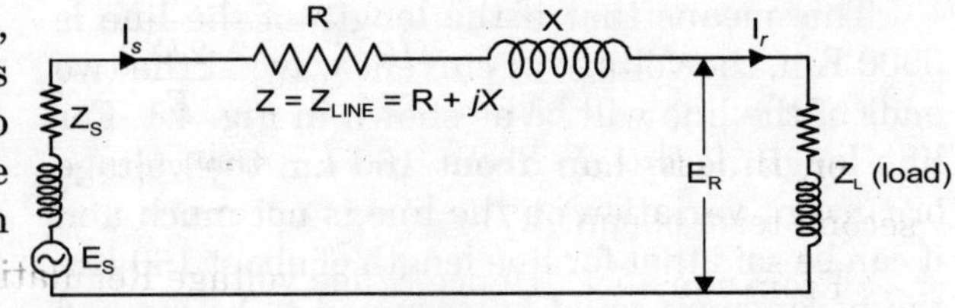

Fig. 4.4

4.2 DEFINITION OF PERFORMANCE FACTORS IN TRANSMISSION LINES

By performance of lines is meant the determination of efficiency and regulation of transmission lines. The efficiency of line is defined as

$$\% \text{ efficiency} = \frac{\text{Power delivered at the receiving end}}{\text{Power sent from the sending end}} \times 100 \quad ...(4.3)$$

or
$$\% \text{ efficiency} = \frac{\text{Power delivered at the receiving end}}{\text{Power delivered at the sending end} + \text{Losses}} \times 100 \quad ...(4.4)$$

The end of the line where load is connected is called the receiving end and where source of supply is connected is called the sending end.

The regulation of a line is defined as the change in the receiving end voltage, expressed in percent of full load voltage, from no load to full load, keeping the sending end voltage and frequency constant.

Hence this is expressed mathematically,

$$\% \text{ regulation} = \frac{E'_r - E_r}{E_r} \times 100 \quad ...(4.5)$$

One E'_r is the Receiving End voltage **under no load condition** and E_r is the receiving end voltage under full load condition. It is to be noted here that E_r' and E_r are the magnitude of voltages.

Note : E_s and E'_r can be also written as V'_s and V_r.

4.2.1 Expression for Voltage Regulation in *R* and *X* of the Line

Let us consider a short lines, where the load is thrown off, there will be no appriciable *IR* and *IX* voltage drops (refer to Fig. 4.4) so that the receiving voltage would be equal to the sending end voltage under these condition.

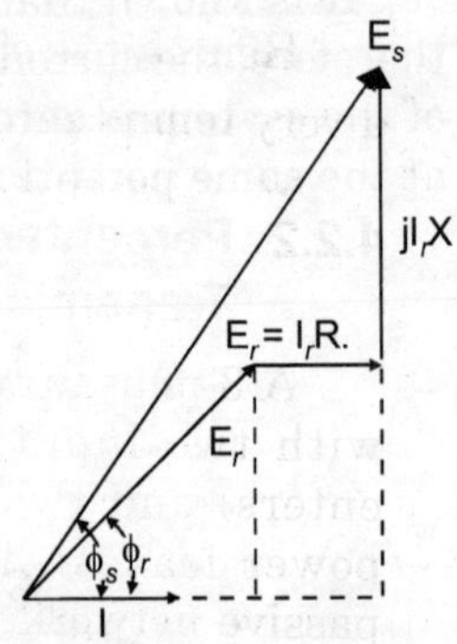

Fig. 4.5. Phasor diagram of a short line.

Hence, for a short line, % Regulation = $\frac{E_S - E_R}{E_R} \times 100\%$. Now from Fig. 4.5, we can write

$$E_S^2 = [E_r \cos \phi_r + IR]^2 + [E_r \sin \phi_r + IX]^2$$

$$= E_r^2 \left[\cos^2 \phi_r + \frac{2\,IR}{E_r} \cos \phi_r + \sin^2 \phi + \frac{2\,IX}{E_r} \sin \phi + I^2 (R^2 + X^2)/E_r^2 \right] \quad ...(4.6)$$

But, considering that $I^2 (R^2 + X^2)/E_r^2$ is pretty small in comparison with other terms of eqn. (4.6).

$$E_s = E_r\sqrt{\left[1+\frac{2IR}{E_r}\cdot\cos\phi_r+\frac{2IX}{E_r}\cdot\sin\phi\right]} = E_R\left[1+\left(\frac{IR}{r}\cos\phi_r+\frac{IX}{E_r}\sin\phi_r\right)\right]^{\frac{1}{2}} \quad ...(4.7)$$

or, $E_s \simeq E_r + I_r\,R\cos\theta_r + I_r \times \sin\phi_r$ (using binomial expansion and limiting on by to second term of eqn. 4.7)

From eqn. (4.7). Hence, %age voltage Regulation

$$= \frac{E_s - E_r}{E_r} = \frac{IR\cos\phi_r + IX\sin\phi_r}{E_r}\times 100\% \quad ...(4.8)$$

In fact, the load is, nomrally lagging in nature and %age voltage regulation is positive as evilent from the eqn. (4.8). However, when the load is of leading power factor, ϕ_r is $-ve$ and thus the expression for %age voltage regulation becomes :

$$\%\text{ regulation} = \frac{IR\cos\phi_r - IX\sin\phi_r}{E_r}\times 100 \quad ...(4.9)$$

Then, voltage regulation becomes zero, when $IR\ \cos\phi_r = I_c \times \sin\phi_r$

$$\therefore \qquad \tan\phi_r = \frac{\sin\phi_r}{\cos\phi_r} \quad ...(4.10)$$

This mean $\qquad \phi_r\,(\text{leading}) = \left(\frac{\pi}{2}-\theta\right)$

where θ is the angle of the line impedance. Voltage regulation many even be negative provided

$$I_X\sin\phi_r > IR\cos\phi_r$$

That means, $\qquad \tan\phi_r\ (\text{leading}) > \frac{R}{X}$

Hence, it is observed that voltage regulation is governed by the load power factor. This is a very important factor to be considered as the characteristics of the transmission line.

4.2.2 Percentage Voltage Regulation with ABCD Constant of Network of a Transmission Line

A 3-phase transmission line as a circuit with two input terminals (where power enters) and two output terminals (where power leaves). In fact in a four terminal passive network, the voltage and current on the receiving and the sanding end are related by the following pairs of equations :

$$V_s = AV_r + BI_r \quad ...(4.11)$$

$$I_s = CV_r + DI_r \quad ...(4.12)$$

(and can be represented in the Fig. 4.7) and use V_s, V_r is equation 4.11 and 4.12).

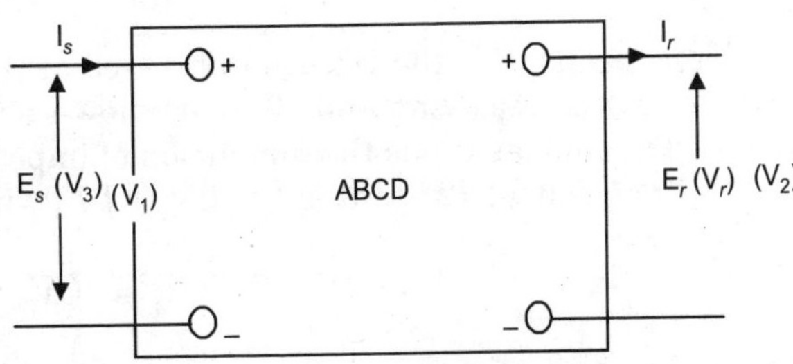

Fig. 4.6. A 2-point (part) representation of a transmission line.

where *A, B, C, D* are called the constants of the network. The transmission line, in fact, is aslo a four terminal network. These constants also can be used for calculation of regulation of the line as discussed below.

Normally, the quantities PI_r and cos ϕ_r at the receiving end are given and of course the *ABCD* constants (In next section we shall discuss what these constants mean in a transmission line). Then, to determine the sending end voltage, E_S (or V_S). We shall use the relation, $V_S = AV_r + BI_r$ as given in eqn. (4.11).

Now to determine V'_r the no load voltage at the receiving end from this equation, this equation can be used.

∴ When, $I_r = 0$ (in load condition), the eqn. (4.11) become $V_s = AV_r'$

That means, $$V'_r = \frac{V_S}{A} \text{ (when } I_r = 0)$$

$$\therefore \text{ \% regulation } = \frac{V_r' - V_r}{V_r} \times 100 \text{ from eqn. (4.5)}$$

as $$V'_r = V_S/A$$

$$\text{As \% regulation } = \frac{(V_S/A) - V_r}{V_r} \times 100.$$

To determine % η of transmission the following relation is made use of % η = % η = Power received at the receiving end)/Power at the receiving and + Losses)] × 100

$$= \frac{P}{P + 3I_r^r R} \times 100$$

4.3 DEFINITION OF CONSTANTS *A, B, C, D* AND VALUES OF *A, B, C, D* FOR SHORT LINE

In this section we shall define these constants to understand what they really mean, before we proceed further.

1. From eqn. (4.11), *i.e.*, $$V_S = AV_r + BI_r, \text{ with } I_r = 0$$

We get, $$V_s = AV_r \;\therefore A = \left.\frac{V_S}{V_r}\right|_{I_r=0}$$

This expression of *A* means that the voltage improved at the sending end per volt at the receiving end when receiving end is open, that is $I_r = 0$. It is **dismensionless.**

2. Again from the eqn. (4.11) with $V_r = 0$, we get

$$V_s = BI_r \;\therefore B = \left.\frac{V_S}{I_r}\right|_{V_r=0}$$

This means, *B* is the voltage impressed at the sending end to receiving end current at the short-circuited receiving end. This is known as transfer impedance in network theory. It is denoted in ohm, as it has the dimension of impedance.

3. From eqn. (4.12), that is $I_b = CV_r + DI_r$, with $I_r = 0$, we get

$$I_s = CV_r \;\therefore C = \left.\frac{I_S}{Vr}\right|_{I_r=0}$$

C is the current in amperes into the sending end per volt on the open circuited receiving end. It has the dimension of admittance.

4. From eqn. (4.12), that is $I_s = CV_r + DI_r$ with $V_r = 0$, we get

$$I_s = DI_r \;\therefore D = \left.\frac{I_S}{I_r}\right|_{V_r=0}$$

D is the current at the sending end per ampere of current at the short-circuited receiving end. This is dimensionless. These constants *A, B, C* and *D* are related for a passive network as below :

$$AD - BC = 1$$

This relation provides a good check on the values of these constants.

The sending end voltage and current can be written from the equivalent network as,

$$V_s = V_r + I_r Z \quad ...(4.13)$$

and $$I_s = I_r \quad ...(4.14)$$

Now comparing the coefficients of eqns. (4.11) and (4.12) and eqn. (4.13) and (4.14) respectively, the constants for **short transmission** line, are,

$$A = 1 \text{ and } B = Z, C = 0 \text{ and } D = 1 \quad ...(4.15)$$

Now checking the values of *A, B, C, D* with the relation

$$AD - BC = 1 \quad ...(4.15)$$

We get, $$1.1 - Z \cdot 0 = 1$$ (by putting the values found for $A = 1$, $B = Z$, $C = 0$ and $D = 1$)

This proves that the values calculated are correct for short transmission line.

Example 4.1. *A 3-phase 50 Hz transmission line has conductors of section 90 mm² and effective diameter of 1 cm and are places at the vertices of an equilateral triangle of side 1 metre. The line is 20 Km long and delivers a load of 10 MW at 33 kV and p.f. 0.8. Neglect capacitance and assume temperature of 20°C. Calculate the regulation and the efficiency of the line.*

Solution : Let us first determine the inductance of the line of Fig. 4.7

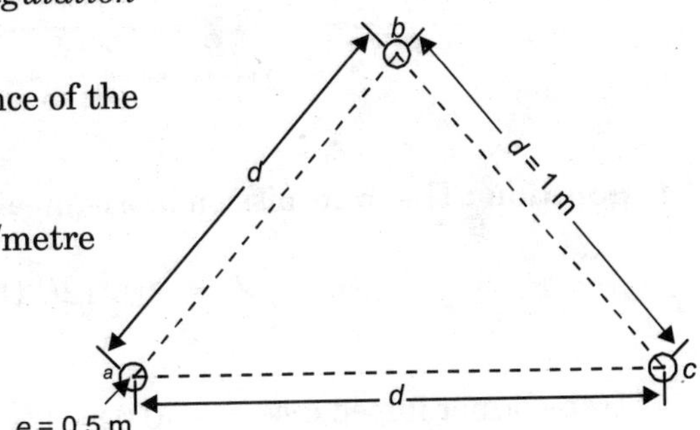

Fig. 4.7

$$L = 2\times10^{-7}\ln\frac{d}{r} = 2\times10^{-7}\ln\frac{100}{0.5} = 10.59\times10^{-7} \text{ H/metre}$$

∴ Total inductance of 20 Km length of line

$$= 20 \times 10.59 \times 10^{-7}$$

$$= 2.119 \times 10^{-2} \text{ H.}$$

∴ X_L = Inductive Reactance = Lw = 6.65 ohm

Now Resistance of copper conductor at 20°C is $\frac{1}{58}$ per metre length when the the section is 1 mm² (from a table).

∴ Resistance $= \frac{1}{58}\times\frac{20\times1000}{90} = 3.83$ ohm

The current $= I = \frac{10\times1000}{\sqrt{3}\times33\times0.8}$ (for a 10 MW, 33 kV, p/0.8 load)

$= 218.68$ nmps

Now as the capacitance is to be neglected, the receiving current is same as the sending end, and hence, we get, the loss on the line

$$= 3 \times (218.68)^2 \times 3.83 \;(= 3 \times I^2 \times R)$$
$$= 0.549 \text{ MW}$$

$\therefore$ The efficiency $\quad \eta = \dfrac{\text{Output}}{\text{Output} + \text{Loss}} = \dfrac{10}{10 + 0.549} = 0.9479$

Now, to determine the voltage regulation we determine the sending end voltage

$$\therefore \quad V_s = (V_r \cos\phi_r + I_r R) + j\,(V_r \sin\phi_r + I_r X)$$
$$= (19052 \times 0.8 + 218.68 \times 3.83 + j\,(19052 \times 0.6 + 218.68 \times 6.65)$$
$$= 16079 + j\,12885$$
$$\therefore \quad V_s = 20605 \text{ Volts}$$

Now as it is a SHORT LINE, the voltage Regulation will be,

% voltage Regulation $= \dfrac{20605 - 19052}{19052} \times 100 = \mathbf{8.15\%}$

Example 4.2. *Determine the voltage at the generating station and the efficiency of the transmission of the following 1-phase system as shown in Fig. 4.8. Transformer Ratio is 2 kV/ 11 kV. Resistance on LV side = 0.04 Ω. Resistance on HV side = 1.3 Ω. Reactance on LV & HV side, is 0.125 and 4.5 Ω respectively.*

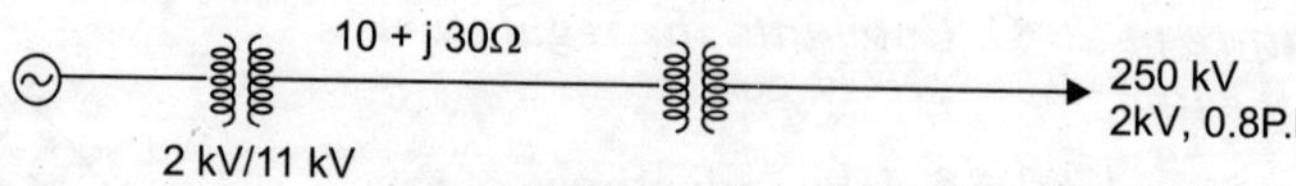

Fig. 4.8

Solution : The transmission line equivalent impedance when referred to L.V. side, will be

$$Z = 10 \times (2/11)^2 + j30 \times \left(\frac{2}{11}\right)^2 = (0.33 + j0.99)$$

Transformer impedance $= 0.04 + 1.3 \times \left(\dfrac{2}{11}\right)^2 + j0.125 + j4.5 \times \left(\dfrac{2}{11}\right)^2$

$$= 0.083 + j\,0.273$$

Then the equivalent circuit for the total system is given in Fig. 4.9. For a load of 250 kVA, 2 kV, 0.8 p.f.

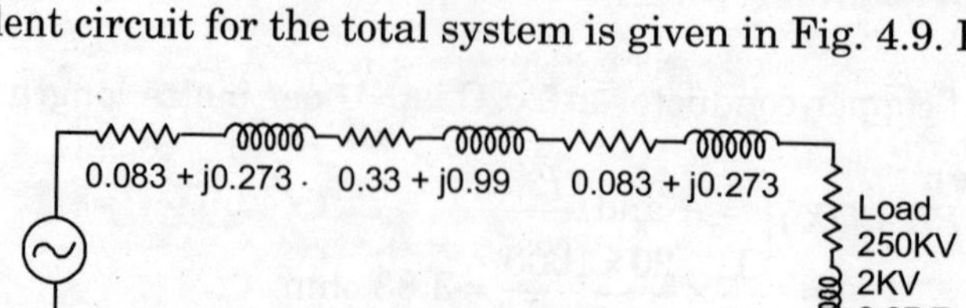

Fig. 4.9

The line current $= \dfrac{250 \times 1000}{2000} = 125$ amps.

The line loss $= I^2 R = 125^2 \times 0.496 = 7.7$ KW

$$\therefore \quad \% \eta = \frac{200}{200 + 7.7} \times 100 = 96.3\%$$

Now taking I_r as the reference, the sending voltage

$$V_s = (V_r \cos \phi_r + IR) + j\,(V_r \sin \phi_r + IX)$$

$$= (2000 \times 0.8 + 125 \times 0.496) + j\,(2000 \times 0.6 + 125 \times 1.536)$$

$$= 2168 \text{ volt}$$

$$\therefore \% \text{ Voltage Regulation} = \frac{V_s - V_r}{V_n} = \frac{2168 - 2000}{2000} \times 100 = \mathbf{98.4\%}$$

4.4 MIXED CONDITIONS FOR A LINE

In a transmission line mixed condition occur when some of the quantities like voltage, current, power and p.f. may be known for the sending end whereas others may be known for the receiving end. Now let us suppose that in a given situation the voltage at the sending end and the p.f. at the receiving end are to be kept fixed, and are known, and it is required to find the receiving end voltage for a given power received.

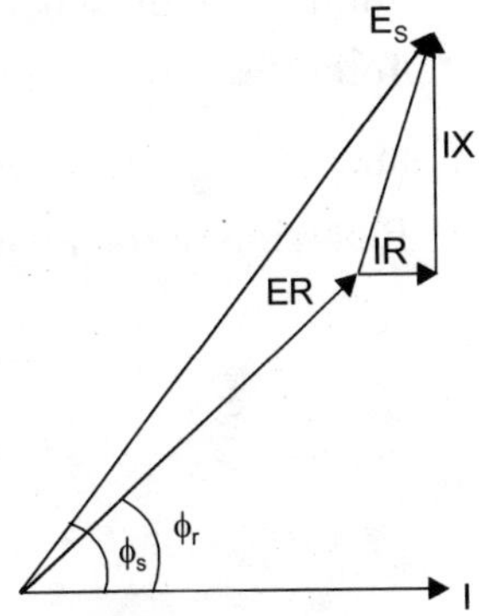

Fig. 4.10. Phasor diagram of a line.

Now let us refer to the Fig. 4.10 from where we can write,

$$E_s^2 = (E_r \cos \phi_r + IR)^2 + (E_r \sin \phi_r + IX)^2$$

$$= E_r^2 + 2IRE_r \cos\phi_r + 2IXE_r \sin\phi_r + I^2(R^2 + X^2)$$

$$\therefore \quad E_s^2 = E_r^2 + 2IRE_r \cos\phi_r + 2IXE_r \sin\phi_r + I^2Z^2 \qquad ...(4.16)$$

Now let, $E_r I \cdot \cos \phi_r$ = real power received = P ...(4.17)

and $E_r I \sin \phi_r$ = VAR received = Q ...(4.18)

Then, from eqn. (4.17), we get, $I = \dfrac{P}{E_r \cos\phi_r}$, and putting this value of I in eqn. (4.16), we get, after simplification,

$$E_s^2 = E_r^2 + 2(PR + QX) + \frac{P^2Z^2}{E_r^2 \cos^2 \phi_r}$$

Multiplying this equation throughout by $E_r^2 \cos^2 \phi_r$ and re-arranging we get,

$$E_r^4 = E_r^2\left[E_s^2 - 2(PR + QX)\right] - \frac{P^2Z^2}{\cos^2\phi_r} \qquad ...(4.19)$$

Now say, $\left[E_s^2 - 2(PR + QX)\right] = A$ and $\dfrac{P^2Z^2}{\cos^2\phi_r} = B$. In eqn. (4.19). Then eqn. (4.19) can be written as,

$$E_r^4 + AE_r^2 + B = 0, \text{ which on solving, we get}$$

$$E_r^2 = \frac{1}{2}\left(1 \pm \sqrt{A^2 - 4B}\right) \text{ or } E_r = \sqrt{\frac{A \pm \sqrt{A^2 - AB}}{2}} \qquad ...(4.20)$$

Note : +ve sign is taken for E_r.

Example 4.3. *A 3-phase, 50 Hz transmission line has the following data :*

1. *Length of the line = 10 Km.*
2. *Sending end voltage = 11 kV.*
3. *Load delivered at receiving end = 1000 kW at 0.8 p.f. log.*
4. *Resistance of each conductor = 0.500 ohm/Km.*
5. *Reactance of conductor = 0.560 ohm/Km.*

Calculate : (6) Line current, Receiving end voltage, a 'Efficiency of Transmission'.

Solution : From the datas given in example,

Total Resistance per phase = 0.5 × 10 = 5, ohm (for 10 Km long conductor)

Total Reactance (to neutial) = 0.56 × 10 = 5.6 ohm (for 10 Km long conductor)

and $E_s = (11/\sqrt{3})$ kV = 6,352 volts phase-to-neutral

Referring to the phasor diagram of Fig. 4.11 (*b*) for this line

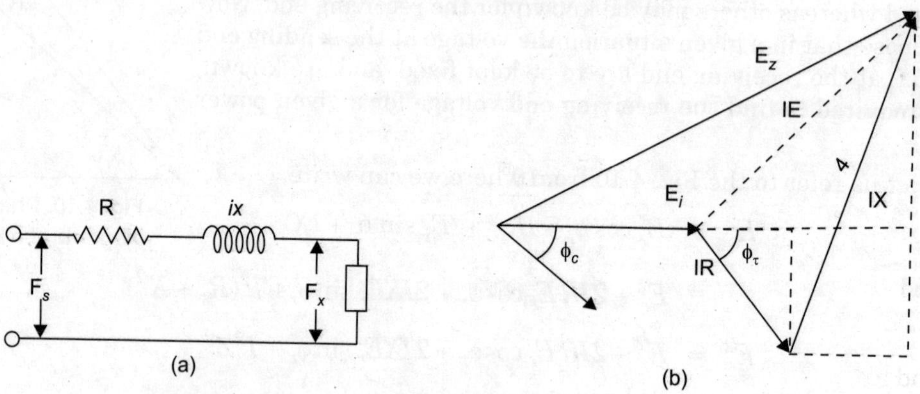

Fig. 4.11

$$E_s = \sqrt{(E_r + IR\cos\phi_r + IX\sin\phi_r)^2 + (IX\cos\phi_r - IR\sin\phi_r)^2}$$

However, considering that (I × cos ϕ_r – IR sin ϕ_r) is small enough to be neglected in comparison with the terms under the square root sign we get,

$$E_s = E_r + IR\cos\phi_r + IX\sin\phi_r$$

Hence, the approximate assumption means that ϕ_r and ϕ_s are equal that E_s and E_r, then, are in phase

We have $|I| = \dfrac{1000\times10^3}{\sqrt{3}\times11\times10^3\times0.8} = 65.6$ (From the load of 1000 KW at 0.8 p.f. long, and 11 kV voltage)

∴ IR = 328 volts $IX_{(\text{Total})}$ = 367.5 volt

So, 6.352 = E_s + 328 × 0.8 + 367.5 × 0.6 = E_s + 262 + 221 = E_r + 483

∴ E_r = 6352 – 483 = 5,869 volts, to neutral

= $\sqrt{3}\times5.869$ = 10.16 kV, line to line volts

Now $I^2 R$ loss in transmission line of 10 Km long = 3 × 65.6² × 5 × 10⁻³ = 6456 kW (for 3-phase).

Therefore, input to the line = 1000 + 6456 = 106456 kW

Hence efficiency of the transmission $= \frac{1000}{1064.56} \times 100 = 93.96\%$

Note, same problem can be solved by MIXED CONDITIONS method (stated above in section 4.4). Let us do it as an useful exercise below. From the datas deduced above.)

We have $Z^2 = (5)^2 + (5.6)^2 = 56.36$ and $P = \frac{1}{3} \times 1000 \times 10^3$ watts/phase

and $$Q = \frac{1}{3} \times \frac{1000 \times 10^3}{0.8} \times 0.6 \text{ (from section 4.9 above)}$$

$$= \frac{1}{3} \times 750 \times 10^3 \text{ var/phase}$$

Now (again for section 4.4), $$A = \frac{1}{3}[(11 \times 10^3)^2 - 2(1000 \times 10^3 \times 5 + 750 \times 10^3 \times 5.6)]$$

$$= \frac{10^6}{3}[121 - 2(5 + 4.2)] = 10^{-6} \times 34.20$$

and $$B = \frac{1}{9} \times \frac{10^6 \times 10^6 \times Z^2}{0.8^2} = \frac{1}{9} \times 10^{12} \times \frac{56.36}{0.64} = 9.78 \times 10^{12}$$

and also, $$4B = 4 \times 9.78 \times 10^{12} = 39.14 \times 10^{12}$$

Then, $$\sqrt{A^2 - 4B} = 10^6 \sqrt{1{,}170 - 39.14} = 10^6 \times 33.63$$

Now taking the +*ve* sign in the $\sqrt{\frac{A \pm \sqrt{A^2 - 4B}}{2}}$

We get, $$E_r = \sqrt{\frac{34.20 + 33.63}{2}} \times 10^3 \text{ volts/phase}$$

$$= 5.823 \text{ kV/ph or } \sqrt{3} \times 5.823 = 10.10 \text{ kV (line to line)}$$

∴ $$\text{Line Current} = \frac{1000 \times 10^3}{\sqrt{3} \times 10.10 \times 0.8} = 71.5 \text{ amps}$$

∴ Total $I^2 R$ loss in transmission line

$$= 3 \times (71.5)^2 \times 5 \times 10^{-3} = 77 \text{ KW}$$

Hence Efficiency of Transmission $= \frac{1000}{1{,}077} = 93.2\%$

Note : Result by both method is almost equal.

4.5 MEDIUM LENGTH LINES

In short length line we neglected the capacitance effect but for the transmission lines with length between 80 Km and 200 Km this capacitance effect (capacitance current of the system) is quite appriciable, and as such cannot be neglected for accurate calculation.

In fact the transmission line with length between 80 Km and 160 Km are categorised as medium length. Here in this medium length lines the capacitance is distributed over the entire length of the line but for a simple treatment its effect can be approximately taken into account by *r* using that it is lumped in the from of capacitors shunted across the line at one or more points.

The shunt capacitance in this medium length lines is either assumed to be concentrated at the middle of the line or half of the total capacitance is concentrated at each end of the line. The two configurations are known as nominal *"T"* and nominal "π" respectively, which are shown in Fig. 4.12 (*a*) and (*b*). However, it is to be noted that the two representations are approximate to the exact representations of the actual line. Also the two representations are not equivalent as can be seen by using star-Data transformations. But it has been found that they are good enough for practical purposes and do not involve much error (approximately with 5% error).

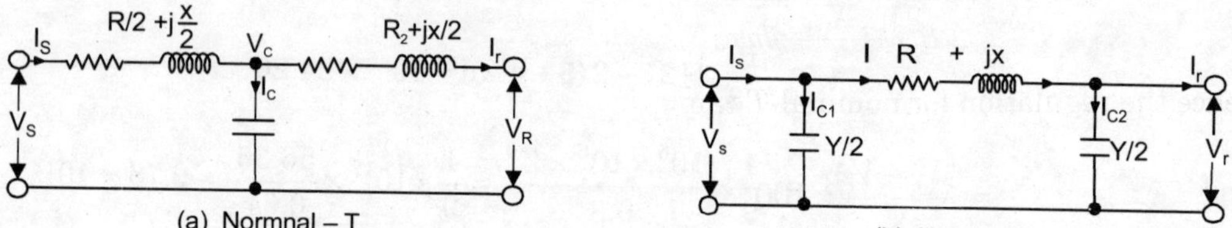

(a) Normnal – T

(b) Normnal – π

Fig. 4.12

4.5.1 Nominal – *T*

Determination of voltage Regulation, efficiency, *ABCD* constants of the medium length line.

1. Regulation and Efficiency

To calculate percentage voltage regulation for the medium length line with reference to Fig. 4.12 (*a*), first objective is to voltage at the receiving and, that is voltage across the capacitance, and also the sending-end voltage V_s.

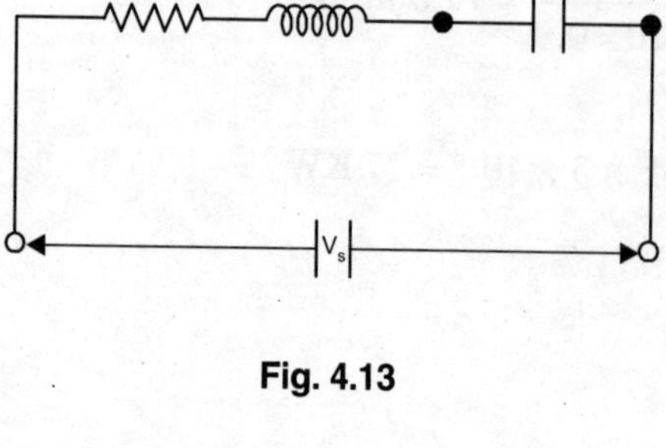

Fig. 4.13

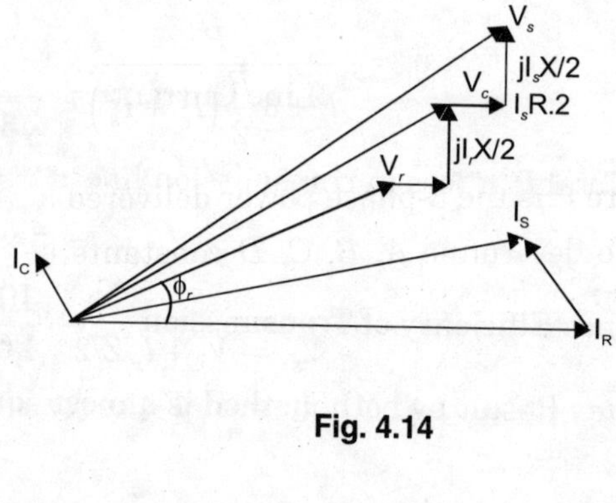

Fig. 4.14

From the phasor diagram of nominal-T, shown in Fig. 4.14, we can write

$$V_c = (|V_r| \cos\phi_r + j\,|V_r| \sin\phi_r) + I_r\left(\frac{R}{2} + j\frac{X}{2}\right) \quad ...(4.21)$$

$$I_c = j\,\omega c V_c \quad ...(4.22)$$

and $$I_s = I_c + I_r = j\,\omega c V_c + I_r \quad ...(4.23)$$

Hence, $$V_s = V_c + I_s\left(\frac{R}{2} + j\frac{X}{2}\right)$$

$$= (|V_r| \cos\phi_r + j\,|V_r| \sin\phi_r) + I_r\left(\frac{R}{2} + j\frac{X}{2}\right) + I_s\left(\frac{R}{2} + j\frac{X}{2}\right)$$

(refer to Fig. 4.13 (a))

Now to calculate regulation it is required to calculate V_r', the receiving end no load voltage V_s as calculated above is fixed in magnitude. The nominal-T circuit for this condition reduces to the cirenit of Fig. (4.15).

∴ From Fig. (4.15) we get

$$V_r = \frac{|V_s|(-j/wc)}{R/2 + j\frac{X}{2} - j/wc}$$

Hence the regulation for nominal-T can be obtained as % voltage regulation

$$= \frac{V_r' - V_r}{V_r} \times 100$$

Now to determine efficiency of the line if η we use the expression, as

$$\%\eta = \frac{\text{Power delivered at the receiving end}}{\text{Power delivered at the receiving end + Loss}} \times 100$$

Another expression used, is

$$\%\eta = \frac{\text{Power delivered at the receiving end}}{\text{Power sent at the sending end}}$$

However, due to small error in determining phase angle between sending end voltage and current (leading to an accurate calculation of efficiency). Hence, it is advised to use the first expression, which can be re-written as

$$\%\eta = \frac{P}{P + 3\frac{R}{2}\left(I_r^2 + I_s^2\right)} \times 100$$

Where P is the 3-phase power delivered at the receiving end, R is the resistance perphase

(b) To determine A, B, C, D constants for nominal-T (Fig. 4.13) we can write following relations :

$$V_s = V_r + I_r Z/2$$

$$I_c = V_c Y$$

But, $$I_s = I_r + I_c = I_r + V_c Y = I_r + \left(V_r + I_r \frac{Z}{2}\right)Y$$

Then, $$V_s = V_c + I_s \frac{Z}{2} = V_r + I_r \frac{Z}{2} + \left[I_r + \left(V_r + I_r \cdot \frac{Z}{2}\right)Y\right] \cdot \frac{Z}{2}$$

$$= V_r\left(1 + \frac{YZ}{2}\right) + I_r\left(\frac{Z}{2} + \frac{Z}{2} + \frac{YZ^2}{4}\right)$$

$$= V_r\left(1 + \frac{YZ}{2}\right) + I_r\left(Z + \frac{YZ^2}{4}\right)$$

$$V_s = V_r\left(1 + \frac{YZ}{2}\right) + I_r Z\left(1 + \frac{YZ}{4}\right) \quad \text{...(4.24)}$$

$$I_s = I_r\left(1 + \frac{YZ}{2}\right) + V_r Y$$

$$= YV_r + \left(1 + \frac{YZ}{2}\right)I_r \quad \text{...(4.25)}$$

Now writing the voltage and current equation with *A, B, C, D* constants of transmission line,

$$V_s = AV_r + BI_r$$
$$I_s = CV_r + DI_r$$

Hence comparing these two equations with the coefficients of eqns. (4.24) and (4.25), we can find

$$A = 1 + \frac{YZ}{2}, B = Z\left(1 + \frac{YZ}{4}\right), C = Y, D\left(1 + \frac{YZ}{2}\right)$$

From above it is clear that, $A = D$

and then $$AD - BC = \left(1 + \frac{YZ}{2}\right)^2 - YZ\left(1 + \frac{YZ}{4}\right) = 1 + \frac{Y^2Z^2}{4} + YZ - YZ - \frac{Y^2Z^2}{4} = 1$$

Hence, from this, the constants as obtained above are correct.

4.5.2 Nominal-π

Determination Regulation, Efficiency of *ABCD* constant of line.

(*a*) The vector diagram of nominal-π circuit of Fig. 4.12 (*b*) is shown in the Fig. 4.15. For nominal-π, it is desirable to take receiving end voltage V_r, as the reference vector.

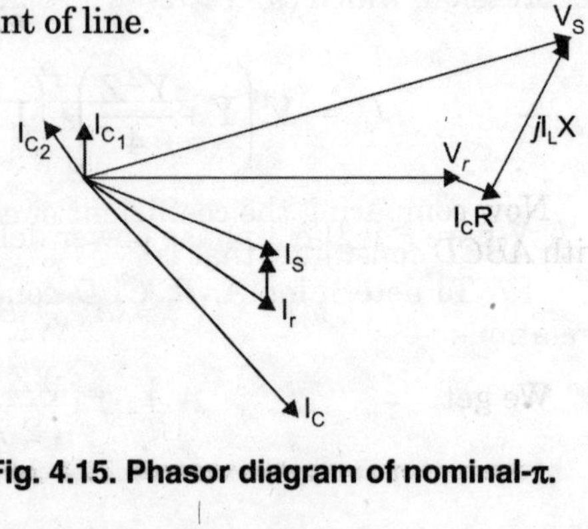

Fig. 4.15. Phasor diagram of nominal-π.

Let us refer to Fig. 4.16 for calculating V_S.

$$I_{c_1} = jV_r \frac{WC}{2}$$

$$I_1 = |I_r| (\cos\phi_r - j\sin\phi_r) + jV_r \frac{\omega C}{2}$$

$$V_s = V_r + \left[|I_r|\ (\cos\phi_r - j\sin\phi_r)\ + jV_r\frac{\omega C}{2}\right](R + jX) \text{ as } V_s = V_r + I_1 Z$$

$$\text{and } I_s = I_1 + I_{c_2} = I_1 + jV_s\frac{\omega C}{2}$$

$$\therefore \quad I_s = |I_r|\ (\cos\phi_r - j\sin\phi_r) + jV_r\frac{WC}{2} + j\frac{WC}{2}\left[V_r + \left(|I_r|\ (\cos\phi_r - j\sin\phi_r) + jV_r\frac{WC}{2}\right)(R + jX)\right]$$

Now having calculated the sending end voltage, it is required to find out no load receiving end voltage for regulation keeping sending end voltage constant in magnitude. The nominal-π circuit for this reduces to Fig. 4.16.

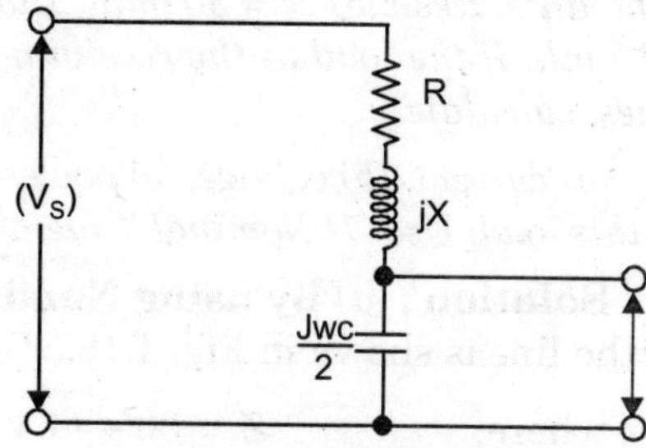

Fig. 4.16. Equation circuit under no load.

$$V_r' = \frac{|V_s|(-2j/wC)}{R + jX - \dfrac{1}{wC/2}}$$

$$\text{Therefore \% regulation} = \frac{V_r' - V_r}{V_r} \times 100$$

$$\text{and} \qquad \%\eta = \frac{P}{P + 3I_r^2 R} \times 100$$

(*b*) To determine *A, B, C, D* constants, let us refer to Fig. 4.12 (*b*).

$$I_{c_1} = V_r\frac{Y}{2}$$

$$I_1 = I_r + I_{c_1} = I_r + V_r\frac{Y}{2}$$

$$V_S = V_r + I_1 Z = V_r + (I_r + V_r Y/2)Z$$

$$V_S = \left(1 + \frac{YZ}{2}\right)V_r + ZI_r \qquad \text{...(4.26)}$$

$$I_s = I_1 + I_{c_2} = I_1 + V_s\frac{Y}{2} = I_r + V_r\frac{Y}{2} + \left[V_r\left(1 + \frac{YZ}{2}\right) + ZI_r\right]\frac{Y}{2}$$

$$\therefore \quad I_s = V_r\left(Y + \frac{Y^2 Z}{4}\right) + \left(1 + \frac{YZ}{2}\right)I_r \qquad \text{..(4.27)}$$

Now comparing the coefficient of eqns. (4.26) and (4.27) with those of equation of V_s and I_s with *ABCD* constants, that is

$$V_s = AV_r + BI_r \qquad I_s = CV_r + DI_r$$

We get

$$A = 1 + \frac{YZ}{2}$$

$$B = Z$$

$$C = Y\left(1+\frac{YZ}{4}\right)$$

$$D = (1 + YZ/2)$$

Hence, we find $A = D$, and $AD - BC = 1+\frac{Y^2Z^2}{4}+YZ-YZ-\frac{Y^2Z^2}{4}=1$

This mean that values of A, B, C, D are correct.

Example 4.4. *A 3-phase, 50 Hz overhead transmission line has the following distributed constants. Resistance = 30 ohm, Inductive Reactance = 65 ohm. Capacitive susceptance = 4.5 × 10^{-4} mh. If the load at the receiving end is 75 MVA at 0.8 p.f. lagging with 132 kV between the lines, calculate.*

(a) current, (b) voltage, (c) power factor at the sending end and (d) Regulation of transmission of this load. Use (1) Nominal T-method and (2) Nominal-π method, with relavent phasor diagram.

Solution : (*a*) **By using Nominal-*T* method.** The circuit diagram for nominal-*T* method of the line is shown in Fig. 4.18.

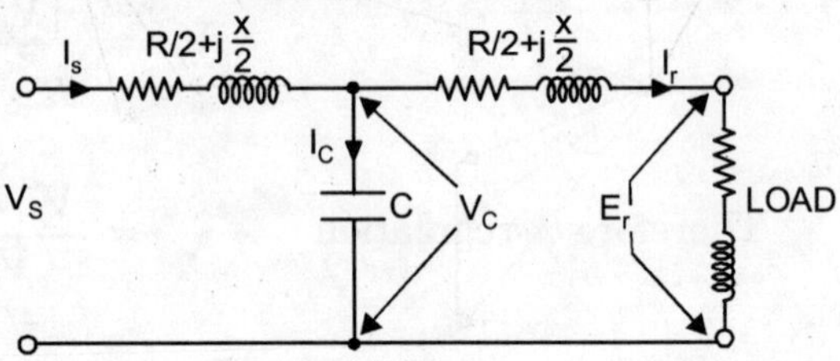

Fig. 4.17

where, $Z = 30 + j\,64$ ohm

$\therefore$ $Z/2 = 15 + j\,32$ ohm

Voltage per phase at Receiving end,

$V_r = \frac{132}{\sqrt{3}} = 76.23$ kV which is taken as the reference phasor in the phasor diagram of the circuit shown in Fig. 4.18. Then for a load of 75 MVA, 50 HZ, 0.8 p.f. (lag) at 132 kV.

The line current at Receiving end

$$I_r = \frac{75\times10^3}{\sqrt{3}\times132} \angle \cos^{-1} 0.8 = 328\,(0.8 - j\,0.6) - j\,197 = 262.5 - j\,197 \text{ amps}$$

Now, V_c (across capacitor C)

$$= V_r + I_r\,(Z/2) = 76.23 + (262 - j\,197)\,(15 + j\,32) \times 10^{-3} = 86 + j\,6 \text{ kV}$$

and $I_a = Y \cdot V_c = j\,4.2\times10^{-4}\,(86 + j\,6)\times10^3 = (j34.5 - 2.3)$ Amps

From the Fig. 4.18, we see

$\therefore$ $I_s = I_r + I_c = 30°\,67' \angle -3161$ amps (by putting values of I_r & I_c)

$\therefore$ $V_s = V_c + I_b\times\left(\frac{R}{2}+j\frac{X}{2}\right) = 95 + j11.8 = 96\angle 6°55 \text{ kV} = 166 \text{ kV}$

(by putting corresponding values)

and $\phi_s = (6^0\,55^1 + 31^0\,61^1) = 38°\,36'$ with I_s lagging V_s

$\therefore$ Sending end p.f. = $\cos\phi_s = \cos\,(38^0\,36^1) = 0.78$ (lagging)

Now refer to phasor Fig. 4.18

Power at sending end is given by, $P_s = \sqrt{3}\ V_s I_s \cos\phi_s = \sqrt{3} \times 166 \times 306.7 \times 78 = 68,542$ KW = **68.54 MW**

Power received, $P_r = 75 \times 0.8 = 60$ MW

Power Transmission Efficiency

$$= \frac{P_r}{P_s} = \frac{60}{68.54} \times 100 = 87.2\%$$

Power Regulation

$$= \frac{V_s - V_r}{V_r} \times 100 = \frac{166 - 132}{132} \times 100 = 25.8\%$$

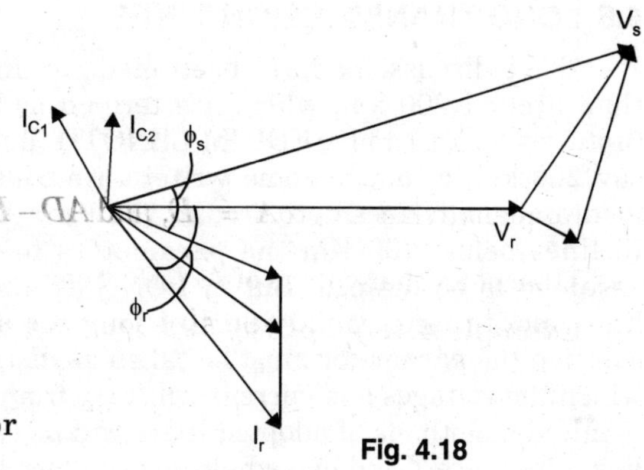

Fig. 4.18

2. By Nominal-π Method

In the Fig. 4.20 the circuit diagram for Nominal π method

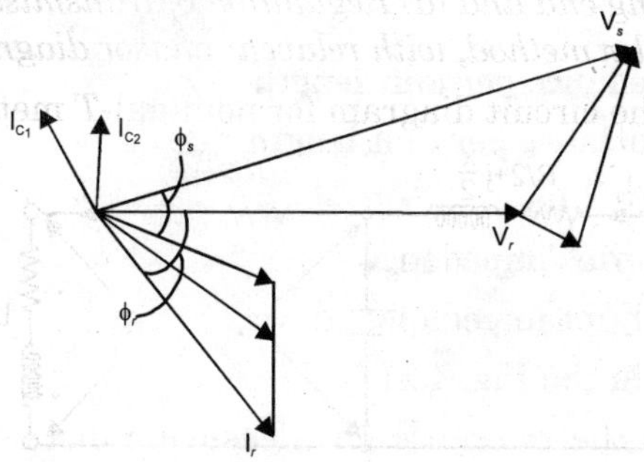

Fig. 4.19

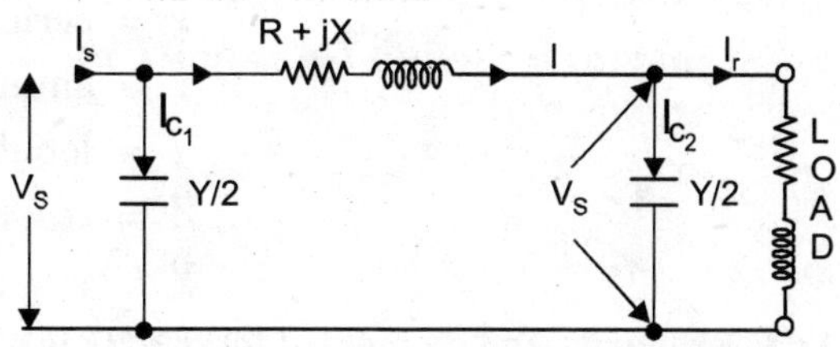

Fig. 4.20

We $\quad V_r = 76.23\,(1 + j\,0)$ (The reference phasor)

$I_r = 262.5 - j\,197$ amps and $Y/2 = j\,2.1 \times 10^{-4}$ mho

$\therefore \quad I_{c_2} = V_r\,(Y/2) = 76.23 \times 10^3 \times j\,2.1 = j\,15.26$ amps

Line current $\quad I = I_r \times IZ = (262.5 - j\,197) + j\,15.26$

$= (262.5 - j\,181.74)$ amp.

Then $\quad V_s = V_r + IZ = 76.23 + (262.5 - j\,181.74)\,(30 + j\,64)$

$= 165.8$ kV (line to line)

$I_{c_1} = V_s\,(Y/2) = (94.97 + j\,11.43) + j\,2.1 \times 10^{-4} = (-2.29 + j\,19)$

Then $\quad I_s = I + I_{c_1} = 260.23 - j\,162.68 = 307 \angle \cdot 32° \, 04'$ amp

(by putting values of I & I_{c_1})

$\therefore$ Sending power factor $= \cos 38° \, 59' \, \overline{0.7784}$ (l_{ag})

Now power at sending end $= P_s = \sqrt{3} \times E_s I_s \cos \phi_s = \sqrt{3} \times 165.8 \times 307 \times 0.7784$

$=$ **68.59 MW**

$$\therefore \text{ Transmission efficiency } = \frac{P_r}{P_s} = \frac{75 \times 0.8}{68.59} = \frac{60}{68.59} \times 100 = 87.46\%$$

$$\text{Then, Percentage Regulation } = \frac{V_s - V_r}{V_r} \times 100 = \frac{165.8 - 132}{132} = 25.47\%$$

The solution of this eqn. (4.32) is :

$$V = A \cdot \exp.(\sqrt{YZ} \cdot x) + B \cdot \exp.(\sqrt{YZ} \cdot x) \quad ...(4.33)$$

Now, from eqn. (4.28), that is, $\frac{dV}{dx} = I \cdot Z$, and eqn. (4.33), let

$$Z_c = \sqrt{\frac{Z}{Y}} \text{ and } Y = \sqrt{YZ} = \alpha + j\beta \quad ...(4.34)$$

where Z_c is known as characteristic impedance and Y propagation constant.

The equations (4.33) and (4.34) can be, hence, rewritten as

$$V = A \cdot e^{Yx} + B \cdot e^{-Yx} \quad ...(4.35)$$

$$I = \frac{1}{Z_c}(A \cdot e^{Yx} + B \cdot e^{-Yx}) \quad ...(4.36)$$

Now two constants A and B are to be determined, hence the boundary condition should be known.

As mentioned previously, the receiving end voltage and current are known.

Therefore, at $x = 0, V = V_r$ and $I = I_s$

Now substituting these values in equations (4.35) & (4.36) we get,

$$V_r = A + B$$

and

$$I_s = \frac{1}{I_c}(A - B)$$

Then from these two equations we get

$$A = \frac{V_r + I_s Z_c}{2}, \text{ and } B = \frac{V_r - I_r Z_c}{2}$$

Now substituting these values of A and B in eqns. (4.35) and (4.36) we get,

$$V = \left(\frac{V_r + I_r Z_c}{2}\right) e^{Yx} + \left(\frac{V_r - I_r Z_c}{2}\right) \cdot e^{-Yx} \quad ...(4.37)$$

and,

$$I = \frac{1}{Z_c}\left[\left(\frac{V_r + I_r Z_c}{2}\right) \cdot e^{Yx} - \left(\frac{V_r - I_r Z_c}{2}\right) \cdot e^{-Yx}\right] \quad ...(4.38)$$

As we know that V and I are the voltage and the current at any point distant x from the receiving end, it can be seen from the above expressions that V and I (in magnitude and phase) are functions of the distance x, receiving end voltage V_r and current I_r and the parameters of the line which means they vary as we move from receiving end towards the sending end of the line.

It is here important to understand to the physical significance of the voltage and current equations (4.37) and (4.38) respectively, where the quantities Z_c and Y are couples, for the reasons below :

$$Z_c = \sqrt{\frac{Z}{y}} = \sqrt{\frac{r + jwL}{g + jwc}}$$

But for a lossless line, $r = 0$, $g = 0$ the expression Z, therefore, becomes, $Z_C = \sqrt{\dfrac{L}{C}}$, and pure resistance and this is known as surge impedance of the line. When dealing with high frequencies on surges normally the losses are neglected and hence, the characteristic impedance become the surge impedance.

In fact, surge impedance loading of a line is the power transmitted when the line is transmitted through a resistance equal to surge impedance. The approximate value of surge impedance for overhead lines is 400 ohm and that for cable about 40 ohm. The phase angle for Z_C for transmission lines is usually between 0° and –15°.

A line terminated in its characteristic impedance is called a flatline on an infinite line. In fact the term, infinite line, arises from the fact that a line of of infinite length cannot have a reflected wave. However, the lower value of surge impedance in case of cables is due to the relatively large capacitance and low inductance of the cables.

In the propagation constant $Y = \alpha + j\,\beta$, the real part is known as attenuation constant and the quadrature component β the phase constant and is measured in radian per unit length.

With these explanation above, the equation (4.37) becomes,

$$V = \frac{V_r + I_r Z_c}{2} \cdot e^{\alpha x} \cdot e^{j\beta x} + \frac{V_r - I_r Z_c}{2} \cdot e^{-\alpha x} \cdot e^{-j\beta x} \quad ...(4.39)$$

In this expression, the first term is called the incident voltage wave as its value increases as x is increased. Now as we are taking receiving end as the reference and as x increases the value of voltage increases which means a voltage wave decreases in magnitude as it travels from the sending end towards the receiving end, and that is why this part of the voltage in the above expression is called incident voltage.

Then for similar reasons the second part is called the reflected voltage. Now, at any point along the line, voltage is the sum total of these two components, that means sum of incident and reflected voltages.

Then as the expression for the current is similar to the voltage the current can also be considered as the sum of incident and reflected current waves.

Then, as explained above, the expressions for voltage and current can be rearranged as below :

$$V = V_r \cdot \frac{e^{yx} + e^{-yx}}{2} + I_r Z_c \frac{e^{yx} - e^{-yx}}{2}$$

$$= V_r \cdot \cosh yx + I_r Z_c \cdot \sinh yx \quad ...(4.40)$$

and, similarly,

$$I = \frac{1}{Z_c}\left[V_r \cdot \frac{e^{yx} - e^{-yx}}{2} + I_r Z_c \frac{e^{yx} + e^{-yx}}{2}\right]$$

$$= \frac{1}{Z_c}[V_r \sinh yx + I_r Z_c \cosh yx]$$

$$= \frac{V_r}{Z_c} \sinh yx + I_r \cosh yx \quad ...(4.41)$$

Now rewriting these equations for $x = l$, where $V = V_s$ and $I = I_s$.

$$\therefore \quad V_s = V_s \cosh yl + I_r Z_c \sinh yl \qquad ...(4.42)$$

$$I_s = V_r \frac{\sinh yl}{Z_c} + I_r \cosh yl \qquad ...(4.43)$$

These two equations, in fact, relate the sending end voltage and current with the receiving end quantities. Previously we have said, in this section at the beginning that V_s and I_s are related by the general equation with *A, B, C, D* constants as :

$$V_s = AV_r + BI_r \qquad ...(4.11)$$

$$I_s = CV_r + DI_r \qquad ...(4.12)$$

where, *A, B, C, D* are such that

$$A = D \text{ and } AD - BC = 1$$

Now comparing the coefficients of the equations (4.42) and (4.41) with equations (4.11) and (4.12) respectively, we get

$$A = \cosh yl, B = Z_c \sinh yl, C = \frac{\sinh yl}{Z_c}, D = \cosh yl$$

Now from this, we also see that

$$A = D = \cosh yl$$

and

$$AD - BC = \cosh^2 yl - Z_c \sinh yl \frac{\sinh \cdot yl}{Z_c} = 1$$

4.6.1 Examples Below on the Above Section will Explain Further Above a Points

Example 4.5. *A single circuit, 50 Hz, 3-phase, transmission line has the following parameters $r = 0.1\ \Omega/km$, $L = 1.4 \times 10^{-3}\ H/km$, $c = 8 \times 10^{-9}\ F/km$, $g = 4 \times 10^{-8}$ mho/km. The voltage at the receiving end is 400 kV. If the line is open at the receiving end find the r.m.s. value and phase angle of the of follows :*

1. *The incident voltage to neutral at the receiving end (take this voltage as reference in computing the phase angles of other voltages.*
2. *The reflected voltage to neutral at the receiving end*
3. *The incident voltage to neutral at 200 km from the received end*
4. *The resultant voltage to neutral and the line voltage at 200 Km from the receiving end.*

Solution : As per example,

$$R = r + j\,wL = (0.1 + j\,1.4 \times 200 \times 50 \times 10^{-3}) = 0.4510\ 4\ \angle 77.19° - 8\ \Omega/\text{Km}$$

$$Y = g + j\,wc = (4 \times 10^{-8} + j \times 2\,\pi \times 50 \times 8 \times 10^{-9}) = (4 + j\,251) \times 10^{-8}\ \text{mho/Km}$$

$$= 251 \times 10^{-8} \angle 89.1^6\ \text{mho/Km}$$

Then, $$Z_c = \sqrt{\frac{Z}{Y}} = 423.9 \angle -5.95°\ \Omega$$

$$y = \sqrt{ZY} = 1.064 \times 10^{-3} \angle 83.14°$$

As $$V(x) = \frac{V_2 - I_2 Z_c}{2} \cdot e^{yx} + \frac{V_2 - I_2 Z_c}{2} \cdot e^{-yx} = V_x^+ + V_x^- \qquad ...(a)$$

where V_x^+ = Incident voltage wave and V_x^- = Reflected voltage wave. As the line is open at the receiving end, hence, $I_2 = O$.

Then,

1. The incident voltage at the receiving end is, referring to eqn. (*a*)

$$V_0^+ = \frac{V_2 + 0}{2} \cdot e^0 = \frac{V_2}{2}$$

But, $$V_2 = |V_2| \angle 0° \text{ and } |V_2| = \frac{400}{\sqrt{3}} = 231\,V$$

$$\therefore \quad V_0^+ = \frac{231\angle 0°}{2} = 115.5\angle 0° \text{ kV}$$

2. The reflected voltage at the receiving and is (again refer to enq. (*a*))

$$V_0^{-1} = \frac{V_2 - 0}{2} \cdot e^0 = \frac{V_2}{2} = 115.5° \angle 0° \text{ kV}$$

3. When $x = l = 200$ km^2. We find $y \times l = 0.213 \angle 83° \cdot 14° = 0.0254 + j\,0.21127$

$$V_1^+ = \frac{V_2}{2} \cdot e^{yl} = 115.5\, e^{(0.0254 - j\,0.21127)}$$

$$= 115.5 \times 1.0257\,(0.978 + j\,0.2097) = (115.561 + j\,2484) \text{ kV}$$

Now at $x = l = 200$, we find,

$$V_2^- = \frac{V_2}{2} \cdot e^{-Yl} = 115.5 \times 0.9749\,(0.978 - j\,0.2097) = 110.126 - j23.6124$$

5. $$V_L = V_L^+ + V_L^- = 115.561 + j\,24.84 + 110.126 - j\,23.6124$$

$$= (222.687 + j\;122.76) \text{ kV to neutral}$$

$$\therefore \quad V_L = 225.69 \text{ kV to neutral}$$

$\therefore$ Line to line voltage = $\sqrt{3} \times 225.69 = 390.9071$ kV

Example 4.6. *Determine A, B, C, D constants for a 3 phase, 50 Hz transmission line 200 Km long having the following distributed parameters.*

$L = 1.20 \times 10^{-3}$ H/km, $c = 8 \times 10^{-9}$ F/km, $r = 0.15\ \Omega$/km, $g = 0$

Solution : From the example,

$$Z = r + j\,wL = (0.15 + j\,0.376)\ \Omega/\text{km}$$

$$Z = Zl = 200 \times (0.15 + j\,0.376)\ \Omega/\text{km} = 81.093 \angle 68° \cdot 28'\ \Omega$$

$$Y = g + j\,wc = j\,2512 \times 10^{-9}\ \Omega/\text{km}$$

$$Y = y \times l = j\,.0005024 \angle 90° \text{ mho}$$

Now, $$yl = \sqrt{ZY} \cdot l = \sqrt{Z \cdot Y} = 0.2018\angle 79.14° = 0.03802 + j\,0.19818$$

$$yl = (\alpha + j\,\beta)\,l = 0.03802 + j\,0.19818 = \alpha\,l + j\,\beta\,l$$

or $\therefore$ $$\alpha l = 0.03802 \text{ radian}, \beta\,l = 0.19818 \text{ radian}$$

$$\cosh yl = \cosh(\alpha\,l) \cdot \cos(\beta\,l) + j \sinh(\alpha\,l) \cdot \sin(\beta\,l)$$

By evaluating values of $\alpha\,L$ and $\beta\,l$, we get, $\cosh yl = 0.98113 \angle 0.43702°$

$$\sinh yl = \sin(\alpha\,l) \cosh(\beta\,l) + j \cos(\alpha\,l) \sinh(\beta\,l) = 0.200515 \angle 79.289°$$

Therefore, we get, $A = D = \cosh \gamma l = 0.98/\angle 0.44°$

$$B = Z_c \sin(\gamma l) = \sqrt{\frac{Z}{Y}} \cdot (\gamma l)$$

$$= 80.59 \angle 68.43° \ \Omega$$

$$C = \frac{1}{Z_c} \sinh \gamma l = 0.49 \times 10^{-3} \angle 90.149° \text{ mho}$$

Example 4.7. *A 132 kV, 3-phase, 50 Hz transmission line of 200 km his the following distributed parameters :*

$$L = 1.3 \times 10^{-3} \ H/km, C = 9 \times 10^{-9} \ F/km, r = 0.2 \ \Omega/km, g = 0$$

Find the sending and voltage, current, power factor and efficiency when delivering 50 MVA at 0.8 p.f. lagging.

Solution : From the parameters of the example, let us first line Z and Y

$$Z = r + j\,wL = 0.2 j\, 2\pi \times 50 \times 1.3 \times 10^{-3} = 0.2 + j\,0.4084 = 0.4547 \angle 63.9° \ \Omega/\text{km}$$

$$Y = j\,wL = j \times 2\,\pi\,50 \times 9 \times 10^{-9} = 2827.433 \angle \times 10^{-9} \text{ mho/km}$$

$$\therefore \ Z_c = \sqrt{\frac{Z}{Y}} = \sqrt{\frac{0.4547 \angle 63.9°}{2827.433 \angle \times 10^{-9} \angle 98}} = 401.04 \angle -13.045 \text{ ohm}$$

$$\gamma = \sqrt{ZY} = 1.1339 \times 10^{-3} \angle 76.95°$$

$$\therefore \ \gamma l = 0.22678 \angle 76.95 = 0.051189 + j\,0.2208 \ (= \alpha + j\,\beta)$$

Now $\cosh \gamma l = \cosh(0.051189) \cdot \cos 0.2208 + j \sinh(0.051189) \cdot \sin 0.2208$

$$= 0.97698 + j\,0.0112158 = \mathbf{0.97704} \angle 0.65770$$

and $\sin \gamma\, l = \sinh(0.97698 + j\,0.0112158)$

$$= \sinh(0.97698) \cos(0.112158) + j \sinh(0.9769) \sin(0.112158)$$

$$= 0.05121 \times 0.9757 + j\,1.0013 \times 0.219$$

$$= 0.0499 + j\,0.2193 = 0.22489 \angle 77.17°$$

Now, $|V_2| = \dfrac{132}{\sqrt{3}} = 76.2 \text{ kV}$

Taking V_2 as reference, $V_2 = |V_2| \angle 0° = 76.2 \angle 0° \text{ kV}$

$\therefore$ $3\,|V_2|\,|I_2| = 50 \times 10^3 \text{ kVA} \therefore |I| = \dfrac{50 \times 10^3}{3 \times 76.2} = 218.7\,A$

With p.f. 0.8 $I_2 = 218.7\,(0.8 - j\,0.6) = 218.7 \angle -36.86\,A$

But $V_1 = V_2 \cosh \gamma l + I_2 Z_c \sinh \gamma l$

By evaluating all the values in this eqn. of V_2, $\cosh \gamma l$, I_2, $Z_c \sinh \gamma l$

We get, $V_1 = 91.9769 + j\,9.89 = 92.507 \angle 6.14° \text{ kV/per ohm}$

$$I_1 = \frac{V_2}{Z_c} \sinh \gamma l + I_2 \cosh \gamma l$$

Again, by evaluating value of V_2, Z_c, $\sinh \gamma l$, I_2, $\cosh \gamma l$, we get in I this equation for I_1, we get

$$I_1 = 172.24 - j\,83.49 = 191.408 \angle -25.86°$$

Now I_2 lags the voltage V_2 by an angle 25.86° but V_1 leads V_2 by an angle 6.1376°, and hence, the phase displacement between V_1 and I_1 is 31.9976°. Here the current I_1 lags behind the voltage V_1.

$\therefore$ $$P_1 = \text{Total sending and power} = 3\,|V_1|\,|I_1| \cos 31.9976 = 45.05 \text{ MW}$$

$$P_2 = \text{Total receiving and power} = 50 \times 0.8 = 40 \text{ MW}$$

$\therefore$ $$\eta = \frac{P_2}{P_1} = \frac{40}{45.05} = 0.888 \times \text{or } 100 = 88.8\%$$

4.7 EQUIVALENT CIRCUIT REPRESENTATION OF LONG LINE

In the above sections for long lines parameters were assumed to be distributed and for which there we hence developed the expressions for voltage and currents. In this section now we shall derive equivalent-π and equivalent-T circuit for such long line. In fact, the nominal-π, and nominal-T circuit do not represent the lines exactly because they do not account for the parameters of the lines being uniformly distributed.

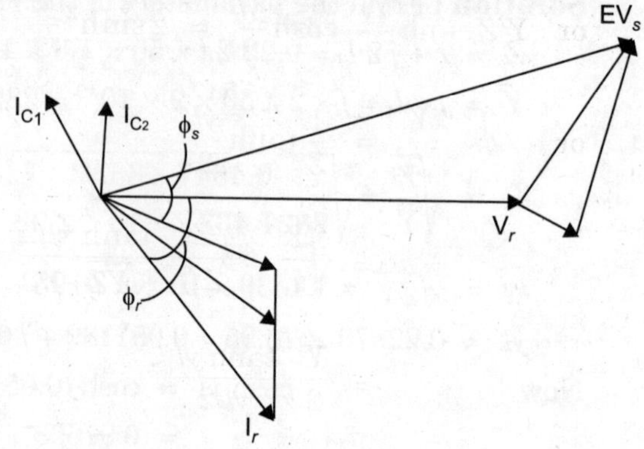

Fig. 4.21. Equation act of a long line.

4.7.1 Equivalent-π of Long Line

In this case let us assume that two terminal conditions, that is the sending end receiving end voltage and current can be related with the following equivalent network of Fig. 4.22. Following relations can be derived from the network of the Fig. 4.22.

$$V_s = \left(1 + \frac{Y'Z'}{2}\right)V_r + Z'I_r \quad \text{...(4.44)}$$

$$I_s = Y'\left(1 + \frac{Y'Z'}{4}\right)V_r + \left(1 + \frac{Y'Z'}{2}\right)I_r \quad \text{...(4.45)}$$

Now, comparing these two equations with equations (4.42) and (4.43) respectively we get

$$\cosh yl = 1 + \frac{Y'Z'}{2}$$

$$Z_c \sinh yl = Z'$$

$$\frac{1}{Z_c}\sinh yl = Y'\left(1 + \frac{Y'Z'}{4}\right)$$

and $$\cosh y = 1 + \frac{YZ'}{2}$$

Then considering the equation $Z' = Z_c \sinh yl$, we can deduced the following relation :

$$Z' = Z_c \sin\text{h}\, yl = \sqrt{\frac{Z}{Y}} \cdot l \cdot \frac{\sinh yl}{\sqrt{YZ} \cdot l} \cdot \sqrt{YZ} = Zl \cdot \frac{\sinh yl}{yl} = Z \cdot \frac{\sinh yl}{yl}$$

This relation means that to get the equivalent series the lumped impedance Z should be multiplied by (sin h yl)/Yl. Then to get the shunt are of the equivalent-π circuit, we shubstitute Z' in equation :

$$\cosh yl = 1+\frac{Y'Z'}{2}$$

$$\therefore \quad \cosh yl = 1+\frac{Y'}{2}\cdot Z_c \sinh yl$$

This can be written as below :

$$\frac{Y'}{2}Z_c\cdot \sin h\, yl = \cosh yl - 1 = \cosh^2\cdot\frac{yl}{2}+\sinh^2\frac{Yl}{2}-\cosh^2\frac{Yl}{2}+\sinh^2\frac{Yl}{2}$$

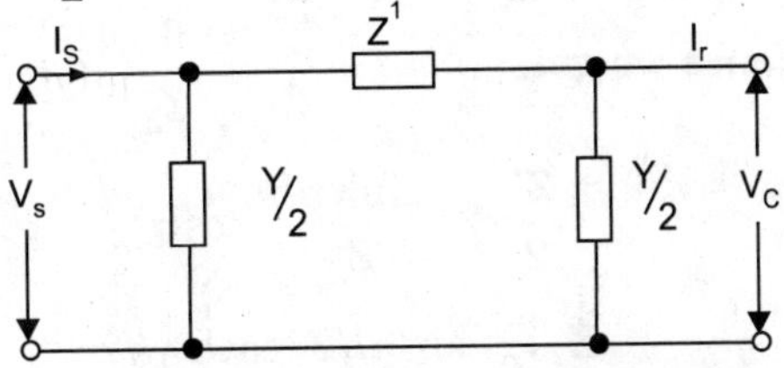

Fig. 4.22. Equivalent-π representation.

$$\text{or}\quad Y'Z_c\sinh\frac{Yl}{2}\cdot\cosh\frac{yl}{2} = 2\sinh^2\frac{yl}{2}$$

$$\text{or}\quad \frac{Y'}{2} = \frac{1}{Z}\tanh\frac{yl}{2}$$

$$= \sqrt{\frac{Y}{Z}}\cdot\frac{\sqrt{YZl}}{2}\cdot\frac{\tanh yl/2}{\sqrt{YZ}\cdot 1/2}$$

$$= \frac{Y}{2}\cdot\frac{\tanh yl}{yl/2}$$

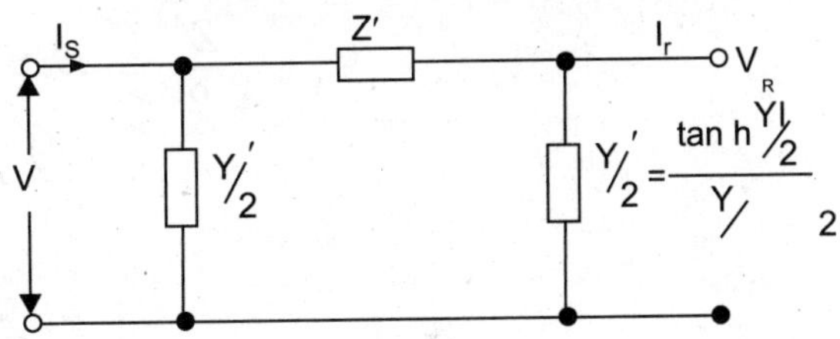

Fig. 4.23

where Y is the shunt admittance. And this known to get the shunt arm of the nominal-π should be multiplied by $\frac{\tanh yl}{yl}$. Hence, the equivalent-π cirenit can be represented by the Fig. 4.23.

4.7.2 Equivalent-*T* Representation of Long Line

Let us refer to the Fig. 4.24 to determine the equivalent-T circuit from this network we can drive the following relation, but with reference to the equations (4.19) & (4.20) for nominal-T network.

$$V_s = \left(1+\frac{Y'Z}{2}\right)V_r + Z'\left(1+\frac{Y'Z'}{4}\right)I_r \qquad ...(4.44)$$

and

$$I_s = Y'V_r + \left(1+\frac{Y'Z'}{2}\right)I_r \qquad ...(4.45)$$

Now company these two equation with equations (4.42) & (4.43). We get

$$1+\frac{Y'Z'}{2} = \cosh yl \qquad ...(4.46)$$

$$Z'\left(1+\frac{Y'Z'}{4}\right) = Z_c \sinh yl \qquad ...(4.47)$$

$$Y' = \frac{\sinh yl}{Z_c} \qquad ...(4.48)$$

and
$$1 + Y' Z'/2 = \cosh y^2 l \qquad ...(4.49)$$

Now to determine the shunt branch of equivalent-T circuit the shunt branch of the nominal-T (lumped shunt admittance) should be multiplied by the factor $\frac{\sinh Yl}{Yl}$.

Now to get the series impedance of the equivalent-T, equation $Y' = \frac{\sinh yl}{Z_c}$ is substituted in equation $1 + \frac{Y'Z'}{2} = \cosh yl$.

Hence we get,
$$1 + \frac{Z'}{2} \cdot \frac{\sinh yl}{Z_c} = \cosh yl$$

$$\therefore \quad \frac{Z'}{2} \cdot 2 \cdot \frac{\sinh yl}{Z_c} \cdot \cosh \frac{yl}{2} = \cosh \frac{yl}{2} + \sinh^2 \frac{yl}{2} - \cosh^2 \frac{yl}{2} + \sinh^2 \frac{yl}{2}$$

$$\therefore \quad \frac{Z'}{2} \cdot 2 \cdot \frac{\sin yl/2 \cdot \cosh yl/2}{Z_c} = 2 \sinh^2 \frac{yl}{2}$$

$$\therefore \quad \frac{Z'}{2} = Z_c \tanh \frac{yl}{2}$$

$$= \sqrt{\frac{Z}{Y}} \cdot \frac{\sqrt{YZ} \cdot l}{2} \cdot \frac{\tanh yl/2}{\sqrt{Y2} \cdot l/2}$$

$$= \frac{Z}{2} \cdot \frac{\tanh (yl/2)}{yl/2} = \frac{Z'}{2} = \frac{Z}{2}\left(\frac{\tan y'/2}{y'/2}\right)$$

Now, from this relationship we find that to get the series branch of the equivalent-T circuit, the series branch of the nominal-T (lumped series impedance) should be multiplied by the factor $\left(\frac{\tanh yl/2}{yl/2}\right)$. Hence the equivalent-$T$ represen-tation is given in the Fig. 4.24.

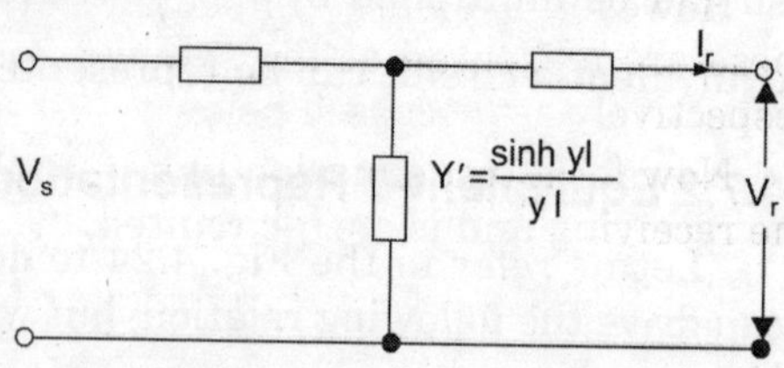

Fig. 4.24. Equivalent T representation.

However here it is to be noted that mind the ratio of tanh $Yl/2$ to $Yl/2$ and sinh Yl to Yl is almost equal to unity for small values of Yl, the nominal circuits, in fact, then represents the medium length lines quite accurately. This fact can be proved by expanding the constants of the equivalent circuit.

$$D = A = \cosh yl = 1 + \frac{YZ}{2} + \frac{Y^2Z^2}{24} + ...$$

$$B = Z_c \sinh yl = \sqrt{\frac{Z}{Y}}\left[\sqrt{YZ} \cdot l + \frac{(YZ)^{3/2}}{6} \cdot l^3 + \frac{(YZ)^{5/2}}{120} \cdot l^5 + ...\right]$$

$$= Z = \left[1 + \frac{YZ}{6} + \frac{Y^2Z^2}{120} + ...\right]$$

Similarly $$C = \frac{\sinh yl}{Z_c} = Y\left[1+\frac{YZ}{6}+\frac{Y^2Z^2}{120}+\ldots\right]$$

$$= Y+\frac{Y^2Z}{6}+\frac{Y^3\cdot Z^2}{120}+\ldots$$

Since, Y^2 is very small $\frac{Y^2Z}{6} \simeq \frac{Y^2Z}{4}$. Therefore, taking into account only the first order term we will get,

$$A = 1 + YZ/2 = D$$

$$B = Z$$

and $$C = Y+\frac{Y^2Z}{4} = Y\left(1+\frac{YZ}{4}\right)$$

which are therefore, *A, B, C, D* parameters of nominal *T*-representation.

4.8 SURGE IMPEDANCE AND SURGE IMPEDANCE LOADING OF TRANSMISSION LINE

Z_c is called the surge impedance. We have seen above that. $Z_c = \sqrt{Z/Y}$, where Z is the series impedance $(R + jX)$ and Y is the shunt admittance $(= G + jB)$ of the transmission line. In fact, it will be seen that characteristic impedance was also defined in the same manner. However the fact is that the characteritic impedance is more correct nomenclature. And Surge Impedance is the characteristic Impedance of a loss less line.

Now, for a line having negligible resistance (and thus will be the case when the conductors are of a large cross-section) and having no shunt leakage and in which case the value of G will be zero $Z_c = \sqrt{L/C}$, which is a pure resistance. It has a value of 400 to 600 (ohm) for an underground cable.

However, the surge impedance of a line may be measured in term of Z_{oc} and Z_{sc} where these are measured at the sending end with Receiving end open-circuted and short circuted respectively as discussed below :

Now for a transmission line $V_s = AV_r + BI_r$ and $I_s = CV_r + DI_r$. Therefore, when the line at the receiving and is open-circuited,

$$I_r = 0, \text{ and } V_s = AV_r, I_s = CV_r$$

$$\therefore \quad Z_{oc} = \frac{V_s}{I_s} = \frac{A}{C}$$

Similarly, when the line at receiving end is short-circuited,

$$V_r = 0, \text{ so that } V_s = BI_r, I_s = DI_r$$

$$\therefore \quad Z_{sc} = \frac{V_s}{I_s} = \frac{A}{C}$$

Now, multiplying, Z_{oc} and Z_{sc}, $Z_{oc} \times Z_{sc} = \frac{A}{C}\times\frac{B}{D} = \frac{B}{C}$

as $A = D$ for a bilateral network which is the case with a transmission line.

But $$B = \sqrt{Z/Y}\sinh\sqrt{YZ}$$

and $$C = \sqrt{Y/Z}\sinh\sqrt{YZ}, \text{ so that } \frac{B}{C} = \frac{Z}{Y}$$

But from the definition of surge impedance, $Z/Y = Z_c^2$

So that, $$Z_{oc} \times Z_{sc} = Z_c^2 \text{ or } Z_c = \sqrt{Z_{oc} \cdot Z_{sc}} \quad \text{...(4.49)}$$

4.8.1 Surge Impedance Loading

Surge Impedence Loading is defined as the load that can be delivered by the line having no resistance, the load being at unity power factor.

Then the power transmitter under these condition is, $P_R = \frac{V_R^2}{Z_0}$ MW where V_R is the receiving end line voltage in kV and Z_0 is the surge impedance of the line in ohm.

P_R, known as the surge impedance loading, also called "Natural Power" of the line.

Now for $$Z_0 = 400\ \Omega, P_R = 2.5\ V_R^2 \text{ kW} \quad \text{...(4.50)}$$

The above equation gives a limit to the maximum power that can be delivered and is useful in the design of transmission lines. This may not necessarily be the maximum loading on the line as discussed elsewhere. Surge impedance loading can be used for the comparison of loads that can be carried on the lines at different voltages.

But in order to increase the power transmitted through a long transmission, lines, can be run in parallel the later method is however very costly. Hence in order to increase P_R either V_R is to be increased, or Z_0 is to be decreased. This is a very important design consideration for transmission.

(*a*) **Increase in Voltage V_R** : In the present transmission system, the trend is for higher and higher voltage as this is the most widely adopted method to increase the power limit for heavily loaded long transmission lines. This system though nowadays has been in use extensively in most of the developed countries and also some developing countries despite it being expensive and having some practical difficulties.

(*b*) **Decrease of Surge Impedance, Z_0**. Considering the fact that spacing between the conductors cannot be discussed much, it (*i.e.*, being dependent or the line voltages and corona etc.), the value of Z_0 cannot be varied as such. But some artificial means are employed to decrease Z_0. For a lossless transmission line $Z_0 = \sqrt{L/C}$ and $y = j\omega\sqrt{LC} = j\beta$. We know that y is the propagation constant and β is the phase shift. The latter determines the torque angle δ between V_S and V_R and has system stability.

But to decrease Z_0 either L is decreased using series capacitors or C is increased using shunt capacitors. However, advantages and disadvantages of these are as under :

Shunt capacitors (in parallel with transmission line) increase the value of C and thus Z_0 is decreased. But as a result angle β is increased. Hence stability conditions because worse expecially when synchronous machines are used as loads. This method is not used for very long transmission lines especially were stability limits are present.

In fact, use of capacitors in series with the lines has been widely used is some countries like U.S.A. and Sweden. This results in a decrease of L so that Z_0 and β get decreased. This improves the stability limits also. But the difficulties arise in short cirenit conditions of the system as no satisfactory protection of capacitors has yet been devised. These capacitors are also sometimes used for smaller voltage lines to reduce line drops and voltage variation.

4.9 FERRANTI EFFECT

When a long line is operating under no load or light load condition, the receiving end voltage is greater than the sending end voltage. This phenomenon is known as Ferranti effect in a transmission line.

This phenomenon, the Ferranti Effect can be explained with following mathematical reasonings.

1. Mathematical Explanation

Let us assume a no load condition the equation

$$V_s = \frac{V_r + I_r Z_c}{2} \cdot e^{\alpha x} \cdot e^{j\beta x} + \frac{V_r + I_r Z_c}{2} \cdot e^{-\alpha x} \cdot e^{-j\beta x}$$

when, $I_r = 0$, and $x = l$. Then this equation reduces to the following.

$$V_s = \frac{V_r}{2} \cdot e^{\alpha l} \cdot e^{j\beta l} + \frac{V_r}{2} \cdot e^{-\alpha l} \cdot e^{-j\beta l}$$

Then, at $l = 0$, this equation again reduced to

$$V_r = \frac{V_r}{2} + \frac{V_r}{2}$$

Now as l increase, the incident component of sending end voltage increases exponentially and turn the vector anti-clockwise through an angle '$\beta\, l$'. Whereas the reflected part of sending end voltage decreases by the same amount and is rotated clock wise through the same angle $\beta\, l$. The sum of these two components of sending-end voltage gives a voltage which is smaller than V_r.

2. Simple Analytical Explanation

A simple explanation of Ferranti effect can be given by approximating the distributed parameters of the line by lumped impedance as shown in Fig 4.25. Since usually the capacitive reactance, under no load or light load condition, the line current is of leading p.f. The phasor diagram of this status is given below for this operating condition. The charging current produces drop in the reactance of the line which is in phase opposition to the receiving end voltage and hence the sending end voltage becomes smaller than the receiving end voltage.

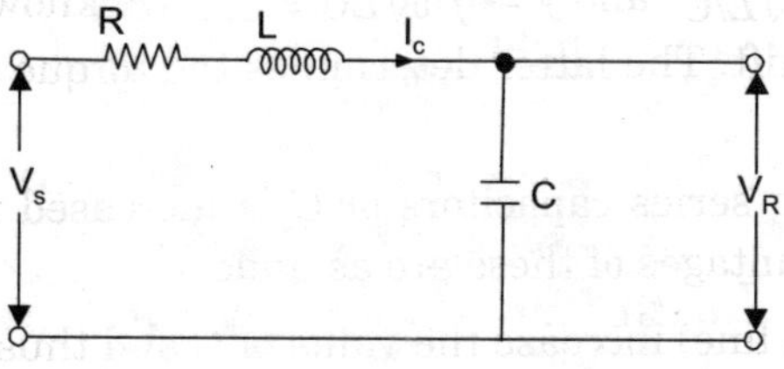

Fig. 4.25

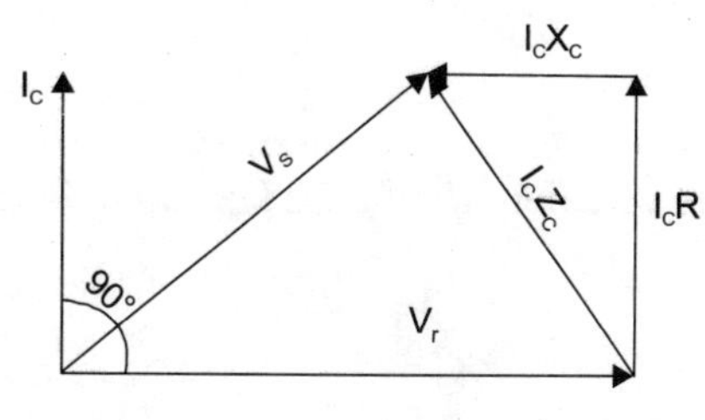

Fig. 4.26

However, another way of explaining the Ferranti effect is based on reactive power flow on the line. It is known that if the reactive power generated at a point is more than the receiver power absorbed, the voltage at the point becomes higher than the normal value and vice-versa. The inductive reactance of the line is 'sink' for the reactive power, whereas the shunt capacitances generate reactive power. In fact, if the loading corresponds to the surge impedance loading, the voltage is same everywhere as the reactive power absorbed then equals the reactive power generated by the line. The SIL, therefore, gives definite meaning to the terms lightly loaded or fully loaded lines. If the loading is less then SIL, the reactive power generated or is more than absorbed, therefore, the receiving end voltage is greater than the sending end voltage.

Hence, this explains the phenomenon due to Ferranti Effect.

4.10 ELECTRICAL LENGTH OF A LINE

We have seen above that a transmission line may be electrically short or long. In fact, eletrically short line indicates that βl is of shorter angle generally in term of degree below 10 electrical degree.

To explain this statement, with an example. Now, say the line is of length 5 electrical degree.

Then $\quad \beta l = (\pi/180) \times 5$ radian

$$\therefore \quad \left(\frac{2\pi f}{v_p}\right) l = \frac{5\pi}{180}, \text{ or } l = \frac{5\pi}{180} \times \frac{v_p}{2\pi f} = \frac{5\pi}{180} \times \frac{3\times 10^8}{2\pi \times 50} = 83 \text{ Km (appex)}$$

Now, when $\quad f = 50$ Hz, and $v_p = 3 \times 10^8$ mt/sec.

Here, v_p is the velocity of travel of Incident waves and reflected waves and is termed as phase velocity. It is defined as the speed with which an observer is to travel along the transmission line to "see" the same phase of the wave. In fact in electricity we don't see it and its many phenomenon but we can "see" and "feel" its phenomenon and its effects.

This v_p is expressed as $\quad v_p = \dfrac{w}{\beta}$

Then, we get, $\quad v_p = \dfrac{w}{\beta} = \dfrac{2\pi f}{\beta}$

Now, as $\quad \lambda = \dfrac{2\pi}{\beta}$, we get, $v_p = \lambda f = \dfrac{\lambda}{T}$

Hence, in lossless lines

$$v_p = \frac{1}{\sqrt{LC}} = \text{velocity of light} = \frac{2\pi f}{\beta} = \lambda f = \frac{\lambda}{T}$$

Hence, $\quad \lambda = \dfrac{v_p}{f} = \dfrac{\text{velocity of light}}{f}$ mt

The wavelength of a 50 Hz. wave is then,

$$\lambda = \frac{3\times 10^8}{50} = 6\times 10^6 \text{ mt}$$

Hence, it is clear from the above brief analysis that electrical degree is equivalent to physical length of line and is a function of frequency and phase velocity v_p. This, then can be said as a line with finite elemental length. Hence by connecting a number of elemental section on electrically long line is feasible.

4.11 ELECTRICAL SHORT LOSSLESS LINES COMPONENTS

Let us consider the π-model of a long line and assuming it to be lossless we find, (refer to the figure of π-model).

4.6 LONG TRANSMISSION LINES

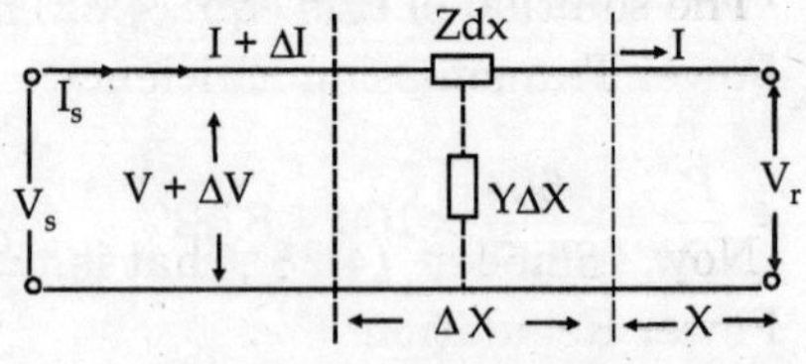

Fig. 4.21. An elemental (point) section of a long transmission line.

Above discussions have been made on lines less than around 200 km, which are termed as SHORT (upto say 80 km) and MEDIUM LENGTH lines (upto say 200 km, although some writers are considering medium length lines upto around 250 Km). Any way, in lines below 200 Km the parameters have been assumed to be lumped. But in LONG Transmission lines more than around 200 Km long for accurate solution the parameter must be taken as distributed uniformly along the length as a result of which the voltages and current will vary from point to point on the line. In fact nominal-T and nominal-π methods, if adopted there will be of considerable error. Hence we shall consider the line impedance Z and line admittance uniformly distributed. The Fig. 4.21 shows a transmission line represented on a single phase basis. Here,

z = series impedance per unit length

y = shunt admittance per unit length

l = length of the line

Z = zl = total series impedance

Y = yl = total shunt impedance

And, for clarity the elemental length dx is shown in the Fig. 4.21.

For analysis, we shall take the receiving end as the reference for measuring distances. Hence we take an elemental length dx of the line at a distance of x from the receiving end.

Now, let us say that the voltage and current at a distance x are V and I and, at a distance $(x + dx)$ are $V + \Delta V$ and $I + \Delta I$ respectively. Hence, $dV = IZ\,\Delta x$ and $dI = Vy\,\Delta x$, are the elemental voltage and currents at the distance x from the receiving end respectively.

Hence, from equations $\Delta V = IZ \cdot \Delta x$, we get $\dfrac{\Delta V}{\Delta x} = IZ$ ...(4.28)

and, from equation $dI = VY \cdot \Delta x$, we get $\dfrac{\Delta I}{\Delta x} = VY$ ...(4.29)

which in the limit, when $\Delta x \to 0$ reduces to reduces to,

$$\frac{dV}{dx} = IZ \quad \text{...(4.30)}$$

and

$$\frac{dI}{dx} = VY \quad \text{...(4.31)}$$

Now differentiating the equn. (4.30), we have

$\dfrac{d^2V}{dx^2} = Z \cdot \dfrac{dI}{dx}$, but, as from eqn. (4.31) we have $\dfrac{dI}{dx} = VY$ this can be written as,

$$\frac{d^2V}{dx^2} = Z \cdot \frac{dI}{dx} = Z \cdot VY = Z \cdot Y \cdot V \cdot$$

$$\frac{d^2V}{dx^2} - ZYV = 0 \quad \text{...(4.32)}$$

$$Z_c = \sqrt{\frac{L}{C}} \qquad \text{[since in a lossless line } r = 0, \text{ and } g = 0]$$

and also, $\quad Z_3 = Z_c \sin \beta l$

and $\quad Y = j\beta = j\omega\sqrt{LC}$ (again in a lossless line, attenuation constant $\alpha = 0$)

Now $\quad Z' = Z_C \sin \beta l$

$$= \sqrt{\frac{L}{C}}.\sin\beta l$$

$$= j\sqrt{\frac{L}{C}}.\sin\beta l$$

As βl is very small $\sin \beta l = \beta$

$$\therefore \quad Z' = j\sqrt{\frac{L}{C}}.\sin\beta l$$

$$= j\sqrt{\frac{L}{C}}\beta l = j\sqrt{\frac{L}{C}}\cdot\omega\sqrt{LC}\cdot l$$

$$= (j\,\omega\,L\,\cdot)\,l = \cdot\, j\,X$$

where, $w\,L$ = reactance per unit length and l = length of the line.

Similarly, $\quad Y_1 = Y_2 = \frac{jwC}{2}\cdot l = \frac{Y}{2}$

Hence, in an electrically short line the distribution effect of the parameter can be neglected and the line parameter can be lumped, as assumed earlier.

4.12 INPUT IMPEDANCE OF A TRANSMISSION LINE

We know that Input Impedance, $\quad Z_{in} = \frac{V_1}{I_1} = \frac{\text{Sending-end voltage}}{\text{Sending-end current}} = Z_{in}$

Hence, representing the input impedance of a transmission line, in lumped from,

$$Z_i = \frac{V_1}{I_1} = \frac{V_R \cosh yl + I_R Z_c \cdot \sinh yl}{(V_R/V_C)\sinh yl + I_R \cosh yl} \qquad ...(4.51)$$

Now dividing numerator and denominator of (4.51) by $I_R \cdot \cosh yl$, we get

$$Z_i = \frac{(V_R/I_R) + Z_c \tanh yl}{(V_R/I_R)\cdot\frac{1}{Z_c}\cdot\tanh yl + 1} = Z_c\frac{Z_{\text{out}} + Z_c \tanh yl}{Z_{\text{out}}\cdot\tanh yl + Z_c}\left(\text{Hence } Z_{\text{out}} = \frac{V_R}{I} = \frac{\text{Receiving end Coltage}}{\text{Receiving end Current}}\right)$$

Then, for an open-circuit line, (when $I_R = 0$)

$$Z_{\text{out}} = \frac{V_R}{I_R} = \frac{V_R}{O} = \text{Infinity}$$

Hence, $\quad Z_{oc} = \left.\frac{V_1}{I_1}\right|_{I_R=0,\ \text{that is, open circuit}}$

By putting $I_R = 0$ in eqn. (4.51), we for O.C. we get,

$$Z_{oc} = \frac{V_1}{I_1}| = Z_c| \tanh \cdot \gamma l = \frac{Z_c}{\tan h\, \gamma l} \text{ (Here } l \text{ is line length)}$$

Thus for short-circuited line, we can write with conductor; $V_R = 0$

$$Z_{out} = \frac{V_R}{I_R} = \frac{O}{I_R} = 0$$

and

$$I_{sc} = \frac{V_1}{I_1}\bigg| = Z_c \cdot \tanh \gamma l$$

Now, writing Z_i is terms of Z_{oc} and Z_{sc}, $Z_{in} = Z_{oc}\dfrac{Z_{out} + Z_{sc}}{Z_{out} + Z_{oc}} = |Z_{in}|e^{j\phi_{in}}$

4.13 MATCHING CONDITION IN A TRANSMISSION LINE

We call a transmission line to be mached, when the terminating load impedance is equal to the characteristic impedance.

This, therefore means that, $Z_{out} = Z_c = \dfrac{V_R}{I_R}$

$$\therefore \quad V_R = I_R Z_C \quad \text{...(4.52)}$$

Now from transmission line equation, we know that

$$A_1 = \frac{1}{2}(V_R + I_R Z_C), A_2 = \frac{1}{2}(V_R - I_R Z_C) \quad \text{...(4.53)}$$

Now substitute $V_R = I_R Z_C$ in eqns. of A_1 & A_2, we get $A_1 = V_R$, $A_2 = 0$. since $A_2 = 0$, it is hence obvious that under condition of perfect matching there will be no reflected wave. So from equations

$$V = \frac{1}{2}(V_R + I_R Z_C)e^{\gamma x} + \frac{1}{2}(V_R - I_R Z_C)e^{-\gamma x}$$

$$I = \frac{1}{2}(V_R/Z_C + I_R)\cdot e^{\gamma x} - \frac{1}{2}(V_R/Z_C - I_R)\cdot e^{-\gamma x}$$

With $V_R = I_R Z_C$, we get,

$$V = V_R \cdot e^{\gamma x}$$

$$I = I_R \cdot e^{\gamma x}$$

Hence, $\dfrac{V}{I} = \dfrac{V_R}{I_R} = \dfrac{V_{out}}{I_{out}} = Z_C$ (from eqn. (4.52))

This proves that at any point on the line, the ratio of the complex voltage and current is equal to the characteristic impedance, Z_c of line. Therefore, any cut section of the line will havethe same characteristic impedance of the entire line. This is another very useful characteristic of the transmission line.

Example 4.8. *A 220 kV, 3-phase overhead transmission line has an impedance per phase of (20 + j 100) ohm, and admittance of j 0.0010 mho. Using π-model, determine the sending voltage and current when the current at the receiving end at 0.9 p.f. lagging, is 300 m Amps.*

Solution : As per the π-model of the Fig. 4.27, we get from the example,

$$|V_2| = \frac{220}{\sqrt{3}} = 127 \text{ kV}$$

and $|I_2| = 300\,A$, and $\cos\phi = 0.9$

and $Z = (20 + j\,100)\,\Omega_1$ and $Y = j\,0.001$ mho

Now taking receiving end voltage as reference phasor, we can write,

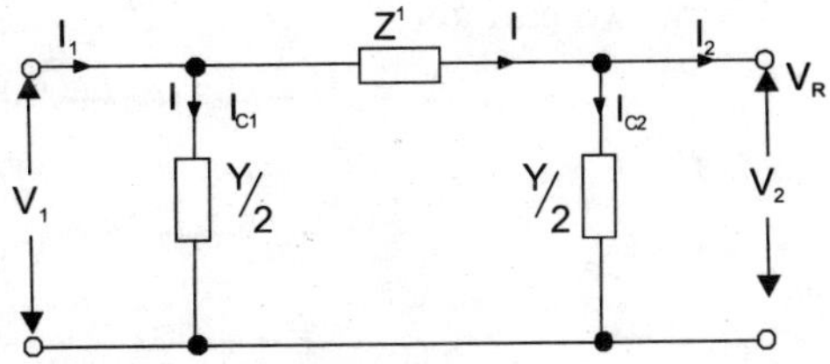

Fig. 4.27

$$V_2 = |V_2| \angle 0° = 127 \angle 0° \text{ kV}$$

and $$I_2 = I_2(\cos\phi_2 - j\sin\phi_2) = 300\,(0.9 - j\,0.436) = 270 - j\,130.8\,A$$

$$I_{C_2} = Y/2\,V_2 = j\,0.5 \times 10^{-3} \times 127 \times 10^3 = j\,63.5\,A$$

$$\therefore \quad I = I_2 + I_{C_2} = (270 - j130.8) + j63.5 = (270 - j67.3)$$

Then, voltage drop in the series impedance, Z

$$IZ = (270 - j\,67.3)\,(20 + j\,100) = (1213 + j\,2565)\,V$$

$$V_1 = V_2 + IZ = 127 + (1213 + j\,2565) \times 10^{-3} \text{ kV} = (139.1 + j\,25.65) \text{ kV}$$

$$= 141.5 \angle 10°.44'$$

Now $$I_{C_1} = Y/2 \cdot V_1 = j\,0.5 \times 10^{-3}\,(139.13 + j\,25.65) \times 10^3\,A = (-\,12.82 + j\,69.56)$$

$$= 70.74 \angle 79°.55'\,A$$

$$\therefore \quad I_1 = I_{C_1} + I$$

$$= (-\,12.825 + j\,69.56) + (270 - j\,130.8)$$

$$= 257.17 - j\,61.235 = 264.364 \angle -\,13.4°\,A$$

Example 4.9. *Determine A, B, C, D constants for a 3-phase, 50 Hz transmission line 200 Km long having the following distributed parameters.*

$$L = 1.20 \times 10^{-3}\,H/Km,\ C = 8 \times 10^{-9}\,F/Km,\ r = 0.15\,\Omega/Km.,\ g = 0$$

Solution : From the parameters given we can find,

$$Z = (r + j\,wL)^{xl} = (0.15 + j\,0.377)\,\Omega/\text{Km}.\ Z = Zl = (0.15 + j\,0.377) \times 200$$

$$= 81.09 \angle 68.28°\,\Omega$$

$$Y = g + wC = 0 + j\,2512 \times 10^{-9}\,\Omega/\text{Km}$$

$$\therefore \quad Y = Yl = j\,0.00005024 \angle 90°$$

$$yl = \sqrt{24} \times l = \sqrt{ZY} = 0.2018 \angle 79.14° = 0.03802 + j0.1982$$

$$yl = (\alpha + j\,\beta) \times l = 0.03802 + j\,0.19818 = \alpha\,l + \beta\,l$$

$$\therefore \quad \alpha\,l = 0.03802 \text{ radian},\ \beta\,l = 0.19818 \text{ radian}$$

Now $$\cosh yl = \cosh\alpha\,l \cdot \cos\beta\,l + j\sin\alpha\,l \cdot \sin\beta\,l$$

$$= 1.000723 \times 0.9807 - j\,0.03801 \times 0.19688 = 0.98113 \angle 0.437°$$

$$\sinh yl = \sinh\alpha\,l\,.\cos\beta\,l + j\cosh\alpha\,l \cdot \sin\beta\,l$$

$$= 0.038 \times 0.980 + j\,1.00072 \times 0.1968 = 0.20051 \angle 79.289°$$

Now, we can get,

$$A = D = \cos h\, yl = 0.981 \angle 0.44°$$

$$B = Z_c \cdot \sin h\, yl = \sqrt{\frac{Z}{Y}} \sin h\, yl = 80.59 \angle 68.43° \,\Omega$$

$$C = \frac{1}{Z_c} \cdot \sin h\, yl = 0.49 \times 10^{-3} \angle 90.149° \text{ mho}$$

4.14 LOSSES IN AN OPEN CIRCUITED LINE

Let us consider a single phase line of length l (metres) and having, a capacitance (charging capacitance) current of i amps per unit length. It is to be noted that the total charging current per phase, $I_c = i \times l$.

Also, $|I_c| = \dfrac{V_s}{(1/wC_0)} = V_s\, wC_0 \cdot l$ where V_s is the sending and voltage (phase-to-neutral) and C_0 is the capacitance to neutral per unit length.

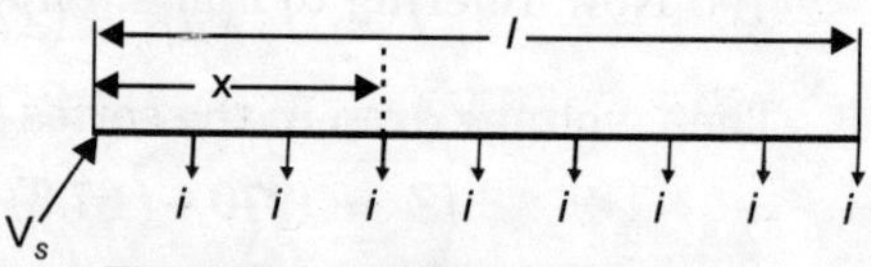

Fig. 4.28. A single phase line of length *l*.

Let us refer to the Fig. 4.28, where the current flowing at a distance x from the sending end is

$$= i\,(l - x) = V_s\, wC_0\,(l - x)$$

Now, if R_0 is the resistance per conductor **per unit length loss in each conductor**

$$= \int_0^l [V_s\, wC_0(l - x)]\, R_0\, dx$$

$$= V_s^2 \cdot w^2 \cdot C_0^2\, R_0 \int_0^l (l^2 - 2lx + x^2)\, dx$$

$$= V_s^2 \cdot w^2 \cdot C_0^2 \cdot R_0 \frac{l^3}{3} = \frac{1}{3} I_2^2\, R \text{ watts}$$

Here, $R = R_0 \times l$, = The total resistance.

4.15 EFFECT OF TRANSFORMERS (AND OR OTHER IMPEDANCES) ON THE PERFORMANCE OF A TRANSMISSION LINE

This is a very important study for transmission lines. Hence to proceed the study let us consider a simplified equivalent circuit of a transformer to either H.V. or L.V. winding is shown in Fig. 4.29, where R and X represent the equivalent resistance and leakage reactance of the winding and R_0 represents resistance responsible for core loss and X_0 represents reactance through which the magneting current flows. Then the Fig. 4.30 shows the equivalent circuit of Fig. 4.29 wherein Z_t and Y_t are the series impedance and shunt admittance of the transformer respectively. However R_0 and X_0 have usually sufficiently high values so that in most cases Y_t can be taken as infinite without much loss of accuracy in which case the equivalent circuit may still further be simplified as shown in Fig. 4.31. It is merely a series impedance Z_t.

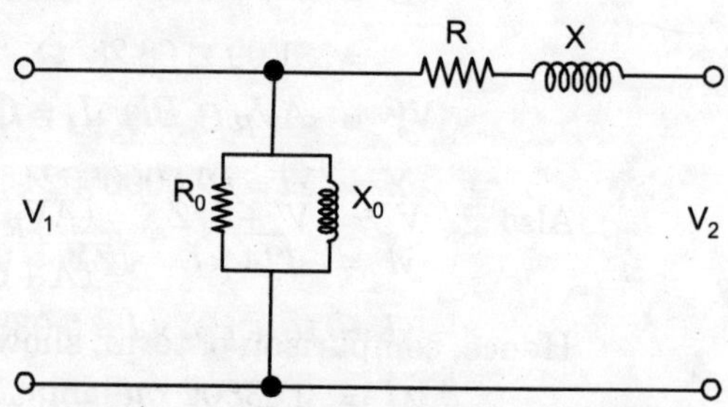

Fig. 4.29. Simplified equivalent circuit at a transformer.

In fact, sometimes a transmission line will have identical transformers at both ends-say a step-up transformers at the generating and a step-down transformer at the load end. Such cases are rather simple since the symmetry of the line is not affected. Very frequently transformers may be considered on a part of the line itself by modifying the *ABCD* constants of the latter.

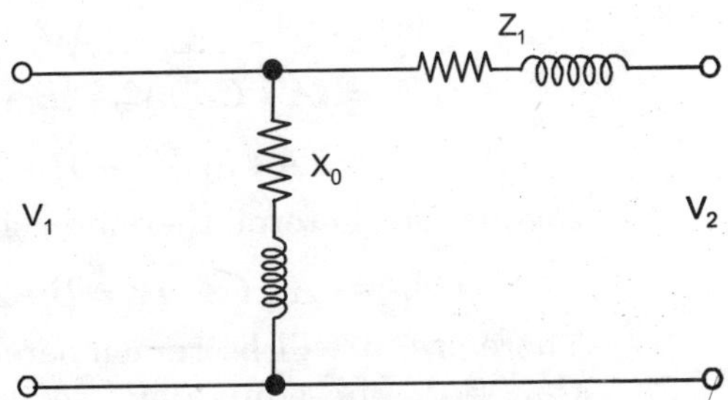

Fig. 4.30. Further simplification of Fig. 4.30.

In fact the inclusion of transformer increases B_0 (which is largely the second impedance) and C_0 (by partly compensating the line charging current with magnetising current). Now let us investigate the problem somewhat in detail :

1. Series Impedance, Z_{sc} at the Receiving End. Now, refering to Fig. 4.32, we get,

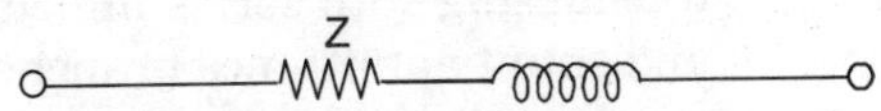

Fig. 4.31 Series impedance representation of transformer equivalence circuit.

$$V_R' = V_{R_1} + I_R \cdot Z_{sc}$$

$$V_s = AV_R' + BI_R = A(V_R + I_R Z_{sc}) + BI_R = AV_R + (B + AZ_{sc})I_R$$

Also, $$I_s = CV_R^1 + DI_R = C(V_R + I_R Z_{sc}) + DI_R = CV_R + (D + CZ_{sc})I_R$$

Comparison of terms, therefore, shown the if Z_{sc} is to be included as a part of the line :

$$A = A_0, B_0 = B + A\,Z_{sc}, C_0 = C, D_0 = D + CZ_{sc} = A + CZ_{sc}$$

2. Series Impedance (Z_{sc}) at the sending End. Let us refer to Fig. 4.33, where

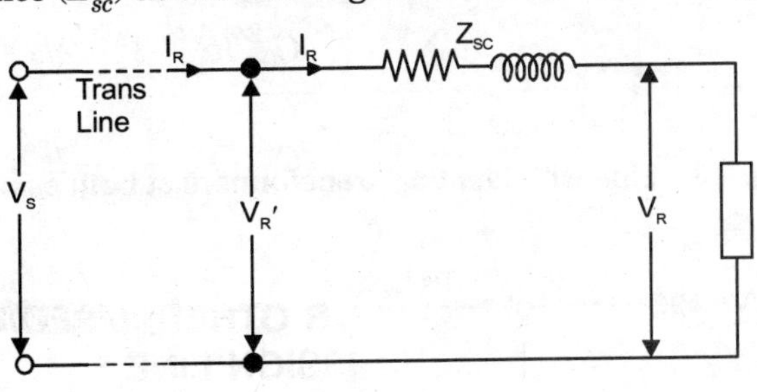

Fig. 4.32

$$V_s' = AV_{R_1} + BI_{R_1}\ I_s = CV_R + DI_R$$

Also $$V_s = V_s' + I_s Z_{sc} = (AV_R + BI_R) + I_s Z_{sc} = AV_R + BI_R + (CV_R + DI_R)Z_{sc}$$

$$= (A + CZ_{sc})V_R + (B + DZ_{sc})I_R$$

Hence, comparison of term, shows that if Z_{sc} is to be included as a part of the line.

$$A_0 = A + CZ_{sc}, B_0 = B + DZ_{sc} = B + AZ_{sc}, C_0 = C \text{ and } D_0 = D = A_0$$

3. Identical transfor at Both Ends ((Magnetising current neglected) from the Fig. 4.34, we find,

$$V_s' = AV_R' + BI_R = A(V_R + I_R Z_t) + DI_R = CV + (D + CZ_t)I_R$$

$$I_s = CV_R' + DI_R = C(V_R + I_R Z_t) + DI_R = CV + (D + CZ_t)I_R$$

$$V_s = V_s^1 + I_s Z_t = [AV_R + (B + AZ_t) I_R] + [CV_R + (D + CZ_t) I_R] Z_t$$
$$= (A + CZ_t) V_R + [B + AZ_t + (D + CZ_t) Z_t] I_R$$
$$= (A + Z_t) V_R + (B + 2AZ_t + CZ_t^2) I_R$$

Comparison of term, therefore, shows that in this case

$$A_0 = A + CZ_t, B_0 = B + 2AZ_t + CZ_t^2, C_0 = C, D_0 = D + CZ_t = A + CZ_t$$

Therefore, it will be remembered that Z_t is the equivalent series impedance of one transformer referred to the line side.

4. Identical transformers at both ends considering both series impedance and shunt admittance branches.

In this case referring to Fig. 4.35 we get.

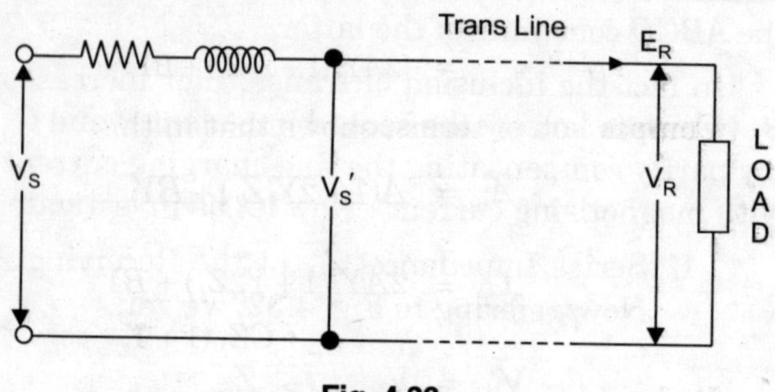

Fig. 4.33.

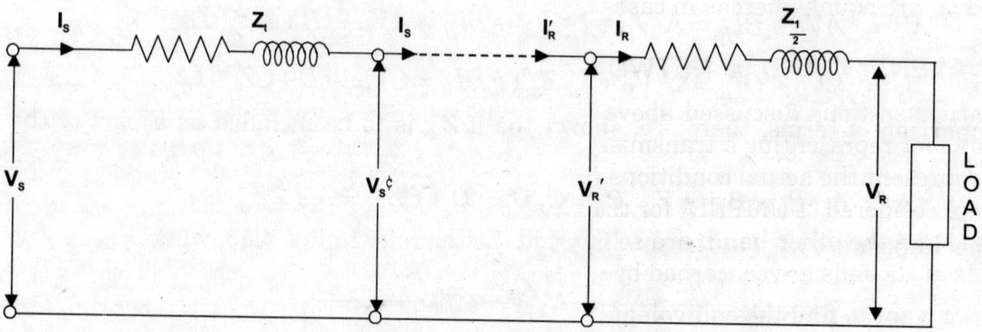

Fig. 4.34. Line with identical transformers at both each.

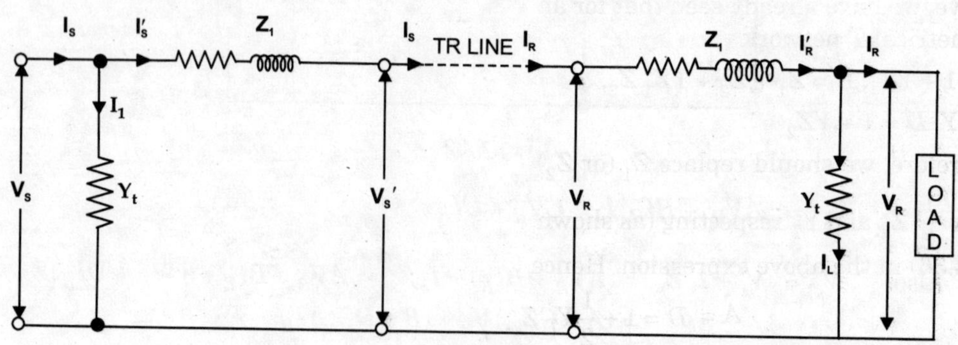

Fig. 4.35

$$V_R' = V_R + I_R' Z_t = V_R + (V_R Y_t + I_R) Z_t$$
$$V_s' = AV_R' + BI_R' = A[V_R + (V_R Y_t + I_R) Z_t) + B(V_R Y_t + I_R)$$

Also $$I_b' = CV_R' + DI_R' = C[V_R + (V_R + (V_R Y_t + I_R) Z_t] + D(V_A Y_t + I_R)$$

So that, $$V_s = V_s' + I_s' Z_t = A[V_c + (V_R Y_t + I_R) Z_t] + B(V_R Y_t + I_R)$$
$$+ CZ_t (V_R + (V_R Y_t + I_R) Z_t) + DZ_t (V_R Y_t + I_R)$$
$$= [A(1 + 2Y_t Z_t) + BY_t + CY_t (1 + Y_t Z_t)] V_R + [2AZ_t + B + CZ_t^2] I_R$$

and $$I_s = I_s' + I_t = C(V_R + (V_R Y_t + I_R) Z_t) + D(V_R Y_t + I_R) + V_s Y_t$$
$$= [2AY_t (1 + Y_t Z_t) + BY_t^2 + C(1 + Y_t Z_t)^2] V_R + [A(1 + 2Y_t Z_t) + BY_t + CZ_t (1 + Y_t Z_t)] I_R$$

Comparison of terms, shows that in this case we get

$$A_0 = A(1 + 2Y_t Z_t) + BY_t + CZ_t (1 + Y_t Z_t), B_0 = 2AZ_t + B + CZ_t^2$$

$$C_0 = 2AY_t (1 + Y_t Z_t) + BY_t^2 + C(1 + Y_t Z_t)^2, D_0 = A(1 + 2 + Y_t Z_t) + BY_t$$
$$+ CZ_t (1 + Y_t Z_t) = A_0$$

It should be noted here for interest that due to symmetry in cases (3) and (4) in this case above, A_0 and D_0 are equal whereas in cases of (1) and (2) it is not so.

4.16 EQUIVALENT 'T' AND 'π' NETWORK OF A LONG LINE

In the earlier sections discussed above in this chapter, we have considered nominal 'T' and 'π' networks for representing a transmission line. We called these methods nominal because they did not represent the actual conditions owing to the fact that actual distributed constants of the line were cosidered "LUMPED" for the purpose of approximate result. The equivalent T and π networks, on the other hand, are so found as to represent the line accurately so far as measurements at its ends are concerned by assuming suitable values of lumped constants.

Now say we want to find the equivalent T (or π) circuit of a long line where $ABCD$ constants are known.

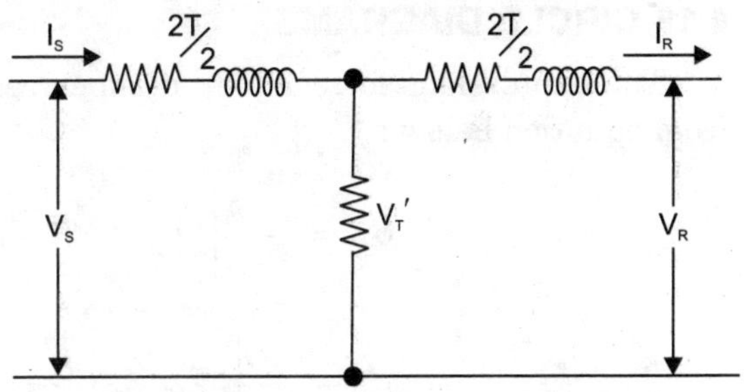

Fig. 4.36

Above, we have already seen that for an unsymmetrical T network

$A = 1 + YZ_1, B = Z_1 + Z_2 + YZ_1 Z_2,$

$C = Y, D = 1 + YZ_2$

Therefore, we should replace Z_1 (or Z_2) and Y by $\frac{1}{2} Z_T$ and Y_T respecting (as shown in Fig. 4.36) in the above expression. Hence for an equivalent symmetrical "T" network.

$$A = D = 1 + \frac{1}{2} Y_T Z_T, B = Z_T + \frac{1}{4} Z_T^2 Y_T, C = Y_T$$

Conversely, if $ABCD$ constants are known and we are to find Y_T and $\frac{1}{2} Z_T$ of the equivalent symmetrical T network we have.

$$Y_T = C \text{ and } \frac{1}{2} Z_T = \frac{A - 1}{C}$$

Similarly in the case of a symmetrical π network we have seen that

$$A = 1 + Y_R Z, B = Z, C = Y_s + Y_R + ZY_s Y_R, D = 1 + Y_s Z$$

Therefore we should replace Z and Y_s (or Y_R) by Z_π and $\frac{1}{2}Y_\pi$ respectively (as seen in Fig. 4.37) giving for an equivalent symmetrical π-network.

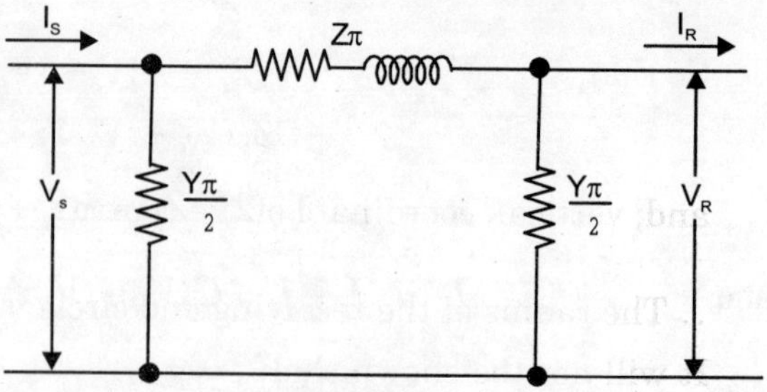

Fig. 4.37

$$A = D = 1 + \frac{1}{2}Y_\pi Z_\pi, B = Z_T, C = Y_\pi + \frac{1}{4}Z_\pi Y_\pi^2$$

4.17 CONSTANT VOLTAGE TRANSMISSION

For constant voltage transmission, a constant voltage drop is always maintained along the line by installing specially designed sychronous motors at the receiving end. The motors change the power factor of the system as the load changes.

This system of transmission has the following advantages :

(1) Some voltage is maintained at both ends of the line for all loads. (2) For long distance heavy power transmission it is possible to carry increased power for a given conductor size. (3) Improvement in power factor is affected at times of heavy loads. (4) A better protection for the line is possible due to the possibility of employing higher terminal reactances.

There are however, some disadvantges also :

1. Risk of interruption of supply becomes more in case of synchronous phase modifiers falling out of synchronism.
2. Short circuit current of the system is increased.

4.18 CIRCLE DIAGRAMS

The expressions for complex number receiving and sending-end powers are reproduced from equation below :

$$S_R = -\left|\frac{A}{B}\right| |V_R|^2 \angle(\beta - \alpha) + \frac{|V_S||V_R|}{|B|}\angle(\beta - \delta) \quad ...(4.53)$$

$$S_S = \left|\frac{D}{B}\right| |V_R|^2 \angle(\beta - \alpha) - \frac{|V_S||V_R|}{|B|}\angle(\beta + \delta) \quad ...(4.54)$$

The units for S_R and S_S are MVA (3-phase) with voltages in KV line. As per the above equations, S_R and S_S are composed of two phasor components – one a constant phasor and the other a phasor of fixed magnitude but variable angle. The loci for S_R and S_S would, therefore, be circles drawn from the tip of constant phasors as centres.

Now, it follows from eqn. (4.543) that the centre of receiving-end circle is located at the tip of the phasor :

$$= -\frac{A}{B}|V_R|^2 \angle(\beta - \alpha) \quad ...(4.55)$$

in polar coordination or in terms of reactangular coordinates. Then, the horizontal coordinate of the centre

$$= -\frac{A}{B}|V_R|^2 \cos(\beta - \alpha) \text{ MW} \quad ...(4.56)$$

and, vertical, coordinate of the centre $= -\frac{A}{B}|V_P|^2 \sin(\beta - \alpha)$ MVAR ...(4.57)

$\therefore$ The radius of the receiving end circle is $= (|V_S|\,|V_R|\,|B|)$ MVA ...(4.58)

It will now be shown that the terms of complex sending and receiving end power is a circle. Since circles are convenient to draw the circle diagram are a useful aid to visualise the load flow problem over a single transmission line.

Now referring to eqn. 4.56, 4.57, 4.58 the receiving-end circle diagram is drawn in Fig. 4.38. The centre is located by drawing OC_R at an angle ($\beta - \alpha$) in the positive direction, from the negative, MW-axis. From the centre C_R the receiving-end circle is drawn with the radius ($|V_S|\,|V_R|/|B|$). The opening point M is located on the circle means of the received real power P_R. The corresponding Q_R (or Q_R) can be immediately read from the circle diagram. The torque angle 'δ' can be read in accordance with the positive direction indicated from the reference line.

For constant $|V_R|$ the centre C_R remains fixed and cencentric circles result for varying $|V_S|$. However, for the case of constant $|V_S|$ and varying $|V_R|$ the centres of circle move along the line OC_R and have radii in accordance to $|V_S|\,|V_R|/|B|$.

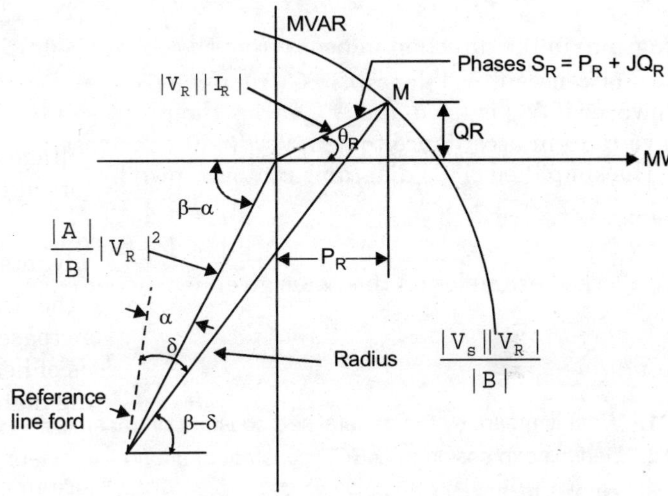

Fig. 4.38. Receiving-end circle diagram.

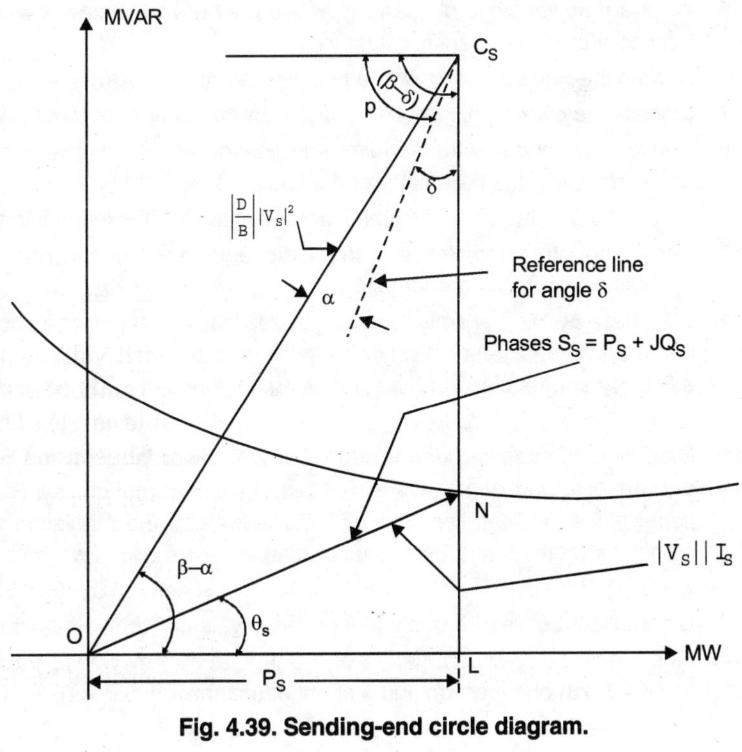

Fig. 4.39. Sending-end circle diagram.

Then similar way, it follows from eq. 4.54 that the centre of the sending-end circle is located at the tip of the phasor in the polar condinations on in terms of rectangular coordinates. The Horizontal coordinate of the centre is $(|D|/|B|)\ |V_S|^2\ \angle(\beta - \alpha)$ MW and vertical coordinate of the centre is $(|D|/|B|)\ |V_S|^2 \sin(\beta - \alpha)$ MVAR. And the radius of sending-end circle is $(|V_S|\ |V_R|)/|B|$. Now the sending-end circle diagram is shown in fig. 4.39. The centre is located by drawing OC_S at angle $(\beta - \alpha)$ from the positive MW axis. From the centre the sending circle is drawn with radius $\dfrac{|V_S||V_R|}{|B|}$, measuring the torque angle δ (as read from the receiving-end circle diagram) in the direction indicate from the reference line.

For constant $|V_S|$ the centre C_S remains fixed and concentric circles result for varying $|V_R|$. However if $|V_R|$ is fixed and $|V_S|$ varies the centres of the circles move along the line OC_S and have radii in accordance to $|V_R|\ |V_S|/|B|$. For the case of a short with a series impedance $|Z| \angle \theta$. The simplified circle diagrams compact own by

$$|A| = |D| = 1, \alpha = 0$$
$$|B| = |Z|, \beta = 0$$

Worked examples on this topic given in Appendix.

QUESTIONS

1. What is meant by the generalised constant of transmission line. What is their significance.
2. Deduce expression for *ABCD* constant of a long line in term of its parameters.
3. What is meant by :
   (1) Surge impedance, (2) Characteristic impedence, (3) Propagation.
4. Draw the equivalent circuit a long line.
5. What are power circle diagram. How are they drawn. Show how receiving end power circle diagram may be drawn for a transmission line?
6. Explain the Ferranti Effect with a phasor diagram.
7. Discuss the classification of lines based on their length of transmission.
8. Derive for a long line the sending end voltage and current relation in term of receiving end voltage and current and the parameter of the line.
9. Differentiate between a nominal-*T* and equivalent-*T* representation of a transmission line.
10. For a short transmission line draw the equivalent circuit and vector diagram and develop the expression for sending end voltage.
11. A 3-phase 50 Hz. transmission line has resistance, impedance and capacitance per phase of .0 own, 0.1 H and 0.9 μF respectively and delivers a load of 35 MW at 132 KV and 0.8 p.f. Determine the efficiency and regulation of the line using, (1) nomina-π, (2) nominal-π.
    (**Ans.** (1) 97.15%, 6.83%, (2) 97.148%, 6.879%)
12. Determine the sending end voltage, current power factor of a 1-phase 50 Hz 76.2 KV transmission delivering a load of 12 MW at 0.8 p.f. The line constant are R = 25 Ω inductance 200 MH, and capacitance between lines 2.5 pF. Also determine the regulation and efficiency of transmission use nominal π-method, and draw phasor diagram.
    (**Ans.** 86.01 KV, 163.82 A, 0.909 p.f. 15.72%, 93.65%)
13. Differentiate between a nominal-*T* and equivalent-*T* representation of a transmission line.
14. Find the *A, B, C, D* parameters of a 3-phase, 80 Km, 50 HZ transmission line series impedance of (0.15 + *j*0.78) ohm per Km and a shunt admittance of *j*5.0 × 10^{-7} mho per Km.

CHAPTER

5

Corona

5.1 CORONA

INSIDE THIS CHAPTER

- Corona
- Critical (Disruptive) voltage
- Corona loss
- Transmission line design based on corona effect
- Inductive interference between power and communication line
- Disadvantages & advantages of corona effect of transmission line

In Transmission Lines Corona effect causes a considerable loss of Power. In fact corona is present where the effective capacitance of the conductor is increased which increase the flow of charging current. Now what is this corona phenomenon?

Corona phenomenon in transmission system is the ionisation of air surrounding the power conductor. Free electrons are normally persent in free space because of radio-activity and cosmic rays. In fact, as the potential between the conductors is increased, the gradient around the surface of the conductor increase. Let us assume that the spacing between the conductors is large as compared with the diameter of the conductors. The free electrons will move with velocity depending upon the field strength. These electron will collide with the molecules of air and or case the speed is large they will dislodge electrons from these molecules thereby the number of electron will increase. The process of ionisation is thus cumulative and ultimately forms an avalanehe. This results in ionisation of the air surrounding the conductor.

In case the ratio of spacing between conductors to the radius of the conductor is more than 15, flash-over can be regarded as impossible.

Therefore, corona phenomenon is defined as a self sustained electric discharge in which field intensifed ionisation is localised only over a portion of the distance between the electrodes.

However, when a voltage higher than the critical voltage is applied between the two parallel polished wires the glow is quite evedent. In fact, after operation for a short line, reddish beads or tufts form along the wire, while around the surface of the wire there is a bluish white glow. If the conductors are examined through a stroboscope, so that one wire is always seen when at a given half of the wave, it is noticed that the reddish tufts or beads are formed when the conductor is negative and a smoother bluish white glow when the conductor is positive. The a.c. corona viewed through a stroboscope, has the same appearing as direct current corona. In fact, as corona phenomenon is initiated a hissing noise heard and ozone gas is formed which can be detected by it characteristic colour and noise.

5.2 CRITICAL (DISRUPTIVE) VOLTAGE

Let us consider a single-phase transmission line as illustrated in Fig. 5.1.

Let 'r' be the radius of each conductor and 'd' the distance of separation such that $d \gg r$. Since a single-phase transmission line has been considered here let q be the charge per unit length on one of the conductor and hence $-q$ on the other. If the operating voltage is V, the potential at conductor A with respect to the neutral plane N will be $V/2$, and that of B will be $-V/2$.

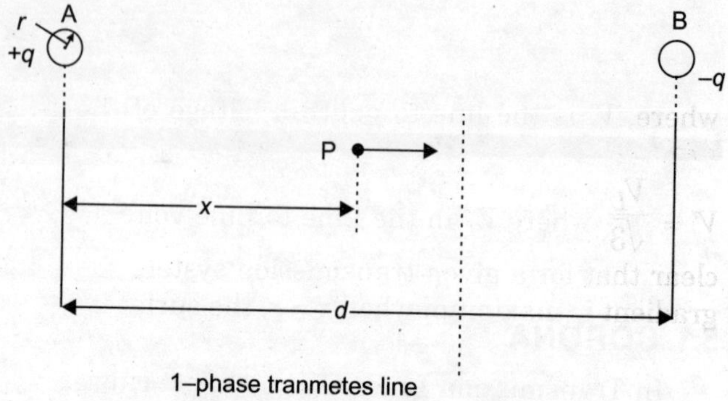

Fig. 5.1. 1-phase transmission line.

Now let us consider a point P at a distance x (in the Fig. 5.1) where we want to find the electric field intensity, with a unit positive charge at P.

Now the field due to conductor 'A' will be repulsive and that due to conductor 'B' will be attractive, thereby the electric field intensity at P due to both the line charges will be additive and it will be attractive, hence the electric field intensity at P due to both the line charges will be additive and it will be,

$$E_x = \frac{q}{2\pi\varepsilon_0 x} + \frac{q}{2\pi\varepsilon_0 (d-x)} = \frac{q}{2\pi\varepsilon_0}\left[\frac{1}{x} + \frac{1}{d-x}\right]$$

The potential difference between the conductor

$$V = -\int_{d-r}^{r} E_x \cdot dx = \int_{r}^{d-r} \frac{q}{2\pi\varepsilon_0}\left[\frac{1}{x} + \frac{1}{d-x}\right] dx$$

$$= \frac{q}{2\pi\varepsilon_0}\left[ln\, x - ln\,(d-x)\right]_r^{d-r}$$

$$= \frac{q}{2\pi\varepsilon_0}\left[ln\, x - ln\,(d-x)\right]_r^{d-r}$$

$$= \frac{q}{2\pi\varepsilon_0} \cdot 2ln\frac{d-r}{r} = \frac{q}{\pi\varepsilon_0} \cdot ln\frac{d-r}{r} \quad \text{...(5.1)}$$

Now, as r is very small as compared to d, $d - r \simeq d$

$$\therefore \quad V = \frac{q}{\pi\varepsilon_0} \cdot ln\frac{d}{r} \quad \text{...(5.2)}$$

Then, the gradient at any point x from the centre of the conductor A is given by :

$$E_x = \frac{q}{2\pi\varepsilon_0}\left[\frac{1}{x} + \frac{1}{d-x}\right] = \frac{q}{2\pi\varepsilon_0} \cdot \frac{d}{x(d-x)} \quad \text{...(5.3)}$$

Now substituting for q from eqn. (5.2) in the expression of eqn. (5.3) we get

$$E_x = \frac{\pi \varepsilon_0 V}{ln\frac{d}{r}} \cdot \frac{1}{2\pi\varepsilon_0} \cdot \frac{d}{x(d-x)}$$

$$= \frac{V}{2ln\frac{d}{r}} \cdot \frac{d}{x(d-x)} = \frac{V'd}{x(d-x)ln\frac{d}{r}} \quad ...(5.4)$$

where, V' is the line to neutral voltage of the system $\left(V' = \frac{V}{2}\right)$ In case of 3-phase system, $V' = \frac{V_L}{\sqrt{3}}$ where V_L in the Line to Line voltage. Now from the expression for the gradient it is clear that for a given transmission system the gradient increases as x decreases, that is, the gradient is maximum when $x = r$, the surface of the conductor and this value is given by,

$$g_{max} = E_n = E_{max} = \frac{V'd}{r(d-r)ln\frac{d}{r}} \simeq \frac{V'}{r\,ln\frac{d}{r}}$$

or

$$V' = r\,g_{max}\,ln\frac{d}{r} \quad ...(5.5)$$

Then the critical disruptive voltage is defined as the voltage at which complete disruption of dielectric occurs. This voltage corresponds to the gradient at the surface equal to the breakdown strength of air. This dielectric strength is normally denoted by g and is equal to 30 kV/cm peak (at 25°C and 76 cm of Hg).

In fact at any other temperature and pressure $g_0' = g_0 \cdot \delta$ where δ is the air density correction factor and is given by

$$\delta = \frac{3.92b}{273+t},$$

where b is the barometric pressure in cm of Hg and t is the temperature of °C

Therefore, the ciritical disruptive voltage is given by

$$V' = r\,g_0\,\delta\,ln\frac{d}{r}\,\text{kV} \quad ...(5.6)$$

While deriving the above expression, an assumption is made that the conductor is solid and the surface is smooth. For higher voltages ACSR conductor are used. The cross-section of such a conductor is a series of arcs of circles each of much smaller diameter than the conductor as a whole. The potential gradient for such a conductor will in consequence be greater than for the equivalent smooth conductor, so that the breakdown voltage for a stranded conductor will be somewhat less than for a smooth conductor. The irregularities on the surface of such a conductor are increased further because of the deposition of dirt and dust on its surface and the breakdown voltage is further reduced. In fact, all average value for the ratio of breakdown voltage for the a conductor and a smooth conductor lies between 0.85 to unity and is denoted by m_0. Suitable values of m_0 are given below :

Polished wires,	= 1.0
Roughened or weathered wires	= 0.98 to 0.93
Seven strand cable	= 0.87 to 0.83
Large cables with more than seven strands	= 0.90 approx.

Then the final expression for the critical disruptive voltage after taking into account the atmosphoric condition and the surface of the conductor is given as :

$$V' = r\, g_0\, \delta\, m_0\, ln \frac{d}{r}\ \text{kV} \qquad ...(5.7)$$

Now when the voltage applied corresponds to the critical disruptive voltage, corona phenomenon starts but it is not visible because the charged ions in the air must receive some finite energy to cause for the ionisation by collisions.

For a radial field, it must reach a gradient g_0 at the surface of the conductor to cause a gradient g_0, a finit distance away from the surface of the conductor.

Then the distance between g_v and g_0 is called the energy distance. This distance, is said to be equal to $(r + 0.01\sqrt{r})$ for two parallel conductors and $(r + 0.308\sqrt{r})$ for co-axial conductors. From this it is clear that g_v, is not constant as g_0 is, and is a function of the size of the conductor.

$$\therefore \qquad g_v = g_0\, \delta \left(1 + \frac{0.3}{\sqrt{r 8}}\right) \text{kV/cm for two conductors in parallel}$$

Also if V_v is the critical visual disruptive voltage, then ...(5.8)

$$V_v = g_v \cdot r\, ln \frac{d}{r}, \text{ then } g_v = \frac{V_v}{r\, ln \frac{d}{r}} = g_0\, \delta \left(1 + \frac{0.3}{\sqrt{r\,\delta}}\right)$$

or
$$V_v = r\, g_0\, \delta \left[1 + \frac{0.3}{\sqrt{r\,\delta}}\right] ln\, \frac{d}{r} kV \qquad ...(5.9)$$

But if the irregularity factor is taken into account

$$V_v = g_0 m \cdot \delta r \left[1 + \frac{0.3}{\sqrt{r\,\delta}}\right] ln \frac{d}{r} = 21.1 m \cdot \delta_r \left[1 + \frac{0.3}{\sqrt{r\,\delta}}\right] ln\, \frac{d}{r}\ \text{kV} \qquad ...(5.10)$$

In this eqn. (5.10), where r is the radius in cms. The irregularity factor m has the following values,

m_v = 1.0 for polished conductors

= 0.98 – 0.93 for rough conductor exposed to atmosphere

= 0.72 for local corona on standard conductor

Now, as the surface of the conductor is irregular the corona does not start simultaneously on the whole surface. It takes place at different points of the conductor which are pointed and this is known as local corona. And for this case m = 0.72 and for decided corona or general corona m_v = 0.82.

5.3 CORONA LOSS

The energy required for the charges to remain in motion is derived from the supply system. In fact that ions produced by the electric field result in space charges which move round the conductor. The space surrounding the conductor is lossy. In order to maintain the flow of energy over the conductor in the filed wherein this additional energy would have been otherwise absent, it is necessary to supply this additional loss from the supply system. This additional power is known as CORONA LOSS.

5.3.1 Factor Affecting Corona Loss

Corona loss on overhead transmission lines are caused due to the following factors

(1) Field around the conductor, (2) Atmospheric factors, (3) Dust, Rain, Snow, Hail effect, (4) Factors connected with the conductor and (5) Electrical Factor.

Let us discuss these factors briefly below :

1. Electrical Factor

Supply frequency and waveform : In fact corona loss is a function of frequency. An expression for corona loss has been developed as $P_{corona} = 241\times10^{-5}\frac{(f+25)}{6\delta}\sqrt{\frac{r}{d}}(V_p - V_0)^2$ kW/Km/phase where f is the supply frequency, δ is the air density. Correction factors V_p is the operating voltage in kV and V_0 is the critical disruptive voltage.

Hence from the expression it can be said that higher the supply frequency, higher are the corona losses. This also says that in d.c. transmission corona loss is less than those in a.c. transmission line. In fact, because of corona phenomenon a.c. third harmonics are always present in the waveform, hence the frequency concerned is not only 50 Hz, but it also contains third harmonic component. As a result the coronal loss is still large compared to d.c. transmitting system.

2. Dust, Rain, Snow and Hail Effect

In the overhead transmission lines, the particles dusts clog to the conductor, thereby the critical voltage for local corona reduces which increases corona loss. Similarly the bad atmospheric conditions such as rain, snow and hailstorm reduce the critical disruptive voltage and hence increase the corona loss.

3. Field Around the Conductor

The configuration of the conductors (that is, whether they are placed in vertical configuration, delta formation etc.) is very much responsible for the field around the conductor in addition to the voltages. Say, if the formation is horizontal the field near the middle conductor is large as compared to the outer conductors, that is the critical disruptive voltage is lowers for the middle conductor and hence the corona loss on the middle conductor is more compared with the two outer conductors. The height of the conductors from the ground has its effect on the corona loss. The smaller the height obviously the greater the corona loss. Then when the line are irregularly spaced, the surface gradient of the conductors and hence the corona losses are unequal.

4. Atmospheric Factor

Again from the expression $P = 241\times10^{-5}\left(\frac{f+25}{\delta}\right)\sqrt{\frac{r}{d}}(V_p - V_0)^2$ KW/Km/Phase we see that it is a function of air density correction factor δ which appears directly in the denominator of the expression and indirectly in the value of critical disruptive voltage. Then, from $V_0 = 21,$ $m_0 \ln\frac{d}{r}$ kV. Hence, lower the value of δ, the lower is the value of V_0 and hence higher the value of $(V - V_0)^2$ (of the expression of P) where V is the operating voltage in kV. This shows that the effect of δ on corona loss is very serious indeed. For lower values the pressure should be low and temperature higher. It is for this reason that the corona loss is more on hilly areas than on plain areas.

5. Factors (Connected with the Conductor) Affecting Corona Loss

The conductor itself (due to its own diameter, number of conductors/phases, surface, profile and last but not the least heating of the conductor due to load current passing through it) is also a major factor contributing to the corona loss. Here we shall discuss these points briefly for the benefit of the reader.

1. Diameter of the Conductor

If we refer to the expression for corona loss-causing power, as stated above, it can be seen that the conductor size appears at two phases and if this remaining constant, we get following two relation.

$$(a)\ P_{(\text{Corona loss})}\ \alpha \sqrt{\frac{r}{d}} \quad \text{and} \quad (b)\ P_{(\text{Corona loss})}\ \alpha\ (V - V_0)^2$$

Hence from (*a*) we find that corona loss is proportional to the square root of the size of the conductor, *i.e.*, larger the diameter of the conductor, larger will be the loss.

But from (*b*) we see that, as V_0 is approximately directly proportional to the size of the conductor, hence larger the size of the conductor larger will be the critical disruptive voltage and hence smaller will be the factor $(V - V_0)^2$.

However, in practice it is found that the effect of the second expression in (*b*) above is much more than the first in (*a*) on the corona losses and hence larger the size of the conductor lower is the corona loss.

2. Profile of the Conductor

This means actually the shape of the conductor whether cylindrical, flat, oval etc. Because of field uniformity in case of cylindrical conductor the corona loss is less in this as compared to any other shape. Hence generally the conductors shape are mostly cylindrical in nature.

3. Surface Conditions

In overhead transmission lines of power system, the conductors are exposed to atmospheric conditions. The surface would have dirt, dust, snow etc., deposited on it, which in fact lower the disruptive voltage and increase corona loss.

4. Number of Conductors/Phases

While operating voltage is 400 kV and above it is found that one conductor per phase gives large corona loss and hence large radio interference level, which interferes with communication lines which normally run parallel to the power lines. This problem of large corona loss is solved by using two, or more than two conductors, bundled together per phase, which is known as bundling of conductors. In fact by bundling the conductors, the self GMD of the conductors is increased, and thus the critical disruptive voltage is increased and hence corona loss is reduced.

5. Heating of the Conductors Due to Load Current

In fact, the heating of the conductor by the load current passing through them has an indirect reducing effect on the corona loss. Without such heating the conductor would tend to have a slightly lower temperature than the surrounding air. In the absence of heating, daw in the form of tiny water drops would form on the conductor in foggy weather or at times of high humidity, which induces additional corona loss. However the heating effect of the load current is large enough to prevent such condensation. In fact, during rains, the heating of the conductor has no influence on the corona loss, but after rain it accelerates the drying of the conductor surface. And the time during which the water drops remain on the surface is reduced and thus the loss is reduced.

5.3.2 Methods of Reducing Corona Loss

Methods of reducing corona losses are (1) by using large diameter conductors, (2) hollow conductors, and (3) bundled conductors. From discussion made in 5.3.1 we can also see the resons of using these methods. However the idea of using the hollow conductors is also have the same reasonings. In fact, using a hollow conductor means to have a large diameter without materially adding to its weight and thus reducing corona loss.

5.4 TRANSMISSION LINE DESIGN BASED ON CORONA EFFECT

One of the major objectives of a Power System design engineer is to use the methods to avoid corona loss on power lines under fair weather conditions. Bad weather conditions, such as rain, sleet greatly increase the corona loss and also lower the critical voltage on the line. On account of this, it is not practical to design high voltage lines which will be corona free at all time. If the lines are designed without considering corona even during bad weather conditions the size of the towers and the conductors will be uneconomical.

In fact, as the bad weather conditions in our country, that also in a particular region, prevail for a shorter period in a year, the average corona loss throughout the year will be comparatively very small. For information, a typical transmission line may have a coronal loss of approximately 1 KW to 1.5 KW per 3-phase, Km, and foul weather corona loss may be approximately 15 to 20 KW per 3 phase Km.

However, while on the design we can mention here that to reduce corona loss, one of the design has one or more layers of copper conductors stranded over a twisted *I*-beam core. Another design consists of tongued and grooved copper segment spiralled together to form a self-supporting hollow tube. This conductor has a smooth surface. Expended steel cored aluminium conductors which incorporate plastic or fibrous spacing material. Lines using the above types of conductors are, however, more expensive then these using the conventional type and the economic limit to the conductor diameter to round 3.75 cm to 5 cm. These special conductors are more effective in reducing corona losses during fair weather condition and there may not be the same degree of improvement during bad weather condition.

5.5 INDUCTIVE INTERFERENCE BETWEEN POWER AND COMMUNICATION LINES

For various practical purposes, it is a common practice to run communication lines along the same route as the power lines the user of electrical energy is also the user of communication system. However recently, say since past few years, using of satelite communication system communication, the communication line sooner or later may become useless. But it is useful for knowledge purpose we should have some brief discussion below on the Inductive interference between power and communication lines.

5.5.1 Electromagnetic & Electrostatic Effect

The transmission lines transmit bulk power at relatively higher voltage. As such these lines give rise to electromagnetic and electrostatic field of sufficient magnitude which induce currents and voltages respectively in the neighbouring communication circuits and also other electrical current. These interfere with communication services performance which may not only be disturbing the hearing and speech but also sometime handling of the telephone receiver may become dangerous.

Due to electromagnetic effects there are mutual inductances M_a, M_b and M_c for conductors, *a*, *b* and *c* lines of 3-phase transmission line respectively. Therefore total mutual inductance, $M = M_a + M_b + M_c$.

where, $M_a = 2\times10^{-7} \cdot \ln \frac{D_{ac}}{D_{ad}}$ H/metre

Similarly, $M_b = 2\times10^{-7} \cdot \ln \frac{D_{bc}}{D_{bd}}$ H/metre

$M_c = 2\times10^{-7} \cdot \ln \frac{D_{ce}}{D_{cd}}$ H/metre

where D_{ac}, D_{ad}, are the distance between conductor and between power conductor and communical line, with conductor a, b, c and communication lines d and e as shown in Fig. 5.2.

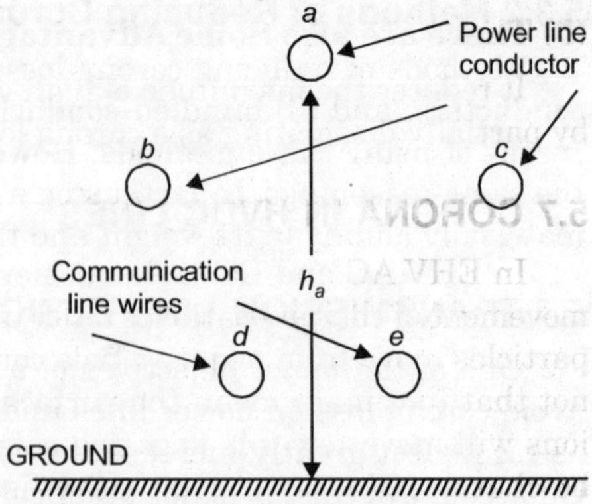

Fig. 5.2. 3-phase lines *a, b, c* and communication *d, e*.

Hence, if I is the current in the power conductors and f is the supply frequency, then the voltage induced in the communication conductors d and e, will be, $V_c = 2\pi f MI$. Volts per m. Here it is to be noted that larger the distance between the power conductors and the communication conductors smaller is the value of mutual inductor and hence smaller the value of induced voltage on communication lines.

It must be also noted that since the current through the power conductors is displaced by 120°, there is appreciable amount of cancellation of the power frequency voltage. But the presence of harmonics and multiples of third hermonics will not cancel as they are in phase in all the power conductors and hence, are dangerous for the communication circuits.

(*b*) As for **Electrostatic effect** again with refer to Fig. 5 above (with conductor a be at a height h_a from the ground. Then the voltage of conductor '*d*' due to charge on conductor '*c*' can be

$$V_{ad} = \frac{q}{2\pi\varepsilon_0}\int_{h_a}^{D_{ad}} \frac{1}{x} + \frac{1}{(2h_a - x)}\, dx = V_a \cdot \frac{\ln(2h_a - D_{ad})/\text{rad}}{\ln 2h_a/r}$$

Similarly we can obtain the potential of conductor d due to conductors b and c and hence potential of conductor d due to conductors a, b and c.

$$V_d = V_{ad} + V_{bd} + V_{cd}$$

Similarly, the potential of conductors e due to conductors a, b and c can be obtained.

5.6 DISADVANTAGES & ADVANTAGES OF CORONA EFFECT OF TRANSMISSION LINE

(*a*) Disadvantage of Corona

Following disadvantages are noticed in transmission system for corona :

1. When corona is present the effective capacitance of the conductors is increased because the effective dia of the conductor is increased. This effect increases the flow of charging current. In fact, because of corona triple frequency currents flow through the ground in case of grounded system and they give rise to a voltage of triple frequency in an ungrounded system.

These triple frequency current and voltages interfere with the communication circuit due to electromagnetic and electrostatic induction effects.

2. There is a definite loss of power even though it is not much during fair weather condition.

(b) There are also Some Advantages of Corona

It reduces the magnitude of high voltage steep fronted waves, lightning and switching effects by partially dissipating as a corona loss. In this way it acts as a safety valve to some extent.

5.7 CORONA IN HVDC LINES

In EHV AC and DC transmission on lines, ionisation being the main cause of corona, the movement of charged particles differ around the conductor. In fact, in DC field negatively charged particles move from negative pole conductor connectively and also steadily. Obviously, they do not that to remain mean the surface of negative pole conductor. Similarly positively charged ions will move towards negative pole conductor. thus the movement of charged particles will be characterised by electric field distribution rather than the surface gradient. However, in AC transmission alternate half cycles resulting in significant charged clouds around the conductors.

In fact, in HVDC lines, the corona loss is lesser due to absence of factor $\sqrt{2}$ in the voltage waveform, corona over insulators and metal surface are negligible. But the corona loss in HVDC system is an important when designing a line. Corona loss in HVDC lines during bad weather condition does not increase abruptly like that in AC lines. Also, the corona loss in monopolar HVDC line is minimum.

The critical line voltage, V_C for monopolar line at which corona will be initiated is given by the expression below :

$$V_c = E_c \cdot r \cdot m \cdot \log \frac{2h}{r} \text{ KV (DC)}$$

Here, E_c = Critical stress kV/cm (in DC, E_c = 14 kv/cm)

m = Surface irregulating factor (0.8 to 0.92)

h = Height of conductor above earth (cm)

and r = Conductor radium in cm.

For bipolar line, V_c becomes as below :

$$V_c = 2E_c\, rm\, ln\, \frac{D}{r}\left[\frac{1}{\sqrt{1-\left(\frac{D}{2h}\right)^2}}\right] \text{kV}$$

Here D = pole to pole spacing

The corona loss in bipolar line is given by

$$P_{dc} = P + a\,(E - E_c)^2 \text{ KW/Km}$$

P = Corona loss at critical gradient (0.3 – 0.5 Kw/Km)

E = Actual surface stress KV/cm

a = Constant

Example 5.1. *Determine the critical disruptive voltage and corona loss for a 3-phase line operating at 110 KV which has conductor of 1.25 cm. dia. arranged in a 3.05 metre delta. Assume air density factor of 1.07 and the dielectric strength of air to be 21 KV/cm.*

Solution : We know, the disruptive critical voltage,

$$V = 21 m\delta r \cdot ln \frac{d}{r}$$

$$= 21 \times 1.07 \times .625\, ln \frac{305}{.625}$$

$$= 21 \times 1.07 \times .625 \times 6.19$$

$$= 87 \text{ K}$$

Here, $m = 1$

$\delta = 1.07$

$r = 0.625$

The line to line voltage is = $\sqrt{3} \times 87 = 150.6$ kV. Since the operating voltage is 110 kV, the corona loss will be absent Corona loss is zero.

Example 5.2. *A conductor with 2.5 cm dia is passed centrally through a porcelain bushing $\varepsilon_r = 4$, having internal and external diameters of 3 cm and 4 cm respectively. The voltage between the conductor and an earthed clamp surrounding the poreclain is 20 kV rms. Determine whether corona will be present in the air space round the conductor.*

Solution : Let $g_{1\,max}$ be the maximum gradient on the surface of the conductor, and $g_{2\,max}$ the maximum gradient on the inner side of the porcelain.

We know then,

$$g_{1\,max} = \frac{q}{2\pi\varepsilon_0 r} \text{ and } g_{2\,max} = \frac{q}{2\pi\varepsilon_0\varepsilon_r r_1}$$

$$\therefore \quad g_{1\,max} r = g_{2\,max}\, \varepsilon_1 r_1$$

$$\therefore \quad g_{1\,max} \times 1.25 = g_{2\,max} \times 4 \times 1.25$$

$$\therefore \quad g_{2\,max} = 4.80\, g_{2\,max} = g_{1\,max}$$

$$\text{or} \quad g_{2\,max} = \frac{g_{1max}}{9.8} = 0.208\, g_{1max}$$

Now, we also know from expression

$$V = g_{1max} r \cdot ln \frac{d_1}{r_1} + g_{2\,max} r \cdot ln \frac{d_1}{r_2}$$

$$= g_{1\,max} \cdot 1.25 \cdot ln \frac{1.5}{1.25} + 0.208\, g_{1\,max} \cdot ln \frac{4.5}{1.5}$$

$$V = 20 \text{ kV } 20 = 0.228 \cdot g_{1\,max} + 0.3427\, g_{1\,max} = 0.570\, g_{1\,max}$$

$$\therefore \quad g_{1\,max} = \frac{20}{0.570} = 35 \text{ kV/cm}$$

As the gradient exceeds 21.1 kV/cm, corona will be present.

Example 5.3. *A single-phase overhead line has two conductors of dia 1 cm with a spacing of 1 metre between centres. If the dielectric strength of is 21 kV/cm, determine the line voltage for which corona will commence on the line.*

Solution : We know from the article above on disruptive critical voltage, the disruptive critical voltage, V_d

As per example $$V_d = 21 m\,\delta \cdot e \cdot ln \frac{d}{r}$$

$$d = 21 \times 1.0 \times 0.5 \cdot ln \frac{100}{0.5} = 21 \times 0.5 \times 5.298 = \mathbf{55.6\ kV}$$

Here, d = 1 metre = 100 cm

r = 0.5 cm

δ = density factor = 1.0

Example 5.4. *Determine the critical disruptive voltage and the artical voltage for local and general corona on a 3-phade overhead transmission line, consisting of three stranded copper conductors spaced 2.5 m apart at the corners of an equilateral triangle. Air temperature and pressure are 21°C and 73.6 cm. Hg respectively. The conductor dia, irregulatory factor and surface factors are 10.4 mm, 0.85 and 0.8 repectively.*

Solution : The critical disrruptive voltage,

$$V_d = 21.1\, m\, \delta\, r\, ln\, (d/r)$$

where $$\delta = \frac{3.92b}{273+t} = \frac{3.92 \times 73.6}{273+21} = 0.9812$$

Now, $$V_d = 21.1 \times 0.85 \times 0.9812 \times 0.52 \times ln \frac{250}{0.52} = 56.5 \text{ KV}$$

$\therefore$ Critical disruptive line-to-line voltage = $\sqrt{3} \times V_d = \mathbf{.98.9\ KV}$

Then, the visual ciritical voltage

$$= 21.1 \times m\ \delta \cdot r \left(1 + \frac{0.3}{\sqrt{rE}}\right) In\ \frac{d}{r}$$

Here, we get, m = 0.7 for local Corona

and m = 0.8 for general Corona

Then, $$\sqrt{r\,\delta} = \sqrt{0.52 \times 0.9812} = 0.7143$$

$$\therefore \quad V_r \text{ for local Corona} = 21.1 \times 0.7 \times 0.9812 \times 0.52\ (1 + 0.42) \ln \frac{d}{r}$$

$$= 10.7 \times 6.175 = 66.66 \text{ KV.}$$

$\therefore$ The line to line voltage will be $\sqrt{3} \times V_d = \sqrt{3} \times 6606 = \mathbf{114.4\ KV}$

Then the visual ciritical voltage for general Corona will be

$$= 114.4 \times \frac{0.8}{0.7} = \mathbf{130.78\ KV}$$

QUESTIONS

1. Explain the Corona phenomenon in power transmission line.
2. State what are the advantages and disadvantages of coronal in the power system.
3. State what are the factors affecting Corona loss.
4. State what is Critical disruptive Voltage. Derive the expression of critical Disruptive Voltage.
5. A single phase transmission line has conductors of diameter 1.25 cm and spaced 2.5 metre apart. Derive an expression for the potential gradient at any point on a line poining the centres of the conductor if the operating voltage of line in 60 kV. Calculate the voltage at which Corona will strart. (**Ans.** 158.KV]
6. A single phase overhead transmission line has two conductors of diameter 1 cm with a spacing of 1 metre between centres. If the disruptive critical voltage for air is 21.1 KV/cm, for what value of the line voltage will Corona commence. (**Ans.** 111.3 kV)
7. A 3-phase transmission line the conductors are equilaterally spaced. The line has a total Corona loss of 53 KW at 106 KV and a loss of 98 KW at 110.1 KV. What is the Critical Disruptive Voltage between the lines? Also determine the Corona loss at 113 KV. (**Ans.** 93.44 KV; 125 KW)
8. Calculate the critical disruptive voltage and Critical Voltages for visual Corona of a 3-phase conductors 7/4.75^2 mm, area 120 mm^2, spaced in a 3.5 metre delta arrangement. Assume, m = 0.85, m_v = 0.72. (**Ans.** V_d = 131 KV, V_r = 150 KW)
9. Determine the distruptive Critical Voltage and the critical voltage for local and general Corona on a 3-phase overhead transmission line consisting of three stranded copper conductors spaced 2.44 metres apart at the corners of an equilateral triangle. Conductor diameter is 1.04 cm. Air density $\delta = \frac{3.92b}{203+t}$ (where b is barametric pressure in cm of Hg, t is temperature in °C). m = 0.85 and surface factors (m_v) for local and general Corona 0.7 and 0.8 respectively. Breakdown strength of air 21.1 KV rms/cm. Air temperature and pressure are 21°C and 73.5 cm of Hg respectively. (**Ans.** V_d = 98 KV, V_r = 115 KV)
10. Explain briefly Corona in HVDC line.

CHAPTER

6

Overhead Insulators and Underground Cables

INSIDE THIS CHAPTER

- Introduction
- Overhead line insulators
- Steps to increase string efficiency
- Configuration of insulators strings (for EHV & HVDC transmission lines)
- Insulators for other electrical engineering application in power system
- Potential distribution over a string of suspension insulators
- Steps for equalising the potential
- Insulated Cables
- Current rating of a cable
- Dielectric loss

6.1 INTRODUCTION

In power system, power is transmitted by (*a*) Overhead lines and (*b*) Underground Cables. In case of overhead lines base Copper or Aluminium conductors are insulated from the ground by Insulators, but in case of underground cables or electrical cables are insulated with insulating materials and tapes etc. In this chapter we shall discuss both the cases.

6.2 OVERHEAD LINE INSULATORS

To maintain safety and necessary clearance between overhead bare conductors and nearest earth point like metal structure of the supports on which insulators are mounted. The insulators provide insulation of high voltage conductor with metal structures and also at the same time provide support to the conductors. Insulators are also used for busbar supports and other high voltage equipment terminals. Here we shall discuss on overhead insulators.

In case of overhead lines of power systems the insulators are connected to the cross arm of the supporting structors and the power conductor passes through the clamp of the insulator. These insulators are mainly used of porcelain and free from defects and of special design for particular use. These insulators should be vitrified, because the presence of pores of air in the porcelain will lower down it dielectric strength. Moreover, any sealed air impurity will also lower the dielectric strength of porcelain. The dielectric strength of porcelain should be 15 kV to 17 kV for every one tenth inch thickness. Poreclain is mechanically strong, less affected by temperature and has minimum leakage problem.

Besides porcelain, toughened glass in also sometimes used for insulators because it has higher dielectric strength (around 35 kV for one tenth inch thickness) which makes it possible to make use of single piece construction whatever be the operating voltage. Glass being transparent it is very easy to detect any flow like air trapped inside, cracks etc.

The major drawback of glass is that moisture condenses very easily on its surface/which is not in the case of porcelain and hence its use is limited to 33 kV.

Synthetic Resin is also used, but these are mostly used in various indoor applications. They contain compounds of silicon rubber resin etc. Synthetic insulators have high strength and lower weight. But leakage current is higher and longetivity is low. In fact these insulators are comparatively cheaper and have applications in bushings mainly for internal application say in switchgear parts.

6.3 TYPES OF INSULATORS

There are three types of insulator for overhead application: (1) Pin type, (2) Suspension Type, (3) Strain Type which will be discussed below :

1. Pin Type Insulators

In the Fig. 6.1 a sketch of a pin type insulator is shown. It consist of a single or multiple shells (rain sheds) adopted to be mounted on a spindle to be fixed to the cross arm of the supporting structure. Multiple shells are provided in order to obtain sufficient length of leakage path so that the flash over voltage between the power conductor and the pin of the insulator is increased. The design of the shell is such that when the upper most shell is wet due to rain but the lower shells are dry and sufficient leakage resistance is provided. This type of insulators are uneconomical beyond 33 kV operating voltage. It is to be noted that the power conductor passes through the groove at the top of the insulator and is tied to the insulator by the annealed wire of the same material as the conductor material. These conductors are not desirable to use beyond 50 kV as the cost of such insulators then increases much faster than the voltage. The cost beyond 50 kV can be given by the relation, cos α V^x where $x > 2$.

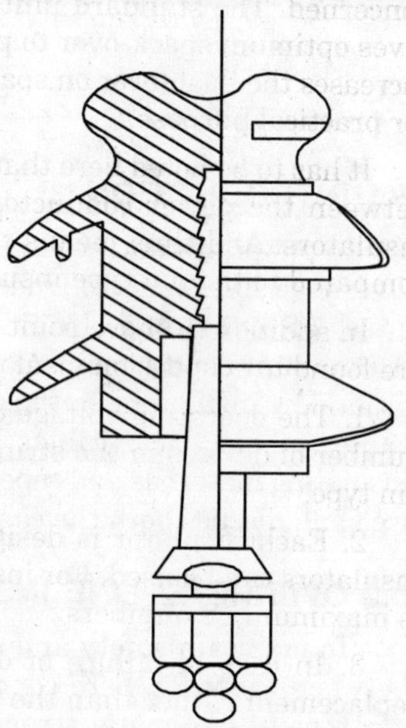

Fig. 6.1. Pin-type insulator.

The insulators and its pin should be sufficiently mechanically strong to withstand the resultant force due to combined effect of the weight of the conductor, wind pressure and ice loading, if any, per span length of the overhead line, while the leakage resistance for each shell is about of the same value, the capacitance for each shell is also nearly constant.

2. Suspension Type Insulators

The Fig. 6.2 shows a sketch of typical suspension type insulator. As the pin type insulators are uneconomical beyond 33 kV operating. Voltage and also replacement of these insulators are expensive. For these reasons for insulating overhead lines against higher voltages suspension type insulators are used.

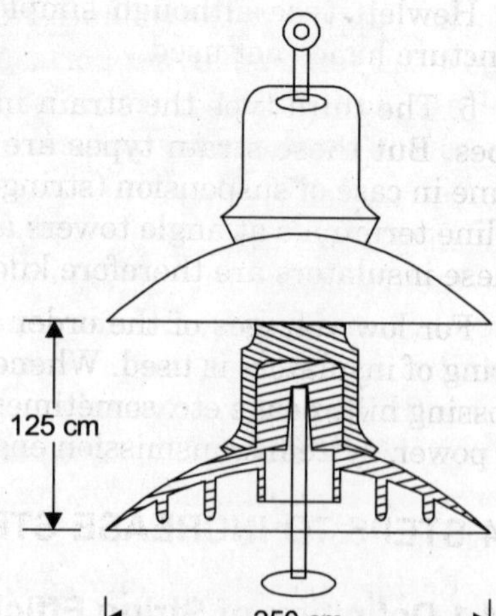

Fig. 6.2. Suspension insulators.

As shown in the Fig. 6.2, these insulators consist of one or more insulators units flexibly connected together and adopted to be hung for the cross-arms of the supporting structure and to carry a power conductor at its lower extremity. Two important types of suspension insulators are (1) Coasted-cap type and (2) Henlell or interlinking type. The coasted-cap type as shown in Fig. 6.2 is the most commonly used type and consists of a single die shaped piece of poreclain grooved on the under surface to increase. The surface leakage path and to a metal cap at the top are to a metal pin under neath. In order to build up string of any number of units the cap increased so as to take the pin of another unit. In this type the cap is fixed to the insulator by cement while the pin may either be cemented or secured by means of springring of steel wire.

This type of insulator has become quite dependable so far as the reliability of service is concerned. The standard unit is 10″ × $5^3/_4$″ in size. The diameter is taken as 1″ (25 cm) as it gives optimum spark-over to puncture voltage ratio. However, increasing the diameter further increases the flash-over on spark-over voltage, but it lowers the above ratio which is undesirable for practical purpose.

It has to be noted here that suspension insulators being free to using the clearances required between the power conductor and supporting structure are more as compared to pin-type insulators. And thus means the length of the cross arm for suspension insulators is more as compared to the pin type insulator.

In addition to above point and also being quite economical, the following further advantages are found in suspension insulators in comparison to the pin type for voltage of 33 kV and more :

1. The operating voltage of the existing transmission can be increased by adding suitable number of devices in the string instead of replacing all the insulators as is necessary in case of pin type.

2. Each insulator is designed for 11 kV and hence for any operating voltage a string of insulators can be used. For instance for 132 kV transmission, the number of insulators required is maximum 12 numbers.

3. In case of failure of one of the units in the string, only that particular unit needs replacement rather than the whole string.

4. As the power conductor and string together in case of wind pressure the mechanical stresses at the point of attachment are reduced as compared with the pin-type ones where because of the rigidity of the attachment fatigue and ultimate brittleness of the wire result.

Hewlett type although simple indesign, but due to its basic structure is more liable to puncture hence not used.

5. The third type the strain insulators are exactly identical in shape with the suspension types. But these strain types are placed in the horizontal plane, rather than in the vertical plane in case of suspension (string) insulators. These are used to take tension of the conductors at line terminals at angle towers at road crossing and at junction of overhead lines with cables. These insulators are therefore known as TENSION OR STRING insulators.

For low voltages of the order 11 kV, shakle insulators are used. But for higher voltages a string of insulators is used. Whenever, the tension in the conductor is very high as at long river crossing hills peaks etc. sometimes two a even three string insulator in parallel have been used by power, system transmission engineers.

6.4 STEPS TO INCREASE STRING EFFICIENCY

6.4.1 Definition of String Efficiency

In fact, owing to the proximity of supporting structure which is at the earth potential, the total system voltage is not equally distributed over the units of the string. The line unit (unit nearest to the conductor) carries the maximum percentage of voltage, the figure progressively decreasing as the unit nearest the tower is approached. The inequality of voltage distribution between individual units is all the more pronounced with larger number of insulator units and also depends on the ratio : (Capacity of insulator/capacity to earth). Then the string efficiency is defined as :

$$\text{String efficiency} = \frac{\text{Flash-over Voltage of } n \text{ units of string}}{n \text{ times flash-over voltage of 1 unit}}$$

6.4.2 Steps to Increase String Efficiency

Following steps are taken for increasing string efficiency :

1. To correct grading of the various capacitance. To make ratio : capacity to earth/capacity per insulator as small as possible.

Now say the neutral capacitance of the lowest unit is increased progressively decreasing this figure as top unit is approached. This will result in reduction of voltage across the line unit since the voltage for a given current is inversely proportional to the capacitance. Then to correct capacitance grading, as the method or step is known is very inconvenient in practice, because it implies all insulators in the string to be different from one another. Alternatively capacity to earth of the bottom unit could be increased by fitting sheet metal cap or rings in intimate contact with the porcelain. The method of capacitance grading is only suitable for very high voltage systems say 200 kV and above.

3. Third Step is: Static Shielding

This step makes use of a guard ring, which is a large metal ring surrounding the bottom unit and connected to the metal work at the bottom of this unit. This unit screens the lower units, thus decreasing their earth caacitances and introduces number of capacitances between the line and the insulator caps. These capacitances are greater for the lower units and thus voltage across these units is reduced. An equal distribution of voltages is not however possible by this method. The guard ring when used along with an arcing horn fixed at the top end of the string also serves the purpose of an arcing shield and protects the insulator string from damage due to overvoltage in the transmission system.

6.5 CONFIGURATION OF INSULATORS STRINGS (FOR EHV & HVDC TRANSMISSION LINES)

Transmission engineers should have the idea of configuration of insulator strings for EHV (Extra High Voltage) and HVDC (High Voltage DC) lines. Below different configurations at given;

There are mainly following configurations : (*a*) vertical string (tangent configuration), (*b*) V-configuration, (*c*) Horizontal configuration (*d*) Special configuration (for EHV requirements).

Above in Art. 6.3 in string insulators we discussed these configurations but here we explain same with figures, 6.3, *a*, *b*, *c*, *d*

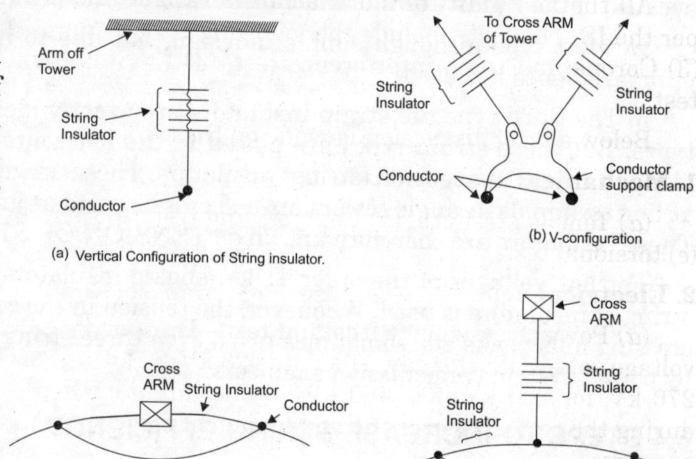

Fig. 6.3.

6.6 INSULATORS FOR OTHER ELECTRICAL ENGINEERING APPLICATION IN POWER SYSTEM

1. Post Insulators

Post insulators are used for supporting bare busbars, isolator switches etc. and are extensively used in busbar panels, switchgear panel boards, in sub-stations. These insulators are similar to a pin type insulator but has a metalic base with a metal cap so that a number of units can be cascaded. These insulators are mostly solid core insulators made of resin. In these insulators porcelain is also used.

2. Bushings

Bushings are used in transformers, Switchgears, Isolators etc. to take out live conductor through earthed tanks or metallic parts. In bushing design the central conductor is in the form of a brass or copper or stranded copper conductors pasing through brass tube or Aluminium tube supporting the bushing core.

The bushings in higher voltage grade bushing are filled with electric insulating oil. While designing bushing the creepage distance for particular voltage is to be maintained. Creepage distance is the distance along the surface of solid insulator across two conduction parts separated by solid insulator. In fact the creepage distance on the outer air-end surface of the insulator is decided by the amount of pollution (dust, dirt, chemicals etc.) in the surroundings.

6.6.1 Insulator Failures

The main causes of an insulator failure are abnormal electrical and mechanical stress, cracks, flash over and ageing. Flow of leakage current through a portion of the insulator may heat the localised portion and may cause its failure. However, the most common cause of insulator failure is due to flash-over, which creates an arc through the air surrounding the insulator. Switching and lightning surge are also responsible for insulator failure and subsequent short circuit of the conductor. Infact arcing horms are used to pass away the impulse voltage in the transmission line.

6.6.2 Testing of Insulators

All the manufacturers of insulators have to carry out tests electrical and mechanical as per the IS. The tests include the following : (1) Mechanical tests, (2) Electrical insulation test, (3) Corona and radio interference test, (4) Environmental test, including temperature cycle test.

Below a brief discussion is made on these tests :

1. Mechanical Tests Include

(*a*) Tensile strength, (*b*) Compressing test, (*c*) Bending test, (*d*) Mechanical vibration test, (*e*) torsional test.

2. Electrical Test

(*a*) Power frequency withstand test—During the power frequency voltage withstand test voltage set by BIS (for say, 11 kV it is 28 kV, 33 k it is 70 kV 132, kV is applied for a period it is 270 kV, for 220 kV it is 460 kV) of one minute. There should be no flash-over or puncture during the test. However the application of test voltage at site is about 75% of the test voltage at works.

3. Environmental Tests

Electrical insulators are subjected to alternate temperature cycles, sudden temperature changes, pollution and other enviornmental stresses. For EHV insulators following test are mandatory : (*a*) Sudden temperature drop test (thermal shock test), (*b*) Extremely low temperature test, (*c*) Pollution test.

4. Corona & Radio Interference Test

The clamps, counectors, metal fittings and the string of insulators the voltage distribution pattern is analysed. During test and inspection by providing suitable voltage grading rings and smooth surface, the possibility of corona discharge and radio interference is elemented for certain voltage ranges.

5. Other Tests Include

(*a*) *Porosity Test :* This is also a destructive test. The specimen sample insulator is broken into pieces are immersed in a 1% solution of dye in alcohol under a pressure of 15×10^3 kg/mt^2. After a specified time the pieces are removed and the peneration of dye is observed. This gives the degree of porosity.

(*b*) *Galvanining Test :* The metal parts of the insulators are galvanised and this test is conducted to assess the strength and thickness of galvanisation.

(*c*) *Power Frequency Puncture Test :* This is also a destructive test. Here the insulator with the pin and conductor in proper position is immersed in an insulating oil and a steadily increasing power frequency voltage is applied. The voltage at which conduction starts is called the puncture voltages.

6.7 POTENTIAL DISTRIBUTION OVER A STRING OF SUSPENSION INSULATORS

String insulators are widely used in transmission line. In this section we shall discuss therefore potential distribution over a string of suspension insulators.

The thumb rule for finding the number of insulator discs for a particular operating voltage is to have one disc for every 11 kV. This however, does not mean that the voltage across the devices of the string is uniformly distributed. This is because of the capacitances formed between the metal parts of the insulators and the tower structure. These capacitances is formed between the metal parts of the insulators and the tower structure. These capacitances could be made negligibly small by increasing the distance between the insulators and the tower structure which requires larger length of cross arms. This will result into bigger size of the towers and hence it becomes uneconomical. Therefore, in practice the insulators are not very fax from the tower structure and hence these capacitances affect the voltage distribution across the strings. The capacitance of each unit is known as mutual capacitance. Fig. 6.4 represents an equivalent circuit for a string of 4 insulator discs. Let

$$m = \frac{\text{Mutual capacitance}}{\text{Capacitance to ground}} = \frac{mC}{C}$$

Here the capacity to ground is the capacitance of metal part of the insulator disc to the tower

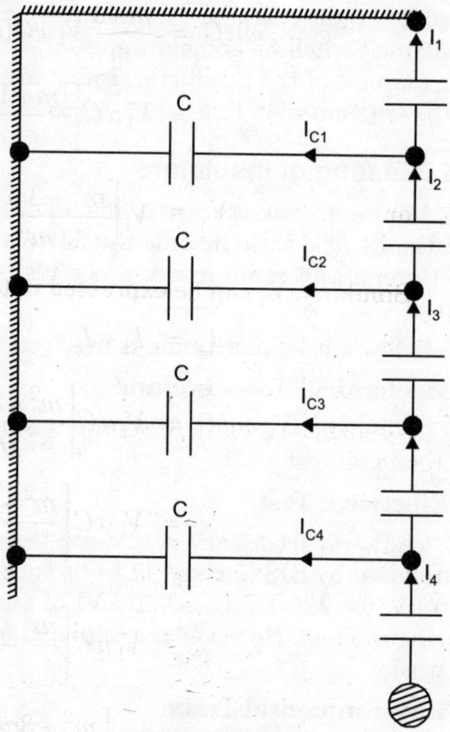

Fig. 6.4. Potential distribution over a string of four insulator.

structure. As the insulator discs are identical each disc is represented by its mutual capacitance mC. Let V be the operating voltage and V_1, V_2, V_3 and V_4 are the voltage drops across the unit starting from the cross arm towards the power conductor,

$$\therefore \quad V = V_1 + V_2 + V_3 + V_4 \quad \text{...(6.1)}$$

Here, our aim is to find out the voltage across each disc as a multiple of the operating voltage and to compare the voltage across each unit. From the diagram it is clear that,

$$I_2 = I_1 + I_{c_1}$$

$$= V_1\, mwC + V_1\, wC \quad \text{...(6.2)}$$

or $$I_2 = V_1\, Cw\,(m + 1)$$

Now, $$I_2 = V_2\, mwC$$

$$\therefore \quad I_2 = V_1\, Cw\,(m + 1) \text{ becomes}$$

$$V_2\, mwC = V_1\, Cw\,(m + 1)$$

$$\therefore \quad V_2 = \frac{V_1}{m}(m+1) = \left(\frac{m+1}{m}\right)V_1 \quad \text{...(6.3)}$$

Also, $$I_3 = I_2 + I_{c_2}$$

$$= V_2\, mwC + (V_1 + V_2)\, wC$$

Now $$I_3 = V_3\, mwC_1$$

$$\therefore \quad V_3\, mwC = V_2\, wC\,(m + 1) + V_1\, wC$$

Now substituting V_2 in terms V_1 from eqn. (6.3), we get

$$V_3\, mwC = \frac{m+1}{m} V_1 wC(m+1) + V_1 wC$$

$$= V_1 wC\left[\frac{(m+1)}{m} + 1\right]$$

or $$V_3 = V_1\left[\frac{m^2+3m+1}{m^2}\right] \quad \text{...(6.4)}$$

Similarly, V_4 can be expressed in terms of V_1 as below :

$$I_4 = I_3 + I_{c_3}$$

$$\therefore \quad V_4\, mwC = V_1 wC\left[\frac{m^2+3m+1}{m}\right] + wC\,[V_1 + V_2 + V_3]$$

$$= V_1 wC\left[\frac{m^2+3m+1}{m}\right] + wC\left[V_1 + \frac{m+1}{m}V_1 + \frac{m^2+3m+1}{m^2}V_1\right]$$

$$= V_1 wC\left[\frac{m^2+3m+1}{m} + \frac{3m^2+4m+1}{m^2}\right]$$

or $$V_4 = V_1\left[\frac{m^2+3m+1}{m^2} + \frac{3m^2+4m+1}{m^3}\right] \quad \text{...(6.5)}$$

Hence we have expressed, V_2, V_3 and V_4 in term of V_1 and the ratio of the capacitances, *i.e.*, *m*.

As, $V = V_1 + V_2 + V_3 + V_4$, and *m* is known, V_1 can be expressed in terms of *V* and then V_2, V_3 and V_4 can be obtained.

However, generally the value of $m > 1$; Now say $m = 5$, With this if there is a string of four insulators as shown in Fig. 6.4.

$$V_2 = \frac{m+1}{m} V_1 - \frac{6}{5} V_1, V_3 = \frac{m^2 + 3m + 1}{m^2} \times V_1 = \frac{41}{25} V_1$$

Similarly,
$$V_4 = \left[\frac{41}{25} + \frac{75+20+1}{125}\right] V_1$$

From these expression of V_2, V_3, V_4 in term of V_1 it is found that

$$V_1 < V_2 < V_3 < V_4$$

which means that the voltage drop across the unit nearest the cross-arm is minimum and it goes on increasing as we move towards power conductor. The voltage drop across the unit nearest the power conductor is maximum. Hence as we move down towards the power conductor, the current goes on increasing being the maximum in the lowest unit, and as stated, the voltage drop is maximum there.

From above analysis, it is clear that the lower most unit in a string of insulators is fully stressed or utilised. As we move towards cross arm the units are less stressed as compared to their capacity and hece they are not utilised fully. In fact, string efficiency is a measure of the utilisation of material in the string and is defined as below.

$$\text{String Efficiency} \quad \eta = \frac{\text{Voltage across the string}}{n \times \text{Voltage across the unit near the power conductor}}$$

$$= \frac{\text{Spark over voltage (S.O.V.) for the string}}{n \times \text{S.O.V. of one disc}}$$

where *n* is the number of insulators in the string.

Now the potential distribution across the insulator string can be obtained as follows in a simple method.

The Fig. 6.5 shows arrangement of string insulators (to keep symmetry in writing mathematical expressions) to determine potential distribution over a string of $(n + 1)$ unit of equal capacitance say (ref. to Fig. 6.5) at junction, *n*.

We find, $I_{n+1} = I_n + I_{c_n}$

$$v_{n+1} \cdot wmc = v_n\, wmC + V_n\, wC$$

$$v_{n+1} = \frac{V_n}{m} + v_n \qquad \text{...(6.6)}$$

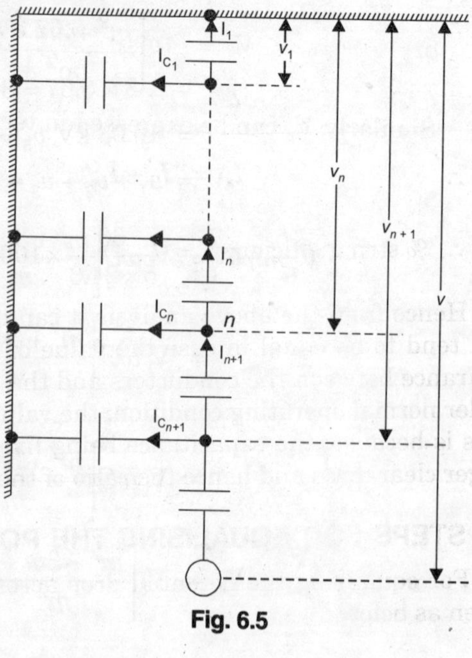

Fig. 6.5

In this expression, V_n is the voltage across n units from the top, and v_n is the drop across nth units. with the expression for v_n, of eqn. (6.6) voltage drop across any unit can be found in term of v_n. As it is known, the total voltage across the string is the sum of voltages across all the units (express in v_n) the value of v_n can be found and so by back substitution the value of voltage across every unit can be calculated. As an example, say m = 5, no of units = 5 and total operating voltage is 66 kV line to ground.

Now, using the relation,

$$v_{n+1} = \frac{V_n}{m} + v_n$$

(for n = 1) $$v_2 = \frac{V_1}{n} + v_1$$

Now as, $V_1 = v_1$, $$v_2 = v_1\left(\frac{1}{m}+1\right) = \left(\frac{1}{5}+1\right)v_1 = 1.2v_1$$

and $m = 5$

(Now, $n = 2$, $V_2 = v_1 + v_2$)

Then, $$v_3 = \frac{V_2}{m} + v_2 = \frac{v_1+v_2}{5} + v_2 = \frac{v_1 + 1.2v_1}{5} + 1.2v_1 = 1.64\,v_1$$

Similarly $$v_4 = \frac{V_3}{m} + v_3 = \frac{V_2+v_3}{m} + v_3 = \frac{v_1+v_2}{m} + \frac{v_3}{m} + v_3 = \frac{2.2v_1 + 1.64v_1}{5} + 1.64v_1 = 2.408\,v_1$$

Then similarly $$v_5 = \frac{V_4}{m} + v_4 = \frac{V_3+v_4}{m} + 2.408v_1 = \frac{3.84\,v_1 + 2.408\,v_1}{5} = 2.408v_1 = 3.657\,v_1$$

$\therefore$ $$V = v_1 + v_2 + v_3 + v_4 + v_5\,(1 + 1.2 + 1.64 + 2.408 + 3.6576) = 9.9\,v_1$$

As $$V = 9.9\,v_1 \text{ as } V = 66 \text{ kV}$$

$\therefore$ $$v_1 = \frac{66}{9.9} = 6.67 \text{ kV}$$

$$v_2 = 1.2 \times 6.67 = 8 \text{ kV}, v_3 = 1.64 \times 6.67 = 10.92 \text{ kV}, v_4 = 2.408 \times 6.67$$
$$= 16.03 \text{ kV}, v_5 = 3.65 \times 6.67 = 24.3 \text{ kV}$$

$\therefore$ $$V = v_1 + v_2 + v_3 + v_4 + v_5 = 66 \text{ kV}$$

$\therefore$ % string efficiency $$= \frac{66}{5 \times 24.3} \times 100 = 54.4\%$$

Hence from the above analysis it can be seen that the voltage drops across various units will tend to be equal in case the value of m is large. In case of high voltage lines since the clearance between the conductors and the tower structure should be more to avoid flash over under normal operating condition, the value of m will go on increasing with operating voltage. This is because, the capacitance being fixed the ground capacitance goes on decreasing with larger clearances and hence the ratio of the two capacitances goes an increasing.

6.8 STEPS FOR EQUALISING THE POTENTIAL

For equalising the potential drop across the various units of the string few steps can be taken as below :

1. Grading of Units

In the Fig. 6.5, it has be seen that unequal distribution of voltage is due to the leakage current from the insulator pin to the lower structure. Infact this current can not be eliminated. Another possibility is that the discs of different capacities culd be used such that the product of their capacitive reactance and the current flowing through the respective unit is same. This requires that the unit nearest the cross arm should have the minimum capacitance and as we move towards the power conductor the capacitance should increase. By this grading it can be seen that the complete equality of voltage can be obtained. In this case let us refer to Fig. 6.6. Here also the capacitance between the metal work and the power conductor is neglected. Here the ground capacitance C are of equal value whereas capacities are different. But as we assume that the equal voltage drop across the various units,

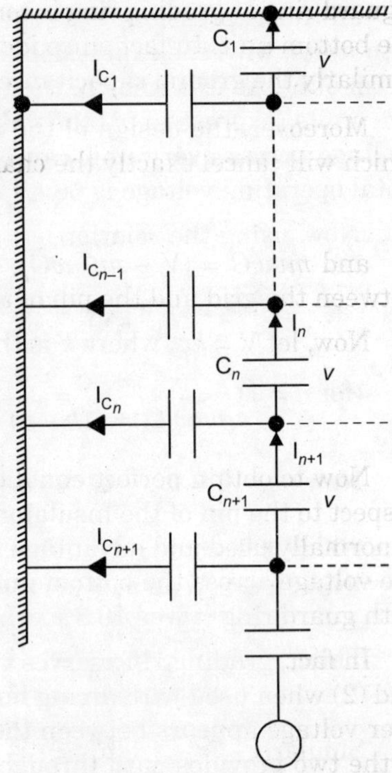

Fig. 6.6. Potential distribution over a string of $n + 1$ insulator unequal capacitance.

$$I_{c_1} = \frac{I_{cn}}{n}$$

But at function in, $$I_{n+1} = I_n + I_{c_n}$$

$\therefore$ $$v\,w\,(C_{n+1}) = v\,w\,C_n + wvwC$$

or $$vw\,C_{n+1} = vw\,(C_n + nC)$$

$\therefore$ $$C_{n+1} = C_n + nC \quad \text{...(6.7)}$$

Hence from this it is clear that in order to carry out unit grading, units of different capacities are required. This obviously requires large stocks of different sized units, which is uneconomical and impractical.

2. Selection of *m*

One of the methods of equalising the potential drop across the various units of the string is to have a larger value of m which as is said earlier needs longer cross arm and hence taller supporting structures and hence it is uneconomical to go beyond certain value of the length of cross arm. It has been found that value of $m = 10$ is about maximum which may be obtained.

3. Static Shielding

In case of unit grading we used units of different capacities, so that the flow of different current through the respective units produces equal voltage drop. In static shielding main idea is to cancel exactly the pin to tower charging current so that the same current flows through the unit of identical capacities to produce equal voltage drops across each unit. This arrangement is shown in Fig. 6.7. In this arrangement

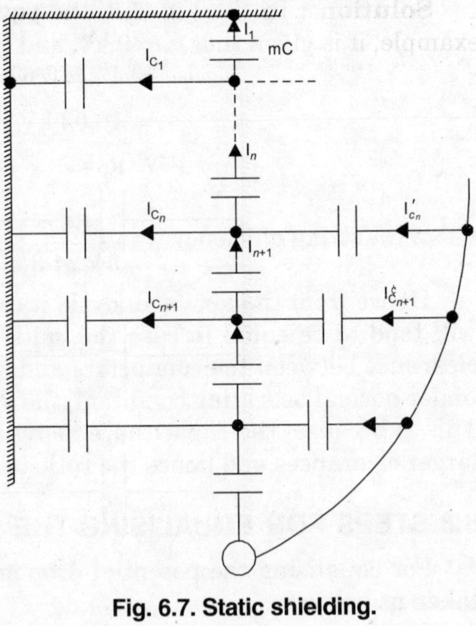

Fig. 6.7. Static shielding.

a guard ring or grading ring is connected round to the power conductor such that this surrounds the bottom unit. In fact since identical units are being used their mutual capacities are equal. Similarly the ground capacitances are equal.

Moreover, the design of the ring should be shuch that this gives rise to the capacitances which will cancel exactly the charging current in that particular section so that

$$I_{n+1} = I_n \text{ and } I_{c_a} = I'_{c_a}$$

and $mvwC = (V - nv)\, wC_n$, where V is the operating voltage and C_n is the capacitance between the grid and the pin of the nth unit.

Now, let $V = kv$, where k is the number of units used than

$$wnv\,C = (k-n)\,vwC_n \text{ or } nC = (k-n)\,C_n \text{ or } C_n = \left(\frac{n}{k-n}\right)C \qquad ...(6.8)$$

Now to obtain perfect equal distribution of voltage the capacitance of the guard ring with respect to the pin of the insulators can be given by the expression above. However this method is normally used and advantage is gained partially. It has been found that for a 14 unit string the voltage across the bottom unit without guard ring was 18.3% of the operating voltage and with guard ring it was 10.8%, which of course is a great improvement.

In fact, grading ring serves two purposes : (1) equalisation of voltage drop across the unit and (2) when used with arcing horn it protects the insulator string from flash over whereas an over voltage appears between the tower structure and the power conductor. The combination of the two provides path through the air medium to discharge the energy. Specially designed insulators are used in transmission lines taken through the areas where there is lot of smoke, Chemical Industry, frequency of lightning strokes in large.

Example 6.1. *A 3-phase overhead transmission line is being supported by three disc insulators. The potentials across top unit and middle unit are 9 kV and 11 kV respectively Calculate (1) the ratio of capacitance between pin and earth to the self capacitance of each unit, (2) the line voltage and (3) string efficiency.*

Solution : In the Fig. 6.8, the equivalent circuit of string insulators is shown. As per example, it is given that $v_1 = 9$ kV, and $v_2 = 11$ kV.

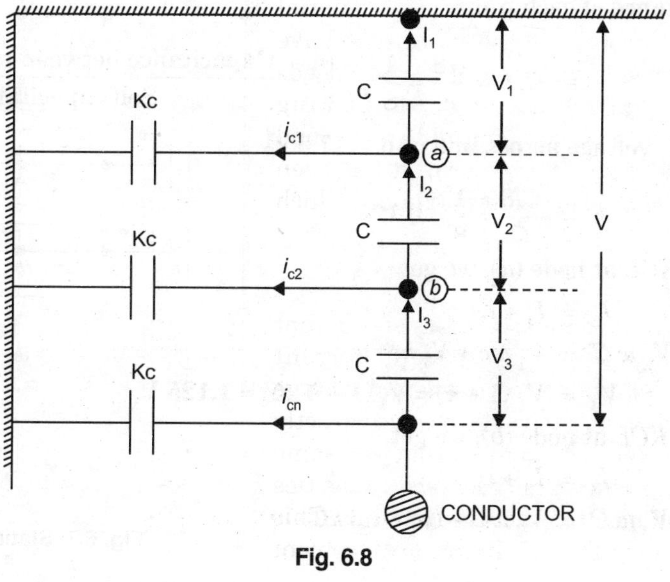

Fig. 6.8

1. Let k bet the ratio of capacitance between pin and earth to self capacitance. If C farad is the self capacitance of each unit, then capacitance between pin and earth = kC.

   Now applying KCL to node (a), we get,

$$I_2 = I_1 + i_1$$

or $$V_2 wC = V_1 wC + V_1\ kwC$$

or $$V_2 = V_1 (1 + k)$$

$\therefore$ $$k = \frac{V_2 - V_1}{V_1} = \frac{11 - 9}{9} = \frac{2}{9} = 0.22$$

2. Again apply KCL to node (b), we get

$$I_3 = I_2 + i_2$$

or $$V_3 wC = V_2 wC + (V_1 + V_2)\ kwC$$

or $$V_3 = V_2 + (V_1 + V_2) K = 11 + (9 + 11) \times .22 = 15.42$$

Voltage between line and earth,

$$V_1 + V_2 + V_3 = 9 + 11 + 15.42 = 35.42 \text{ kV}$$

3. String Efficiency $$= \frac{\text{Voltage across string}}{\text{No of insulators} \times V_3} \times 100 = \frac{35.42}{3 \times 15.42} \times 100 = 76\%$$

Example 6.2. *Each line of a 3-phase system is suspended by a string of 3 similar insulators. If the voltage across the line unit is 17.5 kV; calculate the line to neutral voltage. Assume that the shunt capacitance between each insulator and earth is 1/8th of the capacitance of the insulator itself. Also find the string efficiency.*

Solution : The equivalent circuit of the given line will be same as Fig. 6.8 but with different values given in this example.

Here capacitance between each unit and earth

$$= \frac{1}{8} \cdot C \qquad \text{(as per example)}$$

The self capacitance of each unit

$$= C, k = \frac{c/8}{c} = \frac{1}{8} \qquad \left(k = \frac{\text{Capacitance between pin and earth}}{\text{Self capacitance}} \right)$$

As per example, voltage across line unit = 17.5 kV

$\therefore$ $$k = \frac{C/8}{C} = \frac{1}{8} = 0.125$$

Now applying KCL at node (a), we get

$$I_2 = I_1 + i_1$$

$\therefore$ $$V_2 w C = V_1 wc + V_1 kc \cdot w$$

$\therefore$ $$V_2 = V_1 (1 + k) = V_1 (1 + 0.25) = \mathbf{1.125\ V_1}$$

Then applying KCL at node (b), we get,

$$I_3 = I_2 + i_2$$

or $$V_3 w C = V_2 wC + (v_1 + v_2)\ kC \cdot w$$

or $V_3 = V_2 + (V_1 + V_2)\,k = 1.125\,V_1 + (V_1 + 1.125\,V_1) \times 0.125$

$\therefore \quad V_3 = 1.39\,V_1$

$\therefore$ Voltage across top unit,

$$V_1 = V_3/1.39 = 12.59 \text{ kV}$$

$\therefore$ Voltage across middle unit,

$$V_2 = 1.125 \times V_1 = 1.125 \times 12.59 = 14.16 \text{ kV}$$

$\therefore$ Voltage between line and Earth (*i.e.*, Line to Neutral)

$$V = V_1 + V_2 + V_3 = 12.59 + 14.16 + 17.5 = 44.25 \text{ kV}$$

String efficiency $= \dfrac{44.25}{3 \times 17.5} \times 100$

$= 84.28\%$

Example 6.3. *A string of light suspension insulators is to be fitted with a grading ring. If the pin to earth capacitance are all equal to C find the values of line to pin capacitances that would give a uniform voltage distribution over the string.*

Solution : As per the example in Fig. 6.9 an equivalent circuit of the line is drawn.

Here from the Fig. 6.9, we see, $I_a = I_A$

and $I_b = I_B$ and so on.

Also for uniform voltage distribution

$$V_1 = V_2 = V_3 = \ldots = V_8$$

$\therefore \quad I_a = wC\dfrac{V}{8}$ and $I_A = \dfrac{7V}{8}$

and $\quad wc\dfrac{V}{8} = wc_1 \times \dfrac{7V}{8}$

or $\quad C_1 = \dfrac{C}{7}$

Similarly $\quad I_b = I_B$

and $\quad I_b = \dfrac{2V}{8}$

and $\quad I_B = \dfrac{6V}{8} w C_2$

$\therefore \quad wc = 3\,C_2\,w$

or $\quad C_2 = \dfrac{C}{3}$

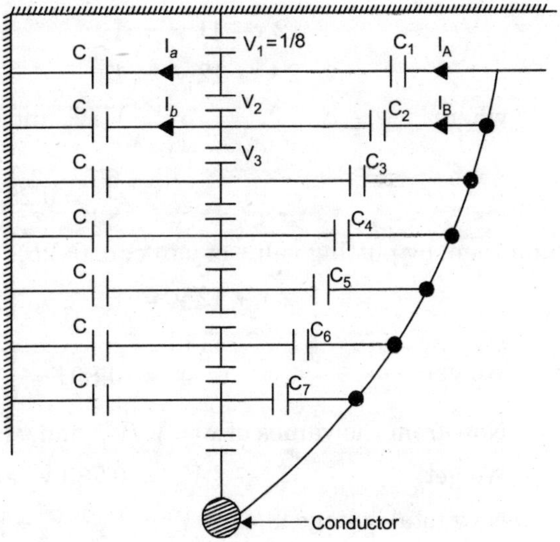

Fig. 6.9

Similarly other results can be obtained, it will be seen that the other values are $\dfrac{3C}{5}, C\dfrac{5C}{3}$, 3C and 7C respectively, for C_3, C_4, C_5, C_6, C_7.

Example 6.4. *Determine the voltage across each disc suspension insulators as a percentage of the line voltage to earth. The self and capacitance to ground of each disc is C and 0.2 respectively. The capacitance between the live pin and the guard ring is 0.1 C. If the capacitance* *to the link of the lower link pin were increased 0.3 C by means of a guard ring, determine the redistribution of voltage. Also determine the string efficiency.*

Solution : In the equivalent circuit of the Fig. 6.10 let V_1, V_2 and V_3 be the voltage drops across the discs as shown in Fig. 6.10. Applying *KCL* in the node (*a*) we get,

$I_2 + I_x = I_1 + I_a$

or $V_2 wC + (V_2 + V_3) \times .1\, C$

$= V_1 wC + 0.2\, V_1 wC = 0$

By simplification,

$$1.2\, V_1 - 1.1\, V_3 - 0.1\, V_3 = 0 \quad ...(6.1)$$

Next, considering *KCL* at the node (*b*), we get

$$I_3 + I_y - I_2 - I_b = 0 \text{ or } V_3$$

$Cw + V_3\, 0.1\, wC - V_2\, wC - (V_1 + V_2)\, 0.2\, C = 0$

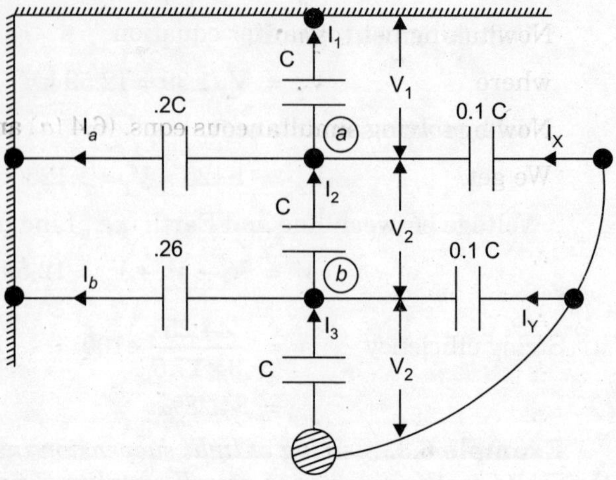

Fig. 6.10

By simplification this equation, we get,

$$0.2\, V_1 + 1.2\, V - 1.1\, V_3 \quad ...(6.2)$$

In the eqns. (1) and & (2) there are three unknowns, V_1, V_2, V_3 with two equations to solve these two equations. Let us divide both by V_3 and can rewrite them as :

$$12\, x - 11\, y = 1 \quad ...(6.2\,(a))$$

$$2\, x + 12\, y = 11 \quad ...(6.2\,(b))$$

where $x = V_1/V_3$ and $y = V_2/V_3$

$$y = \frac{65}{83} = 0.283 = \frac{V_2}{V_3} \quad ...(6.3)$$

and then, by putting value of y in eqn. (6.2*b*)

$$2\, x + 12\, y = 11$$

We get, $x = 0.801 = \dfrac{V_1}{V_0}$ ...(6.4)

Now from the values of x (= V_1/V_3) and y (= V_2/V_3)

We get, $V_2 = 0.783\, V_3$ and $V_1 = 0.801\, V_3$

Now total voltage is $V = V_1 + V_2 + V_3 = 0.801\, V_3 + 0.783\, V_3 + V_3 = 2.585\, V_3$

$$\therefore \quad V_3 = \frac{V}{2.584} \times 100 = 38.7; \; V_2 = 30.3, V_1 = 31$$

$$\therefore \% \text{ String efficiency} = \frac{1}{3 \times 0.387} \times 100 = 86.1\%$$

(*b*) If the capacitance to the line of the lower pin is increased to 0.3 *C* the equation at node (*a*) is unchanged whereas that a (*b*) now becomes

$$0.2\, V_1 + 1.2\, V_2 - 1.3\, V_3 = 0 \quad ...(6.4)$$

Again as above by dividing eqn. (6.4) by $\forall_3$ we can write this (eqn.) as below :

$$2x + 12y = 13 \quad ...(6.4\ (a))$$

Now taking help of earlier equation $\quad 12x - 11y = 1 \quad ...(6.2\ (a))$

where $\quad x = V_1/V_3, y = V_2/V_3$

Now by solving simultaneous eqns. (6.4 (*a*) and (6.2 (*a*)

We get, $\quad y = 0.9277 \text{ and } x = 0.9337$

$$\therefore \quad \frac{V_1}{V_3} = 0.9337, \text{ and } \frac{V_2}{V_3} = .9277$$

$$\therefore \quad V_1 = 0.9337\ V_3 \text{ and } V_2 = .9277\ E_3$$

$$\therefore \quad V = V_1 + V_2 + V_3 = 0.9337\ E_3 + 0.9277\ E_3 + E_4$$

$$\therefore \quad V_3 = 0.3494\ E = 34.94\%$$

$$V_2 = 0.9277 \times 34\% = 32.47\%$$

$$V_1 = 32.64$$

$$\therefore \% \text{ string efficiency} = \frac{1}{3 \times 0.3494} \times 100 = 95.4\%$$

This example proves that by means of guard rings string efficiency is increased considerably.

Note: More solved examples are given in Appendix of this Book.

6.9 UNDERGROUND CABLES

A cable is basically an insulated conductor. External protection against mechanical injury, moisture entry and chemical reaction is provided on the cable. The conductor is usually aluminium or annealed copper while the insulation is mostly PVC or *XLPE*. (Cross Linked Polythelene)

Cables are used for underground transmission and distribution of electrical energy in densely populated areas of town and cities. Cables are also used in Industries through cable tunnels. Some special cables are also used for transmission purpose under the sea or water body.

Obviously, the cables are more expensive than overhead lines particularly at high voltages. Also temperature rise is high in cables. There is limitation of raising the operating voltage. However, in low and medium voltage distribution in urban areas, cables are more widespread.

6.9.1 Types of Cables

For transmission & distribution what type and capacity of cables to be used at any particular location is determined by the mechanical consideration and the voltage and the type of load at which it is required to operate. Usually the operating voltage and the current carrying capacity decide the type of insulations. And cables are placed in various categories depending upon the voltage for which they are designed. Some of the most commonly used and more important type of cables are discussed below :

1. Low-tension Cables

This type of cables are used for voltage below 1000 volts. Various materials are used for insulation. The insulation may consist of paper impregrated with oil. Resin is sometimes used to increase the viscosity and thus to prevent drainage. Where varnished cambric is used for insulation, a coasting of petroleum jelly is provided between layers. This allows for the bonding of the cable without consisting any damage.

Vulcanised bitumen in which refined bitumen is treated with sulphur and vegetable oils is also used as insulation. Under the action of alkalis the bitumen becomes soft and the conductor is displaced from it proper position.

However, vulcanised betumen cables are now a days being replaced by other types mostly by the paper insulated cables.

High Tension Cables Upto 11 kV

This type of cables are usually of the paper insulated type. The 3-case belted type cable has the cores formed together, and each core being insulated. The three core are further insulated with impregrated paper and finally a lead sheath covers the whole to prevent moisture from entering the cable. The cable may be provided with steel armouring if used at places subject to mechanical stress and finally covered by jute covering dressed with preservative compound. However, continuous development is being carried out in design and manufacture of these cables. At present XLPE cable are used.

High Tension Cables for 22 kV and 33 kV

This type (*H*-type) of cable has each core suitably insulated and the insulation is surrounded by metallised paper which is maintained at earth potential. In the SL-separate lead type cables each core is surrounded by it own lead sheath the arrangement is equivalent to three single core cables. The H & SL combined type cable has each core insulated, sheathed with metallised paper or copper tape and then lead sheathed. The three cores are then laid up, filled with filler, armoured and finally served.

Extra High Tension Cables

In this type, cables are normally single core cables. Three cables are used as three core cables which are too large. But there are greater sheath losses in the single core cables and therefore they are not usually armoured. In the EHV cables there is a danger of breakdown of the dielectric on account of presence of voids. The failure of the dielectric, results in ionisation and allied chemical reaction which damage the insulation. In order to minimise the formation of voids two methods are generally employed (*a*) using thin oil under pressure, (*b*) using gas pressure. In the first type, the core of the cable is hollow and is supplied for oil reservoirs suitably placed. It is also called oil-filled cable. In another design the core is of stranded conductor and the oil channels are provided near the sheath. The pressure cables are placed in metal pipes which are gas-tight and filled with nitrogen at pressures of 12 to 15 atmospheres. The pressure of gas does not promit formation of voids on account of compression. As compared to the normal cables of other design a pressure cable can be worked at double the voltage and nearly one a half times the current.

Oil Filled Cables

It require smaller emulation. Here a hollow conductor of soft stranded copper, fed by oil reservoirs, placed at intervals along the lines. Oil is kept under pressure.

Gas Filled Cables

To counter the disadvantages of oil filled cable (Higher Cost, due to laying and maintenance gas filled cables are used. For external pressure cables. Press were is applied externally so that no ionisation takes place. As for interval pressure cables there are (i) High-pressure gas filled, (2) Gas cushion cables, (3) Impregnated pressure cables.

6.9.2 Selection of Cables

The selection of LT cables is easier. Only continuous current rating voltage grade and voltage drop are to be calculated to select the conductor size and type of cable. Obviously thermal conditions and derating factor, (that is, derated capacity of the cable due to temperatures rise from the adjacent cable) are to be checked. However, for HV cables the additional factor being the most important, is short cicuit current rating. It is required because in high voltage systems the fault current is very high. Thus even if the continuous current rating is satisfactory for the selected cable sizes it may not be sufficient from the view point of short circuit current rating. Higher size cable may then be required such that it is capable of dissipating the short circuit fault energy.

Operating problem with cable are to be considered also in selecting cables and particular application. Cables require high charging current and reactive power for operation. This reactive power is fed to the system and is capacitive. This may lead to rising of sending end voltage of the cable due to ferranti effect at lightly loaded condition. Flow of charging current causes heating of cable. This reduces the load current capability. Sometimes higher dielectric loss may further increase the temperature rise.

Moreover, generation of capacitive reactive power may hamper the stability considerations of large generators. Further switching of cable capcitive current may give rise to over-voltage.

6.9.3 Comparison Between Overhead Lines and Underground Cables

The objective of using overhead lines and underground cables the objective of using overhead lines and cables being the same, that is transmission and distribution of electrical energy, use of cables is limited due to large charging current for long length of transmission and high cost of cable. Moreover the cross section of cables is higher than equivalent overhead lines. The insulation requirement of cables is high, particularly at high voltage as the conductors of different phases run side by side. Heat dissipation of cables is less than overhead lines. Though transmission line reaction requires steel towers, still it is cheaper than eraction cost of underground HV cables. However, cable installations are useful for submerge crossings, railway crossing inside of power station and sub-station and in densely populated areas. Underground cables also provide greater safety, less interference with amenities and better outlook.

Cable Laying

Cable laying may be done by: (1) Direct laying, (2) Draw-in system on Duet/pipe buying, (3) Solid system.

6.9.4 Breakdown of Cables

Breakdown of cables may be due to puncture, thermal instability and void formation. A cable punctures when a discharge occurs between the conductor and the sheath at one or more than one points within the dielectric forming a hole in it. This leads to cable breakdown.

Thermal instability occurs in the cable when the power dissipation in the cable is high due to dielectric loss beyond the recommended value. Additional ohmic loss may be cumulatively added such that the rate of heat within the cable is very high. This may lead to thermal breakdown of the cables. However, if the temperature rise can be halted permanent damage can be stopped.

The formation of air space (void) in the dielectric medium due to imperfect impregnation, faulty design or material and successive changing of load cycles may be detrimental for the cable. Ionisation can take place within the void and the voltage gradient may be very high in dielectric. This may produce minute carbon within the insulator and near the conductive surface. In HV cables the formation of carbonised track is gradually extended. This leads to breakdown of cables which at any cost of time and money must be minimised.

6.9.5 HVDC, Cryogenic, Oil-filled Cables

(*a*) Cables for High Voltage Direct Current Transmission (HVDC)

For absence of periodic charging current in HVDC system. Cables play an effective role in dc transmission links. The dielectric loss is very low also. In HVAC system, the electric stress distribution is governed by the dielectric medium.Whereas in HVDC cables, it is determined by the resistance of dielectric. Due to absence of periodic charging current there is practically no limitation regarding distance of transmission by underground cable in "dc" system.

(*b*) Cryogenic Cables

Cryogenic cables have superconductivity. As superconductivity is still not very much feasible in electric power transmission hence a possible alternative to superconductive cables is the resistive operation of conductor at very low-temperature. This utilises the decrease in resistivity with decreasing temperature. However, the cable cost is very high. A coolent is used to keep the temperature within the cable very low. Liquid helium or liquid nitrogen are commonly used as coolent.

(*c*) Oil Filled and Gas-filled Cables a Comparison

Oil-filled cables are used mostly as 1-phase cables and these are common in high voltage system. 3-core oil-filled cables also exit but not commonly used.

A single-phase oil-filled cable consists of a concentric stranded conductor built around an open helical core that serves as a channel for the oil flow. The cable is kept constantly supplied with low viscosity mineral oil. The cable is insulated and sheathed in the same manner as the solid type cables. The ducts within the core are connected to oil reservoirs and hence expansion or contraction of the cable do not produce any void. Oil-filled cables can sustain voltage stress upto 130 kV/cm. In gas filled cables paper dielectric is impregnated with petroleum jelly and the iner space between the dielectric layers is filled up with gas, generally dry Nitrogen at around 1400 kg/m^2. Pressure is retained by pumps and by the lead sheath within the cable. The advantage of gas pressured cables over oil filled cables is that it is not necessary to keep pressure reservoir. A water-proof covering is provided above the sheath. In oil-filled and gas filled cables ionisation passibility is low. SF_6 being of high dielectric strength and good thermal characteristic, is also used in gas-filled cables. Oil-filled and gas-filled cables are used in the range of 132 kV to 400 kV transmission.

The 3-core oil filled cables may be of conventional circular design or even may be flat type. The flat sides are reinforced with metallic tapes and binding wires. However as said in the beginning use of 3-phase oil-filled cables are comparatively much less than 1-phase oil or gas-filled cables. Below in the Fig. 6.11 sketches of gas-filled/oil-filled, shielded cables cross section are given to give an idea of inside structure of cables.

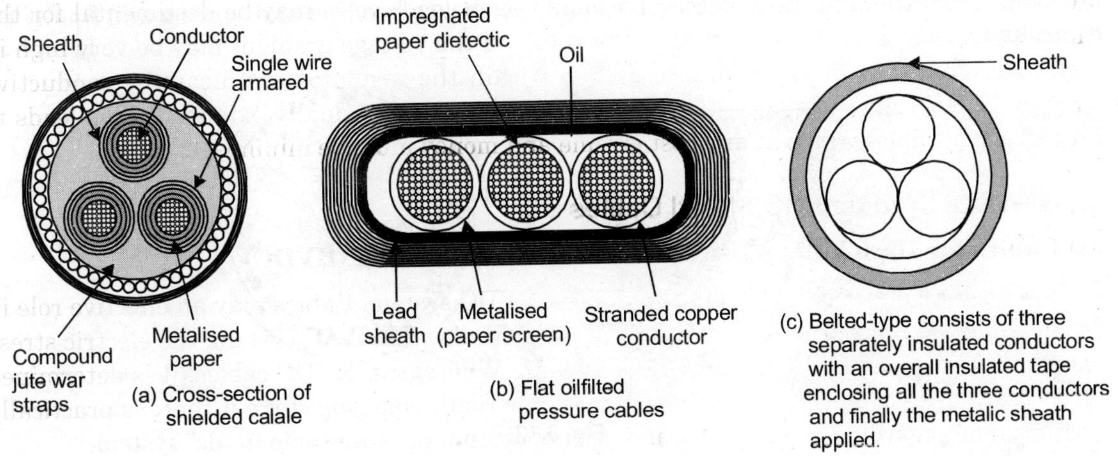

Fig. 6.11.

6.10 CURRENT RATING OF A CABLE

Considering the cost investment on underground transmission particularly it is of great importance to determine the optimum current carrying capacity of the cables. Transmission engineers have to give serious thought in this matter.

We have to give importance to following factors which decide the safe continuous current in cable most economically :

1. The maximum permissible temperature at which the insulation surrounding the conductor can be operated.
2. The method of heat dissipation through the cable on load.
3. The installation conditions and the surrounding and the ambient condition.

In fact, the temperature of a cable rises when the heat is generated when the cable is on load. But the heat generated in the cable are greater than the heat dissipated. Hence insulation of the cable is most important to avoid the premature failure.

1. The allowable temperature values of various types of insulation of the cable are given below :

1. VIR and PVC insulated cables ... 60°C impregnated paper insulated cables, XLPE – 70°C
2. Oil filled and gas pressure cables — 85°C
3. 33 kV solid type armoured cables — 65°C
4. 22 kV sorren cables, armoured — 65°C
5. 11 kV belted cables, armoured — 65°C

Heat, obviously, dissipates through the insulation metal sheath, cable-bedding and servings and finally into the surrounding earth or air. An idea is given in the Fig. 6.12 (*a*) about the heat flow due to a 3-single core cable laid directly into the ground. Similarly in Fig. 6.12 (*b*) the mechanism of heat flow in a 3-phase belted type cable is shown. It is seen that the heat flow in a cable is similar to the leakage current, that means, both flow radially out from the core of the cable through the dielectric national the sheath bedding and serving and finally to the ground.

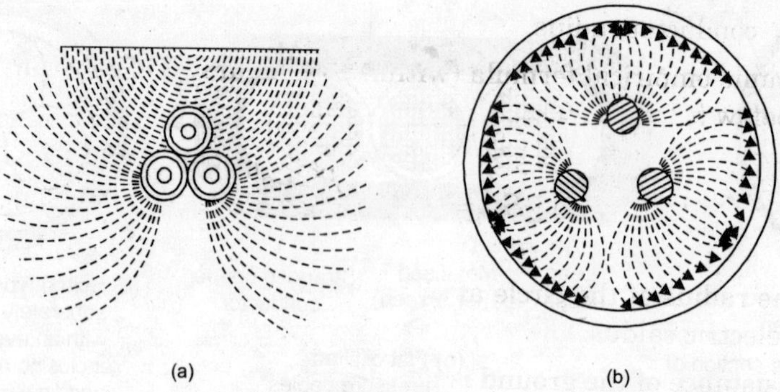

Fig. 6.12

$\therefore$ Heat flow, H (say) is given as $= \dfrac{\text{Temp. difference in °C}}{\text{Thermal resistance in thermal ohms}}$

which is almost same as current (in an electric circuit) given by ohms Law,

$$= \frac{\text{Potential difference in voltage}}{\text{Resistance in ohm}}$$

In case of heat flow, thermal resistance = $g\dfrac{l}{A}$, where g is the thermal resistivity of the material, and Resistance Ohms Law = $\dfrac{l}{A}$, where l is the resistivity.

l = length of the conductor

l = is the length of conductor heat flow paththe material = is the length of Heat flow

A = section through which heat flow

Calculation of Current Rating

To calculate the thermal rating of the cable the thermal resistance of various types of cables: say thermal resistance of 1-core cable, 3-core cable and the ground are to be known for information.

1. Thermal resistance of the 1-core calculation be expressed as

$$S_1 = (g/2\pi)\, ln \frac{R}{r}$$

where 'r' is the radius of the conductor, R is the inner radius of the sheath and g is thermal resistivity of the material. Usual value of g are : for oil filled cable, 4.5 thermal ohm/m for impregnated pressure cable, 5.5 thermal ohm/m.

2. Then thermal resistance of a 3-core cable

$$S_3 = \frac{g}{6\pi}\left(0.85 + \frac{0.2t}{T}\right)\ln\left[\left(4.15 - \frac{1.1t}{T}\right)\left(\frac{T+t}{r}\right)+1\right]$$

where T = thickness of conductor

t = Thickness of belt insulation

r = conductor radius

However, a simple empirical formula (within 10% accuracy) is generally used as below for a 3-core is given below :

$$S_3 = \frac{g}{6\pi} ln \frac{R^6 - a^6}{3R^3 \cdot a^2 r} \text{ thermal ohm/m}$$

where 'a' is the radius of the circle at which the centres of the conductor cross section lie and 'R' is outer dielectric raidus.

3. Thermal resistance of the ground is given by the relation below :

$$G = \frac{g}{2\pi} ln \frac{2h}{R} \text{ thermal ohm/m}$$

where g is the thermal resistivity of the ground, h the depth of cable axis below ground.

However, in practice it is found that the value of g determined in the laboratory must be multiplied by a correction factor of 2/3 to obtain the actual value. Hence the actual formula to be used is :

$$G = \frac{g}{3\pi} ln \frac{2h}{R} \text{ thermal ohm/m}$$

Calculation of current rating is carried out with known thermal resistance of the various factors, as applicable (stated above) under steady state conditions which means that when the current load is continuous or loading which keeps steady state thermal conditions. Neglecting the dielectric losses let T_b be the sheath temperature and T_a the ambient temperature and T_m is maximum permissible temperature.

Then the heat generated in the core of the cable will pass through the dielectric medium whereas through the bedding, serving and the ground the heat flow is sum of the heat generated in the core and the sheath. With these observation the following relations hold good *i.e.*

$$nI^2 R = \frac{T_m - T_s}{S_1} \quad ...(a)$$

where n is the number of cores

I is current in each core

R is resistance of each core

Now say S_1 is the thermal resistance of the dielectric. In case the ratio of sheath loss to core loss is λ, the heat flowing through bedding, serving and the ground will be $(1 + \lambda)\, nI^2 R$ and the following relation will hold good :

$$(1 + \lambda)\, nI^2 R = \frac{T_b - T_a}{S_4 + S_5 + G} \quad ...(b)$$

where S_4, S_5, G are thermal resistances of bedding, serving, grounds

In fact, normally T_b is not known, eliminating T_b from the eqns. (*a*) and (*b*), we get,

$$T_m - T_b = nI^2 Rb_1$$

$$T_b - T_a = (1 + \lambda)\, nI^2 R\, (S_4 + S_5 + G)$$

Now adding these two equations, we get

$$T_m - T_a = \text{Temperature Rise due to Current } I \text{ through the cable}$$

$$= nI^2 R\, [b_1 + (1 + \lambda)\, (b_4 + b_5 + G)]$$

Current rating to be $$I = \sqrt{\frac{T_m - T_b}{nR\left[S_1 + (1+\lambda)(S_4 + S_5 + G)\right]}}$$

6.10.1 Heating of Cables

It is known to us that the temperature rise of a body depends upon the rate of generation and dissipation of heat by the body. If the rate of generation is greater than the rate of dissipation the temperature goes on rising and viceversa.

In case of an underground cable the sources of heat generation are : (1) core loss *i.e.*, copper loss in the core of the cable, (2) the dielectric loss, (3) sheath loss and the heat is dissipated through the dielectric to the ground and finally to the atmosphere.

Core Loss

To find out the core loss the value of the resistance of the cable is calculated as follows :

(*a*) Knowing the resistance of the conductor at ambient temperature. 20°C, the resistant is calculated assuming on operating temperature of say 65°C, $R_{65} = R_{20}\ (1 + \alpha\, t)$.

Where, α is the temperature coefficient of the conductor material and *t* is the difference in temperature (which in this cases 65° – 20° = 45°C).

(*b*) Since the effective area of section of the cable is smaller than the actual physical section, the effective resistance of the cable is larger. A factor of 1.02 is multiplied to get the resistance.

(*c*) The length of the outer most strand is greater than the central strand. The effect of stranding on the resistance is obtained by multiplying the resistance as calculated according to the length of the central strand by a factor of 1.02.

Now having calculated thus the reistance of the cable the core loss is calculated as $I^2 R$ where *I* is the current carried by the cable.

6.10.2 Resistance Capacitance of a Cable

In this section we shall briefly discuss how to calculate insulation Resistance, Capacitance of the cable as below :

6.10.3 Insulation Resistance of a Cable

The usual load current flows through the core of the cable whreas leakage current, *i.e.*, the current which is not useful flows radially, *i.e.*, from the conductor to the sheath through the dielectric material as referred to Fig. 6.13. The flow of leakage current is shown by dotted lines. The resistance of any material is given by :

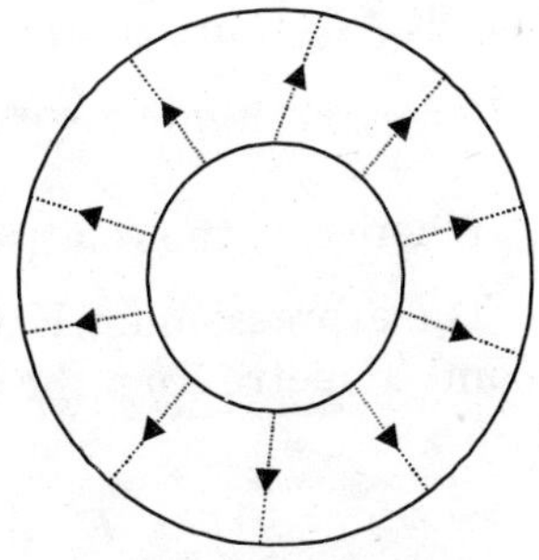

Fig. 6.13. Insulation resistance of a cable.

$$R = \rho \cdot \frac{l}{A}$$

In case of a cable since the area of section increases as we go from the core to the sheath we first write an expression for the insulation resistance of an annular cylinder with radii x and $(x + dx)$ units as measured from the centre of the core.

$$\therefore \quad dR = \rho \frac{dx}{2\pi x l}$$

Here unit in the denominator represent the unit length of the cable, *i.e.*, dR represent the differential leakage resistance for unit length of the cable.

$$\therefore \quad R = \int_r^R dR = \int_r^R \rho/2\pi \cdot \frac{dx}{x} = \frac{\rho}{2\pi}\int_r^R \frac{dx}{x} = \frac{\rho}{2\pi} ln \frac{\rho}{r} \text{ ohm/metre length}$$

In case the length of the cable is l unit the leakage resistance

$$R = \frac{\rho}{2\pi l} ln \frac{R}{r} \text{ ohm}$$

Note : Whereas the resistance of the core of the cable is directly proportional to the length of the cable the leakage resistance is inversely proportional to the length of the cable.

6.10.4 Capacitance of a 1-core Cable

A single-core cable is in effect an electrostatic capacitor because it has two electrodes, the core of the cable and the sheath separated by a dielectric material as shown in Fig. 6.14. Let λ be the charge per unit length. By definition capacitance is the ratio of the charge on one of the electrodes to the potential difference between the electrodes is as we know from basic first principle of Electrostatic stress, the V the potential of the conductor with respect to the sheath, where R is the inner radius of the sheath and r is the radius of the conductor then with λ the charge per unit length, we get voltage,

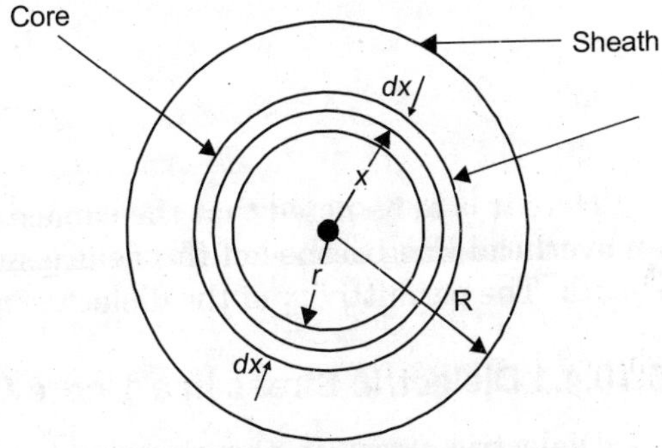

Fig. 6.14. Capacitance of a 1-core cable.

$$V = \frac{\lambda}{2\pi\varepsilon_0} ln \frac{R}{r},$$

or $$V = \frac{\lambda}{2\pi\varepsilon_0\varepsilon_r} \cdot ln\frac{R}{r}$$

Where ε_r is the relative permittivity of dielectric in between the core and the sheath.

The expression for V was obtained, as below w.r.t. Fig. 6.14. The electric field intensity at a point 'x' metre from the centre of the core,

$$E_x = \frac{\lambda}{2\pi\varepsilon_0\varepsilon_r x} = \frac{\lambda}{2\pi\varepsilon_0\varepsilon_r}\frac{1}{x} \; V/m$$

∴ Voltage difference between core and sheat

$$= V = \int_r^R E_x \, dx = \frac{\lambda}{2\pi\varepsilon_0\varepsilon_r} \cdot ln\frac{R}{r} \cdot \text{Volts}$$

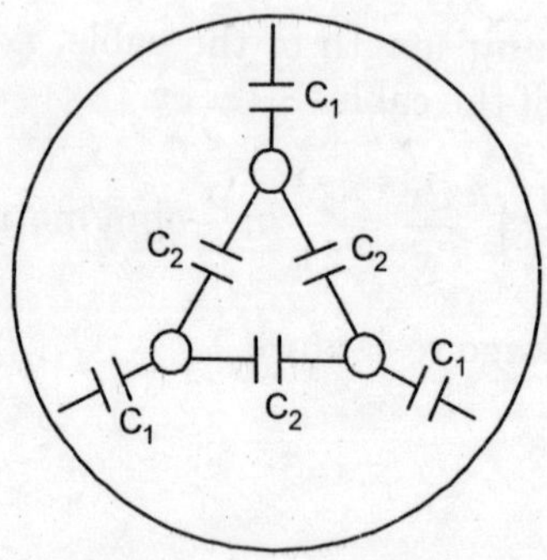

Fig. 6.15 (*a*). Capacitance of a 3-core cable.

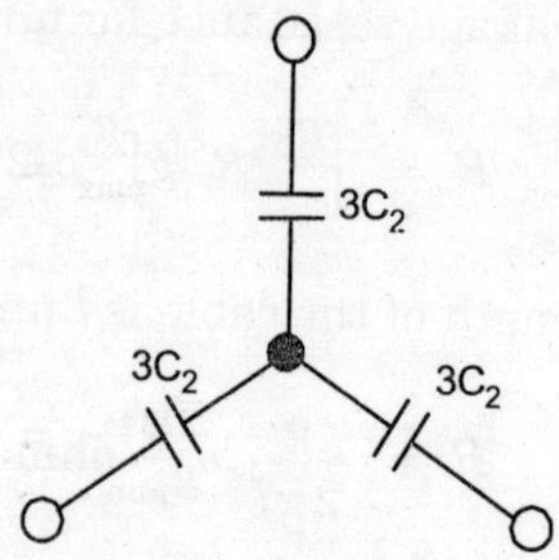

Fig. 6.15 (*b*)

or $$V = \frac{\lambda}{2\,\pi\varepsilon_0\varepsilon_r} ln\, R/r$$

or $$\frac{\lambda}{V} = C = \frac{2\,\pi\,\varepsilon_0\varepsilon_r}{ln\, R/r} \text{ F/metre}$$

Here it is to be noted that the capacitance of a cable is much more important than that of an overhead line because of the nearness of the conductor to one another and to the earthed sheath. The permittivity of the dielectric material is also higher than that of air.

6.10.4.1 Dielectric Stress in a 1-core Cable

Dielectric stress in the electric stress, the insulation of a cable is subjected to under operating conditions. Infact for voltage gradient or potential gradient at any point is known or defined as the rate of increase of potential at that point and is the same as the dielectric stress at the point.

We know that voltage gradient = Electric Intensity

hence, $$g = \frac{\lambda}{2\pi\varepsilon_0\varepsilon_r} \cdot \frac{1}{x} V/m \text{ at a radius } x$$

Now, we also know from above that,

$$V = \frac{\lambda}{2\pi\varepsilon_0\varepsilon_r} \cdot ln\frac{R}{r}\text{d}$$

Hence by putting the value of λ in the expression for

We get, $$g = \frac{2\pi\varepsilon_0\varepsilon_r V}{ln\dfrac{R}{r}} \cdot \frac{1}{2\pi\varepsilon_0\varepsilon_r} \cdot \frac{1}{x} = \frac{V}{x \, ln \, R/r} \cdot V/m$$

As "g" varies inversely as x, 'g' will be maximum when 'x' is minimum, *i.e.*, $x = \frac{d}{2}$ "r" and g will be maximum when

$$x = R\left(= \frac{D}{2}\right)$$

$$\therefore \quad g_{max} = \frac{V}{r \, ln\dfrac{R}{r}} \cdot V/m \text{ and } g_{min} = \frac{V}{R \, ln\dfrac{R}{r}}$$

$$\therefore \quad \frac{g_{max}}{g_{min}} = \frac{R}{r}$$

It is to be noted also that in HV Cables the value of 'g' is taken to the highest possible limit to keep to over all dimeter to minimum. But as the dielectric losses increase rapidly with 'g', the maximum permissible value of 'g' is around *one fifth* of the breakdown value.

6.10.4.2 Conductors of Most Economic Diameter

From the above discussion (Article 6.8.4.1) we know that for fixed values of the voltage V and overall diameter D (radius R) the g_{max} will be minimum when $d \, ln\frac{R}{r}$ is maximum. $d \, ln\frac{R}{r}$ or $d \, ln\frac{D}{d}$ will be maximum if $\frac{d}{dd}\left[d \; ln \, \frac{D}{d}\right] = 0$ or $ln\frac{D}{d_g} = 1$ or when $\frac{D}{d} = e = 2.71828$ or when, $d = \frac{D}{2.71828}$ or $r = \frac{R}{2.71828}$ and $g_{max} = 2V/d$. It is to be noted that for low and medium voltage cables, the diameter of the conductor determined from $d = \frac{2V}{g_{max}}$ is some what less than that determined from the consideration of safe current density. Hence, main important design factor is current carrying capacity of the cable to determine conductor diameter. To have the derived 'D' of conductor as the out increasing cross-section area. We can use (1) At core instead of copper, (2) By stranding copper wires round a hemp centre, (3) stranding copper wires over a lead tube.

6.10.5 Capacitance of a 3-core Cable

In this case of a 3-core cable capacitance determination we could assume that the dielectric is uniform between the core and the sheath it is possible to calculate the capacitance of a 3-core cable. But normally it is not so and therefore, it is desirable to find the capacitance by measurements. In a 3-core cable, sheath is at earth potential and the three conductors at supply terminals. There are six capacitances formed between those systems. In the Fig. 6.15 (*a*) *i.e.*, shown that three capacitances are between the sheath and the conductors and the remaining three capacitances are between the conductors.

Let C_1 be the capacitance between sheath and the conductor and C_2 be the capacitance between each conductor it is desirable to contact this system of capacitors into an equivalent star connection. The equivalent star of a delta connection (by delta star transformation) is shown in Fig. 6.15 (*b*)).

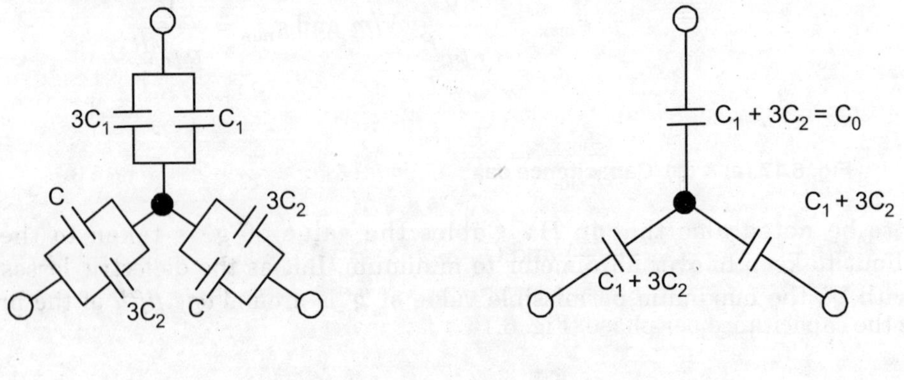

Fig. 6.16 (*a*) & (*b*). Equivalent capacitance of a 3-core cable.

Since the star point is at sheath potential and the other terminals correspond to the conductors of the cable the whole system of capacitors can be reduced to the following star system of capacitors as can be seen in Fig. 6.16 (*a*) and (*b*).

There are two unknown C_1 and C_2 to find out the capacitance per phase of the cable. Here we have to make two measurements. (1) Bunch the three core and measure the capacitance between the bunched conductors and the sheath. Now let this be C_x given by $C_x = 3C_1$ as shown in Fig. 6.17 (*a*) and (2) Connect any two cores to the sheath and measure the capacitance between the remaining conductors and the sheath. Let this be C_y given by Fig. 6.16 (*b*). The equivalent circuit will be $C_y = C_1 + 2C_2$. From these two measurements we get,

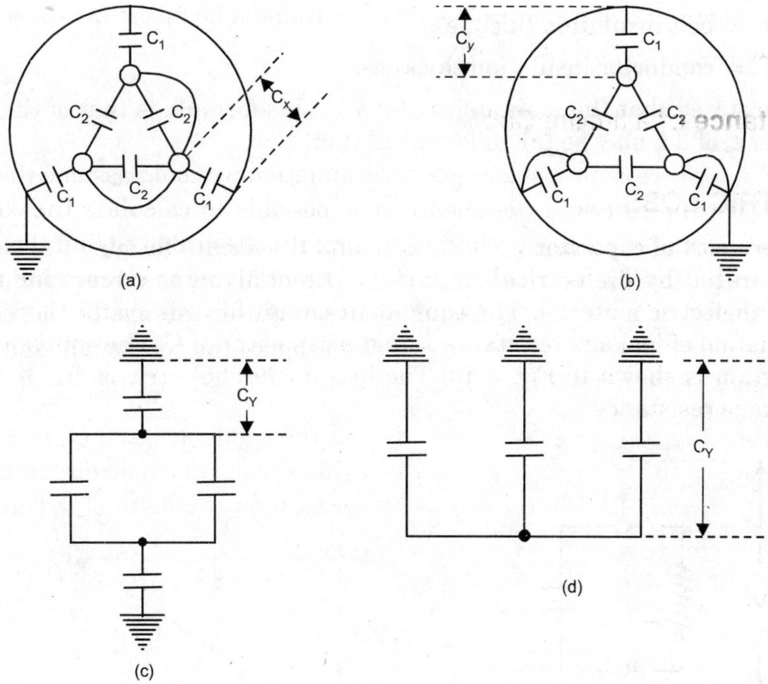

Fig. 6.17 (*a*) & (*b*). Capacitance calculation (*c*) & (*d*). Equivalent of 6.16 (*b*).

$$C_1 = \frac{C_x}{3}, \text{ and } C_2 = \frac{1}{2}(C_y - C_1) = \frac{1}{2}\left[C_y - \frac{C_x}{3}\right]$$

Since the capacitance per phase (Fig. 6.15 (*b*)) is given by

$$C_0 = C_1 + 3C_2 = \frac{C_x}{3} + \frac{3}{2}\left(C_y - \frac{C_x}{3}\right)$$

$$= \frac{C_x}{3} + \frac{3}{2}C_y - \frac{C_x}{2}$$

$$= \frac{3}{2}C_y - \frac{C_x}{6}$$

However, in case the test figures are not available the following emperical formula gives an approximate value of capacitance for circular conductors :

$$C_0 = \frac{0.0299\varepsilon t_r}{In\left[1 + \frac{T+t}{d}\left(3.84 - 1.70\frac{t}{T} + 0.52\frac{t^2}{T^2}\right)\right]}$$

where ε_r = relative permittivity of the dielectric

d = conductor diameter

t = belt insulation thickness

T = conductor insulation thickness

It should be stated that the main uncertainty in this formula is that of the value of ε_r. An average value of ε_r of 3.5 may be taken for calculation.

6.11 DIELECTRIC LOSS

The cable is a sort of capacitor with the core and the sheath forming the two plates of the condenser separated by dielectrical material. The equivalent circuit for this system is represented by dielectric material. The equivalent circuit for this system is represented by a parallel combination of leakage resistance R and a capacitance C. The equivalent circuit with its phasor diagram is shown in Fig. 6.18. The loss in the dielectric is due to the loss in the equivalent leakage resistance :

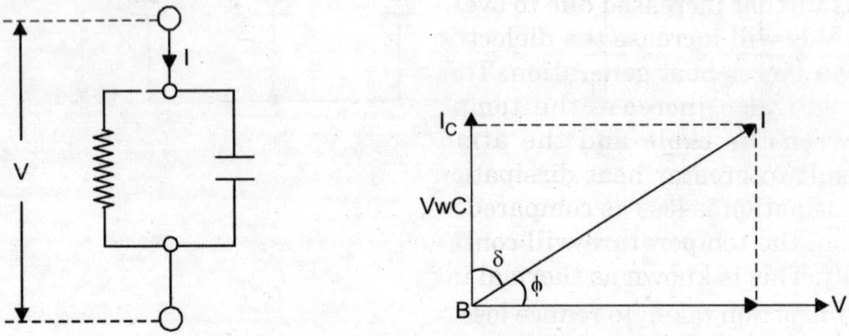

Fig. 6.18 (*a*) Equivalent of a cable. (*b*) Phasor diagram of (*a*).

$$P = \frac{V^2}{R}$$

From phasor diagram,
$$\frac{V/R}{VwC} = \tan\delta$$

or
$$\frac{V}{R} = VwC\tan\delta$$

$$\therefore \quad P = V^2 wC\tan\delta$$

where δ is the dielectric loss angle and w is the power supply frequency.

Since δ is nomrally very small,

$$\therefore \quad \tan\delta = \delta$$

$$\therefore \quad P = V^2 wC\tan\delta$$

Becomes,
$$P = V^2 wC\,\delta \text{ watts (where } \delta \text{ is in radius)}$$

Now from the phasor diagram the power factor angle of dielectric is given by :

$$\phi = 90 - \delta$$

$$\therefore \quad \cos\phi = \cos(90-\delta) = \sin\delta$$

This means, the power factor of a dielectric is a function of the temperature of the dielectric and also depends upon the voltage stress to which the dielectric is stressed :

6.11.1 Variation of Dielectric Power Factor with the Temperature

The variation of dielectric power factor with temperature of a cable operating at a normal voltage is given in Fig. 6.19. The variation roughly follows a *V* shape, it decreases with increase in temperature. The minimum point lies somewhere between 30°C and 60°C depending upon the type of impregnating compound. It is known to us that the operating temperature of a paper insulated cable is about 65°C which is to the right of the minimum point on the *V* curve.

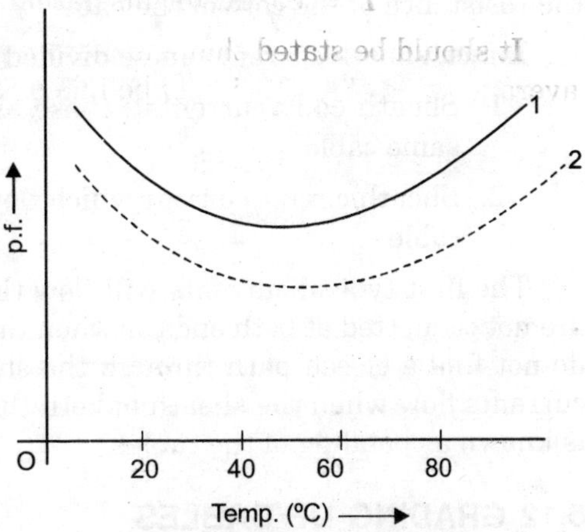

Fig. 6.19. Variation of p.f. with temperature for :
1. Water washed paper.
2. Deionised water washed paper.

Around the operating temperature, if the temperature is further increased due to overloads or other reasons this will increase the dielectric losses further giving a larger heat generation. The rise in temperaturs will also increase the temperature gradient between the cable and the atmosphere which will result in greater heat dissipation if the rate of heat dissipation is less as compared with the heat generation, the temperature will continue to increase until the dielectric overheats and fails electrically. This is known as thermal instability.

However, the action taken to reduce loss angle generally flattess the loss angle/temperature curve (2) of Fig. 6.19 and reduce the tendency towards thermal instability.

In case of solid type of cables when the stresses are high, dielectric loss does not vary directly as square of the voltage rather the losses are more due to the ionisation (corona loss) at weak points in the insulation. As a result there is increase of dielectric power factor. The weak point may be in the form of moisture in the insulation or more generally the presence of void formation. A void is a space which may be between the core papers and the conductor or sheath, or may lie as, more or less, flat film between one layer of paper and another. This space instead of being filled with dielectric strength of air is smaller than the normal working stress of the dielectric. Such a space is liable to ionisations. The Fig. 6.19 shows the variation of dielectric power factor as a function of electric stress. Since the electric stress near the surface of the conductor is maximum the voids near the surface are the first to breakdown and the ionisation then spreads progressively through the whole insulation; the voids near the sheath are the last to break down.

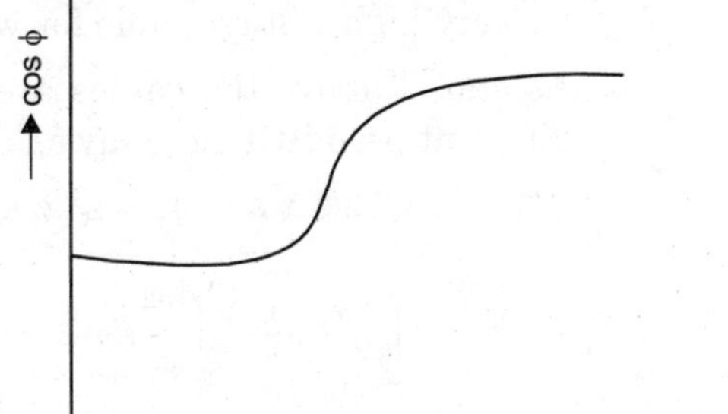

Fig. 6.20

6.11.2 Sheath Loss

While using single core cables for a.c. transmission, the current flowing through the core of the cable gives rise to a pulsating magnetic field which when links with the sheath, induces voltage in it. This induced voltage set up current under certain condition in the sheath and this result in sheath losses. Since the sheath current is proportional to the cable core current the sheath losses are also proportional to the conductor losses. If λ is the sheath loss to the conductor loss, the equivalent a.c. resistance of the cable will be $R_{eq} = R\,(1 + \lambda)$ where R is the resistance of the core of the cable.

Infact, sheath current can be divided into two kinds :

1. Sheath eddy currents : these are the currents which flow entirely in sheath of the same cable.
2. Sheath circuit current which flow for the sheath of one cable to the sheath of another cable.

The first type of currents will flow through the sheath when the sheath of the two cables are not connected at both ends or when they are connected only at one end because the currents do not find a closed path through the sheath of both the cables. Whereas the second type of currents flow when the sheath of both the cables are electrically connected at both ends. This is known as bonding of the cables.

6.12 GRADING OF CABLES

The process of uniformity in dielectric stress is known as grading of cables. Infact, there are two methods of grading as : (1) capacitance grading and (2) intersheath grading.

1. In capacitance grading the uniformity in dielectric stress is obtained by using various layers of different dielectrics in such a manner that the permittivity ε_r of any layer is inversely proportional to its radius. *i.e.*, $\varepsilon_r \propto \dfrac{1}{x}$, or $\varepsilon_r x$ = constant or $g \propto \dfrac{y}{2\pi\varepsilon_r}$ which is constant. In such condition the value of dielectric stress at any point is constant and independent of distance from the centre. This is the ideal grading.
2. Intersheath grading in homogeneous dielectric is used, which is divided into various layers by sustably placing the metallic intersheaths. However grading is only useful for very high voltage cable for which D/d is large.
3. As stated above the cables are graded by having insulation mode of dielectrics of different permittivities, say ε_1, ε_2 for two different dielectrics with stresses,

$$g_1 = Q/2\pi\varepsilon_1 x \text{ and } g_2 = Q/2\pi\varepsilon_2 x \text{ [refer to Fig. 6.14]}$$

$$\therefore\; V = \int_{d/2}^{d_{1/2}} g_1 dx + \int_{d_{1/2}}^{D/2} g_2 dx = \frac{Q}{2\pi}\left[\frac{1}{\varepsilon_1} ln(d_{1/d}) + \frac{1}{\varepsilon_2} ln(D/d_1)\right] \quad \therefore C = \frac{Q}{V}$$

$$\therefore\; Q = \frac{2V\pi}{d\left[\dfrac{1}{\varepsilon_1} ln(d_1/d) + \dfrac{1}{\varepsilon_2} ln(D/d_1)\right]} \quad \therefore g_{1_{\max}} = Q\Big/\left(2\pi\varepsilon_1 \cdot \frac{d}{2}\right) = \frac{2V}{d\left[ln\, d_1/d + \dfrac{\varepsilon_1}{\varepsilon_2} ln\, D/d_1\right]}$$

g_1 will be max. when $x = d/2$ and g_2 will be max when $x = d_1/2$

$$\text{and } g_{2_{\max}} = \frac{Q}{\left(2\pi\varepsilon_2 \cdot \dfrac{d_1}{2}\right)} = \frac{2V}{d_1\left[\dfrac{\varepsilon_2}{\varepsilon_1} ln \dfrac{d_1}{d} + ln\, D/d_1\right]}$$

6.13 CURRENT RATING OF CABLES & MAXIMUM CURRENT CARRYING CAPACITY OF CABLES

6.13.1 Current Rating of Cables

To determine the value of current I which will raise the conductor temperature to the maximum permissible value, the thermal resistance of the cable is to be known. This is the cable rating under the specified conditions and under continuous loading.

Infact, the insulation of the cable will be passing heat through it due to the conductor losses, (neglecting the dielectric losses) while bedding and serving will be passing heat due to the sheath losses in addition.

If the power losses in the "n" number of cores of cable with each core of resistance R Ω per metre are nI^2R with per metre length, when I is the current per core in amperes. If T_m is the maximum temperature allowable, then $nI^2R = (T_m - T_s)/R_{OD}$, where R_{OD} is the thermal resistance of Dielectric. Let us assume that sheath losses are "y" times the conductor power losses and neglecting armour losses, the total losses become $(1 + y)\ nI^2R$ and for ambient temperature T_a, we get, $(1 + y)\ nI^2R = \dfrac{T_s - T_a}{R_{OP} + R_{OG}}$ where R_{OP} and R_{OG} are thermal resistances of protective covering and ground or soil. Now eleminating T_s (as it is not known) from the above equations, we get

$$nI^2R\ R_{OD} + (1 + y)\ nI^2R\ (R_{OP} + R_{OG}) = T_m - T_a \quad \therefore I = \sqrt{\frac{T_m - T_a}{nR\ [R_{OD} + (1 + y)\ (R_{OP} + R_{OG})]}} \quad \textbf{Ans.}$$

6.13.2 Maximum Current Carrying Capacity of Cables

It should be carefully noted that, cables should not be operated too hot for the following reasons : (1) unequal expansion may create void insulation which will lead to ionisation, (2) Expansion of the oil may cause the sheath to burnt in (3) The oil may lose its oversity and it may start draining away for higher levels (4) Thermal instability may rise due the very fast increase of dielectric losses with temperature. In fact the current rating of cables, like that by any other electrical equipment is guided by maximum permissible temperature rise which the insulation used can withstand. Temperature rise depends on. (1) Heat produced (2) surrounding temperature (3) thermal conductivity, (4) no. of cores, (5) spacing.

Example 6.5. *The capacitance of a 3-core load sheathed cable measured between any two of the conductors with sheath earthed is 0.19 μF/kms. Determine the equivalent star connected capacity and the KVA required to keep 16 Kms of the cable charged when connected to 20 kV, 50 Hz supply.*

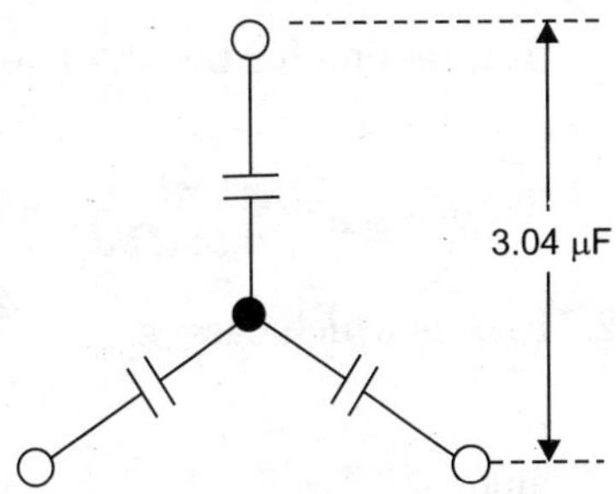

Fig. 6.21

Solution : The equivalent circuit is shown in Fig. 6.21. Since the capacitance measured is 3.04 μF between the conductors, the capacitance per phase will be,

$$= 2 \times 3.04 = 6.08\ \mu F$$

$\therefore$ 3-phase MVA required $= V^2 wC$

$= 20^2 \times 3.4 \times 6.08 \times 10^{-6} = 0.763$ MVA or 763 kVA

Example 6.6. *Calculate the capacitance and charging current of 1-core cable used on a 3-phase 66 kV system. The cable is one Km long having a core diameter 15 cm and sheath diameter 60 cm. the relative permittivity of the insulation may be taken as 3.6 and the supply frequency 50 Hz.*

Solution : Capacitance is given by,

$$C = \frac{\varepsilon_r}{18\,ln\left(\frac{D}{d}\right)} \mu\text{F/Km}$$

$$= \frac{3.6}{18 \times 1.386} = 0.145\ \mu\text{F}$$

The core with respect to the sheath is at $\frac{66}{\sqrt{3}}$ kV.
Therefore charging current

$$= \frac{66 \times 10^3}{\sqrt{3}} \times cw = \frac{66 \times 10^3}{\sqrt{3}} \times 0.145 \times 10^{-6} \times 314 = 1.74 \text{ Amps}$$

Example 6.6. *A 4 kV single-conductor cable has an outside diameter of 8.0 cm. Determine the conductor radius and the electric field strength that must be withstood by the insulating material in the most economical (optimal-rate) configuration.*

Solution : From the equation we get,

$$\frac{D}{d} = r = 2.718$$

We have $\frac{D}{d} = \frac{8/2}{d} = 2.718$

$\therefore$ $d = 1.47$ cm

Maximum value of electric strength,

$$g_m = \frac{V}{r\,ln\,(D/d)} = \frac{11 \times 100}{1.47 \times 10^{-2}\,ln\left(\frac{4}{1.47}\right)} = 747.53 \text{ kV/mt.}$$

Example 6.7. *Find the most economical diameter of single-core cable of a 132 kV 3-phase system. Find also the overall diameter of the insulation if the peak permissible stress is not to exceed 60 kV per cm.*

Solution : As per example, peak value of phase voltage $= \frac{132}{\sqrt{3}} \times 2 = 108$ kV

Expression for the stress is $g = \frac{V}{x\,ln(D/d)}$

Now, $g_{max} = \frac{2V}{d\,ln(D/d)}$ which will be minimum when $I_n\ D/d = 1$. That mean, when $D =$ 2.718 d, is which case, $g_{max} = \frac{2V}{d}$ or $60 = \frac{2 \times 108}{d}$ (as per example $g_m = 60$).

$\therefore$ $d = 3.6$ cm

and $D = 2.718 \times 3.6 = 9.786$ cm

Example 6.8. *Estimate the charging current drawn by a 3-core cable protected by a metal sheath when switched on the an 11 kV, 50 Hz supply. The capacitance between two cores with the third core connected to the sheath is measured to be 3.7 μF.*

Solution : Refering to Fig. 6.22 capacitance is measured between conductors 2 and 3. The value that is measured is

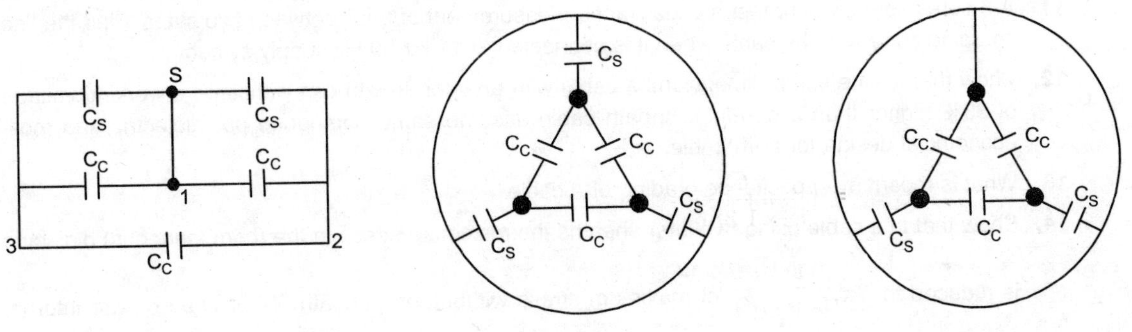

Fig. 6.22

$$C_c + \frac{1}{2}C_c + \frac{1}{2}C_3 = \frac{3}{2}C_c + \frac{1}{2}C_3 = \frac{1}{2}(3C_c + C_b)$$

But $\quad 3C_c + C_b = C_e$

The measured value of capacitance

$$= \frac{1}{2}C_c \text{ or } C_0 = 2 \times 3.7 = 7.4\ \mu\text{F}$$

Therefore charging current perphase $\quad wC_0 V = 314 \times 7.4 \times 10^{-4} \times \frac{11}{\sqrt{3}} \times 10^9 = 14.77$ cm

QUESTIONS

1. A string of six insulator unit has neutral capacitance 10 times the capacitances to ground. Determine the voltage across each unit as a fraction of the operating voltage. Also determine the string efficiency.
2. Each conductor of a 33 kV, 3-phase system is suspended by a string of three similar insulators the capacitance of each insulator. Determine the string efficiency.
3. A string of eight suspension insulators is to be graded to obtain uniform distribution of voltage across the string. If the capacitors of the top unit is 10 times the capacitance to ground of each unit determine the capacitance of the remaining seven units.
4. Define what is string Efficiency.
5. Discuss the methods Equalising the potential drops across the various unit of the string. Explain with sketches of system.
6. State what are the various types of insulators in overhead system with sketches.
7. Derive an expression for the potential Distribution over a string of Suspension Insulators.
8. A string of 5 suspension insulators is to be graded for uniform voltage distribution across the string. If the pin-to-earth capacitance are all equal to C and the mentral capacitance of the top insulator is to 10 C, find the mentral capacitance of each unit in term of C, 6A. A string of 6 suspension insulators is to be fitted with aquatdring. The pin-to-earth capacitance are all equal to C. What should be the values of line-to-line capacitances so as to have uniform voltage distribution over the string.
9. What is the purpose of using inter sheaths in a cable.
10. Show that in a 3-core (bolted-type) cable the neutral capacitance of each. Conductor C_n is equal to C_b + 3 C_c where C_s = capacitance of each conductor to sheath and C_c = capacitor to each of them.

11. A 3-core cable gives or test a capacitance measurement of 2 μF between two cates. Find the line charging current of the cable when it is connected to 11 kV 50 Hz supply system.
12. Show that for the same dimensions a cable with an inter sheath can withstand a working voltage of 33% higher than a non-inter sheath cable assume same homogone our dielectric and most economcial design for both cable.
13. What is meant by capacitance grading of a cable.
14. Show that in a cable using two inter sheaths the maximum stress in the (homogeneous) dielectric is reduced to $\frac{1}{\sqrt{3}(1+\alpha+\alpha^2)}$ of maximum stress without inter sheath. What *D* and *d* are internal discuss core dia d_1 and internal d_2.

$$\alpha = d_1/d = d_2/d_1 = D/d_2$$

CHAPTER

7

Control of Voltage, Load Frequency and Power in Transmission System

INSIDE THIS CHAPTER

- Reactive power
- Relation of active power reactive power at a nodal voltage at a node
- Power system & reactive compensation types
- Voltage control
- Inter-relationship between voltage & reactive power, frequency & active power
- Sending end power circle diagram
- Load frequency control
- Concept of control area
- Effect of voltage and frequency on load

In a power transmission system, Reactive Power is the latent force. Generally, its role being not appreciated, it is very precious in keeping the system voltage stable. However, frequency being the safe indicator of active power balance, voltage is the sole indicator of reactive power balance. Generation and absorption of reactive power in the power system is thus necessary to be studied.

A power system is said to be well designed if it gives a good quality of reliable supply. By good quality is meant the voltage levels within the reasonable limits. Practically all the equipments of the power system are designed to operate satisfactorely only when the voltage levels on the system correspond to the their rated voltages or at the must the variations are within say ±5. In case the voltage variatious is more than a specified value the performance of the equipments suffers and the life of most of the equipment is also sacrificed. Hence the necessity of controlling the voltage on transmission system is much strong. However, let us first look into Reactive Power's case.

7.1 REACTIVE POWER

For generation and absorption of reactive power in the power system the equipments and components of the transmission system are as follows :

(*a*) Alternators

An alternator when over excited generates reactive power while an under excited machine absorb it generally, the alternators are operated in overexcited mode such that they can act as main source of supply of reactive power to the power system. The capability of supplying the reactive power is determined by the short circuit ratio.

(b) Transformers and Overhead Lines

An overhead line when loaded fully, absorbs reactive power, given by I^2X where I is the line current through the line reactance X ohms per phase on light loads or at no load the distributed shunt capacitances throughout the line predominant and the line supplies charging VAR, that is generates reactive power. It is then given by (V^2 wC) where V is the line voltage w the system frequency in radius and C is the capacitance of the line to earth.

Transformers also absorb reactive power considerably, and it is given by,

$$P_{VAR} = I^2 \cdot \frac{V}{I_{rated}} \cdot X_t = I^2 \cdot \frac{V}{I_{rated}} \cdot X_t \text{ VAR/ph.}$$

where I is actual phase current and I_{rated} is the rated phase current, V is the phase voltage, X_t the P.U. reactance of the transformer

7.1.1 Role of Reactive Power on Voltage & Voltage Regulator

In the Fig. 7.0 in a simple transmission network, the sending end voltage (V_S), the receiving end voltage (V_R) and power angle, δ are interrelated a phasor diagram is shown.

∴ From this power diagram it is evident that

$$V_S^2 = (V_R + \Delta V_R)^2 + \delta V_R^2$$

$$= (V_R + V_R I \cos\phi + XI \sin\phi)^2 + (XI \cos\phi - RI \sin\phi)^2$$

$$\therefore \quad V_s^2 = \left(V_R + \frac{RP}{V_R} + \frac{XQ}{V_R}\right)^2 + \left(\frac{XP}{V_R} - \frac{RQ}{V_R}\right)^2 \quad \text{...(7.1)}$$

Now, as $\quad I \cos\phi = \dfrac{P}{V_R}$

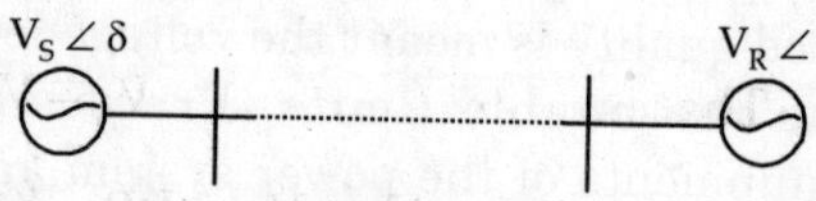

Fig. 7.0

and $\quad \sin\phi = \dfrac{Q}{V_R}$

Then, $$\Delta V_R = \frac{RP + QX^2}{V_R} \quad \text{...(7.2)}$$

and $$\delta V_R = \frac{XP - RQ}{V_R}$$

Now assuming, $\quad \delta V \ll (V_R + \Delta V_R)$

We get, $$V_s^2 = \left(V_R + \frac{RP + QX}{V_R}\right)^2$$

and $$V_s - V_R = \frac{RP + XQ}{V_R} = \Delta V_R \quad \text{...(7.3)}$$

Therefore the arithmetic difference between the voltages is approximately given by $\left(\dfrac{RP + QXQ}{V_R}\right)$.

Now assuming that the line resistance, $\quad R = 0 \quad$ ...(7.4)

Then, we get $$V_s - V_R = \frac{XQ}{V_R} \quad ...(7.5)$$

This therefore, reveals that the voltage depends mainly on reactive power (Q). The angle of transmission with power angle δ is governed by real power (P) and is obtained from

$$\delta = \sin^{-1}(\delta V_R/V) \quad ...(7.6)$$

Now if R is assumed to be zero, and V_s is specified, then from eqn. (7.1), we get,

$$V_R = \sqrt{V_s - \frac{QX}{V_S} + \left(\frac{PX}{V_S}\right)^2} \quad ...(7.7)$$

where V_S = Sending and voltage

V_R = Receiving and voltage

7.2 RELATION OF ACTIVE POWER, REACTIVE POWER AT A NODAL VOLTAGE AT A NODE

In a transmission circuit as any node the phase angle voltage is governed by the variation of load power, P & Q, provided there is no external means to control the bus voltage, V_1.

$$\therefore \quad V_R = f(P, Q)$$

$$\therefore \quad dV = \frac{\delta V_R}{\delta P}\cdot dP + \frac{\delta V_R}{\delta Q}\cdot dQ = \frac{dP}{(\delta P/\delta V)} + \frac{dQ}{(\delta Q/\delta V)} \quad ...(7.8)$$

However, $$\Delta V_R = \frac{RP + XQ}{V_R} \text{ for eqn. (7.4)}$$

That is, $$V_S - V_R = \frac{RP + XQ}{V_R}$$

or $$(V_b - V_R)\cdot V_R - RP - XQ = 0$$

or $$-V_R^2 + V_S\cdot V_R - RP - XQ = 0$$

Now by differenting this equation, we get

$$\frac{\delta P}{\delta V_R} = \frac{V_S - 2V_R}{R} \quad ...(7.9)$$

(Here of course it is assumed that V_S is constant)

Similarly we can get for $$\frac{\delta Q}{\delta V_R} = \frac{V_S - 2V_R}{X} \quad ...(7.10)$$

Now substituting the values of $\frac{\delta P}{\delta V}$ and $\frac{\delta Q}{\delta V}$ from eqns. (7.9) & (7.10) in eqn. (7.8), we get

$$dV = \frac{dP\cdot R + dQ\cdot X}{V_R - 2V_S}$$

If the voltage, V is constant then $dV = 0$. Hence this equation becomes $dP\cdot R + dQ\cdot X = 0$

or $dQ = \left(-\frac{R}{X}\right)dP$, Imfact in the present context is of greater interest. It is then evident from the relation, $\frac{\delta Q}{\delta V} = \frac{V_S - 2V_R}{X}$, that the smaller the reactance associated with the node, the large will be the value of $\left(\frac{dQ}{dV}\right)$ for a given voltage drop. This indicates that the voltage drop becomes inherently small, provided the nodal reactance is low. However, nodal reactance can be made low by use of parallel transmission lines.

In fact, in actual operation, in EHV lines, $R \ll X$. Hence at any short circuit in the node, short circuit current $I_{SC} = \frac{V_{SC}}{X}$ Amp. But, if the system is at no load, from the expression of $\frac{\delta Q}{\delta V} = \frac{V_S - 2V_R}{X}$, with $V_S = V_R$, we get

$$\frac{\delta Q}{\delta V} = -\frac{V_S}{X} \quad ...(7.12)$$

Therefore the magnitude of short circuit becomes numerically equal to the magnitude of ($\delta Q/\delta V$). Infact, it is now to be realised/understood that the factor $\delta Q/\delta V$ (a very important factor) is a measure of determining the short circuit current at a node.

7.3 POWER SYSTEM & REACTIVE COMPENSATION TYPES

It is now evident from the discussion made above that sufficient reactive reserve is required to maintain terminal voltage at the load bus. The devices that supply this reactive power at receiving end, bus in transmission system or load bus in distributing system are called reactive power compensating devices. But, in addition to voltage control the other effects of providing reactive power compensating devices at the load bus are voltage control the other effects of providing reactive power compensating devices at the load bus are : (1) Decrease in KVA loading of alternators, (2) Reduction in investment per kW of load supplied, (3) Reduction in supply side reactive burden and (4) Reduction of system copper loss due to reduction of reactive current. (5) Other benefits include improvement in system power factor which is an important aspect to achieve better economy power distribution system.

Below now, discussions are briefly made on different types of Reactive Compensators :

1. Series Capacitor

Series capacitors work as a powerful compensator by improving the power transfer capability as well as the voltage profile of the line as well as the that of the receiving end bus.

In fact, series capacitors are connected in series with the line conductors and hence reduce the net inductive reactance between the two end of the line. It improves the voltage stable state and supplies reactive power to the receiving end so that the voltage profile is maintained.

Besides this useful benefit, the steady state stability also improves due to net decrease of series inductive reactance. Power transfer also increases and switched series capacitors can be used in order to improve transmit stability. However, its use is still limited except for EHV systems because of the following reason :

(*a*) Series capacitor compensation may lead to mal-operation of distance relays in the transmission line when the degree of compensation is high.

(*b*) Severe over voltage during line fault.

(*c*) Ferro resonance in series compensated lightly loaded line with transformer at the receiving end.

(*d*) Sub-synchronous resonance with electro-mechanical system of the turbine generator with the line causing high terminal stress.

(*e*) Economy is affected and its installation is not just fed when the load reactive power requirement is small.

2. Synchronous Compensators

Synchronous compensators are also known as synchronous Phase Nodifier. In fact it is basically a synchronous motor running at no load. Depending on the value of exitation it can absorb or generate reactive power. But the losses within the motor are more, and wear and tear is high. A great flexibility of operation for all load conditions lagging or leading is thus achieved. Thus, its use in high voltage transmission line helps in either absorption of reactive power or supply of reactive power under light load or heavy load conditions respectively.

However, additional advantage of this device include important in system stability, reduction of the effect of sudden change in load, arresting voltage dips and supply of heavy amount of reactive power during short period. Moreover, the control is continuous. Disadvantages include high cost of installation contribution of more short circuit current in the system, more loss and higher maintenance.

3. Shunt Capacitors

Shunt capacitors are the most useful compensator in the power system. Small banks of static capacitors are usually connected in star delta and used in substation for supplying reactive power, that is Q, to the system. It improves the power factor of the load and hence has got wide application in industrial bus of distribution system and in sub-transmission systems. Now by supplying reactive power to the bus it maintain its voltage profile. In transmission system the static capacitor banks are often connected to the tertiary delta winding of the substation transformer.

Then, the reactive power supplied by capacitors (Q_c) is given by

$$Q_c = VI_c = \frac{V^2}{X_c} = V^2 wC \text{ MVAR/phase}$$

where X_c = shunt capacitance reactance/phase,
V = phase voltage,
C = capacitance/phase.

Shunt connected inductors are frequently used at the load end of the transmission line to control the over voltage of the receiving end at light load condition, the Ferramti Effect. The inductors are usually coreless type and possess linear type characteristics. If X_L be the reactance/phase of the inductors with phase voltage V, the reactive power absorbed by these inductors, Q_c (in star or in Delta connection) is given by

$$Q_l = -VI_l = \frac{V^2}{X_L} = -\frac{V^2}{wL} \text{ MVAR per phase}$$

where, V = phase voltage
X_L = wL ohm
= Reactance of inductor/phase
L = inductance/phase

Note : –*ve* sign indicates reactive power absorption.

7.4 VOLTAGE CONTROL

When power is supplied to a load through a transmission line keeping the sending end voltage constant the receiving end or load voltage undergoes variations depending upon the magnitude of the load and the power factor of the load. The higher the load with smaller power factor, the greater is the voltage variation. The voltage variation at node is an indication of the unbalance between the reactive power generated and consumed by that node. If the reactive power generated is greater than consumed, the voltage goes up and vice-versa. Whenever the voltage level of a particular bus undergoes variation this is due to the unbalance between the two bars at the bus. To under stand this problem let us refer Fig. 7.1, where the node one is a generator node with reference to voltage and node two is the load node with voltage V_2. The two bus bars are inter connected through a short line.

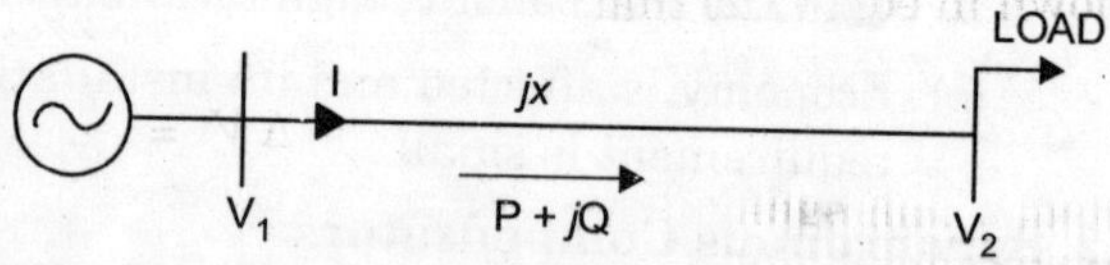

Fig. 7.1. Load connected to the source through a line.

We have explained above the relation between incremental Reactive Power, Active Power & Voltage at node. Below let us now discuss the method of voltage control in transmission system.

7.4.1 Method of Voltage Control in Transmission System

We have discussed above the methods for voltage control by using series capacitors synchronous capacitors, shunt capacitor. But there are other methods of (1) Tap changing transformers and (2) Booster transformers and (3) Static VAR system which are discussed below :

1. By Transformer Tap Changing

The main job of a transformer is to transform electric energy from one voltage level to another. Almost all power transformers on transmission lines are provided with taps for ratio control which means control of secondary voltage. In high voltage or extra high voltage lines on load tap changer (OLTC) transformers are used while ordinary off load tap changers prevail in distribution circuit. It must be stated here that transformer does not generat reactive power (on the other hand it consumes Reactive Power) and only transfers the reactive power from one side to another side altering the in phase component of the system voltage.

Now let us consider two transformers in the transmission system one at the sending end with off nominal tap ratio $1 : t_s$ and another at receiving and with tap ratio $t_r : 1$ such that the product of t_s and t_r is unity. When taps are used the actual sending end receiving end voltages would be $t_s\ E\ (= V_s)$ and the actual receiving end voltage would be $t_r\ V\ (= V_R)$ for nominal sending end receiving end voltage E and V respectively as can be seen in Fig. 7.2.

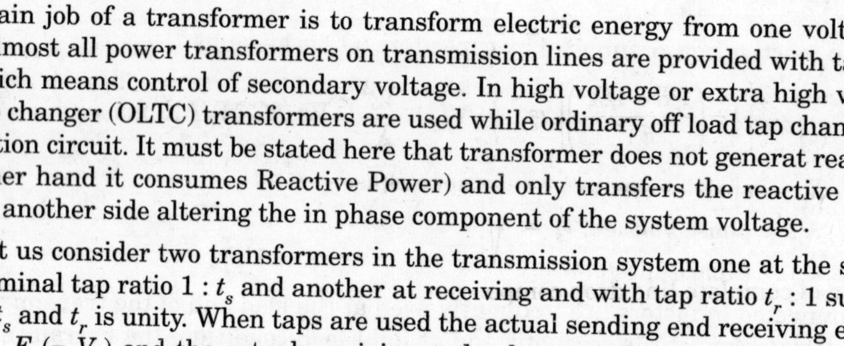

Fig. 7.2. Voltage control by transformer.

Now let ΔV be the voltage drop between the sending end and receiving end It has been shown in eqn. (7.2) that

$$\Delta V = \frac{PR+QX}{V_R} = \frac{PR+QX}{t_r V}$$

or $$V_S - V_R = \frac{PR+QX}{t_r V}$$

or $$t_s E - t_r V = \frac{PR+QX}{t_r V}$$

or $$t_r E = t_r V + \frac{PR+QX}{t_r V} \qquad ...(7.3)$$

Now to ensure same voltage level throughout the line, it is necessary that,

$$t_s \cdot t_r = 1$$

or $$t_s = \frac{1}{t_r}$$

(where t_s = number of turn of Sending End Transformer t_r = number of turn of Receiving End Transformer)

Hence, from enq. (7.13)

$$t_s = \frac{1}{E}\left[\frac{1}{t_s}\cdot V + \frac{PR+QX}{V/t_s}\right]$$

$$\therefore \quad t_s^2 = \frac{V}{E} + \left(\frac{PR+QX}{EV}\right)t_s^2$$

or $$t_s^2\left[1-\left(\frac{PR+QX}{EV}\right)\right] = \frac{V}{E}$$

$$\therefore \quad t_s = \sqrt{\frac{V/Q}{\left[1-\left(\frac{PR+QX}{EV}\right)\right]}} \qquad ...(7.14)$$

Now, in case of complete line drop compensation, $E = V$

Then, $$t_s = \sqrt{\frac{1}{\left[1-\left(\frac{PR+QX}{E^2}\right)\right]}} \qquad ...(7.15)$$

From the above expression we find the tap settings for complete line drop compensation. However, in practice the compensation is limited for 3.7%. Beyond that drop the tap setting requirement is very high and as the transformers do not have any reactive power generation capability hence for high line drops the tap changing transformers do not improve voltage profile. For these reasons use of tap changing transformers to improve system voltage profile is thus limited to smaller voltage variation only.

The tap changing transformers do not control the voltage by regulating the flow of reactive VARs but by changing the transformation ratio, the voltage in the secondary circuit is varied and voltage control is obtained. This method is the most popular as it can be used for controlling voltages at all levels.

2. Booster Transformer

The two-winding load tap changing transformer performs two functions, transforming the voltage and boosting the voltage whereas the booster transformer performs the latter function only. It can be installed it a sub-station as an additional equipment if voltage regulation is further found to be necessary or it can be installed as a separate piece of equipment at any intermediate point in the line. The latter application may be desirable on economical or technical grounds to increase the voltage at an intermediate point in a line rather than at the ends as with tap changing transformer.

However, for small outputs and voltages upto 2000 volts, the simplest booster courts of an auto transformer with necessary tapping whereas for higher voltages and larger sizes it is necessary to utilise on-load tap changing gear and also to perform the switching in an isolated circuit, the voltage of which is only a fraction of the line voltage.

One method is to energise the primaries of the boosting transformer by means of a regulating transformer the secondary of which is provided with tapping along with tap changing gear as shown in Fig. 7.3.

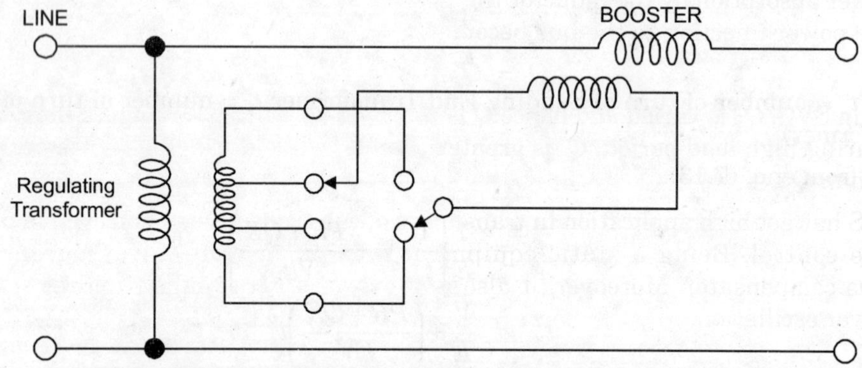

Fig. 7.3. Booster transformer along with regulating transformers.

The voltage changes are made by means of a motor operated controller and arrangements are made to reverse the connection to the primaries of the regulating transformers so that both buck and boost can be obtained. The sensing devices for voltage variation should be sensitive to current rather than voltage as the current varies 100% from no load to full load whereas the voltage varies by 10% or so. Booster transformer method has the following advantages.

1. The rating of the booster is the product of the current and the injected voltage and is hence only about 10% of that of a main transformer.
2. The transformer can be used at any immediate print in the system.
3. When it is used along with a fixed ratio transformer it can be later taken out for inspections or overhaul without affecting much of the system.

There are, however few disadvantages in this method also :

1. This system takes more floor space.
2. This system is less efficient due to the losses in the booster.
3. This method is more expensive than tap-changing method. Infact the booster transformers method are normally used in distribution feeders where the cost of tap changing transformer is very high.

3. Static Var System (SVS)

A static VAR System (SVS) can perform these two tasks together utilising the static VAR compensator. A static VAR compensator is basically a parallel combination of controlled reactor and fixed static capacitor as shown in Fig. 7.4.

It must be stated here that shunt capacitor compensation is required to enhance the system voltage during heavy load condition. As stated SVS can perform these two tasks. In SVS, the reactor control is done by an anti-parallel thysistor switch assembly. The firing angle of the thyristors govern the voltage across the inductor, thus controlling the reactor current. Thus the reactive power absorption by the inductor can be controlled. The capacitor in parallel with the reactor supplies reactive power to the system. Assuming Q_c as the reactive power to the system. Assuming Q_c as the reactive power charging by the capacitor and Q_i is the reactive power absorption by the inductor the net reactive power injection to the bus becomes.

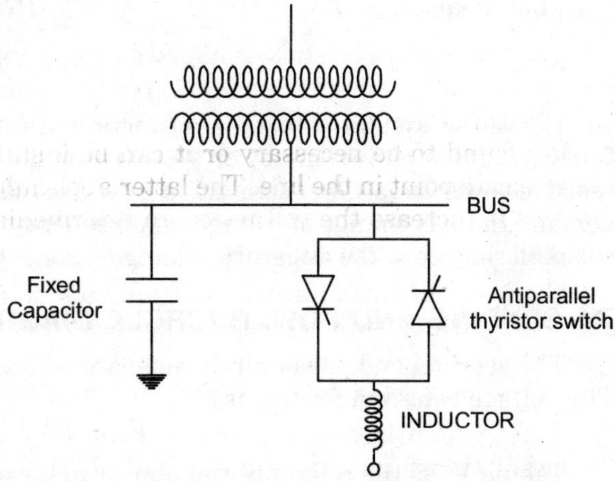

Fig. 7.4. Schematic static VAR compensator.

$$Q = Q_c - Q_i \quad \text{...(7.16)}$$

In fact, in SVC, Q_1 is varied and hence Q is controllable. The bus voltage is thus controllable by SVC. During high load period, Q_i is greater than Q_c while during heavy load condition, Q_c is greater than Q_i.

The SVS has got high application in transmission, sub-transmission and even in distribution bus voltage control. Being a static equipment it is much more advantageous than the synchronous compensator. Moreover, it also improves system stability, voltage stability and reduces power oscillation.

7.5 INTER-RELATIONSHIP BETWEEN VOLTAGE & REACTIVE POWER, FREQUENCY & ACTIVE POWER

We have seen above, in power transmission line the regulation is given by

$$\Delta V = \frac{PR + QX}{V} \quad \text{...(7.17)}$$

In HV or EHV lines normally R is very small and hence can be neglected,

Thus, $$\Delta V = \frac{QX}{V} \quad \text{...(7.18)}$$

That is, $$\Delta V = f_1(Q) \quad \text{...(7.19)}$$

Again, the active power in the transmission line is expressed

where V_S = Sending end values

V_R = Receiving end values

and δ = Power angle between V_S & V_R.

$$P = \frac{V_S \cdot V_R}{X} \cdot \sin\delta \quad \text{...(7.20)}$$

$$\therefore \quad \sin\delta = \frac{PX}{V_S \cdot V_R} \quad \text{...(7.21)}$$

That mean, $$\delta = f_2(P) \quad ...(7.22)$$

Also, $$\frac{d\delta}{dt} = w = 2\pi f = f_2'(P) \quad ...(7.23)$$

Hence the system voltage is a function of reactive power while the system frequency is a function of real power. In power system it is not easy to keep the balance between active and reactive power supply and demand in a precise manner. With change in active and reactive power demand there is change of voltage and frequency and for change on supply voltage or frequency the performance of the apparatus changes. Some permissible limit of variation is allowed.

7.6 SENDING END POWER CIRCLE DIAGRAM

The sending end power circle diagram is developed on the same line as the receiving end. The voltage equation for this is :

$$V_r = DV_s - BI_s \quad ...(7.24)$$

Taking V_s as the reference the phasor diagram as in Fig. 7.5 to express the above equation is as follows.

Now multiplying eqn. (7.24) by $-V_s/B_1$ the equation becomes

$$\frac{-V_r V_s}{B} = \frac{-DV_s^2}{B} + V_s \cdot I_s \quad ...(7.25)$$

This operation result in changing the magnitude of all the phasors in Fig. 7.5 by an amount $|V_s|$ $|B|$ and rotating them through

$$\frac{-V_s}{B} = \frac{|V_s| \angle 180°}{|B| \angle B} = \frac{|V_s|}{|B|} \angle (180 - B)°$$

$(180 - B)°$ clockwise. This result in Fig. 7.6. For the same reasoning as for the receiving end power circle diagram this diagram in Fig. 7.6 is rotated through 180° along the horizontal axis and results to Fig. 7.7. Here O is the centre of the sending and power circle and O_m is the radius of length $(|V_s|\,|V_r|)/|B|$. For a fixed sending end voltage since the location of the centre is fixed there will be concentric circles for different values of Receiving end Voltages. In fact it is possible to find out the maximum power that can be transmitted over the line both at the receiving end and sending end using the receiving end and sending end power circle diagram respectively. The difference in power at the ends is known as transmission loss and is due to the losses on the overhead line.

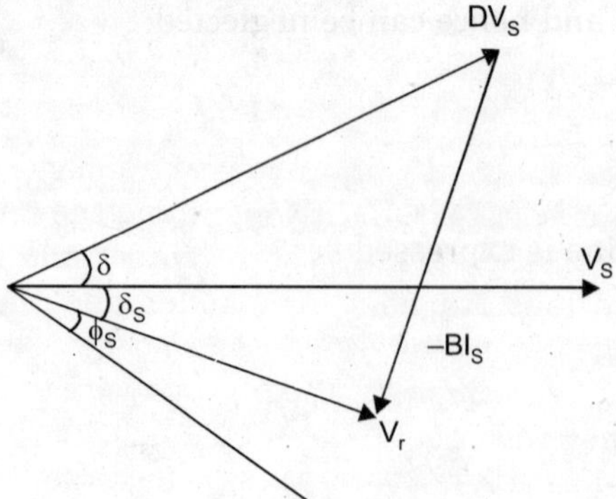

Fig. 7.5. Phasor diagram for eqn. (7.24).

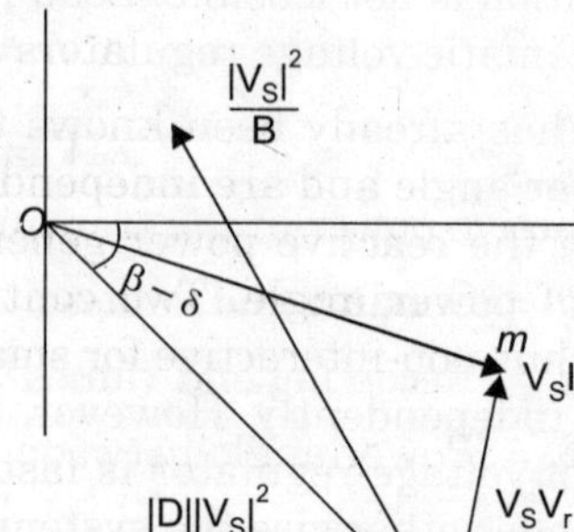

Fig. 7.6. Phasor diagram of Fig. 7.5. multiplied by – V_s/B

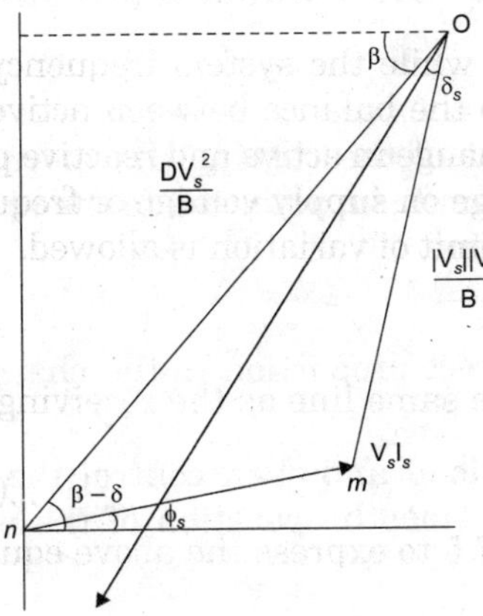

Fig. 7.7. Phasor diagram 7.7 rotated through 180° along horizontal axis.

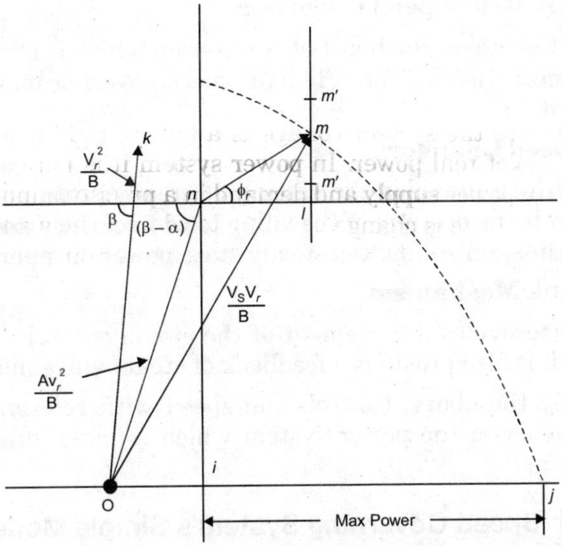

Fig. 7.8. Power circle diagram (receiving end) indicating various operating condition).

To find the maximum power that can be transmitted at the receiving and refer to Fig. 7.8.

7.7 LOAD FREQUENCY CONTROL

The relation between system frequency and speed of AC motors and the reason of governing effect of frequency on electric clock require close regulation of system frequency is the need for synchronous operation of power stations. However, by governments of the prime movers.

In fact, when the load on the turbine increases the decreases change of governor setting is required in order to maintain the nominal speed. If the system is consisting of a single machine supplying load then the speed and frequency change will be in accordance with the governor characteristics. Normal variation of frequency of 3.5% is allowed at different loading condition.

In case the system comprises of two machines running in parallel the system becomes more complex. The machine will share load according to their characteristics. The frequency control may be done by automatic or by nominal regulators. In a large inter-connected system manual regulation is not feasible. Load frequency control equipment is then installed for each generator. Automatic voltage regulators are also installed to control the system voltage.

In fact, it has already been knows to us that small variations in load depend upon the change in power angle and are independent of bus voltage. On the other hand the bus voltage is governed by the reactive power generation (*i.e.*, excitation of the machine) and is almost independent of power angle. Two controls, the load frequency controller and the voltage regulator are thus non-interactive for small changes. In power system control they are modelled and analysed independently. However, the load frequency controller response is slow while the response of voltage regulator is fast. The frequency regulator should be sensitive to fast and small changes otherwise the system will be prone to hunting of the alternator. In a speed governing system of the turbine the following components are of importance :

1. Hydraulic Amplifier

With movement of the speed governor it amplifies the movement which centres in the opening or during of steam value through a properly designed link machanism.

2. "Fly Ball" Speed Governor

It is infact, the heart of the system which sense the change in speed with change in speed the mechanism of the "Fly-ball" speed governor moves upwards and downwards to control the operation.

3. Speed Changer

Which provides the setting of steady state power output of the turbine. Its movement controls the opening of the value to control the amount of steam to be admitted to the turbine and thus controlling the steady state power output of the turbine.

4. Link Mechanism

It provides a movement of the steam control valve in direct proportion to the change in speed. It also provides a feedback of steam valve movement.

By the above, controls the speed with reference to the load and then control the load frequency of the power system which is most important for smooth operation of the whole system.

7.7.1 Speed Governing System's Simple Model

A simplified model of speed governing system is shown in Fig. 7.9 (we shall not go in detail through of the system) in a block diagram.

where K_{bg} = gain of speed governor
T_{sg} = Time constant of the speed governor
R = speed regulation of the governor
ΔD = command signal
ΔY = output movement
ΔF = frequency change signal

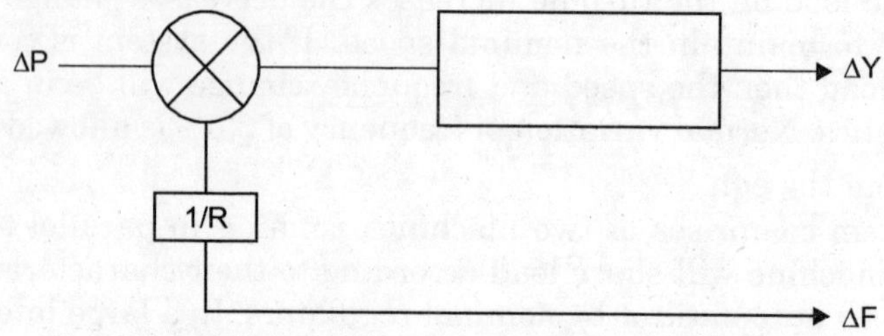

Fig. 7.9. Simplified model of speed governing system.

7.7.2 Turbine Model

In the Fig. 7.10, a simplified steam turbine model (not going in detail again) is shown. Here it is assumed to have a single equivalent time constant. Here in this Fig. 7.10.

K_t = gain of turbine
T_t = time constant of turbine
ΔY = output of governor
ΔP = output of turbine

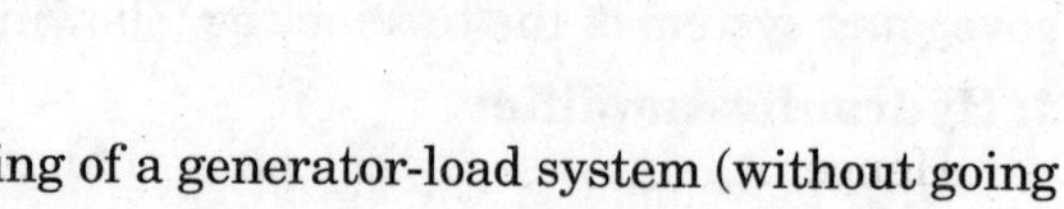

Fig. 7.10. Turbine transfer model.

7.7.3 Generator Load Modelling System

In the Fig. 7.11 again a very simplified modelling of a generator-load system (without going in detail) is shown in Fig. 7.11.

ΔP_g = change in generator output power
ΔP_D = change in load demand
K_{ps} = power system gain
T_{ps} = power system time constant
ΔF = change in frequency

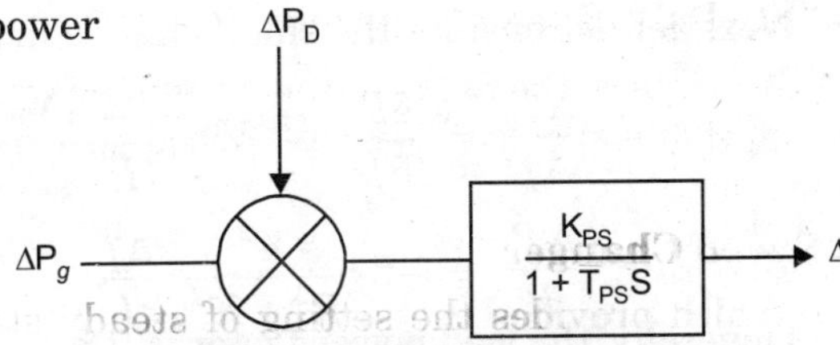

Fig. 7.11. Modelling of generator.

7.8 CONCEPT OF CONTROL AREA

In case of a single generator system supplying an isolated load, the control concept can be easily visualised as point described. In fact, a practical power system is divided into sub-areas in which generators are tightly coupled to respond in unsion that is, generator swing coherently. In a coherent system the speed governors regulation is best achieved when the dynamic frequency response is taken into account. In a proportional control system, the speed changer setting depends on steady state frequency change. This control will be unable to tolerate larger dynamic frequency change with frequency dependent load. Thus if the speed changer setting is adjusted automatically by monitoring the frequency change then the governor setting will be able to cope up with the dynamic variation of frequency. For this purpose a signal from frequency variation sensor is made throughs an integrator to the speed changer. The system hence now becomes a proportional integral controller giving zero steady state error.

7.9 EFFECT OF VOLTAGE AND FREQUENCY ON LOAD

It is known that the performance of impedance type load like heating lighting ironing etc. depend on voltage at their terminals. The complex load power is given by :

$$S = P + jQ = VI = |Z|\,|I|^2 = |Y|\,|V|^2$$

$$= |V|^2 \cdot \frac{1}{R - jX} = |V|^2 \frac{R + jX}{R^2 + X^2}$$

or

$$P + jQ = |V|^2 \cdot \frac{R}{R^2 + X^2} + j\,|V|^2 \frac{X}{R^2 + X^2}$$

Now, we get, from the eqn.

$$P = |V|^2 \cdot \frac{R}{R^2 + X^2} = \text{and}\, Q = |V|^2 \frac{X}{R^2 + X^2}$$

From these expressions we see that real and reactive load are proportional to square of voltage.

Now, we know that X depends upon frequency as $X = 2\pi fL$. Hence, it is evident that the performance of impedance type of load depends upon voltage and frequency.

Now let us see the role of small change of voltage and frequency say for small change of voltage, we get

$$\frac{\Delta P}{\Delta |V|} = \frac{\delta P}{\delta |V|} = 2\,|V| \frac{R}{R^2 + X^2} = \frac{2}{|V|} \cdot P$$

$$\therefore \quad \frac{\Delta P}{P} = 2 \cdot \frac{\Delta |V|}{|V|}$$

Hence, it is now seen that for a change in $\Delta |V|$. The change in real power (ΔP) is twice which means that for one per cent drop in voltage there will be two percent drop in load.

Next, let us consider the small change in frequence and it effect on load,

$$\frac{\Delta P}{\Delta f} = \frac{\delta P}{\delta f} = -|V|^2 \cdot \frac{4RX\pi L}{(R^2 \times X^{22})^2} = -|V|^2 \cdot \frac{2RX^2}{f(R^2+X^2)^2} = -P\frac{2X^2}{f(R^2+X^2)}$$

$$\therefore \quad \frac{\Delta P}{P} = -2 \cdot \frac{X^2}{R^2+X^2} \cdot \frac{\Delta f}{f}$$

Generally, the load power factor is 0.8

Hence $\cos\phi = 0.8$

$$\therefore \quad \frac{X^2}{R^2+X^2} = \sin^2\phi = 0.36$$

$$\therefore \quad \frac{\Delta P}{P} = 0.72\frac{\Delta f}{f}$$

This expression says that for one percent drop in frequency there will 0.72 percent load increase (reactive sign here indicates load increased). The performance of constant power loads, *i.e.*, inductor motor loads also depends on voltage and frequency.

In case of inductor motor the torque is propotional to square of voltage. Hence with drop in voltage torque reduces. But as it is a constant power device the motor current will increase to compensate for drop in voltage. This hence, indicates that with drop in voltage within specified limit the power output of the motor does not change remarkably.

Then for change in frequency the motor reactance changes as well as efficiency its power factor. In case of frequency drop motor reactances reduces. Then the motor takes more current and hence the torque increases. This may of course speed up the motor if the load on it remain constant.

Example 7.1. *A 230 kV line is fed through 33/230 kV transformer, from a constant 33 kV supply. A single line diagram of the 3-phase system is shown in Fig. 7.12. The impedance of the line and transformers at 230 kV is (30 + j 80) ohms. Both the transformers are equipped with tap changing facilities which are so arranged that the product of the two off nominal settings is unity. If the load on the system is 150 MW at 0.9 p.f., determine the settings of the tap changes required to maintain the voltage of the load bus bar at 33 kV.*

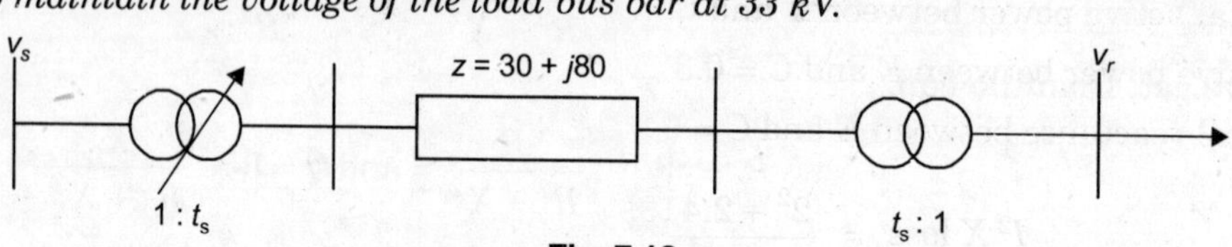

Fig. 7.12

Solution : As per example it is desired to have $V_s = V_r$ and $t_s \cdot t_r = 1.0$. The load is 150 MW and 72.65 MVAR we have the eqn.

$$V_s/V_r = t_s^2\left(1 - \frac{RP+XQ}{V_s V_r}\right) = 1.0$$

Substituting the values (from the example) :

$$r_s^2\left(1 - \frac{3\times\frac{150}{3}\times 10^6 + 80\times\frac{72.65}{3}\times 10^6}{\left(\frac{230}{\sqrt{3}}\right)^2 \times 10^6}\right) = 1.0\left(V_s = V_r = \frac{230}{\sqrt{3}}\times 10^3\right)$$

or $\qquad t_s = $ **1.11 p.u.**

Example 7.2. *In the radial transmission system shown in Fig. 7.13 all p.u. values referred to the voltage bases shown and 100 MVA. Determine the total power active and reactive supplied by the generator and the p.f. at which the generator must operate.*

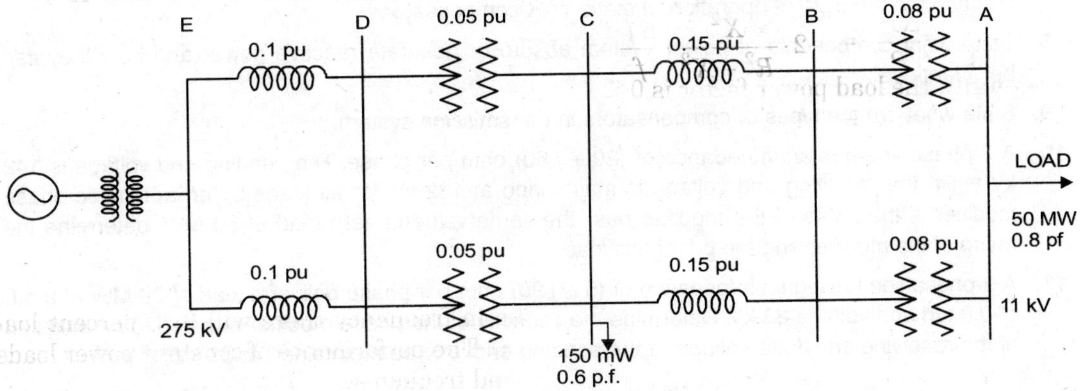

Fig. 7.13

Solution : To solve this problem Nominal Voltage are assumed and voltage drops in the circuit are neglected. Now to find out the total power active and reactive we add up all the active powers connected to the system and reactive power connected to the system plus the reactive power losses in the lines and the transformers. Taking voltage at 1.0 p.u. every where. Now, power at Bus A = Active power 0.5 P.U. and Reactive power 0.375 p.u. The total reactance between bus C and A - 0.575 + 0.04 = 0.115 p.u.

$$\therefore \qquad I^2 X = \frac{P^2 + Q^2}{V^2} \cdot X = \frac{0.5^2 + 0.375^2}{I^2} \times 0.115 = 0.0449 \text{ p.u.}$$

Active power tapped at bus C = 1.5 p.u. and reactive power 2.0 p.u.

$\therefore$ Total active power between E and C = 0.5 + 1.5 = 2.0 p.u.

Reactive power between E and C = 0.375 + 0.449 + 2.0 = 2.4199 p.u.

$\therefore$ Total reactance between E and C = 0.05 + 0.025 = 0.075 p.u.

$$\therefore \qquad I^2 X \text{ loss} = \frac{2^2 + 2.4199^2}{I^2} \times 0.075 = 0.73919 \text{ p.u.}$$

$\therefore$ Total active power supplied by the generator = 200 MW

and

The reactive power supplied = 315.9 MW

The p.f. of the generator = 0.5349

QUESTIONS

1. Establish analytically the role of reactive power on receiving and voltage and voltage regulation of a load bus.
2. Develop the relationship between the incremental changes of Q, P and V in a load node.
3. State what is reactive compensation and what are the benefits.

4. State what is a booster transformer. Where it is used?
5. What is meant by load frequency control?
6. State what is the effect of voltage and frequency change on impedance load.
7. Discuss the principle of operation of static VAR compensator.
8. State what components of power system absorb or generate reactive power and how they can be quantised.
9. State what are the types of compensators in transmission system.
10. A 3-phase line has an impedance of (20 + j 60) ohm per phase. The sending end voltage is 142 kV while the receiving and voltage is maintained at 132 kV for all loads by an automatic phase modifier. If the KVAs of the modifier has the same value of zero load of 50 MW, determine the rating of the modifier and the p.f. of this load.
11. A 3-phase line having an impedance of (5 + j 20) ohm per phase deliver a load of 30 MW at a p.f. of 0.8 lag and voltage 33 kV. Determine the capacity of the phase modifier required to be installed at the receiving and if the voltage at the sending end is to be maintained at 33 kV.

CHAPTER

8

Essentials of Mechanical Design of Transmission Line

Power system perfection requires sound knowledge of essentials of mechanical design of transmission lines. Infact, the transmission line conductors (bare) are supported on transmission line towers with insulators help. These mechanical supports for transmission lines are of the following types. (1) Steel poles, (2) Reinforced concrete poles and (3) Broad-base steel lattice structure towers.

However, for short spans and voltages upto 83 kV, normally the first two type of support are used whereas for long spans and higher voltages the broad-base steel lattice structures are used. In fact, when a perfectly flexible wire of uniform weight is hung between the two horizontal supports, it will form a catenary curve, which will be studied in the following section.

INSIDE THIS CHAPTER

- The catenary curve
- Sag tension
- Effect of wind & ice laoding on overhead lines
- Support at different levels
- Sag template for the span of transmission towers
- Effect of wind pressure
- Stringing chart
- Different types of steel towers line support defined for use in transmission of HVAC & HVDC
- Line supports
- Conductor materials
- Stringing Operations
- Overhead lines vs. underground cables
- Indian electricity rules (1956) for overhead lines

8.1 THE CATENARY CURVE

In the Fig. 8.1, let the conductor be strung between supports A and B. Let l be the distance between the supports, w the weight per unit length of the wire to the tension in the wire at lowest point H of the wire in kg, and OX and OY are the axes drawn from the origin O. The location of O is c units below the lowest point H such that $T_0 = wc$ or $c = T_0/\omega$. In the Fig. 8.1 ψ is the angle between T with the horizontal axis.

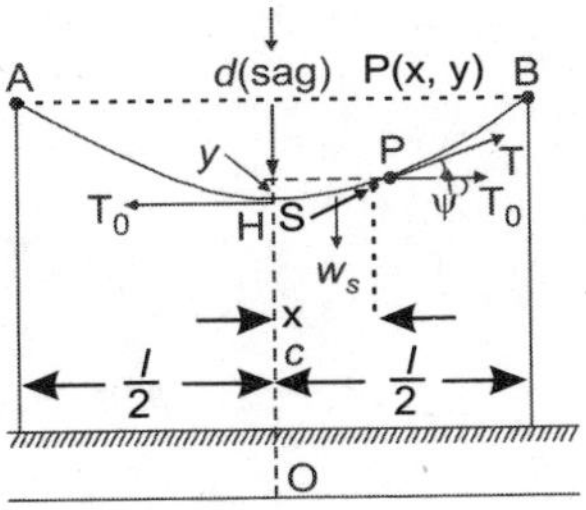

Fig. 8.1.

Now let us consider the equilibrium of the small length s of the wire upto from H point $P(x, y)$. Three forces are acting on this length of the wire :

1. The horizontal tension $T_0 = wc$.
2. The vertical weight ws.
3. The tension T.

Now from the Fig. 8.1, it is clear that

$$T \cos \psi = T_0 = wc \qquad \text{...(8.1)}$$

$$T \sin \psi = ws \qquad \text{...(8.2)}$$

From eqns. (8.1) and (8.2)

$$\tan \psi = \frac{dy}{dx} = \frac{ws}{wc} = \frac{s}{c} \qquad \text{...(8.3)}$$

Then for a differential length

$$ds = \sqrt{dx^2 + dy^2} \qquad \text{...(8.4)}$$

or

$$\frac{ds}{dx} = \sqrt{1 + \left(\frac{dy}{dx}\right)^2} \qquad \text{...(8.5)}$$

Now substituting for dy/dx from eqns. (8.3) to (8.5)

we get,

$$ds/dx = \sqrt{1 + \frac{s^2}{c^2}} = \sqrt{\frac{c^2 + s^2}{c}}$$

or

$$\frac{c \cdot ds}{\sqrt{c^2 + s^2}} = dx \qquad \text{...(8.6)}$$

Now let

$$s = c \sinh \theta, \therefore ds = c \cos h\, \theta \cdot d\, \theta \qquad \text{...(8.7)}$$

∴ Eqn. (8.6) can be written as,

$$\frac{c \cdot c \cosh \theta \cdot d\theta}{c \cosh \theta} = dx$$

or

$$c \cdot d\, \theta = dx$$

or

$$c \cdot \theta = x + A \qquad \text{...(8.8)}$$

Now for $x = 0, s = 0, \therefore \theta = 0$ in eqn. (8.8), we get, $A = 0$

∴ Enq. (8.8) becomes $c\,\theta = x$

or

$$\theta = \frac{x}{c}$$

Now from eqn. (8.7) we get,

$$\theta = \sinh^{-1} \frac{s}{c}$$

or

$$\frac{s}{c} = \sinh \cdot \frac{x}{c}$$

∴

$$s = c \sinh \frac{x}{c} \qquad \text{...(8.9)}$$

From eqns. (8.8) and (8.9) we get,

$$\frac{dy}{dx} = \sinh\frac{x}{c}$$

$$dy = \sinh\frac{x}{c}\cdot dx$$

or $$y = c\cosh\frac{x}{c} + B \qquad ...(8.10)$$

Now from Fig. (8.1), $x = 0, y = c$

Substituting this condition in eqn. (8.10) for evaluaing B

$$c = c + B$$

∴ $$B = 0$$

Hence, eqn. (8.10) becomes,

$$y = c\cosh\frac{x}{c} \qquad ...(8.11)$$

Equations (8.9) and (8.11), represent a catenary and they give the relationships between the length of the wire measured from the lowest point H and the vertical height of any point $P(x, y)$ as measured above the origin O respectively in terms of the distance x as measured from O along OX.

Expanding the term sinh x/c and cosh x/c, the eqns. (8.9) and (8.11) become,

$$s = c\cdot\frac{x}{c} + \frac{x^3}{3!\,c^3} + \frac{x^5}{5!\,c^5} + ... \qquad ...(8.12)$$

$$y = c\cdot 1 + \frac{x^2}{2!\,c^2} + \frac{x^4}{4!\,c^4} + ... \qquad ...(8.13)$$

The expressions can be approximated by taking the first two terms in case the span length is not large as compared to with c. Hence after approximation the equations (8.12) & (8.13) becomes as :

∴ $$s = x + \frac{x^3}{6c^2} \qquad ...(8.14)$$

and $$y = c + \frac{x^2}{2c} \qquad ...(8.15)$$

These two expression are approximate representation of a catenary.

8.2 SAG TENSION

Let T be the sag tension. Hence to calculate tension T at any point $P(x, y)$ on the conductor (refer Fig. 8.1), we again here use eqns. (8.1) & (8.2) by squaring and adding these two equations, we get

$$T^2\cos^2\psi + T^2\sin\psi = w^2c^2 + w^2s^2$$

∴ $$T^2 = w^2(s^2 + c^2) = w^2\left(c^2\sinh^2\frac{x}{c} + c^2\right)\left[\text{as } s = c\sinh\frac{x}{c}\right]$$

$$= w^2c^2\left(1+\sinh^2\left(\frac{x}{c}\right)\right)$$

$$= w^2c^2 \cosh^2 x/c$$

$$\therefore \quad T = wc \cosh x/c \qquad ...(8.16)$$

$$\therefore \quad T = wy \qquad \left[\text{as } y = c\cosh\cdot\frac{x}{c}, \text{ from eqn. (8.11)}\right]$$

Hence from eqn. (8.16), it is clear that the tension in the conductor at any point $P(x, y)$ in the conductor is the product of the y-coordinate of the point and the weight per unit length of the wire/conductor.

8.2.1 Sag "*d*"

At the point $P(x, y)$ the sag d is the vertical distance between the point P (in the Fig. 8.1), and the lowest point H of the centenary curved conductor. To calculate the sag, equation (8.15) is used

$$y = c + \frac{x^2}{2c}$$

$$\therefore \quad y - c = \frac{x^2}{2c} = Sag\ d$$

Now this sag is maximum when $x = l/2$

$$S_{ag}\ 'd' = y - c = x^2/2c = l^2/8c$$

$$\therefore \text{ With } x = \frac{l}{2} \text{ for max sag, i.e.} \quad d = \frac{l^2}{8c} = \frac{wl^2}{8c} = \frac{wl^2}{8T} = \frac{wl^2}{8fA} = \frac{l^2\,\delta}{8f} \qquad ...(8.17)$$

where, f = stress corresponding to tension T,

A = area of cross section of the conductor

and $\delta = \dfrac{w}{A}$, a constant, δ is the density of the conductor material and is therefore, constant for a particular material.

8.2.2 Length *L* of the Conductor

Again we shall use equation (8.14),

$$s = x + \frac{x^3}{6c^2}$$

Substituting $x = l/2$ in this equation, to get the length of the conductor between the point "*H*" and the support end *A* or *B*,

$\therefore$ Eqn. (8.14) becomes,

$$\frac{L}{2} = \frac{l}{2} + \frac{l^3}{48c^2}$$

$$\therefore \quad L = l + \frac{l^3}{24c^2}$$

Now, substituting for $c = f\delta$, in this equation of L, we get,

$$L = l + \frac{l^3 \cdot \delta^2}{24f^2} \qquad \text{...(8.18)}$$

Here we should note that, we have used the relation $T = wc$ instead of $T_0 = wc$. This approximation is true for small value of sag and it can be assumed that when d is small the tension T is approximately uniform throughout the wire/conductors.

After we have derived the basic equation for wire/conductor strung between two supports we are now ready to design the transmission lines. An overhead line must be designed from the viewpoint of worst probable loads rather than the worst possible loads because the cost of the overhead line will become very large if designed on the basis of worst possible condition.

In fact, the sag to be allowed in an overhead conductor at the time of erection, a specified maximum tension for the conductor must not exceed. The problem can be stated in the way that given the maximum tension which must not be exceeded under specific sever conditions of wind, ice or any other kind of loading at a specified temperature to determine the sag and tension at some other conditions of loading and temperature, including the still air and higher termperatures.

8.3 EFFECT OF WIND & ICE LOADING ON OVERHEAD LINES

In overhead lines, the severe conditions are the wind storm and ice loading. Under this condition the per unit length of the wire experiences the following : (1) the weight of the conductor, w acting vertically downwards, (2) the ice loading w_i acting vertically downwards and (3) the wind loading w_a acting horizontally.

8.3.1 Ice Loading

Let us refer to the Fig. 8.2, where a cross section of an overhead conductor is shown. Let r be the radius of the conductor and t the thickness of ice per unit length

$\therefore$ volume of ice per unit length

$$= \pi [(r + t)^2 - r^2] \cdot l$$

$$= \pi(2rt + t)^2 \times l$$

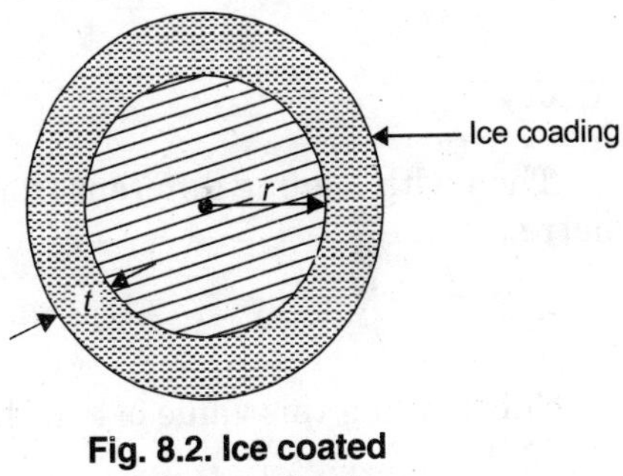

Fig. 8.2. Ice coated conductor of rad *r*.

If p is the density of ice (910 kg/m^3) the weight of ice per unit length of conductor. Then

$$w_i = \pi\rho\,(2rt + t^2) \text{ kg/metre}$$

8.3.2 Wind Loading

Say p is the wind pressure in kg/m^2, assuming the ice coating of thickness t the projected area per unit length on which the wind acting is : $(r + t) \times l$ sq. metres.

Hence, wind loading w_a per unit length will be, (When l = 1m)

$$w_a = 2(r + t) \cdot p \cdot \text{kg/metre}$$

Then, the total vertical loading $= w + w_i$.

Total loading $W = \sqrt{(w + w_i)^2 + w_a^{\,2}}$

Then, the loading factor, q = W/m.

Now let the two conditions, *i.e.*, the severe conditions of wind and ice loading and the still air be characterised by the following :

Wind and Ice Laoding

L_1 = length of the conductor between the supports

W/m = q_1 = loading factor

f_1 = maximum working stress in the conductor in kg/cm^2

t_1 = ambient temperature

Still Air and Higher

L_2 = length of the conductor between its supports

q_2 = loading factor which is unity under these condition

f_2 = maximum working stress in the conductor in kg/cm^2
Corresponding to f_1 under wind and ice loading condition

t_2 = ambient temperature

E = modulus of elasticity in kg/cm^2

$\propto$ = coefficient of linear expansion

Now the problem is : given f_1 let us calculate f_2 which is required for stringing the conductor during fair weather (still air) conditions

Length L_2 from eqn. (8.14) above, $$s = x + \frac{x^3}{6c^2}$$

Now, for $x = l/2$, $$\frac{L_2}{2} = \frac{l}{2} + \frac{l^3}{48c^2}$$

or $$L_2 = l + \frac{l^3}{24c^2} \quad \text{....(8.19)}$$

Then with loading different from w the value of $c = T/W$, where W is the total loading kg/metre.

$$\therefore \quad c = \frac{fA}{W} = \frac{f\,Aw}{Ww} = \frac{f}{\delta q}\left(\text{as } o = \frac{w}{A}, q = \frac{w}{w}\right)$$

Substituting this value of c in the expression of L_2 of eqn. (8.19)

We get, $$L_2 = l + \frac{l^3 \cdot \delta_2^2 q_2^2}{24 f_2^2}$$

As said earlier under this condition of standstill air and higher temperature,

$$q_2 = 1$$

Similarly, $$L_1 = l + \frac{l^3 \delta_1^2 q_1^2}{24 f_1^2}$$

Now to relate f_2 with f_1 one possibility is to find out some relation between the two lengths L_1 and L_2. Now due to higher temperature the length under standstill condition is $l \times \alpha (t_2 - t_1)$ metre more than under ice loading condition but due to increased sag of the overhead conductor at higher temperature the stress in the material is redued from f_1 to f_2 and as such there is contraction of length at higher temperature than at lower temperature.

Therefore, $$L_2 = L_1 + l\,\alpha(t_2 - t_1) - \frac{f_1 - f_2}{E} \times l$$

Here, it is to be noted that little error is introduced if l is taken instead of L for the last term on the right hand side of the above equation.

Now substituting for L_2 and L_1

$$l + \frac{l^3 \cdot \delta_2^2 \cdot q_2^2}{24 f_2^2} = l + \frac{l^3 \cdot \delta_1^2 q_1^2}{24 f_1^2} + l \propto (t_2 - t_1) - \frac{f_1 - f_2}{E} l$$

or $$f_1 - \frac{l^2 \delta^2 q_1^2}{24 f_1^2} E = F_2 - \frac{l^2 \delta^2 q_2^2}{24 f_2^2} \cdot E + \alpha(t_2 - t_1) E$$

The quantities on the left hand of the above expression are known so that putting this equal to K, we get,

$$K = f_2 - \frac{l^2 \delta^2 q_2^2}{24 f_2^2} E + \alpha(t_2 - t_1) E$$

or $$K - \alpha(t_2 - t_1) E = f_2 - \frac{l^2 \cdot \delta^2 q_2^2}{24 f_2^2} \cdot E$$

Again the quantity on the left hand side in known and let this be equal to N we have

$$N = f_2 - \frac{l^2 \cdot \delta^2 \cdot q_2^2}{24 f_2^2} \cdot E$$

or, $$f_2^2 (f_2 - N) = \frac{l^2 \cdot \delta^2 q_2^2}{24} E$$

The quantity on the right hand side of the above expression is known and let this be equal to M, then, we have,

$$f_2^2 (f_2 - N) = M$$

This is a cubic equation in f_2. The procedure for evaluating f_2 can be summarised as follows :

1. Evaluate the loading factors q_1 and q_2 for the two conditions of Load from,

$$q = \frac{\sqrt{(w + w_i)^2 + w_a^2}}{w}$$

2. Calculate K from the expression,

$$K = \left(f_1 = \frac{l^2 \delta^2 q_1^2 E}{24 f_1^2} \right)$$

3. Evaluate N and M from the expressions,

$$N = K - xtE \text{ and } M = \frac{l^2 \delta^2 q_2^2 E}{24}$$

4. Evaluate f_2 from the cubic equation,

$$f_2^2 (\delta_2 - N) = M$$

5. The sag is then evaluated from the expression,

$$d = \frac{l^2 \cdot \delta^2, q_2^2}{8f_2} = (e^2 \cdot \delta^2 \cdot q_2^2/8f_2)$$

8.4 SUPPORT AT DIFFERENT LEVELS

When transmission lines are taken on steep inclines, say hills, the two ends *A* and *B* of towers will not be at the same height. The shape of the wire strung between the support will form a part of the catenary, and as such, the lowest point of the catenary will not lie in the middle of the span. Hence referring to Fig. 8.3 *A* and *B* are the support ends. To have complete catenary, extend the curve *BHA* to *A′* so that *A′* is at the same level as that of *B*. *H* is the mid point, then of *A′ B* and hence is the lowest point. Let *H* be at a horizontal distance of '*a*' units from *A*, and *b* unit from *B*. The difference in levels of supports *A* and *B* is *h* units. The sag *d* is as shown below.

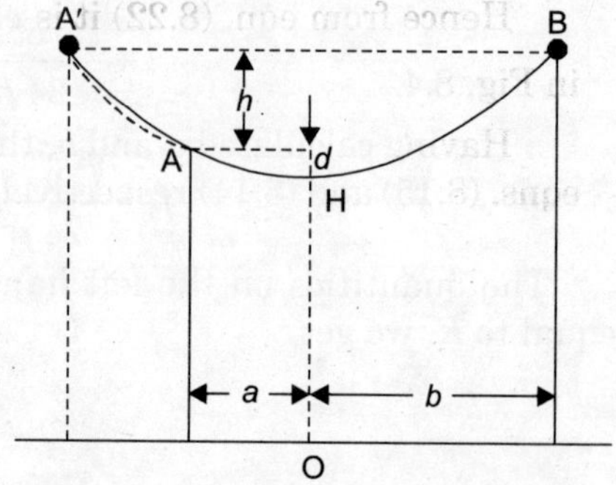

Fig. 8.3. Supports at unequal level.

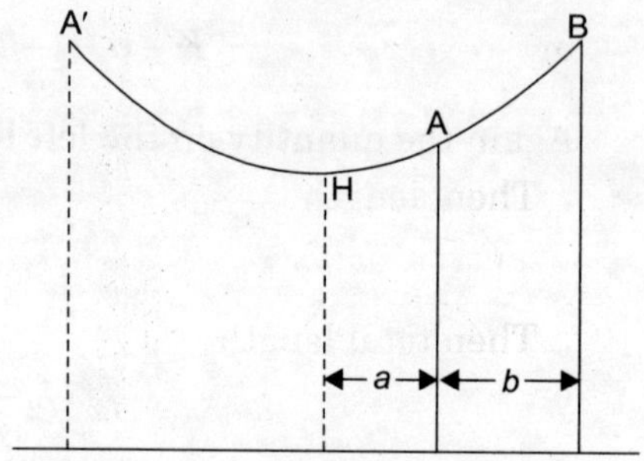

Fig. 8.4. Supports at unequal level, $\frac{ch}{l} > \frac{l}{2}$.

The objective to begin with is to find out *a* and *b* this is done as follows :

From equation (8.15),

$$y = c + \frac{x}{2c}$$

Taking O as the origin,

$$At_1\ A_1\ x = a_1\ y = c + d_1$$

$$At_1\ B_1\ x = b_1\ y = c + d + h$$

Substituting these in equation (8.15)

$$c + d = \frac{a^2}{2c} \qquad \text{...(8.20)}$$

and

$$c + d + h = \frac{b^2}{2c} \qquad \text{...(8.21)}$$

Using these equation,

$$h = \frac{b^2 - a^2}{2c} = \frac{(b+a)(b-a)}{2c} = \frac{l(b-a)}{2c}$$

Now,

$$b - a = (a + b - 2a) = (l - 2a)$$

$\therefore$

$$h = \frac{l(l-2a)}{2c}$$

$$\frac{2ch}{l} = (l - 2a)$$

$$2a = l - \frac{2ch}{l}$$

or,
$$a = \frac{l}{2} - \frac{ch}{l} \quad (8.22)$$

$$b = \frac{l}{2} + \frac{ch}{l} \quad ...(8.23)$$

Hence from eqn. (8.22) it is clear that in case $\frac{ch}{l} > \frac{l}{2}$, a becomes negative which is shown in Fig. 8.4.

Having calculated a and b, the sag and the length of the conductor can be calculated using eqns. (8.15) and (8.14) respectively.

$$Sog\ d = \frac{a^2}{2c} = \left(\frac{l}{2} - \frac{ch}{l}\right)^2 \cdot \frac{1}{2c} = \left(\frac{l}{2} - \frac{fh}{l\delta q}\right)^2 \cdot \frac{\delta q}{2f}$$

$$= \frac{q\delta}{2f}\left[\frac{l^2}{4} + \frac{f^2h^2}{l^2q^2\delta^2} - \frac{lfh}{lq\delta}\right]$$

$$= \frac{q\delta l^2}{8f} + \frac{fh^2}{2q\delta l^2} - \frac{h}{2} \quad ...(8.24)$$

Then, length
$$L_A = a + \frac{a^2}{6c^2} \text{ and } L_B = b + \frac{b^2}{6c^2}$$

Then total length
$$L = L_A + L_B = (a+b) + \frac{a^3 + b^3}{6c^2}$$

$$= (a+b) + \frac{(a+b)(a^2 - ab + b^2)}{6c^2}$$

$$= l\left[1 + \frac{a^2 - ab + b^2}{6c^2}\right] = l + \frac{q^2\delta^2 l^3}{24f^2} + \frac{h^2}{2l}$$

Note: After calculating f_2 from eqn. (8.19), the value of d_2, the sag can be calculated from eqn. (8.17). Various values of f_2 and d_2 are calculated using eqns. (8.19) and (8.17) repeatedly for different temperatures are called the STRINGING CHARTS and are useful while erecting the transmission line conductor for adjusting the sag and tension properly.

8.5 SAG TEMPLATE FOR THE SPAN OF TRANSMISSION TOWERS

In overhead transmission lines, normally there are two types of supports being used in power system : (1) the standard or straight run or intermediate tower. (2) The angle or anchor or tension tower.

The straight run towers are used for straight run and at normal condition. And the angle towers are used at angles, terminal and other point where a considerable amount of unbalanced pull may be thrown the supports. Hence the angle towers are designed to withstand heavy mechanical loading as compared to standard straight run towers.

To locate the position of the towers, the first step is to know a suitable value for the support height and if there are no special guiding factors which dictate this choice, several alternative may be on trial and error basis.

In this case, the next step is to plot a sag template on a piece of transparent paper which consists of a set of curves (Catenesy Curves) as to can be seen in Fig. 8.5. The horizontal and vertical distance represent the span length and sags respectively. Here GHI is (curve 2) the tower footing line. This means this line gives the location of the footing of the tower. Then, DEF, (curve 1), is the ground clearance line. This means that this is the minimum clearence of the power conductor from the ground. Infact this clearance to ground will depend upon the operating voltage, and is given according to Indian Electricity rules in India, which is given in Table 8.1.

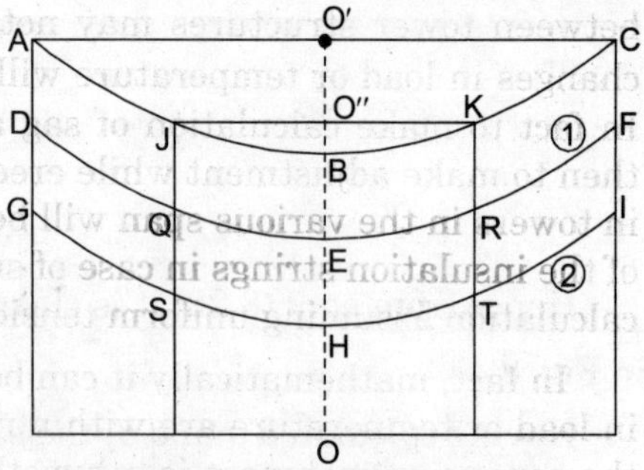

Fig. 8.5. Sag template (1) ground clearance line (2) tower footing line.

Table 8.1. Minimum Clearance to Ground for Overhead Transmission Lines as per IER

Voltage Between Lines	Minimum Height (meter)	Remarks
Less than 650 V. d.c. or 325 V. a.c.	5.8 — —	Across Public Roads
	5.2 — —	Other position
	4.6 — —	Increasible areas to vehicle
Less than 66 kV	6.0	
Between 66 kV and 110 kV	6.3	
Between 110 kV and 165 kV	6.6	
Exceeding 165 kV	6.9	

Note: In practice the height may be a bit more than above figures in the table above, but never under any circumstances less than the minimum height (in metres) specified in the Table 8.1.

Also note that the curve *ABC* (Fig. 8.5) in such that the maximum sag of the conductor would be *O′ B* and with span *JK* the maximum sag is *O″ B*. The curve *DEF* is obtained from *ABC* by subtracting ordinates from *ABC* equal to the minimum ground clearance required and curve GHI is obtained from *ABC* by setting off from *ABC* a distance representing the height of the standard tower from the point of attachment of the lowest conductor to the ground level.

Hence it is now clear that if such a transparent template is placed upon a profile map of the route, as indicated so that the ground clearance line just touches the profile as at *Q* thus points *G* and *I*, where the 'support foot' line cuts the profile will indicate the position of the towers represented by *GA* and *IC*. The curve *ABC* will represent the actual shape and position of the lowest conductor and since *JQ* represent the correct ground clearance, the conductor no where approaches the ground by more than the safe limit.

Above process discussed above for the case where points of conductor supports are on the same horizontal level, but the same process also applies when the route is a steeply sloping one, and the shape and position of the conductor will be always represented by the curve *ABC* (in Fig. 8.5) for transmission lines.

8.5.1 Equivalent Spans of the Overhead Line Towers

From the discussions made above we can see that the location of the towers depends upon the profile of the land along which transmission line is to be run and hence the span length between tower structures may not/will not be equal. When successive spans are unequal, changes in load or temperature will bring unequal changes in tension in the different spans. In fact to make calculation of sag and tension for each and every span is very tedious and then to make adjustment while erecting the transmission line, and in any case the difference in towers in the various span will be automatically equalised by the deflection for the vertical of the insulation strings in case of suspension type of insulators. Hence it is necessary to have calculation assuming uniform tension in the conductor between two tension towers.

In fact, mathematically it can be shown that the variations in the tension with variations in load or temperature are within reasonable limits, the same as those which would occur in the same circumstances is a hypothetical span of one particular length, which is termed as equivalent span expressed in the following expression :

$$\text{Equivalent Span Length,} \qquad L_{span} = \sqrt{\frac{l_1^2 + l_2^2 + l_3^2 + ...}{l_1 + l_2 + l_3 + ...}}$$

where l_1, l_2, l_3 etc. are the length of the individual spans between the two tension towers. To prepare a sag template this equivalent span length value is used.

It should be also stated here that the method of sag template for locating towers should not be used for long spans as well as where the slope of the profile is very steep when it is advisable to make actual calculation for sag and tension of each span.

8.5.2 Stringing Operation

Transmission line towers of suitable height and span are fixed to the ground before the stringing operation conductors are to be started effectively and efficiently. At first the insulator strings are attached to the cross arms of the tower. The next step is the stringing of conductors. The operation of stringing of conductors is divided into following parts :

1. Taking out the overhead line conductors from the drum and hauling them over snatch-blocks hung from the corss-arm on a level with the suspension clamps at the end of insulator strings.
2. Pulling up the conductors, still hanging in the snatch-blocks to the correct tension and adjusting the allowable sag as per design.
3. Transforming the conductors from the snatch blocks to the insulator clamps.

It must be ensured while stringing conductors, the tension in the conductor should be recorded with a dynamometer but it is very much necessary to make the final adjustment by sag measurements, as the tension recorded by the meter may be substantially higher than tension in the conductor at the far end of the section due to friction in the pulling-blocks employed on the cross-arm. During commissioning the transmission lines with towers stringing operation is very important. To ensure the correct stringing of conductors some operators use a 1″ × 2″ batten (painted white) is fixed horizontally to each of the two adjacent towers, at the designed level of the lowest point of one conductor. A climber on one tower keeps his line of sight between the battens and the conductor is slowly let down until he signals that the sag is correct. Only the lowest conductor need be sagged this way, the others then can be adjusted parallel. This method of attaining perfection varies from operator to operator and their long experience on the site doing installation of towers, insulation and stringing of conductors.

Vibration in Conductors and Remedies

The operators also look after the effect of vibrations experienced by overhead transmission lines in the vertical plane by providing vibration dropers. In fact there are two types of vibrations. However, these are not to be confused with the much slower swing of the conductors in the horizontal plane due to simple wind deflection.

Anyway the types of vibrations are : (1) Acoline vibration (2) Galloping or dancing of conductors.

The first one is of high frequency and low amplitude, and the second type are the galloping or dancing of conductors are the low frequency high amplitude vibrations of the conductors. Infact the former are forced vibrations the latter are of the self-excited type. These vibrations are caused due to the wind when it blows past a non-circular conductor. Once this starts it builds within itself and the vibration may become very large and may result, eventually in flexure fatigue in the conductors or there may be flashover between conductors of different phases leading to breakdwon of the transmission.

The virbation problem hence been solved by the use of special devices which fall into two groups : (*a*) reinforcement or armour rods, and (*b*) dampers. In the first group will fall the varnery system of reinforcement, in which the conductor, where it lies in the suspension clamp is surrounded by a spiral layer of small round rods preferably tapered at each end and appreciably larger in diameter than the individual wires of the conductor. These rods in effect give at the point of suspension a stranded cable of much larger diameter than the actual conductor, so that they provide a resistance to bending equal to that of a much larger conductor, whereas the energy which must be absorbed by this bending is only that produced by the wind acting on the much conductor itself. Armour rods also provide excellent protection against flashover and have been used for this purpose in various cases when no vibration was experienced or anticipated.

However, a true dumper is a device specially designed to absorb the vibrational energy, and thereby prevent the occurrence of any vibration at all. Many disign have been put forward among which is the stock-bridge dumper, an extremely simple, neverthless, effective device for suppressing high frequency vibrations. It consists of two hollow weights of special shape fixed at either end of a length of flexible steel cable which is itself fastened to the conductor at its midpoint by means of an aluminium clamp.

In fact two dumpers are used for each point of suspension of the conductor, one or either side, clamped to the conductor at appropriate distances depending on the conductor size. Each span of conductor will thus contain two dumpers and this generally is enough for normal condition. However, for very long span it may be necessary to install additional dumpers on trial error basis till the vibration problem is removed totally.

Dancing of conductors causing mechanical oscillations can be damped by making the conductors circular. For stranded conductor PVC tape is wrapped to make the conductor circular. This method is useful only when steel formation is not there on the conductor. For this situation the sleet or ice coating could be reduced by increasing the I^2R loss on the conductor which of course is not a practical solution to the problem.

8.6 EFFECT OF WIND PRESSURE

The wind pressure on the conductors of the transmission line is considered as acting perpendicular to the conductors and taken as pressure on the projected area of the conductors. The wind pressure is assumed to act on 2/3rd of the projected area for cylindrical surfaces like tubular poles, conductors etc. while it is taken to act on full face area in case of the flat surfaces. In case of towers and lattice type poles the wind pressure is assumed to act on a surface area 1.5 times the projected area of one face to allow for the wind action on both the faces. If d is be diameter of the conductor in metres and p the wind pressure in kg/sq metre then the wind load per metre of the conductor length (assuming the wind pressure to act on 2/3rd of the projected are is $\frac{2}{3}\ p\ (d \times 1)$ kg. When the effect of both ice and wind is to be taken d will be substituted by D *i.e.*, $d + 2t$ where t is the thickness of ice

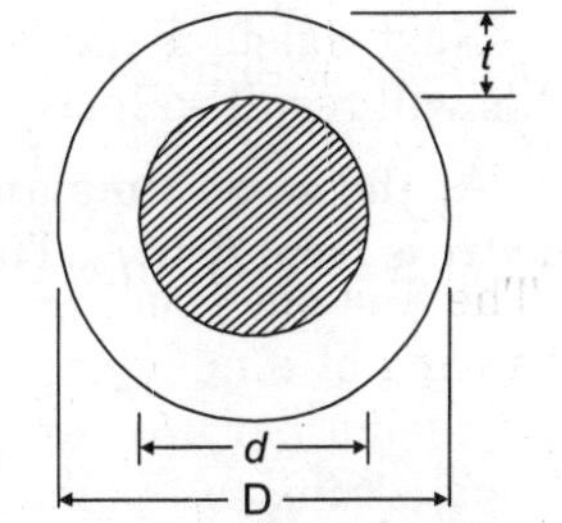

Fig. 8.6. Diameter of the covered conductor.

Because the wind pressure is assumed to act perpendicular to the conductors its effect is only to increase the transverse loading on the conductors. The maximum wind load varies from 100 kg/m^2 to 150 kg/m^2 in this country. The wind load can also be calculated by applying the formula $W_w = 0.006\ V^2$ where W_w is wind loading kg/sq m and V is wind velocity in km/hr.

Thus if W_C is the weight of conductor/metre itself, W_i is the weight of ice per metre and W_w is the weight of wind per metre the resultant weight per metre is given by

$$W = \sqrt{(W_o + W_i)^2 + W_w{}^2}$$

The total sag is worked out for this load on the conductors. To calculate the vertical sag the component of W in the direction of $(W_C + W_i)$ should be taken into consideration (Fig. 8.7).

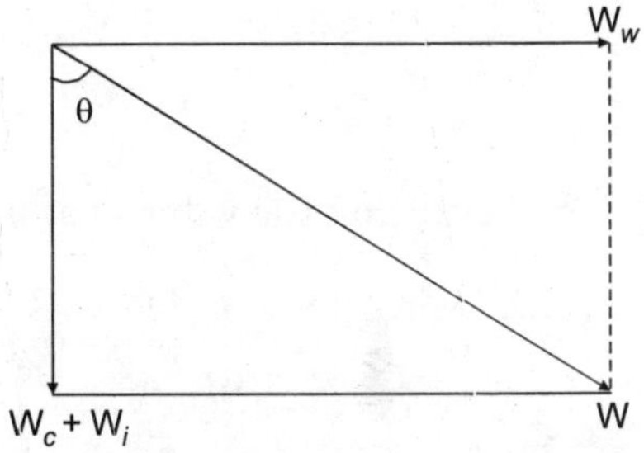

Fig. 8.7. Components of *W*.

8.7 STRINGING CHART

In still air the conductor is acted on by its own weight only. Also if at the same time the temperature is high the resulting sag will give a low tension. We thus find that this is the easiest-condition for a transmission line. Compare this with conditions arising from a low temperature (which reduces the sag) and effect of ice which increases the weight per unit length and also the projected area for wind. A high crosswind will make conditions all the more severe.

Further, the regulations insist on factors of safety based upon a particular tension and sag occurring at specified temperature and ice and wind conditions. Obviously, these conditions are different from those at the time of erection of the line. It is here that a stringing chart is helpful for it enables tension and sag to be determined at any temperature and specified loading conditions once similar figures are known at a given temperature and loading conditions.

It has already been deduced that

$$T = H \quad y = \frac{wx^2}{2T}; \quad D = \frac{wl^2}{2T} \quad \text{and } s = l + \frac{w^2 l^3}{6T^2} \qquad ...(8.24)$$

Let w, f, s D and t represent the load per unit length, the stress, the half span, sag and temperature respectively. Suffix 1 will denote maximum load conditions (with ice and wind and low temperature usually –5.5°C); suffix 2 will similarly denote corresponding values under erection conditions (no ice and wind and actual temperature at the time of erection of line).

Let $\quad a$ = area of cross-section or concern α = coefficient of linear expansion

and $\quad E$ = Young's modulus

Then $\quad f = \dfrac{T}{a}$ so than $s_1 = l + \dfrac{w_1{}^2 l^3}{6 f_1{}^2 a^2}$

Next consider the effect of increase of temperature from t_1 to t_2 over half span length s_1. This will result in increase in length given by $(t_2 - t_1)\,\alpha = s_1\,(t_2 - t_1)\,\alpha\, l$.

At the same time another effect of increase of temperature from t_1 to t_2 is the reduction in stress from f_1 to f_2. The decrease in length on this account is

$$\frac{f_1 - f_2}{E}\, s_1 = \frac{(f_1 - f_2)}{E}\, l$$

The new length s_2 is thus $s_2 = s_1 + (t_2 - t_1)\alpha l - \dfrac{(f_1 - f_2)}{E}\, l$ ...(8.27)

We may now write an expression for s_2 similar to that given by Eq. (8.24) with proper suffixes

$$s_2 = l + \frac{w_2^{\,2}\, l^3}{6 f_2^{\,2}\, a^2} \qquad ...(8.28)$$

Substituting the value of s_1 in Eq. (8.15) and equating with expression given by Eq. (8.28).

$$l + \frac{w_1^{\,2}\, l^3}{6 f_1^{\,2}\, a^2} + (t_2 - t_1)\,\alpha\, l - \frac{(f_1 - f_2)\, l}{E} = l + \frac{w_2^{\,2}\, l^3}{6 f_2^{\,2}\, a^2}$$

or
$$\frac{w_1^{\,2}\, l^2}{6 f_1^{\,2}\, a^2} + (t_2 - t_1)\,\alpha - \frac{(f_1 - f_2)}{E} = \frac{w_2^{\,2}\, l^2}{6 f_2^{\,2}\, a^2}$$

or
$$\frac{w_1^{\,2}\, l^2 E}{6 f_1^{\,2}\, a^2} + (t_2 - t_1)\,\alpha E + (f_2 - f_1) = \frac{w_2^{\,2}\, l^2 E}{6 f_2^{\,2}\, a^2}$$

or
$$f_2^{\,2}\left[f_2 - f_1 + (t_2 - t_1)\,\alpha E - \frac{w_1^{\,2}\, l^2 E}{6 f_1^{\,2}\, a^2}\right] = \frac{w_2^{\,2}\, l^2 E}{6a^2}$$

Eq. (8.17) is a cubic equation in f_2 and knowing the value of f_2, D_2 can be determined since

$$D_2 = \frac{w_2 l^2}{2T_2} = \frac{w_2 l^2}{2 f_2 a}$$

Thus it is possible to plot a graph between temperature (independent variable) and sag (dependent variable). Also f_2 can be plotted (as dependent variable) against temperature (independent variable). Fig. 8.8 shows the plot of tension vs temperature and sag vs temperature. This is called the stringing chart.

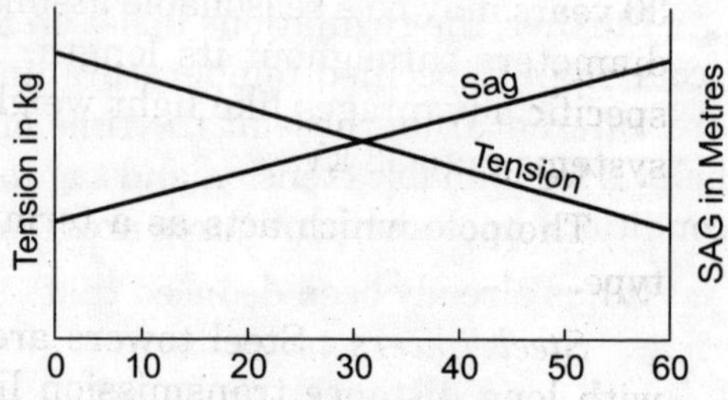

Fig. 8.8. Sag vs temperature chart.

8.8 DIFFERENT TYPES OF STEEL TOWERS LINE SUPPORT DEFINED FOR USE IN TRANSMISSION OF HVAC & HVDC

The designs of different types of steel towers used for transmission of power is given in Fig. 8.9 for the information and knowledge of the reader of this book on power system.

8.9 LINE SUPPORTS

The supports for an overhead transmission line are to be capable of carrying the load due to conductors (including loading of ice deposition and wind thrust) and insulators together with the wind load on the support itself.

The various types of supports include

1. Wooden poles
2. Reinforced concrete poles
3. Tubular steel poles
4. Latticed steel towers

Wooden poles are the simplest and perhaps the cheapest but are useful for voltages upto 132 kV and for spans upto about 60 metres. If properly impregnated against decay these may give a long useful line. Under special conditions single and *H* poles may be employed for lines working at voltages as high as 132 kV and span length of 150 metres. Double pole structures of the *A* or *H* type are often used to obtain a higher transverse strength than what a single pole could provide. Wooden poles are elastic and lines employing this type of support are often designed throughout for the transverse load. At terminals and for anchor supports longitudinal strength is provided by means of guys. Wooden poles are protected by metallic cap at the top to protect it from decay.

The objectionable features of this type of supports are the tendency to rot at the ground level and the fact that their life cannot be predicted with certainty.

Reinforced Concrete Poles : Their use has been more extensive in this country in recent installations. The difficulty with the use of these poles is the high cost of transport owing to their great weight. To overcome this the poles may often be cast on site. The advantages claimed are very long life and exceeding low maintenance charges. They ae used upto 33 kV systems.

Tubular Steel Poles : These are often used as a substitute for wood poles and enable rather longer spans to be used. along the road-side or in developed localities where base area requirement may be an important consideration their use is generally more favoured. This type of support needs to be galvanised which prolongs its life. However a useful life of at least 30 years may be a reasonable assumption. These poles may be swaged type *i.e.*, having different diameters throughout its length. They are costlier than wooden or RCC poles but possess specific advantages like light weight, long life and higher strength. They are widely used for systems upto 33 kV.

The pole which acts as a terminal or carries switchgear or transformers is usually of *H* type.

Steel Towers : Steel towers are extremely useful for system voltage of and above 33 kV with long distance transmission lines running across oepn country where long spans are a decided advantage. Even though they are two to four times costlier than wooden poles yet for tall supports and long spans they prove economical. The steel may be galvanised or painted. These can be classified into (*i*) narrow base lattice type and (*ii*) broad base towers. The former is used for spans not so long while the latter is particularly useful when the span is very large. (With use of a broad base tower it is rather easy to make an adequate foundation). There are separate types of towers designed for single circuit and double-circuit lines. Steel towers are fabricated from galvanised angle iron sections and transported separately to site for assembly and erection. They are finally painted to withstand severe weather conditions. Towers of EHV AC and HVDC transmission lines, are almost similar in appearance although HVDC towers are comparetively simpler. Following types of towers are in use :

1. Self-supporting steel towers

2. Guyed towers
3. Chainnet type towers

Lattice type structure being one of the most common forms of self-supporting tower, H type frames are also used for voltages 345 kV and above. V type and H type frames are used in guyed and chainnet type of towers also. Guys are used to secure more mechanical strength of the tower. Fig. 8.7 ($a - j$) shows lattice type and guyed type of towers.

Any tower consists of (a) Main stem (b) cross-arm (on which conductors with insulators are supported) (c) insulators and (d) Optional guy wire. There should be adequate clearance between conductor to conductor, conductor to ground and conductor to nearest tower member. The configuration of tower is determined by the requirement of various clearances and mechanical loading on structure members.

Though self-supporting structures are most widely used in India, guyed towers are used in USA and Canada. the self-supporting feature of the tower is replaced by guys in the direction of the line and with an increased strength. The guyed towers are mostly supported on hinged foundation.

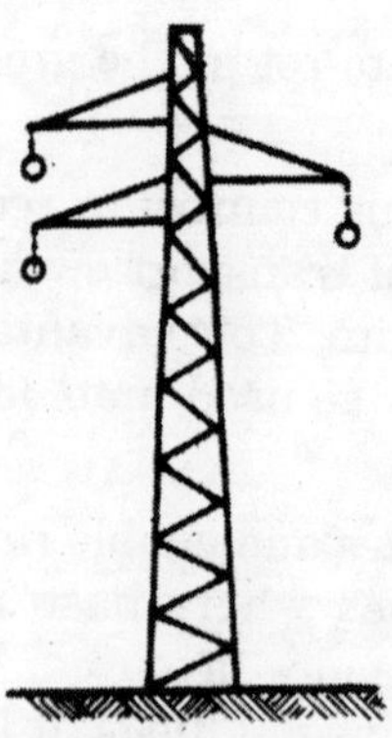

(*a*) Narrow base (33 kV) single circuit.

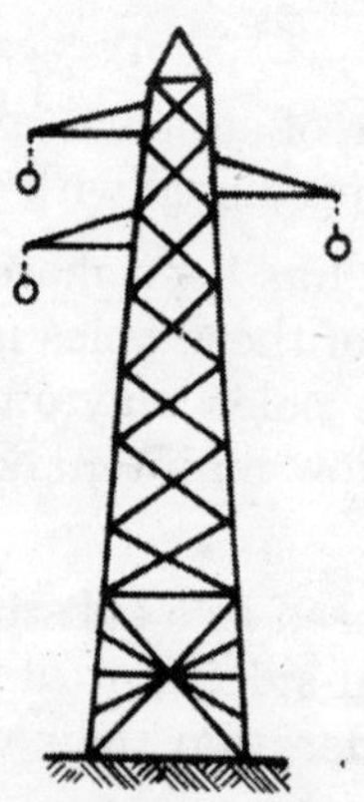

(*b*) Broad base (33 kV/66 kV) single circuit.

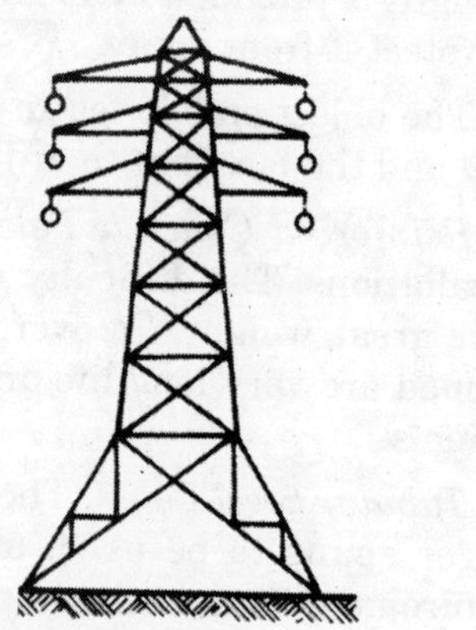

(*c*) Broad base (66 kV/132 kV) double circuit.

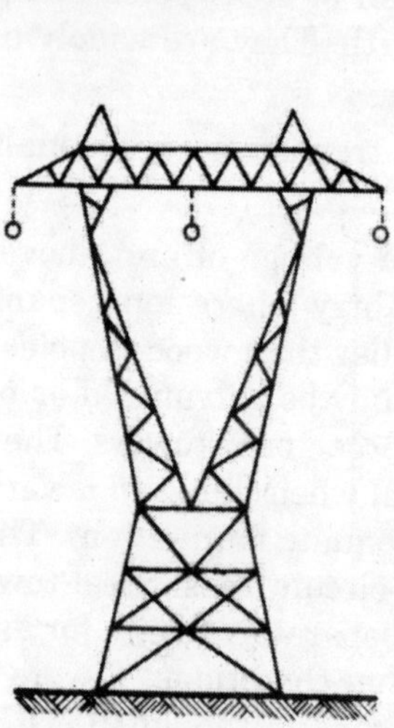

(*d*) Broad base (220 kV/400 kV) single circuit.

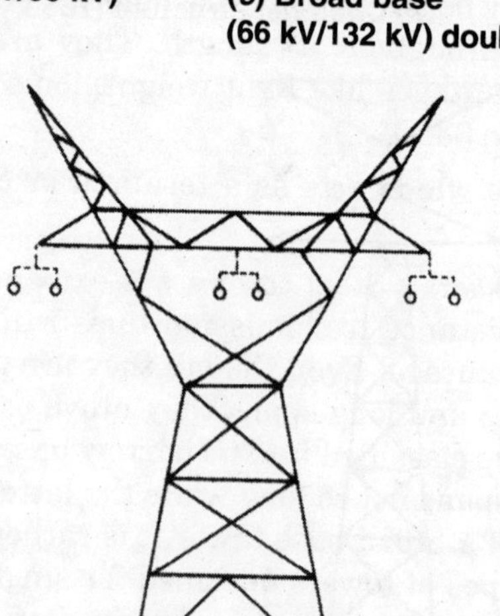

(*e*) Broad base (400 kV multiple conductor) single circuit.

Fig. 8.9

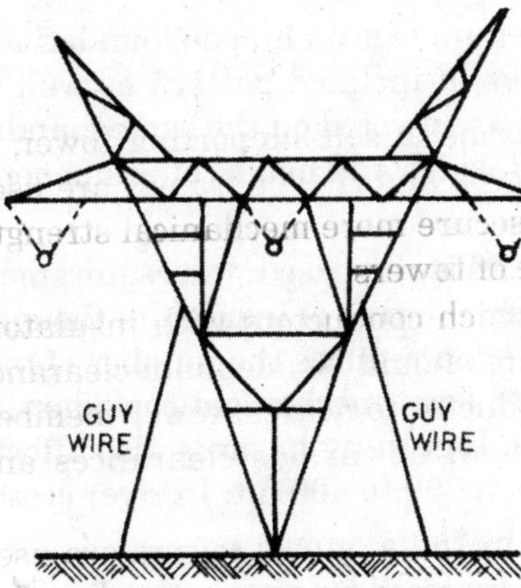

Guyed *V* type tower (500 kV/765 kV) single circuit.

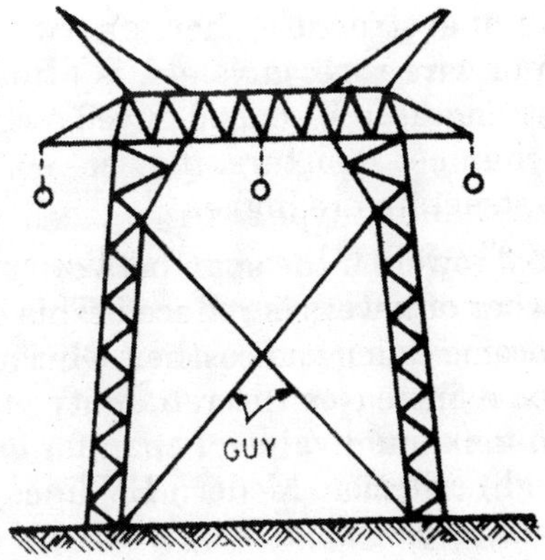

(*g*) Guyed *H* structure (765 kV/800 kV) single circuit.

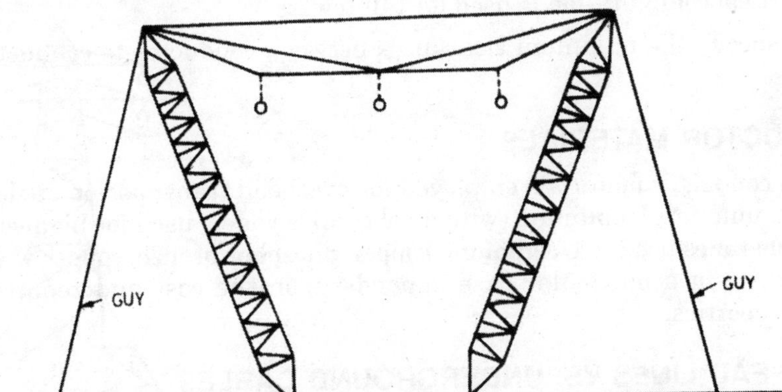

(*h*) Chainnet structure (765 kV/800 kV/1100 kV) single circuit.

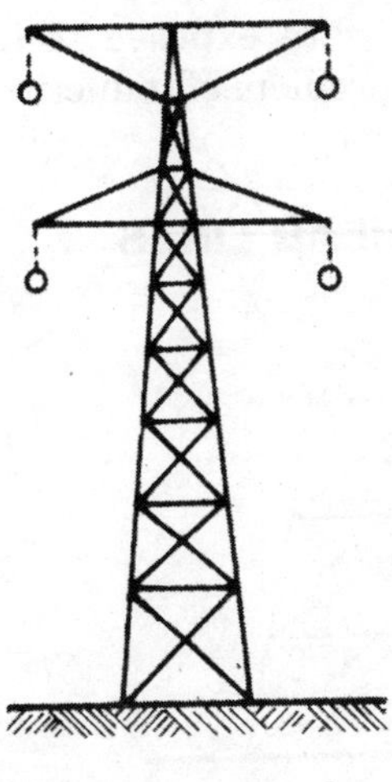

(*i*) Self-supporting tower for HVDC Bipolar line.

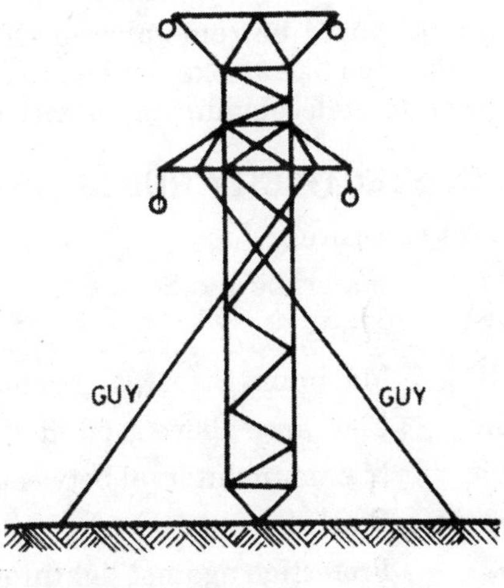

(*j*) Guyed tower for HVDC Bipolar tower.

Fig. 8.9.

In chainnet structure, there are two lattice steel/aluminium legs on hinged foundation and a system of wire rope, guys etc. is utilised to keep the legs in inclined pattern as well as to provide strength. In *V* shaped guyed structure the two legs are pivoted on the same foundation while in chainnet structure, they are spaced apart. Overall width and height of components of chainnet structure are higher.

In steel towers if the span between successive two towers is increased, the requirement of total number of towers is reduced. This enables reduction in the cost of insulators though the towers become taller and costlier. Obviously, use of longer span reduces the number of towers and hence it increases the reliability of the line. However, from mechanical considerations, there is a maximum value of span for each conductor size. Lightning hazards also affect the tower height criterion. Modern HV lines have spans between 200 to 400 mt. In river crossing, hilly terrain, a long span length 800 mt. or more is also utilised.

Regarding conductor configuration, there is no special advantage for symmetrical conductor configuration. Flat horizontal or vertical configurations are frequently used from mechanical considerations. The tower height becomes shorter when horizontal type of conductors are utilised. A vertical configuration is used for tall towers.

IE rules specify the minimum clearances between conductor to conductor and conductor to ground.

8.10 CONDUCTOR MATERIALS

Principal conductor materials employed for overhead transmission of electrical power are copper and aluminium. Aluminium (with steel core) is widely used for high voltage lines. Other materials, sometimes used, are cadmium, copper, phosphor bronze, copper weld and galvanized steel. The choice in a particular case depends upon the cost and required electrical and mechanical properties.

8.11 OVERHEAD LINES VS. UNDERGROUND CABLES

Transmission of power by overhead lines is very much cheaper than by underground cables. Also repairs can be carried out comparatively more easily with overhead transmission or distribution systems than with underground cables.

However it should be remembered that overhead lines are exposed to lightning and atmospheric hazards like smoke, ice etc. Besides there is the problem of inductive interference between power and telecommunication circuits in such systems.

8.12 INDIAN ELECTRICITY RULES (1956) FOR OVERHEAD LINES

Important rules are :

Rule 74	Materials and Strength
Rule 75	Joints
Rule 76	Maximum Stresses : Factor of Safety
Rule 77	Clearance above ground of the lowest conductor
Rule 85	Maximum interval between supports
Rule 90	Earthing
Rule 92	Protection against lightning.

Example 8.1. *A 132-kV transmission line uses A.C.S.R. conductors whose data are : Nominal copper area 110 mm^2, size 30 + 7/2.79 mm, weight 844 kg/km; Ultimate strength 7,950 kg.*

Calculate the height above ground at which the conductors with a span of 300 metres should be supported, the factor of safety being 2. Wind pressure 75 kg/m² of projected area. Ground clearance required is 7 metres.

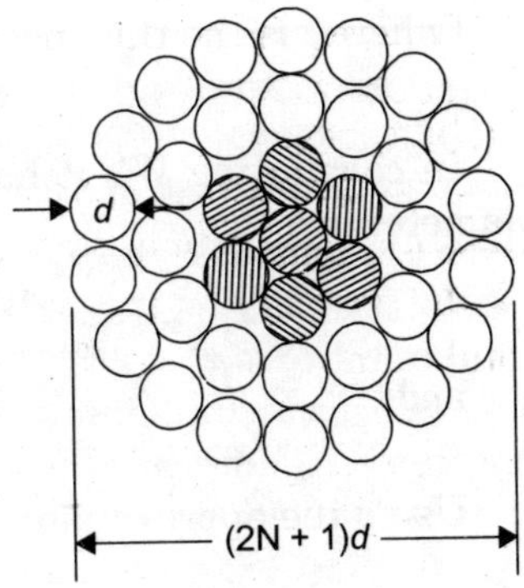

Fig. 8.10.

Solution : Stranded conductors usually have a central wire around which are successive layers of 6, 12, 18, ... wires. The diameter of the cable (stranded conductor) is given by $(2n + 1)\, d$ where d is the dia. of the strand and n is the number of layers around the central strand.

For Fig. 8a shows a 37 strand conductor

We find that $n = 3$, $d = 2.79$ mm, so that

Dia. of conductor = 7 × 2.79 = 19.53 mm.

Let wind pressure per metre acting horizontally be W_W and weight of conductor per metre run acting vertically be W_C. Then

$$W_W = 75 \times 19.53 \times 10^{-3} = 1.465 \text{ kg}$$

and $$W_C = 0.844 \text{ kg}$$

The total force per metre run :

$W = \sqrt{W_C^2 + W_W^2}$ and acts at an angle $\theta = \tan^{-1} \dfrac{W_W}{W_C}$ to the vertical

$$W = \sqrt{(0.844)^2 + 1.465^2} = 1.691 \text{ kg}$$

The approximate formula for sag is $$D = \frac{Wl^2}{2T}$$

Allowable tension $$T = \frac{\text{Ultimate strength}}{2}$$

$$\therefore \quad D = \frac{1.691 \times (150)^2 \times 2}{2 \times 7{,}950} = 4.78 \text{ metres}$$

Ground clearance required being 7 metres height above ground for conductor to be supported is 11.78 metres.

Example 8.2. *Transmission line conductor whose data is given in Example 8.1 is subjected to a horizontal wind pressure of 40 kg/m² of projected area and 1.25 cm radial ice coating. If the maximum permissible sag is 6 metres calculate the permissible span between two supports allowing a factor of safety of 2. Weight of ice is 913.50 kg/m².*

Solution : Let W_C = weight of conductor W_i = Weight of ice on conductor

and W_W = wind pressure on conductor per metre length

Total force on 1 metre of conductor is $W = \sqrt{(W_C + W_i)^2 + W_W^2}$ kg/m and acts at an angle to vertical. We are given

$$W_C = 0.844 \text{ kg/m}$$

Then $$W_i = 913.50 \frac{\pi}{4}\left[(d + 2t)^2 - d^2\right] = 913.50\, \pi\, t(d + t)$$

(where t is the thickness of ice covering)

$$= 913.5\ \pi \times 12.5 \times 32.03 \times 10^{-6} = 1.08 \text{ kg/m}$$

In calculating W_W it has to be remembered that wind now acts on conductor of overall diameter $(d + 2t)$.

$$\therefore \qquad W_W = 40(19.53 + 25) \times 10^{-3} = 1.781 \text{ kg/m}$$

and
$$W = \sqrt{(0.8444 + 1.08)^2 + 1.781^2} = 2.621 \text{ kg/m}$$

Using the approximate formula $D = \dfrac{Wl^2}{2T}$ and putting the values $6 = \dfrac{2.621 l^2}{2 \times (7{,}950/2)}$

or
$$l = \sqrt{\frac{12 \times 3{,}975}{2.622}} = 13.5 \text{ metres}$$

so that span length $= 2 \times 135 = 270$ metres.

Example 8.3. *Calculate max. sag (total and vertical) of a line with the copper conductor 7/0.295 cm size, area 0.484 sq cm. overall dia. 0.889 cm, weight 428 kg/km and breaking strength 1,973 kg. Assume factor of safety 2. Span 200 metres. Level supports :*

(i) Due to weight of the conductor.

(ii) Due to additional weight of ice loading of 1 cm thickness.

(iii) Due to both (i) and (ii) plus wind acting horizontally at a pressure of 39 kg per sq metre.

Solution :

Area of conductor	= 0.484 sq cm	Breaking strength	= 1,973 kg
Overall dia.	= 0.889 cm	Factor of safety	= 2
Weight	= 428 kg/km	Span	= 200 metres

(i)
$$D = \frac{W_C l^2}{2T} = \frac{0.428 \times (100)^2}{2 \times (1{,}973/2)} = 2.18 \text{ metres}$$

(ii) Weight of ice is 913.5 kg/m^3

$$W_i = 931.5\ \pi t\,(d + t) = 2{,}870\ (0.889 + 1) \times 10^{-4} \times 1 = 0.543 \text{ kg/metre}$$

Totat weight of conductor and ice $W = W_C + W_i = 0.428 + 0.543 = 0.971$ kg/m

$$D = \frac{Wl^2}{2T} = \frac{0.971 \times (100)^2}{2 \times (1{,}973/2)} = 4.94 \text{ metres}$$

(iii) Wind acts on conductor of overall dia

$$D = d = 2t = 0.889 + 2 \times 1 = 2.889 \text{ cm}$$

$$W_w = 39 \times 2.889 \times 10^{-2} = 1.113 \text{ kg/metre}$$

$$\therefore \qquad W = \sqrt{(W_C + W_i)^2 + W_w{}^2} = \sqrt{(0.428 + 0.543)^2 + (1.13)^2}\ 1.49 \text{ kg/m}$$

and
$$D = \frac{1.49 \times (100)^2}{2 \times (1973/2)} = 7.55 \text{ metres}$$

The conductor makes an angle θ with the vertical where

$$\theta = \tan^{-1}\left(\frac{W_W}{W_C + W_i}\right) = \tan^{-1} = \frac{1.13}{0.961} = 49.2°$$

∴ Vertical sag $= 7.55 \cos 49.2° = 7.55 \times 0.655 = 4.95$ metres.

Example 8.4. *If in Example 8.3 the supporting towers have a difference in levels of 3 metres determine the point of maximum sag.*

Solution : Referring to Fig. 8.3 let *O* be the point where the sag is maximum. Then

$$x_2 = \frac{l}{2} + \frac{Th}{Wl} \text{ and } x_1 = \frac{l}{2} - \frac{Th}{Wl}$$

Putting the values $x_2 = \frac{200}{2} + \frac{1{,}973 \times 3}{0.428 \times 200 \times 2} = 100 + 34.5 = 134.5$ metres

and $x_1 = 200 - 134.5 = 65.5$ metres

Example 8.5. *A transmission line conductor having dia of 19.5 mm weights 0.85 kg/m. The span is 275 metres. The wind pressure is 39 kg/m² of projected area with the ice coating of 13 mm. The ultimate strength of the conductor is 8000 kg. Calculate the maximum sag if the factor of safety is 2 and ice weighs 910 kg/m².*

Solution : From the datas given by example the overall dia. of conductor with ice-coating $= 1.95 + 2.6 = 4.55$ cm

The projected area per metre length of conductor $= 4.55 \times 1 \times 10^{-2}$ sq. metre

The wind load per metre length $= 4.55 \times 10^{-2} \times 39 = 1.77$ kg/metre. As per example the wind pressure $= 39$ kg/m².

Then the area of section of ice $= \pi[(r + t)^2 - r^2] = \pi[22.75^2 - 9.75^2] = 1.327.10^{-3}$ sq/metre

∴ Ice weight per metre length $= 1.327 \times 10^{-3} \times 910 = 1.207$ kg/metre

The total weight $W = \sqrt{(0.85 + 1.207)^2 + 1.77^2} = 2.71369$ kg/metre

The working tension of the conductor $= \frac{8000}{2} = 4000$ kg

(From the expression above)

$$\therefore \quad C = \frac{T}{W} = \frac{4000}{2.71369} = 1474 \text{ metres}$$

(From the expression above)

∴ The maximum sag $= \frac{l^2}{8C} = \frac{275^2}{8 \times 1474} =$ **6.4 metres**

Example 8.6. *An overhead line has the following data : span length 160 metres, conductor dia 0.95 cm, weight per unit length of the conductor 0.65 kg/metre. Ultimate stress 4250 kg/cm², wind pressure 40 kg/cm² of projected area. Factyor of safety 5 calculate the sag of the line.*

Solution : Factor of safety = 5. The working stress $= \frac{4250}{5} = 850$ kg/cm². The area of section of the conductor $= \frac{\pi}{4} \cdot d^2 = \frac{\pi}{4} \times 0.95^2 = 0.7084$ sq. cm. The wind pressure per unit length of the conductor = $40 \times 0.95 \times 10^{-2} = 0.38$ kg/cm. The weight of conductor per unit length = 0.65 kg/metre.

$$\text{The total effective weight} = \sqrt{0.65^2 + 0.38^2} = \sqrt{0.5669} = 0.753 \text{ kg/metre (W)}$$

$$\text{Working tension} = 850 \times 0.7084 = 602.5 \text{ kg } (= T)$$

Then,
$$C = \frac{T}{W} = \frac{602.5}{20.753} = 800 \text{ metre}$$

$$\therefore \quad \text{sag} = d = \frac{l^2}{8C} = \frac{160 \times 160}{8 \times 800} = \mathbf{4 \text{ metre}}$$

Example 8.7. *An overhead line at a river crossing is supported from two towers of heights 30 metres and 90 metre above water level with a span of 300 metres. The weight of the conductor is 1 kg/metre and the working tension is 2000 kg. Dtermine the clearance between the conductor and the water level midway between the towers.*

Solution : The working tension is 2000 kg and $W = 1$ kg/metre

$$\therefore \quad C = \frac{T}{1} = \frac{2000}{1} = \mathbf{2000 \text{ metre}}$$

Now,
$$h = 90 - 30 = 60 \text{ metres}$$

$$a = l/2 - \frac{ch}{l} = 150 - \frac{2000 \times 60}{300} = 150 - 400 = -250 \text{ m}$$

$$b = \frac{l}{2} + \frac{ch}{l} = 150 + \frac{2000 \times 60}{300} = 550 \text{ metre}$$

$$\therefore \quad d_1 = \frac{a^2}{2C} = \frac{(250)^2}{2 \times 2000} = 15.625 \text{ metres}$$

$$\text{Then at 400 metre} = a,\ d_2 = \frac{400 \times 400}{2 \times 2000} = 40 \text{ metre} \quad \left(d_2 = \frac{a^2}{2c}\right)$$

Therefore, height of the mid-point with respect to A of Fig. 8.4

$$= 40 - 15.625 = 24.375 \text{ metre}$$

QUESTIONS

1. Show how the sag of an overhead line can be claculated in case of supports are at different levels.
2. Show how the effects of wind and ice loading are taken into account while determining the sag and stress of an overhead line conductor.
3. Prove that a transmission line conductor between two support at equal height takes the form a catenary. Define of what is catenary.
4. An overhead transmission line has a span of 220 metres, then conductor weighing 604 kg/km. calculate the maximum sag if the ultimate tensile strength of conductor is 5.758 kg. Assume a factor of safety = 2.
5. State how are the sag and the tension of the conductor affected by the fall of temperature.
6. Explain the necessity of stringing chart for a transmission line and show how such a chart can be made.

CHAPTER

9

Extra High Voltage and High Voltage DC

INSIDE THIS CHAPTER

- Extra high voltage ac
- Importance of EHV transmisision
- Major factors considered in EHV transmission
- Limitations of EHV ac power transmission
- Comparison of ac & dc transmission of high voltage d.c. transmission in power system (HVAC versus HVDC)
- Basic principle of HVDC
- Type of dc links (HVDC links)
- Terminal equipments
- Operational aspects of HVDC systems
- Main Design Consideration of HVDC Systems

9.1 EXTRA HIGH VOLTAGE AC

Since 1950s, a number of developments have taken place in the field of transmission of electric power leading to adoption of over-increasing high voltages. Recent trend is to transmit large amount of power over medium and long lines at very high voltages, above 300 kV, and then rms value of these voltages are termed as EHV, — Extra High Voltage. In our country, transmission voltages range from 66 kV to 400 kV rms, line to line, in three phase bulk power transmission. However, in some sectors 500 kV transmission is being implemented. In highly developed countries like, UK, Europe, USA etc. voltage above 500 kV exist in transmission system ranging even upto 1500 kV. In fact the transmission line voltages above 750 kV are termed as Ultra High Voltages—UHV. In fact, almost all the aspect of characteristics and operations of EHV lines are applicable to UHV lines also.

9.2 IMPORTANCE OF EHV TRANSMISISION

Important reason for the operation of transmission lines in the EHV/UHV ranges are as follows :

1. For a given amount of power to be transmitted over a given distance the transmission efficiency increases as the transmission voltage increases.

2. Power transmitting capacity of the transmission line, at the EHV/UHV range, increases tremendously on the transmission capacity is proportional to the square of the operating voltages. Though the costs of the tower mechanical design's intricacy, insulation and terminal equipments increase but in general these costs are proportional to the voltages rather than square of voltages. As a result, overall capital cost of transmission decreases as the voltage

increases. In fact, transmisision of large blocks of power is thus economically flexible at EHV and UHV.

3. In the transmission lines EHV/UHV lines the P.U. resistance decreases and values of conductors material decrease.

4. With increase in voltage grade/level in transmission lines, the installation cost of transmission line/per km decreases with increase in voltage level.

5. A conventional practice in transmission line is a conventional practice to express the load carrying capability of lines in term of "Surge Impedance Loading" (SIL). At the loading the reactive power common by the inductive part of the line, that is, wLI^2, is theoretically equal to the reactive power charging that is wCV^2, in the line.

For a transmission line, the SIL—Surge impedance loading, P_{SIL}, is given by : $P_{SIL} = \dfrac{V^2}{Z_0}$, where Z_0 is surge impedance of the line (which is equal to $\sqrt{L/C}$, L being the line inductance and C the capacitance per unit length). It is evident that the SIL itself increases which indicates that power transfer increases.

6. Practically, it is not possible to have interconnections of the power system on a large scale without EHV transmission.

7. Many economic considerations have led to the construction of large capacity power station and hence the corresponding transmission lines need to evaluate bulk power. As bulk power transmission is economically flexible at EHV/UHV range, hence with higher voltage range of operation it has become possible to achieve economic in generation also.

9.3 MAJOR FACTORS CONSIDERED IN EHV TRANSMISSION

Following major factors associated with EHV transmission must be considered seriously to ensure trouble free Extra High Voltage transmission system.

1. Line Supports

Heavy mechanical loading is experienced by EHV lines due to bundle conductors large air and ground clearance, considerable dynamic forces due to broken conductors etc. Transmission line towers with fabricated steel members have mostly been used. Statistics indicate that the cost of steel towers varies from 30 to 50 per cent of the total cost of the line, whether of single circuit or double circuit construction for voltages upto 500 kV. Therefore, better and mechanically stronger but cheaper tower designs must be evolved to effect economy. Recent designs for line support include any or many of the following concepts :

(*a*) Suspension tower employing reinforced concrete tubes.

(*b*) H-frame structure supported by steel guy cables.

(*c*) Guyed Aluminium towers (instead of steel).

(*d*) Composite design using standard structural steel for lightly loaded members and high tensile steel for heavily loaded member.

(*e*) Standard wood pole structure with a *V* centre string etc.

The terrific wind pressure during storms and cyclones causes incalculable damage to towers, conductors and insulators. The transmission line must, therefore, be made more wind-resistant. If, however, the amount of metal in the tower is increased the cost goes up appreciably. Therefore, the concepts of streamlining are being applied to the design of structural elements including line conductors.

2. Corona Loss and Radio Interference

In the corona loss, transmission line voltage level being a governing factor, it is more acute in EHV. Infact under bad weather condition, this loss further increases. Since radio and TV have developed greatly in the areas covered by power liine the interference in radio and TV has become a critical factor of communication disturbance.

For reducing corona loss, the spacing between the conductors or the diameter of the conductors can be increased. However, if the spacing is increased, the cost of supports generally becomes very high. The diameter of conductors can be increased by using hollow conductors or steel cored aluminium—ACSR conductors. But the cost of manufacture of such conductors (Hollow or ACSR) is high and their handling is both difficult and expensive. Another difficulty is large wind and snow loading with the use of such conductors. Bundled conductors offer an economical solution for minimising corona loss and radio influence for EHV lines, as the size of the conductors are effective in this regard. These conductors hence filters material or air space in the inside such that the overall diameter increases. Sometimes two or more conductors can be utilised per phase at 400 kV and above line voltage.

3. Insulation Requirement

The magnitude of likely voltage surges determines the required line insulation level. These surges may be due to internal causes like switching operation or due to external causes like atmospheric disturbances including lightning.

4. Erection Problem

Extremely high standard of erection workmanship is a very important factor for EHV system, otherwise wide range of problem due to poor erection quality cause serious problem in EHV transmission lines. Problems of transposition arise as the supporting structures are to be transported across every corner of the country. It also becomes necessary to evolve and employ new erection and conductor handling techniques.

5. Protection from lightning and switching surges.

These two factors must be seriously considered when erecting EHV transmission lines. Protection of EHV lines against lightning in most cases is obtained by the ground wire and rapid auto-reclosing circuit breakers. The ground wire, the line insulation and the tower footing impedances are carefully co-ordinated for adequate lightning protection. Experience with 500 kV line protect from direct lightning strokes by two ground wires with protective angles of 20° to 30° and tower footing impedance of not more than 5-ohms has shown that on such EHV lines lightning is not a serious source of over-voltages and outages. Switching surges (especially those due to arc restriking in circuit breakers) are, however, dangerious as they may cause over-voltages of 2-3 times the normal voltage. With the development in the design of relay breakers systems, however, it is possible to control and minimise switching over-voltages.

In view of the above the level of line insulation for a given voltage rating is showing a downward trend in recent years. For example, inSweden the impulse level of 400 kV lines was reduced from 1,750 kV to 1,425 kV in 1960. Typical present day values for transformer B.I.L. are 1050 kV for 345 kV systems, 1425 kV for 500 kV systems and 2,175 kV for 700 kV system.

6. Power Station and Sub-station Equipment

The design and manufacture of power station equipment has also undergone some exemplary changes. In 1950s turbo-generator units were in the range of 100 – 150 MW whereas turbo generators of 500 MW are now-a-days common in many super power station in India.

Three-phase, two-winding transformers of 700 MW capacity have been in use for quite some time now the capacity is to touch 1,100 MVA soon. Similar is the case with auto-transformers where output will soon exceed 1,200 MVA . For 220 KV circuit, circuit breaker ratings have gone up from 3500 MVA to 20,000 MVA. Oil-less type breakers like *VCB*. SF_6 Breakers' developments have resulted in designs which are more efficient are light in weight and which possess high speed 400 kV and 500 kV breakers are also in use nowadays.

9.4 LIMITATIONS OF EHV AC POWER TRANSMISSION

In the transmission of power by a 3-phase ac EHV system there are certain limitations. These limitations are on account of very important factors, which are given below :

1. Current Carrying Capacity

For overhead transmission line conductor, the maximum power that can be transmitted is never dependent on thermal considerations. However, for underground cables both thermal current limit and charging current are very important while designing the system. Infact due to these considerations a cable is not operated at a loading as high as its natural load. Also since for a given length the cost of cable is 10 to 20 times that of an overhead line, it puts serious limitations on the use of cables (underground on above-ground) in high voltage schemes. For power system engineers overcoming of these problems is a challange in case of cables. However, as said, for overhead conductors designers are less worried regarding current carrying capacities. But there are many other problems for overhead lines which power system designers have to consider seriously and most efficiently, as stated below, briefly.

2. Ferranti Effect

In case the overhead transmission line is loaded with leading reactive-volt-amperes there will be rise of voltage at the receiving end. This rise is of the order of 1.5 percent at 160 km, 15 percent for 500 km and 100 percent for 960 km. Moreover there is also a rise in voltage at the sending end whenever the load on the generator is thrown off suddenly. In AC system, these two aspects are to be taken care of. In fact in many cases. Shunt Reactors at the load end are generally used to control this undesired voltage rise.

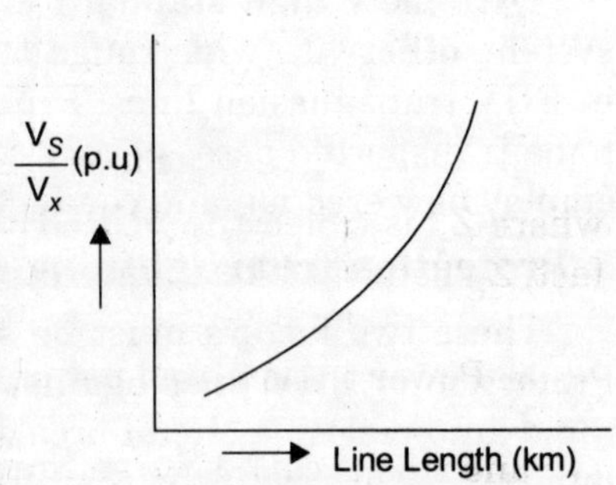

Fig. 9.1. Ferrante effect in power system.

Let us now see below how the voltage rise in lightly loaded EHV lines can be analytically found.

Let the sending end voltage V_s be written in term of *A, B, C, D* constants, receiving end voltage, V_r and current I, is given by :

$$V_s = AV_r + BI \qquad ...(9.1)$$

Now for open ended or lightly loaded line, $I \simeq 0$, then $V_s = AV_v$ from the eqn. (9.1)

$\therefore$ We get $\qquad V_s = AB_r$

or

$$\frac{V_r}{V_s} = \frac{1}{A} = \frac{1}{\cos\frac{2\pi fl}{v}} \qquad ...(9.2)$$

We know from earlier explanation

$$A = D = \cos\beta L = \cos\frac{2\pi l}{\lambda} = \cos\frac{2\pi fl}{v}$$

where l = line length
λ = wavelength
v = velocity of wave in the line
f = system frequency
β = phase constant

Now with length increasing the value of $\cos \frac{2\pi fl}{v}$ in eqn. (9.2) decreases. With decrease of this value the ratio V_r/V_s increases. As sending end voltage remain constant hence V_r increases then. This shows that for a lightly loaded line the receiving voltage increases than the sending end voltage. This is expressed in the curve shown in Fig. 9.1. The rise of receiving and voltage for a lightly loaded transmission line is called Ferranti Effect in Power System.

3. Reactive Losses

In a transmission line, let us consider the performance of the line with negligible resistance, but having an inductance L (Henry) and shunt capacitance C (Farads) per unit length (metre). Now it is derived to supply a load current I (ampere) keeping sending end and receiving end voltages equal then the lagging volt-amperes must be equal to the leading volt-amperes.

That means, $$(IX_L)\cdot I = V\cdot\frac{V}{X_C} \text{ or } I^2Lw = V^2wC$$

$$\therefore \quad I^rLs = \sqrt{CW}$$

$$\therefore \quad \frac{V}{I} = \sqrt{\frac{L}{C}} = Z_C \text{ (ohm)} \qquad ...(9.3)$$

$$\therefore \quad \frac{V^2}{I^2} = \frac{L}{C}$$

where Z_C is called the SURGE IMPEDENCE or the NATURAL IMPEDENCE of the line. In fact, Z_C behaves like a pure resistance.

$\therefore$ Power transmitted per phase is $$VI = V\cdot\frac{V}{Z_C} = \frac{V^2}{Z_C} \text{ wats}$$

and this is called Surge Impedence Loading (SIL or the natural Load, NL) of the line. But in practice a line is not always operated at its natural load. while transmitting power at a lagging power factor there will be a voltage drop along the line but if the power is transmitted at a leading power factor there will be a rise of voltage along the line :

Table 9.1

Voltage, kV	132	220	275	380	500
Surge impedence (Z_C) own	350	3203	315	295	276
Natural load (NL) MW	50	150	240	490	900
Current, amp.	220	385	500	790	1,000

Above typical values of surge impedence and natural load for overhead line are given. The reactive voltage drop (or rise, as the case may be) and the natural load do not in any way place any restriction on the distance over which power may be transmitted. They, however, fix the voltage that may be employed for the transmission of certain amount of power.

4. Stability Consideration

In transmission line if resistance and shunt leakage are neglected the expression of power, developed earlier, $P = \frac{V_s V_r}{X} \sin\delta$ watts per phase, where as stated, V_s = voltage per phase at the sending, V_r = voltage per phase at receiving end, X = reactance in ohm per phase and δ = Load ange in degree. Now for equal sending and receiving end voltages, $V_S = V_R = V$. Hence,

$P = \frac{V_s V_r}{X}$ becomes, $P = \frac{V^2}{X_6}$ sind δ watt per phase. $= \frac{(kV)^2}{X}$ sin δ MW for 3 phases (where kV is voltage in kilo volts) then, for δ = 90, the maximum steady state transfer of power takes place on theoretical consideration. But in actual practice δ is not allowed to vary more than 20° to 30° on full load. This limitation is imposed due to the possibility of losing synchronism during transient disturbances. Now by taking X = 0.3475 ohms per km, and δ = 30°, we get

$$P = \frac{(kV)^2}{X}\sin\delta, P = \frac{(kV)^2}{0.3475 \times l} \times 0.5 = 1.43\frac{(kV)^2}{l} \quad ...(9.4)$$

(where l is the length of the line in kms)

Now, if the transmission line at its Natural Load (NL)

then
$$P = NL = \frac{(kV)^2}{Z_C} \text{ MW}$$

$$\therefore \quad 1.43\frac{(kV)^2}{l} = \frac{(kV)^2}{Z_C}$$

or
$$l = \text{length of line} = 1.43\, Z_C \text{ kilometre} \quad ...(9.3)$$

This means that the theoretical length in kilometres transmission line that can be operated on its natural load and power angle of 30° without loss of stability is only 1.43 times Z_C. This step limit the transmission distance in ac system. The length determined above is, infact, too small for many transmission system. But in EHV system lines, exceeding 200 km in length additional and some costly equipments like series capacitance and also shunt reactors has to be used.

6. Insulation Coordination

This factor perhaps require most serious attention of the Power System engineers engaged in designing and installation and also maintenance of transmission lines. Hence insulation strength of transmission line must be decided by carrying out careful analysis of the different kinds and grades of voltages and over voltages in the line. However, most attention is to be given to over voltages as the system is any way designed for the rated voltage with standard limits set by B.I.S. Hence besides the standard rated voltages insulation strength should be decided for the system for following over voltage causes, mainly :

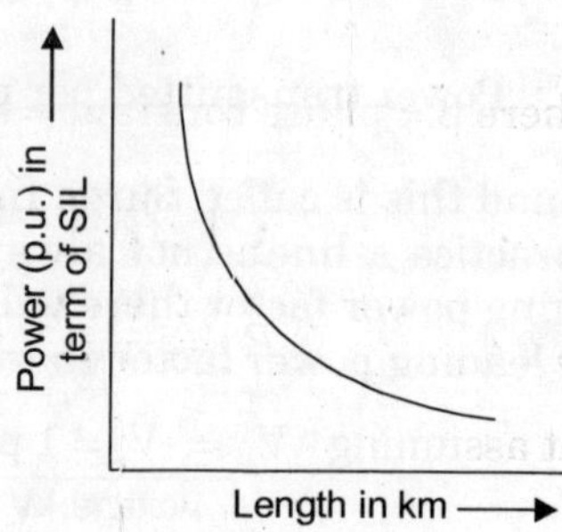

Fig. 9.2.

(a) Over Voltage Due to Lightning

This kind of over voltages is the decesive factor in the choice of insulation level of the line, as over voltages due to switching occurs regularly in the system operation. In fact, to be precise, this occurs due to energisation of a line reclosing, load throw, fault interruption, out of phase switching and also sometime emergency sudden switching off of a line. The switching surges a step frontal wave and of short duration, but its impact is pretty high. As the switching surge is are marginally affected by relative air density humidity rain and wind flow, so it is very much necessary for corrections in the calculation of the insulation coordination of the system.

(b) Temperary Over Voltages

Due to sudden loss of load over voltages of temporary nature in a power frequency are caused. This also occurs due to removal of reactor from a lightly loaded line or sudden connection of capacitors in the line.

(c) Over Voltage Line due to Lightning

In fact over voltages due to lightning is not a decesive factor in insulation coordination. Generally it is most important to assess the probable number of outages due to lightning per 100 km or for more precise study, per 200 km of line length per year. However, as a solution to this problem, Lightning Arrestors are used to protect the system equipment from damage due to lightning arrestors with good earthing.

7. Transmission Lines Loadability

In a transmission line loadability is defined as the optimum power transfer capability of a transmission line under specific operating criteria like prescribed voltages drop, etc.

We know that the power transfer P is given as :

$$P = \frac{V_s V_r}{X} \sin \delta$$

Now assuming 'x' as the reactance per unit length for the line having length l then,

$$X = xl = wL \cdot l \qquad \text{where } L \text{ is the line inductance per unit length}$$

Now, X can re-expressed as :

$$X = wl \cdot L = wl \cdot \sqrt{\frac{L}{C} \times LC} = w \cdot l \times \sqrt{\frac{L}{C}} \cdot \sqrt{LC}$$

(where C is the distributed line capacitance per unit length)

Now, as $X = Z_0 \cdot \beta \cdot l$

(where β = phase constant = $w \sqrt{LC}$ and $Z_0 = \sqrt{\frac{L}{C}}$, surge impedance)

$$= Z_0 \cdot \theta = Z_0 \sin \theta \qquad \text{(at very small value of } \theta, \sin \theta = \theta)$$

So, $P = \frac{V_s \cdot V_r}{X} \sin \delta$ becomes, $P = \frac{V_s \cdot V_r}{Z_0 \cdot \sin \theta} \cdot \sin \delta$

But assuming $V_s = V_r = 1$ p.u.

$$\therefore \quad P = \frac{1}{Z_0} (\sin \delta / \sin \theta)$$

It has been also seen that the power transfer is inversely proportional to the line length (θ) [θ is the line length expressed in parallel].

$\therefore$ The power transfer decreases with increase in length

Now as the surge impedance loading is given by,

$$SIL = \frac{V^2}{Z_0{}^2} \text{ and hence, } V = 1.00 \text{ p.u.}$$

$$SIL = \frac{1}{Z_0} \cdot \text{ then } \frac{P}{SIL} = \frac{\sin \delta}{\sin \theta}$$

Then the power transfer, P_1 with respect to SIL have been showing in Fig. 9.2. Under a specific set of operating criterion of voltage and power angle limiting this power is denoted by loadability.

8. Series Reactive Loss in EHV Transmission Line

It has been observed that the rate of series reactive loss is proportional to the reciprocal of voltage cubed $\left(\frac{dQ_L}{dV} \alpha \frac{1}{|V|^3}\right)$ while the charging rate is proportional to system voltage. Then, in case of voltage, reactive power loss increases while charging capability decreases. This leads to further drop in voltage. This effect is cumulative and if not additional reactive power source is installed this cumulation drop in voltage may lead to voltage instability.

Now to prove, $dQ_L/dV \; \alpha \; \frac{1}{|V|^3}$.

Let *b* be the MVA power being fed at *a* load bus by a radial EHV line of per phase reactance *X* ohm. Line resistance *R* is neglected as in practical EHV lines *R*/*X* ratio is too small.

Let Q_2 be the series reactive loss in MVAR and is given by $|Q_L| = |I^2X|$ where *I* be the line current and for constant MVA load.

$$|I| = \frac{|S|}{|V|}$$

$$\therefore \qquad Q_L = \frac{|S|^2}{|V|^2} \cdot |X|^2$$

Then the change in series reactive loss for change in voltage is given

$$\frac{dQ_L}{dV} = -\frac{2}{(V)^3}|S|^2|X|$$

Now assuming |*S*| and |*X*| constant

we get, $\left|\frac{dQ_L}{dV}\right| \alpha \frac{1}{|V|^3}$. That means the series reactive loss increases sharply with the reciprocal of voltage cubed.

(*a*) Lines and Cables Charging in EHV Lines

In EHV transmission line, let us assume Q_C be the charging MVAR of the lines and the cables :

$$\therefore \qquad |Q_C| = |C|\,|V|^2$$

When, *C* is the distributed line/cable capacitance and *V* the system voltages

Thus, the change in charging due to voltage change is given by

$$\left|\frac{dQ_C}{dV}\right| = 2w\,|C|\,|V|$$

This expression proves that the cable and line charging increments are directly proportional to voltage.

10. Mechanical Vibrations and Osillations

As the number of subconductors increases in bundle conductor transmission hence there may be enormous effect of vibration of conductor. To encounter this problem mechanical design of the system, that is, tower dimension, phase spacings, subconductor spacing, conductor height etc. are to be modified. These modifications affect the electrical design particularly the voltage gradient. When a condcutor is under tension and a steady wind blows across it, small vertics are formed which detach themselves creating minute vertical force on the conductor. The resulting vibrational force causes fatigue of conductor and also in the supporting structures. This is also known as Aeolian Vibration.

Another problem is following, which is a very high amplitude, low frequency conductor vibration occurring mainly in area of flat terrain under freezing rain and icing of conductors. Galloping can be controlled by applying weights at different weight at different location of the span of the line.

11. Environmental and Biological Aspects

Recent studies and research work in this field show that EHV and UHV lines generate electrostatic and electromagnetic fields. These fields can induce current and voltage in animals, human beings and birds. However, these problems do not exceed the satisfactory level upto 500 kV.

12. Compensation of EHV Systems

For compensation of EHV systems, power system engineers have to take various steps. In fact to inject reactive power and control the load end voltage, shunt compensation with capacitive VAR is used. And to encounter Ferranti effect, shunt reactor compensation is used.

Series capacitor compensation is used in order to enhance the power transfer capability of EHV lines. Below commonly used compensators steps are given in brief :

(*a*) Shunt reactors are used so as to drain reactive powers from the line to control the over voltage to nominal values in case the line is surplus of reactive power.

(*b*) Static switchable shunt capacitor banks are used to inject capacitive power at suitable points of the line and to keep the bus voltage to nominal values.

(*c*) Synchronous conductors may be used in order to keep the receiving end voltages to nominal values by adjusting the p.f. of the machine, that is, by adjusting the excitation of it. It is possible to supply both lagging and loading VARs.

(*d*) Static VAR compensators. These are generally used to serve the same purpose of the synchronous condensors but have no rotary part is required. It has got higher speed of response, greater reliability, lower maintenance etc. and is basically a controllable shunt reactor in parallel with static capacitor. Sometimes switched capacitor banks are also provided instead of the fixed capacitor. SVC also reduces power system oscillation and enhances the power handling capacity also.

(*e*) Series compensation is a very important method in improving the performance of EHV lines. It increases the power handling capacity over a line is given by :

$$P = \frac{|V_s|\,|V_r|}{X_L} \sin \delta \qquad \text{...(9.5)}$$

Now, if a series capacitor is installed in the line, having X_C as the capacitive reactance the net reactance is given by $|X_L - X_C|$ and the power transfer in such a line is given by,

$$P' = \frac{|V_s|\,|V_r|}{|X_L - X_C|} \sin \delta \qquad \text{...(9.6)}$$

Now comparing eqns. (9.5) and (9.6), the increase in power transfer is given by :

$$P'/P = \frac{|X_L|}{|X_L - X_C|} = \frac{X_L}{X_L\left[1 - \dfrac{X_L}{X_C}\right]} = \frac{1}{1-k} \quad ...(9.7)$$

where $$k = \frac{X_L}{X_C}$$

But if $X_C < X_L$, the power transfer increases optimum compensation is achieved, with $X_C/X_L = 0.4$ to 0.7. Therefore, it is evident for the power flow equation that for the same amount of power transfer, V_s being equal to V_r, the power angle δ is leader for compensted line than uncompenated line. Lower values of power angle operation improve stability of power system.

Moreover, switch series capacitors also improve transient stability. However, below some major problems accociated with series compensation are stated below for information :

1. High recovery voltage series compensation may produce high recovery voltage across the circuit breaker contact after opening of contact.
2. **Maloperation of line protection :** Series compensation may lead to maloperation of the protection system leading to unnecessary tripping.
3. **Sub-synchronous reconance :** In some cases of series compensation, torsional oscillation may be created causing subsynchronous resonance.
4. **Ferroresonance :** when a lightly loaded transformer is energised through a series compensated line, ferroresonance may occur.

9.5 COMPARISON OF AC & DC TRANSMISSION OF HIGH VOLTAGE AND DC TRANSMISSION IN POWER SYSTEM

The use of transformer for transmitting power over long distances and at higher voltages is possible with alternating current (a.c.) especially where the electric energy is to be generated from water sources or thermal stations (with nuclear fuel or Coal) which are generally situated far away from the load centres (due to many practical reasons). Moreover the the polyphase induction motors which serve the majority of industrial and also some residential and service purposes, these induction motors are simpler and rugged in construction and cheaper also as compared to d.c. machines which were in use much earlier, around 1882. Infact to start with since 1890s, all the electricity supply was of d.c. But it is of interest to know that a.c. almost replaced d.c. lines for part for decades for various practical and economical reasons. But for past few years direct current again is being used for some High Voltage transmission lines. In fact some application, in modern days needed the use of d.c. which has following advantages.

1. The limitation of high voltage AC transmission of bulk power over long distance invite the transmission of high voltage DC for long length of line.
2. DC transmission of high voltage for transmitting bulk power over long distance, say above 550 km is economical.
3. There is greater power transmission for conductor and the DC line is cheaper as it requires only two conductors instead three conductors hence causing less cost for lesser number of towers, insulators and conductors.
4. Each onductor in DC line can be operated as an independent circuit.
5. There is no charging current in DC transmission systems and hence line length limitation does not arise. Neither any corona effect in DC line. In fact, cables in DC

system do not suffer from high dielectric loss. Also in DC system the skin effect is low and hence the current density in the DC transmission line can be higher.

6. The voltage reguation problems are less serious in DC system as in DC line, at steady state, reactance drop does not exit.
7. DC transmission system line does not have any stability problem in itself and hence synchronous operation of transmission link along with the connected machines is possible.
8. In DC system corona loss is such lower compared to AC system. Radio interference HVDC lines is lesser.
9. Line loss is simpler and ground return is possible by using single conductor in HVDC (High Voltage DC) transmission.
10. Line loss is lesser in DC transmission.
11. Low short circuit current is required on HVDC lines.
12. During fault in DC lines grid control of the convertor can drastically reduce the fault current.
13. In AC transmission long line shunt compensation is frequently required.
14. The interconnection of AC grids through AC the lines increase the fault level whereas interconnection of AC grids with DC links does not increase fault level to that extent.
15. There is easy reversibility and controllability of power flow through an HVDC link.
16. HVAC lines and HVDC links can be used in parallel to improve the transient stability network.
17. But besides above advantages, HVDC transmission system has a few disadvantages as well :
    1. Voltage transformation is to be provided on the AC side only.
    2. Maintenance of insulators in HVDC lines is more.
    3. The overall cost increases, particularly for line length below 500 km because of additional requirement of converters filters reactive power compensators.
    4. Circuit breaking in multi-terminal HVDC system is difficult and expensive as well.
    5. Convertor station require considerable reactive power.
    6. Harmonic generation and subsequent filtration in DC are big problem.
    7. Overload capacity of HVDC converters is low.

It is noted that in practice because of these disadvantages (which are very serious in nature causing many practical & technical problem), AC transmission system is still mostly used throughout the world.

9.6 BASIC PRINCIPLE OF HVDC

In a HVDC transmission system there is a rectifier station at the sending end and one inverter at the receiving end. The two stations are connected by a DC transmission line. In the sending end the rectifier station converts AC to DC while the inverter station at the receiving end converts DC to AC to feed the load which are all in ac system.

In fact, by varying the fring angle of the thyristor in the converter the DC output voltage magnitude is controlled. In a rectifier the fring angle is $0° < \alpha < 90°$ and in an inverter the

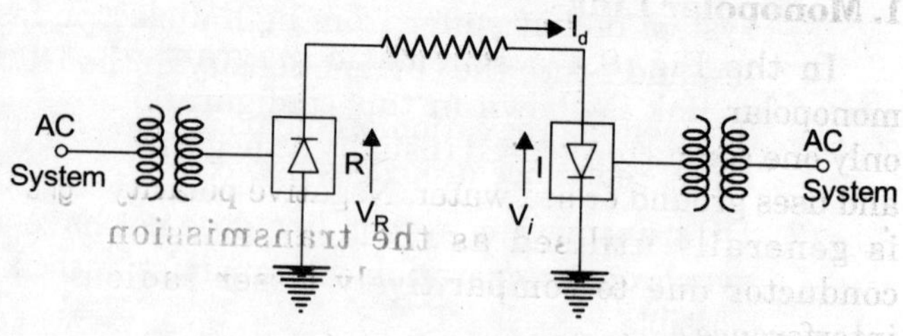

Fig. 9.3

firing angle is (90° < α/180°). As the DC output voltage is a function of cosine of firing angle in convertor hence the convertor voltage becomes negative when α > 90°. This makes the convertor to operate as an inverter. In HVDC converter stations three phase convertors are used both in rectifier and inverters side. By controlling the firing angle reversible operation of converters as well as directional power flow is possible in HVDC link in AC transmission systems the power transfer is governed by voltage difference as well as angular difference between the sending end receiving end whereas in DC system the transmitted power is governed only by the magnitude of terminal DC voltages between the two ends. Thus the controllability of HVDC power is fast and stable. The current flow from higher voltage to lower voltage by proper setting of the rectifier and the inverter.

In a DC transmission system as schematically shown in Fig. 9.3 the line current, I_d is given by,

$$I_d = \frac{V_R - V_i}{R}$$

where V_R is the DC output voltage at the rectifier end and V_i is the DC input voltage at the inverter end, R is the resistance of the entire link.

Again,
$$V_R = \left[\left(\frac{3\sqrt{2}}{\pi} V_{ac_r}\right)\cos\alpha - \frac{3X_{cr}}{\pi} I_d\right]$$

and
$$V_i = \left[\frac{3\sqrt{2}}{\pi} V_{ac_i} \cos y - \frac{3X_{ci}}{\pi} I_d\right]$$

where α = firing angle of rectifier

γ = extinction angle of the inverter

V_{ae_r} = AC side line to line RMS voltage at Rectifier side

V_{ae_i} = AC side line to line RMS voltage at inverter side

X_{cr} and X_{ci} = commutation reactance at the rectifier and inverter

Thus the power transfer is given by

$$P = I_d V_i = \left(\frac{V_R - V_i}{R}\right) V_i \text{ watts}$$

The dc power can be controlled by relative control of DC terminal voltage by controlling angles α and γ. Tap changers in AC side take careof the slow voltage variation on the AC side.

9.7 TYPE OF DC LINKS (HVDC LINKS)

In High Voltage Direct Current (HVDC) System HVDC links play most important role. HVDC links can be classified as below :

1. Monopolar Link

In the Fig. 9.4 a schematic diagram of monopolar link is shown in this configuration only one conductor is used (usually at negative) and uses ground or sea water. Negative polarity is generally utilised as the transmission conductor due to comparitively lesser radio interference.

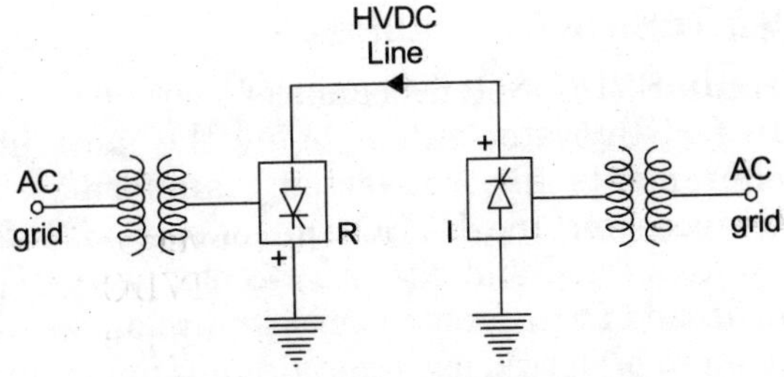

Fig. 9.4. Scheme of monopolar like.

2. Bipolar Link

The Fig 9.5 shows a schematic of a bipolar link. This configuration has two conductors one positive and another negative. In each terminal, two converters of equal rated voltages are connected in series, neutral point being grounded as can be seen in Fig. 9.5. Two poles can operate independently when both the neutrals are grounded. When the current in the two conductors are equal the ground current is zero. In case of a fault in one conductor the remaining conductor along with ground return can supply 50% of the rated load. The rated voltage of a bipolar link is usually expressed as ±Vth.

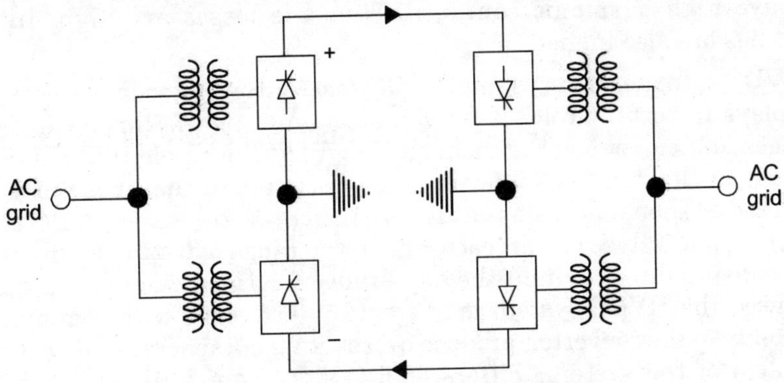

Fig. 9.5. Schematic of bipolar link.

3. Homopolar Link

In Fig. 9.6, a schematic diagram of a homopolar link is shown. It has two conductors but having some polarity, usually negative. This link operates with ground return. In case of a fault in one conductor the converter can be connected such that the heathy conductor can supply power.

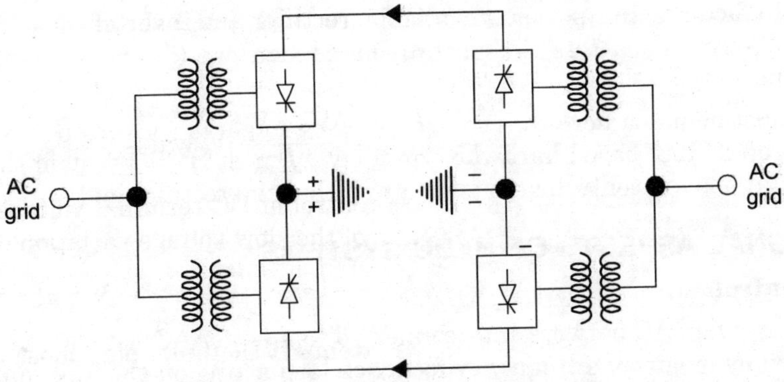

Fig. 9.6. Schematic of a homopolar link.

9.8 TERMINAL EQUIPMENTS

In HVDC systems banks of thyristers are connected and either air cooled or water cooled. In fact thyristor values being the most important terminal equipment, there are other equipments like converted transformer, DC reactor harmonic filters, reactive power compensator and HVDC control system. Infact in practical HVDC system bank of thyristor are connected and either air-cooled or water-cooled thyristor assemblies are protected from *dv/dt* and *DI/dt* protection. Six pulse or twelve pulse converters being in use. The twelve pulse is more preferred as it contain minimum ripple. Two, three phase transformer are conducted in star/star and other two transformers are connected in star/delta in a twelve pulse arrangement.

Main purpose of the converter transformer is to transform AC voltage to the suitable voltage for feeding the converter. It also serves to control the AC side voltage to the suitable voltage for feeding the converter. It also serves to control AC side voltage through tap changing. The converter transformer also offers a suitable reactance thereby controlling the fault level. Twelve pulse operation is feasible due to transformer connections. It also helps in harmonic suppression. However, a convertor transformer has a different design than that of an ordinary transformer. In a convertor transformer the insulation requirement is high and the current flowing through the winding have high harmonic contents. Thus the losses are high. Infact, converter transformer ratings are also higher.

Infact in HVDC transmission system the DC reactor connector in series with each pole of the converter plays important role. It prevents commutation failure in invertor, decreases harmonic voltages and current in the dc line, smoothens the ripple in the line current and it limits the current in the rectifier following a short circuit in the line. But the value of the inductor in the reactor should be such that it does not create resonance in DC circuits at power frequency. Infact normal values of the reactor lie in the range of 0.4 to 1H. In converter station, it is necessary to install filters to limit the amplitude of harmonics, entering the AC system. If filters are not used the HVDC system may create interference with communication system and cause instability of the converter, produce overheating of capacitors and causing some other unwanted problem of the system. Filters on DC side require filtration of 12th and 24th harmonics and in AC side require, 11th, 13th, 23rd and 25th harmonics. Filters in both AC and DC side are required. However, the AC side filters supply the reactive power also and are turned at the frequency for which they are designed.

Infact, HVDC converter station requires reactive power particularly because of converter control and commutation process. Besides, converter transformers consume reactive power. The reactive power drawn is usually in the range of 50% to 60% of the transmitted power. As reactive power is transmitted over the DC line converter station require reactive power compensators. In fact, reactive power is available from : (1) static shunt capacitors, (2) AC side filter, (3) static VAR compensators, (4) synchronous condensors, (5) line charging VAR from AC transmission line.

In fact, it must be noted here that control of firing angle in converters is very important. This is done by optic fibre based hardware circuitry. In most HVDC system the values of the two converter stations are controlled in such a way that the rectifier end controls the voltage.

9.9 OPERATIONAL ASPECTS OF HVDC SYSTEMS

Frequency Control

The frequency of an AC network interconnected to a larger AC network by an HVDC link can be controlled by means of a frequency feedback loop acting on DC link controls such that the small network draws the required power change from the greater network. A DC link is normally insensitive to frequency variations. In case of load throw by a receiving system, a constant power flow though the DC link from sending end can overspeed the receiving end generators causing instability and hence, the incorporation of frequency control is recommended.

Dynamic Stabilization of AC System

A power system is said to be stable if, after a disturbance, the system returns to the condition of stabilization. The power system stabilization from the point of view of dynamic stability is manifested and the power exchanged between them is determined by their relative angular position. After the disturbance, when the stability is attained, the motor positions are in a new state. If the power angle between the machnies increases steadily, the system may face transient instability. When the machines fall out of step after a period of increasing oscillations around the equilibrium point, the power system is dynamically unstable. Dynamic instability rarely occurs in rigid systems. Though large systems, when connected by weak AC transmission lines dynamic instability may occur due to low frequency power swings. This causes negative damping in the system. Damping may be produced by DC power modulation proportional to the frequency difference across the AC interconnector. Thus HVDC systems in parallel to EHV AC intertie is preferred.

Large Signal Modulation

Small signal modulation is suitable for restoring dynamical equilibrium but inadequate for damping of large disturbances. Large signal modulation is then needed to counter large disturbances. Special hardware circuitry is utilized to arrange for large signal modulation.

Controlled Damping in DC Interconnectors

In case an AC tie line is tripped following disturbance, the entire network may face problem in maintaining global stability. On the other hand, for an HVDC link power transfer through it can be increased in order to assist the unhealthy system suffering from power shortage. In this case, the instability of the DC link does not arise. HVDC Link also enables damping of disturbances and, if it has overload capacity, substantial power flow can be provided in addition to providing positive damping.

Torsional interactions result in series capacitor compensated EHV AC lines. This creates torsional modes of vibration leading to sub-synchronous resonance. HVDC systems may also lead to torsional oscillation. However, this problem can be solved easily by providing power modulation control in EHV AC-DC systems.

Trnsient Stabilization of AC Systems

Following large disturbances in a sub-grid, transient instability may occur and transmit through the entire network. HVDC interconnectors, in parallel with AC tie line improve transient stability. In case of a disturbance of one sub-grid, its AC interconnector can be switched off and allowing power transfer from the grid to sub-grid through DC link only. Special purpose controller can be utilized for this purpose.

The thyristor devices, used in converter stations, are rated to withstand sufficient overload and this capability provides the basis for improvement in transient stability of the network.

Voltage and Reactive Power Control

The degree of DC power modulation is restricted by terminal reactive power constraints. It has been already indicated that increase in terminal active power flow is accompanied by a large increase of terminal reactive power requirement. Local reactive power sources like filters, static capacitors, SVC etc. are thus very much essential.

Co-ordination between active and reactive power modulation can be achieved by DC system voltage modulation. An increase in DC voltage will increase DC power transfer as well as power factor and hence decreases reactive power consumption. Usually the sending end (rectifier) is in current control more while the receiving end (inverter) is on voltage control mode. The settings control the power flow and terminal voltages. Small AC side voltage variation can be ecountered by adjusting the on load tap changer in the AC side.

Transient Overvoltages and Insulation Co-ordination

(*a*) disturbance on DC side

(*b*) disturbance on AC side

(*c*) transient in converter itself

(*d*) transient of external origin

Disturbance on DC side that causes overvoltage may be due to short circuit of one pole in a bipolar link and transient overvoltage induced in the healthy pole. However it is not a decisive factor in insulation co-ordination. Re energization of thyristor bridges can cause large transient overvoltage.

The most severe voltage oscillation will occur across the value when an earth fault occurs in the DC reactor or in its bushing. Mostly this overvoltage does not affect insulation.

In AC side, due to resonance between the filter and the system for a low order harmonic, severe overvoltage may occur. Other reasons in AC side that may cause overvoltage are load rejection, AC systems-faults, transformer switching etc. Transformer saturation due to harmonic current may also create overvoltage.

Transient overvoltage may also occur due to converter's own problem like commutation failure or repetitive misfire. Sudden blocking of inverter firing may also create overvoltage across the converter.

Lightning and switching surge are external agents that may cause transient overvoltage in HVDC system. Also switching surges and lightning surges in the AC system can be transfered to the DC side through the transformer.

For co-ordination of insulation, the design aspect should take into account the overvoltage in the plant protection scheme will make the HVDC system design more economic by limiting the overvoltage.

Aspects of Protection

Converter malfunction may occur due to

(*a*) misfire and backfire in converter values.

(*b*) commutation failure, the most common disturbance in inverter operation.

(*c*) short circuit in converter station.

Many of the valves faults are of temporary nature and can be eliminated by temporary absence of conduction. Modern controllers control the firing angles of the converters and the faults can be eliminated by this control action.

In AC side of the HVDC system, faults may occur (short circuits or earth faults) leading to depression of voltage at the converter terminals. The AC side faults must be cleared as soon as possible so that normal DC power transfer can take place. Faults may generate in filter banks and may cause disturbance. Inverse relays and differential protections are utilized to clear the fault in the filter bank.

HVDC Circuit Breaking Principles

The process of arc extinction in AC circuit breakers is easier due to natural current zero phenomenon. However, this does not occur in DC circuit breakers as there is no current zero in direct current. In DC breakers the direct current is reduced to zero by some auxiliary means. A vacuum interrupter is used for circuit breaking in HVDC systems. An auxiliary device is attached with the vacuum interrupter such that the arc current is opposed by the auxiliary device current (generally a discharging capacitor). However, in practice, due to some practical limitation, the fault current is shaped before being interrupted. Commutation of load current from a saturable reactor to a RC circuit is made such that after commutation the current is opposed by the high inductance of the reactor. For successful HVDC circuit breaking, it is necessary breaker and commutation must be initiated as fast as possible.

Researches on HVDC circuit breakers are still in progress and use of these circuit breakers is limited in multiterminal DC systems only. For straight way DC links valve control methodology is still now accepted as feasible solution of DC circuit breaking.

9.10 MAIN DESIGN CONSIDERATION OF HVDC SYSTEMS

Thyristor Converters

The structural design of the recent converters is very simple. Each thyristor module consists of the thyristors and the converter pole is composed of series connection of several thyristor modules. In addition to the thyristor elements, each module contains damping and overvoltage protection circuitry, voltage dividers and firing control units. The modules are mounted in layer form in a box-like structure to form the valve unit

Basically, the converters are of 12 pulse operation in recent days. However, the use of a single 12 pulse converter per pole as well as multiple valves per pole are available in practice. The former design being the simplest in operation and construction, it requires heavy transformer causing transportation and erection problem. Use of single phase transformers has been attempted in the west to overcome the problem. The schematic of a 6 pulse bridge has been shown in Fig. 9.7.

Use of four valve units per phase of a 12 pulse converter configuration is common. These valve unites are stacked vertically to form a quadruple valve per phase. This arrangement results in the most compact and economical layout of the valves and the valve housing. Three quadruple valves constitute a twelve pulse converter (a typical quadruple valve structure has been shown in Fig. 9.8. Valve surge arresters may be directly mounted on the valve structure. In practice, the need of enhancement of reliability leads to the choice of two. 12 pulse converters per pole, connected in series or in parallel (Figs. 9.9 and 9.10). In a parallel connection, a smoothing reactor is required for each 12 pulse converter and current balancing controls must be used for the parallel converters. However, it enhances the reliability of transmission. A parallel configuration provides less line loss during the period when full line currents are not yet established.

Valve Cooling System

As thyristors produce heat of 30-40 w/cm^2 hence effective cooling system is required in practical installation. Each thyristor is generally provided with a double heat sink and the sinks are forced cooled (by air or water). Air cooling serves additional purpose of insulation while water cooling is recommended for high power applications. However, high purity water must be used. For this purpose necessary firing of the thyristors, "Fibre Optic" channels are increasingly in use. Microcomputers are also used in the control room to process the informations from the valves. The auxiliary power needed for firing the thyristors is generally obtained from the voltage across the thyristor.

Valve Protection Circuitry

Following protections are usually employed to protect the thyristors.

(*a*) Forward overvoltage firing

(*b*) dv/dt protection

(*c*) overtemperature protection

(*d*) Forward recovery protection

(*e*) Current sharing protection between two parallel thyristors.

Thyristor Station Layout

The major space is needed by the external plant components, particularly the capacitors used in the form of filters and for VAR support. The valve-hall also consists of surge arresters, and filter reactors in addition to the valves. The transformers and smoothing reactors are placed close to the hall walls with their bushing passing through the wall. The location of the equipments in the substation is designed to combine a low probability of internal flashovers with the best possible utilisation of the space available. The auxiliary power equipment used for cooling is placed just below the valve hall. The building is generally designed in such a way so as to form a "Faraday Cage". This reduces electromagnetic radiation from the valve hall which might cause radio interference. (A typical station layout has been shown in Fig. 9.11).

Convertor Transformers

Any of the following transformer designs can be adopted to achieve a 12 pulse convertor operation :

(*a*) 6 Single-phase 2 winding

(*b*) 3 Single-phase 3 winding

(*c*) 2 three-phase 2 winding.

Star or delta connection can be utilised as per the requirement. The entire winding must be totally and properly insulated. This results in radial leakage flux at the ends of the windings and due to presence of harmonics, converter transformers are subjected to higher losses. Noise due to magnetostriction is higher and may require special tank design to suppress the same. On load tap changing is normally used to reduce the demand of reactive power in steady state operation. The number of tap changer operations is higher in HVDC systems and hence proper quality control of tapchanger components is needed.

DC Reactor

The existence of a DC reactor is an essential part of the HVDC system. While the current waveform in the dc side improves with increasing inductance, the control response slows down and the resonance frequency reduces making the stabilisation of current control more difficult. The limitation of DC line short circuit currents, to prevent valve damage, is normally the decisive factor and leads to smoothing reactances of between 0.5 H to 1.0 H. The DC reactor is generally of air core type and the magnetising characteristics are nonlinear.

DC Power Transmission

The basic principles that determine the size of the overhead lines, towers and the cable, are very much the same for AC and DC transmission. The need for ground wires is less justified in DC lines as only the half of the power is affected by a line fault and the fault current can be controlled from the converter stations. Some utilities prefer ground wires so as to reduce telephonic interference. The choice of conductors also depends on corona and field effect considerations. However, this factor may not be considered for voltages below 1200 kV. The existence of ion currents under the lines is a characteristic of HVDC transmission, but the magnitude of touch potentials appears to be considerably lower under DC lines than those which occur under corresponding AC lines.

Due to the absence of ionic motion in the DC cable insulation, the working stress of paper insulated cables may be as high as 40 KV/mm while operating in DC system while that is only 10 kV/mm in AC system. Thus a thinner insulation may be designed for cables in DC system. Also this improves conductor cooling. Due to absence of any appreciable current in the sheath and reinforcing material; DC cables are subjected to less overcurrent stresses.

Harmonic Elimination

Since the commutation reactance is low in comparison to the DC smoothing reactance, an HVDC converter acts, from the ac point of view, as a source of harmonic currents and from the DC point of view, as a source of harmonic voltage. The appearance of harmonics is an inherent property of the static converters, it is necessary to reduce their penetration into the AC and DC systems.

Increase of pulse number reduces the harmonic order penetrating into both sides of the converter. Beyond the economic range of higher pulse configuration, harmonic elimination is normally done by utilising filters.

AC Filter Design

The role of AC filter is to eliminate all the detrimental effects caused by waveform distortion and particularly the telephone interference. Harmonic voltage is also to be low and for this the following factors are to be considered for assessing the voltage distortion :

(*i*) individual harmonics,

(*ii*) the total voltage distortion and

(*iii*) telephonic interference.

Two basic concepts in filter design are the filter size and its quality. The size of a filter is defined as the reactive power that the filter supplies at fundamental frequency. The quality of a filter expresses the sharpness of tuning which increases with the ratio of its resonance inductance or capacitance to its resistance for resonant filters. In case of high pass filter, the sharpness increases in inverse proportion to that ratio.

The early DC schemes had entirely shunt harmonic tuned filters, each of which consisted of a series PLC circuit tuned to the frequency of a low characteristic harmonic. The filter quality is selected to achieve optimal filter operation *i.e.*, the selected value should inject minimum harmonic current into the network for the assumed operating condition. A larger Q reduces the filter losses and the harmonic voltage, but it increases the risk of parallel resonance between the filters and the network. Fig. 9.12 represents typical AC filters in hvdc installations.

DC Filter Design

On the DC side of HVDC converters the voltage harmonics generate harmonic currents, whose amplitudes depend on the delay, extinction angles and the impedance of the DC circuit itself. Component ratings are considerably different to those for an AC filter, since the harmonic current is reduced to a relatively small value by the large DC smoothing reactor. Q factor is selected in a similar way to the ac Q factor selection. Being the most expensive component, the capacitor is selected first and thus the value of the inductor is determined.

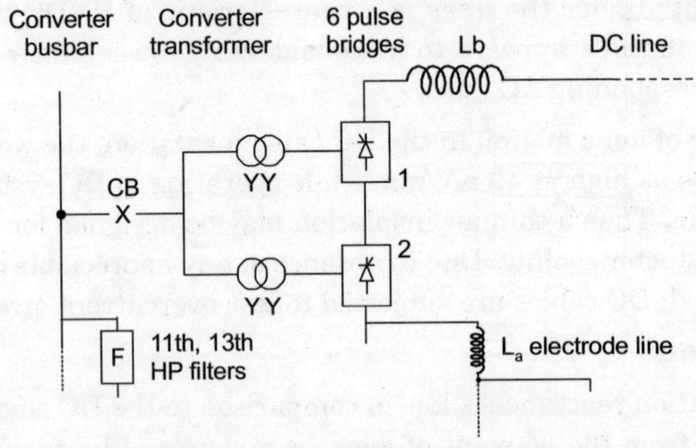

Fig. 9.7. Unit 6-pulse converter.

The recent studies reveal that thyristors being the good alternative of mercury arc valves in recent HVDC installations, proper selection of rating of the thyristors and their protection are the prime requirement in any HVDC schemes design. Different configuration of the thyristor modules can be prepared. Proper HVDC system operation needs thyristor cooling and firing angle control circuitry in addition to conventional protection circuitry of the thyristors.

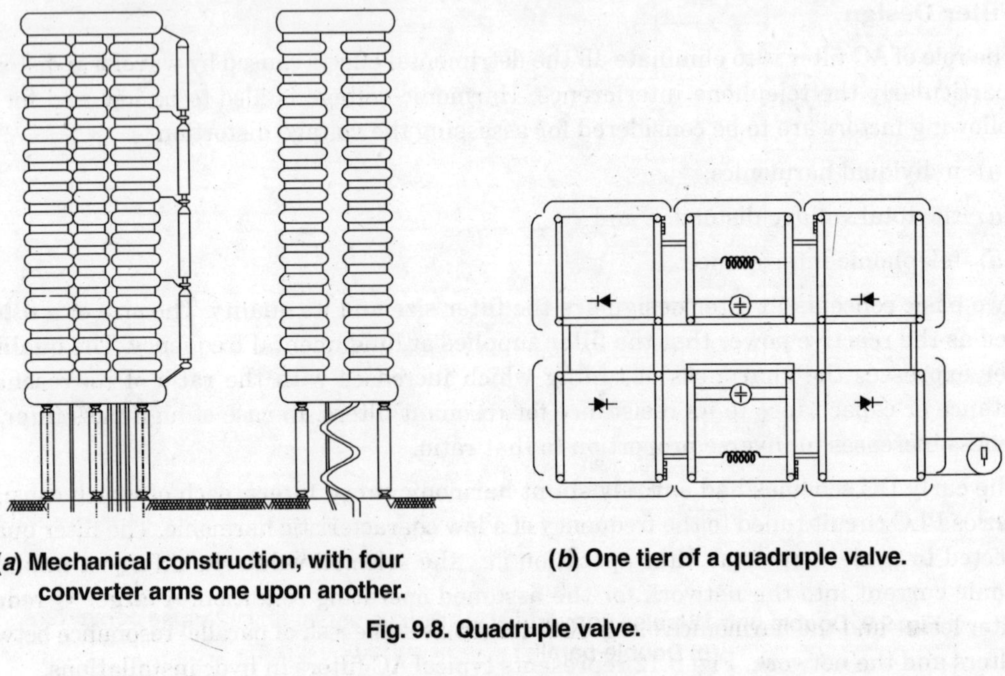

(*a*) Mechanical construction, with four converter arms one upon another.

(*b*) One tier of a quadruple valve.

Fig. 9.8. Quadruple valve.

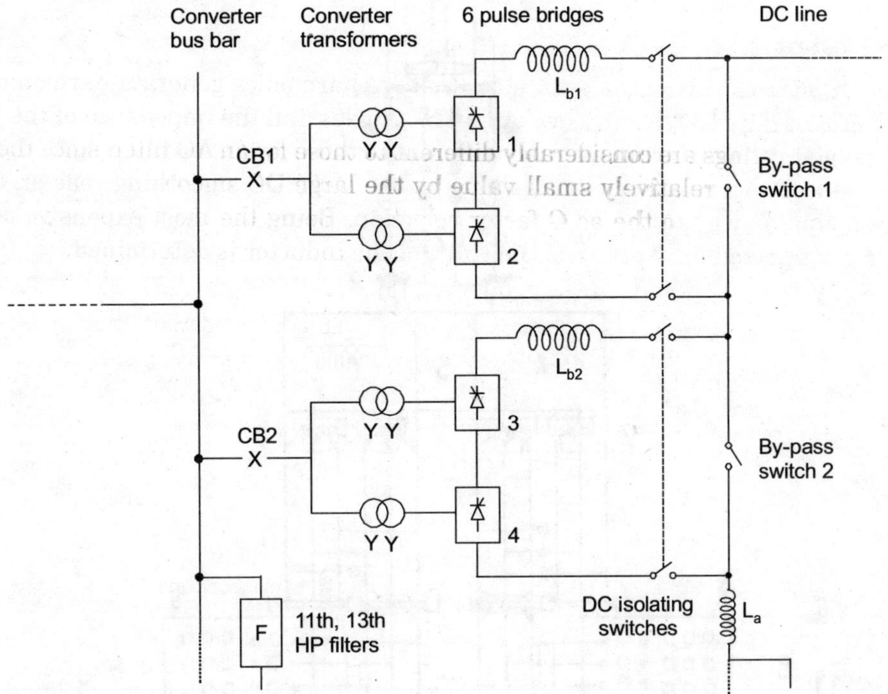

(*a*)

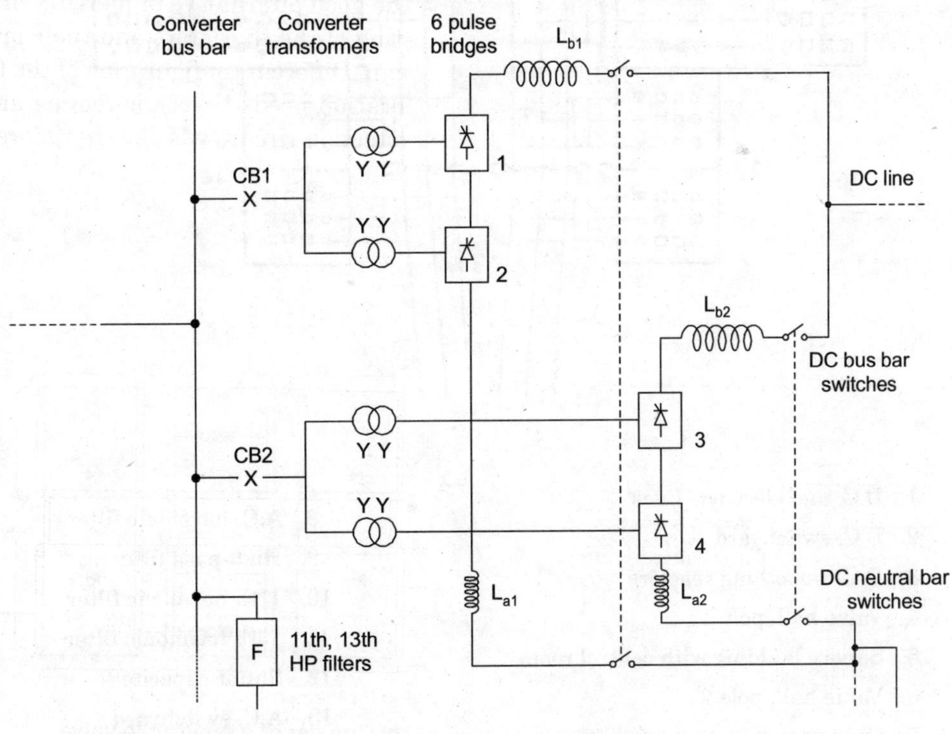

(*b*)

Fig. 9.9. Double unit 12-pulse converters. (*a*) Double-series unit 12-pulse converters. (*b*) Double-parallel unit 12-pulse converters.

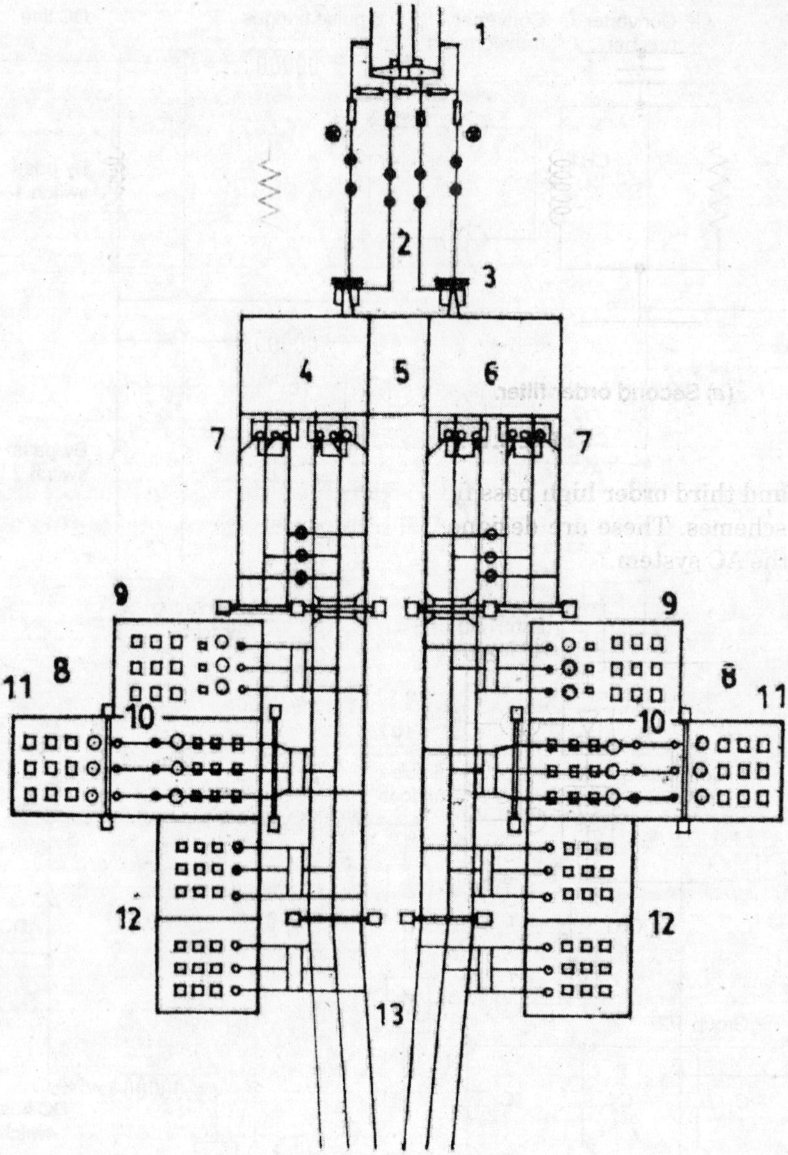

1. D.C. and electrode lines
2. D.C. switchyard
3. D.C. smoothing reactors
4. Valve hall, pole 1
5. Service building with control room
6. Value hall, pole 2
7. Converter transformers
8. A.C. harmonic filters
9. High-pass filter
10. 11th harmonic filter
11. 13th harmonic filter
12. Shunt capacitors
13. A.C. switchyard

Fig. 9.10. Station layout for a bipolar h.v.d.c. station (Reference : Journal).

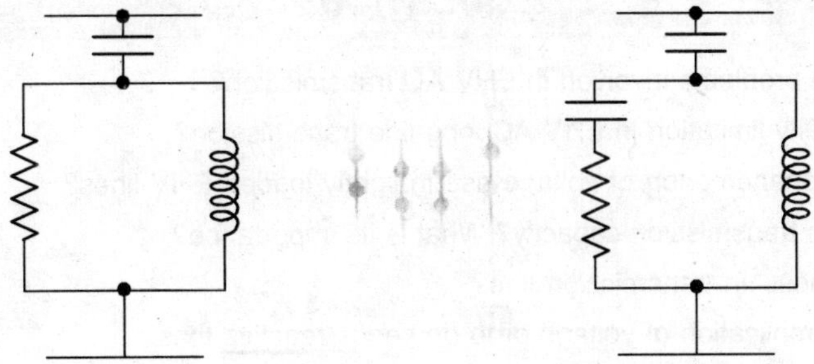

(a) Second order filter. **(b) Third order filter.**

Fig. 9.11. High pass filters.

The second and third order high pass filters shown in Fig. 9.12 (*a*) and (*b*) are extensively used in HVDC schemes. These are designed to reduce the injection of harmonics above the 17th order into the AC system.

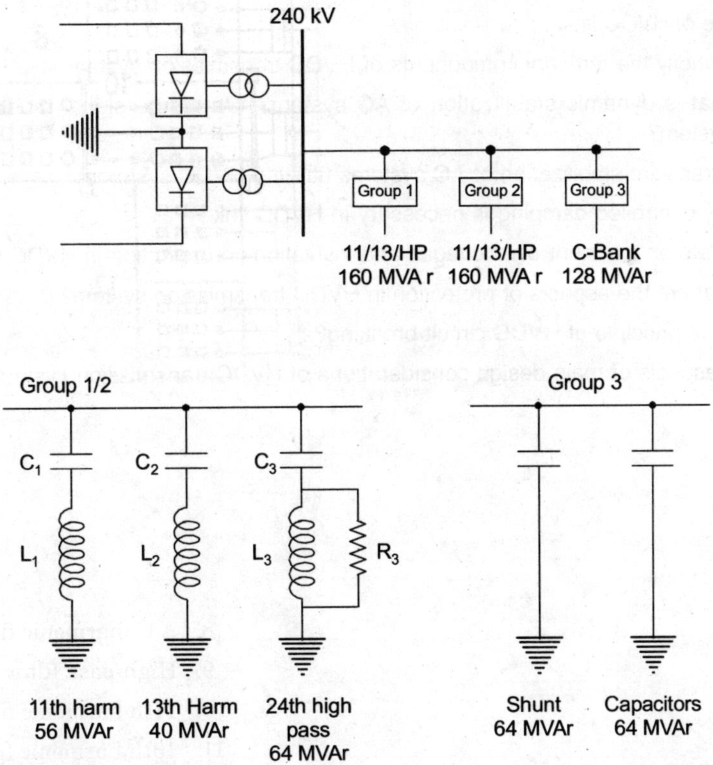

(a) Single line diagram. (b) Detail of filter and capacitor groups.

Fig. 9.12. Filter configuration for a modern 1000 MW converter station.

QUESTIONS

1. What are the problems involved in EHV AC transmission?
2. What is stability limitation in EHV AC long line transmission?
3. Discuss the phenomenon of voltage rise in lightly loaded EHV lines?
4. What is static transmission capacity? What is its importance?
5. Explain 'stiffness' in transmission line?
6. Discuss the implication of voltage drop on series reactive loss.
7. State how voltage is related to reactive charging?
8. What is role of reactive loss on voltage stability?
9. Discuss 'loadability'?
10. State how insulation co-ordination is affected in EHV AC transmission?
11. State the methods of compensation inEHV system.
12. State what are the advantages and disadvantags of HVDC transmission?
13. State the principle of HVDC system operation?
14. What is the control of HVDC transmission systems?
15. State type of HVDC links.
16. Discuss briefly the terminal components of HVDC transmission system.
17. State what is dynamic stabilization of AC system? Why large signal modulation is needed in HVDC system?
18. Discuss transient stabilization of AC systems utilizing HVDC systems.
19. State why controlled damping is necessary in HVDC link?
20. Write a note on transient overvoltages and insulation co-ordination in HVDC systems.
21. State what are the aspects of protection in HVDC transmission system?
22. What is the principle of HVDC circuit breaking?
23. Discuss aspects of main design considerations of HVDC transmission systems.

CHAPTER

10

Transients in Power System

INSIDE THIS CHAPTER

- ❖ Transients in Transmission Lines
- ❖ Travelling waves on transmission lines
- ❖ Line terminate through a capacitance
- ❖ Attenuation of travelling waves in overhead line
- ❖ Capacitance switching
- ❖ Arcing ground causing overvoltage
- ❖ Phenomenon of lightning affecting overhead lines
- ❖ Transmission line design based on lightning

10.1 TRANSIENTS IN TRANSMISSION LINES

The electrical power system can be considered as made up of linear impedance elements of Resistance, Inductance and capacitance. The circuit is normally energised and carries load until a fault suddenly occurs. The fault then corresponds to the closing of a switch or switches, depending upon the type of faults in the electrical circuit. The closing of this switch changes the circuit so that a new distribution of currents and voltages is brought about. This redistribution is accompanied in general by a transient period during which the resultant currents and voltages may momentarily be relatively high. It is very important to realise that this redistribution of currents and voltages cannot take place intantaneously for the following reasons :

1. The electrostatic energy stored by a capacitive C is given by $\frac{1}{2}CV^2$, where V is the instantaneous value of voltage. Assuming capacitance to be constant, the change in energy requires change in voltage, that is for $dt = 0$, the change in current required is infinite which again can not be achieved in practice and so it can be said that change in energy in a capacitor is also gradual.

2. The electromagnetic energy stored by an inductance L is $\frac{1}{2}LI^2$, where I is the instantaneous value of current. Assuming inductance to be constant the change in magnetic energy requires change in current which an inductor is opposed by an emf of magnetic $L \cdot \frac{di}{dt}$. In order to change the current instantaneously $dt = 0$ and therefore $L \cdot \frac{dI}{O}$ is infinity, that means, to bring about instantaneous change in current the emf in the inductor should become infinity which is practically not possible and therefore, it can be said that the change of energy in an inductor is gradual.

Hence, it is known that the two component L and C in an electrical circuit which store energy and it is seen that the change in energy through these components is gradual and therefore, the redistribution of energy following a circuit change take a finite time. The third component, the resistance R, consumes energy.

Moreover, it can be said that (1) the current cannot charge instantaneously through an inductor, (2) the voltage across a capacitor cannot change instantaneously and (3) the law of conservation of energy must hold good. And these three fact are fundamental to the phenomenon of transient in electric power systems.

Hence, now, it can be said that in order to have transients in an electrical system the following requirement should be met :

1. A sudden change in the form of a fault or any switching operation should take place.
2. Higher inductor or capacitor on both should be present.

10.1.1 Transient Components

There are two components of voltages in a power system during transient period : (1) fundamental frequency voltages and (2) natural frequency voltages usually of short duration which are superimposed upon the fundamental frequency voltages. There is third component also known harmonic voltages resulting from unbalanced currents flowing in rotating machines in which the reactances in the direct and quadratures axes are unequal.

However, natural frequency appear immediately after the sudden occurrence of a fault. They simply add to the fundamental frequency voltages. Since resultant voltages are of greater importance from a practical view point it will be preferable to speak of the fundamental frequency and natural frequency components simply as a transient voltage. The transient voltages are affected by the number of connections and arrangements of the circuit transient in which only form of energy storage magnetic or electric is concerned are called single energy transients where both magnetic and electric energies are contained in or accepted by the circuit, double energy transients are involved.

10.2 TRAVELLING WAVES ON TRANSMISSION LINES

In the transmission line the parameters L, C and R are uniformly distributed over the length of the line. For steady state operation of the line, the transmission lines could be represented by lumped parameters, but for the transient behaviour of the lines they must be represented by their actual circuits, that is, distributed parameters.

For a 50 Hz supply and short transmission line the sending end current equals the receiving end current and the change in voltage from sending end to receiving end is smooth. But this is not so when transmission line is subjected to a transient. Now to understand the travelling wave phenomenon over transmission line let us refer to Fig. 10.1 (*a*). The line is assumed to be lossless. Let L and C be the inductance and capacitance respectively per unit length of the line. The line has been represented in Fig. 10.1 (*b*) by a large number of L and C π-section. The closing of the switch is similar to opening the valve at the end of a channel, thereby admitting water to the channel from some reservoir behind. When the valve is opened the channel does not get filled instantaneously. It is observed that the water advancing down the channel. At any instant the channel ahead of the wavefront is dry while that behind is filled with water to the capacity.

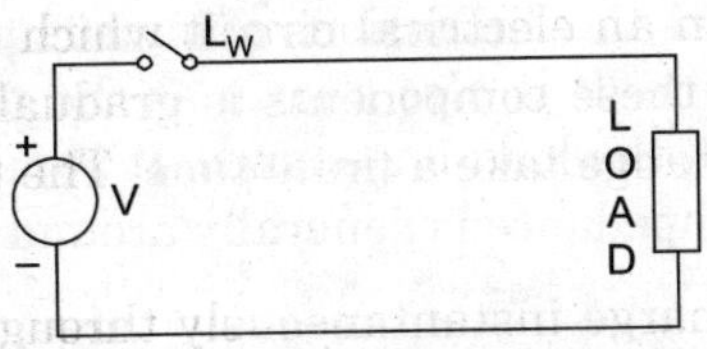

Fig. 10.1 (*a*). Long transmission line.

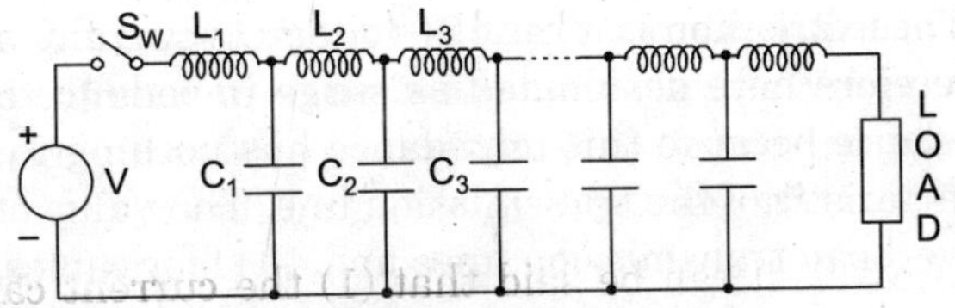

Fig. 10.1 (*b*). Equivalent π-section of a long transmission line.

Similarly, when the switch *b* is closed the inductance *L*, acts as an open circuit and *C*, acts as short circuit instantaneously. At the same instant the next section cannot be charged because the voltage across the capacitor C_1 is zero. So unless the capacitor C_1 is charged to same value whatsoever, charging of the capacitor C_2 through L_2 is not possible which of course will take some finite time. The same argument applies to the third section, fourth section and so on. So we see that the voltage at the successive sections builds up gradually. This gradual build up of voltage over the transmission line conductors can be regarded as though a voltage wave is travelling from one end to the other end and the gradual charging of the capacitances is due to the associated current wave.

Now it is desired to find out expressions for the relation between the voltage and current waves travelling over the transmission lines and their velocity of propagation.

Now let us suppose that the wave after time '*t*' has travelled through a distance '*x*'. Since we have assumed lossless lines whatever is the value of voltage and current waves at the start they remain same throughout the travel. Consider a distance *dx* which is travelled by the waves in time *dt*. The electrostatic flux is associated with the voltage wave and the electromagnetic flux with the current wave. The electrostatic flux which is equal to the charge between the conductors of the line upto a distance *x* is given by

$$q = VC\,x \qquad ...(10.1)$$

The current in the conductor is determined by the rate at which the charge flows into end out of the line

$$I = \frac{dq}{dt} = VC\frac{dx}{dt} \qquad ...(10.2)$$

Here *dx*/*dt* is the velocity of the travelling wave over the line conductor and let this be represented by *v*.

Then,
$$I = VCv \qquad ...(10.3)$$

Similarly the electromagnetic flux linkages erected around the conductors due to the current flowing in them upto a distance of *x* is given by :

$$\psi = IL\,x \qquad ...(10.4)$$

The voltage is the rate at which the flux linkages line around the conductor

$$V = I \cdot L \cdot \frac{dx}{dt} = ILv \qquad ...(10.5)$$

Dividing the eqns. (10.5) by (10.3), we get,

$$\frac{V}{I} = \frac{ILv}{VCv} = \frac{I}{V}\cdot\frac{L}{C}$$

$$\therefore \quad \frac{L}{C} = \frac{V^2}{I^2} \text{ or } \frac{V}{I} = \sqrt{\frac{L}{C}} = Z_0 \qquad ...(10.6)$$

The expression is a ratio of voltage to current which has the dimensions of impedance and is therefore here designated as surge impedance of the line. It is also known as the natural impedance because this impedance has nothing to do with the load impedance. It is purely a characteristic of the transmission line. The value of this impedance is generally around 400 Ω for overhead transmission lines and 40 Ω for cables.

Now, multiplying eqns. (10.3) with (10.5), we get,

$$VI = V \cdot Cv \cdot ILv = VILCv^2$$

or $$v^2 = \frac{1}{LC}$$

or $$v = \frac{1}{\sqrt{LC}} \quad \text{...(10.7)}$$

Now we know, expression for L and C for overhead lines are,

$$L = 2\times10^{-7}\, ln\frac{d}{r}\ \text{H/metre, and } C = \frac{2\pi\varepsilon}{ln\frac{d}{r}}\ \text{F/metre}$$

Now substituting these values in equation (10.7), the velocity of propagation of the wave,

$$v = \frac{1}{\sqrt{LC}} = \frac{1}{\left(2\times10^{-7}\, ln\frac{d}{r}\times\frac{2\pi\varepsilon}{lnd/r}\right)^{1/2}}$$

$$= \frac{1}{\sqrt{4\pi\varepsilon\cdot10^{-7}}} = \frac{1}{\sqrt{4\pi\cdot\frac{1}{36\pi}\times10^{-9}\times10^{-7}}}$$

$$= 3\times10^8 \text{ metres/sec}$$

This is the velocity of light. This hence is , the velocity of propagation of the travelling waves over the overhead transmission lines which equals velocity of light. In reality, because of the resistance and leakage of the lines the velocity of the travelling wave is slightly less than the velocity of light. Normally a velocity of approximately (= 250 m/μ. sec) is assumed. It can be seen from the expression that the velocity of these waves over the cables will be smaller than over the overhead line because of the permittivity term in the denominator. Since, $\varepsilon = \varepsilon_a \varepsilon_r$ for overhead line $\varepsilon_r = 1$. Whereas for cables worth conductor is surrounded by same dielectric material for which $\varepsilon_r > 1$, the term ε is greater for cables than for overhead lines and therefore the velocity of the waves over the cables is smaller than over the overhead lines.

Now let us study the behaviour of these lines to the travelling waves when they reach the other end of the lines or whenever they see a change in the impedance (impedance other than characteristic impedance of the lines).

10.2.1 Open End Line

Let us consider a line with the receiving end open-circuited as shown in Fig. 10.2, where when switch b is closed a voltage and current wave of magnitudes V and I respectively travel towards the open end. These waves are related by the equation : $Z = \frac{V}{I}$, where Z is the characteristics impedance of the line. Consider the line element dx of the line because it is here where the wave is going to see a change in impedance, an impedance different from Z is infinite impedance as the line is open ended.

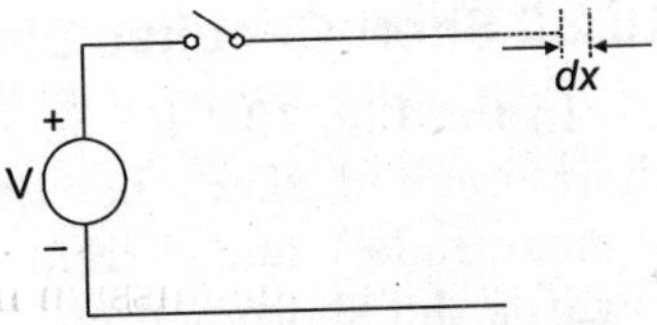

Fig. 10.2. Case of an open end line.

In fact, the electromagnet energy stored by the element dx is given by $\frac{1}{2}$ $L\,dx\,I^2$, and then the electrostatic energy in the element $dx, \frac{1}{2} C \cdot dx \cdot V^2$.

As the current at open end is zero, the electromagnetic to energy. As a result, let the change in voltage be l_1 then,

$$\frac{1}{2} L \cdot dx \cdot I^2 = \frac{1}{2} C \cdot dx \cdot V^2$$

$$\text{or} \quad \left(\frac{l}{I}\right)^2 = \frac{L}{C} \cdot \text{ or } \frac{l}{I} = \sqrt{\frac{L}{C}} \cdot \text{ or } l = I\sqrt{\frac{L}{C}} = I \cdot Z = V \quad \text{...(10.8)}$$

This mean the potential of the open end is raised by V volts, therefore, the total potential of the open end when the wave reaches this end is :

$$V + V = 2V \quad \text{...(10.9)}$$

Now when the switch b is closed, the wave that starts travelling over the line could be considered as the refracted or transmitted wave and is thus,

Refracted wave = Incident wave + Reflected wave

Now, we have seen that for an open end line a travelling wave is reflected back with positive sign and coefficient of reflection as unity.

The Fig. 10.3 (a) shows the variation of voltage.

Now let us see about the current wave : As soon as the incident current wave I reaches the open end, the current at the open end is zero, this could be explained by stating that a current wave of I magnitude travels back over the transmission line. This actually means that for an open end line a current wave is reflected with negative sign and coefficient of reflection unity. In the Fig. 10.3 (b) these variations of current wave over the transmission line is explained. Now after the voltage and current waves are reflected back from the opend end, they reach the source end the voltage over the line becomes $2V$ (as seen in eqn. (10.9) and the current is zero. The voltage at source end cannot be more than the source voltage V therefore a voltage wave of $-V$ and current wave of $-I$ is reflected back in Fig. 10.3. It can, however, be seen that after the waves have travelled through a distance of $4l$ where l is the length of the line, they would have wiped out both the current and voltage waves, leaving the line momentarily in its original state. The above cycle goes on repeating.

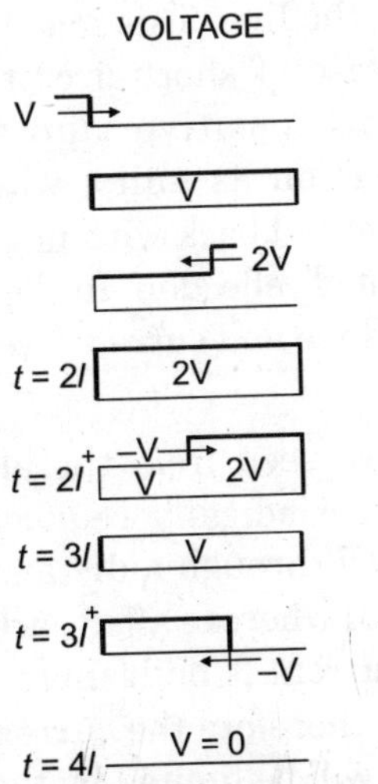

Fig. 10.3 (*a*). Variation of voltage in open ended line.

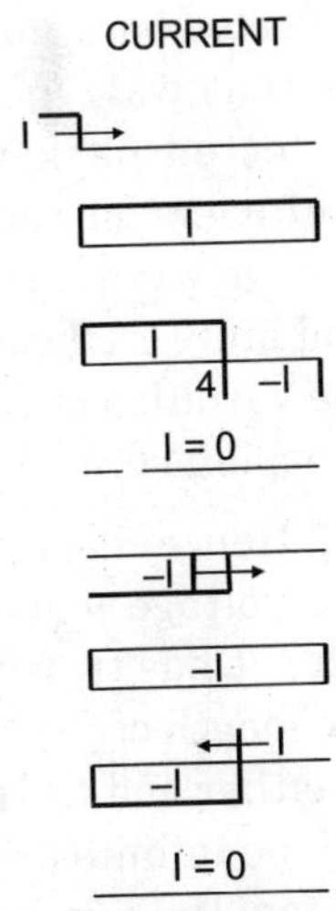

Fig. 10.3 (*b*). Variation of current in open ended line.

10.2.2 Short Circuited Line

In the Fig. 10.4 let us consider the line with receiving end short circuited. Now when the switch 'b' is closed, a voltage wave of magnitude V and current wave of magnitude I start travelling towards the shortened end. Consider again the element dx (as before above) where the electrostatic energy stored by the element is $\frac{1}{2}C \cdot dx \cdot V^2$, and then the electromagnetic energy $\frac{1}{2}L \cdot dx \cdot I^2$.

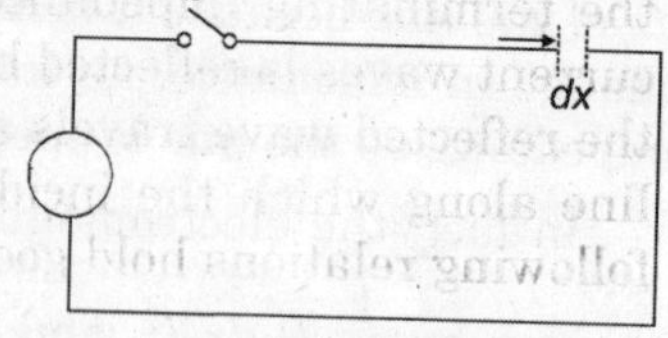

Fig. 10.4. A short circuited in

Since the voltage at the shortened end is zero, the electrostatic energy vanishes and is transformed into electromagnetic energy. As a result, let the change in the current be i, then

$$\frac{1}{2}C \cdot dx \cdot V^2 = \frac{1}{2}L \cdot dx \cdot I^2$$

or
$$V = iZ \text{ (as explained for eqn. (10.7)}$$

or
$$i = \frac{V}{Z} = I$$

And this means the increase in current is I amperes. As result the total current at the shortened end, when the wave reaches the end is $(I + I) = 2I$. This could be considered due to a reflected current wave of magnitude I amperes. Therefore, for a short circuited end the current wave is reflected back with positive sign and coefficient of reflection as unity. Now since the voltage at the shortened end is zero a voltage wave of $-V$ could be considered to have been reflected back into the line. This means that the current wave in case of short-circuited end is reflected back with positive sign and with coefficient of reflection as unity, whereas the voltage wave is reflected back with negative sign and unity coefficient of reflection. In the Fig. 10.5, the variation of voltage and current over the line is explained.

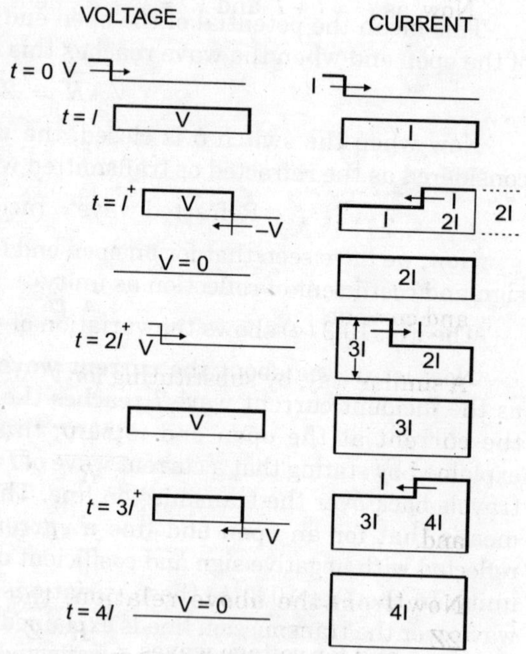

Fig. 10.5. Variation of voltage & current in short ended line.

Hence, it is now seen from the above that the voltage wave periodically reduces to zero after it has travelled through a distance of travel the length of the line whereas after each reflection at either end the current is built up by an amount $I = V/Z_n$. Infact theoretically, the reflection will be infinite and therefore the current will reach infinite value. But practically in an actual system the current will be limited by the resistance of the line end the final value of the current will be $I = V/R$ where R is the resistance of the transmission line.

10.2.3 Line Terminated Through a Resistance

Let Z be the surge impedance of the line terminated through a resistance R as shown in

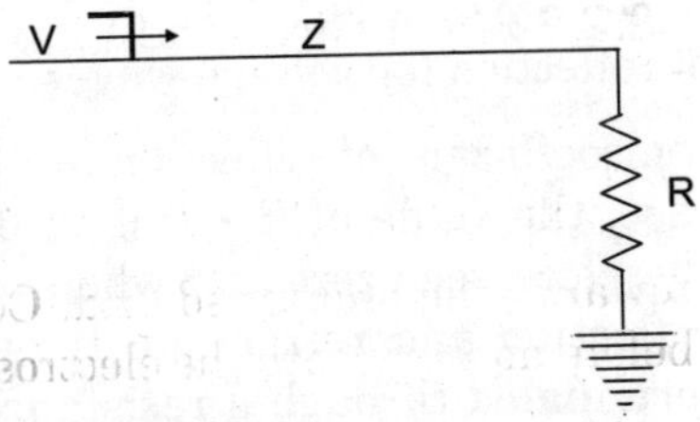

Fig. 10.6. Line terminated through resistance.

Fig. 10.6. We have seen above that whatever be the value of the terminating impedance whether it is open a voltage or current waves is reflected back with negative sign. Also, since the reflected wave travels along the overhead line or over the line along which the incident wave travelled therefore, the following relations hold good for reflected current and voltage waves as well. That is, $I' = -\frac{V'}{Z}$, where V' and I' are the reflected voltage and current waves, and Z is the line impedance.

Refracted on transmitted wave = Incident wave + Reflected wave

Now let V'' and I'' be the refracted voltage and current waves into the resistor R when the incident waves V and I reach the resistance R. The following relation hold good as well,

$$I = \frac{V}{Z},\ I' = -\frac{V'}{Z} \text{ and } I'' = \frac{V''}{R}.$$

Now, as $I'' = I + I'$ and $V'' = V + V'$, hence by using these established relation, we have,

$$\frac{V''}{R} = \frac{V}{Z} - \frac{V'}{Z} \qquad \text{...(10.9)}$$

$$= \frac{V}{Z} - \frac{V'' - V}{Z} \qquad (\text{as } V'' = V + V')$$

$$= \frac{2V}{Z} - \frac{V''}{Z} \text{ or } V'' = \frac{2VR}{Z + R} \qquad \text{...(10.10)}$$

and current,

$$I'' = \frac{2V}{R + Z} = \frac{V}{Z} \cdot \frac{2Z}{R + Z} = I \cdot \frac{2Z}{R + Z} \qquad \text{...(10.11)}$$

A similar way, by substituting for V'' in term of $(V + V'')$, eqn. (10.9) becomes,

$$\frac{V + V'}{R} = \frac{V}{Z} - \frac{V}{Z} \text{ or } V' = V \cdot \frac{R - Z}{R + Z} \qquad \text{...(10.12)}$$

and,

$$I' = \frac{V'}{Z} = -\frac{V}{Z} \cdot \frac{(R - Z)}{R + Z} \qquad \text{...(10.13)}$$

Now from the above relation, the coefficient of refraction for current waves, is $= \frac{2Z}{R + Z}$ and for voltage waves $= +\frac{R - Z}{R + Z}$. Then, the two extreme cases can be derived out of this general expression.

Now for upon end circuit, $R \to \infty$, and hence, coefficient of refraction of current waves $\left(= \frac{2Z}{R + Z}\right)$ becomes, $\frac{2Z}{\alpha + Z} = 0$, and coefficient refraction for voltage waves $\left(= \frac{2R}{R + Z}\right)$ can be rewritten as

$$\frac{2R}{R + Z} = \frac{2}{1 + Z/R} = \frac{1}{1 + Z/\infty} = 2. \qquad (\text{as } R \to \infty \text{ for open end line})$$

Similarly, coefficient of reflection for current waves $= -\frac{R - Z}{R + Z} = \frac{1 - Z/R}{1 + Z/R} = -1$, and coefficient of reflection for voltage waves = $\dfrac{R-Z}{R+Z} = \dfrac{1-Z/R}{1+Z/R} = \dfrac{1-Z/\alpha}{1+Z/\alpha}$. As worked above, now to find out the coefficient of reflection and refraction for current and voltage waves for the short circuit case, the value of $R = 0$ is to be substituted in the corresponding relation as derived. It is, therefore, seen zero that wherever a travelling wave looks into a change in impedance, it suffers reflection and refraction. It is shown below that in case $Z = R$. That means, the line is terminated through a resistance where value equals the surge impedance of the line that is no change in the impedance these will be no reflection and wave will enter fully into the resistance, that means the coefficient of refraction will be unity whereas the coefficient of reflection will be zero.

Now when $R = Z$, substituting this, the coefficient of reflection for current wave

$$= -\frac{R-Z}{R+Z} = \frac{Z-Z}{Z+Z} = 0 \quad (\text{as } R = Z)$$

and for voltage wave, $= \dfrac{R-Z}{R+Z} = 0$

Then the coefficient of refrection for current wave

$$= \frac{2Z}{R+Z} = \frac{2Z}{2Z} = 1$$

and for voltage wave $= \dfrac{2R}{R+Z} = 1$

Therefore, it is seen that when a transmission line is terminated through a resistance equal to its surge impedance the wave does not suffer reflection and hence such lines could be said to be of infinite length. Such lines are also known as infinital lines and the load corresponding to this is known as surge impedence loading or natural impedance loading.

10.2.4 Transmission

Line connected to a cable. In the power system this phenomenon requires study. A wave travels through the line and enters the cable as diagrametically shown in Fig. 10.7. As the wave moves into a different kind of impedence, it obviously suffers reflection and refraction at the junction and the refracted voltage wave is given by,

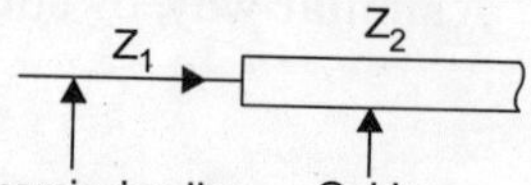

Fig. 10.7. Line connected to cable.

$$V'' = V \cdot \frac{2Z_2}{Z_1 + Z_2} \qquad \text{...(10.14)}$$

The other wave can be obtained by using the reactions in (eqns. (10.11) to (10.13). The impedance of the overhead line and cable are approximately 400Ω and 40 Ω respectively. With these values it can be seen that the voltage entering the cable will be, by using eqn. (10.14) we get, $V' = V \cdot \dfrac{2\times 40}{40+400} = \dfrac{2}{11}V$. That means, it is about 20% of the incident voltage V. It is for this reason that an overhead line is terminated to a substation by connecting the station equipment to the overhead line through a short length of underground cable. Beside the reduction in the magnitude of the voltage wave, the steepness is also reduced because of the capacitance of the cable. The reduction in steepness is very important because this is one of the factors for reducting the voltage distribution along the winding of the equipment. While connecting the overhead line to a station equipment through a cable it is important to note that the length of the cable should not be very short (infact it should not be shorter than the expected of the length of the wave), otherwise successive reflection at the junction may result is piling up of voltage and the voltage at the junction may reach the incident voltage.

10.2.5 Reflection and Refraction at a *T*-Junction

Say a voltage wave V is travelling over the line with surge impedance Z, as shown in Fig. 10.8. When it reaches the junction, it looks a change in impedance and therefore, suffers reflection and refraction. Now let V''_2, I''_2 and V''_3, I''_3 be the voltages and current in the lines having surge impedances Z_2 and Z_1 respectively. Since Z_2 and Z_3 form a parallel path as far as the surge wave is concerned, $V''_2 = V''_3 = V''$. Therefore, the following relation hold good.

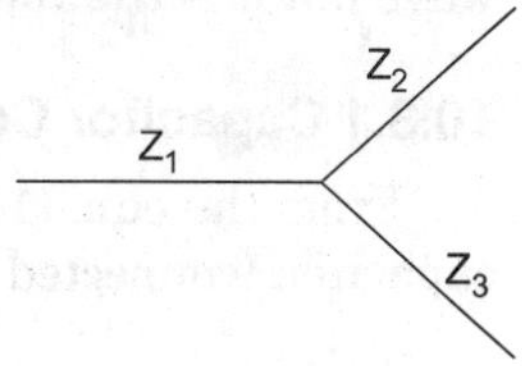

Fig. 10.8. A bifurcated line (*T*) line.

$$V + V' = V'' \text{ and } I = \frac{V}{Z_1};\ I' = -\frac{V'}{Z_1},$$

and, $$I''_2 = \frac{V''}{Z_2} \text{ and, } I''_3 = \frac{V''}{Z_3}$$

and, $$I + I' = I''_2 + I''_3 \qquad ...(10.15)$$

Now, substituting in eqn. (10.15), the values of currents,

$$\frac{V}{Z_1} - \frac{V'}{Z_1} = \frac{V''}{Z_2} + \frac{V''}{Z_3} \qquad ...(10.16)$$

Then substituting for $V' = V'' - V$ (from eqn. $V + V' = V''$) in eqn. (10.16) we get :

$$\frac{V}{Z_1} - \frac{V'' - V}{Z_1} = \frac{V''}{Z_2} + \frac{V''}{Z_3}$$

or $$\frac{2V}{Z_1} = V''\left[\frac{1}{Z_1} + \frac{1}{Z_2} + \frac{1}{Z_3}\right]$$

or $$V'' = \frac{2V/Z_1}{\frac{1}{Z_1} + \frac{1}{Z_2} + \frac{1}{Z_3}} \qquad (10.17)$$

Same way other quantities, as above can be derived (this can be an exercise for the student also).

10.3 LINE TERMINATED THROUGH A CAPACITANCE

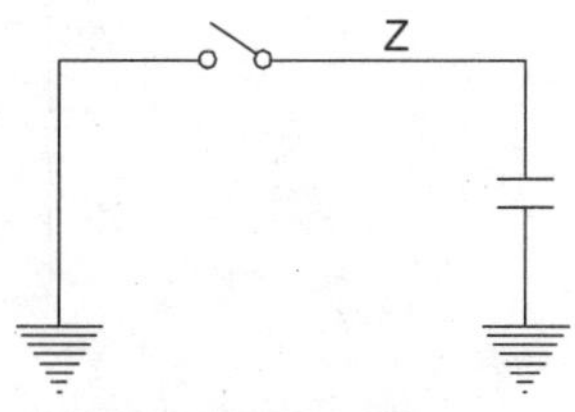

Fig. 10.9. Line terminated through a capacitance.

In this case let us consider that a d.c. surge of infinite length travels cover the line of surge impedance, Z, and is incident on the capacitor as shown in Fig. 10.9. Here our object is to find out the voltage across the capacitor that is, the refracted voltages. The refracted voltage, using eqn. (10.10).

$$\therefore\ V''(S) = \frac{2 \times 1/C_S}{Z + 1/C_S} \times \frac{V}{S} = \frac{2V}{S} \cdot \frac{1}{ZC_S + 1} = \frac{2V}{S} \cdot \frac{1/ZC}{S + 1/ZC} = 2V\left[\frac{1}{S} - \frac{1}{S + 1/ZC}\right] \qquad ...(10.18)$$

By Laplace transform we get,

$$V''(t) = 2V[1 - e^{-i/ZC}] \quad ...(10.19)$$

The variation of voltage is shown in Fig. 10.10 (*b*). Here it is to be noted that, since terminating impedance is not a transmission line, therefore, $V''(S)$ is not a travelling wave but it is the voltage across the capacitor C.

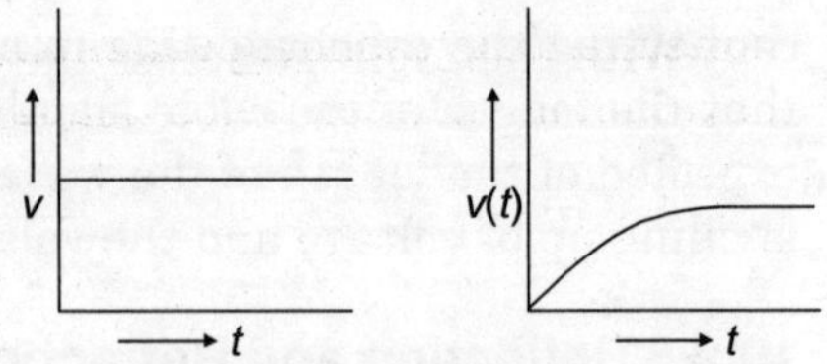

(*a*) Incident voltage (*b*) voltage across capacitor.

Fig. 10.10

10.3.1 Capacitor Connection at *T*

From the eqn. (10.17) we can write the voltage across capacitor, (connected at a *T*) in Laplace from, with reference to the Fig. 10.11.

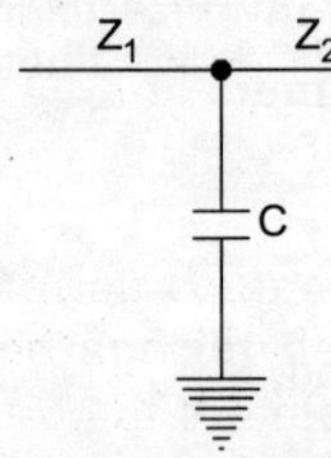

Fig. 10.11. Capacitor connected at *T*.

$$V''(S) = \frac{2V/Z_1 S}{\dfrac{1}{Z_1}+\dfrac{1}{Z_2}+C_S} = \frac{2VZ_2}{S}\cdot\frac{(1/Z_1 Z_2 C)}{\dfrac{(Z_1+Z_2)}{Z_1 Z_2 C}+S}\cdot\frac{2V}{SZ_1C}\cdot\frac{1}{S+\dfrac{Z_1+Z_2}{Z_1 Z_2 C}}$$

Now let $\dfrac{Z_1+Z_2}{Z_1 Z_2 C} = \alpha$

then, $V''(S) = \dfrac{2V}{S}\cdot\dfrac{1/Z_1C}{S+\alpha}$

or $V''(S) = \dfrac{2V}{S}\dfrac{Z_2}{Z_1+Z_2}\cdot\dfrac{\alpha}{S+\alpha} = \dfrac{2VZ_2}{Z_1+Z_2}\cdot\left[\dfrac{1}{S}+\dfrac{1}{S+\alpha}\right]$

or $$v''(t) = \frac{2V\cdot Z_2}{Z_1+Z_2}\left[1-\exp\left(-\frac{Z_1+Z_2}{Z_1 Z_2 C}\right)\right] \quad ...(10.20)$$

The variation at the wave is shwon in Fig. 10.12. Here we have assumed in the derivation of the expression for voltage across the capacitor in the above section, that the travelling surge is of infinite length. Let us now derive the expression when the surge is of finite duration, say here, as shown in Fig. 10.13. Also let the magnitude of this wave be V units. Then the wave could be decomposed into two waves as.

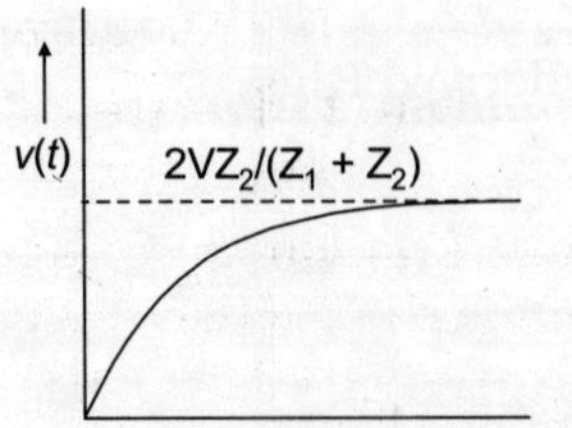

Fig. 10.12. Variation of voltage across capacitor.

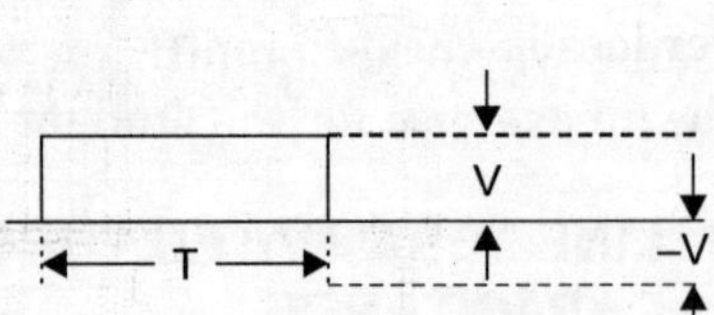

Fig. 10.13. Surge of finite length,

$$f(t) = Vu(t) - Vu(t-r)$$

Here, $$Vu(t-r) = V \text{ for } t \geq r$$
$$= O \text{ for } t < r$$

With this, voltage across the capacitor is given by

$$V''(S) = \alpha f(t) \cdot \frac{2/Z_1}{\dfrac{1}{Z_1} + \dfrac{1}{Z_2} + C_s} = \frac{2V/Z_{1s}}{\dfrac{1}{Z_1} + \dfrac{1}{Z_2} + C_s} = \frac{(2V/Z_1 S) \cdot e^{-Cs}}{\dfrac{1}{Z_1} + \dfrac{1}{Z_2} + C_s}$$

$$v''(t) = 2V \cdot \frac{Z_2}{Z_1 + Z_2}\left[1 - \exp\left(-\frac{Z_1 + Z_2}{Z_1 Z_2 C} t\right)\right] \cdot - \frac{2VZ_2}{Z_1 + Z_2}\left[1 - \exp \cdot \left\{-\frac{Z_1 + Z_2}{Z_1 Z_2 C}(t - \hat{i})\right\}\right]$$

The variation of this voltage is shown in Fig. 10.12. Thus for time $0 < t < \hat{i}$ only the first term in expression is active and for $t \geq \hat{i}$ both the terms are active. The rise in voltage is maximum at $t = \hat{i}$ when the value will be :

$$v''(t) = \frac{2VZ_2}{Z_1 + Z_2}\left[1 - \exp\left(-\frac{Z_1 + Z_2}{Z_1 Z_2 C}\hat{i}\right)\right] - \frac{2VZ_2}{Z_1 + Z_2}[1 - e^{\circ}]$$

$$= \frac{2VZ_2}{Z_1 + Z_2}\left[1 - \exp\left(-\frac{Z_1 + Z_2}{Z_1 Z_2 C}\right)\right] \quad ...(10.18)$$

Therefore, it is clear that attenuation in the magnitude of voltage for a short wave is much more rapid than for a long wave.

It has been seen that the effect of a shunt capacitor is to redue the steepness and magnitude of the wave reaching an equipment. Since an inductor is dual to a capacitor, an inductor in series of the lines should give the same effect.

10.4 ATTENUATION OF TRAVELLING WAVES IN OVERHEAD LINE

Refer to the Fig. 10.14 where a transmission line is represented, where let R, L, C and G be the resistance, inductance, capacitance and conductance respectively per unit length of the line shown in Fig. 10.14. Now let the value of voltage and current wave at $x = 0$, be V_0 and I_0. The object is to find the values of voltage and current waves when they have travelled through a distance of x units overhead line. Let the time taken be t units when voltage and current waves are V and I respectively. Now to travel a distance of dx, let the time taken be dt. Then the equivalent circuit for the differential length Δx of the is shown in Fig. 10.15. Then the power loss in the differential element is

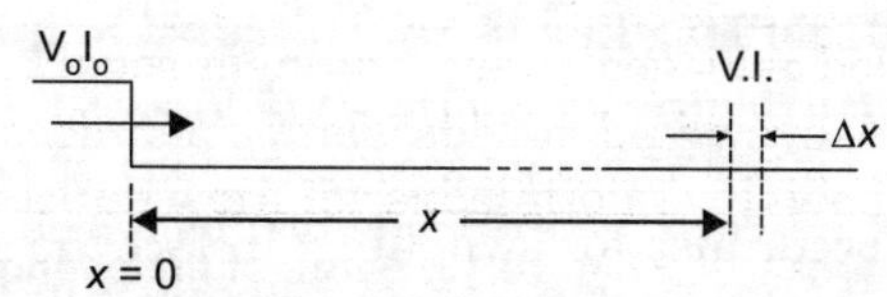

Fig. 10.14. Travelling wave on lossy line.

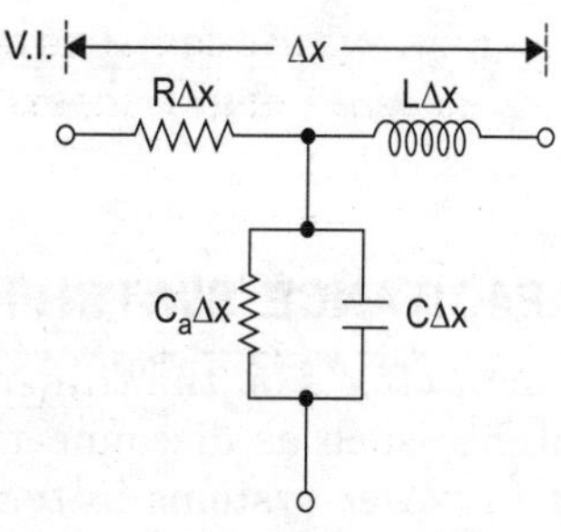

Fig. 10.15. Differential element of transmission.

$$dp = I^2R \cdot dx + V^2G\, dx \quad ...(10.19)$$

Then, the power at a distance x, $V \cdot I = p = I^2Z_n$, that means,

$$p = I^2 Z_n$$

$\therefore$ Differential power, $$dp = -2IZ_n \cdot dI \qquad ...(10.20)$$

where Z_n is the natural impedance of the line. Here negative sign has been assigned as there is reducing in power as the wave travels with line.

Now equating the equations (10.19) and (10.20), we get,

$$-2I \cdot Z_n \cdot dI = I^2 R\, dx + V^2 G\, dx = I^2 R\, dx + I^2 Z^2{}_n \cdot G\, dx$$

$\therefore$ $$dI = -\frac{I(R + GZ_n^2)}{2Z_n} \cdot dx$$

or $$\frac{dI}{I} = -\left(\frac{R + GZ_n^2}{2Z_n}\right) dx$$

then, $$ln\ I = -\left(\frac{R + GZ_n^2}{2Z_n}\right) x + A \qquad ...(10.21)$$

Now, at $x = 0$, we get, $I = I_0 \therefore A = ln\ I_0$

or, now these results are put in eqn. (10.21) to get,

$$ln \frac{I}{I_0} = -\frac{R + GZ_n^2}{2Z_n} x = -ax \text{ where } a = \frac{S + GZ_n^2}{2Z_n}$$

$\therefore$ $$I = I_0 \cdot e^{-ax} \qquad ...(10.22)$$

Exactly, the same way, it can be proved that $V = V_0 \cdot e^{-ax}$. This shows that the current and voltage waves get attenuated exponentially as they travel over the line and the magnitude of attenuation depends upon the parameters of the line. Now since the value of resistance depends not only on the size of the conductors but also on the shape and length of the waves. An empirical relation takes into account the shape and length of the wave for calculating the voltage and current at any point on the line after it has travelled throught a distance x unit, and hence is given as,

$$V = \frac{V_0}{1 + kx V_0} \qquad ...(10.22)$$

where x is the distance in kms, V and V_0 are in kV and k is the attenuation constant of value,

$$k = 0.00037 \text{ for chopped waves}$$
$$= 0.00019 \text{ for short lines,}$$
$$= 0.0001 \text{ for long line}$$

10.5 CAPACITANCE SWITCHING

Transient effects on the transmission lines occur also for switching. In fact the switching of capacitance such as disconnecting a line or a cable or a bank of capacitors poses serious problems in power systems in terms of abnormally high voltages across the circuit breaker contacts. In such situation or condition the current leads the voltage by about 90°, say. The current interruption taken place when it is passing through zero value the capacitor will be charged to maximum voltage. Due to capacitance effect, as the capacitor is at the moment is isolated from the source. It retains its charge as shown in Fig. 10.16 (c) and because of trapping of this charge, half a cycle after the current zero the voltage across the circuit breaker contact is $2V$ which may prove to be dangerous and may result in the circuit breaker strike. This is equivalent to closing the switch suddenly which will result into oscillation in the circuit at the natural frequency,

$$f = \frac{1}{2\pi\sqrt{LC}}$$

In fact the circuit condition corresponds to Fig. 10.16 (*e*) but the only difference between the two circuits is that whereas in Fig. 10.15, the capacitor is charged to a voltage V_1 in Fig. 10.16, it is assumed to be without charge. Therefore, the voltage across the capacitor reaches $3V$. As the source voltage is the voltage across the breaker contacts after another half cycle will be $4V$ which may cause another restrike. This phenomenon may continue indefinitely, theoritically, increasing the voltage by successive icrement of $2V$. This may result into an external flashover, or the failure of the capacitor itself.

This is due to the inability of the circuit breaker to provide sufficient dielectric strength to the contacts to avoid restrikes after they are opened first. This problem is actually solved in practice by using air blast circuit breaker or multibreak circuit breakers.

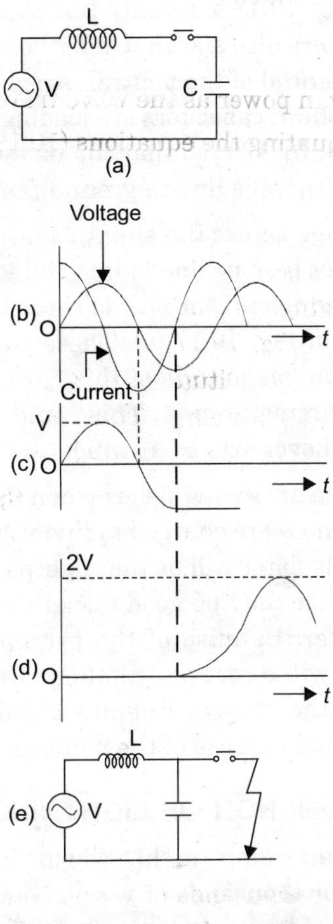

Fig. 10.16. (*a*) **Equivalent circuit for capacitor switching,** (*b*) **system voltage and current,** (*c*) **capacitor voltage,** (*d*) **voltage across the switch.** (*e*) **equipment circuit to determine the restriking voltage.**

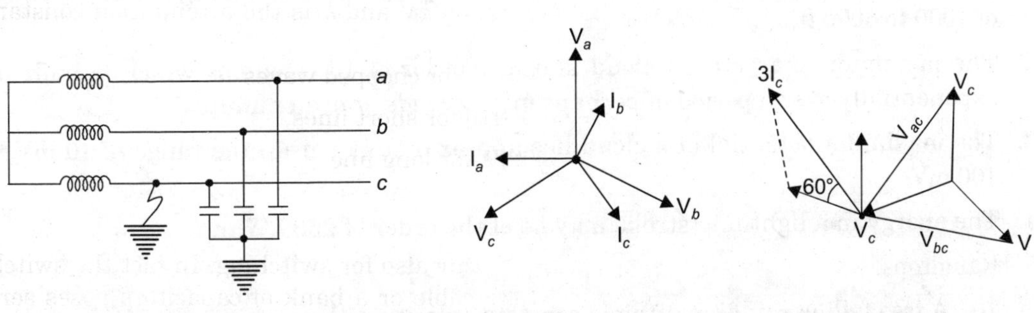

Fig. 10.17. (*a*) **3-phase, line with isolated neutral** (*b*) **Phase diagram under healthy state** (*c*) **phasor diagram at fault.**

10.6 ARCING TO GROUND CAUSING OVERVOLTAGE

In the Fig. 10.17 a 3-phase line system with isolated neutral system is shown. The shunt capacitances are also shown. Under balanced conditions and complete transposed transmission lines, the potential of the neutral is near the ground potential and the current in various phase through the shunt capacitors are leading their corresponding voltages by 90°. They are displaced for each other by 120° so that the net sum of the three current is zero as shown in Fig. 10.17 (*b*). Now, say there is line-to-ground fault on one of the three phases.

The voltage across the shunt capacitor of that phase reduces to zero whereas those of the healthy phases become line-to-line voltages and now they are displaced by 60° rather than 120°. The net charging current now is three times the phase current under balanced conditions, as can be seen in Fig. 10.17 (*c*). These currents flow through the fault and the winding of the alternator. The magnitude of this current is often sufficient to sustain an arc and therefore, we have an arcing ground. This could be due to a flashover of a support insulation. Infact, here this flashover acts as a switch.

Now, if the arc extinguishes when the current is passing through zero value the capacitors in phases *a* and *b* are charged to line voltages. The voltage across the line and grounded points of the point insulator will be the superposition of the capacitor voltage and the generator voltage and this voltage may be good enough to cause flashover which is equivalent to restrike in a circuit breaker. Because of the presence of the inductance of the generator winding, the capacitances will form an oscillatory circuit and these oscillations may build up to still higher voltages and the arc may reignite causing further **transient** disturbances which may finally lead to complete rupture of the point insulators in the transmission line system.

10.7 PHENOMENON OF LIGHTNING AFFECTING OVERHEAD LINES

For human being in this world, lightning has been a source of wonder & sometime destruction for thousands of years. Now what is this phenomenon of lightning which affects the operation of transmission system, if not controlled with preventive measures. To take these precautionary measues, it is desirable to review some of the facts concerning the thunder cloud and the associated phenomenon :

1. The height of the cloud base above the surrounding ground level may vary from 500 to 30,000 ft. The charged centres, which are responsible for lightning are in the range of 1000 to 5000 ft.
2. The maximum charge on a cloud is of the order of 10 coloumbs which is built up exponentially over a period of perhaps may seconds or even minutes.
3. The maximum potential of a cloud lies approximately within the range of 10 mV to 100 mV.
4. The energy in a lightning stroke may be of the order of 250 kWhr.
5. Raindrops.
   (*a*) A free falling raindrop attain a constant velocity with respect to the air depending upon its size. This velocity is 800 cm/sec for drops of the size 0.25 cm dia. and is zero for spray. This mean that in case the air currents are moving upwards with a velocity greater than 800 can/see, no raindrop can fall.

(*b*) Raindrops elongate and become unstable under an electric field, the limiting diameter being 0.3 cm in a field of 10 kV/cm.

(*c*) Falling raindrop greater than 0.5 cm, in dia become unstable and break up into smaller drops.

(*d*) When a drop is broken up by air current the water particle become positively charged and the air negatively charged.

(*e*) When an ice crystal strikes with air currents the ice crystal is negatively charged and the air positively charged.

Mechanism of Lightning Stroke

Lightning phenomenon is the discharge of the cloud to the ground. The cloud and the ground form two plates of a gigantic capacitor and the dielectric medium is air. Since the lower part of the cloud is negatively charged, the earth is positively charged by induction. Lightning discharge will require the puncture of the air between the cloud and the earth. For breakdown of air at STP condition the electric field required is 30 kV/cm, back. But in a cloud where the moisture content in the air is large and also because of the high attitude (at lower pressure) it is seen that for breakdown of air the electric field required is only 10 kV/cm. The mechanism of lightning discharge is best explained with the help of Fig. 10.18.

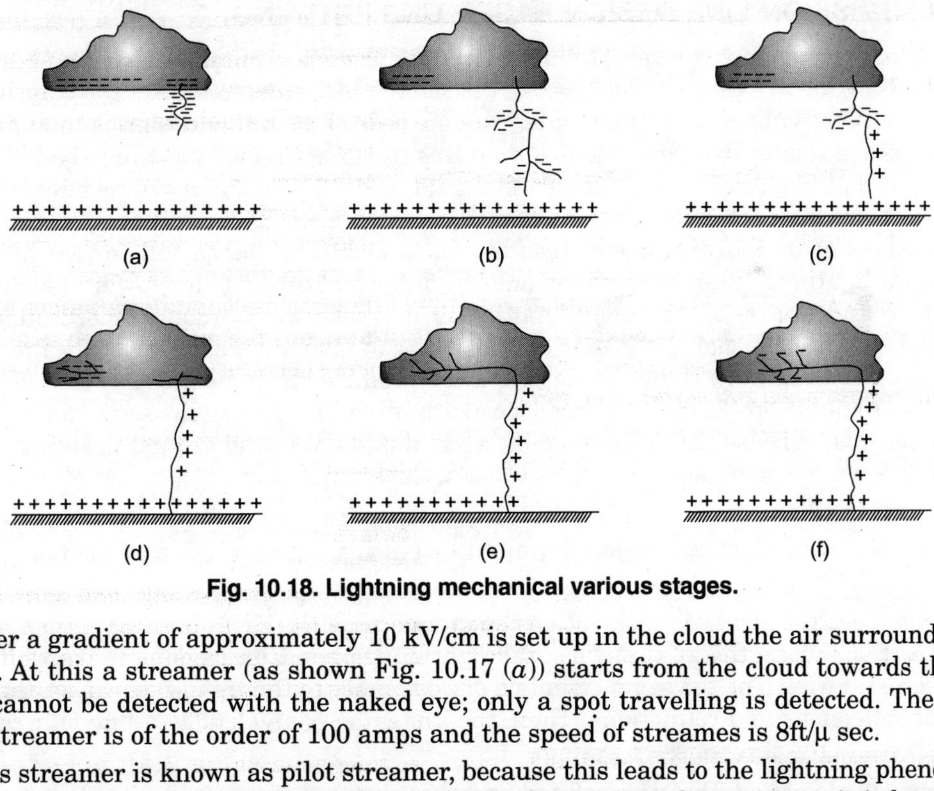

Fig. 10.18. Lightning mechanical various stages.

After a gradient of approximately 10 kV/cm is set up in the cloud the air surrounding gets ionised. At this a streamer (as shown Fig. 10.17 (*a*)) starts from the cloud towards the earth which cannot be detected with the naked eye; only a spot travelling is detected. The current in the streamer is of the order of 100 amps and the speed of streames is 8ft/μ sec.

This streamer is known as pilot streamer, because this leads to the lightning phenomenon. Depending upon the state of ionisation of the air surrounding the streamer, it is branched to several paths and this is known as stepped Ladder, as shown in Fig. 10.17 (*b*). The Laddre steps are of the order of 50 m in length and are accomplished in about a microsecond. The charge is brought from the cloud through the already ionised path to these pauses. The air surrounding these pauses, is again ionised and the leader in this way reaches the earth as seen in Fig. 10.17 (*e*) above.

In fact, once the stepped ladder has made contact with the earth it is believed that a power return strokes, shown in Fig. 10.17 (*c*), moves very fast up towards the cloud through the path already ionised by the leader (Fig. 10.17 (*b*)). This streamer is very intense where the current varies between 1000 amps and 200,000 Amps and the speed is about 10% that of light. It is here where the *–ve* charge of the cloud is being neutralised by the positive induced charge on the earth as shown in Fig. 10.17 (*d*). It is this very instant which gives air to lighning flash which we observe/sec. with out naked eyes. There may be another cell of charges in the cloud near the neutralised charged cell. This charged cell will try to neutralise through this ionised path. This streamer is known as dart ladder as shown in Fig. 10.17 (*e*). The velocity of the dart leader is about 3% of the velocity of light. The effect of the dart on ladder is much more severe than that of the return stroke.

In fact, the discharge current in the return streamer is relatively very large but as it last only for a few micro seconds the energy contained in the streamer is small and hence this streamer is known as cold lightning stroke whereas the dert leader is known as hot lightning stroke because even though the curent in this ladder is relatively smaller but it lasts for some milli seconds and therefore the energy contained in this leader is relatively larger. It is found that each thunder cloud may contain as many as 40 charged cells and a heavy lightning stroke may occur. This is known as multiple stroke.

10.8 TRANSMISSION LINE DESIGN BASED ON LIGHTNING

If comparison is made, the severity of switching surges for voltage 400 kV and above is much more than due to lightning voltages. All the same it is desired to protect the transmission lines against direct lightning strokes. Infact the purpose of good line design is to reduce the number of outage caused by lightning. To obtain this the following actions are to be taken : (1) the incidence of stroke on to power conductor should be minimised, and (2) the effect of these strokes which are incident on the system should be minimised. To obtain the first step stated in (1), we know that lightning normally falls on tall objects, hence tall towers are more vulnerable to lightning than the smaller one. In order to keep smaller towers height for a particular ground clearance, the span length will decrease which require more number of towers and hence the associated accessories like overhead insulators etc. Obviously, the overall cast will go up considerably. Hence, in design, a compromise has to be made so that adequate clearance to ground is provided, by giving special attention in reducing the sag of the overhead lines between two poles so that clearance between the line and the ground is kept at a safe values while keeping longer span and thus lesser number of towers.

In fact another step is, with a particular number of towers the chances of incidence of ligthning on power conductor, can be minimised by placing a ground wire at the top of tower structure.

However, once the stroke is incident on the ground wire the lightning current propagates in both directions along the ground wire, the lightning current propagates in both direction along the ground wire. The tower represents a discontinuity to the travelling waves, therefore they suffer reflections and refractions. Then, the system is equivalent to a line bifurcated at the tower point of the transmission system.

By now, we also know that the voltage and the current transmitted into the tower will depend upon the surge impadence of the tower and the ground impedance that is the tower footing resistant of the tower. Then say if it is low, the wave reflected back up the tower will largely remove the potential existing due to the incident wave. In this way the chance of flash over is eleminated. But, on the other hand, if the incident wave encounters a high ground impedence, positive reflection will take place and the potential on the top of the tower structure will be raised rather than lowered.

Hence, it desired that, for good transmission line design high surge impedences in the ground wire circuits, the tower structures and the tower footing should be avoided there are various methods for lowering the tower footing resistances as stated below. The standard value of this resistance acceptable is approximately 10 Ω (ohm) for 66 kV lines and this increases with the operating voltage. Say, for 400 kV line it is around 80 Ω (ohm). The tower footing resistance is the value of the footing resistance when measured at 50 Hz. The line performance with regard to lightning depends upon the impulse value of the resistance which is a function of the soil resistivity, critical breakdown gradient of the soil, length and type of driven grounds or counterpoises and the magnitude of the surge current. In case the construction of the tower does not give a suitable value of the footing resistance following methods are generally adopted: (1) One possitility could be the chemical treatment of the soil. This method is not practically possible for long length of lines. More over this method requires regular check up. Hence it is generally limited to the substation soils. However, the methods normally used for improving the grounds of transmission tower are the use of (1) ground rods and (2) counterpoises. Briefly these two methods are discussed, below.

1. Grounds Rods to Reduce the Tower Footing Resistance

Ground rods are put into the ground surrounding the tower structure. It must be stated here that the thickness of the rod does not play a major role in reducing the ground resistance as does the length of the rod. Hence, it is better to use thin but long rods or many small rods.

2. Counterpoise

A counterpoise is galvanised steel wire run in parallel or radial or a combination of the two, with respect to the overhead line.

Now one may ask, why we should have a low value of tower foot resistance. It is clear that, whenever a lightning strikes a power line, a current is injected into the power system the voltage to which the system will be raised depends upon what impedences the current encounters. Say if the lightning stroke strikes a tower, the potential of the tower will also be high which will result the flashover of the insulator dries and result in a line-to-ground fault. The flash over will take place from the tower structure to power conductors, and therefore it is known as back flashover surge absorbers are also used in the power system to stop damage of the system. Hence let us now discuss briefly about Surge Asorbers below :

Surge Absorbers

A surge absorber is a device which absorbs energy contained in a travelling wave. It is also known that corona is a means of absorbing energy in the form of corona loss. A short length cable between the equipment and the overhead line absorbs energy in the travelling wave because of its high capacitance and low inductance. Another method is the use of Farranti surge absorber which consists of an air core inductor connected in series with the line and surrounded by an earthed metallic sheet called a dissipator which is insulated from the inductor from the air.

In fact the surge absorber acts like an air cored transformer whose primary is the low inductance inductor, and the dissipator acts as the single turn short circuit secondary. Whenever the travelling wave is incident on the surge absorber a part of the energy contained in the wave is dissipated as heat due to transformer action and by eddy current. Because of the series inductance the steepness of the wave is also reduced. It is, in fact, claimed that the stress in the end turn is reduced by 15% with the helf of surge absorber.

SOLVED PROBLEMS

Example 10.1. *A surge voltage of 100 kV travelling in a line of natural impedance 600 Ω arrives at a junction with two line of impedances 800 Ω and 200 Ω respectively. Find the surge voltages and current transmitted into each branch line.*

Solution : The problem deals with a reflection at a *T*-joint the various natural impedences are : $Z_1 = 600\ \Omega_1\ Z_2 = 800\ \Omega_1\ Z_3 = 200\ \Omega$. The surge magnitude is 100 kV.

The surge as it reaches the joint suffers reflection and here the two lines are in parallel. Therefore, the transmitted voltage will have the same magnitude and is given the relation (developed above).

$$V'' = \frac{2V/Z_1}{\frac{1}{Z_1}+\frac{1}{Z_2}+\frac{1}{Z_3}} = \frac{(2\times100)/600}{\frac{1}{600}+\frac{1}{800}+\frac{1}{200}} = 42.04 \text{ kV}$$

$$\text{The transmitted current in line } Z_2 = \frac{42.04\times1000}{800} = \mathbf{52.55\ A}$$

$$\text{The transmitted current in } Z_1 = \frac{42.04\times1000}{200} = \mathbf{210.2\ A}$$

Example 10.2. *A surge of 15 kV magnitude travels along a cable towards its junction with an overhead line. The inductance and capacitance of the cable and overhead line are respectively 0.3 mH, 0.4 hF and 1.5 mH, 0.012 μF per km. Find the voltage rise at the junction due to the surge.*

Solution : This problem says the surge travels from the cable towards the overhead line and hence there will be positive voltage reflection at the junction.

$$\text{The natural impedance of the cable} = \sqrt{\frac{L_C}{C_L}} = \sqrt{\frac{0.3\times10^{-3}}{0.4\times10^{-6}}} = 27.38\Omega$$

$$\text{The natural impedance of the line} = \sqrt{\frac{L_C}{C_L}} = \sqrt{\frac{1.5\times10^{-3}}{0.012\times10^{-6}}} = 353\Omega$$

The voltage rise at the junction is the voltage transmitted into the overhead line as the voltage is zero before the surge reaches the junction.

$$V'' = \frac{2\times353\times15}{353\times27} = \frac{2\times353\times15}{380} = \mathbf{27.87\ kV}$$

Example 10.3. *A 500 kV, 2 μ sec rectangular surge on a line having a surge impedance of 350 ohms approaches a station at which the concentrated earth capacitance is 3000 pF. Determine the maximum value of the transmitted wave.*

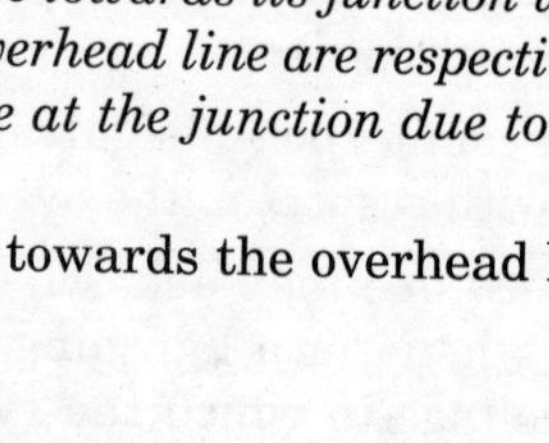

Fig. 1.

Solution : The maximum value of voltage will be,

$$V'' = 2V\left[1-\exp\left(\frac{T}{ZC}\right)\right]$$

$$= 2\times500\left[1-\exp\left(\frac{2\times10^{-6}\times10^{12}}{350\times300}\right)\right]$$

$$= 2 \times 500\,[1 - e^{-19}]$$
$$= 2 \times 500\,[1 - 0.15]$$
$$= 850 \text{ kV}$$

Example 10.4. *A 3-phase transmission line has conductor 1.5 cms in diameter spaced 1 metre apart in equilateral formation. The resistance and leakage reactance ('leakance') are negligible.*

Calculate (1) the natural impedance of the line, (2) the line current if a voltage wave of 11 kV travels along the line (3) the rate of energy, absorption the rate of reflection and the state and the form of reflection of the line is terminated through a star. Connected lead of 1000 ohm per phase, (4) the value of the terminating resistance for no reflection and (5) the amount of reflected and transmitted power if the line is connected to a cable extension with inductance and capacitance per phase per cm. of 0.5×10^{-8} H and 1×10^{-6} μF respectively.

Solution : From the above section on inductance we can write,

$$\text{The inductance per unit length} = 2 \times 10^{-7}\, ln \frac{d}{r} \text{ H/metre}$$

Now with the datas given in the example, we get inductance per unit length

$$= 2 \times 10^{-7}\, ln \frac{100}{0.75} = 2 \times 10^{-7} \ln 133.3$$

The capacitance per phase per unit length

$$L/m = 2 \times 10^{-7} \times 4.89 = 9.78 \times 10^{-7} \text{ H/metre}$$

$$C/m = \frac{2\pi\varepsilon}{ln\, d/r} \text{F/m} = \frac{2\pi \times 10^{-9}}{36\pi\, ln \frac{100}{0.75}} = \frac{2\pi \times 10^{-9}}{36\pi \times 4.89} = 1.136 \times 10^{-11}$$

$$\therefore \text{ The natural impedance} = \sqrt{\frac{L}{C}} = \sqrt{\frac{9.78 \times 10^{-7}}{1.136 \times 10^{-11}}} = \mathbf{294\,\Omega}$$

2. The line current $= \dfrac{11 \times 10^3 (= V)}{\sqrt{3} \times \text{ Natural Impedance of line}} = \dfrac{11{,}000}{\sqrt{3} \times 294} = \mathbf{21.6\,A.}$

3. As the terminating resistance is of higher value as compared to the value of the surge impedance of the line, the reflection is with a positive sign.

   The voltage across the terminating resistance, $E'' = \dfrac{2Z_2}{Z_1 + Z_2} \times E$, where Z_1 = line surge impedance, Z_2 = terminating impedance and E = the incident voltage.

$$\therefore \quad E'' = 2 \times \frac{11000}{\sqrt{3}} \times \frac{1000}{12.94} = 9.8 \text{ kV}$$

$$\therefore \text{ The rate of power consumption} = \frac{3 \cdot E''^2}{R} \text{ MW} = \frac{3 \times 98 \times 9.8}{1000} \times 1000 = 288 \text{ kW}$$

$$\text{The reflected voltage} \quad E' = \frac{Z_2 - Z_1}{Z_2 + Z_1} \times E = \frac{1000 - 2.94}{12.94} \times \frac{11}{\sqrt{3}} = 3.465 \text{ kV.}$$

∴ The rate of reflected energy $= \dfrac{3\times 3.465^2}{294}\times 1000 \text{ kW} = 121.8 \text{ kW}.$

4. So that the incident wave when reaches the terminating resistance, does not suffer reflection, the terminating resistance should be equal to the surge impedance of the line *i.e.*, **294 Ω.**

5. The surge impedance of the cable $= \sqrt{\dfrac{L}{C}} = \sqrt{\dfrac{0.5\times 10^{-8}}{10^{-12}}} = 770.7\,\Omega$

$$V_{\text{refracted}} = \text{The refracted voltage} = \frac{2\times 70.7}{294+70.7}\times\frac{11}{\sqrt{3}}\left(=\frac{2\times\text{surge impedance}}{Z_n+Z_{\text{surge}}}\right)$$

$$V_{\text{reflect}} = \text{The reflected voltage} = \frac{\text{Surge Impedance} - \text{Natural Impedance}}{Z_n+Z_{\text{surge}}}\times\frac{11}{\sqrt{3}}$$

∴ The refracted and reflected powers are respectively

$$= \frac{-223.3}{3647}\times\frac{11}{\sqrt{3}} = \mathbf{-3.9\,kW}$$

(*a*) $\dfrac{3\times V_{\text{refracted}}^2}{Z_{\text{surge}}}\times 1000 = 256 \text{ kW} = \text{The refracted power,}$

(*b*) $\dfrac{3\times V_{\text{reflect}}^2}{Z_{\text{natural}}}\times 1000 = \dfrac{3\times 3.9^2}{294}\times 1000 = \mathbf{155\,kW}$

Example 10.5. *A travelling wave 50 kV outers an overhead line of Surge impedance 400 (ohm) and conductor resistance 6 Ω per km. Determine the value of the voltage wave when it has travelled through a distance of 50 kW.*

Solution : Since the line has same specific resistance, the wave as it travels get attenuated in magnitude.

The magnitude of the wave is given by, $e = e_0 \cdot \in^{-1/2}(R/Z + GZ)\,x$, where e = the value of voltage when travelled through a distance of x kilometers; R, G the resistance and leakage reactance per kilometer length of the line and Z is the surge impedance e_0 = initial magnitude of the surge voltage, ∈ the Naperian base.

In this example, $e_0 = 50$ kV, $x = 50$ Km, $R = 6\,\Omega$, and $Z = 400\,\Omega$ and $G = 0.0\,℧$.

Substituting these values, in the expression,

$$e = e_0 \cdot \in^{-1/2}(R/Z + GZ)\cdot x$$

$$e = 50\cdot \in^{-1/2}\cdot\left(\frac{6}{400}\right)\times 50 = 50\times 0.69 = \mathbf{34.5\,kV}$$

QUESTIONS

1. State what is a travelling wave? Discuss on the development of such a wave on an overhead line.
2. Show that a travelling wave moves with a velocity of light on the overhead line and its speed is proportional to $1/\sqrt{\in_r}$ on a cable with dielectric material of paramillivittvity $\in_r$
3. Explain the variation of current and voltage on an overhead line when one end of the line is (1) short-circuited and (2) Open-circuited and at the other end a source of constant emf V is switched in.
4. What is arcing ground? Explain its effect on the performance of a power system.
5. What is "Capacitance Switching"? Explain its effect on the performance of the circuit breaker.
6. Discuss the mechanism of lightning stroke.
7. Discuss the phenomenon of lightning affecting the performance of overhead lines in power system.
8. Derive the expression for the surge voltage in the line having surge impedances at a *T*-junction of the line.
9. State what is meant by transients in transmission lines. Also state what are the transients components, and causes of transient in transmission lines.
10. State why a short length of cable is connected between the dead end tower and the terminal apparatus in a station.
11. An overhead line with surge impedance 450 ohms, bifurcates into two lines of impendences 400 ohms and 40 ohms respectively. If a surge of 20 kV is incident on the overhead line, calculate the magnitudes of voltages and current entering the bifurcated lines.
12. State what is meant by capacitance Switching Discuss its effect on the performance of the circuit breaker.

CHAPTER

11

Power System Stability

11.1 WHAT IS POWER SYSTEM STABILITY

INSIDE THIS CHAPTER

- What is power system stability
- Power system network & stability
- The power flow in power network
- Transient Stability & swing equation
- Swing Equation
- Multi Machine systems swing equation
- Steady state & transient stability steps
- Equal area criterion
- Transient disturbances
- Various methods to improve system stability
- Solution of swing equation (mathematical calculation)
- Algorithm of derivation in solution at swing equation
- Numerical techniques to solve swing equation (Euler's method, Rungekutta techniques)
- Solution of swing equation (by computation)
- Voltage stability in the power system
- Voltage control in long radial lines
- Effect of grounding on stability transmission system
- Power transfer capability of transmission lines
- Complex operating power contour in a power system

In an interconnected power system the stability of the system is its ability to return to normal or stable operation after having been subjected to some form of disturbance. In fact, instability means a condition denoting loss of synchronism or falling out of step. Stability consideration have been recognised as an essential part of power system planning for a long time. With interconnected systems continually growing in size and extending over vast geographical regions, it is becoming increasingly more difficult to maintain synchronism between various parts of power system. However, the dynamics of a power system are characterised by its basic features as stated below :

1. The system is basically a spring inertia oscillatory system with inertia on the mechanical side and spring action provided by the synchronous tie wherein power transfer is proportional to sin δ (for small δ). δ is the relative internal angle of machines.
2. Synchronous tie exhibits the typical behaviour that a power transfer is gradually increased, a maximum limit is reached beyond which the system cannot stay in synchronism, that means, it falls out of step.

3. As power transfer is being proportional to sin δ, the equation determining system dynamic is nonlinear for disturbances causing large variations in angle δ. Stability phenomenon peculiar to nonlinear system as distinguished from linear systems is therefore exhibited by power system.
4. Power systems are stable upto a certain magnitude of disturbance and unstable for larger disturbances.
5. Accordingly power system stability problems are classified into three basic types like: (*a*) Steady state, (*b*) Dynamic and (*c*) Transient. Briefly these three types are defined or explained below :
   (*a*) Steady state stability is basically concerned with the determination of the upper limit of machine loadings before losing synchronism, provided the loading is increased gradually.
   (*b*) Dynamic instability in power system is more probable than steady state instability which is a system stability problem. Infact small disturbance do continue to occur in power system due to variation in loading, changes in turbine speed, etc. which are though small enough not to cause the system to lose synchronism but do excite the system into the state of natural oscillations. The system is said to be dynamically stable if the oscillation do not acquire more than certain amplitude and die out quickly which means the system is well-damped. In a dynamically unstable system, the oscillation amplitude is large and these persist for a long time, that means, this system is underdamped. This kind of instability behaviour constitutes a serious threat to system security and creates very difficult operating condition.

   Dynamic stability can be significantly improved through the use of power system stabilizers. Computer simulation is the only effective means of studying dynamic stability problem. The same simulation programmes are of course, applicable to transient stability studies.
   (*c*) Transient stability is a very fast phenomenon usually occurring within 1 sec for a generator close to the cause of disturbance. In the previous chapter (chapter 10), we have seen that following a sudden disturbance on a power system rotor speeds, rotor angular differences and power transfer undergo fast changes where magnitudes are dependent upon severity of disturbance. For a large disturbance, changes in angular differences may be so large as to cause the machines to fall out of to step, which is known as the transient instability in power system. Transient instability on a heavily loaded line subjected to a fault requires opening the line to clear the fault is usually of greatest concern. The tripping of a loaded generator on the abrupt droping of a large load may also cause instability. Moreover, the effect of faults like short circuit, the most severe type of disturbances to which a power system is subjected, must be determined in nearly all stability studies. In fact during a fault, electrical power from nearby generators is reduced drastically, while power from remote generators is scarcely affected.

However, in some cases, the system may be stable even with a sustained fault, whereas other systems will be stable only if the fault is cleared with sufficient rapidity. Whether the system is stable on occurrence of a fault depends not only on the system itself but also on the type of fault, location of fault, rapidity of clearing and method of clearing the fault. This means whether the fault is cleared by the sequential opening of two or more breakers or by simultaneous opening and whether or not faulted line is reclosed. The transient stability is almost always lower than the state stall limit, but unlike the latter, it may exhibit different values depending on the nature, location and magnitude of disturbance.

11.2 POWER SYSTEM NETWORK & STABILITY

Power systems complete arrangement have many interconnected generating stations, each with several generators, transmission and distribution system connected to many loads of various types. The Figs. 11.1 & 11.2 shows single line diagrams of a typical power system network. The Fig. 11.1 shows a loop or mesh network and Fig. 11.2 shows a radial network. A power system network, as can be seen in the figures, consist of Generators, Transformers, circuit breakers, transmission lines, loads. Transmission network commonly is termed as grid.

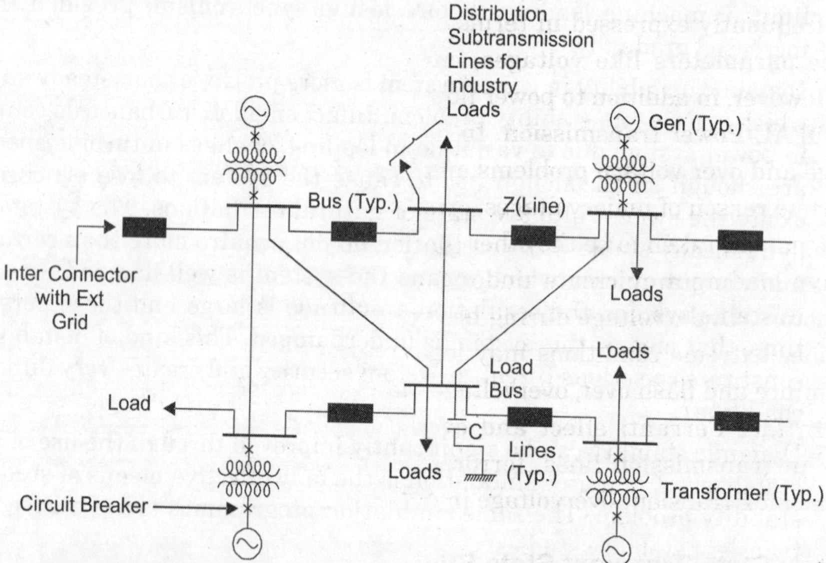

Fig. 11.1. Line diagram of a loop/mesh network of a power system.

A power system is divided into several subsystems based upon operating voltage level. Highest on the voltage level is termed as transmission system/network (or grid), and distribution network voltage levels in descending order to the 11 kV and as low as 440 V (3-phase), 230 V/110 V (1-phase) etc.

The short circuit capacity (Fault Level) is comparatively higher in the loop/mesh network of Fig. 11.1 than the radial network of Fig. 11.2.

However, power system network may vary widely in different utility areas and to meet the load requirement. In fact the voltage level in the transmission as well as in distribution vary within a range. This impairs the uniformity of the nodal

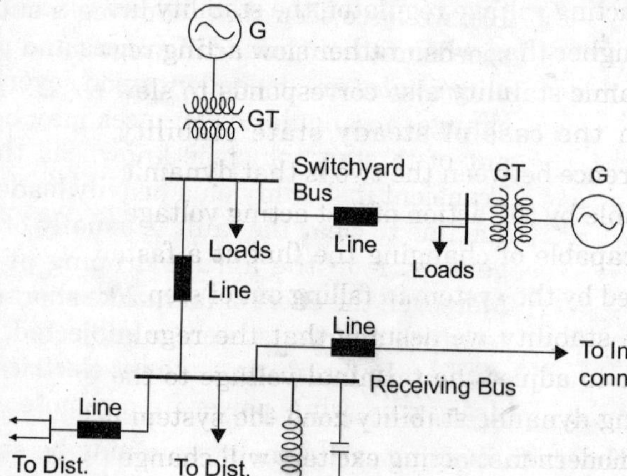

Fig. 11.2. Line diagram of a radial network of a power system.

strength at different place in the subsystem. In most of the power network systems a distribution (sub-transmission) zone usually is provided to distribute bulk power to large industries at a voltage level higher than the distribution voltage level.

Power transmission system is, though, capable of running steadily at or near the steady state, stability limit, still practical power system keeps a margin of power and operates well below this limit to allow for disturbances such as load variation, and faults like short circuit etc. Though this margin is principally determined by the transient stability criterion, another concept known as dynamic stability is useful in determining this margin. It is the capability of the system if it recovers normal operation following a specific minor disturbance. Dynamic stability is frequently expressed in terms of rate of damping of the transient components of performance parameters like voltages, current, load or torque angle of the synchronous machines. However, in addition to power flow stability, voltage stability is also a major criterion for successful AC Power transmission. In fact transmission lines are frequently subjected to undervoltage and over voltage problems and the performance is then affected. Heavy loading being the prime reason of undervoltages, open conductor, high-resistance fault, starting of very large motors may be taken to be the other sources of undervoltages. In reactive power constrain system heavy loading may create undervoltage degrading particularly the performance of induction motors. Undervoltage during heavy loading condition also enhances series reactive loss and under extreme conditions may lead to voltage instability. But owing to the risk of insulation failure and flash over, overvoltage condition is also treated as a dangerous operating condition. In fact Ferranti effect and over compensation being the consumer causes of overvoltage in transmission lines, ferroresonance, switching, faults and lightning arc the conditions causing transient overvoltage in transmission line.

11.2.1 Steady State, Transient State Stability & Dynamic Stability

When synchroneous machines are operated alongwith fast acting voltage regulator, the stability limits of the system are higher than when rather slow acting regulators are used. Dynamic stability also corresponds to slow changes in load as in the case of steady state stability, but the major difference between the two is that dynamic stability is made possible by the action of fast acting voltage regulators which are capable of changing the flux at a faster rate than that caused by the system in falling out of step. Whereas in steady state stability we assume that the regulator act slowly in order to adjust the terminal voltage to the prescribed value. However, it is to be noted that during dynamic stability zone the system does not operation a single power angle curve, but the modern fast-acting exciters will change the operating curve during the period under study. A typical curve is shown in Fig. 11.3. The power systems are usually not designed to operate in the region of dynamic stability as absolute dependance on voltage regulator performance has not been considered useful. However, during disturbance and under emergency conditions, the power system can be operated for realising additional transfer of power by operating in the dynamic stability.

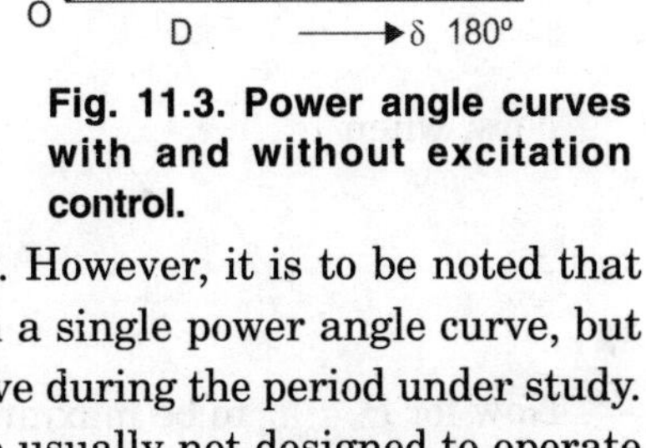

Fig. 11.3. Power angle curves with and without excitation control.

11.3 THE POWER FLOW IN POWER NETWORK

Before we consider the study of stability in power system, let us first consider the power flow calculation in a simple, 1-phase transmission line. Hence for this purpose let us consider a simple 1-phase line in the Fig. 11.4 in its simplest form. In calculation, all the quantities have been expressed in polar form. Here the current $I \angle \phi$ flowing through the transmission line of impedance $Z \angle \theta$ and the sending end voltage is $V_1 \angle \delta$ and receiving and voltage $V_2 \angle 0°$.

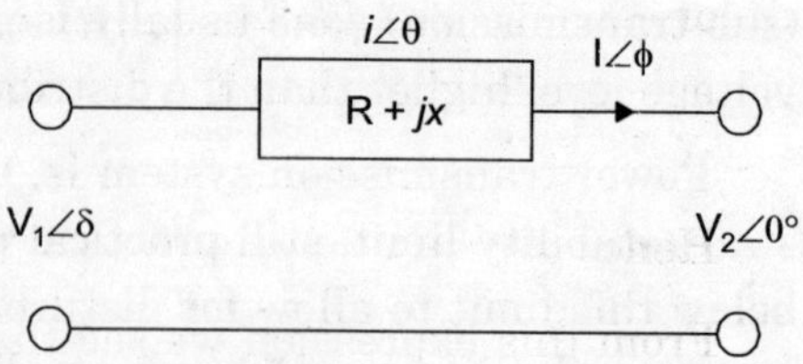

Fig. 11.4. A 1-phase transmission line.

Hence the current, $$I = \frac{V_1 \angle \delta - V_2 \angle 0}{Z \angle \theta} = \frac{V_1}{Z} \angle(\delta - \theta) - \frac{V_2}{Z} \angle -\theta$$

Hence power received, P_2, at the receiving end, is

$$P_2 = R_e\left[V_2 \cdot I\right] = R_e\left\{V_2\left[\frac{V_1}{Z} \angle \delta - \theta) - \frac{V_2}{Z} \angle -\theta\right]\right\}$$

$$= \frac{V_1 V_2}{Z} \cos(\theta - \delta) - \frac{V_2^2}{Z} \cdot \cos\theta \qquad \text{...(11.1)}$$

Now let, $\theta = 90 - \alpha$

(where α is a function of the impedance of the line, Z)

where $Z = R + jX$

$$\therefore \quad P_2 = \frac{V_1 V_2}{Z} \cos(90 - \alpha - \delta) - \frac{V_2^2}{Z} \cos(90 - \alpha) \qquad \text{...(11.2)}$$

Therefore, the power P_2 received is maximum when $\alpha + \delta = 90°$ or, $\delta = (90 - \alpha)$ and the value is therefore given by

$$P_{2\,(\max)} = \frac{V_1 V_2}{Z} - \frac{V_2^2}{Z} \cdot \sin\alpha \qquad \text{...(11.3)}$$

where $$\sin \propto = \frac{R}{Z}$$

$$\therefore \quad P_{2\,(\max)} = \frac{V_1 V_2}{\sqrt{R^2 + X^2}} - \frac{V_2^2}{\sqrt{R^2 + X^2}} \frac{R}{\sqrt{R^2 + X^2}}$$

Now, when $V_1 = V_2$

$$\therefore \quad P_{2\,(\max)} = V_2^2\left[\frac{1}{\sqrt{R^2 + X^2}} - \frac{R}{R^2 + X^2}\right] \qquad \text{...(11.4)}$$

Now for $P_{2\,(\max)}$ to be maximum, $\dfrac{dP_{2(\max)}}{dx} = 0.$

$$\therefore \quad \frac{dP_{2(\max)}}{dx} = 0 = V_2^2\left[\frac{X}{(R^2 + X^2)^{3/2}} - \frac{2XR}{(R^2 + X^2)^2}\right]$$

or $$\frac{V_2^2 X}{(R^2+X^2)^2}\left[\sqrt{(R^2+X^2)}-2R\right] = 0$$

$\therefore$ $$R^2 + X^2 = 4R^2$$

Hence $$X = (\sqrt{3})R = R\sqrt{3} \qquad ...(11.4)$$

From this expression we see that the maximum power can be transferred from one end (1) to end (2) when the reactance of the line is $\sqrt{3}$ times its resistance. Generally, the reactance is quite large as compared to the resistance. The equation shows that it is not necessarily desirable to compensate by series capacitance for all the reactance. Also it is clear that the power can be transferred only if reactance is present. Now in case reactance is zero power cannot be transmitted.

Then for a lossless line, we know, $R = 0$ and hence the power transmitted, $P_2 = \frac{V_1 V_2}{X}$ sin δ. From the expression the transmitted power depends upon the system reactance and the angle between the two buses. The curve between P_2 and δ is known as the power angle curve and is shown in the Fig. 11.5.

Then at δ = 90°, in the Power/Power Angle Curve the maximum Power, P_{max} is given by

$$P_m = \frac{V_1 V_2}{X} \qquad ...(11.6)$$

$\therefore$ P_m for given V_1, V_2 and X occurs at an angle 90°.

The torque angle 'δ' is positive for generator action and negative for motor action. In case of generator action the rotor advances in the direction of rotation, whereas for motor action, the rotor rotates or falls back opposite to the direction of rotation.

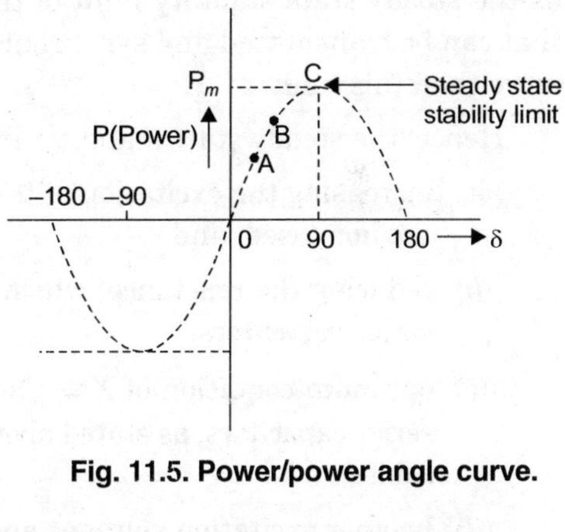

Fig. 11.5. Power/power angle curve.

We should here state that operation of the power grid at steady state is one of the most fundamental requirement of proper operation of a power system. Slight insight into the operational pattern suggests that the steady state operation of the power network is principally governed by the system voltages at the sending and receiving end, the transfer reactance of the line, and the power angle between the two buses. But operational aspects vary depending on the nature of power transmitting network. Hence we have done some analysis on the power related to the power angle and two end voltages and the system reactance. We have seen in our analysis that in the eqn. (11.6). The maximum value of power transmitted can be varied by varying V_1, V_2 and X, the critical Reactance of the Line, at a power angle δ of 90° as shown in Fig. 11.5.

Condition of Stability

Now to analyse further, we see that the power system is stable if, and only if, for an increase in rotor angle δ, the transmitted power also increases. Which mean that ($dp/d\delta$), the rate of increase of power with respect to the increase of δ should be positive. We can see from the

Fig. 11.5 that the angle where $dp/d\delta$ is positive lies between +90° and –90°. But when the tie-line impedance is purely capacitive (negative reactance). The range of angle for delivering power to the system is for 180° to 270° instead of from 0 to 90°.

We also see from the Fig. 11.5, at zero degree with inductive reactance, the power transmitted is zero. Whereas at 180° with capacitive reactance even though the power to be transmitted is zero, but a large wattless current will flow which is not desirable and, therefore, normally over compensation of line (by using series capacitor) is never done.

Now let us study the power angle curve in detail to understand better the stability. Now let P be the mechanical power input to the generator and the mechanical output for the motor assuming negligible frictional and transmission losses say initially this power corresponds to point, A on the power angle curve. If a small increment of shaft load is added to the motor, the output power of the motor increases as the speed does not change monentraily, whereas the input to the motor remain unchanged. Hence, there is a net torque or the motor tending to retard it and its speed decreases temporarily. As a result of reduction in motor speed the rotor angle δ increases and consequently the power input to the motor increases until finally the input and output are again in equilibrium and steady operation takes place at a new point B higher than A on the power angle curve. The gradual addition of load on the motor shaft is possible till the point C is reached on the power angle curve where $P = P_{max}$ and any further addition of load will result in increase in angle δ but reduction in input power to the motor and hence, the motor will decelerate further and it will pull out of step and will probably stall unless it has damper winding which may keep it running as an induction motor. P_m is known as the steady state stability limit of the system which means that it is the maximum power that can be transmitted and synchronism will be lost if an attempt is made to transmit power more than this limit.

Hence, the steady state stability limit can be increased by :

(*a*) increasing the excitation of the motor or generator, or both so that the internal e.m.f., are increased, and

(*b*) reducing the reactance, which is done by either running parallel lines or by using the series capacitors.

(*c*) optimum condition of $X = \sqrt{3} \cdot R$ for maximum power transfer is approached by the series capacitors, as stated above for overhead lines and series reactors for underground cables.

(*d*) heigher excitation voltages and quick response excitation system.

Remarks on Stability in Power System

The value of P_{max} found by the eqn. (11.6) *i.e.*, $P_{max} = \dfrac{V_1 \cdot V_2}{X} = \dfrac{V_S \cdot V_R}{X}$ (where $V_1 = V_S$ = Sending and Voltage, $V_2 = V_R$ = Receiving and Voltage) can be trasmitted over the line provided the load increases are made very gradually and is called steady state limit. Sometimes the load changes are sudden may be due to a fault or switching operation. The stability of a power system under these conditions is called transient stability and the maximum load that the line can transmit under these condition is called transient state stability limit.

It is necessary to have same knowledge of steady state stability and stability limit defined by IEEE as below.

Steady State Stability

When used with reference to a power system is that attributed of the system or part of the system which enables it to develop restoring forces between the element thereof equal to or greater than the disturbing forces so as to restore a state of equilibrium between the element.

Stability Limit

Stability limit is the maximum power flow possible through some particular point in the system when the entire system or part of the system to which the stability limit refers is operating with stability.

Hence it is clear that a power system is stable under certain specified loading and circuit conditions if or being disturbed by some arbitrary forces it tends to come back to original operating position.

In fact, an actual power system would consist of a large number of machines located at different point but multi-machine stability problem are too involved to be discussed level. For such system calculations are in actual practice carried out with computers.

Here it must be stated that the active power P remains affected with the inclusion of the shunt susceptance. Hence the concept of steady state stability limit does not alter with change of line length : Short, Medium or Long line.

It must be also stated that, the equation (11.6) for is an approximate formula for calculating the maximum steady state power since we neglected both Resistance and shunt admitance of the line. Omission of resistance gives a somewhat optimistic result for maximum steady state power limit. In transient stability calculation, however, the result obtained by neglecting resistance are pessimistic since it means damping is neglected. Neglecting shunt capacitance also gives persimistic results. But it must be said that still the eqn. (11.6) is approximately accepted for calculating the maximum study power. But considering Resistance, P_{max} of eqn. (11.6) becomes = $\dfrac{V_1 V_2}{Z} - \dfrac{R \times V_2^2}{Z^2}$, which means deducting I^2R.

Example 11.1. *The sending end and the receiving end voltages of a transmission line at 100 MW load are equal at 132 kV. Per phase line impedance is (4 + j6) ohm. Calculate the maximum steady state power that can be transmitted over the line.*

Solution :

$$V_1 (= V_S) = \frac{132 \times 10^3}{\sqrt{3}} = 76.213 \text{ volts} = 76$$

From

$$Z = 4 + j\,6, R = 4$$

$$V_1 = V_2 = \mathbf{76.213\ kV}$$

We know, consider the resistance of the line

$$P_{max} = \frac{V_1 V_2}{Z} - \frac{RV_2}{Z^2}$$

$$= \frac{(76.213)}{\sqrt{4^2 + 6^2}} - \frac{4 \times (76.213)^2}{(4^2 + 6^2)} \text{ MW/ph.}$$

$$= 358.68 \text{ MW/ph}$$

$$= 107.6 \text{ MW}$$

Example 11.2. *A power system has been shown in Fig. 11.6. Compare the magnitude of maximum power transfer if (a) a shunt inductor is connected at bus (2) and (b) a shunt capacitor is connected at the bus (2).*

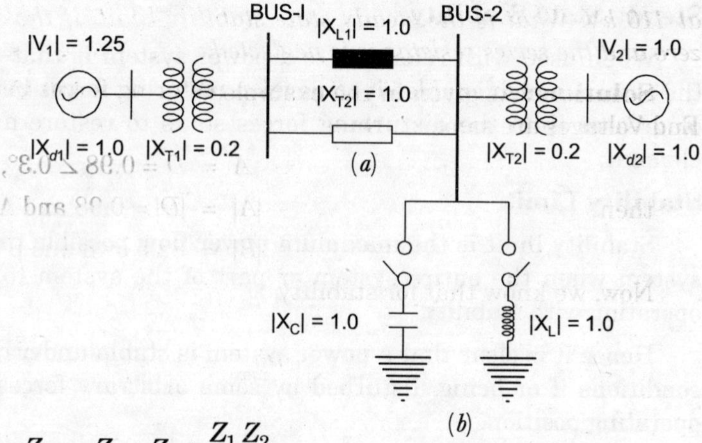

Fig. 11.6

Solution : As per example and reference to Fig. 11.6, (*a*) with inductor (shunt) at *B* 2.

Transfer reactance between the two generating sources is

$$Z = Z_1 + Z_2 + \frac{Z_1 Z_2}{Z_3}$$

where
$$|Z_1| = |X_{d_1}| + |X_{T_1}| + \left|\frac{X_{L_1}}{2}\right| = 1 + 0.2 + \frac{1.0}{2} = 1.7 \text{ p.u.}$$

$$|Z_2| = |X_{d_2}| + |X_{T_2}| = 0.2 + 1.0 = 1.2 \text{ p.u.}$$

$$|Z_3| = |X_L| = 1.0 \text{ p.u.}$$

$$\therefore \quad Z = Z_1 + Z_2 + \frac{Z_1 Z_2}{Z_3} = 4.94 \text{ p.u.}$$

∴ Maximum power transfer under this condition is

$$P_{\max} = \frac{|V_1| \ |V_2|}{|Z|} = \frac{1.25 \times 1.0}{4.94} \text{ p.u.} = 0.253 \text{ p.u.}$$

(*b*) With capacitor (shorted) at *B* 2

Again, transfer reactance between the two generating sources becomes :

$$Z = Z_1 + Z_2 + \frac{Z_1 Z_2}{Z_3}$$

As, $|Z_1|$, $|Z_2|$ remain same as in (*a*), but $|Z_3| = -|X_c| = -1.0$ p.u.

$$\therefore \quad Z = \cos Z_1 + Z_2 + \frac{Z_1 Z_2}{Z_3} = 0.86 \text{ p.u.}$$

∴ Maximum power transfer under this condition is

$$P_{\max} = \frac{|V_1| \ |V_2|}{|Z|} = \frac{1.25 \times 1.0}{0.86} = 1.45 \text{ p.u.}$$

Here it may be observed that the connection of the inductor at BUS-2 reduces the power transfer limit while the capacitor connection improves the power limit. However, inductor connection is needed in order to control the overvoltage at the receiving end when the line is at no load or light loaded.

Example 11.3. *The generalised circuit constants of a nominal π-circuit representing a 3-phase transmission line are : A = D = 0.98 ∠ 0.30°, B = 82.5 ∠ 76° Ω C = 0.0005 ∠ 90° mho. Find the steading state stability limit of the line if the two terminal volages are held constant* *at 110 kV. What is the steady state stability limit, if the shunt admittance is assumed to be zero and the series resistance is neglected.*

Solution : As given in the example, $V_1 = V_2 = 110$ kV (that is sending End & Receiving End Voltages are same)

$$A = D = 0.98 \angle 0.3°,$$

then,
$$|A| = |D| = 0.98 \text{ and } \Delta = 0.3° = \alpha$$
$$|B| = 82.5 \text{ own and } \beta = 76°$$

Now, we know that for stability,

$$P_{max} = \frac{|V_1|\,|V_2|}{|B|} - \frac{|A|\ |V_2|^2}{|B|} \cos(\beta - \alpha)$$

∴ Here
$$P_{max} = \frac{110 \times 110}{82.5} - \frac{0.98 \times 110^2}{82.5} \cdot \cos 75.7° = 111.29 \text{ MW}$$

Now if the shunt admittance is assumed to be zero,

$$Y = 0, \therefore = 1.0 \angle 0. \text{ That is, } \alpha = 0$$

∴
$$P_{max} = \frac{110 \times 110}{82.5} - \frac{110 \times 110}{82.5} \cdot \cos 76° = 111.18 \text{ MW}$$

Now, if the shunt admittance is assumed to be zero as well as the series resistance is neglected

$$Y = 0, R = 0, A = 1 \angle 0°$$

and
$$B = |Z| \sin \beta = 82.5 \sin 76° = 80 \text{ (m)}$$

(as in impedance triangle, $X = Z \sin \theta$ i F $Z = |Z| \angle \theta°$)

However, if only X is present (that is the resistance), then β becomes 90°

∴
$$P_{max} = \frac{110 \times 110}{80} - \frac{(110)^2}{80} \cos 90° = \mathbf{151.24\ MW}.$$

Example 11.4. *Deduce an expression for the maximum steady state power which can be transmitted over a line (neglecting capacitance of the line) if the voltage at each end is kept constant. Show that if the reactance X of the line be varied, the resistance R remaining constant the maximum steady state power that could be transmitted over the line would be greatest when X =* $\sqrt{3}\,R$.

Solution : In the Fig 11.7, V_R and V_S are receving end and Sending End voltages respectively. Since the capacitance is neglected, the line current, and end and Receiving end currents are the same, that is I which is lagging by an angle ϕ_r from V_R and ϕ_S from V_S. Then IR and IX are the component voltage drops due the Resistance R and Reactance X of the impedance Z of the line when a current I flows through it. Now from the phasor diagram of Fig. 11.7, we can write,

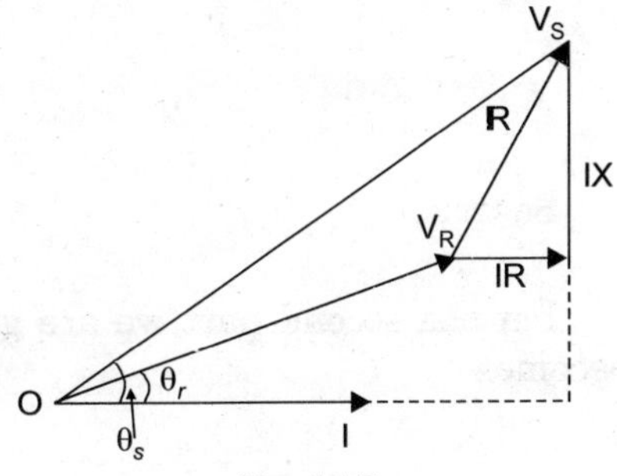

Fig. 11.7

$$V_S \cos \phi_S = V_R \cos \phi_r + IR$$

and
$$V_S \sin \phi_S = V_R \sin \phi_r + IX$$

Now squaring and adding these two equations, we get,

$$V_S^2 = V_R^2 + 2V_R I(R\cos\phi_r + X\sin\phi_r) + I^2(R^2+X^2) \qquad \text{...(11.5)}$$

Now writing, $V_R I\cos\phi_r = P$ = Power received, and $V_R I\sin\phi_r = Q_r$ = Reactive Volt – Amps. in eqn. (11.5) we get,

$$V_S^2 = V_R^2 + 2PR + 2QX + \frac{(P^2+Q^2)(R^2+X^2)}{V_R^2}$$

or rearranging, $$-V_S^2 + V_R^2 + 2PR + 2QX + \frac{(P^2+Q^2)(R^2+X^2)}{V_R^2} = 0 \qquad \text{...(11.6)}$$

(Here in this equation. V_S, V_R, R and X are constants, P and Q are the only two variable)

Now we have to determine the maximum value of P, when Q is varied. To do so by differentiating the eqn. (11.6) with respect to Q and then equating dP/dQ to zero for maximum value of P, we get

$$2R\frac{dP}{dQ} + 2X + \left(\frac{R^2+X^2}{V_R^2}\right)\left(2P\cdot\frac{dP}{dQ} + 2Q\right) = 0$$

So that, $$2X + 2Q\frac{(R^2+X^2)}{V_R^2} = 0, \text{ or } Q = \frac{V_R^2 X}{R^2+X^2}$$

Now substituting the value Q in eqn. (11.4.2) we get

$$-V_S^2 + V_R^2 + 2PR - \frac{2V_R^2X^2}{R^2+X^2} + \frac{R^2+X^2}{V_R^2}\left[P^2 + \frac{V_R^4X^2}{(R^2+X^2)^2}\right] = 0$$

This being a quadrating equation in P, multiplying throughout by $\dfrac{V_R^2}{R^2+X^2}$ and putting $R^2 + X^2 = Z^2$, we get

$$P^2 + 2PR\frac{V_R^2}{Z^2} + \left(V_R^2 - V_S^2\right)\frac{V_R^2}{Z^2} - \frac{V_R^4\cdot X^2}{Z^4} = 0$$

$$\therefore \qquad P = -\frac{RV_R^2}{Z^2} \pm \sqrt{R^2\cdot\frac{V_R^4}{Z^4} + \frac{V_S^2V_R^2}{Z^2} - \frac{V_R^4R^2}{Z^4}}$$

$$= -\frac{RV_R^2}{Z^2} \pm \frac{V_SV_R}{Z} = \frac{V_R^2}{Z^2}\left(-R \pm \frac{V_S}{V_R}\cdot Z\right)$$

So that, $$P_{\max} = \frac{V_R^2}{Z^2}\left[\frac{V_S}{V_R}Z - R\right]$$

For the second part we are given that $V_S = R_R$, so that expression of $P_{\max}$ in this care becomes

$$P_{\max} = \frac{V_R^2}{Z^2}(Z - R)$$

Now, we want to keep R fixed varying X so that P_{max} is greatest this means that $\frac{dP_{max}}{dX}$ should be zero for this condition,

hence,

$$\frac{dP_{max}}{dX} = \frac{d}{dX}\left[\frac{V_R^2}{Z^2}(Z-R)\right] = 0$$

or

$$\frac{d}{dX}\left[\frac{V_R^2}{R^2+X^2}\cdot\sqrt{(R^2+X^2-R)}\right] = 0, \text{ or } \frac{d}{dX}\left[\frac{V_R^2}{\sqrt{(R^2+X^2)}} - \frac{RV_R^2}{(R^2+X^2)}\right] = 0$$

or

$$-\frac{V_R^2\cdot 2X}{2(R^2+X^2)^{3/2}} + \frac{RV_R^2 2X}{(R^2+X^2)^2} = 0, \text{ or } \frac{R}{\sqrt{R^2+X^2}} = \frac{1}{2}.$$

From this, we get $X = \sqrt{3}.R$ **proved**

Example 11.5. *A 3-phase overhead transmission line in which the line voltage at such end is maintained constant at 33 kV has a resistance per phase of 6 (ohm) and a reactance per phase of 15 (ohm). Deduce an expression for the power delivered over the line in terms of the angle between the sending and receiving end voltages plot to scale a curve to a base of angles and determine graphically or otherwise the maximum power this line is capable of transmitting under these condition. Also show that with equal voltage at both ends no power can be transmitted without change of phase between sending and receiving end voltages.*

Solution : In the Fig. 11.8, the phasor diagram for the condition stated in the example.

Phasor diagram for the conditions is shown in Fig. 11.8.

Let $OA = V_R$, the receiving end voltage be a fixed phasor. Then the locus of sending end voltage V_S is a circle of radius OB and centre O.

Let δ be the angle of phase shift between V_R and V_S for conditions considered, AB representing the voltage drop in the line.

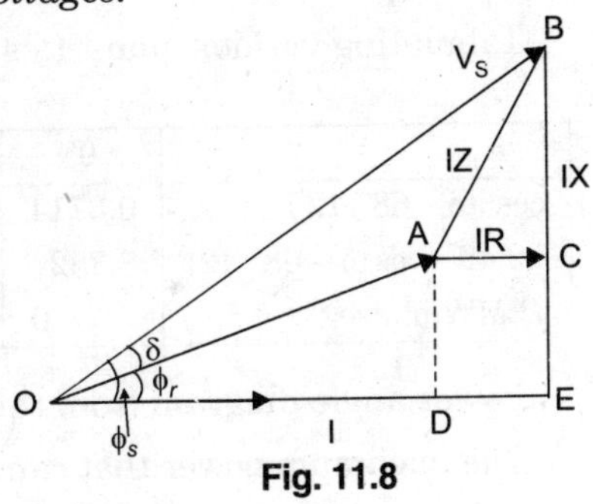

Fig. 11.8

Resolving V_S along OA, $V_S \cos\delta - V_R = IR\cos\phi_r + IX\sin\phi_r = I_aR + I_bX$ ...(11.7)

I_a and I_b being the power and wattless components of load current I.

Similarly resolving V_S along a direction perpendicular to OA

$$V_S \sin\delta = IX\cos\phi_r - IR\sin\phi_r = I_aX - I_bR \quad ...(11.8)$$

Multiplication of (*i*) and (*ii*) by R and X respectively gives

$$V_SR\cos\delta - V_RR = I_aR^2 + I_bRX \quad ...(11.9)$$

and

$$V_SX\sin\delta = I_bX^2 - I_bRX \quad ...(11.16)$$

Adding (*iii*) and (*iv*), $V_S(R\cos\delta + X\sin\delta) - V_RR = I_a(R^2+X^2) = I_aZ^2$

Power received = $V_R \cdot I_a = P$

$\therefore P = (V_R|Z^2)[V_S(R\cos\delta + X\sin\delta) - V_RR] = [V_R/Z_2][V_SZ\cos(\beta-\delta) - V_RR]$ ...(11.11)

where

$$\beta = \tan^{-1}\frac{X}{R}$$

Eq. (v) then is the equation for power delivered in terms of V_S, V_R and δ.

Here R = 6 ohms and X = 15 ohms $\qquad V_S = \dfrac{V_R = 33{,}000}{\sqrt{3}}$ volts

$$\beta = \tan^{-1}\frac{15}{6} = \tan^{-1} 2.5 = 68°12' \qquad Z = \sqrt{6^2 + 15^2} = 16.16 \text{ ohms}$$

The following figures are worked out :

$$V_R/Z^2 = \frac{33 \cdot 10^3}{\sqrt{3} \cdot 261} = 0.1265 \cdot \frac{10^3}{\sqrt{3}}$$

$$V_S Z = \frac{33 \cdot 10^3}{\sqrt{3}} \cdot (16.16) = 533.2 \cdot \left(\frac{10^3}{\sqrt{3}}\right)$$

$$V_R R = \frac{33 \cdot 10^3}{\sqrt{3}} \cdot 6 = 198 \cdot \frac{10}{\sqrt{3}}$$

Substituting these values in Eq. (v), we get

$$P = 0.1265 \cdot \left(\frac{10^3}{\sqrt{3}}\right)\left[(533.2)\frac{10^3}{\sqrt{3}} \cos(\delta - 68°12') - 198\left(\frac{10^3}{\sqrt{3}}\right)\right] \text{ watts/phase}$$

$$= \frac{0.1265}{3}[533.2 \cos(\delta - 68°\,12') - 198] = 22.49 \cos(\delta - 68°\,12') - 8.352 \text{ MW/phase}$$

Tabulating various quantities

δ	0°	30°	60°	68° 12′	90°	120°	156° 24′
cos (δ – 68° 12′)	0.3714	0.7858	0.9898	1	0.9285	0.6064	0.3714
22.49 × cos (δ – 68° 12)	8.352	17.67	22.26	22.49	20.87	13.59	8.352
P MW/phase	0	9.32	13.91	14.14	12.52	5.24	0

Power angle diagram from these readings is plotted and shown in Fig. 11.9.

The maximum power that can be transmitted. P_{max} is found by differentiating the expression for P with respect to δ and equating $\dfrac{dP}{d\delta} = 0$.

We, therefore, have $(V_R/Z^2)\, V_S Z \sin(\beta - \delta) = 0$ giving $\delta = \beta$.

Substitution of this value for δ gives

$$P_{max} = (V_R/Z^2) \cdot (V_S Z - V_R R]$$

$$= 0.1265\,(533.2 - 198) \times \frac{10^6}{3} \text{ watts/phase}$$

= 14.14 MW/phase (the same value having been found from graph).

When δ = 0 (*i.e.*, with no phase shift between V_S and V_R) and $V_S = V_R$ the expression for P reduces to $(V_R/Z^2)\,[V_S Z \cos\beta - V_R R] = (V_R/Z^2)\,[V_S R - V_R R] = 0$ which proves that no power can be transmitted without change of phase if equal voltages are to be maintained at both ends.

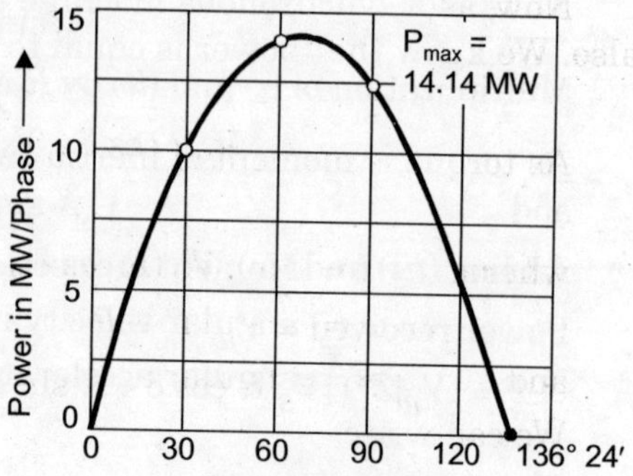

Fig. 11.9

11.4 TRANSIENT INSTABILITY & SWING EQUATION

In a power system, in case of a sudden disturbance, the generator rotor speeds, rotor angular differences and power transfer undergo fast changes where magnitudes are governed by the severity of disturbances.

If the disturbance is large, changes in the angular difference may be well enough to cause the loss of synchronism of the machines. In fact, the types of disturbances being mostly the faults, sudden large load changes may also cause instability of the machine. This type of instability is known as transient instability and is thus defined as the response to large disturbances that may cause instability. The system's response to such a disturbance is usually evident with 1.0 to 3.0 secs.

In fact due to these transient disturbances and sudden large change in loads of that is some load is suddenly added or removed from the shaft of synchronous machine, the rotor will declerate or accelerote respectively with respect to the synchronously rotating stator filed and a relative motion begins. It is said hence, that the rotor is "swinging" with respect to the stator field. The equation describing the relative motion of the rotor, the load angle δ with respect to the stator field as a function of time is known as SWING EQUATIONAL. Now let us develop the swing equation as shown in the section below.

11.4.1 Swing Equation

It is found that the relative position of the rotor axis and the stator magnetic field axis, under normal operation of the system is fixed. The angle between the two is known as the load angle or torque angle denoted by δ and depends upon the loading of the machine. Larger the loading, larger is the value of the torque angle δ. As stated above due to addition or removal of some load from the shaft of the synchronous machine, the rotor will decelerate or accelerate respectively with respect to the synchronous rotating stator field and relative motion begins. In this situation if T_S represent the shaft torque and T_e the electromagnetic torque and if these are assumed positive for a generator the net torque causing acceleration is $T_a = T_S - T_e$, where T_a is positive if shaft torque input is greater than the electromagnetic power output. Now, if for a motor T_e, the electromagnetic torque input is greater than the shaft torque output the motor rotor will accelerate. A similar relation holds good when expressed in term of power (as power is directly related to torque) :

Hence,
$$P_a = P_s - P_e \begin{pmatrix} \text{or, } wT_a = wT_s - wT_e \\ \text{or} \quad P_a = P_s - P_e \end{pmatrix} \quad \text{...(11.2)}$$

where P is the accelerating power

Now, as a synchronous machine has a rotating body, the laws of mechanism apply to this also. We know that power is equal to torque times the angular velocity.

$$P_a = T_a \cdot w \quad \text{...(11.3)}$$

As torque is moment of inertia times the angular acceleration

$$\therefore \quad P_a = T_a \cdot w = I \cdot \alpha \cdot w = M.\ \alpha \text{ (as } M = Iw) \quad \text{...(11.4)}$$

where I = Moment of Inertia in kgm^2

w = angular velocity in rad/sec

and $\propto$ = angular acceleration in rad/sec^2

We can write,

$$w = \frac{2\pi n_s}{60}$$ where n_s is the synchronous speed of the machine is r.p.m.

Now equation (11.3) and (11.4), w should be used in mechanical radians/sec only and not electrical radians/sec.

Now *In* eqn. (11.4), $M = I \cdot w,$

M is in joule sec/mechanical radians

As we are interested in studying the rotor motion in terms of electrical degrees or electrical radians we make use of the following relation :

No. of electrical radians of degrees = No. of mechanical radians or degrees × Number of pairs of poles.

Now, if M is to be expressed in joule-sec/electrical radian when w is in mechanical radians/sec. then,

$$M = I \cdot w/\text{Number of pairs of poles}$$

and if M is to be expressed in joule-sec/electrical degree, then,

$$M = I \cdot w/(\text{Number of pairs of poles} \times 57.32)$$

as 1 rad is = 57.32°

Here M is known as the angular momentum and is expressed in terms of mega-joules-seconds per electrical degree if P_a is expressed in megawatts and α in electrical degrees per second squared. The acceleration α can be expressed in terms of the angular position of the rotor as,

$$\alpha = \frac{d^2\theta}{dt^2} \qquad ...(11.5)$$

Here, the angle 'θ' changes, continuously with respect to time when a sudden change occurs in the system. The value of θ is given by $\theta = w_r t + \delta$ – (11.6) where w_r is the angular velocity of the reference synchronously rotating axis and δ is the angular displacement in electrical degrees from the synchronously rotating reference axis (Fig. 11.9.1).

Taking the derivative the equation (11.6)

we get, $$\frac{d\theta}{dt} = w_r + \frac{d\delta}{dt} \qquad ...(11.6)$$

and $$\frac{d^2\theta}{dt^2} = \frac{d^2\delta}{dt^2} \qquad ...(11.7)$$

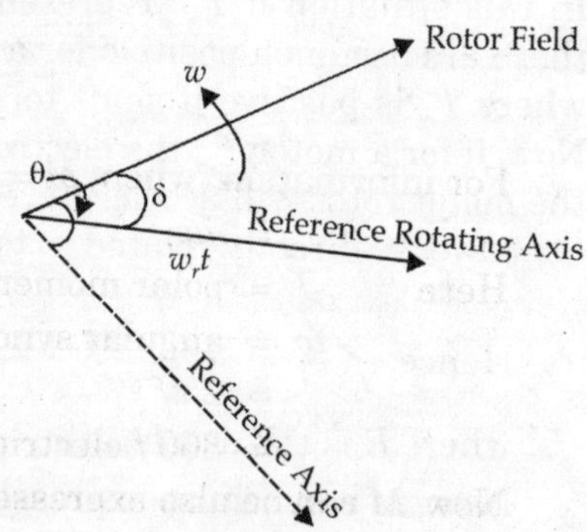

Fig. 11.9

Now from equations (11.2), (11.4) and (11.5) and (11.7), we obtain,

$$M\frac{d^2\theta}{dt} = M\frac{d^2\delta}{dt^2} = P_a = P_S - P_e \qquad ...(11.8)$$

$$\therefore \qquad M\frac{d^2\theta}{dt} = M\alpha = P_a$$

This equation (11.8) is known as the swing equation. The angle δ is the difference between the internal angle of the machine and the angle of the synchronously rotating reference axis which in this case corresponds to the infinite bus. If it is a two machine system two swing equations are required one for each machine, the torque angle between the two machines depends upon the angles between each machine and the synchronously rotating reference.

Now from eqn. (11.4) $M = I \cdot w$ which therefore is not constant but varies somewhat during the swings due to variation in w. In practice, the change in w from the normal system angular velocity is not much during swing except of course when the machine falls out of step and, therefore, almost negligible error is involved by the assumption that M is constant and is equal to the value Iw_n, where w_n is the normal angular velocity of the machine. This value of machine is known as the inertia constant of the machine, and is normally used in calculation for stability studies. The inertia constant is through constant because it is the angular momentum at synchronous speed.

11.4.1.1 Constant, H

In the studies of stability in the power system, there is another very important constant which is very useful in stability studies is denoted by *H*, which is defined as the ratio of the Kinetic Energy at rated speed to the rated apparent power of the machine.

That means,

$$H = \frac{\text{Stored Energy in Mega joules}}{\text{Rating in MVA}}$$
$$= \text{The inertia constant}$$

Relation between *H*, Inertia constant and *M*, the angular momentum. Now let *G* be the rating of the machine in MVA, and *f* the frequency of the system.

Then by the definition

$$G \times H = \text{stored energy in Mega joules} = \frac{1}{2}\text{ M.}\omega. \quad ...(11.9)$$

$$= \frac{1}{2}\text{ M.}\omega. = \frac{1}{2}\text{ M. } 2\pi f = M\pi f$$

or ∴

$$M = \frac{GH}{\pi f} \text{ mega joules = second/radian} \quad ...(11.10)$$

$$= \frac{GH}{180 f} \text{ mega joules = second/electrical degree}$$

∴

$$M = \frac{GH}{180 f} \text{ mega joules = second/electrical degree}$$

For information (when, M = angular momentum)

$= Jw$

Here J = polar moment of inertia of all rotating parts in kg mt^2

w = angular synchronous speed in radians/sec

$= 2\pi f$

$= 360 f$ electrical degree/sec

Now, *M* can be also expressed directly related to H, M as below : from eqn. (11.9)

$$GH = \frac{1}{2}\text{ M.w.}$$

$$M = \frac{2H \cdot G}{w} = \frac{2H}{w_S}$$

where ∵ $w_S = \dfrac{G}{w}$

∴

$$M = \frac{2H}{w_S} \cdot \text{ mega joules/MVA} \quad ...(11.11)$$

$$M = \frac{2H}{2\pi f} = \frac{H}{\pi f} = \frac{H}{180 f} \text{ sec}^2\text{/red} \quad ...(11.12)$$

Now from the above relation between M and H it is clear that M depends upon the size of the machine as well as on its type whereas H does not vary widely with size and has a characteristic value or set of value for each class of machines. In this respect, however, H is similar to per cent reactance of machines. Moreover, whenever value of H is not known a characteristic value may be used. The value of H is lower in case of water wheel generator, as compared to turbo-alternators.

Express the Swing Equation $P_s - P_e = M\dfrac{d^2\delta}{dt^2}$ related to H,

Swing Equation is : $$P_s - P_e = M\cdot\frac{d^2\delta}{dt^2}$$

But from eqn. (11.11), $$M = \frac{2H}{w_s} = \frac{H}{\pi f}$$

where $P_s = P_m$ = Mechanical (Shaft Torque) Power

P_e = Electromagnetic power ($P_m (= P_s) > P_e$)

Now putting the value of M in the swing equation above we get

$$P_s - P_e = \frac{2H}{w_s}\cdot\frac{d^2\delta}{dt^2}\cdot \text{p.u.} \qquad ...(11.13)$$

or

$$P_s - P_e = \frac{H}{180f}\cdot\frac{d^2\delta}{dt^2} \qquad ...(11.14)$$

The swing equation contain information regarding the dynamics and stability of the synchronous machine associated with the connected power system. It may be noticed at this instant, that M and H have been taken as constant in deriving the swing equation. But, in fact, the angular momentum M of the machine is not constant since the angular velocity is changing. But M has been treated constant here since the speeed of the machine does not differ much from the synchronous speed unless the stability limit of the system is succeeded. However, H remains constant as it is a form of representation of M at synchronous speed. Moreover, the damping term proportional to $(d\delta/dt)$ has been neglected, if considered, damping has been found to improve transient stability. Shaft power, P_s or P_m is considered constant throughout the transient stability condition as during this fast period, governors do not act instantaneously.

We know, in a power system, the steady given state electrical power is given by

$$P_e = P_{max}\cdot\sin\delta$$

where, $$P_{max} = \frac{|E_1||E_2'|}{X}$$

Here $|E'_1|$ is emf behind transient reactance of the source $|E_2|'$ the emf behind the transient reactance of the motor and X is transfer reactance.

Hence swing equation can also be written as :

$$M\frac{d^2\delta}{dt^2} = P_m - \frac{|E_1'|\,|E_2'|}{X}\sin\delta \qquad ...(11.15)$$

11.4.2 Multi-Machine Systems Swing Equation

For multi-machine systems solution several swing equations are required. Below the expression for same is deduced.

Say, in a multi-machine system, let,

G_m = Base rating of the machine,

G_{Sys} = System base, and

and H_{Sys} = Inertia constant in system base

$= H_m \times (G_m/G_{Sys})$

where H_m is inertia constant base rating

Hence swing equation, $P_m - P_e = \dfrac{H}{180f} \cdot \dfrac{d^2\delta}{dt^2}$ can be written as below by

Multiplying both sides by

$$\frac{G_m}{G_{sys}};\ (P_m - P_e)\frac{G_m}{G_{Sys}} = \frac{H_m}{180f} \cdot \frac{d^2\delta}{dt^2}\left(\frac{G_m}{G_{Sys}}\right)$$

or

$$P_m - P_e = \frac{H_{Sys}}{180f} \cdot \frac{d^2\delta}{dt^2} \text{ (in p.u. system base)} \qquad ...(10.16)$$

This is the swing equation for a multi-machine system.

However, normally the point by point method is used for solution of the multi-machine system stability problem. In fact even for a single machine system connected to infinite bus with resistance reglected, the format solution of swing equation is possible when with swing equation $P_s = 0$, and by the use of elliptical integrals. In fact the solution of swing equation gives the relation between rotor angle as a function of time and this relation is plotted in terms of curves. If the curves show that the angle between the machine rotors increases without limit, the system is UNSTABLE.

But, on the other hand, if the angles initially increase and then starts reducing, it is probable, though not certain, that the system is STABLE.

However, in a multi-machine system, many a time, one of the machines may stay in step on the first swing and yet go out of step in subsequent swing.

For a two-machine system under the assumption of (1) constant input, (2) no damping and (3) constant voltage behind transient reactance, the machines either fall out of step in the first swing or NEVER. Under this condition the two machines are said to be running at stand still with respect to each other or not. This method is known as Equal Area criterion for stability which we shall discuss later in this chapter in detail with curves. The use of this method eleminates partially or wholly the calculation of swing curves which thus saves a considerable amount of work. The method is applicable to any two machine systems which satisfies the above assumption. This method, infact, is not applicable to a multi-machine system directly. Even if the assumptions made are not strictly true but they do not invalidate the criterion because if these assumption were not true, they would help the system to restore rather than make it fall out of step. Before we go into this in detail, let us establish below the swing equation for. Machines swinging coherently and non-coherently machines.

(a) Swing Equation of Machines Swinging Coherently

Let us consider two machines A & B, on a common system base. Let us refer to the eqn. (10.14) developed for a multi-machine system :

$$P_{m_A} - P_{e_B} = \frac{H_A}{180f} \cdot \frac{d^2\delta_A}{dt^2} \text{ p.u.} \qquad ...(10.17)$$

and, $$P_{m_B} - P_{e_B} = \frac{H_B}{180f} \cdot \frac{d^2\delta_B}{dt^2} \quad ...(10.18)$$

Now, due to coherent swing of the rotors of the two machines say two generators A and B,

$$\delta_A = \delta_B = \delta$$

Then, by adding equations. (11.17) and (11.18) we get,

$$\frac{H_{eq}}{180f} \cdot \frac{d^2\delta_{AB}}{dt^2} = P_{m_{AB}} - P_{e_{AB}} \quad ...(11.19)$$

where $$P_{n_{AB}} = P_{m_A} + P_{m_B}$$

$$P_{e_{AB}} = P_{e_A} + P_{e_B}$$

and $$H_{eq} = H_1 + H_2 = H_{1m}\frac{G_{1m}}{G_{Sys}} + H_{2m}\frac{G_{2m}}{G_{Sys}} \quad ...(11.20)$$

(*b*) Swing Equation for Non-coherent Machines in Power System

In the Power System, machines which swing together are called coherent machines. In fact, two synchronous generators can swing together. Even if their speeds are different, the power angles may be same and their representation of a single base is possible. Then the net effect of such a swing known as coherent swing, is additive as practical. But for a two machine system if one is a synchronous generator and other is a synchronous motor, the swing is non-coherent and the net effect is obtained by subtracting the respective swing equations.

Hence, the swing equation, here, can be represented as :

$$\frac{2H_A}{w_s} \cdot \frac{d^2\delta_n}{dt^2} = P_{m_A} - P_{e_A} \quad \left(\text{Note } \frac{2H}{w_s} = \frac{H}{180} \text{ as } w_s = 360\right)$$

and $$\frac{2H_B}{w_s} \cdot \frac{d^2\delta_B}{dt^2} = P_{m_A} - P_{e_B}$$

Then, subtracting one equation from the other of above two equations we get :

$$\frac{d^2\delta_A}{dt^2} = \frac{d^2\delta_B}{dt^2} = \frac{w_s}{2}\left(\frac{P_{m_A} - P_{e_A}}{H_A} - \frac{P_{m_B} - P_{e_B}}{H_B}\right)$$

Now multiplying both sides by $\frac{H_A H_B}{H_A + H_B}$ rearrangement results,

$$\frac{2}{w_s}\left(\frac{H_A H_B}{H_A + H_B}\right)\frac{d^2(\delta_A - \delta_B)}{dt^2} = \frac{P_{m_A} H_B - P_{m_B} H_A}{H_A + H_B} - \frac{P_{eA} \cdot H_B - P_{eB} \cdot H_A}{H_A + H_B}$$

or $$\frac{2H_{AB}}{w_s} \cdot \frac{d^2\delta_{AB}}{dt^2} = P_{n_{AB}} - P_{e_{AB}} \quad ...(11.21)$$

where, $$P_{m_A} = \frac{P_{m_A} H_B - P_{m_B} H_A}{H_A + H_B}, \; P_{e_{AB}} = \frac{P_{eA} H_B - P_{eB} H_A}{H_A + H_B},$$

$$H_{AB} = \frac{H_A H_B}{H_A + H_B}$$

The equation (11.21) represent the swing equation of a non-coherent system of synchronous machine.

As stated for a multi-machine system involving several swing equations, no formal solution of the equation is feasible. In fact solution of the swing equation gives the values of δ vs. t as shown in Fig. 11.10. Such a graph is known as swing curve. If the swing curve diverges, the system goes to instability while in case of convergent of the swing curve the system stability is maintained.

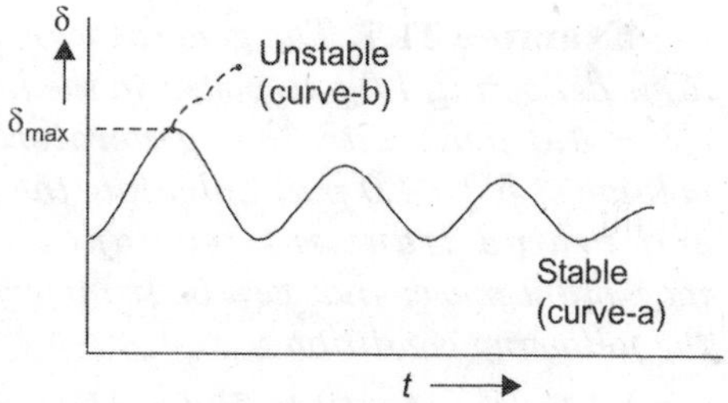

Fig. 11.10. δ/t curve for transient stability profile.

Example 11.6. *In a power system shown in Fig. 11.11 (a), a 3-phase static Generator capacitive reactor of reactance 1 p.h. per phase is connected through a switch at motor bus bar. Calculate the limit of steady state power with and without reactor switch closed. Calculate the power limit with capacitive reactor replaced by an inductive reactor of the same value. Assume the internal voltage of the generator to be 1.2 p.u. and that of the motor to be 1.0 p.u.*

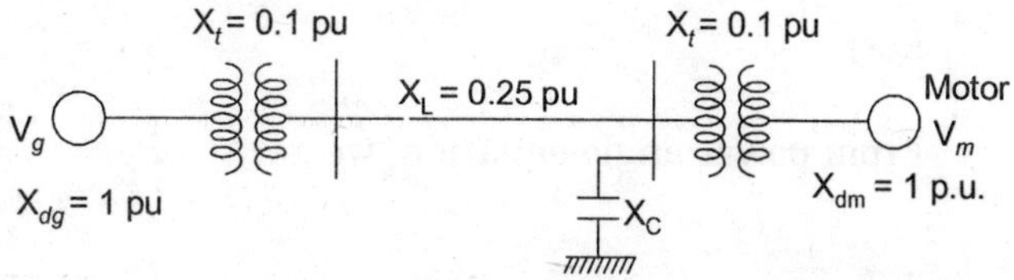

Fig. 11.11 (a)

1. As per example, and for the Fig. 11.11 (*a*) with different values of X with v_g = 1.2 p.u. and V_m = 1.0 p.u. we get. From the expression of steady state power limit,

$$P = \frac{|V_g|\ |V_m|}{X_{(\text{total})}} = \frac{1.2 \times 1}{1 + 0.1 + 0.25 + 0.1 + 1} = 0.49 \text{ p.u.}$$

2. The equivalent circuit with capacitive reactor is shown in Fig. 11.11 (*b*), which is a *T*-circuit (that means a star circuit). Now converting this start to delta, the netwok reduced to Fig. 11.11 (*c*) where,

$$jX_{(\text{transfer})} = \frac{j1.35 \times j1.1 + j1.1 \times (-j1.0) + (-j1.0) \times j135}{-j1.0} = j\,0.965$$

Them steady state power limit $= \dfrac{1.2 \times 1}{0.965} = 1.244$ p.u.

3. Now, with capactive reactance replaced by inductive reactance, we get the equivalent circuit of Fig. 11.11 (*d*). Converting star to delta, we have the transfer reactance of as done in (2),

$$jX_{(\text{transfer})} = \frac{j1.35 \times j1.1 + j1.1 \times j1.0 + j1.0 \times j1.35}{j1.0} = j3.935$$

$$\therefore \text{Steady state power limit} = \frac{|V_8|\ |V_m^j|}{X} = \frac{1.2 \times 1}{3.935} = 0.304 \text{ p.u.}$$

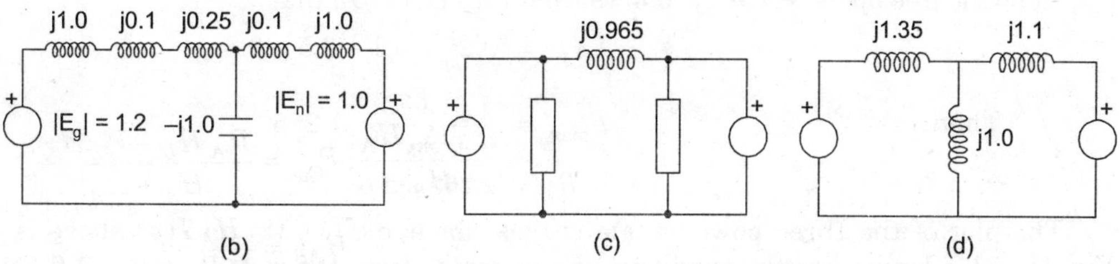

Fig. 11.11

Example 11.7. *The generator of Fig. 11.12 (a) is delivering 1.0 p.u. power to the infinite bus (|V| = 1.0 p.u.) with the generator terminal voltage of $|V_t|$ = 1.0 p.u. Calculate the generator emf behind transient reactance. Find the maximum power that can be transferred under the following condition :*

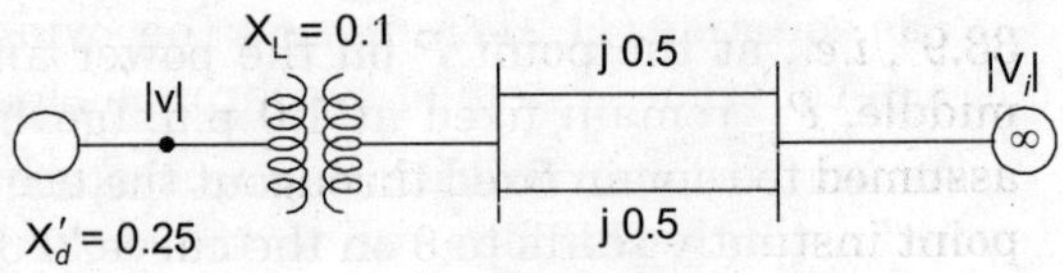

Fig. 11.12 (a)

(a) System healthy, (b) One line shorted (of 3-phase line) in the middle (c) One line open. Draw all the three power angle curves.

Solution : With reference to the power system of Fig. 11.12 (*a*),

Let, $V_t = |V_t| \angle \alpha = 1 \angle \alpha$

From power angle equation, we get, $P_e = \dfrac{|V_1||V_2|}{X} . \sin \alpha$

$\therefore$ $P_e = \dfrac{|V_1||V|}{X} .\sin \alpha$

When power delivered, P_e = 1.0 p.u. (as per example) and X_{total} = 0.25 + 0.1 = 0.35

$\therefore$ $1 = \dfrac{|X|}{0.25+0.1} \cdot \sin \alpha$ or $\alpha = 20.5°$

$\therefore$ Current into infinite bus, $I = \dfrac{|V_t| \angle \alpha - |V| \angle 0°}{jX_{total}}$

$= \dfrac{1\angle 20.5° - 1\angle 0°}{0.35} = 1 + j0.18 = 1.016 \angle 10.3°$

Hence, voltage behind transient reactance

$$E' = 1\angle 0° + j\,0.6 \times (1 + j\,0.18) = 0.892 + j\,0.6 = \mathbf{1.075 \angle 33.9°}$$

(*a*) To find the maximum that can be transferred under the system healthy

$$P_{max} = \frac{|V|\ |E'|}{X_{12}} = \frac{1 \times 1.075}{0.5 + 1.0} = 1.79 \text{ p.u.}$$

$\therefore$ $P_e = \dfrac{|V|\ |E'|}{X_{12}} \cdot \sin \delta = 1.79 \sin \delta$ ...(1)

(*b*) One line shorted in the middle X_{12} = 1.55.

$\therefore$ $P_{max} = \dfrac{1 \times 1.075}{1.55} = 0.694$ p.u.

or $P_e = 0.695 \cdot \sin \delta$ ...(2)

(*c*) One line open : It easily follows from Fig. 11.12 (*b*), that

X_{12} = 0.25 + 0.1 + 0.5 = 0.85

Then, $P_{max} = \dfrac{1 \times 1.075}{0.85} = 1.265$

$\therefore$ $P_e = 1.265 \sin \delta$ ...(3)

The plot of the three power angle curves (for eqns. (1), (2) and (3) above is drawn in Fig. 11.12 (*c*). Under healthy condition, the system is operated with $P_m = P_e$ = 1.0 p.u. and δ_0 =

33.9°, *i.e.*, at the point P on the power angle curve 1.79 sin δ. As one line is shorted in the middle, P_m remain fixed at 1.0 p.u. (governing system acts instantaneously) and is further assumed to remain fixed througout the transient (governing action is slow), while the operating point instantly shifts to θ on the curve, 0.694 sind δ at δ = 33.9°. We note here that because of machine inertia, the rotor angle cannot change suddenly.

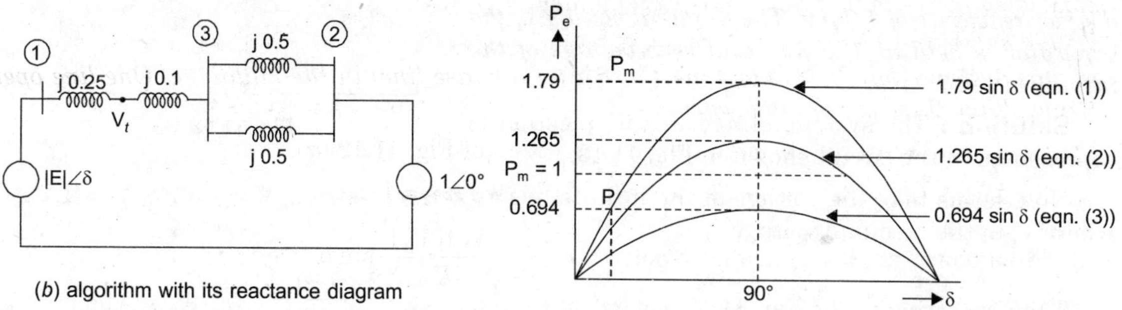

(*b*) algorithm with its reactance diagram

(c)

Fig. 11.12

Example 11.8. *In the above example (example 11.7), if the generator has an inertia constant of 4 MJ/MVA, write the swing equation upon occurrence of the fault. What is the initial angular acceleration? If this acceleration can be assumed to remain constant for Δt = 0.05s find the rotor angle at the end of this time interval and the new acceleration.*

Solution : In transient stability, the swing equation upon occurrence of fault is :

$$P_m - P_e = \frac{H}{180f} \cdot \frac{d^2\delta}{dt^2} \quad \text{(equation developed above)}$$

As per data got from the example : (and the m/result of the above example)

$$1 - 0.694 \sin \delta = \frac{4}{180 \times 50} \cdot \frac{d^2\delta}{dt^2}$$

$$\therefore \qquad \frac{d^2\delta}{dt^2} = 2250\,(1 - 0.694 \sin \delta)$$

Now from the above example, we found initial rotor cycle δ_0 = 33.9 (as calculated in example 11.8).

$$\left.\frac{d^2\delta}{dt^2}\right|_{t=0^+} = 2250\,(1 - 0.694 \sin 33.9°)$$

$$= 1379 \text{ elect degree/s}^2$$

$$\left.\frac{d\delta}{dt}\right|_{t=0^+} = 0, \text{ (rotor speed connect change suddenly)}$$

Then, $$\Delta\delta \text{ (in } \Delta t = 0.05 \text{ sec)} = \frac{1}{2} \times 1379 \times (0.05)^2 = 1.7°$$

$$\therefore \qquad \delta_1 = \delta_{0^+}\ \Delta\delta = 33.9 + 1.7° = 35.6°$$

Then, $$\left.\frac{d^2\delta}{dt^2}\right|_{t=0.05s} = 2250\,(1 - 0.694 \sin 35.6°) = 1341 \text{ elect dg/s}^2$$

It can be observed here that as the rotor angle increases the electrical power output of the generator increases and so the acceleration of the rotor reduces.

Example 11.9. *Find the steady state power limit of a system consisting of a generator of equivalent reactant 0.50 p.u. connected to an infinite bus through a series reactance of 1.0 p.u. The terminal voltage of the generator is held at 120 p.u. and zero voltage of the infinite bus is 1.0 p.u.*

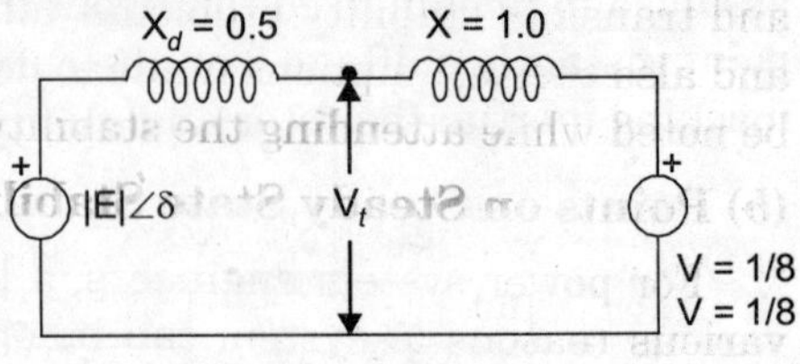

Fig. 11.13

Solution : The system, stated in the question is represented in the circuit shown in Fig. 11.13.

Now, let us take the voltage of the infinite bus as reference then, $V = 1.0 \angle\delta$, $V_t = 1.2 \angle\theta$ where V_t in the terminal voltage.

Then, referring to the Fig. 11.13, we get, $I = \dfrac{V_t - V}{jX} = \dfrac{12\angle\theta - 1.0}{j1}$

But E (as shown in Fig. 11.13) $= V_t + jX_d \cdot I = 1.2\angle\theta + j0.5\left[\dfrac{1.2\angle\theta - 1.0}{j1}\right]$

$$\therefore \quad E = 1.8\angle\theta - 0.5 = (1.8\cos\theta - 0.5) + j\,1.8\sin\theta$$

But steady state power limit is reached when E has an angle of $\delta = 90°$. That means, its real part is zeo. Hence, enq. (11.34a) becomes (by equation with real parts),

$$1.8\cos\theta - 0.5 = 0$$

$$\therefore \quad \theta = 73.87°$$

Now we can write, terminal value $V_t = 1.2\angle\theta = 1.2\angle 73.87° = 0.332 + j\,1.152$

$$\text{As,} \quad I = \frac{(V_t - V)}{X}$$

$$I = \frac{(0.332 + j1.52) - 1}{j1} = 1.152 + j0.668, \text{ and } E = (0.332 + j1.152) + j0.5(1.52 + j0.668), \text{ As, } E = V_t + I \cdot X_d$$

$$\therefore \quad E = (0.332 + j\,1.152) + j\,0.5\,(1.152 + j\,0.668) = -0.002 + j\,1.728 = 1.728\angle 90°$$

Now we know that the steady state power limit, eqn. is given by

$$P_{\max} = \frac{|E|\;|V|}{X_d + X} = \frac{|E|\;|V|}{X_{\text{Total}}} = \frac{1.728 + 1}{1.5} = 1.152 \text{ p.u.}$$

However, if instead, the generator emf is held fixed at a value of 1.2 p.u. the steady state power limit would be :

$$P_{\max} = \frac{1.2 \times 1}{1.5} = 0.8 \text{ p.u.}$$

Hence, it is noted that regulating the generator EMF to hold the terminal generator voltage at 1.2 p.u. raises the power limit from 0.8 p.u. to 1.152 p.u. And this is the method how taken the voltage regulating loop helps in the stability of power system.

11.5 STEADY STATE & TRANSIENT STABILITY STEPS

As a summary of the discussions made along with few worked examples on steady state and transient stability problems and the steps to be taken to improve stability in the system and also the assumptions made to develop the stability circuit equation, following point are to be noted while attending the stability problems in power system.

(*b*) Points on Steady State Stability

For power system engineers, a knowledge of steady state stability limit is important for various reasons. A system can be operated above its transient stability limit, but not above steady state limit. As is clear from the steady state stability equation $P_{nax} = \frac{|E|\ |V|}{X}$, the methods of improving steady state stability limit of a system are to reduce X, (the total reactance of the system) and increase either or both $|E|$ (the sending End Voltage) and $|V|$ (the receiving End Voltage). In case the transmission lines are of sufficiently high reactance, the stability limit can be raised by using two parallel lines, which incidentally also increases the reliability of the system. Then, the series capacitors are sometimes employed in lines to get better voltage regulation and to raise the stability limit by decreasing the line reactance. Also to, improve the stability limit higher oscillation voltages and quick excitation systems are employed.

(*b*) Transient Stability

We have seen above that the dynamics of a single synchronous machine connected to infinite bus bars is governed by the nonlinear differential equation : $M\frac{d^2\delta}{dt^2} = P_m - P_e$ where $P_e = P_{max}$ sin δ. This equation is the Swing Equation.

As stated above in the section related to transient stability no closed form of solution exists for swing equation except for the simple case, $P_m = 0$. (which infact is not a practical case) which involves elliptical integrals. However, for small disturbance like gradual loadings, the equation can be linearised.

However, the practical approach to the transient stability problem is therefore to list all important severe disturbances along with their possible location to which the system is likely to be subjected according to the experience and judgement of the power system analyst. Numerical solution of the swing equation or equation in case of multi-machine case, is then obtained in the presence of such disturbances giving a curve of δ versus t which is known as swing curve. If δ starts to decrease after reaching a maximum value, it is normally assumed that the system is stable and the oscillation of δ around the equilibrium point will decay and finally die out. Infact, important severe disturbances are a short circuit or a sudden loss of load.

In fact in arriving at the swing equation certain assumption and simplifications are made and also for ease of analysis. All the assumptions are listed below along with their justification and consequences upon accuracy of results :

1. Voltage behind transient reactance is assumed to remain constant, *i.e.*, action of voltage regulating loop is ignored. It also leads to pessimistic results.
2. Loads are modelled as constant admittances. This is reasonably accurate representation.
3. Shunt capacitances are not difficult to account for in a stability study. Where ignored, no great significant error is caused.

4. Transmission lines as well as synchronous machine resistance are ignored. This leads to pessimistic results as resistance introduces dumping term in the swing equation which helps stability.
5. Rotor speed is assumed to be synchronous. It fact, it varies insignificantly during the course of the stability transient.
6. Damping term contributed by synchronous machine damper windings is ignored. This also leads to pessimistic result.
7. Mechanical input to machine is assumed to remain constant during the transient, this means that the regulating action of the generator loop is ignored. This leads to pessimistic result.

It should be noted here that the present day power system are so large that even after lumping of machines, the system remains a multi-machine one. Even then, a simple two-machine system greately aids the understanding of the transient stability problem.

Due to occurrence of a severe disturbance, say a short circuit, the power transfer between the machines is greatly reduced, causing the machine torque angles to swing relatively. The circuit breakers near the fault disconnect the unhealthy part of the system so that power transfer can be partially restored, improving the chances of the system remaining stable. The shorter the time to breaker operating called clearing time, the higher is the probability of the system being stable. In fact most of the line fault are transient in nature and get cleared on opening the line. Hence it is a common practice to employ auto reclose breakers which automatically close rapidly after each of the two sequential opening.

The procedure of determining the stability of a system upon occurrence of a disturbance followed by various switching off and on action is called a stability study. Hence, steps to be followed in stability study are outlined below for a single-machine infinite bus bar system. The fault is assumed to be a transient one which is cleared by the time of first recolosure. In the case of a permanent fault, this system completely falls apart. This will not be the case in a multi- machine system. The steps listed, in fact, apply to a system of any size.

Step 1. From the prefault loading determine the voltage behind transient reactance and the torque angle δ_0 of the machine with reference to the infinite bus.

Step 2. For the specified fault, determine the power transfer equation $P_e(\delta)$ during fault. In this system $P_e = 0$ for a three-phase fault.

Step 3. From the swing equation starting with δ_0 as obtained in step 1, calculate δ as a function of time using a numerical technique of solving the non-linear differential equation.

Step 4. After clearance of the fault once again determine P_e (δ) and solve further for $\delta(t)$. In this case $P_e(\delta) = 0$ as when the fault is cleared the system gets disconnected.

Step 5. After transmission line is switched on, again find $P_e(\delta)$ and continue to calculate $\delta(t)$.

Step 6. If $\delta(t)$ goes through a maximum value and starts to reduce the system is regarded as stable. It is unstable if $\delta(t)$ continues to increase calculation is ceased after a suitable length of time.

For the single machine infinite bus bar system, stability can be conveniently determined by the equal area criterion method discussed below :

11.6 EQUAL AREA CRITERION

In a power system where one machine is swinging with respect to an infinite bus, it is possible to study transient stability by mean of a simple criterion, without resorting to the numerical solution of a swing equation.

In a stable system δ (δ describes the motion or swing of the rotor. Infact, when δ increases indefinitely with time, stability of the machine is lost) undedrgoes oscillation which eventually, disout as shown in the curve of δ versus t in the Fig. 11.10 from which it is evident that the system to be stable, ($d\delta/dt$) must be zero at any instant (*X*'on curve (*a*)) on the other hand the profile of instability (*i.e.*, curve *b* in Fig. 11.10) has no $\frac{d\delta}{dt} = 0$. This criterion of stability can be obtained from the swing equation.

The equal area criterion is derived using the swing equation for a machine connected to an infinite bus.

We have seen, that the swing equation for a machine has been developed as :

$$M\frac{d^2\delta}{dt^2} = P_a = P_s - P_e = P_m - P_e$$

P_s = Rotor shaft power
= Mechanical power
= P_m

Now multiplying both the sides of the equation by 2 $d\delta/dt$ (assuming that H is constant and damping is negligible) and then integrating with respect to time, we get,

$$\int 2M\frac{d^2\delta}{dt}\cdot dt = \int 2(P_s - P_e)\cdot\frac{d\delta}{dt}\cdot dt$$

$$M\left(\frac{d\delta}{dt}\right)^2 = 2\int_{\delta}^{\delta}(P_s - P_e)\,d\delta$$

or

$$\frac{d\delta}{dt} = \sqrt{\frac{2}{M}\int_{\delta_0}^{\delta}(P_s - P_e)\,d\delta + C} \qquad \text{...(11.25)}$$

Here δ_0 is the initial torque angle before any disturbance occurs and at this time $d\delta/dt$ = 0. The angle δ will stop changing and the machine will again be operating at synchronous speed after a disturbance when $d\delta/dt = 0$, or when

$$\int_{\delta_0}^{\delta}(P_b - P_e)\,d\delta = 0 = \int_{\delta_0}^{\delta} P_s\,d\delta\left(= \int_{\delta_0}^{\delta} P_s\,d\delta\right)\left(= \int_{\delta_0}^{\delta} P_m\,d\delta\right) \qquad \text{...(11.26)}$$

which means that the area under the curve P_a should be zero which is possible only when P_a is both accelerating and decelerating powers, that is, for a part of the graph $P_s > P_e$ and for the other $P_e > P_s$ as can be seen Fig. 11.14 (*a*). From the condition. $\int_{\delta_0}^{\delta} P_a\,d\delta$ hence states that the area under the accelerating power P_a vs. δ must be zero. That means this is possible if the positive (*i.e.*, accelerating) area under the graph is equal to the negative (*i.e.*, decelerating).

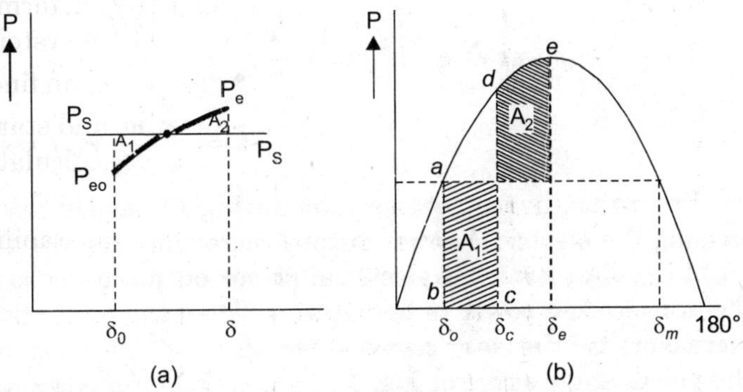

Fig. 11.14. (*a*) Equal area criterion, (*b*) Sudden change of load-equal area criterion.

For a generator action $P_s > P_e$ for positive area A_1, shown in Fig. 11.14 (*a*) and $P_e > P_s$ for negative area A_2 for stable operation. The area represent the Kinetic Energy stored by the rotor during acceleration and the area A_2 represents the Kinetic Energy given up by the rotor to the system and when it is all given up, the machine has returned to its original speed.

The condition, $\int_{\delta_0}^{\delta} P_a \, d\delta = 0$ states that, for stability, the area under the graph of accelerating power P_a vs. δ must be zero. This is possible only if the positive (accelerating) area (A_1) in the graph is equal to the negative (or decelerating) are (A_2). And this criterion is hence known as Equal Area Criterion.

We should note following points with regard to the change in torque angle whenever a disturbance occurs in the system :

1. The angle increases in case of a motor if $P_s > P_e$, *i.e.*, the mechanical output (P_s = the rotor shaft output = the mechanical output) is more than the electrical input and the speed goes down.
2. There is no change in torque angle when the speed of the rotor is the synchronous speed.
3. The angle decreases if the speed is more than the synchronous speed.

With these points with us, Equal Area Criterion can be explained in more details on the $P - \delta$ curve. Let us assume that a generator supplies power to a transmission line. Let us assume that a generator supplies power to a transmission network. Let there be one more short line feeding some load from the generator. A 3-phase fault is assumed on the short line. Before the occurrence of the fault, the generator was operating at synchronous speed with a rotor angle δ_0 while the mechanical power input P_m equals the output electrical power P_e (in Fig. 10.14 (*b*) *"a"* is the operating point of the system in the swing equation curve) due to slow action of the governors. This difference of power results in the speed up of the generator as the kinetic energy is immediately released by the rotor. This causes accelerating power P_a to come into picture. Denoting it for time *t* (*t* is less than time taken to clear the fault, *i.e.*, t_o), the swing equation is given by

$$\frac{2H}{w_s} \cdot \frac{d^2\delta}{dt^2} = P_m - o = P_m \quad ...(11.27)$$

While the fault is ON, the increase in velocity is obtained by integrating the equation (11.27), we get,

$$\frac{d\delta}{dt} = \int_0^t \frac{w_s}{2H} P_m \, dt = \frac{w_s}{2H} P_m \, t \quad ...(11.28)$$

Then the power angle is given by :

$$\delta = \int_0^t \frac{w_s}{2H} P_m \, t = \frac{w_s \, P_m}{4\,H} \cdot t^2 + \delta_0 \quad ...(11.29)$$

From this, hence, we see that during the speed up, the power angle increases abruptly causing the electrical power output increasing rapidly to *d* on the power angle curve of Fig. 10.14 (*b*). But, here, the electrical power output exceeds the mechanical power input causing the acceleration power to be negative. Then the rotor starts slowing down but δ angle goes on increasing till the rotor comes down again at synchronous speed. This point is given by *'e'* on the power angle curve of Fig. 11.14 (*b*). The rotor angle has been advanced to δ_1. Retardation takes place and the rotor speed is falling below synchronous speed, δ is decreasing. The operating point then reaches to point *'a'* where the rotor speed is less than the synchronous speed. In fact, theoretically, the rotor should stop here oscillating if the fault is cleared but due to inertia it slow down further reducing the power angle and thereby reducing the electrical power poutput. However, at 'a', $P_m = P_e$ and there is no accelerating or decelarating power. As for Equal Area Criterion, transient stability is maintained provided $A_1 = A_2$.

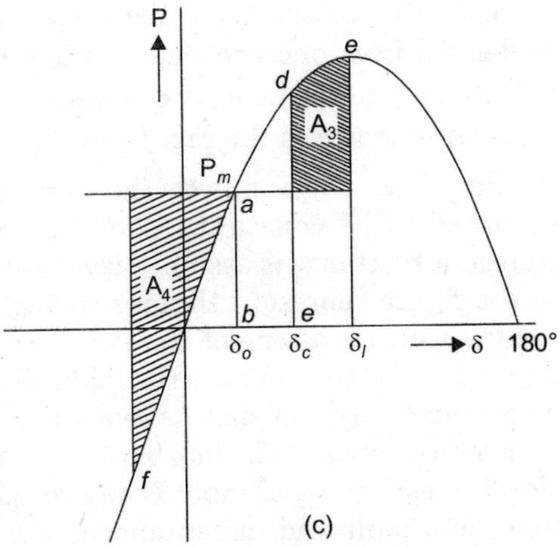

Fig. 11.14 (c). Refer oscillation represented by A_3 & A_4.

Now, in case, the oscillation does not stop at 'a', the rotor slow down and during this time $P_m > P_e$. This causes the rotor to speed up again. The rotor will achieve synchronous speed again at f refer to Fig. 10.14 (c) where the area A_3 and A_4 are shown. The transient stability will be maintained in the system if the area A_3 = area A_4.

In the asbence of damping the rotor continues to oscillate in the sequence $f - a - e$ and $e - a - f$. Synchronous speed is attained at e and f respectively. Now the equal area criterion condition is :

$$\text{Area } A_1 = \text{Area } A_2$$

or
$$\int_{\delta_0}^{\delta_c} (P_m - P_{max} \sin\delta)\, d\delta = \int_{\delta_c}^{\delta_l} (P_{max} \sin\delta - P_m)\, d\delta \quad ...(11.30)$$

However, the left integral, corresponds to the fault period while the right integral corresponds to the past fault period. It is, hence, evident that P_e is zero during the fault period. The same concept is true for the case corresponding to areas A_3 and A_4. Areas A_1 and A_4 are directly proportional to the increase in Kinetic Energy of the rotor, whle it is accelerating, and areas A_2 and A_3 are proportional to the decrease of *KE* while it is decelarating. The physical interpretation of equal area criterion states that if *KE* is added to the rotor, it must be removed from the rotor to restore stability.

However, areas A_1 (or A_4) depends on the line taken to clear the fault. In fact, for any delay in clearing, angle δ_0 increases and A_1 increases. Equal Area Criterion requires that A_2 is to be increased causing maximum swing of δ_l, if there is further delay in clearing the fault, the rotor angle may swing beyond δ_{max}. But this will cause positive acceleration power (as this time $P_m > P_e$) and δ will further increase, without limit causing the loss of stability.

Hence, it is to be noted that the ultimate satisfaction of equal area criterion, that is the transient stability is governed by the critical clearing angle δ_e. The corresponding critical line for remaining the fact is known *H* as the "t_{c^-}" CRITICAL TIME.

11.6.1 Practical Application of Equal Area Critarian

Let us consider two commonly noticed conditions in power system like (*a*) sustained line fault and (*b*) a line fault cleared after sometime by the simultaneous tripping of the breakers at both the ends of the system, and apply Equal Area Criterion on these conditions :

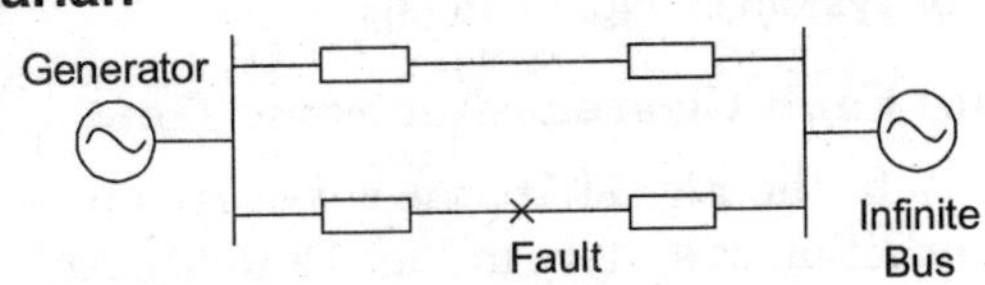

Fig. 11.15 (*a*) Parallel fault in a system.

In the Fig. 11.15 (*a*) in the system it is a parallel feeder fed from one end and at the other end is the infinite bus. In the healthy condition the power angle curve, *A* is shown in the Fig. 11.15 (*b*).

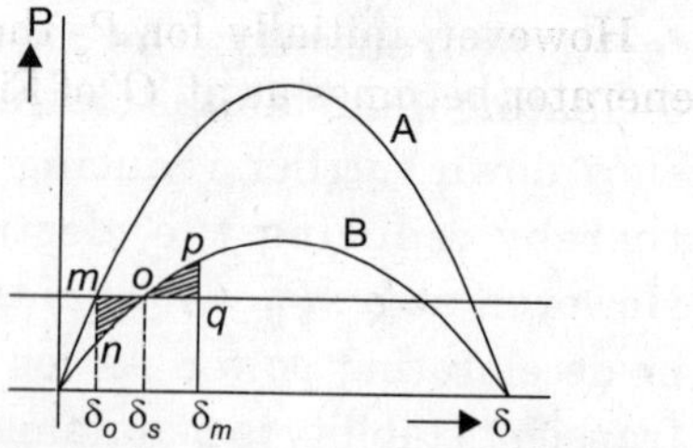

Fig. 11.15 (*b*). Curve "A" – power angle curve in healthy curve 'B' : Power Angle 'B' curve in faulty contain.

Say P_s is the input to the generator which is assumed to be constant. Also the voltage behind transient reactance is assumed constant. The curve 'B' in the figure represent the power angle curve when the fault occurs at one of the two lines shown in the Fig. 11.15 (*a*), and the breakers operate instantaneously and simultaneously at both the ends on that line. As a result the equivalent impedance between the bus bars is increased and hence curve 'B' will be lower than curve 'A'. Now corresponding to input P_s the torque angle of the generator is δ_0 initially. Now as soon as there is a fault and instantaneously it is cleared the output of the generator goes down to point *n* of curve *B* and since input remain constant which is higher than the output, the rotor acceleration and hence the torque angle increases and the operating point moves along curve 'B' towards 'O' from 'n'.

Now as it reaches *O*, the accelerating power becomes zero and the speed of the generator is more than the infinite bus and the speed continues to increase. Then from 'O' to 'P', the rotor experiences declaration but the speed is more than the infinite bus till it reaches point 'P' where the relative speed is zero and the torque angle ceases to increase. At point 'P', the output is more than the input to the generator and hence the rotor decelerates and the speed goes down relative to the infinite bus till it reaches point 0, where the speed is minimum. This torque angle continues to decrease till it reaches point 'n' where again the speed is equal to the speed of infinite bus.

The cycle repeats itself if damping is not present. It is found that in practice because of the damping present the rotor operates at point '0' an curve *B* and the torque angle is δ_0. Now to determine the transient stability limit for this case we should raise the input line P_s such that the area below the line P_s and the curve 'B' and above the line P_s and the curve 'B' at the intersection of P_s andm *B* are equal which is shown in the $P - \delta$ curve of the Fig. 11.15 (*a*). The value of P_s corresponding to this situation is known as the transient stability limit of the power system shown in Fig. 11.15 (*a*).

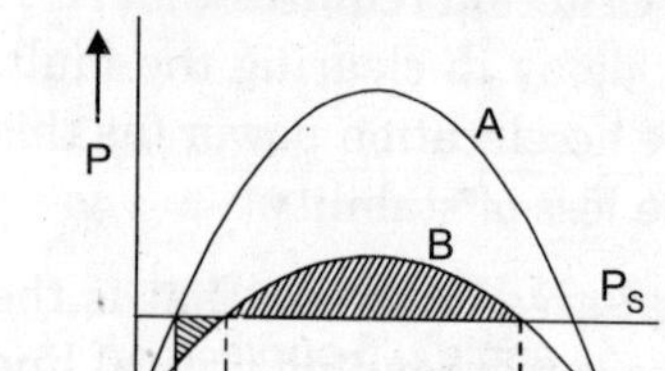

Fig. 11.15 (*c*). Transient stability limit for system of Fig. 11.15 (*a*).

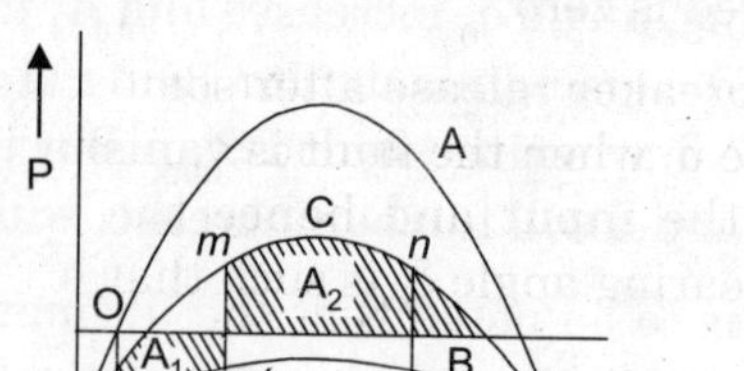

Fig. 11.16. Equal Area criteria applied to the system of Fig. 11.15 (*a*) for a fault cleared at an angle $\delta \angle \delta_0$

(*b*) Fault Cleared After Some Time

In the Fig. 11.16, the curve 'A' represents the power angle curve corresponding to healthy condition of system in Fig. 11.15 (*a*). And curve 'B' represents condition corresponding to fault on one of the two lines and fault allowed to exist for some time. And curve 'C' in Fig. 11.16, corresponds to the condition when the faulted line is cleared of fault.

However, initially for P_s the torque angle is δ_0. At the time of fault, the output of the generator becomes as at 'O' of Fig. 11.16.

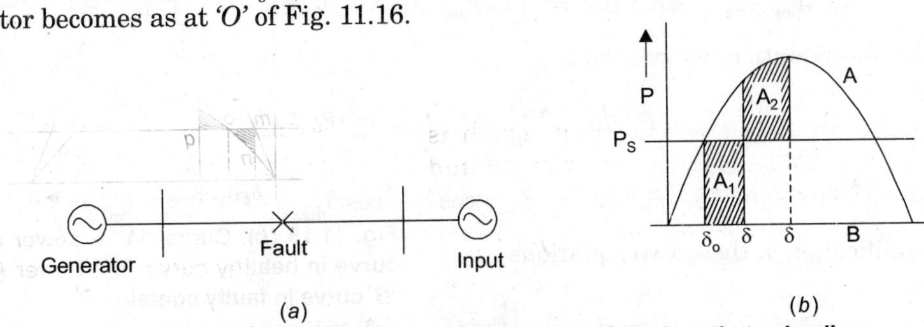

Fig. 11.17 (*a*). Generator bus connected to infinite bus through a line. (*b*). Power angle curve *A*–healthy condition *B* 3-phase faults.

The rotor, hence, accelerates and the rotor accelerates and the rotor accelerates and the rotor moves along the curve 'B' up point "*m*" when the faulted line is removed and the operating point becomes "*m*" on the curve 'c' where the output is more than the input and the rotor decelerates till the speed becomes equal to the speed of the infinite bus and the torque angle ceases to increase at point 'n' in Fig. 11.16.

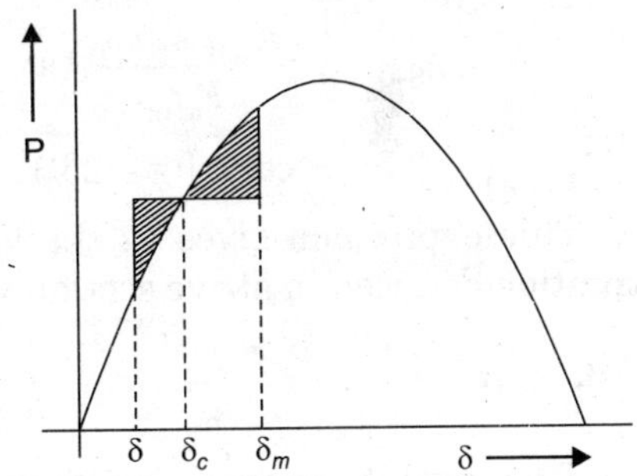

Fig. 11.17 (*b*)

Therefore from this Figure (Fig. 11.16), it is clear that the transient stability limit not only depends upon the type of disturbance, but it also depends upon the clearing time of the breaker. Further the breaker operation smaller will be the area A_1 and hence larger will be the transient stability limit.

Let us study one more practical case as another case study. To do so let us refer to Fig. 11.17 (*a*) when generator is connected to an infinite bus through a single feeder.

Now, suppose there occurs a 3-phase fault on the line temporarily. Let us then refer to Fig. 11.17 (*b*) for power-angle curve which will correspond to the horizontal axis because power transferred is zero.

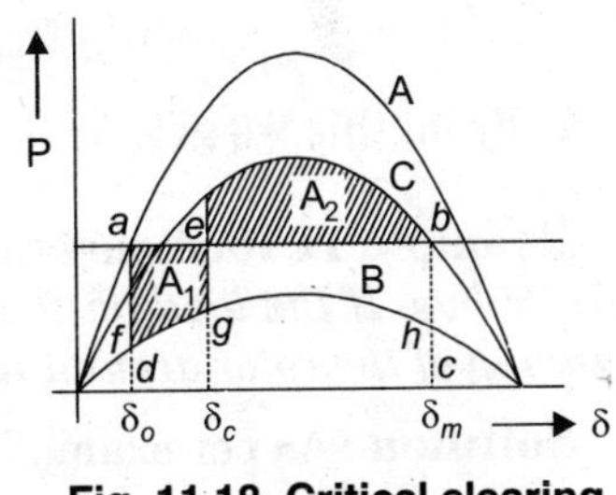

Fig. 11.18. Critical clearing angle *p*-δ-curve.

Now if the breaker release after some time corresponding to clearing angle δ_c when the fault is vanished, the output will be more than the input and hence the rotor decelerates. Finally, if the clearing angle δ_c is such that $A_1 = A_2$, the system becomes stable.

11.6.2 Critical Clearing Angle & Critical Clearing Time

It is noted that for any given initial load there is a critical angle. It is also a fact that the actual clearing angle is greater than the critical value the system is unstable, otherwise it is stable. Now in this section we shall determine the value of critical clearing angle for a given load.

In the section above we have seen that, with equal area criterion principle,

$$\text{Area } A_1 = \text{Area } A_2$$

or $$\int_{\delta_0}^{\delta_c} (P_m - P_{max} \sin\delta)\, d\delta = \int_{\delta_c}^{\delta_1} (P_{max} \sin\delta - P_m)\, d\delta$$

Now from this equation, we can with

$$A_1 \rightarrow \int_{\delta_0}^{\delta_c} (P_m - P_e)\, d\delta = \int_{\delta_0}^{\delta_c} P_m \cdot d\delta = P_m(\delta_0 - \delta_0) \quad ...(11.31)$$

and $$A_2 \rightarrow \int_{\delta_c}^{\delta_{max}} (P_{max} \sin\delta - P_m)\, d\delta = P_{max}(\cos\delta_c - \cos\delta_{max}) - P_m(\delta_{max} - \delta_c) \quad ...(11.32)$$

Now simplification of these two equations

Give us : $$\cos\delta_c = \frac{P_m}{P_{max}}(\delta_{max} - \delta_0) + \cos\delta_{max} \quad ...(11.33)$$

But, $$\delta_{max} = (\pi - \delta_0) \text{ radian} \quad ...(11.34)$$

and $$P_m = P_{max} \sin\delta_0 \quad ...(11.35)$$

Now substitution of equations (11.24) ∪ (11.25) in (11.23) gives us,

$$\cos\delta_c = \frac{P_{max}\sin\delta_0}{P_{max}}(\pi - \delta_0 - \delta_0) + \cos(\pi - \delta_0) = (\pi - 2\delta_0)\sin\delta_0 - \cos\delta_0$$

$$\therefore \quad \delta_c = \cos^{-1}[(\pi - 2\delta_0)\sin\delta_0 - \cos\delta_0] \quad ...(11.36)$$

This expression gives as the value of **CRITICAL CLEARING ANGLE.** Now from the equation deduced in above section we have seen the

$$\delta = \int_0^t \frac{w_s}{2H} P_m t = \frac{w_s P_m}{4H} \cdot t^2 + \delta_0$$

Then in this equation, in the case of critical clearing,

$$\delta_c = \frac{W_s P_m}{4H} t_{cr}^2 + \delta_0 \quad (\text{where } t_{cr} = \text{critical clearing time})$$

∴ From this we get, $$t_{cr} = \sqrt{\frac{411(\delta_c - \delta_0)}{W_s P_m}} \quad ...(11.37)$$

Example 11.10. *A motor is receiving 25% of the power that is capable of receiving from an infinite bus. If the load on the motor is doubled calculate the maximum value of δ during the swinging of the rotor around its new equilibrium position.*

Soltuion : As per example, $\sin\delta_0 = 0.25, \therefore \delta_0 = 14.48° \ (P_m/P_m = .25 = \sin\delta_0)$

and $\sin\delta_c = 0.5$, hence, $\delta_c = 30° \ (P_m/P_m = 0.5 = \sin\delta_c)$

Then δ_m to be found.

Now from the equations. $$(\delta_m - \delta_0)\sin\delta = \int_{\delta_0}^{\delta_m} \sin\delta \cdot d\delta$$

We get $$0.5(\delta_m - 0.2527) = (\cos\delta_0 - \cos\delta_m)$$

$$\therefore \quad 0.5\,\delta_m = \cos\delta_0 - \cos\delta_m + 0.1263$$

$$\therefore \quad 0.5\,\delta_m + \cos\delta_m = 0.96823 + 0.1263 = 1.09435$$

For solution of this equation w_2 as other δ_m such that δ_{mn} should be greater than 30° (as δ_c = 30°). And after some trials δ_m found to be 45°.

Example 11.11. *A 50 Hz. generator is delivering 50% of the power that it is capable of delivering through transmission line to an infinite bus. A fault occurs which increases the reactance between the generator and the infinite bus to 500% of the value before the fault. When the fault is isolated, the maximum power that can be delivered is 75% of the original maximum value. Determine the critical angle for the condition described after developing the expression for δ_c from Equal Area Criterism.*

Solution : We shall take P_m as the maximum power that can be delivered. As per the example we can write,

$$P_m \sin \delta_0 = 0.5\, P_m \;\therefore\; \delta_0 = 30°$$

Now, as per example, during fault, the reactance is 500% of the value before the fault occurred. We go further let no determine the alternate method of determining the critical clear angle, which will be followed in solving this problem for a given load.

In the Fig. 11.18, let these be three power-angle curves, A, B, and C represented as :

$$A = P_m \sin \delta, \text{ before the fault}$$

$$B = r_1 P_m \sin \delta \text{ during the fault}$$

$$C = r_2 P_m \sin \delta \text{ after the fault}$$

where r_1 and r_2 is percentage of the maximum power available. Then for transient stability limit, the two areas $A_1 = A_2$ or equivalently the area under the curve *abcd* (*a* reactangle as seen in Fig. 11.18) should be equal to the area under *dfghbe* that is

$$(\delta_m - \delta_0) P_s = \int_{\delta_0}^{\delta_c} r_1 P_m \sin \delta \cdot d\delta + \int_{\delta_c}^{\delta_m} r_2 P_m \sin \delta\, d\delta$$

$$= r_1 P_m [\cos \delta_0 - \cos \delta_c] + r_2 P_m [\cos \delta_c - \cos \delta_m]$$

Then putting $P_m \sin \delta_0$ for P_s we get

$$(\delta_m - \delta_0) P_m \text{ sind } \delta_0 = r_1 P_m [\cos \delta_0 - \cos \delta_c] + r_2 P_m [\cos \delta_c - \cos \delta_m]$$

Hence after simplifying this we get,

$$\cos \delta_c = \frac{(\delta_m - \delta_0) \sin \delta_0 - r_1 \cos \delta_0 + r_2 \cos \delta_m}{r_2 - r_1} \quad ...(11.38)$$

Then from the curves of the Fig. 11.18,

$$P_s = P_m \sin \delta_0 = r_2 P_m \sin \delta_m = r_2 P_m \sin (\pi - \delta_m)$$

or

$$\sin \delta_0 = r_2 \sin (\pi - \delta_m)$$

$$\therefore \quad \delta_m = \pi - \sin^{-1}\left(\frac{\sin \delta_0}{r_2}\right)$$

Hence, if r_1, r_2 and δ_0 are know then angle δ_c can be obtained from the eqn. (11.38).

Eqn. (11.38) can also be written as with $P_s = P_m \sin \delta_0$

Hence,

$$\sin \delta_0 = P_s / P_m$$

$$\cos \delta_c = \frac{(\delta_m - \delta_0) P_s/P_m + r_2 \cos \delta_m - r_1 \cos \delta_0}{r_2 - r_1} \quad ...(11.39)$$

Now that the expression has been developed for δ_c, referring now to the example given,

We have got, $\delta_0 = 30°$ (from $P_m \sin \delta_0 = 0.5\, P_m$)

Now as per example during fault the reactance is 500% of the value before the fault,

$$\therefore \quad r_1 = 0.2$$

and

$$r_2 = 0.75 \text{ (given)}$$

The critical clearing angle δ_c is given by the equation (11.49), developed above

$$\delta_c = \cos^{-1} \frac{(P_s/P_m)(\delta_m - \delta_0) + r_2 \cos \delta_m = r_1 \cos \delta_0}{r_2 - r_1}$$

Now as per example,

$$0.5\, P_m = 0.75\, P_m \sin \delta_m$$

$$\therefore \quad \delta_m = 41.8° \text{ or } \delta_m = 180 - 41.8 = 138.2° \text{ or } \delta_m = 2.412 \text{ radius}$$

Now substituting these values in the above expression in eqn. (11.50) developed above

$$\therefore \quad \delta_c = \cos^{-1} \frac{0.5(2.412 - 0.5236) - 0.75 \times 0.74547 - 0.2 \times 0.866}{0.55}$$

$$= \cos^{-1} 0.38.36 = 67°\, 44'.$$

Example 11.12. *A synchronous generator is operating at an infinite bus and supplying 45% of its peak power capacity. As soon as a fault occurs, the reactance between the generator and the line becomes four times its value before the fault. The peak power that can be delivered after the fault is cleared is 70% of the original maximum value. Determine the critical clearing angle.*

Solution : From the example, we find,

$$P_{max} \sin \delta_0 = 0.45\, P_{max} \quad \therefore \delta = 26.74°$$

But we know,

$$P_{max} = \frac{EV}{X}$$

During falut 'X', the reactance becomes, as per example, 4 times its value before the fault, we see, 'X' at fault becomes $4X$.

$$\therefore \quad r_1 P_{max} \sin \delta = \frac{EV}{4X} \sin \delta_m = \frac{1}{4} P_{max} \sin \delta_m \text{ (substituting } P_{max} \text{ for } EV/X)$$

$$\therefore \quad r_1 = 0.25$$

After clearance of the fault, $r_2 = 0.7$

$$\therefore \quad r_2 P_{max} \sin \delta_m = P_m = 0.45\, P_{max} \text{ (as per example)}$$

$$\therefore \quad \delta_m = 40°$$

In radian $\therefore$

$$\delta_m = 180° - 40° = 140° = 2.44 \text{ radian}$$

$$\therefore \quad \cos \delta_c = (0.45\,(2.44 - 0.467) + 0.7 \cos 140° - 0.25 \cos 26.74°)$$

$$= \cos 73°$$

$$\therefore \quad \delta_c = 73°$$

Example 11.13. *A 50 Hz, 3-phase synchronous generator delivers 1.00 p.u. power to an infinite bus bar through a network in which resistance is negligible. A fault occurs which reduces the maximum power transferable to 0.40 p.u., whereas, before the fault, this power was 1.8 p.u. and after the occurrence of the fault 1.30 p.u. By the use of the equal area criterion, determine the critical angle.*

Solution : As per example :

$$P_m = 1.00 \text{ p.u. and } P_{max} = 1.80 \text{ p.u.}$$

$$\therefore \quad r_1 P_{max} = 0.40 \text{ p.u., and } r_2 P_{max} = 1.30 \text{ p.u.}$$

Then, $$\delta_0 = \sin^{-1}\left(\frac{P_m}{P_{max}}\right) = \sin^{-1}\left(\frac{1.0}{1.8}\right) = 33.80°$$

$$\delta_f = 180° - \sin^{-1}\left(\frac{P_m}{r_2 P_{max}}\right) = 180° - \sin^{-1}\left(\frac{1.0}{1.30}\right) = 129.6°$$

Also, $$r_1 = \frac{0.40}{1.80} = 0.222 \text{ and } r_2 = \frac{1.30}{1.80} = 0.722$$

We know, $$\delta_c = \cos^{-1}\left[\frac{1}{r_2 - r_1}(\delta_f - \delta_0)\sin\delta_0 + (r_2\cos\delta_f - r_1\cos\delta_0)\right]$$

$$= \cos^{-1}\left[\frac{1}{.722 - .222}\left\{\frac{\pi(129.6° - 33.8°)}{180°}\sin 33.8°\right.\right.$$

$$= [\{1 + 0.93 + (-0.645)\}/0.50] + (0.722 \cos 129.6° - 0.222 \cos 33.8°\}\}]$$

$$= \cos^{-1} 0.57$$

$$\therefore \quad \delta_c = \mathbf{55.20°}$$

11.7 TRANSIENT DISTURBANCES

Transient disturbances require some study regarding transient stability for which we have to consider the effect of (1) Load increases, (2) Switching operation and (3) Fault with subsequent circuit isolation of the system. Infact power system engineers have to regularly face these problems and solve them to bring stability in the system. Briefly we shall discuss these points below with reference to equal area criterion.

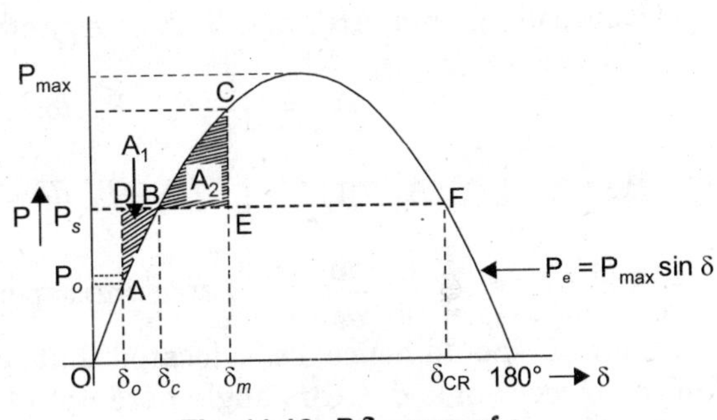

Fig. 11.19. ***P-δ* curve of a synchronous note.**

(a) Load Increases

From the stability point of view load increases result in transient disturbance, if (*a*) the total load exceeds the steady state limit of the system, or (*b*) the total load increase is so sudden as to set up an oscillation beyond the critical point.

To study the case, let us consider a synchronous motor connected to an infinite bus bar. It supplies a load P_0 and is operating at synchronous speed with a torque angle δ_0, corresponding to this output power on the sinusoidal power angle diagram (with negligible resistance in comparison with reactance) being P_c as shown in Fig. 11.19. Now for plotting for power angle diagram, we use the relationship.

$$P_e = \frac{|E_g|\ |E_m|}{|X|}\sin\delta$$

where E_g is the voltage of the infinite bus bar, $|E_m|$ is the voltage behind the transient reactance of the motor X is the transient reactance of the motor including reactance of the transmission line, transformer etc., if any in the system.

Under these conditions, it is clear that, mechanical power input is P_s equal to P_e. Then, if the load on the motor is suddenly increased to P_s the electrical input power will initially be the same. Infact, the sudden excess of load must be met from the Kinetic Energy stored in the rotating system. This is possible only if the motor starts slowing down the torque angle δ increases and with the increase in δ there is an increase in electrical input power. This goes on till P_e and P_s are equal (corresponding to new increased load), at which point the deceleration ceases. The operating point on the power angle curve is the *B*. However, from *A* to *B* the motor has been decelerating at a finite rate and when at *B* its speed is below synchronous speed. The result is that δ continues to increase until point *C* is reached at which point the speed is again synchronous. Torque angle corresponding to *C* is δ_m as show in Fig. 11.19. Then from *B* to *C* electrical power input P_e has been more than P_s and speed continues to increase but δ starts decrease immediately synchronous speed is reached. As δ decreases point *B* is reached again with the speed above synchronous speed.

In fact, at point *B*, the acceleration ceases and the rotor moves towards *A* with a finite deceleration, the cycle of oscillations then repeating itself till the oscillations are damped.

Now as per equal area criterion of stability ignoring losses areas *ABD* and *BCE* must be equal. The decelerating torque at any instant between *A* and *B* is $T_s - T_e$. Kinetic energy lost by the rotor is the area *ABD*. Similarly the accelerating torque at any instant between *B* and *C* is $T_e - T_s$ and the kinetic energy gained by the rotor is this interval is the area *BCE*. Since speeds at *A* and *C* are synchronous the two areas are equal. Then the point *C* corresponding to torque angle δ_m is the point of maximum swing of the rotor. The question which is important from stability consideration is what should be the value of δ_m if stability is to be maintained.

Now calling area, *ABC* and *BCE* as A_1 and A_2 respectively referring to Fig. 11.19.

$$A_1 = \int_{\delta_0}^{\delta} (P_s - P_e)\, d\delta, \text{ and } A_2 = \int_{\delta_s}^{\delta_m} (P_e - P_s)\, d\delta$$

Hence,
$$A_1 - A_2 = \int_{\delta_0}^{\delta_m} (P_s - P_e)\, d\delta = 0 \text{ if equal area criterion is to be satisfied}$$

And
$$\frac{d\delta}{dt} = 0 \text{ at 'e' corresponding to } \delta_m$$

Point *C* should hence, be so located that two area A_1 and A_2 are equal. Now if point *C* lies on the power angle diagram angle here upto *F*, *i.e.*, above the line passing through point P_s it will indicate transient stability.

However, say the condition are as shown in Fig. 11.20. The machine initially supplying a load P_0 corresponding to a torque angle δ and there is a sudden increase in load equal to P_s. Now, let us assume that this corresponds to rotor swing from *A* to *C*. Energy lost during deceleration from *A* to *B* is given by Area A_1. Energy recovered by the rotor due to its acceleration upto point *C* (which it the critical point of swing for transient stability) corresponding to area A_2. In this case $A_2 > A_1$ so that rotor will swing beyond *C*, to (say) *C*′. But it is clear that a point *C* represents input power P_e less than output power P_s and hence an unstable operation. The rotor will therefore, more beyond *C* with the result that synchronism will be lost. It will also be that $d\delta/dt$ is not zero at *C* since δ continues to increase beyond *C* as contrasted with situation shown in Fig. 11.20.

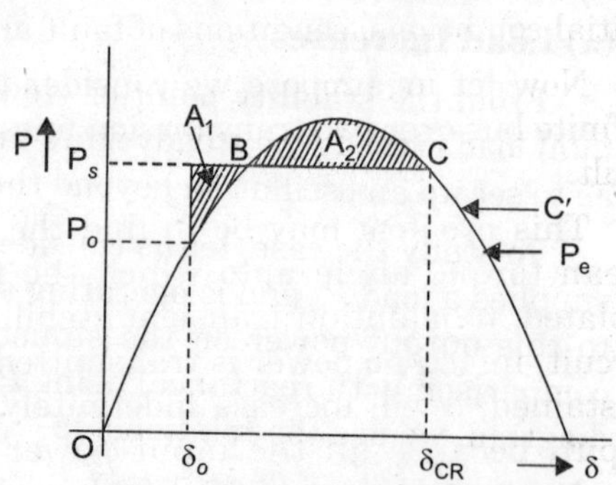

Fig. 11.20. Sudden application of load on motor equal to P_s.

Hence, it is thus clear that if transient stability is to be maintained the rotor swing must not be beyond the critical torque angle δ_{cr}. We will be approaching thus condition if initial operating torque angle, δ_0 was high or sudden increase of power which the system was required to meet was two large in the system.

(b) Switching Operations

In this case again we use equal area criterion for determining the transient stability of the system. We, infact, shall require two power angle diagram : One for initial condition and the other for final condition, that is, after the switching operation. Say, a generator supplying power to infinite busbars over two transmission lines operating in parallel as shown in Fig. 11.21. The mechanical input power is P_s corresponding to point A on power angle curve for 2 lines, the torque angle being δ_0. Suppose one of the lines is opened. the power which can be transmitted over one line with this torque angle is P_1, neglecting losses. Now opening the second line, however, means increased reactance and the torque angle must, therefore, increase for transmission of original power output. The generator is however, accelerated due to the fact that it supplies reduced power output in the transient interval, the torque angle increases because of acceleration. The torque angle δ oscillates between δ_0 and δ_m around equilibrium position δ_s corresponding to new conditions such that areas A_1 and A_2 are equal. It should be noted here that the maximum swing of rotor should not be beyond torque angle δ_{CR} otherwise transient stability will be lost. Hence, it is thus clear that transient stability in this case depends upon : (1) steady stability limit of the remaining system and (2) the difference in initial and final torque angles.

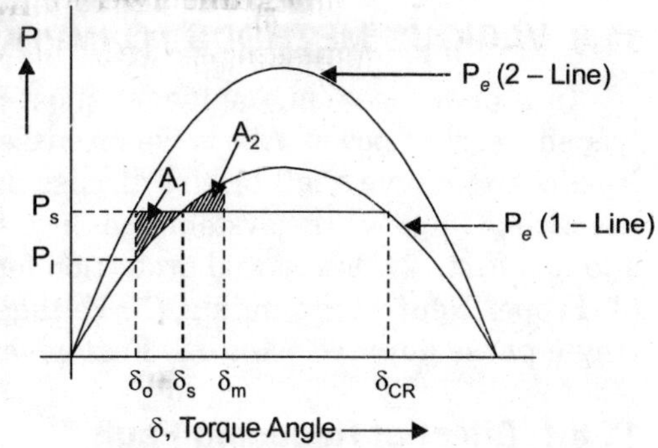

Fig. 11.20 (a). P-δ curves of a generator supply input B_m with 2 line of line.

(c) Fault with Subsequent Circuit Isolation

In the power system this perhaps the most important type of disturbance. The study of problem is, however, made more complicated by behaviour of circuit breakers which may isolates the line in steps, that is sequential switching, high speed or auto-reclosing circuit breakers etc.

In this case, the procedure is similiar to switching operation discussed above, once the initial conditions, conditions of fault are known.

Now let us suppose we consider the case of power being supplied by a generator to an infinite bus over two transmission lines in parallel, one of which develops a 3-phase short circuit fault.

This problem may be to find the critical angle, (that mean torque angle upto which the faulty line must be isolated, to maintain transient stability. During the short circuit, infact, no power is transmitted. Now, if the fault is sustained, δ will increase indefinitely, assuming constant input, because all the input power would be used for acceleration. However, if the fault is cleared at an angle δ_e as can be seen in Fig. 11.21, the rotor will swing upto a torque angle δ_m on the lower power angle curve such that areas A_1 and A_2 are equal. Now, if the point of maximum swing cannot be located above horizontal line through P_s the system will not have transient stability.

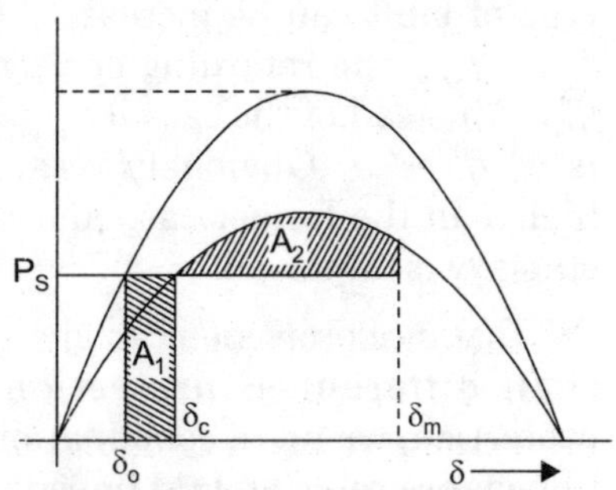

Fig. 11.21 (b). Equal area criterion with fault.

Hence, it is thus clear that larger the time taken for a fault to be cleared then more is the tendency to approach transient stability limit of the system. Also the larger the value of P_s the quicker the fault should be cleared for maintaining stability. Another point to be noted in this if some power is transmitted during the fault it helps in maintaining transient stability of the system.

11.8 VARIOUS METHODS TO IMPROVE SYSTEM STABILITY

In a power system, engineers' most important task is to improve system stability in the transmission of power. Above we discussed on various aspects of stability, mostly theoretically. In this section we shall briefly discuss different methods used by power system engineers in practice to improve the system stability. These methods are : (1) Reducing fault clearance time and its affect, (2) Differential protection for improving transient stability, (3) Automatic reclosing, (4) Proper Neutral grounding, (5) Voltage regulators and its effect, (6) Electrical breaking, (7) single-phase auto reclosing, (8) Fast governor action and (9) other measures.

11.8.1 Effect of Reducing Fault Clearance Time

To improve transient stability, the fault should be cleared as fast as possible. Thus, by reducing the clearance time, the maximum power that could be delivered under a particular disturbance, can be increased (P_e could be improved if the clearance time is reduced). Clearing the fault faster, there is reduction in accelerating energy and increase in the retardation energy. This is evident in Fig. 11.22.

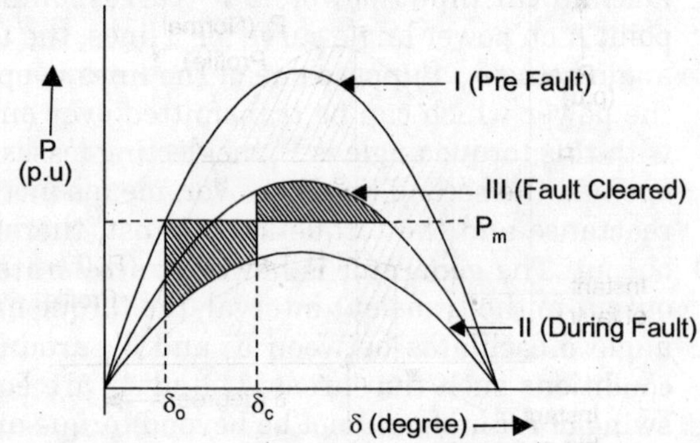

Fig. 11.22

11.8.2 Differential Protection for Improving Transient Stability

To improve transient stability, it is customary to provide the retarding energy always greater than the accelerating energy.

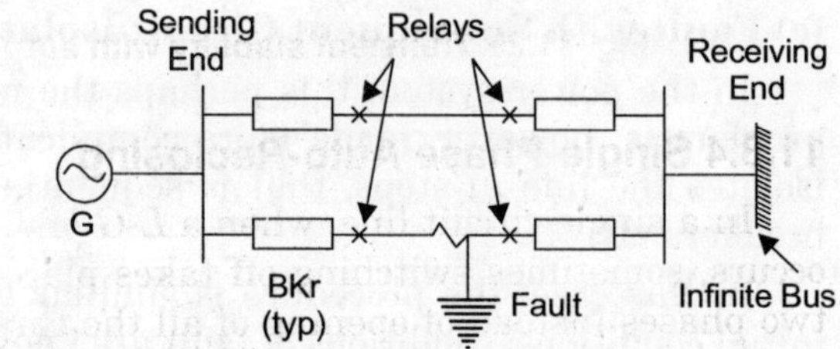

Fig. 11.23 Fault in parallel feeder.

Two circuit breakers shown in Fig. 11.23 should trip simultaneously (in ideal condition) so that the available retardation energy for any type of fault can be increased. In Fig. 11.24 if $t_{cr\,1} = t_{cr\,2}$, the retarding energy is given by $d''\,ff'\,ed'$ instead of the case if $t_{cr\,1} \neq t_{cr\,2}$ when that is $d''\,ff'\,ee'\,d$. Obviously retarding energy is higher in the former case and hence transient stability is improved.

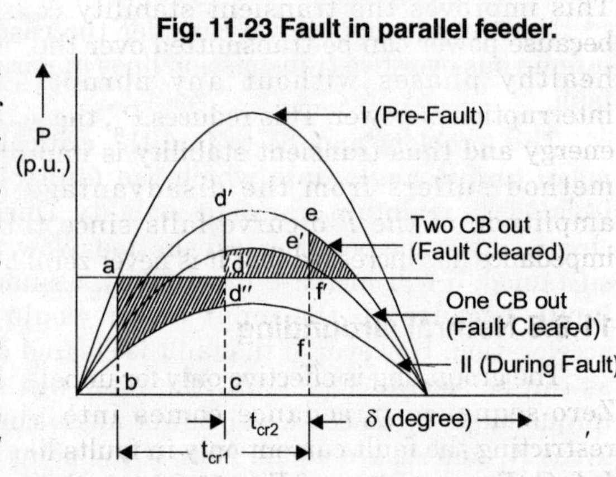

Fig. 11.24 Role of critical clearing times of CBs.

Distance protection can give faster clearance than differential protection. In distance protection, we use a combination of (*a*) definite impedance relay and (*b*) impedance time relay/mho/reactance relay.

11.8.3 Automatic Reclosing

The main type of fault which is quite common is lightning on transmission lines. Higher the operating voltage of the line, the higher is the deionising time. for a 220 kV line, the deionising time is of the order of 15 to 20 cycles (*i.e.*, for $T = 0.02$ sec, t_{dion} is 0.3 to 0.4 sec).

When the fault occurs the relay trips the circuit breaker. The circuit breaker is allowed to close automatically only after the deionising time. This autoreclosing method is more effective in a single circuit line than in a double cirucit line (Fig. 11.25).

ab = relay time to actuate the circuit breaker

cd = deionising time $= t_{dion}$ (bkr. remains off)

The automatic reclosing should be faster enough so that the accelerating energy can be lower and retarding area is increased.

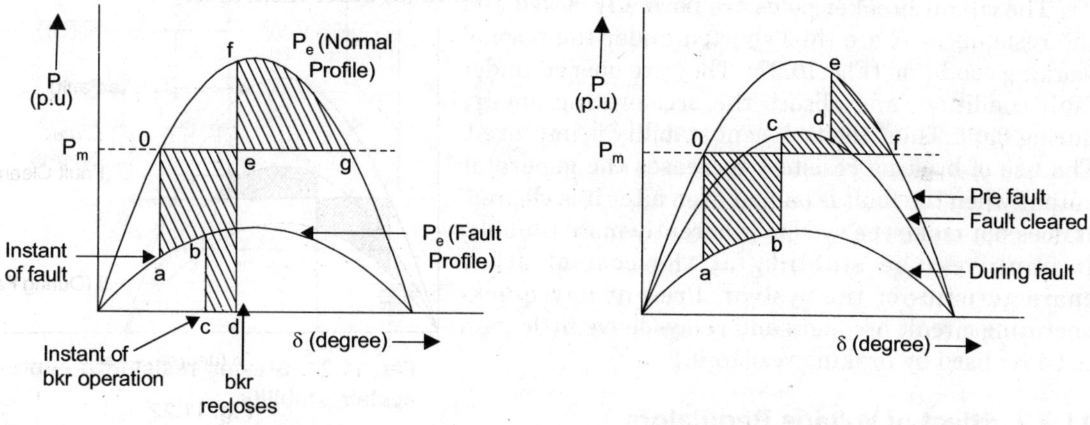

Fig. 11.25. Transient stability with auto reclosure. (*b*). Single phase auto reclosing.

11.8.4 Single-Phase Auto-Reclosing

In a single circuit line, when a *L-G* or *L-L-G* fault occurs, sometimes switching off takes place in one or two phases instead of opening of all the three phases. This improves the transient stability considerably because power can be transmitted over the remaining healthy phases without any abrupt and total interruption of power. This reduces P_a, the accelerating energy and thus transient stability is improved. This method suffers from the disadvantage that the amplitude of the P-δ curve falls since the system impedance has increased (but it is never zero) Fig. 11.25.

11.8.5 Neutral Grounding

The grounding is effective only for unbalanced fault. Zero-sequence impedance comes into picture for restricting the fault current only in faults like *L-G* and *L-L-G*. For a system of Fig. 11.26 (*a*), the equivalent impedance diagram has been shown in Fig. 11.26 (*b*).

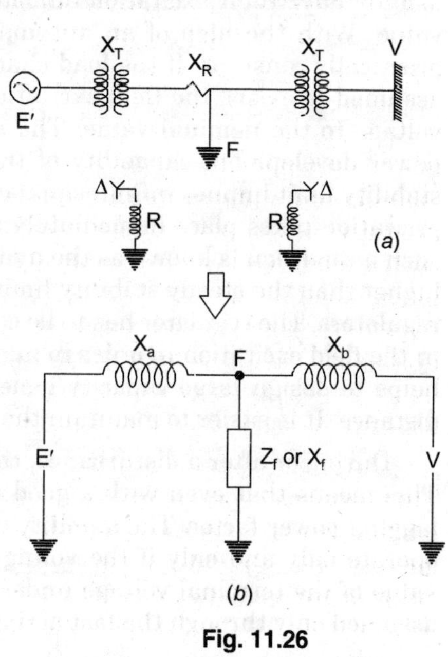

Fig. 11.26

The power transfer is given by

$$P = \left(\frac{EV}{X_a + X_b + \dfrac{X_a X_b}{X_f}} \right) \sin \delta \quad ...(11.40)$$

where X_f = equivalent shunt-fault impedance

Physically the resistance in the neutral of the transformer represents an absorption of electrical energy which in turn decreases the accelerating energy. This improves the transient stability.

11.8.6 Electrical Breaking

The circuit breaker poles are normally closed and the resistances R are thus shorted under the normal working condition (Fig. 10.27). They are opened under fault condition, and absorb the accelerating energy during fault. Thus, the transient stability is improved. The use of braking resistors increases the generator output when the fault is on and even after it is cleared. It does not make the system inherently more reliable. It improves the stability at the cost of other characteristics of the system. Present day quick-operating circuit breakers and relays leave little gain to be realised by braking resistors.

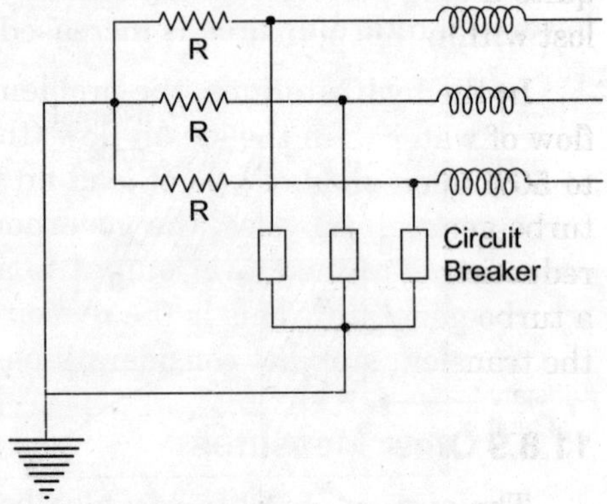

Fig. 11.27. Braking resistor to improve system stability.

11.8.7 Effect of Voltage Regulators

The maximum power which generator can deliver is governed by field excitation of the generator. The excitation again depends upon the system operating conditions. Generators usually have their excitation adjusted to hold their terminal voltage at some predetermined value. With the hlep of an automatic voltage regulator, the terminal voltage can be held practically constant. If the load change is slow (such that the steady-state conditions may be assumed to exist), the field excitation would change slowly to adjust the terminal or system voltage to the nominal value. The steady-state stability limit corresponds to the maximum power development capability of the generator for slow changes of load. Thus steady-state stability limit implies infinitesimal change in load with constant field current. If the change in excitation takes place immediately after each small change in load, the stability limit under such a condition is known as the dynamic stability limit. This dynamic stability limit is always higher than the steady stability limit and is obtainable only with the help of automatic voltage regulators. The regulator has to be capable of making continuous and sufficiently rapid changes in the field excitation in order to maintain the dynamic stability. The use of voltage regulators helps to design large capacity generators when power is to be transmitted over fairly long distance. It is easier to maintain the dynamic stability with the help of regulators.

During or after a disturbance, the terminal voltage usually tends to be lower than normal. This means that even with a good excitation system, the current would be drawn at highly lagging power factor. The stability can be maintained and even the relays may be allowed to operate only and only if the voltage is not allowed to fall appreciably below the prescirbed value of the terminal voltage under the normal condition of operation. This condition can be assumed only through the fast acting voltage regulators.

However, use of automatic voltage regulators increases the transient stability of the system after the first swing but the stability limit during the first swing can be improved only by clearing the faults as early as possible by means of the circuit breakers and the protective devices used in the system.

11.8.8 Fast Governor Action

The fast changes occurring in the power angle during system disturbances can be arrested by varying the prime mover output with the help of the fast acting governor. If the prime mover output can be changed in time, the accelerating energy of the rotor can be decreased. Experiences have revealed that the earliest action of the governor can be brought in the system after about 15 cycles (*i.e.*, 0.30 see for 50 Hz system) of the occurrence of the fault. This is quite a long period from the severity of the faults and the system stability might have been lost within this period.

In the hydrostations, the problem is still more difficult because of difficulty to control the flow of water than the steam flow (the maximum instantaneous overspeed has been about 35 to 50% upon sudden loss of load on the water-wheel generators while it is only 10% for the turbo-generators). Also, the governors and hydraulic conditions do not always permit rapid reduction of prime mover output to a water-wheel generator as is to some extent possible for a turbo-generator. This is the reason why governor action has not been so much advocated by the transient stability considerations.

11.8.9 Other Measures

The system stability can also be improved by inserting a permanent or switched series capacitor, or disconnecting a shunt reactor and connecting a shunt capacitor and by closing throttle or gate valve in the sending area as soon as possible or by opening throttle or gate valve in the receiving area also as early as possible. Use of generators with higher H values also improves transient stability though it is a costly method. Reduction of transfer reactance by reducing the conductor spacing or by increasing conductor diameter also improves transient stability. Use of bundled conductors is also an alternative. Another method to improve system stability is to increase the number of interconnecting EHV lines. Recent trends of improving transient stability includes lowering of x_s (synchronous reactance) of alternator (*i.e.*, enhancement of SCR, short circuit ratio) and use of HVDC links in between regional grids.

11.9 SOLUTION OF SWING EQUATION (MATHEMATICAL CALCULATION)

There are basically two methods of solution of swing equation for power system stability as : (*a*) pre-calculated swing curve and (*b*) step by step solution. Below discussions are made on these two methods :

(*a*) Pre-calculated Swing Curve

In fact, easiest and the simplest way to find the clearing angle corresponding to a given clearing time is to refer to the appropriate pre-calculated swing curve. This curve can be used with system of two finite machines connected through any line or network.

We have seen above that, the swing equation is given by :

$$M \cdot \frac{d^2\delta}{dt^2} = P_m - P_e = P_a$$

Now the equivalent power output is governed by δ and can get it by the power-angle equation for the reactance network as, $P_e = P_{max} \sin \delta$. Then to make this curve generally applicable, we can plot them in terms of a dimensionless variable, that means, the time constant $\hat{i}$, which is

$$\hat{i} = t\sqrt{\frac{\pi}{180} \cdot \frac{P_{max}}{M}} = t\sqrt{(\pi f P_{max})/G_H} \quad \text{...(11.41)}$$

$$\left(\text{as } M = \frac{GH}{180\,F}\right)$$

We know, from above $\quad M\frac{d^2\delta}{dt^2} = P_m - P_{max} \sin \delta \text{ (as } P_e = P_m \sin \delta)$

Now let, $\quad \frac{d^2\delta}{dt^2} = \left(\frac{d\hat{i}}{dt^2}\right)^2 \cdot \frac{d^2\delta}{d\hat{i}^2}$

But, we also, know, from eqn. (11.52)

$$\frac{d\hat{i}}{dt} = \sqrt{\frac{\pi}{180} \cdot \frac{P_{max}}{M}} = \sqrt{\frac{\pi f P_{max}}{G_H}} \quad \text{...(11.42)}$$

Hence, from this equation we can write,

$$M\left(\frac{d\hat{i}}{dt}\right)^2 \cdot \frac{d^2\delta}{dt^2} = P_m - P_{max} \sin \delta$$

or $$M\left(\frac{\pi f}{180} \cdot \frac{P_{max}}{M}\right)\frac{d^2\delta'}{dt^2} = P_m - P_{max} \sin \delta$$

or $$\frac{\pi f}{180} \cdot \frac{d^2\delta'}{d\hat{i}^2} = \frac{P_m}{P_{max}} - \sin \delta$$

But if δ′ is in radian, $\quad \left(\frac{\pi f}{180}\right) = 1.00$

Then, $$\frac{d^2\delta'}{d\hat{i}^2} = P - \sin \delta \quad \text{...(11.43)}$$

where $\quad P = \frac{P_m}{P_{max}}$

Hence, the equation $\frac{d^2\delta'}{d\hat{i}^2} = P - \sin \delta$ is the equation of dimensionless swing curve which is used for the determination of the pre-calculated swing curve, a set of such curves are shown in Fig. 11.28

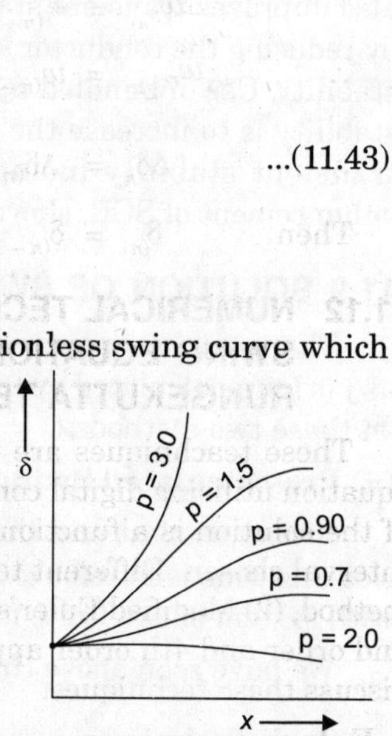

Fig. 11.28. Set of predetermined curves.

11.10 STEP BY STEP SOLUTION

By utilising the step by step procedure, the swing equation can also be solved. In this case we assume that the accelerating power P_a and the rotor angular velocity w_c are constant within each of a succession of intervals (top and middle) as shown above in the Fig. 11.28. These values are used to find the change in the power angle δ during each interval.

In the first step, $P_a(0^+)$ is to be regulated. This is $P_a(0^+) = P_m - P_e(0^+)$. Then the swing equation becomes, $\frac{d^2\delta}{dt^2} = \propto (0^+) = \frac{P_a(0^+)}{M}$ and the change in w_r is given by $\Delta w_r = \alpha(0^+)\,\Delta t$. Then $w_r = w_0^+ \Delta w_r = w_0 + \alpha(0^+)\Delta t$. The change in power angle for the first interval is $\Delta\delta_1 = \Delta w_r\ \Delta t$.

and hence, $\delta_1 = \delta_0 + \Delta\delta_1 = E_0 + \alpha(0^+)\,\Delta t^2$

To Determine P_a

Say, if there is no discontinuity in the swing curve during an iteration interval, $P_a(0^+)$ is equal to half of P_a immediately after the fault. However, if there is a discontinuity at the beginning of the jth interval then,

$$P_{a_{(j-1)}} = \frac{1}{2}\left[P_{a_{(j-1)-}} + P_{a_{(j+1)+}}\right]$$

Hence, $P_{a_{(j-1)-}}$ indicates the accelerating power before the fault is cleared $P_{a_{(j-1)+}}$ is the accelerating power after the clearance of the fault.

Now, if the discontinuity occurs at the middle of the interval, then for that interval

$$P_a = P_m - P_{output} \text{ (during fault)}$$

But as soon as the fault is cleared, $P_a = P_m - P_{output}$ (after clearance of the fault).

11.11 ALGORITHM OF ITERATIONS IN SOLUTION OF SWING EQUATION

Here, $P_{a_{(n-1)}} = P_m - P_{e_{(n-1)}}$ (Similarity to equation $P_a(0^+) = P_m - P_0(0^+)$ can be noticed)

$$\therefore \quad P_{e_{(n-1)}} = \frac{|E|\,|V|}{X_{12}} \sin\delta_{(n-1)} \text{ and } \propto_{(n-1)} = \frac{P_{a_{(n-1)}}}{M} \left(\text{same as equation } (0^+) = \frac{P_{a(0^+)}}{M}\right)$$

$$\Delta w_{r_{(n)}} = \propto_{(n-1)} \cdot \Delta t \text{ (Similar to equation } \Delta w_r = \propto (0^+)\,\Delta t)$$

$$\therefore \quad w_{r_{(n-1)}} = w_{(n-1)} + \propto_{(n-1)} \cdot \Delta t \text{ (Similar to equation } w_0 + \alpha(0^+)\,\Delta t)$$

$$\therefore \quad \Delta\delta_n = \Delta\delta_{(n-1)} + \frac{P_{a_{(n-1)}}}{M} \cdot \Delta t^2, \text{ as } \Delta\delta_n = \Delta\delta_{(n-1)} + w_{r_{(n)}} \times \Delta t$$

Then $\delta_{(n)} = \delta_{(n-1)} + \Delta\delta_n$ (Similar to equation $\delta_{(n)} = \delta_0 + \Delta\delta_1$)

11.12 NUMERICAL TECHNIQUES TO SOLVE SWING EQUATION (EULER'S METHOD, RUNGEKUTTA TECHNIQUES)

These teachniques are necessary for solving swing equation utilising digital computers. Infact, the accuracy of the solution is a function of the teachnique and time interval chosen. Different techniques are as : (1) Euler's method, (2) Modified Euler's method and (3) Rungekutta 2nd order and 4th order approximation. Below we shall discuss these techniques.

1. Euler's Technique

Let us refer to the Fig. 11.29

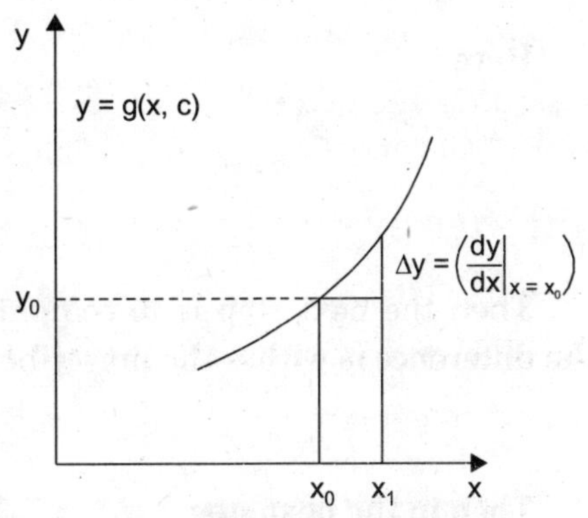

Fig. 11.29. Euler's method.

Now say $$\frac{dy}{dx} = f(x, y) \qquad ...(11.44)$$

where x is the independent variable and y is the dependent variable.

Now to determine, $y = f(x)$, the independent variable is given a set of predetermined values and the dependent variables corresponding values are evaluated. Now, say, the solution is

$$y = g(x, e)$$

Again let us assume that x_0 and y_0 are known.

Then, $$x_1 = x_0 + h$$

where $$h = \Delta t$$

and $$y_1 = y_0 + \left(\left.\frac{dy}{dx}\right|_{x=x_0}\right) \cdot h \qquad ...(11.45)$$

Here $$\left.\frac{dy}{dx}\right|_{x=x_0} = f(x_0, y_0)$$

Then in the 2nd step, $$x_2 = x_1 + h_1$$

and similarly, $$y_2 = y_1 + \left(\left.\frac{dy}{dx}\right|_{x=x_1}\right) \cdot h \qquad ...(11.46)$$

Here, $$\left.\frac{dy}{dx}\right|_{x=x_1} = f(x_1, y_1)$$

There is some inaccuracy in this method and that is that it is assumed a constant slope over the entire interval h as can be seen in Fig. 11.29.

2. Modified Euler's Method

Here let x_1, y_0 be known, $\therefore$ $\left.\frac{dy}{dx}\right|_0$ is known $\left(\text{as } \frac{dy}{dx} = f(x, y)\right)$. Then, in the next step, $x_1 = x_0 + h$. Now let $y_1^{(0)}$ = first approximate value, $\left(\text{Here, } y_1^{(0)} = y_0 + \left(\left.\frac{dy}{dx}\right|_0 \cdot h\right)\right)$. Hence, x_1 and $y_1^{(0)}$ are known. Again, $y_1^{(1)}$ = becomes approximate value.

Here $$\Delta y = \frac{\left(\left.\frac{dy}{dx}\right|_0 + \left.\frac{dy}{dx}\right|_1^{(0)}\right) \cdot h\, y_1^{(0)}}{2}$$

$$= y_0 + \Delta y \left(\text{as } \left.\frac{dy}{dx}\right|_1^{(0)} = f(x_1\, y_1^{(0)} \text{ and is known}\right)$$

Then the next step is to compare $y_1^{(1)}$ and $y_1^{(0)}$ (as expressed above) and check whether the difference is within the prescribed tolerance. If not, then continue the solution.

Then in the next step $$y_1^{(2)} = y_0 + \frac{\left(\left.\frac{dy}{dx}\right|_0 + \left.\frac{dy}{dx}\right|_1^{(1)}\right) \cdot h}{2} \qquad ...(11.47)$$

This process continues till the difference between the last two approximates lies well within n prescribed tolerance upto 10^{-4}.

3. Runge Kutta Second Order Approximation

In Runge Kutta method, the independent variable is given a set of predetermined values and the corresponding dependent variables are evaluated directly from a set of formula, these formula are expressed in some derivatives at predetermined points. The method is self starting and does not require any successive approximation as used in the modified Euler's method.

Here again, let, $\frac{dy}{dx} = f(x, y)$

Now, say, x_0, y_0 are the initial values are known apriori.

Then let us select very small increment to x_1 say h.

Now, $x_1 = x_0 + h$ (given value), $y_1 = y_0 + \Delta y$ (to be evaluated)

Expressing, $y_1 = y_0 + a_1 k_1 + a_2 k_2$, $k_1 = f(x_0, y_0)\, h$, and $k_2 = f(x_0 + b_1 h, y_0 + b_2 k_1)\, h$. Our next step is to evaluate the constant a_1, a_2, b_1 and b_2.

Now, $$k_2 = f(x_0 + b_1 h_1, y_0 + b_2 k_1)\, h$$

Now let us now expand k_2 by Taylor's expansion around (x_0, y_0) as the first approximation.

Then, we get, $$k_2 = \left[f(x_0, y_0) + b_1 h \left.\frac{\delta f}{\delta y}\right|_0 + b_2 k_1 \cdot \left.\frac{\delta f}{\delta y}\right|_0 \right] h$$

$$= f(x_0, y_0)\, h + b_1 h^2 \left.\frac{\delta f}{\delta x}\right|_0 + b_2 k_1 h \left.\frac{\delta f}{\delta y}\right|_0$$

$$y_1 = y_0 + a_1 f(x_0, y_0)\, h + a_2 f(x_0, y_0)\, h + a_2 b_1 h^2 \left.\frac{\delta f}{\delta x}\right|_0^0 + a_2 b_2 f(x_0, y_0)\, h \left.\frac{\delta f}{\delta y}\right|_0 + \quad \text{...(11.48)}$$

$$y_0 = y_0 + f(x_0, y_0)(a_1 + a_2)\, h + a_2 b_1 h^2 \left.\frac{\delta f}{\delta x}\right|_0 + a_2 b_2 f(x_0, y_0)\, h^2 \cdot \left.\frac{\delta f}{\delta y}\right|_0 + \ldots \quad \text{...(11.49)}$$

Then, again expanding, y_1 by Taylor's expansion around (x_0, y_0)

$$y_1 = y_0 + \left.\frac{\delta y}{\delta x}\right|_0 h + \frac{1}{2} \left.\frac{\delta^2 y}{\delta x^2}\right|_0^{h^2} + \ldots$$

But $$\left.\frac{\delta^2 y}{\delta x^2}\right|_0 = \frac{\delta}{\delta x}[f(x, y)] = \left.\frac{\delta f}{\delta x}\right|_0 + \left.\frac{\delta f}{\delta y}\right|_0 \cdot \left.\frac{\delta y}{\delta x}\right|_0 = \left.\frac{\delta f}{\delta x}\right|_0 + \left.\frac{\delta f}{\delta y}\right|_0 f(x_0, y_0)$$

$$y_1 = y_0 + f(x_0, y_0)\, h + \frac{1}{2} h^2 \left[\left.\frac{\delta f}{\delta x}\right|_0 + \left.\frac{\delta f}{\delta y}\right|_0 f(x_0, y_0) \right] \quad \text{...(11.50)}$$

Now comparing eqns. (11.58) and (11.60), we get

$$a_1 + a_2 = 1,\ a_2 b_1 = \frac{1}{2},\ a_2 b_2 = \frac{1}{2}$$

Now let us assume, $$a_1 = \frac{1}{2},\ a_2 = \frac{1}{2},\ b_1 = 1_1\ b_2 = 1$$

$$k_1 = f(x_0, y_0)\,h;\; k_2 = f(x_0 + h, y_0 + k_1)\,h$$

Then, once k_1 and k_2 are known, $y_1 = y_0 + a_1 k_1 + a_2 k_2$, and $\Delta y = \frac{1}{2}(k_1 + k_2)$. In this case, the accuracy of this method is of the order of h^3.

RUNGEKUTTA FOURTH ORDER APPROXIMATION

In this fourth order approximation method the series is to be truncated after the fourth term. Here accuracy is of the order of h^5 (in previous method of 2nd order it was h^3). Here the process used, is almost same as the that adopted for the 2nd order approximation as done above. Here again,

say, $x_1 = x_0 + h$

$$y = y_0 + a_1 k_1 + a_2 k_2 + a_3 k_3 + a_4 k_4$$

where $k_1 = f(x_0, y_0)\,h,\; k_2 = f(x_0 + b_1 h, y_0 + b_2 k_1)\,h$

$$k_3 = f(x_0 + b_3 h, y_0 + b_4 k_1)\,h,\; k_4 = f(x_0 + b_5 - h, y_0 + b_6 k_3)\,h$$

Now choosing, $a_1 = 1/6, a_2 = 2/6, a_3 = 2/6, a_4 = 1/6$

and $b_1 = 1/2, b_2 = \frac{1}{2}, b_3 = \frac{1}{2}, b_4 = \frac{1}{2}, b_5 = 1, b_6 = 1$

Here, Runge Kutta fourth order approximation for solving simultaneous differential equations is :

$$\frac{dy}{dx} = f(x, y, z), \text{ and } \frac{dz}{dx} = g(x, y, z)$$

Now, $y_1 = y_0 + 1/6\,(k_1 + 2k_2 + 2k_3 + k_4)$

$$z_1 = z_0 + 1/6\,(l_1 + 2l_2 + 2l_3 + l_4)$$

$\therefore$ $k_1 = f(x_0, y_0, z_0)\,h$ $\quad l_1 = g[x_0, y_0, z_0]\,h$

$k_2 = f[x_0 + (h/2), y_0 + (k_1/2), z_0 + (l_1/2)]\,h$ $\quad l_2 = g[x_0 + (h/2), y_0 + (k_1/2), z_0 + (l_1/2)]\,h$

$k_3 = f[x_0 + (h/2), y_0 + (k_2/2), z_0 + (l_2/2)]\,h$ $\quad l_3 = g[x_0 + (h/2), y_0 + (k_2/2), z_0 + (l_2/2)]\,h$

$k_4 = f[x_0 + h, y_0 + k_3, z_0 + l_3]\,h$ $\quad l_4 = g[x_0 + h, y_0 + k_3, z_0 + l_3]\,h$

11.13 SOLUTION OF SWING EQUATION (BY COMPUTATION)

The swing equation expressed in terms of state equations can be solved by (*a*) Modified Euler Method and (*b*) Rugne Kutta Method and many other methods. However here we shall use the mostly used two methods discussed above, and as below mathematically and computation.

(*a*) Modified Euler's Method

One object is to solve the equations $x = f(x)$. The independent variable is discretised into time element, $t^0, t^1, \ldots t^k$equidistance maintained. Then the following steps are ;

1. Let us assume an initial solution x^0 and time count is set as $k = 0$.
2. Then compute the derivative $X^k = f(X^k)$.
3. Now compute first estimate of the state-vector.

$$X^{k+1} = X^k + X^k \cdot \Delta t$$

4. Then compute second estimate of the state derivative.

$$X^{k+1} = f(X^{k+1})$$

5. Now compute the average of the state derivatives.

$$X_{av}^{k} = \frac{1}{2}\left[X^{k} + X^{k+1}\right]$$

6. Now let us evaluate the second and first of the stae vector.

$$X^{k+1} = X^{k} + X^{k+1}_{av}\,\Delta t$$

Now print the X^{k+1}

7. Check if t_{max}, if not increment time $t = t + \Delta t$, *i.e.*, $k + k + 1$ and let us go to step (2).
8. Now let us terminate the computation.

In the Fig. 11.30, we can see various steps involved in solution by computation.

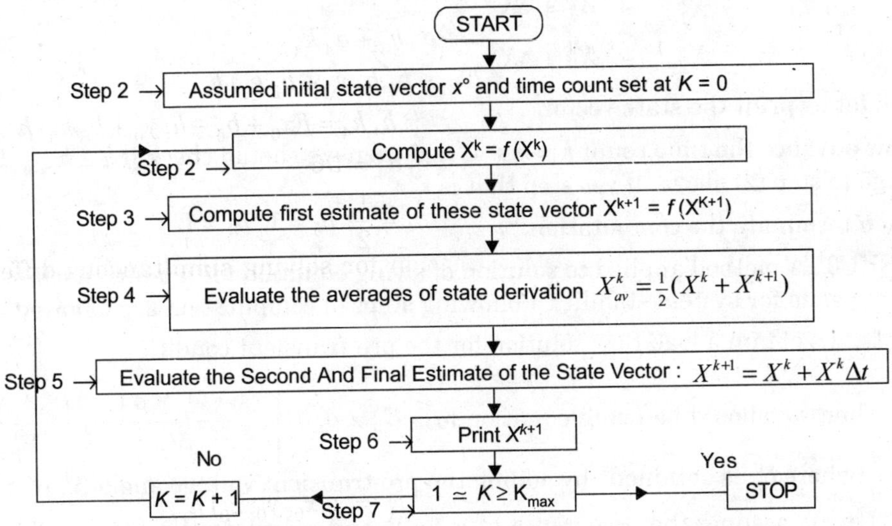

Fig. 11.30. Computing steps for modified Euler's method.

(*b*) Computing of Rungekutta Method

In the case let us consider two first order differential equation in two variables X_1 and X_2, such as

$$X_1 = f_1(x_1, x_2) \text{ and } X_2 = f_2(x_1, x_2)$$

Again as above :

1. Let us assumed x_1^0, x_2^0 and set time count is set at $k = 0$.
2. Now compute the following eight constant $(k_1, l_1 \ldots)$.

$$k_1^k = f_1\left(x_1^k, x_2^k\right)\Delta t$$

$$L_1^k = f_2\left(x_1^k, x_2^k\right)\Delta t$$

$$k_2^k = f_1\left[x_1^k + \frac{1}{2}k_1^k, x_2^k + \frac{1}{2}l_1^k\right]\Delta t$$

$$l_2^k = f_2\left[x_1^k + \frac{1}{2}k_1^k, x_2^k + \frac{1}{2}l_1^k\right]\Delta t$$

$$l_3^k = f_1\left[x_1^k + \frac{1}{2}k_2^k, x_2^k + \frac{1}{2}l_2^k\right]\Delta t$$

$$k_4^k = f_1\left[k_1^k + k_4^k, x_2^k + l_3^k\right]\Delta t$$

$$l_4 = f_2\left[k_1^k + k_3^k, k_2^k + l_4^k\right]$$

3. Then compute the change in state vector.

$$\Delta X_1^k = \frac{1}{6}\left[k_1^k + 2k_2^k + 2k_3^k + k_4^k\right]$$

$$\Delta X_2^k = \frac{1}{6}\left[l_1^k + 2l_2^k + 2l_3^k + l_4^k\right]$$

4. Now evaluate the new state vector.

$$X_1^{k+1} = X_1^k + \Delta X_1^k$$

$$X_2^{k+1} = X_2^k + \Delta X_2^k$$

and let us print the state vector.

5. Now advance the time count $k = k + 1$, and then we should check if $k \geq k_{max}$. If not let us go to step (2) above. If yes step that is

6. Then terminate the computation.

RUNGEKUTTA method applied to solution of swing equation for a multi-machine system in the power system for system stability. Following steps of computation are followed :

Step 1. Let us obtain a load flow solution for the pre-transient condition.

Step 2. Then we should be using equation (*a*), $E_p \angle \delta'_p = \left[V_p^0 + \frac{Q_p^0 \times p}{V_p^0}\right] + j\frac{P_p^0 + X_{ep}}{V_p^0}$ and $\delta_p^0 = \delta'_p + \beta_{pp}$ (where δ_p^0 is obtained) by adding the pre-transient voltage angle β_p^0 to δ'_p.

Step 3. Let us assume the occurrence of a fault and calculate the reduced admittance matrix for this condition and initialise time count $k = 0$, initialise, $j = 0$.

Step 4. Determine the light constant using equation

$$\frac{d^2\delta_p}{dt^2} = (p_s - p_p) \qquad ...(11.51)$$

and $$x_1 = x_2 = \frac{d\delta}{dt} = w - 2\pi f, \text{ and } x_2 = \frac{d^2\delta}{dt^2} = \frac{1}{M}\left[P_s - P_p\right] = W \qquad ...(11.52)$$

Step 5. Then compute the change in state vector

$$\Delta\delta^k = \frac{1}{6}\left[k_1^k + 2k_2^k + 2k_3^k + k_4^k\right] \qquad ...(11.53)$$

and $$\Delta\omega^k = \frac{1}{6}\left[k_1^4 + 2l_2^k + 2l_3^k + l_4^k\right] \qquad ...(11.54)$$

Step 6. Now evaluate the new state vector,

$$\delta^{k+1} = \delta^k + \Delta\delta^k$$

$$\omega^{k+1} = \omega^k + \Delta\omega^k$$ and then print the state vector

Step 7. Now, evaluate the internal voltage behind transient reactance using the equation :

$$E_p^{k+1} = \left|E_p^k\right| \cos\delta_p^{k+1} + j\left|E_p^k\right| \sin\delta_p^{k+1}$$

Step 8. Now check if $t < t_e$ is yes, $k = k + 1$ and go to step (4).

Step 9. Then check if $j = 0$, yes, modify the network data and obtain a new admittance matrix corresponding to point-fault condition. Then set $j = j + 1$.

Step 10. Now set $k = k + 1$.

Step 11. Now check if $k < k_{max}$, yes, go to step (4).

Step 12. Then terminate the process.

Now refer to the Fig. 11.25 for computational. Algorithm using modified Euler method for Power System Problem.

Step 1. Let us obtain a load flow solution for the pre-transient condition.

Step 2. Calculate the generator voltages behind transient reactances using equations used above in step (2). The state vector have finite values whereas all $w^s = 0$ under pre-transient condition.

Step 3. Let us assume the occurrence of fault and initative time $k = 0$. then let us calculate the reduced admittance matrix for this condition. Then set count $j = 0$.

Step 4. Now we determine the state derivatives using the same equation we used above in step (4) and then calculate the first estimate.

$$\delta_p^{k+1} = \delta_p^k + \left.\frac{d\delta_p}{dt}\right|_k \cdot \Delta t, \text{ and } \omega_p^k + \left.\frac{d\omega}{dt}\right|_k \cdot \Delta t$$

Step 5. Then the second estimate of the variables can be obtained if derivatives at $t = t + \Delta t$ and obtained. For this we must have new values of E' at time $t = t + \Delta t$ so that the generated powers can be calculated as : $E_p^{k+1} = |E_p|^k \cos \delta_p^{k+1} + j|E_p|^k \sin \delta_p^{k+1}$ and let use these values in equations same as mention in the step (4).

Step 6. Then we shall determine the average value of state derivative and obtain the second estimate of state variables and the second estimate of the interval voltage angles and machine angular speeds :

$$\delta_p^{k+1} = \delta_p^k + \left[\frac{\dfrac{d\delta_p^k}{dt} + \dfrac{d\delta_p^{k+1}}{dt}}{2}\right] \cdot \Delta t$$

$$\omega_p^{k+1} = \omega_p^k + \left[\frac{d\omega_p/dt) + \dfrac{d\omega_p^{k+1}}{dt}}{2}\right] \cdot \Delta t$$

Step 7. New we compute find interval voltages of the generators at the end of $(t + \Delta t)$, $E_p^{k+1} = |E_p^k| \cos \delta_p^{k+1} + j|E_p^k| \sin \delta_p^{k+1}$ and then print the result.

Step 8. Now we check if $t < t_e$ (clearing time). If yes, advance time by Δt and go to step (4).

Step 9. Then we check if $J = 0$, if yes, the nodal admittance matrix is changed corresponding to the post fault condition and a new reduced admittance matrix is obtained. Then we set, $j = j + 1$.

Step 10. Set $k = k + 1$, and $t = t + \Delta t$.

Step 11. Then we check if $t \le t_{max}$, if yes, then go we go to step (4).

Step 12. Now we terminate the process of computation. Then the relation between δ and t for various generators, are thus obtained and stability or otherwise of the system can be estimated for a particular type of fault. In the Fig. 11.35 all the above computation steps are shown.

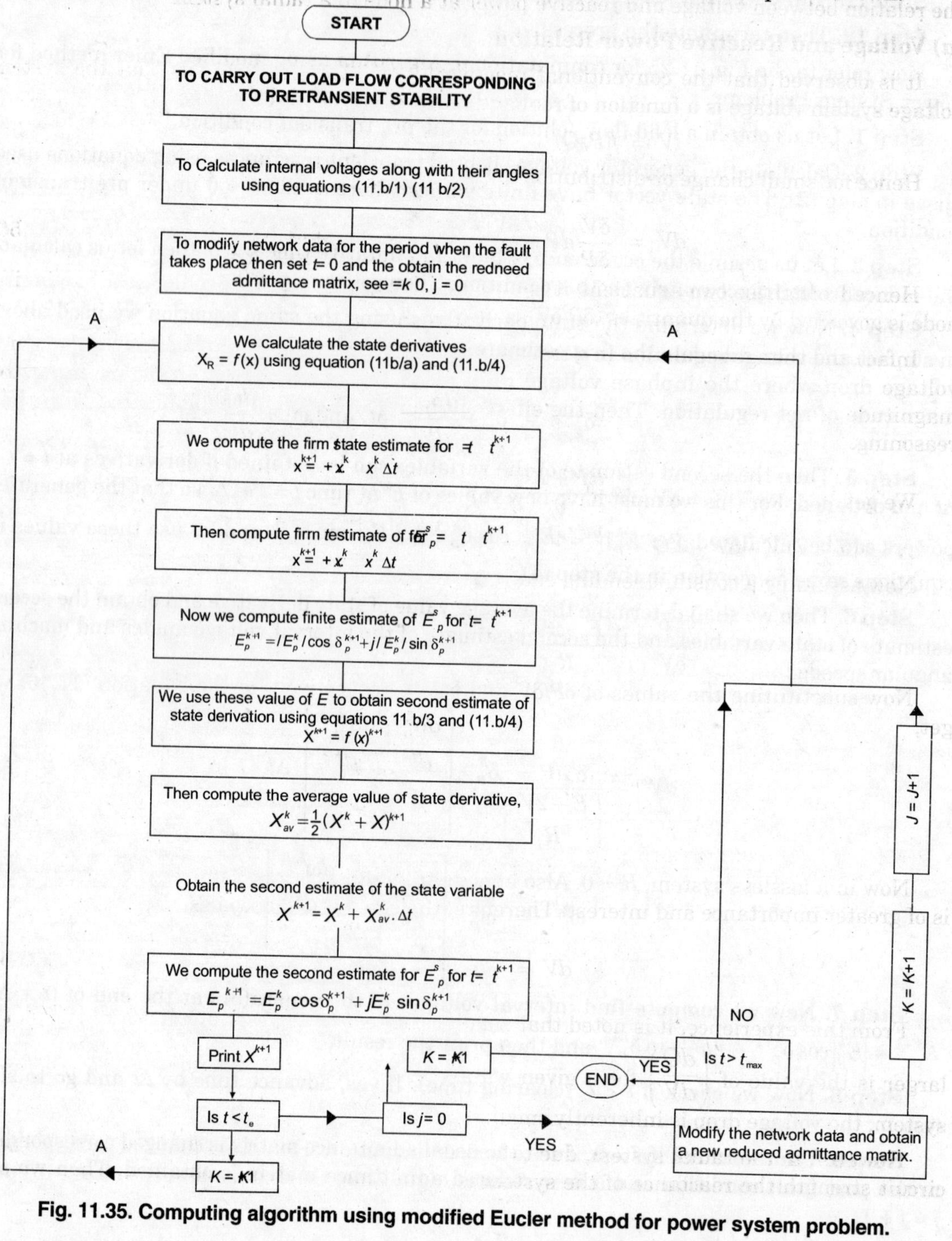

Fig. 11.35. Computing algorithm using modified Eucler method for power system problem.

11.14 VOLTAGE STABILITY IN THE POWER SYSTEM

In this section, discussion will be made on importance of voltage stability. Infact the dependence of voltage on the reactive power status of the corresponding is very important for stability of the system. Moreover voltage stability is essentially an aspect of stability of the load connected to a receiving end bus or a load bus. However, we should have some idea on the relation between voltage and reactive power at a node in a radial system.

(a) Voltage and Reactive Power Relation

It is observed that the conventional reactive power expression reveals that the system voltage system voltage is a function of real and reactive power which means.

$$V = f(P, Q) \qquad ...(11.55)$$

Hence for small change or distribution (or perturbation),

$$dV = \frac{\delta V}{\delta P} dP + \frac{\delta V}{\delta Q} \cdot dQ = \frac{dP}{(\delta P/\delta V)} + \frac{dQ}{(dQ/\delta V)} \qquad ...(11.56)$$

Hence from these two equations it is observed that the change in voltage at the specific node is governed by the quantities $(\delta P/\delta V)$ oand $(\delta Q/\delta V)$.

Infact, the voltage regulation in a transmission line is compared of inphase and quadrature voltage drop where the inphase voltage drop plays the principal role in determining the magnitude of net regulation. Then the effect of quadrature drop is negligible and with then reasoning.

We get, $$\Delta V = \frac{RP + XQ}{V} \text{ or } E - V = \frac{RP + XQ}{V}$$

$$\therefore \quad (E - V)V - RP - XQ = 0 \qquad ...(11.57)$$

Now assuming a constant sending end vcltage from eqn. (11.67)

$$\frac{\delta P}{\delta V} = \frac{E - 2V}{R} \text{ and } \frac{\delta Q}{\delta V} = \frac{E - 2V}{X}$$

Now substituting the values of $\delta P/\delta V$ and $\delta Q/\delta V$ respectively in the equation (11.56) we get,

$$dV = \frac{dP}{\left(\frac{E - 2V}{R}\right)} + \frac{dQ}{\left(\frac{E - 2V}{X}\right)} = \frac{dP \cdot R + dQ \cdot X}{E \cdot 2V} \qquad ...(11.58)$$

Now in a lossless system, $R = 0$. Also applying the concept of decoupling the term $(\delta Q/\delta V)$ is of greater importance and interest. Therefore, the eqn. (11.68) becomes,

$$dV = d\theta \cdot \frac{X}{E - 2V} \text{ or } \frac{dQ}{dV} = \frac{E - 2V}{X} \qquad ...(11.59)$$

From this experience, it is noted that smaller the reactance associated with the node, the larger is the value of $\left(\frac{dQ}{dV}\right)$ for a given voltage drop. This means that for a low reactance system, the voltage drop is inherently small.

However, in the radial system, due to limited number of EHV lines and low system short circuit strength the reactance of the system is high. This, then cause higher voltage drop in the system. This drop, in fact can be compensated by artificial injection of reactive power thus making (dQ/dV) higher. But this obviously enhances the cost of the power system.

Then, in case of light loading, the receiving end voltage is approximately the same as to the sending end voltage and thus equation (11.69) becomes $(dQ/dV) = -(E/X)$. But (E/X) representw the short circuit current flowing through the system. As such it is noted that the magnitude of (dQ/dV) is equal to the magnitude of the short circuit current. Therefore it can be said that the radial system with higher system reactance indicate lower short circuit strength that means obviously lower magnitude of short circuit current.

11.14.1 VOLTAGE CONTROL IN LONG RADIAL LINES

In long transmission lines voltage control is serious problem and infact this problem further aggravates due to high terminal reactances. Hence, it is at this stage it is necessary to establish the reactance power requirement of lines to maintain the system voltage near rated value, which mean that when the receiving end voltage varies from 0.95 p.u. to 1.05 p.u., the rated value being 1.0 p.u.

To determine on this, it is useful to use the generalised line equation for the line including the source reactance and the line transformers.

Hence, $E = AV + BI_R$. Here we assume the phase angle as ϕ_R at the receiving end and taking V as the reference phasor.

$$A = A\angle\alpha, B = B\angle\beta. \text{ Hence from equation } E = AV + BI_R$$

We get, $$E\angle\delta = AV\angle\alpha + BI_R\angle\beta - \phi_R$$

or $$E\angle\delta = AV\cos\alpha + jAV\sin\alpha + BI_R\cos(\beta - \phi_R) + jBI_R\sin(B - \phi_R)$$

Now equating the magnitudes only,

$$E^2 = A^2V^2 + B^2I_R^2 + 2ABVI_R\left[\cos(\alpha-\beta)\cos\phi_R - \sin(\alpha-\beta)\cdot\sin\phi_R\right] \quad ...(11.60)$$

But, we know,

$$\cos\phi_R = \frac{P}{VI_R} \text{ and } \sin\phi_R = \frac{Q}{VI_R} \quad (11.61)$$

This is for the loading of $(P + jQ)$ at receiving end b_m.

Now substituting the values of $\cos\phi_R$ and $\sin\phi_R$ from eqn. (11.57) in equation

$$E = AV + BI_R$$

We get, $$E^2 = A^2B^2 + B^2\left[I_{PR}^2 + I_{QR}^2 + 2ABP\cos(\alpha-\beta) - 2ABQ\sin(\alpha-\beta)\right]$$

(Here I_{PR} = real part of I_R and I_{QR} = reactive part of I_R)

$$\therefore \quad E = \sqrt{A^2V^2 + B^2\left[\frac{P^2}{V^2} + \frac{Q^2}{V^2}\right] + 2ABP\cos(\alpha-\beta) - 2ABQ\sin(\alpha-\beta)}$$

From this expression we can conclude that, for a given system A and B are known. Then Q may be calculated for any specific amount of real power flow to make the receiving end voltage (V) at derired voltage value in term of E (sending and voltage).

11.14.2 Voltage Stability in Power System

By now it is known that the receiving end voltage in transmission line is a function of the sending end voltage surge impedance line length and load along with its power factor.

In the Fig. 11.32 we can see that it represent the profile of the receiving end voltage for a particular line at different loads and power factors. In fact it can also be shown that in a transmission line model there are two roots of system voltage equation and at a particulars state of operation and these roots converge to a a point to denote the singe solution of the system voltage equation. The upper part of the characteristics of this Fig. 11.32 has been termed as STABLE VOLTAGE STATE OR LONE, while the lower part of the curves (where small changes in load curve unusual voltage changes) is terred as UNSTABLE VOLTAGE STATE OR ZONE in the system.

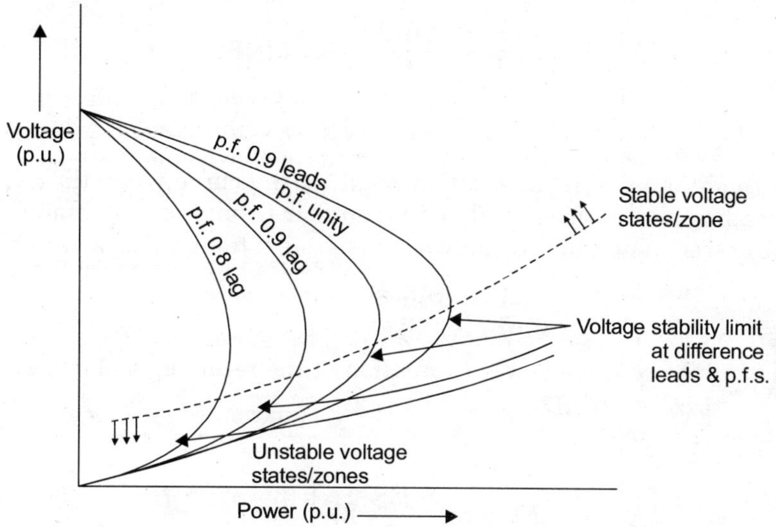

Fig. 11.32 Profile of V_R versus power of a power line at different p.f.

In the Fig. 11.26 it is noted that each of the characteristic curve represent different states of system operation where the "nose" of the characteristics represents the voltage stability limit. The magnitude of voltage at this state of power transfer, for any of the characteristics represent the critical voltage beyond which the system suffers voltage collapse.

Infact, the po ver system loads, particularly the induction motor play vitat role in governing the voltage stability of a transmission system. The nature of the load also affect the stability of voltage in the system.

Below it is shown that the steady state stability limit governs the voltage stability limit. For any consant flow of real power (that means when the generation is limited at a specific value) for any particular line transforming power from the sending end to the receiving end :

$$\sin \delta = \frac{P_x}{EV_R} \quad \text{...(11.71a)}$$

In this condition, $\delta = f(E \text{ or } V_R)$ and the Fig. 11.33 shown profile of δ versus V_R (Receving end voltage). In this we see that at $\delta = 90°$, the receiving end voltage, represented in the x axis is the lowest voltage to allow maximum load demand.

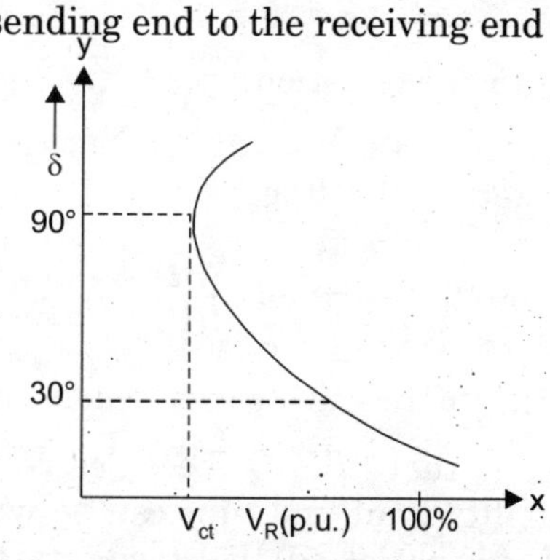

Fig. 11.33. Versus V_R profile.

Now the power expression for a loss less system is represented as $P = \frac{EV}{X} \sin \delta$

$$\therefore \qquad \frac{dP}{dE} = \frac{V}{X}\sin\delta + \frac{EV}{X}\cdot\cos\delta\cdot\frac{d\delta}{dE} \qquad ...(11.62)$$

Now assuming the power generation to be constants

$$\frac{dP}{dE} = 0 \qquad ...(11.63)$$

Hence from equations (11.72), using (11.73)

$$\frac{d\delta}{dE} = \frac{\tan\delta}{E}$$

Now as $\delta \to 90°$, $\dfrac{d\delta}{dE} \to -\infty$

Hence, this condition shown the location point on the curve when the x-axis in the Fig. 11.27 is E, the sending and voltage, E. Then that similar results can be obtained assuming the sending end voltage constant and then analysis the system taking V as a variable parameter.

$$\text{Here,} \qquad \frac{d\delta}{dV} = \frac{\tan\delta}{V} \qquad ...(11.64)$$

Now considering the reactive power expression at the receiving end bus we may write the expression for, $Q = \dfrac{EV}{X}\cdot\cos\delta - \dfrac{V^2}{X}$.

$$\therefore \qquad \frac{dQ}{dV} = \frac{1}{X}\left[E\cos\delta - EV\sin\delta\cdot\frac{d\delta}{dE} - 2V\right]$$

$$= \frac{1}{X}\left[E\cos\delta - EV\sin\delta\left(-\frac{\tan\delta}{V}\right) - 2V\right] \text{ (as from eqn. (11.74)}$$

$$= \frac{1}{X}\left[E\cos^2\delta + E\sin^2\delta - 2V\cos\delta\right]\frac{1}{\cos\delta} = \frac{1}{X}\left[\frac{E}{\cos\delta} - 2V\right] \qquad ...(11.65)$$

Hence, it is seen that eqn. (11.65) represent $\left(\dfrac{dQ}{dV}\right)$ in term of the power angle δ. We have seen above that $\dfrac{dQ}{dV} = \dfrac{E - 2V}{X}$. Which will be same as the equation (11.57) find $\dfrac{dQ}{dV}$ provided $\cos\delta = 1$ which mean $\delta = 0$.

In the equation, $\dfrac{dQ}{dV} = \dfrac{E}{X}$, this has been assumed during developing this expression by neglecting the quadrature voltage drop, which also gives the phase angle difference between the terminal voltage.

Now, at $\delta \to 90°$, $\dfrac{dQ}{dV} \to \delta$. This obviously manifests the fact that at steady state stability limit, the reactive power sensitivity $\left(\dfrac{dq}{dV}\right)$ with voltage becomes infinity which in other way means that (dV/dQ) becomes zero. Hence, this condition obviously represents the voltage stability limit and so it can be well assumed that under same assumptions (like when load effect is neglected) the steady state stability limit consider with voltage stability limit.

11.15 EFFECT OF GROUNDING ON STABILITY TRANSMISSION SYSTEM

In the transmission system Grounding Faults play a very important part in the stability of the power system. In fact the line-to-ground fault is the most common fault in the system and the magnitude of fault current depends upon the zero, positive and negative sequence impedance. Out of these impedances, positive and negative-sequence impedances are practically constant. The zero-sequence impedance depends upon the impedance of the grounding resistor or reactor, number of grounding points, zero-sequence impedance of the lines end of the tower footings. Normally an *L-G* (Line-to-Ground) fault is less serve as compared to other type of fault due to high value of zero-sequence impedance. However, effect of a double line-to-ground fault approaches that of a line-to-line (*L-L*) fault and a line-to-ground (*L-G*) fault approaches not fault when zero-sequence impedance is appreciably large.

Now, from stability point of view, the impedance grounding is better than solid grounding especially during ground fault. Either resistors or reactors are used to decrease the severity of ground fault. Reactors are more commonly used as they are cheaper than the equivalent values of resistors.

11.16 POWER TRANSFER CAPABILITY OF TRANSMISSION LINES

For a longitudinal power supply system, the power transfer capability of transmission lines is substantially affected by nodal power injection and topological changes. It has been generally accepted that the nodal strength, *i.e.*, the short circuit capability of the system is the inverse of the positive sequence equivalent impedance in pwer unit and it indicates the rebustness of the concerned power network. In LPS line, this impedance (which may be regarded as also source reactance) is dominated by the series reactance of the line and terminal reactance when analysed in a lossless frame. The terminal short circuit strength of LPS systems are generally load in comparison to those of robust systems.

In the longitudinal power supply system, this capability can be synonymous to loadability that has been defined earlier as the power transfer capability renders some specific sets of criteria. The power transfer capability can be represented in terms of surge impedance loading (SIL) where SIL = (V^2/Z_0). The operating power being represented as P_b is given by the basic expressions as

$$P_l = \frac{V^2}{X}\sin\delta \qquad \text{...(11.66)}$$

Assuming equal terminal voltages at sending and receiving end of a radial power line.

Now, $X = Ll\omega$, l being the inductance of line per unit length and L the line length

i.e., $X = \sqrt{\frac{l}{c}}\,\omega L\,\sqrt{lc}$

or $X = Z_0 \times \beta L$

Here, Z_0, the surge impedance = $\sqrt{\frac{l}{c}}$, c being the shunt capacitance per unit length, β is the phase constant.

$\therefore$ $X = Z_0 = \theta$ where, θ represents the electrical line length (= βL) in radian

$$= Z_0 \sin\theta, \theta \text{ being a small}$$

Hence, magnitude of power transfer is given by

$$P_l = \frac{V^2}{Z_0 \sin\theta} \sin\delta = (\text{SIL}) \frac{\sin\delta}{\sin\theta}$$

or
$$\frac{Pl}{\text{SIL}} = \frac{\sin\delta}{\sin\theta}$$

Thus the power transfer capability can be conveniently represented in terms of SIL. Fig 11.34 represents its profile for a typical radial system for varying line lengths. Fig. 11.35 shows comparison of loadability profiles (*i.e.*, power transfer capability) of a set of robust and weak power systems.

High source reactance of weak systems plays a vital role in limiting the line loadability. Loadability can be improved by reducing the system reactance (*i.e.*, by reducing reactance of overhead wires and placing series capacitors in the line) and relaxing voltage drop constraint. Reduction of line power loss improves line loadability also. This can be achieved by utilizing low resistance conductors, parallel wires in the transmission system as well as by placing shunt capacitor at the load end. However loadability is seriously impaired by application of shunt reactors (this is particularly evident upto a voltage level of 500 kV).

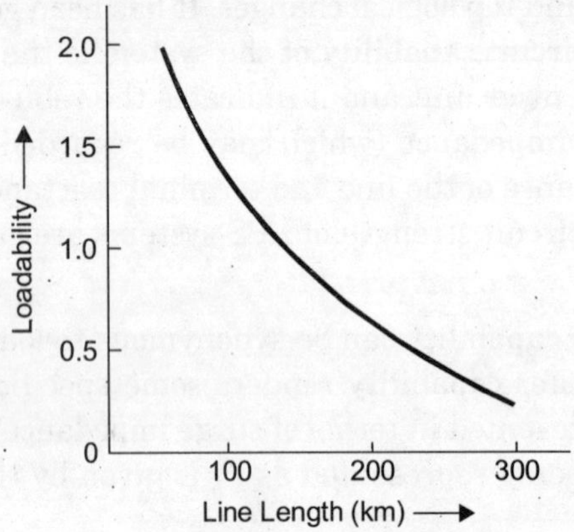

Fig. 11.34. Loadability of power line.

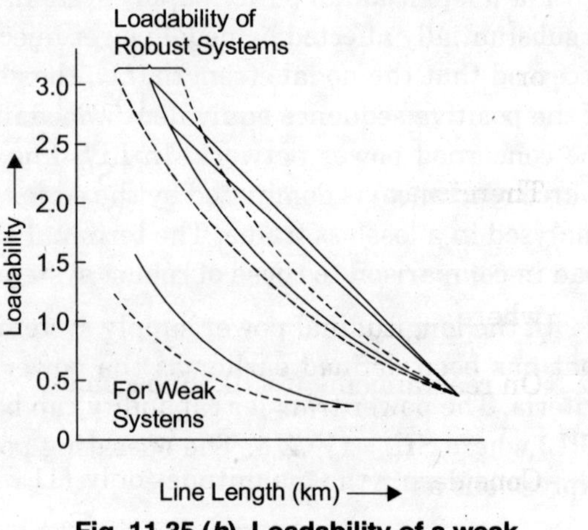

Fig. 11.35 (*b*). Loadability of a weak and robust system.

11.17 COMPLEX OPERATING POWER CONTOUR IN A POWER SYSTEM

Fig. 11.36 represents the equivalent π line model where the shunt capacitances of the line have been assumed to be lumped and connected at the two extreme ends (the sending end and the receiving end) of the line. The entire series reactance of the system has been assumed to be connected between the two buses utilising the lumped parameter concept. The system may be considered as a two area power network with longitudinal EHV transmission line transporting power from the source side to the load side. The terminal reactances assumed to be included in the series reactance between the buses and the mutual induction effect between the conductors have been neglected.

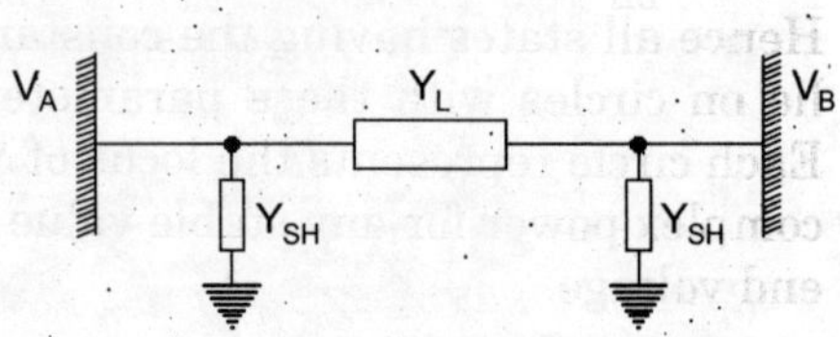

Fig. 11.36. A π line model.

The receiving end current in this equivalent line model is given by

$$I_B = (S/V_B)^* \qquad \text{...(11.67)}$$

Also, the system admittance (Y) is given by

$$[Y] = \begin{bmatrix} Y_{BB} & Y_{BA} \\ Y_{AB} & Y_{AA} \end{bmatrix} = \begin{bmatrix} (Y_L + Y_{SH}) & (-Y_L) \\ (-Y_L) & (Y_L + Y_{SH}) \end{bmatrix}$$

Y_L indicates the transfer while Y_{SH} indicates the shunt admittances. Utilising the conventional expression [I] = [Y] [V], this system can be represented by

$$\begin{bmatrix} I_B \\ I_A \end{bmatrix} = \begin{bmatrix} Y_{BB} & Y_{BA} \\ Y_{AB} & Y_{AA} \end{bmatrix} = \begin{bmatrix} V_B \\ V_A \end{bmatrix} \qquad \text{...(11.68)}$$

i.e.,

$$I_B = Y_{BB} \cdot Y_B + Y_{BA} \cdot V_A \qquad \text{...(11.69)}$$

Comparing equations (11.67) with (11.69)

$$\left(\frac{S}{V_B}\right)^* = Y_{BB} \cdot V_B + Y_{BA} \cdot V_A$$

or

$$S^* = Y_{BB} \cdot V_B^2 + Y_{BA} \cdot V_B^* \cdot V_A$$

or

$$\left(\frac{S^*}{Y_{BB}}\right) = V_B^2 + \left(\frac{Y_{BA}}{Y_{BB}} \cdot V_A\right) \cdot V_B^*$$

Therefore,

$$\left(\frac{S^*}{Y_{BB}}\right) = V_B^2 + vV_B^*$$

where,

$$v = \left(\frac{Y_{BA}}{Y_{BB}} \cdot V_A\right) \qquad \text{...(11.70)}$$

On rearrangement, (10.58) becomes,

$$S - Y_{BB}^* \cdot V_B^2 = v^* \cdot Y_{BB}^* \cdot V_B \qquad \text{...(11.71)}$$

Considering the magnitudes only, (11.81) becomes,

$$|S| - |Y_{BB}^* \cdot V_B^2| = |v^* \cdot Y_{BB}^* \cdot V_B| \qquad \text{...(11.72)}$$

Equation (11.72) represents the equation of a circle with ($Y_{BB}^* \cdot V_B^2$) as centre and $|v^* \cdot Y_{BB}^* \cdot V_B|$ as its radius. Hence all states having the constant amplitudes of (v) lie on circles with these parameters. on the S-plane. Each circle represents the locus of S, the receiving end complex power for any stable value of V_B, the receiving end voltage.

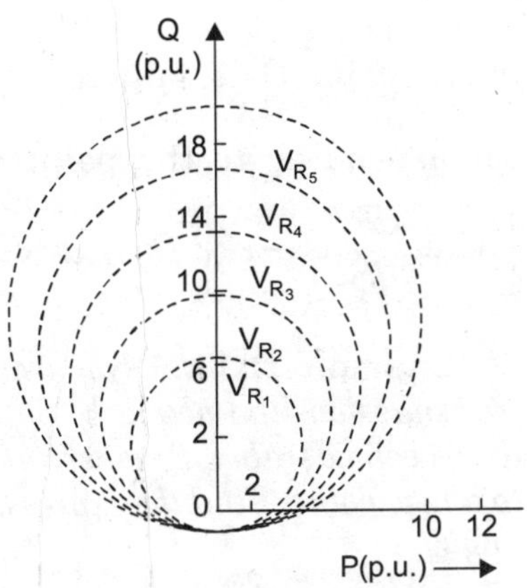

Fig. 11.37. Operating power contour.

It has thus been revealed that complex operating power in a reactive power constrained EHV system is a *circular contour*. Each circular locus represents the maximum magnitude of real and reactive power at a specified sending end voltage an at nay fixed value of receiving end voltage. Any improvement in the receiving end voltage enhances the power transfer capability, particularly the reactive power transfer capability of the line.

Though a large number of circular contours can be plotted utilising the above concept, the operating contours of receiving end voltages varying within the range of 0.95 p.u. to 1.05 p.u. are only feasible in any practical power system. In Fig. 11.38 this feasible portion of the contour has been shaded.

A preliminary concept regarding the operation of a power line within steady state stability limit being included in the basic courses of power system, further discussion has been presented here to clearify the concept. A conventional two mechine system has thus been assumed with no shunt branch and all losses are neglected. The power flow equations are given by

$$P_{AB} = \frac{|V_A| \ |V_B|}{X} \sin\delta = |V_A|\,|V_B|\,|Y_{AB}| \sin\delta = -P_{BA} \quad \text{...(11.73)}$$

and

$$Q_A = |V_A|^2 Y_{AB} - |V_A|\,|V_B|\,Y_{AB} \cos\delta \quad \text{...(11.74)}$$

$$Q_B = -|V_B|^2 Y_{AB} + |V_A|\,|V_B|\,|Y_{AB}| \cos\delta \quad \text{...(11.75)}$$

Setting all the derivatives of these equations to zero, a set of a non-linear algebraic equations is left. Solution of these equations leads to the mathematical representation of "equilibrium" point. for a particular system, when Y_{AB} is known, the equilibrium point will depend on a given set of performance parameters (*i.e.*, P_A, Q_A, P_B and Q_B). However, in addition to these parameters, there is another set of free variables viz. $(V_A - V_B)$ and $(\delta_A - \delta_B)$. Hence there may not be single solution of the power flow equations for any arbitrary value of these parameters and the free variables.

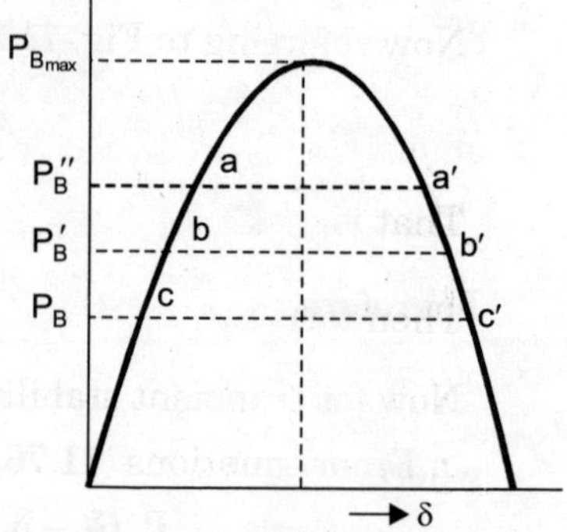

Fig. 11.38

Assuming fixed bus voltages, the *P*-δ characteristic of a power line can be plotted at the receiving end and is shown in Fig. 11.38. It may be observed that there is no load flow solution beyond $P_B = P_{B_{max}}$. Below $P_B = P_{B_{max}}$, there are always two equilibrium points (*i.e.*, two solutions of the load flow equations). Out of these positions, the state of operation corresponding to point *a, b, c* etc., below point *O* is said to be stable $\left[\text{as here, } \frac{dP}{d\delta} \text{ is } +ve\right]$ while the state of operation corresponding to the point *a, b, c,* and any other point beyond point *O* is unstable $\left[\text{here, } \frac{dP}{dS} \text{ is } -ve\right]$. *Critical state of stability* is achieved at point $\left[\text{where } \frac{dP}{d\delta} = 0\right]$.

Example 11.14. *A synchronous generator supplies power to a large grid through a parallel transmission line network. Suddenly a fault occurs at one of the lines and the power output drops considerably. Breakers at both the sides clear the fault and the power transfer is restored to a new value. Find the expression for critical clearing angle, initial angle and maximum power angle.*

Solution : Let us refer to the *P*-δ curve of Fig. 11.39,

Say δ_0 = initial power angle is steady state condition

δ_c = critical clearing angle

δ_m = max. power angle

In the Fig. 11.32 areas A_1 and A_2 are the areas for which equal area criterion is to be applied to check the transient stability and in order to determine the critical clearing angle.

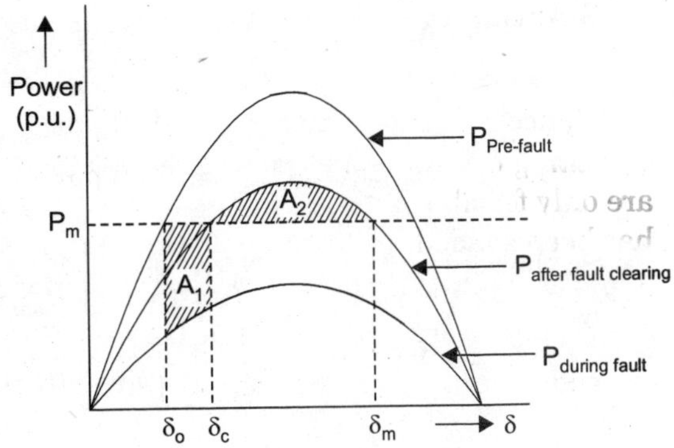

Fig. 11.39

Now let us say that

1. $P_{\text{pre fault}} = P^{(0)} = P_{\max} \sin \delta$
2. $P_{\text{during fault}} = P(f) = Y_1 P_{\max} \sin \delta$
3. $P_{\text{after fault clearing}} = P = Y_2 P_{\max} \sin \delta$

Here, Y_1 and Y_2 are the ratios of maximum power that can be transferred during and after the fault to the maximum power that can be transmitted before the fault occurred in the system.

Now smaller values of Y_1 result in greater disturbance to the system as low Y_1 indicates low power transfer during fault, *i.e.*, higher P_a, causing higher acceleration energy.

Now referring to Fig. 11.39, for areas A_1 and A_2 we get

$$A_1 = P_m(\delta_c - \delta_0) - \int_{\delta_0}^{\delta_c} Y_1 P_{\max} \sin \delta \, d\delta$$

That is,
$$A_1 = P_m(\delta_c - \delta_0) + Y_1 P_{\max} (\cos \delta_c - \cos \delta_0) \quad ...(11.76)$$

Then for,
$$A_2 = \int_{\delta_c}^{\delta_m} Y_2 P_{\max} \sin \delta \cdot d\delta - P_m(\delta_m - \delta_c) \quad ...(11.77)$$

Now for transient stability to be maintained, $A_1 = A_2$

∴ From equations (11.76) and (11.77), we get

$$P_m(\delta_c - \delta_0) = Y_1 P_{\max}(\cos \delta_c - \cos \delta_0) = Y_2 P_{\max}(\cos \delta_c - \delta_m) - P_m(\delta_m - \delta_c)$$

or
$$P_m\delta_c - P_m \delta_0 = Y_1 P_{\max} \cos \delta_c - Y_1 P_{\max} \cos \delta_0 = Y_2 P_{\max} \cos \delta_c - Y_2 P_{\max} \cos \delta_m - P_m \delta_m + P_m \delta_c$$

$$\therefore \quad (Y_1 - Y_2) P_{\max} \cos \delta_c = P_m(\delta_0 - \delta_m) + Y_1 P_{\max} \cos \delta_0 - Y_2 P_{\max} \cos \delta_m$$

Hence, for δ_c, we get,
$$\delta_e = \cos^{-1}\left[\frac{(P_m/P_{\max})(\delta_m - \delta_0) + Y_2 \cos(\delta_m - Y_1)\cos \delta_0}{Y_2 - Y_1}\right]$$

Here,
$$P_m = P_{\max} \sin \delta_c = Y_2 P_{\max} \sin \delta_m$$

$$\therefore \quad \delta_0 = \sin^{-1} \frac{P_m}{P_{\max}} \text{ for } \delta_0 < 90°$$

and
$$\delta_m = \sin^{-1} \frac{P_m}{Y_2 P_{\max}} \text{ for } \delta_m > 90°$$

Example 11.15. *A 50 Hz 3-phase synchronous generator delivers 100 p.u. power to an infinite busbar through a network in which resistance is negligible. A fault occurs which reduces the maximum power to 0.40 p.u. whereas before fault this power was 1.8 p.u. and after the fault clearance 1.30 p.u. By the use of equal area criterion determine the critical angle.*

Solution : From the example, we have

$$P_m = 1.00 \text{ p.u., and } P_{max} = 1.80 \text{ p.u.}$$

Hence, $Y_1 P_{max} = 0.40$ p.u.

and $Y_2 P_{max} = 1.30$ p.u.

So, here,
$$\delta_0 = \sin^{-1}\left(\frac{P_m}{P_{max}}\right) = \sin^{-1}\left(\frac{1.0}{1.8}\right) = 33.8°$$

and
$$\delta_\delta = 180° - \sin^{-1}\left(\frac{P_m}{Y_2 P_{max}}\right) = 180° - \sin^{-1}\left(\frac{1.0}{1.8}\right) = 129°.6$$

Also, we know,
$$Y_1 = \frac{Y_1 P_{max}}{P_{max}} = \frac{0.40}{1.80} = 0.222, \text{ and } Y_2 = \frac{Y_2 P_{max}}{P_{max}} = \frac{1.30}{1.80} = 0.722$$

Now, we know that,
$$\delta_c = \cos^{-1}\left[\frac{1}{Y_2 - Y_1}\{(\delta_f - \delta_0)\sin\delta_0 + (Y_2 \cos\delta_f - Y_1 \cos\delta_0)\}\right]$$

$$\therefore \quad \delta_0 = \cos^{-1}\left[\frac{1}{0.722 - 0.222}\left\{\frac{\pi(1.29.6° - 33.8°)}{1.80}\sin 33.8° + (0.722 \cos 129.6° - 0.222 \cos 33.8°)\right\}\right]$$

$$\therefore \quad \delta_c = \cos^{-1}[\{+0.93 + (-0.695)\}/(0.50)\}] = \cos^{-1} 0.57$$

$$\therefore \quad \delta_c = 55.28.$$

Example 11.16. *A synchronous generator is operating at an infinite bus and supplying 45% of the peak power capacity. As mean as a fault occurs the reactance between the generator and the line becomes four term it value before the fault. The peak power that can be delivered after the fault is cleared is 70% of the original maximum value. Determine the critical clearing angle.*

Solution : As from the example, we get,

$$P_{max} \sin\delta_0 = 0.45 P_{max} (= P_m) \therefore \delta = 26.79°$$

But we know,
$$P_{max} = \frac{EV}{X}\left(= \frac{V_S + V_R}{X}\right)$$

Now during fault "X" become $4X$ as given by the per example

$$\therefore \quad Y_1 P_{max} \sin\delta_m = \frac{EV}{4X}\sin\delta_m = \frac{1}{4}P_{max}\sin\delta_m \left(as\ P_{max} = \frac{EV}{X} \text{ or } \frac{V_s \cdot V_r}{X}\right)$$

$$\therefore \quad Y = 0.25$$

Now after clearance of the fault, $Y_2 = 0.7$

or $Y_2 P_{max} \sin\delta_m = P_m = 0.45 P_{max}$

$$\therefore \quad \delta_m = 40°$$

That is,
$$\delta_m = 180° - 40° = 140° = 2.44 \text{ rad}$$

Now, we know,
$$\cos\delta_c = \frac{1}{0.7 - 0.25}[0.45(2.44 - 0.467) + 0.7 \cos 40° - 0.25 \cos 76.79°]$$

$$= \cos 73° \therefore \delta_c = 73°$$

Example 11.17. *A synchronous motor of negligible resistance is receiving 25% of power that is capable of receiving form an infinite bus. If the load on the motor is suddely doubled calculate the maximum value of power angle during the swinging of the motor around its new equilibrium.*

Solution : As per example, $0.25 P_m = P_m \sin \delta_0$

$\therefore$ $\delta_0 = 14.5° = 0.25$ rad

Now we know, input electrical power to the motor is given by, $P_e = P_m \sin \delta_0$, where P_m is the maximum power which the motor is capable of receiving and say is equal to 1.00 p.u.

Now if the load on the motor is suddenly doubled, as stated in the example, then let δ′ be the new power angle.

$$\therefore \quad 2 \times 0.25 P_m = P_m \sin \delta'_2$$

(as $P_e = 2 \times P_m = 2 \times 0.25 P_m$ as per example)

$$\therefore \quad \delta'_2 = \delta' = 30° = 0.525 \text{ rad}$$

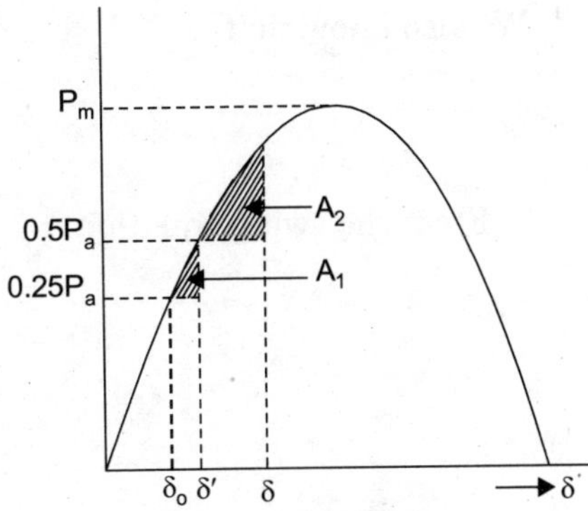

Fig. 11.40

Now let the max power angle be δ_1 as shown in the Fig. 11.33, so that the stability is maintained in the system.

Now applying equal area criterion we get,

$$-A_1 = A_2 \text{ (for motor operation)}$$

Hence, $-A_1 = A_2$

We get, $-\int_{\delta_0}^{\delta'} (0.5 P_{max} - P_m \sin \delta) d\delta = \int_{\delta'}^{\delta} (0.5 P_m - \sin \delta) d\delta$

or $-P_m (0.5\delta + \cos\delta)\Big|_{\delta_0}^{\delta_1} = P_m (0.5\delta + \cos\delta)\Big|_{\delta_1}^{\delta}$

or $-(0.5\delta + \cos\delta)\Big|_{0.25}^{0.525} = 0.5\delta + \cos\delta\Big|_{0.525}^{\delta}$

or $-0.0335 = 0.5\,\delta + \cos\delta - 0.262 - 0.866$

$\therefore$ $\cos \delta_m + 0.5\, \delta_m = 1.090$

This gives (by trial and error method)

$$\delta = 0.81 \text{ elat rad } 465°$$

Example 11.18. *A 50 Hz 4 pole turbo generator of rating 20 MVA, 13.2 kV has an inertia constant of H = 9 KW sec/kVA. Find the K.E. (Kinetic Energy) stored in the rotor at synchronous speed. Find the acceleration if the input less the rotational losses is 28,800 hp and the electric power developed is 16 MW.*

Solution : We know from our knowledge on the subject that stored energy is

$$G_H = 20 \times 9 = 180 \text{ M.W. sec.}$$

where G is the machine rating H is inertia constant.

$$= 180 \text{ M.J.}$$

As per example, the electric power developed,

$$P_m = 16{,}000 \text{ KW.}$$

Now as per example the Mechanical Power,

$$P_m = 26{,}800 \text{ HP} = 26{,}800 \times 0.746 = 20{,}000 \text{ KW.}$$

Hence, the accelerating power, $P_a = 20{,}000 - 16{,}000 = \mathbf{4000\ K.W. = 4\ MW}$

Again, we know, $P_a = M\dfrac{d^2\delta}{dt^2}$ (which is the swing equation)

We also know that, $M = \dfrac{GH}{180f} = \dfrac{180}{180\times 50} = \dfrac{1}{50}$ (as found above

$GH = 20 \times 9 = 180$ MW sec)

∴ From the swing equation, $P_a = M\dfrac{d^2\delta}{dt^2}$/we found $P_a = 4$ MW, $M = \dfrac{1}{50}$

$$\therefore \quad \frac{d^2\delta}{dt^2} = 200°/\text{sec}^2$$

$$= 200\times\frac{\pi}{180}\text{ rad/sec}^2$$

$$= 3.49 \text{ rad/sec}^2$$

Example 11.19. *A 20 MVA generator of the problem of example 11.18, is supplying rated power at 0.8 p.f. (lag) when a fault reduces the electric power output by 50%. Find the accelerating torque at the time the fault occurs. Losses may be neglected and power input to the generator may be assume constant.*

Soltuion : From the examples datas, we find

That, output power before fault, $P_m = 20{,}000 \times 0.8$

$= 16{,}000$ KW

Now also as per example,

Electrical power output, P_e, after fault = 50% of inital power

$$= \frac{1}{2}\times 16000 = 8000 \text{ KW}$$

(Initial electric power before fault was 16000 kW)

Hence, accelerating power, $P_a = P_m - P_e = 8000$ kW

Which mean, $T_a \cdot w_s$ = accelerating power P_a

$$\therefore \quad T_a \cdot w_s = P_a = 8000$$

$$\therefore \quad T_a = \frac{8{,}000}{w_s} = \frac{8000\times 10^3}{2\pi\times\left(\dfrac{120\times 50}{4}\right)/60} \quad \left(\because w_s = \frac{2\pi N_s}{60} \text{ and } N_s = \frac{120f}{P}\right)$$

$$= 50929.52 \; Nw \text{ mt.}$$

Example 11.20. *If W be the weight of the rotating part of an alternator in pounds, R its radius of gyration in fact, the manufactures supply the value of WR^2. Assuming that the rating of the machine to be 5 MVA, find a relationship between H (inertia constant) and WR^2.*

Solution : We know that the *KE* of rotation of the rotor at synchronous speed is

$$KE = KE = \frac{1}{2}\frac{WR^2}{32.2}\left(\frac{2\pi N}{60}\right)^2 \ ft.\ lb$$

where, N = rpm of the rotor

However, $1\ HP$ = 550 ft. lb/sec

= 746 W

550 ft. lb/sec = 746 W

$\therefore$ 1 ft. lb = (746/550) Joule

Hence, the conversion of ft. lb to Mega Joules and dividing the expression of *KE* by the machine rating :

$$H = \frac{K.E.\text{ft. lb.}}{S} = \frac{\left(\frac{1}{2}\frac{WR^2}{32.2}\right)\left(\frac{2\pi N}{80}\right)^2 \times \frac{746}{550} \times 10^{-6}}{S}$$

$$= \frac{2.31\times 10^{-10}\ WR^2 N^2}{S} \text{ MS/MA}$$

Example 11.21. *A two-machine power system delivers a load of 25 MW at 0.8 p.f. lag and has double circuit. The system reactance is 150 percent on 100 MVA base. A sudden symmetrical line-to-ground fault occurs in one of the circuit which reduced the power supplied to 40% which is subsequently cleared by the simultanous act of the circuit breakers on both sides of the faulted line. Calculate in electrical degrees the critical clearing angle.*

Assume that during fault condition the system reactance attains such a value that the maximum power becomes 30% of the normal maximum value when the fault line is isolated the maximum power of the system becomes 60% of the normal maximum power.

Solution : As per the datas given by the example we have first determine then δ_1, we know that, the expression for $\tan\delta_1 = \frac{2mV\cos\phi}{V^2 - mV}$ and from

where $m = \frac{IX}{2} = X_P \frac{VP/\cos\phi}{2(kVA)_b \times 100}$

$$= 150\left[\frac{V\times\frac{25\times 10}{0.8}}{2\times 100\times 10^9 \times 100}\right] = \mathbf{0.2344\,V}$$

$$\therefore \quad \tan\delta_1 = \frac{2\times mV\cos\phi}{V^2 - mV^2} = \frac{2\times 0.2344V^2}{V^2 - 0.0549V^2} = \frac{2\times .234\times 08}{0.9451} = 0.396$$

$$\therefore \quad \delta_1 = \tan^{-1}(0.396) = 21.6° = 0.377 \text{ rad}$$

Now we know, $P_1 = P_{\max_1}\sin\delta_1$

$$\therefore \quad P_{\max_1} = \frac{P_1}{\sin\delta_1} = \frac{25}{0.358} = 68.0 \text{ MW}$$

As per example, during fault, $P_{2(\text{fault})} = 0.4 \times P_1 = 0.4 \times 25 = 10$ MW

Also as per example, $P_{\max_2} = 0.3\, P_{\max_1} = 0.3 \times 68 = 20.4$ MW

$\therefore$ $P_{2(\text{fault})} = P_{\max_2} \cdot \delta_2$

$$\therefore \quad \sin \delta_2 = \frac{P_{2(\text{fault})}}{P_{\max_2}} = \frac{10}{20.4} = 0.49$$

$$\therefore \quad \delta_2 = \sin^{-1} 0.49 = \mathbf{29.4°}$$
$$= 0.5131 \text{ rad}$$

$$P_{\max_2} = 0.5 \times P_{\max_1} = 0.5 \times 68.0 = \mathbf{34\ MV}$$

As, $P_1 > P_{\max_1}$, hence to determine the critical angle considering the equation (for δ_e),

$$\cos \delta_c = \frac{1}{P_{\max_3} - P_{\max_2}} \left[P_1(\pi - \delta_1 - \delta_1) - P_{\max_2} \cos \delta_1 - P_{\max_2} \cos \delta_1\right]$$

$$= \frac{1}{34.0 - 20.4}\left[25(3.14 - 0.377 - 0.377) - 20.4 \times 0.93 - 34 \times .93\right]$$

$$= \frac{1}{13.6}[59.65 - 50.6] = \frac{9.06}{13.65} = 0.665$$

$$\therefore \quad \cos \delta_c = 0.665$$

$$\therefore \quad \delta_c = \cos^{-1}(.665) = \mathbf{48^0\ 19'\ Ans.}$$

QUESTIONS

1. Define steady state stability and transient stability of a power system.
2. State the factors that effect (*a*) steady state stability and (*b*) transient state stability of the power system.
3. State what is meant by critical angle.
4. Derive swing equation and discuss its application in the study of power system stability.
5. State what is equal area criterion discuss its application and limitation in the study of power system.
6. Discuss the application of equal area criterion for the system stability study when (1) a sudden increase in load taken place, and (2) a short circuit on one of the parallel factors takes place which is cleared after certain time.
7. Discuss the effect of neutral grounding on the stability of power system.
8. Develop an expression for the maximum power transfer between two nodes. Prove that the power is maximum when $X = \sqrt{3} \cdot R$, where X is the reactance and R is the resistance of the system.
9. The generalised circuit constants of a nominal circuit representing a 3-phase transmission line are :

   $A = D = 0.975\ \angle 0.3°$, $B = 82.5\ \angle 75.0°\ \Omega$, $C = 0.005\ \angle 90°\ ℧$

   Determine the steady state stability limit of the line if the sending end and receiving end voltages are held at 220 KV.

10. A 4-pole, 50 Hz., 22 kV turbo alternator has a rating of 100 MVA p.f. 0.8 lag. The moment of inertia of rotor is 8000 kgm^2. Determine M and H.

11. The sending end and receiving end voltages of transmission line at a 100 MW load are equal at 115 kW. The per phase line, impedance is (4 + j 7) ohm. Calculate the maximum steady state power limit of the line.

12. The Fig. *Q*.1 shows a radial power system where 3-phase fault occurs is at point *F*. Find the critical clearing angle. The generator may be assumed to deliver, 1.0 p.u. power at the instant preceding the fault.

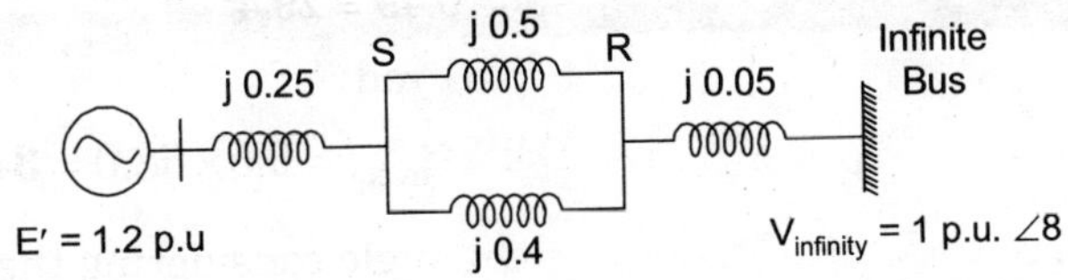

Fig. *Q*. 1

For more worked-one examples on Power System stability see the Appendix.

CHAPTER 12

Load Flow Analysis

12.1 LOAD FLOW

INSIDE THIS CHAPTER

- Load flow
- Type of buses
- Bus (node) admittance matrix
- Development of load flow equations
- Load flow equation & Gauss-Seidel Method
- Load Flow Equation
- Extension of N-R Method (Fast-decoupled load flow)
- What is meant by sparsity matrix
- Solved examples on load flow problem solution

This chapter is devoted to the load flow problem encountered by the power system operation. In fact load flow solution is a solution of the network under steady state condition subject to certain can be in the form of load nodal voltages, reactive power generation of the generators, the tap settings of a tap changing under load transference etc. The load flow solution gives the nodal voltages and phase angles and hence the power injection at all the buses and power flows through interconnecting power transmission lines.

For a power system engineers load flow solution is essential for designing a new power system and for planning extension of the existing one for increased load designe. These analysis require the calculation of numerous load flows under both normal and abnormal operating conditions. Load flow solution also gives the initial conditions of the system when the transient behaviour of the system is to be studied. A load flow solution of the power system requires mainly the following steps : (1) formulation of the network equations and (2) Suitable mathematical technique for solution of the equation by digital computers also for quick and correct solutions.

The load and hence generation are continually changing in a power system. However for no solution, we shall assume in our analysis below that the loads and hence generation are fixed at a particular value over a suitable period of time say half an hour so. But since we are studying the system under steady conditions the network equations will be in the form of simple algebraic equation.

In a power system the bus is associated with four quantities, real and reactive power, bus voltage magnitude and its phase angle. In the solution of load flow problem in the power system, two out of these four quantities are specified and the remaining two are required to be found through the solution of the algebraic equation determined on nodal analysis basis. However, the buses are classified depending upon which the quantities have been specified.

12.2 TYPE OF BUSES

As stated above in power system the buses are of following types, depending on which quantities they have been specified :

Types of Bus	*Specified quantities*	*Quantities to be obtained*
1. Load Bus	P_1	$\lvert V\rvert, \delta$
2. Generator Bus	$P_1 \lvert V\rvert$	Q, δ
3. Slack Bus	$\lvert V\rvert, \delta$	P, Q

We should briefly define these three types of buses :

1. Load Bus

In load bus the real and reactive components of power are specified. Hence, it is desired to find out the voltage magnitude and phase angle through the load flow solution.

2. Generator Bus

In this case voltage magnitude corresponds to the generation voltage and real power P_G corresponding to the rating specified.

3. Slack or Swing or Reference Bus

As stated in a power system there are mainly two type of buses. Load and generator buses. For these buses we have specified the real power P injection. Now $\sum_{i=n}^{\infty} P_i$ real power loss P_L and P_i is the power injection at the buses, which is taken as positive for generator buses and is negative for load buses. Infact, the losses remain unknown until the load flow solution is complete. And it is for this reason that generally one of the generator buses is made to take the additional real and reactive power to supply transmission losses. This is why this type of bus is known as the Slack or Swing bus. At this bus voltage V and phase angle δ are specified whereas the real and reactive powers at P_0 & Q_0 respectively are found by load flow solution.

For load flow solution, the phase angle of the voltage at the slack bus is usually taken as the reference.

And in the analysis below the real and reactive component of voltage at a bus are taken as the independent variables for the load flow equation, that mean, $V_i = \angle\delta_i = e_i + j f_i$, where e_i and f_i are the real and reactive component at voltage or current or both are taken as the independent variables. The load flow equation can be formulated using either the load or bus frame of reference. However, from the viewpoint of computer time and memory, the nodal admittance formulation, using the nodal voltages as the independent variables is the most economic method of load flow solution.

12.3 BUS (NODE) ADMITTANCE MATRIX

In the Fig. 12.1 a three-bus system is shown, for which the load flow equation, using nodal admittance formulation for a three bus system, are developed first and then they are generalised for an m bus system.

Now, to get load flow equation using nodal admittance method, we can write at three nodes, (1), (2) & (3) following equation.

At node of the Fig. 12.1, that at the bus no. (1) by using nodal analysis principle of circuit theory and network analysis.

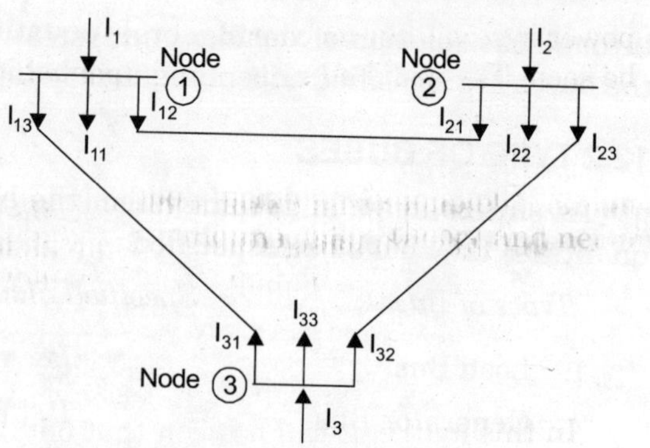

Fig. 12.1. A 3-bus system.

We get,

$$I_1 = I_{11} + I_{12} + I_{13}$$
$$= V_1 Y_{11} + (V_1 - V_2) Y_{12} + (V_1 - V_3) Y_{13}$$
$$I_1 = V_1(Y_{11} + Y_{12} + Y_{13}) - V_2 Y_{12} - V_3 Y_{13}$$
$$= V_1(Y_{11} + Y_{12} + Y_{13}) - V_2 Y_{12} + V_3 Y_{13} = V_1 Y_{11} + V_2 Y_{12} + V_3 Y_{13} \quad ...(12.1)$$

In eqn. (12.1) Y_{11} is the shunt charging admittance at the Bus (1), and

This $\quad Y_{11} = Y_{11} + Y_{12} + Y_{13}, Y_{12} = -Y_{12}$ and $Y_{13} = -Y_{13}$

Similarly by nodal analysis, we find nodal current equation for the nodes (2) and (3), as :

At node (2) $\quad I_2 = V_1 Y_{21} + V_2 Y_{22} + V_3 Y_{23} \quad ...(12.2)$

At node (3), $\quad I_3 = V_1 Y_{31} + V_2 Y_{32} + V_3 Y_{33} \quad ...(12.3)$

Now these equation can be written in a matrix form as below :

$$\begin{bmatrix} I_1 \\ I_2 \\ I_3 \end{bmatrix} = \begin{bmatrix} Y_{11} & Y_{12} & Y_{13} \\ Y_{21} & Y_{22} & Y_{23} \\ Y_{31} & Y_{32} & Y_{33} \end{bmatrix} \begin{bmatrix} V_1 \\ V_2 \\ V_3 \end{bmatrix} \quad ...(12.4)$$

or in the compact mathematical expression these equations can be written as :

$$I_p = \sum_{q=1}^{3} Y_{pq} V_q, \text{ where } p = 1 \text{ to } 3 \quad ...(12.5)$$

Hence, from this equation, we can now write nodal current equation for n-bus system where each node is connected to all other nodes :

$$T_p = \sum_{q=1}^{n} Y_{pq} V_q, \text{ where } p = 1, 2, 3 \ldots n \quad ...(12.6)$$

The same can be expressed in matrix form

$$\begin{bmatrix} I_1 \\ I_2 \\ I_3 \\ I_n \end{bmatrix} = \begin{bmatrix} Y_{11} & Y_{12} \ldots\ldots\ldots Y_{1_n} \\ Y_{21} & Y_{22} \ldots\ldots\ldots Y_{2_n} \\ Y_{31} & Y_{32} \ldots\ldots\ldots Y_{3_n} \\ Y_{n_1} & Y_{n_2} \ldots\ldots\ldots Y_{n_n} \end{bmatrix} \begin{bmatrix} V_1 \\ V_2 \\ V_3 \\ \vdots \\ V_n \end{bmatrix} \quad ...(12.7)$$

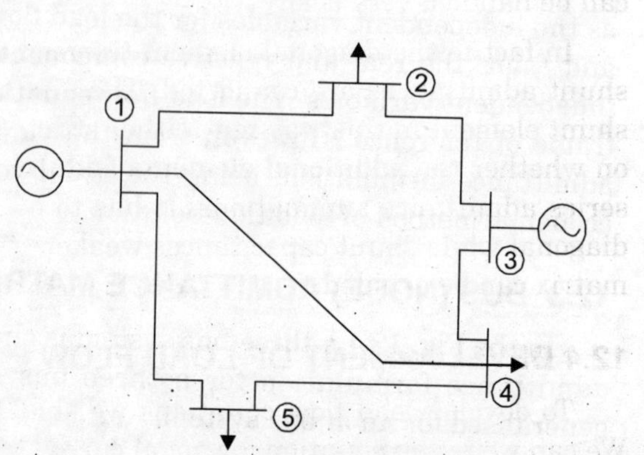

Fig. 12.2. Five bus system

In fact nodal admittance matrix is a "sparse" matrix (which means in this type of matrix a few number of elements are non zero) for an actual power system. Let us consider one system as shown in Fig. 12.2 where a five-bus system can be seen. The nodal admittance matrix for this system can be :

$$Y_{pq} = \begin{bmatrix} Y_{11} & Y_{12} & 0 & Y_{14} & Y_{15} \\ Y_{21} & Y_{22} & Y_{23} & 0 & 0 \\ 0 & Y_{32} & Y_{33} & Y_{34} & 0 \\ Y_{41} & 0 & 0 & Y_{43} & Y_{44} \\ Y_{51} & 0 & 0 & Y_{54} & Y_{55} \end{bmatrix}$$

In this matrix it can be seen that out of 25 elements eight elements are zero, whereas 17 are non-zero. But in a large system these non-zero get less in numbers. For example in a very large system say of 100 nodes there may be non-zero element number as small as 2% to 3% of the total number of elements.

And this is where we see that the computer memory requirements for storing the nodal admittance matrix is very low. It needs to store only a very few non-zero element, it need not store the zeros of the matrix. Again the nodal admittance being a symmetric matrix along the loading diagonal, the computer need store the upper triangular nodal admittance matrix only. As such, the computer memory requirement for storing the nodal admittance is all the more reduced.

However, if the interconnection between the various nodes for a given system, and the admittance value for each interconnecting circuit are known, then the admittance matrix may be assembled as follows : (*a*) The diagonal element of each node is the sum of the admittances connected to it. (*b*) The off-diagonal element is the negated admittance between the nodes.

But it can be seen that the sum of the elements in each column of the admittance matrix addition becomes zero, and this mean that nodal admittance matrix is a singular matrix and as such the rows of the matrix are linearly dependent.

Above it has been stated the necessity of selecting one of the buses as the slack or reference bus based on power balance in the power system. Then, based on admittance matrix approach, it can be said mathematically that one of the buses should be taken as slack otherwise the nodal matrix is singular and cannot be handled.

Now by taking one of the buses as reference, correponding row and column are deleted from the nodal admittance matrix and hence the reduced matrix becomes non-singular, which can be handled very easily.

In fact to the diagonal element (corresponding to the nodes at which they are connected) shunt admittances are connected. The off-diagonal elements are unaffected. The addition of shunt element in this way may either strangthen or weaken the *Y*-matrix diagonal depending on whether the additional elements add to or subtract from the diagonal terms representing series admittance summations. It has to be said that the shunt inductances strengthen the diagonal while shunt capacitances weaken it. With this also the singularity of the admittance matrix can be avoided.

12.4 DEVELOPMENT OF LOAD FLOW EQUATIONS

To develop load flow equations we shall use the nodal current equations derived above. We can write the equation for nodal current for a *n* bus system.

$$I_p = \sum_{q=1}^{n} Y_{pq} V_q \text{ where } p = 1, 2, \ldots n$$

$$I_p = V_{pp} V_p + \sum_{\substack{q=1 \\ q \neq 1}}^{n} Y_{pq} \cdot V_q \qquad \ldots(12.8)$$

or $$V_p = \frac{I_p}{Y_{pp}} - \frac{1}{Y_{pp}} \sum_{\substack{q=1 \\ q \neq p}}^{n} Y_{pq} \cdot V_q \qquad \ldots(12.9)$$

Now, $$V_p^* \cdot I_p = P_p - jQ_p \qquad \ldots(12.10)$$

or $$I_p = \frac{P_p - jQ_p}{V_p^*} \qquad \ldots(12.11)$$

Substituting for I_p in equation (12.9)

$$V_p = \frac{1}{Y_{pp}} \left[\frac{P_p - jQ_p}{V_p^*} - \sum_{\substack{q=1 \\ q \neq p}}^{n} Y_{pq} \cdot V_q \right] \text{ where, } p = 1, 2, \ldots n \qquad \ldots(12.12)$$

The I_p has been substituted by the real and reactive powers because normally in a power system these quantities are specified.

Equation (12.12) are the load flow equation where bus voltages are the variables. It can be seen that the load flow equations are nonlinear and they can be solved by an iterative method which are : (1) Gauss Method and (2) Gauss Seidel Method. Besides these two the load flow problem can be also solved by using another method, the Third Method : (3) Newton-Raphson Method.

Here it must be again stated that load flow analysis is very important when new networks or addition to new networks are considered. Infact the load demand undergoes wide change in a day of 24 hours. This means that at every moment the load demand changes occur throughout the interconnected network. The generation of power at all these moment must equal the demand. For most econonic operation the load changes must be divided in a unique ratio. This mean that load must flow in between the generating station but care must be taken that we may not overaload the interconnecting transmission lines resulting into operation of these lines close to their stability limit.

Below, hence, we shall now discuss the methods mentioned above for solving the Load Flow problem developing different voltages at different buses or nodes of the system network where the admittance voltages of these equations are complex quantities and the number of non-linears equations is $(n - 1)$ where n is the total number of buses (or nodes) in the system.

Let us first look into the Gauss Method as discussed (as easily as possible) below :

12.4.1 Gauss Method (Iterative Procedure)

This method is essentially a trial and error method and provides a sufficiently accurate numerical solution. In fact this solution is followed following way :

1. We guess the initial value of the solution.

2. This guessed solution is used along with the original equation to calculate as correct and better estimate for the solution.
3. Then the second better estimate is used to find the fourth estimate.
4. This process is continued until the difference between successive estimates is extremely negligible.

Now let us see the method as below :

1. Let us assume a flat voltage profile for all nodal voltages except the slack bus (1). Now let the slack bus voltage be $a + j\ 0 \times 0$. Now let us assume a suitable value of convergence criterion $\in$, that is, if the absolute value of the maximum change in voltage between any two consecutive iterations is less than a pre-specified tolerance, $\in$, the convergence is achieved and the iterative procedure is terminated.
2. Set iteration count $k = 0$.
3. Set bus count $p = 1$.
4. Check for the slack bus. If it is not a slack bus go to next step. Since voltage at the slack bus is fixed both in magnitude and phase, it does not vary during iterative procedure and hence go to step (6) if it is a slack bus.
5. Then calculate the bus voltage V_p^{k+1} using equation (12.12) and the difference in bus voltage $\Delta V_p^k = V_p^{k+1} - V_p^k$.
6. Now advance the bus count by 1 to evaluate other values of V_p^{k+1} and ΔV_p^k.
7. Now check if all buses have been taken into account. If yes, then go to the next step.
8. Now determine the largest absolute value of change in voltage $|\Delta V|_{max}$.
9. If $|\Delta V_{max}|$ is less than a specified tolerance $\in_1$ evaluate line flows and print the voltage and line flows. If not, advance the iteration count $k = k + 1$ and go back to step (3).

Let us now look into the computational feature in this method. In fact it is possible to obtain significant reduction in computer time if all arithmetic operation can be performed in advance when they do not change with interation. Now, as P and Q and Y at a bus do not change with iterations, then $\dfrac{P_p - jQ_p}{Y_{pp}}$ can be evaluated in advance.

Now, say, $$A_p = \frac{P_b - jQ}{Y_{pp}} \text{ for all } p = 1, 2, ..., n, p \neq s \quad ...(12.13)$$

Similarly, say, $$B_{pq} = \frac{Y_q}{Y_{pp}} \text{ for all } p = 1, 2, ...n; q \neq p \quad ...(12.14)$$

Then with these simplifications, the voltage equation now becomes

$$V_p^{k+1} = \frac{A_p}{\left(V_p^k\right)^*} - \sum_{q=1, q \neq p}^{n} B_{pq} \cdot V_p^k,\ p = 1, 2, ... n_1 p \neq s \quad ...(12.15)$$

Now let us see how the solutions can be achieved for Load Flow problem by using Gauss-Seidel Technique / Method (Iterative Method) including P.V. Buses.

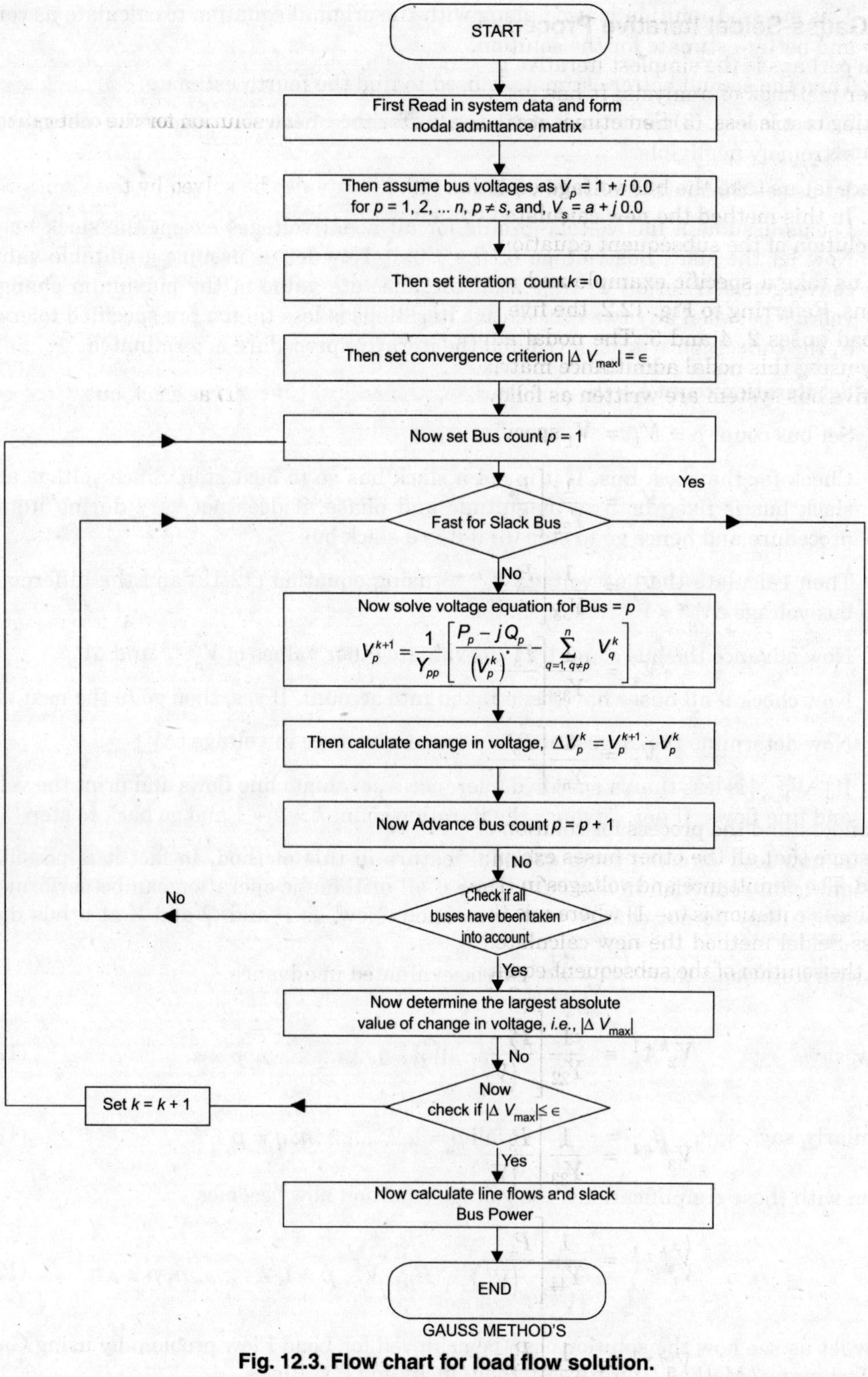

GAUSS METHOD'S

Fig. 12.3. Flow chart for load flow solution.

12.4.2 Gauss-Seidel Iterative Procedure

This perhaps is the simplest iterative method and has been in use since early days of digital computer methods of analysis. It has got the following advantages : (1) It is very simple, (2) Computing cost is less, (3) Sometimes it is used to find the initial solution for the other iterative procedures.

Hence let us take the bus voltage equation (12.15) can also be solved by the Gauss-Seidel method. In this method the new calculated voltage V_i^{k+1} immediately replaces V_i^k and is used in the solution of the subsequent equation.

Let us take a specific example and apply these methods for the solution of the load flow equations. Referring to Fig. 12.2, the five bus system has two generators at buses 1 and 3 and three load buses 2, 4 and 5. The nodal admittance matrix is also given alongwith Fig. 12.2. Then by using this nodal admittance matrix alongwith equation (12.12), the load flow equation for the five bus system are written as follows. Now, assuming bus (1) as slack bus :

$$V_1 = V_1 \text{ specified fixed values}$$

$$V_2 = \frac{1}{Y_{22}}\left[\frac{P_2 - jQ_2}{V_2^*} - (Y_2, Y_1 + Y_{23} \cdot V_3)\right]$$

$$V_3 = \frac{1}{Y_{33}}\left[\frac{P_3 - jQ_3}{V_3^*} - (Y_{32}\, V_2 + Y_{34}\, V_4)\right]$$

$$V_4 = \frac{1}{Y_{33}}\left[\frac{P_3 - jQ_3}{V_3^*} - (Y_{41}\, V_1 + Y_{45}\, V_5)\right] \quad \text{...(12.16)}$$

$$V_5 = \frac{1}{Y_{55}}\left[\frac{P_5 - jQ_5}{V_5^*} - (Y_{51}\, V_1 + Y_{54}\, V_4)\right]$$

To understand the process for solution of these equation we make a simplifying assumption at this stage that all the other buses except bus No. 1, are loaded, *i.e.*, buses where P and Q are specified. The admittance and voltages in these equations are complex quantities and the number of non-linear equation is $(n-1)$ where n is the total number of buses in the system. In the process of Gauss-Seidel method the new calculated voltage, V_i^{k+1} immediately replaces V_i^{k+1} and is used in the solution of the subsequent equation. Hence the equations (12.16) in this case become:

$$V_2^{k+1} = \frac{1}{Y_{22}}\left[\frac{P_2 - jQ_2}{\left(V_2^k\right)*} - (Y_{21}\, V_1 + Y_{23} V_3)*\right]$$

$$V_3^{k+1} = \frac{1}{Y_{33}}\left[\frac{P_3 - jQ_3}{\left(V_3^k\right)*} - \left(Y_{32} V_2^{k+1} + Y_{34} V_4^k\right)\right]$$

$$V_4^{k+1} = \frac{1}{Y_{44}}\left[\frac{P_4 - jQ_4}{\left(V_4^k\right)*} - \left(Y_{41} V_1 + Y_{43} V_3^{k+1} + Y_{45} V_5^k\right)\right]$$

$$V_5^{k+1} = \frac{1}{Y_{55}}\left[\frac{P_5 - jQ_5}{\left(V_5^k\right)*} - \left(Y_{51} V_1 + Y_{54} \cdot V_4^{k+1}\right)\right] \quad \text{...(12.17)}$$

Hence, the general load flow equation resultant for the Gauss-Seidel (*G-S*) method will be given as below :

$$V_p^{k+1} = \frac{1}{Y_{pp}}\left[\frac{P_p - jQ_p}{\left(V_p^k\right)^*} - \sum_{q=1}^{p-1} Y_{pq} V_q^{k+1} - \sum_{q=p+1}^{n} Y_{pq} V_q^k\right] \quad ...(12.18)$$

On the above equation (12.18) the second term of the RHS is clear because the voltage prior to bus *p* should correspond to the value as calculated during the current iteration. However, the procedure for solution of these equation (12.18) being stated below and takes into account the presence of voltage controlled buses in addition to the load buses.

But during the procedure it must be always remembered that : (1) The voltages and admittances are complex quantities, (2) Buses 2, 4 and 5 are loaded buses where *P* and *Q* are known quantities whereas bus 3 is a voltage controlled bus where only *P* is known and *Q* is not known. Therefore, for solving equation corresponding to a voltage. Controlled bus *Q* must finit be calculated and is written as :

$$Q_p = \sum_{q=1}^{n}\left\{f_p\left(e_q G_{pq} + f_q B_{pq}\right) - e_p\left(f_q G_{pq} - e_p B_{pq}\right)\right\} \quad ...(12.19)$$

Here with equation (12.19), $Y_{pq} = G_{pq} - j\,B_{pq}$ and $V_p = e_p + j\,f_p$

(This equation we shall derive later to explain the equations parameter). Now once the *Q* at the voltage controlled bus has been calculated, it should be checked whether this value violates the reactive power generation at the bus, that means, whether it is less than the minimum reactive power generation permissible because of power system stability problem, or it is more than the maximum reactive power generation permissible, because of rotor heating. Now, if it violates, the voltage controlled bus is made to act as a load bus for that iteration only and the reactive power to be substituted in the expression for that iteration will correspond to the limit it has violated. That if it is less than q_{min}, the value to be substituted will be Q_{min}. Then if the calculated value of *Q* lies within the limit the value of *Q* will be substituted in the expression.

In fact, the violation of reative power limit could be due to the specified voltage either being too low, or too high. Since $|V|_{\text{specified (space)}}$ obtained only by controlling reactive power, it is, hence, possible that $|V|$ has been specified beyond the capability of the reactive power generation of the generator. Now if the value of *Q* for a generator bus is calculated after assuming magnitude equal to $|V|_{\text{spec.}}$ for that bus and phase angle corresponding to the value in that iteration and if *Q* so obtained violates the limit, the value of voltage should be taken correspondingly to the voltage in that iteration and not V_{spec}. otherwise the voltage should correspond to specified voltage and phase angle as in that iteration.

3. And the third point to remember in this case is : The number of non-linear equation is $(n - 1)$, where *'n'* is the buses in the number of buses in the system.

12.5 LOAD FLOW EQUATION & GAUSS-SEIDEL METHOD

Now below the Gauss-Seidel iterative procedure steps for solving equation of load flow is given with a flow chart for load flow solution using Gauss-Seidel technique in the Fig. 12.4.

Step 1. Let us assume a flat voltage profile $1 + j\,0{\cdot}0$ for all nodal voltage except the slack bus 1. Now we assume a suitable value of the convergence criterion that is, if the absolute

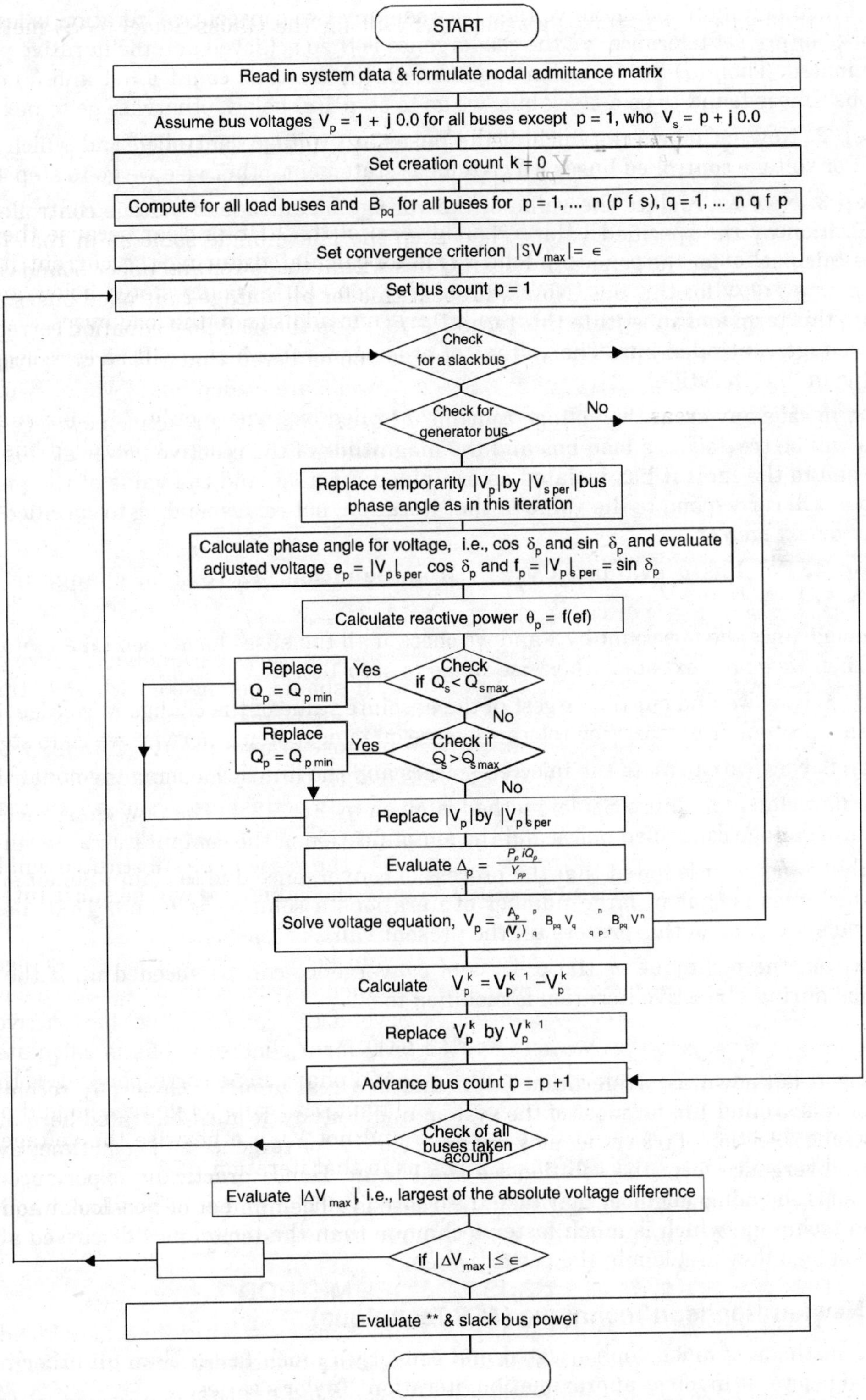

Fig. 12.4. Flow chart for laod flow solution using *G-S* method (voltage controlled buses included).

value of the maximum change in voltage between any two consecutive iteration is less than a prescribed or pre-set tolerance "e", the convergence is then achieved and the iterative procedure is terminated. Then (a) Set iteration count $k = 0$, (b) Sset bus count $p = 1$ and (c) check for slack bus. If it is found to be a slack bus, we go to step 4(a) below, otherwise go to next step.

Step 2. Now let us check which of the buses are voltage controlled and which are load buses. For voltage controlled buses we go to next statement, otherwise we go to step 4 below.

Step 3. Now we replace the value of the voltage magnitude of voltage controlled bus in that iteration by the specified value. Then keep the phase angle some as in that iteration. Now we calculate Q for the generator bus. If Q lies within the lower and upper bounds calculate the term $(P - j\,Q)/V$ for this bus. Now we repeat this for all voltage controlled buses and then calculate this term and substitute this term $(P - j\,Q)/V$ in the load flow equation corresponding to the voltage controlled bus. The voltage of other generator buses will be corresponding to the value in that iteration.

Now in case any or all the voltage controlled buses violate the reactive power generation, the bus will be treated as a load bus and the magnitude of the reactive power at this bus will correspond to the limit it has violated as explained reactive, and the value of the magnitude of voltage will correspond to the value in that iteration, not corresponding to specified voltage, then go to next step.

Step 4. For bus p evaluates V_p^{k+1} from equation (12.18) and change in voltage $\Delta V_p^K = V_p^{k+1} - V_p^k$.

Then advance the bus count by 1 and we check if all the buses have been taken into account if yes. Then we go to next step otherwise we go to step 1(c).

Step 5. Now we find out the largest of the absolute value of the change in voltage. If this is less than a present or prespecified tolerance more on to next step otherwise we go to step (2).

Step 6. Now we calculate the injected powers and the line flows using the nodal voltages.

The flow chart for Gauss-Seidel method is given by Fig. 12.4. The flow chart includes the presence of voltage controlled buses and the simplification of the computational technique.

In this practice, it is found that the process of convergence due to Gauss-Seidel method is slow, which means that in large number of iterations a solution is found. That means two many trials are done in this process till the present value is reached.

However, the technique or the process of convergence can be speeded up if the voltage correction during corrective iteration is modified to

$$V_{p(\text{spec.})}^{k+1} = V_p^k + \alpha(V_p^{k+1} - V_p^k) \qquad ...(12.20)$$

where α is known as acceleration factor and is a real number. Generally recommended value for α is around 1.6, for most of the system under study. It must be stated here, however, that a wrong selection of α's value, may result in slower convergence and sometimes even may result in divergence form the solution we shall want. Hence practically experienced power system laod controlling engineer may take the desired value of 'α'. Let us now look into Newton-Raphson technique which is much faster technique than the techniques discussed above for solution of load flow problem in the power system.

12.5.1 Newton-Raphson Technique (*N-R* Technique)

This methods is much sophisticated and converges much faster than all other methods discussed before. It involves approximation, iteration, Taylor's series.

In this *N-R* technique, the load flow problem can also be solved by deriving the equations for the method as follows :

From above discussion we know that at any bus p;

$$P_p - j\,Q_p = V_p^{\,*} \cdot I_p = V_p^{\,*} \sum_{q=1}^{n} Y_{pq}\, V_q \qquad ...(12.21)$$

Now let, $V_p = e_p + j\,f_p$

and $Y_{pq} = G_{pq} - j\,B_{pq}$

∴ Now putting these values for of V_p and Y_{pq} in the eqn. (12.21)

We get, $$P_p - j\,Q_p = (e_p + jf_p)^* \sum_{q=1}^{n} (G_{pq} - jB_{pq})(e_q + jf_q)$$

$$= (e_p - jf_q) * \sum_{q=1}^{n} (G_{pq} - jB_{pq})(e_p + jf_q) \qquad ...(12.22)$$

Now separating the real and imaginary parts from eqn. (12.22)

We get, $$P_p = \sum_{q=1}^{n} \{e_p(e_q\,G_{pq} + f_q\,B_{pq}) + f_p(f_p\,G_{pq} - e_q\,B_{pq})\} \qquad ...(12.23)$$

and $$Q_p = \sum_{q=1}^{n} \{f_p(e_q\,G_{pq} + f_q\,B_{pq}) - e_p(f_p\,G_{pq} - e_q\,B_{pq})\} \qquad ...(12.24)$$

Also $$|V_p|^2 = e_p^2 + f_p^2 \qquad ...(12.25)$$

In fact these three equations (12.23), (12.24), (12.25) are the load flow equation for *N-R* technique, and it can be noted that they are non-linear equations in terms of real end imaginary component of nodal voltages. The left-hand quantities of the equation above; P_p, Q_p for a load bus; and P_p and $|V_p|$ for generator bus are specified and e_p and f_p are unknown quantities. Now for an 'n'-bus system the number of unknowns are $2(n-1)$ as the voltage at the slack bus is known and is kept fixed both in magnitude and phase. Hence, if bus 1 is taken as the slack, the unknown variables are (e_2, e_3 ..... e_{n-1}, e_n, f_2, f_3, ... f_{n-1}, f_n). Thus, to solve the problem for $2(n-1)$ variables we need to solve $2(n-1)$ sets of equation.

In fact, *N-R* method is an iterative method which approximates the set of non-linear simultaneous equations to a set of linear simultaneous equation using Taylor's series expansion and the term expansion and the term are limited to first approximation. Now the mathematical process of this method can be explained as below.

Say, the unknown variables be (x_1, x_2, x_3, ... x_n) and the specified quantities y_1, y_2, y_3, ... y_n. These are related by the set of non-linear equations as below :

$$\begin{aligned} y_1 &= f_1(x_1, x_2, x_3, \ldots\ldots, x_n) \\ y_2 &= f_2(x_1, x_2, x, \ldots\ldots, x_n) \\ y_n &= f_n(x_1, x_2, x_3, \ldots\ldots, x_n) \end{aligned} \qquad ...(12.26)$$

Now, we solve these equations by initially having an approximate solution of say (x_1^0, x_{2f}^0, ..., x_n^0); where the superscript zero means the zeroth iteration in the process of solving the above non-linear equation (12.26) above.

Here it is to be also noted that the initial solution for the euqation should not be very for from the actual solution, otherwise there are changes of leading to unexpected solution, and hence it may not be possible to achieve a solution, as expected, whatever may be the computer time utilised. Although, at finit gland, it may appears to be a great drawback to the *N-R* method of solution, but with experience and knowledge on the field of load flow, the problem of initial quess for a power system is not difficult. In fact, a flat voltage profile, that is, $V_p = 1.0 + j0.0$, for $p = 1, 2, 3, ..., n$ except the slack bus has been found to be satisfactory for all most all practical system of load flow.

We should, hence, not that the equations are linearised about the initial guess. We shall expand first equation (of 12.26), $y_1 = f_1$ and the result for the other equation will follow.

Now, let us assume, $\Delta x_1^0, \Delta x_2^0, \Delta x_3^0, ..., \Delta x_n^0$ are the correction are required for, $x_1^0, x_2^0, x_3^0, ..., x_n^0$ respectively for the next better solution. Then the equation $y_1 = f_1$ will be expressed as :

$$y_1 = f_1(x_1^0 + \Delta x_1^0, x_2^0 + \Delta x_2^0 + \Delta x_2^0, x_3^0 + \Delta x_3^0, ..., x_n^0 + \Delta x_n^0)$$

$$= f_1\left(x_1^0, x_2^0, x_3^0, ..., x_n^0\right) + \Delta x_1^0 \left.\frac{\delta f_1}{\delta x_1}\right|_{x^0} + \Delta x_2^0 \left.\frac{\delta f_1}{\Delta f_2}\right|_{x^0} + \Delta x_3 \left.\frac{\delta f_1}{\Delta x_3}\right|_{x^0} + ... + \Delta x_n^0 \left.\frac{\delta f_1}{\delta x_0}\right|_{x^0} + \phi_1$$

Here in this equation ϕ_1 in the function higher order of Δx^0 and the higher derivatives which are neglected according to *N-R* technique. It is this assumption which needs the initial solution to be close to the find solution. Now, if all the all the equation (12.26), we get :

$$\begin{bmatrix} y_1 - f_1(x_1^0, x_2^0, x_3^0, ..., x_n^0) \\ y_2 - f_2(x_1^0, x_2^0, x_3^0, ..., x_n^0) \\ \vdots \\ y_n - f_n(x_1^0, x_2^0, x_3^0, ..., x_n^0) \end{bmatrix} = \begin{bmatrix} \frac{\delta f_1}{\delta x_1} \frac{\delta f_1}{\delta x_2} \frac{\delta f_1}{\delta x_3}, ..., \frac{\delta f_1}{\delta x_n} \\ \frac{\delta f_2}{\delta x_1} \frac{\delta f_2}{\delta x_2} \frac{\delta f_2}{\delta x_3}, ..., \frac{\delta f_2}{\delta x_n} \\ \vdots \\ \frac{\delta f_n}{\delta x_1} \frac{\delta f_n}{\delta x_2} \frac{\delta f_{n1}}{\delta x_3}, ..., \frac{\delta f_n}{\delta x_n} \end{bmatrix} \begin{bmatrix} \Delta x_1^0 \\ \Delta x_2^0 \\ \vdots \\ \Delta x_n^0 \end{bmatrix} \quad ...(12.27)$$

$$B = J \cdot C$$

where J is the first derivative matrix known as the Jacobian matrix. The solution of the equations requires calculation of left hand Vector, which is the difference of the specified quantities at $(x_1^0, x_2^0, ..., x_n^0)$. Similarly J is calculated at this guess. Then, solution of the matrix equation gives $(\Delta x_1^0, \Delta x_2^0, \Delta x_3^0, ... \delta x_n^0)$ and the next better solution is obtained as below :

$$x_1^1 = x_1^0 + \Delta x_1^0; \; x_2^1 = x_2^0 + \Delta x_2^0, ... x_n^1 = x_n^0 + \Delta x_n^0$$

Now, the better solution is available and is,

$$(x_1^1, x_2^1, x_3^1, ..., x_n^1)$$

Then, with these values the process is repeated till (a) the largest element, in magnitude, in the left column of the equation is less than a prespecified value on (b) the largest element in the column vector (Δx_1, Δx_2, Δx_3, ..., Δx_n) is less than a prespecified value.

Now, say, when we refer to a power system problem the above set of linearised equation become (assuming there is only one generator bus which is taken as slack bus and all other buses are load buses) as given below :

$$\begin{bmatrix} \Delta P_2 \\ \Delta P_3 \\ \vdots \\ \Delta P_n \\ \Delta Q_2 \\ \Delta Q_3 \\ \vdots \\ \Delta Q_n \end{bmatrix} = \begin{bmatrix} \frac{\delta P_2}{\delta e_2} & \frac{\delta P_2}{\delta e_3} & \cdots & \frac{\delta P_2}{\delta e_n} & \frac{\delta P_2}{\delta f_2} & \frac{\delta P_2}{\delta f_3} & \cdots & \frac{\delta P_2}{\delta f_n} \\ \frac{\delta P_3}{\delta e_2} & \frac{\delta P_3}{\delta e_3} & \cdots & \frac{\delta P_3}{\delta e_n} & \frac{\delta P_3}{\delta f_2} & \frac{\delta P_3}{\delta f_3} & \cdots & \frac{\delta P_3}{\delta f_n} \\ \vdots & \vdots & & \vdots & \vdots & \vdots & & \vdots \\ \frac{\delta P_n}{\delta e_2} & \frac{\delta P_n}{\delta e_3} & \cdots & \frac{\delta P_n}{\delta_n} & \frac{\delta P_n}{\delta f_2} & \frac{\delta P_n}{\delta f_3} & \cdots & \frac{\delta P_n}{\delta f_n} \\ \frac{\delta Q_2}{\delta e_2} & \frac{\delta Q_2}{\delta e_3} & \cdots & \frac{\delta Q_2}{\delta e_n} & \frac{\delta Q_2}{\delta f_2} & \frac{\delta Q_2}{\delta f_3} & \cdots & \frac{\delta Q_2}{\delta f_n} \\ \frac{\delta Q_3}{\delta e_2} & \frac{\delta Q_3}{\delta e_3} & \cdots & \frac{\delta Q_3}{\delta e_n} & \frac{\delta Q_3}{\delta f_2} & \frac{\delta Q_3}{\delta f_3} & \cdots & \frac{\delta Q_3}{\delta f_n} \\ \vdots & \vdots & & \vdots & \vdots & \vdots & & \vdots \\ \frac{\delta Q_n}{\delta e_2} & \frac{\delta Q_n}{\delta e_3} & \cdots & \frac{\delta Q_n}{\delta e_n} & \frac{\delta Q_n}{\delta f_2} & \frac{\delta Q_n}{\delta f_3} & \cdots & \frac{\delta Q_n}{\delta f_n} \end{bmatrix} \begin{bmatrix} \Delta e_2 \\ \Delta e_3 \\ \vdots \\ \Delta e_n \\ \Delta f_2 \\ \Delta f_3 \\ \vdots \\ \Delta f_n \end{bmatrix} \quad \text{...(12.28)}$$

$2(n-1) \times 1$ $\qquad\qquad$ $2(n-1) \times 1$

This equation (12.27) can be expressed is short form as below :

$$\begin{bmatrix} \Delta P \\ \hdashline \Delta Q \end{bmatrix} = \begin{bmatrix} J_1 & \vdots & J_2 \\ \hdashline J_3 & \vdots & J_4 \end{bmatrix} \begin{bmatrix} \Delta e \\ \hdashline \Delta f \end{bmatrix}$$

However, in case the system contain all types of buses, the set of equation can be expressed as below :

$$\begin{bmatrix} \Delta P \\ \hdashline \Delta Q \\ \hdashline |\Delta V_p|^2 \end{bmatrix} = \begin{bmatrix} J_1 & \vdots & J_2 \\ \hdashline J_3 & \vdots & J_4 \\ \hdashline J_5 & \vdots & J_6 \end{bmatrix} \begin{bmatrix} \Delta e \\ \cdots \\ \Delta f \end{bmatrix}$$

Now, the element of the Jacobian matrix can be derived from the load flow equations, (12.29), (12.36), (12.31).

Then the off diagonal elements of J_1 are,

$$\frac{\delta P_p}{\delta e_q} = e_p\, G_{pq} - f_p\, B_{pq},\ q \neq p \qquad \text{...(12.29)}$$

Now, the diogonal element of J_1 are

$$\frac{\delta P_p}{\Delta e_p} = 2e_p \cdot G_{pp} + f_p\, B_{pp} - f_p B_{pp} + \sum_{\substack{q=1 \\ q \neq p}}^{n} (e_p\, G_{pq} + f_q\, B_{pq}) \qquad \text{...(12.30)}$$

Now, the off-diagonal elements of J_2 are

$$\frac{\delta P_p}{\delta f_p} = e_p\, B_{pq} + f_p\, G_{pq},\ q \neq p \qquad \text{...(12.31)}$$

We get the diagonal element of J_2, as :

$$\frac{\delta P_p}{\delta f_p} = 2f_p G_{pp} + \sum_{\substack{q=1 \\ q \neq p}}^{n} (f_p G_{pp} - e_q B_{pq}) \quad \text{...(12.32)}$$

and the off-diagonal element of J_2 are

$$\frac{\delta Q_p}{\delta e_q} = e_p B_{pq} + f_p G_{pq}, q \neq p \quad \text{...(12.33)}$$

The diagonal elements are,

$$\frac{\delta Q_p}{\delta e_p} = 2e_p B_{pp} - \sum_{\substack{q=1 \\ q \neq p}}^{n} (f_q G_{pq} - e_q B_{pq}) \quad \text{...(12.34)}$$

Similarly we can wrtie the expression for the off-diagonal and diagonal element of J_4 respectively as :

$$\frac{\delta Q_p}{\delta e_p} = -e_p G_{pq} + f_p B_{pq} \pi q \neq p \quad \text{...(12.35)}$$

$$\frac{\delta Q_p}{\delta f_p} = 2f_p \cdot B_{pp} + \sum_{\substack{q=1 \\ q \neq p}}^{n} (e_p G_{pq} + f_q B_{pq}) \quad \text{...(12.36)}$$

Then the off-diagonal and diagonal elements of J_5 are

$$\frac{\delta |V_p|^2}{\delta e_p} = 0, g \neq p \quad \text{...(12.37)}$$

and,

$$\frac{\delta |V_p|^2}{\delta e_p} = 2e_p \quad \text{...(12.38)}$$

The off-diagonal and diagonal element of J_6 are

$$\frac{\delta |V_p|^2}{\delta f_q} = 0, q \neq p \quad \text{...(12.39)}$$

and,

$$\frac{\delta |V_p|^2}{\delta f_p} = 2f_p \quad \text{...(12.40)}$$

Then, we calculate the residual column vector consisting of ΔP, ΔQ and $|\Delta V|^2$. Now, let P_{sp}, Q_{sp} and $|V_{sp}|$ be the specified quantities at the bus p. Now, assuming a suitable value of the solution (flat voltage profile in this case) the value of P, Q and $|V|$ at the various buses are calculated.

Hence,

$$\Delta P_p = P_{sp} - P_p^0$$

$$\Delta Q_p = Q_{sp} - Q_p^0$$

$$|\Delta V_p|^2 = |V_{sp}|^2 - |V_p|^2 \quad \text{...(12.41)}$$

Here in this equation (12.41) the subscript zero means the value calculated corresponding to the guess, this is, zeroth iteration, that is the initial guess.

Now, after calculating the Jacobian matrix and the residual column vector corresponding to initial guess that is the initial solution the desired increment voltage vector $\begin{bmatrix} \Delta e \\ \Delta f \end{bmatrix}$ can be calculated by using any standard technique, preferably gauss elimination with sparsity techniques. The next solution will be :

$$e_p^1 = e_p^0 + \Delta e_p^0$$

$$f_p^1 = f_p^0 + \Delta f_p^0$$

These values of voltager will be used in

START

Read in system death and formula nodal admittance matrix.

Now assume bus voltage an $V_p = 1 + j\,0.0$ for $p = 1, \ldots n, p \neq s$, $V_s = a + j\,0.0$.

Set convergence criterions = $\in$

$K = 0$

Set bus count $p = 1$

Check for slack bus

Calculate P and Q_p using equation 12 and 12

Check for generator bus

is $Q_p < Q_{p\,min}$

is $Q_p > Q_{p\,max}$

Sat $Q_p = Q_{p\,max}$ — — Evaluate $|\Delta V_p|^2 = |V_{ps}|^2 = |V_p|^2$

Evaluate $\Delta Q_p = Q_{ps} - Q_p$

advance bus count $p = p + 1$

$p > n$

Fig. 12.5. Determine the largest of the absolute of the residue

These values of voltages will be used in the next iteration. The process will be repeated and in general the now better estimate for bus voltages will be :

$$e_p^{k+1} = e_p^k + \Delta e_p^k$$

$$f_p^{k+1} = f_p^k + \Delta f_p^k$$

The process is repeated till the magnitude of the largest element in the residual column vector is less than the prescribed value. Then the sequence of steps for the solution of load flow problem using *N-R* technique is explained as follows, with reference to the flow chart in Fig. 12.5. This steps of the *N-R* method :

1. First assume a suitable solution for all buses except the slack bus. And let, $V_p = 1 + j0.0$, for $p = 1, 2, ... n, p \neq b, V_s = a + j\,0.0$.
2. Then set convergence criterion = 6, if the largest of absolute of the residue exceeds $\in$ the process is repeated otherwise it is terminated.
3. Set iteration count $k = 0$.
4. Set bus count $p = 1$.
5. Check if p is a slack bus. if yes we go to 10.
6. Then calculate the real and reactive powers P_p and Q_p respectively using equation (12.18) and (12.19).
7. Then evaluate $\Delta P_p^k = P_{sp} - P_p^k$.
8. Now check if the bus in question is a generator bus. If yes, compare the Q_p^k with the limits. Then if it exceeds the limit fix the reactive power generation to the corresponding limit and then treat the bus as a load bus for that iteration and go to next step. Then if the lower is violated set $Q_{p_{\text{spec.}}} = Q_{p_{\max}}$. If the limit is not violated evaluate the voltage residue. $|\Delta V_p|^2 = |V_p|^2_{\text{spec.}} - |V_p^k|^2_k$ and we go to step.
9. Now evaluate $\Delta Q_p^k = Q_{ps} - Q_p^k$.
10. Advance the bus count by 1, *i.e.*, $p = p + 1$ and then we check if all the buses have been accounted. If not, go to step 5.
11. Now we determine the largest of the absolute value of the residue.
12. Then if the largest of the absolute value of the residue is less than $\in$, we go to step (17).
13. Now evaluate element for Jacobian matrix.
14. Calculate voltage element Δe_p^k and Δf_p^k.
15. Then we calculate new bus voltages, $e_p^{k+1} = e^k + \Delta e_p^k$ and $f_p^{k+1} = f_p^k - \Delta f_p^k$.

16. Then advance iteration count $k = k + 1$ and then we go to step (4).

17. Now we evaluate bus and line power and print the result.

12.5.2 *N-R* Method's Approximation in Load Flow Problem Solution

We know from one basic knowledge on the power system operation that a small change in phase angle changes the flow of active power, and in fact does not affect much the flow of reactive power. And similarly a small change in nodal voltage affects the flow of reactive power whereas active power practically does not change. Now, with above these two basic facts in mind and using the polar coordinates, the set of linear load flow equations can be written in matrix form as follows :

$$\begin{bmatrix} \Delta P \\ \Delta Q \end{bmatrix} = \begin{bmatrix} J_1 & 0 \\ 0 & J_4 \end{bmatrix} \begin{bmatrix} \Delta\delta \\ |\Delta V| \end{bmatrix} \qquad ...(12.42)$$

where J_1 corresponds to the element $\frac{\partial P}{\partial \delta}$ which exist.

J_2 corresponds to the elements $\frac{\partial Q}{\partial \delta}$ which do not exist and hence are zero.

then J_3 corresponds to the elements $\frac{\partial Q}{\partial \delta}$ which do not exist and are zero.

and J_4 corresponds to the elements $\frac{\partial Q}{\partial |V|}$ which exist.

In fact this approximation to Newton-Raphson method simplifies the calculation and results in smaller computation time for solution of load flow problem. Now let us make some comparison of solution by several methods discussed above :

12.5.3 Comparison of three Methods of Solution of Load Flow Problems

In the above discusions we have seen that *G-S* (Gauss-Seidel) method is superior to Gauess method, but then we have also seen that *N-R* (Newton-Raphson) method is superior to *G-S* (Guess-Seidel) method. Hence we find comparison can be made only between *G-S* method and *N-R* method, and that two when *Y* bus matrix is used for problem formulation.

However, from the computer–memory requirements, poles coordinates are preferred for solution based on *N-R* technique and the rectangular coordinates for the *G-S* technique. However if we look into the steps taken *G-S* and *N-R* techniques are find :

1. The time taken to perform one iteration of the computation is relatively smaller in *G-S* technique in comparison to that of *N-R* technique.
2. But the number of iritation required by *G-S* method for a particular system are greater as compared to *N-R* method and they increase with increase in the size of the system.
3. In case of *N-R* technique the number of iteration is more or less independent of the size of the system and vary between 3 to 5 iterations.
4. The convergence characteristics of *N-R* method are not affected by the solution of a

slack bus whereas that of *G-S* method is sometimes very seriously affected and the solution of a particular bus may result in poor convergence.

5. The main advantage of *G-S* method as compared to *N-R* method is its ease in programming and most efficient use of core memory.
6. However, for large power system, *N-R* method is found to be more efficient and practical from the viewpoint of computing time and convergence characteristics.
7. Although *N-R* method can solve most of the practical problems, it may fail in respect of some ill-conditioned problem where other advanced mathematical programming technique like the non-linear programming techniques can be used.

12.6 LOAD FLOW EQUATION

In above methods we have seen how iterative methods help solution of bus voltages. Hence after the iterative solution of bus voltages is completed , the current at bus p as shown in Fig. 12.6 the line connecting bus p to q is

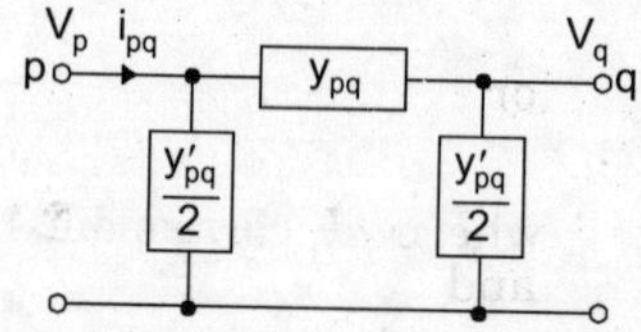

Fig. 12.6. Equivalent π-circuit of a transmission link for evaluating line flow the reactive power flow.

$$i_{pq} = \left(V_p - V_q\right) y_{pq} + V_p \frac{y'_{pq}}{2} \qquad ...(12.42)$$

We get from the relationship between, $P_{pq}, Q_{pq}, V^*_{pq} \cdot i_{pq}$

We get,
$$P_{pq} - j\,Q_{pq} = V^*_{pq} \cdot i_{pq}$$

or
$$P_{pq} - j\,Q_{pq} = V^*_p \left[\left(V_p - V_q\right) y_{pq} + V_p \frac{y'_{pq}}{2}\right] \qquad ...(12.44)$$

where P_{pq} is the real power from bus p to q and Q_{pq} is the reactive power flow from bus p to q.

However, similarly, at bus q, the power flow from bus q to p n :

$$P_{qp} - j\,Q_{qp} = V^*_q \left(V_q - V_p\right) y_{pq} + V^*_{q\,y_{pq}} + V^*_q V_q \frac{y_{pq}}{2} \qquad ...(12.45)$$

Than the power loss in line pq is the algebraic sum of the power flows

$$(P_{qp} - j\,Q_{pq}) \text{ and } (P_{pq} - j\,Q_{qp})$$

12.7 EXTENSION OF *N-R* METHOD (FAST-DECOUPLED LOAD FLOW)

Extension of *N-R* method is also known as Fast-Decoupled load flow which is formulated in polar coordinates with certain approximation which result into a fast algorithm for load flow solution. We have to first derive the load flow equations in polar coordination then we shall discuss the method of fast developed load flow an extension of *N-R* method.

To do so, we know from above discussion

$$P_p - j\,Q_p = V^*_p I_p \text{ and } I_p = \sum_{q=1}^{n} Y_{pq} \cdot V_q$$

$$\therefore \quad P_p - j\,Q_p = V^*_p \sum_{q=1}^{n} Y_{pq} V_q \qquad ...(12.46)$$

The voltage and admittance in polar coordinates are expressed as :

$$V_p = |V_p| \exp(j\,\delta_p) \text{ and } Y_{pq} = |Y_{pq}| \exp(-j\,\delta_q)$$

Then the voltage and admittance in polar coordinates are expressed as :

$$V_p = |V_p| \exp.(j\,\delta_p) \text{ and als } Y_{pq} \exp(-j\,\delta_q)$$

Now substituting values of V_p and Y_{pq} in the eqn. (12.46) we get

$$P_p - j\,Q_p = |V_p| \exp(-j\delta_p) \sum_{q=1}^{n} Y_{pq} \Big| \exp(-jQ) |V_p| \exp(j_q^{\delta})$$

$$= \sum_{q=1}^{n} |V_p||V_q| Y_{pq} \exp\{-j(Q_{pq} + \delta_p - \delta_q)\}$$

or

$$P_p = \sum_{q=}^{n} |V_p||V_q| Y_{pq}| \exp(-j(Q_{pq} + \delta_p - \delta_q)) \quad ...(12.47)$$

and

$$Q_p = \sum_{q=1}^{n} |V_p, V_q Y_{pq}| \cos(Q_{pq} + \delta_p - \delta_q) \quad ...(12.48)$$

where $p = 1, 2, 3, ..., n$

But the equations, (12.47) and (12.48) above can be written also as :

$$P_p = |V_p V_p Y_{pp}| \cos Q_{pp} + \sum_{\substack{q=1 \\ q \neq p}}^{n} |V_p V_q Y_{pq}| \cos(Q_{pq} + \delta_p - \delta_q) \quad ...(12.49)$$

and

$$Q_p = |V_p V_p Y_{pp}|_{\sin} Q_{pp} + \sum_{\substack{q=1 \\ q \neq p}}^{n} |V_p V_q Y_{pq}| \sin(Q_{pq} + \delta_p - \delta_q) \quad ...(12.50)$$

We can now write these equations after linearisation can be rewritten in matrix from as :

$$\begin{bmatrix} \Delta P \\ \Delta Q \end{bmatrix} = \begin{bmatrix} H & N \\ M & L \end{bmatrix} \begin{bmatrix} \Delta\delta \\ \Delta|E|/|E| \end{bmatrix} \quad ...(12.51)$$

Here in this equations H, N, M and L are elements of Jacobian matrix. In decouple load flow method (extension of N-R method) the first assumption is that real power changes (ΔP) are less sensitive to changes in voltage magnitude and are mainly sensitive to angle. Similarly the reactive power changes are less sensitive to change in angle but mainly sensitive to change in voltage magnitude. With these assumption the equation (12.51) is changed to :

$$\begin{bmatrix} \Delta P \\ \Delta Q \end{bmatrix} = \begin{bmatrix} H & O \\ O & L \end{bmatrix} \begin{bmatrix} \Delta\delta \\ \Delta|E|/|E| \end{bmatrix} \quad ...(12.52)$$

The equation (12.52) is decoupled equation which can be expanded as

$$|\Delta P| = |H|\,|\Delta\delta| \quad ...(12.53)$$

and

$$|\Delta Q| = |L|\,|\Delta|\,|E|/|E| \quad ...(12.54)$$

Now by using equations (12.49) and (12.50) the element of the Jacobin matrices. H and I are obtained as below :

1. **Off Diagonal Element of *H***

$$H_{pq} = \frac{\partial P_p}{\partial \delta_q} = |V_p V_q Y_{pq}| \sin[Q_{pq} + \delta_p - \delta_q]$$

$$= |V_p V_q Y_{pq}|[\sin Q_{pq} \cos(\delta_p - \delta_q) + \sin(\delta_p - \delta_q)\cos Q_{pq}]$$

$$= |V_p V_q| [\,|Y_{pq}| \sin Q_{pq} \cos(\delta_p - \delta_q) + |Y_{pq}| \cos Q_{pq} \sin(\delta_p - \delta_q)]$$

$$= |V_p V_q| [-B_{pq} \cos(\delta_p - \delta_q) + G_{pq} \sin(\delta_p - \delta_q)] \quad ...(12.55)$$

2. Some way, we get off diagonal element of *L* as below :

$$L_{pq} = \frac{\partial Q_p |V_q|}{\partial V_q} = |V_p V_q Y_{pq}| \sin(Q_{pq} + \delta_p - \delta_q)$$

$$= |V_p V_q| [-B_{pq} \cos(\delta_p - \delta_q) + G_{pq} \sin(\delta_p - \delta_q)] \quad ...(12.56)$$

Now from equations (12.55) and (12.56), it can be observed that

$$H_{pq} = |V_p V_q| [G_{pq} \sin(\delta_p - \delta_q) - B_{pq} \cos(\delta_p - \delta_q)]$$

3. Then, the diagonal elements of the *H* are :

$$H_{pp} = \frac{\partial P_p}{\partial \delta_p} = \sum_{\substack{q=1 \\ q \neq p}}^{n} |V_p V_q Y_{pq}| \sin(Q_{pq} + \delta_p - \delta_q)$$

$$= -Q_p + |V_p V_p \cdot Y_{pp}| \sin Q_{pp} = -Q_p - V_p^2 B_{pp} \quad ...(12.57)$$

4. Some way, diagonal elements for the matrix *L* are :

$$L_{pp} = \frac{\partial Q_p |V_p|}{\partial V_p} = |2V_p^2 Y_{pp}| \sin Q_{pp} + \sum_{\substack{q=1 \\ q \neq p}}^{n} |V_p V_q Y_{pq}| \sin(Q_{pq} + \delta_p - \delta_q)$$

$$= |2V_p^2 Y_{pp}| \sin Q_{pp} + Q_p - |V_p^2 Y_{pp}| \sin Q_{pp}$$

$$= Q_p + |V_p^2 Y_{pp}| \sin Q_{pp} = Q_p - V_p^2 B_{pp} \quad ...(12.58)$$

5. **Approximation**

Now in the case of fast decoupled load flow method following approximations are further made for evaluating jacobin element.

$$\cos(\delta_p - \delta_q) \simeq 1$$

$$G_p \sin(\delta_p - \delta_q) \leq B_{pq}$$

and $$Q_p \ll B_{pp} V_p^2$$

∴ The Jaeobian elements now become

$$L_{pq} = H_{pq} = -|V_p V_q| B_{pq} \text{ for } q \neq p$$

and $$L_{pp} = H_{pp} = -B_{pp} |V_p|^2$$

With these Jacobian elements equations (12.51 *a*) and (12.51 *b*) become

$$[\Delta P_p] = [V_p][V_q][B'_{pq}][\Delta\delta_q] \quad ...(12.59)$$

and $$[\Delta Q_p] = [V_p][V_q][B''_{pq}]\frac{\Delta|E_q|}{E_q} \quad ...(12.60)$$

where, B'_{pq} and B''_{pq} are the elements of $[-B_{pq}]$ matrix

Then further decoupling is obtained as below :

1. Omit from B'' the angle shifting effect of phase shifters.
2. Omit B' the representation of these network elements that affect *MVAr* flows, *i.e.*, shunt reactor and off-nominal in phase transformer taps.
3. Divide eqns. (12.60) and (12.61) by V_p and assuming V_g = 1 p.u. and also neglecting the series resistance in calculating the element of B' with these assumption equations (12.60) and (12.61) for the load flow solution take the form.

$$\left[\frac{\Delta P_0}{E_p}\right] = [B'][\Delta\delta] \qquad \text{...(12.61)}$$

and

$$\left[\frac{\Delta Q_p}{E_p}\right] = [B''][\Delta E] \qquad \text{...(12.62)}$$

Here it is to be noted that $[B']$ and $[B'']$ are real and sparx and having similar strnetime as there of H and L respectively. Since the two matrien any constant and do not change during successive irritations for solution of the load flow problem, these are to evaluated only once and inverted once during the first iteration and then used in all successive iteration.

In fact, it must be stated here that this method is fast only because of the nature of Jacobian matries $[B']$ and $[B'']$ and the spareity of these matrices.

12.8 WHAT IS MEANT BY SPARSITY MATRIX

We must define what is meant by Sparsity Matrix. Let us briefly discuss the matter with an example say, in realistic power systems there may be 100 buses. In such a system each bus may be connected with only two or three of the remaining buses. This means that only three or four of the elements ın each row or column are non-zero obviously all the diagonal elements are non-zero. Such a matrix were there are a large number of non-zero elements present it is a "SPARSE MATRIX". Say if the matrix is $(n \times n)$ and there are Z zero element then sparsity is defined as $S = \frac{Z}{n}$.

In fact, in realistic system, the sparsity of a Y_{Bus} system may be as high as 95 – 97%. In sparce matrix, there is enough saving of commutational time.

12.9 SOLVED EXAMPLES ON LOAD FLOW PROBLEM

Example 12.1. *By using Gauss-Seidel method determine the voltages at the end of first iteration of the following load system for a load flow solution : (Take α = 1.6).*

The line admittances are given to the following bus-code :

Bus Code	*Admittance*
1 – 2	*2 – j 8.0*
1 – 3	*1 – j 4.0*
2 – 3	*0.666 – j 2.664*
1 – 4	*1 – j 4.0*
3 – 4	*2 – j 8.0*

The shedule of active and reactive powers :

Bus Code	*P*	*Q*	*V*	*Remarks*
1	–	–	*1.06*	*SLACK*
2	*0.5*	*0.2*	*1 + j 0.0*	*PQ*
3	*0.4*	*0.3*	*1 + j 0.0*	*PQ*
4	*0.3*	*0.1*	*1 + j 0.0*	*PQ*

Solution : From the data given above in the example, the admittance matrix will be as below :

$$Y_{pq} = \begin{bmatrix} 3-j1.20 & -2+j8.0 & -1+j4.0 & 0.0 \\ -2+j8.0 & 3.666-j14.664 & -0.666+j2.664 & -1+j4.0 \\ -1+j4.0 & -0.666+j2.664 & 3.666-j14.664 & -2+j8.0 \\ 0.0 & -1+j4.0 & -2+j8.0 & 3-j12.0 \end{bmatrix}$$

Then the power for load buses are to be taken as negative and that for generator buses as positive.

Then for the stated system by using the equation for voltage.

$$V'_2 = \frac{1}{Y_{2_2}}\left|\left[\frac{P_2 - jQ_2}{V_2^*} - Y_{2_1} V_1^0 - Y_{2_3} V_3^0 - Y_{24} V_4^0\right]\right.$$

∴ For Bus Code 1 – 2

$$V_2' = \frac{1}{(3.666 - j14.664)}\left|\frac{-0.5 + j0.2}{1 - j0.0} - 1.06(-2 + j8) - 1.0[(-0.666 + j2.664) - (-1 + j0.4)1.0]\right.$$

$$= (1.01187 - j\,0.02888)$$

Now, voltage correction during consecutive iteration is modified to

$$V_{p(\text{ace})}^{k+1} = V_p^k + \alpha(V_p^{k+1} - V_p^k) \text{ (from eqn. 12. 16 } a)$$

Hence by using this equation

$$V_{2_{\text{ace}}}' = (1.0 + j\,0.0) + 1.6\,(1.01187 - j\,0.02888 - 1.0 - j\,0.0)$$

$$= 1.01899 - j\,0.046208 \quad \textbf{Ans.} \text{ (for Bus Code 1.2)}$$

Similarly for Bus Code 1-3

$$V_3' = \frac{1}{Y_{3_3}}\left[\frac{P_3 - jQ_3}{V_3^*} - Y_{3_1} V_1 - Y_{3_2} V_2' - Y_{3_4} V_4^0\right]$$

$$= \frac{1}{3.666 - j14.664}\left[\begin{array}{l}\dfrac{-0.4 + j0.3}{1 - j0.0} - (-1 + j4.0)1.06 - (-0.666 + j2.664)(1.0189 - j.0462) \\ -(-0.606 + j2.664)(1.01899 - j\,0.046208) - (-2 + j8)(1 + j0.0)\end{array}\right]$$

$$= 0.994119 - j\,0.029248$$

Again by using the eqn. (12.76 *a*)

$$V'_{3_{ace}} = (1 + j\,0.0) + 1.6\,(0.994119 - j\,0.02944 - 1 - j\,0.0) = 0.99059 - j\,0.0467968 \textbf{ Ans.}$$

Then Similarly for Bus Code 1 – 4

$$V'_4 = \frac{1}{Y_{44}}\left[\frac{P_4 - jQ_4}{V_4^{0*}} - Y_{42}\,V'_4 - V_{43}\,V_3^1\right]$$

$$= \frac{1}{(3-j12)}\left[\frac{-0.3+j0.1}{1-j0.0} - (-1+j4.0)(1.01899-j0.046208) - (-2+j8)(0.99059-j0.0467968)\right.$$

$$= 0.9716032 - j\,0.064684$$

Then as above

$$V'_{4_{ace}} = 1.0 + j\,0.0 + 1.6\,[(0.9716032 - j\,0.064684) - 1 - j\,0.0]$$

$$= 0.954565 - j\,0.1034944 \textbf{ Ans.}$$

Example 12.2. *The laod flow data for the sample power system are given below. The voltage magnitude at bus 2 is to be maintained at 1.04 p.u. The maximum and minimum reactive power limit of the generator at bus 2 are 0.35 and 0.0 p.u respectively. Determine the set of load flow equation at the end of first iteration by using Newton-Raphkson method. Impedance for sample system are given in the following Bus codes :*

Bus Code	*Impedance*	*Line Changing Admittance*
1 – 2	*0.08 + j 0.24*	*0.0*
1 – 3	*0.02 + j 0.06*	*0.0*
2 – 3	*0.06 + j 0.18*	*0.0*

Schedule of generation and loads is given below :

Bus Code	*Assumed Voltages*	*Generation MW*	*Generation MVAR*	*Load MW*	*Load MVAR*
1	*1.06 + j 0.0*	*0*	*0*	*0*	*0*
2	*1.0 + j 0.0*	*0.2*	*0.0*	*0.0*	*0.0*
3	*1.0 + j 0.0*	*0*	*0*	*0.06*	*0.25*

Solution : With the datas given in the above example

$$y_{12} = \frac{1}{z_{12}} = \frac{(0.08 - j0.24)}{(0.08 + j0.24)(0.08 - j0.24)} = 1.25 - j_0 3.75$$

Similarly we can get y_{13}, y_{23} and the nodal admittance matrix below :

$$y_{13} = 5 - j\,15 \text{ and } y_{23} = 1.6667 - j\,5.0$$

and the nodal admittance matrix, Y is :

$$Y = \begin{bmatrix} 6.25 - j18.75 & -1.25 + j3.75 & -5 + j15 \\ -1.25 + j3.75 & 2.916 - j8.75 & -1.666 + j5.0 \\ -5 + j1.5 & -1.666 + j5.0 & 6.666 - j20 \end{bmatrix}$$

Now let us **assume** a flat voltage profile for bus 2 and 3 and for bus 1.

$$V_1 = 1.06 + j\,0.0$$

Then from the nodal admittance matrix and the assumed voltage solution :

$G_{11} = 6.25$	$B_{11} = 18.75$	$e_1 = 1.06$	$f_1 = 0.0$
$G_{12} = -1.25$	$B_{12} = -3.75$	$e_2 = 1.0$	$f_2 = 0.0$
$G_{13} = -5.0$	$B_{13} = -15.0$	$e_3 = 1.0$	$f_3 = 0.0$
$G_{22} = 2.966$	$B_{22} = 8.75$		
$G_{23} = -1.666$	$B_{23} = -5.0$		
$G_{33} = 6.666$	$B_{33} = 20.0$		

With roter to the expression developed for Real Power P buy N-R method discussed above in such 1.24 for NR method.

Now,
$$P_2 = e_2(e_1 G_{21} + f_1 B_{21}) + f_2(f_1(G_{21} - e_1 B_{21}) + e_2(e_2 G_{22} + f_2 B_{22})$$
$$+ f_2(f_2 G_{22} - e_2 B_{22}) + e_2(e_3 G_{23} + f_3 B_{23}) + f_2(f_3 G_{23} - e_3 B_{23})$$
$$= 1.01\,\{1.06\,(-1.25) + 0.0\,(-3.75)\} + 0.0\,\{0.0\,(-1.25) - 1.06\,(-3.75)\}$$
$$+ 1.0\,\{1.0(2.916)\} + 0.0\,(\;) + 1.0\,\{1.0(1.0(-1.666)\} + 0.0\,(\;)$$
$$= -1.325 + 2.916 - 1.666 = -0.075$$

Then, the same way, $P_3 = -0.3$

Now, to determine, Q, from the expression deduced for reactive form above in MR method.

$$Q_2 = f_2\,(\;) - e_2(f_1 G_{21} - e_1 B_{21}) + f_2\,(\;) - e_2(f_2(G_{21} - e_2 B_{22}) + f_2\,(\;) - e_2(f_3 G_{23} - e_3 B_{23})$$

Now putting the values of parameters in this equation, we get

$$Q_2 = -0.225$$

The same way, we can find Q_3 as :

$$Q_3 = -0.9$$

$$\therefore \quad \Delta P_2 = P_{2_{\text{spec.}}} - P_{2_{\text{calculated}}} = 0.2 - (-0.075) = 0.275$$

Similarly,
$$\Delta P_3 = P_{3_{\text{spec.}}} - P_{3_{\text{calculated}}} = -0.6 - (-0.3) = -0.3$$

Now, as the lower limit on Q_2 is 0.0 and the value of Q_2 as calculated above violates this limit, bus 2 is treated as a load were $Q_{2_{\text{spec.}}} = 0.0$.

$$\therefore \quad \Delta Q_2 = 0.0 - (-0.225) = 0.225$$

and
$$\Delta Q_3 = -0.25 - (-0.9) = 0.65$$

1. In the N-R method we have seen above that we have to find diagonal and off-diagonal element as below :

Diagonal elements for real power : (By use expressions of the N-R method discussed above in detail.

$$\frac{\partial P_p}{\partial e_p} = 2e_p G_{pp} + \sum_{\substack{q=1 \\ q \neq p}}^{n} (e_p G_{pp} + f_q B_{pq})$$

Here for Bus (2) *i.e.*, (p = 2) and (q = 1)

$$\frac{\partial P_2}{\partial e_2} = 2e_2 G_{22} + e_1 G_{21} + f_1 B_{21} + e_3 G_{23} + f_3 B_{23}$$

Now by putting the values of the parameters in the above equation we get,

$$\frac{\partial P_2}{\partial e_2} = 2 \times 1.0 \times 2.916 + 1.06 (-1.25) + 0.0 (-3.75) + 1.0 (-1.666) + 0.0 (-5.0)$$

$$= 2.848$$

In the same way,

$$\frac{\partial P_3}{\partial e_3} = 2e_3 G_{33} + e_1 G_{31} + f_1 B_{31} + f_1 B_{31} + e_2 G_{32} + f_2 B_{32}$$

$$= 2(1.0)(6.666) + 1.06 (-5.0) + 0.0(-15.0) + 1.0(-1.666) + 0.0(-5.6)$$

$$= 6.3666$$

Now,

$$\frac{\partial P_p}{\partial f_p} = 2f_p G_{pp} + \sum_{\substack{q=1 \\ q \neq p}}^{n} (f_q G_{pq} - e_q B_{pq})$$

$$\frac{\partial P_2}{\partial f_2} = 2f_2 G_{22} + f_1 G_{21} - e_1 B_{21} + f_3 G_{23} - e_3 B_{23}$$

$$= 2(0.0)(2.916) + (0.0)(-1.25) - 1.06(-3.75) - 1.0(-5.0)$$

$$= 8.975$$

Then,

$$\frac{\partial P_3}{\partial f_3} = 20.90$$

1. **Off-diagonal elements :** The same was as above using the expression for off-diagonal elements we get :

$$\frac{\partial P_p}{\partial e_q} = e_p G_{pq} - f_p B_{pq}$$

$$\therefore \quad \frac{\partial P_2}{\partial e_3} = e_2 G_{23} - f_2 B_{23} = -1.666$$

Similarly,

$$\frac{\partial P_3}{\partial e_2} = -1.666$$

Then,

$$\frac{\partial P_p}{\partial f_q} = e_p B_{pq} + f_p G_{pq}$$

$$\therefore \quad \frac{\partial P_2}{\partial f_3} = -5.0$$

and

$$\frac{\partial P_3}{\partial f_2} = 5.0$$

Now, in the same way we can find the partial derivatives of the reactive power for both diagonal element and off diagonal elements as below :

2. Diagonal Element for Reactive Power, Q

We use the expression for diagonal element for Q deduced in N-R method section above as below :

$$\frac{\partial Q_p}{\partial e_p} = 2e_p B_{pp} - \sum_{\substack{q=1\\q\neq p}}^{n} (f_p G_{pp} - e_q B_{pq})$$

$$\frac{\partial Q_2}{\partial e_2} = 2e_2 B_{22} - f_1 G_{21} + e_1 B_{21} - f_2 G_{23} + e_2 B_{23}$$

$$= 2(1.0)(0.75) + 1.06(-3.75) + 1.0(-5.0) = 8.525$$

Similarly, $$\frac{\partial Q_3}{\partial e_3} = 19.1$$

Then, $$\frac{\partial Q_p}{\partial f_p} = 2f B_{pp} + \sum_{\substack{q=1\\q\neq p}}^{n} (e_p G_{pq} + f_q B_{pq})$$

$\therefore$ again the same way as done above to find $\frac{\partial Q_2}{\partial e_2}$ & $\frac{\partial Q_3}{\partial e_3}$, we find

$$\frac{\partial Q_2}{\partial f_2} = -2.991 \text{ and } \frac{\partial Q_3}{\partial f_3} = -6.966$$

Similarly off diagonal elements are calculated and the final set of linear equations at the end of iteration 1 are by using the expression below.

$$\begin{bmatrix} \Delta P_2 \\ \Delta P_3 \\ \Delta Q_2 \\ \Delta Q_3 \end{bmatrix} = \begin{bmatrix} \frac{\partial P_2}{\partial e_2} & \frac{\partial P_3}{\partial e_2} & \frac{\partial P_2}{\partial f_2} & \frac{\partial P_2}{\partial f_3} \\ \frac{\partial P_3}{\partial e_2} & \frac{\partial P_3}{\partial e_3} & \frac{\partial P_3}{\partial f_2} & \partial P_3/\partial f_3 \\ \frac{\partial Q_2}{\partial e_2} & \frac{\partial P_2}{\partial f_3} & \frac{\partial Q_2}{\partial f_2} & \partial P_2/\partial f_3 \\ \frac{\partial Q_2}{\partial f_3} & \frac{\partial Q_3}{\partial e_3} & \partial P_3/\partial e_2 & \partial Q_3/\partial f_3 \end{bmatrix} \begin{bmatrix} \Delta e_2 \\ \Delta e_3 \\ \Delta f_2 \\ \Delta f_3 \end{bmatrix}$$

Now evaluating the values calculated above we find :

$$\begin{bmatrix} 0.275 \\ -0.3 \\ 0.225 \\ 0.65 \end{bmatrix} = \begin{bmatrix} 2.846 & -1.666 & 8.975 & -5.0 \\ -1.666 & 6.366 & -5.0 & 20.90 \\ 8.525 & -5.0 & -2.991 & -1.666 \\ -5.0 & 19.1 & -1.666 & -6.966 \end{bmatrix} \begin{bmatrix} \Delta e_2 \\ \Delta e_3 \\ \Delta f_2 \\ \Delta f_3 \end{bmatrix}$$

QUESTIONS

1. State what are the various types of buses in a power system for load flow studies. State what are the reasons for the classification.
2. Explain the significance of load flow solution in power system after defining the load flow solution.
3. Develop load flow equations suitable for solution by (*a*) *G-S* (Gauss-Seidel) method and (*b*) *N-R* (Newton-Raphson) method by using nodal admittance approach.
4. Discuss and explain with a flow-chart the computational procedure for load flow solution using (*a*) Gauss-Seidel method and (*b*) Newton-Raphson method when the system contains all types of buses.
5. Compare the performance of *G-S* (Gauss-Seidel) method and *N-R* (Newton-Raphson) method for load flow solution using nodal admittance apporoach for the formulation of load flow equation.
6. Discuss the modified Newton-Raphson or fast-decoupled load flow method.
7. State what is meant by Sparsity Matrix.
8. Find the bus admittance matrix for the system shown in the Fig. Q. 1, taking ground as reference. It is given that all the lines are characterised by a series

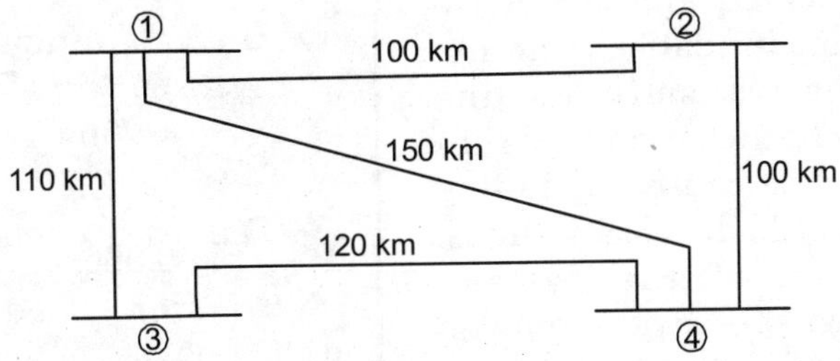

impedance of $(0.1 + j\,0.7)$ Ω/km and a shunt amittance of $(j\,0.35 \times 10^{-5})$ ℧/km. Lines are rated at 220 KV. Use the base values 220 KV and 100 MVA. Express all impedances and admittances in per unit.

9. For the system of Fig. Q. 2 find the voltage at the receiving bus at the end of the first iteriation. Load in $2 + j\,0.8$ p.u. Voltage at the sending end (black) is $1 + j\,0$ pu. Line admittance is $1.0 - j\,9.0$ pu. Transformer reactance is $j\,0.4$ pu off-nominal turns ratio is 1/1.04. use the GS technique. Assume $V_R = 1\angle 0°$.

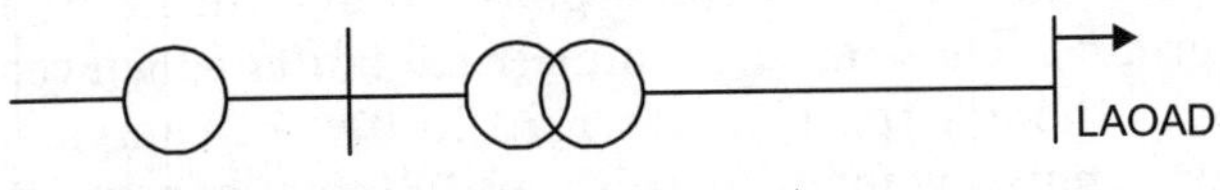

Note: For solved examples see Appendix.

CHAPTER

13

Symmetrical and Unsymmetrical Fault Calculation

13.1 SYMMETRICAL FAULT

During the operating power system, fault occurs because of insulation failure and mechanical damage. Insulation failure in the system may be caused by a system over-voltage such as switching surges or a lightning stroke or may be due to damaged and sometime insufficient insulation and insulation clearance to the ground for the transmission line conductors and mechanical damage in transmission lines may be caused by falling trees and also due to very high velocity winds. In fact for protection of the system, determination of settings of relays is done on the expected knowledge of fault current due to unsymmetrical fault condition too for which knowledge of symmetrical component etc. is needed. In this section at first we shall discuss symmetrical fault calculation and later the unsymmetrical fault calculations.

INSIDE THIS CHAPTER

13.1.1 Type of Faults

(*a*) Most of the system faults are not three phase faults but faults involving one line to ground or two lines to ground. These are unsymmetrical faults which require special study of symmetrical components though the symmetrical faults are comparitively rare. But the symmetrical fault analysis must be made as this type of faults generally leads to most severe fault current few against which the system has to be protected. In fact design of protection system depends largely on the analysis of expected symmetrixal and unsymmetrical faults current.

Infact in this section discussion will be made on the symmetrical fault analysis. A power network comprises synchronous generators, transformers, lines and loads. However, the operating conditions at the time of fault are important the loads can be neglected during fault as voltages dip very low so that current drawn by loads can be neglected in comparison to fault currents.

It has been found that the synchronous generators short circuit has a characteristic time-varying behaviour. During short circuit, the flux per pole undergoes dynamic changes with associated transient in damper and field windings. The reactance of the circuit model of the machine changes in the first few cycles from a low substransient reactance to a higher transient value, finally settling at a still higher synchronous (that is steady state) value. Depending upon the arc interruption time of circuit breakers, a suitable reactance value is used for the circuit model of synchronous generator for short circuit analysis.

Modern large interconnected power system experience heavy current flowing during a fault must be interrupted much before the steady state conditions are established. Moreover, if we consider mechanical forces which act on fault-clearing circuit breaker and their mechanical component. The maximum that a breaker has to carry momentarily must also be determined (to design a circuit breaker). Hence for selecting a circuit breaker we must, therefore determine the initial current that flows on occurrence of short circuit and also the current in that transient that flows at the time of circuit interruption on which we have discussed on transient in power system in the chapter 10. Below we shall now discuss the type of faults, sources of fault power, limitations to fault current, percentage reactance, etc.

(*b*) Below a Table 13.1 shows Types of fault and their percentage in the system :

Table 13.1. Type of Faults

Type of Faults	Percentage of Cause of Failure	Probability of Failures
1. Insulation Failure due to :	45%	
(*a*) Lightning	5%	
(*b*) Switching fault	10%	
(*c*) Insufficient insulation design (poor quality insulation distance between line ground)	10%	
(*d*) Relay (protective system) failure	15%	
2. Mechanical Failure		
(*a*) Sleet, wind tree falling etc.	15%	
(*b*) Jumpers, conductors fixing poor quality mechanical component in circuit breaker, isolators	20%	
3. These faults can result into one of the following forms :		
(*a*) Single phase to ground faults		10%
(*b*) Phase to phase faults		15%
(*c*) Two-phase to ground faults		10%
(*d*) Three-phase faults		5%

These faults cause heavy currents, called short circuit currents, to flow in the system. The determination of the values of such currents enable us to make proper selection of circuit breakers protective relays and also helps to ensure that the associated apparatus, *e.g.*, bus bars connections, current transformers will withstand the forces which arise due to the fault currents during the period prior to the interruptin device clearing the fault.

In fact single phase to ground faults are most common whereas the 3-phase short-circuits are found to have most severe fault current (which we shall calculate later in this chapter), as these high currents involve symmetrical conditions.

13.1.2 Factors Effecting Symmetrical Short-Circuit Current in a System

From the basic relation between the current, voltage and impedance we can determine the factors on which depends the symmetrical short circuit current. I_{sc}. First it is necessary to know the reactance (X) of each phase up to the fault point and if V is the phase voltage,

then, $$I_{sc} = \frac{V}{X_{sc}} \text{ amperes} \qquad ...(13.1)$$

From this eqn. (13.1) we can see that the short circuit current due to the fault depends on the Short Reactance developed at the moment of short circuit fault. It must be mentioned that the line resistance is so small that in comparison to the short circuit reactance X_{sc} is negligibly small.

In fact in the power system, the reactance of line, synchronous machines, transformers, reactors etc. are usually expressed as a percentage.

Now let us see below what is meant by Percentage Reactance :

Say, a reactance X percent will have a reactive voltage drop of X percent of normal voltage when carrying full load current corresponding to normal rating.

$$\text{(percentage } X) \ \% X = \frac{IX}{V} \times 100 \qquad ...(13.2)$$

1. Percentage Reactance

This can be written, hence, as, where X is the reactance, V the rated voltage and I is full load current.

Now, if X is the only reactance in the circuit, the short circuit current is given by V/X which by equation (13.2) is,

$$I_{sh} = \frac{V}{X} = I \times \left(\frac{100}{\% X}\right) \qquad ...(13.3)$$

As, from eqn. 13.2, we get,

$$\frac{V}{X} = I \times \left(\frac{100}{\% X}\right)$$

Therefore it means, that if the secondary of a transformer having a reactance of, say, 5% were short circuited, the applied voltage need only be 5% of the normal line to neutral voltage in order to circulate full-load current through its windings. Alternatively, with second short-circuited and with normal voltage applied to the primary the symmetrical short-circuit current will be 20 times the normal full load current.

It must be stated here that in the case of cables and over-load lines, the rectance values in most cases are given in ohms for a fixed length or per unit of length and it may become necessary to circuit such values into percentage values. This we can do by using the equation (13.2) as discussed below :

From eqn. (13.2) we can write $$X = \frac{(\% X) V}{100 \cdot I}$$

Which can be written as, $$X = \frac{(\% X) \cdot V \cdot V}{100 I \cdot V} = \frac{(\% X) V^2}{100 VI}$$

We can aslo write, then, $$X = \frac{(\% X)\left(\frac{V}{1{,}000}\right) \cdot \frac{V}{1000} \times 1000}{100\left(\frac{V}{1000}\right) \cdot I}$$

$$= \frac{(\% X)(KV)^2 \cdot 10}{KVA} \quad ...(13.4)$$

From eqn. (13.4) once we get, (% X), percentage Reactance

$$= \frac{(KVA)X}{10(KV)^2} \quad ...(13.5)$$

Now let us develop the expression **Base *KVA*.**

2. Base *KVA* Related % *X*

It is an established fact, that since different plants in any power system may have different *KVA* ratings. Now as the percentage reactance quoted by the manufacturers are of the respective *KVA* ratings, hence it is necessary to convert all the reactances in the system to a common *KVA* rating base. The value of this, however is quite unimportant and may be : (*a*) that of the largest plant, (*b*) that of the total plant capacity and (*c*) any arbitrary value. The common *KVA* is known as the base *KVA* of the system.

The conversion can be affected by using the expression,

We get, % Reactance at bus *KVA* $= \frac{\text{Base } KVA}{\text{Plant } KVA} \times \%$ Reactance Plant *KVA* ...(13.6)

As an example, to explain further, a transformer has a rating of 10,000 KVA with 5% percentage reactance will have a (KVA) of 25,000 then a percentage reactance will be as percentage eqn. (13.6).

$$= \frac{25{,}000}{10{,}000} \times 5 = 12.5\%$$

3. Voltage Base & Reactance

There are cases where it will be more convenient to work in ohmic values of the various reactances. The method would be simple if all the reactances relate to same voltage but if step-up or step-down transformers or other equipments which operate at different voltages are also included then all the ohmic values will hence to be reduced to a common voltage base. The following expression may be used for this conversion.

$$X_2 = \left(\frac{V_2}{V_1}\right) X_1$$

where X_1 = Reactance in ohms at system voltage V_1

X_2 = Reactance in ohm at base voltage V_2

4. Transient in *R-L* Series Circuit & Fault (Switching Operation)

Study on fault in the system cannot be completed without study of the transient in a series *R-L* circuit. In fact to perform a complete study on faults it is necessary to determine the initial value of current when a fault occurs in a system, when a generator shorted at its terminal.

From our study of basic electrical engineering, we know that when a sinusoidal voltage, $e = E_{max} \sin wt$ is suddenly (by siwtching on) applied in a series *R-L* circuit we get voltage equation as :

$$L(di/dt) + R_i = E_{max} \sin wt \quad ...(13.7)$$

We get from this differential equation after solving it, the expression of

$$i = I_{max} \sin(wt - \phi) + A \cdot e^{-(R/L)t} \quad ...(13.8)$$

where $I_{max} = (E_{max})/\left(\sqrt{R^2 + X^2}\right)$, $\phi = \tan^{-1}(WL/R)$ and A is a constant of integration which can be determined by applying the condition existing when the circuit is closed.

Now with initial conditions, in this equation of i, with $t = 0$, when $e = 0_1$ and let the circuit be closed when than any value, to then eqn. (13.8) becomes

$$0 = I_{max} \sin(wt_0 - \phi) + Ae^{-(R/L)t_0}$$

$$A = -I_{max} \sin(wt_0 - \phi)\, e^{-(R/L)t_0}$$

Hence the final equation of i becomes after putting the expression of A

$$i = -I_{max} \sin(wt_0 - \phi) - I_{max} \sin(wt_0 - \phi)\, e^{-(R/L)(t-t_0)} \quad ...(13.9)$$

When, R is small compare to L $\phi = \frac{\pi}{2}$, then the eqn. (13.9) becomes

$$i = I_{max} \sin\left(wt - \frac{\pi}{2}\right) - I_{max} \sin\left(wt_0 - \frac{\pi}{2}\right) e^{-(R/L)(t-t_0)} \quad ...(13.10)$$

Now after studying the eqn. (13.10) of i, it is noticed that it contains two component: (1) steady state and (2) transient component exponentially decaying, as below :

1. Steady state component : $$i_{steady\ state} = I_{max} \sin\left(wt - \frac{\pi}{2}\right) \quad ...(13.11)$$

2. Transient component : $$i_{tansient} = I_{max} \sin\left(wt - \frac{\pi}{2}\right) \cdot e^{(-R/L)(t-t_0)} \quad ...(13.12)$$

From this expression, we find the amplitude of the transient component will depend on t_0; that is, the time at which the switch is closed. Now for switching operation, we shall consider the following two instant of closing the switch :

1. Switch closed when e is passing through maximum values as can be seen in Fig. 13.1

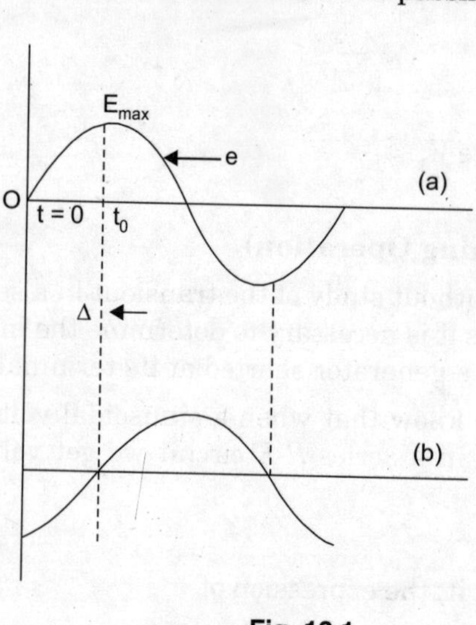

Fig. 13.1

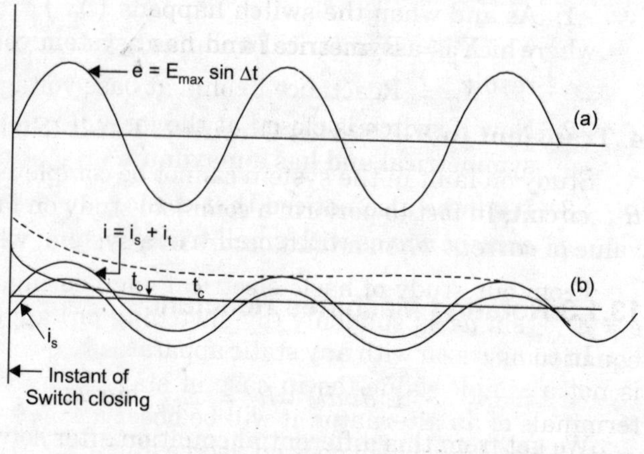

Fig. 13.2

When, $\theta = wt_0 = \pi/2,$

and
$$i = I_{max} \cdot \sin\left(wt - \frac{\pi}{2}\right) - I_{max} \sin\left(\pi/2 - \frac{\pi}{2}\right) \cdot e^{-(R/L)(t-t_0)}$$

$$= I_{max} \sin\left(wt - \frac{\pi}{2}\right) \quad ...(13.13)$$

This equation shows that there is no transient and the current through the circuit is sinusoidal right from the instant of switch as can be seen Fig. 13.1 (*b*). It is noticed that at the time of closing the switch the current is zero.

(*b*) When "*e*" is passing through zero towards positive when the switch is closed as can be seen in Fig. 13.2 (*a*)

$\therefore \quad t_0 = 0, wt_0 = 0$

$$\therefore \quad i = I_{max} \sin\left(wt - \frac{\pi}{2} - I_{max} \sin\left(0 - \frac{\pi}{2}\right)\right) e^{-(R/L)t}$$

$$= I_{max} \sin\left(wt - \frac{\pi}{2}\right) + I_{max}\, e^{-(R/L)t} \quad ...(13.14)$$

Hence, the current contains both the steady state and the transient component.

In fact, during the early period after switching $e^{-(R/L)\,t}$ does not decay rapidly from the value of unity and the current is therefore approximate

$$i = I_{max} \sin\left(wt - \frac{\pi}{2}\right) + I_{max} = I_{max}\,(1 - \cos wt)$$

The maximum value of $i = 2I_{max}$ and it is reached at $wt = \pi$.

The maximum value of $i = 2I_{max}$ and it is reached at $Q = wt = \pi$. Hence we see that in this case the current rises to a value very nearly equal to $2I_{max}$ soon after the switch is closed. This is known as the DOUBLING EFFECT. It may also be seen from Fig. 13.2 (*b*) that the presence of the transient mark is the current wave ASYMMETRICAL. Soon after the transient effect dies out the current settles down to the steady state value and becomes sinusoidal. Hence we can summerise the whole incedence as below :

1. As and when the switch happens to be closed at instant when $e = 0$, a current wave which is assymetrical and has a maximum instantaneous value of roughly to $2I_{max}$ result.
2. Now if switch is closed at the instant where '*e*' is maximum a current wave which is symmetrical and has a maximum instantaneous value equal to I_{max}.
3. It can be also noticed that closing the switch at other instants will give asymmetric current waves with small transients.

13.1.3 Rotating Machines Reactance

In comparison with any static apparatus like transformer, the reactance of rotating machine is not a simple value (as in case of static apparatus). Say, if a short circuit occurs at the terminals of an alternator, it will be observed that the short circuit current will be maximum in the beginning which goes on decreasing until it reaches a steady state value as can be seen in Fig. 13.3. Thus it will be observed that the reactance of rotating machine is a complex quantity which depends on time. In fact three types of reactances are assigned to rotating machine :

1. Sub-transient Reactance (X''_d)
2. Transient Reactance (X'_d)
3. Synchronous Reactance (X_s)

Below briefly we shall discuss the above types of reactances is given :

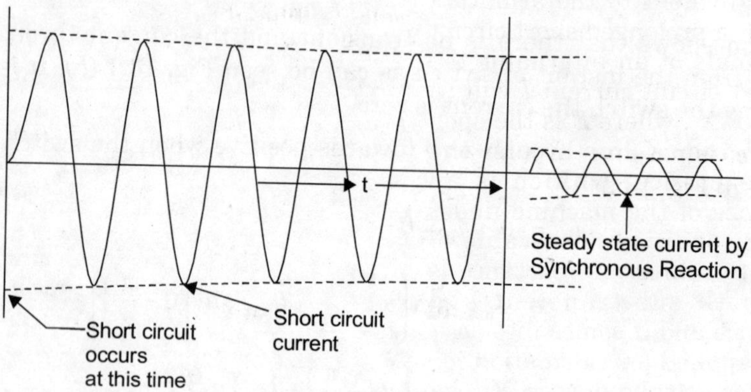

Fig. 13.3. Representation of short circuit at the terminals of an Alternator.

1. Sub-transient Reactance

Sub-transient Reactance is the apparent reactance of the stator winding just at the instant the short circuit occurs and it causes high current to flow during the first few cycles. If the machine is fitted with damper windings or if the machine has solid pole faces as in the turbo-alternators, the air gap flux at the first instant of short circuit is prevented from changing to any great extent. This results from the fact that there is very little leakage flux. As a result initial short circuit current in such machines are greater. Then, the subtransient current is, $I'' = \frac{E}{X_d''}$, where X''_d is the sub-transient reactance which in fact approaches the armature leakage differing from the quantity only by the leakage of the damper winding. In fact for all practical purpose $X'''_d = X_L$.

2. Transient Reactance

When sudden short circuit occurs the terminals of an alternator, the machine has a much lower effective value of reactance than the steady-state synchronous reactance or the synchronous reactance.

When a short circuit occur across the terminals of an alternator the load current will increase to an enormous value, and as it resistance is very low as compared to the reactance this current will lost almost by 90° from the voltage. This will result in an increased demangnetising action caused by armature reaction resulting into a decreased generator voltage per phase say E. Hence the short circuit current is given as : $I' = \frac{E'}{X_L}$, where X_L is called the leakage reactance, then the equation, $I' = \frac{E'}{X_L}$ can also be represented as $I' = \frac{E'}{X'_d}$. Here E is the normal phase voltage and X'_d is taken as the transient reactance which takes into as the transient Reactance which includes armature and leakage reactance.

3. Synchronous Reactance

Synchronous is the value commonly used for all problems involving steady state calculation of machine performance and includes armature leakage reactance and the reactance equivalent to the armature reaction. Thus for a prolonged short circuit across the terminals of an alternator the steady state short circuit current I will be given by E/X_d or E/X_s, where E is the open circuit e.m.f., and X_d or X_s is the synchronous reactance, whhich is an apparent reactance of the machine under normal working condition. The value of either X_d or X_s. Synchronous reactance is greater than both sub-transient and transient reactance and it comes into action only a few seconds after the short circuit occurs. Hence, it is not applied for calculation for the solution of circuit breakers, fuses, contactors, but useful for relay settling purposes.

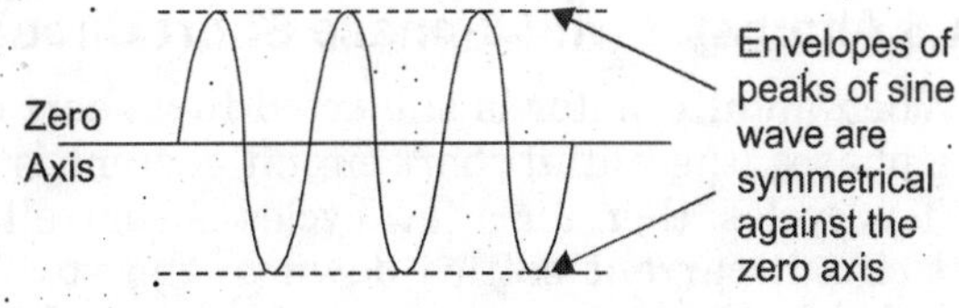

Fig. 13.4 (*a*). Symmetrical wave envelopes.

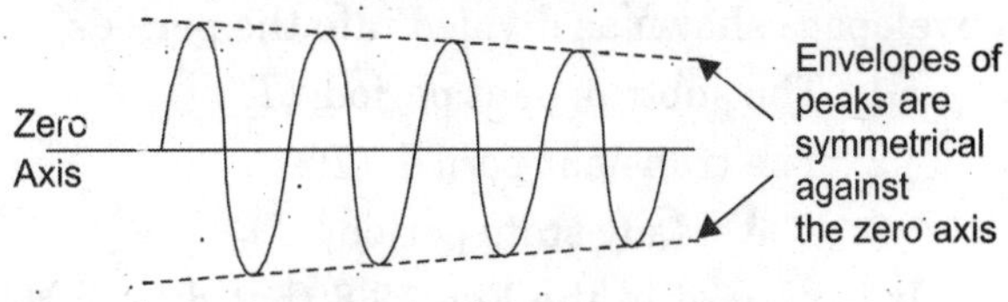

Fig. 13.4 (b). Symmetrical short circuit current wave envelopes.

13.1.4 Symmetrical & Asymmetrical Short Circuit Current

What we mean by a symmetrical & assymmetrical current. In fact a current is said to be symmetrical if the peaks of the envelopes of the current wave are symmetrical about the zero axis as we can see in Fig. 13.4 (*a*) and (*b*). Shows symmetrical short circuit current wave.

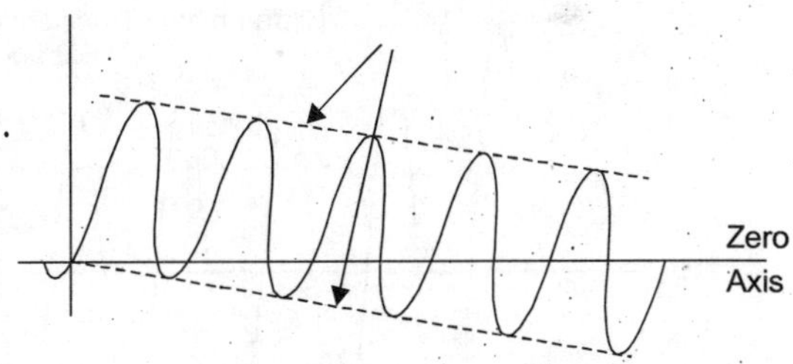

Fig. 13.5. Asymmetrical short circuit current envelopes.

But if the envelops of the peaks of the current are not symmetrical about the zero axis, it is then called assymetrical current as we can see in the Fig. 13.5. In fact, it should be mentioned here that it is very hard to interpret the asymmetrical short circuit current. These currents are arbitrarily divided into simple a.c., symmetrical and d.c., component the sum of these two components at any instant will be equal to the magnitude of that asymmetrical a.c., wave form as can be seen in Fig. 13.5.

It is to be noted here that the magnitude of the d.c. component depends on the instant the short circuit occurs. In the Fig. 13.6 it can be observed that at the time of short circuit, both a.c., and d.c. components are equal as a result, the resultant current is zero at this instant. However, as a matter of fact, as there is no source of d.c. e.m.f. in the system it does not continue to flow at a constant value but the energy represented by the direct component is dissipated as I^2R LOSS. The whole exercise is shown in Fig. 13.7.

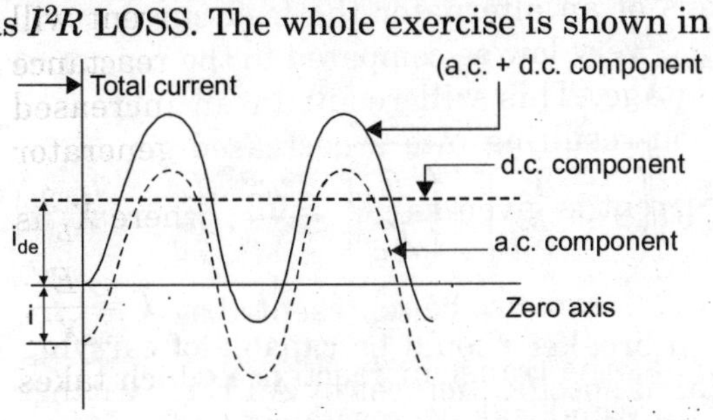

Fig. 13.6. Asymmetrical a.c. waveform.

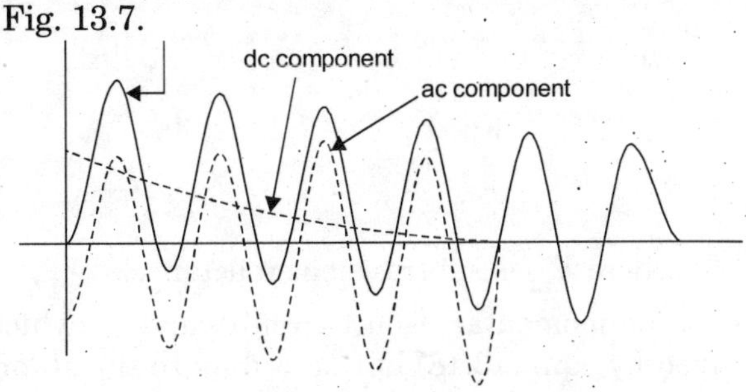

Fig. 13.7. Showing decay of d.c. component.

13.1.5 Alternator with 3-phase Short Circuit

When an alternator is subjected to a short circuit across its terminals, that is on all the three phases, the initial short circuit current is limited by the subtransient reactance for the first few cycles, then for a few cycles is controlled by the transient reactance and finally the short circuit current settles down to the steady state short circuit current which is then controlled by synchronous reactance. In the Fig. 13.8 a representation of the oscillogram of the symmetrical short circuited current at the armature terminals of the alternator. The wave envelope is shown as divided into the periods :

1. The subtransient period, (T'')
2. The transient period, (T')
3. The steady state period, (T).

It is noticed in the Fig. 13.8 that during the subtransient period the current decrease is very fast and this period, as stated last for few cycles only. Then the transient period lasts relating longer time during which the current decrease is moderate that is not as fast as in case of sub-transient period.

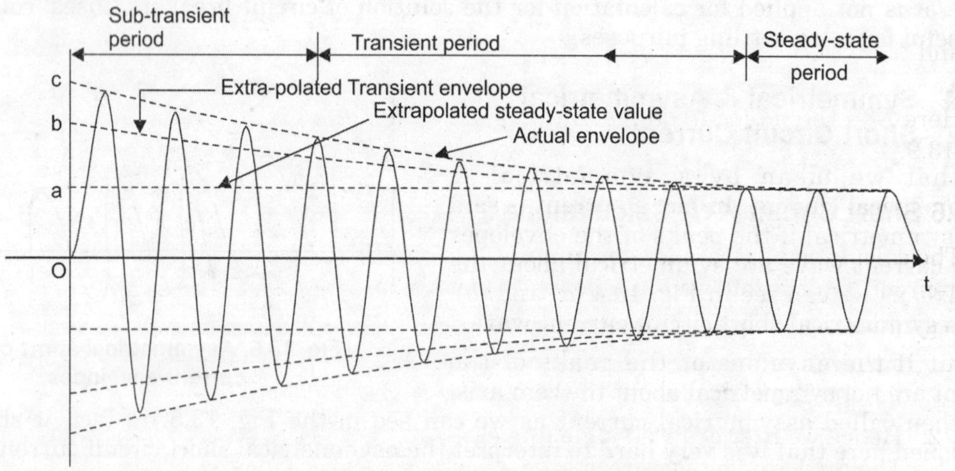

Fig. 13.8. Representation of oscillogram of symmetrical short circuit current.

The RMS value of armature short circuit current for 3-phase short circuit, is given by

$$i_{\text{r.m.s.}} = \sqrt{i_{\text{d.c.}}^2 + i_{\text{a.c.}}^2} \quad ...(13.15)$$

where a maximum d.c. current $\quad i_{\text{d.c.(max)}} = \sqrt{2} \cdot \dfrac{E}{X''_d} \quad ...(13.16)$

and $\quad i_{\text{ac(max)}} = \dfrac{E}{X''_d} \quad ...(13.17)$

$$\therefore \quad i_{\text{r.m.s(max)}} = \sqrt{\left(\sqrt{2} \cdot \frac{E}{X''_d}\right)^2 + \left(\frac{E}{X''_d}\right)^2} = \sqrt{3} \cdot \frac{E}{X''_d} \quad ..(13.18)$$

where X''_d is subtransient reactance.

The momentary short circuit current, which a breaker should be capable of carrying is given by eqn. (13.18) but according to the standard specification set by A.I.E.I. Switchgear committee, the multiplying factor instead of being $\sqrt{3}$ should be 1.6 which can be further reduced to 1.5 for 5 kV breaker or less. The multiplying factor for normal operating capacity depends on the speed of the breaker, for slower breaker it is (8 cycle) it is 1.0, for 5 cycle breaker it is 1.1 and for 2 cycle breaker it is 1.4. Now referring to the Fig. 13.8, the currents during three periods mentioned above are limited by various reactances of alternator, neglecting the armature resistance) and can be defined by the following equation assuming that the alternator is operating at no load in the pre fault period.

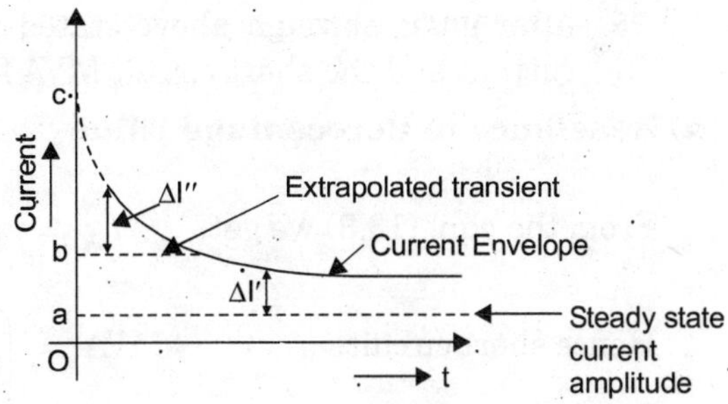

Fig. 13.9

$$\therefore \quad |I''| = \frac{Oc}{\sqrt{2}} = \frac{|E|}{X''_d}, \text{ for subtransient period } (T'')$$

$$|I'| = \frac{Ob}{\sqrt{2}} = \frac{|E|}{X'_d}, \text{ for transient period } (T')$$

and

$$|I| = \frac{Oa}{\sqrt{2}} = \frac{|E|}{X_d}, \text{ for steady state period } (T).$$

Here E is the no load voltage of the alternator and the r.m.s. currents are drawn in the Fig. 13.9.

13.1.6 Short Circuit KVA Calculations Process

The calculation process for determining the short circuit current resulted due to a 3-phase short-circuit fault is followed in following steps :

1. First we shall consider that the 3-phase short circuit fault result in symmetrical fault currents in the three phases and as such problems involving such fault can be solved for one phase the result for the other two phases being the same.
2. Hence we first draw a single line diagram of the complete network indicate on it : (1) The capacity, voltages, reactance, resistance of all synchronous plants, (2) The capacity, type, reactance, resistance and length of all cables and overhead lines with spacings where necessary, (3) The capacity and reactance of transformers, (4) The values of any artificial reactance reactors, introduced in the system.
3. After the single-line diagram as stated in (1) is drawn, then, corresponding to this line diagram we draw a single-line reactance or impedance diagram, showing one phase of the system and the neutral.
   3.1. Then in this diagram we write down the reactance or impedances either in percentage values on a common KVA base or ohmic values on a common KV base.
   3.2. However, while drawing the single-line reactance, the following datas of the network must be noted : (*a*) The resistance of generators or transformers, being small in comparing to reactors being small, can be neglected, (*b*) The transformer is represented by a reactance in series, (*c*) Resistance in the case of long overhead transmission lines and cables, assumes appreciable value and cannot be ignored. A transmission line is represented by series reactance and resistance.
4. Now after drawing the reactance diagram, we calculate the total reactance or impedance upto the point of fault. This max involve series parallel combinations and delta/star or star/delta conversion of reactances.

5. After going through above stated processes in three steps above. The short circuit current and the short circuit MVA for the system can be calculated as :

(a) Reactance in Perecentage Value

From the eqn. (13.3) we get, I_{sh} = (Short Circuit Current) $= \dfrac{I \times 100}{\% X\ \% X} \cdot I$

Hence short circuit MVA $= \left(\dfrac{100}{\% \text{ Reactance}} \times \text{base MVA} \right)$

$$= \left(\frac{100}{\% \text{ Reactance}} \times \text{base KVA} \right) \times 10^{-3} \quad \text{...(13.19)}$$

Short circuit current (r.m.s) in terms of short circuit MVA and voltage

$$= \frac{\text{Short Circuit MVA} \times 10^3}{\text{Line-to-line (KV)} \times \sqrt{3}}, \text{ amps.} \quad \text{...(13.20)}$$

(b) Reactance in Ohms

In this case, short circuit current (rms) $= \dfrac{V}{\sqrt{3} \times \text{reactance}}$ amps. ...(13.20)

$$\text{Short Circuit MVA} = \frac{\text{Short Circuit Current} \times V \times \sqrt{3}}{10^6} \quad \text{...(13.21)}$$

It must be noted that in these two equations, V is the line voltage in **Volts**.

Below we shall now discuss on per unit system in calculations.

13.2 PER UNIT SYSTEM

In the network we know there are two quantities : (1) Basic quantities and (2). Per Unit quantities. Below we shall discuss these quantities which will help the calculation for short circuit fault analysis.

1. Base Quantities

(a) For single-phase circuit, (1) $I_b = \dfrac{KVA_b}{KV_b}$, Here suffix '$b$' indicates base quantities

$$\text{(2) } Z_b = \frac{V_b}{I_b} = \frac{KV_b}{I_b} \times 10^3 = \frac{KV_b^2}{KVA_b} \times 10^3 = \frac{KV_b^2}{MVA_b} \quad \text{...(13.22)}$$

(b) For three-phase circuit, (1) $I_b = \dfrac{KVA_{b(3\phi)}}{\sqrt{3}\, KV_{b(L-L)}}$,

where $V_{b(L-L)}$ = Line to line base voltage

$$\text{(2) } Z_b = \frac{V_{b_{L-L}}}{\sqrt{3} \cdot I_b} = \frac{KV_{v(L-L)}}{\sqrt{3} \cdot I_b} \times 10^3 = \frac{KV_{b_{L-L}}^2}{KVA_{b(3\phi)}} \times 10^3 = \frac{KV_{b(L-L)}^2}{MVA_{b-b(3\phi)}}$$

...(13.23)

2. Per Unit Quantities

In this case say, **1. Base Voltage** = 1. p.u.

2. Base Current = 1. p.u.

3. Base Impedance = 1. p.u.

Then, *V, I, Z* (Voltage, Current, Impedance) of a given system can be expressed in p.u., values as stated below :

(*a*) For single-phase circuit, $V_{P.U.} = \frac{V}{V_b}$, $I_{P.U.} = \frac{I}{I_b}$

and $$Z_{P.U.} = \frac{Z}{Z_b} = Z\left[\frac{KVA_b}{KV_b^2} \times 10^{-3}\right] = Z\left(\frac{MVA_b}{KV_b^2}\right) \quad ...(13.24)$$

(*b*) For three-phase circuit, $$Z_{P.U.} = \frac{Z}{Zb} = Z\left[\frac{KVA_{b(3\phi)}}{KV_{b(L-L)}^2} \times 10^{-3}\right] = Z\left(\frac{MVA_{b(3\phi)}}{KV_{b(L-L)}^2}\right) \quad ...(13.25)$$

We can see that the final p.u. expression for $Z_{p.u.}$ is single-phase circuit, and three-phase circuit are some.

3. Change of p.u. Values on One Base to p.u. Values on Different Base

However, sometimes, it becomes necessary to change per unit values on one base to per unit values on a different base. This can be easily done with by using the formula below :

1. $$Z_{P.U._{(new)}} = Z_{P.U._{(old)}} \times \frac{MVA_{b(new)}}{MVA_{b(old)}} \times \left(\frac{KV_{b(old)}}{KV_{b(new)}}\right)^2 \quad ...(13.26)$$

Example 13.1. *Two 11 KV, generating stations are connected through two transfers and a transmission line as shown in the line diagram of Fig. 13.10. The sub-transient reactances of generators and transformers are being marked in the Fig. 13.10. The reactance of the line is 60 Ω. If a 3-phase short circuit occurs the bus bars of one of the stations, calculate the short circuit current. Throughout, carry out calculation on per unit base. The system of network is shown in Fig. 13.10.*

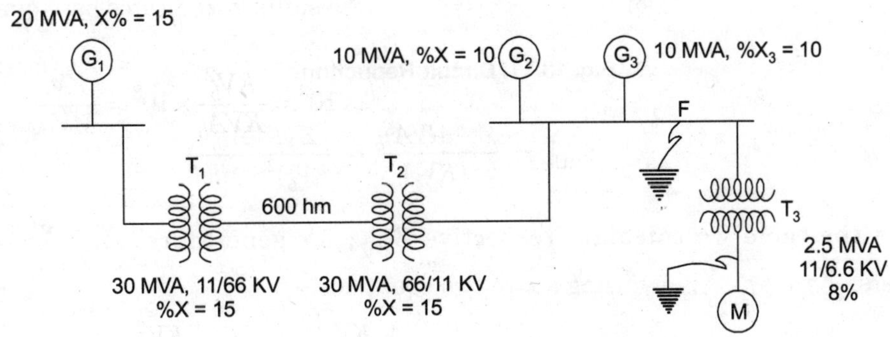

Fig. 13.10

Solution : A single line diagram for the system (stated for example 13.1) is drawn (in p.u.) in Fig. 13.11 (*a*) representing in a circuit of the system network of Fig. 13.11. Now let the KVA rating of the Generator G_1 (of the Fig. 13.11) 20,000, and 11 KV be the base KVA, and the base volts respectively. Now let us convert the reactance of three transformers to the same base KVA and obtain the actual value of reactance in view of equation.

$$\left(X = \frac{X_p\,(\text{Voltages})^2}{(\text{Output in } VA) \times 100} \cdot \Omega, \text{ where } X_p \text{ (percentage reactance) } = \frac{IX}{V} \times 100 \right), \text{ and we convert}$$

these reactances into p.u. basis by equation

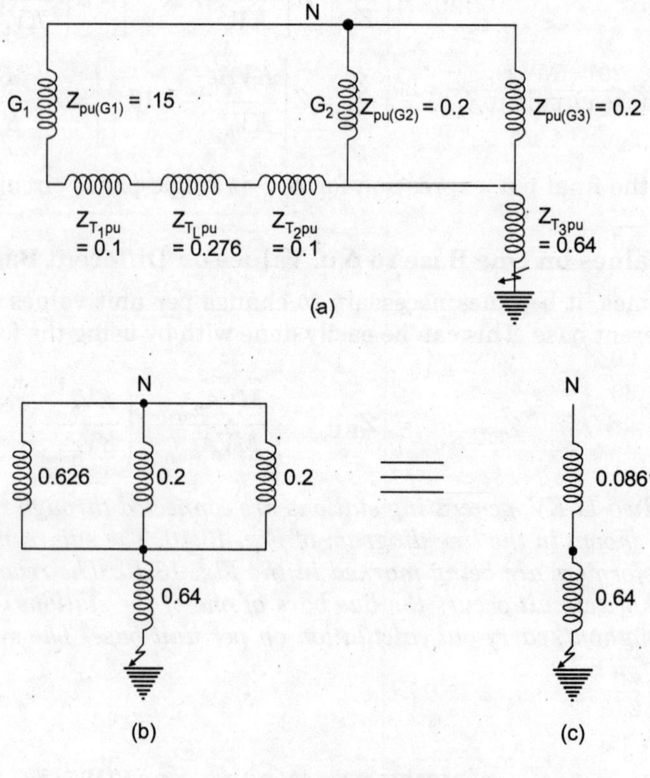

Fig. 13.11. Circuit Reduction.

$$Z_{\text{P.U.}} = \frac{Z \cdot (MVA)_b}{(KV_b)} = \frac{Z\,(KVA)_b}{(KV_b)^2 \times 1000}$$

Now in the table we calculate respective $Z_{\text{P.U.}}$, for generators, $X_{G_1}, X_{G_2}, X_{G_3}$ and transformers, $X_{T_1}, X_{T_2}, X_{T_3}$ by using above two equations.

System component	*% Reactance (X_p) referred to 20 MVA base (Given in the Line diag.)*	$Z = \frac{X_p (KV)_b^2 \times 10}{(KVA)_b}$	$Z_{P.U.} = \frac{Z \times (KVA)_b}{(KV)_b^2 \times 100}$
Generator (20 MVA) G_1 (base KVA) $(X_{G_1} = 15\%)$	15	$Z_{G_1} = 15 \times \frac{(11)^2 \times 10}{20{,}000}$ $= 15 \times .605$ $= .908$	$Z_{P.U.(G_1)} = .908 \times \frac{20{,}000}{(11)^2 \times 1000}$ $= .908 \times 0.165 = 0.15$ $= 0.15$
Generator G_2 (10 MVA) $(X_{G_2} = 10\%)$	$10 \times \frac{20\,(= (\text{MVA})_{\text{base}})}{10 (= \text{En. MVA})} = 20$	$Z_{G_2} = 20 \times .605 = 1.21$	$Z_{P.U.(G_2)} = 1.21 \times .165 = 0.2$
Generator G_3 (10 MVA) $(X_{G_3} = 10\%)$	$10 \times \frac{20}{10} = 20$	$Z_{G_D} = 20 \times .605 = 1.21$	$Z_{P.U.(G_3)} = 1.21 \times .165 = 0.2$
Transformer, T_1 (30 MVA) $(X_{T_1} = 15\%)$	$15 \times \frac{20}{30} = 10$	$Z_{T_1} = 10 \times .605 = .605$	$Z_{P.U.(T_1)} = .605 \times .165 = 0.1$
Transform T_2 (30 MVA) $(X_{T_2} = 15\%)$	$15 \times \frac{20}{30} = 10$	$Z_{T_2} = 10 \times .605 = .605$	$Z_{P.U.(T_2)} = .605 \times .165 = 0.1$
Transformer T_3 (2.5 MVA) $(X_{T_3} = 8\%)$	$5 \times \frac{20}{2.5} = 64$	$Z_{T_3} = 64 \times .605 = 38.72$	$Z_{P.U.(T_3)} = 38.72 \times .165 = 0.64$

$$\text{Now base current} = \frac{\text{base KVA}}{\sqrt{3} \times (\text{KV})_b} = \frac{20{,}000}{\sqrt{3} \times (\text{KV})_b} = 1{,}050 \text{ Amp.}$$

∴ Base reactance of transmission line,

$$X_{TL_{b_p}} = \frac{(\text{Base Voltages, KV})^2 \times 1000}{\text{Base KVA}} = \frac{(66)^2 \times 1000}{20{,}000} = 217.8\,\Omega$$

$$\therefore \quad X_{TL_{P.U.}} = \frac{\text{Actual Reactance}}{\text{Unit Reactance}} = \frac{60}{217.8} = 0.276\,\Omega$$

Hence in the circuit of Fig. 13.11 (*a*) the $Z_{TL_{P.U.}}$ (Reactance (p.u.)) is shown between $Z_{T_{1(b)}}$ and $Z_{T_{1(b)}}$ and $Z_{T_{2(P.U.)}}$. Then by circuit theory 13.11 (*a*) is reduced to that of 13.10 (*b*), which finally is reduced to the circuit of Fig. 13.11 (*c*).

∴ Now from Fig. 13.11 (*c*) we can find Fault Reactance

$$\therefore \quad X_{Fault_{(P.U)}} = 0.0861 + 0.64 = 0.7261 \text{ ohm.}$$

$$\therefore \text{ Per unit fault current} = \frac{\text{Per Unit Voltage (base)}}{\text{Fault Reactance P.U.}} = \frac{1}{0.7261} = \mathbf{1.377}$$

$$\therefore \text{ Actual fault current} = \text{per unit fault current} \times \text{Base current}$$

$$\therefore \text{ Base current} = \frac{20{,}000}{\sqrt{3} \times 11} = 1.377 \times 1050 = \mathbf{1{,}446\ Ans.}$$

$$= 1{,}050\ A$$

$$\left(\text{Hence Base Current} = \frac{\text{Base MVA.}}{\sqrt{3} \times \text{Base KV.}}\right)$$

Example 13.2. *An alternator of 11 KV, 10 MVA rating has 10% subtransient reactance (X''_d), 15% transient Reactance (X'_d) and 80% synchronous reactance (X_s or X_d). It is connected to a load system through a circuit breaker and a transformer. If under no load conditions a 3-phase fault occurs between the transformer and the circuit breaker calculate, (a) the initial symmetrical short circuit current, (b) the sustained short circuit, (c) the maximum d.c. components of the short circuit current, (d) the current interrupting capacity of the breaker, (e) the momentary current rating of the breaker, (f) the interrupting KVA. Assume that circuit breaker is a 2-cycle breaker.*

Solution :

(*a*) In study of symmetrical short circuit analysis. We find the initial symmetrical short circuit current will be controlled by sub transient react once, X''_d. Hence, above we developed, $X''_d = \dfrac{X''_{pd}\,(\text{KV})^2 \times 10}{(\text{KVA})_b}$ where X''_{pd} is the percentage sub transient reactance.

∴ As per example, $\% X''_{pd} = 10$

$$\therefore \quad X''_d = \frac{10 \cdot (11)^2 \times 10}{10{,}000} = 1.21 \text{ ohm.}$$

$$\therefore \quad I''_d = \frac{11{,}000}{\sqrt{3} \times X''_d} = \frac{11{,}000}{\sqrt{3} \times 1.21} = \mathbf{5{,}249\ Amp.}$$

(*b*) The sustained short circuit current is controlled by the synchronous reactance X_d or X_s.

$$\text{Hence,} \quad X_s \text{ (or } X_d) = \frac{X_{ps}\,(\text{KV})^2 \times 10}{(\text{KVA})_b}$$

where X_{ps} = percentage synchronous reactor or (X_{dd})

$$= \frac{X_{pd}\,(\text{KV})^2 \times 10}{(\text{KVA})_b} = \frac{80 \cdot (11)^2 \cdot 10}{10{,}000} = 9.68\,\Omega$$

As per example, % $X_{ps}\ (= X_{ps}) = 80$

∴ The sustained short circuit (at synchronous reactance)

$$I_b = \frac{11{,}000}{\sqrt{3}\times 9.68} = 656.1 \text{ amp.}$$

(c) $$I_{\text{d.c.(max)}} = \sqrt{2}\frac{E}{X''_d} = \sqrt{2}\times I''_d = \sqrt{2}\times 5{,}249 = 7{,}423 \text{ amp.}$$

(d) Current interrupting capacity of the breaker (for a 2 cycle breaker, multiplying factor is 1.4).

= 1.4 × initial symmetrical r.m.s. current

= 1.4 × 5, 249 = 7,348.6 Amp.

(e) Momentary interrupting capacity of the breaker is given as in this case of momentary interrupting capacity, multiplying factor, is 1.6

= 1.6 × 5249 = 8,398.4 amps.

In this case of percentage interrupting capacity multiplying factor is 1.6

Example 13.3. *A 3-phase transmission line operating at 33 kV having resistance and reactance of 50 ohm and 20 ohm respectively is connected to the generating station bus bar through a 5000 KVA step-up transformer of 6 percent reactance. Two alternators are connected to the bus bars, one 10,000 KVA having 10% reactance, and another 5000 KVA having 7.5% reactance. Calculate the KVA a short circuit fault between phases occuring (a) at the high voltage terminals of the transformer, (b) at load end of transmission line.*

Solution : The system of the example is shown in the single line diagram of the Fig. 13.12 (*a*).

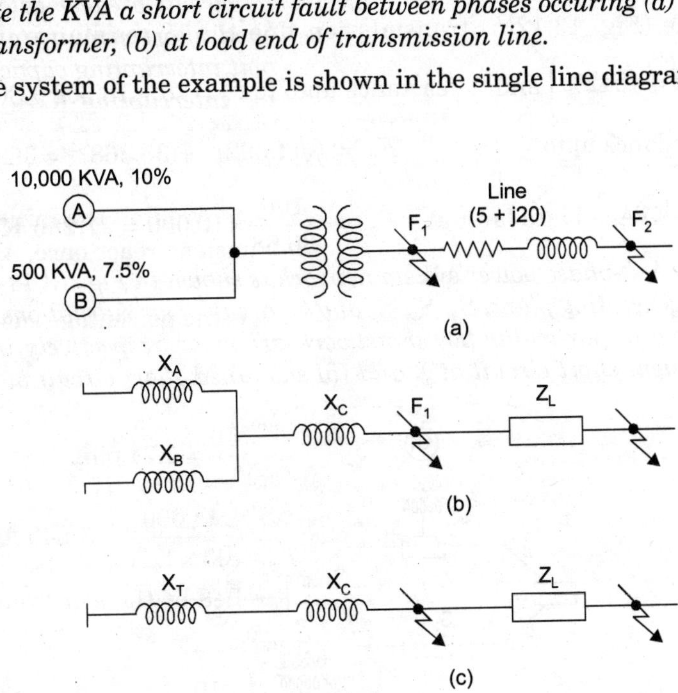

Fig. 13.12.

We shall choose the higher KVA in the system as the base KVA.

∴ Reactance of alternator (A) = 10%.

∴ With 10,000 KVA as the base KVA.

Reactance of alternator $(B) = \dfrac{7.5 \times 10{,}000}{5000} = 15\%$

Reactance of transformer $= \dfrac{6 \times 10{,}000}{5000} = 12\%$

The line impedance in ohm can be connected to percentage impedance using percentage reactance equation,

Here

$$X = 20\ \Omega$$

$$\% X = \frac{X \cdot (\text{KVA}) \cdot X}{10 \cdot (\text{KV})^3} = \frac{10{,}000 \times 20}{10 \times (33)^2} = 18.368\%$$

$$R = \mathbf{5\ \Omega}$$

$$\% R = \frac{\text{KVA} \cdot R}{10 \times (33)^2} = \frac{10{,}000 \times 5}{10 \times (33)^2} = 4.592$$

The single line impedance diagram of Fig. 13.2 (*a*) cannot be drawn is Fig. 13.12 (*b*) which again simplified to Fig. 13.12 (*c*).

∴ Hence, for a 3-phase fault at the high voltage terminal F_1 (in Fig. 13.12 *a*) of the transformer, the impedance upto the fault $= \dfrac{15 \times 10}{15 + 10} + 12 = 18\%$.

∴ Short circuit KVA fed into the falut $= 10{,}000 \times \dfrac{100}{18} = 55{,}560$ KVA. Now for a fault at the end, F_2, in the (Fig. 13.12 *b*) transmission line the % resistance upto $F_2 = 4 \times 592$ $\left(\%R = \dfrac{10{,}000}{10 \times 332} \times 5 = 4542\%\right)$ and % reactance upto to $F_2 = 18 + 18.368 = 36.368\%$.

∴ Total % impedance upto $F_2 = \sqrt{(4.592)^2 + (36.368)^2} = 36.68\%$

∴ Short circuit KVA fed into fault at $F_2 = \dfrac{100}{36.68} \times 10{,}000 = \mathbf{27{,}260\ KVA}$

Example 13.4. *A, 3-phase power system network is shown in Fig. 13.13 a. In the figure G_1 and G_2 are the two generating plant, S_1, S_2, S_3 and S_4 are the four substations. The % reactances based on KVA rating of particular power network are given respectively in the table below. Determine (a) transient short circuit at X and (b) sustained short circuit at X where the fault occurd table.*

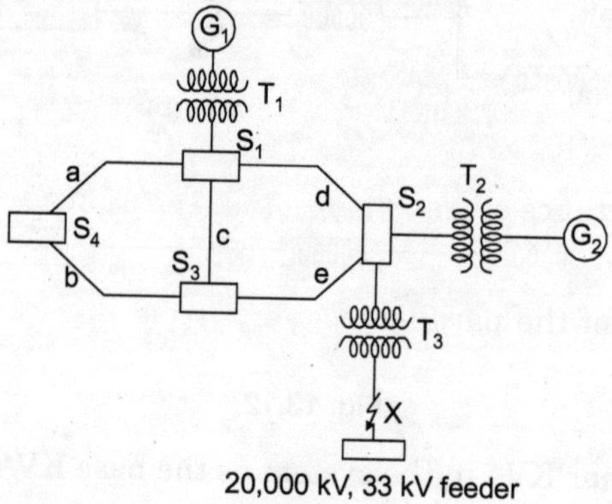

(*a*)

Fig. 13.13

Circuit components	*Capacity (KVA)*	*% Reactance*
Transmission Line		
a	40,000	*10*
b	20,000	*5*
c	50,000	*20*
d	30,000	*15*
e	10,000	*6*
Transformers		
T_1	1,50,000	*10*
T_2	50,000	*8*
T_3	20,000	*5*
Generator		
G_1	1,50,000	$90 = X_{s_1}$ (*or* X_a) $30 = X'_d$ *(Transient)*
G_2	50,000	$50 = X_{s_1}$ (X_d) $175 = X'_d$ *(Transient)*

Solution : We here choose a base KVA as 2,00,000 KVA, the various % reactances are shown on the line reactance diagram shown in Fig. 13.13 (*b*), which refer to Fig. 13.13 (*a*) of one-line diagram of the network.

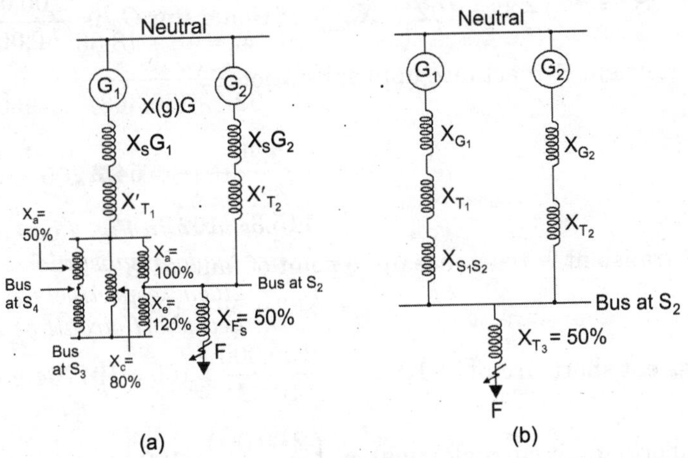

Fig. 13.13

Hence, equivalent % reactance of transimission lines '*a*' and '*b*' (connected is series).

$$= 50 + 50$$

Equivalent % reactance of the parallel combination of line '*c*' with series combination of lines '*a*' and '*b*'

$$= \frac{1}{\dfrac{1}{100} + \dfrac{1}{8c}} = 44.5$$

Now a study of the reactance diagram will show that this equivalent % reactance 44.5 is in series with X_e and that this combination in parallel with X_d gives the total reactance between substations S_1 and S_2.

The total % reactance between sub-stations S_1 and S_2.

$$\therefore \quad X_{S_1 S_2} = \frac{(44.5+120)100}{44.5+120+100} = 62\%$$

The reactance diagram many now be redrawn in Fig. 13.13 (*c*).

1. Transient Short Circuit Current

Equivalent transient % reactance from generator G_1 to substation S_2.

$$= 40 + 40/3 + 62$$

$$= 115.3$$

$$X_{tr(Gen)} = \frac{2{,}00{,}000}{1{,}50{,}000} \times 30 = 40\%$$

$$X_{T_{1(tr)}} = X_{T_1} = \frac{2{,}00{,}000}{1{,}50{,}000} \times 10 = \frac{40}{3}\%$$

$$X_{S_1 S_{2(tr)}} = X_{S_1 S_2} = 62\%$$

Similarly, equivalent transient % reactance from generator G_2 to substation

$$S_2 = 70 + 32 = 102 \left[\because X_{tr(gen)} \text{ of Generator } G_2 \text{ is } \frac{2{,}00{,}000}{50{,}000} \times 17.5 = 70\%\right]$$

$\therefore$ Equivalent transient % reactance upto substation S_2

$$= \frac{1}{\dfrac{1}{115.3} + \dfrac{1}{102}} = 54.2$$

and equivalent transient % reactance up to point of fault at '*X*' – '*F*'.

$$= 54.2 + 50 = 104.2$$

Therefore, transient short circuit $\quad \text{KVA} = \dfrac{2{,}00{,}000}{104.2} \times 100 = 191754.55$ **KVA**

and transient short circuit current (rms) $= \left(\dfrac{2{,}00{,}000}{104.3} \times 100\right) \times \dfrac{1}{\sqrt{3} \times 33}$

$$= 3360 \text{ amps.}$$

The peak value of the transient short circuit current would be $\sqrt{2} \times 3360 = 4751$ amps.

It should be noted that this value of short-circuit current is based on the assumption of no d.c. component.

2. Sustained Short Circuit Current at the Fault Point, *F*

We shall first find equivalent steady state reactance from G_1 to substation S_2

$$= 120 + \frac{40}{3} + 62 = 195\%$$

$$\left[\text{As, } X_{S(gen)} = \frac{2,00,000}{1,50,000} \times 90 = 120\%,\ X_{T_1} = \frac{40}{3}\%,\ X_{S_1 S_2} = 62\%\right]$$

Similarly, equivalent steady state % reactance from G_2 to S_2,

$$= 200 + 32 = 232\%$$

$$\therefore \text{ Equivalent reactance upto } S_2 = \frac{1}{\frac{1}{195} + \frac{1}{232}} = 106\%.$$

∴ Total steady state reactance upto fault point '*F*'

$$= \frac{2,00,000 \times 100}{156} \times (3 \times 33) = 2242.3 \text{ amps.}$$

The peak value of sustained short circuit current

$$= \sqrt{2} \times 2243.3 \text{ amp}$$

$$= 3162 \text{ Amps.}$$

Example 13.5. *A 625 KVA, 480 V alternator supplies a purely resistive load of 500 KW at 480 V. The sub-transient reactance of the generator is 8%. Assuming the load to be directly connected across the generator terminals, find the initial symmetrical rms current in P.U. at the generator terminal for a 3-phase deat short at its terminals.*

Solution : Say, in this case, 625 KVA be the base KVA and 400 V be the base voltage.

$$\therefore \quad I_{\text{Load}} = \frac{500}{625} \text{(with p.f. = Unity)}$$

$$= 0.8 \text{ P.U at } \angle 0^\circ$$

Now let the terminal voltage (480 V) $= 1 \angle 0^\circ$ P.U. reference

$$\therefore \quad E'' = V + j \cdot I_L X''_d$$

$$\therefore \quad E'' = (1 + j_0) + 0.8\,(j\,0.8)\ \{X''_d = 8\% = 0.08 \text{ p.u.}\}$$

$$= (1 + j0.0064) \text{ P.U.}$$

$$\therefore \quad I'' = \frac{E''}{X''_d} = \frac{1 + j0.064}{j0.08} = (0.8 - j12.5) \text{ p.u.}$$

13.3 UNSYMMETRICAL FAULTS IN POWER SYSTEMS

In above section 13.1 we discussed on the symmetrical faults calculation (in power system), but, in fact, the possibility of such a fault in a power system is quite rare. The actual fault may occur on one line and ground or two lines and ground, or there may occur fault between of the three lines and ground, such fault are called unsymmetrical faults. To detect such faults it is vary much essential that the fault current and voltages should be determined. Hence the types of unsymmetrical fault can be tabled as below :

(1) Three-phase fault, (2) Single-line to ground fault, (3) Double line fault, (4) Double line to ground fault.

Infact proper analysis of these faults are necessary for correct assessment of the fault current and voltage of the power system.

13.3.1 Symmetrical Component and Sequence Components

Before we go into the calculation of unsymmetrical fault calculation in a power system, we shall look back briefly on know to determine the sequence components :

Let us first, consider the symmetrical component of the positive phase sequence (refer to Fig. 13.14. Then these can be expressed as :

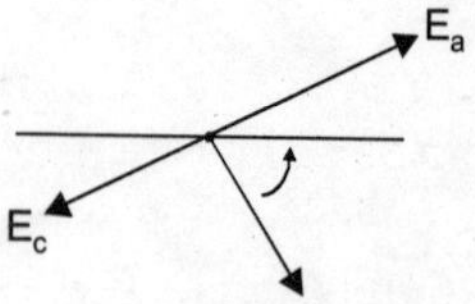

Fig. 13.14. Unbalanced vectors.

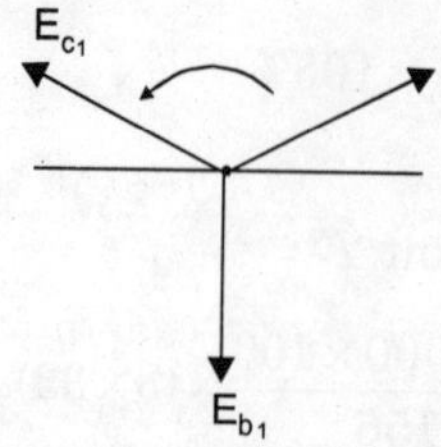

Fig. 13.15 (*a*). Positive segment components of the vectors of 13.15.

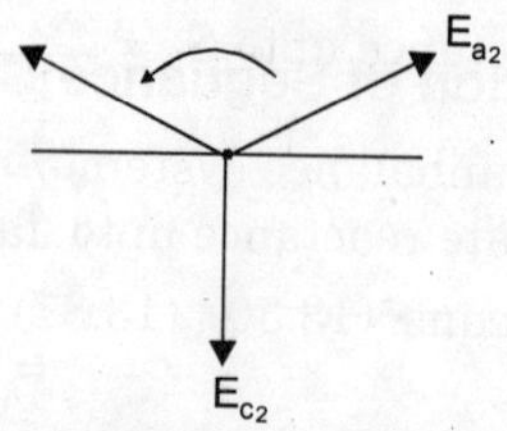

Fig. 13.15 (*b*). Negative sequence components of Fig. 13.15.

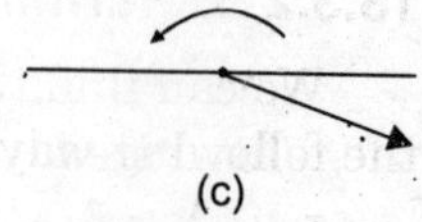

Fig. 13.15 (*c*). Zero-sequence components of the vectors of Fig. 13.15.

$$E_{a_1} = E_{a_1}, E_{b_1} = a^2 E_{a_1}, E_{c_1} = aE_{a_1} \quad ...(13.27)$$

$$a = e^{j\,120°} = -0.5\,j + 0.866$$

$$a^2 = e^{j\,240} = -0.5 - j\,0.866, a^3 = e^{j\,360} = 1,$$

$$a^4 = a^3 \times a = -j\,0.5 + j\,0.866,\ a^5 = a^3 \times a^2 = -0.5 - j\,0.866$$

$$\therefore 1 + a + a^2 = 0 = 0 + j0,\ a - a^2 = j\sqrt{3} = \sqrt{3}\angle 90° = 0 - j\,1.732$$

$$a^2 - a = -j\sqrt{3} = \sqrt{3}\angle -90° = 0 - j1.732;\ a + a^2 = 1\angle 180° = -1 - j\,1.0$$

$$1 + a = 1\angle 120° = 0.5 + j0.866 = -a^2, 1 - a = 1 - (-0.5 + j0.866) = 1.5 - j0.866 = \sqrt{3}\angle -30°$$

We shall need these operators in calculations below :

$\therefore$ Similarly the symmetrical components of the negative phase sequence (referring to Fig. 13.15 (*b*), can be expressed as,

$$E_{a_2} = E_{a_2}, E_{b_2} = aE_{a_2} \quad E_{c_2} = a^2 E_{a_2} \quad ...(13.28)$$

Similarly the zero phase sequence component (as referred to Fig. 13.15 *c*), can be expressed as below :

$$E_{a_0} = E_{a_0}, E_{b_0} = E_{a_0}, E_{c_0} = E_{a_0} \quad ...(13.29)$$

The equations, (13.27) to (13.29) that for can be now written to together as below.

$$E_a = E_{a_1} + E_{a_2} + E_{a_0} \quad ...(13.30)$$

(which means E_a is the addition of positive, negative and zero sequence phase voltages).

Similarly,
$$E_b = E_{b_1} + E_{b_2} + E_{b_0}$$

This expression of E_b can be now expressed in terms of E_a by putting the expression of E_b, E_{b_2} and E_{b_0} from the equations (13.27) to (13.29) and hence we get,

$$E_b = a^2 E_{a_1} + a\,E_{a_2} + E_{a_0} \quad ...(13.31)$$

Similarly, for E_c we can get, from $E_c = E_{c_1} + E_{c_2} + E_{c_0}$ as,

$$E_c = a\,E_{a_1} + a^2\,E_{a_2} + E_{a_0} \qquad ...(13.32)$$

Therefore, in the eqns. (13.30), (13.31), (13.32) on the right hand side, the first term on the right-hand side constitute the positive sequence system, the second terms constitute the negative sequence system and the third term constitute the zero sequence system.

13.3.2 Determination of Sequance Component

We can find the unbalance system of voltages and current in term of the phase current in the following way :

Let us first add eqns. (13.30), (13.31) and (13.32) and then after simplification of same we get :

$$\therefore \quad E_a + E_b + E_c = (1 + a^2 + a)\,E_{a_1} + (1 + a + a^2)\,E_{a_2} + 3\,E_{a_0} = 3E_{a_0} \text{ [as } 1 + a + a^2 = 0]$$

$$\therefore \quad E_{a_0} = \frac{1}{3}(E_a + E_b + E_c)$$

Then multiplying (13.30) by a, and (13.32) by a^2 and add these equation to eqn. (13.30), we get,

$$E_a + aE_b + a^2E_c = E_{a_1}(1 + a^2 + a^3) + E_{a_2}(1 + a^2 + a^4) + E_{a_0}(1 + a + a^2) = 3a_1$$

$$\therefore \quad E_{a_1} = \frac{1}{3}(E_a + a\,E_b + a^2\,E_c)$$

Then, multiplying 13.31 by a^2 and 13.32 by a and then adding these equation to 13.30 (a) we get :

$$E_a + a^2E_b + aE_c = E_{a_1}(1 + a^4 + a^2) + E_{a_2}(1 + a^3 + a^3) + E_{a_0}(1 + a^2 + a) = 3E_{a_2}$$

$$\therefore \quad E_{a_2} = \frac{1}{3}(E_b + a^2\,E_b + a\,E_c)$$

Thus, we find for any unbalanced system of voltages

$$E_{a_1} = \frac{1}{3}(E_a + aE_b + a^2\,E_c)$$

$$E_{a_2} = \frac{1}{3}(E_a + a^2\,E_b + a\,E_c) \qquad ...(13.33)$$

$$E_{a_0} = \frac{1}{3}(E_a + E_b + E_c)$$

The same way as we found unbalanced system voltages, we also for any unbalanced system of current, we get,

$$I_{a_1} = \frac{1}{3}(I_a + aI_b + a^2\,I_c)$$

$$I_{a_2} = \frac{1}{3}(I_a + a^2\,I_b + aI_c) \qquad ...(13.34)$$

$$I_{a_0} = \frac{1}{3}(I_a + I_b + I_c)$$

In matrix form symmetrical component of voltages can be as below from eqns. (13.30), (13.31) and (13.32).

$$\begin{bmatrix} E_a \\ E_b \\ E_c \end{bmatrix} = \begin{bmatrix} 1 & 1 & 1 \\ 1 & a^2 & a \\ 1 & a & a^2 \end{bmatrix} \begin{bmatrix} E_{a_0} \\ E_{a_1} \\ E_{a_2} \end{bmatrix}$$

or $$\begin{bmatrix} E_a \\ E_b \\ E_c \end{bmatrix} = [A] \begin{bmatrix} E_{a_0} \\ E_{a_1} \\ E_{a_2} \end{bmatrix}$$

where $$[A] = \begin{bmatrix} 1 & 1 & 1 \\ 1 & a^2 & a \\ 1 & a & a^2 \end{bmatrix}$$

$$\therefore \quad \begin{bmatrix} E_{a_0} \\ E_{a_1} \\ E_{a_2} \end{bmatrix} = [A]^{-1} \begin{bmatrix} E_a \\ E_b \\ E_c \end{bmatrix}$$

where $$[A]^{-1} = \frac{1}{3} \begin{bmatrix} 1 & 1 & 1 \\ 1 & a & a^2 \\ 1 & a^2 & a \end{bmatrix}$$

13.3.3 Zero Sequence Component Analysis

From the equations above the equations (13.33) and (13.34) are the equations for unbalanced system of voltages and currents respectively. After studying these equations it can be found that there will be no zero-sequence components of voltages when the phasor sum $E_a + E_b + E_c = 0$, we find that there will be no zero sequence component, when the phasor sum $I_a + I_b + I_c = 0$.

Infact, as the phasor sum of the line-to-line voltage in a 3-phase system is always zero, there will be no zero-sequence component in the line voltage of a 3-phase system. Similarly as the phasor sum of the three line current in a 3-phase, 3-wire system is zero, there will be no zero sequence component in the line current of a 3-phase 3-wire system.

However, zero sequence components may be present in the phase voltages of star circuit on phase current of a delta circuit. In a star circuit the zero sequence component of the phase voltages are all in phase. The line voltage being the vector difference of two phase voltages the zero sequence component cancel out when the line voltage is determined. Moreover in a delta circuit the zero-sequence components of the phase currents are all in phase and. consequently from a local circulating current inside the mesh. Hence, there can be no zero-sequence components in the line currents.

13.3.4 3-phase Power in Term of Symmetrical Components

We can find the power conserved in a 3-phase circuit directly from the symmetrical components provided the currents and voltages are known.

Now, let S be the power consumed in a 3-phase circuit, and it is the total complex power flowing into a 3-phase circuit through its phases R, Y, B. While P and Q are the real and reactive powers in three phases. E_R, E_Y and E_B indicate phase voltages at terminals and I_R, I_Y and I_B the currents flowing into the circuit.

Here,
$$S = P + jQ = E_R I_R^* + E_y I_Y^* + E_B I_B^* \qquad ...(13.35)$$

In matrix notation,
$$S = [E_R \, E_Y \, E_B] \begin{bmatrix} I_R \\ I_Y \\ I_B \end{bmatrix} = \begin{bmatrix} E_R \\ E_Y \\ E_B \end{bmatrix}_t \begin{bmatrix} I_R \\ I_Y \\ I_B \end{bmatrix} \qquad ...(13.36)$$

However,
$$\begin{bmatrix} E_R \\ E_Y \\ E_B \end{bmatrix} = [A] \begin{bmatrix} E_{R_0} \\ E_{R_1} \\ E_{R_r} \end{bmatrix} \qquad ...(13.37)$$

where
$$[A] = \begin{bmatrix} 1 & 1 & 1 \\ 1 & a^2 & a \\ 1 & a & a^2 \end{bmatrix}$$

and
$$\begin{bmatrix} I_R \\ I_Y \\ I_B \end{bmatrix} = [A] \begin{bmatrix} I_{R_0} \\ I_{R_1} \\ I_{R_2} \end{bmatrix} \qquad ...(13.38)$$

Now using eqns. (13.33) and (13.34) in matrix eqn. (13.32),
$$S = [AE]_t \, [AI]^* \qquad ...(13.39)$$

where
$$[E] = \begin{bmatrix} E_{R_0} \\ E_{R_1} \\ E_{R_2} \end{bmatrix} \text{ and } [I] = \begin{bmatrix} I_{R_0} \\ I_{R_1} \\ I_{R_2} \end{bmatrix}$$

But in matrix algebra, the transposes of the matrices in reverse order, that means,
$$[AV]_t = A_t \, V_t$$
$$\therefore \quad S = A_t A_t \, [AI]^*$$

Noting that $A_t = A$ and that a and a^2 are conjugate, we obtain
$$S = \left[E_{R_0} \cdot E_{R_1} \cdot E_{R_2}\right] \begin{bmatrix} 1 & 1 & 1 \\ 1 & a^2 & a \\ 1 & a & a^2 \end{bmatrix} \begin{bmatrix} 1 & 1 & 1 \\ 1 & a & a^2 \\ 1 & a^2 & a \end{bmatrix} \begin{bmatrix} I_{R_0} \\ I_{R_1} \\ I_{R_2} \end{bmatrix} \qquad ...(13.40)$$

However, $A_t A^* = 3U$, u is the unity matrix,
$$U = \begin{bmatrix} 1 & 0 & 0 \\ 0 & 1 & 0 \\ 0 & 0 & 1 \end{bmatrix}$$

Then, $$S = 3\left[E_{R_0}\ E_{R_1}\ E_{R_2}\right]\begin{bmatrix}E_{R_0}\\E_{R_1}\\E_{R_2}\end{bmatrix}^* = 3\left(E_{R_0}\ I^*_{R_0} + E_{R_1}\ I^*_{R_1} + E_{R_2}\ I^*_{R_2}\right) \quad ...(13.41)$$

That means, the sequence power is one-third the power in terms of phase quantities.

Example 13.6. *A short-circuit to earth occurs near the terminals of phase. A of a 3-phase alternator, star connected with neutral point earthed, the current to earth being 1000 amps. If the alternator is not supplying any normal current, calculate the positive, negative and zero sequence components of current of all phases.*

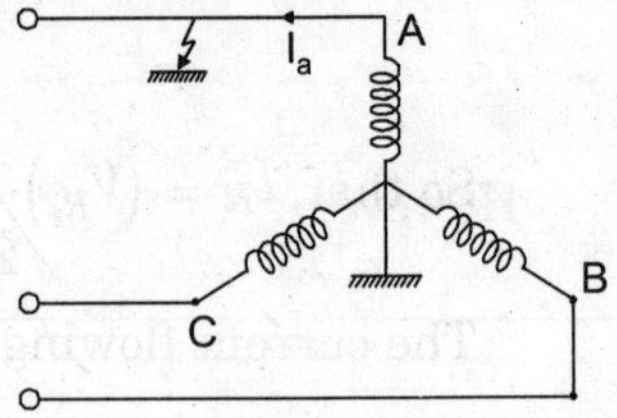

Fig. 13.16

Solution : With reference to the above example we refer to Fig. 13.16. As per example we are given,

$$I_a = 1000\ \angle 0°, I_b = 0, I_c = 0$$

(taking current in phase 'a' as reference)

Zero sequence component,

$$I_0 = I_{a_0} = I_{b_0} = I_{c_0} = \frac{1}{3}[I_a + I_b + I_c] = \frac{1}{3}[1000\ \angle 0°] = 333.3\ \angle 0° \text{ amps.}$$

Positive sequence component are,

$$I_{a_1} = \frac{1}{3}\left[I_a + aI_b + a^2 I_c\right] = \frac{1}{3}[1000\ \angle 0° + 0 + 0] = 333.3\ \angle 0° \text{ amps.}$$

Negative sequence components are,

$$I_{a_2} = \frac{1}{3}\left[I_a + a^2 I_b + aI_c\right] = \frac{1}{3}[1000\ \angle 0° + 0 + 0] = 333.3\ \angle 0° \text{ amps.}$$

$$I_{b_2} = aI_{a_2} = 333.3\ \angle 120° \text{ amps.}$$

$$I_{c_2} = a^2\ I_{a_2} = 333.3\ \angle 240° \text{ amps.}$$

Example 13.7. *An unbalanced star connected 3-phase supply having a negligible internal impedance is connected to three identical voltmeters in star connection. Each voltmeter has a resistance of 20,000 ohm and negligible reactance. The supply voltages of the 3-phases from line-to-neutral are respectively,*

$$E_R = 100\ \angle 0°, E_Y = 200\ \angle 278, E_B = 100\ \angle 120°$$

Using the method of symmetrical component determine the reading of the voltmeter connected to the yellow line. Also find the current through the voltmeters.

Solution : As per example, it is given,

$$E_{R_N} = (100 + j\,0) \text{ volts, } E_{Y_N} = (0 - j\,200) \text{ volts}$$

and $$E_{B_N} = (-\,50 + j\,86.6) \text{ volts}$$

Now $$V_{R_0} = \frac{1}{3}\left[E_{R_N} + E_{Y_N} + E_{B_N}\right] = \frac{1}{3}[50 - j113.4] = (16.67 - j\,37.8) \text{ volts}$$

$$V_{R_1} = \left[E_{B_N} + aE_{Y_N} + a^2 E_{B_N}\right] = \frac{1}{3}[100\angle 0° + 200(270° + 120°) + 100\angle(120° + 240)]$$

$$\text{(after simplification)} = \frac{1}{3}[373 + j100] = 124.3 + j33.3 = 128.7\angle 15° \text{ volts}$$

$$V_{R_2} = \frac{1}{3}\left[E_{R_N} + a^2 E_{Y_N} + a E_{B_N}\right] = \frac{1}{3}[100\angle 0° + 200\angle(270° + 240°) + 100\angle(120° + 120°)]$$

$$\text{(after simplification)} = \frac{1}{3}[-123 + j13.4] = -41 + j4.5 = 41.2\angle 173.7° \text{ volts}$$

So that, $I_{R_1} = (V_{R_1})/20{,}000$; $I_{R_2} = (V_{R_2})/20{,}000$ and $I_R = I_{R_1} + I_{R_2}$

The current flowing in voltmeter connected in yellow phase

$$I_Y = I_{Y_1} + I_{Y_2} = a^2 I_{R_1} + aI_{R_2} = a^2\frac{V_{R_1}}{20{,}000} + a\frac{V_{R_2}}{20{,}000} = \frac{V_{Y_1}}{20{,}000} + \frac{V_{Y_2}}{20{,}000}$$

Reading of voltmeter connected in yellow phase is, V_{Y_1}

where, $V_Y = I_Y \times 20000 = V_{Y_1} + V_{Y_2}$

We have $V_{Y_1} = a^2 V_{R_1} = 128.7(15° + 240°) = (-33.3 - j124)$ volts

$$V_{Y_2} = aV_{R_2} = 41.2\angle(173.7° + 120°) = 41.2(0.42 - j0.916)$$
$$= (16.6 - j37.8)^2$$

Therefore, $V_Y = V_{Y_1} + V_{Y_2} = (-16.7 - j161.8)$

$\therefore$ $|V_Y| = 162.5$ V

$\therefore I_Y$ voltmeter current $= \dfrac{162.5}{20{,}000} = 8.125$ **Ans.**

Example 13.8. *An unbalanced delta connected load is connected to the terminals a, b, c of a 400 volt balanced three-phase supply. The load consists of ab : a resistor of 20 ohms : bc : a capacitor of 40 ohms reactance, ca : an inductor of 10 ohms reactance and 5 ohms resistance.*

Find the three line currents and the reading of a watt meter connected with its current coil in line 'a' and pressure coil between 'b' and 'c', if the phase sequence is V_{ab}, V_{bc}, V_{ca}.

Solution : As per the example, we have

$V_{ab} = (400 + j0)$ volts.

$V_{bc} = 400\angle 120° = (-200 - j346.4)$ volts, $V_{c_a} = 400\angle -120°\ (-200 + j346.4)$ volts.

Hence, $I_{ab} = V_{ab}/Z_{ab} = 400/20 = 20\angle 0° = 20 + j0$ amps.

$$I_{bc} = V_{bc}/Z_{bc} = \frac{400\angle -120°}{-j40} = \frac{400\angle -120°}{40\angle -98} = 10(0.865 - j0.5) = 8.66 - j5 \text{ amps.}$$

$$I_{c_a} = V_{ca}/Z_{ca} = \frac{400\angle 120°}{5+j10} = \frac{400\angle 120°}{11.2\angle 63.5°} = 35.7\angle 56.5° = 35.7(0.552 + j0.834)$$

$$= (19.70 + j29.8) \text{ amps.}$$

In the Δ load, it is clear that line current I_a, I_b, I_c are given as :

$$I_b = I_{bc} = I_{ca} = -11.34 - j5 = 12.4\angle -23.68$$

$$\therefore \quad |I_b| = 12.4 \text{ Amp.}$$

Now,
$$I_c = I_{ca} - I_{bc} = 11.04 - j29.8 = 27.14\angle 66°$$

$$\therefore \quad |I_c| = 27.14 \text{ Amp.}$$

We know, the wattmeter reading, would be,

$$P = I_a V_{b_c} \cos\phi$$

Now
$$\phi = 120° - (-89.4°) = -120° + 99.4° = -30.6°$$

$$\therefore \quad P = I_a V_{b_c} \cos\phi = 29.8 \times 400 \times 0.435 = \mathbf{10.26\ KW}$$

13.3.5 Unsymmetrical Faults in Power System

In order to detect the unsymmetrical faults in the form of say one line to ground or two lines or there may occur fault between two lines and ground etc., it is essential that the fault current and voltage should be known. In the following section we shall see how we can determine the fault current and voltages and to do so we have to determine the unsymmetrical series impedances, that is the sequence impedances :

13.3.6 Sequence Impedances

Let us consider the circuit of Fig. 13.17 is which the three impedances Z_a, Z_b and Z_c are different phase differences and Z_n is the neutral impedance. We will take this circuit and the sequence impedances, Z_a, Z_b, Z_c, Z_n and the sequence current, I_a, I_b, I_c and I_n to develop expressions. It should be noted here that in an unsymmetrical system positive sequence voltage caused by only positive sequence currents to flow. The impedance to the *+ve* sequence current resulted in the circuit, being termed as positive sequence impedance. Then the negative sequence voltages cause only negative sequence current to flow, as a result, the impedance to *–ve* sequence current being termed as negative sequence impedance. Then the zero sequence voltages cause only zero sequence currents to flow. The impedance to the zero sequence current (being termed as zero sequence) as zero sequence impedance.

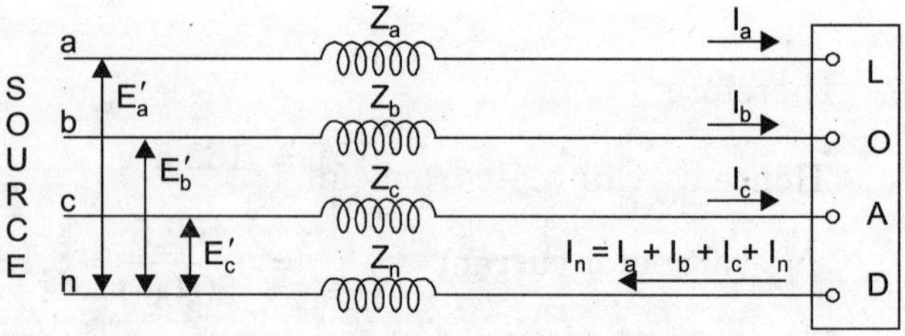

Fig. 13.17

13.3.7 Sequence Networks

A sequence network of the Fig. 13.17 is considered to determine the expressions for $E'_{a_0}, E'_{a_1}, E'_{a_2}$ below :

By applying Kirchhaff's law in the circuit of Fig. 13.19, we get

$$E'_a = Z_a I_a + E_a + Z_a (I_a + I_b + I_c) \qquad \text{...(13.42)}$$

$$E'_b = Z_b I_b + E_b + Z_n (I_a + I_b + I_c) \quad ...(13.43)$$

$$E'_c = Z_c I_c + E_c + Z_n (I_a + I_b + I_c) \quad ...(13.44)$$

Writing the expression for E'_{a_0} and expressing the current in the symmetrical components where necessary, we get :

$$E'_{a_0} = \frac{1}{3}(E'_a + E'_b + E'_c) = \frac{1}{3} Z_a \left(I_{a_0} + I_{a_1} + I_{a_2}\right) + \frac{1}{3} Z_b \left(I_{a_0} + a^2 I_{a_1} + a I_{a_2}\right)$$
$$+ \frac{1}{3} Z_c \left(I_{a_0} + a I_{a_1} + a^2 I_{a_2}\right) + \frac{E_a + E_b + E_c}{3} + 3Z_n \frac{I_a + I_b + I_c}{3}$$

$$E'_{a_0} = I_{a_0}(Z_a + Z_b + Z_c)/3 + I_{a_1}(Z_a + aZ_b + aZ_c)/3 + I_{a_2}(Z_a + a^2 b + a^2 Z_c) + E_{a_0} + 3Z_n \cdot I_{b0} \quad ...(13.45)$$

In this expression there is similarity between the the three Z combination (that is for zero segment, positive sequence and negative sequence impedance) and the three I Combinations (that is zero sequence, positive sequence and negative sequence expressed in eqns. (13.42, 13.43, 13.44).

Hence, we may similarly write,

$Z_{a_0} = \frac{1}{3}(Z_a + Z_b + Z_c)$, zero sequence component of the three unequal impedance, Z_a, Z_b, Z_c

$Z_{a_1} = \frac{1}{3}(Z_a + aZ_b + a^2 Z_c)$, positive sequence component of the three equal unequal impedance, Z_a, Z_b and Z_c

$Z_{a_2} = \frac{1}{3}(Z_a + a^2 Z_b + aZ_c)$, negative sequence component of three unequal impedance, Z_a, Z_b and Z_c

Hence we can write from eqn. (13.45), impedance, Z_a, Z_b and Z_c

$$\therefore \quad E'_{a_0} = E_{a_0} + I_{a_0} Z_0 + I_{a_1} Z_{a_2} + I_{a_2} Z_{a_1} + 3Z_n \cdot I_{a_0} \quad ...(13.46)$$

$$\therefore \quad E'_{a_0} - E_{a_0} = I_{a_0} Z_{a_0} + Z_{a_1} Z_{a_2} + I_{a_2} Z_{a_1} + 3I_{a_0} Z_n \quad ...(13.47)$$

Similarly as done to determine expression for E'_{a_0}

$$E'_{a_1} = \frac{1}{3}(E'_a + a E'_b + a^2 E'_c)$$
$$= \frac{1}{3}\{Z_a I_a + E_a + Z_n (I_a + I_b + I_c)\} + \frac{1}{3} \cdot a\{I_b I_b + E_b + Z_n (I_a + I_b + I_c)\} \frac{1}{3} \cdot a^2 \{Z_c I_c + E_c + Z_n (I_a + I_b + I_c)\}$$

$$= \frac{Z_a}{3}\{I_{a_0} + I_{a_1} + I_{a_2}\} + \frac{Z_b}{3}\{aI_a + a^3 I_{a_1} + a^2 I_a\} + \frac{Z_c}{3}\{a^2 I_{a_0} + a^3 I_{a_1} + aI_{a_2}\}$$
$$+ \frac{E_a + aE_b + a^2 E_c}{3}$$

$$= I_{a_0}\left(\frac{Z_a + aZ_b + a^2 Z_c}{3}\right) + I_a\left(\frac{Z_a - I_b + Z_c}{3}\right) + I_{a_2}\left(\frac{Z_a + a^2 Z_b + aZ_c}{3}\right) + E_{a_1} + 0$$

$$\frac{Z_n (I_a + I_b + I_c)(1 + a + a^2)}{3}$$

$$\therefore \quad E'_{a_1} = E_{a_1} + I_{a_0} Z_{a_1} + I_{a_1} Z_{a_0} + I_{a_2} Z_{a_2} \quad ...(13.48)$$

$$E'_{a_1} - E_{a_1} = I_{a_0} Z_{a_1} + I_{a_1} Z_{a_0} + I_{a_2} Z_{a_2} \quad ...(13.49)$$

Then, similarly,

$$E'_{a_2} = \frac{1}{3}(E'_a + a^2 E'_b + a E'_c)$$

$$= \frac{1}{3}\{Z_a I_a + E_a + Z_n (I_a + I_b + I_c)\} + \frac{a^2}{3}\{Z_b I_b + E_b + Z_n (I_a + I_b + I_c)\} + \frac{a}{3}\{I_c I_c + E_c + Z_n (I_a + I_b + I_c)\}$$

$$= \frac{Z_a}{3}\{I_{a_0} + I_{a_1} + I_{a_2}\} + \frac{Z_b}{3}\{a^2 I_{a_1} + a I_{a_1} + a^3 I_{a_2}\} + \frac{Z_c}{3}\{a I_{a_0} + a^2 I_{a_1} + a^3 I_{a_2}\}$$
$$+\left(\frac{E_a + a^2 E_b + aE_c}{3}\right) + I_{a_1}\left(\frac{Z_a + a Z_b + a^2 Z_c}{3}\right) + I_{a_1}\left(\frac{Z_a + Z_b + Z_c}{3}\right) + E_{a_2}$$

$$= I_a\left(\frac{Z_a + a^2 Z_b + aZ_c}{3}\right) + I_{a_1}\left(\frac{Z_a + Z_b + a^2 Z_c}{3}\right) + I_{a_2}\left(\frac{Z_a + Z_b + Z_c}{3}\right) + E_{a_2}$$

$$\therefore \quad E'_{a_2} = E_{a_2} + I_{a_0} Z_{a_2} + I_{a_1} Z_{a_1} + I_{a_2} Z_{a_0} \quad ...(13.50)$$

$$E'_{a_2} - E_{a_2} = I_{a_0} Z_{a_2} + I_{a_1} Z_{a_1} + I_{a_2} Z_{a_0} \quad ...(13.51)$$

The above three equations can be rewritten as eqn. below :

$$\left.\begin{aligned} E'_{a_0} - E_{a_0} &= I_a (Z_{z_0} + 3Z_n) + I_{a_1} Z_{a_2} + I_{a_2} Z_{a_1} \\ E'_{a_1} - E_{a_1} &= I_{a_0} Z_{a_1} + I_{a_1} Z_{a_0} + I_{a_2} Z_{a_2} \\ E'_{a_2} - E_{a_2} &= I_{a_0} Z_{a_2} + I_{a_1} Z_{a_1} + I_{a_2} Z_{a_0} \end{aligned}\right\} \quad ...(13.52)$$

In fact, in reality, we shall be concerned with circuit that are normally balanced only upon the occurrence of an unsymmetrical fault.

Hence, let us, therefore, refer to the eqn. (13.52), when the three impedances are equal, *i.e.*, $Z_a = Z_b = Z_c = Z$ (say). Hence, under the condition these equation (13.52) become respectively as below :

$$\left.\begin{aligned} E'_{a_0} - E_{a_0} &= I_{a_0}(Z + 3Z_n) \\ E'_{a_1} - E_{a_1} &= I_{a_1} \cdot Z \\ E'_{a_2} - E_{a_2} &= I_{a_2} \cdot Z \end{aligned}\right\} \quad ...(13.53)$$

Now from the eqn. (13.53) it can be concluded that the symmetrical components of the unbalanced currents flowing in balanced impedances produces a voltage drop of zero sequence type only and so on. In other words, as said in the beginning the zero sequence, impedances positive sequence impedances and negative sequences are produced due to zero sequences positive sequence, positive sequence and negative sequence voltage in the network.

In the following section we shall discuss on sequence impedance of transmission lines in a power system.

13.3.8 Transmission Lines Sequence Impedances

In transmissions lines of the power system network sequence impedances play a very important role is calculating the fault current of transmission lines. Transformers and alternators are the most important to parts of the power systems.

Infact, the positive and negative sequence impedance of linear symmetrical static impedances circuit do not depend on the phase sequence and hence the positive and negative sequence impedances of the transmission line are some. But the zero sequence impedance, particularly the reactance is about 2 to 4 times the positive sequence reactance.

1. Transformers Sequence Impedances

It is known that in practice any power system network consists of a large number of transformers for stapping up and down the voltage levels. The positive sequence impedance of a transformer is equal to its leakage reactance, the resistance of the winding is usually small in comparison to the leakage reactance. Transformer basically is a static equipment, as such the positive and negative sequence impedance are identical.

However, the zero sequence impedance of a three-phase transformer is slightly different from positive sequence impedance. The flow of the zero sequence current through the transformer depends greatly on the type of connections. In fact it can flow only through the winding connected in star and the star point is earthed.

In case of delta winding or in star winding with neutral isolated, the zero sequence currents cannot flow. In the Fig. 13.18 some transformer connection and corresponding zero-sequence network of synchronous generators and transformers are shown.

(*a*) Generators

(1)

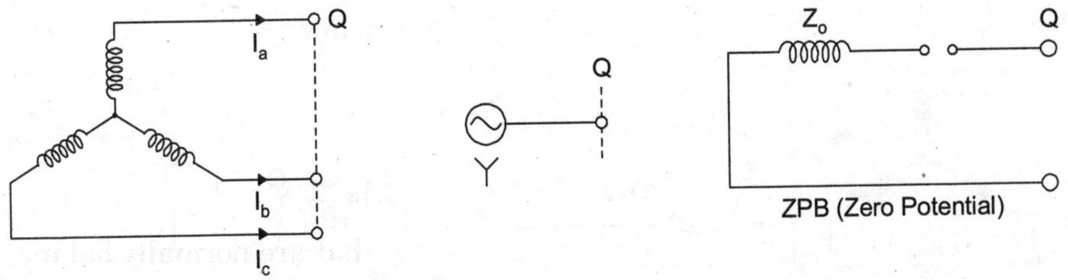

(2)

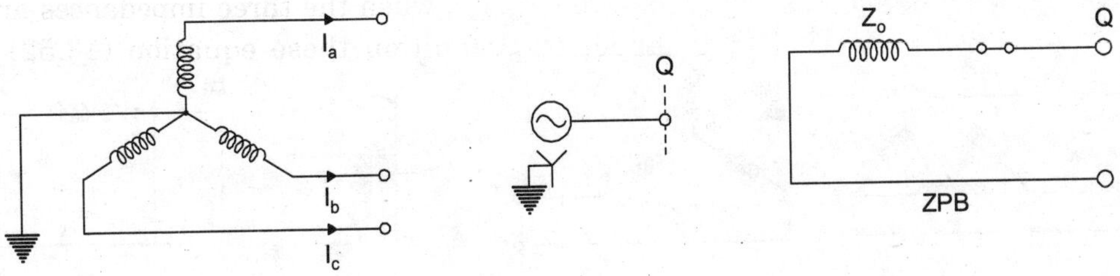

(3) 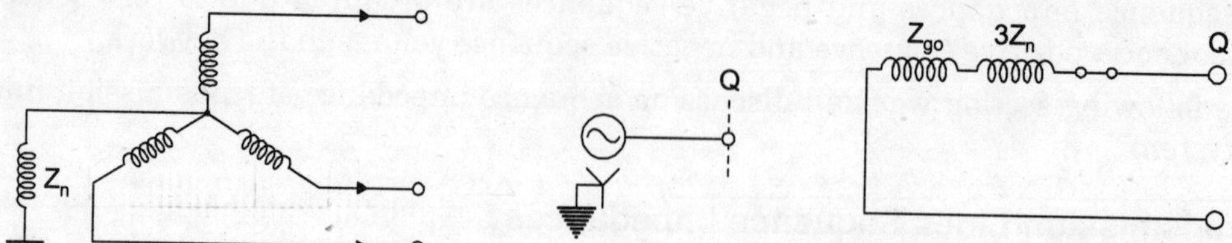

(4) 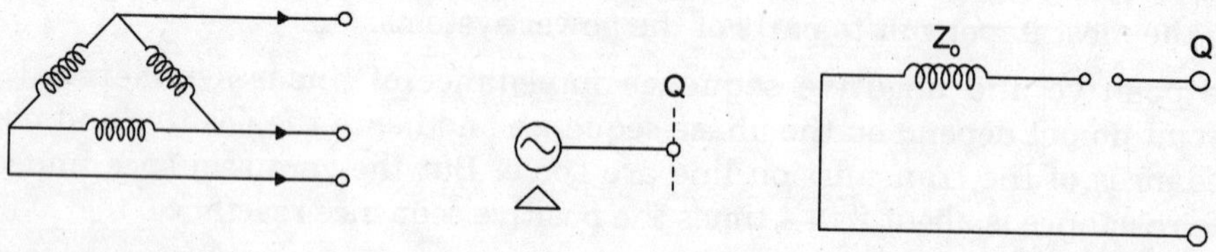

(b) Transformers

(5) 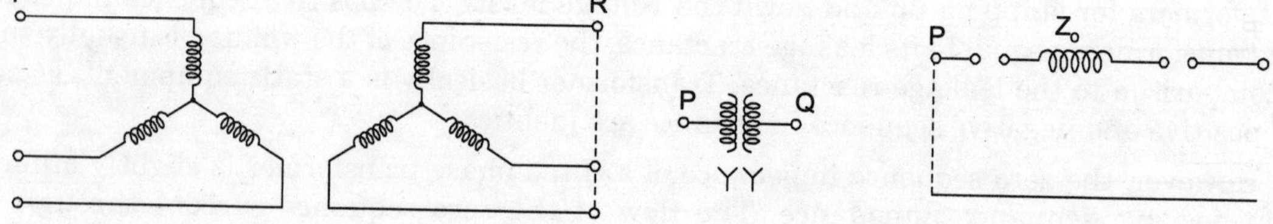

(6) 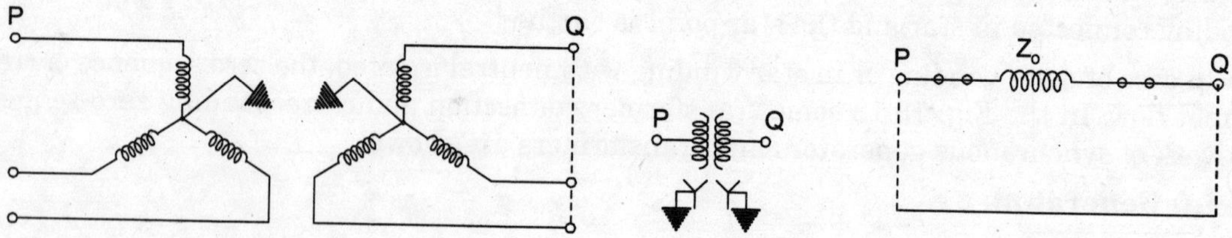

(7) 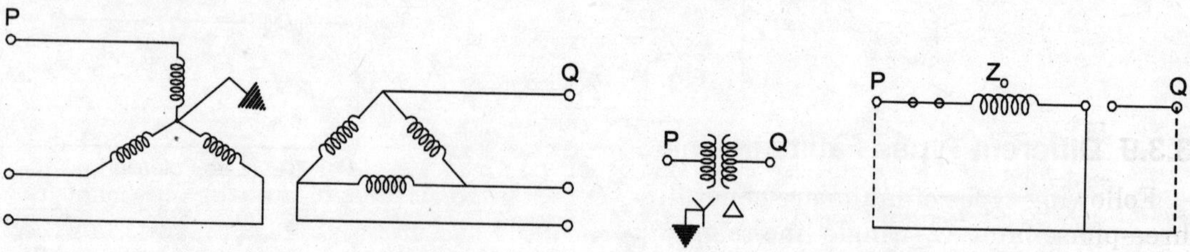

(8) 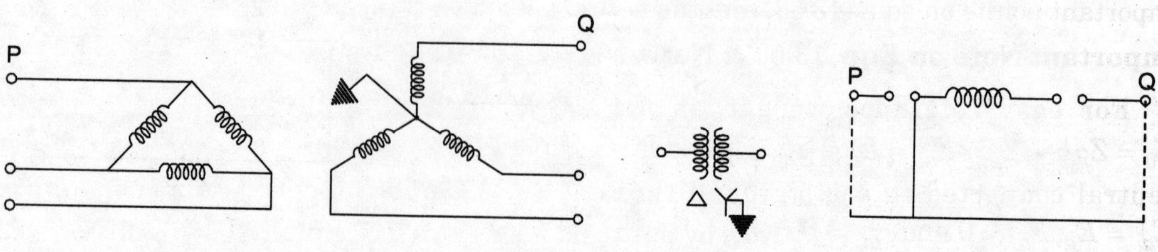

(9)

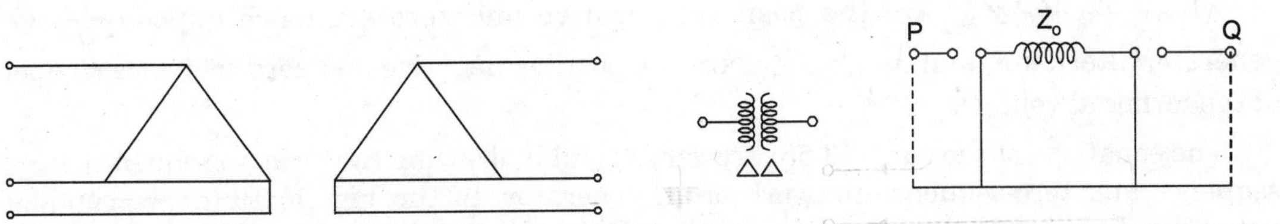

(10)

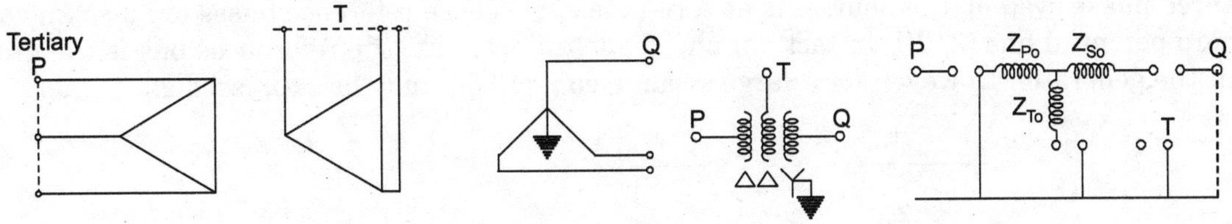

(11)

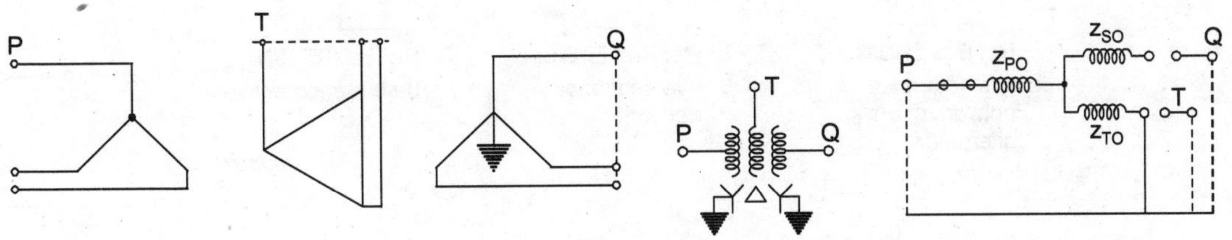

(12)

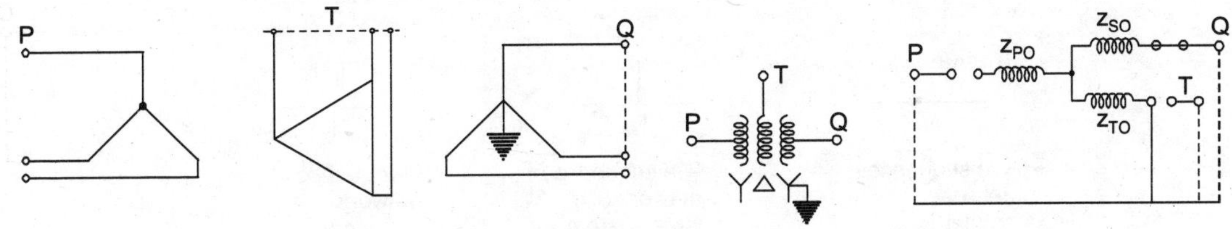

Fig. 13.18. Different.

13.3.9 Different Types Faults in Unlaoded Alternator

Following types of faults are generally occur at the terminals of unloaded alternators : (1) Three-phase fault, (2) Single-line-to ground fault and (3) Line-to-line/Double-line fault which are discussed below. However, before we start discussion on the fault we to consider the important points on eqn. (13.53) and the network bond on this equation

Important Note on Eqn. 13.53 & Network Based on the Eqns.

For easy reference we get, in the equation (13.53) : $E_{a_0} = E'_a - I_{a_0} \cdot Z_0$, where $Z_0 = Z_{g_0} + 3Z_{g_n} \cdot E'_{a_1} = E_{a_1} + I_{a_1} \cdot Z_1$ and $E_{a_2} = E'_{a_2} + I_{a_2} \cdot Z_2$. Since synchronous machine (with neutral connected to earth/ground through Z_n) generate only *+ve* sequence voltages, $E_{a_1} = E_g, E'_{a_2} = 0$ and $E'_{a_0} = 0$. Now by putting these values in the eqns. (13.53) we get,

$$V_{a_1} = E_g - I_{a_1} Z_{g_1}, V_{a_2} = 0 - I_{a_2} Z_{g_2}, V_{a_0} = 0 - I_{a_0} Z_{g_0} \quad ...(13.54)$$

where $Z_{g_1}, Z_{g_2}, Z_{g_0}$ are the positive, negative and zero sequence impedances of the generator/alternator, and $V_{a_1}, V_{a_2}, V_{a_0}$ are the positive negative and zero sequence component of the terminal voltages.

The equations of the eqn. (13.55) are very useful in drawing the positive sequence, negative sequence and zero-sequence network of an generator. In the Fig. 13.19 (*a*) we can see the networks of single-phase equivalent circuits of the balanced three phase circuit through which the symmetrical component of the unbalanced current are considered to flow. In a balanced three-phase system, the neutral is at zero-potential. These reference buses are also knwon as zero-potential bus (*ZPB*). In fact, for the sequence network, the references bus is the ground at the generator. However, for a large system, eqn. (13.54) may be expressed as

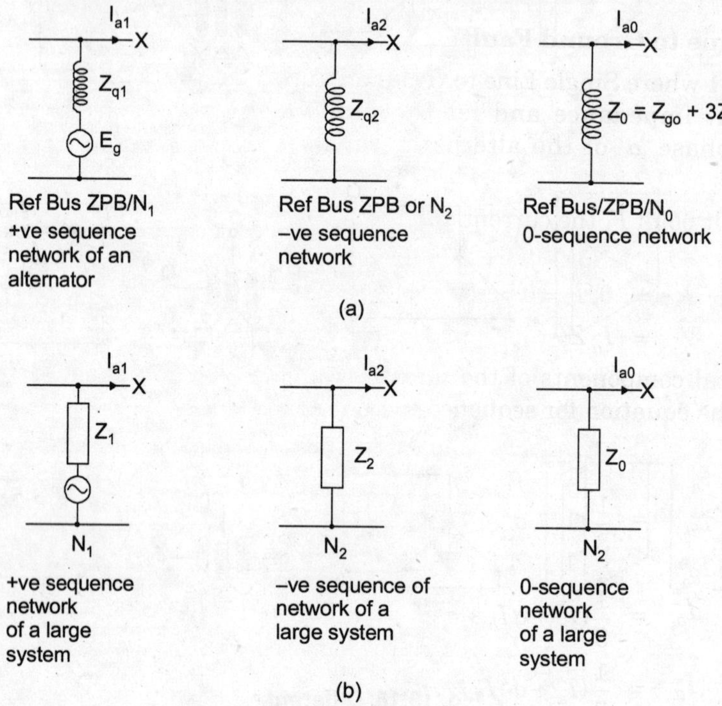

Fig. 13.19

$$V_{a_1} = E_g - I_{a_1} Z_1, V_{a_2} = 0 - I_{a_2} \cdot Z_2, V_{a_0} = 0 - I_{a_0} Z_0 \qquad \text{...(13.55)}$$

Then, the sequence networks are drawn past in the same way of Fig. 13.19 (*a*), but the sequence impedance Z_1, Z_2, Z_0 are the equivalent system sequence impedance and the corresponding networks of *+ve*, *–ve* and 0-sequence are shown in Fig. 13.19 (*b*).

1. Three-phase Faults

A 3-phase alternator earthed or grounded through an impedance, Z_{g_n} in represented in the Fig. 13.20 where the generator is assumed to be at no load when a 3-phase fault strikes its terminals and hand current, I_a, I_b, I_c flow through phases *a, b, c* respectively. In this condition, as the generator develops only positive sequence voltage hence with reference to eqn. (13.55) developed above, we have got :

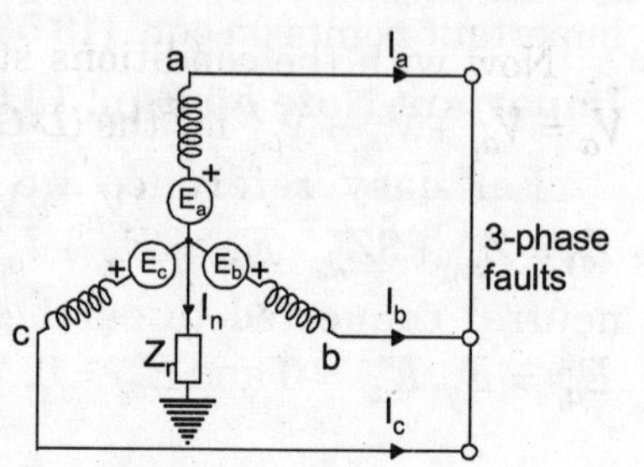

Fig. 13.20

$$V_{a_1} = E_g - I_{a_1} \cdot Z_{g_1}, V_{a_2} = 0 \cdot I_{a_2} \cdot Z_{g_2}, V_{a_0} = 0 - I_{a_0} \cdot Z_{g_0},$$

where $\quad Z_0 = Z_{g_0} + 3Z_{g_n}$

which in matrix from we can write this equation as below :

$$\begin{bmatrix} V_{a_1} \\ V_{a_2} \\ V_{a_0} \end{bmatrix} = \begin{bmatrix} E_g & 0 & 0 \\ 0 & Z_{g_2} & 0 \\ 0 & 0 & Z_0 \end{bmatrix} \quad ...(13.56)$$

The sequence network is same as shown in the Fig. 13.19 (*a*).

2. (*L-G*), Single Line to Ground Fault

In the Fig. 13.21 where Single Line to Ground Fault, let Z_1 be the fault impedance and let the fault be occurring at the phase '*a*' of the alternator which is unloaded.

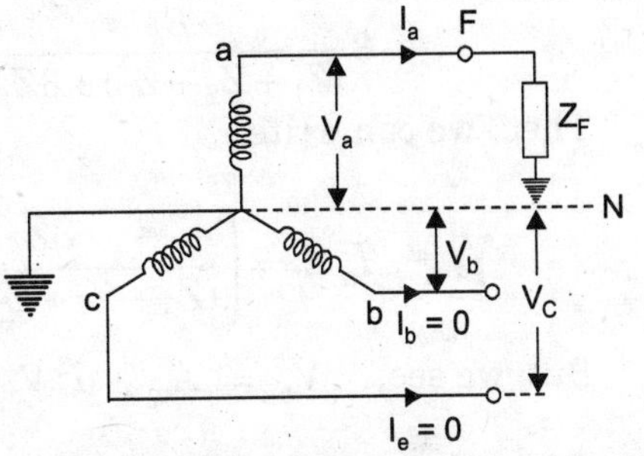

Fig. 13.21. (a) Single line to ground (*L-G*) fault at *F*.

Now, at the fault point *F*, the current and voltage are given by :

$$I_b = 0, I_c = 0 \quad ...(13.57)$$

and

$$V_a = I_a Z_f \quad ...(13.58)$$

The symmetrical components of the fault current are given by as below in matrix from with reference to the equation for sequence current in the eqn. (13.34).

$$\begin{bmatrix} I_{a_1} \\ I_{a_2} \\ I_{a_0} \end{bmatrix} = \frac{1}{3}\begin{bmatrix} 1 & a & a^2 \\ 1 & a^2 & a \\ 1 & 1 & 1 \end{bmatrix}\begin{bmatrix} I_a \\ 0 \\ 0 \end{bmatrix} \quad ...(13.59)$$

$$I_{a_1} = \frac{1}{3}(I_a + aI_b + a^2 I_c)$$

$$I_{a_2} = \frac{1}{3}(I_a + a^2 I_b + a I_c)$$

$$I_{a_0} = \frac{1}{3}(I_a + I_b + I_c)$$

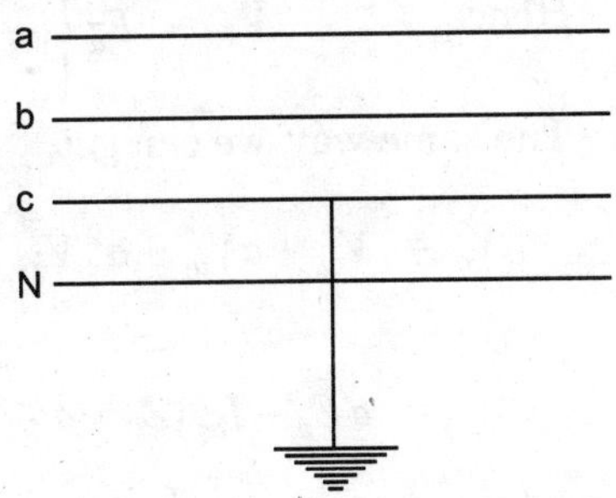

Fig. 13.21(b). Alternate line diagram of L-G Fault

As $I_b = I_c = 0$ in eqn. (13.58)

Hence, $\quad I_{a_1} = I_{a_2} = I_{a_0} = \frac{1}{3} I_a \quad ...(13.60)$

Now with the conditions state in eqn. of (13.58) and the eqn. (13.59) in the expression, $V_a = V_{a_1} + V_{a_2} + V_{a_0}$ for the (*L-G*) fault network in the Fig. 13.21, we get,

$$V_{a_1} + V_{a_2} + V_{a_0} = V_a I_a Z_f \text{ (as } V_a = I_a Z_f \text{ of eqn. (13.58)}$$

$$= \left(\text{as from, } I_{a_1} = \frac{1}{3} I_a \therefore I_a = 3I_{a_1}\right)$$

$$= 3I_{a_1} Z_f \text{ (from in eqn. (13.58)}$$

$$\therefore \quad V_{a_1} + V_{a_2} + V_{a_0} = 3I_{a_1} Z_f \quad ..(13.61)$$

Now as equations (13.60) and (13.61), it is noticed that all sequence current are equal and the sum of sequence voltages $(3Z_f I_{a_1})$. This logic hence suggests a-series connection of sequence network through an impedance of Fig. 13.22.

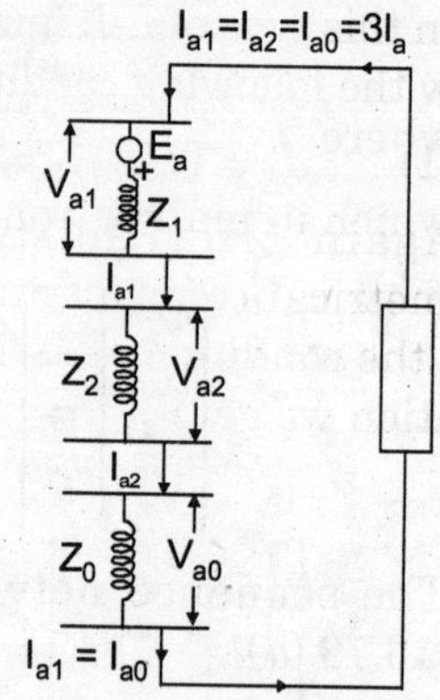

Fig. 13.22. Sequence network representation of *L-G* fault.

Hence, here, $I_{a_1} = \dfrac{E_a}{(Z_1+Z_2+Z_0)+3Z_f}$

and the fault current, $I_a = 3I_{a_1}$

$$= 3\frac{E_a}{(Z_1+Z_2+Z_0)+3Z_f} \qquad ..(13.61a)$$

Then, we can write,

$$V_a = I_a Z_f = \left(\frac{3Z_f}{(Z_0+Z_1+Z_2+3Z_f)}\right)E_a \qquad ...(13.61b)$$

But, we see, $V_b = A_{a_0} + a^2 V_{a_1} + aV_{a_2} = (-I_{a_0} Z_0) + a^2(E_a - I_{a_1} Z_1) + a(-I_{a_2} Z_2)$

$$= a^2 E_a - I_{a_1}(Z_0 + a^2 Z_1 + aZ_2)$$

$$\therefore \quad V_b = E_a\left\{\frac{a^2(Z_0+Z_1+Z_2+3Z_f)\cdot(Z_0+a^2 Z_1+aZ_2)}{(Z_0+Z_1+Z_2+3Z_f)}\right\}$$

Hence, $$V_b = E_a\left\{\frac{(a^2-1)Z_0+(a^2-a)Z_2+3a^2 Z_f}{(Z_1+Z_2+Z_0+3Z_f)}\right\} \qquad ...(13.62)$$

The same way, we can get,

$$V_c = V_{a_0} + aV_{a_1} + a^2 V_{a_2} = (-I_{a_0} Z_0 + a(E_a - I_{a_1} Z_1) + a^2(-I_{a_2} Z_2)$$

$$= aE_a - I_{a_1}(Z_0 + aZ_1 + a^2 Z_2) = E_a\left\{\frac{a(Z_0+Z_1+Z_2+3Z_f)-(Z_0+aZ_1+a^2 Z_2)}{Z_0+Z_1+Z_2+3Z_f}\right\}$$

$$= E_a\left\{\frac{(a-1)Z_0+(a-0^2)Z_2+3aZ_f}{Z_0+Z_1+Z_2+3Z_f}\right\} \qquad ...(13.63)$$

It has to be stated here that if the generator was supplying balanced load during pre-fault condition the load current was of positive sequence only. Infact, a generator generated positive sequence voltage and no negative and zero sequence voltages, and so during pre-fault condition there was no negative or zero sequence component of current to the load. However, under the fault condition, the positive sequence current to the fault is to be added to the pre-fault current. But only the positive sequence current flows into the fault provided the pre-fault current into the fault being zero.

3. Line to Line (*L-L*) or Double Line Fault

A network with a Line-to-Line fault at F of an unloaded alternator involving phases 'b' and 'c' through a fault impedance Z_f is represented in the Fig. 13.23.

In this type of *L-L* fault, the current and voltages at the fault follow the following conditions :

(1) $I_a = 0$, (2) $I_c = -I_b$, (3) $V_b - V_c = I_b Z_f$ ...(13.64a)

Again referring to eqn. (13.63) we can write the symmetrical components of the fault current in matrix form with the condition of *L-L* fault as stated in eqn. 13.63 (1) & (2) equation we get,

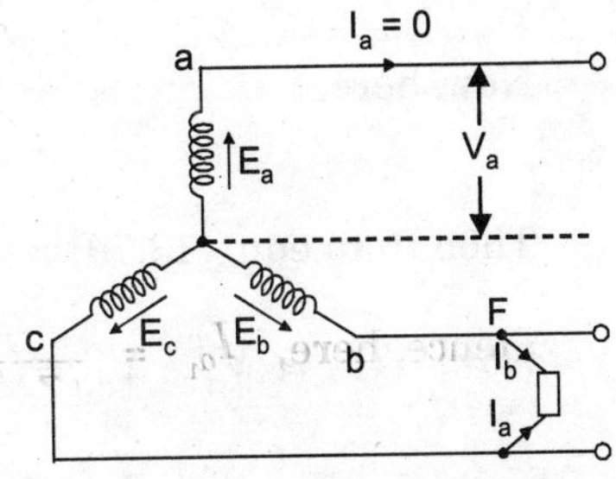

Fig. 13.23(a). *L-L* or double-line fault.

$$\begin{bmatrix} I_{a_1} \\ I_{a_2} \\ I_{a_0} \end{bmatrix} = \frac{1}{3}\begin{bmatrix} 1 & a & a^2 \\ 1 & a^2 & a \\ 1 & 1 & 1 \end{bmatrix}\begin{bmatrix} 0 \\ I_b \\ -I_b \end{bmatrix} \quad ...(13.64b)$$

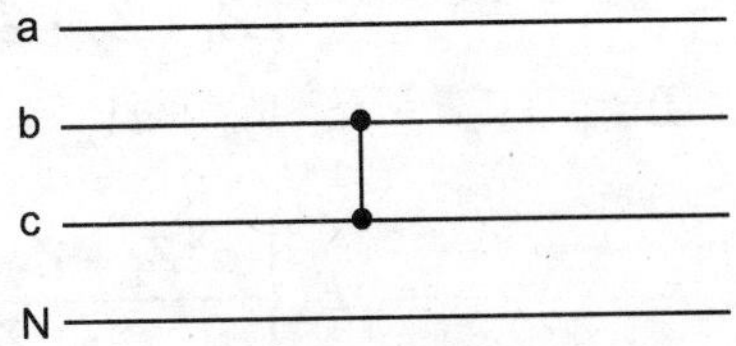

Fig. 13.23(b) Alternate dgm of L-L Fault.

As from eqn. (13.64), (1) & (2) eqn. $I_a = 0$, and $I_c = -I_b$

Hence, we get,
$$\left.\begin{aligned} I_{a_2} &= -I_{a_1} \\ I_{a_0} &= 0 \end{aligned}\right\} \quad ...(13.65)$$

Then, the symmetrical components of the voltages at *F* referring to eqn. (13.65), can be expressed in matrix form as :

$$\begin{bmatrix} V_{a_1} \\ V_{a_2} \\ V_{a_0} \end{bmatrix} = \frac{1}{3}\begin{bmatrix} 1 & a & a^2 \\ 1 & a^2 & a \\ 1 & 1 & 1 \end{bmatrix}\begin{bmatrix} V_a \\ V_b \\ V_c = V_b - I_b Z_f \end{bmatrix} \quad ...(13.66)$$

∴ From eqn. (13.63)

$$V_c = V_b - I_b Z_f$$

∴ From the matrix equation (13.66) we get

$$3V_{a_1} = V_a + (a + a^2) V_b - a^2 Z_f \cdot I_b \quad ...(13.67)$$

$$3V_{a_2} = V_a + (a + a^2) V_b - a Z_f I_b \quad ...(13.68)$$

Hence, from these two equations (13.67) & (13.68), we get,

$$3(V_{a_1} - V_{a_2}) = (a - a^2) Z_f I_b = j\sqrt{3}\, I_b Z_f \quad ...(13.69)$$

However, as $I_{a_2} = -I_{a_1}$ and $I_{a_0} = 0$

We get, $I_b = (a^2 - a) I_{a_1} = -j\sqrt{3}\, I_{a_1}$

$$\therefore \quad I_b = (a^2 - a) I_{a_1} = -j\sqrt{3} \cdot I_{a_1} \quad ...(13.70)$$

Now substituting the value of I_b from eqn. (13.70) in eqn. (13.69), we get,

$$\therefore \quad V_{a_1} - V_{a_2} = I_{a_1} Z_f \quad ...(13.71)$$

Now the equations (13.65) and (13.71) suggest parallel connection of positive and negative sequence network through a series impedance Z_f as can be seen in Fig. 13.20. Now as $I_{a_0} = 0$, there is no connection of the zero sequence network.

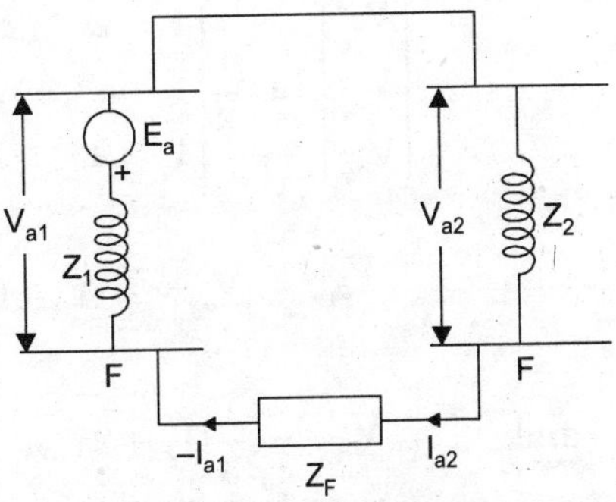

Fig. 13.24. Sequence network for *L-L* fault.

Now, here, $$I_{a_1} = \frac{E_2}{Z_1+Z_2+Z_f}$$

Then from eqn. (13.70) we get, $$I_b = I_c = -\frac{j\sqrt{3}\cdot E_g}{Z_1+Z_2+Z_f} \qquad ...(13.71)$$

Now, $$V_a = V_{a_1}+V_{a_2} = (E_a - E_{a_2}\cdot Z_1)+(-I_{a_2}\cdot Z_2) = E_{a_0} - I_{a_1}(Z_1 - Z_2)$$

$$= E_a\left\{\frac{(Z_1+Z_2+Z_3)-(Z_1-Z_2)}{Z_1+Z_2+Z_f}\right\}$$

$$\therefore \quad V_a = E_a\left(\frac{Z_f+2Z_2}{Z_1+Z_2+Z_f}\right) \qquad ...(13.72)$$

But, $$V_b = a^2 V_{a_1} + aV_{a_2} = a^2(E_a - I_{a_1} Z_1) + a(-I_{c_2} Z_2) = a^2 E_a - I_{a_1}(a^2 Z_1 - aZ_2)$$

$$= E_a\left\{\frac{(a^2 Z_1 + Z_2 + Z_f)-(a^2 Z_1 - a Z_2)}{Z_1+Z_2+Z_f}\right\}$$

$$\therefore \quad V_b = E_a\left\{\frac{-Z_2+aZ_f}{Z_1+Z_2+Z_f}\right\} \qquad ...(13.73)$$

Similar way, we get, $$V_c = E_a\left\{\frac{-Z_2+aZ_f}{Z_1+Z_2+Z_f}\right\} \qquad ...(13.74)$$

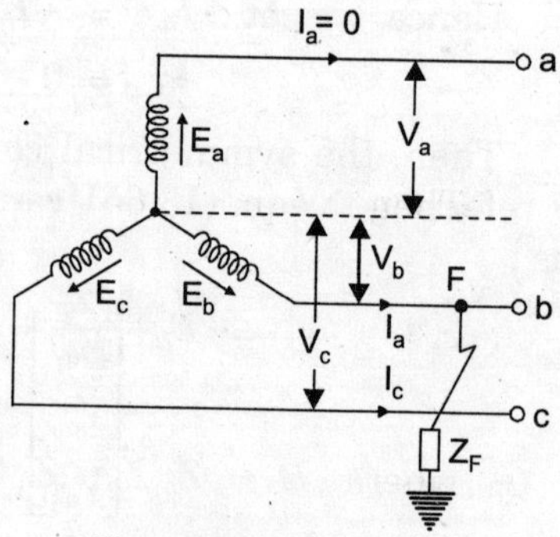

Fig. 13.25(a). *L-L-G* fault representation.

4. *L-L-G* Double Line to Ground Fault

A double-line to Ground (*L-L-G*) fault at *F* of an unloaded alternator is shown in Fig. 13.25 where the fault impedance is Z_f. In this fault condition, the current and voltages are as below :

$$I_a = 0 \qquad ...(13.75)$$

and $$V_b + V_c = (I_b + I_c)Z_f = 3I_{a_0} Z_f$$

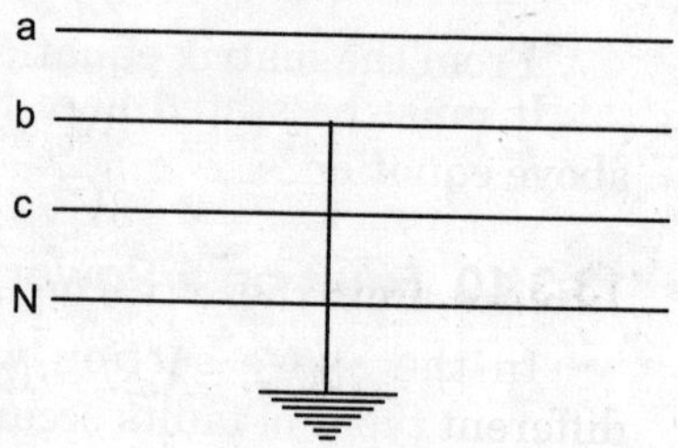

Fig. 13.25(b)

The symmetrical components of voltages are given in matrix form by :

$$\begin{bmatrix} V_{a_1} \\ V_{a_2} \\ V_{a_0} \end{bmatrix} = \frac{1}{3}\begin{bmatrix} 1 & a & a^2 \\ 1 & a^2 & a \\ 1 & 1 & 1 \end{bmatrix}\begin{bmatrix} V_a \\ V_b \\ V_c \end{bmatrix}$$

$$\therefore \quad V_{a_1} = V_{a_2} = \frac{1}{3}(V_a + (a+a^2)V_b) \qquad ...(13.76)$$

and $$V_{a_0} = \frac{1}{3}(V_a + 2V_b) \qquad ...(13.77)$$

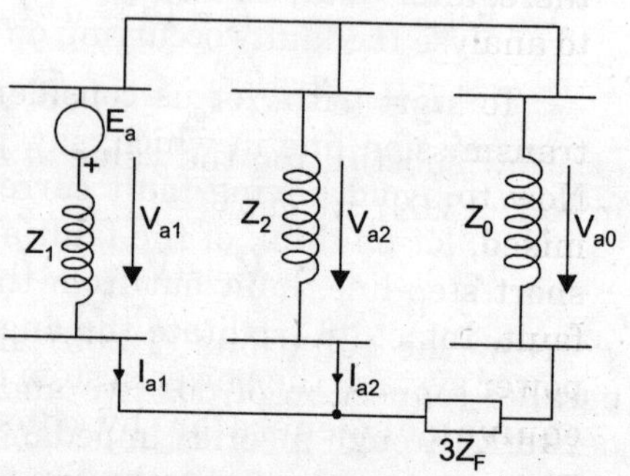

Fig. 13.26. Equivalent sequence network

Now, subtracting eqn. (13.76) from eqn. (13.77), we get,

$$V_{a_b} - V_{a_1} = \frac{1}{3}(2 - a - a^2)V_b = V_b = 3I_{a_0} Z_f \quad \text{...(13.78)}$$

$$\therefore \quad V_{a0} = V_{a_1} + 3Z_f I_{a_0}$$

By studying the equations (13.75), (13.76) and (13.78) is we find that sequence network drawn in Fig. 13.26 where with respect to fault, F, the total circuit total impedance is $(Z_1 + Z_2) \parallel (Z_0 + 3Z_f)$

$$\therefore \quad I_{a_1} = \frac{E}{(Z_1 + Z_2) \parallel (Z_0 + 3Z_f)}$$

$$= \frac{E_a}{[(Z_1 + Z_2)(Z_0 + 3Z_f)/(Z_1 + Z_2 + Z_0 + 3Z_f)]}$$

$$= \frac{E_a}{Z_1 + Z_2 \dfrac{Z_0 + 3Z_f}{Z_0 + Z_2 + 3Z_f}} \quad \text{...(13.79)}$$

$$\text{Then,} \quad V_a = V_{a_0} + V_{a_1} + V_{a_2} = -I_{a_0} Z_0 + (E_a - I_{a_1} Z_1) - I_{a_2} Z_1$$

$$\therefore \quad V_a = E_a \frac{Z_0 Z_2}{D} + E_a - \left(E_a Z_1 \frac{Z_0 + Z_2 + 3Z_f}{D} \right) + \frac{E_a Z_2 (Z_0 + 3Z_f)}{D} \quad \text{(Using eqn. (13.79))}$$

$$\text{where} \quad D = Z_1 Z_2 + Z_2 Z_0 + Z_0 Z_1 + 3Z_f (Z_1 + Z_2)$$

$$\therefore \quad V_a = 3E_a \left\{ \frac{Z_2 Z_0 + 2Z_f Z_2}{Z_1 Z_2 + Z_2 Z_0 + Z_0 Z_1 + 3Z_f (Z_1 + Z_2)} \right\} \quad \text{...(13.80)}$$

It must be stated here that in case of a solid/bolted fault, $Z_f = 0$, which is to be put in above equation.

13.3.10 Fault on a Power System

In the above section we have discussed on the different types of faults occurring at the terminals of an unloaded alternator. Infact, the very result obtained there, after analysis and calculations, can be also used to analyse the faults occurring on a power system.

To start with, let us consider a short section of a transmission line in which, say, phase '*a*' is grounded. Now to avoid having fault current and load currents mixed, let us think of the fault as being occurred on a short step-line, built out from the line at the point of fault. Infact, to facilitate the analysis of the fault on a power system, the power system may be replaced by an equivalent generator, by considering the laws of Thevenin's. Hence, say, the e.m.fs, E_{a_1}, E_{a_2} and E_{a_0} are the symmetrical components of the open-circuit voltage

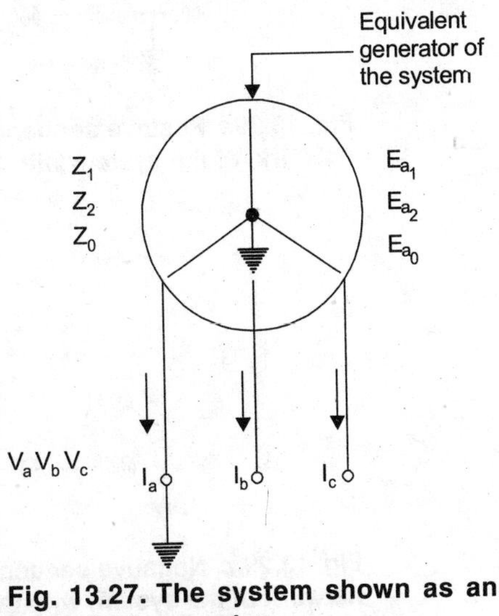

Fig. 13.27. The system shown as an equivalent generator.

to ground before the fault actually occurred in the system. The assumption can be represented in the Fig. 13.27. from clearing our conception of the complex nature of fault on the system. In the Fig. 13.27, the impedances Z_1, Z_2 and Z_0 are the impedances of positive, negative and zero sequence currents respectively from the point of fault.

Now, the basic equation applied here, and $I_{a_1}, I_{a_2}, I_{a_0}$ and $V_{a_1}, V_{a_2}, V_{a_0}$ can be determined the same way as we did in above section. Then after these sequence currents and voltages are known the corresponding sequence currents and voltages through out the network can be calculated by applying Kerchhoff's laws.

To explain or analyse the subject further, let us consider that a fault has occurred at the point F in power system shown in a single line diagram in the Fig. 13.28.

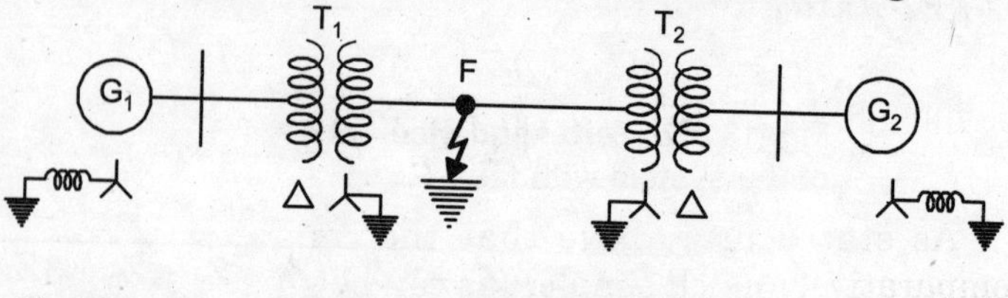

Fig. 13.28. A single Line Simplified Diagram (SLD) of a power system under fault at point *F*

To analyse the system at fault as shown in Fig. 13.28, we have to draw the corresponding system networks for positive, negative and zero sequence equances. Although the construction of positive and sequence network is not a very hard work as shown in the Fig. 13.29*a* and 13.29*b*, but the same for zero sequence needs some explanation in view of a transformer having been interposed in the zero-sequence circuit of Fig. 13.29*e*.

The networks for positive, negative and zero sequence networks along with their corresponding Thevenin equivalent circuit are shown in Fig. 13.29*b*, 13.29*d* and 13.29*f* respectively below.

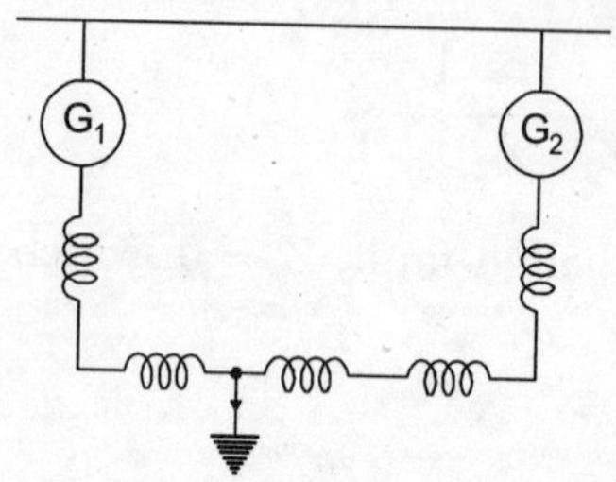

Fig. 13.29*a*. Positive sequence network of the system with fault *F*.

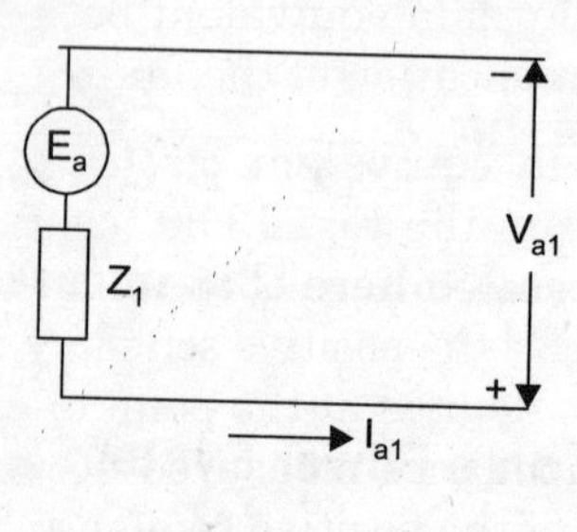

Fig. 13.29 *b*. Thevenin equivalent circuit of the positive sequence network of Fig. 13.19*a*.

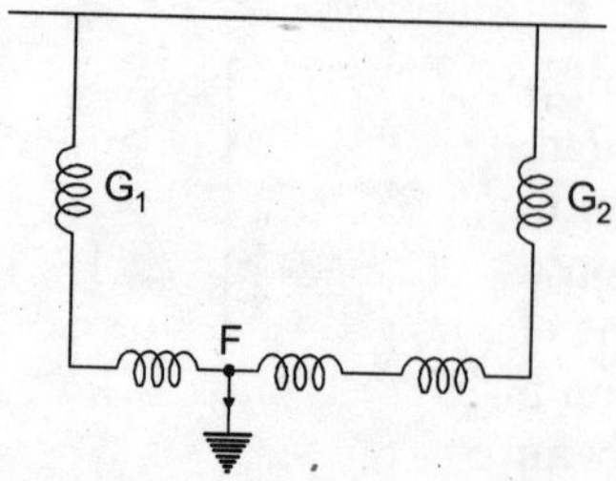

Fig. 13.29 *c*. Negative sequence network of the system with fault *F*.

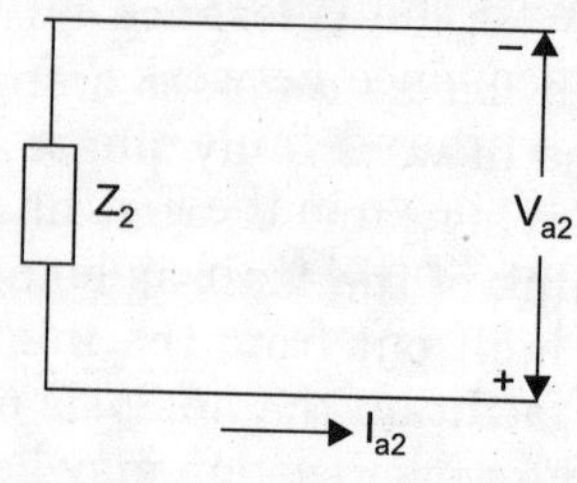

Fig. 13.29 *d*. Thevenin equivalent circuit of the negative sequence of the network of Fig. 13.29*c*.

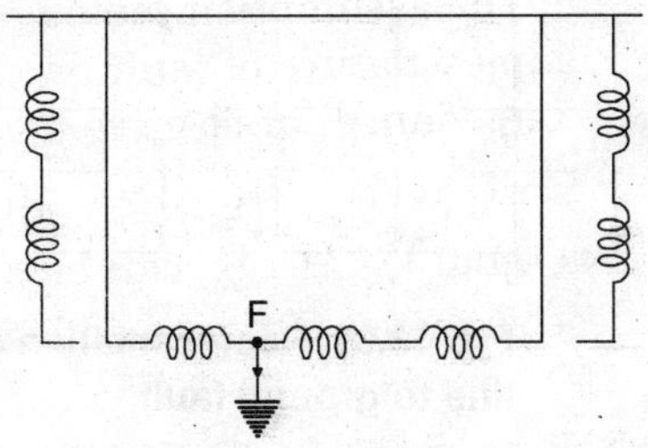

Fig. 13.29*e*. Zero sequence network of the system with fault *F*.

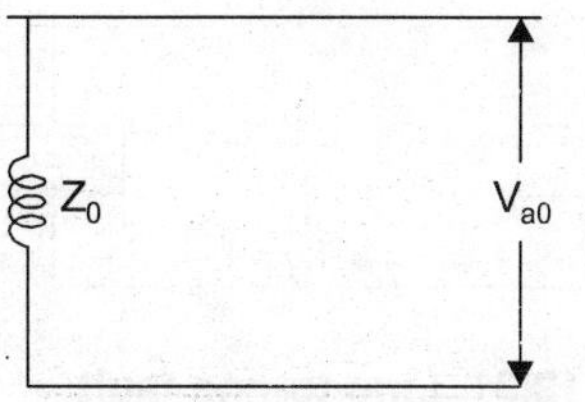

Fig. 13.29*f*. Thevenin equivalent of the zero sequence network of Fig. 13.19*e*.

As stated just above that the construction of positive and sequence network are comparatively much simpler (as can be seen in the Fig. 13.29*a*, 13.29*b*) than that of the zero-sequence network of Fig. 13.29*e*, in view of a transformer having been interposed in the circuit of Fig. 13.29*e*. In fact, for an earth fault on the earthed side, there is a path for the zero sequence current to flow. The corresponding compensating current can also be circulating to the delta. The equivalent circuit therefore must provide for a path for the line on star side through the equivalent impedance of the transformer to the reference bus and an open circuit must exist between the line and the reference bus on the delta. This is shown in the Fig. 13.29*f*. Now, if the convection from star point on the star side, to ground contained an impedance of say Z_n, the zero sequence equivalent circuit would have contained on impedance of $3Z_n$ in series with the equivalent impedance of the transformer to connect the line on the star side to ground that is, the reference bus.

In the Fig. 13.29 each of the networks (of positive, negative and zero sequence) can be replaced by its Thevenin equivalent between the two terminals composed of its reference bus and the point of application of the fault.

Then Thevenin equivalent circuit of each sequence network is shown in the Fig. 13.30 for the corresponding network. The internal voltage of the single generator of the equivalent circuit for the positive sequence network, E_a, the prefault voltage to neutral at the point of application of the fault Z_1 is the impedance measured between point *F* and the reference bus of the positive sequence network with all the internal e.m.f. short circuited.

Now, as no negative or zero sequence currents are flowing befors the fault occurs, the pre-fault voltage between point *F* and the reference bus is zero in the negative and zero sequence networks. With these points, for a single line-to-ground fault the connection of the sequence networks is shown in the Fig. 13.31, while for line to line fault the connection of the sequence network is now shown in Fig. 13.32.

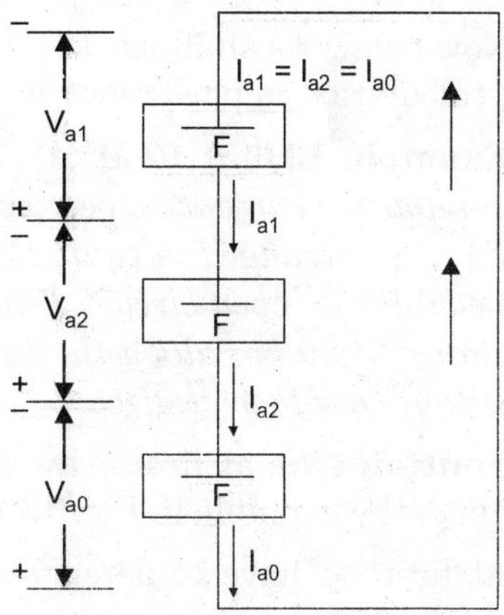

Fig. 13.30. Single line ground fault thevenin equivalent circuit of system.

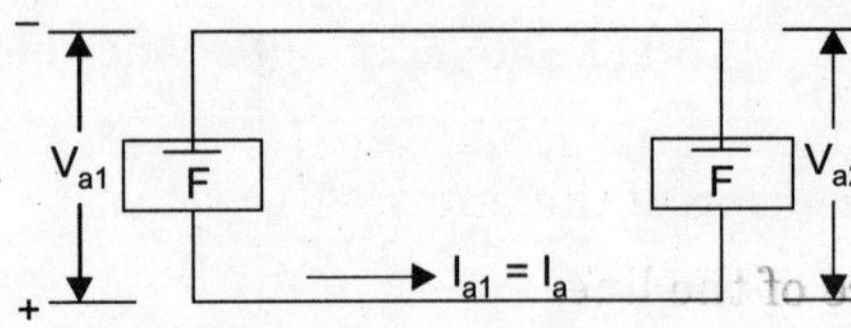

Fig. 13.31. Line-to-line fault.

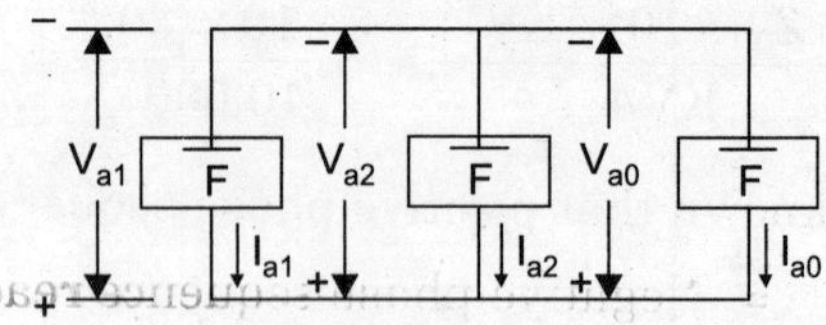

Fig. 13.32. Representation of double line to ground fault.

13.3.11 Fault Calculation Procedure

Discussions made above can be summed up below for the procedure to be followed for calculation of fault current etc. :

1. We should first draw by inspection of the positive, negative and zero sequence impedance diagram. We should remember that zero sequence impedance is needed only if there is an earth fault also.
2. Next we have to set the value of sequence impedances in each network.
3. Then reduce the networks into their equivalent sequence impedance Z_1, Z_2 and Z_0 (positive, negative and zero sequence impedances).
4. Then draw the equivalent network.
5. Then obtain the sequence component of current.
6. Then obtain the current is each phase from these sequence component using concerned equation.
7. If the voltage are needed they can be obtained from knowledge of impedances and sequence current by using concerned equations developed above.

Now below we shall use the procedures and equations developed above for unsymmetrical fault calculation in power system by solving few practical examples below.

Example 13.6. *A 10 MVA, 50 Hz generator has reactances of 30%, 10%, 5% to +ve, –ve and o-sequence current respectively. It is connected to a line comprising three conductors of 1 cm diameter arranged in equivalateral triangular spacing of 5 m side. The generator is excited to give 30 KV on open-circuit. Find the current in the line when two lines are short-circuited at a distance of 20 km along the line. Resistance and capacitance are to be neglected.*

Soution : We at first draw the diagram of the system of the example in Fig. 13.33.

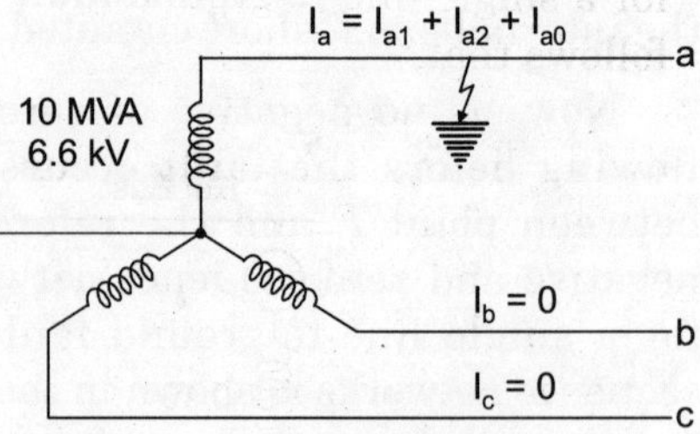

Fig. 13.33

At first we have to determine, *+ve*, *–ve* and *o*-sequences Z_{g_1}, Z_{g_2} and Z_{g_0} (of the generator) respectively from the data given in the example by using equation for Z_{g_1} Z_{g_2} and Z_{g_0}, as below

$$Z_{g_1} = \frac{(\% Z_1)\times 10\times(\text{KV})^2}{\text{KVZ}} = \frac{30\times 10\times(30)^2}{10{,}000} = 27\ \Omega \text{ Positive sequence impedance of generator}$$

$$Z_{g_2} = \frac{(\% Z_2)\times 10\times(\text{KV})^2}{\text{KVA}} = \frac{10\times 10\times(30)^2}{10{,}000} = 27\times\frac{10}{30} = 9\ \Omega = -ve \text{ sequence impedance of generator}$$

$$Z_{g_0} = \frac{(\%\ Z_0)\times 10\times (\text{KV})^2}{\text{KVA}} = \frac{5\times 10\times (30)^2}{10{,}000} = 27\times\frac{5}{30} = 4.5\ \Omega$$ *o*-sequence impedance of generator

It is known that positive-phase-sequence reactance of the line

= Negative-phase-sequence reactance of the line

$$= 2\pi f(L_S - L_M) \text{ which can be } = 2\pi f\left(\frac{1}{2} + 4.606\log\frac{D}{r}\right)\times 10^{-7} \text{ ohm per metre}$$

Now in this example, $D = 5\ m$, $r = 0.005\ m$

$\therefore$ positive-phase-sequence reactance of line 20 km long, negative sequence reactance = $Z_{l_2} = 9\Omega$.

$\therefore Z_1$ upto the point of fault = $Z_{g_1} + Z_{f_1} = j(27+9) = j36\Omega$

Z_2 upto the point of fault = $Z_{g_2} + Z_{l_2} = j(9+9) = j18\Omega$

$$\therefore \quad I_{a_1} = \frac{E_a}{Z_1+Z_2} + \frac{30{,}000/\sqrt{3}}{j(36+18)} = -j320\,A$$

$$I_{a_2} = -I_{a_1} = j320 \text{ Amps.}$$

But, $I_{a_0} = 0, I_a = 0$ as per the fault condition

$$I_b = I_{b_0} + I_{b_1} + I_{b_2} = a^2 I_{a_1} + a I_{a_2} = a^2 I_{a_1} + a(-I_{a_1}) = (a^2 - a) I_{a_1} = -j\sqrt{3}(-j320) = -555 \text{ amps.}$$

$$I_c = -I_b = 555 \text{ amps.}$$

Example 13.7. *A 6.9 KV, 10 MVA alternator has $X'' = X_2 = 0.15$ P.U. and $X_0 = 5\%$, its neutral being grounded through a current limiting reactor of 8% (based on the generator rating). The generator is connected to alternator connected step-up transformer with a reactance of 6%, its neutral being solidly grounded. When the terminal voltage of the generator is 13.8 KV a single-line-to-ground fault occurs on the open circuited high voltage side of the transformers. Find the initial symmetrical rms, current in all phase generator.*

Solution : With reference to the example we find in the Fig. 13.34 (*a*), (*b*) & (*c*) positive, negative and zero sequence networks respectively for the system. It should be stated here that for a single-line-to ground fault these sequence networks will be connected in series. Hence it follows that,

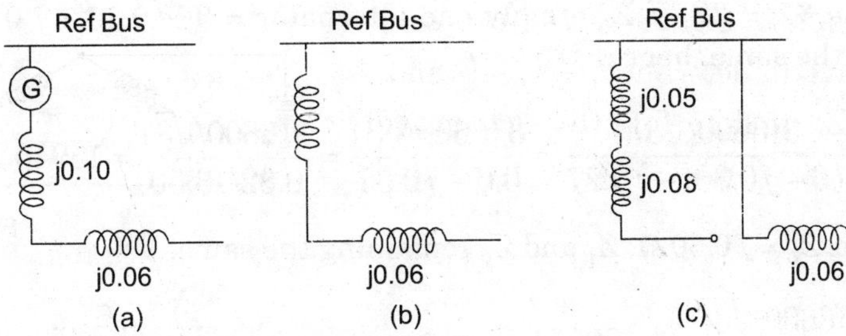

Fig. 13.34

$$Z_1 = 0.16 \text{ P.U.}, Z_2 = 0.16 \text{ P.U.}, Z_0 = 0.06 \text{ P.U.}$$

$$I_{a_1} = \frac{E_0}{Z_1 + Z_2 + Z_0} = \frac{1}{j0.38} = -j2.63 \text{ P.U.}$$

(as the generator is connected to Δ side of the transformer hence zero sequence current will not flow through Δ side of the transformer and generator and will flow only through star side for a $h \cdot v$, side fault).

Now, denoting the currents in the generator by suffixes A, B, C we have,

$$I_{A_1} = jI_{a_1} = +2.63 \text{ P.U.}, I_{A_2} = -jI_{a_2} = -2.63 \text{ p.u.}$$

This means, $I_A = I_{A_1} + I_{A_2} = 0, I_{B_1} = a^2 I_{A_1} = 2.63 \angle 240°, I_{B_2} = a I_{A_2} = 2.63 \angle -60°$

$$I_B = I_{B_1} + I_{B_2} = 2.63(--0.5 - j0.866 + 0.5 - j0.866) = 4.56 \angle -90° \text{ p.u.}$$

and $I_{C_1} = a I_{A_1} = 2.63 \angle 120°, I_{C_2} = a^2 I_{A_2} = 2.63 \angle 60°$

$$I_C = I_{a_1} + I_{C_2} = 2.63(-0.5 + j0.866 + 0.5 + j0.866) = 4.56 \angle 90° \text{ p.u.}$$

But we have, as per example, $I_{\text{base}} = \dfrac{10{,}000}{\sqrt{3} \times 13.8} = 418.5$ Amps.

So that, $I_A = 0, I_B = 4.56 \angle -90° \times 418.5 = 1908 \angle -90°$ Amps

$$I_C = 4.56 \angle 90° \times 418.5 = 1908 \angle 90° \text{ Amps.}$$

Example 13.8. *A total of six numbers of 6.6 KV, 3-phase alternators are connected to a common set of bus bars. Each has positive and zero sequence reactances of 0.90 Ω, 0.72 ohm and 0.30 ohm respectively. An earth fault occurs on one bus bar. Determine the value of the fault current if, (a) all alternator neutrals are solidly grounded. (b) If one alternator neutral is grounded through a resistance of a 0.2 Ω, the ohms being isolated. (c) If only one alternator neutral is solidly grounded, the other being isolated.*

Solution :

(*a*) When the alternators are connected in parallel so are their reactances, $Z_1 = j\,0.15\,\Omega$, $Z_2 = j\,0.122\,\Omega$, and $Z_0 = j\,0.05\,\Omega$. Hence, $Z_1 + Z_2 + Z_0 = j\,0.32\,\Omega$.

Now, assuming '*a*' phase to be at fault

$$I_a = \frac{3E_a}{Z_1 + Z_2 + Z_0} = \frac{3(6600/\sqrt{3})}{j0.32} = 35{,}700 \text{ Amps.}$$

(*b*) In this case, $Z_0 = 3Z_n + Z_0$ of only one alternator $= 3 \times 0.20 + j\,0.30\,\Omega$, Z_1 and Z_2 remaining the same, hence,

$$I_a = \frac{3(6600/\sqrt{3})}{(0.6 + j0.30) + j0.27} = \frac{3(6600/\sqrt{3})}{0.6 + j0.57} = \frac{3(6600/\sqrt{3})}{0.82/43.50} = 13940 \angle -43.5 \text{ amps.}$$

(*c*) In this case, $Z_0 = j\,0.30\,\Omega$, Z_1 and Z_2 remaining the same.

$$I_a = \frac{3(6600/\sqrt{3})}{j0.57} = -j20054.74 \text{ Amps.}$$

Example 13.9. *A 30 MVA, 3-phase, 11 KV, 50 Hz alternator, having these neutral solidly grounded, is operating at no-laod. It has a reactance to positive sequence reactance of 2.5 ohm. The reactances to negative and zero sequence current are 80% and 30% of the positive sequence value respectively. For a double line-to-ground fault, determine : (a) The current in the faulted phase, (b) The current through ground, (c) The voltage of healthy phase to neutral.*

Solution :

(*a*) Here in this example, we have got :

$$Z_1 = j\,2.5\,\Omega, Z_2 = j\,0.8 \times 2.5 = j\,2\,\Omega, Z_0 = j\,0.3 \times 2.5 = j\,0.75\,\Omega$$

$$\text{Now total impedance} = \frac{Z_0 Z_2}{Z_0 + Z_2} = \frac{(j0.75)(j2)}{j(0.75+2)} = \frac{-1.50}{j2.75} = j0.545\,\Omega$$

Now, assuming phases *b* and *c* to be under fault

$$I_{a_1} = \frac{E_a}{Z_1 + (Z_0 Z_2 / Z_0 + Z_2)} = (\text{with } Z_f = 0)$$

$$= \frac{11{,}000/\sqrt{3}}{j(0.5 + 0.545)} = -j\frac{11{,}000}{\sqrt{3} \times 3.045} = -j\,2090 \text{ amps.}$$

Then, $$I_{a_2} = \frac{-Z_0}{(Z_0 + Z_2)} \cdot I_{a_1} = \frac{-j\,0.75}{j\,2.75} \times (-j\,2090) = j\,570 \text{ amps.}$$

$$I_{a_0} = \frac{-Z_2}{Z_0 + Z_2} \cdot I_{a_1} = \frac{-j\,2}{j2.75} \times (-j\,2090) = j\,1.520 \text{ amps.}$$

$$\therefore \quad I_a = I_{a_0} + I_{a_1} + I_{a_2} = 0$$

and $I_b = I_0 + I_{b_1} + I_{b_2} = I_0 + a^2 I_{a_1} + a I_{a_2} = 1520\angle 90° + 2090\angle 150° + 570\angle 210°$

$= 1520(0 + j1) + 2090(-0.866 + j0.5) + 570(-0.866 - j0.5)$

$= j1520 - 1820 + j1045 - 495 - j285 = 23.5 + j2280 = 3250\angle 135.9°$ amps.

Now from the equation deduced in above discussion on the subject, we get,

$$I_{a_2} = \frac{E_a (Z_0 + 3Z_f)}{Z_1 Z_2 + Z_2 Z_0 + Z_0 Z_1 + 3Z_f}(Z_1 + Z_2) \text{ with } Z_f = 0, I_{a_2} = I_{a_1} \cdot \frac{-Z_0}{Z_0 + Z_2}$$

$$I_{a_0} = -E \frac{Z_2}{Z_1 Z_2 + Z_2 Z_0 + Z_0 Z_1} + 3Z_f (Z_1 + Z_2) \text{ with } Z_f = 0, I_{a_0} = -I_{a_1} \frac{Z_2}{Z_0 + Z_2}$$

Also $I_c = I_0 + I_{c_1} + I_{c_2} = I_0 + a\,I_{a_1} + a^2\,I_{a_2} = 1520\angle 90° + 2090\angle 30° + 570\angle -30°$

$= 1520\,(0 + j\,1) + 2090\,(-0.866 + j\,0.5) + 570\,(0.866 - j\,0.5)$

$= j1520 + 1820 + j1045 + 495 - j\,285 = 2315 + j\,2280 = 3150\angle 44.6°$ amps.

(*b*) Current through ground, $I_n = 3I_0 = 3 \times 1520 = 4560$ amps (we know that ground serves as the return path of zero sequence currents).

(*c*) In this case, we know that voltage of healthy phase '*a*' is given by $V_a = V_{a_1} + V_{a_2} + V_{a_0}$ from the discussion on the subject above. However,

$$V_{a_1} = V_{a_2} = V_{a_0} \text{ and therefore, } V_a = 3V_{a_2}, \text{ where}$$

$$V_{a_2} = \frac{Z_0 Z_2}{Z_0 + Z_2} \times I_{a_1} = (j0.545)(-j2090) = 1140 \text{ volts.}$$

$$\text{As} \quad V_a = 3V_{a_2} = 3420 \text{ volts (as a fault conductor } V_h = V_c = C)$$

Example 13.10. *Three 6,600 volts, 10,000 KVA, 3-phase alternators are connected to a common set of busbars. Each has a reactance to +ve sequence current of 5%. The reactance to zero sequence currents are 75% of the +ve sequence and the reactance to zero-sequence currents are 30% of the +ve sequence value. If an earth fault occurs on one busbar, determine the fault current (a) if all the alternator neutrals are solidly earthed, (b) if only one of the neutrals of the neutralise is solidly earthed and others are isolated (c) if one of the neutrals is earthed through a resistance a resistance of 0.3 ohm and the other are isolated.*

Solution :

(*a*) When all the three alternators are solidly earthed as shown in Fig. 13.35, *a* to find the fault current at let us follow the process below :

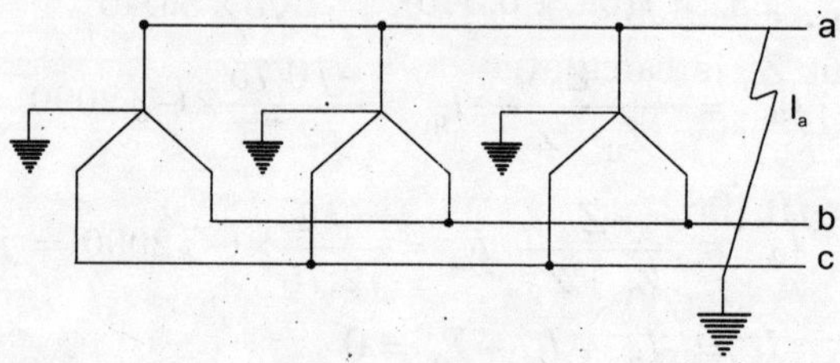

Fig. 13.35

$$\text{Full load currents of each alternator} = \frac{10{,}000 \times 1000}{\sqrt{3} \times 6{,}600} = 875 \text{ amps.}$$

Now, let the value of positive phase-sequence reactance of one alternator be X, then

$$\frac{X, I}{6{,}600 \times \sqrt{3}} = \frac{15}{100} = \frac{3 \times 5}{100} \text{ (5\% is the +ve sequence phase reactance)}$$

$$\therefore \quad X = \frac{15}{100} \times \frac{6{,}600}{\sqrt{3}} \times \frac{1}{875} = 0.654\,\Omega$$

$$\therefore \quad Z_{g_1} = j\,0.654 \text{ (+ve sequence impedance)}$$

$$Z_{g_2} = 0.75 \times j\,0.654 = j\,0.492\,\Omega \text{ (–ve sequence, 75\% of +ve sequence reactance)}$$

$$Z_{g_3} = 0.3 \times j\,0.654 = j\,0.1962\,\Omega \text{ (zero sequence reactance, = 30\% +ve sequence reactance)}$$

When the three alternators are running in parallel, the resultant reactances will be $\frac{1}{3}$rd, *i.e.*,

$$Z_1 = j\,0.218,\ Z_2 = j\,0.614,\ Z_0 = j\,0.065$$

∴ The fault current,

$$I_a = \frac{3Z_g}{Z_1+Z_2+Z_0} = \frac{3\times(6600/\sqrt{3})}{j(0.218+0.164+0.065)} = j\,25{,}600 \text{ amps.}$$

(*b*) In this case, as per example, one of the three alternators is being solidly earthed and others are isolated, which is represented in the Fig. 13.36 and the fault occurs at *F* to earth.

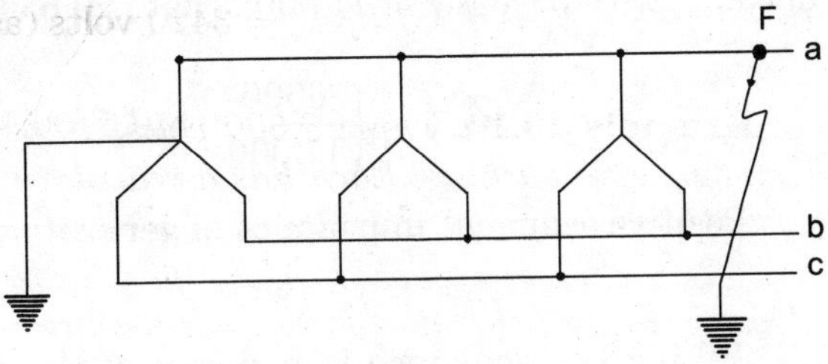

Fig. 13.36

In this case from the calculation is (*a*)

$$Z_0 = Z_{g_0} = j\,0.1962$$

The fault current

$$I_a = \frac{E_a}{Z_1+Z_2+Z_0}$$

In this case, Z_2 are calculated same as in (*a*) above : but Z_0 is calculated in this case,

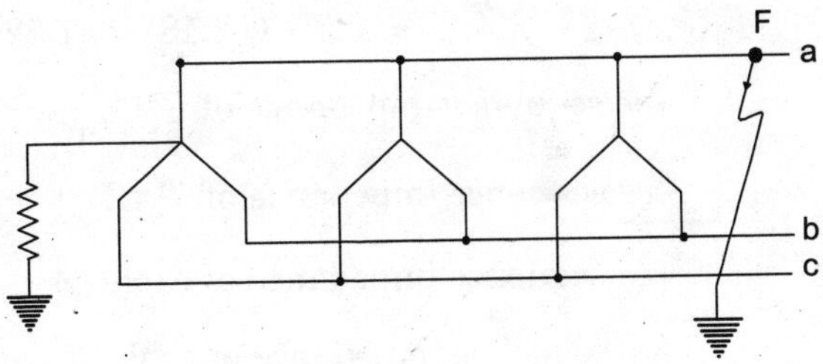

Fig. 13.37

$$Z_0 = Z_{g_0} = j\,0.1962$$

$$\therefore \quad I_a = \frac{\sqrt{3}\times 6{,}600/\sqrt{3}}{j(0.218+0.164+0.1962)} = -j\,19{,}800 \text{ amps.}$$

(*c*) In this case one of the alternators neutrals is earthed through a resistance of 0.3 Ω and other two are isolated as shown in Fig. 13.37 *c*. In this case,

$$Z_0 = 0.3\times 3 + j\,0.1962 = (0.9 + 0.1962)\ \Omega$$

$$\therefore \text{ Fault current} = \frac{\sqrt{3}\times 6{,}600}{j0.218 + j0.164 + (0.9 + j0.1962)} = \sqrt{3}\times 6600\,(0.795 - j0.5) \text{ Amps.}$$

$$= \sqrt{3}\times 6600\times\sqrt{0.795^2 + 0.5^2} = 10700 \text{ Amps.}$$

Example 13.11. *In the power system network of Fig. 13.38 (a) a fault occurs at F. Find the fault current for (a) Line fault, (b) Line to ground fault. The technical data about the system is as follows :*

Generators : G = 2000 KVA, 6.6/11 KV, reactance 5%, solidly earthed on 11 KV side.

T_2 = 2000 KVA, 11/6.6 KV resistance 5%, solidly earthed on 6.6 KV side.

Cable : jX = 0.5 Ω at 11 KV.

Solution : Let us take the base KVA as 2000, and the base KV = 6.6.

$$Z_{g_1} = 10\% \text{ reactance at 6.6 KV}$$

$$= \frac{\%\,\text{Reactance} \times 10\times(\text{KV})^2}{\text{KVA}}$$

(By using the equation for positive sequence impedance of generator)

$$= \frac{10 \times 10 \times (6.6)^2}{2000} = 2.18\,\Omega$$

Transformer 5% reactance at 66 KV when refered to = 1.09 Ω. Positive sequence impedance, of cable, with base KV of 11,000. That is 0.5 ohm at 11 KV when referred to 6.6 KV. We get,

$$Z_{c_1} = \left(\frac{6600}{11,000}\right)^2 \times 0.5 = 0.18\,\Omega$$

Positive sequence impedance of generator,

$$Z_{g_1} = j\,2.18\,\Omega$$

Taking –*ve* sequence impedance of the generator equal to 70% of the +*ve* sequence impedance,

$$Z_{g_2} = 0.7 \times (j\,2.18) = j\,1.526\,\Omega$$

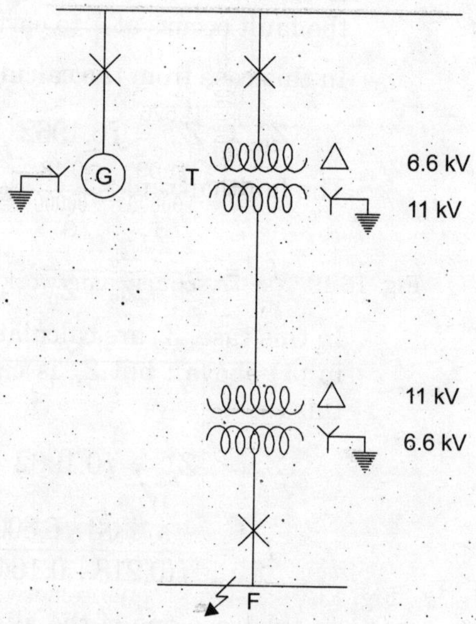

Fig. 13.38

$\therefore$ +*ve* sequence impedance of $T_1 (Z_{T_1})_1 = j1.09\,\Omega.$

–*ve* sequence impedance of $T_2 \left(Z_{T_2}\right)_2 = j1.09\,\Omega.$

+*ve* sequence impedance of cable $Z_{c_1} = j0.18\,\Omega.$

–*ve* sequence impedance of cable $Z_{c_2} = j0.18\,\Omega.$

+*ve* sequence impedance of $T_2 \left(Z_{T_2}\right)_2 = j1.09\,\Omega.$

–*ve* sequence impedance of $T_2 \left(Z_{T_2}\right)_2 = j1.09\,\Omega.$

The sequence networks of the system and their equivalent are shown in Fig. 13.39 (*b*), 13.39 (*c*) +*ve* sequence; –*ve* sequence and zero sequence respectively. Hence from these figures :

$$\therefore \quad Z_1 = j\,2.18 + j\,1.09 + j\,0.18 + j\,1.09 = j\,4.54\,\Omega$$

$$Z_2 = j\,1.526 + j\,1.09 + j\,0.18 + j\,1.09 = j\,3.886\,\Omega$$

$$Z_0 = j\,1.09\,\Omega$$

Based on above calculation now we can find :

(*a*) Line to line fault, $I_{a_1} = -I_{a_2} = \dfrac{E_a}{Z_1 + Z_2} = \dfrac{6.600/\sqrt{3}}{j0.454 + j3.886}$

$$I_{a_0} = 0, I_a = 0, I_b, I_{b_0} + I_{b_1} + I_{b_2} = I_{a_0} + a^2\,I_{a_1} + a\,I_{a_2}$$

$$\therefore \quad I_b = I_{a_0} + a^2\,I_{a_1} + a\,(-I_{a_1}) = (a^2 - a)\,I_{a_1}$$

as $\quad I_a = 0, I_{a_1} = -I_{a_2}$

$$\therefore \quad I_b = (a^2 - a) \times \frac{6600/\sqrt{3}}{j8.426} = \frac{(-j/\sqrt{3})\,6600/\sqrt{3}}{j8.426} = -785 \text{ amps.}$$

$$I_c = -I_b = 785 \text{ amp. (shown in Fig. 13.39 } d)$$

(*b*) Line-to-line Ground Fault (refer to Fig. 13.39 *e*).

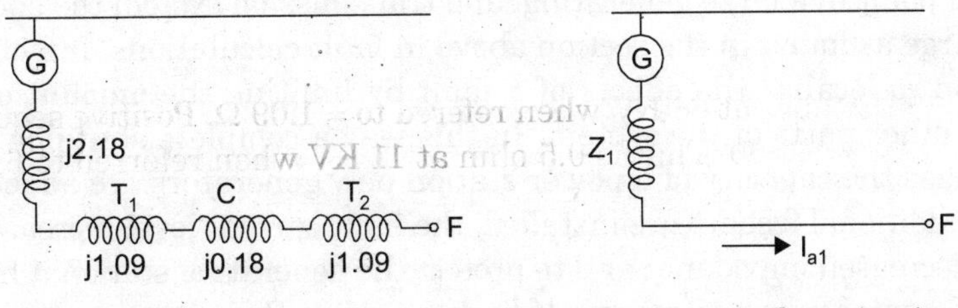

Fig. 13.39 (*a*). Positive network of thevenin's equivalent circuit

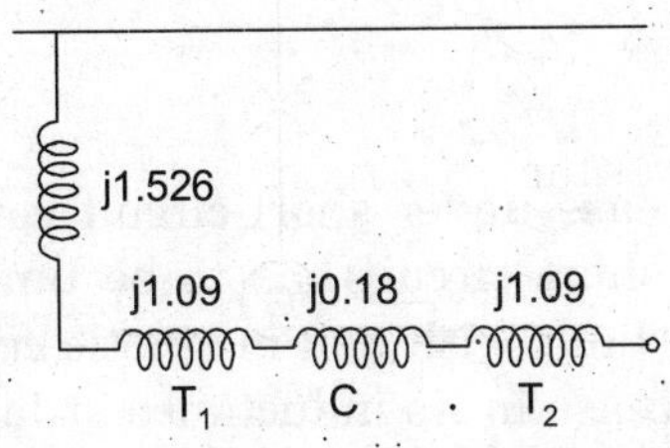

Fig. 13.19 (*b*). –*ve* sequence network of the Fig. 13.38

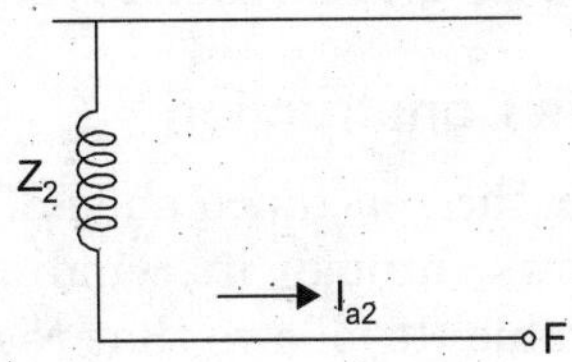

Thevenin equivalent circuit of –*ve* sequence (negative sequence)

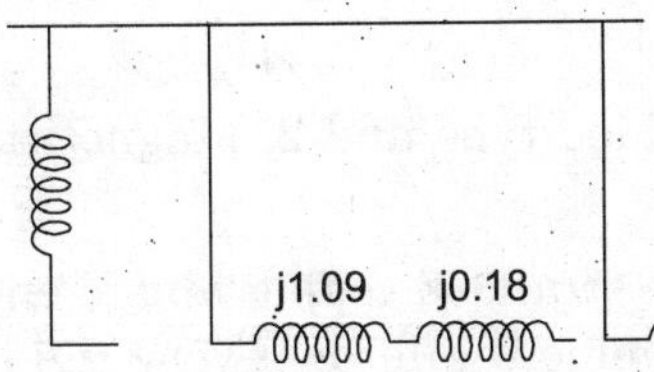

Fig. 13.19 (*c*). Zero sequence network of Fig. 13.38

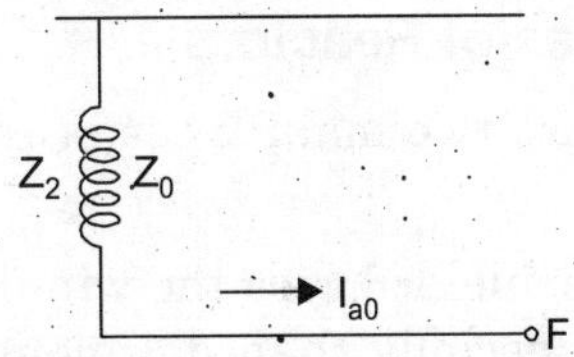

Thevenin equivalent circuit of 0 sequence

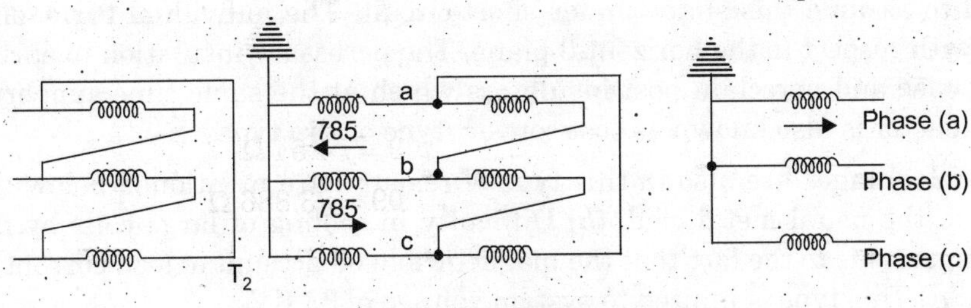

Fig. 13.39 (*d*)

$$I_a = \frac{3\,E_a}{Z_1 + Z_2 + Z_0}$$

$$= \frac{3 \times (6600/\sqrt{3})}{j(4.54 + 3.886 + 1.09)}$$

$$\therefore \quad |I_a| = 1{,}200 \text{ amp.}$$

13.4 FAULT LIMITING REACTORS

To limit the short-circuit current flowing to a safe value and for the purpose of protecting the plant and the system current limiting reactors which are large coils wound for high self-inductance and very low resistance are used. As their resistance is kept small in comparison to reactance no appreciable amount of energy is wasted in them. It is known that when a short-circuit occurs at a point in a large generating and transmission system the short-circuit current flowing is very large as found in the section above in fault calculations. Infact, current limiting reactors also tend to localise the effects of a fault by limiting the amount of power that can flow into it from other parts of the system. In this way a complete shut-down for the plant is avoided. To increase the capacity of a power station new generators are added in parallel with the original and additional feeders are installed, the fault current level increases due to increase KVA rating and decreased impedance and to protect the generating station it becomes necessary to replace the old switchgear by new one of higher rating. Switchgear equipment of the same rating can, infact, be used if reactors are fitted.

13.4.1 Reactor's Construction

For a reactor the reactance should not decrease under short-circuit conditions, when a large current flows through its windings, if a short circuit has to be limited to a correct predetermined value. It follows that reactors be built with non-magnetic cores (that means, saturation of magnetic core should load to a decrease in its inductance at large currents and thereby increased short-circuit current) or alternatively with an iron core but with an air gap included in magnetic circuit.

13.4.2 Types of Reactors

There are two main types of reactors : 1. Unshielded type and 2. Magnetically shielded types.

1. In unshielded type the winding is usally of bare stranded copper and is embedded in a number of specially shaped cement slabs. This arrangement affords a very rigid mechanical support against the mechanical forces which are released when a short-circuit occurs. These forces load to compress the coil axially and expand it radially. A circular coil is employed since all coils tend to assume this shape under short-circuit. The individual turns of the winding are inclined with respect to the horizontal plane. The necessary insulation to earth is provided by a convert base and poreclain post insulators which at the same time support the reactor. This type of reactor is also known as cost convert type or dry type.

Some disadvantages are also in this type of reactors are mentioned below : (*a*) These do not conform to the usual metal clad. (*b*) Difficulty in cooling of large coils by fan. (*c*) Large space requirement due to the fact that the magnetic fold on account of load current is practically unrestricted, (*d*) This type is limited to system voltage of 33 KV.

2. Magnetically Shielded Type Reactor

In this type the coil is placed in a standard transformer tank and is oil insulated. Thus the coil can be self or fan-cooled. The objectively of shielding (to present the magnetic flux from entering the transormer tank walls) is to avoid considerable losses and heating under normal loads when would be prohibitive under short-circuit conditions. Out method makers use of iron laminations, which provide path for the stray magnetic field outside the winding. Unfortunately this result in a decrease of reactance under abnormal current conditions. In this arrangement the coil has no iron inside. In another method the shields are in the form of short circuited rings at earth potential and are supported from non-magnetic end plates. Induced currents in these rings setup mm.f.s., which continue the flux to follow path on the inside of the shield. With this arrangement the reactance does not decrease under heavy current condition. In large reactors these shielding sings may be in the form of cooling and thus perform both functions of magnetic shielding and cooling.

Magnetically shielded type reactors are also known as oil immersed type. These types of reactors can be applied to a circuit of any voltage level for either indoor or outdoor installation. Other advantages of these reactors are : (1) A high factor of safety against flashover, (2) No magnetic field outside the tank to cause heating or magnetic forces in adjacent reactors or metal structures during short-circuit, (3) High thermal capacity.

13.4.3 Methods of Locating Reactors Positions

The reactors may be located in the following positions :

1. Generator Reactors

In this system reactors may be insulated in series with each generator as can be seen in Fig. 13.40.

(It must be noted here that a reactor may be located in such away that while there is no large voltage drop across it under normal conditions, it will prevent a short circuit being fed by the generators into the fault, thereby limiting the breaking current of the circuit-breakers).

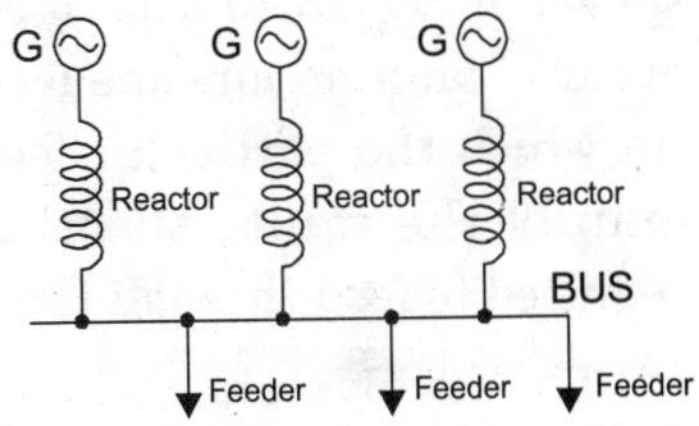

Fig. 13.40. Generator reactor.

The disadvantage of this method is that in the event of a fault occurring on a feeder the voltage on the common busbar drops to a low value with the result that the remaining feeders, which are healthy, may also fall out of step. The whole system is interrupted and will need resynchronising after cutting out of the faulty feeder. Another disadvantage is that full load current is always passing through the reactor causing voltage drop and a constant less under normal operating condition.

Then considering that modern alternators have large transient reactances (which protect the machine against short-circuit current) separate reactors in series with machines are employed only in the case of old machines having low reactance.

2. Feeder Reactors

The main disadvantage of the first method is avoided by putting the reactors in series with various feeders instead of generators, this arrangement hence is called Feeder Reactors, which is shown in Fig. 13.41. In the case of a fault on any one feeder the main voltage drop is in its reactor only so that the bus-bar voltage is not affected much. It will be seen that so protection is offered against a fault occurring on the bus bars. Considering the fact that bus bars faults are rare. This drawback is not of much importance. The disadvantage of constant voltage drop and losses is a quality present in this method also.

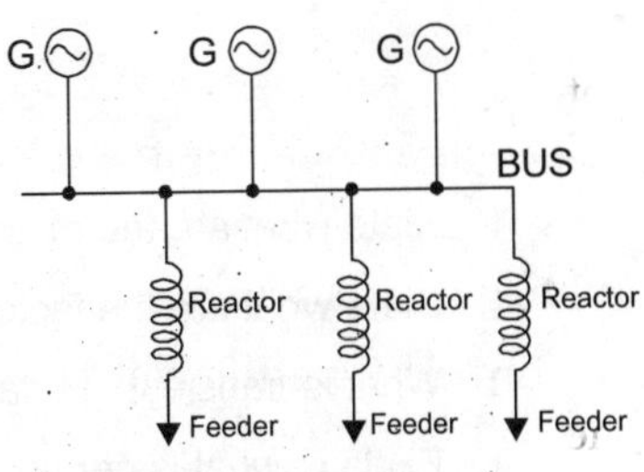

Fig. 13.41. Feeder reactor.

As reactance in the feeders will tend to affect adversely the voltage regulation of the system when operating under normal condition, feeder reactors are only employed in special cases such as short feeders of large cross-section fed from a large power station or in case of parallel feeders.

3. Bus Bar Reactors

Reactors located in both the positions mentioned above suffer from the disadvantage that they carry the full load current of the circuit with which they are associated, thus causing voltage drop and copper loss. This advantage may be removed by disadvantage the bus bar into sections and inter connecting the section through reactors as below :

(*a*) Ring System

In this case each feeder is normally fed by one generator. As a very small amount of power flash across the reactors, voltage drop, caused under operating conditions is small. This arrangement is shown in Fig. 13.40 what it can be noted that in the event of fault on any one feeder only one generator (to which the particular feeder is conected) feeds mainly the fault, all other generators being required to feed through the reactors.

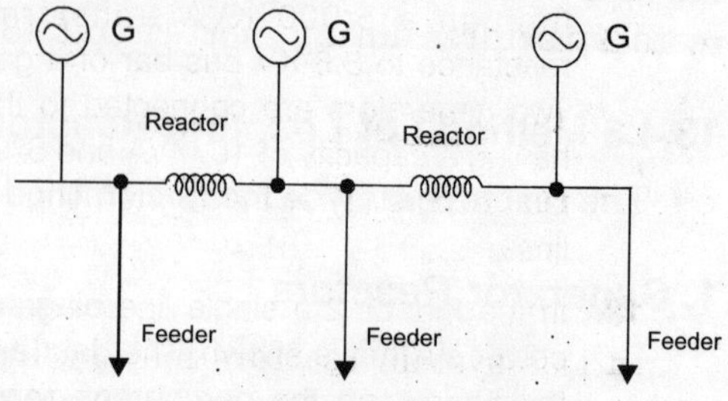

Fig. 13.42. Ring system.

(*b*) Tie-bar System

This system also works in a manner similar to ring system and is particularly suited to a system where large extensions are likely to be made. For a fixed size of individual reactors, the current fed into a fault does not exceed a certain value with increase in number of sections. In the Fig. 13.43 arrangement of a tie-bar system is shown. In this system, in fact, no modification of the switchgear is required when the system is extended.

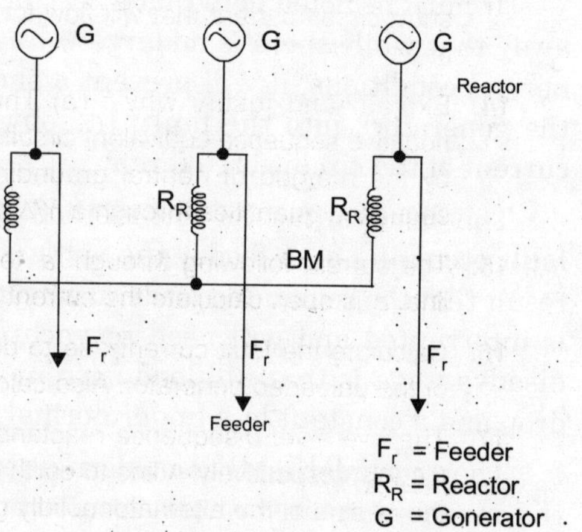

Fig. 13.43. Tie-bar system.

QUESTIONS

1. State what are the major type of faults in power system.
2. State what are the factors effecting symmetrical short-circuit current in a system.
3. What is translent reactance? Explain.
4. Explain, what is per unit system. How per unit quantities are determined in a system.
5. Define symmetrical and asymmetrical current in a system with wave diagrams.
6. State the short circuit KVA calculations in a system.
7. State what is meant by unsymmetrical fault in power system.
8. Explain what is positive sequence, negative sequence and zero sequence component with vector diagram.
9. What is meant by sequence impedance in a system?

10. Discuss (*a*) 3-phase fault, (*b*) Line to ground fault expression for the corresponding voltage and current in the system.

11. A 3-phase dead short circuit occurs at a point *F* in a power system shown in the Fig. Q. 1.

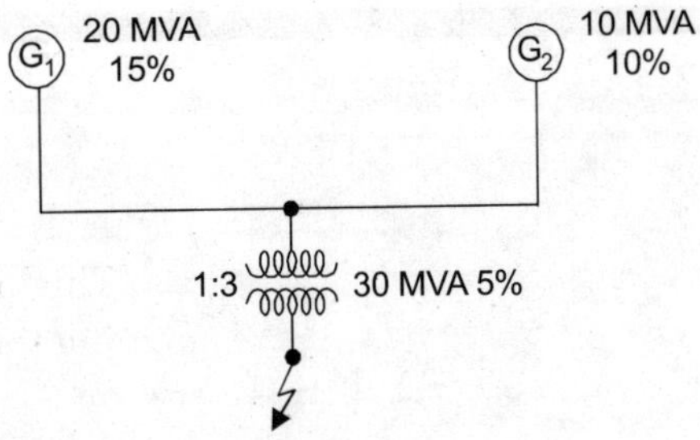

Q. 13.1

12. A short 3-phase transmission line operating at 11 KV has an impedance of $(1.8 + j\,2.1)\ \Omega$ and is connected through a 5,000 KVA transformer having 5% reactance to 5.6 KV bus bar of a generating station. Two generators are connected to the bus bar, each having a capacity of 10 MVA and 5% reactance calculate the short circuit out rent, if a symmetrical fault occurs (1) at the *h.t.* terminals of the transformer and (2) at the fatend of the transmission line.

13. In the Fig. Q. 2 a single line diagram of a 3-phase power system is shown. The data are indicated can the figure and the percentage reactance of each alternator is based on its own capacity. Find the short circuit current that will flow for a 3-phase fault at point *"F"*.

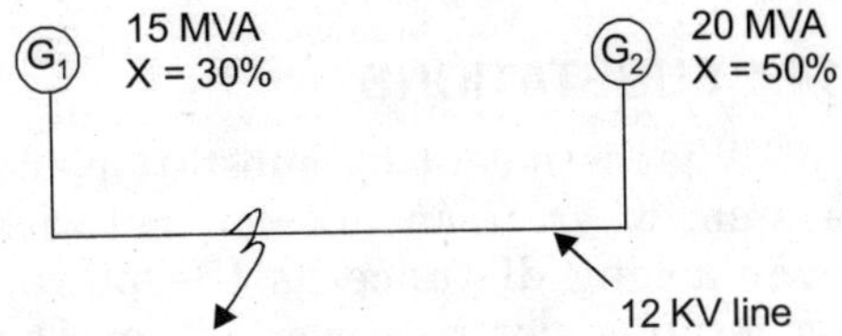

Q. 13.2

14. Explain and justify why : (*a*) The positive and negative sequence equivalent circuit are not affected by the method of neutral grounding. (*b*) The phase shifts in positive sequence and negative sequence quantities through a *Y*/Δ transformer are opposite to each other.

15. The current following through *'a'* terminal of a delta-connected load is 100 amps. Assuming the line *'c'* is open calculate the current in all the lines and their symmetrical components.

16. Calculate the fault current due to double line to ground fault occurring at the generator terminals of the unloaded generator. Also calculate line voltages. Generator of 20 MVA, 13.2 KV.

17. The *+ve*, *−ve*, *o*-sequence reactances of a 20 MVA, 13.2 KV alternator are 0.3 p.u., 0.2 P.U. and 0.1 P.U. respectively. A line to earth fault develops on phase *'a'* of the unloaded generator. Assume the neutral of the alternator solidly grounded. Determine the fault current and fault voltage in p.u.

18. State what is Fault Limiting Reactor. State what are the types of reactors.

CHAPTER

14

Substations

14.1 SUBSTATIONS

INSIDE THIS CHAPTER

- Substations
- Various other types of substations
- Construction of substation
- Planning objective & philosophy of substation-distribution networks
- Bus Bar Arrangements
- Common arrangement of bus-bars
- Network configuration

What is meant by substation in a power system where transmission network spread over a long distance as the power houses (generating electric power) either Thermal, or Nuclear or Hydel power stations generally situated far away from the load centres due to various economical and geographical reasons. In fact long and high voltage transmission lines are necessary to transmit large blocks of power for the sources of generation to the load centres, to interconnected power houses for increased reliability of supply, greater system stability and lesser stand-by power plant and helping consumers to get cheaper electric energy. Now in between the power house and ultimate consumers a number of transformation and switch-stations with proper protection scheme have to be created, and these stations are called substations from where the domestic and Industrial power distribution system is planned considering proper substation equipment for primary and secondary substations which are to be economically designed.

14.1.1 Total Electrical Distribution System Planning

For planning a system following steps are to be considered :

1. At first we should select the distribution voltage based on : (*a*) Load demand (present and future), (*b*) standard equipment rating, (*c*) nature and physical location of loads, (*d*) economy.
2. Arrange loads in logical groupings : (*a*) Tentatively designate individual 'large' loads-like large industrical loads (large motors, furnaces etc.). (*b*) Group small loads (domestic and service sectors lighting load etc.) in economical substation sizes to fit plant layout. (*c*) Then to devide substation loads into feeder groups.
3. Consider special load requirements : (*a*) starting requirements of "large" motors, (*b*) operation of arc furnaces, welders, and other cyclic loads, (*c*) critical or vital loads (Hospitals, Railways etc.).

4. Select plant distribution system from the following types : (*a*) Radial (old type), (*b*) Radial (modern), (*c*) Secondary selective, (*d*) Primary selective, (*e*) Looped primary, (*f*) Secondary network, (*g*) Combination of above.
5. Select system components for : (*a*) Main-supply substation. (*b*) Secondary distribution system, (*c*) Distribution/utilisation circuits.
6. Comparison of various alternates, taking into account : (*a*) Initial cost, (*b*) Safety, (*c*) System reliability, (*d*) Operation and maintenance—safety, simplicity and economics, (*e*) Flexibility for future expansion.

Most of the above points require detail discussion which is beyond the scope of this book. But above points are mentioned here for distribution engineers in power system to have idea of the major steps to be considered while planning an overall power distribution system.

However, here we shall briefly discuss the points mentioned in (4), (5) & (6).

For example, let us look at the Fig. 14.1 showing three segments of distribution system (of a particular power consumer) comprising, primary, load center and secondary. This figure will indicate that while load-center systems do not comprise the important heart of any well-designed, efficient and economical system, they are the basic principle of modern distribution system.

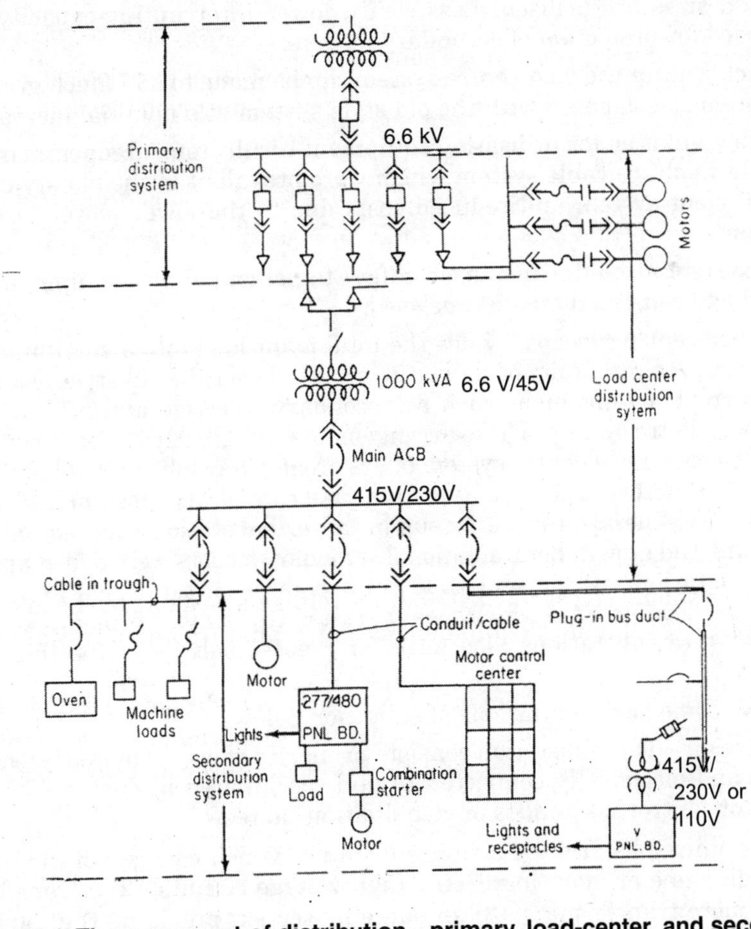

Fig. 14.1. Three segment of distribution—primary, load-center, and secondary.

These principles apply equally well to both primary and secondary unit substations, but this section primarily covers load-centre systems serving loads at 415 V or below since these are predominant in the typical industrial plant.

14.1.2 Load Centre System—Concept and Advantages

The basic concept of load centre systems is to distribute power at highest economical voltage level (upto 12 KV) to areas of concentrated electrical load (*i.e.*, the load centres) where voltage is transformed down to the lower utilisation voltage and delivered to utilisation equipment via relatively short secondary feeders.

Where total plant load is such that it exceeds the most economically size substation (750 to 1500 KVA), the load is served by two or more smaller substations rather than by one large substation at the point of primary service entrance; again the secondary feeders are relatively short. In fact, in the old-style radial system, the total load was served by a single substation at the point of primary service entrance. This arrangement resulted in long and heavy secondary feeder cable. The Fig. 14.2 shows the difference between the old-style radial and the modern load-centre systems.

Many advantages accrue to the consumer from the use of the load-centre concept. Initial costs are lower. Carrying power to load centres via smaller high-voltage cables rather than large multiple low-voltage cables result in a substantial saving in cable costs. While transformer costs are somewhat higher for the load-centre system but this cost is offset by the fact that smaller substation sizes are utilised. Less costly, lower interrupting-capacity breakers can therefore be utilised for protection of secondary feeders.

Primary switchgear for the load-centre system can be made to add much greater selectivity to the overall system as composed with the old style system at a nominal increase in cost.

Better voltage regulation for utilisation equipment result from a reduction of voltage drops inherent in long low-voltage cable system which were prevalent in the old-style systems. This means a more efficient system and reduced costs due to the lower power losses found in a load-centre system.

Load centre system also offer increased safety to personnel and equipment since all live parts are isolated and completely metal-enclosed.

Because the load-centre concept divides the total plant load into a small number of widely dispersed substations, the continuity of service is increased over the old-style system. In addition many different current arrangement such as secondary selective and primary selective are available in the load-centre system. These arrangements can greatly increase service continuity at a minimum extra cost. Load centre system offers greater flexibility, since small units may be added whenever and wherever they are needed to take care of increased plant load. In the old-type it system, it was general practice to put in larger substation then actually required in anticipation of future load growth here, additional secondary feeders were added upto the capacity of the original substation and beyond that point a large substation was added again in anticipation of future load growth, which meant wasted investment if plant expansion did not come about load centre substations also have for greater salvage value than the old-style substation.

Now let us give some basic definitions of subtation below :

(*a*) Unit substation—This substation consists primarily of one or more transformers which are mechanically and electrically connected to and coordinated in design with one or more switchgear or motor control assemblies or combination thereof.

(*b*) Articulated unit substation is a unit substation which consists of : (1) an increasing line section providing one or more increasing high voltage circuits, (2) a transformer section consisting one or more transformers, (3) an outgoing section providing the connection of one or more outgoing feeders, each of which provided with interrupting devices, (4) subassenblies manufactured outside for assembly and connection in the field.

(*c*) Primary unit substation is a unit substation which has a voltage rating of 1000 V or above, (*d*) secondary unit substation having the low voltage rating of less than 1000 V.

In the Fig. 14.2 an outline diagram of a mobile unit substation is shown.

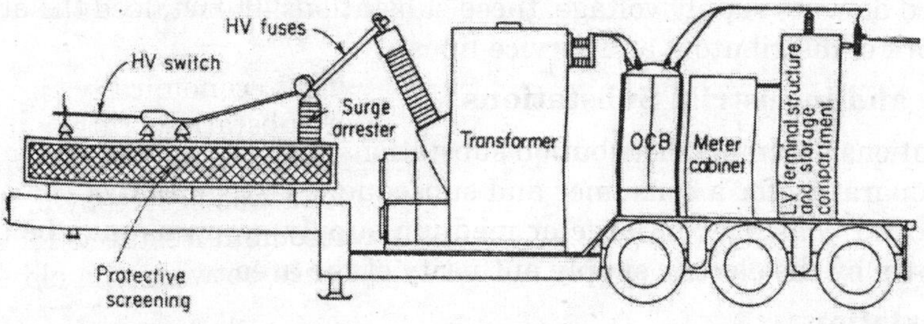

Fig. 14.2. Mobile substation. Primary protection provided by fuses while secondary protection is provided by oil circuit breakers with meter/relay cabinet.

(*d*) Secondary unit substation consists of (*a*) Incoming-line section the incoming-line section of secondary unit substation are generally rated at 15 KV. (*b*) Transformer section for secondary unit substation is available in four basic designs : (1) Oil immersed for outdoor use (or indoor vault). (2) Open-ventilated dry or cost dry or oil type for indoor or outdoor use. (3) Non-flammable-liquid-immersed for indoor or outdoor use. (4) Sealed dry (gas-filled) for indoor or outdoor use. It is to be noted here that as in the case of primary unit substations, the bushings utilised for secondary unit substation transformers are selected to have an insulation level equal or greater than the insulation load of the winding to which they are connected.

(*c*) Outgoing section of secondary unit substation is available in the following options : (1) Metal-enclosed switchgear with stationary or drawout low-voltage power circuit breakers, (2) Metal-enclosed switchgear or switchboard construction with moulded-case circuit breaker, (3) Metal-enclosed switch board construction with switch and fuse combination, (4) Motor-control-centre assemblies, (5) Combination of the above points.

14.2 VARIOUS OTHER TYPES OF SUBSTATIONS

We have discussed above in general the planning of substations and basic principle of same with discussion on mobile unit substation consisting of primary and secondary substation. Here we shall discuss various type of substations generally used in power system. We shall discuss on important points on civil and electrical work requirement in a substation including busbar arrangementsd. In fact, the substations in the power system can be classified depending on the purpose of the substation starting from the generating stations to the consumer's end.

1. Step-up of Substations at Generating Stations

Three types of substations are constructed at the power stations to step up the voltage generated by the generators at power houses, which are generally of 6 KV level stepped up by specially designed transformers for transmission of power at much higher value of 132 KV to 400 KV range through different stages. These substations thus enable transmission of huge blocks of power over long distances to the load centre's substations most economically.

2. Primary Grid Substations & Secondary Substations

The primary grid substations are created/constructed at suitable load centres along the primary transmission lines. In these substations the primary transmission voltage (for the power station) received through transmission line down to a suitable secondary voltages to be fed to the secondary substations. That means from these grid substations the secondary transmission lines are carried over to the secondary substations which are constructed near the actual load points, where the voltage is further stepped down to sub-transmission and distribution voltage for the consumer-domestic and Industrial-refer to Fig. 14.1.

3. Distribution Substations

These substations are constructed where the sub-transmission voltage/primary distribution voltage is stepped down to supply voltage, these substations, in fact, feed the actual consumers through a network of distributors and service lines.

4. Bulk Supply and Industrial Substations

These substations are infact distribution substations with the difference that the substation is constructed separately for a consumer and subsequent distribution of the electric power is left to the particular consumer of large or medium supply group as may be designed by the standard laid down by the electric supply authority of the area.

5. Mining Sub-station

These substations are of special types and equipped with special substation equipments meeting the special flame proof specifications. These substations need special design considerations to arrange extra precautions for safety needed in the operation of the electric supply. Substation equpments like transformers, switchgear and the panel have to be flame proof.

6. Mobile Substations

Refer to Fig. 14.2 showing a mobile substation which are of special design and for special use say, at open cast mine, on highways and places where a permanent substation is not required.

7. Pole-Mounting, Open or Kiosk Type

These substations are used mostly in the rural areas where the load density is low and the overhead lines are used. Generally for MV/LV transformers below 100 KVA it is possible to bolt the transformer onto a single pole or place it on a metal frame attached to the pole. For larger-sized units a platform supported by two poles is used to hold the transformer up. Typical arrangement are shown in Fig. 14.3.

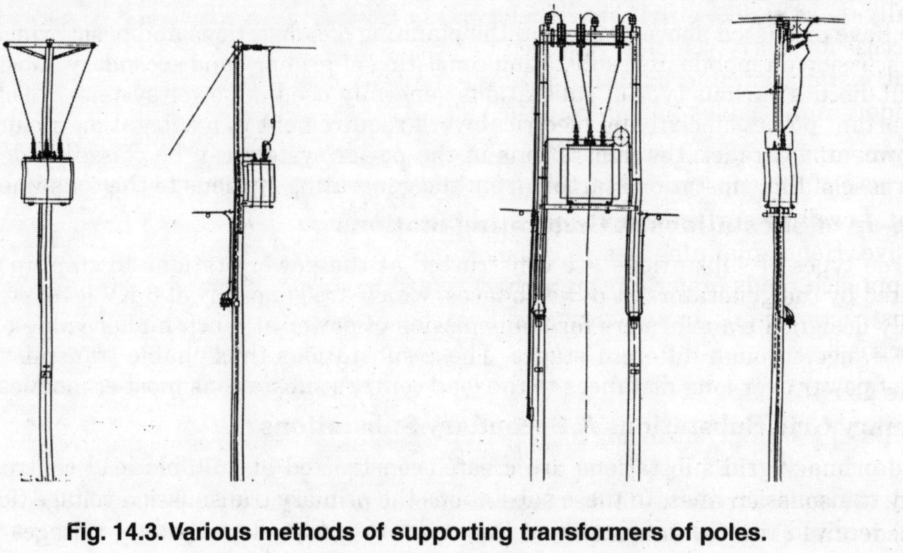

Fig. 14.3. Various methods of supporting transformers of poles.

For protection the use of individual MV fuses is not recommended since they frequently operate modes storm or transient fault conditions. One 3-phase disconnector/isolator (drop-out-type) can cover up a group small pole-mounted substations connected to a section of the MV network. Larger transformers are equipped with individual disconnectors as shown in Fig. 14.3*b*. On the low voltage side each feeder is usually equipped with its own protective fuses, or mini circuit breakers are also used in place of fuses in many places nowadays.

For overvoltage protection spark gaps are in common use on the MV side. A prequisite for this arrangement is that the transformer-winding construction is not oversensitive to the rapid collapse of single voltages. Depending on the average level of lightning strikes in the area, it may be economic to protect the larger rented transformers say over 200 KVA, by lightning arrestors.

Pole mounted mine substations also provide an economic arrangement in areas where MV overhead lines (in rural areas) and underground cables (in urban areas also) are used.

8. Outdoor Substations

Considering cost and safety and operation it is common to have outdoor substation for 33 KV voltage level. These substations distribute electric power both at 11 KV or 6.6 KV (for industrial purpose) and 415V 3-phase (and then to 230 V-1-phase supply for domestic purpose. These substations are equipped with transformer, isolators, switchgear panels with relay control panels and circuit breakers and busbar arrangement. Proper earthing and lightning arrestors are also arranged in these substations for safety.

9. Indoor Type Substations

These substations are created at large consumers premises.

14.3 CONSTRUCTION OF SUBSTATION

A brief discussion is made here an construction of substation to have some idea on the means are requirement for constructing outdoor substations.

In fact, a modern substation is a complex structure as it requires numerous item of equipment and allied services. First step for constructing a substation is to make civil and electrical survey of the area thoroughly to choose the site. The ideal site for a substation is naturally the centre of gravity of the electrical load but in practice the cost and availability of land access for heavy vehicles (required to bring transformer, switchgear etc.), considerations of plant installation and handling future expansion of the substation, laying overhead lines and underground cables to and from the substation etc. are important.

However, where land is cheap the need for economy in area is not always appreciated. In fact, it is his only the cost of land but also that of its treatment which must be taken into account. Levelling surfacing and enclosing the land are usually both time consuming and also expensive. Hence planning has to consider both time and cost. Substations obviously has to have approach roads preferably on highways. After arranging the planned land for substation, the construction work for substation is divided primarily into two groups : (*a*) Civil work and (*b*) Electrical works.

(*a*) The Civil Works

(1) Land levelling, surfacing as generally most of the substations are built on farm lands or low lands near the highway or main roads in rural areas. Fencing around the land to be created. (2) Roads, trenches to be made, (3) Building for, (*a*) office, stores, control rooms etc., (*b*) resedential flats for substation staffs. (4) Construction of foundations for switchyards,

Transformers, (5) Cable trenches, (6) If large substation constructed near the rail stations then Railway siding and railway tracks are to be constructed. (7) Water supply system, (8) Drainage and Sewage system, (9) Gardening.

(*b*) Electrical Works

1. Earthing, cabling, oil handling system, illumination system, compressed air system, arranging 1-ph. 230 V and 3-ph. 415 V line for erection work, d.c. auxiliary supply, telecommunication system.
2. Choice of busbar arrangement, preparation of key diagrams and layout showing proposed bus bar arrangements, the relative positions of various equipments as per the layout giving distance between various equipments, phase segregation distances of both horizontals and vertical arrangements as per voltage grade and as per standard specification.
3. Selection of Isolators, circuit Breakers, lightning arrestors, power transformers, Instrument Transformers, protective relaying schemes, control and relay boards, voltage regulating equipment, protection against voltage surges.

From the above brief discussions on construction of substation it is noted that design of both civil construction and electrical system involves mainly the selection of proper location, proper civil work and proper electrical equipments. This obviously requires a fair knowledge and understanding of the principle, operation of working and the types of the various equipments as per the planning and design of electricity distribution networks.

14.4 PLANNING, OBJECTIVE & PHILOSOPHY OF SUBSTATION DISTRIBUTION NETWORKS

In industrialised countries the existing supply system covers virtually all the inhabited areas. In these areas the existing network usually provides a good starting point for planning future system arrangements. The need for further investment is usually to cater for load growth, or to replace ageing assets on the network. Uprating of existing overhead lines and other plant where technically feasible can often be more economic than installing new circuits or equipment when costs of obtaining new line routes or new substation sites are taken into account. However, the longer-term cost of developing the network must also be examined. However, as the line passes these objectives may be modified. In addition, it will be necessary to consider the effect of future changes in such factors as the growth of load in different areas, variations in the relative levels of the costs of materials and energy, as well as the increasing use of technical innovation.

The planning activity often includes not only determining major future network requirements, *e.g.*, additional primary substations, but also producing planning guidelines.

Infact the objective of planning of distribute network a large variety of regulations, guidelines and recommendations are produced to ensure that the electricity supply system can be properly operated for the benefit of society. In most countries regulation concerning safety, the supply arrangement and the proper use of electricity from part of governmental legislation. In fact planning objectives concerning the quality of supply, tariff price levels and stable employment are in common use in industrialised countries.

14.5 SUBSTATION NETWORK

14.5.1 Layout Drawing for the Substation Network

Layout drawings are made to show the actual physical position of each equipment. Complete work of constructing a substation is based on this layout drawing which gives the complete picture of the substation including indicating the boundaries of the land. This layout is prepared based on the initial layout prepared on the land plan where space is marked for civil works and electrical works and also the roads and paths for movement of men and material. The electrical work starts generally after all the major civil work has been completed as per the final layout drawing giving the physical position of the electrical equipment for which some civil works will be required during erection of electrical equipment including the outdoor bus bar arrangements to supports etc.

In the final layout for electrical works a detailed electrical layout is prepared indicating the placing of various equipment in the switchyard and their connections with each equipment starting from the High Voltage (or EHV) incoming to the outgoing MV (or LV). The number and types of columns and beans to support the outdoor busbars inward/outward connection will obvious depend on the system voltage and the bus-bar arrangement selected. However, there are very many numbers of bus bar arrangements, some of which generally used are discussed briefly below in the next section. For proper earthing bars obviously the civil work's help is to be taken at the time of preparing and levelling substations switchyeard ground as per the design instruction generally used internationally. However, on the detailed electrical layout overhead ground wire, if decided to be used for protection against the direct lightning strokes, is marked. This hence, determines the position and number of columns which require peaks for fixing or binding the ground wire rigidly.

Besides the main columns and beams, supporting structures are required for mounting individual items such as lightning arrestors, voltage transformers, current transformers, Isolators, Circuit Breakers, post insulators. The design of such very important structures can be finalised with the type of equipments to be placed on then with the equipments drawings giving the overall discussions and fixing details. This is a very detail design work in the design stage. The supporting structures column and beam are generally of steel (with antirust coating and weather-proof anti-corrosive point) or even reinforced coment conorete. The selection is made at the design stage for considering the cost, availability of times and quality. R.C.C. structures are cheaper than the steel structure due to high price rise of the steel structures. Hence designers are inclined to use the RCC structures, of course without compromising with the quality. Main problem of using RCC structures is the quality of the concrete cost, as due to lowest bidding many manufactures of these RCC structures do not give attention to the mechanical strength of these structures (to ensure their profile) by using either low quality cement or less quantity cement. This is a point all electrical authorities must not over look at any cost, as any breakage of structures in the switch yeard may dangerously reduce the gap between the lines and ground causing serious fault in the substation. Below we shall briefly discuss some of the most commonly used Bus bar arrangement in the substations.

14.5.2 Bus Bar Arrangement

In a substation there are two types of Bus bar : (1) Outdoor and (2) Indoor, outdoor bus bar arrangement is meant for switch yeard where all the electrical equipments as stated above, are erected as per substation layout. Indoor bus bar arrangements are generally used in the switchgear panels having 11 KV and also in the *LT* panel boards. Let us first look into the outdoor bus bar arrangement in the switchyard which are of various types for various applications :

In fact the outdoor bus bars are either of (*a*) the rigid or of (*b*) the strain type.

(*a*) Rigid Type Bus Bars

In the rigid type, pipes are used for bus-bars and also for making connections among the various equipments wherever required. The bus bars and the connections are supported on pedestal insulators. This obviously results to a low type of switchyard wherein equipments as well as the bus bars are spread out thus requiring a large space. As the bus-bars are rigid the clearances remain constant. Moreover, in this type of arrangement, the maintenance job is pretty easy, since the bus-bars and connections are not very high from the ground considering the required insulation levels of the operating voltage in the switchyard. Also due to large diameter of the pipes the corona loss is substatially less.

However, due to large area for the outside switchyard in the substation, this type of bus bars, despite all the advantages, is comparitively costlier. But it must be stated that despite being costlier, rigid type outdoor bus-bars are more reliable than strain type.

(*b*) Strain Type Bus Bars

As for the strain-type bus-bars, this type is a overhead system of wires between two supporting structures and supported by strain type insulators. The stringing tension may be limited to 500 – 900 kg depending upon the size of the conductor used. This arrangement is more economical than the rigid bus bar arrangement. However, in case of shortage of aluminium pipes in the country only strain type bus bars are recommended.

(*c*) Rigid-type Bus Bar Materials

Generally aluminium pipes are used. The sizes of pipes commonly used for various voltages are given in the Table 14.1. However, as aluminium pipes being exposed to all kind of weathers aluminium oxides are developed in these pipes hence periodical attention must be given in the connections particularly. In case of long spans expansion joints should be provided to avoid strain on the supporting insulators due to thermal expansion or contraction of the pipe.

Table 14.1

Voltage Level	*Diameter of Aluminium Bus bar Pipes*
33 kV	40 mm.
66 kV	65 mm.
132 kV	80 mm.
220 kV	80 mm.
400 kV	100 mm.

(*d*) Strain-type Bus Bar Material

In the strain type bus bars materials used for bus bars and connections of the strain type are ACSR and all are aluminium conductors. Bundle conductors are used where high ratings for bus-bars are required. The sizes of the corresponding ACSR conductors are given in the Table 14.2 for various voltage levels.

Table 14.2

Voltage Level	*Diameter of Strain Type Bus Bar Conductors*	*Type of Conductors*
66 kV	37/2.79 mm.	ACSR
132 kV	37/4.27 mm.	ACSR
220 kV	61/3.99 mm.	ACSR
400 kV	61/4.27 mm.	ACSR Duplex

14.6 BUS BAR ARRANGEMENTS

As stated above there one various types of bus-bar arrangements The choice of a particular arrangement depends on many factors like system voltage, position of the substation in the system reliability, flexibility and the cost, and the stability of supply. Moreover following technical points must be considered while deciding any bus-bar arrangement.

1. Simplicity of the system. It must be remembered that the simplest design is the best design, and economical also.
2. Arrangement should be ensured that periodical (say every two or even, one week in some cases) maintenance should be possible without interruption of power supply on any possible danger to the maintenance and operational personnel.
3. Every piece of equipment is a source of possible danger due to various reasons like : design, manufacturing defect, erection defect etc. Therefore, spare equipment has to be kept ready in stock to meet the contingencies to avoid any disruption of supply.
4. The installation should be as economical as possible considering the requirement and continuity of supply of power.
5. The layout should be such that there should be all necessary means available in the event of expansion and/or augmentation of bus-bars.

14.7 COMMON ARRANGEMENT OF BUS-BARS

There are commonly used bus bar arrangements as stated below :

(*a*) Single Bus-bar Arrangement

This single bus-bar arrangement is the simplest and cheapest. In the Fig. 14.4 this arrangement is shown in a line diagram. However, this arrangement is subjected to two major defects : (*a*) maintenance of the line is not possible without interruption of supply, (*b*) without shut-down, extension is not possible. But despite these two disadvantages, in this arrangement, the equipment connections are very simple and operation of the system is very convenient.

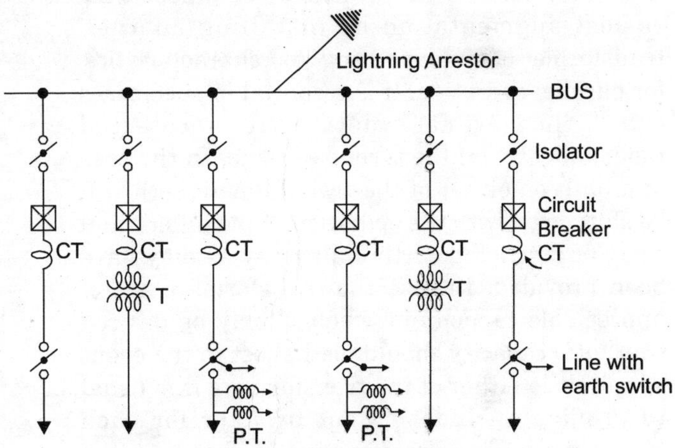

Fig. 14.4. Single bus bar arrangement.

Generally the indoor 1 kV switch boards have quite often single bus-bar arrangement. However, as it is, the scheme is not very popular for system voltages of 33 kV and above except where the relative importance of the sub-station is less or the position of the substation does not justify elaborate schemes. In such case it is used in slightly modified form as discussed below.

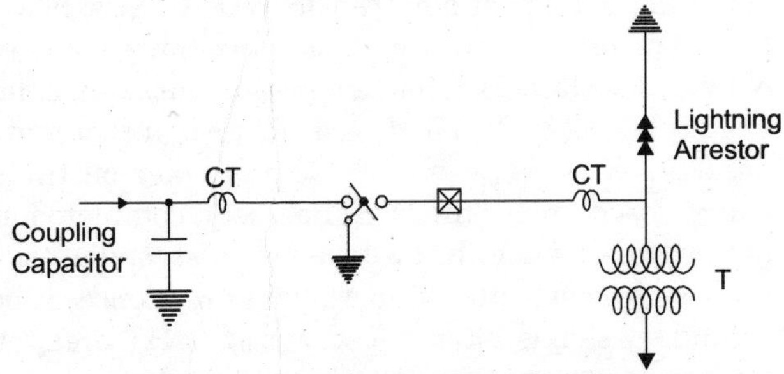

Fig. 14.5. Terminal station arrangement

In the Fig. 14.5, an arrangement is shown for a terminal station with a single incoming line and a single power transformer. Power fed is on one side only and no line protection is necessary. The potential transformers or capacitive voltage transformers are therefore not necessary for protection. However, a voltage indication may be required for which one single phase potential transformer or capacitive voltage transformer would do. Capacitive voltage transformers may be preferable in case carrier communication is desired. If the voltage indication on the secondary side is considered adequate, which should be the case under normal circumstances, only a coupling capacitor for carrier communication would do. The wave trap is, in fact, required only when it is desired to establish a carrier communication link. Considering lightning, this is a dangerous exposition and every wave reaching the transformer will reflect with about double the magnitude. Hence for protection against lightning, lightning arrestors should be located as close to the transformer as possible. A slight modification to this arrangement is shown in Fig. 14.6. In this arrangement power transformers are installed. It is intended to have a common breaker for the control of both the transformers as shown in the Fig. 14.6. Two additional isolators are needed, one with each transformer for easy manovarity. In developing stages conversion of the arrangement in Fig. 14.5 may be required to be changed to the arrangement in Fig. 14.6 quite often as the load develops. This is the quickest, cheapest and easiest augmentation by installing another transformer of similar rating and characteristics for parallel operation. It is essential in such cases that the transformers are protected independently on the secondary side. In the case of a fault on either of the transformers both will be shut down with the operation of the differential relay or restricted earth fault relay as may have been provided. In case the transformers are of appreciable capacity or when it may be desired that total capacity should not be lost in the event of a fault in either of the transformers it is usual to provide an independent breaker for each transformer. This arrangement is shown in the Fig. 14.7. The provision of transformer isolator between the breaker and transformer is advisable for extra precaution. Some scheme however does not provide it. Cost can be reduced by eliminating these isolators on bus schemes other than the mesh scheme but adequate measures should be taken to avoid back feed. If maintenance work has to be done on one of the HT breakers it is necessary that the breaker on the secondary side is open and is incapable of being closed unless the work has been completed and HT breaker or isolator closed. However, in particular, when the switchgear on secondary side is installed indoors and is out of visibility the risk becomes all the more general. Hence, it becomes necessary that the breakers on the secondary side is inter locked with the HT breaker or bus isolator on the HT side so that the former cannot be closed unless the latter is closed. The HT breaker is primarily for the protection of the transformer. It is a normal feature, therefore, that if the HT breaker trips under instructions from the differential or restricted earth fault relay it trips the breaker on the secondary side as well.

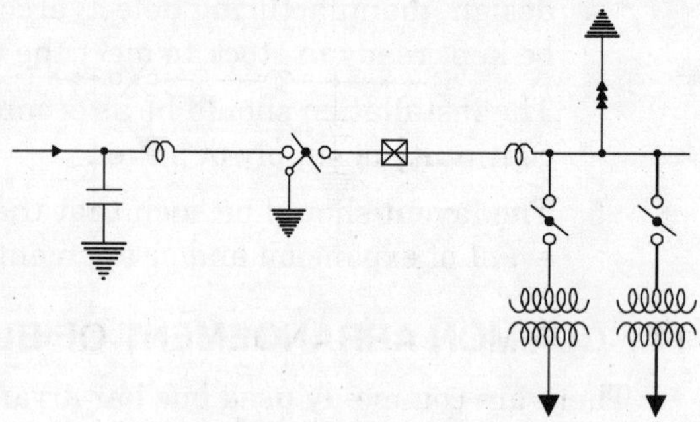

Fig. 14.6. Modified terminal station arrangement.

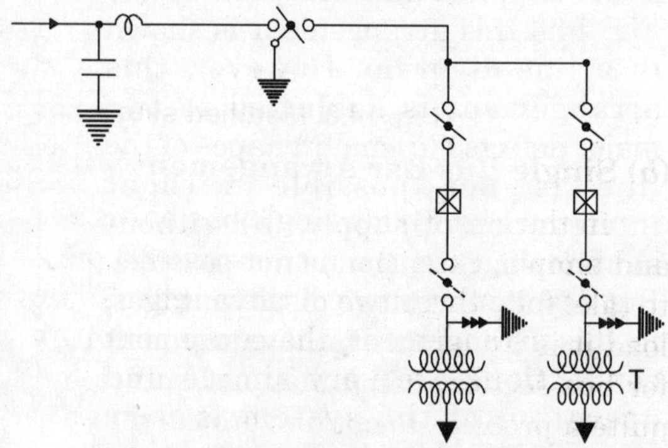

Fig. 14.7. Further modified terminal station scheme.

In the Fig. 14.8, we see a modification of the single bus arrangement, when the line circuits are limited to two. A bypass, normally open isolator, is provided to continue the supply if the line breaker has to be pulled out for maintenance. As is quite obvious the scheme envisages use of two additional isolators. Then extension to the scheme for another incoming or outgoing circuit will present difficulties. For location of lightning arrestors the same considerations as for the terminal substation discussed above is required. However, under normal working conditions when both the incoming and outgoing circuits are in the reflected wave then it will not have an amplitude greater than that of the incident wave. However, it is quite possible that under stormy conditions when chances of lightning surges are greater, one of the circuits may be out in which case the arrangement will act exactly like a transformer terminal station.

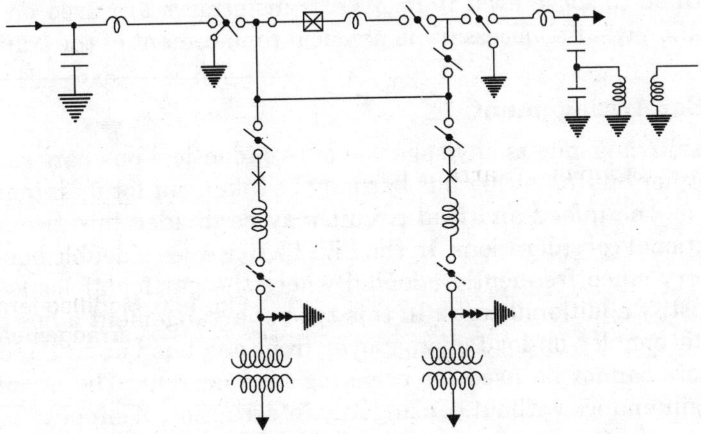

Fig. 14.8. Modified single bus arrangement through station scheme.

(*b*) Single Bus Bar Arrangement with Bus Sectionaliser

In the Fig. 14.9, a single bus bar is adopted with sectionalising arrangement for cheapness and simplicity. If the increasing and outgoing circuits are distributed evenly on the sections to take full advantage of these scheme. However, where double feed is provided for any single load it is preferable to have one circuit from each section of the bus bar. The isolators alone for bus sectionalising may not suffice for they have to be operated off circuit and it may be quite a problem to obtain zero current across the section. For fuller benefits a circuit breaker should be used. This will enable bus differential protection to trip only the faulty section and will keep the healthy section running. Another interesting and useful application of this arrangement is found where the two winding transformers are used. Say two transformers 66 KV/11 KV each of 4 MVA are used. 11 KV from each is brought on either section of the bus.

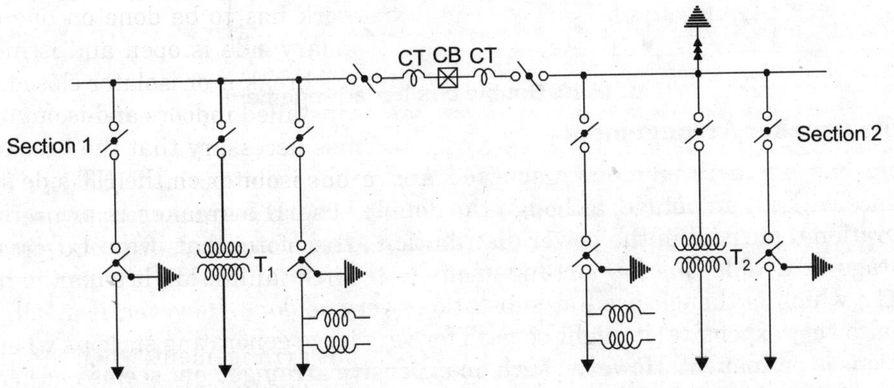

Fig. 14.9. Single bus bar arrangement with sectionaliser.

The bus sectionaliser and the two incoming breakers are so inter looked that only two breakers can be closed at a time. Normally each section continues to it own load. If the incoming circuit on any section has to be taken out of the sectionaliser, breaker can be closed and supply restored to the outgoing circuits of the section. The gears installed generally have a normal rating of 800 amps and rupturing capacity of 150 MVA only.

Now, say, if we assume a normal percentage impedance of 10%, the entire gear would need replacement if the load develops and transformers have to be replaced with 7.5 MVA or larger rating with the inter looking arrangement between incoming and bus sectionaliser breaker the gear can continue to serve even if 15 MVA transformers are used on either section. In such an augmentation what is necessary is physical replacement of the transformers with the bigger ones.

(*c*) Double Bus Bar Arrangement

Double bus bar arrangements envisage use of two identical bus bars so that (*a*) each load may be fed from either bus (*b*), either bus bar may be taken out for maintenance and cleaning of insulators and (*c*) the infeed and load circuit may be divided into two separate groups if needed from operational considerations. In the Fig. 14.10 we see a double bus bar arrnagement, which has been very much frequently adopted where the particular loads and continuity of supply of power justify additional costs. In this type of arrangement a bus coupler breaker is to be provided as it enables on-load change over from one bus bar to the other. The normal bus section isolators cannot be used for breaking load current. The arrangement does not permit breaker maintenance without causing any interruption of supply.

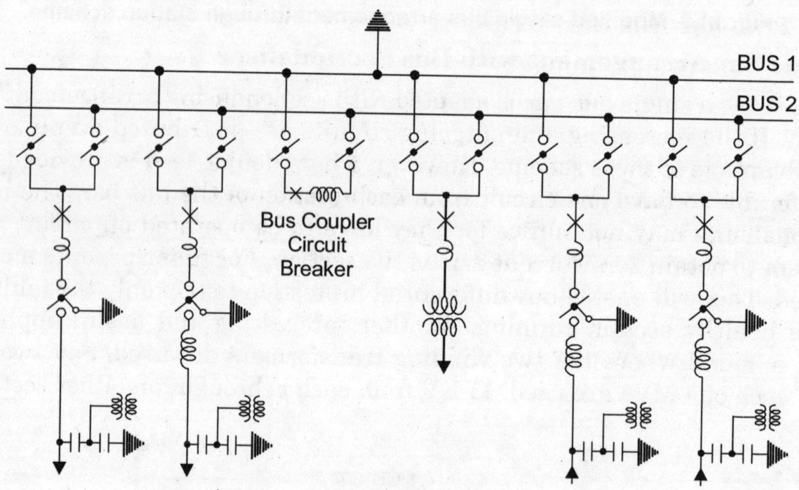

Fig. 14.10. Double bus bar arrangement.

(*d*) Double Breaker Arrangement

Double bus bar arrangement discussed above does not persuit breaker or isolator maintenance without an outage, although the double bus bar arrangement is useful for most of the operational purpose in the power distribution system. A simple way of overcoming this disadvantage of double bus bar arrangement is the use of double breaker arrangement (Fig. 14.11), which is though used at substations very seldom. However, if at all, it is used (which is also very expensive) it might be used for very large generating stations where security of connections is permanent. However, such an expensive arramgement seems neither desirable nor-necessary, since any well-planned and developed system when the number of power station and inter-connections have increased outage of any power station and much less an individual circuit causes negligible effect on the supply position anywhere.

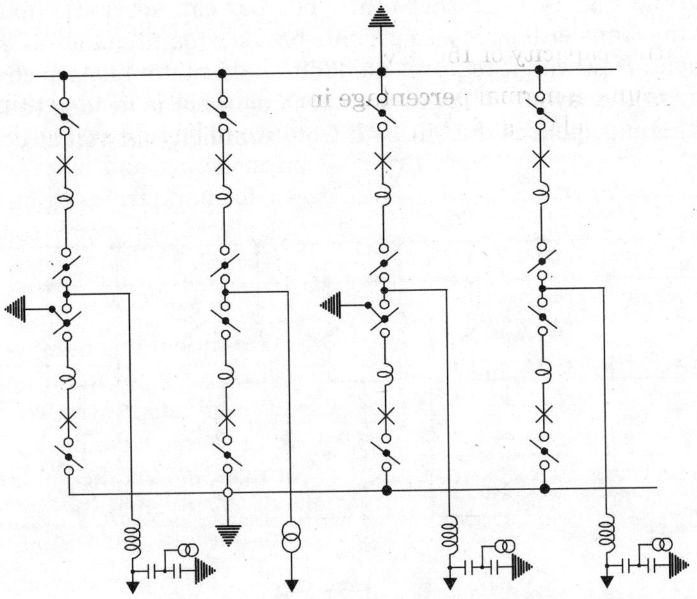

Fig. 4.11. Double breaker arrangement.

(e) Main and Transfer Bus Bar Arrangement

This arrangement is in fact an alternative to double-bus bar arrangement. The Fig. 14.12, shows this alternative arrangement of double bas bar arrangement discussed above in (c) above where it provides facility to change over to either bus to carry out maintenance on the other but provides no facility to carry out breaker maintenance. But the main and the transfer bus works the other way round. Infact it provides facility for carrying out breaker maintenance but does not permit bus maintenance. Whenever maintenance is required on any breaker the circuit is changed over to the transfer bus and controlled through bus coupler breaker. As would be seen in Fig. 14.12, which envisages use of an additional isolator on each circuit and this adds to the cost. However, as the connection to the transfer bus is through isolators careful intere-locking is very much important and necessary with the bus coupler.

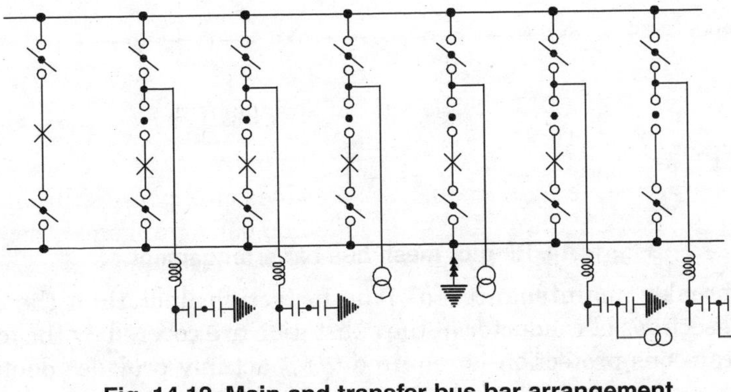

Fig. 14.12. Main and transfer bus bar arrangement.

(f) Double Bus Bar with bypass Isolators

This arrangement is a combination of the double bus and main and transfer bus schemes which is shown in the Fig. 14.13. Either of the bus bar can act as the main bus, the second bus being used as the transfer bus. It also permits breaker maintenance as envisaged by double breaker arrangement. However, it requires an additional isolator on each circuit and introduces slight complication of the layout. Whenever something more is needed than a double bus bar arrangement this arrangement is the but both from simplicity as well as economy.

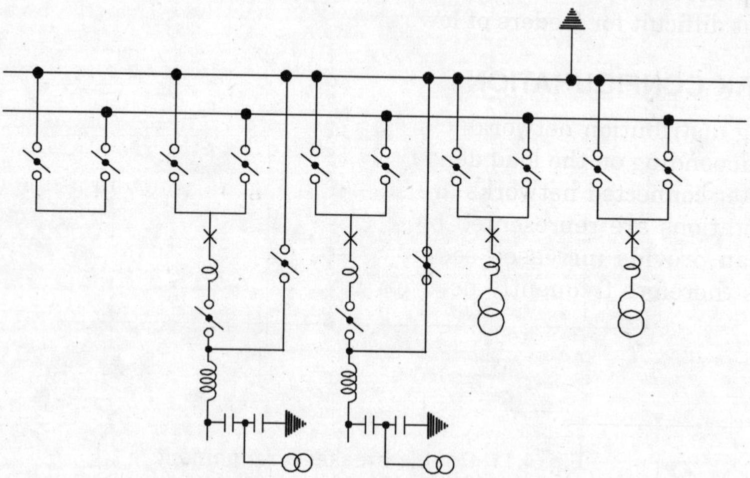

Fig. 14.13. Double bus bar arrangement with bypass isolators.

(g) Ring Bus Bar or Mesh Bus Bar Arrangement

This bus bar arrangement is also known as Ring Bus or Ring main arrangement. The Fig. 14.14 shows one such bus bar arrangement, which has following good and useful features :

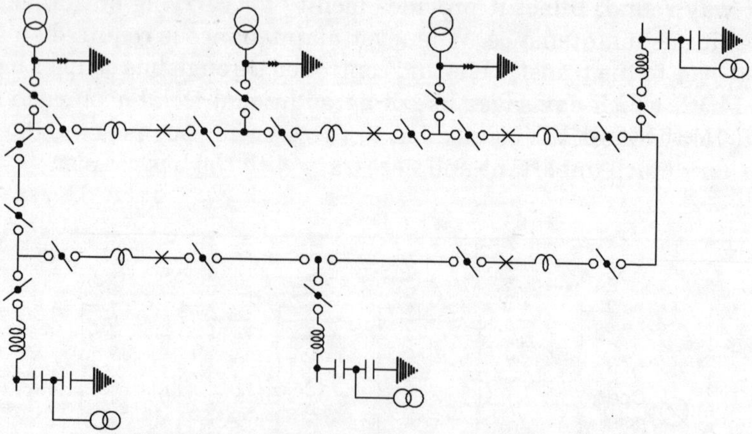

Fig. 14.14. Ring or mesh bus bar arrangement.

(*a*) It permits breaker maintenance. (*b*) It is in fact cheaper than the double bus bar arrangement, (*c*) All sections of conductor in the substation are covered by the feeder protection and hence no separate bus protection is required (*d*) it actualy provides double feed to each circuit and as such the opening of one breaker under maintenance or otherwise does not interrupt the power supply any circuit.

However this arrangement has certain operational and safety deficiencies and hence is not suitable particularly in developing systems. In fact, as long as the mesh or ring is closed each circuit has duplicate feed. Any breaker can be taken out for maintenance or inspection or change. But if a circuit trips under these circumstances supply to a number of circuits in the substation supply area may be lost. To measure the current passing to an individual feeder, it is necessary to compare the current from two sets of current transformers which must be capable of carrying the current passing round the mesh or ring while the feeders current represented by differential current may be small. This phenomenon makes a getting accurate values of current difficult for feeders of low rating.

14.8 NETWORK CONFIGURATION

In designing distribution networks can be provided to different areas of the system in a variety of way depending on the load density and system voltage level. Various arrangements of mesh and inter connected networks are shown in Fig. 14.15 (*a*) in a line diagram. Where individual substations are represented by circles. The interconnecting circuits of the mesh arrangement can provide increased security of supplies to individual substation and this arrangement is therefore frequently used in HV systems. While this arrangement requires

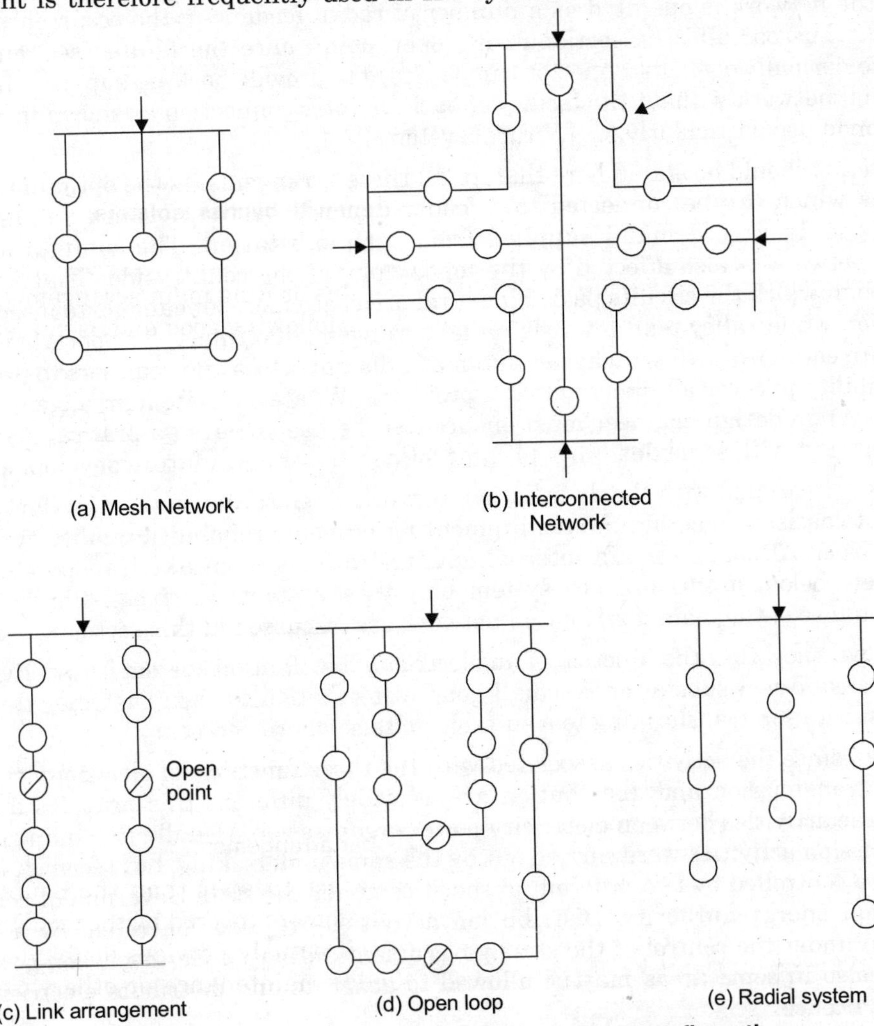

Fig. 14.15. Various types of distribution network configuration.

more substation equipment overall, for example switchgear and electrical connection, it is usually more efficient in terms of total circuit costs. The mesh arrangement is easier to extend and has a higher utilisation of circuits when fully developed than systems may be operated split with normally open points, or closed to improve security, although the latter require more circuit breakers and more sophisticated protection.

In fact, by inter connecting and operating a number of infeed substation in parallel as shown in Fig. 14.15*b*, it is possible to reduce the total transformer capacity into the group. Subject to satisfactory network circuit loadings such an arrangement can accept the loss of one infeed without interruptions of supplies within the network. Care must be taken that the fault levels within the network are acceptable. Voltage levels and reactive power flows throughout the network may be a problem on extended interconnected systems and parallel operation of the infeed point can result in reverse power flows the infeed transformers under outage conditions on the higher-voltage system. A similar interconnection between infeed point can be provided by the link arrangement as shown in Fig. 14.15*c*. However, by opening interconnectors as indicated the system can operate as radial feeders with closures of the open point restoring supplies if one of the infeed substation is out of service. A common option at MV and LV is the open loop arrangement shown in Fig. 14.15*d*. Under normal operating condition the network is operated as a number of radial feeders on the occurrence of a fault between the bus bar and the sectionalising open point, once the faulted section has been isolated the normally open disconnector can be closed to provide back-up supplies. In addition, purely radial network without the facility of back-up inter connection as shown in Fig. 14.15*e* are in common use particularly for LV rural systems.

However, it should be stated here that, in all these arrangements the optimum number of substations which can be connected to a feeder depends on the network voltage and the distances and average demand supplied from each substation. The configuration of an individual network is also affected by the topography of the countryside. That is the rural areas, through which the circuits pass. Flat rural areas are more ameanable to interconnected configuration while valley tend towards radial arrangements. Infact, for operational reasons networks are equipped with switchgear, automatic disconnectors and realesers to provide day-to-day flexibility to cover all likely network problems. Whatever system arrangement is used the power system design engineer must ensure that he has achieved a pratical arrangement at the lowest cost with some flexibility to cater for as yet unknown future developments.

However, associated with the distribution networks themselves, various auxiliary systems are needed to assist in meeting the requirement for economy reliability, quality and safety of supply of power. Although certain internal administrative system like transport specialised machinery etc. belong in this auxiliary system, only, these systems which assist in the operation and maintenance of the main distribution network are discussed in this section.

In fact, we know that the function of an electricity distribution system is to deliver energy from the transmission substation or small generating station to each customer through the secondary substation transforming to a suitable voltage where necessary.

However, since the activities associated with the three functions of electricity supply, *i.e.*, generation transmission and distribution are as widely different this provides a basis for splitting these activities between electricity power organisation. Usually the main generation and transmission activities were carried out by the same undertaking, but recently, these two activities are controlled by two different authorities under the state Government control and on a national energy authority. Distribution activities are also controlled by a separate organisation under the control of the state government authority. However, non-government authorities also in some areas may be allowed to generate and distribute electricity under government licence.

14.9 SYMBOLS FOR CIRCUIT ELEMENTS/EQUIPMENT IN SUB-STATIONS

Symbols of important circuit elements/equipment in sub-station are given below in the Fig. 14.16

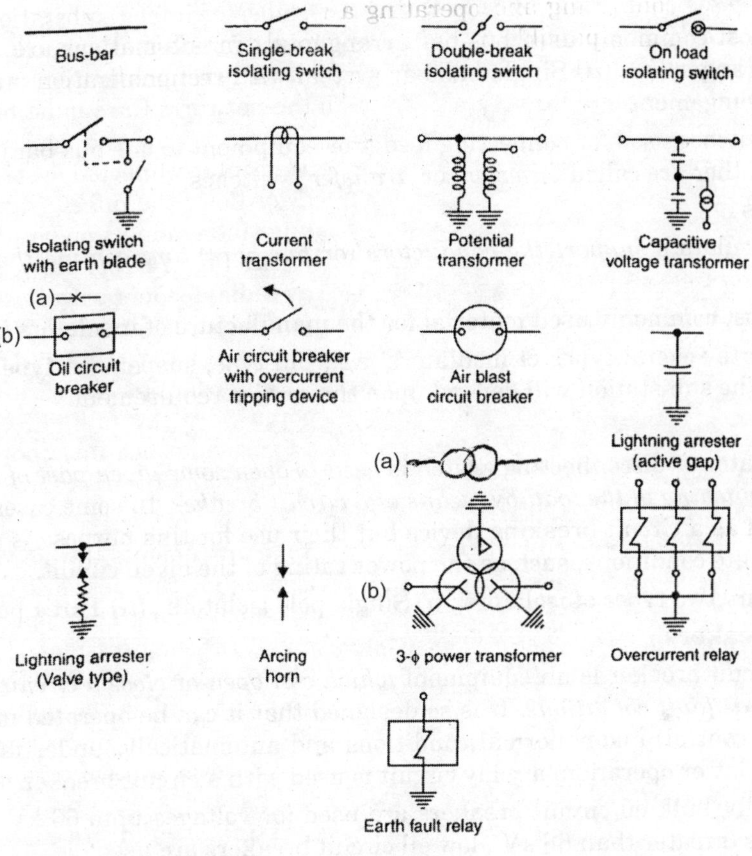

Fig. 14.16. Symbols of important circuit elements/equipment in sub-station.

14.10 EQUIPMENT FOR SUB-STATIONS AND SWITCHGEAR INSTALLATIONS

The main equipment required for sub-stations and switchgear installation is given below :

1. Main bus-bars
2. Insulators
3. Isolators
4. Circuit breakers
5. Load-interrupter switches
6. Fuses
7. Power transformers
8. Current and potential transformers
9. Indicating and metering instruments
10. Protective relays
11. Carrier-current equipment
12. Control cables.

1. **Main bus-bars**
   - When several lines operating at the same voltage have to be directly connected electrically, bus-bars are used as the common electrical component. Bus-bars are copper or aluminium bars, generally of rectangular cross-section, and operate at constant voltage. The incoming and outgoing lines in a sub-station are connected to the bus-bars.
   - The most commonly used bus-bar arrangement in sub-stations are : (*i*) Single bus-bar arrangement; (*ii*) Single bus-bar system with sectionalisation ; (*iii*) Double bus-bar arrangement.
   - The switches used for connecting feeder or equipment to one bus-bar (or bus in short) or the other are called *"selector"* or *"transfer"* switches.
2. **Insulators**
   - The insulators *support the conductors (or bus-bars) and confine the current to the conductors.*
   - The most commonly used material for the manufacture of insulators *is porcelain.*
   - There are several types of insulators (*e.g.,* pin type, suspension type etc.) and their use in the sub-station will depend upon the service requirement.
3. **Isolators**
   - An isolator or disconnecting switch is *used to open some given part of a power circuit after switching of the load by means of a circuit breaker.* In some cases isolators may be used as a circuit breaking device but their use for this purpose is strictly limited by definite conditions, such as the power rating of the given circuit.
   - There are two types of isolators : (*i*) Single-pole isolators ; (*ii*) Three-pole isolators.
4. **Circuit breakers**
   - The circuit breaker is an equipment *which can open* or *close a circuit under normal as well as fault conditions.* It is so designed that it can be operated manually (or by remote control) under normal conditions and automatically under fault conditions. For the latter operation, a relay circuit is used with a circuit breaker.
   - Generally, bulk oil circuit breakers are used for voltages upto 66 kV while for high voltages (greater than 66 kV), low oil circuit breakers are used.
5. **Load-interrupter switches**
   - The switches are designed and used to close and open high voltage circuits under normal working conditions (at normal load). The arc extinguishing device of the load interrupter is made in the form of a split, moulded plastic chute fitted with organic glass inserts. This chute surrounds the moving knife of the arc extinguishing system. The stationary acting contact is located in the lower part of the chute.

     When the switch is opened, the working contacts between the arc are drawn separate. Acted upon by the high temperature of arc, the walls of the organic material inserts generate gases (mainly hydrogen), which create a longitudinal blast serving to extinguish the arc.
   - For closing and opening the load interrupter switches lever-arm *manually operating mechnisms* are employed.
6. **Fuse :** Fuses have already been discussed in the earlier part of this book.
7. **Power transformers**
   - Those transformers are used for stepping-up the voltage for transmissidrl at generating

stations and for stepping-down voltage for further distribution at main step-down transformer sub-stations. Usually naturally cooled, oil immersed, known as 'ON' type, two winding, 3-ϕ transformers, are used upto the rating of 10 MVA. The transformers of rating higher than 10 MVA, are usually air blast cooled. For very high rating, the forced oil, water cooling and air blast cooling may be used.

- For regulating the voltage the transformers used are provided with on load tap changer.
- The transformers are generally installed upon lengths of rails fixed on concrete slabs having foundation 1 m to 1.5 m deep.

8. Current and potential transformers (Instrument transformers)

Current transformer (C.T.)

- It is essentially a step-up transformer which steps down the current to a known ratio. The primary of this transformer consists of one or more turns of thick wire connected in series with the line. The secondary consists of a large number of turns of fine wire and provides for the measuring instruments and relays a current which is a constant fraction of the current in the line.
- The current transformers are *rated* for rated voltage of the installation, the rated currents of the primary and secondary windings and the accuracy class (the accuracy class indicates the 1 mit of the error in the percentage of the rated turn-ratio of the given current transformers).

Potential transformer (P.T.):

- It is essentially a step-down transformer and steps down the voltage to a known ratio. The primary of this transformer consists of a large number of turns of fine wire connected across the line. The secondary winding consists of a few turns and provides for measuring instruments and relays a voltage which is a known fraction of the line voltage.
- These instruments make the ordinary low voltage instruments suitable for measurement of high voltage and isolate them for high voltage.
- The potential transformers are rated for primary and secondary rated voltage, accuracy class, number of phases, system of cooling.

9. Indicating and metering instruments : A large number of indicating and metering instruments struments (*e.g.*, ammeters, voltmeters, energy meters etc.) are installed in a sub-station to maintain watch over the circuit quantities. The instrument transformers are invariably used with them for satisfactory operation.

10. Protective relays : These have already been discussed earlier.

11. Carrier-current equipment : Such equipment is installed in the sub-stations for communication, relaying, telemetering or for supervisory control. This equipment is suitably mounted in a room known as *carrier room* and connected to the high voltage power circuit.

12 Control cables : For affecting automatic controls, the control cables and conduit system are required. The control system generally operates at 110 V or 220 V and the cables employed for this purpose are multi-core cables having 10 or 37 or 61 conductors according to requirement. In order to lay these cables generally ducts are run from control room basement to centrally located junction box from where the conduits are run to the required points.

14.11 KEY DIAGRAM OF 11 KV/400 V INDOOR SUB-STATION

The key diagram of 11 kV/400 V indoor station is shown in Fig. 14.17 which is self explanatory.

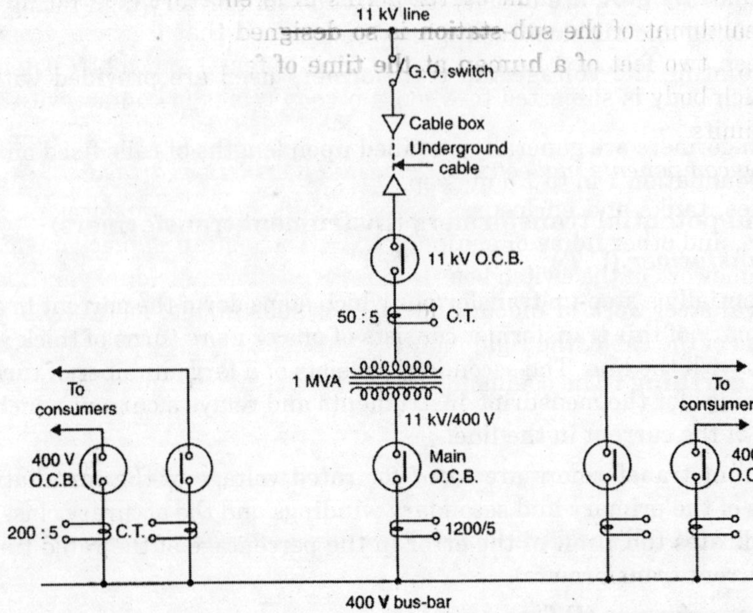

Fig. 14.17. Key diagram of 11 kV/400 V indoor station.

14.12 EARTHING

"Earthing" or **"grounding"** is *the connecting of an electrical equipment or apparatus to the earth with the help of a connecting wire of negligible resistance.*

If a metallic part of an electric appliance in an electric installation comes in direct contact with a base or live wire, the metal being a good conductor of electricity is charged and static charge on it will accumulate. Now if any person comes in contact with this charged metal part, he will get a severe shock. But if the metallic parts of the equipment/apparatus are earthed, the charge will be given to earth immediately as the metallic part comes in direct contact with a bare or live wire or breakdown occurs. As discharge takes place to the earth, the impedance of the path of the current being low, a heavy current flows to the earth and the protective device (usually a fuse) operates (the instant the current exceeds the limiting value) and isolates the equipment/apparatus from the supply. Thus it is very essential to provide earthing to ensure safety to the users as well as to the operating and maintenance personnel.

The earthing can be categorised in the following two parts way :

1. System earthing.
2. Safety earthing.

1. System earthing : It is required to provide low fault impedance to the ground fault currents for proper operation of the protective relays and for improving the system requirements by an effectively earthed system.

2. Safety earthing : It is required to provide protection to the operating staff working in the yard and sub-station from any injury during fault conditions by keeping the voltage gradient within safe limits.

The above two categories of earthing have common earthmat from which flat iron risers are taken out, to connect all the non-current carrying metal parts of the equipment. At the time of fault the earthmat conductors rise to a voltage, which is equal to the resistance of the earthmat multiplied by ground fault current. This difference in potential results in voltage gradients. The earthmat of the sub-station is so designed that the *step potential* (potential difference between two feet of a human at the time of fault) and *touch* potential (potential difference to which body is subjected to when a person comes in contact with an earth body) are within safe limits.

The *following components in electrical installations must be earthed :*

(i) The frames, tanks and enclosures of electric machines, transformers, and apparatus lightning fittings, and other items of equipment; *(ii)* The operating mechanisms of the switch gear; *(iii)* The framework of the switch boards, control boards, individual panel boards, cubicles; *(iv)* The structural steel work of indoor and outdoor substations, metal cable jointing boxes, the metal sheaths of the cable, the rigid metal conduit runs, and similar metal work.

Methods of earthing : The commonly used methods of earthing are :

1. Pipe earthing
2. Plate earthing.

1. Pipe earthing

Pipe earthing with full specifications as per Indian Standard is described below. Refer to Fig. 14.18.

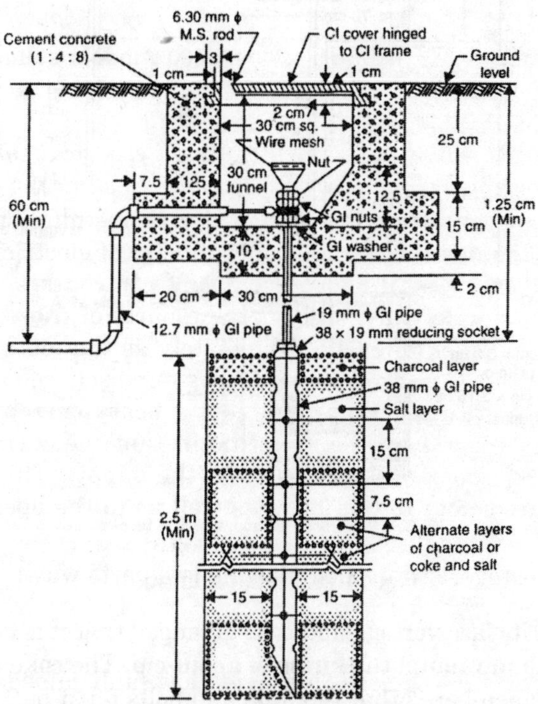

Fig 14. 18. Pipe earthing.

- In this system of earthing a G.I. pipe of 38 mm dia. and 2 metres length is embedded vertically in ground to work as earth electrode but the depth depends upon the soil conditions ; there is no hard and fast rule for this.

- The earth wires are fastened to the top section of the pipe with nut bolts.
- The pit area around G.I. pipes is filled with *salt and coal mixture* for *improving the soil condition and efficiency of the earthing system.*

The contact surface area of G.I. pipe with soil is more in comparison to *theplate* because of its circular section and hence can *take up heavy leakage current for the same electrode size.* The earth wire connection with the G.I. pipe being above the ground level can be checked for carrying out continuity test as and when desired (while in plate earthing it is difficult).

2. Plate earthing

- In plate earthing, the looping earth wire is bolted effectively with the earth plate made up of copper size 60 cm × 60 cm × 3.18 mm (2′ × 2′ × 1/4″) and embedded 3 metres in ground. Copper plates are found to be most effective earth electrodes and are not affected by the soil moisture, *i.e.,* these do not get rusted. But on account of its high material cost, galvanized iron plates are preferred and usually used for normal work.
- In case the soil resistivity is high, the plate should be placed vertically in ground at a higher depth. A line diagram showing the plate earthing system is shown in Fig. 14.19.

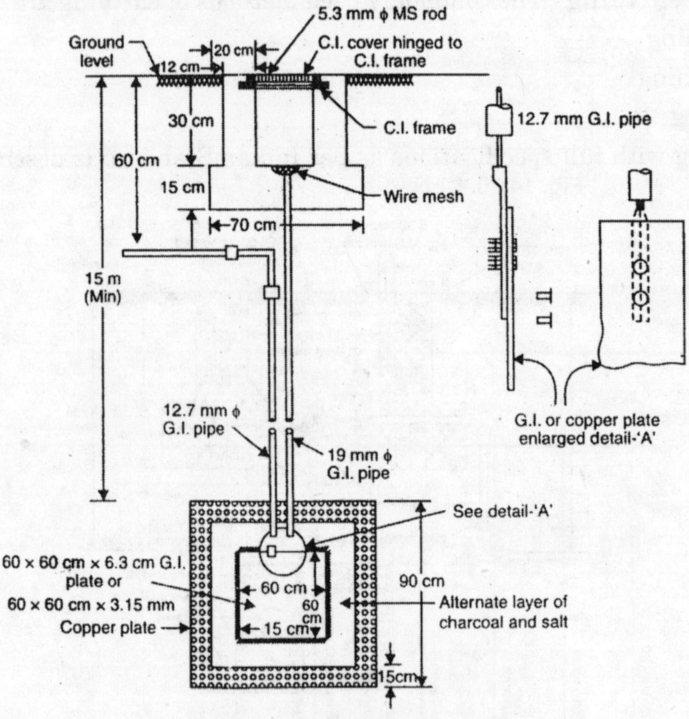

Fig. 14.19. Plate earthing.

The plate is kept with its face vertical and is so arranged that it is embedded in an alternate layer of coke and salt for a minimum thickness of about cm. The coke salt decreases the earth resistance. It should be remembered that the nuts and bolts must be of copper for copper plate and should be of G.I. for G.I. plates. Usually, the earth wire is drawn through a G.I. pipe fitted with a tunnel on the top through which salty water is poured in the pit of earth plate from time to time in summer sea on when the moisture of the soil will decrease to a larger extant which will increase the earth resistance.

14.13 CIRCUIT (LINE) DIAGRAM OF 66 KV (FIG 14.21) & 220 KV (FIG. 14.22) SUBSTATIONS

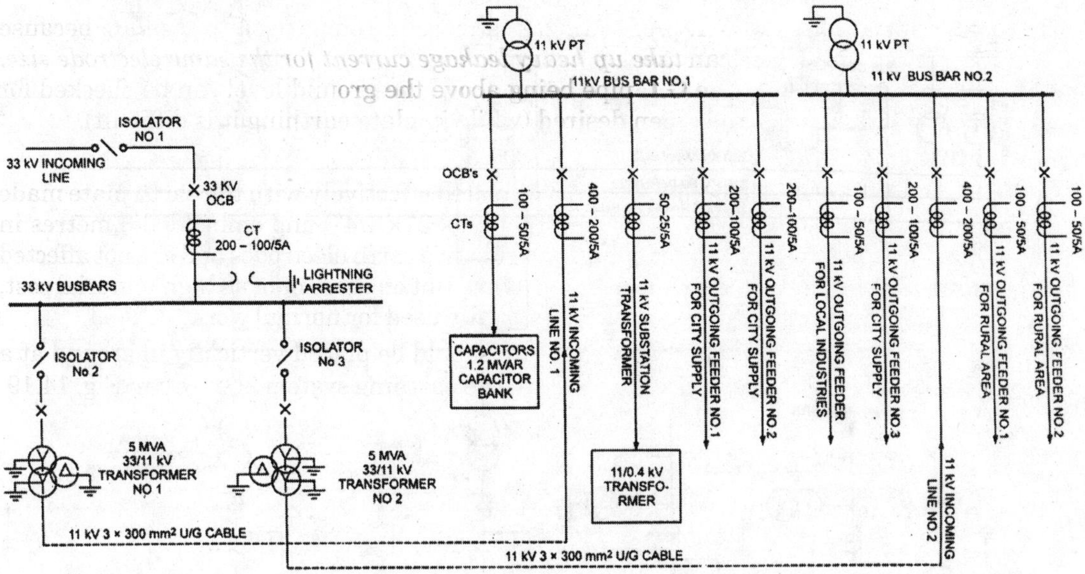

Fig. 14.20. Circuit diagram for 33 kV substation.

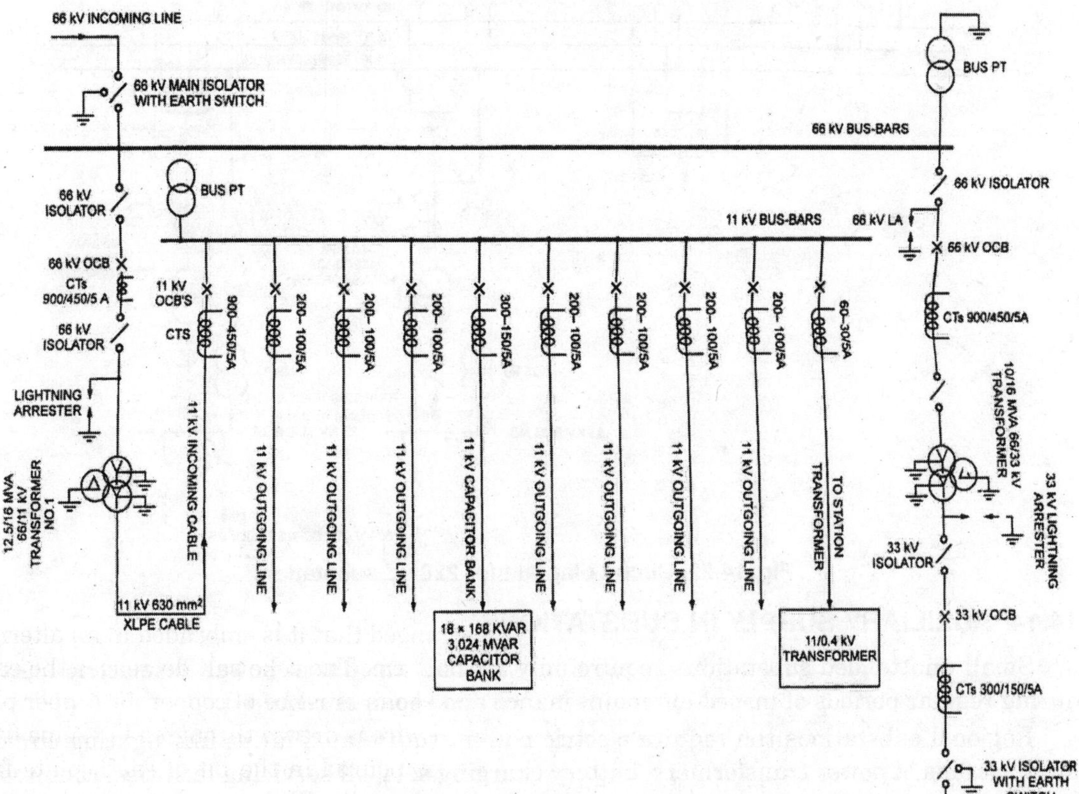

Fig. 14.21. Circuit diagram for 66 kV substation.

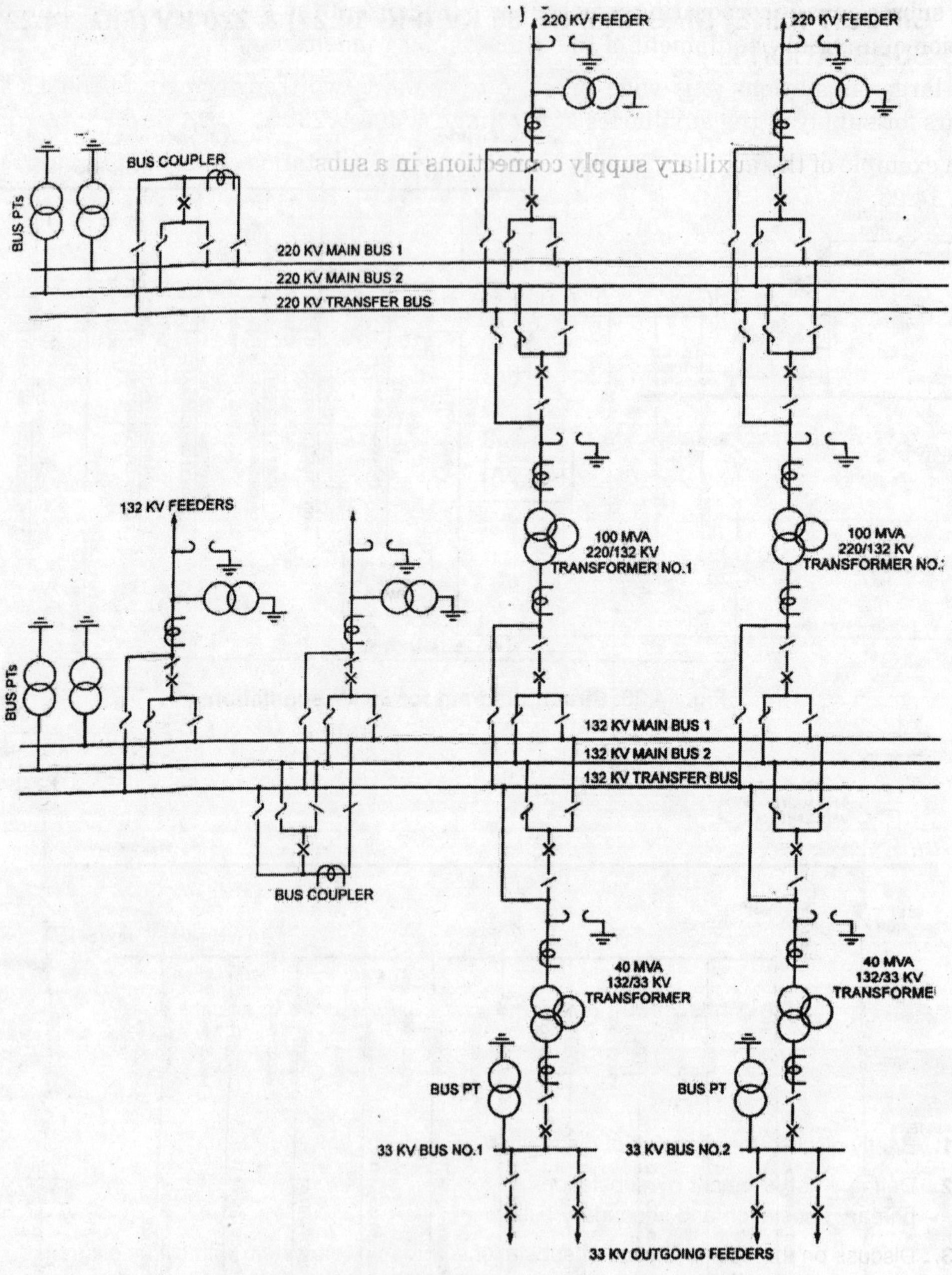

Fig. 14.22. Circuit diagram for 220 kV substation.

14.14 AUXILIARY SUPPLY IN SUBSTATIONS

Small unattended substations require only a small amount of power for electric lighting during regular periods of inspection maintenance and repair is required.

Regional substations the require electric power require auxiliaries like lighting circuits, air blast fans of power transformers, battery charging sets, oil servicing facilities, compressor units in case of air blast circuit breakers, ventilating fans of the substation buildings, water supply and heating system equipment etc.

In substations incorporating synchronous condensers the supply is also required for the operation of auxiliary equipment of the synchronous condensors.

In large substations it is wide practice to connect two transformers to the 11 kV main bus-bars for supply of the auxiliaries at a voltage of 400 V/230 V.

An example of the auxiliary supply connections in a substation of medium capacity is shown in Fig. 14.23.

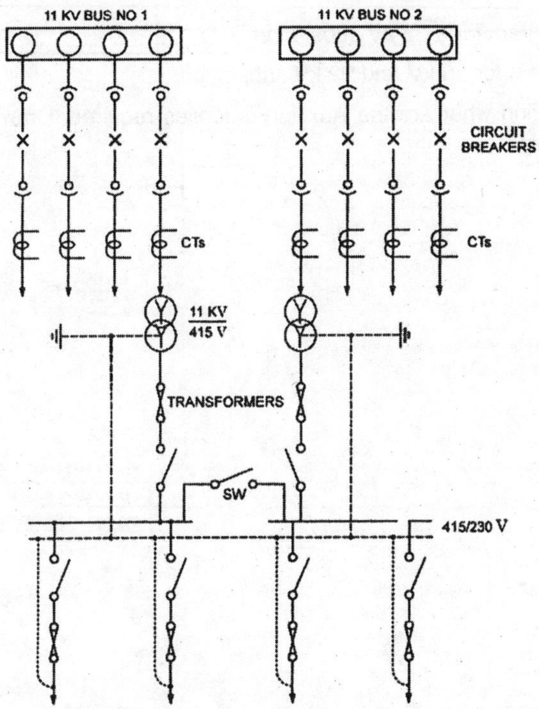

Fig. 14.23. circuit diagram for auxiliary supply in substation.

QUESTIONS

1. Briefly discuss the planning of distribution networks in power system.
2. Define what is meant by substations in a power system. State what is the difference between a primary substation and secondary substation.
3. Discuss on the different types of substations in the power system with circuit diagram.
4. State which type of substations is most useful for distribution of electrical powers in rural area.
5. Write a brief note on distribution Network Configuration.
6. What are the different types of Bus Bar arrangements.
7. Write note on :

   (*a*) Double Bus bar arrangements.

   (*b*) Main and transfer Bus Bar arrangement.

   (*c*) Ring or Mesh Bus Bar arrangements.

8. Write what are the basic requirements for constructing a substation in (*a*) Urban Area and (*b*) Rural Area.
9. Prepare a layout a for construction of a substation.
10. State how the capacity of a substation is determined to meet the demand of the area covered by the substation.
11. State what are the main equipment required in a sub-station.
12. State why earthing or grounding is necessary and what are the major methods employed in a substation.
13. Draw a circuit diagram for a 33 kV substation.
14. Draw circuit diagram for 66 kV and 22 kV substation.
15. State in a substation what are the Auxiliary supplies required. Draw a main connection diagram for Auxiliary Supply.

CHAPTER

15

Protection in Power System

15.1 INTRODUCTION

INSIDE THIS CHAPTER

- Introduction
- Grounding in Power Systems
- Protection Components and Power System Protection
- HRC Fuses
- Circuit Breakers : OCB, VCB, SF_6
- Relay for Protecting Power System
- Static Relays

Protection of power system circuits will be discussed on the following points which constitute full protection of the system from various types of faults. These points are :

1. Grounding in Power System
2. Protection Components : Fuses, Circuit Breakers Relays
3. Protection of Power System Equipments and Bus Bars Line.

15.2 GROUNDING IN POWER SYSTEMS

Proper Grounding of Power System & Substations is very important not only for protection system, but also for safety to human life. In fact modern system operate with their neutral points properly grounded. The advantages of the grounded neutral are : (1) persistent arcing grounds are eliminated, (2) earth faults can be utilised to operate protective relay to disconnect the fault.

However Grounding of neutral does not have any effect on the operations of a poly phase system modern steady state and balanced condition by the type of grounding neutral.

Below we shall discuss both ungrounded and grounded neutral system.

15.2.1 Ungrounded Neutral System

(*a*) In the Fig. 15.1 (*a*) a simple ungrounded-neutral system is shown, where the line conductors have capacitances between one another and to ground as represented by the delta and star connected capacitors. The delta connected capacitances in fact, hence has little influence on the grounding characteristics of the system and hence will not be considered in the discussion here. In a perfectly transposed line each conductor will have the same capacitance to ground. As such in normal conditions, the line to neutral charging currents I_{CR}, I_{CY}, I_{CB} will form a balanced set of currents as shown in Fig. 15.1 (*b*), where phasors V_{RN}, V_{YN}, V_{BN} represent the phase to neutral voltage of each phase. The charging currents I_{CR}, I_{CY}, I_{CB} load this respective phase voltages by 90°. The magnitute of each phase currents is = V_{ph}/X_C, where X_C is the capacitance of the line to ground. As these phasor currents are balanced and hence no current can flow to earth obviously.

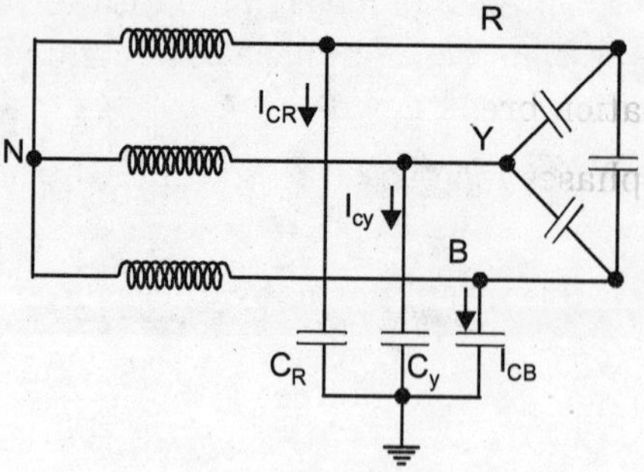

Fig. 15.1 (*a*)

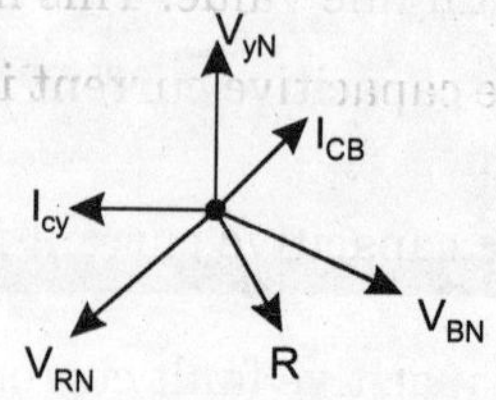

Fig. 15.1 (*b*)

(*b*) Now let us consider a system as shown in Fig. 15.2 (*a*) where a phase to Earth fault at point *F* in line *B* occurred. We shall now analyse this faulty line.

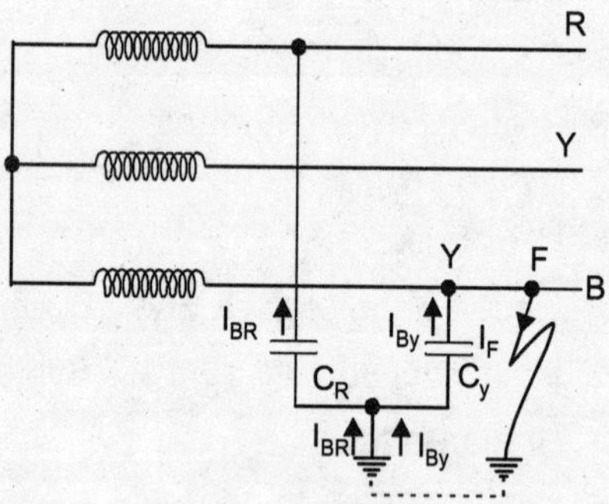

(*a*)

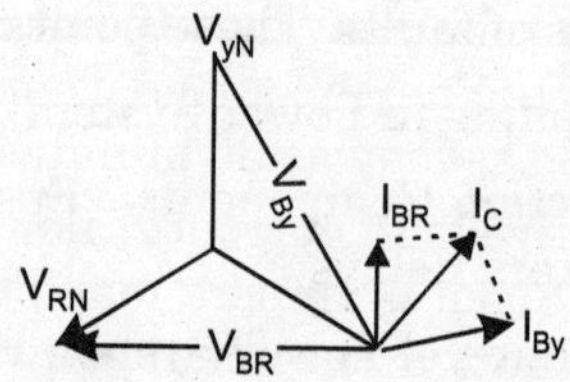

(*b*)

Fig. 15.2 (*b*). Phasor diagram for a ground fault in phase *B* as shown in Fig. 15.2 (*a*).

From the Figure we see that the current in phase *B*, has two components : (1) I_{BR}, through phase *B*, and capacitance C_R to the phase *R* conductor and (2) I_{BY}, through phase *B*, the fault current and capacitance C_Y to the phase '*Y*' conductor. The voltages driving these currents are V_{BR} and V_{BY} respectively and since the impedance of the circuit traversed by these components is predominently capacitance they lead their respective voltages by nearly 90°, as shown in the phasor diagram in Fig. 15.2 (*b*) as I_{BR} and I_{BY}, value of which will be :

$$I_{BR} = \frac{V_{BR}}{X_C} = \frac{\sqrt{3}\,V_{PH}}{X_C} \text{ and } I_{BY} = \frac{\sqrt{3}\,V_{PH}}{X_C}$$

And the fault current I_F is the vector sum of I_{BR} and I_{BY}

$$\therefore \qquad I_{BY} = \sqrt{3}\cdot I_{BR} = \sqrt{3}\times\sqrt{3}\cdot\frac{V_{PH}}{X_C} = 3\,\frac{V_{PH}}{X_C}$$

From the above analysis the following conclusions can be drawn :

(*a*) It is not a practice to operate system with ungrounded neutrals as :

1. Such a system cannot be adequately protected for fault to earth.
2. The insulation of such a system is likely to be overstressed by the over-voltages.

3. Insulation overstress may give rise to insulation failure on other parts of the system which they lead to heavy phase-to-phase fault conditions.

(b) 1. In an undergrounded neutral system under a single line to ground fault the voltage to earth of the two healthy phases rises from their normal phase to neutral voltage to full line value. This may result in insulation break down.

2. The capacitive current in the two healthy phases increases to $\sqrt{3}$ times the normal value.

3. The capacitive current in the fautly phase is 3 times in normal value.

4. A capacitive fault current I_F. For system made of overhead lines $I_F = \dfrac{\text{kv} \times \text{km}}{260}$ Amp.

5. For system made of H type of cables, $I_F = \dfrac{\text{kv} \times \text{km} \times 25}{280}$ km.

For system made of belted type cables $I_F = \dfrac{\text{kv} \times \text{km} \times 25}{280}$ amp.

In case of overhead lines the fault current may be so small as to render automatic isolation by protective relay may be difficult. Also the fault current being small, there will be very little influence on the neighbouring communication circuits. However, the line duration of the arc which may be as long as 30 minutes may offset this advantage.

A capacitive fault current I_F flows into the earth. Experience shows that capacitive currents in excess of about 4 ampere may be sufficient to maintain an arc in the ionised path of this fault even though the medium which actually caused the fault has cleared itself. The persistance of the are due to the flow of capacitive current gives rise to a condition known as "arcing ground" in which cycle charging and discharging of the system capacity through the fault result in high frequency oscillations being superimposed on the whole system and a build up of very high voltages can occur. It is observed that under ground conditions the phase voltage of the system may rise to 5 to 6 times its normal value. The build up of high voltages may result in insulation break-down.

Now below we shall discuss on Grounded neutral system and analyse the system accordingly.

15.2.2 Grounded Neutral System

From own experience, we know that the modern power system operate with their points grounded. The advantages of the grounded neutral are : (a) persistant arcing grounds are eliminated, (b) Earth fault can be utilised to operate protective relays to disconnect the fault. However, the power system neutral can be grounded effectively on non-effectively. In effectively grounded systems, the neutral point is grounded directly, also called solid grounding, to the earth. On the other hand for non-effectively grounded systems, any of the following these methods can be adopted : (1) Resistance grounding, (b) Reactor grounding, (c) Resonent grounding by arc supression coil. Below we shall discuss these types of groundings one by one as knowledge on this subject is very important for power system engineers.

15.2.3 Direct Grounding (Solid Grounding)

As discussed above in solid grounding or direct grounding, a direct metallic (conductor) connection is made as shown in Fig. 15.3 from the system neutral to one on more each electrode consisting of plate rods or pipes buried in or driven into the ground.

A three phase system with its neutral directly or solidly grounded and an earth fault at point F in line B is shown in Fig. 15.4 (*a*). The phase diagram of this fault condition is shown in Fig. 15.4 (*b*). The current in phase 'B', has in fact, three components : (1) I_{NR} through phase B the fault and capacitance C_R to the phase R conductor, (2) I_{NY} through phase B the fault and capacitance C_Y to the phase-Y conductor, (3) I_F through phase B, the fault and the ground.

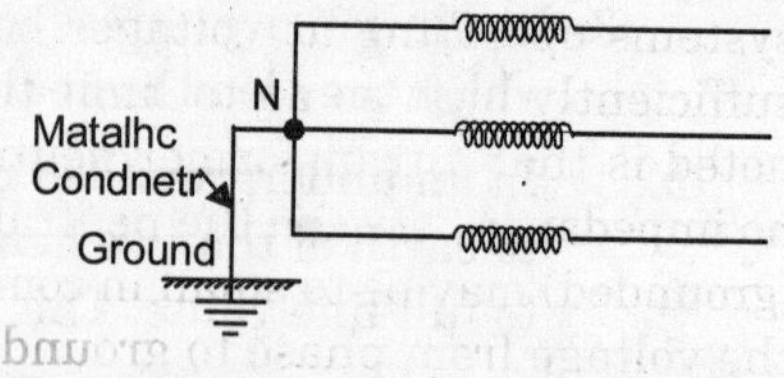

Fig. 15.3. Solid/direct grounding of neutral.

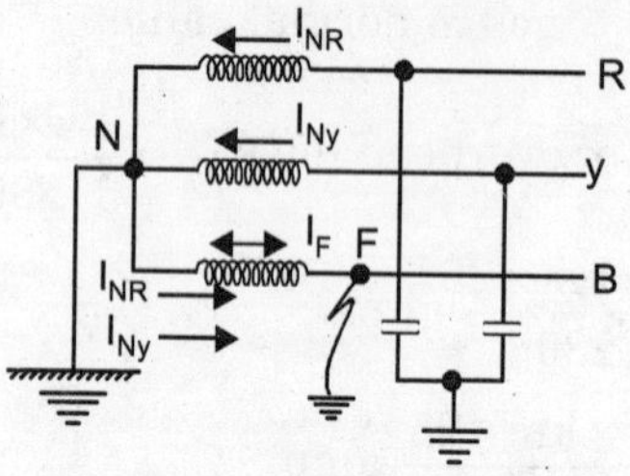

Fig. 15.4 (*a*). Fault in a 3-phase solidly grounded system.

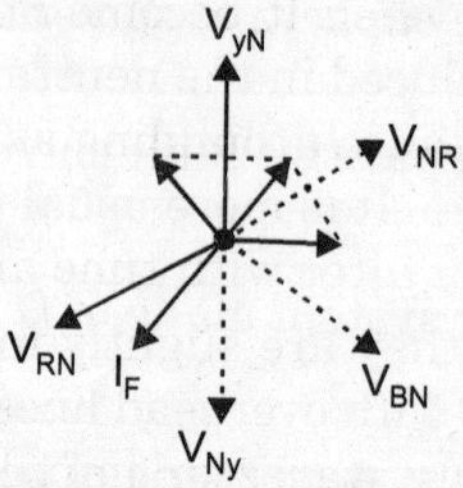

Fig. 15.4 (*b*). Phasor diagram of a fault of solidly 3-phase system.

An analysis of the fault by symmetrical components gives $I_F = \dfrac{3\,V_{Ph}}{Z_1 + Z_2 + Z_0}$ since $Z_1 + Z_2 + Z_0$ is pre-dominently inductive, I_F behind the phase to neutral voltage of the faulted phase by nearly 90^6. This is shown in Fig. 15.4 (*b*). The voltage driving the currents I_{NR} and I_{NY} are V_{NR} and V_{NY} respectively and since the impedance of the circuits traversed by these current in pre-dominently capacitive, they lead their respective voltages by 90° as shown by phasors I_{NR} and I_{NY}. I_{CF} is the resultant of I_{NR} and I_{NY} is in phase opposition to I_F.

In case of solidly grounded system, we should note the following points :

1. Whenever, a fault to earth occurs on anyone phase of the system the voltage to earth of the faulty B phase becomes zero, but the healthy phases, in general, remain at their normal phase value. As such lightning arresters rated for phase voltage can be employed. Also, the equipment should be insulated for phase voltage. Thus a saving in cost of the equipment results.
2. The flow of heavy fault current I_F will completely multify the effect of the capacitive current I_{CF} at the fault and so no arcing ground phenomenon or over-voltage can occur.
3. The flow of heavy fault current permits the use of discriminative protective gear.
4. Increased earth fault current result in greater influence on neighbouring communication circuits.
5. Since the fault has to be cleared by the circuit-breakers heavy earth fault current may result in burning of contacts.
6. The adverse effects of heavy earth fault current listed under (4) and (5) have been overcome by employing high-speed and higher capacity breakers.

It should be also noted that the use of solid or effective grounding is limited to system where the normal circuit impedance is sufficient to prevent very high value of fault current being reached. This is necessary in order to avoid excessive damage at the fault location. Our experience shows that combined impedance of the apparatus circuit and earth return path in systems operating at voltages below 2.2 kv and those operating at voltages above 33 kv is sufficiently high so as to limit the value of the current to a safe value. Another point to be noted is that a transformer neutral may be solidly/effectively grounded so that there may be no impedance between the neutral and earth. But, the transformer capacity thus solidly earthed (grounded) may be to small in comparison with the sizes of the system to be effective in stabiling the voltage from phase to ground when ground or earth fault occurs in the system.

15.2.4 Resistance Grounding

At a time when it become necessary to limit the earth fault current, a current limiting device is introduced in the neutral and the earth. One method of introducing a current limiting device is resistance grounding as shown in Fig. 15.5. The resistance may comprise of metallic resistance units. It is more usual to use liquid resistors if the voltage is 6.6 kv or more. Metallic resistors do not alter with time and little or no maintenance is required.

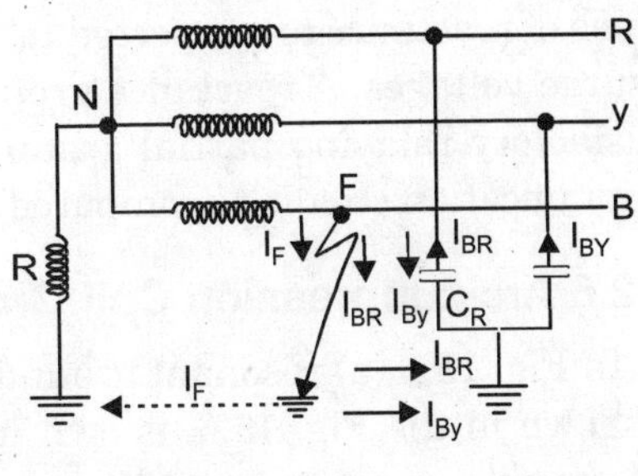

Fig. 15.5. Resistance grounding system.

However, they are slightly inductive and this is a disadvantages with overhead lines exposed to lightning, since travelling wares or impulses are subjected to positive reflection and this may sudenly stress the insulation of the equipment and cause breakdown. However, liquid resistors are free from this advantages and are normally of robust construction. These may be mounted outdoors as well. We must state here that an important consideration in resistance grounded systems is the power loss in the resistor during line to line to the ground faults. In fact, it is, in general, common to fix a value which will limit the earth fault current to the full rating of largest generator or transformer. Based on the practice the value of resistance to be inserted in the neutral connection to earth is : $R = \dfrac{V_L}{\sqrt{3} \cdot I}$, where R = Resistance in ohm, V_L = Line voltage in volts. I = Full load current of largest machine of transformer in amps.

Resistance grounding is normally employed on system operating at voltages between 2.2 kv and 33 kv when the total power source capacity exceeds 5,000 kVA as the circuit characteristics of such systems usually give rise to excessive currents under ground fault conditions. Neutral Grounding Resistors NGR are normally designed to carry their rated current for a short period, usually 30 seconds.

We can below note the merits and demerits of a resistance grounding :

1. It minimises the hazard of arcing grounds.
2. It permits the use of discriminitive protective gear.
3. In general a resistance grounded system will have materially lower ground-fault current than the solidly grounded system and hence will have less inductive influence on neighbouring communication circuits.
4. The system neutral will almost invariably be displaced in the case of a ground fault thereby necessary for the use of lightning arrestors at increased cost and sacrifice in protective performance.
5. This a system a cost less than the solidly grounded system.
6. There is enormous energy loss in NGR for dissipation of fault energy.

15.2.5 Reactance Grounding

Reactance grounding implies insertion of a reactance in the ground connection infact a system is said to be reactance grounding if X_G/X_L exceeds 30 but is less than the value necessary for resonant grounding.

There is no idea in purposely increasing the grounding reactance of a system beyond that required to keep current within non-destructive range except of curve for the special case of ground fault neutralisers used in resonant grounding. The transient voltages, resulting from arcing increase as the reactance is increased. Similarly during switching operation higher value of reactance is increased. Similarly during switching operation higher values of reactance are expected to result in higher value of surge voltages. A reactance grounded system ensures satisfactory relaying partial grading of the apparatus insulation, reduced interference with communication circuit as compared with that in solidly grounded system.

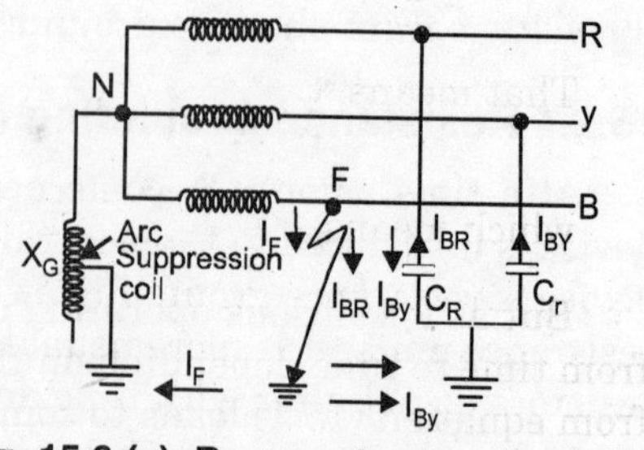

Fig. 15.6 (*a*). Resonant grounding with fault *F*.

15.2.6 Arc-suppression Coil Grounding (Resonant Grounds)

In Fig. 15.6 (a) resonant grounding system with a fault is shown. An arc-suppression coil as shown in the Fig. 15.6, is own iron cored reactor mounted in the neutral earthing circuit and capable of being turned to resonate with the capacitance of the system when one line becomes earthed. In fact the function of the are suppression coil is to make arcing earth fault self-extinguishing and in the case of sustained fault to reduce the earth current to lower value so that the system can be kept in operation with one line earthed. The are suppression coil is sometimes referred to as a "Peterson Coil" or "Ground fault neutraliser" while the grounding so achieved is referred to as a Resonant grounding in the system The Fig. 15.6 shows the B-phase earthed by single line to earth fault on arc-suppression coil neutral ground system. Due to the fault the resultant capacity current is three times the normal line to neutral charging current of one phase as $I_{CF} = \sqrt{3} \times I_{BR} = \sqrt{3}(V/X_C) = \sqrt{3}\dfrac{\sqrt{3}\,V_{ph}}{X_C} = 3 \cdot \dfrac{V_{ph}}{X_C} = 3 \times$ line to Neutral charging current.

Under these conditions the voltage of the faulty phase is impressed across the arc-suppression coil and a current I_F flows lagging by approximately 90° and in phase opposition to I_{CF} as can be seen in the phasor diagram of Fig. 15.6 (*b*). Infact by adjusting the tapping on the coil I_F can be made to neutralise I_{CF} so that the resultant current in the fault is limited to practically flow through the fault. The system can also operate with a sustained earth fault on phase without harmful results and no arcing phenomenon can occur. In practice there will be a small resultant current present in fault since absolute turning between the inductance of the coil and the capacitance of the system may not be possible. However we know from our practical experience that the small resultant currents due to deviations of the order of 20% for system voltage upto 66 kV and 10% for higher voltage from resonance cannot maintain the are.

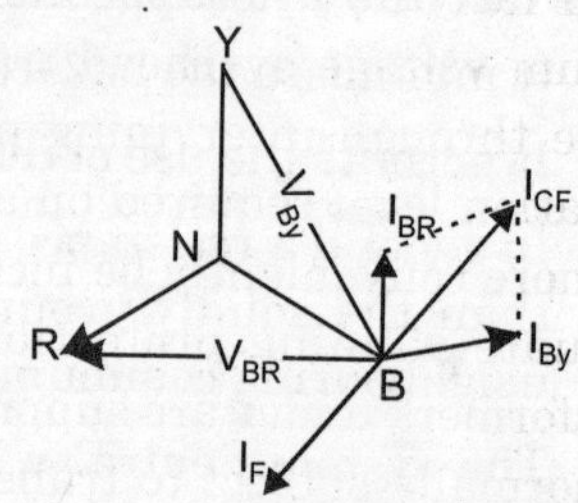

Fig. 15.6 (*b*). Phasor dgm of a resonant dgm grounded system.

The inductance of the coil can be determined following way : $I_{CF} = 3 \cdot \frac{V_{ph}}{X_C}$ and also $I_F = \frac{V_{ph}}{X_L}$, where X_L is the inductance of the coil. But at resonance, $I_F = I_{CF}$.

That means, $$\frac{V_{ph}}{X_L} = \frac{3\,V_{ph}}{X_C} \therefore X_L = \frac{X_C}{3} \quad \therefore wL = \frac{1}{3wC}$$

which means, $$L = \frac{1}{3w^2C} = \frac{1}{3w^2C} \qquad ...(15.1)$$

But as for various operating conditions the capacitance of the system network varies also from time to time depending on the operating conditions. And this is why the value of L found from equation (15.1) leads to some problem.

But this problem can be overcome by using a tapped coil for which the proper tapping is being used for each change in network conditions.

Then, the current rating of the coil is given by

$$I_F = \frac{3\,V_{ph}}{X_C} \qquad ...(15.2)$$

In fact, the normal rating of coils used on systems on which permanent earth fault can be located and removed promptly in ten seconds, whereas continuous time-rated coils are used on all other systems. In such cases, it is also usual to provide automatic means as shown in Fig. 15.6 (*c*) to by pass the are suppression coil after some time. This allows fault current to flow and the normal system protective gear then operates to disconnect the faulty line from supply. However, it may be noted that during the period the arc suppression coil is functioning and the voltage to earth of the healthy phases rises to $\sqrt{3}$ times the phase to neutral value and so the insulation must be capable of withstanding this voltage increase. This also necessitates the use of lightning arresters which are rated for full voltage.

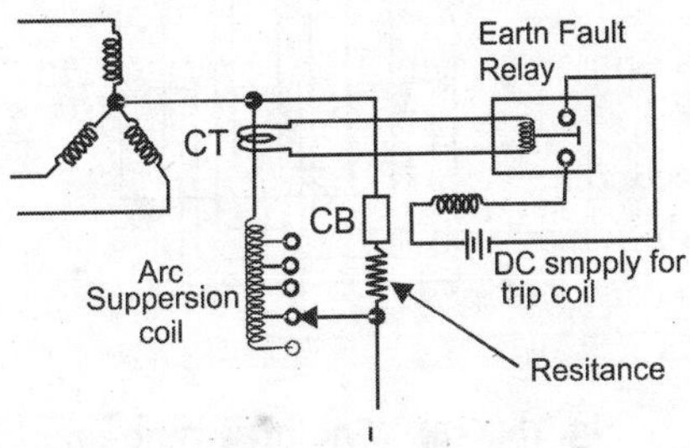

Fig. 15.6 (*c*). Arc-suppression coil application.

In fact the arc-suppression coil method of neutral grounding is normally confirmed to medium voltage overhead transmission lines which are connected to the system generating source through intervining power-transformers. The reasons arc stated as : (1) The higher insulation level required on apparatus associated with are suppression coil grounded system can more conveniently be incorporated in power transformers than in generators. In addition adequate winding protection cannot so readily be obtained for generators as for power transformers under arc suppression coil operating conditions. (2) Overhead transmission lines are normally subject to transient earth fault due to lightning, birdage etc. of short duration. Moreover, it has been found that ground-fault neutralisers decrease the line outages from fault to 20 or 30 per cent of these obtainable with some from of grounded operations. Hence this type of arc-suppressing grounding is used on radial lines with many advantages, as this method avoids the construction of parallel circuits for service continuity and hence capital cost is reduced.

15.2.7 Non-effecting Grounding

Regarding non-effective type of grounding we shall discuss the following two types :

1. Grounding Through an Earthing Transformer

Where the neutral points of a three phase system are not accessible or where the transformers or generators are delta connected an artificial neutral earthing point may be created with the help of an interconnected star earthing transformer generally called neutral earthing Compensator as can be seen in Fig. 15.7 (*a*). Earthing transformer is a 3-limbed type transformer having two equally proportional winding on each limb. One set of windings is connected in star to provide the neutral point. The other ends of this set of winding are connected in the second set of windings as shown that the directions of the currents in the two wings on each limb are opposite to each other. With this arrangement of windings the transformer offers a low impedance path to the flow of zero phase sequence currents under system fault conditions since the only magnetic flux which results from the zero sequence currents is the leakage flux about each winding section. Under normal operational conditions only small exciting current circulates in the windings of the earthing transformer.

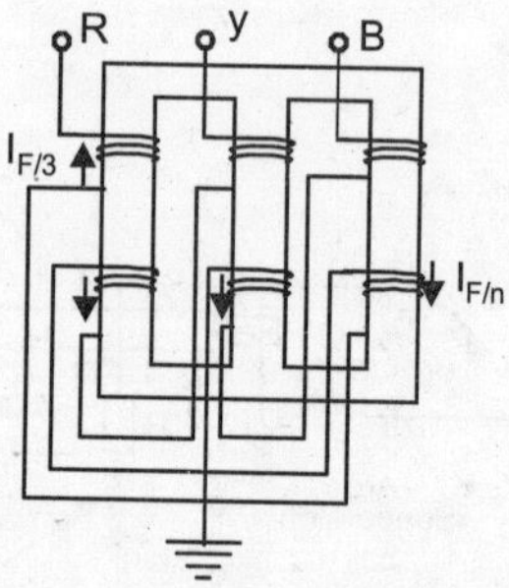

Fig. 15.7 (*a*). Winding arrangement of an inter-connected star earthing transformer.

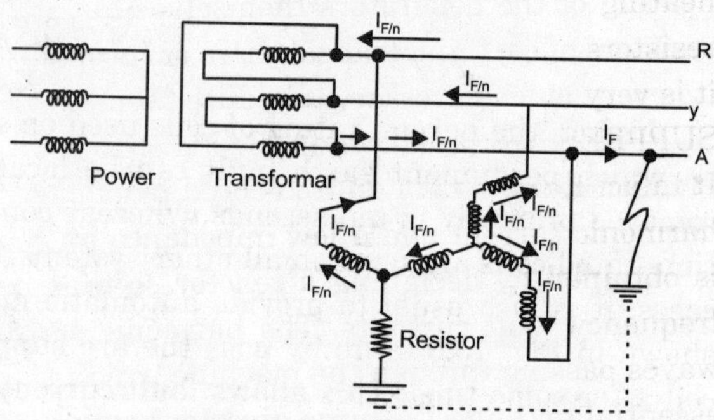

Fig. 15.7 (*b*). System neutral grounded through an earthing transformer.

However, due to the very low impedance of the earthing transformer windings under fault condition, it is necessary sometimes to limit the value of the fault current by the addition of a resistor in series with the neutral earthing connection which is shown in Fig. 15.7 (*b*). In fact, this is necessary in systems operating at voltage between 2.2 kv and 3.3 kv where the total power capacity exceeds 5 MVA.

2. Voltage Transformer Earthing

Any system with its neutral point earthed through a single phase voltage transformer operates virtually as an insulated neutral system. The voltage transformer acts as a very high reactance earthing device. The over voltage conditions are associated with insulated neutral systems. Its presence in the neutral connection as shown in Fig. 15.8 serves only as a voltage measuring device to indicate a fault to earth on the system. The voltage transformer acts as a reflection point for travelling waves passing through the machine winding and in order to prevent high voltage build-up it is necessary to connect a surge diverter between the machine

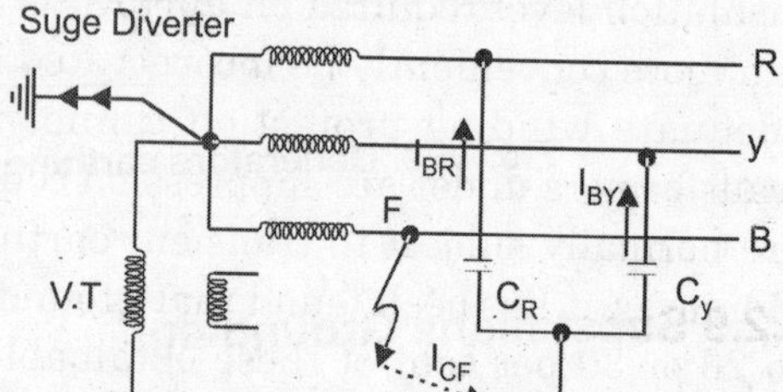

Fig. 15.8. Voltage transformer earthing of neutral.

neutral point and the earth as shown in the Fig. 15.8. The application of the voltage transformer in the neutral earthing circuit normally confined to generator equipment which are directly connected to step-up power transformers. The generator circuits of these equipments are physically isolated for the main distribution system and as the interconnecting cables between the generator and transformer windings are usually very short the electrostatic capacity of the circuit hence will be negligible and consequently, there will be no risk of over voltage conditions arising due to arcing ground phenomenon.

15.2.8 Harmonic Suppressions

In the case when the voltage wave of the generator is non-sinusoidal the 3rd harmonic voltage may cause third harmonic current to flow from the generator line terminals through the system and returns by way of comparitively low zero sequence reactance path of the system capactiance or the windings of the transformer or other generators to its neutral terminals. These 3rd harmonic currents which are continuous in nature do not give rise to injurious over-heating of the generators themselves but they can cause dangerous overheating of neutral resistors of earthing transformers or incorrect operation of earth fault protective gear. Hence, it is very much necessary to reduce them to safe value. This is done by the use of HARMONIC SUPPRESSOR (as shown in the Fig. 15.9) in the neutral earthing connection of the generator. It infact it consists of an iron-cored reactor and present a high impedance path to the flow of harmonic current and a low impedance path to the flow of fault current. This characteristic is obtained by designing a reactor so that its core becomes saturated at very low system frequency fault currents. The harmonic suppressors act as a reflection point for travelling waves passing through the machine winding and then to prevent a high voltage build-up it is essential to connect a surge-diverter between the machine neutral point and earth.

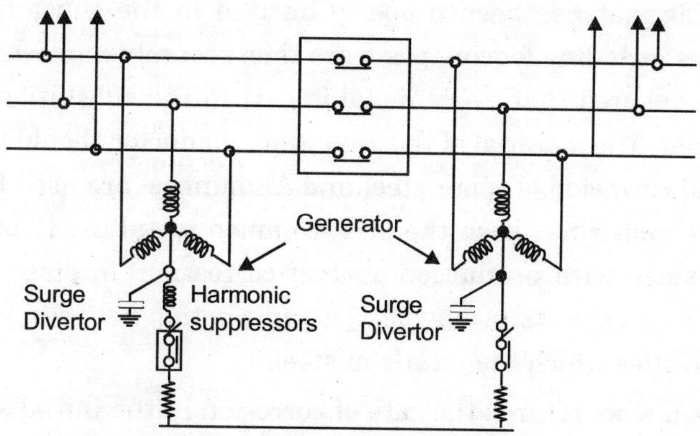

Fig. 15.9. Generators earthing through individual resistors and harmonic suppressors.

15.2.9 Substations Grounding

In the power system, the grounding at substations plays a vital role in the protection scheme and the safety in the system.

In fact, for any particular value of the fault current, the potential V_C and the potential gradient over the surface out from the electrode in the case considered above will depend upon the ground resistance. In systems where maximum ground fault current may be very high, it is seldom possible to obtain a ground resistance so low as to ensure that the total rise of the grounding system potential will not reach values unsafe for the human contact. In fact, this being the case the hazard can be corrected only by control of local potentials. A mat is usually the most practical way to do this at sub-station yeards. The mat (which is also called Earthing Grid) comprises a number of meshes as shown in Fig. 15.10. A typical size of mesh is of 4m × 5m of earthing conductor. These meshes are connected to several earth electrodes of size 1.9 cm × 3 m driven at intervals. The total number of electrodes may be found from the following relationship. No. of electrodes = $\frac{\text{Maximum Fault Ciremit}}{500}$. Grounding Resistance R_g for the earthing (Grounding) mat can be found by : $R_g = \frac{P_s}{4r} + \frac{P_s}{L}$, where L is the total length of the burried conductor in metres and 'r' radius of circular plate in metres, P_s is the resistivity of soil in $\Omega - m$ having an area equivalent to that covered by the mat.

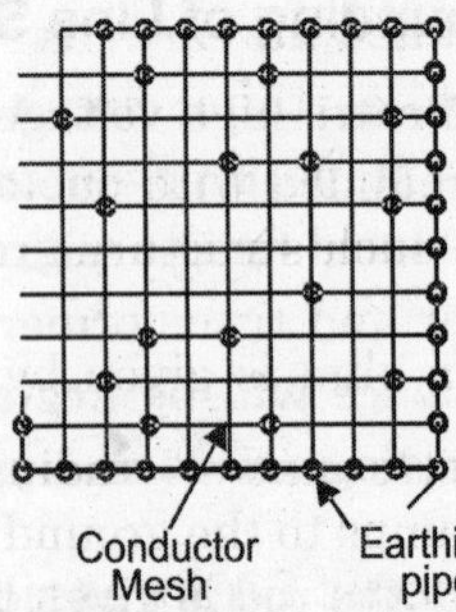

Fig. 15.10. A grounding molor grid.

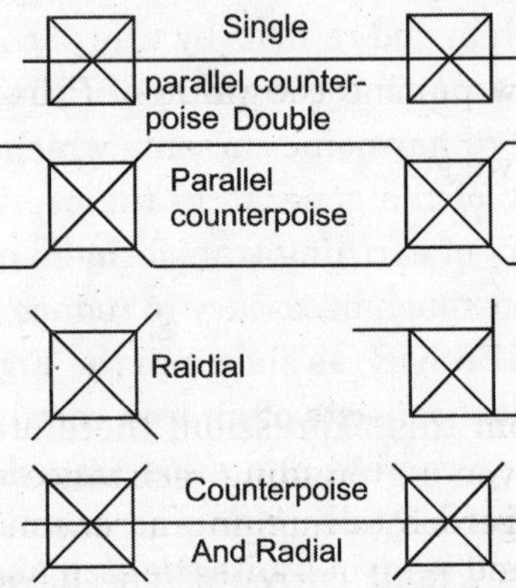

Fig. 15.11. Arrangement of counterpoises.

After the earthing mat has been properly burried in the substation ground all metal structures and frames including fencing posts are then securely connected to the earthing mat or grid. It should be ensured that every metallic part in the substation has two connections with the grounding mat. The material of the grounding conductor should have high conductivity and low underground corrosion. Copper steel and Aluminium are used for grounding system. Although in the past copper has been the most common metal used but the modern trend is to use high quality steel (with precuation against corrossion) in place of copper considering the cost saving and moreover its use avoids galvanic action between grounding system and other undergound utilities which are mostly of steel.

In fact, galvanised steel retards the rate of corrosion in the initial stages but as the time passes Zinc (for Galvanising) coating destroys the rate of corrosion increases. In fact depending on the corrosivity of the soil, Zinc coating of the steel pipe and mesh may be destroyed in 3 to 10 years, also depending on the quality of Galvanrising. Hence it is most important that good quality steel pipes with best quality zinc coating should be used, as not only the protection system even the safety of the people engaged in the substation operation depends on good earthing. Therefore a proper size and quality of the conductor with best quality zinc coating should be used, so that (1) it is mechanically sound and (2) it has propers thermal stability to ground fault current.

15.2.10 Grounding of Line Structure

In case of extra high voltage transmission lines, these are carried on lattice steel structures which are grounded with one or more grounding rod driven vertically at the surface. Let us consider one such structure grounded through one grounding rod. The grounding resistance of this rod in ohm is given by, $R = \frac{P}{2rl} ln\left(\frac{2l}{a}\right)$ where, l = length of the rod driven in the ground in metres and a = radius of the rod in metres. Now when the grounding rod discharges a current I_F amps to the ground, the voltage gradient on the surface of the ground at a distance x metres from the rod is given by

$$\frac{dv}{dx} = \frac{pI_F}{2r\sqrt{(l^2 + x^2)}} \cdot \text{V/m}$$

Now putting the value of P from the equation $R = \frac{p}{2rl} ln\left(\frac{2l}{a}\right)^{dx}$ in the equation of $\frac{dv}{dx}$ above, we get

$$\frac{dv}{dx} = \frac{Rl\, I_f}{ln\left(\frac{2l}{a}\right)\sqrt{l^2 + x^2}}$$

From this expression, therefore, it is clear that when the ground fault current is very high, even a grounding resistance may produce potential gradient on the ground which may be dangerous to human and animal life. Hence for providing proper safety it is essential to check and keep in limits the step and touch potentials by decreasing the resistance still further or by using more than one grounding rod, when the earth resistivity is high and the driven rods are not adequate, the remedy is to buy a wire in the earth and connect it to the tower footing.

In fact, the wire may be run parallel to or at some angle to the line conductor themselves. This is called counterpoise. The parallel counterpoise, as compared to one at right angles given a little more coupling with the line conductor. Some of the arrangements of counterpoises that have been used in actual construction are shown in Fig. 15.11 for the system.

However overhead earth wire interconnecting the earth electrodes at the various towers are also used for this provides a number of parallel path to the ground fault current. This also provides a shield against lightning strokes. In shuch cases, an effective earthing bond is necessary on each structure that is lower structure at two points firstly, between the ground and the structures, secondly, between the structure and the continuous earth wire. Since the latter is normally supported on suspension clamps to avoid damage due to vibrations, a flexible bond is necessary between the earthwire and the structure so as to be able to take the vibrations. In fact generally used common form of electrode is G.I. pipe, 2.5 m to 3 m long, and 25-30 mm internal diameter, provided with a suitable driving point at the bottom end and at the top with a cup.

15.2.11 Earth Wire Size

Where the earth resistance of individual earth pipes is relatively high a high conductivity earthwire may be necessary to provide a low resistance multiple earth. In fact, the size of the continuous earth wire may have to be selected not only from the point of view of mechanical consideration but also keeping in view of conductivity, as the earth wire provides a multiple earth connection in conjunction with the individual earth pipes at each tower location.

In the Table 15.2 some of the earth wire sizes actually used on lower line construction are given.

Table 15.2

System Voltage KV	Earth Wire Aluminium	Size (Metre) Steel
220	—	19/0.104
220	12/.1151	7/0.1151
220	—	19/2.64
132	18/0.102	19/0.102
132	11/0.093	11/0.093
132	—	7/0.16
66	12/0.1214	7/0.126
66	—	7/0.11

15.2.12 Earthing & Safety

When a fault occurs in a system, then under fault conditions the non-current carrying metal parts of an electrical installation like frames, enclosures, supports, fencing etc. may attain high potentials with respect to the ground so that any person or strong animal touching these or approching these will be subjected to potential difference which may result in the flow of a current through the body of the person or the animal of such a value may prove fatal.

However, experimentally, it has been found that the safe value of current in amps (in rms) which a human body can tolerate is, $I = \dfrac{0.165}{\sqrt{E}}$ for $t < 3$ secs and $I = 9$ mA for $t < 3$ secs, where t is time duration in seconds of the flow of current.

The object of grounding or earthing is to ensure safety of the operating personnel and other human or animal bodies by discharging the electrical energy to the earth and this is done by connecting the non-current carrying metallic parts to the general mass of earth by means of a grounding or earthing system which comprise electrical conductors to conduct fault currents, and the connecting leads. The conductors may be in the form, as stated above, of Mat Grid or multiple electrodes or even a single electrode in the form of rods, plates pipes etc.

15.2.13 Grounding or Earthing Practice

In the power system there are various types of Earthing practice as stated below :

1. Grounding/Earthing at each voltage level shown in Fig. 15.12 (*a*). In the power system there are various voltage levels and hence for protection and safety each voltage is to be grounded at the neutral lead of a generator power transformer bank or grounding transformer. Any generator or transformer used for grounding should be one as for as possible which is always connected to the system. Alternatively a sufficient number of generators or transformers should be grounded to ensure at least one ground on the system at all times. Grounding at each voltage level is shown in fig. 15.12 (*a*).

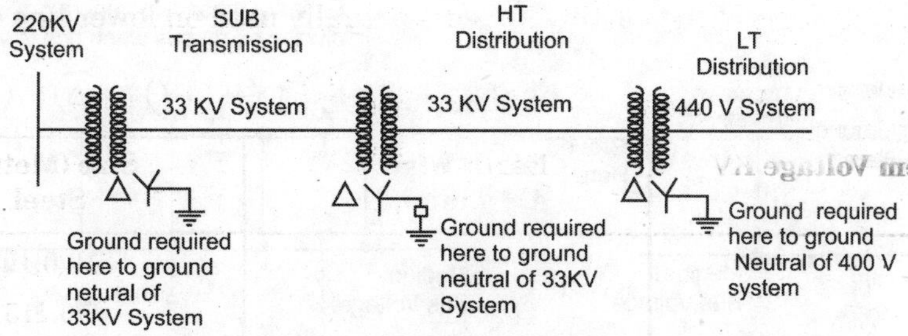

Fig. 15.12 (*a*). Each voltage level is grounded independently.

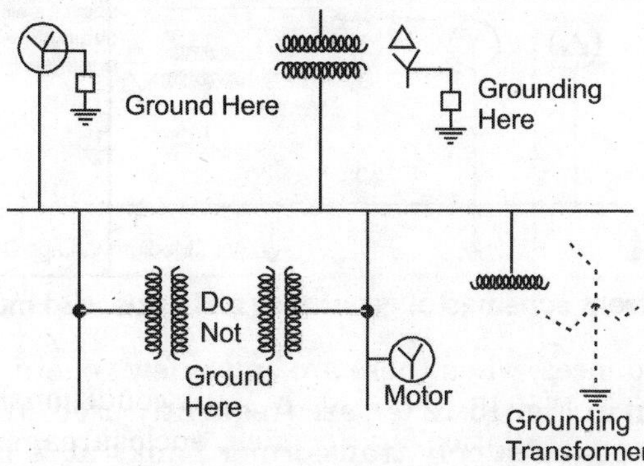

Fig. 15.12 (*b*). Grounding at the power house.

2. Grounding is to be done at power source and not at the load as shown in Fig. 15.12 (*b*). If power sources are delta connected add grounding transformer rather than ground at load.
3. Grounding is to be provided at each major source bus section as shown in Fig. 15.12 (*c*).

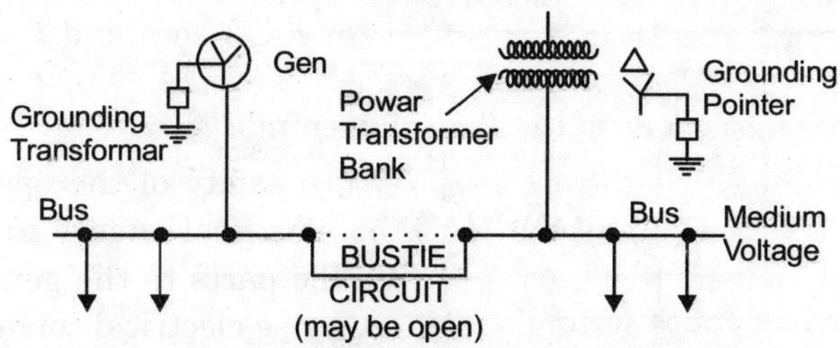

Fig. 15.12 (*c*). Grounding of bus section.

4. Neutral circuits are shown for grounding of following sources.
   (*a*) Single power source shown by Fig. 15.12 (*d*)
   (*b*) Multiple power source shown by Fig. 15.12 (*e*).

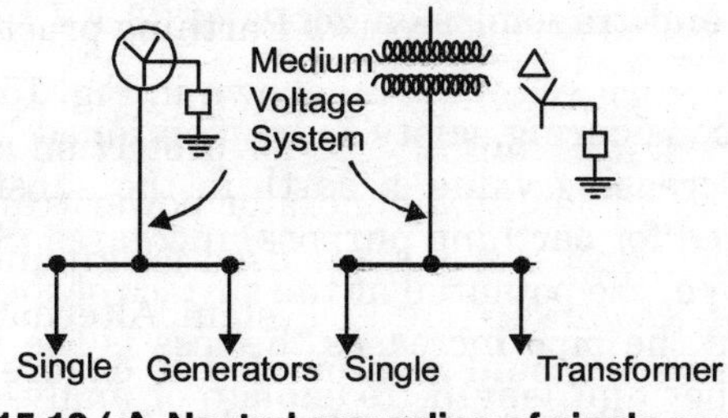

Fig. 15.12 (*d*). Neutral grounding of single sources.

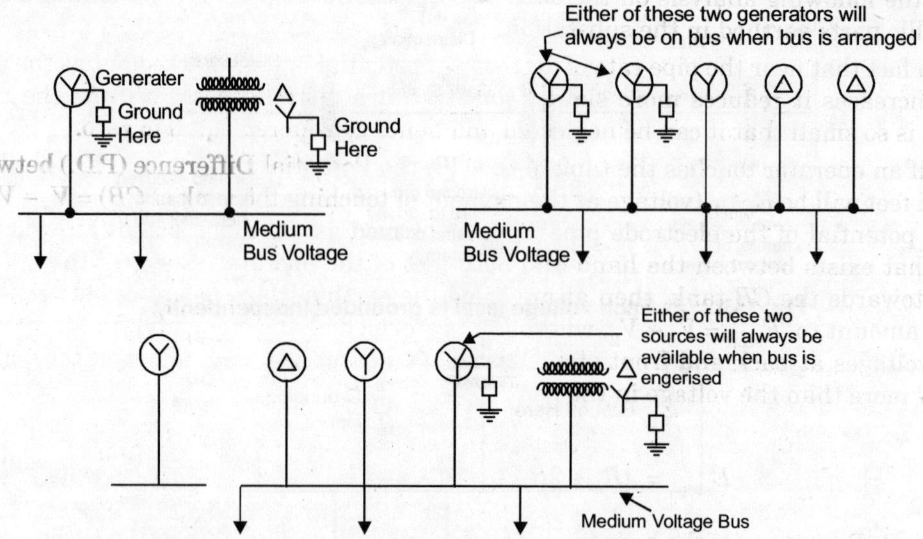

Fig. 15.12 (*e*). Different schemes of grounding a MV bus with multiple sources.

Individual neutral impedances, when there are only a few generator or power transformers banks at a station, as shown in Fig. 15.12 (*e*), are frequently used. When there are more than two or three generators on power supply transformer banks at it is commonly desirable to use one resistor. Each power sources is then connected to the resistor through a neutral bus and neutral switching equipment as shown in Fig. 15.12 (*f*). This keeps the ground fault current to a practical minimum since the ground current from the station is never greaters than can that be supplied through a single resistor. It also ensures the same value of ground current regardless of the number of generators or transformers in use and simplifies ground relaying.

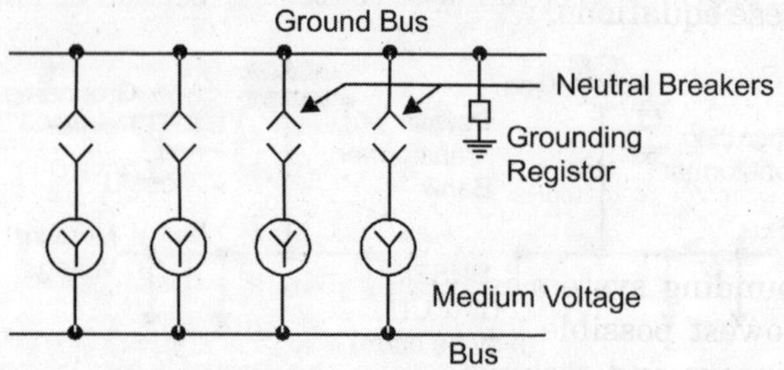

Fig. 15.12 (*f*). Neutral grounding by means of neutral bus and generator.

However, as a thumb rule, sources of all L.T. supplies of 230 V/440 V 3-phase should be solidly grounded while for medium voltage distribution of 3.3 kV/6.6 kV/11 kV. The grounding of neutral at source should earthed or grounded non-effectively. But for all voltage levels relating to sub-transmission and transmission zones of 33 kV and above solidly grounded neutrals are arranged.

Another important point considering safety to be mentioned here. In fact as the current flows through a constantly increasing value of earth in the substation its density drops as the distance from the pipe (used for earthing purpose) increases. Since the current density is highest at the surface of the pipe, the potential at the surface of the earth will be higher round the pipe. As the distance from the pipe increases the less is the difference in earth surface potential between two points per unit length. To explain or analyse the subject further let us consider the following analysis on the surface potential developed on the earthing pipes and the metallic parts earthed in the substation.

It is a fact that near the pipe rate of decrease in potential is very marked but as the distance further increases it reduces more slowly. Now say at a distance of 20 m from the pipe the potential is so small that it can be neglected and hence considered equal to zero.

Say, if an operator touches the tank of the *CB*, the Potential Difference (P.D.) between his hand and feet will be E_{touch} (voltage at the moment of touching the tank of *CB*) = $V_e - V$, where V_e is the potential of the electrode pipe E_{touch} is termed as the touch potential, which is the voltage that exists between the hand and both feet of the operator. Now say the operator is walking towards the *CB*-tank, then at any time the earth surface potential between his foot step will amount to, $E_{step} = V_1 - V_2$, where E_{step} is termed as the step potential and V_2 and V_1 are the voltages at back and front step ($V_2 < V_1$ as nearer the tank (earthed tank of *CB*) is obviously more than the voltage in the back step).

But $$E_{step} = (R_k + 2R_f)\, I_k \text{ volts, and } E_{touch} = \left(R_k + \frac{R_f}{2}\right) I_k \text{ volts}$$

where R_k Resistance of the body in ohms, generally taken to be 1000 Ω, R_F is the grounding resistance of one foot in ohms (taken for all practical purposes to be equal to 6 P_s, where P_s is the resistivity of the soil near the surface of the ground in ohm-meter), I_k is the current in amps (rms) following through the body.

Now substituting the tolarable limits of current, we get

$$E_{step} = (1000 + 6\,P_s)\,0.165/\sqrt{t} \text{ for } t < 3 \text{ sees}$$

and $$= (1000 + 6\,P_s)\,.009 \text{ for } t > 3 \text{ sees}$$

$$E_{touch} = (1000 + 1.5\,P_s)\,0.165/\sqrt{t} \text{ for } t < 3 \text{ sees}$$

and $$= (1000 + 1.5\,P_s)\,.009, \text{ for } t > 3 \text{ sees}$$

Now simplifying these equations, we get

$$E_{step} = (165 + P_s)/\sqrt{t} \text{ for } t < 3 \text{ sees}$$

and $$= (9 + 0.054\,P_s) \text{ for } t > 3 \text{ sees}$$

Then, $$E_{touch} = (165 + 0.25\,P_s)/\sqrt{t} \text{ for } t < 3 \text{ sees}$$

and $$= (9 + 0.0135\,P_s) \text{ for } t > 3 \text{ sees}$$

Hence, when a grounding system is installed in a substation of the power system, the objective is to obtain lowest possible values of E_{step} and E_{touch} to ensure full safety for the operators and the engineers and technicians in the installation. Thus values should never exceed the value obtained from above equation of E_{step} & E_{touch}.

15.3 PROTECTION COMPONENTS AND POWER SYSTEM PROTECTION

In a power system, Grounding or Earthing is certainly very important, for safety and protection, but an electric power system should ensure the availability of electrical energy without interruption to every load connected to the system. When the electric power supply is connected to remote villages the power system would consist of several thousand kilometers of distribution lines. The high voltage transmission lines carrying back power could extend over several hundred kilometres, since all these lines are generally overhead lines are exposed there are many chances of their breakdown due to storms, falling of external object, damage to insulators, short circuit and Earth fault. To save the power system and the system equipments a protection scheme is designed. To make a protection scheme for the system major component like fuses, Isolators, Circuit Breakers and Relays are used. In this section we shall discuss briefly these components.

15.3.1 Fuses

1. What is a Fuse?

A fuse is the very first stage of protection in the electrical power system. Fuse is essentially a short piece of a fusible metal like copper wire of very low diameter of proper rating inserted in a circuit which melts when a pre-determined value of current flows through and thus breaks the circuit by dislocating the faulty section of the circuit.

2. Now what is the Rated Carrying Current of a Fuse?

For a fuse the rated carrying current is the maximum current which it can carry without any excessive heating and melting. The rated carrying current of the fuse depends upon the permissible temperature rise of the contacts of the fuse holder, fuse and upon the deterioration of fuses caused by oxidation of the fusing material.

3. What is Fusing Current?

This is the minimum current at which a fuse element shall melt. In fact it is defined as the minimum current to fuse the wire in such a time interval as shall be necessary for the wire to hence attain its steady temperature.

In fact for round wire the approximate value of fusing current is given by $I = k\sqrt{d^3}$, where k is a constant depending on the property of metal of the wire and 'd' is the diameter of the wire.

This fusing current, in fact depends upon mainly for following factors : **(1)** Material, **(2)** Length, **(3)** Diameter, **(4)** Shape of cross-section of the fuse element, **(5)** Size and location of the terminal, **(6)** Type of enclosure employed, **(7)** whether the fuse is stranded or not. The fusing current for a stranded fuse will be less than the product of the fusing current for one strand and the number of strands as shown in the table no. 15.2.1.

No. of wire	1	2	3	4	7
Fusing Current	1	$1\frac{2}{3}$	$2\frac{1}{4}$	$2\frac{3}{4}$	4

If the strands are twisted together the fusing current will be reduced further on account of the local heat produced which will cause melting of the fuse earlier.

4. Fusing Factor

Fusing factor is defined as the ratio between the minimum fusing current and the rated carrying current. Its value is always more than 1.

Fusing factor of a fuse can be determined from the time-current characteristic of the fuse. This characteristic defines the operating time as a function of the fusing current. In the Fig. 15.13 characteristic for a fuse for different current ratings is shown. From this it can be seen that

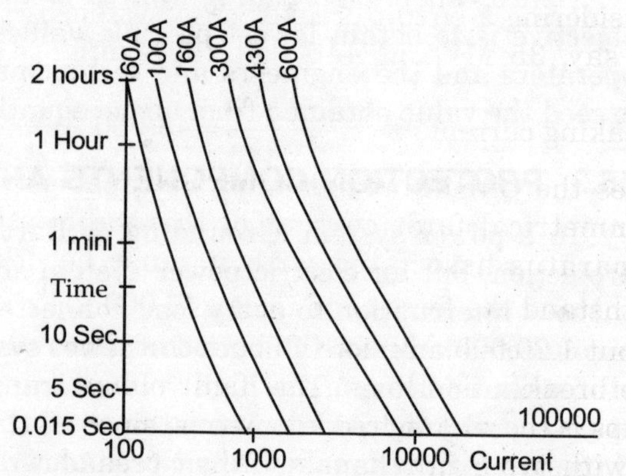

Fig. 15.13. Time-fusing current charac-teristic of a fuse.

the characteristics are inverse type, that means, the time of operation is inversely proportional to the fusing current. The minimum fusing current is the asymptotic value of the time-current curve. For example, the minimum fusing current for a 60 A rated fuse is 100 A and therefore to the fusing factor is 100/60, *i.e.*, 1.66. The inverse character of the time-current curve makes fuse very suited for over-current protection.

5. Prospective Current and Cut-off Current

The Fig. 15.14 shows how ac current is cut-off by a fuse. The fault current normally has a very fast loop, but since it generates sufficient energy, the fuse melts well before the peak of this loop is reached. The rms value of the first loop of the fault current is called the prospective current which may hence be defined as that current which would flow under fault condition when the fuse is replaced by a link of negligible impedance. The cut-off current is the maximum value of current actually achieved. Fig. 15.15 shows the typical fuse characteristic "cut-off" curves for one design of cartridge fuse. In fact, since the cut-off current is largely determined by the amount of energy needed to melt the element the cut-off will vary according to the prospective current.

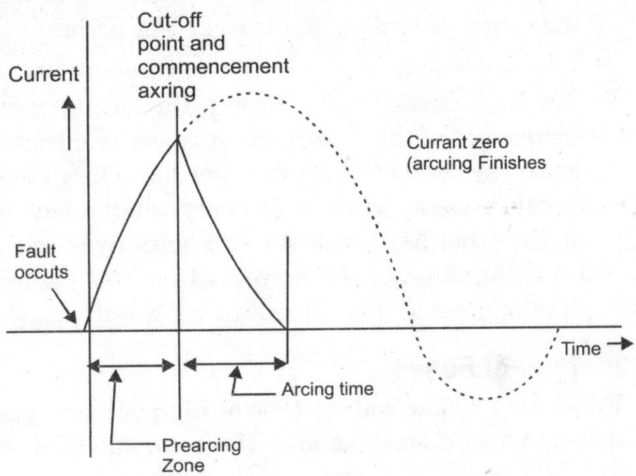

Fig. 15.14. Cut-off of a.c. current by fuse.

Let us take an example. Say a 600 Ampere rated fuse cuts off a prospective fault current of 30 kA (rms) at 14.4 kA (peak) whereas a prospective current of 20 kA (rms) is cut-off at 12.3 kA by the sames fuse. It should be noted that lower the normal rating of the fuse lower is the cut-off current. Infact cut-off is a valuable feature when a fuse is used for providing back up protection to circuit-breakers or other apparatus as considerable reduction in the electro-mechanical forces thrown on the system is achieved. This can be seen by considering a circuit-breaker on a system. With a fault of, say, 35 MVA at 440 volt the symmetrical (rms) breaking currents is : $\frac{35\times10^6}{\sqrt{3}\times440} = 46{,}000$ Amps. Without fuses the Circuit-breaker must be capable of breaking symmetrical fault current of this value and all the apparatus associated with it must be designed to withstand the forces due to say 46,000 × 1.8 × $\sqrt{2}$, *i.e.*, about 1,20,000 amp for the period of time during which the breaker is clearing the fault (where, the 1,20,000 amps is the making current which the breaker may have to withstand if the breaker gets closed on to a fault. If fuses are now inserted and assuming the rating of the fuses to be 300 amps and also assuming these fuses to

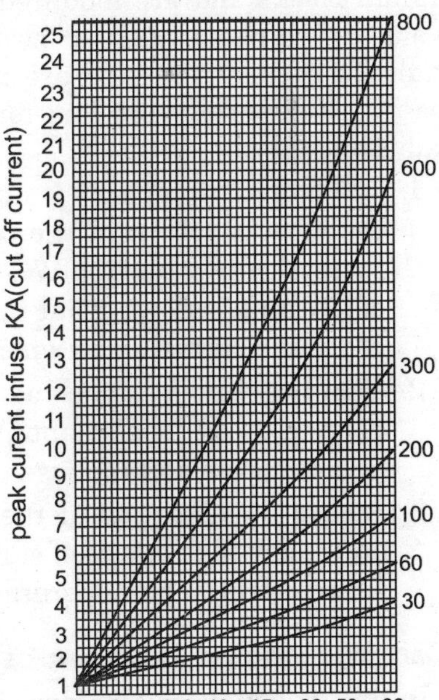

Fig. 15.15. Characteristic "cut-off" curves for a typical HRC cartedge type fuse.

give a cut-off at 29,200 amps then these fuses would operate to clear the fault at the cut-off value of 29,200 amps. This represents a considerable basemant because the circuit breaker has to withstand forces due to this current, the electromagnetic forces being proportional to the sequence of the current.

6. Pre-arcing time (melting time) is the time between the commenciment of the current which tends to cause a break in a fuse element and the instant that the arc is initiated (refer to Fig. 15.14).
7. Arcing time is the time between the end of the pre-arcing time and the instant when the circuit is broken and the current becomes permanently zero (refer to Fig. 15.14).
8. Total operating time is the sum of the pre-arcing and arcing time (Refer to Fig. 15.14).
9. Breaking capacity of a fuse is the rating corresponding to the rms value of the a.c. component of the maximum prospective current and the system voltages. For example, when say that a Circuit Breaker has a breaking capacity of 50 MVA, we expect that the circuit breaker will operate only when a current equivalent to 50 MVA passes. A fuse on the other hard almost never actually passes a current equivalent to its rupturing or breaking capacity. We however take this higher value to a fuse because it prevents the development of this current by its cut-off actions.

15.3.2 Type of Fuses

Fuses may be low voltage type or high voltage type. Low voltage fuses can be further sub-divided into two classes namely the semi-enclosed rewireable fuses and the cartridge type fuses.

A. Low Voltage Fuses : Rewirable fuses

Semi-enclosed rewirable fuses have one or more strands of fuse wire stretched between terminal blocks and are mounted on a moulded or poreclain handle. These types of fuses have a limited use, usually at power points where very low value of fault currents are to be handled. Fuses of this type are made upto about 500 A rated current but their breaking capacity is generally too low for modern needs. These fuses suffer from the following disadvantages :

1. The protective capacity is uncertain. This means that no dependance can be placed upon the wire to interrupt the circuit when a given current flows. For instance although theoretically a No. 155 WG copper wire fuse should interrupt a circuit when a current of 197 amp is flowing but the circuit may be interrupted when a lower current flows or at other circumstances when only a much higher current flows.
2. The fuse wire is subjected to deteroration due to oxidation through the continuous heating up of the element. After a relatively short period the current causes the metal to deteriorate and the fuse operates at a lower current.
3. Accurate calibration of the fuse wire is impossible. It is not possible to determine accurately the amount of current which a fuse will carry before it operates. For instance, a larger length fuse operates earlier than one of shorter length.

C. Cartridge Type (or Totally Enclosed) Fuses

In this type of fuses, the fuse element is encosed in a totally enclosed casing (generally of porcelain) and provided with metal contact on both sides. These fuses are classified as (1) D-Type and (2) Link Type. (1) D-Type Cartridge Fuses are non-interchangeable fuse comprining a fuse base, adaptor ring, cartridge and a fuse cap. The cartridge is pushed in the fuse cap and the cap in served on the fuse cap touching the conductor to complete the circuit between two terminals through the fuse link. In this type standard ratings are 6, 16, 32 and 64 amps. Breaking capacity is 4 KV for 2 to 4 Amps and 16 KA for 6 to 63 Amps rated fuses. These fuses are reliable but obviously requires more time to replace than Rewirable fuse.

C. HRC Cartridge Fuse (Link Type)

HRC means High Rupturing Capacity. In fact, intensive research by manufacturers has developed HRC cartridge fuse. Where large concentrations of power are concerned as in the modern distribution system, it is essential that fuses should have a definite known breaking capacity and also that the breaking capacity should have a high value.

In the Fig. 15.16 HRC cartridge fuse is shown in its simplest from. It consists of a ceramic body having metal end-caps to which are welded fusible silver, or bimetallic current carrying element. The space within the body surrounding the element is completely packed with a filling powder usually quartz (packed in pressure to have hardly any air pockets), which has excellent arc extinguisting properties. Rated values of HRC fuses are of rated voltage, rated frequency, rated current, rated breaking current symmetrical and/or Assymetrical, rated let through current, rated impulse current, rated natural frequency rated amplitude factor. The rated let through current is the largest momentary value of the current during the breaking time.

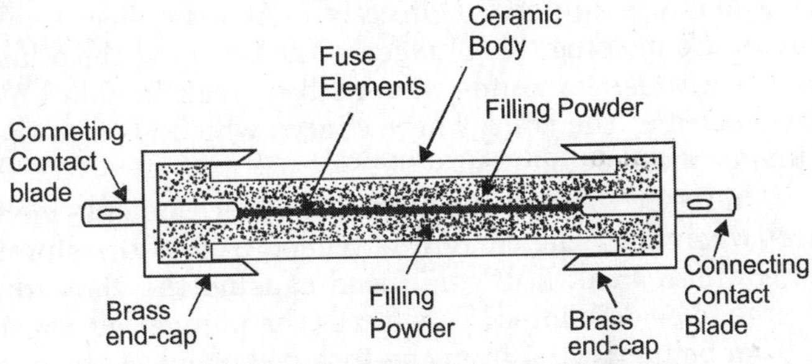

Fig. 15.16. HRC Fuse Cross-section.

In fact HV HRC fuses are switching devices in which the current path is interrupted by means of the fusing of certain parts under the influence of its own current-heat, if the current exceeds the current limit during a certain time corresponding to the fusing time characteristic.

The operation of the fuse consists of the following steps : **(1)** Melting of the silver (fusing) element, **(2)** Vaporisation of the element, **(3)** Fusion of silver vapour and the filling powder-leading to powder 'roping', **(4)** extinction of the arc then the electrical phenomenons associated with the operation of the fuse amp : (*a*) creation of a high resistance as a result of the chemical reaction between the silver vapour and the filling powder, (*b*) change-over of the high resistance into an insulator as the current is cut-off, (*c*) creation of a transient voltage within the fuse at the instant of breaking fault current due to sudden change of energy.

Then the physical phenomena associated with the fuse operation include a sudden rise of temperature and the generation of a high internal pressure derived from the release of energy, a consequent on the interruption of fault current. In the power system, to achieve relialibity or consistantly good performance it is necessary to obtain effective control of both the physical and electrical performance and phenomena which means that the arc should be quenched in a minimum of time. The shorter this is, the higher the breaking capacity provided always that the transient voltage is effectively controlled. Also since arcing also creates a high temperature which rapidly diminishes with arc extinction. The rate of temperature decline should be such as to reduce the risk of gas pressure tending to shatter the ceramic body. In fact, both rapid arc extinction and proper attainment of temperature fall are achieved with a good quality filling powder.

Standard rating are 2, 4, 6, 10, 16, 25, 30, 50, 62, 80, 100, 125, 160, 200, 320, 400, 500, 630, 800 and 1000, 1250 Amp.

15.3.2 (*b*) Liquid Type Fuses

Liquid type fuse has the widest range of application to H.V. systems. They may be used for potential transformer protection or for circuits about 400 a rated current on system upto 132 kV or higher and may have breaking capacities of 6,100 A (350 MVA) at 33 kV.

Another type of H.R.C. cartridge fuse incorporating tripping device is shown in Fig. 15.16 (*a*). The body of the fuse is made of ceramic material with a cap rigidly fixed at either end. These are connected by a number of silver fuse elements. At one end is a plunger which acts as a tripping device for the associated circuit breaker. This may operate either directly or it may close relay circuit. Connecting this plunger with the cap at the other end is a tungsten shunt wire with a weak fusible link introduced at the point where charge which experts the plunger when detonated.

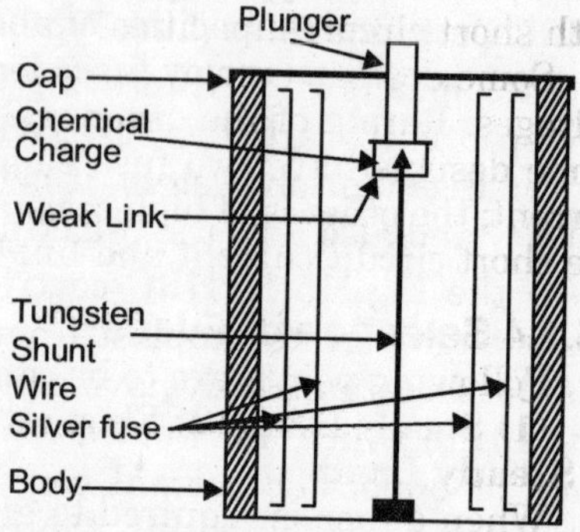

Fig. 15.16 (*a*). H.R.C. Fuse with tripping device.

On the occurrence of a fault, the silver elements melt first whereafter the current is transferred to the shunt wire whose weak link gets fused causing the chemical charge to be detonated. This forces the plunger outwards. A limit is set to its travel to prevent it from being ejected from the fuse body, and at the same time to trap it so that it cannot be reset. It is claimed that the use of the high resistance shunt wire also serves to limit the restriking voltage after fault clearance, acting in a manner similar to that of resistance switching as employed in air-blast circuit breakers.

H.R.C. fuse with a tripping device has the following advantage over a H.R.C. fuse without the tripping device.

In a 3-phase system if a fuse on one phase operates, due to an earth fault, the motors supplied by the system will continue to run on the single phase supply. As a result, the cable and the motor windings may be overloaded to dangerous limit. So when one phase operates it must be ensured that the other two lines are also disconnected. This is achieved with a fuse incorporated with the tripping device since the plunger at the end of its travel closes a trip coil circuit which operates the three phase switch. The plunger also gives a visual indication when a fuse has blown.

Low voltage H.R.C. fuses may be built with a breaking capacity of 16,500 to 33,000 amps at 440 V. (Note that this is the r.m.s. value of a.c. component of the maximum prospective current).

(*a*) Motor Loads

In case of motor loads where starting current is several times (in case of b_q-cage motors it is about 1.6 – 1.8) the normal current, HRC fuses protect motors, the cables and the starter itself from damage due to short circuit and the high starting current. However, protection against overloads should be provided by devices like overload tripping system in the circuit. The use of separate overloads take care of the relatively more frequent occurrences without frequent fuse failure and renewal.

(*b*) Switching of Transformer

The fuse selected for transformer switching should have such a rating as will withstand the transient current surge on switching in. In this case the general practice is to select a fuse which has a rating of 50% to 70% above the normal full load current of the apparatus to be protected along with the circuit.

(*c*) Switching Capacitor Circuit

The same type of selection of protection arrangement as made in case of (*b*) is carried out but here tendency should always be towards the higher percentage. This type of circuit is particularly difficult because on switching in there is heavy current surge of a highly transient and of high natural frequency.

15.3.3 High Voltage H.R.C. Fuses

(*a*) *Cartridge type*, similar in general construction to the 1-v designs, are used upto 33 kV, with short circuit capacities of about 8,760 amps at the voltage (500 MVA three phase).

Some designs employ fuse elements wound to a helix, so as to avoid corona effect at higher voltages. Rating of the order of 200 A at 6.6 kV and 11 kV, and 50 A at 33 kV are available. Some designs have two fuse elements in parallel; one, of low resistance, carries the normal current; the other is of high resistance and, after the low-resistance element has blown, reduces the short circuit current and finally breaks the circuit.

15.3.4 Selection of Fuses

Following points are to be considered for selection of fuse in a power system circuit.

(1) Steady Load, **(2)** Fluctuating Load

1. Steady Load

When a fuse is required to give both overhead and short-circuit protection then it should have a steady current rating not greatly in excess of the full load current of the circuit in order to give close protection. But if the fuse is required to give only short circuit protection then fuse of much greater rating than the normal load current can be selected. Such a fuse will run cooler, hence a larger life and will obviously have a good margin in hand for some temporary overload with one risk of deterioration or serious damage.

2. Fluctuating Load

There are various types of fluctuating loads from which system needs protection which are stated below :

Table 15.2

Metal	Sp. Resistance in Microhm-cm	Melting Point
Tin	11	231.85°C
Zinc	6	419°C
Lead	21.7	327°C
Copper	1.66	1084°C
Silver	1.557	960.5°C
Aluminium	2.788	658.7°C

Tables for Selecting Fuse

(*a*) In the Tables 15.3 and 15.4 fuse wire table for various type fuse wire element timed copper respective are given for ready reference :

Table 15.3. Fuse Wire Table.

(Figures are Approximate and for Commercial use Only)

Diameter in Inches

Fusing Current in amperes	Copper	Aluminium	Tin	Allotin	Lead
1	0.0020	0.0026	0.0076	0.0084	0.0084
2	0.0036	0.0040	0.0116	0.0136	0.0124
3	0.0044	0.0052	0.0148	0.018	0.0164
4	0.0052	0.0068	0.018	0.022	0.020
5	0.0060	0.0076	0.022	0.024	0.024

Fusing Current in amperes	Copper	Aluminium	Tin	Allotin	Lead
10	0.0100	0.0124	0.036	0.040	0.036
15	0.0124	0.0156	0.044	0.048	0.048
20	0.0156	0.0190	0.052	0.064	0.060
25	0.018	0.0220	0.064	0.072	0.072
30	0.020	0.024	0.072	0.080	0.078
35	0.023	0.028	0.076	0.092	0.084
40	0.024	0.030	0.084	0.096	0.096
45	0.026	0.032	0.092	0.104	0.104
50	0.028	0.036	0.096	0.119	0.108
60	0.032	0.040	0.110	0.128	0.124
70	0.036	0.044	0.122	0.144	0.136
80	0.040	0.048	0.134	0.160	0.150
90	0.044	0.052	0.144	0.168	0.162
100	0.048	0.056	0.152	0.180	0.174
120	0.052	0.064	0.176	0.202	0.196

(Fusing current is approximately 1.5 to 2 times the normal rated current)

Table 15.4. Fuse Wire Table for Tinned Copper Wires

S.W.G.	Dia. in inches	Current rating of fuse in amps.	Approximate fusing Current
40	0.0048	1.5	3
39	0.0052	2.5	4
38	0.0060	3.0	5
37	0.0068	3.5	6
36	0.0076	4.5	7
35	0.0084	5.0	8
34	0.0092	5.5	9
33	0.0100	6.0	10
32	0.0108	7.0	11
31	0.0116	8.0	12
30	0.0124	8.5	13
29	0.0136	10.0	16
28	0.0148	12.0	18
27	0.0164	13.0	23
26	0.0180	14.0	28
25	0.020	15.0	30
24	0.022	17.0	33
23	0.024	20.0	38

S.W.G.	Dia. in inches	Current rating of fuse in amps.	Approximate fusing Current
22	0.028	24.0	48
21	0.032	29.0	58
20	0.036	34.0	70
19	0.040	38.0	81
—	0.044	43.0	98
18	0.048	45.0	106
17	0.056	65.0	135
16	0.064	73.0	166
15	0.072	78.0	197
14	0.080	102.2	230
13	0.092	130.0	295

(*b*) In Table 15.5 is given the rated rupturing current of HV HRC fuse inserts related to the appropriate rated currents and rated voltages.

Table 15.5 Rated Rupturing Currents of HV HRC Fuse-wire inserts* Correlated to the Appropriate Rated Currents and Rated Voltages.

Rated current	Rated rupturing current (cos φ = 0.15) in kA ruptering At rated voltages of				
A	3 kV	6 kV	10 kV	20 kV	30 kV
6	75	75	75	30	30
10	75	75	75	30	30
16	75	75	75	30	30
25	75	75	75	30	30
40	75	75	75	30	30
63	75	65	60	—	—
100	75	65	60	—	—
150	75	65	—	—	—
200	75	30	—	—	—

(*c*) In Table 15.6 is given the selection of HV HRC fuse wire inserts for the short-circuit protection of 3-phase transformers.

Table 15.6 Selection of HV HRC Fuse-wire Inserts for the Short-circuit Protection of Three-phase Transformers, when There are no Discrimination* Requirements

Rated power of the transformer (three-phase)	Operating voltage					
	3 kV	6 kV	10 kV	15 kV	20 kV	30 kV
	Rated current of the HV HRC fuse-wire insert					
kVA	A	A	A	A	A	A
20	16	10	6	6	6	6
30	25	10	10	6	6	6

Rated power of the transformer (three-phase) kVA	Operating voltage / Rated current of the HV HRC fuse-wire insert					
	3 kV	6 kV	10 kV	15 kV	20 kV	30 kV
	A	A	A	A	A	A
50	25	16	10	10	6	6
75	40	25	16	10	10	6
100	40	25	25	16	10	10
125	63	40	25	16	16	10
160	63	40	25	25	16	10
200	100	40	40	25	25	16
250	100	63	40	25	25	16
315	100	63	40	40	25	25
400	150	63	63	40	40	25
500	150	100	63	40	40	25
630	200	100	63	40	40	25
800	2 × 150	150	100	2 × 25	40	25
1000	2 × 150	150	100	2 × 40	2 × 25	40

(*d*) In the Table 15.7 is given discriminatively operating HV HRC and LV HRC fuse-wire inserts on the high and low voltage.

Table 15.7 Discriminatively Operating HV HRC and LV HRC Fuse-wire Inserts on the High and Low Voltage Sides of Transformers

Rated power of transformer (three-phase) kVA	High voltage (HV) Rated current of the HV HRC fuse-wire inserts						Low voltage (LV) Rated current of the LV HRC fuse-wire inserts		
	3 kV	6 kV	10 kV	15 kV	20 kV	30 kV	231 V	400 V	525 V
	A	A	A	A	A	A	A	A	A
20	16	10	6	6	6	50	25	25	25
30	25	10	10	6	6	6	80	50	36
50	40	16	10	10	6	6	100	63	63
75	40	25	16	16	10	10	200	100	80
100	63	40	25	16	16	10	224	160	100
125	63	40	25	25	16	10	250	200	125
160	100	40	40	25	16	16	425	224	160
200	100	63	40	25	25	16	500	250	224
250	100	63	40	25	25	16	600	355	250
315	150	100	63	40	40	25	—	425	355
400	200	100	100		40	25	—	500	425

15.3.5 Application of HRC Fuses

The general application of low voltage HRC fuses is on the voltage distribution systems against over-load and short-circuit conditions. Another useful application of HRC fuses is for the back up protection to circuit breakers whose short circuit capacity has been increased owing to extension of generating plant or inter-connection to a value beyond their rated MVA. The fuses are arranged to deal with short circuits exceeding a certain magnitude while the circuit breakers continue to interrupt short circuit within their breaking capacity. The correct operation is answered by adjusting suitably the time current characteristic of the fuses to that of the circuit breakers. This arrangement eliminates the necessity of replacing the Circuit Breaker.

15.3.6 Advantages and Disadvantages of Fuse

There are many advantages in using fuses in the circuit but there are some disadvantages also. Below both these points are discussed.

(*a*) Advantages of HRC Fuses

1. It is the cheapest and "First defence" for the equipment and the Circuit from the effect of overload and other faults.
2. It requires no maintenance except replacement of the burnt fuse wire (in case re-wirable fuses) or replacement by a new one in case of HRC fuses.
3. The increase time current characteristic of a fuse enables it to be used for overload protection. Then the minimum time of operation can be made much smaller than that with Circuit Breakers.
4. It affords current limiting effect under short circuit condition due to cut-off. It interrupts enormous short circuit currents without noise also without flame or gas, smoke as all are suppressed by the quartz powder in the HRC fuses.

Disadvantages

HRC fuses of course has limitations. It cannot do the Circuit Breaker's job. Hence on heavy short circuits discrimination between fuses in series cannot be obtained unless there is considerable difference in the relative size of the fuses concerned. Discrimination between two protective devices (such as two fuses, a fuse and an over current relay or fuse and a circuit breaker) is said to occur when on incidence of a short circuit or an overcurrent only the desired protective device operates.

In fact it is not possible to employ secondary protection to fuses if we cannot use secondary overload relays, earth leakage relays along with the fuses.

It was said in the beginning fuse is a primary defence or primary protection. Hence in all electrical circuits fuse is the very first stage of protection and in the secondary protection as we said above we use switchgear consisting circuit breakers with all kind of relay, which we shall discuss in the following discussions.

15.4 CIRCUIT BREAKERS

Circuit Breakers consist of current carrying contacts called electrodes. These are normally engaged but under pre-determined conditions, separate to interrupt the circuits. When contacts are separated an arc is struck between them. This arc plays an important part in the interruption process as it provides for the gradual transition from the current carrying to the voltage withstanding states of contacts, but it is dangerous on account of the energy generated in it in the from of heat which may result in expensive forces. The main problem in a circuit breaker is, therefore, to extinguish the arc shortly after it has started and before the energy generated by it has reached a dangerous value.

The presence of this arc is at the same time, an advantage and a disadvantage since it is necessary to create a contact gap large enough to withstand the restriking voltage before an attempt is made to extinguish the arc and the extinction should occur at a current zero in an a.c. circuit to avoid creating over voltage problem, some means of prolonging current flow after the contacts have parted is evolved. On the other hand a high current electric arc creates a large quantity of ionsised gas which has to be dispursed in an extremely short time. In all circuit-breaking devices the arc burns in a gas. This gas will be mainly composed of nitrogen in air circuit breakers, hydrogen from vaporised and dissociated hydrocarbon oil in oil circuit breakers, vaporised metal in vacumm circuit breaker, and SF_6 in SF_6 circuit breakers.

Gases have the property that all the molecules and other atomic particles move freely under the influence of forces derived from the environment. The velocity with which these particles move is a function of the temperature, and quite high temperature are encountered in an arc discharge.

15.4.1 Arc Interruption

Since the arc path is purely resistive the arc voltage is alway in phase with the arc current. The magnitude of the arc voltage increases in each successive current loop, since the contacts of the Circuit Breakers are assumed to be separating thereby increasing the arc length and therefore the voltage.

However, there are two methods of interrupting the arc current :

(*a*) High resistance interruption or Rheostatic interruption.

(*b*) Low Resistance interruption of current-zero interruption.

Let us first discuss the method (*a*), in this method the arc is so controlled that its effective resistance increases with time so that the current is reduced to a value insufficient to maintain it with an alternating current the total resistance of the resistance of the circuit also tends to bring the current in phase with the circuit voltage to that at the current zero the voltage appearing across the contact know as restriking voltage is relatively low. The energy dissipated in the arc is however, high and this limits the application of high resistance interruption to low and medium power a.c. switches and circuit-breaking, it is also used for all present d.c. breakers. The reistance may be increased by : (1) Lengthening the arc, (2) Cooling the arc the voltage required to maintain ionization increases with a decrease of temperature so that cooling effectively increases the resistance. (3) Splitting the arc— An appreciable voltage is absorbed at the contact surfaces so that if the arc can be split into a number of small arcs in series, the voltage available for the actual arc column is reduced. (4) Constraining the arc if the arc can be considered into two, a very narrow channel the voltage necessary to maintain it is increased.

(*b*) ***Low resistance interruption of circuit zero interrruption :*** This method is, at present applicable only to alternating current. In this method the arc resistance is kept low until the current zero where the arc extinguishes itself naturally and is prevented from re-striking in spite of the high re-striking voltage. Low resistance interruption also called current zero interruption is used for all high power a.c. circuit breakers.

15.4.2 Current Zero-interruption

There are two main theories explaining current-zero interruption process as : (1) The recovery rate theory (known as Sleepian's theory), (2) The energy balance theory (known as Cassie's Theory).

1. Recovery Role Theory

In this theory the rate at which the ions and electrons recombine to form an arc are replaced by neutral molecules. That is, the rate at which the gap recovers its dielectric strength is compared with the rate at which the restriking voltage rises more rapidly than the dielectric strength the space breaks down and the arc persists.

However, there is an alternative explanation which is afforded by the energy balance theory. Immediately after the current zero the contact space contains ionised gas and therefore has a first part-zero resistance. The rising restriking voltage causes a current to flow between the contacts so that a power is dissipated, as heat, in the contact space. Initially when the restriking voltage is zero, the power is zero and it is again, zero when the space has become fully de-ionised and the resistance is infinitely high. Between these two limits the power dissipated rises to a maximum value. Now if the heat so generated exceeds the rate at which heat can be dissipated from the contact space ionization will persist and the breakdown will occur, giving an arc for another half-cycle.

A rapid increase of dielectric strength or post-zero resistance is necessary for successful interruption and this can be achieved by :

(*a*) *Lengthening of the Gap :* The dielectric strength or post-zero resistance is proportional to length or gap so that lengthening by rapid opening of the contact is an obvious provision. the permissible arc length is limited however, by other consideration *e.g.*, are energy and possibility of transient voltages due to current chopping.

(*b*) *Cooling :* Natural combination of ions and electrons takes place more rapidly if they are allowed tc cool. Cooling by conduction to adjacent parts.

(*c*) *Blast Effect :* Cool non-ionised gas or liquid can be forced into the contact space thereby cooling the ions and electrons and sweeping then away.

15.4.3 Circuit Breaking Terms

There are following Circuit Breaking Terms about which we must have clear knowledge. Hence below discussions are made on following terms associated with Circuit Breaking.

1. Recovery Voltage and Restriking Voltage

At the very point of arc extinction at current zero, the voltages across the breaker terminals which uptil now equalled the arc voltage becomes the voltage of the system. In fact this does not occur instantaneously but occurs under transient conditions, the system acts as an oscillating circuit, there is superposition of transient state, with some damping effect oscillating at the natural frequency of the system on steady state at power frequency.

In fact, at the normal frequency r.m.s. voltage that appears across the breaker poles after final arc extinction has occured is termed as the recovery voltage. Two multiple circuit-breakers, it is expressed as the voltage between the phases of the circuit. In such cases another term, recovery voltage component is used to express the voltage across the breaker of each pole as shown in Fig. 15.17 (*a*). The instantaneous value of the recovery voltage at the instant of arc extinction is called the active recovery voltage as shown in Fig. 15.17 (*b*). The transient voltage that appears across the contact at the instant of arc extinction is called the re-striking voltage. In three-phase circuits the restriking voltage refers to the voltage across the first pole to clear because this voltage is generally higher than that which appears across each of the other two.

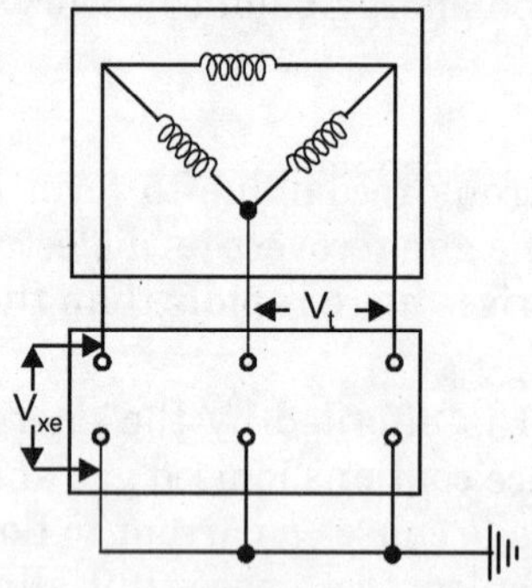

Fig. 15.17 (a). Recovery across poles.

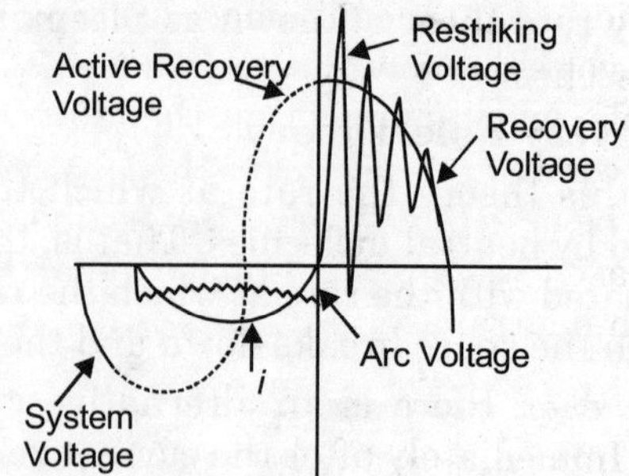

Fig. 15.17 (b). Arc extinction curve.

However, the active recovery depends on the factors : (1) Effect of power factor, (2) The armature reaction, (3) The circuit condition. These factors we shall discuss briefly, below :

1. Effect of Power Factor on Recovery Voltage

When the ratio of reactance to resistance of the circuit is high the active recovery voltage will be high resulting low p.f. But with high p.f., that is when the resistance is high as compared to the reactance the active recovery voltage would be correspondingly lower. Now if (*a*) the p.f. is zero at the instant of arc extinction, the active recovery voltage is at peak value equalling AB (as can be seen Fig. 15.18 (*a*). But in Fig. 15.18 (*b*) the p.f. is 0.5 A and the active recovery voltage at the instant of arc extinction is CD which is less than AB. However, in general the active recovery voltage equal the maximum value of the system voltage multiplied by sin ϕ where, is the power factor angle.

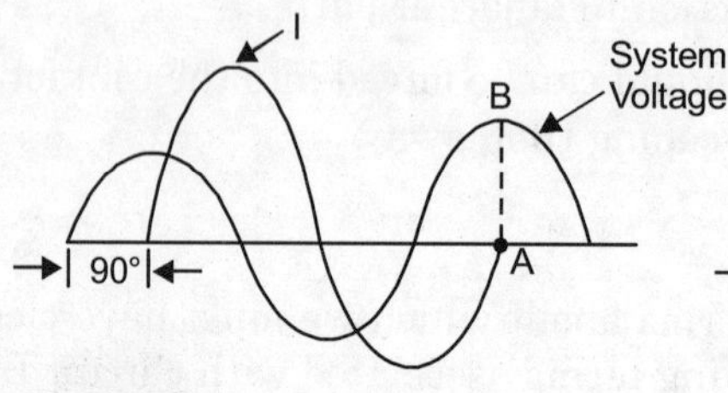

Fig. 15.18 (a). At cos ϕ = 0.

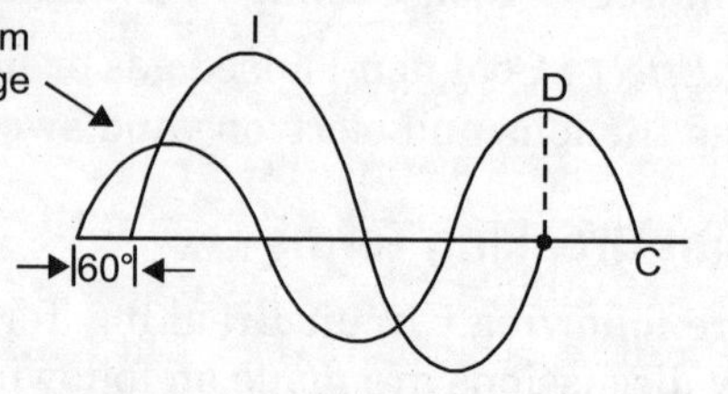

Fig. 15.18 (b). When p.f. cos ϕ = 0.5.

2. Effect of Armature Reaction on Recovery Voltage

Because of demagnitining effect of armature reaction the recovery voltage is less than the normal system voltage. The fault currents flowing in the generator windings are of lagging power factor. They have demagnetising armature reaction, hence they reduce the terminal voltage.

3. Effect of circuit conditions on recovery voltage is another factor that influences the recovery voltage. That means three phase faults that are insulated from earth either at neutral or at fault produce recovery voltages in the first phase to clear which are normally more severe than those produced by single-phase or three-phase fault on system with earthed neutrals which is explained in the following section :

Let us consider an unearthed three phase fault on a three-phase system with the neutral earthed being cleared by a circuit breaker. Now when the breaker opens, it draws out an arc in each phase, assume that the arc in R-phase is the first to be cleared as shown in Fig. 15.19 (a). Then at the instant of extinction of this are the Y and B phases are still arcing and hence the same instantaneous phase voltage, –0.5 V where V is the phase to neutral value of the system voltage. Now the resistances of the ares in the Y and B phases at this instant are neglizible, which means that the fault itself is momentarily at the potential –0.5 V. since the fault is common to all three phases the momentary value of the voltage component in phase R must be (V + 0.5 V) *i.e.*, 1.5 V as shown in Fig. 15.19 (b) recovery voltage component in the first phase to clear on a 3-phase unearthed fault is 1.5 times that on an earth fault assuming an earth neutral as.

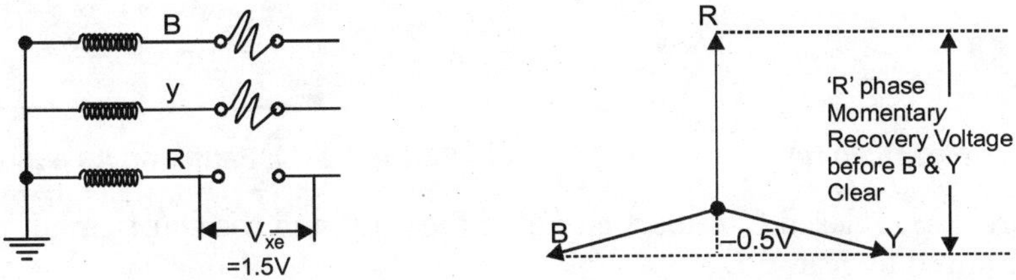

Fig. 15.19. Effect of fault on recovery voltage.

15.4.4 Expression for Active Recovery Voltage in a 3-phase Circuit

To develop the expression for active Recovery Voltage in a 3-phase circuit let the phase-to-neutral (rms) value of system voltage = V_{volts}. Hence maximum value of the phase to neutral system voltage is given by, $V_{max} = \sqrt{2}$ V.

Hence, phase to neutral active recovery voltage = $V_{max} \sin \phi$, where ϕ is the p.f. angle.

Now, if the recovery voltage is less than the system voltage due to the demagnetining effect the phase to neutral value of the active recovery would be $K_3 \cdot V_{max} \cdot \sin \phi$. Where

$$K_3 = \frac{\text{Recovery voltage}}{\text{System voltage}}$$

Then taking into account the circuit condition the phase-to-neutral value of the active recovery voltage becomes $K_2\, K_3\, V_{max} \sin \phi$ where K_2 is called the phase factor.

As explained above, the phase factor has a value of 1.5 in the case of three phase ungrounded fault, k_2, therefore, is = 1 or 1.5.

$\therefore\ v = k_2\, k_3\, V_{max} \sin \phi$ where v is the phase to neutral value of the active recovery voltage.

Now if phase-to-phase value of the active recovery voltage is required then another factor $k_1 = \sqrt{3}$, will be used, and then we get, $v_r = k_1\, k_2\, k_3\, V_{max} \sin \phi$, which is the phase to phase value of the active recovery voltage.

15.4.5 Rate of Rise of Re-striking Voltage (*RRRV*)

Rate of Rise of Restriking Voltage is defined as the ratio of peak re-striking voltage and time taken between voltage zero, and peak voltage.

$$\text{That means, } RRRV = \frac{\textbf{Peak Re-striking Voltage}}{\textbf{Time between Voltage Zero and Peak Voltage}}$$

Here the peak re-striking voltage is the maximum instantaneous voltage obtained by the re-striking voltage. Usually the voltage is in kV and time in micro-seconds so that R.R.R.V. is in kV/μ-see.

In fact, the expression for the active recovery voltage,

$$v_s = v\left(1-\cos\frac{1}{\sqrt{LC}}\right) = \text{V}\,(1-\cos 2\pi\, p_n^t)$$

Where C is the capacitance to the earth and L is the reactance per phase of the system upto the point of short-circuit occurring on a feeder close to the bus-bar. The Fig. 15.20 (*a*) represent a short-circuit occurring on a feeder close to the bus-bars. Fig. 15.20 (*b*) shows the equivalent electric circuit where L and C are defined above. Here R is the resistance per phase of the system upto the point of short-circuit and C is the capacitance to earth of the circuit breaker bushing.

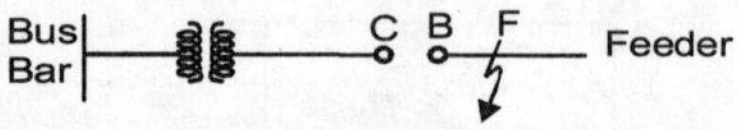

Fig. 15.20 (*a*)

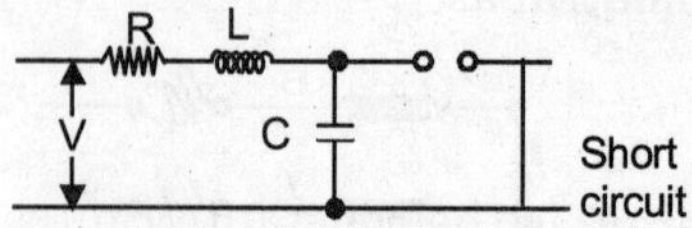

Fig. 15.20 (*b*). Fault and its equivalent circuit.

When the breaker is closed C is short-circuited the fault and the short-circuit current through the breaker is limited by R and L.

If R is negligible compared to L, the short circuit current will lag the voltage by 90°. This condition is shown in Fig. 15.20 (*c*), where i represents the short circuit and l_a represents the arc voltage.

Now with the contact opened and the arc broken, current '*e*' is diverted through '*c*' so that the voltage, which has so far been effective only across the inductance, is suddenly applied to the inductance and capacitance in series which form an oscillatry circuit having a neutral frequency.

$$f_n = \frac{1}{2\pi}\sqrt{\frac{1}{LC}}\ \text{Hz}$$

The initial charging current surge tends to carry the voltage across the condenser, and therefore across the circuit breaker contact to double its equilibrium value. This is the restriking voltage transient which tends to re-establish the arc in the circuit-breaker we can express this transient mathematically analysed below :

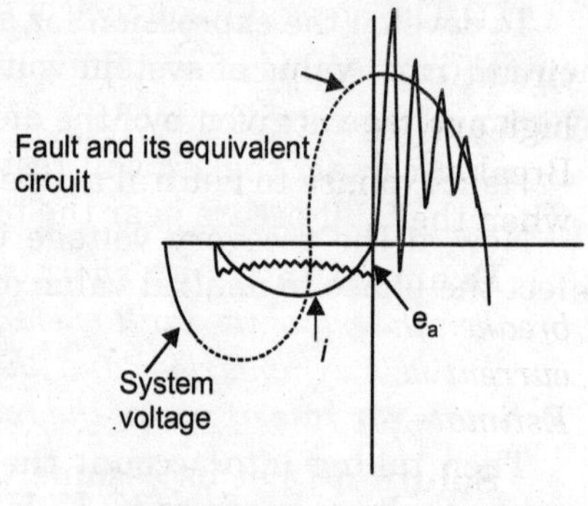

Fig. 15.20 (*c*). Circuit breaker operation.

When CB open and the arcs broken the voltage equation for the circuit, neglecting resistance, is,

$$L\frac{di}{dt}+\frac{1}{C}\int idt = v \qquad \text{...(15.4.1)}$$

where v is the intentaneous value of the recovery voltage (line-to-neutral) which for the short period under consideration can be regarded as constant. Then, differentiating equation (15.4.1) we get,

$$L\frac{d^2i}{dt^2}+\frac{i}{C} = 0 \qquad \text{...(15.4.2)}$$

Solution of this equation is, $\quad i = A\sin(wt+E)$ ...(15.4.3)

amd $\quad \dfrac{di}{dt} = Aw\cos(wt+E)$ ...(15.4.4)

Substituting the initial conditions of $i = 0$ at $t = 0$ in equation we get, $E = 0$. Hence from equation (15.4.4), we get, $LAw = v$.

$$\therefore \quad A = \frac{v}{Lw} = \frac{v}{L}\sqrt{LC} = v\sqrt{C/L} \left(\because w = 2\pi f_n = 2\pi \times \frac{1}{2\pi\sqrt{LC}} = \frac{1}{\sqrt{LC}} \right)$$

$\therefore$ Finally equation (15.4.3) may be written as, $i = v\sqrt{C/L} \cdot \sin(t\sqrt{LC})$.

Again the voltage across the capacitor and therefore across the circuit breaker contact *i.e.*, the restriking voltage (phase-to-neutral) is

$$v_6 - \frac{1}{C}\int idt = \frac{v}{C}\int \sqrt{C/L}\sin\left(\frac{1}{\sqrt{L/C}}\right)dt = \frac{-v}{C}\sqrt{C/L}\cdot\sqrt{LC}\cos\left(\frac{1}{\sqrt{LC}}\right) + k = -v\cos\left(\frac{1}{\sqrt{LC}}\right) + k \quad ...(15.4.5)$$

Where k is the constant of integration which can be evaluated from the initial condition as $v_s = 0$ at $t = 0$ thus, $k = v$.

$\therefore$ From equation (15.4.5) we get,

$$v_s = -v\cos\frac{1}{\sqrt{LC}} + v = v\left(1 - \cos\frac{t}{\sqrt{LC}}\right) = v\left(1 - \cos 2\pi f_n^t\right) \quad ...(15.4.6)$$

is the expression of *RRRV*. The *RRRV* depends on the following factors with reference to equation (15.4.6).

(1) The neutral frequency of oscillation (2) the active recovery voltage. However, for a short circuit taking place near the power station busbars. C being very low, f_n $(= 1/2\sqrt{LC})$ will be high and the rate of rise of the dielectric strength of the insulating medium between the Circuit Breaker contact the arc will restrike. Thus worst conditions for a circuit-breaker would arise when the fault occurs near the bus of the system.

Example 15.1. *In a short circuit test with earthed neutral on a 132 kV, 3-phase circuit-breaker, the p.f. of the fault was 0.3, the recovery voltage was 0.95 of full line value, the breaking current was symmetrical and the re-striking transient had a natural frequency of 16,000 Hz. Estimate the rate of rise of the restriking voltage.*

Solution : Let us assume that the fault is grounded. System voltage line-to-line = 132 KV (rms) then the system voltage, line to neutral $= \frac{132}{\sqrt{3}}$ kV (r.m.s.)

$\therefore$ System voltage, line-to-neutral maximum value $= \frac{132}{\sqrt{3}} \times \sqrt{2}$

Hence the active recovery voltage (line-to-neutral)

$$= k_2\, k_3\, V_{max} \sin\phi, \text{ where } k_2 = 1, k_3 = 0.95$$

Hence, by evaluating value of k_2, k_3, $\sin\phi$

and $\quad \sin\phi = \sin(\cos^{-1} 0.3) = 0.95$

We get $\quad v = k_2\, k_3 \times \sin\phi = 0.95 \times \frac{132}{\sqrt{3}} \times \sqrt{3} \times 0.954$

Maximum re-striking voltage $\quad = 2v = 2 \times 0.95 \times \frac{132}{\sqrt{3}} \times \sqrt{2} \times 0.954$

Now this maximum restriking value is reached in time 't' sec. This maximum restriking value is reached in time 't' $= \frac{132}{2f_n}$ secs.

$$\therefore \qquad \text{R.R.R.V.} = \frac{2\times 0.95\times \dfrac{132}{\sqrt{3}}\times\sqrt{2}\times 0.0957}{\dfrac{1}{16,000\times 2}\times 10^6} = 6.25 \text{ kV}/\mu. \text{ sees}$$

Example 15.2. *The Fig. 15.21 shows one phase of 3-phase 132 kV, 50 Hz circuit breaker situated in a power system. For purpose of analysis resistance is neglected. A fault occurs point at F which is short distance along the line. Develop the expression for the under damped restriking voltage immediately following the current zero.*

Solution : This is an example of double frequency transient. The circuit restriking transient will be due to both the circuit due to the circuit, left of the C.B. and the circuit, right of the C.B., because these two circuits will behave independently.

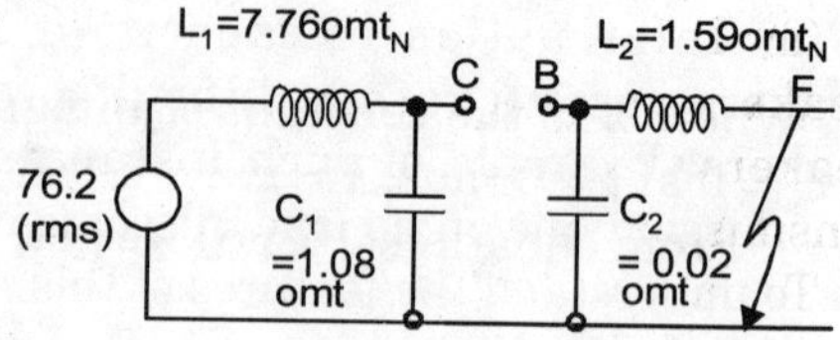

Fig. 15.21.

Before the switch opening the 50 Hz voltage will be divided in proportion to the inductances that is to a close approximation the (peak) voltage of the capacitors will be : $\dfrac{L_2}{L_1+L_2}(76.2\times\sqrt{2}) = 18$ kV (peak). The circuit restriking transient will therefore be $v_s = (76.2\times\sqrt{2}-18)\,(1-\cos w_1 t)+18(1-\cos w_2 t)$

$$\text{where} \quad w_1 = \frac{1}{\sqrt{L_1 C_1}} = 0.397\times 10^{-5} \text{ radian/sec} = 90\,(1-\cos w_1\, t) + 18(1-\cos w_2\, t)$$

$$w_2 = \frac{1}{\sqrt{L_2 C_2}} = 1.77\times 10^{-5} \text{ radian/sec}$$

Example 15.3. *From the following data of a 50 Hz generator : e.m.f. to neutral 7.5 kV (rms), reactance of generator and connected system 4 ohm, distributed capacitance to neutral 0.01 μF, resistance negligible, find, (a) the maximum voltage across the contact of circuit breaker when it breaks a short-circuit current at zero current, (b) the frequency of the transient oscillation and (c) the average rate of rise of voltage up to the first peak of oscillation.*

Solution :

(*a*) Active recovery voltage-phase-to-neutral

$$v = \sqrt{2}\times 7.5 \text{ kV}$$

∴ Maximum re-striking voltage (phase-to-neutral) = $(7.5\times\sqrt{2})\times 2 = 21.2$ kV

(*b*)
$$f_n = \frac{1}{2\pi\sqrt{LC}} = \frac{1}{2\pi\sqrt{\dfrac{4}{314}\times\dfrac{0.01}{106}}} = 14.1 \text{ kHz}$$

(*c*) Since
$$v_s = v\left(1-\cos\frac{t}{\sqrt{LC}}\right)$$

∴ Maximum restriking voltage would occur when

$$\cos\frac{1}{\sqrt{LC}} = -1 \text{ or } \cos\frac{1}{\sqrt{LC}} = \cos\pi$$

or $$t = \pi\sqrt{LC} = \frac{\pi}{2\pi f_n} = \frac{2}{2f_n}$$

$\therefore$ Average rate of rise of restriking voltage $= \dfrac{21.2}{2\times 14.2\times 10^3} = 600.00$ kV/s

15.4.6 Interruption by Circuit Breakers

Circuit Breakers are subjected to (*a*) Capacitive current and (*b*) Low inductive current pose difficult problem for the Circuit Breakers. Below we shall discuss this problem faced by CB and also discuss the Resistance switching to avoid these problems.

(*a*) Interruption of capacitive currents by Circuit Breakers poses a rather difficult problem for the circuit breakers. Example of such instances are opening a long transmission line on no load, disconnecting a capacitor bank etc. To understand the features of this process let us consider the simple circuit shown in Fig. 15.22. The value of capacitance C_L (for example a capacitor bank) is much more than C.

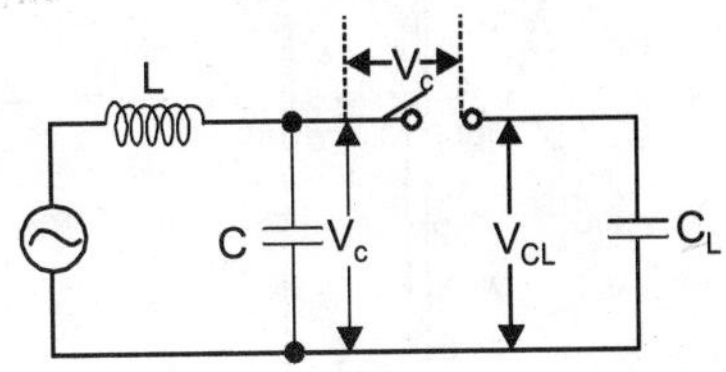

Fig. 15.22.

Now let us refer to Fig. 15.23 which is explaning interruption of capacitive currents. Here in this figure we see that the charging current is easily interrupted near its passing through zero (in the figure, point '*a*') because the current is low and the rate of rise of recovery voltage at the beginning is also low and increase relatively slowly. The openings of the circuit leaves a charge trappen in the capacitance C_L. Then the voltage V_{CL} across the capacitance (indicated in Fig. 15.22) remain at the value it had when the circuit is open which is the peak value of the supply voltage V.

Now after the circuit is opened the voltage V_C across the capacitance C oscillates and approaches the new steady state value. But as the capacitance C is quite small the new value of voltage across this particular capacitance remain at the value of voltage across it will be assumed to be approximately equal the supply voltage V.

Now we find that the recovery voltage V_r is equal to the difference between V_C and V_{C_L}. But at point '*a*' of the Fig. 15.23 its initial value is zero and increases slowly at the start. But half a cycle later when the V_C reverses the V_r (recovery voltage) reaches twice the normal peak value at the point '*b*' of the Fig. 15.23. Then at this instant, the point '*b*' (in the Fig. 15.23) on even earlier it is possible the arc restrikes because the electrical strength between the contact of the circuit breaker may not have attained sufficiently high value. This will reduce the circuit and V_{C_L} will oscillate with a high frequency about the supply voltage V, now, at this instant, the supply voltage is at its negative peak (point 2 in the Fig. 15.23) therefore a high frequency oscillation occurs between + 1 and –3 times the peak value of supply voltage.

However, at the instant when the arc is restruck recovery voltage V_r becomes zero (the point *b* in Fig. 15.23). Now if we assume that at the first zero (at point '*c*') of the high frequency current arc extinction takes place, the voltage across the capacitance C_L attain –3 times the peak value of normal voltage and remain there. Then due to this recovery, voltage reaches point C and starts to increase slowly upto point '*d*'. If at this instant a restrike occurs again a high frequency oscillation of V_{CL} about the supply voltage (shown at point 3 in Fig. 15.23) will take place between –3 and +5 times its peak value. Several repetitious of the restriking cycle can theoretically increase the voltage across capacitance C_L to exceedingly high values (at a rate of nearly 2 times normal peak value of supply voltage per half cycle). But in practical cases, over voltages are kept within 4 times the peak of the supplies voltage.

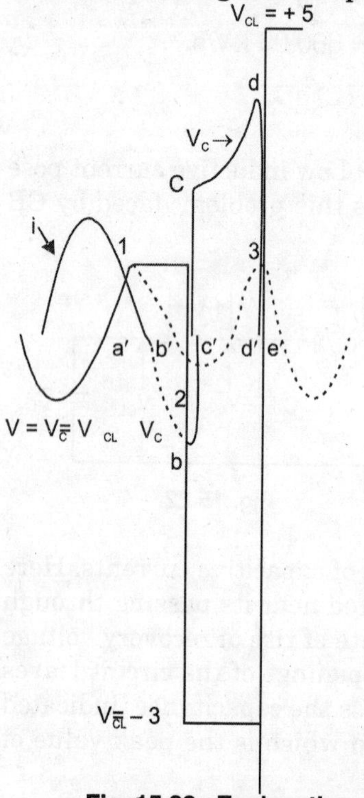

Fig. 15.23. Explanation of interruption of capacitive current.

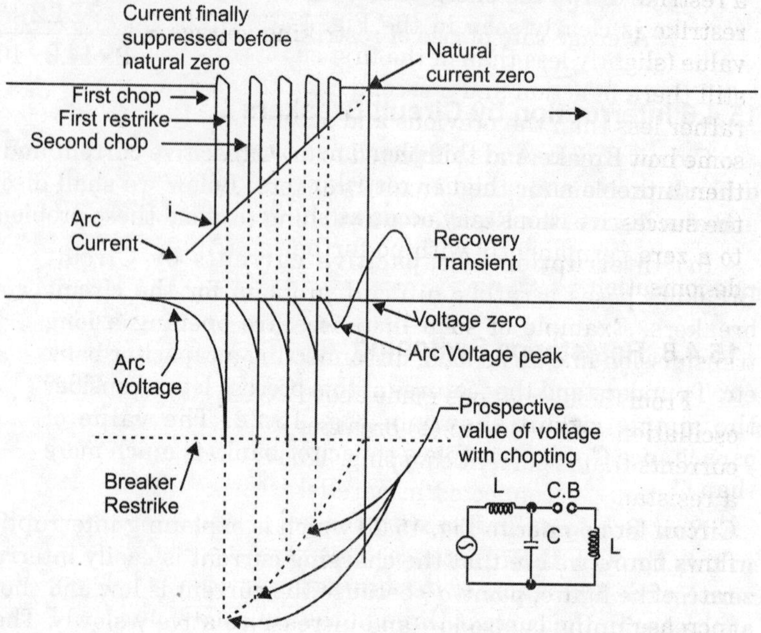

Fig. 15.24. Current chopping.

15.4.7 Interrupting Low Inductive Current

Let us consider the Circuit of Fig. 15.24 (*a*). In such a circuit when interrupting a low inductive current for example, that of a shunt reactor on the magnetining current of a transformer the rapid de-ionising and blast effect may cause the current to fall abruptly to zero well before the natural current zero. This phenomena during interrupting a low inductive current is turned current chopping. However due to this very serious voltage oscillations is caused by current chopping.

Again let us consider the circuit of Fig. 15.24 (*a*). Here let the arc current be *i* when it is chopped down to zero value, as shown by point in the Fig. 15.24 (*b*). As such the stored energy in the inductor, $\frac{Li^2}{2}$ will discharge into the stray capacitance charging the latter to a prospective voltage *e*. Hence, $\frac{Li^2}{2} = \frac{Ce^2}{2}$. That means, $e = i\sqrt{L/C}$ volts and which appears across the *CB* and this prospective voltage may be extremely high in relation to the normal system voltage.

However, the breaker is usually able to relieve the situation by re-striking at some point on the rising chop voltage. In fact, just how far the voltage may rise before restrike depends upon various factors. In fact, the lower the rate of rise of this voltage the more time this is for de-ionisation of the breaker gap and correspondingly high over voltage may be reached. Similarly the effectiveness of the deionising means will influence the restriking voltage. Such a restrike draws the energy out from the capacitance and the voltage there collapses; the first restrike is clearly seen in the Fig. 15.24 (*b*) both as a sudden rise of current to its normal value (slightly less than of the first chop and as a voltage collapse. But, the de-ionising force is still there in action and a second chop takes place. However, the chopped current this time is rather less than the previous and accordingly the rise of voltage (we see it from $dc/dt = i/C$) is some how slower. And this phenomena may give the gap a chance to become further de-ionised then hitherto and a higher restriking voltage may result. This is not shown in the Fig. Then the successive chops may occur as shown in Fig. 15.24 (*b*) until a final chop brings the current to a zero permanently with no further restriking since the gap is now in an advance stage of de-ionisation.

15.4.8 Resistance Switching

From the discussions made above in Articles 15.4.5 and 15.4.6, it is clear that severe voltage oscillations occur due to (*a*) breaking of capacitive currents and (*b*) breaking of low inductive currents that is current chopping. This can be avoided by resistance switching in which process a resistance is connected in parallel with the arc that is in parallel with the contacts of the Circuit Breakers. In fact with the arc so shunted by the resistance a part of the arc current flows through the resistance. This results in decrease in arc current and an increase in the rate of de-ionisation of the are path and the resistance of the arc. This leads to a further increase in the current through the shunt resistances. This build up process continues until the current path through the arc is substituted by that through the resistance either wholly or in greater part. In the latter case, the small value of the current remaining in the arc path becomes so unstable that it is easily extinguished.

However, alternatively the resistance may be automatically switched in by transference of the arc from the main contacts to the aricing contact as in the case of an axial type air-blast circuit breaker, the time required for this action is very short of less than one half-cycle of the current wave. Infact having substituted a metalling path for the arc path the current in its limited value follows through the resistance and is then easily broken.

Resistors are used to ensure dumping of high frequency restriking voltage transient.

It has been established that if the resistance is given a value equal to or less than $\frac{1}{2}\sqrt{L/C}$, the oscillatory nature of the transient will vanish and the R.R.R.V. will be kept within the rating capacity of the breaker. Hence for critical dumping $R = \frac{1}{2}\sqrt{L/C}$.

A resistor in parallel with the gap (between the fixed and moving contact) may be used for the following purposes : (1) To avoid damage when switching capacitive current (2000 – 5000) Ω required for the purpose) in oil circuit Breakers, (2). To suppress Over-voltage due to current chopping (10,000 – 20,000 ohm required) in air blast Circuit Breakers. Air blast CBs are comparatively restrike free or capacitor current switching, as dielectric strength builds up faster. (3) To reduce Circuit damage by reducing/controlling the rate of RRV.

Example 15.4. *In a 132 kV power system, the circuit phase to ground capacitance is 0.01 μF, the inductance is 6 H. Determine the voltage appearing across the pole of a C.B. if a magneting current 5 amp (instantaneous value i) is interrupted. Also calculate the value of resistance to be used across the contact to eleminate the RRRV.*

Solution : As stated above, prospective value of the voltage

$$e = i\sqrt{L/C} = 5\sqrt{\frac{6}{0.01}\times 10^{-6}} = 122.5 \text{ kV}.$$

$$2.\ R = \frac{1}{2}\sqrt{L/C} = 0.5\sqrt{6\times 10^{8}} = 12.25 \text{ ohms}.$$

In the following Articles discussion will be on Circuit Breakers and the types of Circuit Breakers.

15.4.9 Type of Circuit Breakers

In the power system Circuit Breaker is the most important equipment for reliable operation of the system. High voltage Circuit Breakers used in the power system may be classified under two max groups : (1) Oil Circuit Breaker and (2) Oil less circuit Breakers. In fact although the earliest Circuit Breaking devices employed air as the arc-extinguishing medium, but it was not long belfore a better dielectric was needed for better design of insulation and insulating oil soon appeared on the circuit breaking technology. In fact it was initially used in small quantities surrounding the contact. In 1885, Ferranti designed a 500 A 2000 V oil switch. As larger power and voltages were required the oil was used also as an insulating medium in an earthed metal tank and the bulk oil circuit breaker was then born in the electrical circuit breaking technology.

Oil Circuit Breakers can be subdivided into *(a)* Bulk oil circuit Breakers, using a large quantity of oil, *(b)* Minimum Oil Circuit Breakers which are both discussed in the articles below. But before that a brief discussion is being carried out below, on circuit breaking.

15.4.9.1 Circuit Breaking

In all forms of Circuit breaking devices an arc is formed when the contacts part on load or on fault condition. The presence of this arc is at the same time, has an advantage and a disadvantage. Since it is necessary to create a contact gap large enough to withstand the restriking voltage before or attempt is made to extenguish the arc and the extinction should occur at a current zero in an a.c. circuit in order to avoid creating over voltage problems some means of prolonging current flow after the contact have parted is essential. On the other hand a high current electric arc creates a large quantity of ionised gas which has to be dispersed in an extremely short time.

Here it is not the intention to go deep into the physics of arcs, but a brief summary of the main theories relating to the process of arc extinction are useful and helpful in understanding the reasons behind the constructions of the various type of Circuit breaking devices. In all Circuit breaking devices the arc burns in a gas, this gas will be mainly composed of nitrogen in air circuit breakers, hydrogen from vaporised and dissociated hydrocarbon oil in oil circuit breakers vaporised metal in vacuum circuit breakers, SF_6 in SF_6 Circuit Breakers. Gases have the property that all molecules and other atomic particles move freely under the influence of forces derived from the environment. The ability of a gas to conduct is a function of the mobility of electric charge carriers amongst its particles. The velocity with which these particles move is function of the temperature and quite high temperatures arc encountered in an arc discharge. The concept of temperature in a gas discharge is more a measure of particle velocity than of heat in the usual sense, and it is stated in Kelvin which is the Celsius scale starting at absolute zero (–273°C).

In fact, the higher the pressure in the gas the higher will have to be temperature and/or the voltage gradient to ensure that the electrons can acquire the necessary ionising energy in the shorter mean free path. This explains why an increase in any gas leads to an increase in its dielectric strength.

But, at the other extreme the reduction in the gas pressure first of all leads to a reduction in voltage gradient needed for ionisation as the mean free path increases. However, as the space between the electrodes becomes less and less crowded there comes a time when the mean free path exceeds the distance between the electrodes. Now there will be a proportion of electrodes, which travel to the anode and become neutralised there without striking another particle on the way. Also there will be particles which travel out of the electrode space and mix with cooler particles and again play no part in ionisation. So there comes a point in the relationship between voltage gradient and gas pressure, when the voltage gradient has to rise with decreasing pressure in order to maintain the "plasma", as the ionised and conducting body of gas between the electrodes is called.

The Fig. 15.25 illustrate some of the process which take place inside the plasma and also shows typically the arc voltage between the anode and cathode. The numbers refer to the following phenomena.

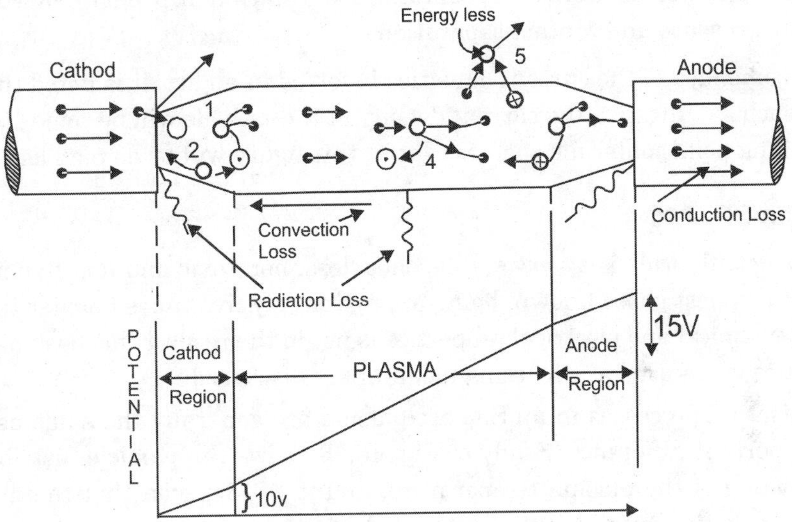

Fig. 15.25. Physical process in the electric arc in CB.

1. A positive ion reaches the cathode is neutralised by collecting an electron and returns as an unchanged particle into the plasma. The energy released by the return of the electron heats the cathode.
2. A simple elastic collision of a neutral particle with the cathode.
3. An electron released from the cathode collides with a neutral atom with sufficient energy to release another electron, resulting in the existence of two free electrons and a positive ion.
4. This shows a similar event to that described in (3), but within the plasma. This is probably the most common event in the hot core of the plasma.
5. This shows an electron and a positive ion being attracted to each other and combining to form a neutral atom outside the plasma. This creates an energy loss from the plasma.

6. Near the anode an ionising collision may not lead to an increase in the electron population as the electron can be absorbed by the anode, or the negative space charge which forms there.

The figure also shows the thermal losses due to radiation, convection and conduction through the contact structure. In practice, conduction plays only a small part in the energy loss from the arc.

The voltage drop along the arc path is in three parts. There is a small voltage drop at each electrode of 10 – 15 V, that as the anode being higher than that at the cathode, and there is a voltage gradient along the arc which has a much stronger relationship to the pressure and thermal conditions in the plasma than it has to the current flowing.

It will be appreciated that as the current being carried by the ionised plasma is nearly always a.c., it will be continually changing in magnitude and the number of charged particles needed for conduction will need to change accordingly. Thus, conditions in the arc will be transient conditions, as well as a complicated relationship between voltage gradient, temperature, pressure and contact separation.

To give some idea of the current densities involved in an arc, it is noted that for refractory electrodes, such as tungsten, the current density at the cathode will be some 1000 A/cm^2, whilst with low boiling point materials such as copper, this figure will be as high as 10^6 A/cm^2.

15.4.9.2 The SF_6 Arc

Sulphur hexafluoride is a very stable, odourless, non-toxic and non-flammable gas which is one of the heaviest gases known, being approximately five times heavier than air. Study of its physical, chemical and electrical properties is not in the respect this book here its particular characteristics when used as an arcing medium are discussed.

The ionisation processes in an SF_6 arc plasma are generally the same as in other gases, with one important difference. Study of Fig. 15.26 shows the particle distribution in an SF_6 arc as a function of the plasma temperature, and it will be seen that there is a substantial population of negative fluorine ions. As expected, with increasing temperature the SF_6 molecule is quickly dissociated into its constituent atoms and then, at a temperature of about 3000 K, ionisation begins with the formation of positive sulphur ions.

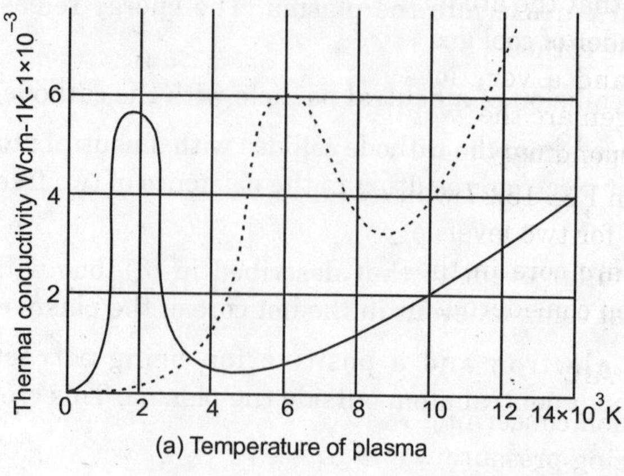

(a) Temperature of plasma

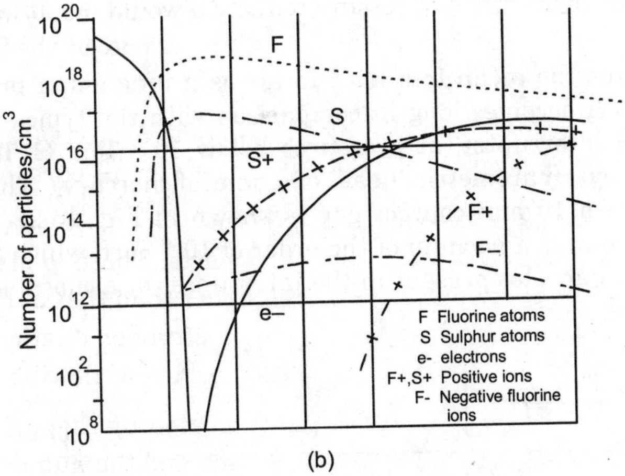

(b)

Fig. 15.26. Characteristics of SF_6.

Contrary to expectation, there is no corresponding appearance of electrons : instead the electrons are captured by the fluorine atoms to form negative ions.

Only when the arc temperature exceeds about 4000 K are there any substantial numbers of free electrons for conductivity to increase significantly. The mass of the negative fluorine ion is 185 times that of an electron and it moves that much more slowly, so that each replacement of an electron by a negative ion is equivalent to a reduction in current flow rate by a factor of 185.

An important characteristic of the arc plasma when considering the arc extinction processes is its thermal conductivity. The upper part of Fig. 15.27 compares the thermal conductivities of SF_6 and nitrogen (N_2). The peaks in the curves correspond to the molecular dissociation temperatures, which can be seen in the lower part of the figure as 2000 K for SF_6, and is 6500 K for N_2. The result of this is that the arc plasma in SF_6 is surrounded by a cylinder of cool gas with a high thermal conductivity and a very low electrical conductivity. In a nitrogen arc the *'mantle'*, as the outer sheath is known, is at a much higher temperature and correspondingly more conductive.

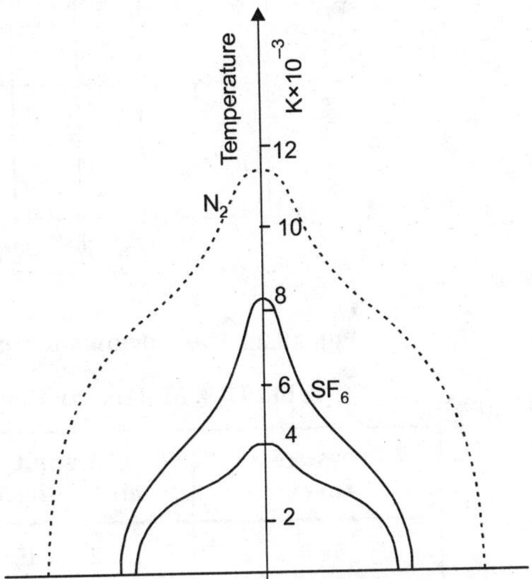

Fig. 15.27. SF_6 and N_2 arc temperature profiles.

This is illustrated in Fig. 15.27 which shows temperature profiles for both gases. For SF_6, two profiles are shown, for two levels of current flow. The result of the cooling mantle and the small diameter of the arc core in the SF_6 arc gives it a very short thermal time constant, which is important when considering the extinction process.

15.4.9.3 The Vacuum Arc

The earlier discussion concerning voltage breakdown as a function of pressure led to the conclusion that a reducing pressure would first lead to a reduction in voltage strength, but a point would be reached where further pressure reduction would result in an increase in voltage strength.

Table 15.9 compares the mean free path in air as a function of pressure and illustrates that the mean free path becomes long in comparison with the typical contact gap of 8 – 12 mm found in vacuum interrupter, at pressures below 10^{-3} Torr (1 Torr is equivalent to a pressure represented by a barometric 'head' of 1 mm of mercury). The effect of this on the breakdown strength of a 10 mm contact gap is shown in Fig. 15.28. To make an effective vacuum interrupter requires a vacuum of the order of 10^{-6} Torr, which represents a condition where for every billion particles present in the interrupter at atmospheric pressure, only one is left after evacuation!

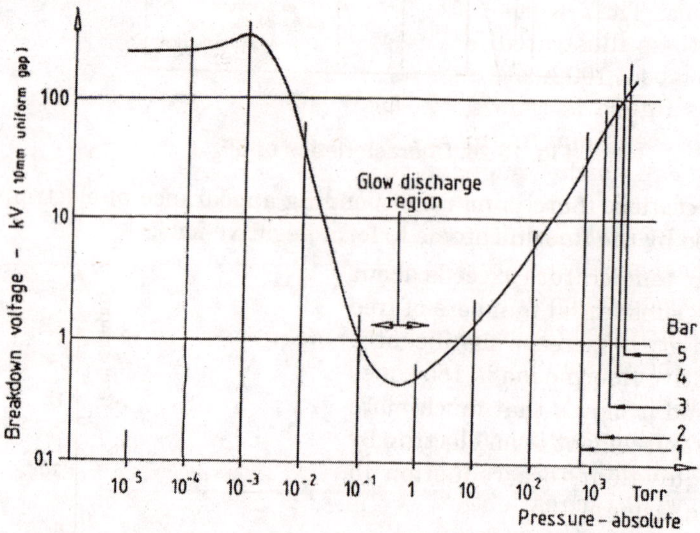

Fig. 15.28. Breakdwon strength of air at different pressures (10 mm gap).

Table 15.9. Molecular Density and Mean Free Path in N_2 at 25°C

Pressure Torr	Density molecules/cm³	Mean free path cm
760	2.5×10^{19}	6.3×10^{-6}
100	3.3×10^{18}	4.8×10^{-5}
10	3.3×10^{17}	4.8×10^{-4}
1	3.3×10^{16}	4.8×10^{-3}
10^{-1}	3.3×10^{15}	4.8×10^{-2}
10^{-3}	3.3×10^{-13}	4.8
10^{-5}	3.3×10^{11}	4.8×10^{2}

Achieving this degree of vacuum economically is just one of the problems that had to be solved before vacuum interrupters could become a commercial proposition.

The vacuum arc started to become important in circuit-breaking in the 1950s when serious efforts were made to obtain worthwhile breaking capacities from vacuum devices. Since the very existence of an arc depends on the presence of gas molecules, it seems at first sight to be a contradiction to talk of an arc in a vacuum. However, the phenomenon that has already been described for the creation of ionisation in normal gaseious arcs serves equally well to initiate an arc in vacuum. In this case the heat created as the contacts finally part generates metal vapour by 'boiling off' some of the contact material. the presence of an electric field in a very small gap quickly generates free electrons and a localised plasma is created in the ionised metal vapour. Although vacuum interrupter contacts are usually made of alloys (or mixtures) of copper and other metals, the copper has the lowest boiling point and the vapour is thus largely composed of copper atoms.

The fact that the arc medium is dependent on the material of the contacts has many implications for the interrupter performance. The arc generated from a single cathode root, or 'spot', as illustrated in Fig. 15.29, will carry currents of up to about 100 A, at which point it splits into two spots. This figure is true for copper, other materials having different maximum currents. Each individual arc is conical, as the magnetic pinch effect concentrates the plasma into a high pressure region. The positive ions produced close to the cathode by the ionisation are ejected from the magnetically constricted high pressure region towards the anode together with the electron stream which is magnetically attracted to the anode. This results in a neutralisation of the plasma resulting in a very low voltage gradient in the arc.

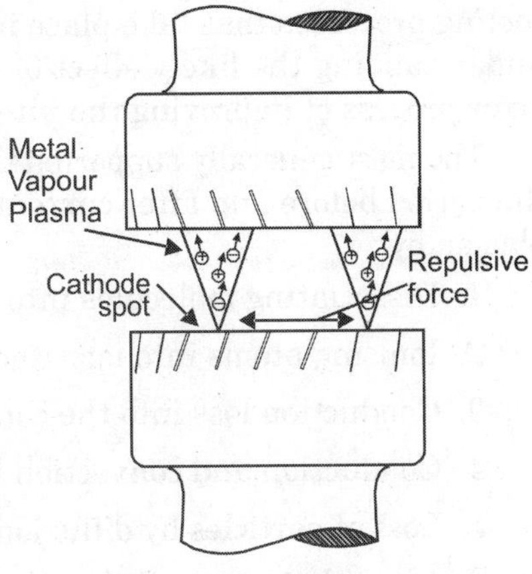

Fig. 15.29. a vacuum arc.

Vacuum arcs also exhibit another unique characteristic in that the separate arc spots repel each other, even though the arcs are carrying currents in the same direction and would normally be expected to attract each other. Whatever the explanation of this behaviour, it is of considerable importance in the process of arc extinction. The individual cathode spots are very small (of the order of 10^{-5} cm^2) and the resulting very high current density (10^5 – 10^7 A/cm) leads to continue 1 melting and evaporation of the contact material, thus maintaining the supply of vapour for continued arcing. These cathode spots are continually in high speed motion with the result that the depth of erosion from each spot is very shallow, of the order of a few tens of microns. This type of arcing is referred to as the 'diffuse phase' and is recognisable by the very low arc voltage (of the order of 30 V) and a very short thermal time constant of a microsecond or less.

However, when the total current through the interrupter exceeds about 1000 A, in a simple butt contact design, the vapour pressure becomes such that normal Maxwellian laws apply, and the arc becomes magnetically constricted. When this happens, in the absence of any external influences, the arc roots become much larger and more deeply heated, so that the conductivity continues through the current zero, and the extinction is prevented. This type of arcing is referred to as the 'constricted phase', and is recognisable by a much higher arc voltage, in the region of 300 V.

The whole development of satisfactory vacuum interrupters has rested on the creation of contact designs and geometries which prevent the excessive heating of the cathode roots, and thus maintain arc time constants in the microsecond region.

ARC INTERRUPTION THEORY

In most fault interruption conditions in an a.c. system, a very rapidly rising voltage is impressed between the contacts of a circuit-breaker as current flow ceases at a current zero.

For successful arc interruption the circuit-breaker designer has to ensure that the contact gap can withstand that voltage so that the arc does not resume as the current reappears with the opposite polarity.

Over the many years since circuit-breakers were first built, researchers have sought to explain theoretically how the extinction process works so that circuit-breakers could be designed from first principles. But the phenomena are so complex, and influenced by so many variables, frequently of a transient nature. that this has proved impossible. Research people working in this field have derived complex equations to describe the ionisation, heating and cooling processes that take place in the arc, and these are considerable help to the designer in understanding the likely effect of design changes and thus guide him in the largely trial and error process of improving the interrupting performance of a circuit-breaker.

The most generally supported theory considers the thermal balance of the arc column during the period before and after current zero. During this period energy is being removed from the plasma by :

1. Dissociating molecules into atoms.
2. Ionising atoms into ions and electrons.
3. Conduction loss into the contacts.
4. Conduction and convection loss from the arc column.
5. Loss of particles by diffusion.

Energy is being supplied to the arc by :

1. The external circuit by current flow or voltage application.
2. By recombination of charged particles.

These phenomena can be represented by assuming that the energy losses can be equated to a rising arc resistance. Study of the form of the energy losses leads to the conclusion that this rise is probably exponential. The application of the transient recovery voltage (TRV) then leads to the appearance of a current after the zero point, known as the *post-arc current*, which initially rises and then falls as the resistance rapidly increases. This is illustrated in Fig. 15.30, which shows the effect of impressing TRV waves (V_1) of increasing frequency across a contact gap where the effective post-arc resistance is increasing exponentially, as shown in Fig. 15.30.

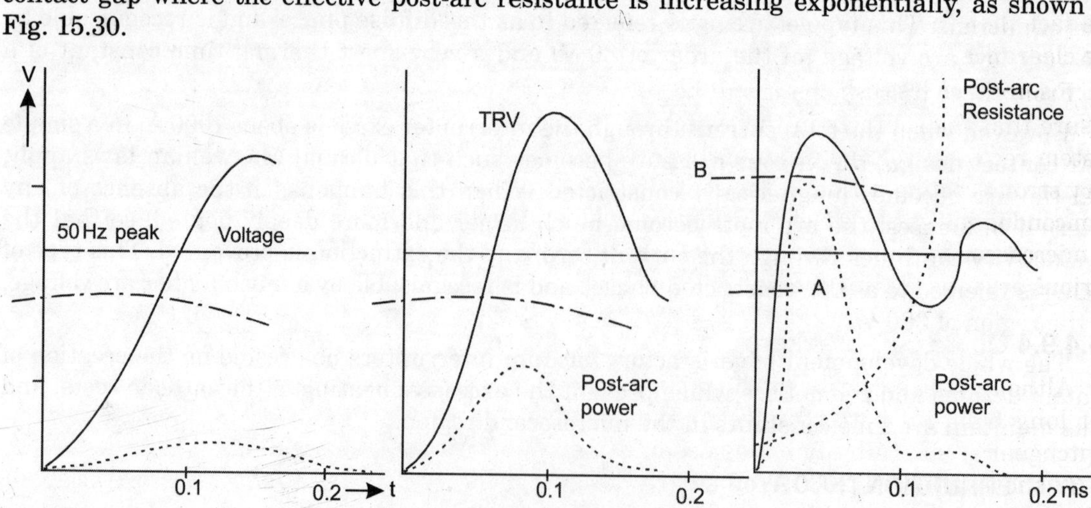

Fig. 15.30. Post-arc conductivity.

The result is plotted as the power developed in the contact zone as a function of time after current zero. If the power removal from this zone by the arc extinguishing means provided by the designer can be represented by a characteristic such as the chain-dotted line shown in Fig. 15.30, then extinction could take place in the first two examples. But the higher power peak in the third case would lead, at point *A*, to a reversal of the resistance rise, a collapse of the recovery voltage and resumption of fault current flow. This would be a failure to clear at that current zero and would indicate a limiting value of TRV for that design. Depending on the type of circuit-breaker under consideration, the next current zero might exhibit a stronger power removal characteristic (as shown at B in Fig. 15.30) such that clearance would follow, even at the higher frequency TRV.

The alternative theory was advanced at an earlier period than the energy balance theory just described and was derived from a presumed dielectric recovery based on breaking the discharge path of the arc by a 'wedge' of clean dielectric. Depending on the effectiveness of the arc extinguishing device the rate of dielectric recovery would exceed the rate of rise of the transeint recovery voltage, and clearance would be effected. Fig. 15.31 shows a typical recovery voltage transient, superimposed on which are two possible gap dielectric recovery curves. Following the 'dielectric race' theory, if the gap recovery was as curve (1) clearance would occur, but if it followed curve (2) the gap would restrike and another half period of arcing would ensue.

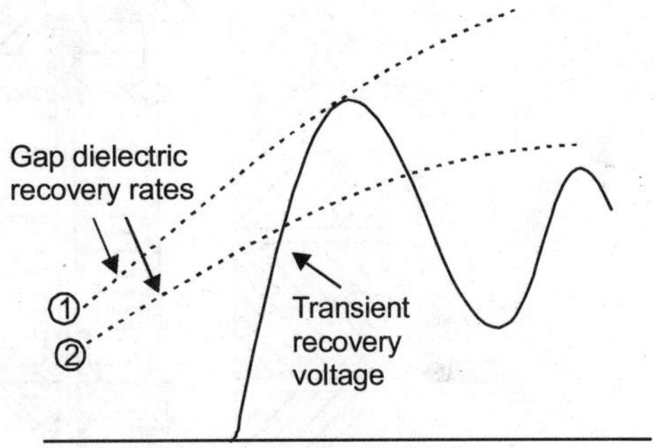

Fig. 15.31. Spark breakdwon.

In practice, calculated rates of increase of dielectric strength fell short of the values demonstrated by the ability of test models to clear, by an order of magnitude or more, which led to greater credence being given to the energy balance theory. In practice, occasions can arise where both theories can play a part in explaining the interruption phenomena, since cases occur where late breakdown of the contact gap takes place, long after deionisation of the gap should have occurred under the energy balance theory. This can only be explained by the fact that the contact gap has not reached the necessary dielectric strength to maintain the clearance.

From these discussions it will be seen that the task facing the switchgear designer is to ensure that the contacts in the switching device establish a gap capable of withstanding the system recovery voltage at an early current zero following contact separation, arrange for a very strong cooling action in the contact gap region and/or arrange for the replacement of the semiconducting gas by a clean, insulating medium at current zero. As the time scale is measured in microseconds, this is not an easy task! How it is achieved in typical circuit-breakers, using various arc interrupting media, will now be discussed.

15.4.9.4 Oil Circuit-breakers

Although the earliest switching devices employed air as the extinguishing medium, it was not long before a better dielectric was needed and insulating oil soon appeared on the switchgear scene. Initially it was used in small quantities just surrounding the contacts. Fig. 15.32 shows a 500 A, 2000 V oil switch designed by Ferranti in 1885. As larger powers and voltages were required, the oil was used also as an insulating medium in an earthed metal tank and the bulk oil circuit-breaker was born.

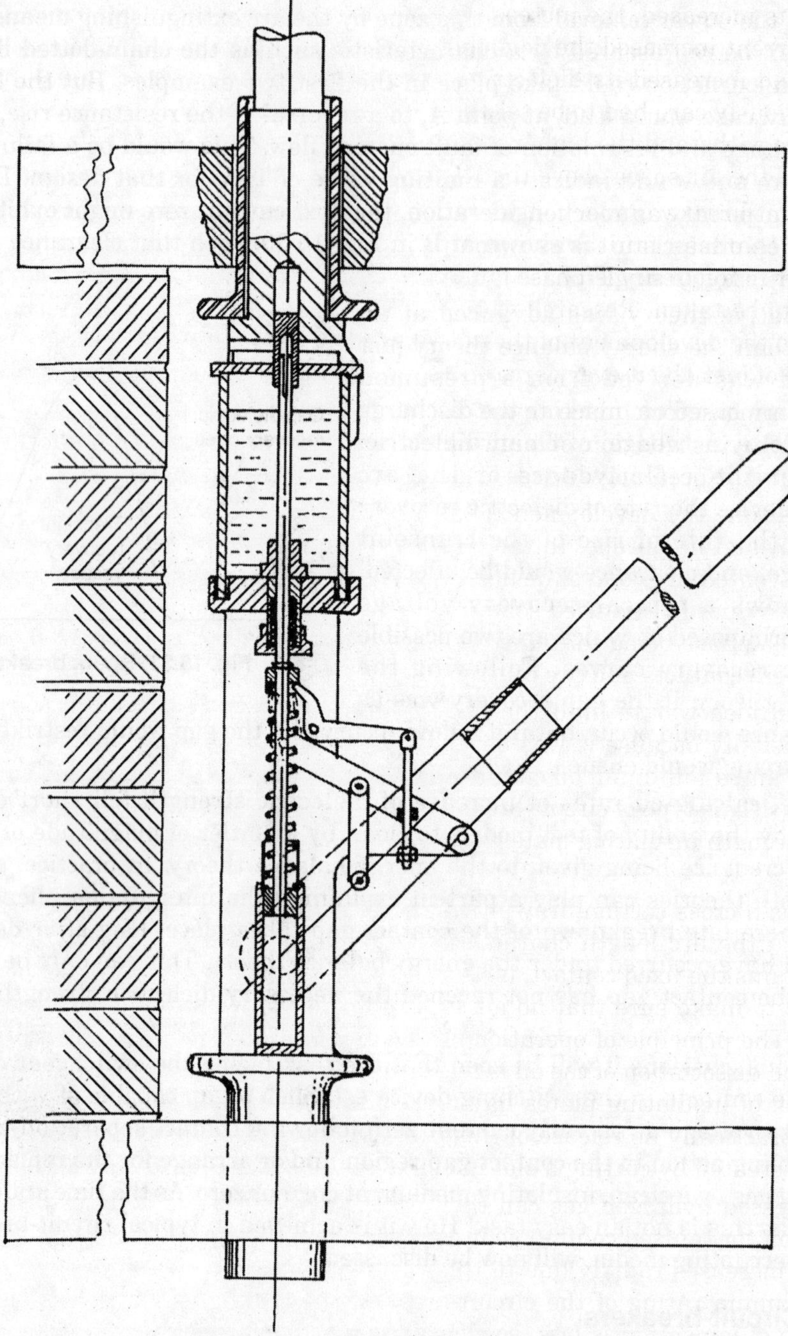

Fig. 15.32. Extract from british Patent 13091/1895: the first oil-break switch. Ferranti 500 A, 2000 V.

Intially, the contacts were parted in the oil bath and the general movement of the oil, the cooling effect of the hydrogen created from the oil by the arc, and the distance which separated the contacts, were the factors that cooled the arc path and created the dielectric strength needed to interrupt the relatively small fault currents in the early days of power distribution. As fault currents increased, the increase in pressure due to the vaporisation of the oil by the short-circuit current increased the deionising properties of the hydrogen bubble surrounding the arc path, and increased its dielectric strength, thus preserving the breaking capacity of the device. Obviously care had to be taken to ensure the mechanical strength of the container was sufficient to withstand the higher pressures as circuit-breaker ratings increased, and the proportioning of the air space above the oil became a critical design parameter.

Eventually a point was reached when the requirement to restrain the pressure with a fully rated three-phase fault was incompatible with the requirements for reliable and consistent interruption of single-phase fault currents of moderate value, and more sophisticated measures had to be taken. Research work carried out at the Electrical Research Association in the UK led to the development of the side-vented 'explosion pot' or the 'cross jet pot', which forms the basis of just about every arc control device used in oil circuit-breakers today.

This development took place towards the end of the 1920s and was followed shortly afterwards by the installation of high power testing stations so that the fault-breaking capabilities of circuit-breaking devices could be scientifically studied.

Also at this time the development of oil circuit-breaking techniques took quite different paths in the UK and in Europe. In the UK the metalclad bulk oil design was refined by the use of the side-vented arc control device, whilst in Europe the 'minimum oil' type of circuit-breaker was introduced. This used solid insulation as the oil container and introduced phase segregation, each phase being contained in a separate, cylindrical, insulating housing. Within that housing were contacts and arc control devices very similar to the ERA pattern.

The original idea was to limit the fire and damage caused if a circuit-breaker failed to clear a fault, usually because it was beyond its rating, but anyone who has seen a bulk oil circuit-breaker and a minimum oil circuit-breaker fail on a short-circuit test plant knows that this is erroneous! This type of circuit-breaker is still produced in large numbers, but now using modern high strength insulating materials for the housings. The earliest models had porcelain bodies.

Fig. 15.33 is a cross section drawing of a typical side-vented arc control device on which is superimposed the arc length characteristic related to fault current. The chamber at the top, which contains the fixed contact, is usually made of metal and is closed, except for a small hole at the top to make sure that no air is trapped inside when the circuit-breaker is being filled with oil. The principle of operation of this device is that the pressure developed by the vaporisation and dissociation of the oil is retained in the pot by withdrawing the moving contact through a stack of insulating plates having a minimum radial clearance round the contact. Thus there is practically no release of this pressure until the moving contact uncovers one of the side vents, created by cutting a slot in one of the plates.

The compressed hydrogen gas can then escape across the arc path, exerting a powerful cooling action on the ionised column as it does so. When the current zero is reached, the post-arc resistance increases rapidly due to this cooling action and clearance occurs. At currents below the maximum rating of the circuit-breaker, the cooling action is less vigorous, 1 but also the degree of ionisation is less, so clearance is still achieved. it may take a little longer to develop the necessary pressure, however, and clearance may be delayed to a later current zero. Design of an effective, wide range, arc control device requires a careful balance of vent areas, vent spacing and contact speed to ensure consistent performance over the full range of currents.

Fig. 15.33 illustrates one of the refinements of the design to improve the interruption of small currents, in that a supplementary oil chamber is shown below the side vents. This is known as a *compensating chamber* and provides a fresh source of oil to be vaporised to feed more clean gas back across the arc path when clearing low currents, such as the load current of the circuit-breaker. The arc length characteristic shows that this feature is brought into use at currents below about 4 kA.

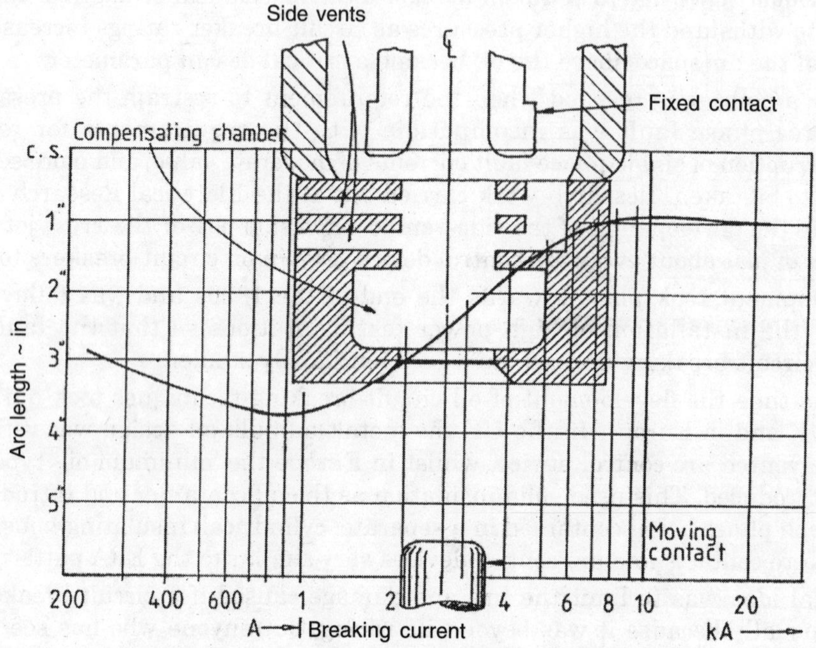

Fig. 15.33. Oil circuit-breaker arc characteristic.

One of the self-compensating features of such a device at the highest fault currents is the fact that the internal pressure which can reach 30 – 40 bar, acts on the moving contact to increase its speed and thus open the side vents sooner. As the diagram shows, once the arc has been extinguished, the moving contact leaves a respectable gap in clean oil before coming to rest. One of the desirable features of such a device is that the arc shall always be extinguished before the contact leaves the 'pot'. Other variations include the addition of small vents in the side of the top chamber, situated so that they relieve excess pressure at full short-circuit currents, but do not release too much pressure when interrupting light currents.

As with all types of circuit-breaker which use the power of the arc to provide the means for its own extinction, there exists a 'critical zone' in the breaking current range. In Fig. 15.33 this occurs at currents in the 900 A region. This arises when the range of operation of the arc control device is being extended. As already indicated, steps have to be taken to cater for the very high pressures that occur at 100% fault currents, yet at low currents the design has to ensure the provision of adequate arc extinguishing effort. The art of the design is to make sure that there is not a 'blind spot' along the way between these two extremes. International standards that lay down high power test procedures recognise this problem by ensuring that tests are arranged to discover the point of maximum arc length and prove the performance accordingly.

Bulk oil circuit-breakers usually have a metal top dome, through which six bushings pass to connect into the system, and the arc control devices, generally of the form illustrated, are attached to the bottom ends of those bushings. The moving contacts are then arranged on a 'bridge', which in turn is fastened to an insulated drive rod. A removable tank, containing the insulating oil (complying with BS 148) is bolted to the dome. Earlier designs had two sets of contacts and arc control devices in series in each phase, and this type of Oil Circuit-Breaker (O.C.B.) was referred to as 'double break'. However, consideration of the voltage division between these two breaks during the recovery period led to some manufacturers producing single-break designs, particularly at the lower normal current and short-circuit current ratings.

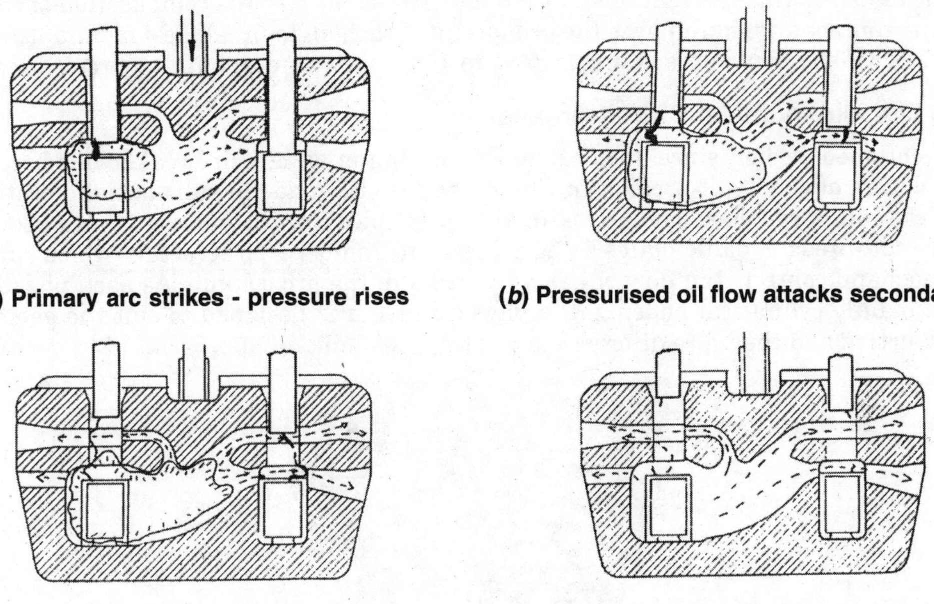

(*a*) Primary arc strikes - pressure rises (*b*) Pressurised oil flow attacks secondary arc

(*c*) Near current zero oil and gas flows cool arcs (*d*) Arc trap purgad as pressure dissipated

Fig. 15.34. The operation of the Caton arc trap.

These o.c.b.s. are generally the same as the double-break variety except that the moving contact is hinged to the end of one of the phase bushings and the fixed contact with its are control device is attached to the bottom of the other bushing. The argument for this comes from the fact that a short circuit usually involves earth, and this means that the bushing and associated fixed contact on the fault side of the O.C.B. are earthed. With the close proximity of the earthed metal tank, there are appreciable leakage capacitances from the conductors to earth and whilst under normal service conditions these are symmetrical, under short-circuit conditions some of these capacitances are shorted, with the result that the division of recovery voltage between the two breaks of a double-break design is far from equal. Meaurements have put this ratio nearer to 80:20 instead of the ideal 50:50 for both devices to be equally effective. So the argument is that if one arc control device can do 80% of the work, it is not too difficult to make it do 100% and save some space and cost as a result.

The only possible weakness of this argument is under capacitance current switching conditions. The leakage capacitance conditions in the O.C.B. are normal when performing this switching duty, so two breaks will give nearly twice the voltage strength as one, for the same gap. Under the rather onerous voltage conditions when switching capacitance loads, the single-break breaker is likely to be more prone to re-ignitions or restrikes.

Many and varied have been the ingenious arc control devices designed over the years to try to extract the maximum rating out of the oil circuit-breaker, and one of the most effective deserves a brief mention. It is always helfpul to use a pressure generating auxiliary arc, if this can be done without sacrificing dielectric strength in the contact zone. The device shown in Fig. 15.34 is quite well known, under the name of its inventor, as the *Caton arc trap*. When the device moves downwards, one of the pair of contacts, which are in series, parts before the other (*a*), and the gas pressure which that arc generates creates a strong extinguishing action on the second pair of contacts (*b*). After arc extinction (*c*) the gas channels are purged by the residual pressure (*d*). The effectiveness of the device is shown by its flat arc characteristic.

At the higher normal current or short-circuit current ratings, the complication of the hinge joint in the contact system removes the economics of the single-break design, and heavy duty O.C.B.s. are usually double-break. Figs. 15.35 to 15.37 show typical bulk oil circuit-breakers.

15.4.9.5 The Minimum Oil Circuit-breaker

As mentioned earlier, since about 1930 the minimum oil circuit-breaker has become the common design of oil circuit-breaker in Europe and is still made in substantial quantities for use in distribution systems. The arc extinction principles are the same as those used in the bulk oil circuit-breaker, baffle plates being placed into an insulated structure which surrounds the contacts and controls the flow of gas in relation to the arc column. As each phase is in a separate, mainly cylindrical housing, the physical layout is designed to suit the geometry of the construction and therefore differs in detail from the bulk oil approach.

Fig. 15.35. Typical bulk oil circuit-breaker. (*Courtesy GEC Distribution Switchchear Ltd.*).

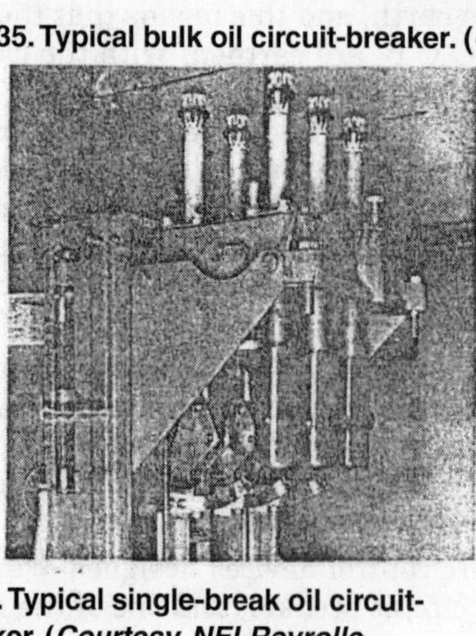

Fig. 15.36. Typical single-break oil circuit-breaker (*Courtesy NEI-Reyrolle Distribution Ltd.*).

Fig. 15.37. Bulk oil circuit-breaker with arc trap. (*Courtesy Yorkshire Switchgear Ltd.*).

One essential difference is that the resultant circuit-breaker has its terminals more easily accessible from the side than from the top, and this leads to considerable differences in the associated cubicle design when used in metal-enclosed switchgear. It also means that all minimum oil circuit-breaker designs are single break. Several ingenious designs of transfer contact for the moving contact have been developed over the years.

The drive to the moving contact is usually a simple crank mechanism with a drive shaft being taken through the housing, sometimes above oil level, but more often below, requiring oil seals. By dividing the tubular body into two compartments, usually in the area of the transfer contact, it can be made convenient for the change in oil volume of the lower compartment due to the movement of the contact rod into it to result in the pumping of the displaced oil into the arc control region.

This will assist in the rate of growth of the dielectric strength of the gap when interrupting small currents and improve performance in that area. This is particularly valuable when switching capacitive currents, and goes a long way to offset the disadvantage of single-break o.c.b.s for that duty. Fig. 15.38 and Fig. 15.39 show typical designs of minimum oil circuit-breaker.

Fig. 15.38. Typical minimum oil circuit-breaker (*Courtesy Calor-Emag Elektrizitats-Aktiengesellschaft*).

Fig. 15.39. Traditional minimum oil circuit-breaker. (*Courtesy AEG-Telefunken*).

AIR CIRCUIT-BREAKERS

Now a days, SF_6, VCB and the oil circuit-breaker developments described above led to the almost total replacement of the earlier designs of air circuit-breaker for service in medium voltage distribution systems by oil circuit-breakers in many countries. Even the oil circuit Breakers are gradually replaced by VCB and SF_6 circuit Breakers for less maintenance and better service than OCBs and ACBs.

Test of C.B.s: All CiBs are to be type tested for short-circuit test, High Voltage and impulse Test.

15.5 RELAYS FOR PROTECTING POWER SYSTEM

15.5.1 Introduction

For power system protection there are various types of relays used for power system protection. Infact, the actuating quantity is an electrical signal, although sometimes the actuating quantity may be pressure or temperature. The electrical type of protective relays can be classified in a number of ways. Relays are accordingly of following types :

Infact the application of a particular relay is decided by its characteristics and other factors like accuracy, operating time burden, method of setting adjustments etc. But above all the relay must be reliable. Moreover, the relay in general is expected to sense the change between healthy and faulty conditions and send a signal when fault occurs. This, in general is achieved by comparing two quantities either in amplitude or in phase. The amplitude or phase relation depends on the system conditions and for a pre-determined value of this relation indicative of a particulars type and location of fault the relay operates. Except in relays such as over current relays, where only one electrical quantity overcomes a mechanical quantity such as the restraint from a spring, it is usual to compare to electrical quantities. As such the device performing the comparison part is the heart of the protective relay and is known as a comparator which devides the operating characteristics of the relay.

Electromagnetic relays that work inherently as amplitude comparators are : hinged armature structure plunger structure, balanced beam structure, induction disc structure with shaded pole driving magnet, those working inherently as phase comparators are : induction cup structure induction, disc structure with watemetric type driving magnets, induction dynamometer structure. Below brief discussions are made on types of protection and relays mostly used for protecting equipments and the power system.

15.5.2 Types of Protection and Relays

In the power system, there are generally two types of protection and these are.

Primary and Secondary (or back-up) Protection

The primary protection is the first line of defence which consists fuses and primary relays clearing the faults in the protected section as fast as possible. But complete reliability or safety can not be ensured this way (100% safety) not only for the protective scheme but also of the associated C.T.s, P.T.s and C.B.s and transformers. Hence some form of back-up, that is secondary protection must be provided. This back-up secondary relays operate if the primary relays fail and cover not only the local section but the next one also and have a time delay long enough for the primary relays to operate if they can.

Let us now explain this with a protection scheme shown in Fig. 15.37 (*a*) for remote back-up protection provided by simple time graded relays. The Fig. 15.37 (*a*) shows a fault at *F* which would normally be seen/sensed by relay R_1 and isolated by the CB_1 at *C*. In the event of failure of this relay, or associated equipment at *C* the fault would be isolated by the operation of the relay R_2 and the $C{\cdot}B_2$ at *B*. Hence here R_2 is back up relay of R_1, and characteristic of this relay is thus shifted upwards as shown in Fig. 15.37 (*b*). Similarly we can explain how R_3 acts as the back-up for R_2 and the CB_2 in the system.

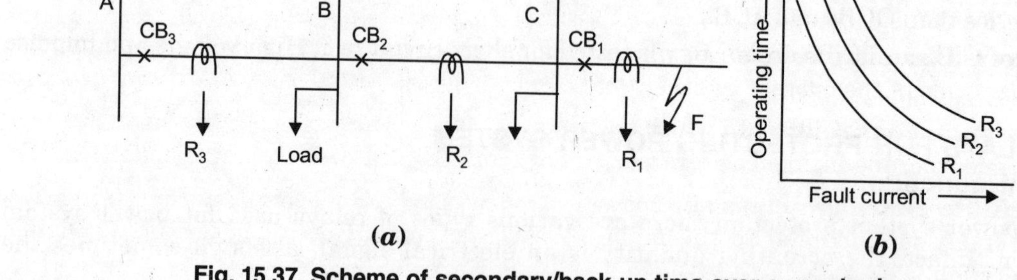

Fig. 15.37. Scheme of secondary/back-up time over current relay.

Hence types of protective relays can be classified depending on their construction and principle of operation as stated below :

1. **Electromagnetic Induction Relays (Induction Relays) :** This type of relays use the principle of the induction motor where torque is developed by induction in a rotor in their operation. This type of relays are energied or acturated by a.c. supply.
2. **Ordinary Electromagnetic Relays :** In this type of relays we find moving plunger, moving iron, attracted armature hinged and balanced beam type of relays are working as ordinary electromagnetic relays.
3. **Electro-thermal Relay :** These relays are overload protection using bimetalic strip.
4. **Electro-dynamic Relays :** These relays operate on the same principle as the moving coil instrument.
5. **Static Relays :** These relays use transisterind chips, or magnetic amplifiers to obain the required operation characteristics.
6. **Physio-electric Relays :** Buchholz's relay is an example of this type relay.

Relays may be also classified based an their application as below :

1. **Over voltage, over current and over-power (Over load) Relays :** In these type of Relays operation takes place when the voltage, current or power rises above a specified value which are mostly instantaneous or induction relays.
2. **Unver-voltage, Over-current and Over-power Relays :** This type of relays operation takes place when the voltage current or power falls below specified value which are mostly instantaneous.
3. **Directionals or Reverse current relays :** In which operation occurs when the applied current assumes a specific phase displacement with respect to the applied voltage and the relay is compensated for fault in voltage-induction current relays.
4. **Directional or Reverse power relays :** in which operation occurs when the applied current and voltage assume a specific phase displacement and no compensation is allowed for fall in voltage-power Relays.
5. **Distance Relays :** In which the operation depends upon the ratio of the voltage to current.
6. **Differential Relays :** In which operation takes place at same specific phase or magnitude difference between two or more electrical quantities.

Relays classified based on their special application :

1. **Inverse Definite Minimum time long-IDMT relays :** In this type of relays the time of operation is approximately inversely proportional to the smaller values of current or other quantities causing operation and tends a definite minimum time as the value increases without limit.
2. **Inverse time-lag relays :** In which the time of operation is quite independent of the magnitude of the current or other quantity causing operation.
3. **Instantaneous relays :** In which operation takes place after a negligibly small interval of time from the incidence of the current or other quantity which causes operation.
4. **Definite Time-lag Relays :** In this type of relays the time of operation is approximately inversely proportional to the magnitude of the current or other quantity causing operation.

15.5.3 Different Useful Electromagnetic Relays (Construction & Operation)

In this section brief discussions will be made on the most useful. Induction type relays for protecting the power system. Here we shall discuss briefly the construction of Relays but certainly discussion will be made on the operation of these relays only.

(a) Earth-leakage (Induction-type) Relay (Non-Directional Over-current Relay)

This type of relay consists of a metallic disc which is free to rotate between the poles of two electromagnete as can be seen in the Fig. 15.38. Here we also see that the spindle of the disc carries a moving contact which bridges two fixed contacts when the disc rotates through an angle which is adjustable between 0° and 360°. By adjusting this angle the travel of the moving contact can be adjusted so that the relay can be given any desired time setting which is indicated by a pointer on a time setting dial, which is calibrated from 0 to 1. These figures do not represent the actual operating times but are multipliers to be used to convert the time known from the relay name plate curve into the actual operating time.

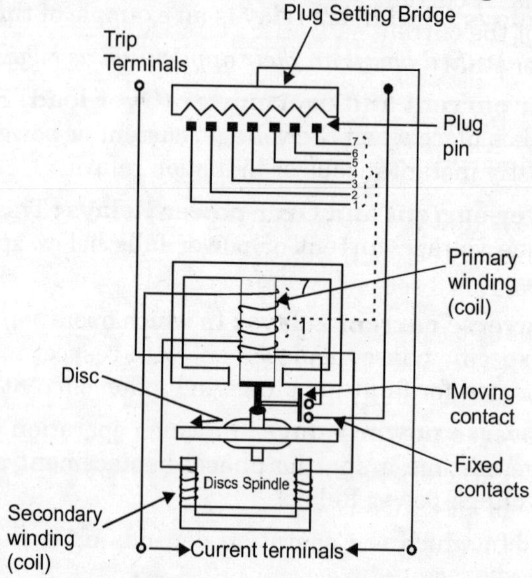

Fig. 15.38. Non-directional induction relay.

As shown in the figure, the upper electromagnet has a primary and a secondary winding. To the secondary of a C.T. (in the line to be protected and is provided) the primary of the upper electromagnet is connected. The primary wind is provided with the tappings say for taps 1 to 7 which are connected to a plug setting bridge which is usually arranged to give a seven section of tapping, the over current range bring 50 percent to 200 percent in steps of 25%, and the earth fault 10% to 70% to 80% in steps of 10%. These values are percentages of the current rating of the relay.

In fact, these values are percentages of the current rating of the relay. Thus a relay may have a current rating of 5A, indicating that it is suitable for use on CT having a secondary current rating of 5A but with a setting of 50% the relay would start to operate at 2.5 A. Similarly if set at 200%, it would start to operate at any desired tapping and therefore, current setting can be selected by inserting a pin between the spring loaded jaws of the bridge type socket at the appropriate tap value. Then when the pin is withdrawn for the purpose of changing the setting while the relay is in service the relay automatically adopts a high setting, thus ensuring that the C.T. secondary is not open circuited and that the relay remain operative for fault during the process of changing the settings. The secondary winding surrounds the limbs of the lower electromagnet as well. The torque exerted on the disc is due to the interaction of eddy current produced there in by means of the leakage flux from the upper electromagnet and the flux from the lower electromagnet and these two fluxes having a phase displacement between turn.

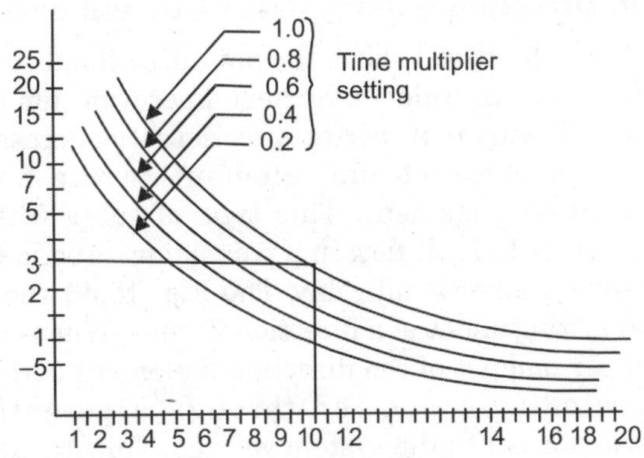

Fig. 15.39.

A set of typical time current characteristic curves of this type of relay is shown in Fig. 15.39. The curve shows the relation between the operating current in terms of current setting multipliers along the x-axis and operating time in seconds along the y-axis. A current setting multiplier indicates the number of times the relay current is in excess of the current setting. The current setting multiplier is also referred to as plug setting multiplier P.SM.

$$\text{Hence} \quad \text{P.S.M.} = \frac{\text{Primary Current}}{\text{Primary Setting Current}} = \frac{\text{Primary Current}}{\text{Relays Current Setting} \times \text{C.T. Ratio}}$$

In fact, usually the case is that the rated current of the relay is equal to the rated secondary current of C.T. From the figure the operating line when current-setting multiplier is 10 and the time multiplier is set at 1, is 3 seconds. This is sometimes called the basic 3/10 curve.

Hence, it is evident that at the same current setting but the time multiplier set at 0.8, the time of operation is 2.4 seconds. Thus to get the actual time of operation against any particular time-multiplier setting multiply the times of operation of the basic curve by the multiplier setting. Thus in this example the time of operation is, 3 × 0.8 = 2.4 sees.

Then the time current characteristics of Fig. 15.39 are the Inverse Definite Minimum Time (IDMT) type since the time of operation is approximately inversely proportional to the smaller values of current and tends to a definite minimum time as the current increases with limit. We can obtain the D.M.T. characteristic by saturating the iron in the upper magnet so that there is practically no increase in flux after current has reached a certain value. This results in the flattening out of the current-time curve.

Example 15.5. *An IDMT over current relay has a current setting of 150% and has a time multiplier setting of 0.5. The relay is connected in the circuit through a C.T. having ratio 500 : 5 amps. Calculate the time operation of the relay if the circuit carries a fault current of 6000 A. The relay characteristic is shown in Fig. 15.39.*

Solution : Fault current is $6000 \times \frac{5}{500} = 60$ A

$$\text{Plug Setting Multiplier (P.S.M.)} \; \frac{\text{Actual ams current in Relay}}{\text{Setting current}} = 60 \times \frac{1}{5 \times 1.5} = 8$$

Times from Fig. 15.39 against this multiplier of 8 = 3.15 see

∴ operating time = 3 × 15 × 0.5 = 1.575 see.

(b) Directional Relay (Over Current or Earth Fault Relay)

As discussed above the non-directional relay discussed above can operate for fault in either direction. In order to achieve operation for the fault flowing in a specific direction, it is necesary to add a directional element to the non-directional element. This type of relay which responds to fault flow in a particular direction is called a directional rélay. The Fig. 15.40 shows an arrangement of a directional relay. Where the upper magnet of the directional element carries a winding connected through a potential transformer to the system voltage. This is called the voltage coil of the relay. The lower magnet of the same element carries another winding (current coil of the relay) which is energised-through a C.T. by the fault current. The contacts of the directional element are connected in series with another winding over the lower magnet of the non-directional element.

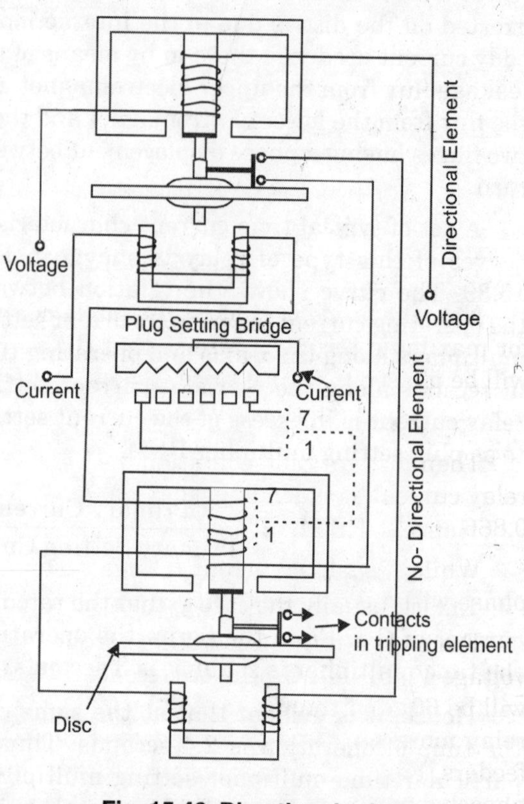

Fig. 15.40. Directional relay.

Now, when a fault takes place the fault current flows throughs current coil of the relay which produces a flux in the lower magnet of the directional element while the current in the voltage coil produces another flux in this upper-magnet of the directional element. The two fluxes produce a torque tending to close its contacts (directional element contacts). However, the relay current also flows through the winding over the upper magnet of the non-directional element and therefore produces a flux in this magnet. This flux induces an e.m.f. in the winding over the lower magnet of the non-directional element. Since this winding provides a closed path, the induced emf circulates a current which, therefore, produces another flux. The two fluxes then produce a torque on the disc of the non-directional element to close the contacts in the trip circuit. Infact the directional element must operate first to operate the non-directional element. In the phasor diagram of Fig. 15.41 (*a*), V is the relay voltage I be the relay current which is leading the relay voltage by an angle θ. V and I produce flux Φ_V and Φ_I, Φ_V lagging by angle ϕ and Φ being in phase with I.

(c) Directional Relay Connections

Regarding directional relay connections there are two types of relay connections in use : (1) The 30°-relay connection with a maximum torque angle of 0° and (2) The 90°-relay with a maximum torque angle of 45°. What is meant by the relay angle. The relay angle is defined as the angle between the voltage and current supplied to the relay under balanced three phase unity p.f. condition. In the Fig. 15.41 (*b*) phasors I_R, I_Y, I_B and V_R, V_Y, V_B represent the phase-currents and voltages of a 3-phase balanced system with unity p.f. conditions. Phasor RB shows the system red phase to blue phase. It is lagging I_R, V_R phasor by 30°. Let the relay element be supplied with current I_R and voltage V_{RB} in phase relation as shown, that is, I_R leading V_{R-B}. This should be together separate by 30°. The connection will therefore be referred to as

30° relay.

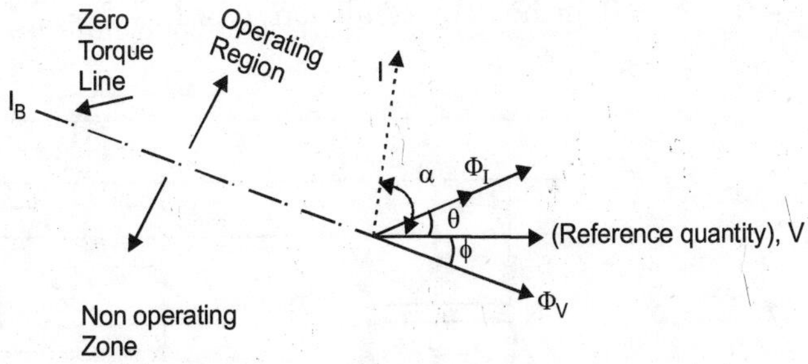

Fig. 15.41 (*a*). Directional element phasor diagram.

Now the angle α, that is, the angle between the current and voltage supplied to the relay for maximum torque is zero so that the position of relay current phasor for maximum torque will be phasor V_{RB}. Also since, $T = kVI$ can $(\theta - \alpha)$, where, $\alpha = 0°$, so that

$$T = KVI \cos \theta \quad ...(1)$$

Then for unity p.f. condition, *i.e.*, system current I_R in phase with system voltage V_R, the relay current I_R is leading relay voltage V_{RB} by 30°, *i.e.*, $\theta = 30°$, so that $T = kVI \cos \theta = kVI \times 0.866$.

While if system current I_R lags system voltage V_R by 30°, the relay current I_R will be in phase with relay voltage V_{RB} so that $\theta = 0$ and $T = Vk_1 = T_{max}$.

From (1) it may be seen that torque is maximum when system current I_R lags the system voltage V_R by 30°. On the occurrence of fault, I_R may lag V_R by, let us say 90° in which case θ will be 60° and the torque will be 0.5 T_{max}. We thus see that the initial lead to 30° makes the relay more sensitive at low power factors. The 30° relay is in general satisfactory for plain feeders. For transformer feeder 90° relay having 45° maximum torque angle is preferred. The phasor diagram of 90° connected directional relay with 45° maximum torque angle is shown in Fig. 15.41 (*c*). Torque T_{max} = kVI cos (45 ° – 45°).

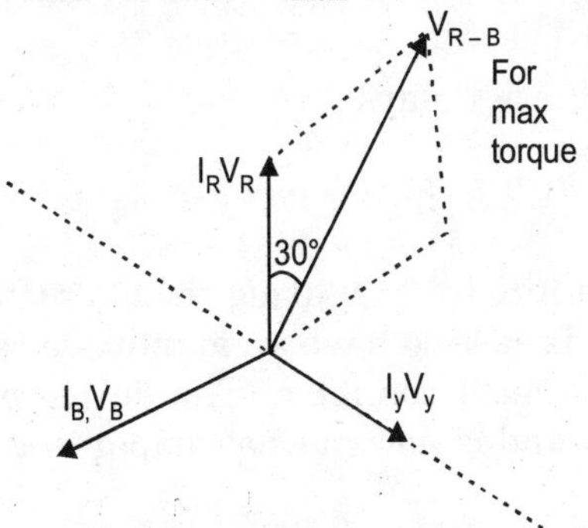

Fig. 15.41 (*b*). Phasor diagram of 0° directional relay with zero maximum torque angle.

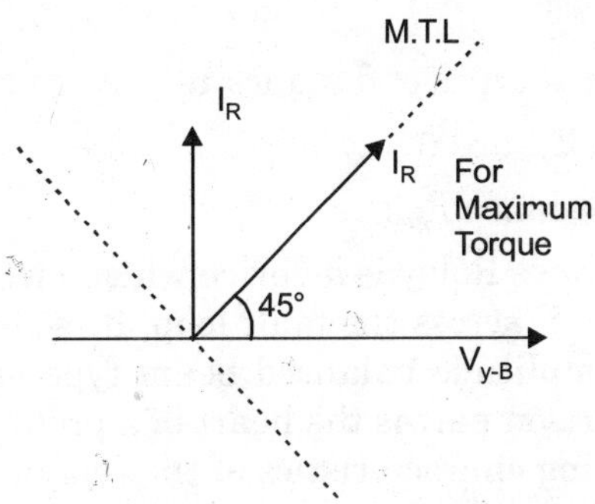

Fig. 15.41 (*c*). Phasor diagram of 45° connection of directional relay.

For θ = 45°, α = 45°, *i.e.*, T_{max} = kVI and *T* for 90° connection = kVI cos (90 – 45) = 0.707.

$$\therefore \quad kVI = 0.707\, T_{max}$$

Then the Fig. 15.42 shows 30° connection directional relay controlling earth fault on all three phases of a three phase circuit is energised by residual current and residual voltage. The residual current is obtained by summing the current in the three phases, using 3 lines

CTs or a core balance CT (CBCT). The former is more usual. Then if all the three line currents are. I_A, I_B, I_C, then $I_A + I_B + I_C = 0$ for healthy condition. $I_A + I_B + I_C = I_{res}$ for an earth fault on one phase.

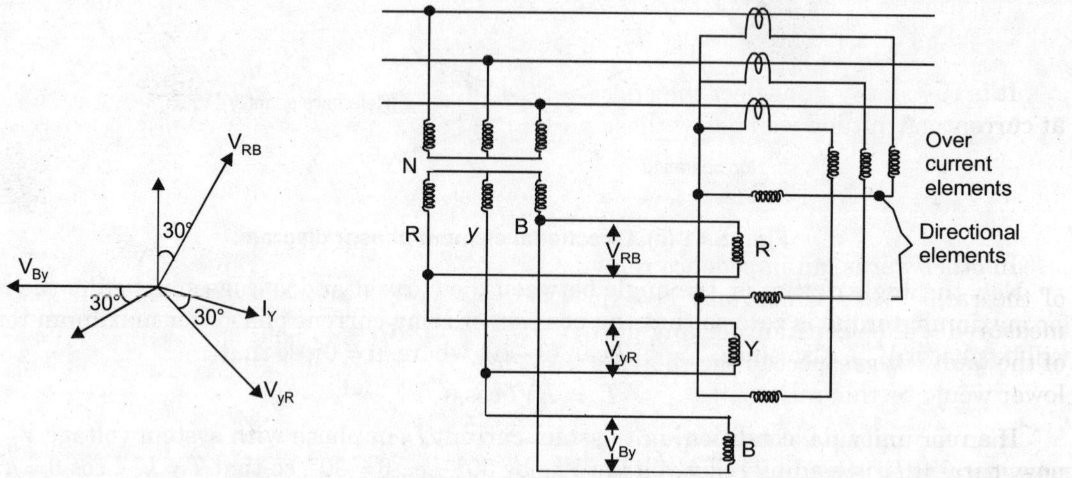

Fig. 15.42. 30° Connection diagram.

(*d*) Distance Relays

In fact, in principle Distance relays differ in principle for other form of protection as their performance is not governed by magnitude of the current or the voltage in the protected circuit but rather on the ratio of these two quantities. In distance relays, there balance between voltage and current the ratio of which can be expressed in terms of impedance. Impedance is an electrical measure of distance along a transmission line. The relay operates when the ratio *V/I* is less than a predetermined value and is also termed as Ratio relays. However, in designing distance relays it is necessary to think in terms of the volts amperes and phase angle to which it must respond because these are the quantities which actually operate the contact actuating parts.

Principal types of distance relays are : (1) impedance Relays

1. Admittance Relay

2. Impedance Relays

Impedance Relay is a device which measures distance by comparing the fault current with the voltage *V* across the fault loop. It is simple in this case to have an amplitude comparator and the amplitude balanced beam type structure is most common. (The device performing the comparison part is the heart of a protective relay and is known as a comparator. It decides the operating characteristics of the relay).

In an impedance relay, the torque produced by a current element is balanced against the torque of a voltage element. The current element produces positive (pick up) torque proportional to I^2 whereas voltage element produces negative (Reset) torque proportional to V^2.

The torque equation infact, is

$$T = K_1 I^2 - K_2 V^2 - K_3$$

where K_1 and K_2 are constants and K_3 is spring constant. At the balance point when the relay is on the verge of operating, the net torque is zero, hence we can say at this moment.

$$K_2 V^2 = K_1 I^2 - K_3 \quad ...(15.1)$$

Now deviding this equation by $K_2 I^2$, we get,

$$\frac{V^2}{I^2} = \frac{K_1}{K_2} - \frac{K_3}{K^2 I^2} \text{ or } \frac{V}{I} = Z = \sqrt{\frac{K_1}{K_2} - \frac{K_3}{K_2 I^2}} \quad ...(15.2)$$

It is customary to neglect the effect of the control spring since its effect is noticeable only at current magnitude well below those normally encountered. Therefore with $K_3 = 0$

$$\therefore \quad Z = \sqrt{\frac{K_1}{K_2}} = \text{constant} \quad ...(15.3)$$

In other words, an impedence relay is on the verge of operating at a given constant value of the ratio V to I which may be expressed as impedence. Theoretically, the V/I ratio as measured will be constant for any particular fault position and will only vary if the position of the fault with respect to the relaying point varies. Thus nearer the fault is to the relay the lower would be this ratio.

If a relay capable of measuring this ratio is now installed at the supply end of a line, its V/I setting can be adjusted so that the relay operates for faults anywhere within a given section of line and remain in operative for any fault beyond this section. The usual way of expressing the operating characteristics of an impedance relay is on R-X diagram as shown in Fig. 15.43 (*a*). It is a circle with its centre at the intersection of the X and R axis, that is where X and R are zero. The equation for a circle is $a^2 + b^2 = r^2$ where a and b are the vertical and horizontal ordinates and r is the radius. Hence, it follows that the locus of the impedance $Z = \sqrt{R^2 + X^2}$ will also be a circle. Since $Z^2 = R^2 + X^2$ with Z as radius. Any value of impedance less than Z, the radius of the circle will cause the relay to operate, whereas any value greater than this will cause the relay to retain. Thus the impedance relay is non-directional, that is, it will operate irrespective of the direction of the current.

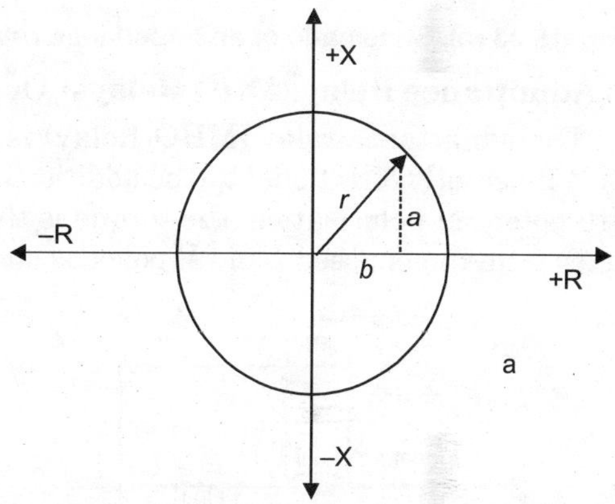

Fig. 15.43 (*a*). Characteristic of an impedance relay on *R-X* diagram.

Now, if such an impedance relay is applied to a transmission line where the voltage element is fed from a voltage transformer and the current element from a current transformer as shown in Fig. 15.43 (*b*). The two quantities supplied to the relay will be proportional to the line current I and the system voltage V. Now considering a fault as shown in Fig. 15.43(b) the relay will be supplied with a voltage equal to $I \times Z_F$. Thus, the ratio of the voltage and the current supplied to the relay will be equal to Z_F, the impedance between the relaying point and the point of fault. As above, the impedance relay being non-directional it would also operate for all fault position within section. At of the line in the figure for which the impedance presented to the relay is less then Z_F. To avoid this unwanted operation it is necessary to use a directional relay in conjunction with the impedance relay, the combined characteristic then being the shaded part of Fig. 15.43 (*c*) in which AB represents the impedance of the line in font of the relay and AC the impedance of the line behind the relaying point. Since the directional unit permits tripping only in the +*ve* torque region the active position of the impedance unit characteristic shown shaded. The net result is that tripping will occur only for points that are both within the circle and above directional unit characteristic.

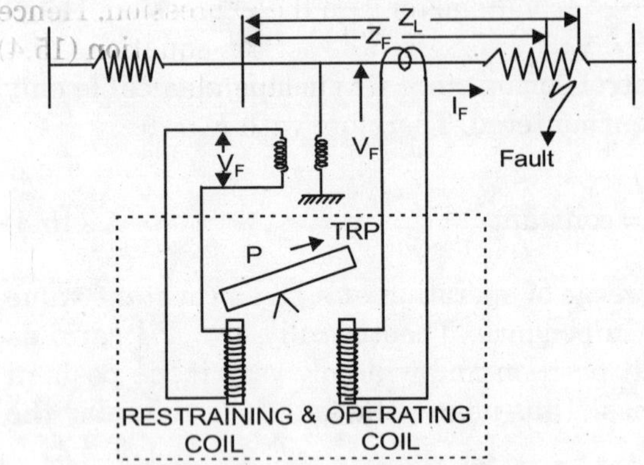

Fig. 15.43 (*b*). Shcematic of an impedance relay.

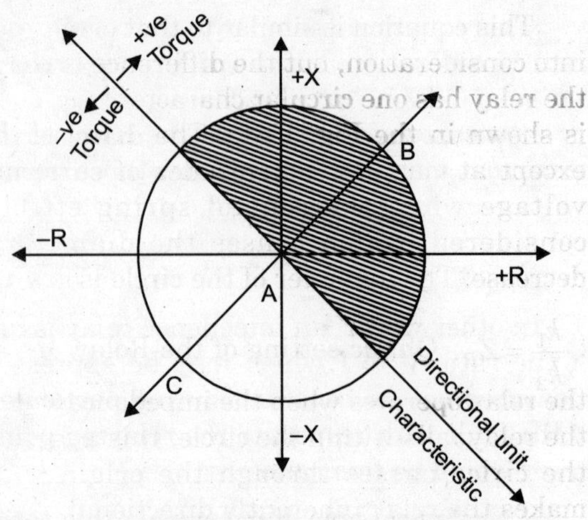

Fig. 15.43 (*c*). Impedance relay characteristic.

(*e*) Admittance Relay (MHO Relays) Or MHO Type Distance Relay

The admittance relay (MHO Relay) is similar in principle to an impedance relay but is made inherently directional by the addition of a voltage winding known as the polarising winding with potential polarisation, the torque is the product of the potential polarising flux times the fluxes from the opposed *I* and *V* poles as shown in Fig. 15.44 (*a*).

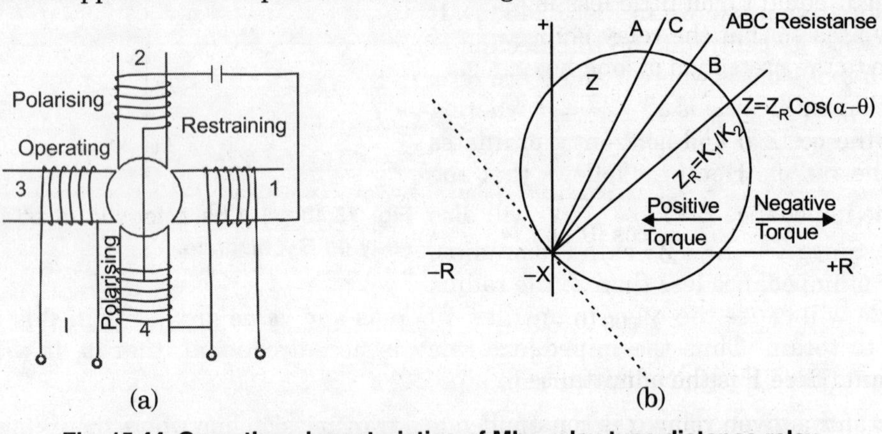

Fig. 15.44. Operating characteristics of Mho relay type distance relay.

Hence the torque equation of such a unit is

$$T = kVI\cos(\theta-\alpha) - k_2V^2 - k_3$$

where θ and α are positive when I lags V.

At balance point, $T = 0$

Hence $k_1 VI\cos(\theta-\alpha) - k_2V^2 - k_3 = 0$

or, $$\frac{k_1}{k_2}\cos(\theta-\alpha) - \frac{k_3}{k_2 VI} = \frac{V}{I} = Z$$

of $(k_3/k_2 VI)$ the spring control effect is neglected

Then, $$Z = \frac{k_1}{k_2} \cos(\theta - \alpha) \qquad \text{...(15.4)}$$

This equation is similar to that of a directional relay when the control spring effect is taken into consideration, but the difference is that there is no voltage term in the expression. Hence the relay has one circular characteristic. The operating characteristic as per the equation (15.4) is shown in the Fig. 15.45. The diameter of this circle is practically independent of V or I_2 except at very low magnitudes of current or voltage when the control spring effect is considered, which causes the diameter to decrease. The diameter of the circle is given as : $\frac{k_1}{k_2} = Z_R$. Ohmic setting of the Relay. In fact, the relay operates when the impedance seen by the relay falls within the circle. It is a fact that the cirlce passes through the origin which makes the relay inherently directional. Due to this characteristic the relay needs only one pair of contacts which makes it fast tripping for fault clearance and reduces the VA burden on the CTs. Infact, the impedance angle of the protected line is normally 60° to 70° which is shown by line OC in Fig. 15.45 (*c*). The arc resistance R, represented by AB, which is horizontal to OC from the extrimity of the chord Z. By making α equal to, on little less lagging than θ, the circle made to fit very closely around the fautly area so that the relay is inccuritive to power swings, and therefore is particularly applicable to the protection of long heavily loaded lines.

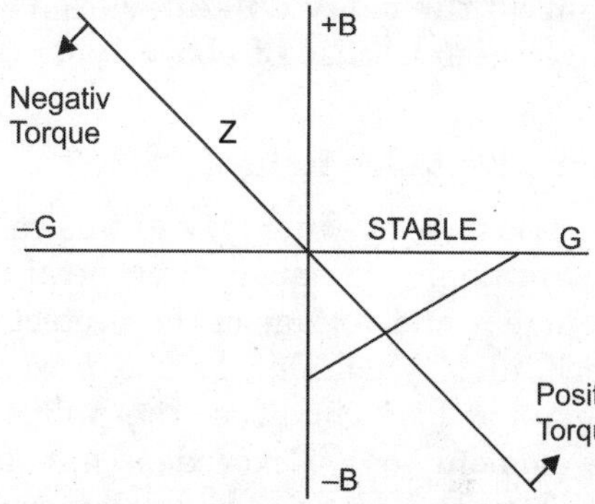

Fig. 15.45. Characteristic of mho type relay on admittance diagram.

Now the equ. $Z = \frac{k_1}{k_2} \cos(\theta - \alpha)$ can also be written as

$$\frac{1}{Z} \cos(\theta - \alpha) = \frac{k_2}{k_1}$$

or $$Y \cos(\theta - \theta) = \frac{k_2}{k_1} = a$$

constant; Here Y is the admittance in mho.

In fact, for a given relay α is constant, and the heads of the admittance phasor Y will lie on a straight line. The characteristic of mho relay on admittance diagram is hence a straight line as shown in Fig. 15.45.

In power system, Mho or Admittance relay suitable for long EHV/UHV heavily loaded transmission lines as its threshold characteristic in Z-plane is a circle passing through the origin and its diameter is Z_R, *i.e.*, k_2/k. Due to this, the threshold characteristic is quite compact, enclosing fault area compactly and hence there is lesser chance to operate during power swing and also it is directional.

(*f*) Distance Relays Types

In Power system Cicuits Distance Relays used for its protection are generally divided into two groups : (1) Definite distance Relay and (2) time-distance Relays and these are defined as below :

1. Definite Time Distance Relays

These relays operate instantaneously when the impedance (reactance) or admittance falls below a specified set value, and these are of impedance, reactance and mho type.

2. Time-distance Relays

These realys have the time of operation dependent upon the value of impedance – reactance or admittance. This means that upon the distance of the fault from the relay point. In fact a fault **nearest the relay will operate it earlier than a fault further away from the relay point.** Again as in case of above type, these relays also may be of impedance reactance or mho type.

1. Definite Distance Relay

A simple form of beam type of this relay is shown in Fig. 15.46. It consists essentially of a balanced beam pivoted about its central point and two electromagnets energised respectively by the current and voltage of the protected circuit. The armatures of the two electromagnets are mechanically coupled to the pivoted beam. Under normal load condition the pull due to the voltage element ($\alpha\ V^2$) exceeds. That due to the current element ($\alpha\ I^2$) and the relay contact thus remain open. On the occurrence of a fault in the protected circuit the voltage applied to the relay decreases and the current increases, the ratio of the two being such that the pull exerted by the current coil now exceeds that due to the voltage coil and the relay closes its contacts to trip the circuit. In fact such a relay operates for a certain value of impedance and for this reason it is known as impedance type measuring relay. In fact if this relay is connected in a circuit such as shown in Fig. 15.46, and the impedance Z adjusted such that the current flowing in the circuit is negligible. Then the relay beam will be held against the voltage energised coil and current would increase until at a certain vlaue of current depending upon the relative number of A.T.s on the two electromagnets. The force due to the current coil will overcome that due to the voltage coil and the relay beam will be tilted over to the other side and close its contacts. Such a relay thus operates for a certain value of impedance and for this reason it is known as impedance type measuring relay.

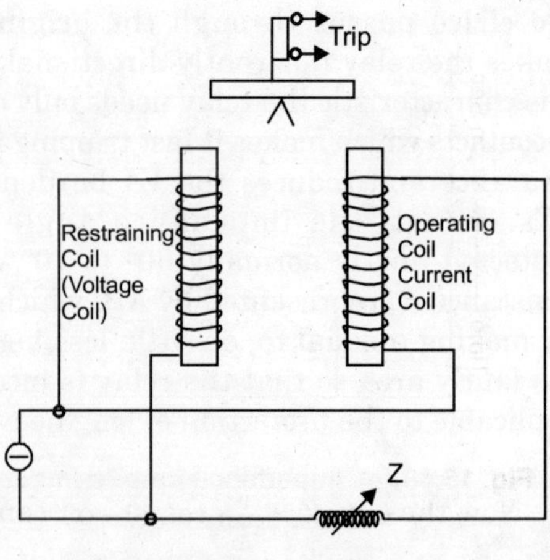

Fig. 15.46. Define distance relay.

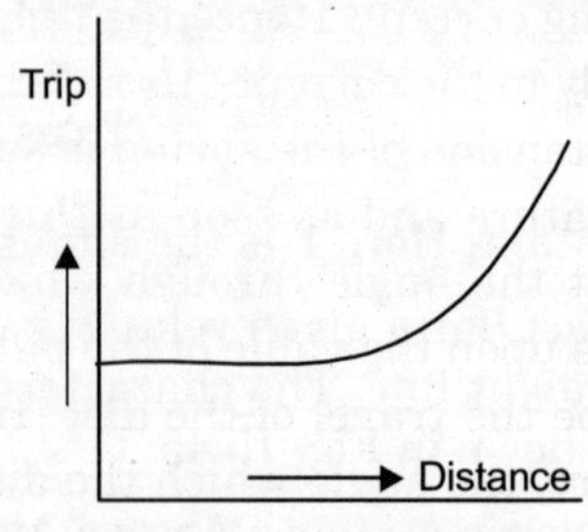

Fig. 15.47. Time *Vs* distance impedance relay characteristic.

The Fig. 15.47 shows the time *Vs* impedance characteristic of a definite distance impedance relay.

2. Distance Time (Impedance)

In the Fig. 15.48 (*b*) a distance time impedance type relay is shown. It consists of a current driven induction element similar in all respects to the induction over current relay. The spindle which carries the disc of the element is shown connected by means of a spiral coupling to a second spindle which carries the bridging piece of the moving contacts of the relay trip contact. This bridge is normally held in the open position by an armature held against the pole face of an electromagnet energised by the voltage of the circuit.

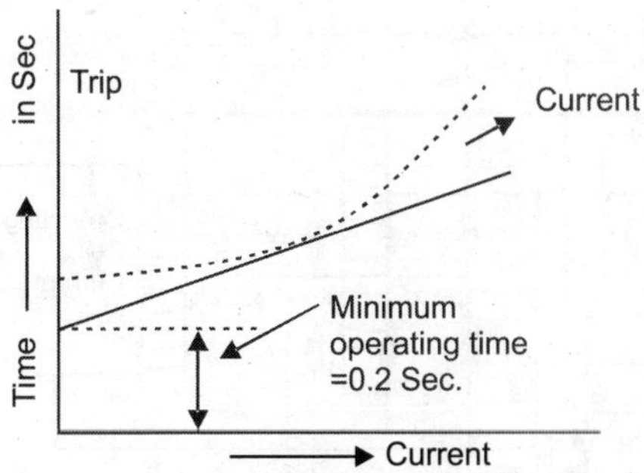

Fig. 15.48 (*a*). Impedance timeVs impedance characteristic.

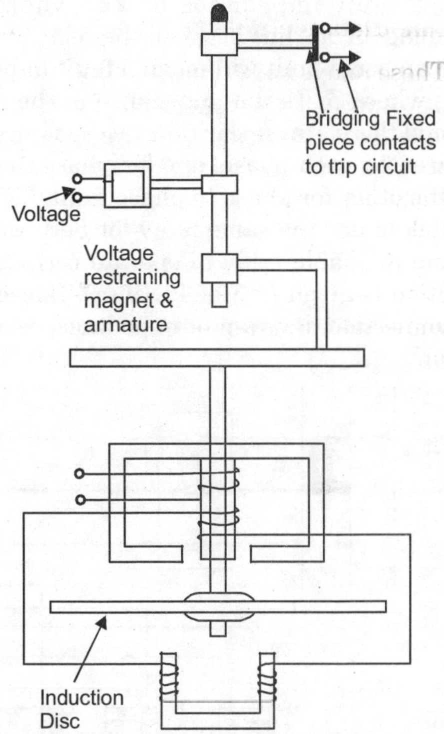

Fig. 15.48 (*b*). Impedance Time Realy.

In fact, on the occurrence of a fault, the induction disc starts to rotate. The speed of rotation of the disc neglecting the effect of the control spring, is approximately proportional to the operating current. Hence the time which the disc requires to turn through a given angle varies inversely to the current. Also as the disc rotates the spring is wound. The disc continues rotating till the tension of the spring is sufficient to overcome the pull of voltage restrain magnet over its armature and as soon as this armature is released the trip contacts are closed. Thus, we see that the angle through which the induction disc must rotate before the relay operates depends upon the value of the pull on the restraint armature. The greater this pull, the greater would be the travel of the disc. However, this pull is also roughly proportional to the voltage, therefore, the angle which the disc must travel to operate the relay is directly proportional to V. Now as the time which the disc takes to travel through a certain angle varies inversely as the current the time of operation of this relay would vary as (V/I), *i.e.*, as Z or distance.

The approximate distance characteristic of distance time impedance relay is shown in Fig. 15.48 (a). The exact time distance characteristic is curved as shown dotted in the figure. This may be explained as that the torque on the disc tends to be proportional to I^2 while the pull of holding or restraint magnet tends to be proportional to V^2 so that the time tends to be proportional to Z^2 thus giving a relation other than the linear one shown by the firm line, in the Fig. 15.48 (*a*)

Impedance Relays Connection for Earth Fault and Phase Fault

In a system, a short-circuit between two lines will have a fault impedance of $2Z$, where Z is the impedance of one line between the relay and the fault, while an earth fault will mean a fault impedance is $Z + Z_m$, where Z_m is the impedance of the earth path. It would therefore appear that two separate relays are required on each phase, one for phase to earth fault and the other for phase to phase fault. However, it is possible to use the same relay for both conditions. A scheme to enable relay to operate correctly for both condition is given in Fig. 15.49 (a). Based on the principle of operation, in the Fig. 15.49 (b) the connection diagram of impedance relays arranged for phase and earth fault protection is shown.

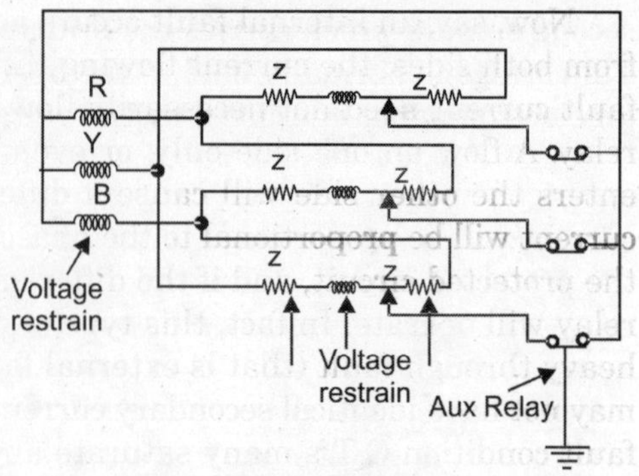

Fig. 15.49 (a)

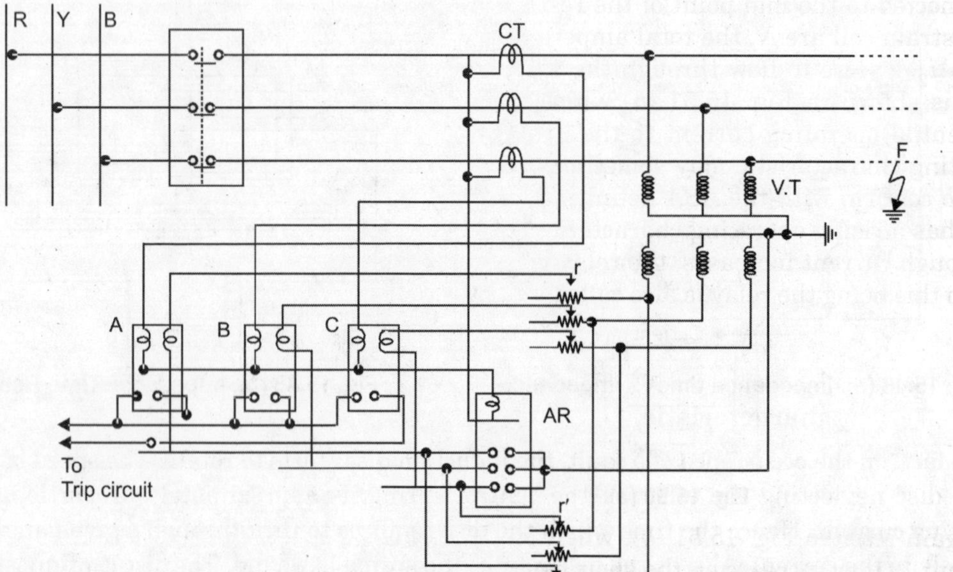

Fig. 15.49 (b). Connection diagram of impedance relay for phase and earth fault protection.

The relays are normally connected for phase to phase operation with their voltage restraint coils energised through r'. When an earth fault occurs the auxiliary relay AR operates and changes the voltage restrain coils from phase-to-phase connection to phase-to-phase without disconnecting the voltage supply to the voltage restraint coils.

Differential Relays

The relay that operates when the phasor difference of two or more similar electrical quantities exceeds a predetermined amount is called a differential relay. Infact, any type of relay, when connected in a certain way, can be made to operate as differential relay. Most differential relays are of the "Current differential" type. The Fig. 15.50 (a) shows the arrangement of an over-current relay connected to work on the current different principle.

Now, say, the current I flows through the primary circuit to external fault. If the two current transformers have the same ratio and are connected as shown no current will flow through the relay and it will remain inoperative.

Now, say, an internal fault occurs at X is the Fig. 15.50 (b) and if current flows to the fault from both sides, the current flowing through the relay will be $i_1 + i_2$. It may be noted that the fault current need not necessarily flow the fault from both sides to cause current flow in the relay. A flow on one side only, or even some current flowing out of one side a large current enters the other side will cause a differential current. In other words the differential relay current will be proportional to the phasor difference between the current entering and lowering the protected circuit, and if the differential current exceeds the pick up value of the relay, the relay will operate. In fact, this type of differential relay is likely to operate inaccurately with heavy through fault (that is external fault) since the supposed identical current transformers may not have identical secondary currents due to constructional errors or under service external fault condition C.T.'s many saturate and cause unequal secondary current and the difference of the secondary current may approach the pick-up value of the relay. This disadvantage is overcome in the percentage differential type relay. Fig. 15.51 (a) shows the simple circuit connection for such a relay. Here, the differential current in the operating coil, C of the Fig. 15.51 (a) is $i_1 - i_2$ while the current in the restraning coil R $(i_1 + i_2)/2$, since the operating coil is connected to the mid point of the restrining coil. In other words if the number of turns on the restrain coil are N, the total ampere turn are $i_1 \times (N/2) + i_2 \times (N/2)$, which is the same as if $(i_1 + i_2)/2$ were to flow through the whole coil. The operating characteristic of this type of relay is shown in Fig. 15.51 (b) which shows that except at low current the ratio of the differential operating current to the average restraining current is a fixed percentage. The operating characteristic may sometimes be plotted using through current I_2 instead of $(I_1 + I_2)/2$ to confirm with the *ASA* definition for a percentage differential relay. However, as this relay has a rising operating characteristic, *i.e.*, the pick up value increases on the magnitude of through current increases, the relay is restrained or biased against operating inaccurately. Due to this being the relay is also called a BIASED RELAY.

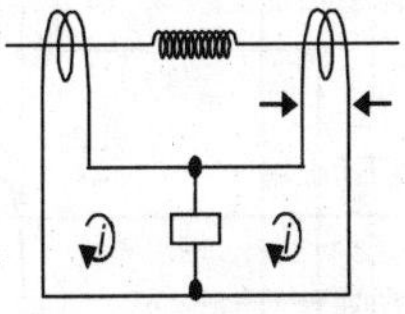

Fig. 15.50 (*a*)

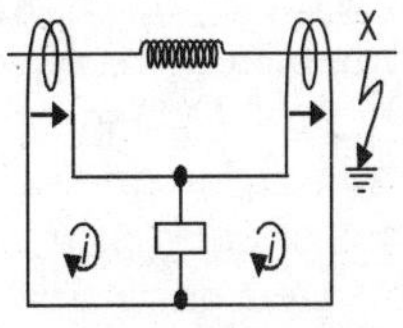

Fig. 15.50 (*b*)

Again refer to Fig. 15.51 (b), where instead of the slope of the characteristic remaining constant, it may increase on the short circuit current increases. Such a characteristic is shown in Fig. 15.51 (c). The relays are accordingly referred to as constant slope and increasing slope relays of the percentage differential biased type. The important difference between them is the fact that the increasing slope type, while costing slightly more requires less accuracy in the performance of its current transformers than the constant slope type does.

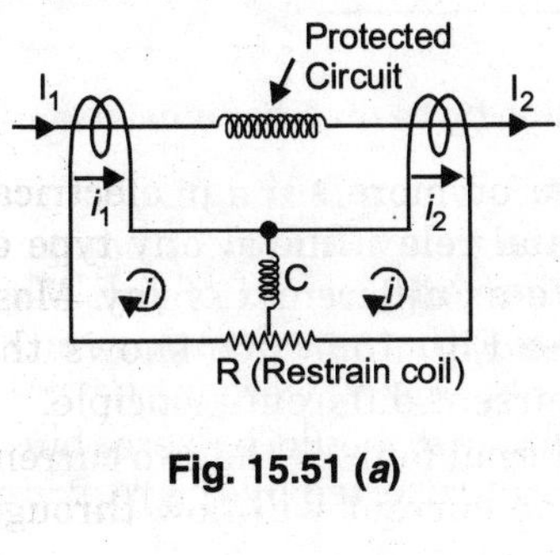

Fig. 15.51 (*a*)

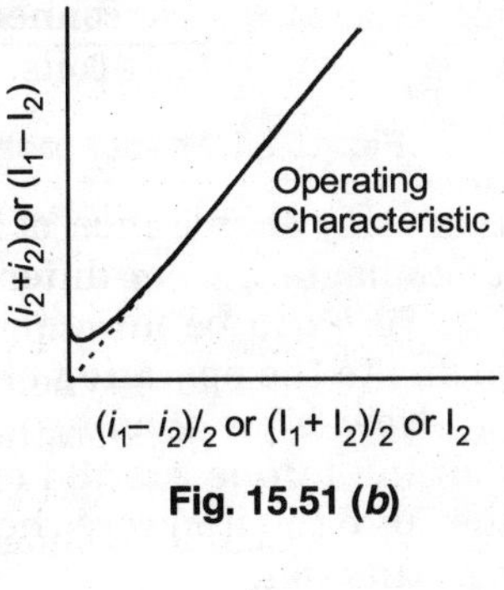

Fig. 15.51 (*b*)

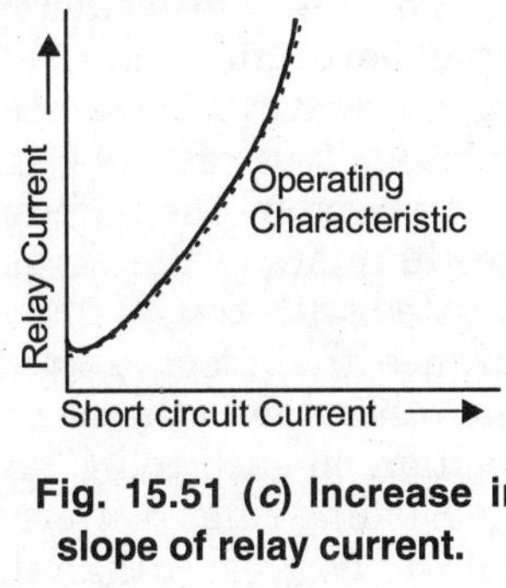

Fig. 15.51 (*c*) Increase in slope of relay current.

PERCENTAGE DIFFERENTIAL RELAY

In the Fig. 15.52 shows a percentage differential or biased relay of the beam type. The restraining coil produces a pull preventing the contacts from closing while the operating coil produces a pull in the closing direction. The percentage restraing current is determined by the relative turns on the restraining and operating coils. However, sometimes a mechanical bias is also given by shifting the fulcrum nearer towards the operating coil.

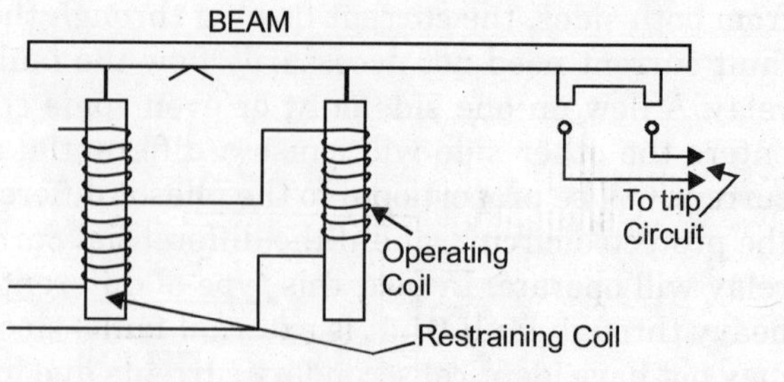

Fig. 15.52. Beam type percentage differential relay.

BIASED BEAM DIFFERENTIAL REALY

In the Fig. 15.53, we can see the connection diagram of a biased beam differential relay protecting an alternator or transformer winding.

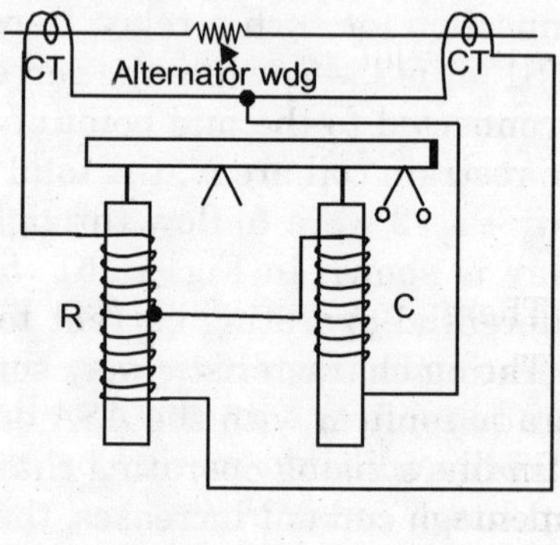

Fig. 15.53. Biased beam differential relay.

In Fig. 15.54 the connection diagram for the differential protection for the differential protection of a feeder using biased beam relays, and the operation is explained as below. Under fault condition such an earth fault shown at F_1 the secondary current in CT_1 at the sending end exceeds that in CT_2 and this difference divides between two alternative paths : One path through circuit D_1 and C_1 at the sending end and the other via R_1, P_1, D_2, C_2, R_2 and P_2, D_1 and D_2 being diverter resistances the magnitude of each being equal to that of each pilot. The resistances of the two path are thus in the ratio 1 : 3. That means, the resistance of the second path is 3 times that of the first path. Consequently the current through the first path and therefore, through C_1 will be 75% of the fault current and if this is sufficiently large the relay will operate.

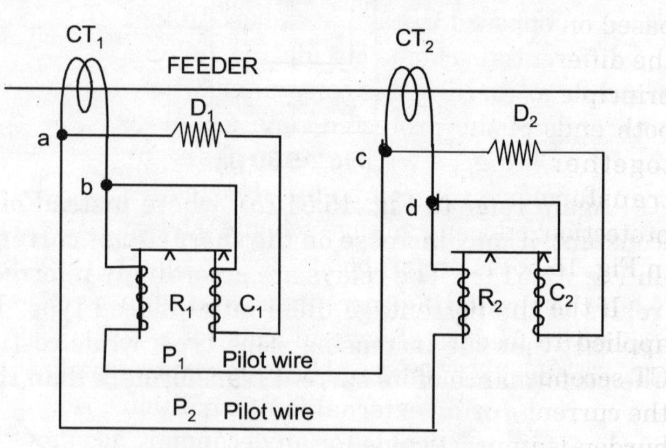

Fig. 15.54. Feeder protection by beam differential relay

Then, if the feeder is supplied from both ends, the relay at the remote end will also operate. Now the beam type relay of Fig. 15.53 may be replaced by the differential induction relay shown in Fig. 15.55, where D is a pivoted disc free to rotate in the air gap of two electromagnet, a portion of each pole of the electromagnet is fitted with copper shading ring. This ring can be moved further into, or out of pole. Two torques are in fact exerted on disc-one due to the operating element, that is, operating coil C, and other due to the restrainting coil R. Now, if the shading rings S_1 and S_2 were in the same position on each coil C and R, the resulting torque exerted on the disc would be zero, but if the shading rings S_2 were moved further into the iron core the torque exerted by the restraint coil will be greater than that of the operating coil.

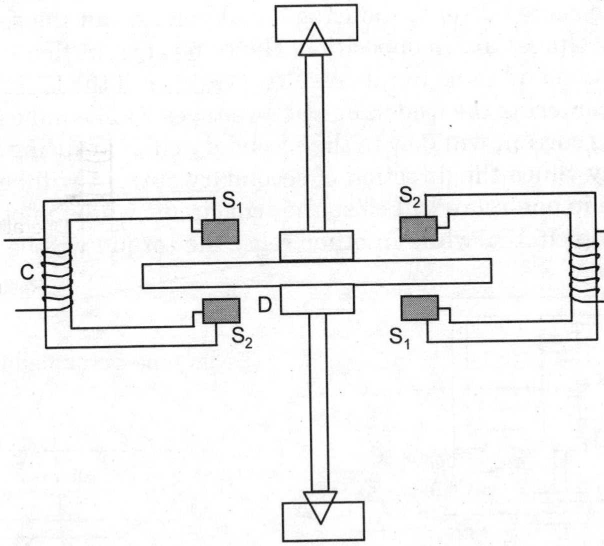

Fig. 15.55. Induction type biased differential relay.

There are some advantages in this type of relay of the beam type as steated below :

The induction element in this type of differential induction relay, is not susceptible to operation due to transients a slight time delay can be obtained and the biasing feature can be finally adjusted merely by altering the position of the shading rings on either or both elements or coils.

Transverse Differential Relay (Opposed Voltage Differential Scheme)

In the Fig. 15.56 a differential scheme based on opposed voltage is shown. In fact the differential scheme circulating current principle as discussed is convenient where both ends of the protected zone are close together, *e.g.*, with generator or transformer protection, but for feeder protection, opposite voltage scheme (shown in Fig. 15.56) is preferable.

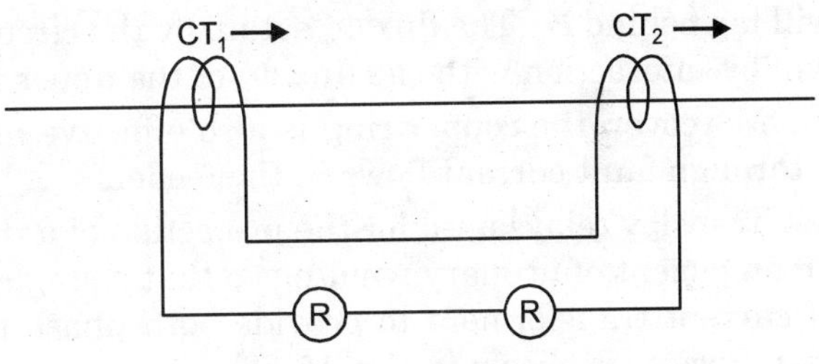

Fig. 15.56

If the circulating current scheme were applied to feeders of several kms long, the CT secondary e.m.fs would be required to circulate about 5A or 1A at full load or several times the current during external fault conditions, round a pilot loop of fairly high impedance such a burden is impracticable for any economic design of C.T. on external faults no current flows in the pilot wires, the CTs being connected in opposition, but on internal faults current flows through both relays which can be located in the substations at each end of the feeder.

Translay Relay

A schematic representation of the translay relay – a differential relay, is shown in the Fig. 15.57 (*a*). It is, infact, similar to an over-current relay except that secondary winding is not closed on itself and also that the central limb is provided with copper rings. Such a relay is extensively used in the transely protection applied to feeders. The secondary winding is connected to pilot wires to which is again connected the secondary winding of a similar relay as the other end of the feeder. When the current at the two ends is the same, the primaries of the two relays carry the same currents inducing equal voltages in the secondaries which are so connected that their voltages are in opposition. Hence no current flows in the two secondary circuits under the condition and so no torque results. However, if the current leaving the feeder differs from the current entering the feeder, unequal voltages will be induced in the secondaries and as such a circulating current will flow in the secondary circuit causing torque to be exterted on the disc of each relay. Since the direction of secondary current will be opposite in the two relays, therefore torque in one relay will close the trip circuit while, in other relay the torque will tend to close the trip circuit, while in other relay the torque will be in the movement in the unoperated direction.

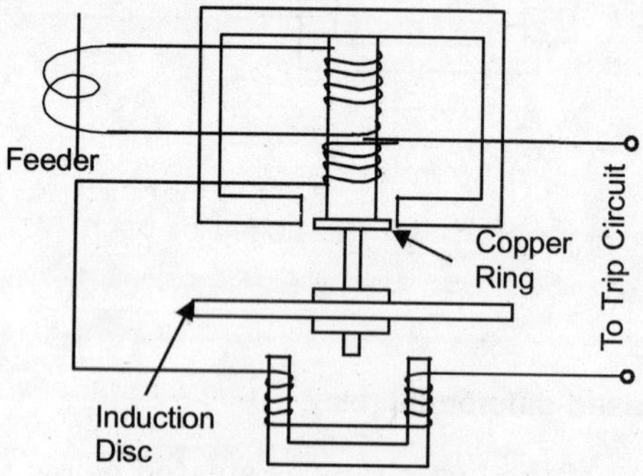

Fig. 15.57 (*a*)

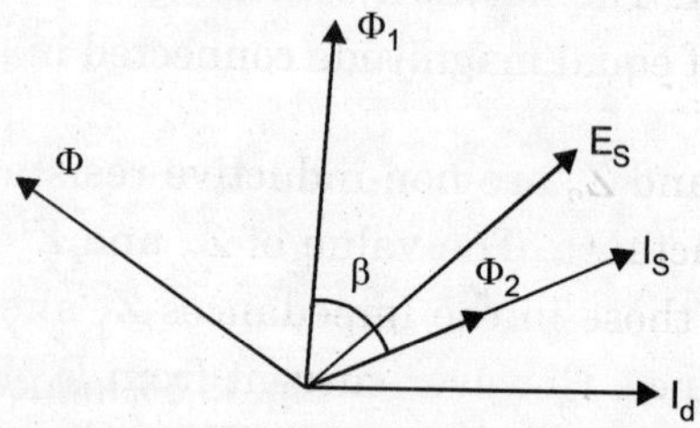

Fig. 15.57 (*b*). Phase diagram of transley relay.

The Fig. 15.57 (*b*) shows the phasor diagram of this type of relay, Φ is the main flux which induces E_s in the secondary winding Φ_1 is the flux which enters the disc from the electromagnetic and result in an induced current I_d in the disc lagging Φ_1 by 90°. Φ_1 lags behind Φ because of the copper ring. I_s is the secondary current and result in the flux Φ_2. The secondary current due to the difference in secondary voltage of two inter connected relays will lag behind E_s. The flux Φ_2 set up by the circulating current in the lower magnetic winding will by interaction with the flux Φ_1 of the upper magnetic element cause the relay to operate.

Moreover, the copper ring is also effective in preventing the relays from operating when a through fault current flows in the feeder.

Transley relays used for the protection of a three-phase circuit are provided with a special arrangement of primary winding so that a single relay can be used with each three-phase set of current transformers to provide both phase and earth fault production. A relay used for this purpose is shown in Fig. 15.58.

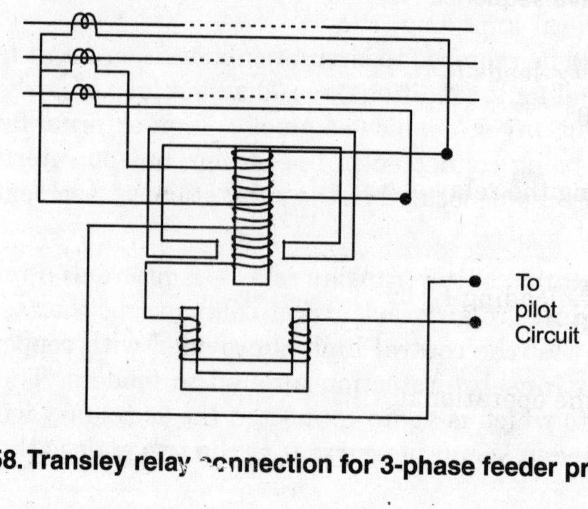

Fig. 15.58. Transley relay connection for 3-phase feeder protection.

Negative Sequence Relays

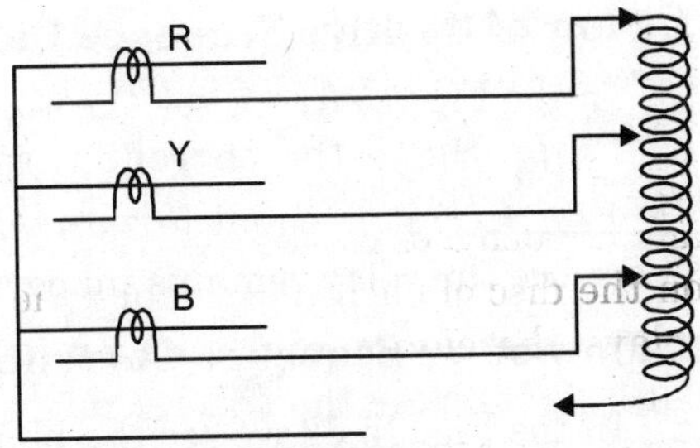

Fig. 15.59. Simplified connection for the primary relay of Fig. 15.66.

A negative sequence relay provides protection to generators and motors against unbalanced loading that result from phase to fault. Fig. 15.59 shows the phase unbalance protective equipment developed by the AEI · (*uk*). The equipment consists of network energised from three current transformers and a single pole relay having an inverse time characteristic connected across the network. The network consists of four impedance Z_1, Z_2, Z_3, Z_4 of equal magnitude connected in a bridge formation. as shown in Fig. 15.60.

Z_1 and Z_2 are non-inductive resistances while Z_3 and Z_4 are composed of both resistance and reactance. The value of Z_2 and Z_4 are so adjusted that the current flowing in these lag behind those in the impedances Z_1 and Z_2 by 60°. The relay is assumed to have a negligible impedance. However current from R phase divide into two equal components, I_{Z_1} and I_{Z_3} at A, I_{Z_1} being of I_{Z_3} by 60°. We study in the Fig. 15.60 (*a*) from this phasor diagram, I_1 = $I_{Z_3} = \frac{I_R}{\sqrt{3}}$. Similarly the current from phase B divide into two equal components I_{Z_2} and I_{Z_4} each equal to $I_B/\sqrt{3}$ and I_{Z_2} leading I_{Z_4} by 60°.

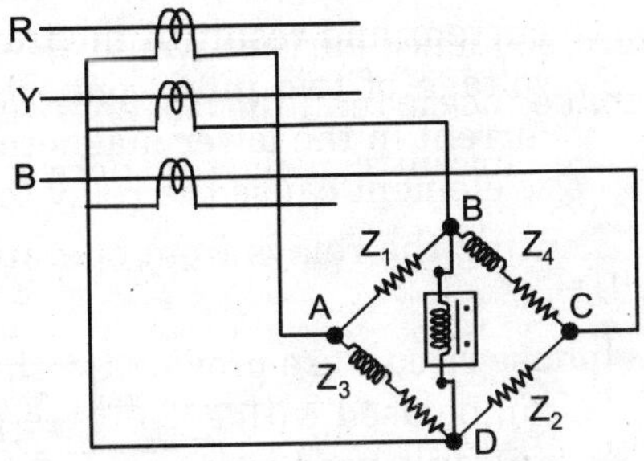

Fig. 15.60. Negative sequence relay.

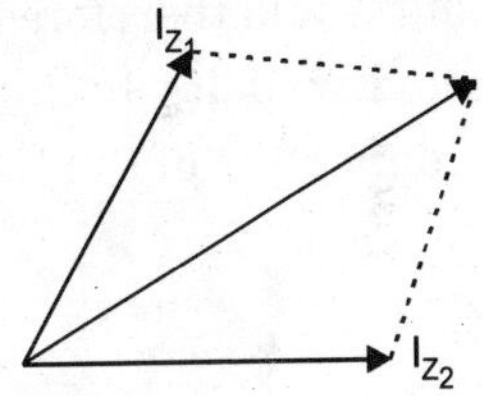

Fig. 15.60 (*a*). Division of currents in *R*-phase.

It is to be noted that I_{Z_1} leads I_R by 30°. While I_{Z_3} lags I_R by 30°. Similarly I_{Z_2} leads I_B by 30°, when I_{Z_4} lags I_b by 30°.

∴ The current entering the relay at point $B = I_{Z_1} + I_{Z_4} + I_Y$.

$$= \left[\frac{I_R}{\sqrt{3}} \text{ leading } I_R \text{ by } 30° + \frac{I_B}{\sqrt{3}} \text{ lagging } I_B \text{ by } 30° + I_r\right]$$

Now we shall study the operation during the flow of +*ve*, –*ve* and zero sequence currents.

1. Flow of Positive Sequence Currents

In the Fig. 15.61 we see the *+ve* sequence current I_R, I_Y, I_B. Since the current through the relay is $I_{Z_1} + I_{Z_4} + I_Y$ it is equal to zero when $I_1 + I_{Z_4} = -I_Y$. Therefore the relay remains inoperative.

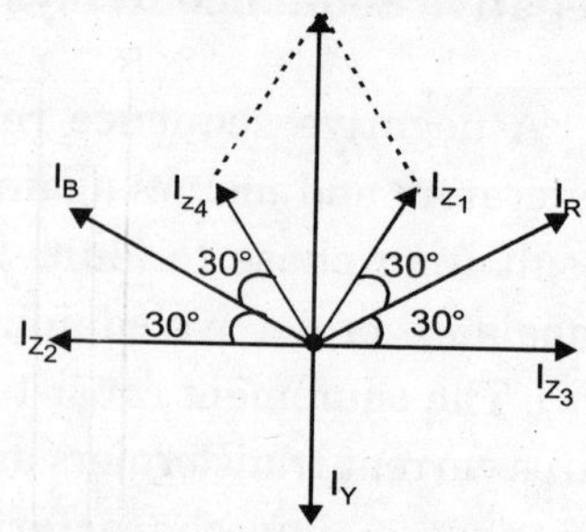

Fig. 15.61. Phasor diagram of *+ve* sequence current in normal condition.

2. Flow of *–ve* Sequence Currents

In the Fig. 15.62 we see the phase representation of the *–ve* sequence current in the 3-phases. Resolving I_R and I_B into their component. $I_{Z_1}, I_{Z_2}, I_{Z_4}$ we find that the current flowing in the relay = I_Y since I_{Z_4} cancels out I_{Z_1}. Thus the relay is operative under the influence of I_Y. A low setting value well below the normal full-load rating of the machine is provided since comparatively small values of unbalance current produce a great danger.

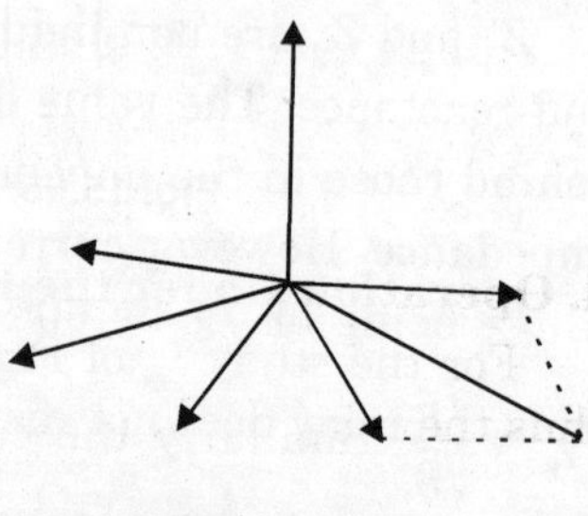

Fig. 15.62. Phasor diagram of *–ve* sequence currents.

3. Flow of Zero Sequence Current

If we study the phasor diagrams of Fig. 15.63, we observe that $I_{Z_1} + I_{Z_4} = I_Y$, so that a total current of twice the zero sequence current would flow through the relay and would therefore cause its operation. To mark the relay inoperative under the influence of the zero sequence current would flow through the relay and would therefore cause its operation. to make the current, the current transformers are connected in delta because then no-zero sequence current can flow in network circuit. The delta connection of the CTs is shown in Fig. 15.64.

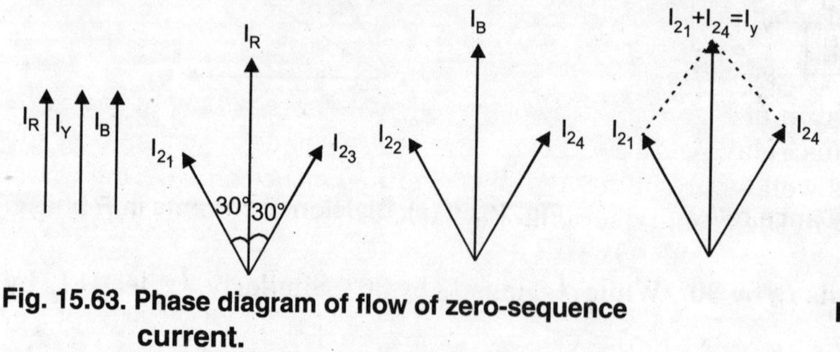

Fig. 15.63. Phase diagram of flow of zero-sequence current.

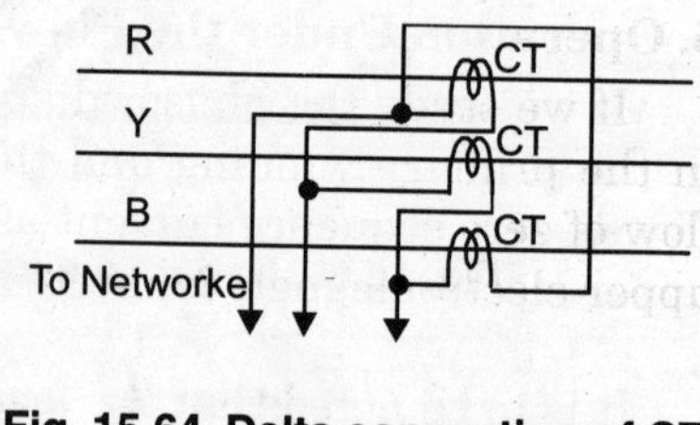

Fig. 15.64. Delta connection of CT.

AEI-U.K. developed another form of negative sequence relay and the associated equipment as shown in Fig. 15.65. The primary winding is provided with a centra tap one half of the winding is energised direct from the *Y*-phase while the other half in energised from the *R* phase through an auxiliary transformer. The auxiliary transformer is provided with an air gap in its magnetic circuit and thus it is possible to adjust the phase displacement between I_R and I_{R_1} to any angle (*i.e.*, other than the usual phase shift of 180°). The adjustment is to make that I_{R_1} tags behind I_R by 120°. The relay primary current is therefore the phasor difference to I_{R_1} and I_Y.

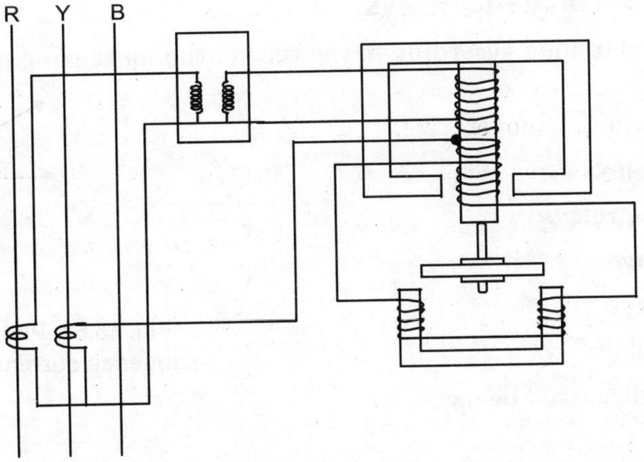

Fig. 15.65. Schematic of AEI-UK negative phase sequence relay.

4. Operation Under the Flow of +*ve* Sequence Current

For the study I_B of Fig. 15.66, it is clear that the relay primary winding current is zero thus the relay does not respond to the flow of +*ve* sequence current.

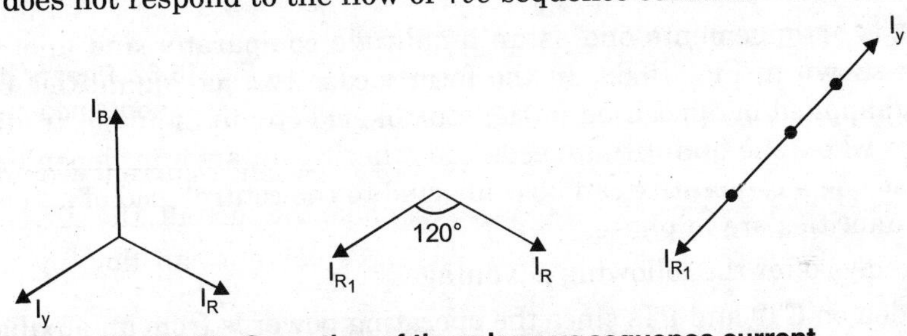

Fig. 15.66. Operation of the relay +*ve* sequence current.

5. Operation Under the Flow of –*ve* Sequence Current

If we study the phasor diagram of Fig. 15.67 it is observed that there is flow of current I in the primary winding and thus relay operates. This relay can be made responsive to the flow of zero sequence current as well by providing another winding on the central limb of the upper electromagnet, connected in the residual circuit of 3 line CTs a.

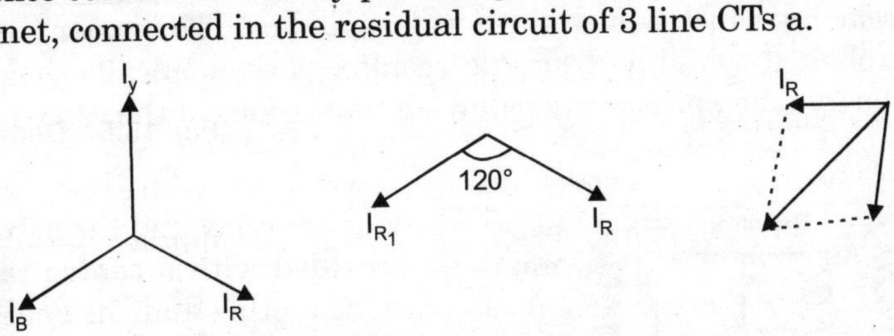

Fig. 15.67. Flow of –*ve* sequence current.

Consequently restricted by factors such as the conflicting requirements of sensitivity and mechanical robustness. Experience shows that there are more exacting requirements can readily be met using new static relays.

15.5.4 Static Relays

15.5.4.1 Classification of Static Relays

Static relays are classified according to the type of the measuring unit or the comparator as follows :

(*i*) Electronic relays
(*ii*) Transductor relays
(*iii*) Rectifier bridge relays
(*iv*) Transistor relays
(*v*) Hall effect relays
(*vi*) Gauss effect relays

These relays are discussed below :

(*i*) Electronic Relays

These were the first to be developed in the series of static relays. They date back to early 1928 when Fitzgerald presented a carrier current pilot relaying which constituted unit protection of the transmission line. Subsequently a series of electronic circuits for most of the common types of protective gear relays were developed. The components used were electronic valves for measuring unit.

The two basic arrangements one as an amplitude comparator and another as a phase comparator are shown in Fig. 15.68. In the former case two a.c. quantities to be compared are rectified and applied in opposition in the control grid circuit of an electronic tube, so that operation occurs when one quantity exceeds the other by an amount depending on the bias. In the latter case one a.c. quantity can be connected to the control grid of an electronic occurs when the two quantities are in phase.

Electronic relays offer the following advantages :

(*a*) Low burden on CTs and PTs since the operating power is from an auxiliary d.c. supply.
(*b*) Absence of mechanical inertia and bouncing contacts.
(*c*) Fast operation.
(*d*) Low maintenance, owing to the absence of moving parts.

In spite of all these however electronic relays have not met with general success except for carrier current relaying which has attained widespread use. The major reason is that carrier relaying has accomplished a highly desirable result not economically feasible by any other method. This popularity is in spite of the following limitations of the electronic circuits.

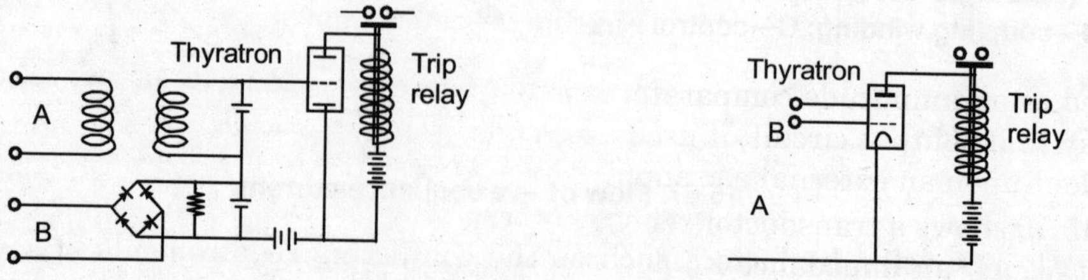

(*a*) Electronic amplitude comparator **(*b*) Electronic phase comparator.**

Fig. 15.68.

(a) Presence of incandescent filament and necessary low-voltage power supply to heat them.

(b) Short life of the electronic valves.

(c) High power consumption.

(d) Requirement of high-tension supply.

(e) High cost for simple relays such as overcurrent relay.

(ii) Transductor (Magnetic Amplifier) Relays

Since relays now have to perform much more complicated functions, many types tend to become very complex mechanically, and hence costly to make, and difficult to test and to maintain. The transductor controlling a simple slave relay, makes it possible to reduce a complex electromechanical device which operates in accordance with the interaction of fluxes and currents in a fully predictable way.

A transductor comprises essentially a magnetic core on which there are two groups of windings, usually known respectively as the operating windings and control windings. Each group may comprise only one winding but if there is more than one winding in a group all those windings are magnetically linked. On the other hand the windings of the different groups are not magnetically linked. The operating windings are energized with a.c. and control windings are energized with d.c. A transductor operates so as to present a variable impedance to currents flowing in the operating windings the magnitude of this impedance being varied by the current in the control winding. Fig. 15.69 shows a single core transductor.

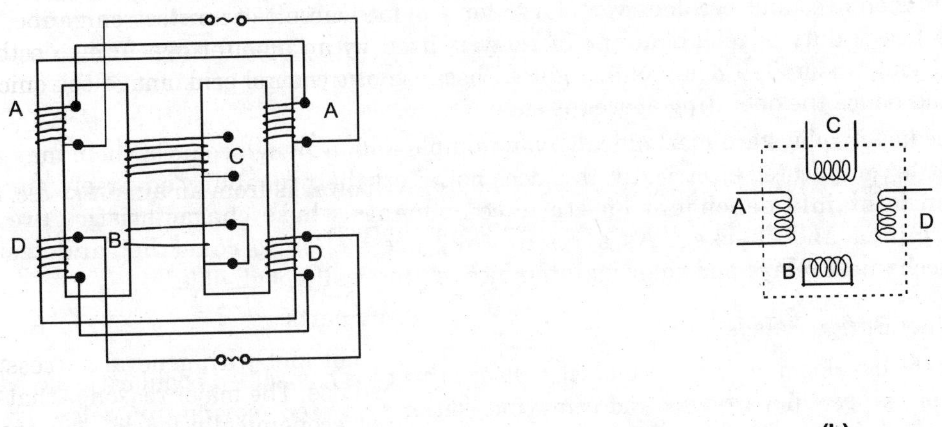

Fig. 15.69 (a). Single core transductor; (b) diagramatic representation of transductor. A—operating winding; B—coupling winding; C—control winding; D—output winding.

If used as an amplitude comparator it is limited in its sensitivity by the sensitivity of the slave relay in its output circuit; if used as a phase comparator to obtain greater sensitivity it is dependent upon an external a.c. supply which is sometimes difficult to arrange.

Fig. 15.70 shows a transductor relay where the restraining voltage is applied to the control winding through an impedance Z_R and a rectifier. The restraining current thus obtained is rectified and partially smoothed by the short-circuit effect of the rectifier on control winding and also due to the short circuited coupling winding. The effect of this is to drive both the limbs on which the operating winding is wound into saturation.

Now the operating winding is wound in such a way that the operating ampere-turns oppose the restraining ampere-turns in one limb and reinforce them in the other. Neglecting the finiteness of permeability and assuming equality of restraining and operating turns, the operating current would drive out of saturation, the limb in which operating current is against the restraining ampere-turns, when the peak value of the operating current exceed the magnitude of restraining current. When this happens a voltage is induced in the output winding and the relay is energized. The same operation is repeated on the other limb during the next half cycle.

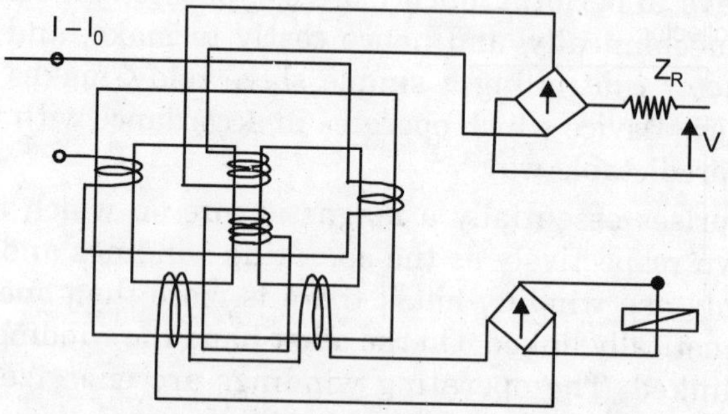

Fig. 15.70. Transductor relay.

The slave relay current is very low until the above mentioned condition exists. Then it suddently increases and becomes very large for a small value of operating current. For this reason a heavy duty of telephone tpe of relay is used with large setting. The to smoothing and rectifying a signal, a delay is introduced because of the time constant of the smoothing circuit and hence the operating spaced is slow.

Transductor relays are mechanically very simple and although some of them may appear to be a little complicated electrically, this does not affect their reliability adversely, since their operation is mainly dependent on static components whose characteristics are easily predetermined and checked. As a result they are easier to construct and test than electromechanical relays and their maintenance is practically negligible.

(*iii*) Rectifier Bridge Relays

With the development in semiconductor diodes this type has gained popularity and basically consists of two rectifier bridges and a moving coil or polarized moving iron relay. The most common are relay comparators based on rectifier bridges, which can be arranged as either phase or amplitude comparators. These will be discussed later in the chapter.

(*iv*) Transistor Relays

These are the most widely accepted type of static relays, so much so that when we say static relays, we can safely infer transistor relays only. The transistor which acts like an electronic valve can overcome most of the limitations posed by the electronic valves and thus has made possible to develop the *electronic relays* more commonly known as *static relays*.

The characteristics of modern transistors are such that they can replace the functional elements which are used in electromechanical relay to give necessary characteristics. The technical features particularly suited to the design of functional units include amplifying and switching characteristics, sensitivity and high speed. Experience has shown that transistor circuits cannot only perform the essential functions of a relay such as summation, comparison of inputs and integrating them, but they also provide necessary flexibility to suit the various relay requirements.

Two basic arrangements of relays based on transistor comparators are shown in Fig. 15.71. In either of these circuits, current of constant magnitude flows in the collector circuit only when the input a.c. quantities are simultaneously negative; a relay in the collector circuit will pick up when the overlap angle exceeds a certain value, *i.e.*, when the mean d.c. level in the collector circuit exceeds the relay pickup as a result of phase coincidence.

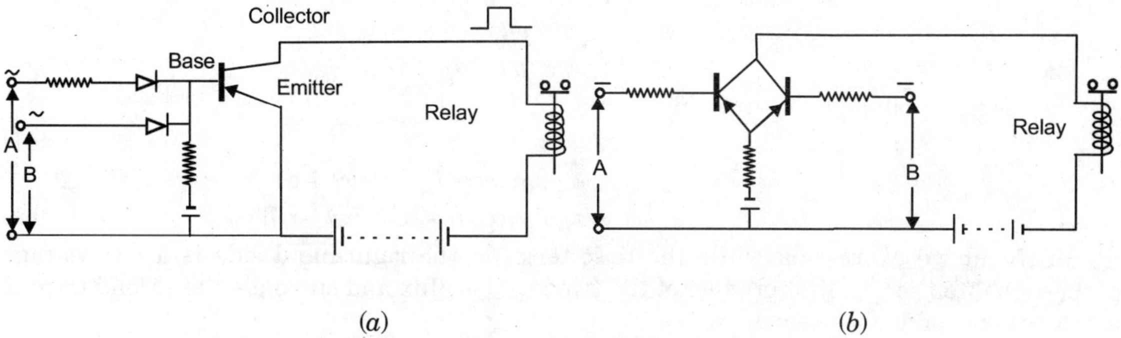

Fig. 15.71. Static comparators : (*a*) multiple inputs to transistor base; (*b*) two back-to-back transistors.

The advnatages of transistorized relays may be enumerated as follow :

(*a*) Quick response, long life, high resistance to shock and vibration.

(*b*) Quick reset action—a high reset value and absence of overshoot is easily achieved because of the absence of mechanical inertia and thermal storage.

(*c*) No bearing fricti n or contact troubles (no corrosion, bouncing or wear), so minimum maintenance required.

(*d*) Ease of providing amplification enables greater sensitivity to be obtained.

(*e*) The basic building blocks of semiconductor circuity permit a greater degree of sophistication in the shaping of operating characteristics, enabling the practical realization of relays with threshold characteristics more closely approaching the ideal requirements.

(*f*) The low energy levels required in the measuring circuits permit miniaturization and minimize current transformer inaccuracies.

(*g*) Use of printed (or integrated circuits) to avoid wiring errors and to facilitate rationalization of batch production.

Static relays employing transistors have their limitations which are :

(*a*) Variation of characteristics with temperature and age.

(*b*) Dependence of reliability on a large number of small components and their electrical connections.

(*c*) Low short time overload capacity compared with electromagnetic relays.

It has now become possibie to compensate for all these factors. For example, the use of thermistors eliminate temperature error, whilst ageing may be minimized by presoaking for several hours at a relatively high temperature. Factors (*b*) and (*c*) above are the design features of the circuit and careful design can compensate fo: these limitations.

(v) Hall Effect Relays

Hall effect is utilized to give a phase comparator. The effect is very predominant in certain semiconductors such as indium arsenide, indium antimonide, indium phosphate, etc. Fig. 15.72 illustrates the basic Hall generator where the Hall crystal in the form of a slab carries current in the X-direction and is placed in a magnetic field in the Y-direction, a voltage V_H known as Hall voltage is induced in the Z-direction across the edges of the crystal.

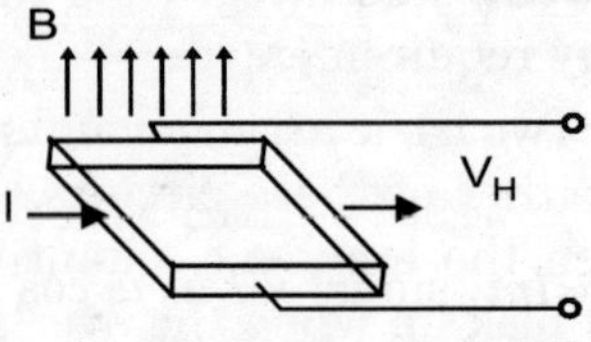

Fig. 15.72. Basic Hall generator.

Let $\phi = \phi_m \sin wt$

and $i = I_m \sin (wt - \alpha)$

the the Hall voltage V_H is given by

$$V_H \propto \phi_i \quad ...(9.1)$$

$$\propto \phi_m \sin wt \times I_m \sin (wt - \alpha)$$

$$\propto \phi_m I_m [\cos \alpha - \cos (2wt - \alpha)] \quad ...(9.2)$$

In the above expression while the first term on the right-hand side is a d.c. voltage proportional to the vectorial product of the two inputs—flux and current—the second term is an a.c. voltage of double frequency.

The double frequency term can be eliminated by cross connecting two Hall generators as shown in Fig. 15.73 where the two input signals are the sinsusoidal a.c. currents I_1 and I_2.

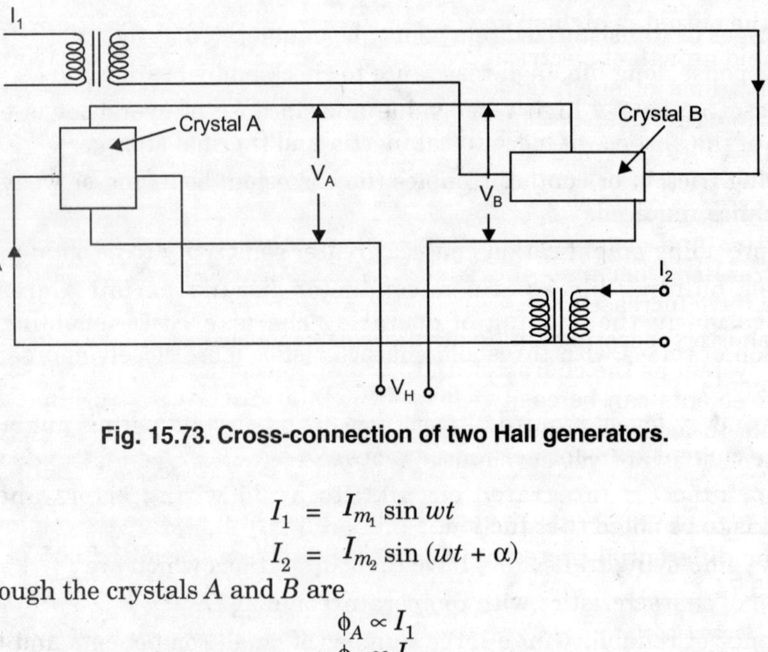

Fig. 15.73. Cross-connection of two Hall generators.

Assuming

$$I_1 = I_{m_1} \sin wt$$

and $$I_2 = I_{m_2} \sin (wt + \alpha) \quad ...(9.3)$$

Fluxes through the crystals A and B are

$$\phi_A \propto I_1$$
$$\phi_B \propto I_2$$

and current through the crystals are

$$\left.\begin{aligned} I_A &\propto \frac{DI_2}{dt} \\ \text{and} \quad I_B &\propto \frac{DI_1}{dt} \end{aligned}\right\} \quad ...(9.4)$$

The two crystals are connected such that their output voltage oppose each other, and the resultant output of the combination is, therefore

$$V_H \propto V_A - V_B \quad ...(9.5)$$

$$\propto I_1 \frac{dI_2}{dt} - I_2 \frac{dI_1}{dt} \quad ...(9.6)$$

$$\propto Im_1 \sin wt \times Im_2\, w \cos (wt + \alpha) - Im_2 \sin (wt + \alpha)\, w\, Im_1 \cos wt$$

or $$V_H \propto Im_1\, Im_2 \sin \alpha \quad ...(9.7)$$

The device therefore now acts as a pure phase comparator. A telephone type relay can be directly operated from the Hall voltage. The Hall effect device does not require an auxiliary d.c. source while the transistorized circuits need about 10 V d.c. supply. The total volt-ampere consumption of fall device will be higher than trnsistor circuits. It s construction will be much simpler. Because of the high cost of the Hall crystal, large temperature error and low output, they have not yet been used in practical relays except in the U.S.S.R.

(*vi*) Gauss Effect Relays

Some semiconductors have the property where resistance varies when magnetic field is applied. This effect is known as *magneto resistivity* or *Gauss effect*. If one a.c. voltage V_1 produces the magnetic field through the crystal, which for the best results should be in the form of a disc having a large diameter and another voltage V_2 sends a current radially through the disc, then the current will be proportional to $V_1\ V_2 \cos \theta$, where θ is the angle between the two voltages. This can thus act as a phase comparator and used as a static relay. This is however better than Hall effect devices because polarizing current is not required, the crystal is simpler and the output is higher. The cost of the crystal has limited its use in static relays.

With the rapid growth of electrical transmission and distribution system during last forty years and with adjust of much larger power station and interconnected systems, the duty imposed upon protective gear become are and more severe. Thus functions of relays because more and more complicated and many types of relays load to become vary complex mechanically and hence costly and difficult to cost and also maintain.

The basis of the static relay operation is the use of the circuits and component to obtain a variety of function and operating characteristics. The need of fuse and reliable protective schemes was realised because short-circuit levels, circuit ratings and complexity of interconnection have increased.

Moreover, shorter operating times have become more essential for preserving dynamic stability of the system as the character and loading approach design limits. Experience says that such requirements can be easily set by using static relays. Infact the transistors have made it possible to achieve greater sensitivity and simultaneously excellent mechanical stability. However, it is not yet economical to replace the existing electromechanical relays with static relays.

Moreover it is to be noted that the static relays have first been made commercially for the distance and the differential protective schemes. Simplified block diagram of static relay is shown in Fig. 15.74.

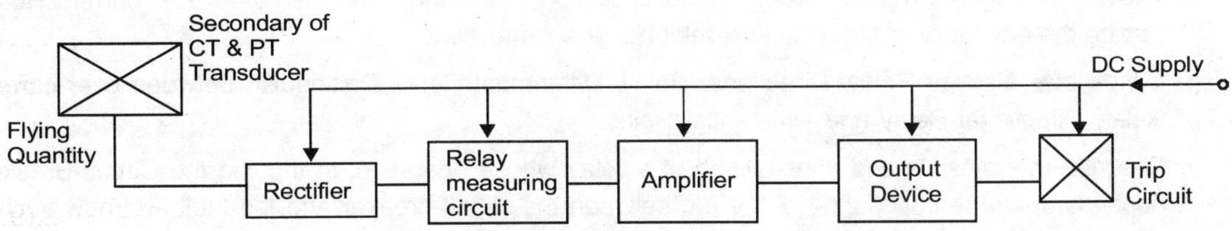

Fig. 15.74. Simplified block diagram of a static relay.

In a static relay the measurement is carried out by static circuits consisting of comparators level detectors, filters etc., while in a conventional electro magnetic relay, it is done by comparing operating torque (or force) with restraining torque (or force).

Static relays can be arranged to respond electrical inputs. The other types of inputs such as heat, light, magnetic field, travelling waves can be suitably converted into equivalent analogue or digital signals and then supplied to the static relay.

However, a multi-input static relay can accept several inputs. The logic circuit in the multi-input digital static relay can determine the condition for relay response and the sequence of events in it.

Advantages of Static Relays

1. The power consumption in case of static relay is usually much lower than that in case of their electro-mechanical equipments. Hence burden on CTs & PTs is reduced and their accuracy is increased. Problem arising out of CT saturation are eliminated and there is an overall reduction in the CTs & PTs cost.
2. Quick response, long life, shock-proof, high reliability, high degree of accuracy, fewer problem of maintenance in static relay in comparison with the induction relays.
3. These are compact. A single static relay can perform several functions. A single microprocessor based system can substitute several independent protection and control relay units.
4. The risk of wanted operation is less with static relay.
5. Quick reset action.
6. A static protection control and monitoring system can perform several functions such as protection, monitoring data acquisition, measurement, memory indication etc.
7. The static relays are provided with self monitoring, easy lasting and servings. Detective module can be replaced easily.

However, the characteristics of static relays are influenced by ambient temperature and ageing. Static relays are sensitive to voltage spikes or voltage transient. Filter circuits in relays are used as remedy. Static relays have low short-time overload capacity. Static relays are costlier, for simple and single function in comparison with electromagnetic relays for same purpose.

QUESTIONS

1. Describe a directional over current Relay and state where and when it is used.
2. State what are the fundamental requirements of protective relaying.
3. Explain with figures the construction and operation of the induction disc type over current Relay. Device the equation for the torque developed by such a relay.
4. Define over Current Relay, Directional Relay, Differential Relay. Distinguish between over current Relay, Directional Relay and Differential Relay.
5. Describe the principle and construction of a relay whose operation, in tripping the Circuit Breaked depends upon the impedance of the arc between the circuit-breaker and the fault and how such a relay can be used to give the selective protection in a network.

6. Explain the principle of distance relays early the difference between impedance relay, reactance relay and mho relay.

7. Write a note on Buchholz's relay protection for transformers.

8. Explain what is meant by positive, negative and zero phase sequence. Illustrate with diagram a system of protection making use of negative-sequence current for the operation of relays.

9. An over-current relay of current rating 5A and setting 150 per cent is connected to the secondary of C.T. of ratio 500 : 5A. Calculate the current in lines for which the relay picks up.

10. An earth fault relay has a setting of 20%, current rating 5A. It is connected to a CT of ratio 500 : 5. Calculate the pick-up current in primary for which the earth relay operates.

11. A 100/5A bar primary current transformer supplies an over current relay set at 25% pick up audit has a burden of 5VA. Determine the knee-point voltage and cross section of the core if the CT has 50 turns on its secondary and the fault current is 15 times the relay setting. Assume the flux density as 1.4 wb/m^2.

Ans. Hint : the operating current of the relay = 5 × 0.25 = 1.25 Amp.

Universal Torque Equation

∴ The secondary voltage = $\dfrac{5\text{VA}}{1.25} = 4$ volts.

The CT secondary voltage when current is 15 times the relay setting = 15 × 4 = 60 volts. Hence the knee voltage must be slightly greater than 60 volts.

Now we know, $E = 4.44\ B_m$ AfN

$$\therefore \quad A = \frac{60}{4.44 \times 1.4 \times 50 \times 50} = 3.86 \times 10^{-3} \text{ sq. m.}$$

= 38.6 sq. cms.

12. Determine the *VA* output of a current transformer having a ratio of 100.5 and secondary resistance of 0.1 ohm. The resistance of connecting lead is 0.1 ohm and the relay burden in 5 *VA*.

Ans. Hints : The output *VA* is $I_S^2 \times (r_s + r_{low}) + VA(\text{Burden})$

= $5^2 \times (0.1 + 0.1) + 5 = 10$ VA

13. State what are the methods of voltage control.

Ans. The methods for voltage control are the use of (1) shunt capacitors (2) shynchtranous capacitors, (3) Tap changing transformers, (4) Borsler transformers.

14. A 230 kV line is fed through 33 kV/230 kV transformer from a constant 33 kV supply. A single line diagram.

15. (*a*) Explain successive restrikes, and current chopping as applied to the interruption of capacitive and low inductive currents.

(*b*) How does resistance switching help in such condition.

(*c*) Show how the restriking transients can be dumped by connecting a resistance across the contact of a Circuit Breaker.

16. Describe the expression for resistance to be provided across circuit breaker contact for critical damping when

(*a*) Circuit breaker is located at the substation and followed by a transmission line.

(*b*) The circuit breaker is located at the end of a long transmission line and fault on.

17. Illustrate with the help of sketches the process of are interruption in multibreak oil C.B., with shunt resistors.

18. Explain the following terms : (*a*) symmetrical breaking circuit, (*b*) Asymmetrical breaking current, (*c*) Making current.

19. Write a brief note on HVDC Circuit Breakers and state this application.

20. An OCB is rated for 1000 MVA, 2 kA, 66 kV, 3-phase, 3 see state what are its : (*a*) rated operating voltage, (*b*) rated operating current, (*c*) rated symmetrical breaking current.

21. Draw schematic diagrams for the (*a*) Impedance Relay, (*b*) Reactance Relay, (*c*) Mho Relay.

22. Explain clearly the basic principle of operation of a differential relay. Explain the working of this type of relay for (*i*) an internal fault, (*ii*) a through fault.

23. Give a scheme of protection for a ring main having three substation and fed from one end. Explain whether the same scheme could be used if the ring main were fed from mere those one end.

24. Explain the principle of Merz-Price system of protection used for power transformers. What are the limitations of this scheme and how are they overcome?

25. What is an IDMT over current relay. Explain its characteristics.

26. What is meant by per cent bias? How is this achieved in practice in a differential relay? Under what circumstances is a percentage differential relay preferred over the differential relay?

27. Explain the basic principle of operation of a percentage differential relay for (1) internal fault and (2) through fault.

28. Explain how the selection of current and time setting is done in a time current graded system.

29. State what are the abnormal condition in a large alternators against which protection is necessary.

30. What is an HRC fuse? Compare an HRC fuse with a circuit breaker as interrupting device.

31. Describe the construction, principle of operation and application of an HRC fuse.

32. Explain the terms (1) Symmetrical breaking current (2) Asymmetrical breaking current, (3) Making current. Explain clearly how these currents can be determined from oscillogram taken during short circuit test on a 3-phase CB.

33. Explain the terms (1) restriking voltage, (2) Recovery voltage and (3) RRRV. Derive the expression for the restriking voltage in terms of system voltage, inductance end capacitance across a C.B. contact when a 3-phase fault takes place assuming the neutral of the system to be solidly grounded.

34. Discuss the principle of are interruption in an oil OCB.

35. Explain the performance of SF_6 gas and vacuum when used for Circuit Breaking in power system.

36. A Circuit Breaker is rated as 2500 A, 1500 mVA, 33 kV, see. 3-phase oil Circuit Breaker. Determine the symmetrical breaking current, rated making current, short time rating and rated service voltage.

37. An 11 kV, 50 Hz, 3-phase alternator is connected a system which has inductance and capacitance per phase of 10 mH and 0.001 μF respectively. Determine (1) the maximum voltage across the breaker contacts, (2) frequency of transient oscillation, (3) the average RRRV, (4) the maximum RRRV.

38. What is a static relay and what is the basis for its development? In what way has it been successful in replacing the conventional electromagnetic relays?

39. Name different types of static relays. Discuss the use of transistors and Hall crystals as static relays.

40. Discuss the advantages and disadvantages of static relays as protective devices.

CHAPTER

16

Protection of Equipments in Power System

INSIDE THIS CHAPTER

- Introduction
- Protection of Transformer
- Generator Protection
- Prime Mover Protection
- Overspeed Protection
- Overvoltage Protection
- Motor Protection
- Bus Zone Protection
- Protection Scheme Regarding a Direct Connected Generator
- Transformer Protection Scheme
- Static Relay

16.1 INTRODUCTION

In power system the two major items of equipments are the transformers and the Generators. Infact on these equipments faults occur less frequently than on lines, but the damage caused by the faults usually take much more time and money to repair than are required to repair the damage done by fault on lines. Rapid operation of circuit breakers equipped with all necessary protection scheme is feasible on lines and it helps in saving the amount of damage.

A fault in a generator or transformer always needs careful attention of the technical supervisory personnels in the system. Clearance of Faults very fast helps in minimizing the damage to the equipments and also reduces the interruption to power service due to system instability and reduced voltage.

Now below discussions will be made on the transformer, generator protection of Motor etc.

16.2 PROTECTION OF TRANSFORMER

Transformers are static equipments and totally enclosed with core and coil oil immersed. As such transformers develop faults rarely, but when the fault occurs in a transformer it may be serious, unless the transformer is quickly disconnected from the system.

However, faults developed in the transformer can be divided into three main categories, as below :

(1) Faults in the auxiliary equipments which are parts of the transformer, (2) Faults in the transformer windings and connections (3) Overloads and external short circuit.

16.2.1 Auxiliary Equipments Subjected to Faults

In auxiliary equipments the detection of faults is necessary to prevent ultimate failure of the main transformer windings. The following can be considered as auxiliary equipments mediums :

1. Transformer Oil

Less quantity of oil (lower level of oil) is a dangerous condition in a transform because live parts and the leads to the bushings, tap changer etc. which have to be well under oil but these get exposed if the oil (inside the tank) drops below the specified oil level indication with alarm contacts available to give indication for immediate attention.

2. Gas Constitution

Transformer oil is most important insulation medium for transformers. Hence maintaining the quality of transformer is very important for transformer protection.

Infact deterioration of transformer oil and insulation is minimised if oxygen and moisture are excluded from the gas space. As the normal operating pressure within the transformer tank varies rather widely, so sealing of the tank is not always recommended. A pressure vacuum gauge can be used on a sealed tank to provide a visual indication of pressure in the tank, conservator tank is mounted on the main tank to take the expansion and contractions of the oil. Whereas silica gel is placed in the breathing vent pipe to prevent moisture from entering. Sometimes a nitrogen cylinder is connected to the gas space, with regulating devices to maintain the pressure between 0.5 and 0.8 atmosphere. This scheme maintains an inert atmosphere in spite of small leaks and provides some indication of presence of leaks by the rate of expansion of nitrogen in the cylinder. Alarm contacts can be provided in the system to indicate pressure inside the tank above or below predetermined limits.

3. Oil Pump and Forced Air Fans

The top oil temperature normally gives indication of the load on the transformer. Increased oil temperature might be an indication of an overload or it might be due to a fault in the cooling system, such as failure of the oil pumps or the blocking of a radiator valve, or non operation of fans. A thermometer with alarm contacts will indicate rise in the oil temperature due to any of these faults. An oil flow indicator is commonly used to indicate proper operation of oil pumps.

4. Core and Winding Insulators

Incipient faults may occur initially which may develop into major faults if not taken care of at the initial stages. Insulation failure may develop because of the following reasons :

(*a*) The insulation of the lamination and core belts may be of poor quality or has been damaged accidentaly during erection.

(*b*) The insulation between the windings, between winding and the core and the conductor insulation may be of poor quality, may have been damaged mechanically, may be brittle because of ageing or overloading.

(*c*) Badly made joint or conjunction.

(*d*) The incipient faults mentioned above need not be attended to immediately and as such do not cause any interruption to the supply service, but must be taken care of as soon as possible, before it becomes a major problem.

16.2.2 Winding Faults

Electrical faults which cause immediate sections damage and are detected by unbalance current or voltage may be divided into the following classes :

1. Faults between adjacent turns or parts of coils such as phase-to-phase faults on the HV and LV external terminals or on the winding itself or short circuits between turns of HV and LV winding.
2. Faults to ground or across complete winding such as phase-to-earth fault on the HV and LV external terminals or on the winding.

In fact a short circuit between turns can start with a point control resulting from mechanical forces of insulation deterioration due to excessive over load or a loose connection, breakdown of transformer insulation by an impulse voltage. The puncture of the turn insulation by an inpulse is supposed to cause a path of destruction, through which the normal frequency voltage can maintain an arc. But if the turn voltage is insufficient to maintain the arc it will be quenched by the oil at the first current zero.

Then faults to ground or across a large part of the winding will result in large value of fault currents as well as the emitting of large amount of gas due the decomposition of oil. This type of fault is not difficult to detect but rapid clearance of fault is essential to avoid excessive damage.

16.2.3 Overload and External Short Circuits

Overload can be sustained for long periods, being limited only by the permitted temperature rise in the windings and cooling medium. Excessive overloading will result in the deterioration of insulation and subsequent failure. In fact, it is usual to monitor the winding and oil temperature condition and an alarm is initiated when the permitted temperature limits are exceeded. External short circuits may only be limited by the transformer reactance and where this low fault current may be excessive.

Differential protectional of transformers is the most important type of protection used for internal phase-to-phase and phase-to-earth faults and is generally applied to transformers having ratings of 3 MVA and above.

In a star delta transformer, the load currents in the two windings are not in direct phase opposition, but are displaced by 30° and to allow for this the CT secondaries are connected in delta on the star side and in star on the delta side. In the Fig. 16.1 we see an ordinary differential protection for a 3-phase star-delta transformer.

The differential scheme described suffers from the following drawbacks : (1) Unmatched characteristics of CTs, (2) Ratio change as a result of tapping, (3) Magnetising inrush current we shall briefly discuss these points below :

16.2.4 CT Characteristics

Unless saturation is avoided the difference in CT characteristics due to different ratios being required in circuits of different voltages may cause appreciable difference in the respective secondary current whenever through-faults occur. This trouble is aggravated in the case of transformers due to unequal ratio CTs being employed on either side of the protected transformer. A source of ratio error which results in circulating currents under through fault condition is the unequal burden imposed on the CTs due to unequal load lengths.

16.2.5 Ratio Change as a Result of Tapping

Tap changing equipment is a common feature of a power transformer which effectively alters the turns ratio. Compensating for this effect by varying the tappings on differential-protection CTs impractable. Biased or percentage differential relays ensure stability with the amount of unbalance occurring at the extremition of the tap-change range. Biased relays are better suited to the overall protection of variable-ratio transformers.

16.2.6 Magnetising Inrush Current

When the transformer is energised, the transient inrush of magnetising current flowing into the transformer may be as great as ten times full-load current and it delays relatively slowly. This is bound to operate the differential protection of the transformer falsely. The magnitude of the magnetising inrush current is a function of the permanent flux trapped in the transformer core and the instant on the voltage cycle when it is switched on.

There are a number of ways of ensuring immunity from operation by magnetising surges. Firstly, the relay may be given a setting higher than the maximum inrush current; secondly, the time setting may be made long enough for the magnetising current to fall to a value below the primary operating current before the relay operates. These simple remedies are incompatible with high-speed and low primary operating current. In the third method the hormonic content of the current flowing in the operating circuit is filtered out and passed through a restraining oil.

16.2.7 Biased Differential Relays

To avoil undesirable operation on heavy external fault due to CT errors and ratio change as a result of tap changing, a restraining winding is provided which is energised by the through current. In other words the operating winding is biased and operates by some percentage of the through current then the relay becomes more sensitive at low current with tripping for external fault under these conditions and if the restraing and operating magnets are similar with the ratio of turns as T, the criterion for operation for a static comparator is $\frac{I_1 - I_2}{T} > \left|\frac{I_1 + I_2}{2}\right|$ and for an electromagnetic comparator is $\left|\frac{I_1 - I_2}{T}\right|^2 > \left|\frac{I_1 + I_2}{2}\right|^2$. Value of T is usually of 0.05 for generators and 0.1 – 0.4 for transformers, the higher values being used if the transformer ratio is varied by tap changer.

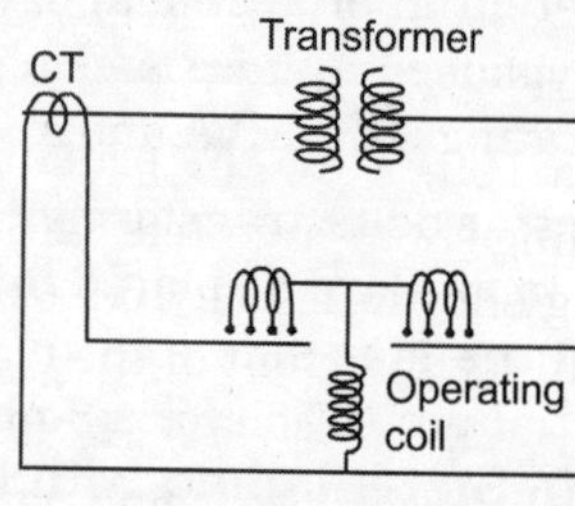

Fig. 16.1 (*a*). Biased differential protection for a transformer.

The Fig. 16.1 (*a*) shows the single line diagram of the biased differential scheme for the transformer. Bias is adjustable by tapping on the restraining winding and setting by spring restrain typical values are 10 to 40% for bias and 50 to 100% for setting. The time characteristics of induction relays are such that magnetising inrush current do not cause operation. The Fig. 16.1 (*b*) shows the bias characteristics for a typical relay.

16.2.8 Earthing Fault Current Affected Due to Influence of Winding Connection

Earth-fault current's magnitude for a given fault position within a winding depends upon the winding connections and the method of neutral earthing. The two conditions for earth current flow in the case of a winding fault to earth are :

(*a*) That a path exists for current to flow into and out of the winding.

(*b*) That ampere-turns balance is maintained between the windings. From the first condition it is clear that it is not necessary for the winding itself to be earthed for fault current to flow only that the system connected to the winding is earthed. The delta side of a transformer is usually earthed through an earthing transformer.

16.2.9 With Resistance Earthed Neutral Star Winding

The earth fault current in this case depends on the value of earthing resistor and is proportional to the distance of the fault from the neutral end of the winding. Considering the case of a delta-star transformer, neutral earthed through a resistor on the secondary side, for a 1 : 1 voltage ratio the primary to secondary turn ratio is $\sqrt{3}$: 1. For an earth fault I_F at 100% of the winding on the secondary side the corresponding value on the primary side is $1/\sqrt{3}$ times the fault current on the secondary $(1/\sqrt{3})I_F$. As the fault position on the secondary side varies, the effective ratio of transformation between the primary and the faulted portion of the secondary winding also varies. Hence for an earth fault at x% of the winding on the secondary side [$(x/100) \times I_F$) the corresponding current on the primary side will be $(x/100)^2 \times I_F/\sqrt{3}$ as the effective turn ratio of primary to secondary now is $\sqrt{3}: x/100$. Thus the primary fault current is proportional to the square of the percentage of winding short-circuited. The variation of fault current both in the primary and the secondary winding shown in Fig. 16.1 (*b*) & 16.1 (*c*) for a 1 : 1 voltage ratio the resistor is assumed to pass full-load current of the transformer with rated line-neutral voltage.

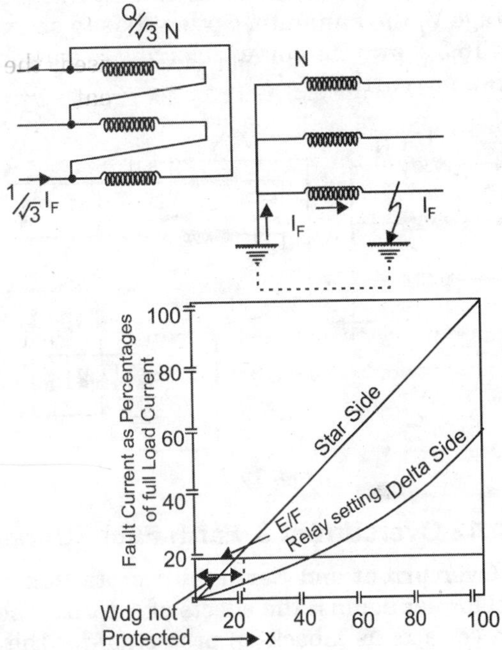

Fig. 16.1 (*b*) Distance *x* of fault from neutral (%).

16.2.10 Star Winding with Neutral Solidly Earthed

It is a fact that the earth fault current is no longer proportional to the position of the fault. The leakage reactance of the faulted winding in terms of reactance per turn increases the nearer the fault is to the star point, but the reactance of the other winding is effectively reduced owing to the change in transformation ratio so that the fault current reaches a minimum at some point near the middle of the winding. Fig. 16.1 (*c*) shows the transformer earth fault current on a solidly earthed star winding.

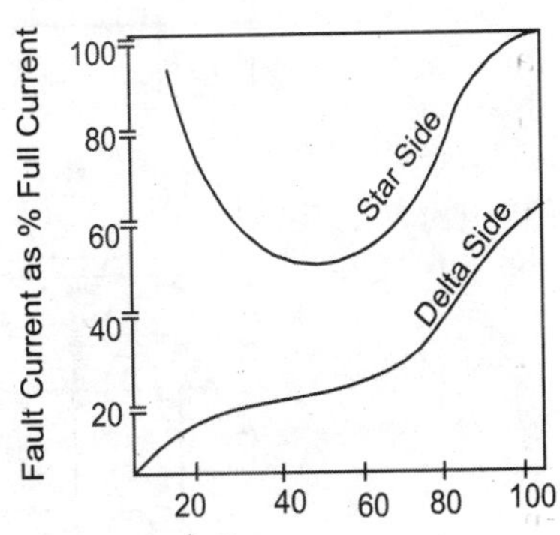

Fig. 16.1 (*c*) Distance *x* of fault from neutral (%).

16.2.11 Delta Winding

The Fig. 16.2 the fault in Delta winding is shown. The minimum voltage on the delta winding is only half the phase-to-earth voltage and occurs on the midpoint of phase. The high leakage reactance 25 to 50% at midpoint gives fault currents with a solidly-earthed system of 100-200% of full load; this divides equally between two phases. If the system is resistance earthed the resistance R should be added vectorially to the high reactance in determining the minimum earth-fault current. However, if the resistance is rated phase-neutral voltage V_1 the minimum current has to be expressed as $V/2R$ for all practical purposes. In the Fig. 16.3 shown the curves referring to transformer earth-fault for delta winding system system resistance earthed.

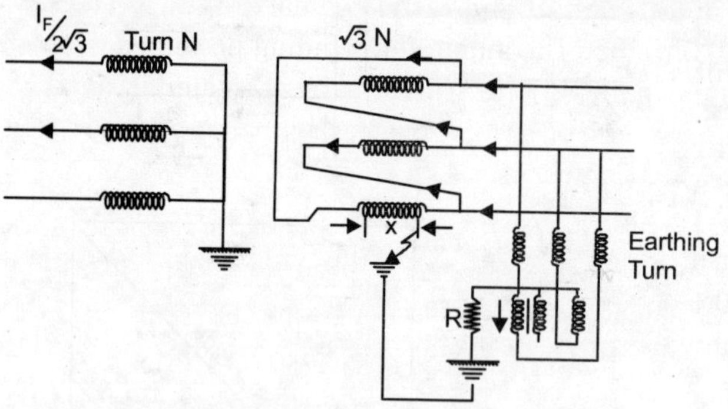

Fig. 16.2

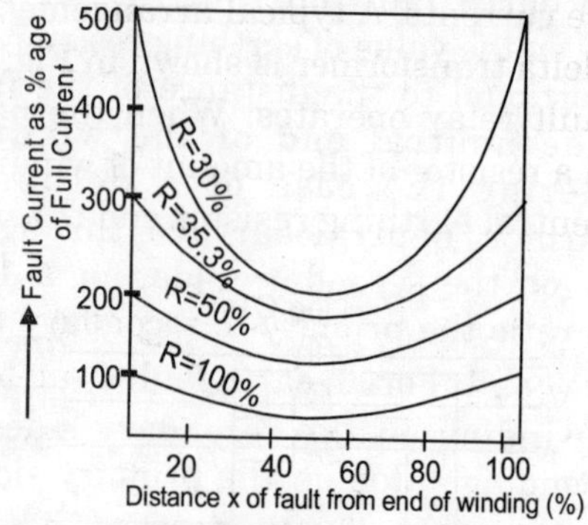

Fig. 16.3

16.2.12 Overcurrent & Earth Fault (Unrestricted)

Overcurrent and Earth-fault protection using IDMT relays is used mainly to protect the transformer against the effects of external short circuits and excess overloads. This protection however acts as a back-up protection for the transformer with sittings properly chosen. The current setting must be above the permitted sustained overload allowance and below the minimum short circuit current. Extremely increase characteristic is the ideal as it closely resembles the thermal cause of transformer. Also the time setting may have to be high in order to grade with other overcurrent relays on the system. The protection is located on the supply side of the transformer and is arranged to trip both the HV and LV circuit breakers.

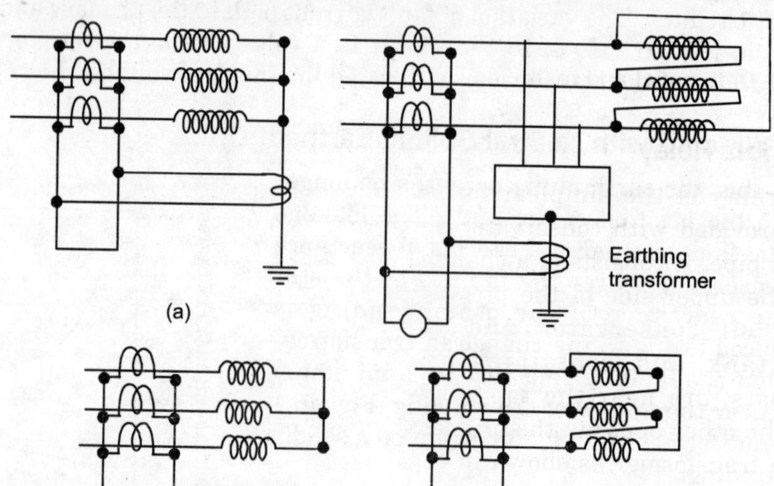

Fig. 16.4. Restricted earth-fault protection : (*a*) neutral earthed within the protected zone; (*b*) neutral not earthed within the protected zone.

16.2.13 Restricted Earth-Fault Protection

This type of protection is provided to detect earth-faults within the protected zone of transformer. ACT is fitted in each connection to the protected zone of the transformer. ACT is fitted in each connection to the protected winding and the CT secondary are connected in parallel to a relay. Fig. 16.4 illustrates the connection of restricted earth-fault protection for star and delta windings. Ideally, the output of the CTs is proportional to the sum of the zero sequence currents. A typical arrangement of differential and restricted earth-fault protection of star/delta transformer is shown in Fig. 16.5. Infact, the amount of winding protected against earth-fault relay operates. When the neutral point of the star of a transformer is earthed through a resistor of the amount of winding protection therefore varies according to the rating of the neutral earthing resistor end the relay setting.

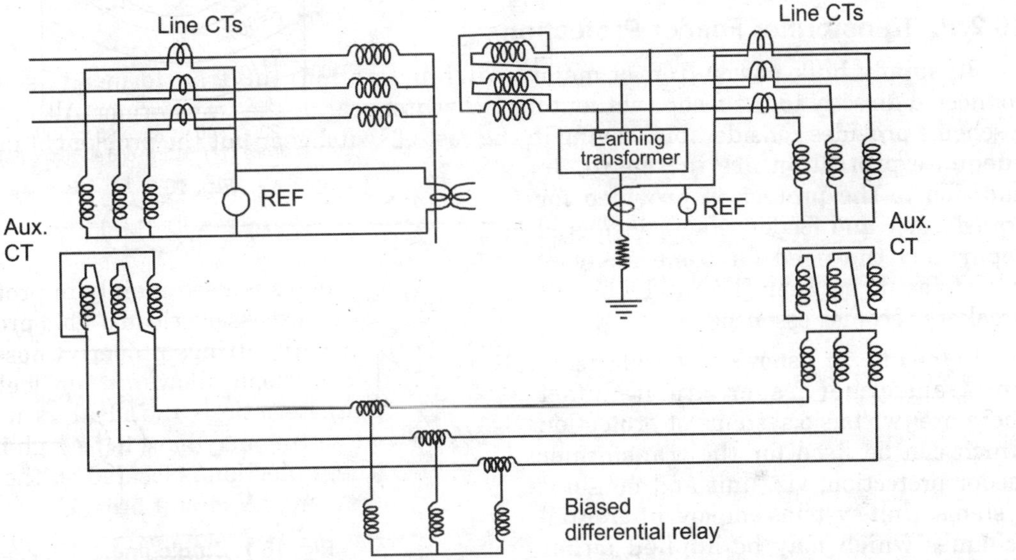

Fig. 16.5. Differential and restricted earth-fault protection of a star/delta transformer.

16.2.14 Buchholz Relay

Buchholz relay is the simplest form of protection which is commonly used in all transformers provided with conservator and tank through a pipe. It consists of a chamber connected in the upper side of the pipe run between the oil conservator and the transformer tank and containing two cylindrical floats, are near the top of the chamber and the other opposite the orifice of the pipe to the transformer as shown in Fig. 16.6.

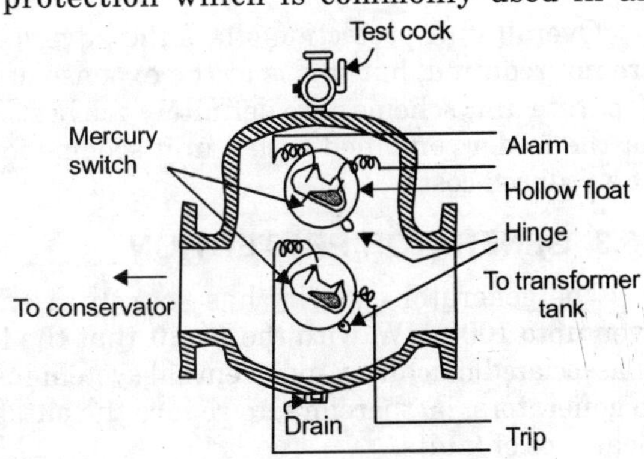

Fig. 16.6. Buchholz relay.

In fact under normal condition the floats are up, but on the occurrence of inter-turn turns fault in the transformer gas bubbles produced (due to the breakdown of oil) flow out of the transformer tank in the direction of the conservator. These gas bubbles on reaching the Buchholz relay are trapped and thereby reduce the oil level in the Relay chamber and cause the upper float to fall and when the float has fallen through a predetermined distance a pair of contacts which are controlled by the float, close and give an alarm bell's audible signal and tripping the Circuit Breaker, if the fault is heavy, closing the normally open contact of the Buchholz Relay.

The main advantage of Buchholz relays is that they indicate incipient fault, for example, between turns faults or core heating and so may enable a transformer to be taken out of service before serious damage occurs. One important limitation, however is that they do not project the connecting cables, which must therefore have a separate protection.

However, for hermetically sealed transformers, these are sometime fitted with hydraulic relays which respond to the rate of change of pressure in the gas cushion above the oil.

16.2.15 Transformer Feeder Protection

To supply bulk power from a major switching station, the transformer is sometimes connected directly to the feeder without an HV switchgear at the transformer. Although such a scheme provides considerably saving in the cost of switchgear but the problem of providing adequate protection are increased. In addition to the protection provided for transformer and feeder when considered separately the need for some mean of inter-tripping between HV and LV circuit breakers becomes essential.

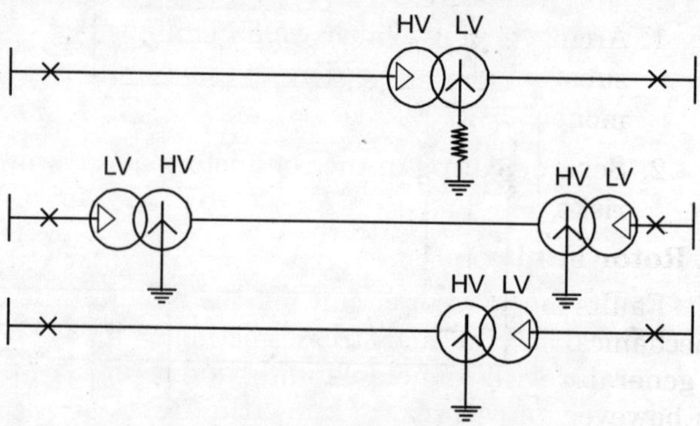

Fig. 16.7. Single-line.

In the Fig. 16.7 shows a typical single-line transformer feeder circuit. Infact there are two basic systems of protection which can be used for the transformer feeder protection, viz. unit and non-unit systems. Unit systems employ differential schemes which may be applied either separately for the transformer and the feeder or as an overall unit protection, combined for both the transformer and the feeder. In both the cases a pilot wire channel is required. In the former to provide a means of inter-tripping and in the latter as an integrand part of the scheme. Non-unit systems provide back up protection for faults out side the protected zone in addition to clearing the in-zone fault.

Overall unit protection offers the advantage that CTs on the HV side of the transformer are not required, but this is at the expense of unit schemes unless additional relays are used. Separate unit schemes are definitely the best form of protection but usually a non unit scheme for the feeder combined with a unit scheme for the transformer, provides adequate protection at a reduced cost.

16.3 GENERATOR PROTECTION

The generator capacity has sharply risen in recent times from 30 MW to 750 MW and even upto 1000 MW, with the result that the loss of even a single unit may cause overloading of associated machines and eventual system instability. The basic function of protection applied to generators, is, therefore, to reduce the outage period to a minimum by rapid discriminative clearance of faults.

Moreover, the increased size of generators and even greater increase in their outputs by more efficient methods of cooling make it imperative to protect them against fault.

In fact, unlike other apparatus, opening a breaker to isolate to the faulty generator is not sufficient to prevent further damage, since the generator will continue to supply power to a stator winding fault until its field excitation is suppressed. It is therefore necessary to open the field, stop the fuel supply to the prime mover and in some cases apply breaking.

It is known, that all large modern generator units are invariably transformer connected owing to interconnection. Before considering the types of protection fitted to generators it is desirable to consider the origin and effects of fault as stated below briefly.

(*a*) Generator Faults

Generator faults can be considered under the following heads :

1. Stator Faults 2. Rotor Faults 3. Abnormal Running Condition

1. Stator Faults

These faults include the following :

(*i*) Phase-to-earth faults, (*ii*) Phase-to-phase faults, (*iii*) Inter-turn fault. Most faults occur in the stator winding and their connection and majority of these are earth faults. Phase faults and inter-turn faults are less common, these usually develop into an earth fault. The effect of earth fault in the stator is two fold :

1. Arcing to core which welds leminations together, causing eddy current hot spots on subsequent use. Repairs to this condition involve considerable expenditure of time and money.
2. Severe heating in the conductors damaging them and the insulation, with possible fire risks.

2. Rotor Faults

Faults in the rotor circuit may be either earth faults or between turns, which result severe mechanical and thermal stresses acting on the winding insulation. Modern practice is to operate a generator with its field winding and rotor body to insulation breakdown can be tolerated. It is however, important, to know the existence of such a fault so that the generator may be taken out of service at leisure, since the incidence of a second fault will short circuit some part of the field winding resulting vibration of the rotor with possible damage to the bearing.

Another point to note is that rotor open circuits, though rare, can cause arcing and thus result in serious condition.

3. Abnormal Running Conditions

In a generator, the abnormal running conditions which are likely to occur are (*a*) loss of excitation, (*b*) unbalanced loading, (*c*) Overloading, (*d*) Failure of prime mover, (*e*) Overspeeding and (*f*) Over-voltage. Let us discuss briefly the matter as below :

Field failure may occur due to a faulty field breaker on failure of the exciter. When a generator loses its field excitation it speeds up slightly and acts as an induction generator deriving excitation from the system and supplying power at a loading power factor. A fall in voltage will also occur owing to loss of excitation which may result in loss of synchronism and system in stability. There is also the possibility of overheating of the rotor due to induced current in the rotor and dumping windings. Unbalance may occur due to single-phase faults unbalance loading open circuits due to broken lines, or due to the failure of one pole of a circuit breaker to close. However, short-circuits will normally be cleared by circuit protection, uncleared faults and cases of unbalanced current condition to be maintained on the generator. The unbalance gives rise to negative sequence currents which produce an armature reaction field which rotates in a direction opposite that of the rotor hence produce a flux which sweeps through the rotor with twice the rotational speed. Hence spurious current of twice the machine frequency are induced in the rotor body, in the field windings and the damper windings. These currents cause to overheating if the degree of unbalance is appreciable. The amount of negative frequency current which can be carried by a generator is usually well below its rated current.

When the generator is overloaded, it will overheat the stator which can further damage the insulation and thus create serious problems in the generator and eventually damaging the machine.

1. When the prime mover fails, it may cause the generator to "motor" by taking power from the system in case it is not a single generator system. The seriousness of this condition depends on the type of drive used for the generator.
2. In case of a sudden loss of load it may cause the machine to overspeed which is more likely to occur with hydraulic generators because the water flow cannot be stopped quickly. Then due to overspeeding or due to defective voltage regulator overvoltages are likely to occur.

16.3.1 Stator Protection

The best way to protect stator from phase-to-phase and phase-to-ground (iron lamination of the stator core) is provided by the longitudinal differential relay. The relay recommended for this application is an instantaneous attracted armature type setting 10 to 40% which is immune to a.c. transients and has the high speed feature is suitable if the CTs are reasonably matched. As in well known that if the CTs on the line side have dissimilar characteristic from those mounted at the neutral end of the winding a relatively large spell current will flow through the relay operating coil. A percentage bias differential relay should be used under such circumstances. Longitudinal percentage biased differential protection for generator stator or trans is shown in the Fig. 16.8.

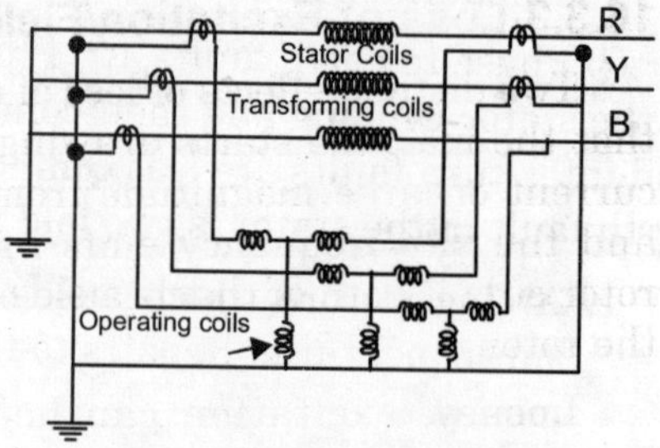

Fig. 16.8. Percentage biased differential relay for stator or Transe coils.

2. Stator Inter-turn Fault Protection

Stator inter-turn fault on the same phase of the stator winding does not disturb the balance between the current in the neutral and the high voltage CTs, as a result such a fault can not be detected coil by longitudinal differential protection. Transverse differential protection can detect the unbalance between normally identical windings caused by non inter-turn fault when the generator has two windings per phase. Bias is used because the sharing will never be exact. Biased transverse differential protection to protect against inter-turn faults is shown in Fig. 16.9.

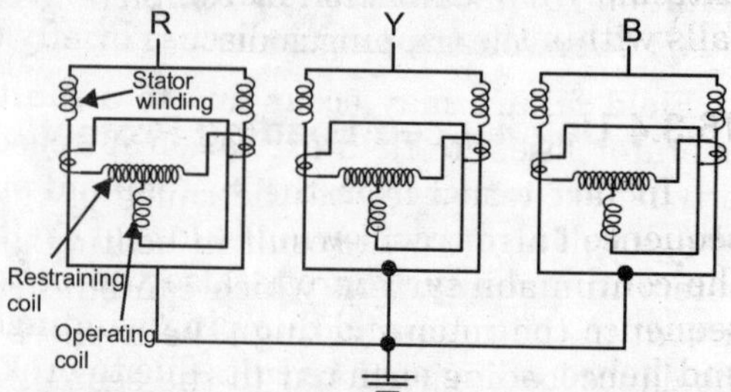

Fig. 16.9. Biased transverse differential protection.

In fact generators having single winding per-phase to those generators whose parallel windings are not accessible can be protected by using zero sequence component of voltage caused by the reduction of emf in the faulted phase. In the Fig. 16.10 one such circuit is shown where the zero sequence voltage appears across the tertiary winding of the voltage transformer which is connected to the operating winding of the three element directional Fig. 16.10 relay. The winding in quadrature to this operating winding of the relay is energised by the secondary of V.T. Inter turn fault.

16.3.2 Rotor Protection

It has been noticed that the rotor windings may be damaged by earth faults or open circuits. In the Fig. 16.10 we see a modern method of rotor earth fault detection. The field is biased by a d.c. voltage which causes current to flow through the relay R for an earth-fault any where on the field system. Detection of rotor open circuits is the same as that used for detecting loss of excitation discussed below.

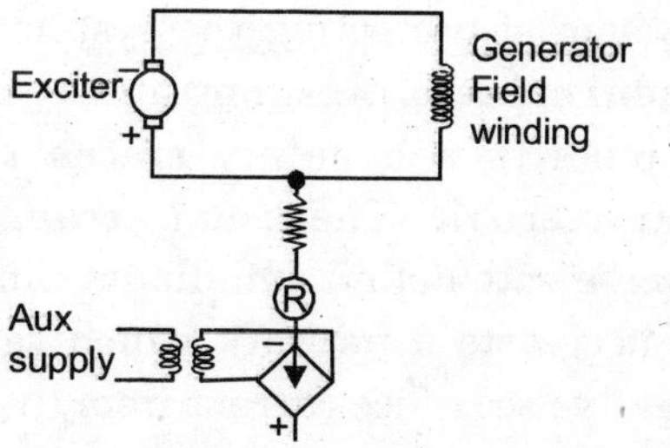

Fig. 16.10. Refer earth-fault protection.

16.3.3 Loss of Excitation/Field Failure Protection

Two distinct effects of loss of excitation are that the machine starts drawing magnetising current of large magnitude from the system, and the slip frequency emfs induced in the rotor circuit both of them cause overheating of the rotor.

Loss of excitation can be detected by increasing the reactive component of stator current, an excessive value of *VAR* component indicates either actual or prospective loss of synchronism. To allow for system transients which may cause a momentary reversal of *VAR* component, it is usual to incorporate a fixed time delay of between one and five seconds in the tripping sequence of the relay. It can be detected by an under current moving coil relay in the field circuit, but some large generators operate over a very wide range of field excitation and such a relay may be an embarassment. Furthermore, field failure due to exciter failure may not be detected because the under current relay may be held in a.c. induced from the stator. An under current relay is fast enough to drop out on a.c. cannot be used because it would be affected by a.c. induced during synchronising and during external fault. An alternative solution is to apply an offset impedance or mho measuring relay at the generator terminals. Its operating characteristic is arranged as shown in Fig. 16.11 so that during conditions of extremely low excitation or complete loss of excitation, the equivalent generator impedance falls within the tripping zone.

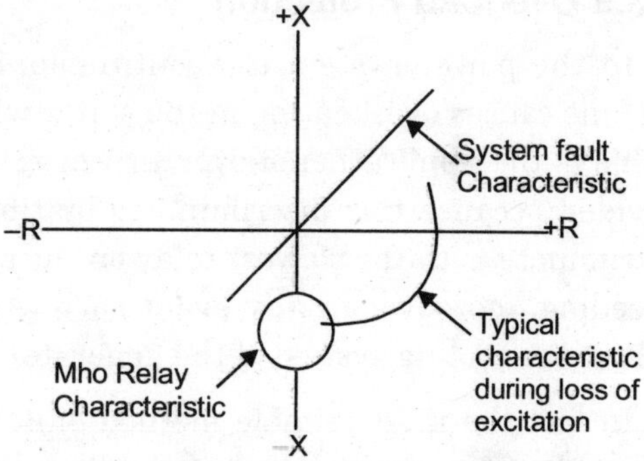

Fig. 16.11. Loss of excitation characteristic.

16.3.4 Unbalanced Loading Protection

In fact whatever be the cause of the unbalance it is obvious that the negative phase sequence current will result in heating of the stator. In case of high-speed turbo-generators, the continuous current which can be carried is usually between 10 and 15% of the positive sequence continuous rating. The negative sequence heating follows a normal resistance law and hence is proportional to the sequence of the current. The heating time constant of the machine is largely a function of the cooling system employed. This is expressed by a rating equation : $I_2^2 t = k$ where I_2 is the negative sequence current expressed on a per unit basis of continuous maximum rating (CMR), t is the duration of current in seconds and k is a constant which for turbo type machine will normally have a value between 3 and 20. However, the problem of protecting the system against this condition lies in obtaining a relay characteristic which will accurately match this heating characteristic. The usual arrangement is an inverse with definite minimum time delay relay connected to a network which segregates the negative sequence current from the positive and zero sequence current as can be seen in Fig. 16.12. The relay has a long operating time and has a range of setting to allow its characteristic to be accurately matched to those of the machine.

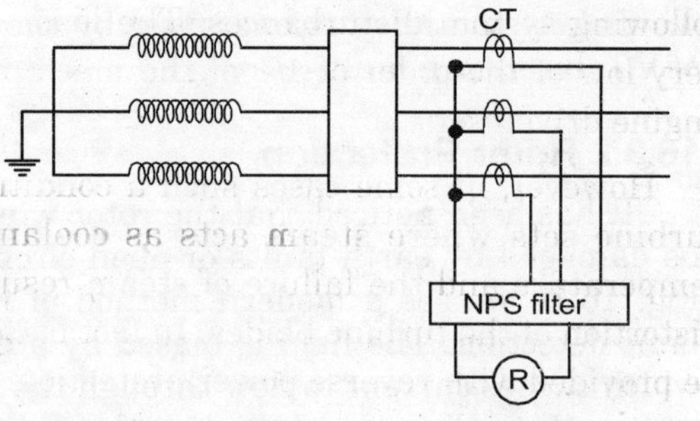

Fig. 16.12. Protection against unbalanced load using negative sequence filter. NPS = Negative Phase Sequence, R = Overcurrent IDMT relay.

16.3.5 Overload Protection

In the power system the continuous balanced overloading of a machine causes overheating in the stator winding. An obvious solution to this is the application of overcurrent relays but this is not usually provided because this discriminates by time, it must be arranged to discriminate with the slowest relay on the system which the generator is feeding, moreover in any event such a protection cannot detect a fault in the cooling system of the generator.

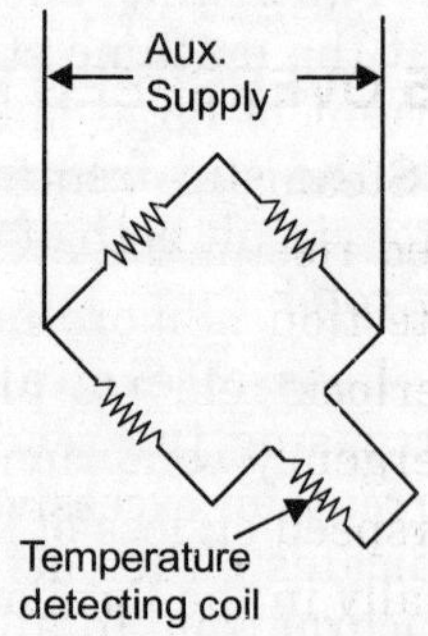

Fig. 16.13.

In fact the most reliable method of detecting such a condition is by means of temperature detector coils embedded at various points in the stator winding. Temperature detectors may be either Thermocouple, Thermistors or resistance temperature detectors (RTD). The temperature detectors coil form one arm of the Wheatstone Bridge as shown in Fig. 16.13. Generally, the generating sets below 30 MW rating are not normally provided with embedded temperature detectors but are usually provided with thermal relays. This type of relay has a bimetal strip heated by secondary current from the stator the housing of the bimetallic strip is designed to have a heating and cooling characteristic similar to that of machine. Such a relay will however will not provide protection against overheating due to the failure of the cooling system.

16.4 PRIME MOVER PROTECTION

In the event of prime-mover failure the machine starts motoring, which means that it draws electrical power from the system and drives the prime mover. This condition imposes a balanced load on the system, which can be detected by a power relay with a directional characteristic, so as shown in Fig. 16.14. It would detect this condition over the full power factor range. An auxiliary time-lag arrangement must be provided to prevent operation by range. An auxiliary time-lag must be provided to prevent operation by synchronising surges and power oscillations

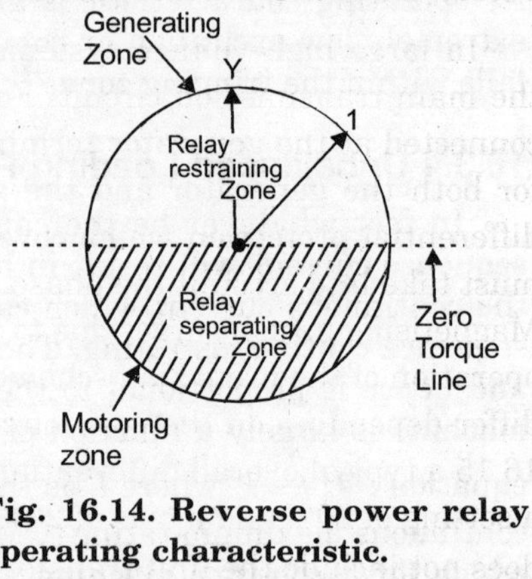

Fig. 16.14. Reverse power relay operating characteristic.

following system disturbances. The load coming on the machine under such conditions are very low of the order of 1% of the machine rating in case of steam sets and 25% in case of engine driven sets.

However, in some cases such a condition could be very harmful as in the case of steam turbine sets where steam acts as coolant maintaining the turbine blades at a constant temperature and the failure of steam results in overheating due to friction with subsequent distortion of the turbine blades. In fact back pressure turbine sets and engine driven sets must be provided with reverse power protection.

However, in condensing turbines the rate of temperature rise is low and such a protection is not necessary. Hydroelectric generators are usually fitted with mechanical devices to disconnect the generator from the system show the water flow to drop to a value likely to cause cavitation. Moreover Back-up reverse-power protection may be provided in unmanned stations by sensitive directional relays with timing relay.

16.5 OVERSPEED PROTECTION

Steam sets which are controlled by fast acting governor gear respond rapidly to the initial speed rise in the event of load being suddenly lost, unlike hydraulic sets. Hence overspeed protection is more necessary in case of hydraulic sets, which is a reverse or under power interlock relay to prevent the main generator circuit breaker being tripped under non-emergency conditions until the set output has dropped to a value low enough a prevent overspeed on loss of load. This is supplementary to the mechanical overspeed device which is usually in the form of centrifugally operated rings, mounted on the rotor shaft, which fly out and close the stop valves if the speed of the set increases by more than 10%.

16.6 OVERVOLTAGE PROTECTION

Overvoltage protection is provided on machines which are subjected to overspeed on loss of load. The protection is provided with an overvoltage relay. The relay is energised from a single-phase voltage transformer and is usually of the induction pattern with IDMT, characteristics. The relay opens the main Circuit Breaker and the field switch if the over voltage persists.

16.6.1 Generator Transformer Units Protection

In large high voltage systems generators are connected through step up transformer to the main transmission circuit. For supplying the generator auxiliaries a unit transformer is connected at the generator terminals. It is usual to provide a common differential protection for both the generator and the step up transformer and the intervening connection. This differential protection which covers a zone from generator neutral to the HV circuit breaker must take into account the phase shift and current transformation in the step-up transformer. Magnetising inrush surges due to restoration of voltage on fault clearance may cause unwanted operation of the protection scheme unless precautious are taken. Also the effective setting will differ depending on the position and type of fault, *i.e.*, H.V. or LV phase to earth. In the Fig. 16.15 a typical overall differential protection scheme is shown. A biased differential relay with a setting of 20% and a bias of 20% is generally satisfactory. This overall differential protection does not include the unit transformer which has a separate differential protection.

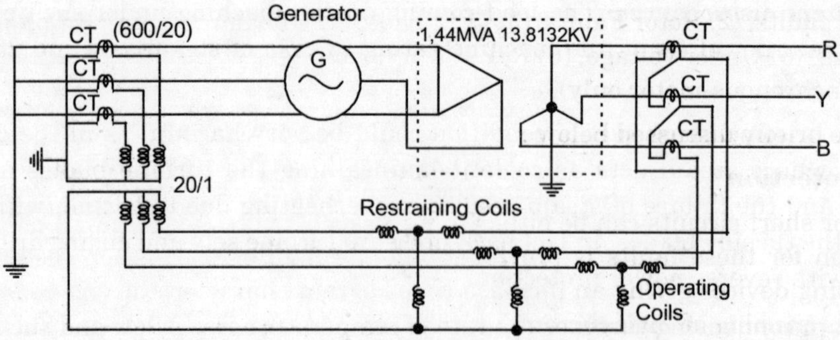

Fig. 16.15. Generator-transformer overall biased differential protection scheme.

In fact for protection against earth-fault an earth-fault relay can be put in the generator neutral : or in the secondary winding of the step up transformer. The differential protection is ineffective against earth-fault in case the HV star side of the generator-transformer is resistance earthed and thus restricted earth-fault protection is provided.

However, in case, it is solidly earthed the differential protection is quite adequate, but still a separate restricted earth-fault protection of the instantaneous type is preferred and the differential protection under such a case acts as a back-up protection to the restricted earth-fault protection. A scheme of restricted *E*/*F* protection along with the differential protection for a resistance earthed transform on the HV side is shown in Fig. 16.16.

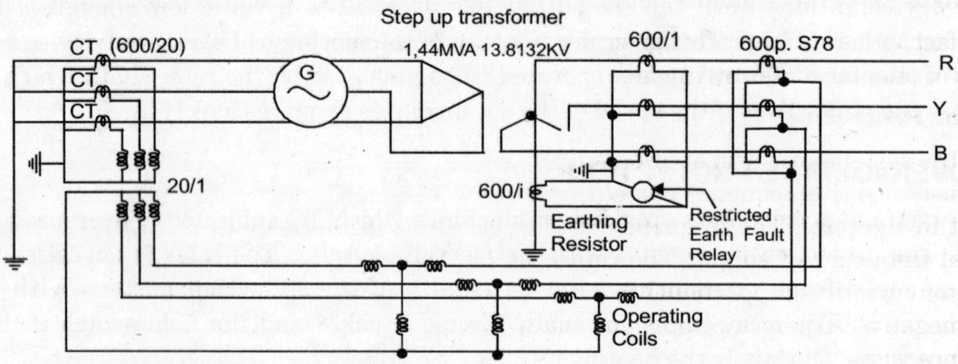

Fig. 16.16. HV restricted earth-fault protection on generator transformer units.

16.7 MOTOR PROTECTION

In the system there are wide ranges of motors for various purposes. But the fundamental problems affecting the choice of protection are independent of the type of motor and the type of load to which it is connected. The motors under discussion here are a.c. motors which include synchronous motors and induction motors.

The types of electrical fault in motors are similar to those of generators. Motors therefore in general are protected against the following faults :

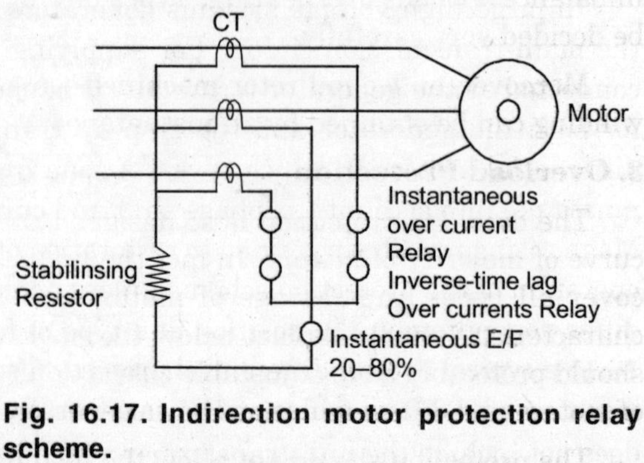

Fig. 16.17. Indirection motor protection relay scheme.

(1) Stator faults, (2) Rotor Faults, (3) Overloads, (4) Unbalanced Supply Voltages including single-phasing, (5) Under voltage, (6) Reverse or open-phase starting, (7) Loss of synchronism (in case of synchronous-motor only).

These are briefly discussed below :

1. Stator Protection

The stator short circuits can be either to Earth or between phases or all phases to Earth. The protection for these faults is provided with the help of thermal or dash-pot type over current tripping devices giving an increase time current characteristic and usually providing instantaneous tripping at high current.

Instantaneous over current relays supplied from CTs are provided for motors of larger rating (usually more than 50 HP).

Phase-fault protection is provided by two high-set instantaneous relay elements, the setting is so chosen that it is well above the maximum starting current. Earth fault protection for a motor operating on an earthed neutral system is provided by means of a simple instantaneous relay having a setting of approximately 30% of the motor full load current in the residual circuit of three CTs. Operation of relay due to CT saturation during initial high starting current should be avoided. This is usually achieved by increasing the voltage setting of the relay of inserting a stabilising resistance in series with it. Details of one such scheme is shown in Fig. 16.17. When a motor operates on an unearthed neutral system. Earth Fault relay as shown in the figure is of no use and a neutral displacement equipment has to be applied.

In fact, differential protection scheme is sometime used on very large and important duty motors of unearthed neutral system.

2. Rotor Protection

In the system any form of unbalance either in the supply voltage or in the loading pattern will cause negative sequence current to flow in the stator which will induce high frequency current in the rotor. The frequency of these currents in the rotor in fact is $(2 - s)$ times the nominal frequency of supply. The rotor heating due to the positive sequence components of the stator current is proportional to d.c. resistance value while the effect on the rator windings of the negative sequence component is proportional to $(2 - s) f$ (approximately 100 Hz) a.c. resistance value. Obviously the heating effect of negative phase sequence current greater than that of the positive phase sequence current. Motor protection therefore must take this into accounts if it is to decide correctly what load the motor can star a given degree of voltage unbalance without overheating. Types of protections provided for unbalanced voltages are to be decided very carefully.

Moreover, on wound rotor machines some degree of protection against faults in the rotor winding can be obtained by an instantaneous overcurrent relay measuring the stator current.

3. Overload Protection

The overload protection is so designed that it matches as closely as possible the heating curve of majority of motors. In fact the wide variety, rather wide diversity of motor duties to cover all types and ratings of motor with a given characteristic curve. The protection characteristic should lie fast below the heating curve of the motor protected. The protection should preferably have adjustable characteristics so that it may be adopted to different designs of motors and different duties they are subjected to.

The protection should not allow the motor to be restarted after tripping while the winding temperature is still high as this may have dangerous consequences. To have an effective safeguard an ideal protection should therefore not only match the heating characteristic of the motor but also its cooling characteristic. It must also be ensured that the relay must not operate under heavy starting current up to, say, 6-8-times full load current which last a few seconds. Generally for half a minute or so or even long or in exceptional cases.

The thermal constant of most types of motors is of the order of 15 to 20 minutes, hence for protection from overloads the relay should have a time constant slightly lower than this.

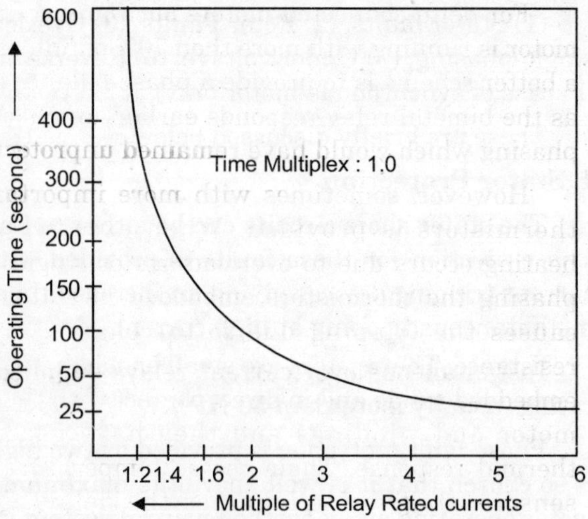

Fig. 16.18. Operation time characteristic of IDMT relay induction type.

In fact when a motor stalls, a current equal to the starting current flows and serious damages occurs if it persists for a time longer than the starting time. Hence, the closer the characteristic of the overload relay relay matches the starting current the better is the motor protected against such damage.

In the Fig. 16.18 is shown that induction overcurrent relays having characteristic of the type are best suited for such purposes as discussed above. A typical setting required for overload protection is 120% of full load current. It can be seen from the Fig. 16.18 that the current setting is 120% of full load, but a starting current of 6 times full load current for 30 seconds will not cause tripping with the help of the time multiplier setting the operating time at high values of overcurrent can be adjusted to match the motor starting characteristic without changing the current-setting.

In practice one phase-connected relay element is sufficient for overload protection but two provides single phase protection as well, hence this is preferred mostly.

4. Unbalanced and Single Phasing Protection

In the system the unbalanced three-phase causes negative sequence current to flow in the motor which is likely to cause overheating of the machine windings. Unbalanced loads or accidental opening of one phase of one phase of the 3-phase supply (called single-phasing) depending on the load still keeps the motor running although such a condition also causes the negative sequence current to flow in the motor.

In star-connected motors, complete overload and single-phasing protection can be provided by fitting two overload elements as shown in Fig. 16.19. The characteristic of the overload elements are such that the motor is permitted to run with supply on two phases only until such time as there is a risk of thermal damage.

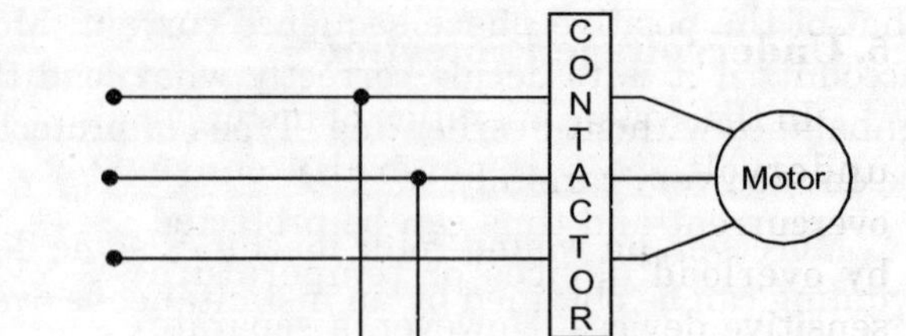

Fig. 16.19. Temperature-sensing operated scheme.

For delta-connected motors such an arrangement gives satisfactory protection when the motor is running with more then 70% of full load. For detecting the condition of single-phasing a better scheme is to provide a phase-balance relay on bimetal relays which protect the motor as the bimetal relay responds earlier, and also in the critical range of 40 – 60% when single-phasing which would have remained unprotected by a conventional overload relay.

However, sometimes with more important and large motors thermal protection with thermistors is provided. When excessive heating occurs due to overloadings or single-phasing the thermistors embedded in stator causes the tripping due to the change in resistance. Thermistors are small enough to be embedded inside and in direct contact with the motor and windings and they have good thermal response. Their use as temperature sensors, therefore eliminates the delay in transferring heat to actual sensing elements.

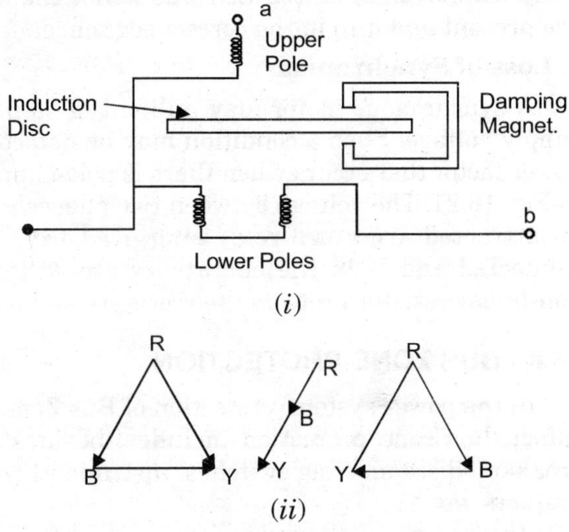

Fig. 16.20. Open and reversed phase relay.

In over-temperature protective scheme for a three-phase direct-on-line start type squirrel cage induction motor, three negative thermal coefficient (NTC) thermistors are located at the hot spots at the surfaces of the three stator winding phase (due on each phase) and are electrically connected into the temperature sensing circuit. The signal from the temperature sensing circuit is fed to a switching amplifier which causes a relay to operate when the signal equal on exceeds a present level. A normally closed (N/C) contact of this relay is connected in the control circuit.

During normal loads the signal from the temperature sensors is below the present value and the relay is non-operative. Its N/C contact keeps the control circuit energised and contactor closed. When the temperature reaches the upper limit the signal from the temperature sensing circuit cause an output in the amplifier and the relay operates to upon its contact. The control circuit gets deenergised and the contactor opens to disconnect the motor from supply.

5. Undervoltage Protection

In the system operation of motor on undervoltage will generally cause overcurrent and thus can be protected by overload devices or temperature sensitive devices. However, a separate single-element under voltage relay energised with phase-earth or phase-phase voltage can be provided to protect against a three-phase drop in voltage or an attempt to start with low voltage on all phases. A time-delay arrangement is usually incorporated to prevent tripping by a transient voltage drop.

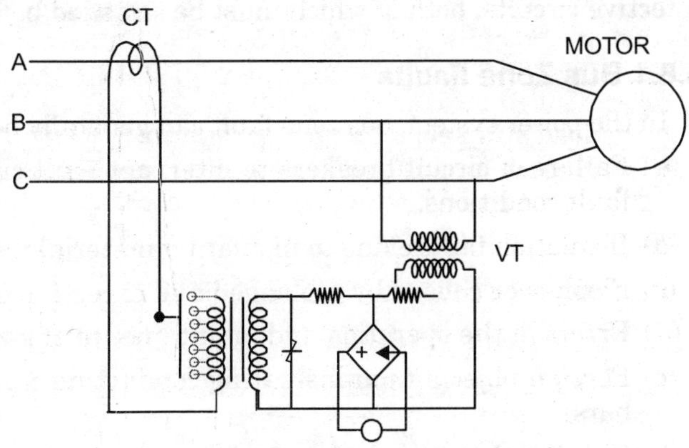

Fig. 16.21. Out-of-step protection relay.

6. Reverse Phase Protection

The direction of rotation of the motor changes if the phase sequence is changed. In some motor application this type of protection does become an essential feature for motor protection scheme. An induction disc poly phase voltage relay is used to protect motors from starting without phase open or with reversed phase sequence. The connections of such a relay are shown in Fig. 16.20, and its torque is proportional to the product of two line-to-line voltages. The relay will not close its contacts and hence the motor will not start unless all the three phases are present and also in the correct sequence.

7. Loss of Synchronism

A synchronous motor may pull out of step due to severe overload or due to reduction in supply voltage. Such a condition may be detected by a relay which responds to the change in power factor that occurs when there is pole-slipping. A typical circuit of this type relay is shown in Fig. 16.21. The voltage between two phases is compared with the current in the third phase; an attracted armature relay energised from a full-wave rectifier bridge is differentially connected and is in the operated state so long as the motor is in synchronous (step). A non-linear resistor protects the rectifiers and extends the operating range of the relay.

16.8 BUS ZONE PROTECTION

In the power system, protection of Bus Zone is a very important task for system engineers. Infact, bus zone protection includes, besides the bus itself the apparatus such as circuit breakers, disconnecting switches, instrument transformers and bus sectionising, reactors, bus couplers, etc.

Although bus-zone fault are rare, but experience shows that bus zone protection is highly desirable in large and important stations such as Grid-substations.

In fact, bus zone is more vulnerable to the effects of faults once it occurs for the reasons : (1) concentration of fault MVA increases risk of considerable damage, (2) Loss of bus zone would result in widespread supply interruptions.

Hence, it is of great importance that bus zone protection should be fast, stable and most reliable. Periodic testing is necessary to check the pick up of the relay on internal faults. It is the opinion of some experts that local bus protection should not be provided and the bus faults cleared by the back-up relays at the neighbouring stations, as the provision of local bus protection would certainly increase the risk of inadvertment tripping.

However, where local bus protection is provided care is taken by providing two independent protective circuits, both of which must be satisfied before tripping can occur.

16.8.1 Bus Zone Faults

In the power system, bus zone fault can generally be classified under one of the followings :

(*a*) Failure of circuit breakers to interrupt fault current or failure to clear under through fault conditions.

(*b*) Insulation failure due to insulating material deterioration.

(*c*) Flash-over caused by prolonged and excessive overvoltage.

(*d*) Errors in the operation and maintainence of switchgear.

(*e*) Foreign objects (non-insulating/conducting materials) accidentally falling across bus bars.

In fact, the clearing of a bus fault requires the opening of all the circuits branching from the fault bus or bus section (except circuit having no back feed).

16.8.2 Bus Back up Protection

In fact when no separate bus protection is provided but distance protection is provided for the feeders connected to the bus, it is possible to cover the bus bars within zone reach of the distance relays. It may however be noted that for small switchgear installations such a scheme may be quite satisfactory, whereas for large and important installations a separate bus zone protection may be provided. Bus back up protection may also mean that in case the breaker fails to operate for a fault on the feeder then it must be regarded as a bus fault. It should then open all other breakers on that bus. Such a back up can be provided with appropriate time delay through a time.

16.8.3 Differential Schemes of Bus Bar Protection

Differential schemes of bus bar protection is based on simple circulating current principle that during normal load condition or external fault condition the sum of current entering bus equals the sum of these leaving. If the sum of these currents (for a given conductor) is not zero, it must be due to a short circuit either a ground fault or a phase-to-phase fault. Hence differential schemes can be applied for both these types of fault the essential difference in the schemes being the method of connecting the CTs.

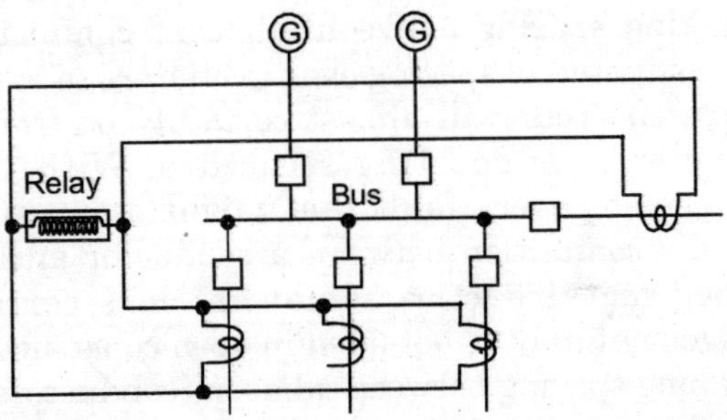

Fig. 16.22. Differential protection of a bus section.

A simple differential protection of a bus section is shown in Fig. 16.22. A current in the relay indicates a fault within the protected region and initiates opening of the generator breaker and each of the line breakers.

However, the main drawback with this type of differential protection is the difference in the magnetic conditions of the iron cored CTs which may result in false operation of the relay at the time of an external fault. Even with identical CTs having large iron cores to avoid the saturation with maximum fault current d.c. transient component present differently because of its slow decay. Biasing of differential relays improves the stability considerably but is not a complete solution.

Moreover, it can be seen that a high impedance bus differential relay can discriminate between internal and external fault better than the usual low-impedance relay. In other words, the ratio of the relay current during an internal fault to relay current during an external fault is greater if the impedance of the relay is higher.

A special type of CT having no iron core also known as the linear coupler is sometimes employed to overcome the difficulties of an iron cored CT. In the case of a linear coupler the secondary voltage is proportional to the primary current, and the secondary windings of all couplers on the same bus section are connected in series to the relay as shown in Fig. 16.23. The sum of their voltage outputs is equal to the vector sum of the voltages in the circuit connected to the bus bars. Under normal and external

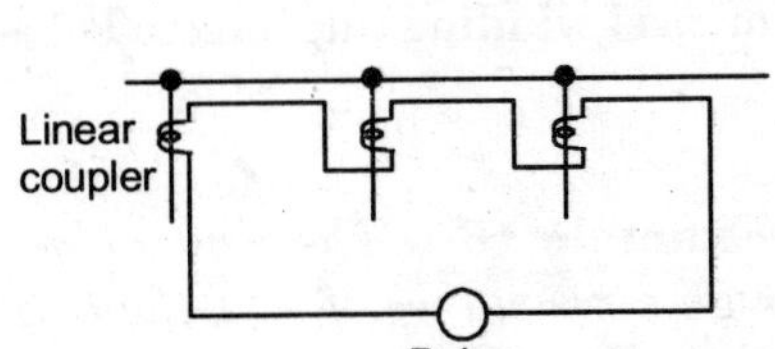

Fig. 16.23. Linear coupler scheme.

fault condition the voltages add upto zero, whereas for an internal fault there is a resultant voltage in the secondary circuit and thus the relay operates.

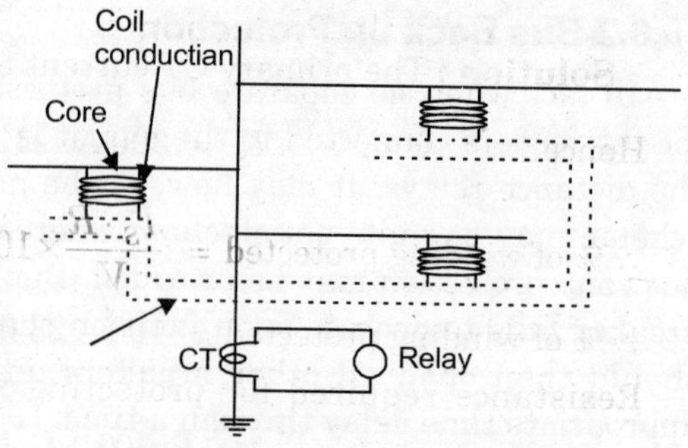

Fig. 16.24. Frame leakage protection.

16.8.4 Frame Leakage Protection

It is possible to design a station so that the faults which develop, are mostly earth faults, by providing earthed metal barrier surrounding each conductor in the bus structure. With this arrangement every fault that might occur must involve a connection between a conductor and an earthed metal part. A somewhat less certain arrangement may be obtained in open construction by making spacing between adjacent conductors great compared to spacing over insulators to earth. Again that occur will almost certainly be from a conductor to supporting structure. With this arrangement every fault that might occur must involve a connection between a conductor and an earthed metal part. a somewhat less certain arrangement may be obtained in open construction by making spacing between adjacent conductors great compared to spacing over insulators to earth. Again faults that occur will almost certainly be from a conductor to supporting structure. However, in each of the preceding cases the metal supporting structure also known a fault bus is earthed through a CT as can be seen in Fig. 16.24. A fault involving a connection between a conductor and the earth, *i.e.*, the supporting structure will result in current flow through this CT and actuation of the relay connected to its secondary. Operation of this relay will result in tripping all breakers connecting equipment to the bus. Sometimes an impedance Z is connected in the earthing connection to limit the value of the short circuit current during line to earth fault. The value of the neutral earthing resistor and the amount of winding to be protected determines the relay fault setting. If the R is the earthing resistor I_F is the earth fault current then, $I_F = (1 - x)\dfrac{V}{R}$ where x% of winding is protected as shown in Fig. 16.25.

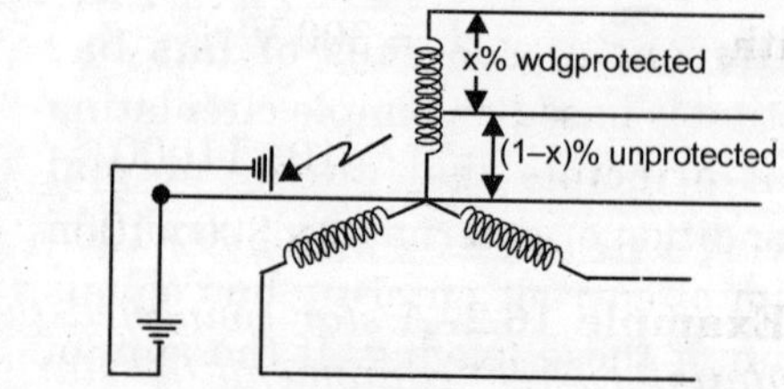

Fig. 16.25.

This current, I_F must be equal to the primary fault setting of the differential protection I_S for minimum operating current. Hence, $I_S = (1 - x)\dfrac{V}{R}$ and $I_R = \dfrac{V}{R}$.

Hence,
$$x = \left(1 - \frac{I_S}{I_R}\right)$$

where I_S = primary fault setting current

I_R = Earthing resistor current setting

or % of winding unprotected $(1 - x) > I_S \times 100$

$$= \frac{I_S \cdot R}{V} \times 100 \cdot I_R$$

Example 16.1. *The neutral point of a three-phase 20 MVA, 11 kV alternator is earthed through a resistance of 5 Ω, the relay is set to operate when there is an out of balance current of 1.5 A. The CTs hence a ratio 1000/5. What percentage of winding is protected against an earth fault and what should be the minimum value of earthing resistance to protect 905 of the winding.*

Solution : The primary CT current for relay operation = $1.5 \times \frac{1000}{5} = 300$ A

Hence, $I_S = 300$ A

$\therefore$ % of winding protected = $\frac{I_S \cdot R}{V} \times 100 = \frac{300 \times 5 \times \sqrt{3}}{11,000} = 23.6\%$

$\therefore$ % of winding protected = 100 – 23.6 = 76.4%

Resistance required for protecting 90% of the winding in given by $10 = \frac{I_S \cdot R \times 100}{V} = \frac{I_6 \cdot R \times 100}{11,000}$.

With, $I_S = 300$ V $\therefore$

$$R = \frac{10 \times 11000}{\sqrt{3} \times 300 \times 100} = \mathbf{2.12\ \Omega}$$

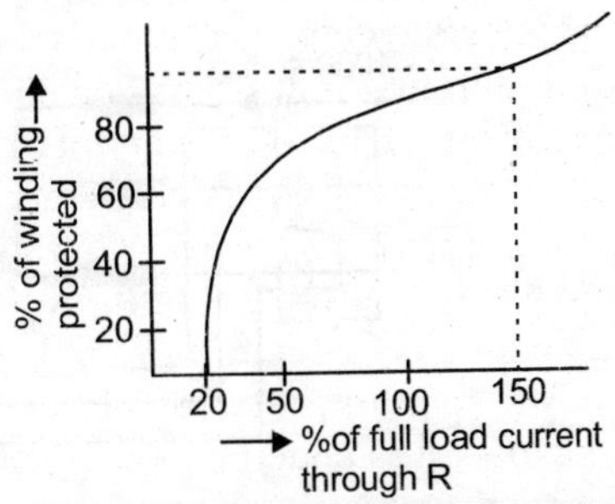

Fig. 16.26. Earth-fault setting characteristic.

Example 16.2. *A star connected three-phase 15 MVA 6.6 kV alternator is protected by circulating current equipment, the star point being earthed via a resistor R. Plot a curve to a base of current up to 150% of full-load current through R showing the percentage of winding protected for an earth-fault setting of 20%. Find the value of earthing resistor if 85 of the winding is to be protected.*

Solution : A fault setting of 20% means that, if a terminal fault has its current limited to the full load rating them 20% of the winding from the neutral and will be unprotected. Now for different values of current through *R* under terminal fault condition by simple proportion it is possible to find the percentage of winding protected. This is shown in the Table 16.1 for a 20% setting. Thus the curve can also be drawn in Fig. 16.26.

Table 16.1

% of full load current through *R* for a terminal	Unprotected % of winding unprotected	Effect of earthing resistance on the winding protection % of winding protected
20	100	0.
40	50	50.
60	33.33	66.66
80	25	75.0
100	20	80
120	16.66	83.33
150	13.33	86.66

For protecting 85% of the winding the current through *R* comes out to be 33.33%.

Full-load current = $I_F = \frac{15 \times 10^4}{\sqrt{3} \times 6.6 \times 10^3} = 1310$ A (100 – 85) = 15, 20% setting, $I_S = 0.2 \times I_F$

Now, $1 - x = \frac{I_S \times R}{V} \times 100$ or $15 = \frac{0.2 \times 1310 \times R}{6600/\sqrt{3}} \times 100$ $R = \mathbf{2.2\Omega}$.

Example 16.3. *A 3-phase transformer having a line voltage ratio of 400V/33,000V is connected in Star-Delta. The CTs on the 400V side have a current ratio of 1000/5. What should be the ratio of CTs on the 33,000 V side.*

Solution : In Fig. 16.27 is shown the connection of CTs. Let us say that the line current on the secondary side I_{L_2} can be found by the equation.

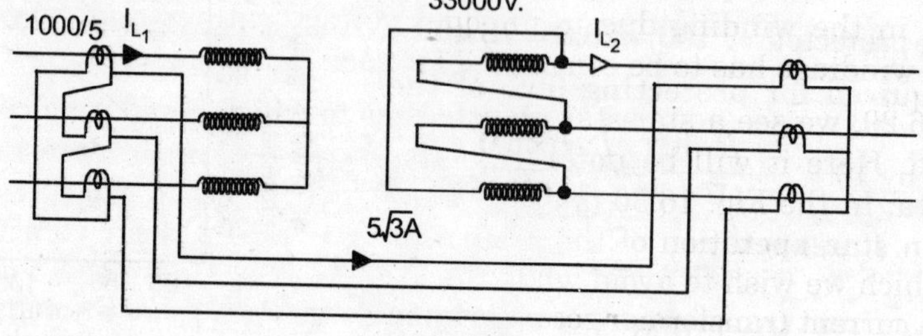

Fig. 16.27

$$\sqrt{3}\times 400\times 1000 = \sqrt{3}\times 33{,}000\times I_{L_2}$$

$$I_{L_1} = 1000\text{A} \qquad \therefore\ I_{L_2} = (400/33)\text{A}$$

The current through the secondary of the CT on the primary side will be 5A. The CTs on the primary side being delta connected, the current through its pilot wire (wires connected between primary side & secondary side CTs) will be $(5/\sqrt{3})$A. The CTs on secondary side being star connected will have a current of $(5/\sqrt{3})$A in the secondary. Hence CT ratio on the 33,000 V side $= \dfrac{400}{33\times 4\sqrt{3}} = 7/5.$

16.9 PROTECTION SCHEME REGARDING A DIRECT CONNECTED GENERATOR

For a Direct connected generator a basic protection scheme is shown in Fig. 16.28, where it comprises (*a*) Biased differential protection, (*b*) Back-up overcurrent and earth fault relay protection, (*c*) Negative Phase Sequence (NPS) protection and (*d*) Stand by Earth fault protection. Voltage restrained over-current relay may also be given in this scheme instead of the back up ordinary overcurrent relay unit.

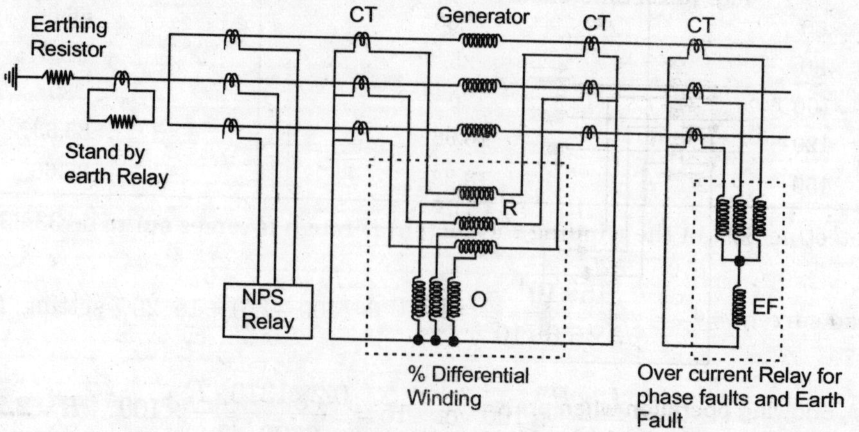

Fig. 16.28. Basic protection scheme direct connected generator.

16.10 TRANSFORMER PROTECTION SCHEME

This is most important protection scheme in power system. Here, differential protection operating on the circulating current system is most widely used for the protection of insulation faults in power transformer winding on their associated connecting cables.

The principle of operation of this type of scheme must be understood properly. However in applying the scheme of the protection of transformers the disparity of the current magnitude and phase angle in the winding due to unequal voltage ratings and varying phase inter-connection of the windings has to be considered for each individual case.

In the Fig. 16.29, we see a star-star transformer to which circulating current protection has been applied. Here it will be noted that the current transformers in both sides are connected as delta. In the Fig. 16.30 (*a*), it is shown that how had the current transformers been connected in star, operation of the protective relay would occur on a fault outside the protected zone which we wish to avoid while the Fig. 16.30 (*b*) shows how this can be avoided by connecting the current transformer secondaries in delta. Because of the delta-star connection there is a phase displacement between the primary and secondary line current and to compensate for this, the CTs on the delta side of the power transformers are connected in star and those on the star side of the transformers are connected in delta. But, this alone is not sufficient. Due to different voltages on the input and the output sides of the transformer the magnitudes of the line current will also be different and unless suitable ratios for the CTs on the input and the output are selected there may be unwanted relay operation. In choosing the ratio of the current transformers the primary current are those based on the kVA of the transformer and the appropriate primary and secondary voltage.

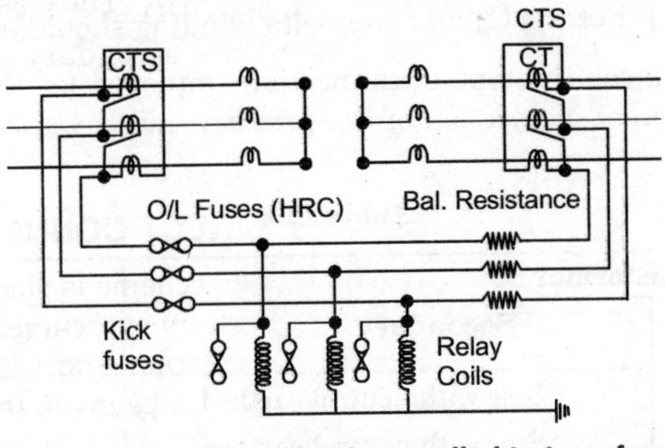

Fig. 16.29. Differential protection applied to transformers.

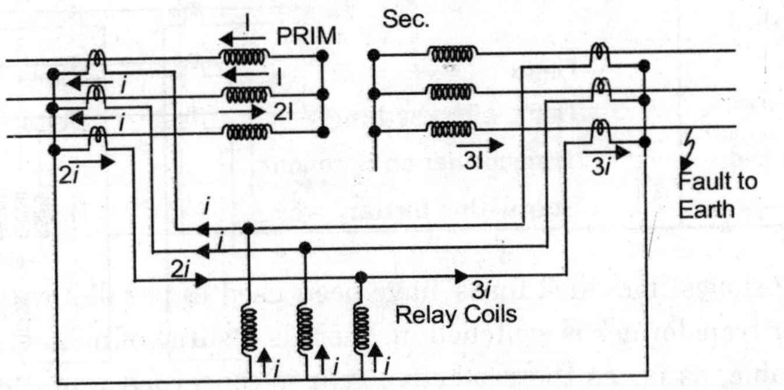

Fig. 16.30 (*a*). Showing operation when protective CTs are connected in star and fault occurs external to the power transformer of ratio 1 : 1.

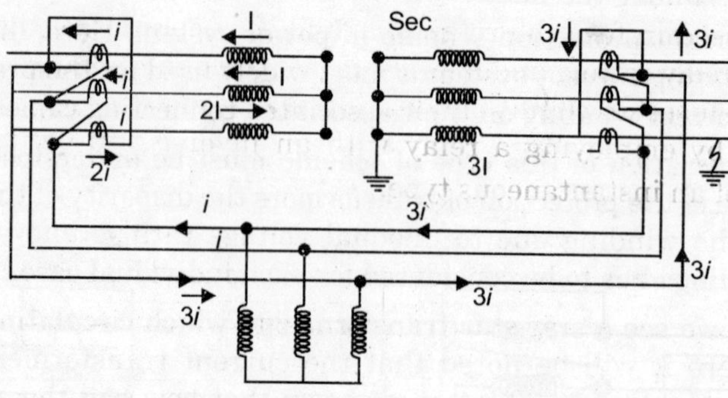

Fig. 16.30 (*b*). Showing operation when protective CTs are connected in Delta and a fault occurs external to the power transformer of ratio 1 : 1.

If the secondary current of the star-connected current transformers are 5 amperes, then these for the delta connected CTs are $5\sqrt{3}$ amps.

In fact, there are two basic requirements that the relay connections must satisfy are :

1. The relay must not operate for load or external.
2. The relay must operate for internal operate.

However, a general rule is that the CTs on any star winding of a power transformer should be connected in delta and that the CTs on any delta winding should be connected in STAR.

In the Table 16.2, we see the type of connection employed for the CTs on the input and the output sides for different connection of the primary and the secondary winding of power transformer.

Table 16.2

Sl. No.	Power Transformer Primary	Secondary	Current Transformer On the side	On the side
1.	Star	Star with neutral earthed	Delta	Delta
2.	Delta	Star with neutral earthed	Star	Delta
3.	Star with Neutral Earthed	Delta	Delta	Star
4.	Delta	Delta	Star	Star
5.	Star with Neutral Earthed	Delta with earthing Transformer on Secondary	Delta	Star
6.	Star	Star with a tertiary way	Delta	Delta

The Fig. 16.29 shows that 'kick fuses' have been used in parallel with the relay coils. In fact, when a power transformer is switched in, there is a surge of magnetising current which is not distinguishable, as far as the protective gear is concerned from the fault condition to present operation due to this the 'kick fuses' are used to shunt the relay coils. These fuses are of the time-limit type with an inverse characteristic and do not operate in the short time of the switching in surge. Under the sustained fault condition however, fuses do operate and the current is then passed in fuse the relay coils and the operation of the relay takes place. This scheme affect the relay setting and in order to overcome this arrangement are sometimes made to have the 'kick-fuses' in circuit only during the period of switching in. The difficulty can also be overcome by employing a relay with an inverse and definite minimum time characteristic instead of an instantaneous type.

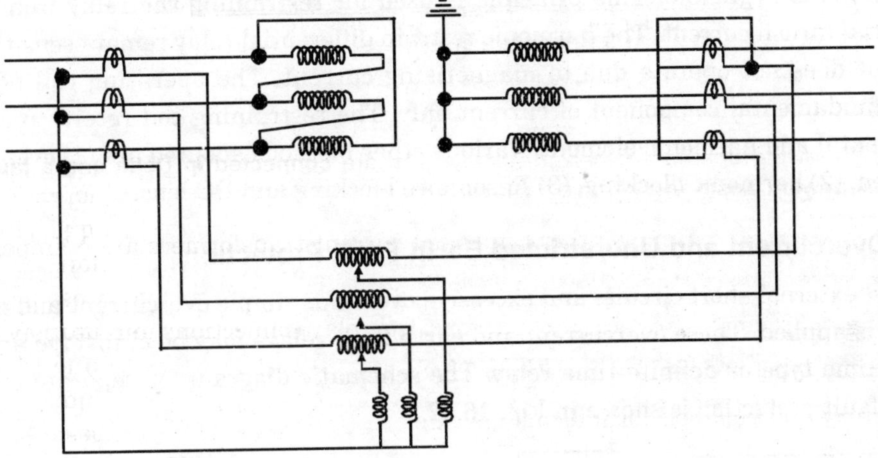

Fig. 16.31

In the Fig. 16.29 the O/L fuses shown provide a form of back-up protection. In the event of a sustained through fault, damage may be caused to the transformer and it is then that one or more of the O/L fuses will operate leaving the relay to be fed from one set of CTs and thus causing the relay to operate.

In fact, circulating current protection provides protection against earth fault and short-circuit between adjacent phases. Protection is also provided for short-circuits between turns on the same phase since such short-circuits alter the turn ratio of the transformer and consequently cause unbalance between current transfomer pairs. The change in the transformer ratio must, however, be considerable in order to pass sufficient current into the relay to cause operation. Such short-circuits are dealt with more effeciently by Buchholz relay.

However due to the various known difficulties (like magnetising current surges, fixed current transformer ratios on variable ratio power transformer due to tapping and the unavoidably different characteristic of CTs on the high and low voltage sides of the transformer) several new protection schemes hence been developed. There are biased Differential Protection, Translay Protection etc. Biased Differential Protection can be used for transformer protection with due consideration to the ratios and connection of the current transformers. A typical diagram of connection showing such a protection applied to a delta-star power transformer is shown in Fig. 16.31.

16.10.1 Harmonic Restraint

We have discussed above that operation of the relays due to the magnetising inrush current can be avoided either by using kick fuses across the relay coils or by using relays with inverse and definite minimum time characteristics. But with the larger and more important transformers, the relay current and time rating necessary to ensure stability on the magnetining inrush current produced by switching on the transformer are indequate to provide high speed protection.

A high speed baised differential relay incorporating a harmonic restraint feature will present relay operation under magnetising inrush current condition. The magnetising inrush current have a high component of even and odd harmonics. Harmonic component of short circuit current is negligible. This principle is used for restraining the relay from operation during initial inrush current. The harmonic restrain differential relay remain sensitive to fault current but does not operate due to magnetising current. The operating coil of the relay receivers fundamental component of current only. The restraining coil receive rectified sum of fundamental and harmonic element. Various other methods adopted are (1) even harmonic cancellation, (2) harmonic blocking, (3) Resonance blocking and DC bias.

16.10.2 Overcurrent and Unerstricted Earth Fault Protection

Against external short circuits and excessive overloads simple overcurrent and earth fault protection is applied. These overcurrent and earth fault relays may be of inverse and definite minimum time type or definite time relay. The schematic diagram for separate overcurrent and earth fault protection is shown in Fig. 16.32.

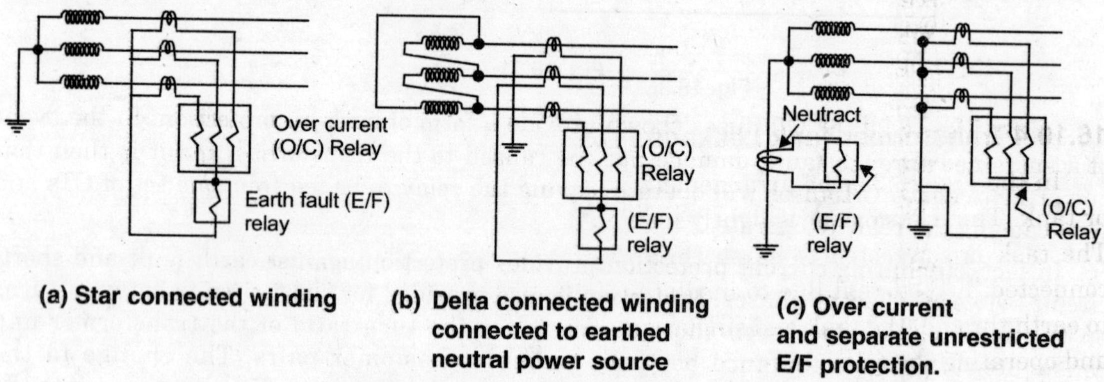

Fig. 16.32. Combined over-current and earth fault relay protection.

In the O/C relay there are three element-one for each phase and earth fault has a single element. The normal range of current settings available on IDMT over current relay is 50% to 200% and on earth fault element 20 to 80%. On the latter another range 10 to 40% is available and may be selected where the earth fault current is restricted due to insertion of impedance in the neutral grounding. In the case of transformer winding with neutral earthed unrestricted earth fault protection is obtained by connecting an ordinary earth fault across a neutral CT as can be seen in Fig. 16.32 (*c*).

16.10.3 Restricted (Balanced) Earth Fault Protection

In the Fig. 16.33 is shown the connection for balanced earth fault protection of one winding (primary, secondary or tertiary) of a transformer. Each transformer winding has to be provided with a separate similar arrangement.

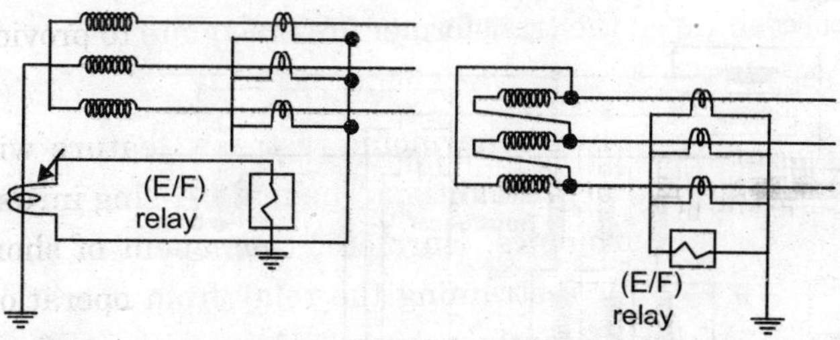

Fig. 16.33. Restricted E/F protection scheme.

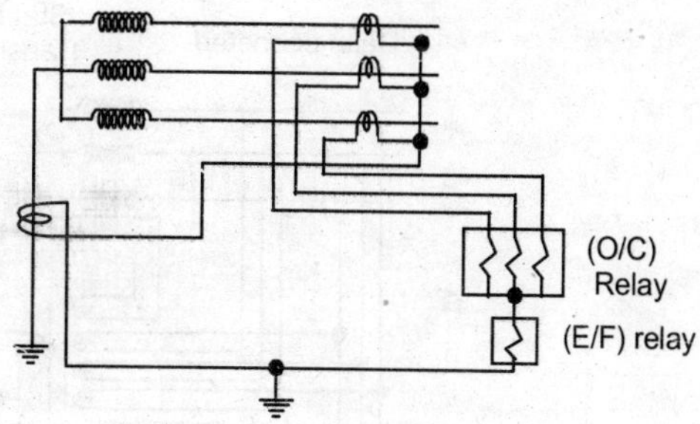

Fig. 16.34. Over-current protection with E/F protection.

In the Fig. 16.34, we can see the overcurrent protection as well as restricted earth fault protection.

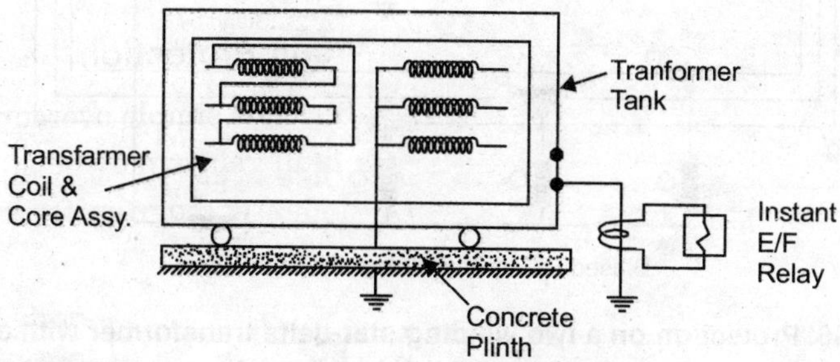

Fig. 16.35. Frame leakage protection.

16.10.4 Transformer Tank Leakage Protection

In the Fig. 16.35, the arrangement is shown for leakage protection of transformer frame or tank. The transformer is lightly insulated from earth by mounting it on a concrete plinth. The task is connected to earth through CT across which an instantaneous E/F relay is connected. E/F current due to insulation failure in any winding of the transformer will flow to earth through the tank and single earth connecting thus energising the current transformer and operating the relay.

16.10.5 Transformer Protection Scheme in a 132 kV Grid Power System

In the Fig. 16.36 a typical 132 kV Grid system transformer protection arrangement is shown for the reference the figure is self explained.

16.10.6 Generator Transformer Protection

In the power system, in high voltage transmission circuits, the bus bars are operated at higher voltages than that of the generators. in fact, it is common practice to connect the generator directly to step-up transformers. The protection of a generator and its step up transformer as a unit is affect by means of circulating current principle of protection. Because of the occurrence of magnetising current transient at any moment the relay setting in this protection must be considerably increased as compared with those protecting a generator only. The circuit used is shown in Fig. 16.37 and consists of a star connected machine connected directly to the delta winding of a delta/star transformer.

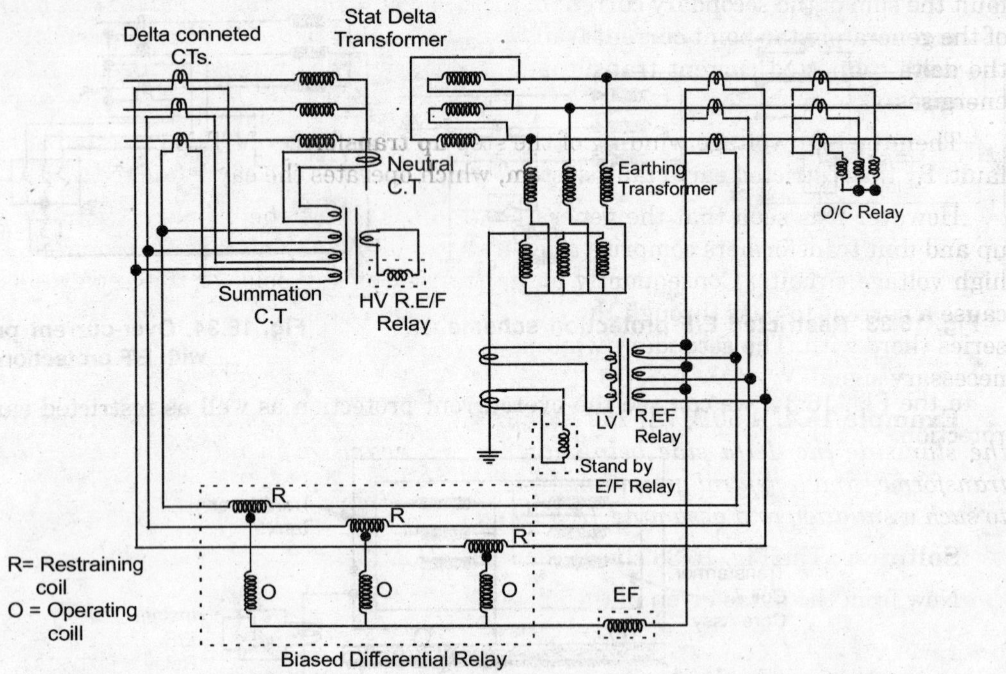

Fig. 16.36. Protection on a two winding star-delta transformer with earthing transformer on the delta side.

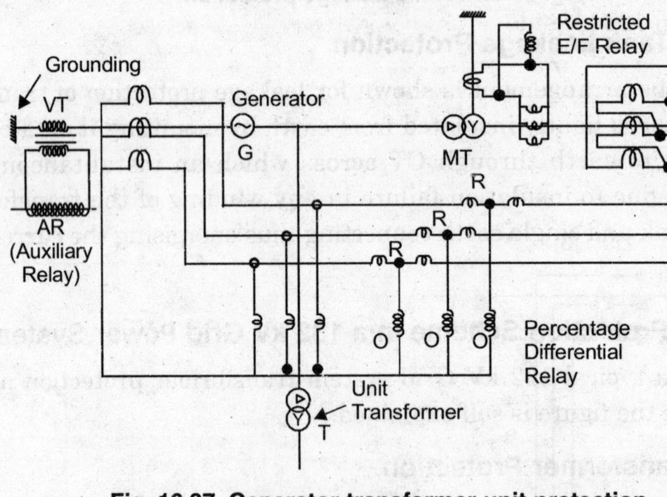

Fig. 16.37. Generator-transformer unit protection.

The zone of differential protection includes the stator winding, the step up transformer and the intervening connection current transformers on the high voltage side of the step up transformer (M.T.—Main Transform) are delta-connected, and those located near the star point of the generator are star-connected so as to cancel the 30° displacement between line current introduced by the delta connected primary of the M.T. The unbalance that would be caused between current transformer pair owing to the load of unit transformer *T* is avoided by initiating a further set of current transformers in the primary leads of the latter. The connections of current transformers is Delta as the primary of unit transformer is star. While there is no fault the sum of the secondary current of these current transformers and the secondary current of the generator star-point current transformers is equal to the current in the pilot wires from the delta connected current transformer secondary. When a fault occurs the relay A.R. are energises.

Then the high voltage winding of the step up transformer M.T. is protected against earth fault. By the restricted earth fault system, which operates the earth fault relay ER.

However it is seen that the generator windings *G* and the primary windings of the step up and unit transformers comprise a separate circuit having no electrical connection with the high voltage circuits. Consequently an earth fault at any point of this separate circuit will cause a current to pass through the earth connection and through voltage transformer VT in series there with. The secondary winding of VT energises an alarm relay AR which gives the necessary signal.

Example 16.4. *A 50 MVA, 132 kV Star-Delta buck supply transformer is earth solidly on the star side the delta side being earthed by means of an interconnected-star earthing transformer and a neutral grounding resistor. Show how Merz-Price protection may be applied to such a situation and assuming 1A secondary CT ratings, suggest suitable CT ratios.*

Soltuion : The Fig. 16.38 shows the connection of the Merz-Price protection.

Now from the datas given in the example,

$$I_{FL} = \frac{50 \times 10^6}{\sqrt{3} \times 33 \times 10^7} = 876 \text{ A}$$

$\therefore$ CT ratio of 33 kV side $= \dfrac{876}{1}$

Hence, CT ratio on 132 kV side

$$= \left(876 \times \frac{33}{132}\right) / \frac{1}{\sqrt{3}}$$

$$= 219/\left(\frac{1}{\sqrt{3}}\right)$$

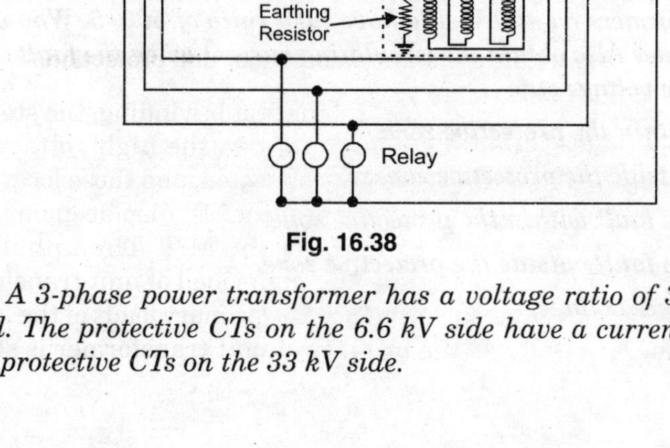

Fig. 16.38

Example 16.5. *A 3-phase power transformer has a voltage ratio of 33 kV/6.6 kV and is star delta connected. The protective CTs on the 6.6 kV side have a current ratio of 100. What must be the ratio of protective CTs on the 33 kV side.*

Solution : This is a star-delta connected transformer, so that protective CTs on the star side must be in delta and these on the delta side must be in star.

Now let us calculate the line current on H.V. side ($I_{H.V.}$) for a line current of (say) 100 amps, or the 2 V side.

$$\therefore \quad \sqrt{3}\times 6.6\times 100 = \sqrt{3}\times 33\times I_{H.V.} \qquad \therefore I_{H.V.} = 20 \text{ Amps}$$

The protective CTs on the LV side are in star and have a secondary current of 1 amps (given) so that current in the pilot wires must be the same, *i.e.*, 1 amps. For balance the secondary current of protective CTs connected in delta on the HV side on the main transformer must therefore, be $1/\sqrt{3}$ amps.

Hence, this means that CTs on the HV side have a primary current of 20 amps for a secondary current of $1/\sqrt{3}$ amps. Therefore the ratio of protective CTs on the HV side (33 kV) = $20 : (1/\sqrt{3}) = 20\times\sqrt{3} : 1$ = **35 : 1 amps.**

Example 16.6. *A three phase 66 kV/11 kV power transformer is connected in star/delta. The transformer is protected by Mertz-Price circulating current method. Protection current transformers on the low voltage side have a ratio of 250/5. Find the ratio of the current transformers on high voltage side.*

Solution : Windings of the low voltage side of the main transformer are connected in delta.

∴ Voltage for phase = 11,000 V = Line Voltage

However, current transformers in low voltage side (*i.e.*, 11 kV side) are connected in star and their ratio is 250/5.

Now in HV side of the transformer, the main windings are connected in star and ratio of the CTs is 250/5 (= 50). But on the high voltage side of the transformer, the main winding are connected in star.

$$\therefore \text{ Phase voltage} = \frac{66{,}000}{\sqrt{3}} = 38{,}000 \text{ volts}$$

$$\text{The ratio of the main transform} = \frac{38{,}200}{11{,}000} = 3.48$$

Also current transformer on to high voltage side are connected in delta

∴ High voltage side current to transformer ratio

$$= \frac{50}{1}\times\frac{1}{348} = \mathbf{14.4}$$

Example 16.7. *A 3-phase, 200 kVA, 11000/400 V transformer is connected as delta/star. The protective transformers on 400 V side have turn ratio of 500/5. Who will be the Ct ratios on the high voltage side? Also obtain the circulating current when the fault of 750 A of following types occur on the low voltage side :*

1. *Earth fault within the protective zone.*
2. *Earth fault outside the protective zone.*
3. *Phase to phase fault within the protective zone.*
4. *Phase to phase fault outside the protective zone.*

Assumke balance voltage.

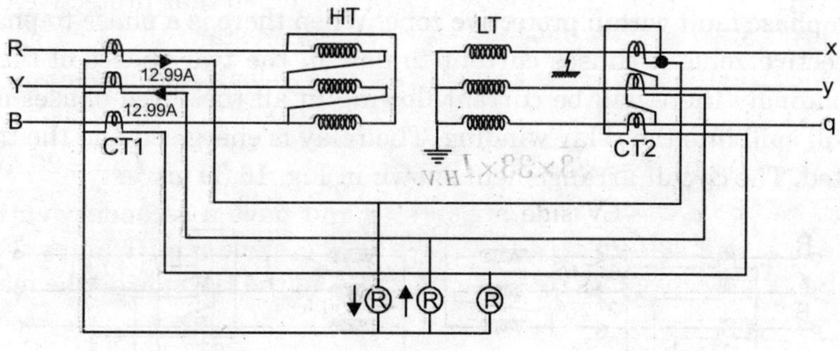

Fig. 16.39 (*a*)

Solution : In the transformer

$$I_1\text{(phase value)} = \frac{E_2 I_2}{E_1} = \frac{(0.4/\sqrt{3})\times 500}{11} = 10.5 \text{ A}$$

Equivalent line current on H.V. side = 10.5 × $\sqrt{3}$ = 18.18 A (As HV is Delta connected)

A current of 500 A on HV side will cause a current of 5 A in the secondary of CT_2.

However the pilot current on HV side = 5 × $\sqrt{3}$ = 866 A

∴ The ratio of CT_1 (HV side) should be 18.18/8.66.

Now,

1. The earth fault is within the protective zone and since the primary isolated, so, no zero sequence current will flow in the primary.

   The phase current in the HV side corresponding to 750 A is given as

   $\frac{750\times(0.4/\sqrt{3})}{11}$ = 15.75 amps. The current ratio of CT_1 is 18/18/8.66, so corresponding current through the relay will be 12.9 A as shown in Fig. 16.39 (*a*).

2. When there is an earth fault outside the protective zone, the distribution of current of the HV side remains the same as in case (1). The secondary current in CT_2 is given as

   $\frac{750\times 5}{500}$ = 7.5 amps.

   The current in pilot wires = 7.5 × $\sqrt{3}$ = 12.99 amp. The direction of which is shown in Fig. 16.39 (*b*). Thus there will not be any current flowing into the relay, so, the circuit breakers for through fault does not isolate the transformers.

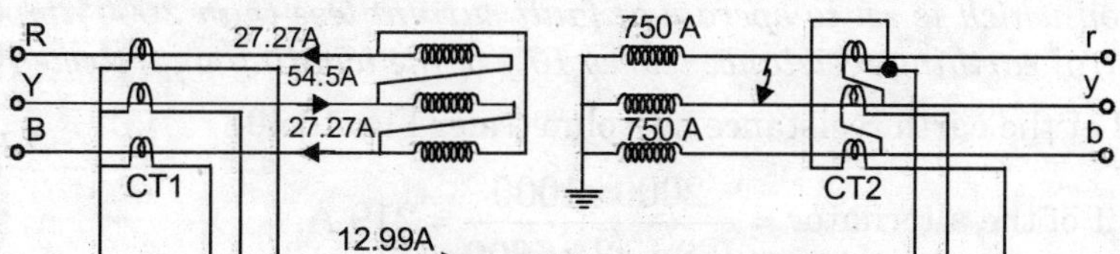

Fig. 16.39 (*b*)

3. Phase-to-phase fault within protective zone. When there is a phase-to-phase fault within the protective zone, it causes current to flow in the two phases of the L.V. windings correspondingly there will be current flowing in all the three phases of H.V. winding which will spill into the relay winding. The relay is energised and the transformer will be isolated. The circuit arrangement shown in Fig. 16.39 (*c*).

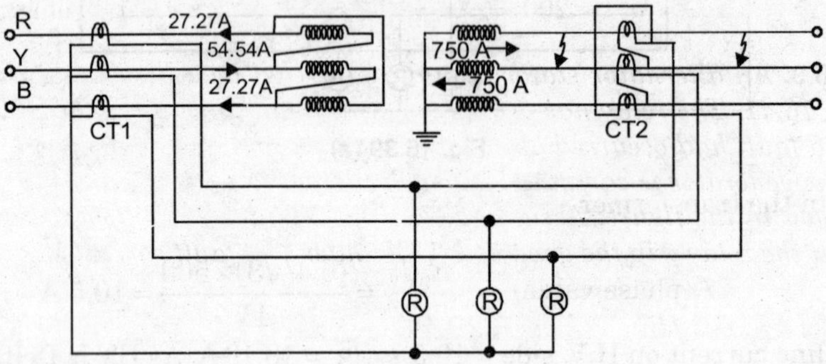

Fig. 16.39 (*c*)

4. Phase-to-phase fault outside the protective zone, when there is a phase-to-phase fault outside the protective zone, the current distribution changes to that is shown in Fig. 16.39 (*d*). It will be observed from this that the relay is not energised, then the transformer is not isolated.

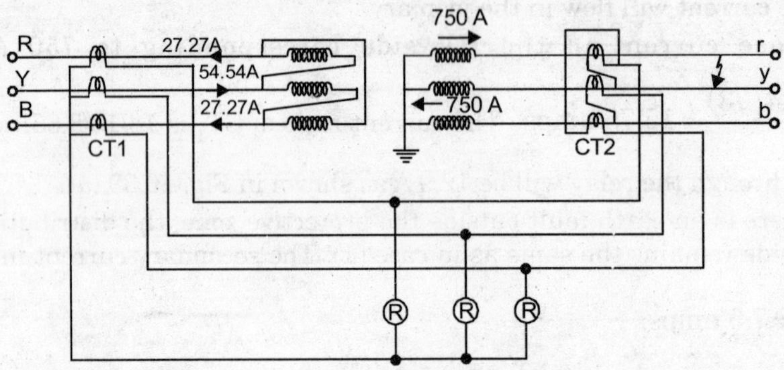

Fig. 16.39 (*d*)

Example 16.8. *A 6,600 volt 3-phase turbo-alternator has a maximum continuous rating of 2000 kW at 0.8 p.f. and its reactances 12.5%. It is equipped with Merz-Price circulating current protection which is set to operate at fault current less than 2000 amperes. Find what value of the neutral earthing resistance leaves 10% of the winding unprotected?*

Solution : Let the earth resistance be r ohm (refer Fig. 16.40).

Full load current of the alternator $= \dfrac{2000\times1000}{0.8\times\sqrt{3}\times6600} = 219$ A.

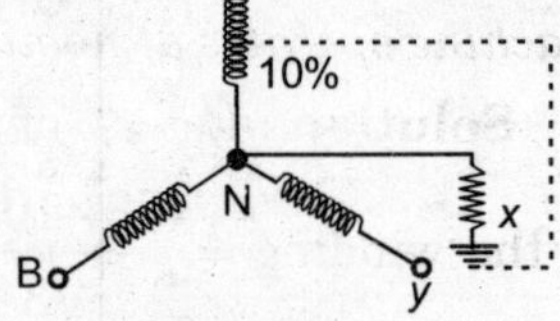

Fig. 16.40

Let the reactance per phase of the alternator be x ohms.

$$\therefore\ 12.5 = \frac{\sqrt{3}\times x\times 219}{6,600}\times100 \quad \therefore\ x = 2.19 \text{ ohm.}$$

∴ Reactance of 10% of the winding = 0.219 ohm.

Voltage induced in winding NA = $\dfrac{6,600}{\sqrt{3}\times 16} = 381$ volts

The protection operates at a current = 200 amps.

$$\therefore \qquad 200 = \frac{381}{\sqrt{r^2 + 0.219^2}} \qquad \therefore r = 1.89 \text{ ohm.}$$

Example 16.9. *An alternator stator winding protected by a percentage differential relay is shown in Fig. 16.41. The relay has a 0.15 amp minimum pick up and a 12% slope. A high resistance ground fault had occurred as shown near the grounded neutral end of the generator winding while the generator is carrying load. The current flowing at each end of the generator winding are shown in the figure. Assuming that the CTs have a 400/5 ampere ratio and no inaccuracies, will the relay trip the generator CTs under this fault condition.*

Solution :
$$i_1 = 960\times\frac{5}{400} = 4.5 \text{ A}$$

$$i_2 = 300\times\frac{5}{400} = 3.75 \text{ A}$$

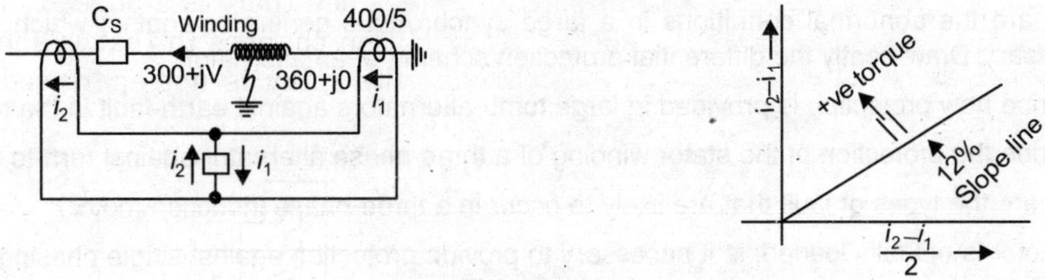

Fig. 16.41

$$i_1 - i_2 = 4.50 - 3.75 = 0.75 \text{ A}$$

$$\frac{i_1 + i_2}{2} = \frac{4.5+3.75}{2} = \frac{8.25}{2} = 4.125 \text{ A}$$

$$\frac{i_1 - i_2}{\frac{i_1 + i_2}{2}} = \frac{0.75}{4.125} = .18 = 18\%$$

∴ This point would lic above 12% slope characteristic, *i.e.*, it would lie on the +ve torque regime and therefore the relay would operate.

Example 16.10. *A 50 MVA, 3-phase 33 kV alternator is being protected by the use of circulating current balance scheme using 2000/m amp CT. The neutral of the generator is earthed through a NGR of 7.5 Ω. It the pick up current for the relay is first above 0.5 ampere, determine what percentage of the winding of each phase unprotected against earth when the machine operates at nominal voltage.*

Solution : Let x% of the winding be unprotected during nominal. The emf induced in x% of the winding = $\dfrac{33\times 1000}{\sqrt{3}}\times\dfrac{x}{100} = 190.5\,x$ volts. In case of ground fault only the limiting factor NGR.

$\therefore$ The fault current due to this voltage is $= \dfrac{190.5x}{7.5}$ amp.

$\therefore$ The limiting condition of not operation will be valid till $\dfrac{190.5\,x}{7.5} = 200 \therefore x = 7.88\%$.

Thus 7.88% of the winding remains unprotected.

QUESTIONS

1. Describe the protection scheme for internal fault in a 3-phase delta-star connected power transformer. Also draw a neat sketch and explain clearly why the CTs are to be connected in particular fashion only.
2. (*i*) Describe a Buchholz relay and discuss its merits and drawbacks.

   (*ii*) Discuss what special factor are to be taken into account while designing the protection of a large Star/Delta Transformer.
3. Describe the principle of Merz-Price system of protection for a power transformer. What are the short coming of this scheme and low are they overcome.
4. Why restricted earth-fault protection is provided to alternators through it does not provide protection against earth-fault to the complete winding. State what is the justification of providing the protection.
5. What are the abnormal conditions in a large synchronous generator against which protection is necessary. Draw neatly the differential protection scheme of an alternator.
6. Describe how protection is provided in large tumb-alternators against earth-fault in the rotor.
7. Describe the protection of the stator winding of a three-phase alternator against turn-to turn faults.
8. What are the types of fault that are likely to occur in a three-phase induction motor?

   If a motor is not fully loaded, is it necessary to provide protection against single phasing. Explain the reason.
9. Give a scheme of protection for a large three-phase induction motor.
10. Describe the method of protecting bus bars by differential relaying. What are the limitation of this method and to what extent these can be overcome.
11. Describe a simple method of bus bar earth-fault protection applicable to metal-clad switch board on distribution system.

CHAPTER

17

D.C., A.C. Distribution System

17.1 DISTRIBUTION OF DC POWER

Transmission and distribution of electric power means its conveyance from the control station where it is generated to places where it is demanded by the consumers like mills, factories, residential and commerical buildings, pumping stations, etc. Electric power may be transmitted by two methods (*a*) by overhead system or (*b*) by underground.

The second method of transmission is suited for densely populated areas generally the urban areas. However the underground method is costlier than the first method.

In overhead system power is conveyed by bare conductors of copper or aluminium which are strong between steel poles and also wooden poses, which are erected at convenient distances along the selected towers. The bare copper or aluminium conductors are fixed to an insulator which is itself fixed on to a cross-arm on the pole. The number of cross-arm carried by a pole depends on the number of conductors it has to carry. Line support, as discussed in earlier chapters of this book, consist of (*a*) pole structures and (*b*) Tower. Poles are made of wood, reinforced concrete or steel are used upto 33 kV whereas only steel towers are used for higher voltages.

The second method of the underground system employs insulated cables which may be single, double or triple or four core etc. Also discussed earlier.

However, a good system whether overhead or underground should fulfil the following requirements, as per Indian or International standard :

INSIDE THIS CHAPTER

1. The voltage at the consumer's premises must be maintained within ± 4% to +16% of the declared voltage. However, the actual value depends on the type of load mainly. In fact, as per India Electricity Act and Rules, voltage fluctuation should not exceed ± 5% of normal voltage for LT supply and ± 12 1/2% for HT supply.
2. The loss of power in the system should be a small percentage :- 10% of the power transmitted.
3. The transmission cost should not be unduly excessive.
4. The maximum current passing through the conductor should be limited to such a value as not to overheat the conductor or injure its insulation.
5. The insulation resistance of the whole system should be very large so that there is no undue leakage or danger to human life.

In fact, at present (and since past few decades) all production of power is a.c. power and almost all d.c. power required for various application is obtained from large a.c. power system by using converting machinary like synchronous or rotary converters, solid state converters and motor generator sets etc. There are many economical and technical reason for producing power in the from of alternating current rather than direct current. In this chapter d.c. distribution will be discussed. In other chapters a.c. system in discussed indetail.

Some of the reasons are stated below :

1. It is possible in practice, to construct large high-speed a.c. generators of capacitor upto 750/1000 MW. Such generators are economical both in the matter of cost for kwh of electric energy produced as well as in operation unfortunately d.c. generators cannot be built of rating higher than 5 MW because of commutation trouble. Moreover, since they must operate at low speeds it necessitates large and heavy machines.
2. A.C. voltages can be efficiently and conveniently raised or lowered for economic transmission and distribution of electric power. On the other hand d.c. power has to be generated at comparatively low voltages by units of relatively low power rating.

17.2 AC POWER TO DC POWER CHANGE OVER

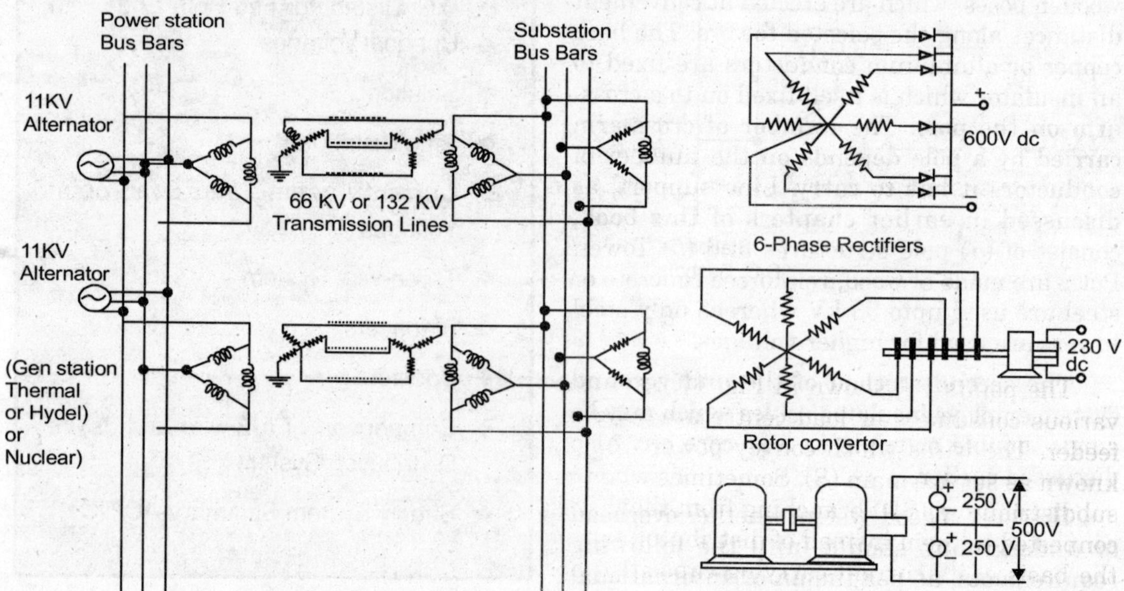

Fig. 17.1

As there is no economical method of raising the d.c. voltage for transmission and lowering it for distribution the Fig. 17.1 shows a typical power system for obtaining d.c. power from a.c. power. In this system other details like instruments, switches, circuit breakers etc. have been omitted.

In the Fig. 17.1, there are two 11 kV alternators run in parallel and supply power to the Power Station Bus Bars. The voltage is stepped up by 3-phase transform to 60 kV or 132 kV or 400 kV for transmission purposes and again stepped down to 11 kV at the sub-station for transforming a.c. power to d.c. power with the help of 6-phase Rectifier, Rotor converters Motor-Generator but as shwon in Fig. 17.1 for d.c. distribution to the consumers.

17.2.1 Distribution System for DC Power

In the Fig. 17.2 a simple schematic diagram of low tension distribution system for d.c. power. The whole system consists of a network of cables or conductors which convey power from control station to the consumer's premises. The station Bus-bars are fed by a number of generators (only two shown in figure) running in parallel. From the bus bar the power is carried by many feeders which are laid to various parts of the locality and Industrial areas. These feeders deliver power at certain points to a distributor which runs along the various streets or routes.

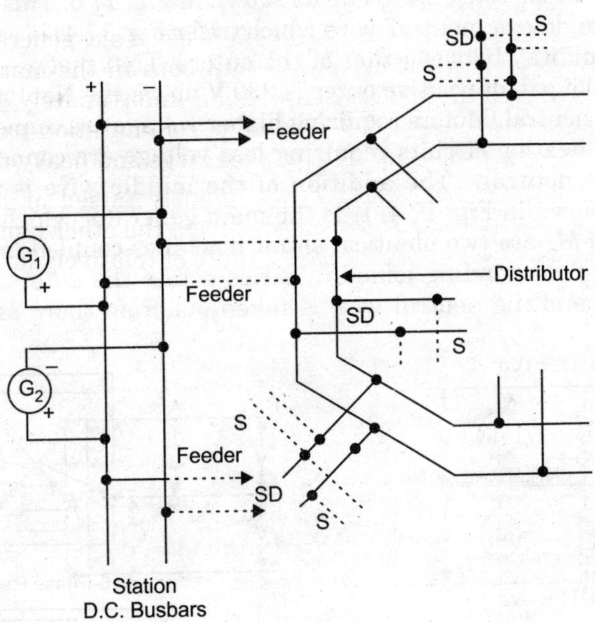

Fig. 17.2

The points FF, shown in Fig. 17.2 are known as feeding points power connection to the various consumers or load centre are given from this distributor and not directly from the feeder. The wires which convey power from the distributor to the consumer's premises are known as service main (*S*). Sometimes when there is only one distributor in a locality, several subdistributors (SD) branching from the distributor are employed and service main are then connected to them instead of distributor as shown in Fig. 17.2. Hence a feeder is designed on the basis of its current-carrying capacity whereas the design of distributor is based on the voltage drop occurring in it.

17.3 TWO-WIRE AND THREE-WIRE DISTRIBUTION

Power, in d.c. distribution system may be fed and distributed either by (1) 2-wire system or (2) 3-wire system.

1. 2-wire System

In this system, one is the outgoing or positive wire and the other is the return or negative wire. In the case of a 2-wire distributor, lamps, motor and other electrical apparatus are connected is parallel between the two wires as shown in Fig. 17.3. As seen the potential difference and current hence their maximum values at feeding points F_1 and F_2. The standard voltage between the conductors is 220 V.

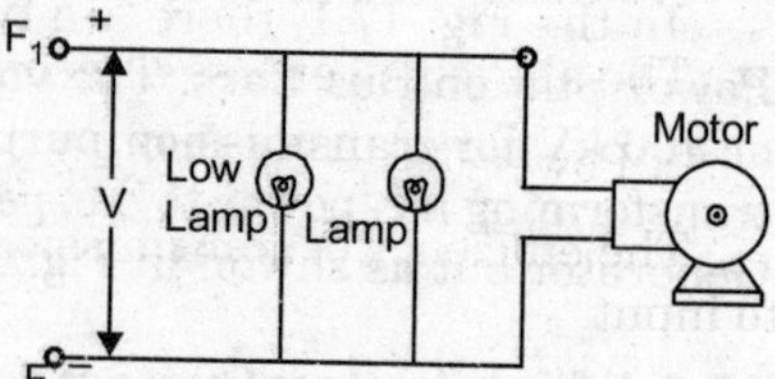

Fig. 17.3

The two-wire system when used for transmission purposes, less much lower efficiency and economy as compared to the 3-wire system.

2. 3-wire System

In the Fig. 17.4 is shown a 3-wire system for d.c. distribution. A 3-wire system has not only a higher efficiency of transmission, but when used for distribution purposes, it makes available two voltages at the consumer's end as shown in Fig. 17.5. This 3-wire system consists of two "outers" and a middle or 'neutral' wire which (refer Fig. 17.4) is earthed at the generator end. Its potential is midway between that of the outers. That means if the p.d. between the outers is 460 V then the p.d. of positive outer is 230 V above the Neutral and that of negative outer 230 V below the neutral. Motors requiring higher voltage are connected across the outers, whereas lighting and heating circuits requiring less voltage are connected between any one of the outers and the neutral. The addition of the middle wire is made possible by the connection diagram shown in Fig. 17.4. It is the main generator which supplies power to the whole system. M_1 and M_2 are two identical shunt machines coupled mechanically with their armatures and shunt field winding joined in series across the outers. The junction of their armatures is earthed and the neutral wire is taken out from there as shown N in the Fig. 17.4.

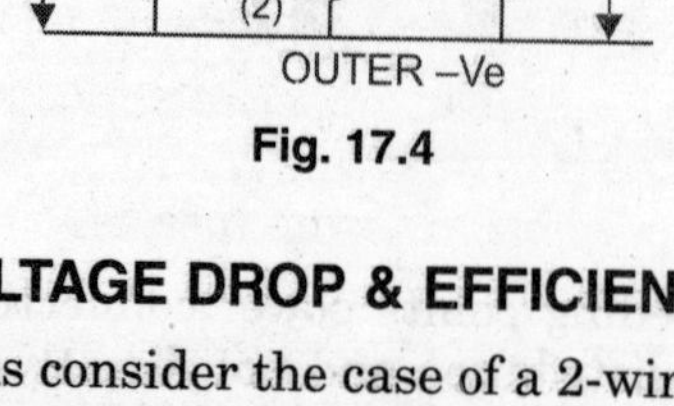

Fig. 17.4

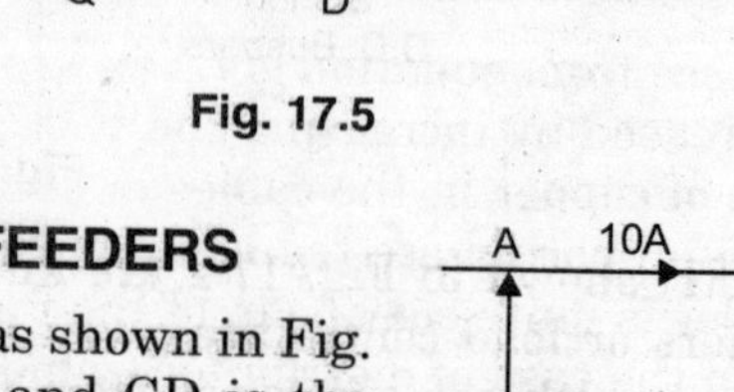

Fig. 17.5

17.4 VOLTAGE DROP & EFFICIENCY OF FEEDERS

Let us consider the case of a 2-wire feeder as shown in Fig. 17.6, where AB is the sending end voltage, and CD is the receiving end. Obviously the p.d. at AB is higher than that at CD. The difference in potential at the two ends is the potential drop or 'dro' in the cable. Suppose the transmitting voltage is 250 V, current in AC is 10 Amp. and resistance of each feeder is 0.5 Ω then drop in both feeder-conductor is 10 × 0.5 = 5 Volts and drop in both feeder conductors is 5 × 2 = 10.

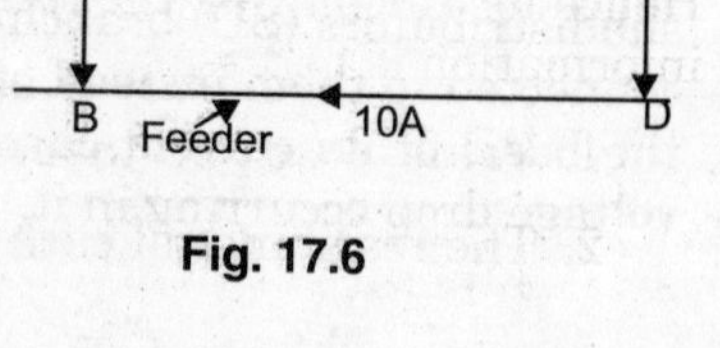

Fig. 17.6

Hence, the p.d. at the receiving and CD = 250 – 10 = 240 V.

Now input power at AB = 250 × 10 = 2,500 W

Output power at CD = 240 × 10 = 2,400 W

∴ Power loss in two feeder = 2,500 – 2,400 = 100 W

The above power loss alternative can be found by using the formula.

∴ Power loss = $2 \times I^2 R = 2 \times 10^2 \times 0.5 = 100$ W

The efficiency of transmission is defined, like any other efficiency, as the ratio of the output to input.

$$\therefore \quad \text{Efficiency of transmission} = \frac{\text{Power delivered by the line}}{\text{Power received by the line}}$$

In present case, power delivered by the feeder is = 2400 W (OUTPUT) and the power received by the feeder from the system is 2500 (INPUT)

$$\therefore \quad \eta = \frac{2400 \times 100}{250} = 96\%$$

In general, if V_1 is the voltage at the sending end and V_2 at the receiving end and I the current delivered, then

$$\text{Input} = V_1 I \qquad \text{and} \qquad \text{Output} = V_2 I, \qquad \therefore \eta = \frac{V_2 I}{V_1 I} = \frac{V_2}{V_1}$$

Now, $V_2 = V_1$ – drop in both the feeders (Conductor or distributors)

$= V_1 - IR$, where R is the total resistance of both the conductors

$$\therefore \quad \eta = \frac{V_2}{V_1} = \frac{V_1 - IR_1}{V_1} = 1 - \frac{IR_1}{V_1} \qquad ...(17.1)$$

$$\therefore \quad \%\ \text{efficiency} = 100 \times \left(1 - \frac{IR_1}{V_1}\right) = 100 - \left(\frac{IR}{V_1} \times 100\right)$$

Now $(IR/V_1) \times 100$ represent the voltage drop in both conductors as a percentage of the voltage at the sending. It is known as percentage drop.

$$\therefore \quad \%\ \eta = 100 - \%\ \text{drop}$$

In the present example, the total drop is 10 volt, which, expressed as percentage becomes (10/250) × 100 = 4%. Hence % η = 100 – 4 = 96% (as found to start as it in the section).

Now, we can see from equation (17.1) that for a given drop, transmission efficiency can be considerably increased by increasing the voltage at the transmitting end, *i.e.*, V_1. More over, the cross-section of copper in the cables is decreased in proportion to the increase in voltage which result in a proportionate reduction of the cost of copperate the cables. The calculation of drop, in a feeder is quite easy bean of the fact that current is constant throught its length. But it is not so in the case of distributors which are tapped off at various points/places along their entire length. Hence their different sections carry different current over different length. Hence for calculating the total voltage drop along the entire length of a distributor, following information is necessary.

1. Value of current tapped at each load point.
2. The resistance of each section of the distribution between the tapped points.

17.5 METHODS OF FEEDING A DISTRIBUTOR

Different methods of feeding a distributor are given below :

1. Feeding at one ends
2. Feeding at both ends with equal voltages
3. Feeding at some intermediate point
4. In addition the nature of loading also varies such as

(*a*) concentrated loading, (*b*) uniform loading, (*c*) combination of (*a*) & (*b*) below some of the important cases are discussed below separately :

17.6 D.C. DISTRIBUTOR FED AT ONE END

One conductor AB of a d.c. distributor with concentrated loads and fed at one end is shown in Fig. 17.7. Now let i_1, i_2, i_3 etc. be the current tapped off at points *C, D, E* and *F* and I_1, I_2, I_3, I_4 be the current in the various section of the distributor. Let r_1, r_2, r_3 etc. be the ohmic resistances of these various section and R_1, R_2, R_3, R_4 be the total resistance from the feeding end *A* to the successive tapping points.

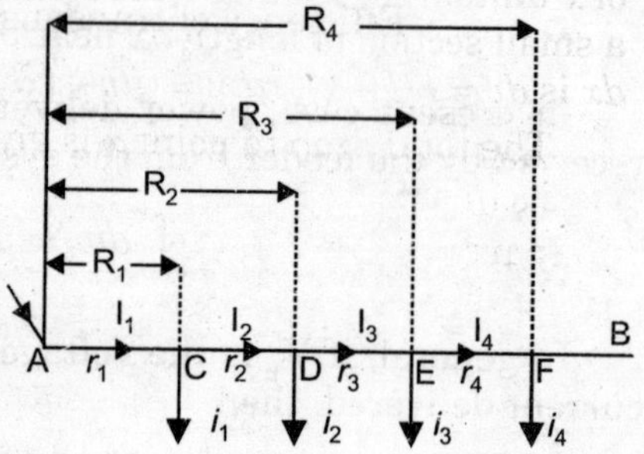

Fig. 17.7

Then, total drop in the distributor is,

$$v = r_1I_1 + r_2I_2 + r_3I_3 + ... \quad (1)$$

But $\quad I_1 = I_2 = i_2 + i_3 + i_4 + ...$

Now putting the values of I_1, I_2, I_3 ... etc. in equation (1), we get,

$$\therefore \quad v = r_1 (i_1 + i_2 + i_3 + i_4 + ...) + r_2 (i_2 + i_3 + i_4 + ...) + r_3 (i_3 + i_4 + ...)$$

$$= i_1r_1 + i_2 (r_1 + r_2) + i_2 (r_1 + r_2 + r_3) + ... = (i_1R_1 + i_2R_2 + i_3R_3 + ...)$$

= sum of the moment of each local current about feeding point *A*.

1. Hence, the drop at the far end of a distributor fed at one end is given by the sum of the moments of various tapped currents about the feeding point *i.e.*, $v = \Sigma iR$.
2. It follows from this that the total voltage drop is the same as that produced by a single load equal to the sum of various concentrated loads, acting at the centre of gravity of the load system.
3. Let us find the drop at any intermediate point like *E*. The value of this drop is

$$= i_1R_1 + i_2R_2 + i_3R_3 + R_3(r_4 + r_5 + ...)$$

= (sum of moment upto *E*) + (moment of whole load beyond *E* assumed acting at *E*)

In fact, in general the drop at any intermediate point is equal to the sum of moments of various tapped current upto that point plus the moment of all the load currents beyond that point assumed to be acting at that point. The total drop of both conductors would obviously be twice the value calculated above.

17.7 UNIFORMLY LOADED DISTRIBUTOR

One conductor AB of a d.c. distributor fed at one end, and uniformly loaded with *i* amperes per unit length, as shown in Fig. 17.8 (*a*). Any convenient unit of length may be chosen *i.e.*, a metre or 10 metres but at every such unit length, the load tapped is the same.

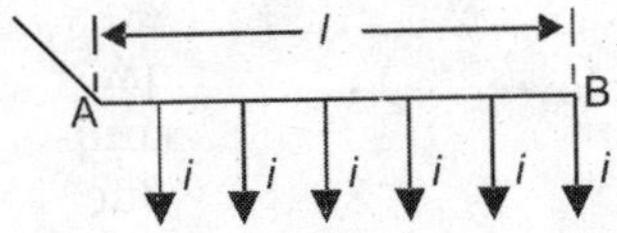

Fig. 17.8 (a)

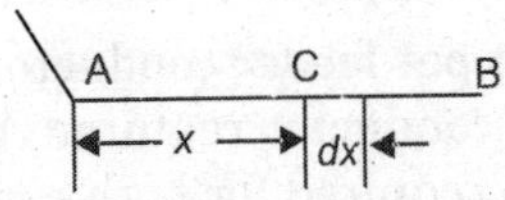

Fig. 17.8 (b)

Hence, let i = current tapped off per unit length,

r = resistance per unti length of the distributor

l = total length of the distributor

Now, let us find the voltage drop at a point C as shown in Fig. 17.8 (b), which is at a distance of x units from feeding end A. The current at point C is $(il - ix) = i\,(l - x)$. Then let us consider a small section of length dx near point C. Its resistance is (rdx). Hence voltage drop over length dx is $dv = i\,(l - x)\,(rdx) = (ilr - ixr)\,dx$.

The total drop to point x is given by integrating the above quantity between proper limits.

$$\therefore \quad \int_0^x dv = \int_0^x (ilr - ixr)\,dx \qquad \therefore v = ilrx - \frac{1}{2}irx^2$$

$$= ir\left(lx - \frac{x^2}{2}\right) \qquad = ir\left(lx - \frac{x^2}{2}\right) \quad ...(17.2)$$

The drop at point B can be obtained by putting $x = l$ in the equation (17.2), we get,

$$\therefore \text{Drop at point } B \quad v = ir\left(l^2 - \frac{l^2}{2}\right) = \frac{irl^2}{2} = \frac{(i\times l)\times(r\times l)}{2} = \frac{I}{2}R = I\times\frac{1}{2}R \quad \begin{pmatrix}\text{as } i\times l = I\\ \text{and } r\cdot l = R\end{pmatrix}$$

where $i \times l = I$ – total current entering at point A

$r \times l = R$ – the total resistance of distribution AB.

Total drop in the distributor $AB = \frac{1}{2}IR.$

1. It follows that in a uniformly load distributor, total drop is equal to that produced by the whole of the load assumed concentrated at the middle point.
2. Suppose that such a distributor is fed at both ends A and B with equal voltages. the resistance of each is $R/2$ and total current fed into each distributor is $I/2$.

Hence drop at the middle point is $= \frac{1}{2}\times\left(\frac{I}{2}\right)\times\left(\frac{R}{2}\right) = \frac{1}{8}IR.$

It is then $\frac{1}{4}$th of that of a distributor fed at one only. Hence the advantage of feeding a distributor at both ends, insted of at one, is obvious.

The equation of drop at point C distant x units from feeding point $A = irlx - \frac{1}{2}irx^2$, shows that the diagram of drop of a uniformly-loaded distributor fed at one end is a parabola.

Example 17.1. *A dc 2-wire feeder supplies a constant load with a sending-end voltage of 220 V. Calculate the saving of copper if thus voltage is doubled with power transmitted remaining the same.*

Solution : Let l = length of each conductor in metre

σ = current density in A/m^2

p = power supplied in watts

1. At 220 V Supply

Current per feeder conductor $I_1 = p/220$.
Area of conductor required $A_1 = I_1/\sigma = P/220\ \sigma$.
Volume required for both conductors is

$$V_i = 2A_1 l = 2pl/220\ \sigma = pl/110\ \sigma.$$

2. 440 V Supply

$$V_2 = pl/220\ \sigma$$

$$\therefore \text{\%age saving in copper} = \frac{V_1 - V_2}{V_1} \times 100 = \frac{(pl/100\sigma) - \dfrac{pl}{220\ \sigma}}{pl/100\ \sigma} \times 100 = 50\%$$

Example 17.2. *A uniform 2 wire d.c. distributor 200 metres long is loaded with 2 amperes/ metre. Resistance of single wire is 0.3 ohm/Km calculate the maximum voltage drop if the distributor is fed (a) from one end, (b) from both ends with equal voltages.*

Solution : (*a*) Total resistance of the distributor is, $R = 0.3 \times \left(\frac{200}{100}\right) = 0.06\ \Omega$. Total current entering the distributor is $I = 2 \times 200 = 400$ A. Total drop on the whole length of the distributor is (refer article 17.2).

$$= \frac{1}{2} IR = \frac{1}{2} \times 400 \times 0.06 = 12 \text{ V}$$

(*b*) Since distributor is fed from both ends, total voltage drop is

$$= \frac{1}{8} IR = \frac{1}{8} \times 400 \times 0.06 = 3 \text{ V}.$$

As explained in Article 17.2, the voltage drop in this case is one-fourth of that in case (*a*) above in this example solution.

Example 17.3. *A 2-wire d.c. distributor AB is 300 metres long. It is fed at point A. The various loads and their positions are given below :*

At point	*Tapped Load*	*Distance from A in metre*
C	*30*	*40*
D	*40*	*100*
E	*100*	*150*
F	*150*	*250*

If the maximum permisible voltage drop is not to exceed 10 V, find the cross-sectional are of the distributor. Take $\rho = 1.78 \times 10^{-8}\ \Omega - m$.

Solution : In the Fig. 17.9, the distributor along with its tapped current is shown.

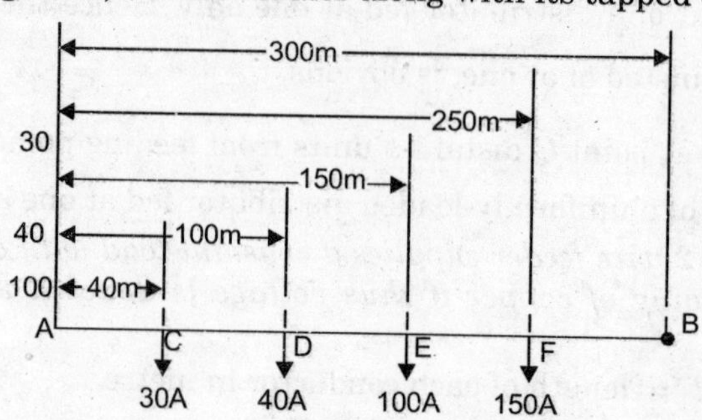

Fig. 17.9

As explained in the above article, the total drop over the distributor is

$$v = i_1R_1 + i_2R_2 + i_3R_3 + i_4R_4 \quad \text{for one conductors}$$
$$= 2(i_1R_1 + i_2R_2 + i_3R_3 + i_4R_4) \quad \text{for two conductors}$$

Here, R_1 = resistance of $AC = \rho l_{AC/A} = (1.78 \times 10^{-8} \times 40)/A$ where A = area of conductor

R_2 = resistance of $AD = \rho l_{AD/A} = (1.78 \times 10^{-8} \times 100)/A$

R_3 = resistance of $AE = \rho l_{AE/A} = (1.78 \times 10^{-8} \times 150)/A$

R_4 = resistance of $AF = \rho l_{AF/A} = (1.78 \times 10^{-8} \times 250)/A$

As v = 10 V = voltage drop between AB

$$\therefore \quad 10 = \frac{2\times1.78\times10^{-8}}{A}(30\times40+40\times100+100\times150+50\times250)$$

$$= \frac{2\times1.78\times10^{-8}}{A}\times32{,}700 \qquad \therefore A = 3.58\times327\times10^{-7}m^2 = 1{,}1653 \text{ cm}^2$$

17.8 DISTRIBUTOR FED AT BOTH ENDS WITH EQUAL VOLTAGE

In this type of d.c. distribution system, it should be noted that in such case :

1. The maximum voltage drop must always occur at one of the load points.
2. If both feeding ends are at the same potential, then the voltage drop between each end and this point must be the same, which in other words, means that the sum of the moment about ends must be equal.

In Fig. 17.10 (*a*) it is shown that a distributor fed at two points F_1 and F_2 with equal voltages. The potential of the conductor will gradually fall from F_1 onwards, reach a minimum value at one of the tapping, say A and then rise again at the other feeding point F_2 Fig. 17.10 (*b*) is approached. All the currents tapped of between point F_1 and A will be supplied from F_1 while these tapped off between F_2 and A will be supplied from F_2. The current tapped at point A itself will, in general, be partly supplied by F_1 and partly by F_2.

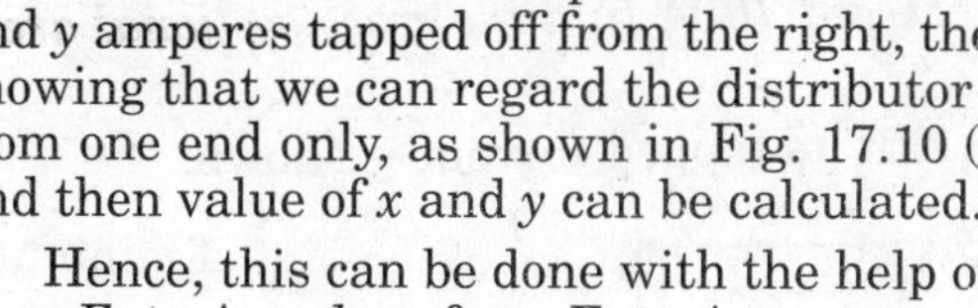

Fig. 17.10 (*a*)

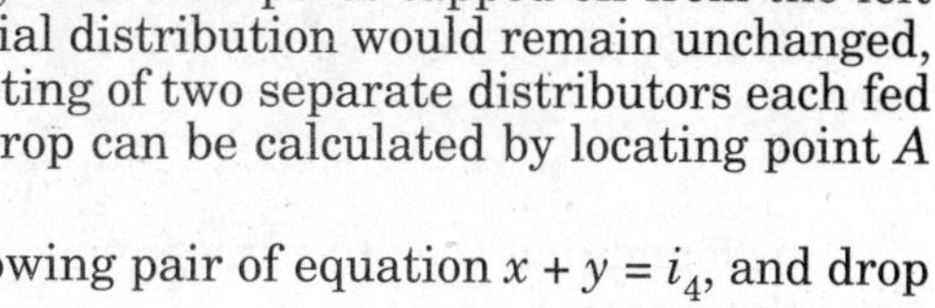

Fig. 17.10 (*b*)

Now let the values of these current be x and y respectively. If the distributor wire actually cut off into two at A the point of minimum voltage, with x amperes tapped off from the left and y amperes tapped off from the right, then potential distribution would remain unchanged, showing that we can regard the distributor as consisting of two separate distributors each fed from one end only, as shown in Fig. 17.10 (*b*). The drop can be calculated by locating point A and then value of x and y can be calculated.

Hence, this can be done with the help of the following pair of equation $x + y = i_4$, and drop from F_1 to A_1 = drop from F_2 to A_2.

Alternative solution to solve the above case :

The alternative method is to take the total current fed at one end, by F_1 as x and then to find the current distribution as shown in Fig. 17.10 (*c*). The drop over the whole distributor is equal to the sum of the products of the product of currents in various section and their resistances. Say for a distributor fed at both ends with equal voltages, this drop equals zero. And thus, the value of x can be found and obviously the current in each section can be found and thus the voltage drop at all the tapped points, and thes the minimum voltage drop point in the distributor.

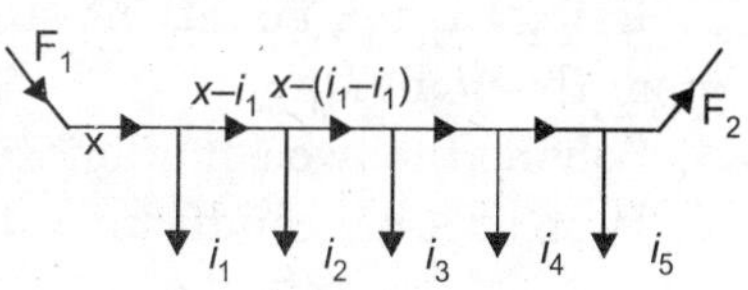

Fig. 17.10 (*c*)

The above points can be easily explained by the following example by using both the methods.

Example 17.4. *A 2-wire d.c. distributor F_1, F_2, 1000 metres long is loaded as under :*

Tapped Point	*Distance from F_1 in Metres*	*Tapped Load / Current at Tapped Point*
A	100	20
B	250	80
C	500	50
D	600	70
E	700	40
F	800	30
G	850	10
H	920	15

So above statement can be also written as below :

Tapped Point :	A	B	C	D	E	F	G	H
Distance From F_1 (in metre) :	100	250	500	600	700	800	850	920
Tapped Load at the Distance : (in Amp.)	20	80	50	70	40	30	10	15

The feeding point at point F_1 and F_2 are maintained at the some potential. Find which point in the distributor will have the minimum potential and what will be the drop at this point? Take the cross section of the conductors as 0.35 cm² and specific resistance of the distribute material; copper on 1.764×10^{-8} Ω cm.

Solution : The distributor with the tapped load current at the tapped load points is shown in Fig. 17.11. The simple method of locating the point of minimum potential is to take the moment about the two ends and then by comparing the two runs make a guess at the possible point. The process is as below :

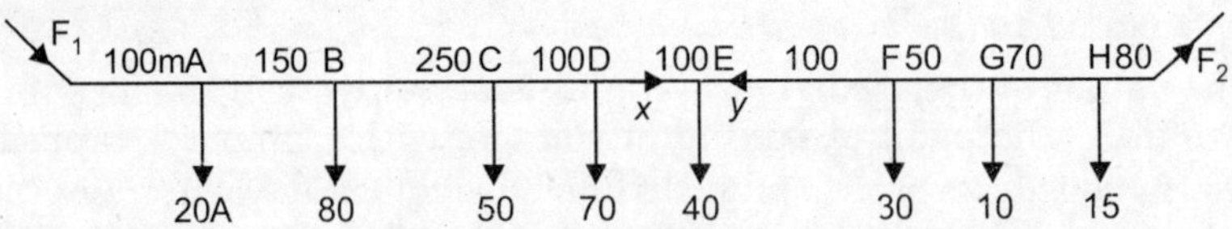

Fig. 17.11

It is found that the 4th point from F_1 that is the required point, *i.e.*, the point of minimum potential, because from

Using the two equations stated in the article 17.3, we have :

at the point D_1 $\quad x + y = 70$ and $47{,}000 + 600x = 20{,}700 + 400y$

Solving the two equations, we get $x = 1.7$ A, $y = 70 - 1.7 = 68.3$ A

Drop at A per conductor = 47,000 + 600 × 1.7 = 48,020 ampere-metre, (V)

$$\text{Resistance} = \frac{pl}{A} = \frac{1.764\times10^{-8}\times1}{0.35\times10^{-4}} = 50.4\times10^{-5}\ \Omega/\text{m}.$$

Hence, drop per conductor = 48,020 × 50.4 × 10^{-5} = 24.2 V

Reckoning/considering both conductors, the drop at A is = **48.4 V**

Moment about F_1 in ampere-metres	Sum	Moment about F_2 in ampere-metres	Sum
20 × 100 = 2000	2000	15 × 80 = 1,200	1,200
80 × 250 = 20,000	22,00	10 × 150 = 1,500	2,700
50 × 500 = 25,000	47,000	30 × 200 = 6,000	8,700
		40 × 300 = 12,000	20,700
		70 × 400 = 28,000	48,700

Alternative Solution Method

But earliest method perhaps is the alternative method for the above example is shown below with reference to the Fig. 17.12 (*a*). The drop over the whole distributor is equal to the sum of the products of current in the various rections and their resistance. For a distributor fed at both end with equal voltages, this drop equals zero. In this way value of x can be found as below :

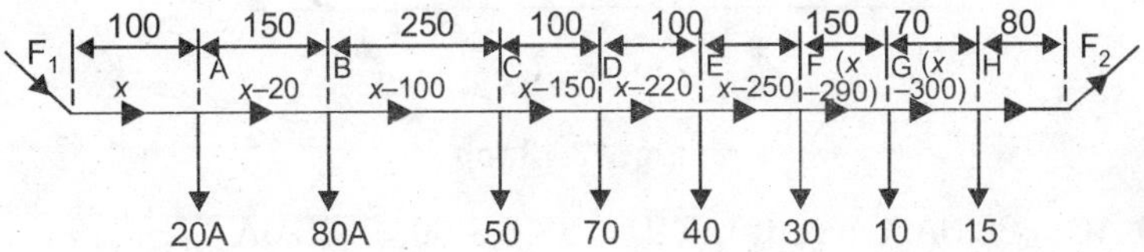

Fig. 17.12 (*a*)

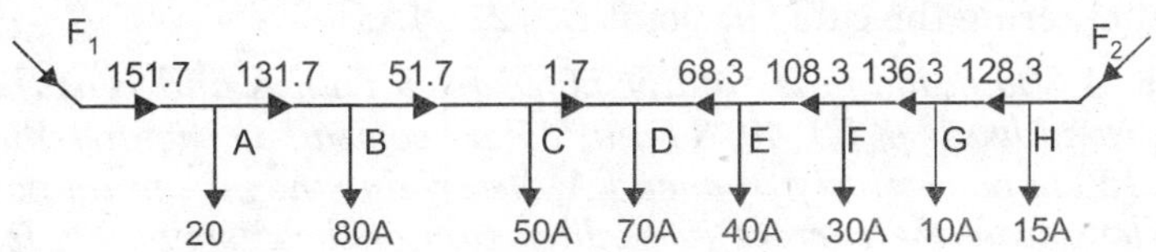

Fig. 17.12 (*b*)

Resistance per metre (single distributor) = $5.04 \times 10^{-4}\ \Omega$

Resistance per metre (double distributor) = $10.08 \times 10^{-4}\ \Omega$

$$\therefore\ 10.08 \times 10^{-4}\,[100x + 15(x-20) + 250(x-100) + 100(x-150) + 100(x-290) + 100(x-260)$$
$$+50(x-290) + 70(x-300) + 50(x-315) = 0$$

or, $1000x = 151,700 \quad \therefore \quad x = 151.7\text{ A}$

This gives a current distribution as shown in Fig. 17.12 (*b*). Obviously point D is the point of minimum potential.

Drop at D (considering both conductor) is

$$= 10.084 \times 10^{-4}\,(100 \times 151.7 + 150 \times 131.7 + 250 \times 51.7 + 100 \times 1.7)$$
$$= 10.08 \times 10^{-4} \times 48,020 = 48.4\text{ V as before}$$

Example 17.5. *The resistance of a cable is 0.1 Ω per 1000 metre for both conductors. It is loaded as shown in Fig. 17.13 (a). Find the current supplied at A and B, if both the ends are supplied as 200 V.*

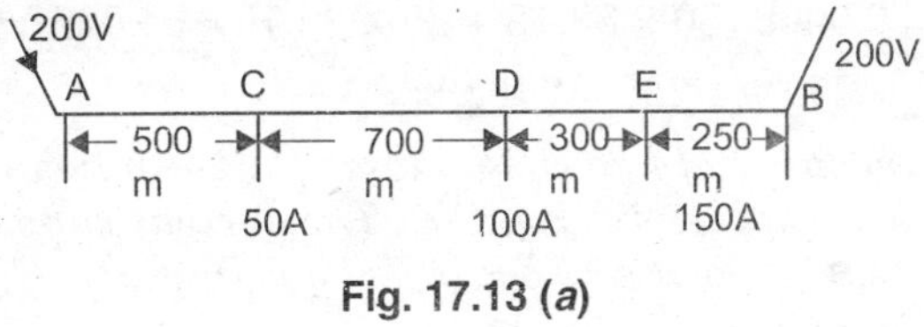

Fig. 17.13 (*a*)

Solution : Let the current distribution be as shown in Fig. 17.13 (*b*). Resistance for both conductors = 0.1/1000 = 10^{-4} Ω/m. The total drop over the whole cable is zero because it is fed at both ends by equal voltages.

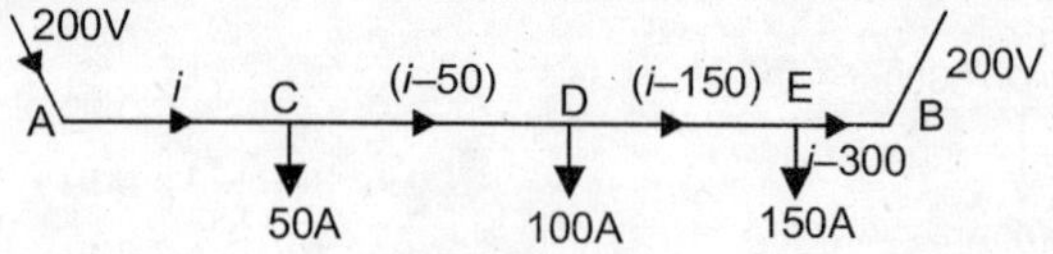

Fig. 17.13 (*b*)

$\therefore$ $10^{-4}[500i + 700(i - 50) + 300(i - 150) + 250(i - 300) = 0$

$\therefore$ $$i = 88.6 \text{ A}$$

This gives the current distribution as shown in Fig. 17.12 (*c*).

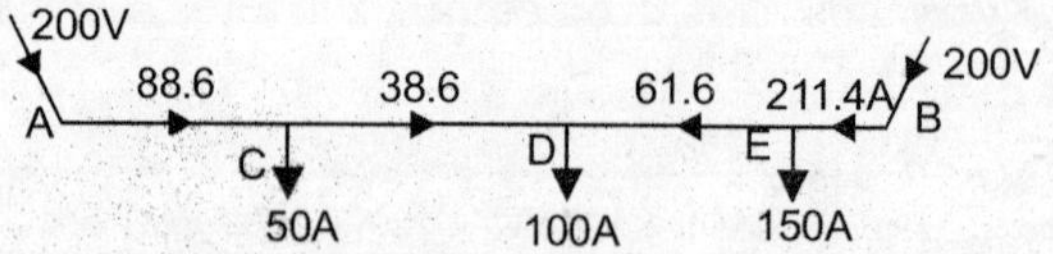

Fig. 17.13 (*c*)

$\therefore$ Current in AC = 88.6A, current in CD = 88.6 – 50 = –38.6A

$\therefore$ Current in DE = (38.6 – 100) = –61.4A

Hence, current entering the cable at point B = 211.4A.

Example 17.6. *A 200 m long distributor is fed from both A and B at the same voltage of 250 V. The concentrated loads of 50, 40, 30 and 25 are coming on the distributor at a distance of 50, 75, 100 and 150 m respectively from end A. Determine the minimum potential and locate its position. Also determine the current in each section of the distributor. It is given that the resistance of the distributor conductor is 0.08 Ω per 100 metres for go and returns.*

Solution : In the Fig. 17.14 (*a*), let current fed at point A be i. The currents in various section is shown in the Fig. 17.14 (*a*). Resistance per unit metre the distributor (go and return) is 0.08/100*r* = 0.008 Ω.

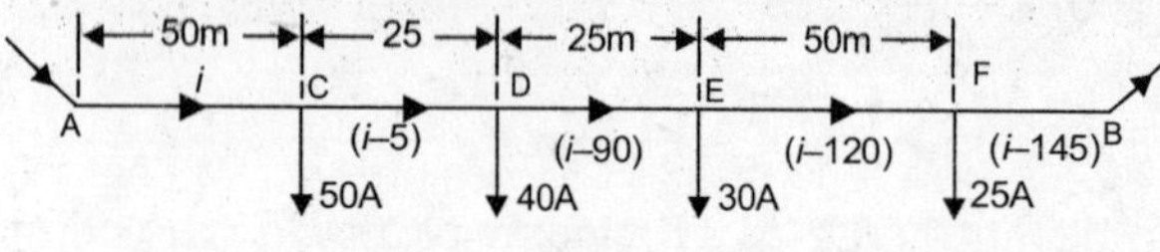

Fig. 17.14 (*a*)

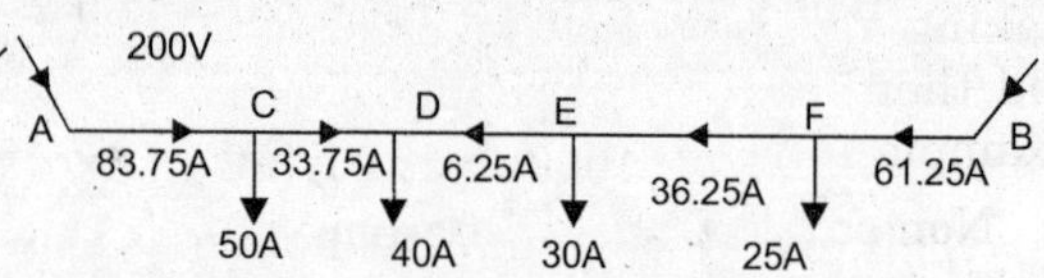

Fig. 17.14 (*b*)

Since the distributor is fed at both ends with equal voltages, total drop over it is zero.

Hence, $0.008/50i + 25(i - 50) + 25(i - 90) + 50(i - 120) + 50(i - 145) = 0$

or $$200i = 16{,}750 \qquad \text{or} \qquad i = 83.75 \text{ A}$$

The to actual current distribution is shown in Fig. 17.13 (*b*). Hence obviously it is noted that point D is the point of minimum potential.

$\therefore$ Drop at point D is = 0.0008 (50 × 83.75 + 25 × 33.75) = 4.025 V

$\therefore$ Minimum potential = 250 – 4.025 = 245.98 V

Example 17.7. *The resistance of two conductors of a 2-conductor shown in Fig. 17.15 (a) is 0.1 Ω per 1000 m for both conductors. Find (a) The current supplied at A, (b) The current supplied at B, (c) The current in each section, (d) The voltages at C, D, E. Both A and B are maintained at 200 V.*

Solution : The distribution along with its tapped current is shown in Fig. 17.15 (*a*). Let the current distribution be as shown in the Fig. 17.15.

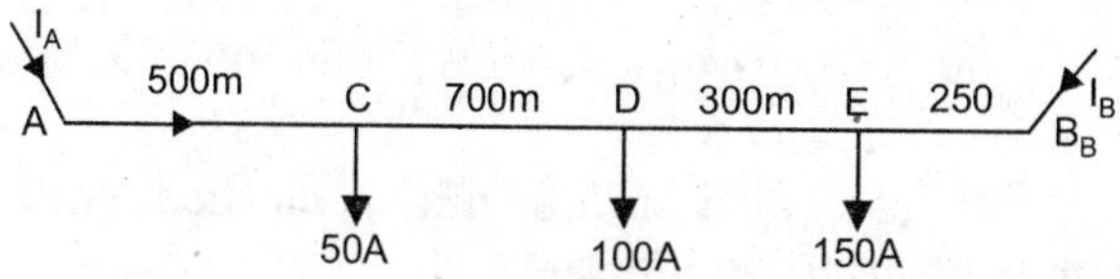

Fig. 17.15 (*a*)

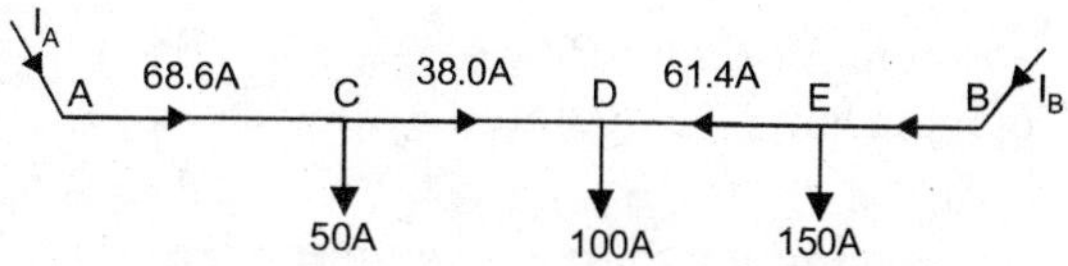

Fig. 17.15 (*b*)

Resistance per metre double is = 0.1/1000 = 10^{-4} Ω. The total drop over the whole distributor is zero as at both the end sare at same voltage.

$\therefore$ $10^{-4}[500i + 700(i - 50) + 300(i - 150) + 250(i - 300)] = 0$

or $\quad 1750i = 155{,}000 \qquad \therefore\ i = 88.6$ A

This gives the current distribution is shown in Fig. 17.14 (*b*).

$\therefore$ (*a*) $= I_a = 88.6$A, (*b*) $I_b = -211.4$A

(*b*) (–*ve* sign says that direction of current will be is opposite direction).

(*c*) Current in section $AC = 88.6\,A$. Current in $CD = (88.6 - 50) = 38.6$ A.

Current in section $DE = (31.6 - 100) = -61.4$ A.

Current in EB = (–61.4 – 150) = –211.4 A.

(*d*) Drop of over $AC = 10^{-4} \times 500 \times 88.6 = 4.43$ V.

Drop over $CD = 10^{-4} \times 700 \times 38.6 = 2.7$ V.

Drop over $DE = 10^{-4} \times 300 \times -61.4 = -1.84$ V.

1. Voltage at $C = 200 - 4.3 = 195.52$ V.
2. Voltage at $D = 195.57 - 2.7 = 192.87$ V.
3. Voltage at $E = 192.87 - (-1.84) =$ **194.71 V.**

17.9 D.C. DISTRIBUTOR FED AT BOTH ENDS WITH UNEQUAL VOLTAGES

This can be dealt with either by taking moment (in amp. m) about the two ends and then making a guess about the point of minimum potential or by assuming a current *x* fed at one and then finding the actual current distribution. Consider the case already discussed in Example 17.4.

Non say, there is a difference of *v* volts into ampere-metres with the help of known value of resistance/metre since F_2 is at a lower potential these ampere-metres appear in the column for F_2 as an initial drop.

If for example '*v*' is 4, then since resistance/metre is 5.04×10^{-4} Ω initial ampere-metres for F_2 are = $4 \times 10^{-4}/5.04$ = 7,938 **amp/metres.**

Then the table of respective moment will be as follows :

Moment About F_2	Sum	Moment About F_2	Sum
20 × 100 = 2000	2000	Initial = 7,938	—
80 × 200 = 20,000	22,000	15 × 80 = 1,200	9,138 $\left(\frac{7938+1200}{7,938}\right)$
50 × 500 = 25,000	47,000	10 × 150 = 1500	10,538
		30 × 200 = 6,000	16,638
		40 × 300 = 12,000	28,638
		70 × 400 = 28,000	56,638

As seen, the dividing point is the same as before :

$$x + y = 70 \text{ and } 47000 + 600x = 28{,}638 + 400y$$

Solving for x and y we get, $x = 9.64$ A, and $y = 60.36$ A. After knowing the value of x, the drop at A can be calculated as before.

But, as in previous case discussed in Art 17.3, there is also an alternative method which is, much simpler than above method, and known as alternative method.

In this method the total drop in the total length of the distributor for a current passing through it with uniform resistance/metre, will be equal to the difference of voltage between the sending end and the receiving end. This alternative method is explained below with the example no. 17.8 below.

Example 17.8. *A two-wire, d.c. distributor P_Q, 800 metre long is loaded a under :*

Distance from P (metres) :	100	250	500	600	700
Load in amperes :	20	80	50	70	40

The feeding point at P is maintained at 248 V and at Q at 245 V total resistance of the distributor (load and return) is 0.1 Ω. Find (a) the current supplied at P and Q and (b) The power dissipated in the distributor.

Solution : As shown in Fig. 17.16 (*a*), let x be the current supplied from and P (sending End). The other current in the various sections of the distributor are as shown in the Fig. 17.16 (*a*). Now the total drop over PQ should be equal to the potential difference between ends P and Q (that is between Sending End and Receiving End points).

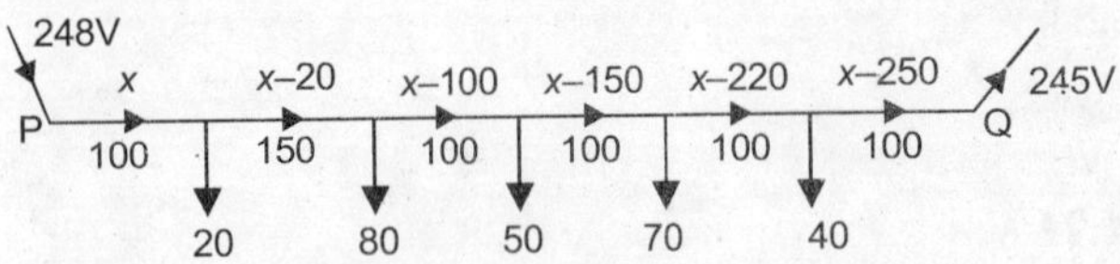

Fig. 17.16 (*a*)

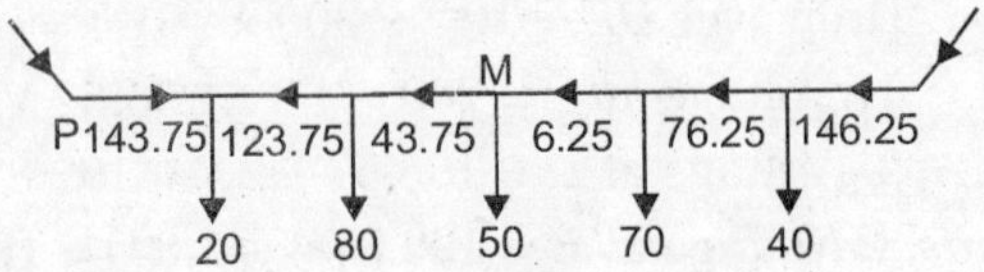

Fig. 17.16 (*b*)

$$= 248 - 245 = 3 \text{ V}$$

Now, resistance/metre of both the conductors of the distributor = 0.1/800

$$= (1/8000)\ \Omega$$

$$\therefore \quad \frac{1}{8000}[100x + 150(x-20) + 250(x-100) + 100(x-150) + 100(x-220) + 100(x-246) = 3$$

$$800x = 115{,}000 \qquad \therefore \qquad x = \text{the current} = 143.75 \text{ A}$$

The actual distribution is shown in Fig. 17.15 (*b*) from where it is seen that point has the minimum potential.

$$\therefore \qquad I_p = 143.75 \text{ A}, \qquad I_Q = 116.25 \text{ A}$$

$$\text{Power Loss} \qquad \Sigma E^2R = \frac{1}{800}[143.75^2 \times 100 + 123.75^2 \times 150 + 43.75^2 \times 250 + 6.25^2 \times 100 + 76.25 \times 100 + 116.25^2 \times 100]$$

$$= 847.3 \text{ W}$$

Example 17.9. *A dc two were distributor AB is 450m long and in fed at both ends at 250 volt. It is loaded as follows : 20A at 60m from A. 40A at 100m from A and a uniform loading of 1.5 A/m from 200 to 450m from A. The resistance of each conductor 0.05 Ω/km. Find the point of minimum potential and its potential.*

Solution : With reference to the example in the Fig. 17.17, let D be the point of minimum potential and let i bc the current in distributor section CD. Then current 250V supplied to load D from end B is $(40 - i)$. If r is the resistance of the distributor/m (for both go and return), then drop at over AD $= (20 + i) \times 60r + i \times 40r$. Drop over BD = drop due to concentrated load + drop due to distributed load.

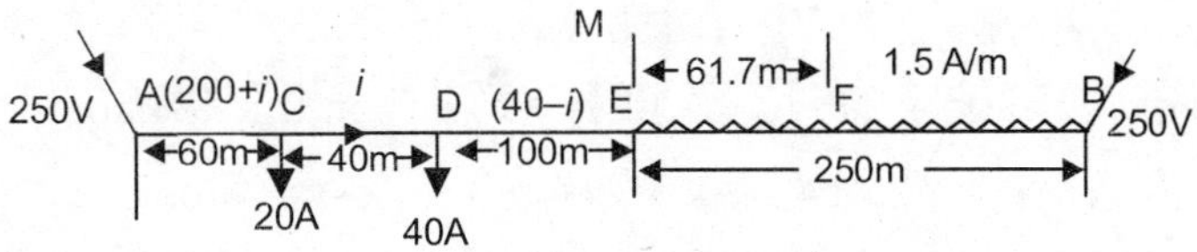

Fig. 17.17

$$= (40 - i) \times 350r + 1.5 \times r \times 250^2/2.$$

Since the two feeding points A and B are at the same potential of 250V, the two drops must be equal.

$$\therefore \quad (20 + i) \times 60r + i \times 40r = (40 - i) \times 350r + 1.5 \times r \times 250^2/2.$$

$$\therefore \quad i = 132.6\text{A}$$

Now as $(40 - i)$ comes out to be negative, it means that D is not the point of minimum potential. The required point is somewhat nearer the other end B. Let it be F. Obviously, current in section DF = 132.6 – 40 = 92.6A. Hence, distance of minimum potential point F from end A is = 60 + 40 + 100 + 92.6/1.5 = 261.7 m. Total voltage drop over section AF is

$$= 2 \times 0.05 \times 10^{-3}(152.6 \times 60 + 132.6 \times 40 + 92.6 \times 100 + 92.6 \times 61.7/4) = 2.65\text{V}$$

$\therefore$ Potential of point $\quad F = 250 - 2.65 = 247.35\text{V}$

Example 17.10. *A two-wire dc distributor AB, 1000 metres long, is supplied from both ends, 240 V at A and 242 V at B. There is a concentrated load of 200 A at a distance of 400 metre from A and a uniformly distributed load of 1.0 A/m between the mid-point and end B.*

Determine (1) the currents fed at A and B (2) the point of minimum potential and (3) voltage at this point. Take cable resistance as 0.005 Ω per 100 metre each core.

Solution : The resistance per 100 metres of both cores = **0.005 × 2 = 0.01 Ω**

Now let us take 100 m as the unit of length. Let current at end B bc I_B as shown in the Fig. 17.18. Now consider any element dx anywhere within distance BD, *i.e.*, distance over which uniformly distributed load is applied. Say it is at a distance of x units (of 100 metre each). The current through dx is $I = (I_B - 100 \times 1.0 \times x) = (I_B - 100\,x)$.

$\therefore$ Voltage drop over $dx = (I_B - 100\,x) \times 0.01 \times dx$

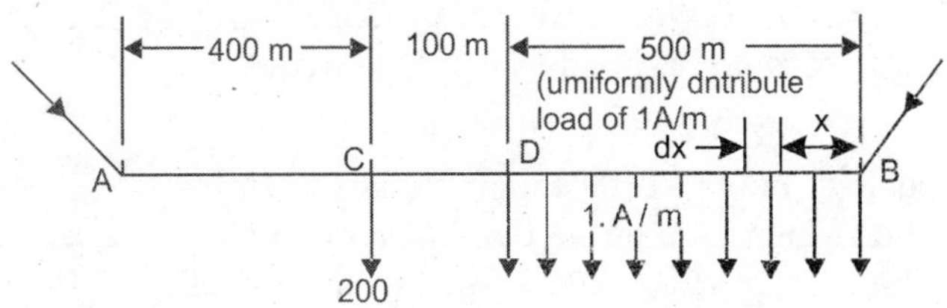

Fig. 17.18

Then voltage drop over distance BD

$$= \int_0^5 (I_B - 100x) \times 0.01 \times dx = 0.01 \int_0^5 (I_B - 100x)\, dx$$

$$= 0.01 \left| (I_B \cdot x - 50 \cdot x^2) \right|_0^5 = (0.05\, I_B - 12.5)\text{ V}$$

Voltage drop over DC $= (I_B - 500) \times 0.01 =$ **$0.01\, I_B - 5$**

Volt drop over CA = $(I_B - 700) \times 0.01 \times 4 = \mathbf{0.04\, I_B - 28}$

Since total drop over AB = 242 – 240 = 2 volts.

$\therefore 0.05I_B - 12.5 + 0.01I_B - 5 + 0.04I_B - 28 = 2V$

(Volt drop over BD + volt drop on CD + voltage drop over AC = 2V).

$$\therefore \quad I_B = 455 \text{ A}$$

Now total current $500 + 200 \text{ A} = 700 \text{ A} = I_A + I_B$ in the distributor is

$$\therefore \quad I_A + I_B = 700 \text{ A}$$

Now as $I_B = 455$ A, $\therefore$ $I_A = 700 - 455 = 345$ A

Now, it is obvious that (245 – 200) = 45 A is fed into the distributed load at D. Hence, point of minimum potential M will be 45/1.0 = 45 metres from D. Its distance from B = 500 – 45 = 455 metre or 455 units of length.

Voltage drop upto this point from end $B = \int_0^{4.55} (455 - 100x) \times 0.01 \times dx = 10.35$ V

$\therefore$ Potential of $M = 242 - 10.35 = \mathbf{231.65\ V}$

17.10 RING DISTRIBUTOR

A ring distributor is a distributor which is arranged to form a closed circuit and which is fed at one or more than one point. For the purpose of calculating voltage distribution, it can be looked upon as consisting of a series of open distribution fed at both ends. by using a ring distributor fed properly, we can gain by saving substantial amount of cost in using copper in the distribution system.

Now, if the ring distributor is fed at one point, then for the purposes of calculation, it is equivalent to a straight distributor fed at both ends with equal voltages as discussed above and explained in the examples no example 17.11.

17.11 CURRENT LOADING AND LOAD-POINT VOLTAGES

To explain this system let us consider a 3-wire 500/250 V distributor shown in Fig. 17.19. The motor requring 500 V is connected across the outers whereas other loads requiring lower voltage of 250 V are connected on both sides of the neutral.

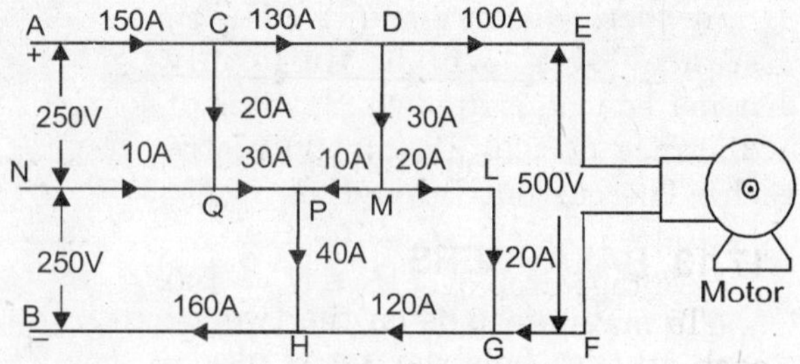

Fig. 17.19

In the Fig. 17.19, the current in positive outer AE flows for the left to right and in negative outer FB for right to left. The current through the various section of the neutral wire NL may flow in either direction depending on the load current on the positive side and negative side but is independent of the load current on the positive side and negative side, but is indepedent of the load connected outers. The Fig. 17.19 is self explanatry. Hence knowing the values of currents in the various conductors, voltage drops can be calculated provided resistances are known. After that voltages at different load points can be calculated as shown in Fig. 17.19.

17.12 THREE-WIRE SYSTEM

Three-wire system consists of the outer conductors (between which the voltage is twice the normal values of lighting) and the third wire which is called the middle or neutral wire. It is of half the cross section as compared to any one of the two outers and is earthed at the generator end. The voltage of the neutral is thus approximately half way between that of the outers.

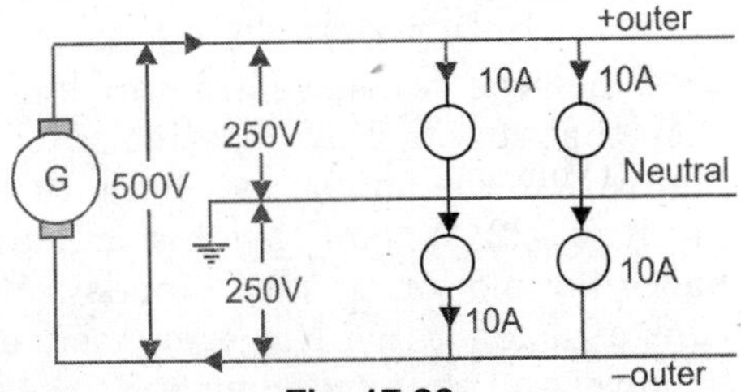

Fig. 17.20

Say if the total voltage between the two outers is 500 V, then the positive outer is 250 V above the neutral and the negative outer is 250 V be can the neutral. Hence the load (say a motor) requiring higher voltages are connected across the outers directly, whereas lighting and heaters load etc. requiring lesser voltage (say 250 V) are connected between any one of the outers and the neutral.

If the loads on both sides of the neutral are equal, that is balanced load, as shown in Fig. 17.21, then there is no current in the middle wire, and the effect is as if the different loads are connected in series across the outers.

However, is practice, although effort is made to distribute the various loads equally on the two sides of the neutral, yet it is difficult to achieve exact balance with the result that some out-of-balance current does flow in the neutral as shown in Fig. 17.26(*a*). In the present case as vienced from generator end, the load is equivalent to a resistance of 25/3 in series with a resistance of 25/2 ohm as shown in Fig. 17.21 (*b*). Then the voltages obviously, across the two will becomes unequal. The voltage across the positive side will full to $\dfrac{500 \times 25/3}{(25/2 + 25/3)} = 200$ V and on the negative sides it will become rise to) $\dfrac{500 \times 25/2}{(25/2 + 25/3)} = 300$ V.

This difficulty of unequal voltages with unbalanced loads is removed by using balancers discussed in the following article.

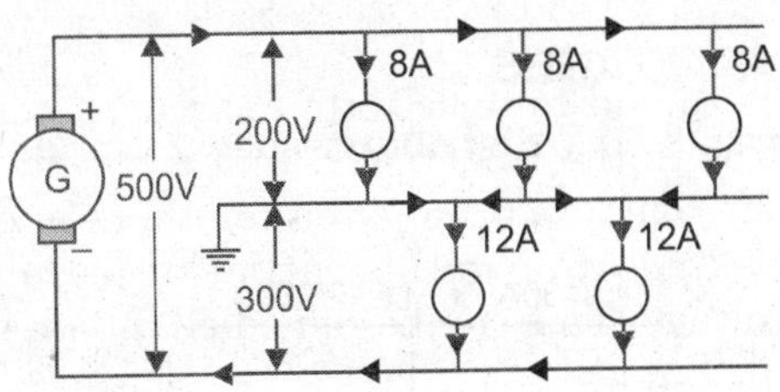

Fig. 17.21 (*a*)

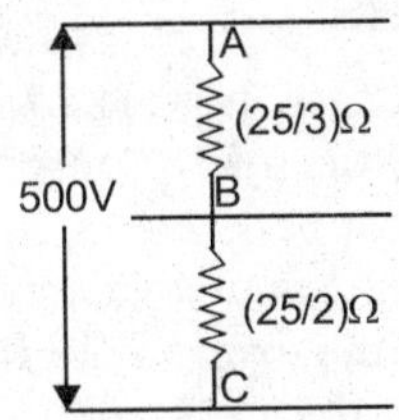

Fig. 17.21 (*b*)

17.13 BALANCERS

To maintain p.ds on the two sides of neutral equal to each other, a balancer set is used. The comment form of balancer consist of two identical shunt-wound machines which are not only coupled mechanically but have their armatures and field circuits joined in series across the outers. The neutral is connected to the function of the armatures as shown in Fig. 17.22. When the system is unloaded or when the loads on the two sides are balanced then :

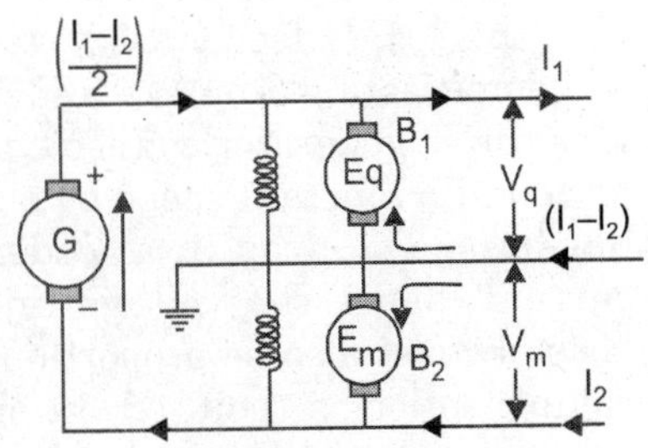

Fig. 17.22

1. Both machines run as unloaded motors and
2. Since their speeds and field-currents are equal, their back e.m.f.s are the same.

When the two sides are unbalanced *i.e.*, when the load supplied by *+ve* outer is different from that supplied by negative outer, then out-of-balance current $(I_1 - I_2)$ flows through the mid-wire to the balancers where it divides into two halvers each equal to $(I_1 - I_2)/2$ as shown in Fig. 17.22. When I_1 is greater than I_2 then *+ve* side is more heavily loaded (than *–ve* side) hence the p.d. on this side tends to fall below the e.m.f. of the balancers on this so that B_1 runs as a generator. However, the p.d. on the lightly-loaded side rises above the e.m.f. of the balancers on that side, hence B_2 runs as a motor. In this way energy is transferred through balancers from the lightly-loaded side the heavily-loaded side. In the Fig. 17.22, machine B_2 is running as a motor end is driving machine B_1 as a generator.

Then, $$V_g = E_g - \frac{(I_1 - I_2)}{2} R_a \text{ and } V_m = E_m + \frac{(I_1 - I_2)}{2} R_a$$

$$\therefore \quad V_m - V_g = (E_m - E_g) + (I_1 - I_2) R_a \quad \text{...(1)}$$

Since the speed and excitation of both the two machines are equal

$$\therefore \quad E_g = E_m, \quad \therefore \quad V_m = (I + I_2)/R_a \quad \text{...(2)}$$

Here, R_a = armature resistance of each machine, V_g = terminal p.d. of machine running as a generator *i.e.*, B_1. E_g = induced emf of B_1, V_m = terminal p.d. of metering machine, B_2, E_m = 1 induced emf of B_2.

Hence, from the two equations (1) & (2), we find that the difference of voltages between the two sides of the system is proportional to : (1) out of balance current $(I_1 - I_2)$ and (2) armature resistance of the balancer.

In fact for this reason, R_a is kept very small and effort is made to arrange the loads on the two sides that out-of-balances current is as small as possible.

17.14 BOOSTERS

Booster is, in fact, a generator where function is to add to inject into a circuit a certain voltage that is sufficient to compensate for the *IR* drop in the feeders etc. In a d.c. system, it may sometimes happen that a certain feeder is much longer as compared to others and the power supplied by it is also larger. In that case, the voltage drop in this particular feeder will exceed the allowable drop of 6% from the declared voltage. This can be remedied in two ways (1) by increasing the cross-section of the feeder so that its resistance and its *IR* drop is decreased, (2) or by increasing the voltage of the station bus-bars.

However, the usual practice is to install a booster in series with this longer feeder as shown in Fig. 17.23. Since it is used for compensating drop in a feeder, it is known as feeder booster. It is series generator connected in series with the feeder and driven at a constant speed by a shunt motor working from the bus-bars. The drop in a feeder is proportional to the load current, hence the voltage injected into the feeder by the booster must also be proportional to the load current if exact campensation is required. In other words, the booster must work on the straight or linear portion of its voltage characteristic.

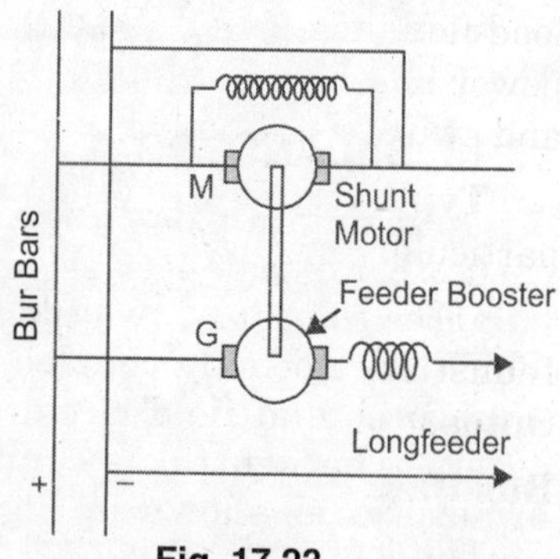

Fig. 17.23

17.15 COMPARISION OF A 2-WIRE AND 3-WIRE DISTRIBUTION SYSTEM

Let us now compare the 2-wire and 3-wire system from the point of view economy of conductor material. For this comparison, it will be assumed that : (1) The amount of power transmitted is the same in both cases, (2) The distance of transmission is the same, (3) The efficiency of transmission (and hence losses) is the same, (4) Voltage at consumer's terminals in the same, (5) The 3-wire system is balanced and (6) In the 3-wire system the mid-wire is of half the cross-section of each outer.

Let W be the transmitted power in watts and V the voltage at the consumer's terminals. Also let R_2 = resistance in ohm of each wire of 2-wire system, R_3 = resistance in ohm of each outer in 3-wire system.

The current in 2-wire system is W/V and the losses are $2(W/V)^2 R_2$. In the case of 3-wire system, voltage between outers is 2 V, so that current through outers is $(W/2V)$, because there is no current in the neutral according to our assumption (5) made above. Total losses is the two outers are $2(W/2V) R_2$.

Since efficiencies are the same, it means the losses are also the same,

$$\therefore \quad 2(W/V)^2 R_2 = 2(W/2V)^2 \times R_2 \text{ or } R_2/R_3 = 4/1$$

Since the cross-section and hence the volume of a conductor of given length is inversely proportional to its resistance,

$$\therefore \quad \frac{\text{Volume of each 3-wire conductor}}{\text{Volume of each 2-wire conductor}} = \frac{1}{4}$$

Now let us represent the volume of copper in the 2-wire system by 100 so that the volume of each conductors is 50. Then, volume of each outer in 3-wire system = 50/4 = 12.5.

Volume of neutral wire 3 tire system = 12.5/2 = 6.2

$\therefore$ Total volume of copper in 3-wire system = 12.5 + 6.25 + 12.5

$$\therefore \quad \frac{\text{Total copper volume in 3-wire feeder}}{\text{Total copper volume in 2-wire feeder}} = \frac{31.25}{100} = \frac{5}{16}$$

Hence, a 3-wire system requires only 5/16th (or 31.25%) as much copper as a 2-wire system.

17.16 SUBSTATION IN DISTRIBUTION SYSTEM

The transmission network in a power system high voltage transmission lines are necessary to transient huge quantity (block) of power from the sources of generation to the load centres to inter connect power houses for increased reliability of supply greater system stability of load flow, and lesser stand by power plant and hence cheaper electric energy. In between the power house and ultimate consumer a number of transformation (by step-down transformer) and switchy station hence to be constructed or creats. These station are called substations.

Types : There are different types of dependency on purpose and application of energy for particular laods as below :

(1) Step up s/sh., (2) Primary grid s/sh., (3) Secondary s/sh., (4) Distribution s/sh., (5) Industrial & Back power s/s, (6) Mining s/sh, (7) Metal s/sh, (8) Cinemetographs s/sh, (9) Outdoor type, (10) Under type, (11) Pole-mounting substation.

Bus Bars

Outdoor Bus-bars are of the rigid type. Bus bars for type and outdoor are generally pipe supported on insulates.

In the rigid bus arrangement Al pipes and steel pipes are used of following diameter for various system voltages:

33 kV	40 mm	132 kV = 80 mm	220 kV	80 mm
66 kV	65 mm		400 kV	100 mm

BUS ARRANGEMENT

1. Single Line Bus bar Arrangement

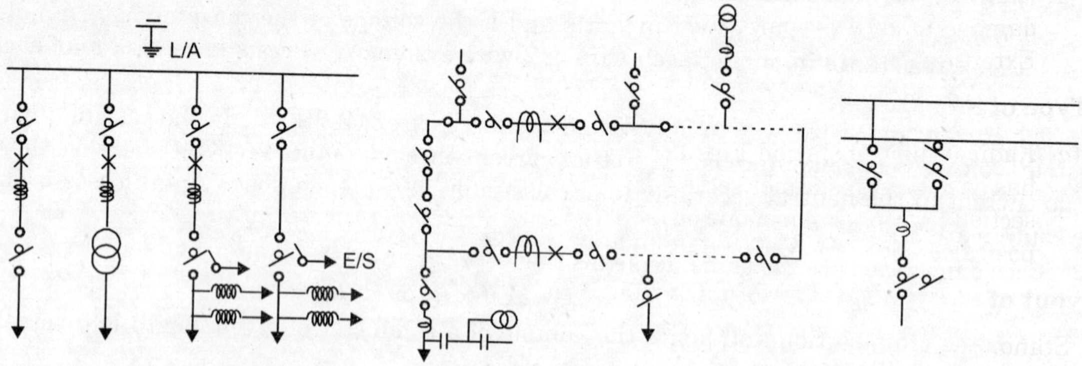

Fig. 17.24

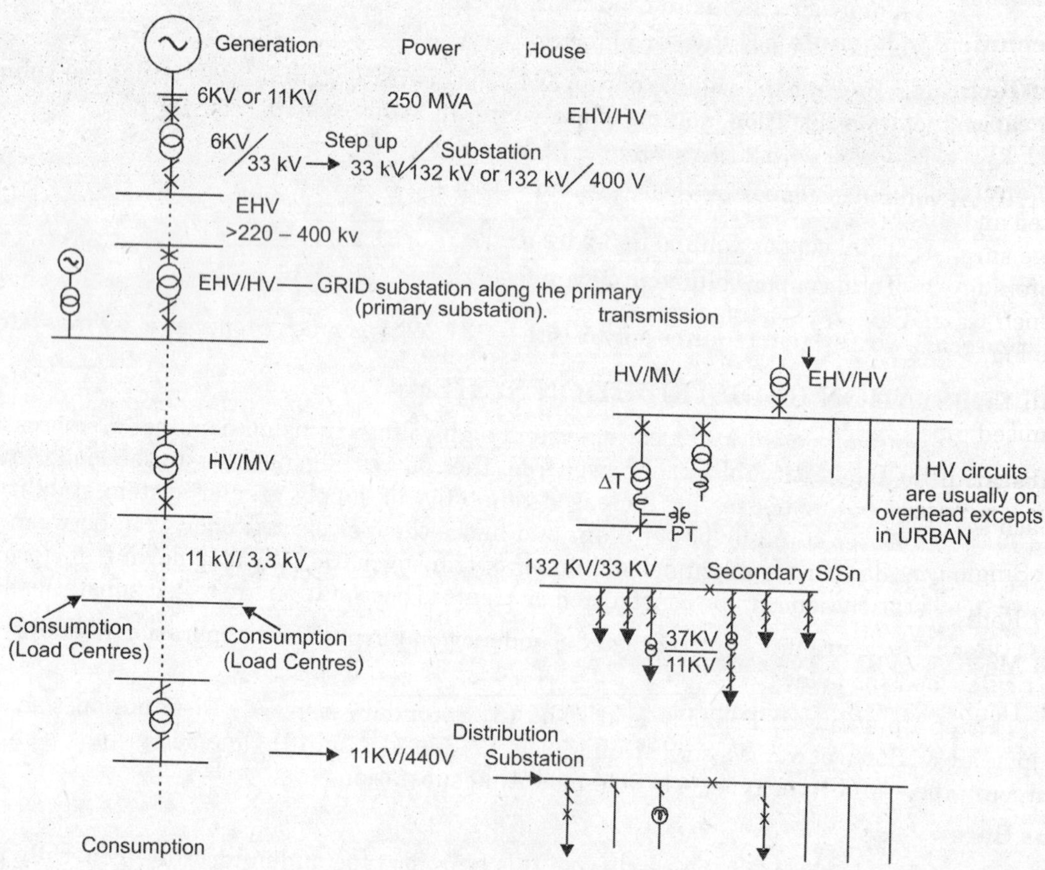

Fig. 17.25

Substations are necessary to

1. Establish Technical and Economical factors including environmental condition.
2. For high voltage/EHV Level, s/s it is common practice to install outdoor.
3. Substations are at convenient points for control and protection of the transmission distribution network.
4. Arrange easy maintenance possible, without damage to staff prepare. Earthing and answer Extension posibility.

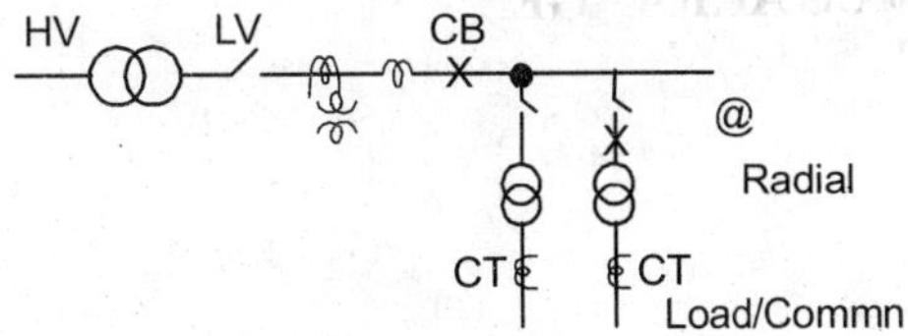

Fig. 17.26

3. Type of Substation

(*a*) Radial Unit substation Fig. 17.26 is one which has a single transformer section and an outgoing section for connection of one or more outgoing Feeder. It may be provided with secondary position.

Layout of Substation

Standardised Layouts have been adopted for substations. These include proper location, communication system. Also following factors are to be considered to locate the site of the substations.

(*a*) Technical & Economical Factors

In Technical Factors : 1. Size of the substation depending on the power capacity of the substation including High Voltage and MVA Rating. Say 11 kV, 33 kV, 132 kV, 220 kV, 440 kV. And power varying on the load requirement. Accordingly transformers and switchgears are to be selected and commissioned. (*b*) Lightning Arrestors are used to curbs high voltage surges. (2) Perfect grounding of the substation is a must considering the safety of the operators for protection system also. Grounding of the substation is done by an Earthing net on which the ground of the S/Stn. is furnished. The Earthing arrangement of a s/s is shown in the Fig. 17.27.

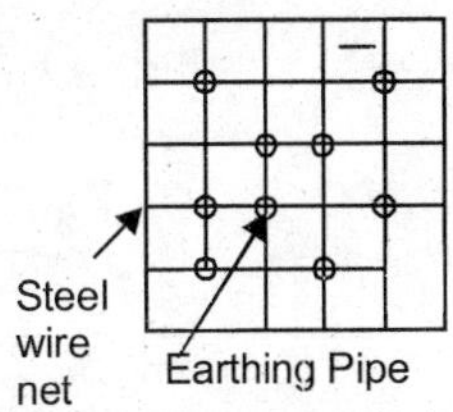

Fig. 17.27

All the equipments are to be Earthed by connecting with Earthing pipes which are galvanised pipe (*Zn* plated) to avoid rusting.

4. Substation Networks

Each s/s has a bus bar network arrangement of following types :

1. Single Bus Bar arrangement shown in Fig. 17.28.

(*a*) Radial Networks

(*b*) Mesh Type/Ring Type shown in Fig. 17.29(*a*)

2. Double Bus bar Arrangements shown in Fig. 17.27.

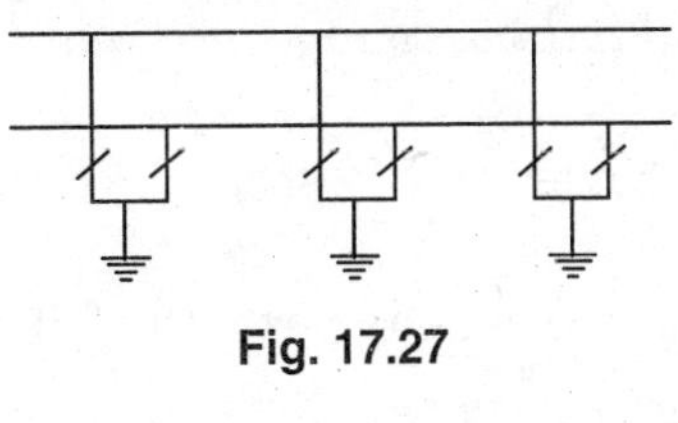

Fig. 17.27

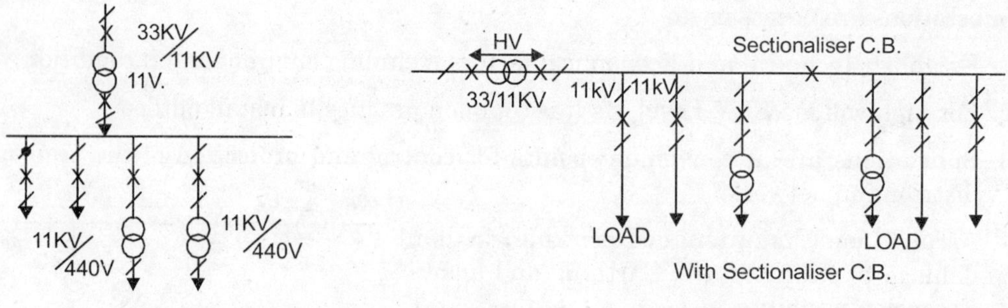

(*a*) (*b*)

Fig. 17.28

Example 17.11. *A 400-metre ring distributor as shown in Fig. 17.29 (a) where distances are in metres. The resistance of each conductor is 0.2 Ω per 1000 metres and the loads tapped of at point B, C and D are as shown. If the distributor is fed at A find voltages at B, C and D.*

Solution : Let us assume a current of I in section AD, shown in Fig. 17.29 (*a*) and then find the total drop which should be equated to zero.

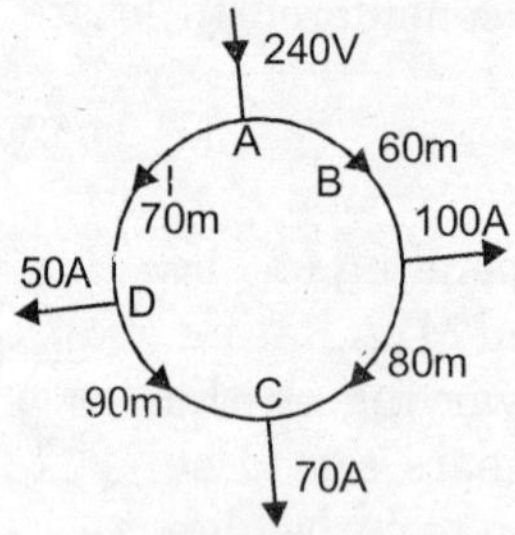

Fig. 17.29 (*a*)

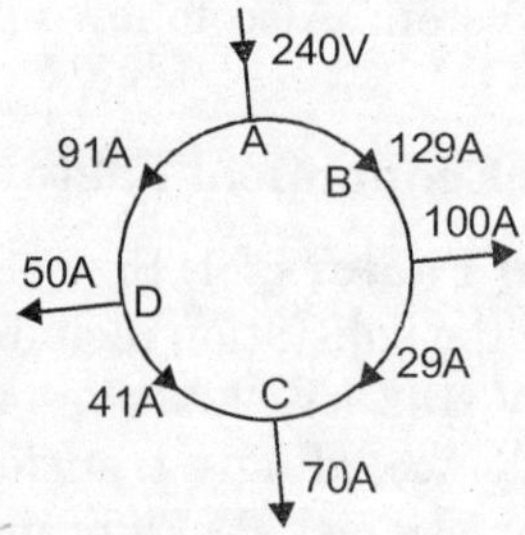

Fig. 17.29 (*b*)

$$\therefore \quad 70I + 90(I - 50) + 80(I - 120) + 60(I - 220) = 0$$

$$\therefore \quad 300I = 27{,}300 \text{ or } I = 91 \text{ A}$$

The distribution becomes as showing in Fig 17.29 (*b*) from where it is seen that C is the point of minimum potential, by putting the value of I.

Drop in AD = 2(91 × 70 × 0.2/1000) = 2.55 V, Drop in DC = 2(41 × 90 × 0.2/1000) = 1.48 V

Drop in CB = 2(29 × 80 × 0.2/1000) = 0.93 V, Drop in BA = 2(129 × 60 × 0.2/1000) = 3.1 V

Voltage at D = 240 – 2.55 = 237.45 V, Voltage C = 237.45 – 1.48 = **235.97 V**

Voltage at B = 240 – 3.1 = 236.9 V

Example 17.12. *In a direct current ring main, a voltage of 400 V is maintained at. At B, 50 metres away from A a load of 150 A is taken and at C 300 metres from B, in load of 200 A is taken. The distance between A and C is 700 metre's. The resistance of each conductor of the mains is 0.03 Ω per 1000 metres. Find the voltage at B and C and also find the current in the section BC.*

Solution : Let us assume a current of I in section AB then find the total drop round the ring and equal it to zero.

∴ Now, as seen from Fig. 17.30 (*a*), we can write the following equation :

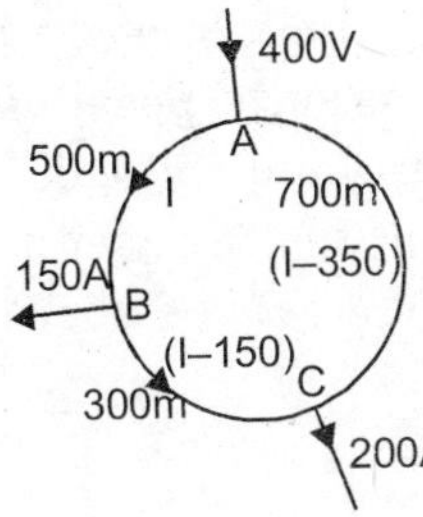

Fig. 17.30 (*a*)

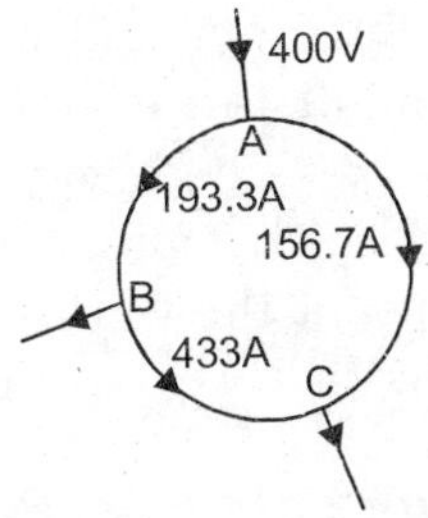

Fig. 17.30 (*b*)

$$500I + 300\,(I - 150) + 700(I - 350) = 0 \qquad \therefore \; I = 193.3 \text{ A}$$

Ten current distribution becomes of the Ring main as shown in Fig. 17.30 (*b*) from where it is seen that C is the point of minimum potential.

Drop over AB = 2(193.3 × 50 × (0.03)/1000)] = 5.8 V

Drop over BC = 2(43.3 × 300 × 0.03/1000) = 0.78 V

Voltage at B = 400 – (drop over AB) = 400 – 5.8 = 394.2 V

Voltage at C = 394.2 – Drop over BC = 394.2 – 0.78 = **393.4 V**

The current in section BC is shown in Fig. 17.30 (*b*).

Example 17.13. *In a 3-wire distribution system, the supply voltage is 250 V on each side. The load on one side is a 3 on a resistance and on the other, a 4 Ω. The resistance of each of the 3 conductors is 0.05 Ω. Find the load voltages.*

Solution : Let the assumed direction of the unknown current in the three conductors be as shown in Fig. 17.31. Applying KVL to closed circuit $ABCDA$, we get,

$$250 - 3.05x - 0.05\,(x - y) = 0$$

$$\therefore \qquad 310x - 5y = 25{,}000 \qquad \ldots(1)$$

Similarly, circuit $DCEFD$ yields

$$0.05\,(x - y) - 4.05y + 250 = 0$$

$$\therefore \qquad 5x - 410y = -25000 \qquad \ldots(2)$$

Now equations (1) & (2), we get x = 81.64 A, y = 61.97 A. Since both currents come out to be positive, it means that their assumed directions of flow are correct.

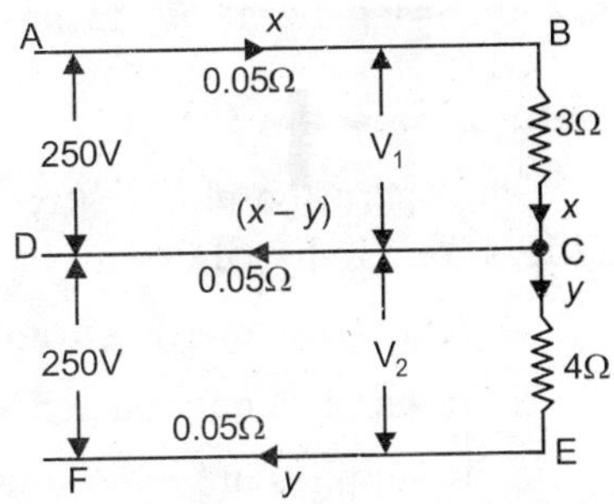

Fig. 17.31

$$\therefore \qquad V_1 = 250 - 0.05 \times 81.64 - 0.05\,(81.64 - 61.97) = 244.9 \text{ V}$$

$$V_2 = 250 + 0.05\,(81.64 - 61.97) - 0.05 \times 61.97 = 247.9 \text{ V}$$

17.17 SUPPLY SYSTEMS

17.17.1 Introduction

Large amount of power is generated at the generating stations. The present trend is to install bigger size of alternators to generate large amount of power to cater the required increasing demand. The site of the power station depends upton the type of power station. The new thermal power stations are being constructed. Station sites are governed by the availability of water and coal resources. The nuclear plants are also situated remote from the centres of consumption due to safety reasons. Thus, the difficulty of getting power station sites near the consuming centres make it inevitable to transfer bulk of electrical energy through longer distances. Long-distance. bulk power transfer is only possibile by high voltage transmission systems. Extra high voltage (EHV) and ultra high voltage (UHV) transmission systems have been developed in most of the countries for transporing energy from remote power stations. By EHV is meant the voltage above 220 kV. The voltages above 760 kV are called ultra high voltages (UHV).

17.17.2 Basic Structure of an AC Power System

Electrical energy generated at generating stations is transported to remote load centres. Between a generating station and a consumer we have transmission, sub trnsmission and distribution levels of voltage. Since the long distance transmission at high voltages is cheap and, low voltages are required for utility purposes, the voltage level goes on decreasing from the transmission system to the distribution system. An electrical power system may be divided into three main components, namely, the generating system, the transmissin system and the distribution system.

Distribution Voltage Level

The component of an electrical power system connecting all the consumers in an area to the bulk power sources is called a *distribution system*. Bulk power stations are connected to generating stations by transmission lines. They feed a number of substations which are usually situated at convenient points near the load centres. A substation distributes the power to domestic, commercial and relatively small consumers. The consumers requiring larger blocks of power are usually supplied at subtransmission or even transmission levels. A bulk power station or a generating station usually supplies power to a subtransmission system.

Subtranmission Level

The operation of a subtransmission system is similar to that of a distribution system. It differs from a distribution system in the following manner :

1. A subtransmission system has a higher voltage level than a distribution system.
2. It supplies only bigger loads.
3. It supplies only few substations as compared to a distribution system which supplies a number of loads.

Transmission Level

A transmission system is quite different from either a subtransmission or a distribution system. it has a higher voltage level than a subtransmission system. It supplies only large blocks of power to bulk power station or very big consumers. The third and the main consideration in which a transmission system differs from either a subtransmission or a distribution system is that it interconnects the neighbouring generating stations into a *power pool*. Thus, a transmission line performs two functions : First, the transport of energy from generating stations to bulk receiving stations and second the interconnection of two or more generating stations. The interconnection of two neighbouring subtransmission systems is also done by transmission lines.

LAYOUT OF A POWER SUPPLY NETWORK

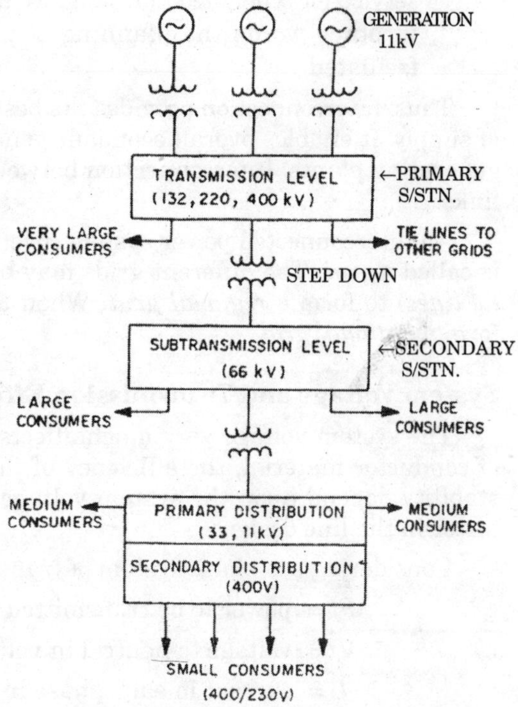

Fig. 17.32. A power system network showing various voltage levels.

Fig. 17.32 shows the layout of a typical ac power supply network by a single-line diagram. Electrical energy is generated by three-phase synchronous generators (alternators). The generation voltages are usually 11 kV and 33 kV. This voltage is too low for transmission over long distances. it is therefore stepped upto 132, 220, 400 kV, of more by means of step-up transformers. At that voltage, the electrical energy is transmitted to a bulk power substation (receiving station) where energy is supplied from several power stations. The voltage at these substations is stepped down to 66 kV in India and fed to the subtransmission system for onward transmission to distribution sub-stations. These substations are located in the vicinity of the load centres. The voltage is further stepped down to 33 kV and 11 kV. Large industrial consumers are supplied at the primary distribution level of 33 kV while smaller industrial consumers are supplied at 11 kV. The voltage is stepped down further by distribution transformers located in residential and commercial areas where in residential and commercial areas where it is supplied to these consumers at the secondary distribution level of 400 V three phase or 230 V single phase.

It is to be noted that it is not necessary that every power system network should have all the stages shown in Fig. 17.32. For example, in a particular network there may be no secondary transmission. Similarly for a smaller power network there is only distribution and no transmission.

System Interconnection

Two or more generating stations are interconnected by tie lines. Interconnection of generating stations has the following important advantages :

1. It enables the mutual transfer of energy from surplus zone to deficit zone economically.
2. Lesser overall installed capacity to meet the peak demand.
3. Lesser standby reserve generating capacity.
4. The size of the biggest unit is not related to the peak load of an individual system but it relates to the peak load of the interconnected system. Thus, fewer but larger machines of greater efficiency are to be installed.
5. It permits the generation of energy at the most efficient and cheapest stations at every time.
6. It reduces the capital cost operating cost and cost of energy generated.
7. If there is a major breakdown of a generating system unit in an interconnected system there is no interruption of power supply. Similarly, when a machine is taken out of

service for scheduled maintenance and inspection the continuity of supply is maintained. In other words the planning of plant outages for maintenance and repair work is faciliated.

Thus, interconnection provides the best use of power resources and ensures greater security of supply. It enables overall economic generation by optimum use of high capacity economical generating plants. Interconnection between networks is done either by HVAC links or HVDC links.

An interconnected power system covering a major portion of a country's territory (or state) is called a *grid*. The different grids may be interconnected through transmission lines (called *tie lines*) to form a *regional grid*. When the different regional grids are interconnected they form a *national grid*.

System Voltage and Transmission Efficiency

The system voltage very much affects the capital cost of a transmission line. The weight of conductor material, the efficiency of the line, the voltage drop in the line and the system stability depend upon the system voltage. The choice of voltage, therefore, becomes a major factor in the line design.

Considering a n-phase system of transmission, let

P = power to be transmitted per phase in watts,

V = voltage to neutral in volts,

I = current in each phase in amperes,

l = length of the line in metres,

A = cross-sectional area of each conductor in m^2,

ρ = specific resistance of the conductor material in ohm-m,

R = resistance of each conductor in ohms.

α = current density in A/m^2, and

$\cos\varphi$ = power factor of the load.

Then, $P = VI\cos\varphi$; $\quad I = \dfrac{P}{V\cos\varphi}$; $\quad A = \dfrac{I}{\alpha} = \dfrac{P}{\alpha V\cos\varphi}$

The resistance of each conductor is given by

$$R = \rho\frac{l}{A} = \frac{\rho\, l\, \alpha\, V\cos\varphi}{P}$$

(*a*) Power Loss

The power loss in the line per phase, $P_L = I^2R = \left(\dfrac{P}{V\cos\varphi}\right)^2 \cdot \dfrac{1}{P}(\rho\, l\, \alpha\, V\cos\varphi) = \dfrac{\alpha\,\rho\, l P}{V\cos\varphi}$ W

which shows that the power loss in the line is inversely proportional to both the system voltage and the power factor.

(*b*) Voltage Drop in Resistance

The voltage drop in resistance per phase $= IR = A\alpha\rho\dfrac{l}{A} = \alpha\rho l$

The percentage voltage drop in resistance in the line $= \dfrac{IR}{V}\times 100\% = \dfrac{\alpha\rho l}{V}\times 100\%$

The voltage drop in reistance per phase is, therefore, constant for a given value of current density and the percentage voltage drop in resistance decreases with the increase in the system voltage.

(c) Weight of Conductor Material

Let σ be the density of the material. The weight of conductor material required for each phase

$$= Al\alpha \frac{Pl\sigma}{\alpha V \cos\varphi}$$

This indicates that the weight of the conductor material required for the line will decrease with the increase in supply voltage and power factor.

The efficiency of transmission, $\eta_T = \dfrac{\text{line output}}{\text{line output + line loss}} = \dfrac{P}{P + \dfrac{\alpha\rho l P}{V\cos\varphi}} = \dfrac{1}{1 + \dfrac{\alpha\rho l}{V\cos\varphi}}$

$$= \left(1 + \frac{\alpha\rho l}{V\cos\varphi}\right)^{-1} = 1 - \frac{\alpha\rho l}{V\cos\varphi} \text{(approx.)}$$

The above relation show that the efficiency of transmission also increases with the increase of supply voltage and power factor. Of more reason for using higher voltages is the enhancement of system stability.

In the above discussion, for the effect of system voltages and power factor, $P\,l$, ρ and α are assumed to be constants. The corona and leakage losses are neglected.

Working Voltage

The above considerations indicate the desirability of using high voltages if the power is to be transmitted over a longer distance. It is also necessary that with the ac systems the load power factor should be as near to unity as possible for maximum economy. It is to be noted that heavy currents are more difficult to handle than high voltages. Considerable saving in the cost of the conductor material is possible when the voltage is high. Although much economy can be effected in the conductor material by employing extra high voltages, the cost of insulation of conductors, whether overhead or underground, increases appreciably. By adopting high voltages the electrical separation or clearances between the conductors is to be increased to avoid electrical discharge. The problems of mechanical supporting structures and right-of-way acquisition become more difficult and expensive.

The other problems encountered with high voltages are the insulation of the equipment, corona, radio and television interference. The insulation costs of transformers, switchgear and other terminal equipment increase tremendously. Corona, radio and television interference become very serious at extra high voltages. Thus, higher the voltage the more costly is the line. The voltage level of a system is, therefore, governed by the amount of power to be transmitted along the length of the line.

Choice of the Next High Voltage

The choice of the highest system voltage for a country is a matter of great significance. It is not merely the economic considerations to choose the next higher voltage but the site of power station, location and density of the load, and the technological developments of the times are also to be kept in view.

The next voltage level should be chosen on the basis of future requirements of load also. To determine the best system voltage the alternative plans should be compared by taking into account the power development over a period of aboul ten years. The long- term forecast should also include the development of power resources during that period.

The interval between the existing and the proposed level should not be too low as it will have a short life. At the same time loo large an interval would lead to heavy expenses. It is, therefore, desirable to adopt the *skip-step basis* in which the next voltage chosen should be at least two steps higher than the existing one.

It is very necessary that a limited number of future EHV levels should be adopted and standardized so as to achieve the advantages of standardization more effectively without sacrificing the necessary flexibility of the existing systems and the proposed schemes.

Standardization of Transmission Voltages

There is much variation in transmission voltages in different countries. A country adopts a voltage or a system of voltage levels to suit its requirements of load. Where long distance transmission is involved, the use of EHV becomes inevitable. Earlier, individual attempts were made to fix voltage levels for high power transmission but such an adoption of individual voltage levels resulted in waste of time. The designs due to their varied nature were costly. It was realized to standardize the transmission voltages for the following reasons :

(*a*) Standardization provides better facilities for research and development.

(*b*) The equipment can be manufactured economically with greater reliability.

(*c*) The maximum possible use of EHV for transmission and interconnection of EHV systems can be made.

The standardization would, therefore, avoid independent attacks to tackle EHV problems in different countries to find the optimum economic voltage and, thus, a lot of time is saved. By standardizing the equipment, the higher voltage can be adopted for a reasonable period of time before next change and, thus, the number of lines will also be lesser.

One of the difficulties encountered with in the slandardization was to adopt the maximum continuous voltage or the nominal voltage to fix up a standard of system voltage. Again, there is a discrepancy in the matter of margin between these values in European and American practices. In the former case, the maximum value is ten per cent higher than the nominal value, while in the lalter this difference is only five per cent. It was proposed to compensate the nominal values in such a way as to have the same maximum value in both the systems in order to avoid this margin. It was preferred to have the criterion of maximum value rather than the nominal value for the standardization purpose.

Earlier, standardization was very successful upto a voltage level of 230 kV. This voltage was the standard voltage used in many countries. Different opinions were put forward to standardize the system voltage above 230 kV and the international slandardization was not so successful. The various voltages adopted by different countries above 230 kV arc *275*. 287. 345. 380, 400. 500, 735, 1100, 1200 kV, etc. The highest voltage used in India is 400 kV. The voltaees above 765 kV are called ullra high voltages (UHV).

CLASSIFICATION OF LINES

Lines may be classifined either electrically or physically.

Electrical Classification

1. Direct Current (DC) System

(*a*) DC 2-wire, 1-wire earthed.

(*b*) DC 1-wire, earth return.

(*c*) DC 2-wire, mid point earthed

(*d*) DC 3-wire.

2. Alternating Current (AC) System

(*a*) Single-phase, 2-wire, 1-wire earthed.

(*b*) Single-phase, 1-wire, earth return.

(*c*) Single-phase, 2-wire, mid-point earthed

(*d*) Single-phase, 3-wire.

(*e*) Two-phase, 3-wire

(*f*) Two-phase, 4-wire.

(*g*) Three-phase, 3-wire

(*h*) Three-phase, 4-wire

All these systems are used in practice. Every system has got its own field of application. The line economics varies with the individual system. Direct current 2-wire, mid-point earthed system is used for transmitting large amounts of power over long distances.

Single-phase a.c. system is rarely used for transmission purposes. Three-phase 3-wire a.c. system finds its greatest used for transmission and primary distribution. For secondary distribution work three-phase, 4-wire system is common. The industrial and other big loads are supplied by three-phase, 4-wire system. The domestic and other small load consumers are supplied with single-phase power obtained from the individual phases of a three-phase supply.

Physical Classification

The other criterion of classification of lines is their physical configuration. The lines may be overhead, underground or underwater. Again, overhead lines may either have bare conductors or insulated conductors.

Most of the high voltage and extra high voltage lines have bare conductors since these arc most economical. The conductors on overhead lines are sometimes insulated for safety or other considerations. The insulated aerial cables although costlier than the bare conductors are far cheaper than (he underground cables. They are used in densely populated urban areas at relatively low voltages or for rural distribution work at medium voltages. Overhead lines and underground systems have been compared in Chapter 4.

For long river or sea crossing, it is not possible to erect overhead lines due to their longer spans. In such cases the need of underwater cables becomes inevitable. Such cables are very costly to manufacture and install. However, with advanced technique over head lines over used of river-crossing.

Comparison of Conductor Costs in Various Systems

The conductors in a line are insulated from each other and from the earth. The cost of insulation varies with the system. Since the conductor cost forms the bulk of expenditure of the line, it is necessary to compare it in various systems from that point of view. A fixed maximum voltage for the same insulation stress should always be specified to make such a comparison. With bare conductor overhead lines and single-core cables, the maximum voltage to earth is taken as the basis of comparing the conductor costs. With multicore belted type cables, it is the maximum voltage between the line conductors which forms the criterion of comparison.

Conductor efficacy is defined as the ratio of conductor material required for a given system to that required for the dc, 2-wire system with one wire earthed.

Let us now compare the volume of conductor material required for a given system to that required for the dc. 2-wire system with one wire earthed.

Criterion of Equal Maximum Voltage to Earth

This criterion is used for overhead lines and single-wire cables. The following assumptions are made for all the systems to be compared regarding the relative amount of conductor material required :

1. The same maximum voltage to earth, V.
2. The same power P to be transmitted.
3. The same route length l of the line.
4. The same efficiency of transmission, that is, the same I^2R losses for all the systems. Corona and leakage losses are neglected.
5. Balanced load so that there is no power loss in the neutral conductor.

General Procedure

1. Draw the circuit diagram of the given system. mark the direction of currents and polarities of voltages. In ac systems the voltages and currents should be expressed in terms of their r.m.s. values.
2. Calculate the I^2R loss in all the conductors of the system and denote it by PL.
3. Equate the I^2R loss of the system under consideration to the I^2R loss in the d.c. 2-wiere system which is taken as the reference system.
4. Determine the volume of conductor material used in all the conductors of the system.
5. Determine the ratio of the volume of conductor found in step 4 to the volume of the conductor used in the 2-wire system with one-wire earthed.

Let us consider the following cases :

(*a*) Direct current 2-wire system with one wire earthed

The dc 2-wire system with one wire earthed is shown in Fig. 17.33.

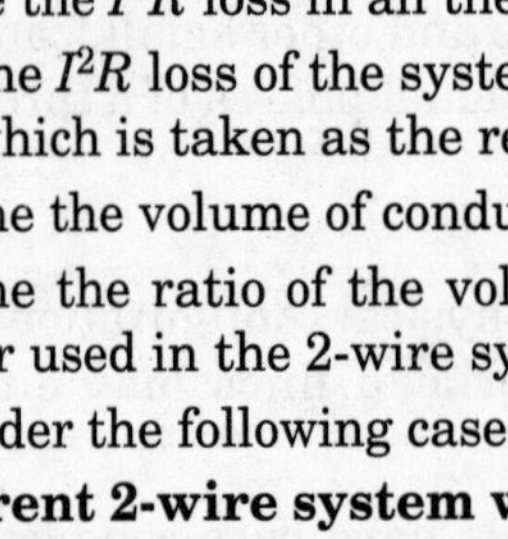

Fig. 17.33. Dired current 2-wire system with one wire earthed.

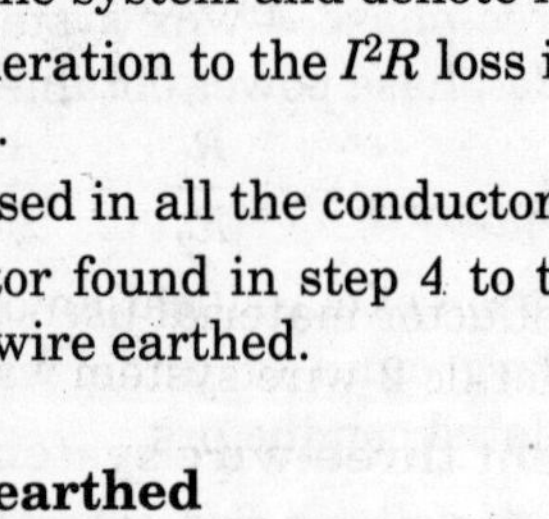

Fig. 17.34. Direct current 2-wire system with mid point earthed.

Let I_1 = line current

R_1 = resistance of each conductor

A_1 = area of corss-section of each conductor

Power transmitted, $P = VI_1$

Line curret $I_1 = \frac{P}{V}$

Total I^2R loss for both the conductors, $p_{L1} = 2I_1^2 R_1 = 2\left(\frac{P}{V}\right)^2 R_1$

Volume of conductor material used in both the conductors $v_1 = 2lA_1$

(b) Direct current 2-wire system with mid-point earthed

The dc 2-wire system with mid-point earthed is called dc 2-wire balanced system. Such a system is shown in Fig. 17.34.

line voltage $= 2V$

line current $I_2 = \frac{P}{2V}$

If R_2 be the resistance per conductor, then the total I^2R loss for both conductors is

$$p_{L2} = 2I_2^2 R_2 = 2\left(\frac{P}{2V}\right)^2 R_2$$

Volume of conductor material used, $v_2 = 2lA_2$

Equating the l^2R loss for cases (*a*) and (*b*) for the same efficiency of transmission, we get

$$p_{L1} = p_{L2}$$

$$2\left(\frac{P}{V}\right)^2 R_1 = 2\left(\frac{P}{2V}\right)^2 R_2; \frac{R_1}{R_2} = \frac{1}{4}$$

But $R_1 = \rho\frac{l}{A_1} \cdot R_2 = \rho\frac{l}{A_2}$

$$\therefore \quad \frac{R_1}{R_2} = \frac{A_2}{A_1} = \frac{1}{4}; \frac{v_2}{v_1} = \frac{2lA_2}{2lA_1} = \frac{A_2}{A_1} = \frac{1}{4} = 0.25$$

Thus, the conductor material used with dc 2-wire mid point earthed system is 25 per cent of that required for dc 2-wire system with one wire earthed.

(c) Direct current three-wire system

Direct current three-wire system is shown in Fig. 17.35. In this system there are three conductors namely, two outers and one middle conductor. The middle conductor is earthed at the supply end. When the load is balanced, the current in the middle (neutral) conductor is zero.

Let the cross-section of any outer conductor be A_3 and that of the neutral wire be half tht of any outer, that is $0.5A_3$.

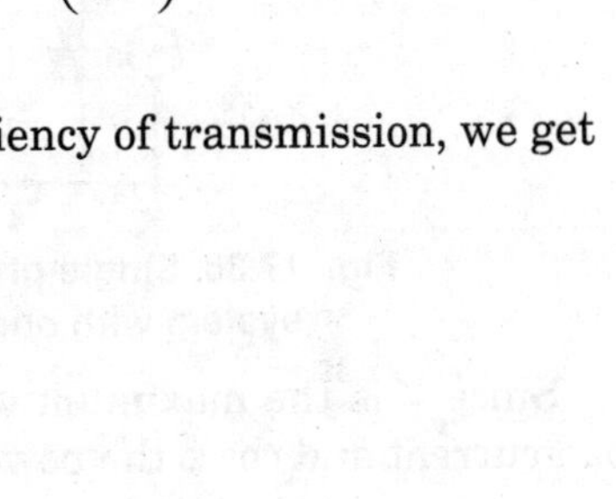

Fig. 17.35. Direct current 3-wire system.

Line voltage = $2V$; Line current, $I_3 = \frac{P}{2V}$

I^2R loss in the system = $2I_3^2R_3 = 2\left(\frac{P}{2V}\right)^2 R_3$

Volume of conductor material required, $v_3 = 2lA_3 + 0.5lA_3 = 2.5lA_3$

For the same efficiency of transmission in cases (*a*) and (*c*)

$$p_{L1} = p_{L3}$$

$$2\left(\frac{P}{V}\right)^2 R_1 = 2\left(\frac{P}{2V}\right)^2 R_3; R_1 = \frac{1}{4}R_3$$

$$\rho\frac{l}{A_1} = \frac{1}{4}\rho\frac{l}{A_3}; \frac{A_3}{A_1} = \frac{1}{4}$$

$$\frac{v_3}{v_1} = \frac{2.5lA_3}{2lA_1} = \frac{2.5}{2\times 4} = 0.3125$$

Thus, the conductor material used with dc 3-wire system is 31.25 per unit of the material required for dc 2-wire system with one wire earthed.

(*d*) Single-phases 2-wire system with one wire earthed

This system is shown in Fig. 17.36.

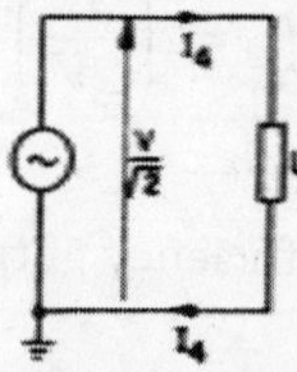

Fig. 17.36. Single-phase, 2-wire system with one wire earthed.

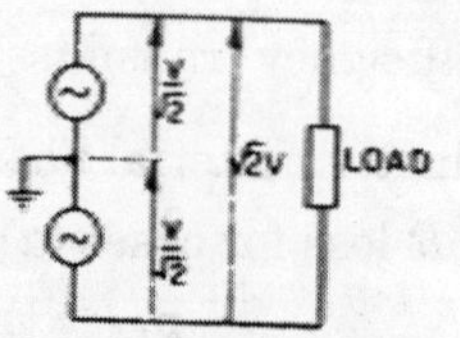

Fig. 17.37. Single-phase, 2 wire mid-point earthed system.

Since V is the maximum voltage to earth, the rms voltage to earth is $V/\sqrt{2}$. If I_4 is the load current and cos φ the power factor of the load.

$$P = \frac{V}{\sqrt{2}} I_4 \cos\varphi; I_4 = \frac{\sqrt{2}\,P}{V\cos\varphi}$$

Let R_4 be the resistance of each conductor of cross section As.

Line loss, $pLA = 2I_4^2R_4 = 2\left(\frac{\sqrt{2}\,P}{V\cos\varphi}\right)^2 R_4$

Volume of conductor material used, $v_4 = 2lA_4$

For the same efficiency of transmissions as in case (*a*)

$$p_{L1} = p_{LA}$$

$$2\left(\frac{P}{V}\right)^2 R_1 = 2\left(\frac{\sqrt{2}\,P}{V\cos\varphi}\right)^2 R_4; \frac{R_1}{R_4} = \frac{2}{\cos^2\varphi}$$

But $$\frac{R_1}{R_4} = \frac{A_4}{A_1}\quad\left(R = \rho\frac{1}{A}\right)$$

$$\frac{v_4}{v_1} = \frac{2lA_4}{2lA_1} = \frac{2}{\cos^2 \varphi}$$

(e) Single-phase, 2-wire mid-point earthed system

A single-phase, 2-wire mid-point earthed system is shown in Fig. 17.34

Maximum voltage between any conductor and earth = V

Maximum voltage between two live conductors = $2V$

Therefore, r.m.s. voltage between the live conductors $= \frac{2V}{\sqrt{2}} = \sqrt{2}\,V$

If I_5 be the line current, R_5 the resistance of each conductor of cross-section A_5, and $\cos \varphi$ the power factor of the load

$$P = (\sqrt{2}V)\, I_5 \cos \varphi;\ I_5 = \frac{P}{\sqrt{2}\, V \cos \varphi}$$

Line loss, $$p_{L5} = 2I_5^2 R_5 = 2\left(\frac{P}{\sqrt{2}\cos \varphi}\right)^2 R_5$$

Volume of conductor material used, $v_5 = 2lA_5$

For the same efficiency transmission as in case (a)

$$p_{L1} = p_{L5}$$

$$2\left(\frac{P}{V}\right)^2 R_1 = 2\left(\frac{P}{\sqrt{2}\, V \cos \varphi}\right)^2 R_5;\ \frac{R_1}{R_5} = \frac{0.5}{\cos^2 \varphi} = \frac{A_5}{A_1}$$

$$\frac{v_5}{v_1} = \frac{2lA_5}{2lA_1} = \frac{0.5}{\cos^2 \varphi}$$

(f) Three-phase, 3-wire star-connected system

This system is shown in Fig. 17.38

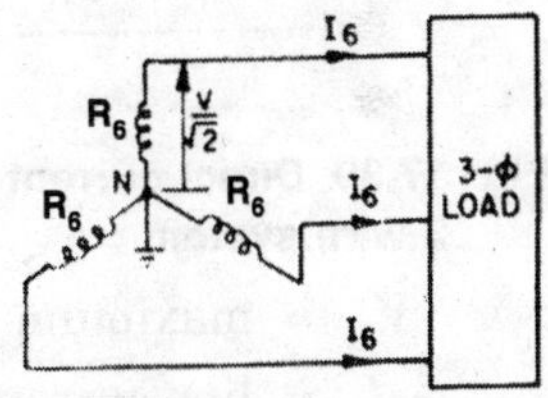

Fig. 17.38. Three-phase, 3-wire star connected system.

Maximum voltage to earth $= V$

Therefore, r.m.s. value of voltage per phase $= \frac{V}{\sqrt{2}}$

If I_6 is the rms phase current, R_6 the resistance of each conductor of cross section A_6 and $\cos \varphi$ the power factor of the load

$$P = 3V_{ph}I_{ph} \cos \varphi = 3 \times \frac{V}{\sqrt{2}} I_6 \cos \varphi$$

$$I_6 = \frac{\sqrt{2}\,P}{3V \cos \varphi}$$

Line loss, $$p_{L6} = 3I_6^2 R_6 = 3\left(\frac{\sqrt{2}\,P}{3V \cos \varphi}\right)^2 R_6$$

Volume of conductor material used $\qquad v_6 = 3lA_6$

For the same efficiency of transmissions as in case (*a*)

$$p_{L1} = p_{L6}$$

$$2\left(\frac{P}{V}\right)^2 R_1 = 3\left(\frac{\sqrt{2}\,P}{3V\cos\varphi}\right)^2 R_6;\ \frac{R_1}{R_6} = \frac{1}{3\cos^2\varphi} = \frac{A_6}{A_1}$$

$$\frac{v_6}{v_1} = \frac{3lA_6}{2lA_1} = \frac{0.5}{\cos^2\varphi}$$

Criterion of Equal Maximum Voltage between Line Conductors

We shall now compare the systems on the basis of the same maximum voltage between two line conductors. Such a criterion is important with multicore cables. Again, it shall be assumed that the the amount of power transmitted route length and efficiency of transmission remain the same in all the systems. Corona and leakage losses are neglected and the system is assumed to be balanced so that there is no power loss in the neutral conductor. The ratio of conductor material required in any system to that required for the dc 2-wire system with one-wire earthed is found in the following cases. We shall use the same general procedure discussed in section 2.13.1.

(*a*) DC 2-wire system

This system is shown in Fig. 17.39.

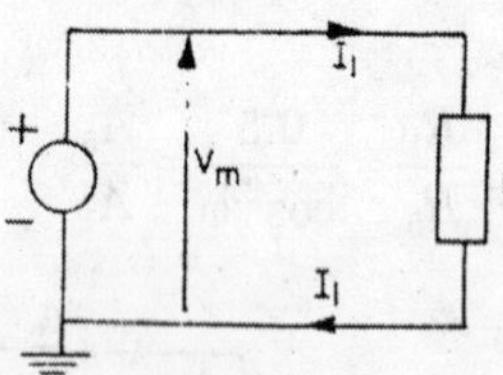

Fig. 17.39. Direct current 2-wire system.

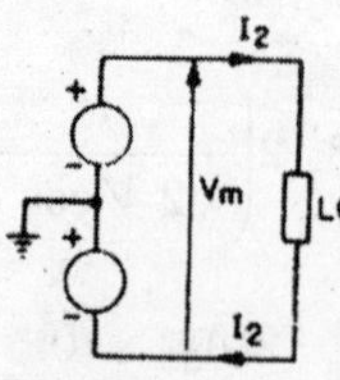

Fig. 17.40. Direct current, 2-wire mid-point earthed system.

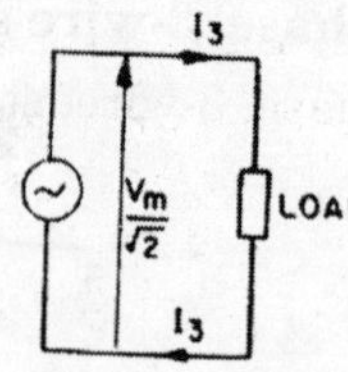

Fig. 17.41. Single-phase, 2-wire system.

Let $\quad V_m$ = maximum voltage between the conductors

I_1 = line current

R_1 = resistance of each conductor

A_1 = area of corss-section of each conductor

Power transmitted, $\qquad P = V_m I_1, I_1 = P/V_m$

Line losses, $\qquad p_{L1} = 2I_1^2 R_1 = 2\left(\frac{P}{V_m}\right)^2 R_1$

Volume of conductor material used, $\qquad v_1 = 2lA_1$

(*b*) Direct-current, 2-wire mid-point earthed system

This system is shown in Fig. 17.40. Since this system is the same as (*a*) the conductor material used is also the same.

(*c*) Single-phase, 2-wire system

This system is shown in Fig. 17.40.

Maximum voltage between line conductors $= V_m$

Elective (rms) line voltage $= V_m/\sqrt{2}$

If I_3 is the r.m.s. line current, R_3 the resistance of each conductor of cross section A_3 and $\cos \phi$ the power factor of the load

$$P = \frac{V_m}{\sqrt{2}} I_3 \cos \varphi, \; I_3 = \frac{\sqrt{2}\,P}{V_m \cos \varphi}$$

Line loss, $$p_{L3} = 2I_3^2 R_3 = 2\left(\frac{\sqrt{2}\,P}{V_m \cos \varphi}\right)^2 R_3$$

Volume of conductor material used, $V_3 = 2lA_3$

For the same efficiency of transmission as in case (*a*),

$$p_{L1} = p_{L3}$$

$$2\left(\frac{P}{V_m}\right)^2 R_1 = 2\left(\frac{\sqrt{2}\,P}{V_m \cos \varphi}\right)^2 R_3$$

$$\frac{R_1}{R_3} = \frac{2}{\cos^2 \varphi} = \frac{A_3}{A_1}; \; \frac{v_3}{v_1} = \frac{2lA_3}{2lA_1} = \frac{2}{\cos^2 \varphi}$$

(*d*) Three-phase, 3-wire star-connected system

Three-phase, 3-wire, star-connected system is shown in Fig. 17.42

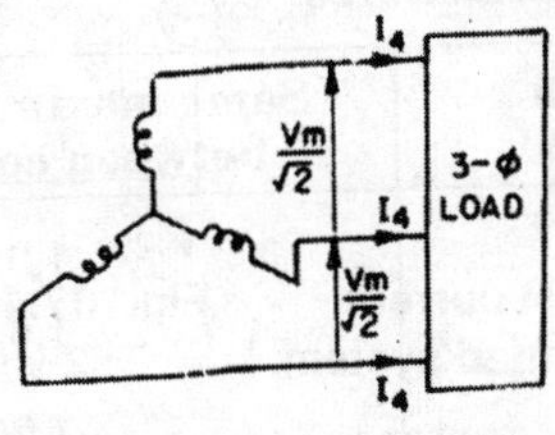

Fig. 17.42. Three-phase, 3-wire, star-connected system.

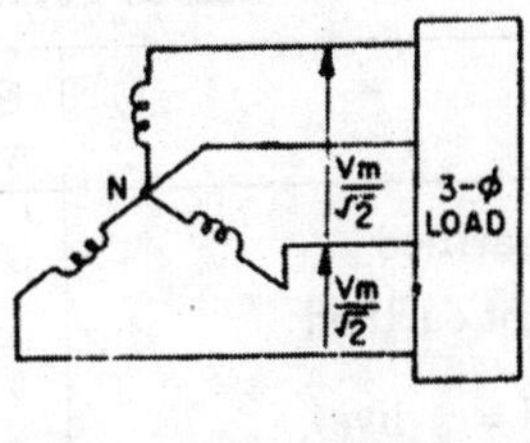

Fig. 17.43. Three-phase, 4-wire system.

Maximum line voltage $= V_m$

Effective (rms) line voltage $= V_m/\sqrt{2}$

If I_4 is the r.m.s. line current, R_4 the resistance of each conductor of cross section A_4 and $\cos \varphi$ the power factor of the load

$$P = \sqrt{3}\, V_l I_l \cos \varphi = \sqrt{3} \times \frac{V_m}{\sqrt{2}} I_4 \cos \varphi$$

$$I_4 = \frac{\sqrt{2}\,P}{\sqrt{3}\, V_m \cos \varphi}$$

Line loss, $$p_{L4} = 3I_4^2 R_4 = 3\left(\frac{\sqrt{2}\,P}{\sqrt{3}\, V_m \cos \varphi}\right)^2 R_4$$

Volume of conductor material used, $v_4 = 3lA_4$

For the same efficiency of transmission as in case (*a*)

$$p_{L1} = p_{L4}$$

$$2\left(\frac{P}{V_m}\right)^2 R_1 = 3\left(\frac{\sqrt{2}\,P}{\sqrt{3}\,V_m \cos\varphi}\right)^2 R_4$$

$$\frac{R_1}{R_4} = \frac{1}{\cos^2\varphi} = \frac{A_4}{A_1}; \quad \frac{v_4}{v_1} = \frac{3lA_4}{2lA_1} = \frac{1.5}{\cos^2\varphi}$$

(*c*) Three-phase, 4-wire system

This system is shown in Fig. 17.42. For a balanced load this system is the same as the three-phase, 3-wire system except that there is an additional neutral wire.

If the cross section of the neutral wire is half that of any line conductor, the ratio of the amount of conductor material required in this case and the d.c., 2-wire one write earthed system is

$$\frac{3.5lA_4}{2lA_1} = \frac{1.75}{\cos^2\varphi}$$

In case the cross section of the neutral is taken to be equal to that of any line conductor, the ratio becomes

$$\frac{4lA_4}{2lA_1} = \frac{2}{\cos^2\varphi}$$

Table 17.1 gives the comparison of conductor material in various systems.

Table 17.1. Conductor Efficacies

System	Same maximum voltage to earth	Same maximum voltage between conductors
DC 2-wire, 1-wire earthed	1.0	1.0
DC 2-wire, mid-point earthed	0.25	1.0
DC 3-wire, (neutral = $\frac{1}{2}$ live)	0.3125	1.25
Single-phase, 2-wire	$2/\cos^2\varphi$	$2/\cos^2\varphi$
Single-phase, 2-wire, mid-point earthed	$0.5/\cos^2\varphi$	$2/\cos^2\varphi$
Single-phase, 2-wire, (neutral = $\frac{1}{2}$ live)	$0.625/\cos^2\varphi$	$2.5/\cos^2\varphi$
Three-phase, 3-wire	$0.5/\cos^2\varphi$	$1.5/\cos^2\varphi$
Three-phase, 4-wire, (neutral = live)	$0.667/\cos^2\varphi$	$2/\cos^2\varphi$

From the Table 17.1, it is observed that on the basis of equal maximum voltage to earth the dc 2-wire mid-point earthed system is the cheapest on the basis of conductor cost alone. The conductor efficacies are the same in single-phase 2-wire and three- phase 3-wire systems. If equal maximum voltage between the conductors is made the basis of comparison, it is found that the dc balanced 2-wire system is the cheapest again and in ac three- phase, 3-wire star-connected system is more economical than single-phase system. It is also observed that the criterion of equal maximum voltage to earth is more economical and, therefore, the overhead lines will be cheaper.

Example 17.14. *In a dc 2-wire system a feeder is working on 250 V supplying a constant load. If the supply voltage is increased to 400 V with the same power transmitted, calculate the percentage saving in conductor material.*

Solution : Let l = length of each conductor in metres

α = current density in A/m²

P = power transmitted in watts

(a) 250 V supply

Line current, $$I_1 = \frac{P}{V_1} = \frac{P}{250}$$

Cross-sectional area of conductor required, $$A_1 = \frac{I_1}{\alpha} = \frac{P}{250\alpha} \text{ m}^2$$

Volume of conductor material required, $$v_1 = 2lA_1 = \frac{2lP}{250\alpha} = \frac{lP}{125\alpha} \text{ m}^3$$

(b) 400 V supply

Line current $$I_2 = \frac{P}{V_2} = \frac{P}{400}$$

Cross-sectional area of conductor required, $$A_2 = \frac{I_2}{\alpha} = \frac{P}{400\alpha} \text{ m}^2$$

Volume of conductor material required, $$v_2 = 2lA_2 = \frac{2lP}{400\alpha} = \frac{lP}{200\alpha} \text{ m}^3$$

$$\frac{v_2}{v_1} = \frac{lP}{200\alpha} + \frac{lP}{125\alpha} = \frac{125}{200} = 0.625$$

Percentage saving in conductor material $$= \frac{v_1 - v_2}{v_1} \times 100 = \left(1 - \frac{v_2}{v_1}\right) \times 100$$

$$= (1 - 0.625) \times 100 = \mathbf{37.5} \text{ per cent}$$

Example 17.15. *A three-phase 4-wire system is used for lighting. Compare the amount of conductor material required with that needed for a 2-wire dc system with same lamp voltage. Assume the same losses and balanced load. The neutral wire has half the cross-section of the outers.*

Solution :

(a) DC 2-wire system

Let V = voltage between conductors

P = power delivered

I_1 = line current

R_1 = resistance of each conductor

A_1 = conductor corss-section

Line current $$I_1 = \frac{P}{V}; \qquad \text{line loss} = 2I_1^2 R_1 = 2\frac{P^2}{V^2} R_1$$

Volume of conductor material required, $v_1 = 2lA_1$

(*b*) AC three phase, 4-wire system

Effective (rms) voltage between line and neutral $= V$

If I_2 = line current

$\cos \varphi$ = power factor

R_2 = resistance of each conductor

A_2 = area of corss-section of each phase conductor

$$P = 3VI_2 \cos \varphi$$

For lighting load, $\cos \varphi = 1$

$$\therefore \quad I_2 = \frac{P}{3V}$$

Total line loss $= 3I_2^2 R_2 = 3\left(\frac{P}{3V}\right)^2 R_2 = \frac{P^2 R_2}{3V^2}$

For equal line loss in both the systems

$$\frac{2P^2}{V^2} R_1 = \frac{P^2 R_2}{3V^2} : \frac{R_1}{R_2} = \frac{1}{6}$$

$$\begin{matrix} R_1 & A_2 & A_2 & 1 \\ R_2 & A_1 & A_1 & 0 \end{matrix}$$

Volume of conductor material required,

$$v_2 = 3lA_2 + 0.5lA_2 = 3.5lA_2$$

$$\frac{v_2}{v_1} = \frac{3.5lA_2}{2lA_1} = \frac{3.5}{2 \times 6} = 0.292$$

Example 17.16. *A 50 km long transmission line supplies a load of 5 mVA at 0.8 power factor lagging at 33 kV. The efficiency of transmission is 90 per cent. Calculate the volume of conductor aluminium required for the line when*

(a) single-phase, 2-wire system is used.

(b) three-phase, 3-wire system is used.

Take the resistivity of aluminium as 2.85×10^{-8} *Ω-m*

Solution : Power transmitted = MVA $\cos \varphi = 5 \times 0.8$ MW $= 4 \times 10^6$ W

Efficiency of transmission = 90

$\therefore$ line loss, p_L = 10 per cent of power transmitted $= \frac{10}{100} \times 4 \times 10^6 = 4 \times 10^5$ W

(*a*) Single-phase system

$$\text{MVA} = VI \times 10^{-6}$$

$$5 = 33 \times 10^3 \times I \times 10^{-6};\ I = 151.5 \text{ A}$$

Let A_1 be the area of cross-section of each conductor.

Line loss, $p_L = 2I^2R = 2I^2\rho\dfrac{l}{A_1}$

$$A_1 = \frac{2I^2\rho l}{p_L} = \frac{2\times(151.5)^2\times2.85\times10^{-8}\times50\times10^3}{4\times10^5} = 1.635\times10^{-4}\text{ m}^2$$

Volume of conductor required, $v_1 = 2lA_1 = 2 \times 50 \times 10^3 \times 1.635 \times 10^{-4} = 16.35\text{ m}^3$

(*b*) Three-phase 3-wire ac system

$$\text{MVA} = \sqrt{3}\,V_L I_L \times 10^{-6}$$

$$5 = \sqrt{3}\times33\times10^3\times I_L\times10^{-6}$$

Line current, $I_L = \dfrac{5\times10^6}{\sqrt{3}\times33\times10^3} = 87.5\text{ A}$

Let the area of each phase conductor be A_2

Total line loss, $p_L = 3I_L^2R_2 = 3I_L^2\rho\dfrac{l}{A_2}$

$$A_2 = \frac{3I_L^2\rho l}{p_L} = \frac{3\times(87.5)^2\times2.85\times10^{-8}\times50\times10^3}{4\times10^5} = 0.818\times10^{-4}\text{ m}^2$$

Volume of aluminium required, $v_2 = 3lA_2 = 3 \times 50 \times 10^3 \times 0.818 \times 10^{-4} = 12.27\text{ m}^3$

Example 17.17. *An existing single phase a.c. system comprising of two overhead conductors is to be converted into a three-phase, 3-wire system by providing an additional similar conductor. Calculate the percentage of additional load that can be transmitted by the three-phase system if the operating line voltage and percentage line losses remain the same in both the systems.*

Solution : Let I_1 = single-phase current

I_3 = three-phase current

R = resistance of each conductor

$\cos\varphi$ = power factor in both the system

(*a*) Single-phase system

Power transmitted, $P_1 = VI_1\cos\varphi$

Line loss $= 2I_1^2R$

Percentage line loss $= \dfrac{2I_1^2R}{VI_1\cos\varphi}\times100$

(*b*) Three-phase system

Power transmitted, $P_3 = \sqrt{3}\,VI_3\cos\varphi$

Line loss $= 3I_3^2R$

Percentage line loss $= \dfrac{3I_3^2R}{\sqrt{3}\,VI_3\cos\varphi}\times100$

For the percentage line losses to be the same in both the system, we have

$$\frac{2I_1^2R}{VI_1\cos\varphi}\times 100 = \frac{3I_3^2R}{\sqrt{3}\,vI_3\cos\varphi}\times 100$$

which gives

$$I_3 = \frac{2}{\sqrt{3}}I_1$$

$\therefore$

$$P_3 = \sqrt{3}\,V\left(\frac{2}{\sqrt{3}}I_1\right)\cos\varphi = 2VI_1\cos\varphi$$

Additional load transmitted $= P_3 - P_1 = 2VI_1\cos\varphi - VI_1\cos\varphi = VI_1\cos\varphi$

Percentage additional load $= \dfrac{P_3 - P_1}{P_1}\times 100 = \dfrac{VI_1\cos\varphi}{VI_1\cos\varphi}\times 100 = 100$

Thus, 100 per cent additional load can be transmitted by converting single phase line to three-phase line.

Example 17.18. *A given amount of power is to be transmitted by an overhead line. Compare the diameter and weight of aluminium conductor with those of copper for the same power loss in the line. The following data may be assumed ;*

Specific resistance of aluminium = 2.85 μΩ-cm

Specific resistance of copper = 1.70 μΩ-cm

Specific gravity of aluminium = 2.71

Specific gravity of copper = 8.89

Solution : For the same power loss in the line, the resistances of aluminium and copper will be the same. If the symbols have their usual meanings and the suffixes a and c be used for aluminium and copper.

$$R_a = R_e; \quad p_a\cdot\frac{l}{A_a} = p_c\frac{l}{A_c} \quad \therefore \quad \frac{A_a}{A_c} = \frac{p_a}{p_c}$$

$$\frac{\frac{\pi}{4}d_a^2}{\frac{\pi}{4}d_c^2} = \frac{2.85\times 10^{-6}}{1.70\times 10^{-6}}$$

$$\frac{d_a}{d_c} = \sqrt{\frac{2.85}{1.70}} = 1.293$$

where d_a = diameter of aluminium conductor, and

d_c = diameter of copper conductor

If σ denotes the specific gravity,

$$\frac{\text{weight of aluminium}}{\text{weight of copper}} = \frac{\sigma_a A_c l}{\sigma_c A_a l} = \frac{2.71}{8.89}\times\frac{2.85}{1.70} = 0.51$$

This relation shows that the weight of an aluminium conductor is 51.1 per cent of the copper conductor of equal resistance.

QUESTIONS

1. A 2-wire dc distributor AB, 300 m long is fed at both ends *A* and *B*. The distributor AB, 300 m long is fed at both ends *A* and B. The distributor supplies a uniformly distributed load of 0.25 A per m together with concentrated load of 40 A at *C* and 60 A at *D*, AC and BD being 120 m each. *A* and *B* are maintained at 300 V, the loop resistance of the distributor is 0.1 Ω/100 m. Determine the current at *A* and *B* and also the potential of points *C* and *D*. (**Ans.** 85.5 A, 85.5 A, 291.54 V, 291.05 V)
2. A ring distributor is supplied through a feeder *AB* and is loaded as shown in Fig. 17.43. Calculate (*a*) cross section of *AB*, and (*b*) that of the ring main for minimum volume of copper. Assume that the maximum voltage drop form *A* to the point of minimum potential is 15 volts, take $f = 1.73\ \mu\Omega$ per can cube. (**Ans.** (*a*) 31.6 m^2, (*b*) 1.3736 m^2)

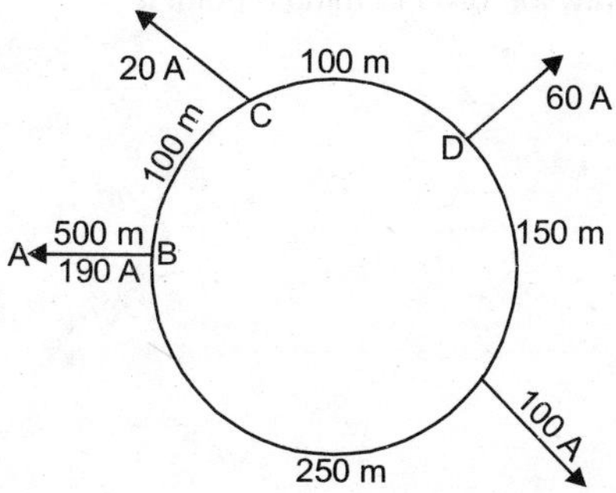

3. A two-conductor street much *A* 500 metres in length in fed from both ends at 250 volts. Loads of 50 A, 60 A, 40 A and 30 A are tapped at distances of 100 m, 250 m, 350 m and 400 m from *A* respectively as shown in Fig. 17.43. If the *x*-seection of the conductors be 1 cm and specific resistance of the material of the conductors is 1.07×10^{-8} ohm-m determine the minimum consumer voltage. (**Ans.** 245 of volts)

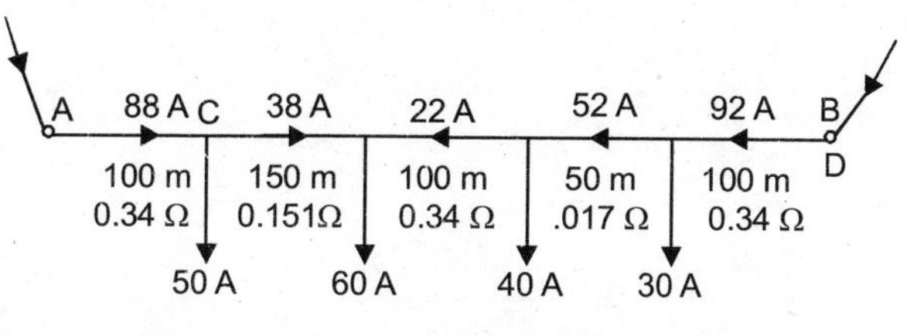

4. Explain different types of distribution system with the half of sketch.
5. State what is (*a*) radial distribution system, (*b*) ring on loop distribution system. Explain with diagram.
6. State what are the advantages of a doubly fed distribution, over singly fed distribution.
7. Explain with diagram a 2-wire dc system of distribution and ring-main distribution system.
8. What is meant of (*a*) balancer, (*b*) Booster, explain.
9. Explain how balances are used to maintain equal voltages on two sides of neutral in a 3-wire dc distribution system.
10. With suitable diagram explain the location, action and use of a booster.
11. State what is meant by primary distribution system and secondary distribution system.
12. State what is the difference between a feeder and a distributor.
13. State why the voltage drop of primary concern in the design of the distributors.
14. State on what factors does the primary distribution voltage depend.

15. As per indian Electricity Act/rules. what is the declared consumer voltage (*a*) for domestic, (*b*) Commercial, (*c*) Industrial light and heaving.

16. State if it is correct to say a feeder is a line which supplies a point of distribution network without being tapped at any intermediate point.

17. State if it is correct to say that a distribution is line from which tappings are taken along its length for providing, supply to the consumers.

18. State if it is correct to say that because of poor efficiency a two-wire system is not used for transmission purpose.

19. State if it is correct to say that radial system of distribution is not used for poorest reliability.

20. A uniformly-loaded dc distribution is fed at both ends with equal voltages. As compared to a similar distributor fed at one end only, the drop at middle point is :

(*a*) one half

(*b*) one fourth

(*c*) one third

(*d*) twice

Ans. (*b*)

CHAPTER

18

Tariffs and Electricity Energy Audit

18.1 INTRODUCTION

INSIDE THIS CHAPTER

- Introduction
- Terms Commonly used in Supply System Operation
- Factors effecting the Cost of Generation
- Economics of Generation
- Choice of Size and Number of Generator Units
- Tariffs
- Energy audit
- Energy conservation

For any area, where electricity is to be made available it is very much necessary to have a correct idea of the load requirement of the area once this has been done for the said area, the next job is how to get this power. It is known that there are three important types of power stations like : Thermal (based on coal mainly), Hydro-electric and Nuclear (besides such as gas-turbines, solar, wind etc).

In choosing the type of generation a number of important basic points have to be considered like; kind of fuel available and its cost, suitable sites, nature of load to be supplied etc.

In fact the nature of load to be supplied affects the choice of plant to a considerable extent. The load which a power system has to supply is, infact never constant because of variable demands at different times of the day. A load curve predicts this variation of load having a peak demand. The minimum capacity of generating staion must be such as to meet the maximum demand of the load but at the same time power system has to maintain reliability and continuity of power supply to consumers at all times. Hence the use of a single set to supply the whole of a variable load with complete reliability of supply is neither practical, not economical. Hence, it is necessary to decide the size and number of generating sets to meet the load curve demand as clearly as possible. In that case the reserve required would be only one unit of the largest size chosen. But when a large number of generating sets is chosen the operational problem and maintainence cost increase. hence the capital cost of a larger number of sets is more than that of the same capacity of plant in a smaller number of units of larger size. As such there has to be a compromise in the selection of size and number of generator units in the station. Hence, here discussion is briefly made in studying the economic aspect of power plant and its selecting also elements which constitute the cost of power.

18.2 ECONOMIC ASPECTS OF GENERATION

It is very necessary to have a correct idea of the load requirements of the area where electricity is to be made available. Once this has been done for a particular area the next problem is how to get this power. We have seen that there are three important types of stations, namely thermal, hydro-electric and nuclear (besides others such as gas turbines etc). In choosing the type of generation a number of points have to be considered. Some of these are: the kind of fuel available and its cost, availability of suitable sites for a hydro station, the nature of load to be supplied etc. (We have dealt with these points in detail in earlier chapters).

The nature of load to be supplied affects the choice of plant to a considerable extent. The load which a power system has to supply is never constant because of variable demands at different times of the day. The variations can be seen from the predicted load curve. The minimum capacity of generating station must be such as to meet the maximum demand. At the same time it is essential for the power system to maintain reliability and continuity of power supply at all times. The use of a single set to supply the whole of a variable load with complete reliability of supply is neither practical nor economical. An alternative way of deciding the size and number of generating sets is to choose a large number of sets to fit the load curve as closely as possible. In that case the reserve required would be only one unit of the largest size chosen. However when a larger number of generating sets is chosen the operational problems and maintenance cost increase; the capital cost of a larger number of sets is more than that of the same capacity of plant in a smaller number of units of larger size. Therefore there has to be a compromise in the selection of size and number of generator units in the station.

In the present chapter we are interested in studying the economic aspects of power plant and its selection, also elements which constitute the cost of power-the fixed and operating cost, depreciation of plant and other factors affecting economics of generation and distribution of power etc. Costs of supply of electricity to consumers (tariffs) are also considered.

18.2.1 Terms Commonly Used in System Operation

Connected load is the sum of the continuous ratings of the load-consuming apparatus connected to the system.

Firm power is the power intended to be always available (even under emergency conditions).

Cold reserve is that reserve generating capacity which is available for service but is not in operation.

Hot reserve is that reserve generating capacity which is in operation but not in service.

Spinning reserve is that generating capacity which s connected to the bus and is ready to take load.

18.2.2 Factors Affecting the Cost of Generation

When considering the installed capacity of a generating station it is seldom necessary to choose the rating of plant to meet all the load at the same time for all the load connected to the station does not come simultaneously. A term commonly used in this connection is Diversity Factor which is being defined as

$$\text{Diversity Factor} = \frac{\text{Sum of consumer' maximum demands}}{\text{Maximum load on the station}}$$

The effect of diversity factor is to reduce the simultaneous maximum demand on the station for the same individual demands. The effect of diversity is reduced capital cost for the station and consequently a lower overall rate for generation of electricity.

Load Curves

The load on a station is seldom steady; it varies from time to time. The daily variations in load from time to time-hourly or half-hourly are given on load curve which shows graphically the variation of load with respect to time in chronological order. Much useful information can be obtained from a daily load curve *e.g.*

(*i*) The Area under the curve gives the number of units generated in the day.

(*ii*) The ratio of the area under the curve to the total area of rectangle in which it is contained gives the load factor for the day.

(*iii*) The peak of the curve gives the simultaneous maximum demand on the station on that day.

(*iv*) The area under the curve divided by the number of hours gives the average demand on the station.

It may be mentioned that the daily load curve for a station is not the same for all days. It differs from day-to-day and season-to-season. It is in fact usual to draw two typical load curves – one for summer and another for winter – and base the annual output figures for the stations on these curves.

Load Factor

The area under the load curve gives the number of units generated and the total area of the rectangle in which the load curve is contained the total number of units which could have been generated had the maximum demand remained at the highest ordinate of the load cur e throughout this period. The ratio of the number of units actually generated in a given period to the number of units which could have been generated with the same maximum demand is called the **load factor** for the station. Alternatively

$$\text{Load factor} = \frac{\text{Average demand}}{\text{Maximum demand}} = \frac{\text{Energy generated in a given period}}{\text{(Max. demand) (Hours of operation in the given period)}}$$

If the period considered is a day the factor is the daily load factor; if period is a month is the monthly load factor and similarly for the yearly load factor, the period is a year.

The load factor as calculated above corresponds to the supplier and may sometimes be referred to as the undertaking load factor.

Two other factors mentioned frequently are

Note : If the period considered is a day the load factor is the daily load factor, if period is a month it is the monthly load factor and for year, the yearly load factor.

The Load Curves

The load curve shown in Fig. 18.1 bears a great importance in generation and power system. The load curve bears a great importance also for power-tariff engineers who after studying this curve can plan the tariff from economic consideration view-point.

In fact load on the power station is seldom constant. It varies during the whole 24 hours of the use of electricity. The daily variation in load on the system (power station) can be plotted hourly or half-hourly on a graph taking load on Y-axis and time on X-axis. The curve so obtained is called Daily Load Curve. For a typical metropoltan area one such load curve is shown for reference. The load curve gives following information for necessary plan :

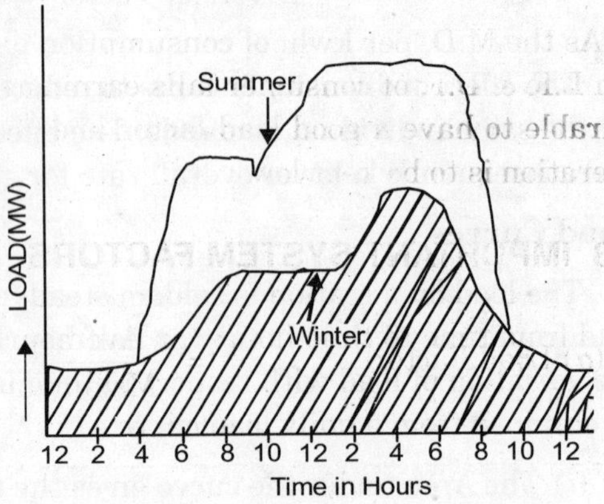

Fig. 18.1

1. The variation of the load during different hours of the day.
2. The area under the curve represent the total number of units generated in a day.
3. The peak of the curve represents the maximum demand on the station on the particular day.
4. The area under the load curve divided by the number of hours represents the average load on the power station.
5. The ratio of the area under the load curve to the total area of the rectanga in which it is contained gives the load factor.

As fixed charges include cost of plants, buildings, transmission and distribution system expenses etc. and Runing charges include whole or major part of fuel, water, wages etc. However it should be emphasised that the fixed charges do not depend upon the number of units generated, where running charges are a function of units generated (or sold).

In case the total cost of electricity supply are to be shared equitably by various consumers each consumer should pay a fixed sum, proportional to the maximum demand which he puts on the station in addition to a sum (of energy consumed) proportional to the units consumed by him. This is the basis of what are called two part tariff. The maximum demand which consumer imposes upon the station is not his actual demand but actual demand. Diversity Factor of the group which is a number. Infact the plant capacity is determined by simultaneous maximum demand which is equal to

$$(\text{Plant Capacity}) = \frac{\text{Arithmatic Sum of M.D.s of all Consumers}}{\text{Diversity Factor for the Feeder}}$$

And,

$$(\text{Diversity Factor}) = \frac{\text{Sum of Consumers Max Demand}}{\text{Max Load at Station}}$$

Hence, it is clear that higher the D.F. less will be Max load at station and more of M.D. and less the Fixed expences the consumer will have to pay.

As the M.D. per kwh. of consumption depends upon load factor it follows that if either or both L.F. & D.F. of consumer falls earning the increase in the cost of energy. Hence it is very desirable to have a good load factor and good diversity factor of the station if overall cost of generation is to be kept low.

18.3 IMPORTANT SYSTEM FACTORS

(*a*) Load Factor $= \dfrac{\text{Average Demand}}{\text{Maximum Demand}} = \dfrac{\text{Energy Generated in a given Period}}{\text{(Max demand) Hours of Operation in the given Period)}}$

(*b*) Diversity Factor $= \dfrac{\text{Sum of Consumers M.D.}}{\text{Max. Load on the Station}}$

1. Firm Power : is the power intended to be always available (even under energy may condition).
2. Cold Reserve : is the reserve generating capacity which is available for service but it is not in operation.
3. Hot Reserve : is the reserve generating capacity which in operation but is not in service.
4. Spinning Reserve : is the generating capacity which is connected to the bus and is ready to take load.

(*c*) Plant Capacity Factor $= \dfrac{\text{Actual Energy Produced}}{\text{Max. Possible Energy that Could have been Produced (installed plant capacity)}}$

$= \dfrac{\text{Average Demand}}{\text{Installed capacity}}$ (during a given period)

(*d*) Plant Use Factor $= \dfrac{\text{Actual Energy Produced}}{\text{Plant Capacity} \times \text{Time (hours) the plant inoperation}}$

(*i*) Plant Capacity Factor $= \dfrac{\text{Actual energy produced}}{\text{Maximum possible energy that could have been produced (based on installed plant capacity)}}$

(*ii*) Plant Use factor $= \dfrac{\text{Acual energy produced}}{\text{Plant capacity} \times \text{Time (hours) the plant has been in operation}}$

We may make two important observations by saying that ideal conditions for cheap electric supply exist when.

(*i*) the installed capacity and hence the corresponding capital cost of the generating plant is kept low (this can be achieved by a good diversity factor) and (*ii*) the daily output of each generator unit is large (which means a good load factor for the station). The reduction in costs with good load factor is due to the fact that overall working cost per unit becomes low, the fixed charges having been distributed over more units of energy generated.

18.3.1 Reduction of Costs by Interconnection of Stations

For simplicity we may assume no spare plant for the generating station but in an isolated station the practice is to provide reserve plant (usually a generator equal to the largest size in the station) to facilitate maintenance and repairs and to meet demand of energy when a unit has a forced outage.

(*i*) If every station worked in isolation the total reserve plant requirement of all stations which is normally (or for most part of it) lying idle would amount to considerable capacity. With an interconnected system the total reserve capacity is considerably reduced.

(*ii*) In an interconnected system it is possible to have larger generator ratings which means reduced capital cost per kW.

(*iii*) It is possible to arrange the sharing of load between two or more interconnected stations in such a manner that the more efficient station can work at a high load factor the inefficient station being made use of only for peak load hours.

(*iv*) It is very unlikely that load curves of two different stations or systems would be identical. (In an extreme case the two peaks may occur at times differing by a few minutes.) On account of maximum demands of individual stations not occurring simultaneously it is possible to work with lesser installed capacity with integrated working of the stations and thus effect a saving in the installed capacity.

(Interconnection of stations has to be considered on its own merits for it involves expense in providing tie lines between stations and at the same time increased switchgear ratings to be employed at different points of the system.)

18.3.2 Curves Useful in System Operation

Chronological curve (*load curve*) is plotted by placing the ordinates (kW) in their proper time sequence.

Load duration curve represents the load vs time with the ordinates rearranged in magnitude sequence *i.e.* with the greatest load on the left, lesser loads towards the right and the least load at the extreme right. Suppose we consider a daily load duration curve of a system. The absciassa corresponding to any kW ordinate of this curve will give the number of hours during which that amount of load (kW) has been on the plant. The area of either of these two curves gives the total units for the period considered.

Integrated duration curve shows the total energy included in all loads upto a given maximum demand, and plotted as percentage quantity, the curve is also known as *Peak Percentage Curve*.

It is known that the area under a load duration curve gives the total kWh.

Integrated duration curve can therefore be derived from the load duration curve remembering that the abscissa corresponding each ordinate should be equal to the area of the duration curve upto the value of that ordinate.

Such curves are useful in that with a given number of kWh available. The kW of the load can be the base or peak load. Use of such curves lies in studying the variations between the rate of water inflow in hydel station and that of the electrical load.

18.3.3 Choice of Size and Number of Generator Units

The study of load curves of a power system reveals that the load on a power station is never constant and that it varies at different times of the day. The peak load occurs for only a short duration. The capacity of the generator units must be selected so as to meet this peak load. The minimum number of generating units chosen could be one having a capacity equal to the maximum demand on the system. The drawbacks in selecting a single unit to meet maximum demand are :

1. As the load on the system is variable and the load factor is less than 100% there is a considerable period during the day when the load on the station would be much less than the maximum load. During this period the generating set may be running at half the load or even practically on no load. Hence it is not running at all times under conditions best suited for its operation to give maximum efficiency. It is also uneconomical to run the set under low load condition as fuel consumption would be more.

2. Reliability of service is a very important factor. Cheaper power without reliability of supply is a costlier power. If it is an isolated individual plant and the requirement of reliability and continuity of supply is to be satisfied another set of equal capacity must be installed for use when the first set is out of order and is under repairs. This results in increased capital cost. Thus it is neither practical nor economical to use a single unit to supply the whole of the variable load where reliability of supply is essential.

The alternative way of deciding the size and number of generator units is to choose the number of sets in such a way that they fit the load curve as closely as possible. Then each unit can be made to operate in such a manner that it runs almost at full load or at a load at which it gives maximum efficiency. The reserve required in this case would only be one unit of largest size chosen and would be much smaller than the maximum reserve capacity required in the first case. As the reserve capacity required would be lesser it would result in better plant capacity factor and plant use factor. This type of selection however involves the following difficulties :

(*i*) With increased number of sets the floor area required increases and so also the building cost.

(*ii*) Maintenance cost increases.

(*iii*) More sets involve more starting, stopping and parallel operation of the equipments. This also results in increased personnel for handling the equipment thus increasing the cost.

(*iv*) Capital cost of the larger number of sets of lower sizes is greater than that of the same capacity of plant in a smaller number of units of larger size.

A study of above factors shows that there should neither be one unit nor a large number of smaller units of different sizes. There has to be a compromise in the selection of units. The aim should be to have a small number of units and to fit them as properly on the load curve. The choice is also governed by the best compromise between plant capacity factor and plant use factor.

18.4 TARIFFS

We know that the rating of plant in a generating or receiving station must be equal to the simultaneous maximum demand which it has to meet (assuming no spares). This means that a fixed sum proportional to the demand to be met has to be invested for the plant required.

Also fuel and maintenance expenses, which vary almost in direct proportion to the units of energy generated, have to be incurred. The first type of expenses are called **Fixed Charges** and include the cost of plant, buildings, transmission and distribution expenses, part of staff salaries etc. while the second type of expenses called Running Charges include whole or major part of fuel, oil, water and wages etc. It may be emphasized that the fixed charges do not depend upon the number of units generated while running charges are a function of units generated (or sold).

If the total costs of electricity supply are to be shared equitably by various consumers each consumer should pay a fixed sum proportional to the maximum demand which he puts on the station in addition to a sum (for energy consumed) proportional to the units consumed by him. This is the basis of what are called *Two Part Tariffs*. In some cases expenses of a station may be worked out in a form Rs. $(A + B.\ \text{kW} + C\ \text{kWh})$ where A, B and C are constants. In this case B is fixed cost and can be explained as (say) salaries of managing agents, taxation and, in some cases, cost of land etc. which are assumed to be fixed expenses irrespective of plant or station size. It is however more usual to have the total expenses in a two-part form-a fixed sum per kW of maximum demand plus a running charge per unit of energy. We can say that in this case constant A of the above expression has been merged in the fixed sum per kW or kVA of M.D.

It has been mentioned above that each consumer should pay a fixed sum based upon his maximum demand. The fixed expenses of a station are determined by the size of the station (or in other words the simultaneous maximum demand which it has to meet) and not by the arithmetical sum of maximum demand of various consumers. The maximum demand which the consumer imposes upon the station is not his actual demand but actual demand/diversity factor of the group of which he is a member. A second thought over the matter will confirm this statement for, as said before, the plant capacity is determined by simultaneous maximum demand which is equal to $\dfrac{\text{Arithmetic sum of M.Ds of all consumers}}{\text{Diversity Factor for the feeders}}$

Where, as stated, $\text{Diversity Factor} = \dfrac{\text{Sum of consumer M.D.}}{\text{Maximum load at Power Station}}$

$$\text{Load Factor} = \frac{\text{Average Demand}}{\text{Maximum Demand}}$$

It is clear that higher the diversity the less will each kW of consumer's M.D. increase the combined maximum and less the fixed expenses he will have to pay.

As the maximum demand per kWh of consumption depends upon the load factor it follows that as either or both the load factor and the diversity factor of a consumer increase the cost of energy per kWh falls.

(Thus it is very desirable to have good load factor and good diversity factor for the station if overall cost of generation is to be kept low.)

Flat Rate Tariffs

If we can estimate the load factor and diversity factor for various uses of electricity (*e.g.* lighting, industrial etc.) we may merge the equitable contribution to the fixed charges of a particular class of consumers with each kWh consumption. The resultant tariff is a flat rate tariff. This flat rate generally varies according to the manner in which the supply is used and different rates are applicable for different uses. This type of tariff has the advantage of being easily understood by consumers but the proper estimation of load and diversity factors is not so simple. Besides separate metres are required for different classes of supply.

Two Part Tariffs

The toal charge under this kind of tariff is split into two components-a fixed charge dependent upon the maximum demand (this maximum demand is estimated on the rateable value of the premises or may be assessed on the basis of connected load or in any similar manner) and a variable charge based on the energy consumption.

$$\text{Load Factor} = \frac{\text{Average demand}}{\text{Maximum demand}} = \left(\frac{\text{Average Demand}}{\text{Sum of M.D.s}}\right)$$

$$\text{Diversity Factor} = \frac{\text{Sum of M.Ds}}{\text{M.D. at power station}}$$

$$\therefore \quad \text{M.D. at Station} = \frac{\text{Sum of M.D.}}{\text{Diversity}}$$

The disadvantage of this tariff from viewpoint of consumer is that he has to pay his fixed charge irrespective of consumption. For example a person may be making sparing use of electricity due to, say, his stay in his premises for a limited period in the whole year in which case he will have to pay a disproportionately high overall price per unit of energy consumption.

Block Rate Tariffs

Under this tariff the fixed charge is merged into the unit charge for one or two blocks of consumption, all units in excess being charged for at low unit rate. The size of each block in terms of units of consumption is fixed by the size of the house, higher block sizes being made applicable to bigger size of house (rateble value or bigger connected load). A typical tariff under this head may be-say for lighting load (connected load upto 1 kW).

First 15 units per month at 30 paise per unit

Next 25 units per month at 15 paise per unit

All additional units per month at 8 paise per unit

This kind of tariff meets the objection of two-part tariff.

Maximum Demand Tariffs

These tariffs are similar to two-part tariffs and differ only in the sense that the maximum demand is actually measured by a demand indicator instead of assessing it merely on the basis of rateable value.

Power Factor Tariffs

In a.c. systems the size of plant is determined not only by the kW capacity required but also by the power factor at which the plant is to work. Power factor tariffs are devised to make a distinction between overall charge per unit to be recovered from two types of consumers, one having a good power factor and the other a poor power factor. Three classes are:

***(i)* kVA Maximum Demand Tariffss :** Instead of measuring the maximum demand in kW it may be measured in kVA and a demand charge per kVA may be recovered in addition to a charge proportional to the kWh consumption.

***(ii)* Sliding Scale or Average Power Factor Tariffs :** In this case an average p.f., say 0.80 lagging, may be assumed as reference and a surcharge for each 0.01 by which the average p.f., falls below this figure may be made. Conversely, if the average p.f. is above this reference a discount may be allowed in a similar manner.

***(iii)* KWh and kVARh Tariffss :** Under this kind of tariff both active (kWh) and reactive (kVARh) units are separately charged, the latter usually as a fraction of the kWh rate. A low p.f. means more kVARh component for each kWh and consequently more overall charge per unit of energy consumed.

EXAMPLES OF TARIFF

Example 18.1. *A generating station has a connected load of 450 mw and a maximum demand of 250 mw; the units generated being 615 × 10^6 per anuum. Calculate (1) the demand factor and (2) load factor.*

Solution : 1. Demand Factor $= \dfrac{\text{Maximum Demand}}{\text{Connected Load}} = \dfrac{250}{450} = 0.556$

2. Average Demand $= \dfrac{\text{Units Generated/annum.}}{\text{Hours is a Year}} = \dfrac{615\times10^6}{8760} = 70{,}205.5$ kw

$\therefore$ Load Factor $= \dfrac{\text{Average Demand}}{\text{Maximum Demand}} = \dfrac{70205.5}{250\times10^3} = 0.281$

That is load factor = 28.1%

Example 18.2. *The Maximum Demand on a steam power station is 480 mw. if the annual load factor is 40%, calculate the total energy generated annually.*

Solution : Energy generated/annum = Maximum demand × Load Factor × Hours in a year

$$= (480\times10^3)\times(0.4)\times(24\times365) = 16{,}819.2\times10^5 \text{ kwh}$$

Example 18.3. *A generating station has a maximum demand of 500 mw. The annual load factor is 50% and capacity factor is 40%. Find the reserve capacity of the plant.*

Solution : Energy generated/anuum = Maximum demand × Load Factor × Hours in a year.

$$= \left(\underset{\text{M.D.}}{500}\right)\times\left(\underset{\text{L.F.}}{0.5}\right)\times(24\times365) = 2.19\times10^6 \textbf{ mwh.}$$

$$\text{Capacity Factor} = \frac{\text{Units generated/anuum}}{\text{Plant capacity}\times\text{Hours in a year}}$$

$$\therefore \quad \text{Plant Capacity} = \frac{\text{Units generated per annum}}{\text{Capacity factor}\times\text{Hours in a year}}$$

$$= \frac{2.19\times10^6}{0.4\times8760} = 625 \text{ mw}$$

$$\therefore \quad \text{Reserve Capacity} = \text{Plant Capacity} - \text{Maximum Demand}$$

$$= 625 - 500 = 125 \text{ mw}$$

Example 18.4. *An electric supply system has a maximum load of 70 mw. annual expenses of the system are :*

1. *Generation – Rs. 8,50,000*
2. *Fuel – Rs. 28,00,000*
3. *Transmission Rs. 3,45,000*
4. *Distribution – Rs. 27,50,000*
5. *Repairs etc. Rs. 3,00,000*

The number of units generated per year are 600 × 10^6 kwh. The consumers have an aggregate maximum demand of 75 mw. Evaluate a two part tariff to be charged from the consumers. Assume that the fixed charges of generation, find, transmission, distribution, repairs etc. are 90%, 15% 85%, 95% and 50% respectively. Losses in transmission and distribution are 20%.

Solution : In order to evaluate the two part tariff, it is necessary to find out the fixed and running charges separately, which are tabulated as follows :

Sl. No.	Item	Total charges	Fixed charges		Running charges	
			Percentage	Value (Rs.)	Percentage	Value (Rs.)
1.	Generation	–8,50,000	90%	765,000	10% (100–90)	85,000
2.	Fuel	–28,00,000	15%	420,000	85%	23,80,000
3.	Transmission	–3,45,000	85%	293,250	15%	51750
4.	Distribution	–27,50,000	95%	261,2500	5%	13,7500
5.	Repairs	–3,00,000	50%	150,000	50%	150,000
		Total		4240750		280,4250

Maximum Demand $= 75 \times 10^3$ kw

Total fixed charges = Rs. 42,40750

∴ Fixed charges per unit $= \dfrac{42,40,750}{75 \times 10^3} =$ Rs. 56,544

Total number of units distributed $= 600 \times 10^6 \times 0.8$

(as losses in T & D = 20%)

Actual Bower distributor = (100 – 20) × Total Units generated

(after generation) = **48 × 10^7 kwh.**

Total running charges shown in

the above table = 2,80,4250 = Rs. **2,8043 × 10^6**

∴ **Running Charges** $=$ Rs.$\dfrac{2,8043 \times 10^6}{48 \times 10^7} \times 100 = 0.584$ paise

∴ Hence two part tariff is Rs. 56,544 per kw (Fixed change) pens 0.584 **paise per** kwh (Running charges).

Example 18.5. *A generating station has a maximum demand 500 mw the annual load factor is 50% and capacity factor is 40%. Find the reserve capacity of the plant.*

Solution : Energy generated/annum

$$= \text{Maximum Demand} \times \text{Load Factor}$$

$$= \left(\text{Max Demand} \times \left(\frac{\text{Average Demand}}{\text{Maximum Demand}}\right) \times \text{Hours in a year}\right)$$

$$= 500 \times (0.5) \times 8760 \text{ mwh} = 2.19 \times 10^6 \text{ mwh}$$

$$\text{Capacity Factor} = \frac{\text{Units Generator/Annum}}{\text{Plant Capacity} \times \text{Hours is a Year}}$$

$$\therefore \text{Plant capacity} = \frac{2.19 \times 10^6}{0.4 \times 8760} = 6750 \text{ mw}$$

$$\therefore \text{Reserve capacity} = \text{Plant capacity} - \text{Max. Demand}$$

$$= 625 - 500 = \mathbf{125 \text{ mw}}$$

Example 18.6. *A generating station supplied the following loads : 150 mw 120 mw, 85 mw, 60 mw and 5 mw. The annual load factor of the station is 48%. Calculate (a) Number of units supplied annually (b) the demand factor. The station has maximum Demand =* ***220 mw***

Solution : Load Factor $= \dfrac{\text{Number of Units Supplied Annually}}{\text{Max. Generating Capacity} \times 8760}$

(*a*) Load Factor $= 0.48$

Number of Units Supplied Annually $=$ L.F. × Maximum Generating top × 8760

$$= 0.48 \times 220 \times 8760 = 9.25 \times 10^6 \text{ units}$$

$$(b) \text{ Diversity Factor} = \frac{\text{Sum of Max. Demands of Individual Consumers}}{\text{Station Max. Demand}}$$

$$= \frac{150 + 120 + 85 + 60 + 5}{220} = \frac{420}{220} = \mathbf{1.909}$$

$$(c) \text{ Station Demand Factor} = \frac{\text{Maximum Demand}}{\text{Connected Load}} = \frac{\text{Maximum Demand}}{\text{Total Load Requirement of each Consumer}} = \frac{220}{420}$$

$$= .524 \text{ or } \mathbf{52.4}$$

Example 18.7. *A base load station having a capacity of 400 mw and a stand by station having a capacity of* ***50 mw*** *share a common load. Find the load factor and capacity factors of two power stations from the following data.*

1. *Annual stand by station output = 87.35 × 10^6 kwh.*
2. *Annual base load station output = 101.0 × 10^6 kwh.*
3. *Peak load on stand by station = 12.0 mw.*
4. *Hour of use of stand by station / year = 3000 hrs.*

Solution :

1. Stand by station

Annual Load Factor $= \dfrac{\text{kwh generator/Annum}}{\text{Max. Demand} \times \text{Annual working hours in use}}$

$= \dfrac{87.35 \times 10^6 \times 100}{(120 \times 10^3) \times 3000} = 24.26\%$

Annual Capacity Factor $= \dfrac{\text{kwh Output/annum}}{\text{Installed Capacity} \times \text{Hours in a Year of Stand by Stand}}$

kwh generated/annum

= kwh output/annum $= \dfrac{87.35 \times 10^6 \times 100}{50 \times 10^3 \times 8760} = 19.90\%$

50 mw = **50 × 10³**

2. Base Station

Annual Load Factor $= \dfrac{\text{kwh generated/annum, by base}}{\text{Max. Demand} \times \text{Annual working hour in use}}$

$= \dfrac{1010 \times 10^6}{(120 \times 10^3) \times 3000}$

Annual Capacity Factor $= \dfrac{101 \times 10^6 \times 100}{400 \times 10^3 \times 876}$

Example 18.8. *A generating station supplies the following loads to various consumers :*

Industrial consumer = 750 mw, Commerical establishment = 350 mw

Domestic Power = 10 mw, Domestic light for = 50 mw

*if the maximum demand on the station is 1000 mw and the number of kwh generated per year is **50 × 10⁵** determine.*

1. The Diversity Factor, 2. Annual Load Factor

Solution : 1. Diversity Factor $= \dfrac{750 + 350 + 10 + 50}{1000} = \dfrac{\text{total Load Connect}}{\text{Max. Demand}}$

= **1.16**

2. Average Demand $= \dfrac{\text{mwh generated/annum}}{\text{Hours in a Year}}$

$= \dfrac{50 \times 10^6}{8760} = \mathbf{570.78\ mw}$

Load Factor $= \dfrac{\text{Average Load}}{\text{Max. Demand}} = \dfrac{570.78}{1000} = .57078 = \mathbf{57\%}$

Example 18.9. *A power supply is having the following loads :*

Type of Load	*Max. Demand (kw)*	*Diversity of Factor of Group*	*Demand Factor*
Domestic	*15,000*	*1.25*	*0.7*
Commercial	*25,000*	*1.2*	*0.9*
Industrial	*50,000*	*1.3*	*0.98*

If the overall system diversity factor is 1.5, determine (a) The maximum demand, (b) Connected load of each type.

Solution : (*a*) The sum of maximum demands of the three loads

$$= 15,000 + 25,000 + 50,000 = 90,000 \text{ kw}$$

Since the system diversity factor is 1.5

$$= \left(\frac{\text{Sum of total maximum loads D.F.}}{\text{Maximum Demand}}\right)$$

∴ Maximum Demand on Supply System $= \dfrac{\text{Sum of the Total Max. Load}}{\text{Diversity Factor}} = \dfrac{90,000}{1.5} = \mathbf{60,000\ mW}$

(*b*) Each type of load has its own diversity factor among its consumers

Max. Domestic Load Demand $=$ Max. Domestic Demand × Diversity Factor

$$= 15000 \times 1.25 = 18,750 \text{ kw}$$

∴ Connected Domestic Load $= \dfrac{\text{Max. Domestic Load Demand}}{\text{Demand Factors}} = \dfrac{18,750}{0.7} = \mathbf{26785.7\ kW}$

Conn. Comercial Load $= \dfrac{25000 \times 1.2}{0.9} = 33333.3$ kw

$$= \frac{\text{Max. Demand} \times \text{Diversity Factor}}{\text{Demand Factor}}$$

Example 18.10. *A generating station has a connected load of 450 MW and a maximum demand of 250 MW: the units generated being 615 × 10⁶ per annum. Calculate (i) the demand factor and (ii) load factor.*

Solution : (*i*) Demand Factor $= \dfrac{\text{Maximum Demand}}{\text{Connected Load}} = \dfrac{250}{450} = 0.556$

(*ii*) Average Demand $= \dfrac{\text{Units Generated/annum.}}{\text{Hours is a Year}} = \dfrac{615 \times 10^6}{8760} = \mathbf{70,205.5\ kW}$

∴ Load Factor $= \dfrac{\text{Average Demand}}{\text{Maximum Demand}} = \dfrac{70205.5}{250 \times 10^3} = 0.281$

Example 18.11. *The maximum demand on a steam power station is 480 MW. If the annual load factor is 40%, calculate the total energy generated annually.*

Solution : Energy generated/annum = Maximum demand × L.F. × Hours in a year

$$= (480 \times 10^3) \times (0.4) \times (24 \times 365) \text{ kWh}$$

$$= 16,819.2 \times 10^5 \text{ kWh}$$

Example 18.12. *A base load station having a capacity of 400 MW and a standby station having a capacity of 50 MW share a common load. Find the annual load factors and capacity factors of two power stations from the following data.*

Annual standby station output $= 87.35 \times 10^6$ *kWh*

Annual base load station output $= 101.0 \times 10^6$ *kWh*

Peak load on standby station $=$ *120 MW*

Hours of use by standby station/year $=$ *3000 hrs.*

Solution : Installed capacity of standby = 50 MW = 50 × 10^3 kW

Installed capacity of base load plant = 400 MW = 400 × 10^3 kW

(*i*) Standby station

$$\text{Annual load factor} = \frac{\text{kWh generated/annum}}{\text{Maximum demand} \times \text{Annual working hours}} \times 100$$

$$= \frac{87.35 \times 10^6}{(120 \times 10^3) \times 3000} \times 100 = 24.26\%$$

$$\text{Annual capacity factor} = \frac{\text{kWh output/annum}}{\text{Installed capacity} \times \text{Hours in a year}} \times 100$$

$$= \frac{87.35 \times 10^6}{50 \times 10^3 \times 8760} \times 100 = 19.94\%$$

In a similar way, annual load factor and annual capacity factor can be calculated for the other station.

Example 18.13. *A generating station has a maximum demand of 500 MW. The annual load factor is 50% and capacity factor is 40%. Find the reserve capacity of the plant.*

Solution : Energy generated/annum = Maximum demand × L.F. × Hours in a year

= (500) × (0.5) × (8760) MWh = 2.19 × 10^6 MWh

$$\text{Capacity factor} = \frac{\text{Units generated/annum}}{\text{Plant capacity} \times \text{Hours in a year}}$$

$$\therefore \text{Plant capacity} = \frac{2.19 \times 10^6}{0.4 \times 8760} = 625 \text{ MW}$$

Reserve capacity = Plant capacity – max. demand = 625 – 500 = 125 MW

Example 18.14. *A generating station supplied the following loads: 150 MW, 120 MW, 85 MW, 60 MW and 5 MW. The station has a maximum demand of 220 MW. The annual factor of the station is 48 percent. Calculate (a) the number of units supplied annually (b) the diversity factor, and (c) the demand factor.*

Solution : $$\text{Load factor} = \frac{\text{Number of units supplied annually}}{\text{Maximum generating capacity} \times 8760}$$

$$0.48 = \frac{\text{Number of units supplied annually}}{220 \times 8760}$$

∴ (*a*) Number of units supplied annually = 0.48 × 220 × 8760 = 9.25 × 10^5 units

(*b*) Diversity factor

$$= \frac{\text{Sum of maximum demands of individual consumers}}{\text{Station maximum demand}}$$

$$= \frac{150 + 120 + 85 + 60 + 5}{220} = \frac{420}{220} = \mathbf{1.909}$$

(*c*) Station demand factor

$$= \frac{\text{Maximum demand}}{\text{Connected load}} = \frac{\text{Maximum demand}}{\text{Load requirement of each consumer}} = \frac{220}{420} = 0.524 \text{ or } 52.4\%$$

Example 18.15. *A 1000 MW power station delivers 1000 MW for 2 hours, 500 MW for 6 hours and is shut down for the rest of each day. It is also shut down for maintenance for 60 days annually. Calculate its annual load factor.*

Solution : Energy supplied for each working day

$$= (1000 \times 2) = (500 \times 6) = 5000 \text{ MWh}$$

Station operates for $= 365 - 60 = 305$ days in a year

$\therefore$ Energy supplied/year $= 5000 \times 305 = 15,2,000$ MWh

$$\text{Annual load factor} = \frac{\text{MWh supplied per annum}}{\text{Max. gen. capacity} \times \text{Year of working in ave.}} \times 100$$

$$= \frac{15,25,000}{(1000) \times (305 \times 24)} \times 100 = 20.8\%$$

Example 18.16. *A generating station supplies the following loads to various consumers :*

Industrial consumer = 750 MW; Commercial establishment = 350 MW

Domestic power = 10 MW; Domestic light = 50 MW

If the maximum demand on the station is 1000 MW and the number of kWh generated per year is 50 × 10^5, determine (i) the diversity factor and (ii) annual load factor.

Solution : (*i*) Diversity factor

$$= \frac{750 + 350 + 10 + 50}{1000} = \mathbf{1.16}$$

(*ii*) Average demand

$$= \frac{\text{MWh generated/annum}}{\text{Hours in a year}}$$

$$= \frac{50 \times 10^5}{8760} = 570.78 \text{ MW}$$

$$\therefore \text{Load factor} = \frac{\text{Average load}}{\text{Max. demand}} = \frac{570.78}{1000} = 0.57078 = 57\% \text{ (app.)}$$

Example 18.17. *A power supply is having the following loads:*

Type of load	*Maximum demand (kW)*	*Diversity of group*	*Demand factor*
Domestic	*15,000*	*1.25*	*0.7*
Commercial	*25,000*	*1.2*	*0.9*
Industrial	*50,000*	*1.3*	*0.98*

If the overall system diversity factor is 1.5, determine (a) the maximum demand and (b) connected load of each type.

Solution : (*a*) The sum of maximum demand of three types of loads

$$= 15000 + 25000 + 50000 = 90000 \text{ kW}$$

Since the system diversity factor is 1.5

$$\therefore \text{Maximum demand on supply system} = \frac{90000}{1.5} = 60000 \text{ kW}$$

(*b*) Each type of load has its own diversity factor among its consumer

Maximum domestic load demand = Maximum domestic demand × diversity factor

$= 15000 \times 1.25 = 18750$ kW

$\therefore$ Connected domestic load $= \dfrac{18750}{0.7} = 26785.7$ kW $\left[\therefore \text{Conducted load} = \dfrac{\text{Max. demand}}{\text{Demand factor}}\right]$

Similarly connected commercial $= \dfrac{25000 \times 1.2}{0.9} = 33333.3$ kW

load and connected industrial load $= 50000 \times \dfrac{1.3}{0.98} = 66326.531$ kW

Hence connected domestic load = 26785.7 kW

Connected connercial load = 33333.3 kW

Connected industrial load = 66326.5 kW

Example 18.18. *A generating station has a maximum demand of 10 MW and the daily load on the station is as follows :*

kW	*Time*
11 PM – 6 AM	*2000*
6 AM – 8 AM	*3500*
8 AM – 12 Noon	*8000*
12 Noon – 1 PM	*3000*
1 PM – 5 PM	*7500*
5 PM – 7 PM	*8500*
7 PM – 9 PM	*10000*
9 PM – 11 PM	*4500*

Choose the size and number of generator units. What reserve plant would be necessary? Find the load factor, plant factor and plant use factor of the station. Also draw up the operating schedule by drawing a load curve.

Solution : Fig. E1.1 shows the load curve plotted from given data. Ma. Demand = 10,000 kW.

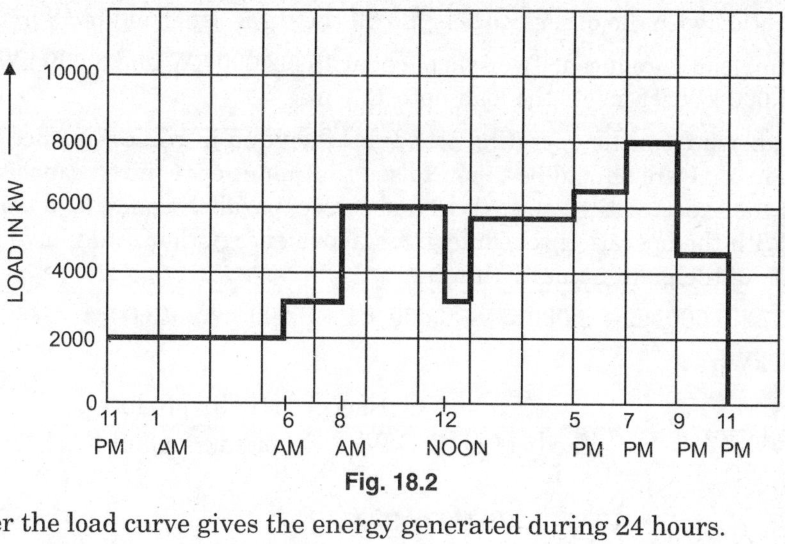

Fig. 18.2

Area under the load curve gives the energy generated during 24 hours.

Therefore, energy generated in 24 hours

= 2,000 × 7 + 3,500 × 2 + 8,000 × 4 + 3,000 × 1 + 7,500 × 2 + 10,000 × 2 + 4,500 × 2

= 132,000 Kwh

$$\text{Load Factor} = \frac{\text{Energy generated during 24 hrs}}{\text{Maximum Demand} \times \text{24 hrs}} = \frac{132,000}{10,000 \times 24} = 0.55(\text{or } 55\%)$$

From the nature of the load curve it can be seen that three generating sets of capacities 5,000, 3,000 and 2,000 kW will suffice and can be best fitted into the load curve as explained in the operating schedule.

Reserve capacity = Largest size of unit in the station

Therefore, reserve capacity = 5,000 kW

and installed capacity = 5,5000 + 3,000 + 2,000 + 5,000 = 15,000 kW

$$\text{Plant Capacity Factor} = \frac{\text{Energy produced during 24 hrs}}{\text{Installed capacity (kW)} \times \text{hrs}} = \frac{132,000}{15,000 \times 24} = 0.3667 \text{ or } 36.67\%$$

Operating Schedule

1. From 11 p.m. to 6 a.m. only 2,000 kW set is run.
2. At 6 a.m. the load increases and therefore 3,000 kW is put in operation in parallel with 2, kW set to meet this increased load.
3. At 8 a.m. the load further increases, 5,000 kW set is started and put in parallel with other two sets already operating. All the load is transferred to 5,000 and 3,000 kW seets and the the 2,000 kW set is stopped.
4. At 12 noon the load decreases and 5,000 kW set is taken out of service. Only 3,000 kW set takes the load.
5. At 1 p.m. the load is expected to increase. 5,000 kW set is started and put in parallel with 3,000 kW set.
6. Between 5 p.m. and 7 p.m. the load further increases. 2,000 kW set is started and put in operation with the other two sets. So all the three sets run upto 9 p.m.
7. At 9 p.m. load is reduced. Two sets of capacities 2,000 kW and 3,000 kW are taken out. Only 5,000 kW set meets the load upto 11 p.m.

Thus we known from the operating schedule when and in what sequence the sets should be started and run. Care should be taken to see that plant of required capacity is kept ready for loading at the expected time of the load and correct paralleling and load transfer operation takes place. With the operating schedule given above energy that could have been generated by the capacity of the plant actually running

= 2,000 × 7 + 5,000 × 2 + 8,000 × 4 + 3,000 × 1 + 8,000 × 4 + 10,000 × 4 + 5,000 × 2

= 141,000 kWh

$$\text{Plant use factor} = \frac{\text{Energy actually produced}}{\text{Capacity of plant (kW)} \times \text{No. of hours plant has been in operation}}$$

$$= \frac{132,000}{141,000} = 0.936 \text{ or } 93.6\%$$

Example 18.19. *A power station has an installed capacity of 210 MW. The capital cost of station is Rs. 1000/MW. The fixed cost is 13% of the cost of investment. On full load at 100% load factor the variable cost of the station per year is 1.3 times the fixed cost. Assume no reserve capacity and variable cost to be proportional to the energy produced, find the cost of generation per kWh at load factor of 100% and 50% Comment on the results. [The cost figure is indicative only].*

Solution : Capital cost of station

$$= 1000 \times 210\ 10^3 = 2100 \times 10^5$$

(*i*) At 100% load factor

Fixed cost Rs. $2100 \times 10^5 \times \frac{13}{100}$ = Rs. $1.3 \times 273 \times 10^5$

Variable cost = Rs. $1.3 \times 273 \times 10^5$ = Rs. 354.9×10^5

Total operating cost/annum = Rs. $(273 + 354.9) \times 10^5$ = Rs. 627.9×10^5

Total units generated = load factor × Maximum demand × 8760

$= 1 \times 210 \times 10^3 \times 8760$ kWh $= 1837.5\ 5\ 10^6$ kWh

Cost of generation/kWh $= \dfrac{627.9 \times 10^5 \times 100}{1837.5 \times 10^6} = 3.42$ Paise

(*ii*) At 50% load factor

Fixed cost = Rs. 273×10^5

∴ Total units generated $= 0.5 \times 210 \times 10^3 \times 8760$ units $= 919.8 \times 10^6$ units

As variable cost is proportional to kWh generated, variable cost at 50% load factor

$$= \text{Rs. } 354.9 \times 10^5 \times \frac{919.8 \times 10^6}{1837.5 \times 10^6} = \text{Rs. } 177.65 \times 10^5$$

Total operating cost = Rs. $(273 + 177.65) \times 10^5$ = Rs. 450.65×10^5

Cost of generation/unit $= \dfrac{450.65 \times 10^5 \times 100}{919.8 \times 10^6} = 4.899$ or 4.9 paise

Comment: As the load factor is reduced, cost of generation is increased.

Example 18.20. *A generating station has a maximum demand of 100 MW. Calculate the cost per unit generated from the following data :*

Capital cost = *Rs.* 200×10^6
Annual load factor = *40%*
Annual cost of fuel and oil = *Rs.* 15×10^6
Taxes, wages and salares etc. = *Rs.* 10×10^6
Interest and depreciation = *15%*

Solution : Units generated/annum = Maximum demand × L.F. × Hours in a year

$= 100 \times 10^3 \times (0.4) \times 8760$ kWh = Rs. 35.04×10^7 kWh

Annual fixed charges: Annual interest and depreciation = 15%

= Rs. $0.15 \times 200 \times 10^6$ = Rs. 30×10^6

Again annual running charges = Annual cost of fuel and oil + Taxes, wages etc.

$= \text{Rs. } (15 \times 10^6 + 10 \times 10^6) = \text{Rs. } 25 \times 10^6$

Total annual charges $= \text{Rs. } (30 \times 10^6 + 25 \times 10^6) = \text{Rs. } 55 \times 10^6$

$\therefore$ Cost per unit $= \text{Rs. } \dfrac{55\times10^6}{35.04\times10^7} = \text{Rs. } 0.16 = 16 \text{ paise}$

Example 18.21. *A generating station has the following data. Installed capacity = 500 MW; Capacity factor = 45% Annual load factor = 60%; Annual cost of fuel, oil etc. = Rs. 10 × 10⁷; Capital cost = Rs. 10⁹; Annual interest and depreciation = 15% Calculate (i) the minimum reserve capacity of the station and (ii) the cost per kWh generated.*

Solution : Capacity factor (C.F.) $= \dfrac{\text{Average demand}}{\text{Installed capacity}}$ ...(a)

Load factor (L.F.) $= \dfrac{\text{Average demand}}{\text{Maximum demand}}$ ...(b)

Dividing (a) by (b), we get, $\dfrac{\text{C.F.}}{\text{L.F.}} = \dfrac{\text{Maximum demand}}{\text{Installed capacity}}$

or Maximum demand = Installed capacity × C.F./L.F. = 500 × 0.45/0.6 = 375 MW

$\therefore$ Reserve capacity = 500 – 375 = 125 MW

(*ii*) Units generated/annum = Maximum demand × L.F. × Hours in a year

$= (375 \times 10^3) \times (0.6) \times 8760 = 1971 \times 10^6$ kWh

Annual fixed charges = Annual depreciation and interest

$= \text{Rs. } 0.15 \times 10^9 = \text{Rs. } 1.5 \times 10^8$

Annual running charges $= \text{Rs. } 10 \times 10^7$

$\therefore$ Total annual charges $= \text{Rs. } (1.5 \times 10^8 + 10 \times 10^7) = \text{Rs. } 25 \times 10^7$

$\therefore$ Cost per kWh $= \text{Rs. } \dfrac{25\times10^7}{1977\times10^6} = \text{Rs. } 0.126 = 13 \text{ paise}$

Example 18.22. A electric supply system has a maximum load of 70 MW. The annual expenses of the system are :

(i) Generation Rs. 8,50,000

(ii) Fuel Rs. 28,00,000

(iii) Transmission Rs. 3,45,000

(iv) Distribution Rs. 27,50,000

(v) Reparis etc. Rs. 3,00,000

The number of units generated per year are 600 × 10⁶ kWh. The consumers have an aggregate maximum demand of 75 MW. Evaluate a two part tariff to be charged from the consumers. Assume that the fixed charges for generation, fuel, transmission, distribution, repairs etc. are 90%, 15%, 85%, 95%, and 50% respectively. Losses in transmission and distribution are 20%.

Solution : In order to evaluate the two part tariff, it is necessary to find out the fixed and running charges separately, which are tabulated as follows :

Sl. No.	Item	Total Charges	Fixed charges		Running charges	
		Rs.	Percentage	Value	Percentage	Value
(i)	Generation	850,000	90	765,000	10	85,000
(ii)	Fuel	2,800,000	15	420000	85	23,80,000
(iii)	Transmission	345,000	85	293250	15	51750
(iv)	Distribution	2,750,000	95	2612500	5	137500
(v)	Repairs	300,000	50	150,000	50	150000
		Total		4240750		2804250

Maximum demand $= 75 \times 10^3$ kW

Total fixed charges = Rs. 4240750

Fixed charges per unit $= \dfrac{42.408\times10^5}{75\times10^3} =$ Rs. 56.544

Total number of units distributed $= 600 \times 10^6 \times 0.8 = 48 \times 10^7$ kWh

Total running charges = Rs. 2804250 = Rs. 2.8043×10^6

Running charges $=$ Rs. $\dfrac{2.8043\times10^6}{48\times10^7}\times100 = 0.584$ paise

Hence two part tariff is Rs. 56.544 per kW of maximum demand plus 0.854 paise per kWh.

Example 18.23. *The capital cost of a hydro-power station of 100 MW capacity is Rs. 1000 per kW. The annual depreciation charges are 15% of the capital cost. A royalty of Rs. 2 per kW per year and Re 0.03 per kWh generated is to be paid for using the river water for generation of power. The maximum demand on the power station is 70 MW and annual load factor is 60%. Annual cost of salaries, maintenance charges etc. is Rs. 10,00,000. If 20% of this expense is also chargeable as fixed charges, calculate the generation cost in two part form.*

Solution : Units generated/annum $= (70 \times 10^3) \times (0.6) \times 8760 = 367.92 \times 10^6$ kWh

Capital cost of plant = Rs. $100 \times 10^3 \times 1000$ = Rs. 100×10^6

Annual fixed charges :

Depreciation = Rs. $0.15 \times 100 \times 10^6$ = Rs. 15×10^6

Salaries, maintenance etc. = Rs. $0.2 \times 10{,}00{,}000$ = Rs. 2×10^5

Total annual fixed charges = Rs. $(15 \times 10^6 + 2 \times 10^5)$ = Rs. 152×10^5

Cost per kW = Cost per kW due to fixed charges + Royalty

$$= \text{Rs.}\left(\frac{152\times10^5}{70\times10^3}+2\right) = \text{Rs. } 219.14$$

Annual running charges :

Salaries, maintenance etc. $= \text{Rs. } 0.8 \times 10{,}00{,}000 = \text{Rs. } 8 \times 10^5$

Cost per kWh = Cost/kWh due to running charges + Royalty

$$= \text{Rs.}\left(\frac{8\times10^5}{367.92\times10^6}+0.03\right)=\text{Rs. }(0.00217+0.03)=\text{Rs. }0.032$$

∴ Total generation cost in two part form is given by Rs. (219.14 × kW + 0.032 × kWh)

Example 18.24. *A 100 MW station has the following data :*

Generating cost (inclusive of station capital costs) per annum	*Rs. 30 per kW*
Fixed cost per annum	*Rs. 4,000,000*
Cost of fuel	*Rs. 60 per tonne*
Calorific value of fuel	*5,700 kcal/kg*
Plant heat rate	*2,900 kcal/kWh at 100% capacity factor and 4,050 kcal/kWh at 50% capacity factor*

Find the overall generating cost per unit at 50% and 100% capacity factor.

Solution : Generating cost per kW=Rs. 30

$$\text{Fixed cost per kW} = \frac{4{,}000{,}000}{100\times10^3} = \text{Rs. } 40$$

∴ Fixed cost per kW capacity = Rs 70

$$\text{Capacity factor} = \frac{\text{Average demand}}{\text{Rating of equipment}}$$

With 100% C.F. average demand = 100,000 kW

With 50% C.F. average demand = 50,000 kW

With average demand of 1 kW energy generated per annum is 8,760 kWh. Similarly when the average demand is 0.5 kW energy generated per annum is 8,760/2 or 4,830 kWh.

∴ Cost per kWh due to fixed charges :

(*a*) With 100% C.F. $\text{Rs. }\dfrac{70}{8{,}760} = \dfrac{70\times100}{8{,}760} = 0.799 \text{ p}$

(*b*) With 50% C.F. $\text{Rs. }\dfrac{70}{4{,}380} = 1.598 \text{ p}$

Coal per kWh

(*a*) With 100% C.F. $\dfrac{2{,}900}{5{,}700} = 0.51 \text{ kg}$

(*b*) With 50% C.F. $\dfrac{4{,}050}{5{,}700} = 0.71 \text{ kg}$

Cost due to coal

(*a*) With 100% C.F. $0.51 \times \frac{60 \times 100}{1,000}$ = 3.06 p/kWh

(*b*) With 50% C.F. $0.71 \times \frac{60 \times 100}{1,000}$ = 4.26 p/kWh

∴ Total cost per unit

(*a*) With 100% C.F. 0.799 + 3.06 = 3.859 p

(*b*) With 50% C.F. 1.598 + 4.26 = 5.858 p

18.5 ENERGY AUDIT

Energy is fundamental to living. Thus, reliable energy supplies in useable forms as provided by existing technology are essential. When energy availability becomes tight or threatened, it is a matter of prime concern. The economic role of energy is basic - energy is needed for production. Therefore, energy policy invariably involves consideration of economic goals. Further, use of energy always affects the environment, so that consideration of environmental goals is required. Finally, energy policy also involves consideration of political goals – *e.g.*, the Arab oil embargo of 1973 had strong political determinants.

Although these were earlier signs, it was the Arab oil embargo and the attendant dramatic price increases that thrust the problem of energy into world consciousness. The energy crisis triggered a search for energy policies by consuming countries, which would enable them to adapt to the market power of **OPEC** and to face a long-term oil shortage. Producing countries also found themselves in need of systematic energy policies.

The growing inertia of the energy sector and the considerable size of financial commitments made long-term energy forecasting increasingly mandatory, both at national or regional basis as well as on a global basis. Improper or untimely decisions will come to bear more heavily upon the community as a whole. Yet forecasts are difficult to make because of many major uncertainties in the various domains :

- For the energy demand, major uncertainties are related to the possible absolute level of the world population, absolute and relative levels of energy consumption per capita, and the structure of energy demand (Energy mix and possible role of secondary energies).
- For energy resources, only solar energy is really infinite and inexhaustible, but we do not know yet how to domesticate it on a very large scale. The extent of availability of uranium as well as fossil resources is uncertain; there is hardly enough for the next decades, certainly not enough to make serious long-term forecasts on relative costs and/or comparative availability
- **Of** the **"constraints"** (pollution, health effect, effect climate, capital requirement, etc.) — the last but not least part of the triangle— our understanding is still very poor, and much effort has to be devoted to this most recent aspect of the energy problem.

A seemingly vital aspect of all major consuming nations' energy policies t is Conservation. The **Energy/GNP ratio** is normally used as a measure of I Conservation. In developed countries, the Government makes extensive use of high taxes on energy and energy consuming equipment. In countries like Switzerland, energy conservation means pricing energy to reflect its social costs, tacitly assuming this promotes the optional level of energy conservation.

This section considers three major questions

- Is energy conservation, defined as a given percentage reduction in current consumption, acceptable or even feasible given our desire to live in a world of improving standards of living?
- "The Rationale for Energy conservation", is whether energy conservation is desirable. In particular, we consider whether energy conservation may result in welfare losses, which more than offset any associated advantages.
- Assuming that a given conservation targets to be met, what is the most efficient policy instruments for its achievement? The alternative policy instruments considered include both price and non-price methods of rationing.

The Industrial Revolution, beginning in the Seventeenth Century, marked the large-scale application of inanimate energy forms through new I technologies. Without technological breakthroughs, coupled with existence i of inanimate energy supplies, the Industrial Revolution would have been of I trivial economic significance. Just as capital and labour are key inputs, energy 1 has been and will continue to be a vital input in the production process. The I present trend of increase in prices of fossil fuel is causing severe strain on ! operating costs of most industries. This is causing worry for management for (cost control and cost reduction of energy input in all Industrial Activities. ! Energy cost could be controlled and reduced either by consumption control : or by price control or both. Consumption of Energy for a particular process is dependant upon the technology adopted, and maintenance of plant and machineries. This calls for annunciation of appropriate policy for Energy management and Energy Conservation.

18.5.1 Energy Intensity

Energy intensity refers to the quantity of energy consumed per unit of real GDP created. Analysis m the trend of energy intensity calls for energy intensity changes because of structural and behavioral changes and separately for improved technical efficiency. The energy intensity is closely linked to economies of scale, technology, instrumentation and controls, structural changes, Government policy and end users' taste. The energy intensive industries include iron & Steel, Fertilizer, Textiles, cement, paper and pulp Aluminum, glass, sugar and refineries. Energy subsidy, particularly those encouraging energy consumption by keeping the energy prices low, impose a burden on economic efficiency, environmental pollution, Government budget and increase in energy intensities. Conservation of energy leads to sustainable development.

On May26, 2003 the Electricity Act, 2003 was passed by the Govt of India to consolidate the laws relating to generation, transmission, distribution, trading and use of electricity and generally for taking measures conductive to development of electricity industry, promoting competition there-in, protecting consumers interest and supply of electricity to all areas, rationalization of electricity tariff, ensuring transparent policies regarding subsidies, promotion of efficient and eco-friendly policies, constitution of central Electricity Authority, Regulatory commissions and establishment of Appellate Tribunal and for matters connected therewith or incidental thereto.

Energy Intensity in Indian Industries

Indian Industry today is unquestionably more energy intensive than those abroad. The intensity of energy use in steel for instance is more than double that of Japan or Italy. Table 18.1 below iliustrates clearly the large gap in energy utilization patterns in key industrial sectors, between India and other developed nations.

Table 18.1. Comparison of Energy use in Selected Industries in India and Abroad

(Million KCal / tonne)

Country	Steel	Cement	Pulp & Paper	
India	9.50	2.00	11.13	11.25
Italy	4.03	0.89	—	9.92
Japan	4.18	1.20	—	—
Sweden	5.02	1.40	7.56	—
UK	6.07	1.30	7.62	12.23
USA	6.06	0.95	9.70	11.32
W Germany	5.21	0.82	—	—

Source : Energy Policy issues, Workshop Proceedings July' 84

This intensiveness in energy use focuses on the need to upgrade and modernize production technologies and provide a measure of integration of energy management practices and concepts/technologies in value added manufacture. It may be pertinent to point out here that energy intensity is not only a function of the production technology; other factors like the quality of raw material and the fuel, location of the industry, capacity installed and utilized, degree of mechanization, operational and maintenance practices, etc. also play a role. Even after appropriately discounting for these factors it appears that the scope for energy conservation in Indian Industries is very large.

Table 18.2 below shows the energy intensities in the industrial sector between 1972-73 and 1982-83.

Table 18.2. Intensity of Energy Consumption in Industrial Sector

1972-73	1975-76	1979-80	1982-83
1.3	1.285	1.33	1.23

Units: MTCR per Rs 100 crores at 1970-71 prices

Interestingly, industry has maintained a reasonably steady record of energy intensity as measured in terms of MTCR per Rs 100 crores at 1970-71 prices.

The elasticity of energy consumption between 1972-73 and 1982-83 as shown in Table 18.3 appear to be erratic in different periods. This is quite usual, because in the short run, various adjustments are possible in response to sharp price changes or physical restriction in the supply of energy; these may differ from long run patterns. The elasticity of consumption for the eleven-year period (1972-73 to 1982-83) is 0.83. The industrial Sector shows some sharp changes, with a particular Low elasticity of consumption.

Table 18.3. Elasticity of Energy Consumption in the Industrial Sector

Year	Elasticity of consumption
1972-73/75-76	0-89
1975-76/79-80	1-23
1979-80/82-83	0-25
1972-73/82-83	0-83

during the period 1979-80 to 1982-83, which was undoubtedly induced by the significant increase industrial production and higher energy prices as will as by restriction in the supply of energy. On earth continuous depletion of Fossil fuel reserves, high capital cost of hydro power, increased population, and concern for environmental pollution have become a matter of concern to every nation whether it is developed or underdeveloped. As such conservation of energy and energy audit are the most important two elements in the process of socio-economic development on earth.

Here a brief introductory discussion is mode below on Energy audit which has become on important task for modern power system engineers.

18.5.2 Energy Audit Management

Energy audit is an integral part of Energy Management. This refers to Methodical examination and review of entire energy system of an Organization. Energy audit is likely to generate sufficient data both in qualitative term and in quantitative term for management action for reduction and/or control of energy cost. This will conserve energy.

Energy audit is carried out through inspection of facilities that provide breakdown of energy usages within an organization.

Energy audit is a tool for Energy conservation. Energy Audit is a vital link in the entire management chain. It is an effective tool in defining and pursuing comprehensive energy management programmes. Energy Audit acts as an instrument in energy cost optimization, pollution reduction, and safety and identifies methods for improving operating performance and maintenance practices in an organization. The objectives of Energy Audit could be mentioned as follows :

- Identifying the quality and cost of various energy inputs
- Assessing current pattern of energy consumption in different cost centers of operation
- Relating energy inputs and production output
- Identification of potential areas of thermal and electrical energy economy
- Highlighting wastages in major areas
- Fixing energy saving potential targets for individual cost centers
- Implementation of measures for energy conservation and realization of saving
- Energy Audit has been broadly classified into two categories
- Preliminary Energy Audit
- Detail Energy Audit

Energy demand is not an exception to the economic theory of limited means and unlimited demand. Energy cost is a significant factor in economic activity at par with land, labour and capital. The demand for energy is increasing .On the other hand energy resources are becoming scarce and costlier. This steady increase in gap calls for energy conservation. This is possible through systematic approach towards present trend of energy consumption through Energy Audit and application of modern techniques and methods for minimizing energy wastages, identifying areas requiring energy cost optimization, pollution reduction and safety. Energy Audit also identifies methods for Energy conservation. It is an integral part of Energy Management.

An Industrial Undertaking is likely to consume energy in various forms as follows :

- **Water**
- **Compressed Air**
- **Coal**
- **Furnace Oil**
- **Diesel Oil**
- **Petroleum**
- **Natural Gas**
- **Steam**
- **Electricity**

Each type of energy as stated above may be used in different forms and/or may be transformed to a different type of energy for effective utilization for commercial gain.

American Society of Heating, Refrigeration and Air-Conditioning Engineers (ASHRAE) identify three levels of Energy Audit as follows :

Level I

The Level I cover an energy report that records information on energy use patterns, design characteristics, equipment systems and modes of operation and establishes one or more energy use indices.

Level II

This level of Energy Audit includes a cursory on the site audit that identifies and estimates energy cost savings for no cost/low cost measures like TOM (Tuning, Operation and Maintenance) and indicates the opportunity for capital-intensive measures. Level II audits include all Level I requirements. The Remedial measures suggested herein constitute-sequencing operation of various sources of Energy to meet the System demand and to maintain the systems losses at optimum level. This would reduce Energy losses and increase the life of Energy supplying and utilizing plants.

TOM requires Knowledge on the existing energy system pattern and type of Energy demand, knowledge of Manufacturing processes and availability / inventory of the type of Energy and quality of available form of Energy. Such knowledge would enable the Engineers and Supervisors to draw a programme to identify and use the optimum energy sources from time to time during operating hours. This calls for higher degree and higher level of supervision in the actual operating period. TOM produces almost immediate results without investment of any capital.

TOM in an existing energy consuming plant requires preparation oJ Model, Simulation of the designed Soffvvare and application of the same in normal operating condition. Thereafter the saving in energy losses could be worked out. It is considered necessary to involve the existing operating personnel during application of TOM for building confidence amongst the existing managers Engineers and Supervisors. Success of this process would demand a fair level of Instrumentation and Controls for better supervision, control and optimum use of precious energy.

Level III

This level includes a detail on-site energy Audit that identifies and qualifies the cost effectiveness of TOM and Capital Intensive Energy measures. It includes all of Level II requirements. This process involves engineering and design for modification of existing form of transportation (transmission), distribution and utililization of energy in the plant. The method is Capital intensive and often calls for investment. Subsequently, this method demands study of economics of such proposal and a cost-benefit analysis before decision-making. The process requires study of existing hardware constituting the energy supply and delivery systems and identification of the particular hardware(s) adding to energy loss. The process also requires identification of System/part of the energy system constituting to additional loss. The next step is to develop an alternative hardware(s) and/or System(s) to re-design the energy delivery systems for improved operating efficiency. Subsequently, a cost-benefit analysis could be done to examine the economic feasibility of such re-engineering. The entire exercise requires detail study of the existing energy delivery systems and its historical operating performances over a period of time.

After field study and identification of possible energy .saving areas as referred to above, a detailed calculation is required to be prepared for estimation of possible savings in each type and form of energy and from each plant/machines/activities. This is to be followed by cost-benefit analysis for assessment of economics.

As mentioned above, the Energy Audit as at level I above include preparation of a report indicating the followings in particular :

- List of all energy generating, transporting, distributing and consuming equipment. The list will indicate detail specification of the equipment, name of manufacturer, and year of manufacture,
- Energy usages pattern of all equipment as above.
- Historical operating data, which are generally available from the Logbook.

This report with the above information will form an Energy Information System of an Organization. This will indicate the type and quantity of Energy, which an organization was using, and the pattern of energy usages during the day/week/month/year.

This report will indicate if an Organization is energy intensive or not.

The Energy Audit at Level II includes study of the report prepared at Energy Audit Level I, analysis of the same and preparation of recommendation on the followings :

- Identification of the equipment and/or process which may be tuned, maintained or adjusted without involving capital investment for giving energy saving and cost saving. The recommendations include details of tuning/adjustment and corresponding type and quantity of energy saving potential.

- Identification of area(s) requiring modification/modernization/refurbishing of the existing equipment/system, which need capital investment for energy savings. Detail Project Report **(DPR)** is required to be prepared for the recommended modification/ modernization as above. The **DPR** should include detail Engineering including identification of the new balancing equipment if any being installed, preparation of **PERT** network for completion of the modification, financial appraisal of the Project Report, identification of the sources of finance and pay back period arising out of saving in cost of energy.

As an example, drawl of electrical energy at low power factor may be considered. Low Power Factor arises from operation of large number of small induction motors at partial load. The supply company charges penalty if power factor is below 0.98 (Lagging). Power factor can be improved by installation of shunt capacitors. Thus, if an Industrial Undertaking is drawing 3-phase, 415 volts, 50 HZ, electrical power at power factor 0.70 (lagging), it will be required to pay penalty to the supply company in every billing month. Such penalty could be saved if shunt capacitors of adequate capacity and class are installed to improve the power factor to 0.98 (lagging). The purchase and installation of the shunt capacitor can be financed by bank borrowings. The borrowed capital along with accrued interest could be paid back from the savings made by improving the power factor from 0.70 (lag) to 0.98 (lag).

The scope of work of Energy Audit could be generally summarized as follows :

18.5.3 Preliminary Energy Audit

This requires the following jobs in particular :

- Collection of historical data
- Quick or walk through assessment of Energy performance of the

System under study

- Comparison of actual energy consumption to standard/design energy
- Consumption and analysis of the variance
- Identifying of potential areas for future investigation for energy saving
- Preparation of Report
- Presentation of the Report

An Industrial Undertaking may be consuming energy of various types and in various forms for conversion of raw materials into finished goods. The types and forms of energy may be purchased Electrical Energy, Own generated electricity, low pressure steam, compressed air, oil fired furnaces and water. This shows that the Undertaking was using purchased electricity, diesel fuel, furnace oil and coal for boiler. Statistical data for consumption of each type of energy for the last 2/3 financial years and the corresponding production output of the Organization would help the Energy Auditor to understand the consumption process characteristics of the Organization. This is likely to show a correlation amongst product mix, production output and consumption of each type of energy. Subsequently, the usages of the type of energy could be ranked in order of their priority in the process. Accordingly, the quantity and type of energy required for such type of Industries as per engineering norms could be worked out and compared with the figure actually consumed. The variance, if any, could be analyzed to identify the potential areas requiring further study for possible energy savings. Thus, preliminary Energy Audit generates initial data, which are essential for study, analysis and identification of the particular process and/or equipment which needs to be modified for possible energy saving.

18.5.4 Detail Energy Audit

The preliminary Energy Audit will indicate whether further study will yield favorable result It is also likely to identify the area(s) needing further study and analysis for possible modification. The following work is generally covered by the scope of Detail Energy Audit

- Analysis of present consumption and past trend
- Comparison of actual performance with target and/or designed

Parameters and analysis of the variances.

- Inter company comparison of various energy parameters
- Checking of capacities and efficiencies of equipment/systems
- Review of fuel handling and storage
- Costing of plant utilities
- Checking of instrumentation and control system
- Developing of energy use indices
- Preparation of energy balance diagram
- Reviewing of Management Information System
- Preparing of Programme for bringing awareness for energy savings amongst employees
- Preparation of recommendation for energy saving and energy Conservation

Study of the logbook of the Energy consuming industry will indicate the past energy consumption data in various forms and corresponding output of the unit on daily/weekly/ monthly basis.

A Production unit is likely to consume the following types of final forms of Energy

- Purchased Electricity
- Own Generated Electricity
- Steam
- Compressed Air
- Water
- Diesel/Petroleum for running of own transport

If, diesel operated low pressure Boilers are used for producing Steam, then only the following type of primary Energy sources/Final form of Energy are in use in the Industry

- Purchased Electricity
- Diesel /Petroleum

Pattern of usages of these types of Energy sources could be obtained from the Logbook. These historical data's could be compared with corresponding physical output of the Production unit and a relation could be established. Analysis of such relation is likely to generate data that guides the management m identifying the best Energy mix for optimizing usages of Energy resources. Such historical data shows the type and source of Energy and Application of Energy.

As for example in a Coal based Thermal Power Plant, the input consists of Coal, Water, Fuel oil, Compressed Air, Auxiliary Power and some Chemicals .The Output consists of Electrical Energy, Heat Energy, Sound Energy and Systems Josses. Using these Historical data of input and output an Energy Balance Diagram Can be prepared to understand the entire conversion process from the stage of Raw Coal input to the Stage of Electrical Power Transmission as shown in Fig.18.3 below. This diagram identifies the losses in each stage of the conversion process. Such losses thereafter can be divided into Controlable losses and Un-Controllable losses as shown in Fig.18.4 below. The foctor(s) responsible for such losses could be identified and corrective actions could be taken.

The Fig. 18.3 below shows the Energy Balance Diagram of a 210MW Coal based Thermal Power Station with Boiler Efficiency of 89 -percent and TG Heat Rate of 2022 K Calories per KWh. The Diagram indicates percentages of losses in each stage of Conversion Process. Fig. 18.4 indicates the controllable and Un-Controllable losses and the factors contributing to such losses. Fig. 18.4 also indicates the suggested Remedial Measures.

Based on above level of losses, a target can be prepared for possible reduction of controllable losses. The actual performance can be obtained from the Logbook .The variance (Actual Loss-Targeted Loss) can be calculated and analyzed for designing further action. Similar studies in other coal based thermal Power plants may develop data for inter Plant comparison. Continuance of such Audit is likely to generate a Standard that can be utilized by all Plant Operators as Bench markings for attainment of Goal. This will bring awareness at all levels regarding improving operating performances, improving quality of maintenance, reduction of cost of conversion and conservation of Energy.

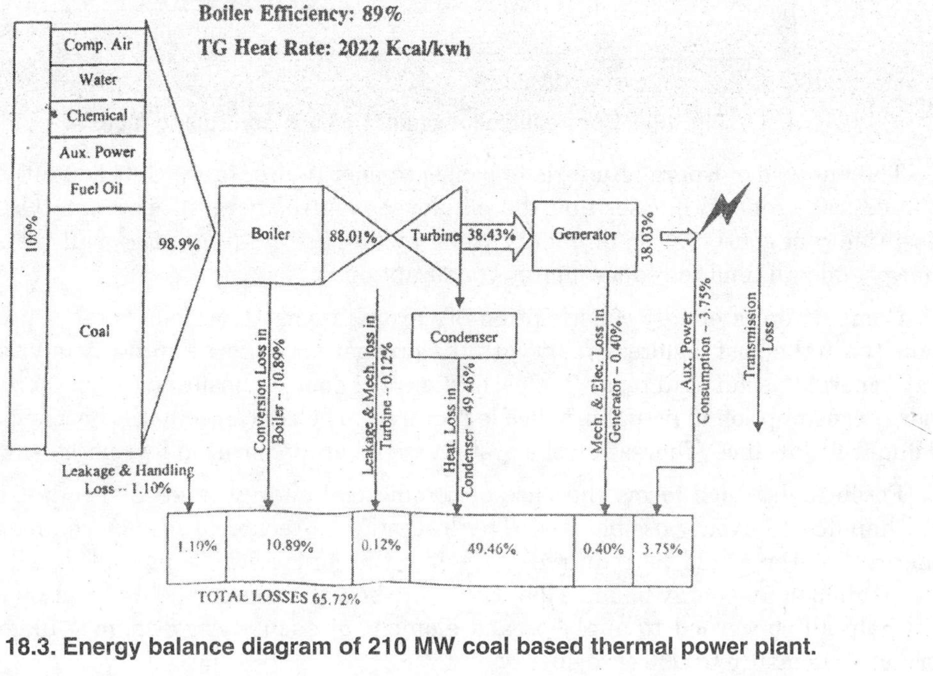

Fig. 18.3. Energy balance diagram of 210 MW coal based thermal power plant.

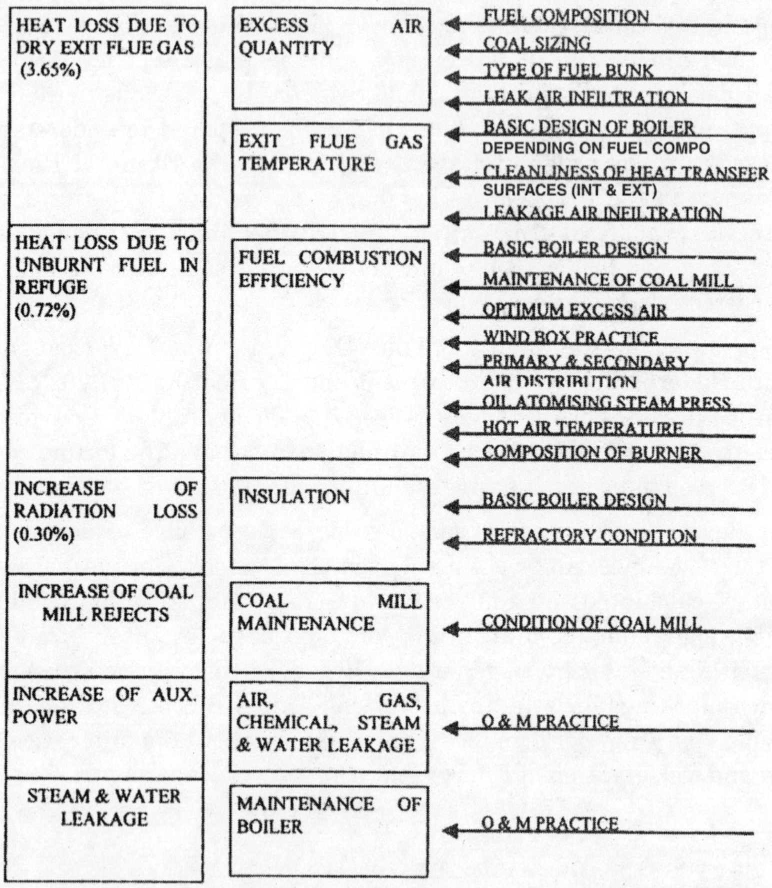

Fig. 18.4. Controllable losses of boiler & controlling factors.

The purpose of Energy Audit is to conserve energy profitably. Subsequently, the first step is to educate general masses and the employees of an Energy Consuming Organization the essentiality of conservation of energy. Such awareness creation drive will reduce wastage of energy and will tend to reduce energy consumption.

Comparison of energy consumption of various types at various level of production of an Industry in the past (before 5-year) with the present may show a trend. Analysis of such trend may generate useful data regarding performance of energy consuming equipment/process. Such energy consumption in actual practice is required to be compared with the designed parameter to quantify variance. Thereafter, the system/engineer modernized for energy saving profitable.

Fossil fuel, which forms the base of commercial energy, is needed to be stored properly and handled to avoid possible loss. The loss may be technical loss or commercial loss. The Energy Auditor would be required to review this process as a part of detail energy audit. Preparation of an energy balance diagram will show the source and application of energy. This will help all concerned to analyze each element of such energy balance diagram and take corrective measure to conserve energy.

18.5.5 Equipment/Systems to be Studied

For preparation of an energy management information system and conducting an energy audit on an Energy consuming Industry the following equipment systems are needed to be studied.

- Electrical distribution network and load management
- Electrical motors above 10 KW rating
- Group drive systems
- Transmission of electric power to motive gear
- Compressed Air generator and distribution network
- Boiler
- Steam generation and distribution
- Power operated material handling equipment
- Lighting systems
- Diesel Generating equipment
- Refrigeration and Air conditioning equipment
- Water supply systems

It is considered that electricity is the largest commercial energy consumed in an Organization. The details of the electrical distribution system inside the Industrial Establishment are needed to be studied. The details would include technical particulars of the substation including capacity, type and rating of the transformers, switchgear, shunt capacitors if any and measuring instruments. The nameplate details of this electrical equipment would indicate various important parameters including internal loss components at various loads in operation. The number and type of outgoing feeders, their areas of distribution, connected load and capacity of each feeder is a matter of necessity to understand the possible losses in the feeder and the type of possible unbalanced loading of the transformer. Reading of all the instruments in the substation could be available from the Logbook. Analysis of such data from the Logbook for past periods is necessary to understand loading details of the distribution network. Such details would be necessary for preparation of an optimum load management system.

Nameplate details of all motors above 10KW are necessary. In preparing such details, the machine the motor is connected to and name and designation of the distribution feeder, which is supplying the concerned motor is required to be mentioned. These details would identify the motors, which are under loaded/overloaded/optimum loaded.

In many Industrial Companies, there are individual drives and also group drives. In-group drive, one large induction motor drives an overhead line-shaft. The motive power for the machines is taken from the overhead rotating line shaft through pulley and belt. Such belt drive is a source of loss of energy. Again, if the line shaft remains operative for a long time under loaded, then the driving motor runs under loaded. This phenomenon results in drawal of low power factor electrical energy from the distribution system, besides under loaded running of overhead line shaft.

Compressed Air constitutes one of the main sources of energy in an Industrial Establishment requiring pneumatic operation of machines, tools, instrumentation and control systems. Compressed Air Generating Plant consists of motor driven rotary/reciprocating compressor, air-receivers and associated pipelines. Both unit type compressor plant and common compressed Air Generating Plants are used. Generally compressed air is distributed at a pressure of 5 kg per square cm. In case of common compressed Air Generating Plant, large pipelines are laid out for distribution of compressed air to various equipment. Study of industrial plant has shown leakage in pipelines and consequent loss of air. The compressed air plants along with the associated pipelines; valves and instruments are needed to be studied to identify leakage area (s).

Most of the process Industries requires low-pressure steam. Low-pressure steam is generated in boiler. The boiler may be coal fired/oil fired/gas fired. The generated steam is distributed through insulated pipelines to various utility points. The energy auditor is required to survey the pipeline to find out possible leakage/loss of heat. In addition, the combustion process is required to be analyzed to see that there is economy in fuel consumption.

Lighting, water supply systems, refrigeration and air-conditioning system constitute large energy consuming area in an Industrial organization. Utility of such services, mode of use and attitude of employees and management on such services are required to be studied for design and implementation of any energy conservation plan.

Most of the industrial companies are equipped with stand-by diesel generating set(s) of adequate capacity and class to avoid possible stoppage of production because of sudden mains power failure. Generally, cost of production of such power is high. It is therefore not advisable to use such power for all loads. A study of such stand-by power plants is likely to give an energy conservation plan.

18.5.6 Processes to be Studied

Detail energy audit requires study and analysis of the followings to identify possible energy wastages for initiating conservation measures.

Flue Gas

Sampling of the flue gas in the boiler will indicate it there was optimum combustion. Optimization of fuel-air ratio is a necessity for attainment of optimum combustion. If the air is less, there will be unburnt fuel. On the other hand, more air will allow the heat to go out of the chimney.

Exhaust Gas

Measurement of volume and exit temperature of the exhaust gas indicates the quantity of heat energy wasted in the form of exhaust gas. Such exhaust gas is also creating air pollution, which ultimately gives rise to health hazard. In high capacity, high speed, four-stroke diesel engine, the exhaust gas temperature was found to be of the order of 550° c-600°C. The volume of exhaust gas is also high. In such cases, a suitably designed boiler/heat engine may be installed to profitably utilize such large heat energy in exhaust gas.

Fossil Fuel

The processes of ordering and receipt of fossil fuel are required to be studied. The solid fuels like coal and ignites are received either through Railway wagons or Road transport. Such transport loaded with fuel enters the plant premises, when they are weighed in a weighbridge, which generates printed weights. After unloading, the empty wagon/transport is again weighed in the same weighbridge and print out taken. The difference in weight is the net material received. Such details are required to be entered into the bin-card and the stores ledger. The received store is now sent to the laboratory for determination of its chemical composition and ash content.

Liquid fossil fuels like diesel, petroleum and furnace oil are received at the point of consumption through railway wagons or road transport. The received liquid fuel is now pumped into the storage tank. A properly calibrated recording type flow meter attached with the input pipeline measures the quantity of liquid fossil fuel received. A sample of such fuel is now sent to laboratory for assessment of chemical properties and water content in the received fuel.

Storage of Fossil Fuel

Proper storage and stores accounting of purchased fossil fuel is a matter of necessity for ascertaining the actual consumption of fuel by the energy conversion plant. Normally, solid fossil fuel like coal is stored in open yard. Such storage may result in pilferage of coal. Liquid fossil fuels are either stored in under ground fuel reservoir or in overhead fuel tank. Possibility of pilferage of such costly mineral oil is required to be avoided. It is also, necessary to ensure that such stored oil is not allowed to mix with water or other liquid. Strict maintenance of logbook showing particulars of receipt of fuel vis-a-vis stock of fuel and issue of fuel is a matter of necessity.

18.5.7 Audit Systems

The task of Energy Audit systems could be summarized as follows :

- Identification of Energy consuming equipment/apparatus and; codification of same for ease of computerization
- Study of electrical/mechanical/thermal characteristics of energy consuming equipment/ apparatus
- Type of energy consumed, as for example water, coal, mineral oil, electricity, heat, air or natural gas
- Comparison between rated performance and actual performance of each energy consuming equipment
- Analysis of quality of energy viz. coal, mineral oil and electricity which an Industrial Undertaking receives for use and application
- Preparation of a standard for all types of input energy being received by an Industry
- Preparation of a standard for performance of each energy consuming equipment
- Installation of an Internal Audit System based on above

Quality of Energy

Energy Audit demands analysis of consumption pattern of energy. Such analysis will indicate the controllable variances, which could be minimized/ stopped only by TOM. The uncontrollable variances could be minimized/ stopped by additional investments in plants and equipments. Such investment decisions would require a detail project report including cost-benefit analysis.

Analysis of energy consumption requires a structural study for determination of the quality of energy being consumed. The following methods are normally used for maintaining of quality of various forms of energy.

Electricity

The fluctuations in voltage and frequency determine the quality of electrical energy. Reliability of power supply is the second component of quality of power. Reduction in voltage increases current for the same power. This increases load losses. Increase in supply frequency increases no-load loss and increases inductive reactance thereby worsening power factor of the system. Low power factor increases reactive power.

Chemical property, ash content and calorific value determines the quality of coal. The calorific value and the ash content determine the quantity of air that will be required for full combustion of the fuel. The chemical property indicates the type of pollution, which could be liberated by the gases generated by burning of coal.

Fuel Oil

Chemical property, ash content and calorific value determines the quality of fuel oil. The water content of fuel oil is required to be determined to assess the quality of fuel.

Steam

The degree of saturation of Steam, its temperature and pressure determines the quality of steam energy.

Compressed Air

The pressure of compressed air and moisture content determines the quality of compressed air.

Whereas coal and fuel oil is original sources of energy, which is preserved in the form of potential energy, the electrical energy, steam and compressed air are all derived energy.

Fig. 18.5 below shows the diagrammatic representation of Energy Quality in various types and classes of Energy Resources. It could be observe^ from above that the quality of Energy in various forms play a significant role in usages and conservation of Energy. Subsequently, the Quality of Energy is a significant component in selection of the type of fuel. Thus the Quality of fuel is an important component in Energy Economics. It is, therefore essential that a proper Quality Management System be installed in Energy Consuming Industries.

18.5.8 Methodology

The Methodology for the Audit will include field study of manufacturing processes, daily and yearly operations and type & pattern of energy consumption. Historical data on cardinal parameters like volume of production and consumption of energy of various forms and types are required to be collected and collated. An Analysis is required to identify correlation between actual output of the manufacturing unit/manufacturing process/Equipment and usage of various forms of Energy. This Analysis will establish usage pattern of various forms of energy for the unit/process/equipment as shown in the chart below;

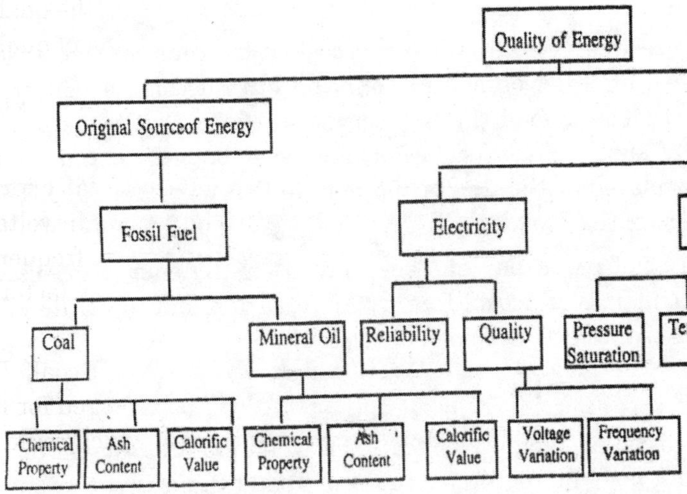

Fig. 18.5. Energy Quality representation Block diagram

The manufacturing processes and the manufacturing cycles are required to be studied to understand and identify the type of energy utilized/consumed as the raw materials are converted into finished goods. This will reflect upon the diversity factor of energy being used in various forms and in various stages of conversion process.

The past Electricity bills log books for consumption of diesel oil and Coal and log books of purchased Electricity and own generated Electricity for past few years is required to be studied and analyzed to identify the high consumption periods and areas.

The energy consuming machines/plants/areas are required to be studied during actual operation. The actual operating conditions are to be recorded and analyzed to compare the actual energy consumption with manufacturers recommended ratings. Thereafter, possible energy losses from each machines/plants/areas are identified. The estimated energy losses from the system as a whole when all the plants are in running condition is studied and analyzed. These losses are now classified under controllable losses and uncontrollable losses. The controllable losses are to be analyzed to prepare a prescription of measures to be taken to minimize the losses. In identifying the controllable losses and preparing recommendations for its elimination, care is essential to ensure that the existing volume and quality of production remained undisturbed. Care is also necessary to be taken to ensure that there was no effect on environmental pollution resulting from such energy wastage minimization programme.

18.5.9 Petrolium Conservation Research Association (PCRA)

PCRA is an independent organization created for conservation of Energy in India.

The objectives of PCRA include the following :

- To create awareness among people
- To promote R & D by use of efficient technologies, appliances, additives and devices
- To promote interfuel substitution
- To promote environmental protection
- To provide training and technical advisory services.

Examplery Case Study

Continues depletion of Fossil fuel reserves, high capital cost of hydropower, increased population on the Earth and concern for Environmental pollution have become a matter of concern to all people on this planet. It is opined that conservation of Energy is one of the important element in the process of sustainable Socio-economic development. In India, the Energy Conservation Bill, 2000 was placed before the Parliament on February 9 of 2000. This has since been promulgated as an Act of the Parliament.

In the following paragraphs, a brief particular has been given for the energy intensive Cement Industry in India. Brief particulars of Actual Energy Audit carried out in a Jute Mill in West Bengal and a Tea estate in Assam have also been discussed.

Cement Industries

The cement industry in India has come a long way since 1914, when the first cement plant was commissioned with a production level of 1000 tons/ annum. Today India is the **second largest** cement producer in the world with a production level of about 120 million tones (about 5% of world production of about 2000 million tonnes). The installed capacity is about 119 million tonnes and at an expected 10 % growth rate the production is likely to grow to about 300 million tones at the end of 2014-2015.

However, cement consumption per capita in our country at about 99-kg/ capita is one of the **lowest.** The world average is about 267 kg/capita, while that of China is 450 kg/capita. Similarly in Japan it is 631 kg/capita while in France it is 447 kg/capita.

Over the years, the growth of the industry has been uneven with traditionally cement deficit regions covering the most of the major growth centers of the country.

Cement plants in our country have mostly changed from the wet process to the energy efficient dry process. Out of 157 kilns, 117 are dry process based, 32 are based on wet process and 8 on semi dry.

Though the best of our industry matches quite well with world standards in terms of energy (thermal energy Kcal/kg of clinker"- India 665 against 690 of Japan) and pollution norms SPM of 40 in India against 20 of Japan) but the average performance of the Indian industry is lagging behind.

In the coming years, in order to survive and grow in the globalize market, rapid modernization and adoption of cost effective energy efficient and environment friendly technologies will be the prime mover for the viability of the industry in the global canvas.

The industry should increasingly look for other cheaper fuel options like sludge from paper plants, sugar cane trash, bagasse, jute dust, textile dust, biogas, refinery waste like pet coke etc. The industry should be known in future as savior of the country for sustainable development by consuming most of the industrial wastes.

India is the second largest cement producer in the world next to China with both cement production and consumption at 99 million tones during 2000. This books is closed by briefly discussning the sources of energy which in distributed consectivation and controlled by Power System.

18.6 ENERGY SOURCES & CONSERVATION OF ENERGY

Thus the energy sources could be broadly divided as shown below in Fig. 18.6 :

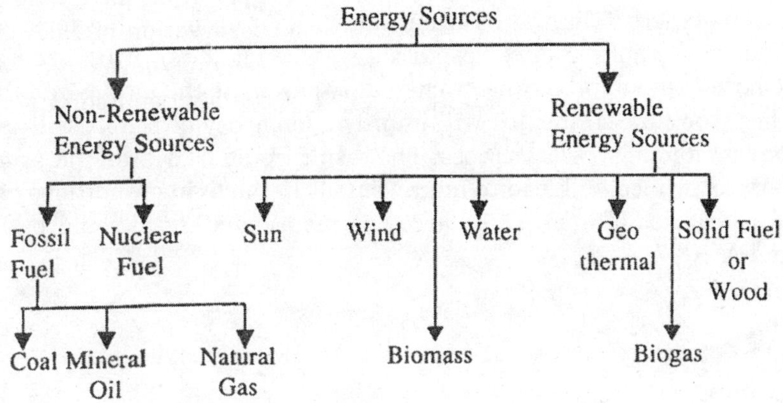

Fig. 18.6

Some Researchers and academicians have preferred to categorize energy resources into following two broad divisions :

(i) Commercial Conventional Energy Resources

(ii) Non-Conventional Energy Resources

The **Commercial Conventional Energy Sources** have been identified are those which are used in large scale for general masses. The term commercial refers to availability of such energy resources easily and at affordable prices. The term conventional refers to some code of practices being followed for production and usage of such type of energy resources. As these resources of energy follows a pre-determined set of procedures for its commercial application, this part of energy resources is well developed technologically, economically and commercially. The **non-conventional energy resources** as the name implies are yet to follow a set standard. Their development is still in the earlier stage. Subsequently, their usage for human consumption is limited. The limitation arose from technological backwardness and inability to compete with conventional form of energy both in terms of quantity, quality and prices. However, as these types of energy resources are Renewable in nature, the mankind has to depend upon such energy resources with days to come. The world population in general and the population of the Asian countries in particular are increasing rapidly. The present day human considers use of energy as an essential element for sustenance. This calls for increased use of energy. Even if per capita energy consumption were reduced by 5% from the present level by improvement in technology, the quanta of energy consumption would continue to increase because of population pressure. Use of conventional form of energy resources would create environmental hazards, which is unhealthy for living being. On the other hand such type of energy being non-renewable in nature, their reserves would reduce at faster rate and is likely to exhaust in the near future.

These phenomena calls for following actions and strategies for development of energy policies on a totality :

(*i*) Technology up-gradation in the field of conventional non-renewable energy resources such that converted energy for usage by human being per unit weight and/or volume of non-renewable fuel is increased without pollution.

(ii) Technology up gradation in the field of non-conventional renewable energy resources such that large-scale energy could be produced from such renewable resources for commercial application.

It is to be remembered that generation and utilization of usable form of energy from non-conventional renewable energy resources would not deteriorate ecology and pollute environment. On the other hand il can enrich quality of soil. As the nature and the environment are not disturbed, the human .health will improve, longevity of people will take an upward trend and labour productivity will increase. The result would be overall increase in production of goods and services, which will lead to improvement in the living conditions of human being. The two broad categories of energy resources as mentioned above could be represented as shown below in Fig. 18.7.

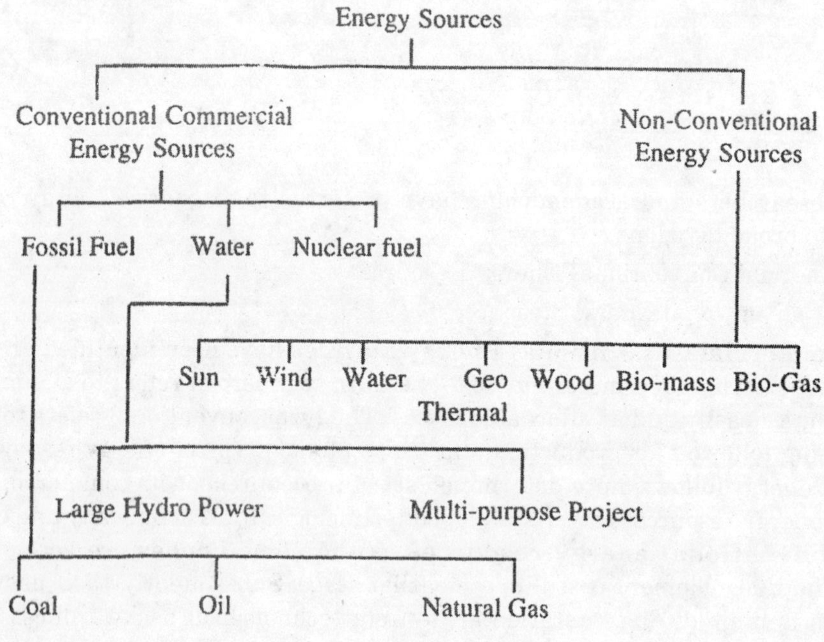

Fig. 18.7

It could be observed from the above that water has been grouped under conventional commercial energy sources and also in non-conventional energy sources. Water is basically a renewable energy resource. However, depending upon quantity of stored energy in water, it is grouped under conventional and non-conventional energy source. When the energy stored in a mass of water body is high so that it could be transformed into large quantity of useable form of energy, then such water energy sources are grouped under conventional commercial energy type. Conversion of large quantity of usable form of energy results in economies of scale of production and/or conversion. Subsequently, such converted energy form could be made available for human consumption for improving their living condition at affordable price. A water body with large volume of water or high head of water or both is a source of large potential energy. Conversion of such large volume of potential energy into useable form of kinetic energy could be undertaken for improving living condition of people only if there are economies of scale of such conversion. Such economies of scale are used as an indicator if such water resources could be utilized for generating electricity or it could be used for a multipurpose project. There are water bodies whose energy reserves are low. A water body may have low volume of water and / or low head such that the stored potential energy is low. When economies of scale as said above is used as an indicator, conversion of such water resources into usable form of energy sounds un-commercial for human consumption. These types of renewable energy sources are termed as mini-hydro power and / or micro-hydro power and are grouped under non-conventional energy resources. Again TIDAL WAVES in water body has kinetic energy. Technically, such energy could be converted into useable form of energy, but economies of scale of conversion would not warrant development of it since it could not be made available for human consumption and in consumable form at affordable price. Subsequently this type of renewable energy is grouped under non-conventional energy resources.

As said earlier the SUN **(Aditya)** is the eternal source of energy in the universe. This renewable source of energy has been converted into usable form of energy. Even in landing artificial satellites in space, this was used as the main source of energy in data collection, data acquisition and control systems. The Scientists have developed technology for conversion of solar energy into useable form of electrical energy and heat energy. Solar photovoltaic Cells have been developed for conversion of solar energy into electrical energy. Solar energy has also been utilized as solar thermal energy for water heating and also for cooking. However, building of plants for conversion of solar energy into usable form of energy is capital intensive. Again large-scale conversion of soiar energy is yet to be done for human consumption. Subsequently, Solar energy is grouped under non-conventional energy resources.

The wind is a source of renewable energy. The wind power systems involve the conversion of the earth's winds, to rotating shaft power and/or Electrical power output. The overall system that accomplishes this energy conversion process is generally called a Wind Turbine Generator (WTG). In wind power systems the actual conversion process uses the basic aerodynamic forces of lift or drag to produce a net positive force on a rotating shaft, resulting in production of mechanical power. In addition to large scale Mega Watt (MW) size wind power generator for production of Electrical energy, wind.

Power systems can be applied for smaller applications in developing countries as well as for energy supply to remote areas or for specialized applications. These categories can include Water Pumping, battery charging and heating and use as component of hybrid energy (Wind/ diesel, wind/photo.

18.6.1 Conservation of Energy

All the nations on earth today are introducing on practising energy policies where the vital aspect is conservation without offsetting any social welfare and economic activities. Below some of the major methods of conservation of energy (1) The choice of electrical or any alternative form of equipment for performing a given service is normally dependent on the relative costs of the possible below and a study of the economics of utilising electrical energy is of importance.

As we discussed earlier the cost of owing and operating any piece of equipment can be split into two parts : (1) Capital or Fixed charges and (2) Running on variable charges. The Fixed charges made to any consumer bear some relation to the capacity of plant, and this is usually regarded as being proportional to the consumer's maximum load or maximum demand. This part of tariff is expressed generally as a certain price per kW of maximum demand. And the running cost depends upon the amount of energy consumed and is charged for at a price per unit energy as kWhr.

To avoid unnecessary use of energy price per unit rate can be increased and thus a cost-consious consumer will use energy saving methods like (1) improving power factor, Instead of charging per kW maximum demand a fixed charge per kVA can be enforced. This penalises for bad power-factor only at the time of consumer's peak load. Also power factor penalty down can be enforced. (2) Improvement of load factor by reaching Maximum laod. This is mostly for Industrial Consumption. For household consumption an idea of energy consumption of various items used is given in Table 18.4, which shows that a substantial quantity of energy is consumed in domestic area. hence target should be prepared for conservation of such energy by proper control and surveliance as about 20% of these loads are spent unwisely by the household people.

Next most important point is about the electric supply company who can take various measures necessary for demand management and hence for conservation of energy. Below in a chart in the Table 18.4 it is explained clearly.

Then in the Fig. 18.8 a chart is shown about implementing conservation programme by the supply company.

Table 18.4. Energy Consumption by Various Items used in Household

Brief details of Applicances	**Estimated Hours of use**	**Consumption in kWh per day**	
		Per one-Hour	**Per Month**
Window Type Air Conditioner 1-Ton:	10	1.45	435
Window Type air conditioner 1.5 Ton	10	1.95	585
Refrigerators operating 14 hours a day (165L)	Full Day	0.12	51
Refrigerators operating 14 hours a day (180L)	Full Day	0.162	68
Electric Iron, 1000W	0.5	1.0	15
Geyser, 10L capacity	0.5	1.5	23
Semi-automatic 5-KG Washing Machine	1	0.29	9
Automatic 5-KG Washing Machine	1	0.37	11
21-Inches CTV	4	0.1	12
Microwave Oven	0.5	1.0	15
½ HP Water PmP	3	0.37	33
1 HP Water Pump	3	0.75	68
9-Watt CFL	10	0.01	3
18-Watt CFL	10	0.02	6
40-Watt Incandescent Lamp	10	0.04	12
100-Watt Incandescent Lamp	10	0.10	30
40-Watt, 4-FT Tube Light with Electronic Choke	10	0.04	12
48-Inch Sweep Ceiling Fan	18	0.075	41
56-Inch Sweep Ceiling Fan	18	0.085	46
9-Inch Sweep Exhaust Fan	18	0.05	27

NB : Months have been considered for 30-days

Units have been rounded off

(Source : CESC Ltd., Kolkata)

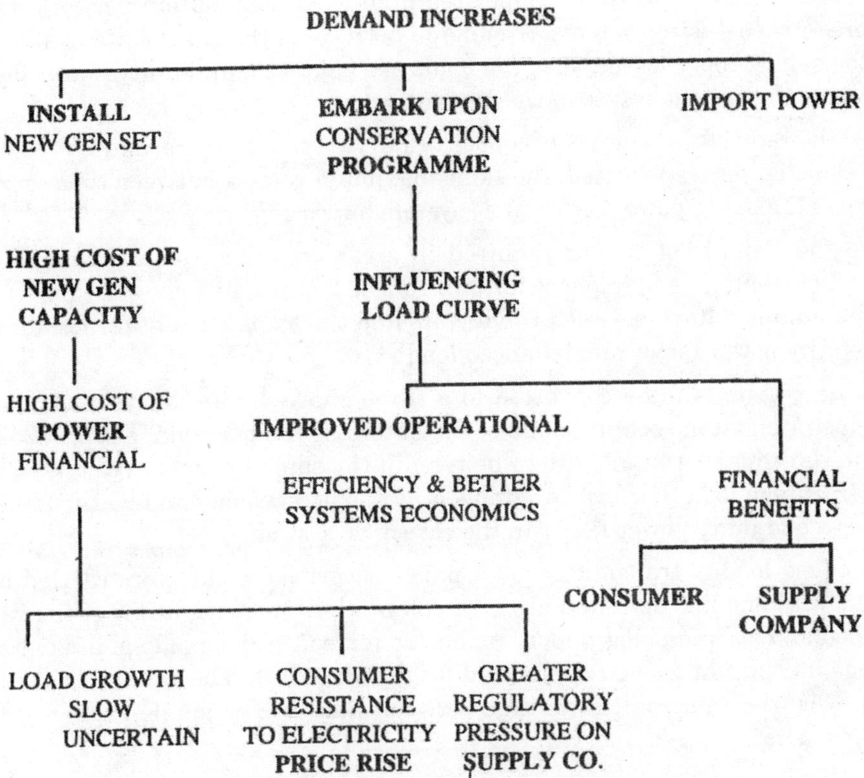

Fig. 18.8. Demand management and energy conservation.

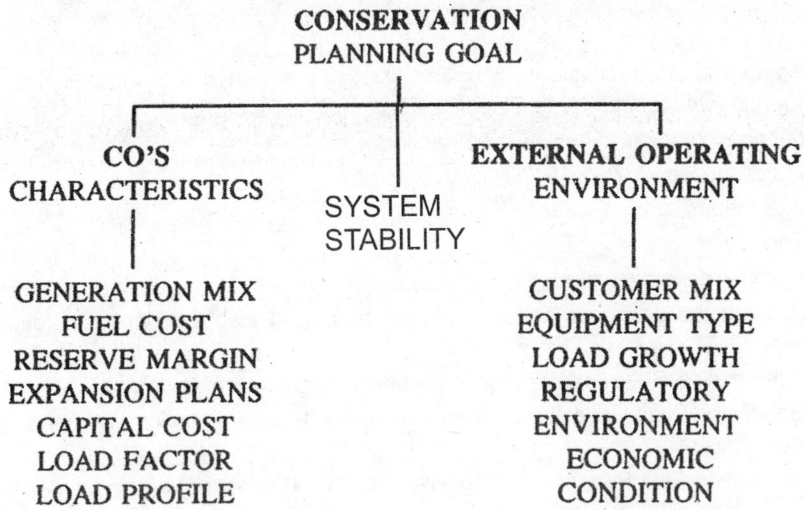

Fig. 18.9. Implementing conservation programmme.

A 2-wire dc transmission line, 0.25 km long supplies a power of 125 kW to a distribution network. Calculate the volume of conductor copper required for the line when the potential difference (pd) maintained between the distributors is (*a*) 250 V and (*b*) 400 V, the current density in the copper in each case being $3 A/mm^2$. Calculate also in each case the power loss in the line, taking the resistivity of copper as $1.7 \times 10^{-}$ Ω-m.

Compare the relative weights of copper required for a distribution network on the dc 3-wire, and three-phase 4-wire systems. Assume in both cases the same voltage at consumers's terminals, the same copper losses, that the loads are balanced, and unity power factor in the three-phase case Neglect the losses in the neutrals.

Calculate the ratio of the weights of copper use in two-phase 4-wire, and 2-wire dc systems. Assume the same power transmitted, the same maximum voltage between conductors and the same power loss. Take the power factor in ac system as cos ϕ.

Compare the weights of copper required in a cable transmission system utilizing (*a*) constant direct current, (*b*) single-phase (*c*) three-phase 3-wires. Assume the same transmitted power and maximum voltage between conductors and the same percentage loss in each case. Assume also unity power factor and balanced load.

A 3-wire dc system is to be converted to a three-phase 4-wire dc system by addition of another conductor equal in section to one of the outers. If the percentage copper loss and the voltage at the consumer terminals are to be remain the same in the two cases, calculate the additional percentage load that can be supplied by the ac system. Assume the balanced load in the two cases and unity power factor in the case of ac system.

A single-phase load is transmitted by a pair of overhead conductors carried on similar insulators but with one line operated at earth potential. a third conductor of the same cross-section is added and a three-phase supply connected instead of the single-phase one. Calculate the percentage increase in power transmitted for the same loss. The voltage to earth for each conductor is to be the same as the voltage between lines in the single-phase case. Assume constant power factor.

QUESTIONS

1. Electric power of 50 MW at 0.85 power factor lagging is to be transmitted over a 220 kV, three-phase 3-wire 200 km transmission line. The efficiency of transmission is 0.9. Calculate weight of conductor material required for the line in the following conductors :
   (*a*) copper conductors, (*b*) aluminium conductors.
   Resistivity of copper = 1.70×10^{-8} Ω-m
   Resistivity of aluminium = 2.85×10^{-8} Ω-m
   Specific gravity of copper = 8.89
   Specific gravity of aluminium = 2.71
2. State what are the economic aspects of generation and also the Factors affecting the cost of generation.
3. What is load curves. State what are the important information we get from it.
4. Define load factor, Plant Capacity, Diversity, factor, Plant Factor Maximum Demand. Also state what are various types of Tariffs.
5. What is meant by Two Part Tariffs. And also state what is meant by cold reserve, Hot reserve, Spinning Reserve. What is Tariff.
6. What is Energy Audit, and the Energy Audit Management and its usefulness in electricity supply chain?
7. From Energy audit management what are the equipments and systems to be studied.
8. A workshop having a number of induction motors has a M.D. of 750 KW with a p.f. of 0.75 and a load factor of 35%. If the tariff is Rs. 0.75 per annum per kVA of M.D. Plan 3 paise per unit extimate what expenditure would it pay to inner to raise the power factor to 0.9.

CHAPTER

19

Economic Load Despatch, Load Frequency Control , Compensation

INSIDE THIS CHAPTER

- Economic Load Dispatch
- Compensation in Power System Fact, Station etc.
- Load Frequency Control
- Insulation Coordination
- Surge, Diverter
- Automatic Voltage Control
- LFC with GRC

19.1 ECONOMIC LOAD DISPATCH

Problem in economic load dispatch involves the solution of two different problems. For first is the unit commitment or predispatch problem wherein it is required to select optimally out of the available generating sources to operate, to meet the expected load and provide a specific margin of operating reserve over a specified period of time. The second aspect of economic dispatch is the on-line economic dispatch when it is required to distribute the load among the generating units actually parallel with the system in such a manner as to minimise the total cost of supplying the minute-to-minute require-ments of the system.

In fact with large intereconnection of the electric networks, the energy crisis in the world and continuous rise in prices, it is extremely important and essential to reduce the running charges of the electric energy that means to reduce the fuel consumption for meeting a particular load demand.

In the earlier chapters in this book we discussed load flow study for the power system, in that for a particular load demand the generation at all generator buses are fixed except at all generator bus known as slack, reference or swing bus where the generation is allowed to take value within certain limits. In case of load dispatch the generations are not fixed but they are allowed to take values again within certain limits so as to meet a particular load demand with minimum fuel consumption. This means economic load dispatch problem is really the solution of a large number of load flow problems and choosing the one which is optimal in the sense that it needs minimum cost of generation.

Hence it is clear form this that since total cost of generation is a function of the individual generation of the sources which can take values within certain constraints, the cost of generation will depend on the system constraints for a particular load demand. which means that the cost of generation is not fixed for a particular load demand but depends upon the operating constraints of the sources.

The modern power system has to operate under various operational and network constraints. Hence it is best to understand the various constraints before actually taking up the economic load dispatching problem.

19.1.1 Constraints in Power Systems

There are two types of constraints : (*a*) Equality and Inequality. Inequality constraints of two types. Hard Type and Soft Type. The Hard Type are these which are definite and specific like the tapping range of an on-load tap-changing transformer whereas soft type are those which have some plexibility associated with them like the nodal voltages and phase angles between the nodal voltages, etc. In fact soft inequality constraints have been very Efficiently handled by penalty function method of Equality Constraints.

The equality constraints are the basic load flow equation give by

$$P_p = \sum_{q=1}^{n} \left| e_p(G_{pq} + f_q B_{pq}) + f_p(f_q G_{pq} + e_q B_{pq}) \right|$$

$$Q_p = \sum_{q=1}^{n} \left| f_p(e_p G_{pq} - f_q B_{pq}) - e_p(f_q G_{pq} - e_q B_{pq}) \right|$$

where $p = 1, 2, 3, ..., n$

Here in above equation, e_p and f_p are the real and imaginary component of voltage at the pth node and G_{pq} and B_{pq} are nodal conductance.

1. Inequality Constraints

(*a*) Generator Constraints : The kVA loading on a generator is given by $\sqrt{P_p^2 + Q_p^2}$ and this should not exceed a prespecified value of C_p because of the temperature rise condition, *i.e.*, $P_p^2 + Q_p^2 \le C_p^2$.

The maximum active power generation of source is limited again by the thermal consideration and also minimum power generation is limited by the flame instability of a boiler. If the power output of a generator for optimum operation of the system is less than a prespcefied value $P_{\min}$ the unit is not put on the bus because it is not possible to generate that low value of power from that unit. As such the Generator power P_p can not be outside the range stated by inequality, *i.e.*, $P_{p_{\min}} \le P_p \le P_{p_{\max}}$. Similarly the maximum and minimum reactive power is limited. The max Reactive power is limited because of overheating of the rotor and minimum is limited because of the stability limit of the machine. So the generator reactive power Q_p can not be outside the range stated by the inequality, *i.e.*, $\theta_{p_{\min}} \le \theta_p \le \theta_{p_{\max}}$.

(*b*) Voltage Constraints : It is most important to ensure that voltage magnitude and phase angles at various nodes should vary within certain limits. The voltage magnitudes should vary within certain limits because otherwise most of the equipment connected to the system will not operate satisfactrily or additional use of voltage regulating devices will make the system uneconomical.

Thus, $$\left|V_{p_{\min}}\right| = \left|V_p\right| \le \left|V_{p_{\max}}\right|$$

and $$\delta_{p_{\min}} = \delta_p \le \delta_{p_{\max}}$$

where $|V_p|$ and δ_p stand for the voltage magnitude and phase angle at the pth node.

The normal operating angle of transmission line lies between 30° to 45° for transient stability reason. Hence 30° to 45° for transient stability reason. Hence a higher limit is imposed on angle δ. A lower limit of δ assures proper utilisation of transmission facility.

(c) Running Spare Capacity Constraints : These constraints are required to meet : (*a*) The forced outages of one or more alternators on the system, and (*b*) the unexpected load on the system. The total generation should be such that in addition selecting the load demand and losses a minimum spare capacity should be available, *i.e.*

$$G \geq P_p + P_{SO}$$

where G is the total generation and P_{SO} is some prespecfied power. A well planned system is one in which this spare capacity P_{SO} is minimum.

(d) Transformer Tap Settings : When an antotransfer is used, the minimum tap setting could be zero and the maximum one, *i.e.*, $0 \leq t \leq 1.0$. Exactly same way for a two-winding transformer if tappings are provided on the secondary side, the $0 \leq t \leq n$, here 'n' is the ratio of transformation. Now, phase shift limits of the phase shifting transfer, $\theta \leq \theta_p \leq \theta_{p_{max}}$.

(e) Transmission Line Constraints : In this case the flow of active and reactive power through the transmission line circuit is limited by the thermal capability of the circuit and is expressed as $C_p \leq C_{p_{max}}$ where $C_{p_{max}}$ is the maximum loading capacity of the pth line.

(f) Network Security Constraints : When a system in operating initially, satisfactorily and there is an outage may be scheduled or forced-one, it is only obvious or natural that some of the constraints of the system will be not violated.

The complexity of these constraints (in terms of number of constraints) is increased when a large system in under study or consideration. A study is to be made with outage of one branch at a time and then more than one branch at time. The nature of constrains are same as voltage and transmission constraints.

Above we have briefly stated the various constraints. Let us now look into the actual problem of economic load dispatch: As we are considering a system which is already in existence, the fixed loss/costs like salaries, and capital costs are neglected from the cost criterion. We also consider here only the thermal plants and therefore we are concerned only with the cost of fuel input to different generating station. Here we shall only discuss problem of optimisation using the conventional B-coefficient method for loss evaluation. In the past the practice was to load the machines such that the incremental cost of production for all the machines was same and thus transmission line was nent considered an integral part of the system. It was later realised that a plant far from the load bus to supply in addition to the load, transmission losses also and, therefore, this plant must be penalised. These transmission line losses are calculated by making use of B-coefficients.

A simplified performance curve of a boiler turbine-generator unit is shown in Fig. 19.1. From the curve, heat rate is defined as the ratio of fuel input to the corresponding power output and hence the units are million BTU per MW hr. From the heat rate curve another characteristic known as incremental fuel curve can be obtained.

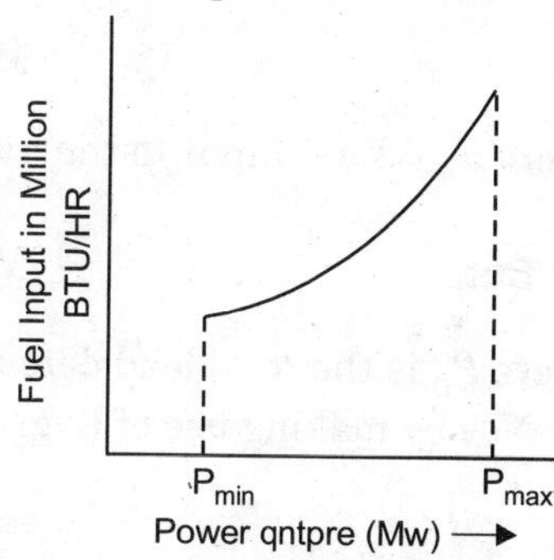

Fig. 19.1. Input-ouput power curve.

$$\text{Incremental fuel rate} = \frac{\Delta \text{ Input}}{\Delta \text{ Output}}$$

This means it is a ratio equal to a small charge in input to the corresponding small change in input. Now as the incremental quantities tend to zero; incremental quantities tend to zero, incremental fuel rate tends to

$$\text{Incremental fuel rate} = \frac{d(\text{Input})}{d(\text{Output})} = \frac{dF}{dP}$$

Where F is the fuel input in mission BTU per hr. and P is the power output in MW. The units associated with incremental fuel rate are therefore million BTU per MW hr and the incremental fuel cost is expressed in terms of Rs. per MW hr. which is obtained by multiplying the incremental fuel rate by the fuel cost in Rs. per BTU.

It should be noted that incremental cost consists of the incremental fuel rate cost plus the incremental cost of labour, supplies maintenance and water. Since it is difficult to express exactly these costs as a fuction of output and also since they form generally a small fraction of the incremental cost of fuel, the incremental cost of production will be considered equal to the incremental cost of fuel. The incremental efficiency is defined as the reciprocal of incremental fuel rate or incremental production cost and is given as below:

$$\text{Incremental efficiency} = \frac{dP}{dF} = \frac{\text{Incremental Output}}{\text{Incremental Input}}$$

As and when the Merit Order Scheduling method is applied to economic scheduling of power plants, it assumes that the incremental cost of all the generators is constant over the full range or over successive discrete portions within the range. The economical way of meeting a load will be to load the machines in order of highest incremental efficiency. This method hence, needs forming of a table which could be looked into for any load condition and does not need any complicated calculation.

Normally the incremental cost curves are not constant, hence, this Merit Order Scheduling scheme is not used and will not be discussed here.

19.1.2 Economic Dispatch Neglecting Losses

Let us first define the economic dispatch problem as below :

Minimum Total Fuel, E_T input to the system, F_T is expressed as :

$$\text{Min } F_T = \sum_{n=1}^{n} F_T \qquad \text{...(19.1)}$$

where F_n = Fuel Input to the nth limit.

But
$$\text{Min } F_T = \sum_{n=1}^{n} P_n \text{ is subject ot } P_D = \sum_{n=1}^{n} P_n \qquad \text{...(19.2)}$$

where P_D is the total load demand.

Nowby making use of Lagrangian multiplier the auxiliary function is obtained as :

$$F = F_1 + \lambda\left(P_D - \sum_{n=1}^{n} P_n\right) \qquad \text{...(19.3)}$$

where λ is the Lagrangian multiplier.

Now differentiating F with respect to the generation P_n and equating it to zero (gives the condition for optimum operation of the system), we get :

$$\frac{dF}{dP_n} = \frac{dF_T}{dP_n} + \lambda(0-1) = 0$$

$$= \frac{dF_T}{dP_n} - \lambda = 0$$

As,

$$F_T = F_1 + F_2 + \dots F_n$$

$$\therefore \quad \frac{dF_T}{dP_n} = \frac{dF_n}{dP_n} = \lambda$$

Hence the condition for optimum operation is

$$\frac{dF_1}{dP_1} = \frac{dF_2}{dP_2} = \qquad \frac{dF_n}{dP_n} = \lambda \qquad \dots(19.4)$$

But, here, $\frac{dF_n}{dP_n}$ = incremental production cost of plant n is **Rs/Mw hr.** Then the incremental production cost of a given plant over a limited range is represented by :

$$\frac{dF_n}{dP_n} = F_{nn} \cdot P_n + f_n$$

Where F_{nn} = Slope of incremental production cost curve

f_n = Intercept of incremental production cost curve

Infact the equation (19.4) means that the machines are to be so loaded that the incremental cost of production cost of each machine is same. However, it is to be noted that the active power generation connstraints are taken into account while solving the equations which are derived above. if these constraints are violated for any generator it is tied to the corresponding limit and the rest of the load is distributed to the remaining generator units according to the equal incremental cost of production. The simultaneous solution of equations (19.2) and (19.4) gives the economic operating schedule. Therefore, any good technique for solving a set of linear equation can be used but since the inequality constrains have also to be taken into account the following iteractive method is used.

(*a*) To assume a suitable value of $\lambda(0)$. This value should be more than the largest intercept of the incremental cost characteristics of the various generators.

(*b*) To compute the individual generations $P_1, P_2, \dots, P_n$ corresponding to incremental cost of production from equation (19.4). Then in case generation at any of the buses is violated the generation of that generator is fixed at the limit violated during that iteration and the remaining load is distributed among the remaining generators.

(*c*) Then to check if the equality $\sum_{n=1}^{n} P_n = P_D$ is satisfied.

(*d*) Then if not, make a second guess of λ' and repeat the above steps. The selection of λ' in this step must be guided by the result in the step in (*c*).

Let us take an example to explain the point further.

Say, if we find that the total generation is less than P_D then correct value of λ is to be selected would be, $\lambda > \lambda°$. Now, if the equality is satisfied, the generations as calculated in the step (*b*) above will give the optimum operating strategy.

Say, there two generators of 100 MW each with incremental characteristics :

$$\frac{dF_1}{dP_1} = 2 + 0.012\,P_1$$

and

$$\frac{dF_2}{dP_2} = 1.5 + 0.015\,P_2$$

Minimum load on each unit is 10 MW,

Total load to be supplied is 150 MW

Determine the economic operating shedule.

Solution of this problem is given as below :

Say, $\lambda = 2.8$ then to determine generation P_1 and P_2 as :

$$P_1 = \frac{2.8-2.0}{0.012} = 66.6 \text{ MW}, \quad P_2 = \frac{2.8-1.3}{0.015} = 86.6 \text{ MW}$$

$$\therefore \quad P_1 + P_2 = 66.6 + 86.6 = 153.2$$

Hence it is noted that the generation is greater than the demand, therefore next approximation of λ should be less than the 2.8. As load demand is also 150 MW, the difference between P_D and P_n is not such. Therefore, value of λ should not be very much different from 2.8. Let it be 2.7.

Then, with this, the generation are 58.4 + 80 = 138.4 MW, which is less than the demand. Therefore, next approximation λ should be greater than 2.7, say now it is 2.75. Hence with then the generation are 62.5 + 83.4 = 145.9 MW.

Let us take the next approximation of λ = 2.78, which gives generation as 65 + 85.4 = 150.4 MW. Now, it is to be seen that the generations are converging as better approximations to λ chosen and finally a value between 2.77 and 2.78 will give the optimal generation shedule.

But in this iterative process there is some limitation, as this method does not sense the location of changes in the loads. As long as the total load is fixed, irrespective of the location of loads, the solution will always be the same and in fact for this reason the solution may not be feasible in the sense that the load voltages may not be within specified limits, the reactive power generations required also may not be within limits. But in case of an urban area where the load density is very high and the transmission distances are very small, the transmission loss could be neglected and the optimum strategy of generation could be based on the equal incremental production cost as outlined above. But where the energy is transported over relatively larger distances with low load density, the transmission losses in some cases may amount to 20% to 30% of the total load and it then becomes very essential to lake these losses into account when developing the strategy for economic load dispatch. Now below we shall discuss on optimum Load Dispatch including Transmission Losses.

19.1.3 Optimum Load Dispatch with Transmission Losses

To derive an expression or strategy for optimum load dispatched let us consider the Fig. 19.2 to consider the inclusion of Losses in the lines. In the Fig. 19.1 there two identical generators, *i.e.*, generators with identical incremental production cost. The total load must be shared equally by both the generators, that means, each generator should supply half of the total load. The common sense tells us that it is more economical to let generator no. 2 supply most of the local load because generator no. 1 has to supply in addition to the load, the transmission losses also. Therefore, the criterion of sharing load by equal incremental production cost does not hold good under such situation and a strategy must be evolved which takes into account the transmission losses also. Hence the optimal load dispatch problem including transmission losses is defined as

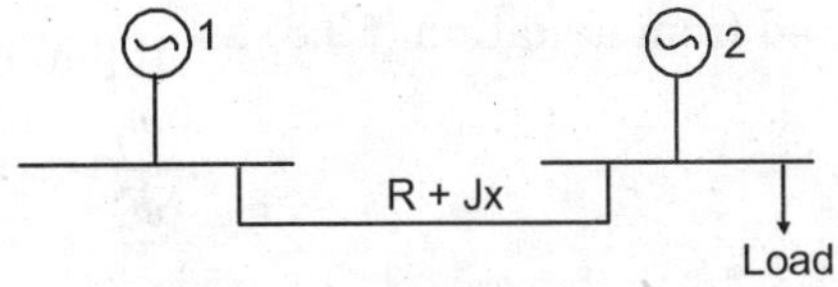

Fig. 19.2. Two identical generators connected through a transmission link.

$$\text{Min } F_T = \sum_{n=1}^{n} F_n \quad \text{...(19.5)}$$

Subject to
$$P_D + P_L - \sum_{n=1}^{n} P_n = 0 \quad \text{...(19.6)}$$

Where P_L is the total system loss which is assumed to be a function of generation and the other term have their usual significance. Making use of the Lagrangian multiplier λ, the auxiliary function is given by

$$F = F_T + \lambda\left(P_D + P_L - \sum P_n\right)$$

The partial differential of this expression when equated to zero gives the condition for optimal load dispatch :

$$\frac{DF}{dP_n} = \frac{\partial F_T}{\partial P_n} + \lambda\left(\frac{\partial P_L}{\partial P_n} - 1\right) = 0$$

or
$$\frac{dF_n}{dP_n} + \lambda\frac{dP_L}{dP_n} = \lambda \quad \text{...(19.7)}$$

Here $\frac{\partial P_L}{\partial P_n}$ is known as the incremental transmission loss at plant n and λ is known as the incremental cost of of received power in Rs. per MW hr.

The equation (19.6) is a set of n equations with $(n + 1)$ unknwons where n generations are unknown and λ is also unknown. These equations are known as coordination equations because they coordinate the incremental transmission losses with the incremental cost of production. Now to solve these equations the loss formula equation below is expressed in terms of generation and is approximately expressed as :

$$P_L = \sum_m \sum_n P_m B_{mn} P_n \quad \text{...(19.8)}$$

Here in this equation (19.8) P_m and P_n are the source loadings, B_{mn} the transmission loss coefficients. The formula is derived under the following assumptions : (1) The equivalent load current at any bus remains a constant complex fraction of the total equivalent laod current (2) the generator bus voltage magnitudes and angles are constant, (3) The power factor of each source is constant.

The solution of coordination equation (19.7) requires the calcuation of $\delta P_L/\delta P_n$ which is obtained from equation (19.8) as

$$\frac{\delta P_L}{\delta F_n} = 2\sum B_{m'} P_m$$

Also, $$\frac{dF_n}{dP_n} = F_{nn} P_n + f_n \qquad ...(19.9)$$

∴ The coordination equation can be written as

$$F_{nn}P_n + f_n + \lambda \sum 2B_{mn}P_m) = 1 \qquad ...(19.10)$$

Now collection the coeffts. of P_n in this equation for P_n we obtain

$$P_n(F_{nn} + 2\lambda B_{nn}) = -\lambda\left(\sum_{n=1} -f_n + \lambda\right)$$

Then solving for P_n we obtain,

$$P_n = \frac{1 - \frac{f_n}{\lambda}\sum_{n=1} 2B_{mn}\, P_n}{(F_{nn}/\lambda) + 2B_{nn}} \qquad ...(19.11)$$

Now to arrive at an optimal load dispatching solution, the simultaneous solution of the coordination equation along with the equality constraint should be sufficient and any standard matrix inversion subroutine could be used.

However, as the plants might go beyond their loading conditions, it is required to solve a new set of equations and thus by the process of elimination this could be done but as this could be very time consuming in a large interconnected system, an iterative process can be used. Following steps are taken as earlier.

(*a*) We assume a suitable value of λ°. This value should be moe than the largest intercept of the incremental production cost of the various generators.

(*b*) Calculate the generation based on equal incremental production cost.

(*c*) Calculate the generation at all the buses using the equation.

$$P_n = \frac{1 - \frac{f_n}{\lambda} - \sum_{m \neq n} 2B_{mn}P_n}{\frac{F_{nn}}{\lambda} + 2B_{nn}}$$

Hence it is to be noted that the power to be substituted on the right hand side during zeroth iteration correspond to the values as calculated in step (*b*). For subsequent iterations the values of powers to be substituted correspond to the powers as calculated in the previous iteration. In case any of the generation violates the limit the generation of that generator is fixed at the limit violated.

(*d*) Check if the difference in power at all generator buses between two consecutive iteration is less than a prespecified value. If not, we shall go back to step (*c*).

(*e*) Now calculate losses using the relation

$$P_L = \sum_m \sum_n P_n B_{mn} P_m$$

and then calculate, $$\Delta P = \left|\sum P_G - P_L - P_D\right|$$

(*f*) Then if ΔP is less than E, stop calculation and calculate cost of generation with these values of powers.

(*g*) Now let us update value of λ and we go back to step (*c*)

Example 19.1. *The fuel inputs per hour of plants 1 and 2 are given as*

$$F_1 = 0.2P_1^2 + 40P_1 + 120 \text{ Rs. per hr.}$$
$$F_2 = 0.25P_2^2 + 30P_2 + 150 \text{ Rs. per hr.}$$

Determine the economic operating schedule and the corresponding cost of generation if the maximum and minimum loading on each unit is 100 MW and 25 MW, the demand is 180 MW, and transmission losses are neglected. If the load is equally shared by both the units, determine. the saving obtained by loading the units as per equal incremental production cost.

Solution : (*a*) The incremental production cost of both the units

are $$\frac{dF_1}{dP_1} = 0.4P_1 + 40 \text{ Rs. per MW hr.}$$

and $$\frac{dF_2}{dP_2} = 0.5P_2 + 30 \text{ Rs. per MW hr.}$$

Now for economic load dispatch operation of the units,

$$\frac{dF_1}{dP_1} = \frac{dF_2}{dP_2}$$

That means, $$0.4\,P_1 + 40 = 0.5P_2 + 30 \quad ...(a)$$

and $$P_1 + P_2 = 180 \text{ MW} \quad ...(b)$$

Solution of these two equation, give,

$$P_1 = 88.89, P_2 = 91.11 \text{ MW}$$

Now cost of generation $= F_1 + F_2$

$$F_1 = 0.2P_1^2 + 40P_1 + 120 + 5255.88 \text{ Rs./hr.}$$
$$F_2 = 0.25P_2^2 + 30P_2 + 150 = 4958.55 \text{ Rs./hr.}$$

Total cost $= F_1 + F_2 +$ Rs. 10214.43/hr.

(*b*) If the load on each unit is 90 MW the cost of generation will be :

$$F_1 = \text{Rs. } 5340/\text{hr.},\quad F_2 = \text{Rs. } 4875/\text{hr.}$$

∴ Total cost $= F_1 + F_2 =$ Rs. 10215/hr.

∴ Saving will be, Rs. 10215 – Rs. 10214.43 = Rs. 0.57/hr.

Example 19.2. *Determine the incremental cost of received power and the penalty factor of the plant shown in Fig. 19.2a below, if the incremental cost of production is:*

$$\frac{dF_1}{dP_1} = 0.1\,P_1 + 3.0 \text{ Rs./MW hr.}$$

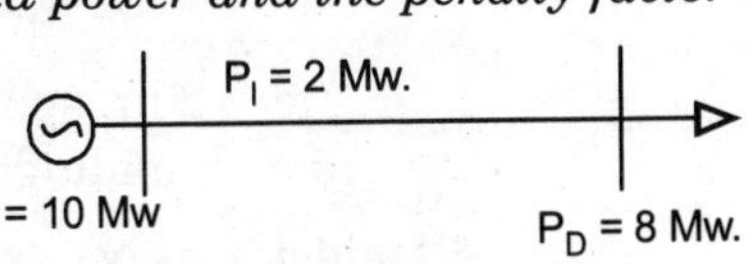

Fig. 19.2a

Solution : The penalty factor $= \dfrac{10}{8}$

$\therefore$ Cost of received power $= \dfrac{dF_1}{dP_l} L_1 = (0.1 \times d + 3) \cdot \dfrac{10}{8} = \text{Rs. } 5/\text{MW hr.}$

Example 19.3. *Assuming that the fuel input in BTU per hour for units 1 and 2 are given by*

$$F_1 = (8P_1 + 0.024P_1^2 + 80)\, 10^6$$
$$F_2 = (6P_2 + 0.04P_2^2 + 120)\, 10^6$$

The maximum and minimum loads on the units are 100 MW and 10 MW respectively. Determine the minimum cost of generation when the following load, (as shown in the Figure 19.2b below) is supplied. The cost of fuel is Rs. 2 per million BTU.

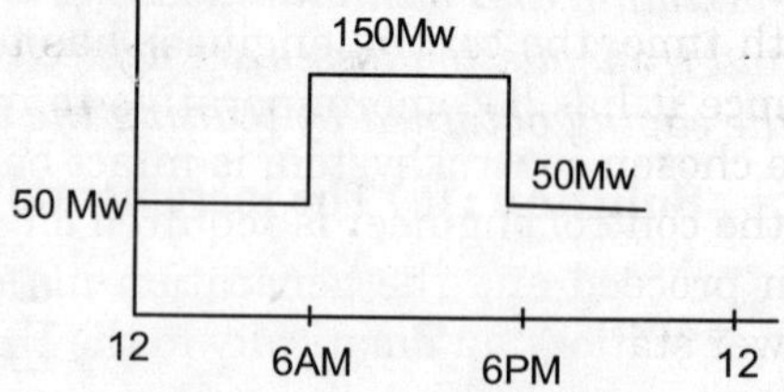

Fig. 19.2b

Solution : From the fuel input characteristics

$$\frac{dF_1}{dP_1} = 0.048P_1 + 8$$

and

$$\frac{dF_2}{dP_2} = 0.08P_2 + 6$$

1. When load is 50 MW, for economic loading the conditions are

$$\frac{dF_1}{dP_1} = \frac{dF_2}{dP_2} \quad \text{or,} \quad 0.048P_1 + 8 = 0.08P_2 + 6 \qquad ...(1)$$

and $\quad P_1 + P_2 = 50 \qquad ...(2)$

i.e., from these equation, $\quad P_1 = 15.625$ MW and $P_2 = 34.375$ MW

By putting the values $\therefore$ P_1 and P_2 in equation of F_1 and F_2 $\quad F_1 = 210.868$ million BTU per MW

2. When load is 150 MW, the equations are

$$0.048P_1 + 8 = 0.08P_2 + 6 \qquad \left[\text{As } \frac{dF_1}{dP_1} = \frac{dF_2}{dP_2}\right]$$

and $\quad P_1 + P_2 = 150$

$\therefore \quad P_1 = 78.126$ MW and $P_2 = 71.874$ MW

$\therefore$ By putting values of P_1 and P_2 in the given equation of F_1 and F_2, we get,

$$F_1 = 851.496 \text{ million BTU/hr.}$$
$$F_2 = 857.87 \text{ million BTU/h.}$$

Total cost $= \text{Rs. } (F_1 + F_2) + (F_1 + F_2)$
(for 12 hr. operation from 12 midnight and 12 mid day as per example) $\quad$ for 50 MW. $\quad$ for 150 MW

$= \text{Rs. } [(210.868 + 373.5) + (851.496 + 757.87)] \times 2 \times 12$

$= \text{Rs. } 52649.64$

19.1.4 Automatic Load Dispatching

By now we understand that economic load dispatching is that aspect of power system operation wherein it is required to distribute the load among the generating units actually paralleled with the system in such a manner as to minimise the cost of supplying the minute-to-minute requirements of the system. In a large interconnected (as in found is present system) system it is humanly impossible to calculate and adjust such generations and hence the help of digial computer system along with analogue devices is sought and the whole process is carried out automatically, hence the term automatic load dispatching is used for the purpose. In fact an energy system is normally controlled by a central load control centre which coordinates the operations of several area centres. The objective is to minimise the cost of supplying electricity to the load points while ensuring security of supply against loss of generation and transmission capacity and also maintaining the voltage and frequency of the system within specified limits. Since the interconnection is growing bigger and bigger in size with time, the control engineer has to make adjustment to various parameter in the system. Hence it has become imperative to make use of automatic control for load dispatch problem the chosen control system is infact based on a digital computer working on-line. The presence of the control engineer is required only to enter certain data. The rest of the control calculation can proceed and the generation instruction caculated, transmitted and implemented at the power stations automatically in the Fig. 19.5 is shown the block diagram of an automatic control system used for load dispatching.

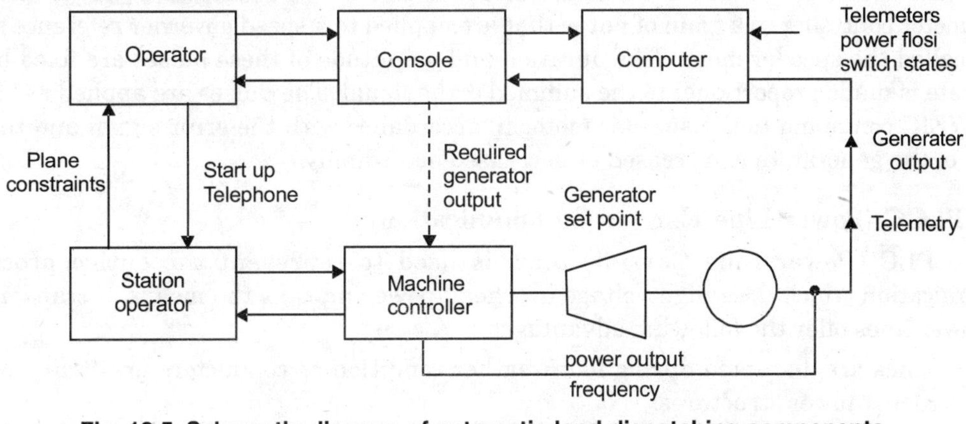

Fig. 19.5. Schematic diagram of automatic load dispatching components.

A brief discussion is made below on the operational function of various components for automatic load dispatching.

Computer

In fact, the computer predicts the load and suggests economic loading. Computer actually transmists information to machine controllers.

Data Input

The computer receives a lot of essential data from the telemetering system and from the paper tape. Telemetering data comes to the computer either as analogue signals representing line power flows, plant outputs or as signal bits indicating switch or isolater position. The system is entirely automatic. Paper tape stores all the basic data required, *e.g.*, the system parameters, load prediction security constraints etc., and finally asking the stations to switch the machine controllers to automatic control.

Console

In this system, the console is the component through which the operator can converse with the computer. The operator can obtain certain information required for some action to be taken under emergency condition in case. Such a condition do arise in the system, or the data can be put by the operator if needed. The instructions being issued and the state of the energy system (tie-line loading, plant outputs, nodal voltages, etc.) can be examined on cathode ray displays associated with the console.

In fact the console has the facilities of security checking and load flows for the network calculation. Such a facility can be called manually at almost any instant of time and the results may be recorded automatically after a specified interval of time, if required. The security check (like loss of generator or loss of one or more than one transmission links) reveals and displays that if one or more than one particular branches suffer outage, at last one of the remaining branches will be overloaded.

Machine Controller

The computer sends instructions regarding the optimal generation to the machine controller at regular intervals which in turn implements them. Control on each machine is applied by a closed loop system which uses a measure of actual power generated and which operates through a conventional speeder motor. These are referred to as controller power loops. However, in the power frequency loop an error signal is ferred from these two component and is converted in the motor controller to a train of put is that are applied to a speed governor reference setting motor called the speeder motor. The duration and amplitude of these pulses are fixed but the pulse rate is made proportional to the summed error signal. The pulses are applied as "RAISE or LOWER" command to the speeder motor in accordance with the error signal and thus the output of the generators is increased or decreased accordingly.

19.1.5 PLCC (Power Line Carrier Communication)

The PLC (Power Line Carrier) term is used to represent the entire process of communication which uses high voltage overhead power lines as the means of transmission. The power lines offer the following advantages.

1. Lines are disrupted during foot weather condition as conductors are strung on very robust tower structures.
2. Leakance are disrupted during foul weather condition as conductors are strung on very robust tower structures.
3. Lines have thicker cross section of wire and hence attenation of signals is not much.
4. As the phases are separated from each other quite appreciably cross-linkage between lines is practically avoided.
5. The cost of providing extra lines for the purpose is avoided.

But, there are certain difficulties associated with the transmission of communication signal over the power lines.

(*a*) Since the power is being transmitted at relatively high voltages, the operation on these lines may prove to be dangerous to human lives and also to telephone apparatus.

(*b*) During transient operation of system, *e.g.*, switching transients, surge voltages or corona phenomenon etc. the presence of higher hormonics in power currents may interfere with the communication signal.

In fact these difficulties have been overcome by selecting suitable coupling capacitors and carrier frequencies. The power line carrier communication finds application in telemetering power line protection, telecontrol etc. We know that the power lines may have higher hormonics due to switching surges and corona loss on overhead lines. These frequencies generally line between 100 Hz to 50 kHz and therefore, if carrier frequencies are chosen in this region noise introduced in carrier signal would be very large and hence carrier frequency in the range of 30 kHz to 500 kHz are used. In order that these carrier signals do not interfere into the adjoining section of lines also that these carrier signals are not lost by being shorted by low impedance of transformer or generator at the end of the line, line traps are used which offer very high in impedance to carrier frequency and low impedance to power frequency but low to carrier frequencies. The two most commonly employed coupling methods are centre-phase to ground and adjacent to ground and adjacent-phase to phase to ground and to phase as shown in Fig. 19.6.

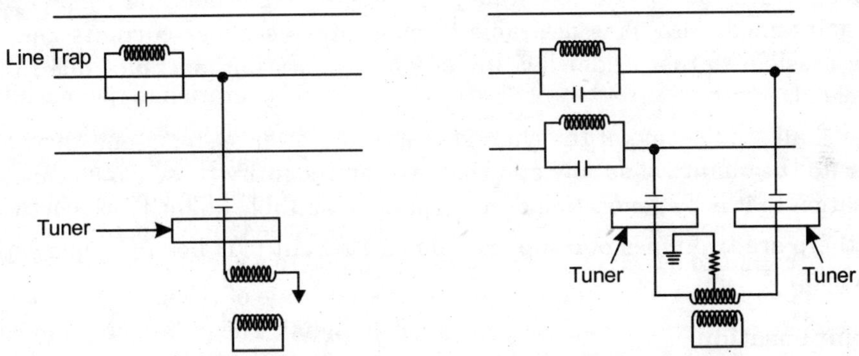

Fig. 19.6. Coupling circuits (*a*) centre-phase to ground and (*b*) adjacent phase to phase.

Modern power system has grown quite complex. The awareness of power system engineer for a good quality of reliable supply forced to adopt some automatic means so that he not only operates his system reliably but most economically. For this he has to know the state of the system so that he can regulate the system in that direction. The state (frequency, voltage, active power, Reactive power flows, position of C.P., Switches etc.) of such a complex system can be obtained by telemetering only. No manual method can handle such a complex job.

To conveniently handle the situation, the whole system is divided into a number of areas or groups, each area having an area (or group) control station and finally these stations are centrally controlled (or coordinated) by what are known as master-controlled station. These different control stations are located at suitable point in the network and they control such item as voltage, power, frequency on getting information from distant points regarding condition of the system by telemetering. There are certain informations to be sent to master control station only yet there are other to be sent to both.

Information telemetered to group control stations areas follows :

1. Generated active and reactive power of each station in the group.
2. Net active and reactive power transfer from the group.
3. Active and Reactive power flow through the tie lines, alongwith direction of flow and overload alarm.

Then information to be telemetered to control station as follows :

1. Voltage and frequency from selected points.
2. Total generated active and reactive power of system.

3. Tie-line loading (active and reactive powers) from selected points.

Then information to be telemetered to both group and central control stations :

1. System frequency and rate of change of frequency.
2. Generated active and reactive power.
3. Area-interconnection active and reactive power.

19.2 INTRODUCTION TO COMPENSATION AND FACTS

19.2.1 Introduction to Compensation and Facts

A power system is expected to operate under widely varying conditions from no-load to over loading to short circuits and it is desired that the quality of supply is maintained under all conditions. Which means that constancy of voltage magnitude and frequence of system should be maintained. Also, it is desirable to maintain the phase currents and voltages as balanced as possible so that under heating of various rotating machines due to unbalances could be avoided.

Here we shall study some of the characteristics of power systems and their loads which can detoriarate the quality of supply and then we can indentify these which can be corrected by compensation that is by generation or absorption of suitable quantity of reactive power.

In fact there are two types of compensation carried out : (1) Load Compensation and (2) Line Compensation

1. Load Compensation

Load compensation is the Management of Reactive Power to improve the quality of supply expecially the voltage and p.f. levels. Here the reactive power is adjusted with respect to an individual load and the compensation device is connected to the load itself. Hence, there are three main objectives.

1. Better voltage profile.
2. p.f. correction.
3. Load balancing

In power system, the voltage profile must remain within ±5% of the rated value for better and efficient operation of various electrical equipments.

Infact, voltage variation is due to unbalance in the generation and consumption of reactive power in the systems.

If the generated reactive power is more than what is being consumed, voltage levels goes up and vice versa. But if the two are equal (*i.e.*, $P = Q$) the voltage profile becomes flat. Unfortunately, the reactive power in a system keeps on varying and hence if the reactive power generation is also simultaneously controlled, a more or less near flat voltage profile could be maintained.

One of the obvious methods would be to have the system of large strength *i.e.*, it interconnects large size machines and a large number of lines so that the effective impedance as seen from any point into the network is negligibly small and hence the voltage profile could be improved. But this would result in high fault levels and would require switch gears of relatively large capacity and hence is uneconomical from view point of improving voltages profile.

In fact, the network should be designed based on active power transfer capability and the reactive power transfer capability and the reactive power should be met locally by installing shunt compensating elements (capacitor and Inductors). Infact, it is to be noted that these elements do not contribute to the fault level end yet provide and maintain proper balance between the generated and consumed reactive power.

2. It is desirable that both economically and technically, to operate the system at near p.f. usually p.f. correction means to general reactive power as close as possible to the load as this result not only in large conductor size but also in increased losses. In fact in order to operate the system at near unity p.f. some of the electric utilities impose certain penalty on account of operating loads at low p.f.

3. A very important concept of load compensation is load balancing. It is desirable to operate the three phase system under balanced condition as unbalanced operating highly dangerous especially for the rotating machines.

Infact, IDEAL Load compensator would be for following function. (1) It will provide controllable and variable reactive power almost instantaneously as required by the load, (2) It should operate independently three phases. (3) It would maintain constant voltage at its terminals. Typical loads requiring compensation are : Arc Fuses Inductor Furnace, Steel Rolling Mills, Large action motors, Arc Welders, inductors.

19.2.2 SIL (Surge Impedance Loading)

SIL (Surge Impedance Loading) is the power delivered at Rated voltage by a lossless line to a load resistance whose value is equal to the surge impedance, $Z_C = \sqrt{L/C}$ for a lossless line Z_C is purely resistive.

$$\text{SIL} = V^2_{R(rated)}/Z_C$$

19.2.3 Reactive Compensation

Considering the voltage profiles of a long transmission line under different load conditions, it is to be observed that at light loads which are marked less than the (SURGE IMPEDANCE LOADING) the voltage at the receiving end is more than the sending end voltage.

Conversely, when the loads are several times more than the SIL the voltage at the receiving end plunges downward of course. When the load is equal to SIL (Surge Impedance Loading), there is no flow of reactive power and the voltage is constant along the line. From the forgoing discussion, it is understood that when the line is lightly loaded and also when the line carriers a heavy load the voltage has to be controlled at both the ends of the line.

Hence to keep the voltage controlled the following type of compensations are employed in Power System.

1. Inductive Shunt Reactor Compensation

Under light load or open line conditions, Shunt Reactor Compensation is employed. The Fig. 19.5 shows an equivalent circuit with an inductive shunt Reactor having a reactance of jX_{Lsh} connected at the receiving end.

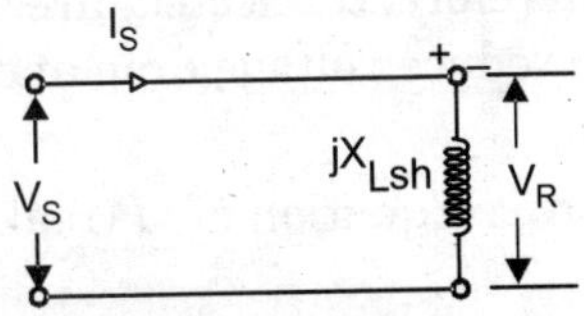

Fig. 19.5. Shunt receive compensation of Long line.

Now if V_R is the receiving end voltage, then the receiving end current is given by,

$$I_R = \frac{V_R}{jX_{L_{Sh}}} \quad ...(19.12)$$

Substituting this value of I_R in the equation of V_S below :

$$V_S = \cos(\beta l)\, V_R + jZ_C \sin(\beta l)\, I_R$$

Now putting the value of I_R in this equation of V_R we get the V_S in terms of V_R.

$$V_S = \cos(\beta l)\, V_R + j\,Z_C \sin(\beta l)\, \frac{V_R}{j\,X_{L_{Sh}}} \quad ...(19.13)$$

On simplification (to get expression for $X_{L_{Sh}}$)

We get,

$$X_{L_{Sh}} = \frac{Z_C \sin(\beta l)}{\left\{\frac{V_S}{V_R} - \cos(\beta l)\right\}} \quad ...(19.14)$$

This equation provides the value of the Inductive Reactance to be connected at the receiving end. It may also be noted that since no real power is flowing along the line the sending end and receiving end Voltage's are in phase. Then sending end current for such a long line can be expressed as,

$$I_S = \left[j\frac{1}{Z_C} \sin(\beta l) \right] V_R + \cos(\beta l)\, I_R \quad ...(19.15)$$

and substituting for V_R from euation (19.12) in equation (19.15), we get

$$I_S = \left[-\frac{X_{L_{Sh}}}{Z_C} \sin(\beta l) + \cos(\beta l) \right] I_R \quad ...(19.16)$$

If the value of $X_{L_{Sh}}$ is substitute by assuming $V_S/V_R = 1$ and $l = 0$ in equation then it is obtained that $I_S = I_R$.

19.2.4 Shunt Capacitor Compensation

When the power factor at the receiving end load is low, then either the magnitude of the sending end voltage is required to be increased, or reactive power is necessary to be supplied at the load bus, in order to maintain the voltage at the consumer terminals. The latter technique for improving the power factor has been found to be more effective. Shunt capacitors are, therefore, connected directly across the load bus or in the tertiary winding of the transformer to reduce voltage drop and line losses.

$\therefore$ Equation of P(real Power)

$$= -\frac{|V_R|^2}{|Z|} \sin\theta + \frac{|V_R|\,|V_S|}{|Z|} \cos(\theta - \delta)$$

and Q_R (Reactive Power) or Imaginary Part

$$= -\frac{|V_R|^2}{|Z|} \sin\theta + \frac{|V_R|\,|V_S|}{|Z|} \sin(\theta - \delta)$$

Express real(P) and reactive (Q_R) Receiving end Power for a nominal π-equivalent circuit and can be used to compute the reactive Volt-Amp (VAR) rating of the shunt capacitance reactor $X_{C_{Sh}}$ required to be connected for a given load.

19.2.5 Capacitive Series Reactor Compensation

When capacitive Reactors are connected on series with long line will help raising both the steedy state and transient stability limit, in addition to minimising the voltage clip at the consumer's terminal during heavy loading of a line. The Fig. 19.6 shows series and shunt capacitors connected in a long power transmission line.

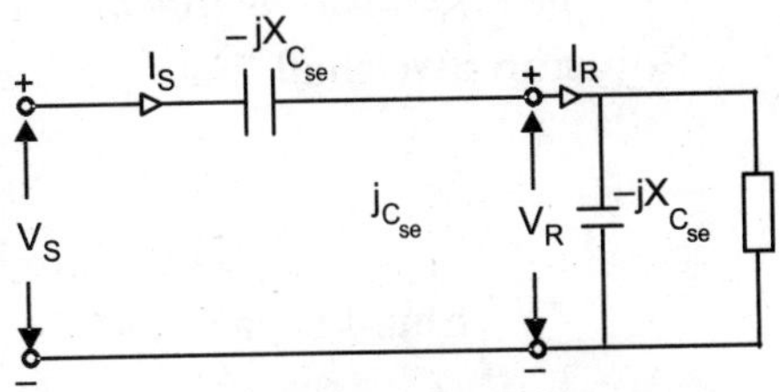

Fig. 19.6. Shunt & series compensation for long line.

The per-phase active power transferred over a loss per long line for the equivalent circuit shown in Fig. 19.6, can be written as :

$$P = \frac{|V_S|\,|V_R|}{(X' - X_{C_{Se}})} \sin\delta \qquad \text{...(19.16)}$$

Here $\quad X_{C_{Se}}$ = Reactance of the capacitor connected in series with line.

This ratio $X_{C_{Se}}/X'$ is defined as the compensation in this case.

From equation (19.16) it may be noted that the theoretical capability of the line to transmit power has been enhanced. Experience has shown that by incurring an additional cost of series capacitors (which is a fraction of the cost of a new line), the transmission capability of extra high voltage lines can be more than doubled.

Another advantage of the use of series capacitors is that their reactive volt-ampere rating varies with the line loading.

The disadvantage of series compensation is that under short circuit conditions, the series capacitors have to be protected against the resultant high current. Special protective devices, which bypass the short circuit current, are employed for this purpose. A second disadvantage is that when a disturbance occurs in the system, a resonant circuit, which resonates at a frequency below the normal synchronous frequency, is set up. The **sub-synchronous frequency** f_r is

$$(SSR) = f_r = f_{S_e}\sqrt{\frac{1}{L'C_{Se}}} \qquad \text{...(19.17)}$$

Where L' is the inductance in line of the π-equivalent circuit, C_{S_e} is the series capacitance in farads, and f_S is the synchronous frequency in Hertz. The sub-synchronous resonance (*SSR*) as it is called, is likely to cause damage to the turbo-generator, if it occurs.

Let us now consider an example to explain the principle of compensation :

Example 19.3. *A 3-phase, 400 kV, 50 Hz transmission line has a series inductive reactance of 0.3 Ω/km and shunt admitance of 3.75 × 10^{-6} S/km. The sending end rated voltage of the 3-phase transmission line is 400 kV.*

1. *Compute the reactance and rating of an inductive shunt reactor to be connected at the receiving and maintained at 400 kV on no load.*

2. *A three-phase load of 800 MVA at 0.8 p.f. lagging a is connected at the receiving end of the line. It is required to maintain a voltage of 400 KV at the receiving end when the line is supplying the load. Find the MVAR and capacitance of shunt capacitor for compensation purpose.*

3. *Calculate the sending end voltage and regulation if a series capacitor connected in the middle of the line provides 50% compensation.*

Solution : We shall first find the following characteristics, value of the line :

$$L = \frac{0.30}{2\times\pi\times 50} = 9.5493\times10^{-4} \text{ H/km.}$$

Shunt capacitance $C = \dfrac{3.75\times10^{-6}}{2\times\pi\times50} = 1.1937\times10^{-6}$ F/km

1. Surge impedance $Z_C = \sqrt{\dfrac{9.5493\times10^{-4}}{1.193\times10^{-8}}} = \left(Z_C = \sqrt{\dfrac{L}{C}}\right) = 282.8427\ \Omega$

Phase constant $\beta = 2\times\pi\times50\sqrt{(9.5493\times10^{-4})\times(1.1937\times10^{-8})}$

$= 0.0011$ rad/Km $\left(= 2\pi f\sqrt{LC}\right)$

2. Propagation constant $j\beta = j\,0.0011$ per Km.

$\therefore Z_C$ = Surge Impedance $= 282.8727\ \Omega$

phase constant, $\beta = 0.0011$ rad/Km, $\beta l = 18.9076°$

per phase sending voltage $V_S = 400/\sqrt{3} = 230.9401$ kV

Then, SIL (Surge Impedance Loading) $= \dfrac{V_R^2\text{(rated)}}{Z_C}$.

Before solving the problem let us first know how to find expression SIL $= \dfrac{V_R^2}{Z_{C_1}}$

Surge Impedance Loading (SIL) is the power delivered at rated voltage by a loss less line to a load of resistance whose value is equal to the surge impedance, $Z_C = \sqrt{L/C}$. for a loss less line Z_C, is purely resistive. To prove that SIL $= \dfrac{V_R^2}{Z_C}$. Let us refer to Fig. 19.7 which shows a lossless line of length l terminated in a resistance equal to surge impedance.

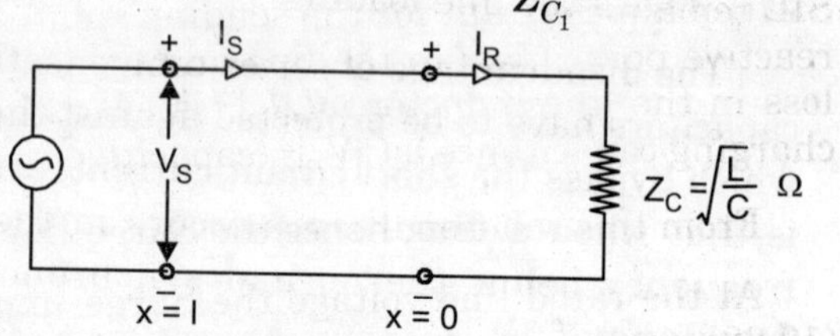

x is the length of line
L = Inductance of line
C = Capacitance

Fig. 19.7

We know from own previous experience that the propagation constant $Y = j\beta$, for a lossless line, the *ABCD* onstant are written as :

$$A = D = \cos h(j\beta l) = \frac{e\,j\beta l + e^{-j\beta l}}{2} = \cos\beta l$$

$$B = Z_C \sin h\,(j\,\beta\,l) = j\,Z_C \sin(\beta\,l)\ \Omega$$

$$C = \frac{1}{Z_C}\sin h(j\beta l) = j\frac{1}{Z_C}\sin(\beta l),\ S\text{(Siemen)}$$

We also know that the voltage at a from the distance x receiving end is written as :

$$V(x) = \cos(\beta x) V_R + j Z_C \sin(\beta x) I_R$$
$$= \cos(\beta x) V_R + j Z_C \sin(\beta x) \left(\frac{V_R}{Z_C}\right)$$
$$= [\cos(\beta x) + j \sin(\beta x)] V_R$$
$$= e^{j\beta x} V_R = V_R \angle \beta x \quad ...(19.18)$$

or
$$|V(x)| = |V_R| \quad ...(19.19)$$

Now equation shows that in a lossless line under surge impedance loading the magnitude of voltage at any point x is constant and is equal to the sending end values but has different phase position.

Simimilarly regarding current,
$$I(x) = j\frac{1}{2} \sin(\beta x) \cdot V_R + \cos(\beta x) \left(\frac{V_R}{Z_C}\right)$$
$$= [\cos \beta x + j \sin(\beta x)] \frac{V_R}{Z_C}$$
$$= e^{-j\beta x} V_R/Z_C \quad ...(19.20)$$

$\therefore$
$$|I(x)| = |I_R| \quad ...(19.21)$$

Equation (19.21) shows that in a lossless line under surge impedance loading the magnitude of current at any point along the line is constant, and is equal to the sending end value.

The complex power $S(x)$ at any point x along the line may be obtained by using equations (19.18) and (19.20) as

$$S(x) = P(x) + jQ(x) = V(x) \cdot I(x)$$
$$= \left(e^{j\beta x} V_R\right)\left(e^{e\beta x} \cdot \frac{V_R}{Z_C}\right) = \frac{|V_R|^2}{Z_C} \quad ...(19.22)$$

Now from equation (19.22) it is observed that the real power flow along a lossless line at SIL remain constant for the sending end to the receiving end since Z_C is resistive, there is no reactive power flow in the line and $Q_S = Q_S = 0$. This indicates that under SIL, the reactive loss in the line inductor $wL|I_R|^2$ is exactly offset by the reactive power supplied by the line charging capacitance $wC|V_R|^2$ (capacitive charge power), this means that, $wL|I_P|^2 = wC|V_R|^2$.

From this relation, hence the Surge Impedance may be obtained as $Z_C = V_R/I_R = \sqrt{L/C}$.

At the rated line voltage the Surge Impedance Load of the line is given by the equation (19.22), as dervied above is

$$\text{SIL} = \frac{V_R^2 \text{ (rated)}}{Z_C}$$

Where rated voltage is used for a single-phase and rated line-to-line voltage is used for a three-phase line. The surge impedance loading of the line in itself is not a measure of the maximum power that can be delivered over the line.

Now from equation (19.19), it is to be noted that the magnitude of the voltage at the sending end is equal to the magnitude of the voltage at the receiving end. Thus there is no voltage drop and the voltage profile along the line is flat as shown in Fig. 19.9.

But, in power engineering practice the load on lines can vary significantly and is different from SIL. It may only be a fraction of SIL, when the line is lightly loaded and it may be several times SIL under heavy load depending on the length of the line and reaactive compensation. For loads much higher than SIL. Shunt capacities may be needed to minimize voltage drop along the line.

On the other hand, for light load conditions, *i.e.*, loads much loss or than the SIL shunt inductors be need to compensate the line charging current.

Hence let us refer to the Fig. 19.8 where the voltage profile is explained in the transmission lines at various loading condition.

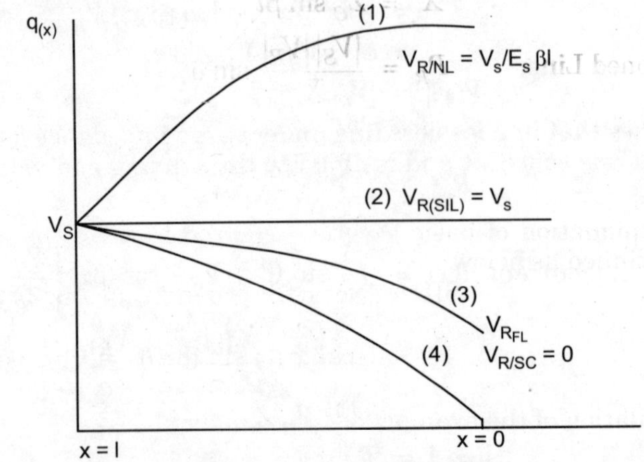

Fig. 19.8. Voltage profile along a transmission line for various loading conditions.

When a transmission line does not carry a load equal to SIL, the voltage profile, along the line is no longer horizontal. Fig. 19.8 shows a plot of the voltage profile for a transmission line of length 1500 Km (quarter wavelength) for different loading condition. The voltage at the sending end is maintained constant for all loading conditions.

For a lossless line, the voltage at any point x of the line is given by equation, $V(x) = \cos(\beta x)\, V_R + jZ_C \sin(\beta x)\, I_R$. The following four cases may be considered for this relation and shown graphically in Fig. 19.7.

1. In the case when the line carries no load (NL) the $I_{R_{(NL)}} = 0$, then voltage at any point x from the receiving end is

$$V_{(NL)}(x) = \cos(\beta x)\, V_{R_{NL}}$$

   where subcript NL stands for no load condition values. Then the sending end voltage $V_{S_{(NL)}} = \cos\cos(\beta l)\, V_{R_{(NL)}}$, and the no load voltage increases for $U_{S_{(NL)}}$ at the sending end to $V_{R_{(NL)}}$ at the receiving end where $x = 0$.
2. When the line carries a load equal to the surge impedance Z_C, the voltage profile at the receiving end $V_{R_{(SIL)}}$ is flat as discussed above shown in curve (3).
3. When there is a short circuit at the receiving end, then the receiving end, then the receiving end voltage $V_{R_{(SC)}} = 0$, where the subscript (SC) stands for short-circuit values. The equation $V(x) = \cos(\beta x)\; V_{R_{SC}} + jZ_C \sin(\beta x)\, I_{R_{SC}}$ becomes $V_{S_{(SC)}} = Z_C \sin(\beta x) I_{R_{SC}}\ (\because V_{R_{SC}} = 0)$, and the voltage decreases from this value at the sending end $V_{R_{(SC)}} = 0$ at the receiving end (curve 4).
4. Then for full load condition, the voltage profile $V_{R_{(FL)}}$. Will depend on the magnitude of load current and the power factor. The voltage profile will lie above the short circuit voltage profile (curve 3) and below the $V_{R_{(SIL)}} = V_R$ that is voltage at Surge Impedance Loading shown in Fig. 19.9. Now let us continue the example :

We also have seen earlier that for steady state stability limit, is

$$P_R = \frac{|V_S||V_R|}{X}\sin\delta$$

where $X' = Z_C \sin\beta l$

But Compensationed Line; $P_R = \frac{|V_S||V_R|}{X'}\sin\delta$

and P_{max} is when $\delta = 90°$

$$\therefore \quad P_{max} = \frac{|V_S||V_R|}{X'}$$

Now with the explanation of basic theories required to solve the process of solving the Example no. 1 is continued as below :

$\therefore$ SIL = Surge Impedance Loading of the Line

$$= \frac{V_R^2}{Z_C} = \frac{(400)^2}{22.8427} = \mathbf{565.68\ MW}$$

From above calculation of the example we have, inductance,

$$L = \frac{0.30}{2\times\pi\times50} = 9.55\times10^{-4}\ \text{H/Km.}$$

Shunt capacitance, pe km $C = \frac{3.75\times10^{-6}}{2\times\pi\times50} = 1.2\times10^{-8}$ F/Km

$\therefore$ Surge Impedance, $Z_C = \sqrt{\frac{L}{C}} = \sqrt{\frac{9.55\times10^{-4}}{1.2\times10^{-8}}} = 282.84\ \Omega$

Propagation constant $= j\beta = j\,0.0011$/Km.

Phase constant $\beta = 2\pi f\sqrt{LC} = 2\pi\ 50\sqrt{(9.55\times10^{-4})\times(1.2\times10^{-8})}$

$\therefore$ βl = (.0011 rad/km) x 300 rad = 0.0011 rad/km.)
= 18.19

per phase sending end voltage $V_S = \frac{400\ \text{KV}}{\sqrt{3}} = \mathbf{230.95\ KV}$

and the per phase receiving end
Voltage on no load, V_R = **243.146 KV**

$\therefore$ From the equation developed above.

For compensating shunt Reactance $X_{L_{SC}}$ as :

$$X_{L_{SC}} = \frac{Z_C\sin(\beta l)}{\left[\frac{V_S}{V_R} - \cos(\beta l)\right]} = \frac{287.8427\times\sin 18.9°}{[1-\cos(18.9°)]}$$

$$= \mathbf{1698.6\ \Omega}$$

Capacity of Shunt Reactor (for Compensation Capacity for Compensation)

as $V_R = V_S$ for a loss less line $= 3\times\frac{V_R\times V_S}{X_{L_{SC}}}$ $[\beta l = 18.9°]$

(2) Second part of Ex. 1 = **94.194 MVAR**

As 3-phase load of 800 MVa at 0.8 p.f. lagging is connected of the receiving end of the line. It is required to maintain a voltage of 400 KV at Receiving end when the line is supplying the load. Then find the value of Shunt capacitance for copensation purpose in the system. As line is lossless, sending end Real Power, P_S is equal to the Receiving End Real Power, P_R. Which means, $P_R = P_S$. Which we know from previous discussion that, $P_R = P_S = \frac{|V_R||V_S|}{X'}$. sin δ = Real Power supplied at 0.8 p.f. lagging where $X' = Z_C \sin \beta l = Z_C \beta \cdot l = Z_C \cdot \sqrt{LC \times l}$.

∴ To find the power angle for supplying real power at the receiving end of 640 MW. We can use the following relationship.

$$\frac{|V_R||V_S|}{X'} \sin \delta = P_{real} = 640$$

∴ As $|V_R| = |V_S|$, we get

$$X' = Z_C \sin \beta l = Z_C \cdot \beta l = Z_C \cdot \sqrt{LC} \cdot l \text{ for short line}$$

$$Z_C = 282.8427 \; X' = Z_C \cdot l = 282.842 \times .0011 \times 300 \text{ for short line} = 91.65\ \Omega$$

$$\beta = 0.0011 \text{ rad/km}$$

$$l = 300 \text{ km or } X' = Z_C \sin \beta l \text{ (for long line) } Z_C = \sqrt{\frac{L}{C}}$$

We also know $\sin \beta l = 18.9076$

For a long line $X' = 282.842 \times \sin 18.907$

$= 91.6532$

∴ From the relationship of Real Power

$$P_{R_e} = \frac{|V_R||V_S|}{X'} \sin \delta$$

$$= \frac{400 \times 400}{Z_C \sin \beta l (= X')} \cdot \sin \delta$$

$$= \frac{400 \times 400}{91.6532} \cdot \sin \delta$$

As $P_{R_e} = 640$ MW

Then get δ from the expression below :

$$\frac{400 \times 400}{91.6532} \cdot \sin \delta_{12} = 640$$

$$\therefore \quad \delta_{12} = \sin^{-1}\left(\frac{640 \times 91.6532}{400 \times 400}\right) = 21.5069^\circ$$

Now for a π equivalent transmission line, the parameters

$$A = \cos \beta l, \qquad A = D$$

and

$$B = j Z \sin \beta l \qquad C = j \sin (\beta l)/Z_C$$

∴ We can write the expression for sending end power and is obtained as follows :

$$Q_{12} = \frac{V_1^2}{X'} \cos (\beta l) - \frac{V_1 V_2}{X'} \cos(\delta_{12})$$

Where $X' = Z_C \sin \beta l$

Here, V_1 is the sending voltage per phase

V_2 Receiving end voltage per phase

Now substituting the values of $V_1, V_2 (V_1 = V_2)$,

$\cos \beta l$, $\cos \delta_{12}$, in the equation for Q_1^2 we get the value of Q_{12}

$$Q_{12} = \frac{(290 \cdot 9401)}{91.6532} \cdot \cos (18.9076°) - \frac{(230.94)^2}{91.6532} \cdot \sin 215°$$

$$= 9.1178 \text{ MVAR/phase}$$

$P_{\text{reactive power}}$ = Three-phase **reactive** power of the load

$$= 800 \cdot \sin \phi = 800 \times 0.6 \text{ (as } \cos \phi = 0.8)$$

$$= 480 \text{ MVAR}$$

∴ Three-phase rating of the shunt capacitor in MVAR

$$= jQ_{12} - jP_{\text{reactive power}}$$

$$= j(3 \times 9.1178) - j\,480$$

$$= -j\,452.6466$$

∴ Magnitude of shunt capacitance can be found after finding capacitive reactance from,

$$X_C = \frac{V^2}{\text{MVAR}} \qquad [\text{as MVAR} = V^2 X_C]$$

$$= \frac{V^2}{j452.6466} = \frac{(400)}{j452.6466}$$

$$= \frac{(400)^2}{j452-6466} = -j353.4766\ \Omega$$

Now as $X_C = \dfrac{1}{C\omega}$ $\quad \therefore C = \dfrac{1}{X_C\,\omega}$

Magnitude of the capacitor required for the compensation purpose is

$$C = \frac{10^{-6}}{353.4766 \times 2 \times \pi \times 56} = 9.0051\ \mu F$$

3. To find the sending end voltage and regulation, if a series capacitor is connected in the middle of the line providing 50% compensation.

Now from above calculations we get,

$$X' = Z_C \sin \beta l = 91.6532\ \Omega$$

∴ Capacitive reactance to be connected in series for 50% of compensation by X',

We get, $X_{C_{Se}} = 0.5 \times (X') = 0.5 \times 91.6532\ \Omega$

$$= \mathbf{45.8266\ \Omega}$$

∴ Receiving end current, $I_R = \dfrac{\text{Three Phase Load}}{\sqrt{3} \times V \cos \phi}$

$$= \frac{800}{\sqrt{3}\times 400\times 0.8} = \mathbf{1.443\ KA}$$

The constant of the π-equivalent circuit with line for 50% compensation are

$$Z' = j\left(X' - X_{C_{Se}}\right) = j(91.6532 - 45.8266) = \boldsymbol{j\,45.8266\ \Omega}$$

$$\therefore \quad Y' = j\frac{2}{Z}\tan\left(\frac{\beta l}{2}\right) = j\frac{2}{282.8427}\tan\left(\frac{18.9076}{2}\right) = \boldsymbol{j\,0.00215}$$

$$\therefore \quad A = \left(1+\frac{Y'Z'}{2}\right) = \left[1+\frac{j0.0021\times j45.8266}{2}\right] = \mathbf{0.9519}$$

and $\quad B = Z' = \boldsymbol{j\,45.8266\ \Omega}$

Now assuming that the per-phase Receiving End voltage V_R is the reference Vector.

We get, $\quad I_R = 1.4434\,(0.8 - 0.6) = (1.1547 - j\,0.8660)$ IA

and $\quad V_S = 0.9519 \times 230.9401 + \left(j X_{C_{Se}}\right) \times (1.1547 - j\,0.8660)$

$$= 0.9519 \times 230.9401 + (j\,45.8266) \times (1.1547 - j\,0.8660)$$

$$= (259.52 + j\,52.91)\text{ kV} = 264.8596\ \angle 11.5248^\circ\text{ KV}$$

Then, sending end line voltage = $\sqrt{3} \times 264.8596\ \angle 11.5248^\circ = 458.75\ \angle 11.52^\circ$ KV

$$\therefore \text{Regulation }\% = \frac{V_S - V_R}{V_R} = \frac{458.7503 - 400}{400}\times 100 = \mathbf{14.6876\%}$$

19.2.6 Important Point to note Regarding Compensation

1. The corrective measures are taken in two cases (where the loading in system is less than the SIL of the line and also the loading is more than the SIL) by connecting suitable compensating Reactive network at suitable P locations in the power system.
2. The basic equation describing a trasmission line is given as :

$$\frac{\partial V}{\partial x} = I_Z,\ \frac{\delta^2 V}{\delta x^2} = (r + j\omega L)\,(g + j\omega C)\,V = Y^2 \cdot V$$

and
$$\frac{\delta^2 I}{\delta x^2} = (r + j\,\omega L)\,(g + j\,\omega C)\ \text{II} = Y^2 \cdot I$$

and
$$Z = (r + j\,\omega L),\ Y = (g + j\,\omega\,C)$$

where x is the distance from the sending end l is the length of the line, the solution of the above equation for a lossless line is obtained with assumptions :

$$r = 0,\text{ and } g = 0$$

Now, we know $\quad V(x) = V_r \cos\,(l - x)\,\beta + j\,Z_C\,I_r\,\sin\beta\,(l - x) \qquad \text{...(19.23)}$

Here β is the phase constant

That is, $\gamma = \propto + j\,\beta$ but for a lossless line $\propto = 0$

$$\therefore \quad \gamma = j\,\beta$$

Now $\quad \gamma^2 = (r + j\,\omega L)\,(g + j\,\omega C)$

or $$r = g = 0, \quad \therefore \gamma^2 = -\omega^2 LC, \text{ and } \gamma = j\beta$$

Here $$\beta = \omega\sqrt{LC} = \frac{2\pi f}{v} = \frac{2\pi}{\lambda}$$

Now from equation $\delta V/\delta x = IZ$ and equation (19.23) above we get

$$I(x) = I_r \cos\beta(l-x) + j\frac{V_r}{Z_C}\cdot\sin\beta(l-x)$$

β is also known as wave number (the number of waves per unit of line length).

3. Note at 50 Hz. The wavelength $\lambda = \frac{3\times10^8}{50} = 6000$ Km and β can be expressed as one wavelength per 6000 kms, *i.e.*, 360° per 6000 Km or $\frac{360}{6000} = 0.06°$ per Km or 1.02166 $\times 10^{-3}$ rad per Km.
4. Z_C in equation is known as the surge impedance on natural impedance its value depends upon the line design. Its value decreases with increase in operating voltage. Also with vertical configuration the surge impedance is lower as compared to horizontal or delta configuration of the conductors. The usual value of Z_C lies in the range of 200 to 400 Ω.
5. The surge impedance of a line is defined as the apparent impedance of an infinite long line and is given by the ratio of voltage to the current at any point along it. Now an infinite line can be simulated by terminating any line by its surge impedance. In that case $V_r = I_r Z_C$ and the apparent impedance of the line at every point is

$$Z(x) = \frac{V(x)}{I(x)} = \frac{V_r\cos(l-x)\beta + jI_rZ_C\sin\beta(l-x)}{I_r\cos\beta(l-x) + j\frac{V_r}{Z_C}\sin\beta(l-x)}$$

$$= \frac{I_rZ_C[\cos(l-x)\beta + j\sin\beta(l-x)]}{I_r[\cos\beta(l-x) + j\sin\beta(l-x)]} = Z_C \quad \text{...(19.24)}$$

$\therefore$ It can be seen that when the line is terminated through Z_C, the voltage and current magnitude remain same all along the length. Therefore, the line is said to have flat voltage magnitude but displaced in phase by β, the sending end voltage leading the receiving voltage.

From equation (19.24) it is clear that the ratio of V/I every which is a real number Z_C and therefore, voltage and current are in phase at all points of the line, *i.e.*, power factor is unity at all point. This also means that at every point of the line, reactive power generated equals the reactive power absorbed. Therefore, whereas the natural active power loading of the line is given by V^2/Z_C the natural reactive power loading is zero.

One of the major advantages of operating the line corresponding to natural loading is that because of flat voltage profile, the insulation all along the line is uniformly stressed.

Suppose the line is open circuited at the receiving end, *i.e.*, $I_r = 0$. Then,

we get, $$V(x) = V_r\cos\beta(l-x)$$

and $$I(x) = j\frac{V_r}{Z_C}\sin\beta(l-x)$$

Now at $$x = 0,\ V(x) = V_S \text{ and } I(x) = I_S$$

$$V_S = V_r\cos\beta l$$

and $$I_S = j\frac{V_r}{Z_c}\sin \beta l = j\frac{V_S}{Z_C}\tan \beta l$$

Now suppose the length of the line is 100 km, cos/(β *l*) = cos (0.06 × 100) = cos 6° ≅ 1.0, and therefore no Ferrent effect is experience whereas for say 300 km,cos (β *l*) =cos (0.06 × 300) = cos 18° = 0.95

i.e., V_r/V_S = 1.05 a rise of 5% of voltage at the receiving end. This could be tolerated. However if the length is 800 km, cos 48° = 0.66, and V_r/V_S = 1.494 a rise of 49.4% in the receiving end voltage as compared to the sending end voltage. Such a high rise in receiving end voltage is unacceptable and some suitable compensating devices must be used to avoid then situation. In practical case, that is, in actual case, the receiving end voltage rise more than what is calculated above, as the sending end voltage will increase with the increase in the charging current due to rise in open circuit voltage at the sending end generator.

In fact the generator should be able to supply so much of leading VARs or absorb lagging VARs. Under-excited operation of the generator below a certain minimum excitation is not desirable for two reasons :

1. Lower excitation results in reduced internal e.mf. of the generator and hence reduces stability limits.
2. Lower excitation increases heating of the ends of the stator core.

Therefore, it is the under-excited operation of the alternator rather than voltage rise due to Ferranti effect that stringent limit is imposed on the length of uncompensated lines.

19.2.7 To Prove $X' = Z_C \sin \beta l$

Assuming the line to be compensated is to be lossless let us analyse the system when a load of $P_L + jQ_L$ is connected. Using the basic equation (1) (in this section)

$$V(x) = V_r \cos (l - x)\beta + j Z_C I_r \sin \beta (l - x)$$

with $x = 0$, we get

$$V(S) = V_r \cos \beta l + j Z_C I_r \sin \beta l$$

Here $$I_r = \frac{P_L + jQ_L}{V_R}$$

$$\therefore \quad V_S = V_R \cos \beta l + j Z_C \sin \beta l \cdot I_r$$

$$V_S = V_R \cos \beta l + j Z_C \frac{P_L + jQ_L}{V_R} \cdot \sin \beta l$$

This equation holds good irrespective of whether the load is synchronous or asynchronous. Assuming the load is synchronous and taking V_r as reference, and say V_S leads V_r by an angle δ

$$\therefore \quad V_S = |V_S| (\cos \delta + j \sin \delta)$$

As $$V_S = V_r \cos \beta l + j Z_C \frac{P_L + jQ_L}{V_r} \sin \beta l = |V_S| [\cos \delta + j \sin \delta]$$

$$\therefore \quad |V_S| [\cos \delta + j \sin \delta] = V_r \cos \beta l + j Z_C \frac{P_L + jQ_L}{V_R} \sin \beta L$$

$$= V_r \cos \beta l + Z_C \frac{Q_L}{V_r} \sin\beta l + j Z_C \frac{P_L}{V_r} \sin \beta l$$

Now, equating the real and imaginary parts on the two sides, we have,

$$V_S \cos \delta = V_r \cos \beta l + Z_C \frac{Q_L}{V_C} \cdot \sin\beta l$$

and $$V_S \sin \delta = Z_C \frac{P_L}{V_r} \sin \beta l$$

or $$P_L = \frac{V_S V_r}{Z_C \sin \beta l} \cdot \sin \delta$$

if the length of the line is short, $\sin \beta l = \beta l$

and $$Z_C \sin \beta l = \sqrt{\frac{L}{C}} \cdot (w\sqrt{LC})l = (\angle w \times l) = (w\sqrt{LC}) \times l = X' = \text{ Total}$$

reactance of line. Thus it is proved that $X' = Z_C \sin \beta l$

$\therefore$ equation $$P_L = \frac{V_S V_r}{Z_C \sin \beta l} \cdot \sin \delta \text{ can be}$$

written, now, as below

$$P_L = \frac{V_S V_r}{X'} \sin \delta$$

Now if $V_S = V_r$, then this equation becomes,

$$P_L = \frac{V_r^2}{Z_C \sin \beta l} \cdot \sin\delta \text{ or } = \frac{V_r^2}{X'} \sin\delta$$

$$\therefore \quad X' = Z_C \sin\beta l$$

or if $$P_C = V_r^2/Z_C, \text{ then } P_L = \frac{V_r^2}{Z_c \sin \beta l} \cdot \sin\delta \text{ can be written as :}$$

$$P_L = \frac{(V_r^2/Z_C)}{\sin\beta l} \cdot \sin\delta$$

$$= \frac{P_C}{\sin \beta l} \cdot \sin \delta$$

For certain and operating voltage of the line P_L is maximum. When $\delta = 90°$ and this maximum depends upon the length of the line. Lower the length, the smaller is the maximum value of P_L. The ratio $P_{L\,max}/P_C$ for a few large is given in the table below for information.

Length in km	**βl/in degrees**	**$(P_{L_{max}}/P_c)$**
300	18	9.236
600	36	1.701
900	54	1.231
1200	72	1.051
1500	90	1.00

19.2.8 Important Aspects of Copensation

(*a*) Important Aspect of Compensation is to increase power transfer capability of the system which is achieved by inserting capacitor at suitable location in series with the line so that the inductive reactance of the line is reduced which is equivalent to reducing the effective length of the line. This method of compensation is known as line-length compendsation.

(*b*) To assess the effectinous of a compensated system a performance index in term of product of length of line and the power to be transmitted is evaluated. This criterion is selected as we known that for longer length lines it is not even possible to load lines to their natural loading.

(*c*) Flat voltage profile on the line can be achieved if the loading of the line corresponds to its surge impedance loading. Hence for this. Compensating device should be no chosen that effective or virtual surge impedance Z_C of the line should give virtual natural loading equal to actual load. Since actually the actual load keeps charging with times the compensating devices should also vary without delay so that every time the effective surge impedance, matches with the actual loading Z at is $P_C = V_r^2/Z_C = P_L$. Where P_L is actual load.

The compensating devices are the suitable connection of capacitors and/or inductor on to the lines. Compensation where carried out with the sole objective of modifying the surge impedance of the line is known as Surge Impdence Compensation or Z_C Compesnation.

(*d*) Compensations are classified as passive and active compensators. Shunt capacitors and reactors and series capacitors are the passive compensators and synchronous capacitors and thyristorised controlled capacitors and reactors are the active compensator.

19.2.9 Types of Compensators

We discussed above series and shunt capacitors as compensators below we shall discuss briefly other compensators used in power system.

1. Synchronous Condensers

Synchronous condensers are the active shunt compensators and have been used in both at transmission and subtransmission levels to improve voltage profile and system stability whereas in static capacitors the reactive power supplied decreases with decrease in system voltages the synchronous capacitors regulate reactive power independent of system voltage. This is the main reason for the supriority of synchronous capacitors over static capacitors in power system transmission lines.

Yet another application of synchronous capacitors is with HVDC transmission where these apply a portion of the convertor reactive power requirement and provide necessary reactive power support to on a.c. network with low short-circuit capacity what is a synchronous capacitor.

A synchronous capacitor is a synchronous machine synchronised to the power grid and controlled to absorb or generate lagging VAR on the system. Most of the synchronous capacitors are outdoor and remain almost unattended. These are automatically controlled that means, these are associated with automatic controls for start up, shut down and on line monitoring. The efficiency of a synchronous capacitor including its auxiliary is 99%.

In fact, synchronous condensers under voltage regulator control automatically charge their output in a direction to minimise voltage variations. Hence the synchronous condensers are used for (*a*) Power System Voltage Control both under normal and emergency operating condition and (*b*) HVDC application. It must be noted here that, the emergency operating condition result from a system fault on a sudden loss of a major transmission link or a major generating station break down such may result in a breakdown of the system or "islanding". By ISLANDING of the system is meant that in a large interconnected system if there is a major breakdown a part of the system may be isolated from the system and keep running before total breakdown lakes place. The system which is isolated from the rest of the unhealty system is known as islanded system and the process is known as Islanding is which condition voltage extremes in either direction may occur depending upon the imbalance in the reactive power in the islanded system.

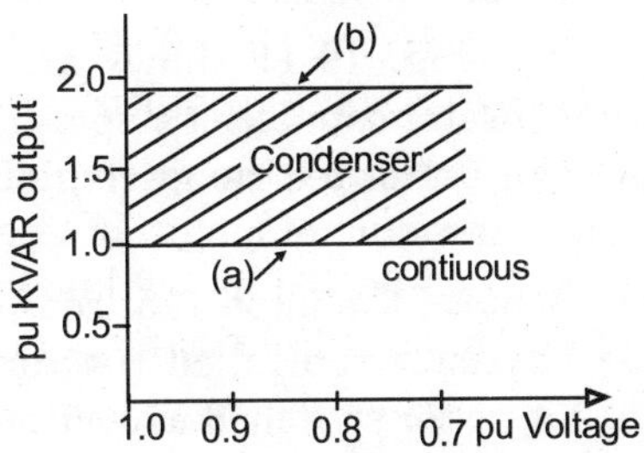

Fig. 19.9. Condenser output : (*a*) Continuous condition, (*b*) Energy condition

From the Fig. 19.9 it is clear that condensers have short-time capabilities well in excess of their wave plate ratings.

The short-time rating may be 150% of the rated value for one minute. In normal practice the condenser output depends upon the system voltage and the excitation wiling. The higher the excitation wiling the faster is the response of the condenser for a particular reactive power.

In Fig. 19.10, we set the application of a synchronous condenser at the receiving and of an HVDC line. The condenser series the purposes as below :

(1) A part of the reactive power requirement of the convertor. (2) A system which low hart circuit capacity is subjected to large voltage dipx is case of shunt fault : For such system the condenser allows d.c. convertor control to maintain acceptable control of the a.c. when the a.c. system has low short circuit capacity, the synchronous condenser increases the short circuit capacity of the system besides providing a portion of the reactive power requirement of inverting process. of the available sources of reactive power (static VAR compensation and filters) the synchronous condenser has the unique property of providing short time voltage stability in the subtransient range.

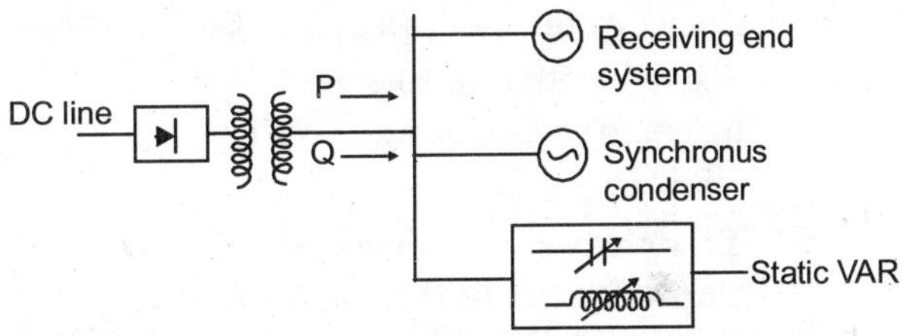

Fig. 19.10. Use of condenser for HVDC line.

Infact all the synchronous condensers are automated with remote short-stop and voltage control device. The protective relaying requirement are more or less same as for a turbo-altermotor except a few special requirement. Since a near zero field current in a normal operation for a synchronous condenser a different type of loss field relay is required.

19.2.10 Static Compensators

Compensators which have no rotating parts and are used for surge impedance compensation and for compensation by sectioning a long transmission line. These are also used for load compensation wherein these maintain constant voltage (1) under slowly varying load changes (2) load rejection, outages of generator and line (3) under rapidly varying load *e.g.*, are for mean. These improve system stability and system power factor. Since these regulate the flow of reactive power, these are used for minimisation of transmission losses and control a.c. voltage near HVDC convertor terminal.

The Fig. 19.11 shows an ideal static reactive power compensator capable of stepless adjustment of reactive power over on unlimited range (lagging and loading) without any time delay.

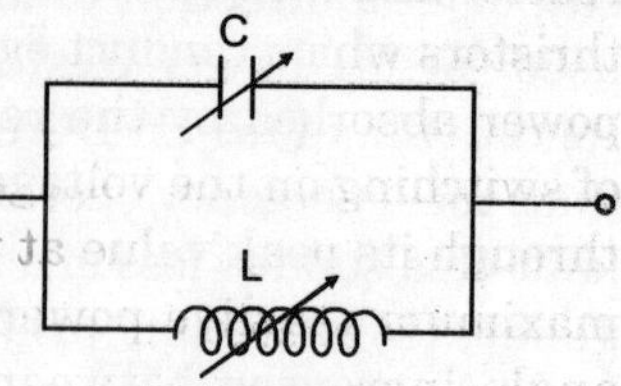

Fig. 19.11

Some of the important compensators used in transmission and distribution networks are:
(*a*) Thyristor controlled reactor (TCR),
(*b*) Thyristor switched capacitors (TSC) and (*c*) Saturated Reactors. Fig. 19.12 shows single line diagram of the above compensators.

(*a*) **Thyristor Controlled Reactor (TCR) :** A fixed shunt capacitor in parallel with the controlled susceptance. The fixed capacitors are usually turned with small reactors to hormonic frequencies to absorb hormonics generated by the controlled sweeptance or (TCR or SR) or to avoid V_{rf} harmful resonances (Fig. 19.12 (*a*)).

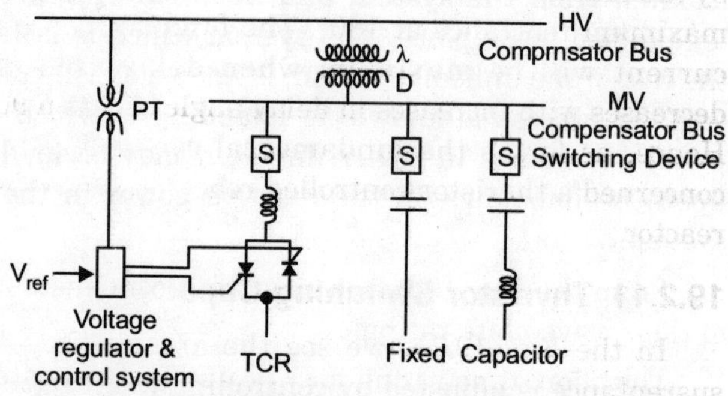

Fig. 19.12 (*a*). TCR-Thyristor controlled reactor.

(*b*) **Thyristor Switched Capacitors :** In the Fig.19.12 (*b*) a step down transformer which significantly affects the performance of the transformer expecially with respect to losses, hormonics and

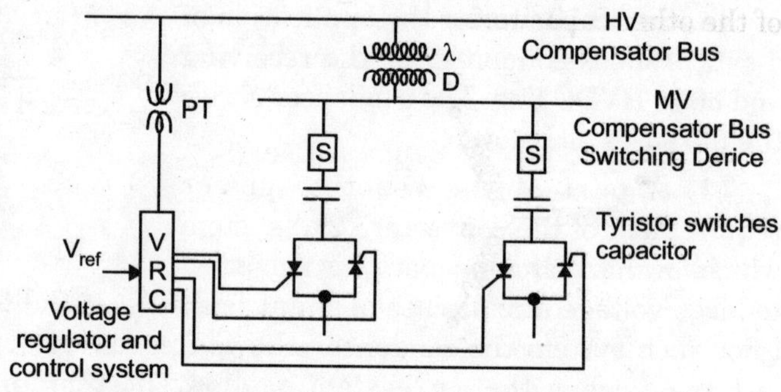

Fig. 19.12 (*b*) TSC-thyristor switched capacitors.

(*c*) **Saturated Reactors :**

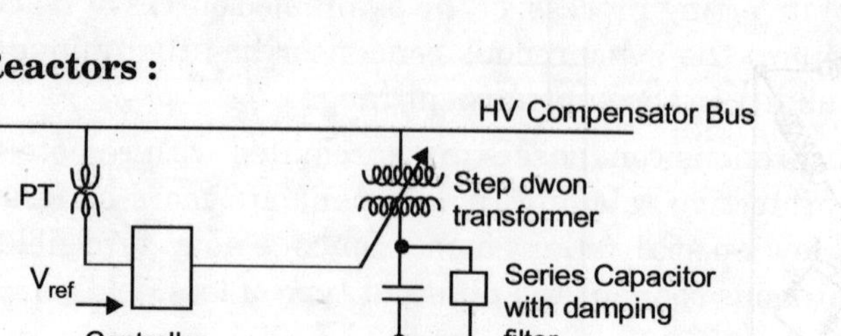

Fig. 19.12 (*c*). Standard reactors.

The Fig. 19.13 shows a basic thyristor controlled Reactor. The controlled element is the reactor and the controlling element is the thyristor controller consisting of two of positely poled thristors which conduct every alternate half cycles of the supply frequency. In fact, the reactive power absorbed by the reactor will depend upon the instant of switching on the voltage wave. When the voltage is passing through its peak value at the instant of switching of thyristor, maximum reactive power will be absorbed and as the delay angle increases between 90° to 180°, the reactive power absorbed decreases which really means as the conductor angle increases between 90° to 180°, the effective reactance offered by the reactor increases with minimum angle being at 90° and maximum reactance at 180°. The fundamental component of current will be maximum when delay angle is 90° and decreases with increases in delay angle between 90° and 180°. Hence, as for as the fundamental component of current is concerned a thyristor controlled reactor acts as a controllable reactor.

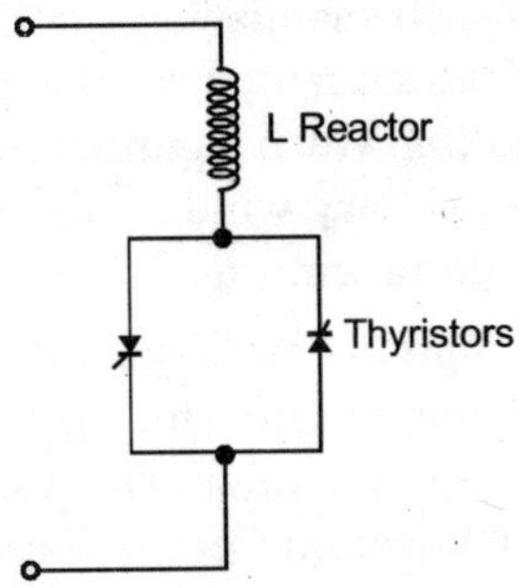

Fig. 19.13. Basic thyrisotor controlled reactor.

19.2.11 Thyristor Switching Capacitors

In the Fig. 19.14, we see the arrangment of thyristor switched capacitor, where the susreptance is adjusted by controlling the number of parallel capacitors in conduction. Each capacitor always conducts for an integral number of cydes. Normally the capacitors used are of the same capacity except one of the capacitors which has its sweeptance half the sweeptance of the other capacitors.

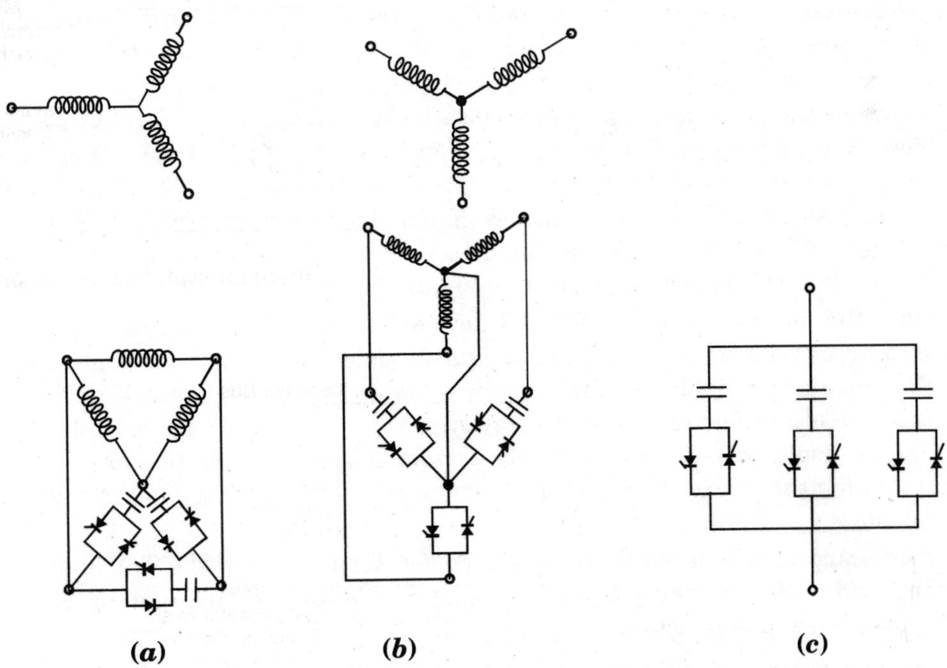

Fig. 19.14. Thyristor switched capacitor (*a*) Delta connected secondary, delta connected TSC, (*b*) Star-connected secondary & TSC (4-wire system), (*c*) Each phase of (*a*) and (*b*).

19.2.12 Saturater-Reactor Compensator

As the voltage or the current contain lot of hormonics the plain saturated reactor is unsuitable for use in transmission system. In the Fig. 19.15 shows a 3-phse saturated reactor having a short-circuited delta winding which eleminates third hormonic current from the primary winding as these current are supplied by the delta winding.

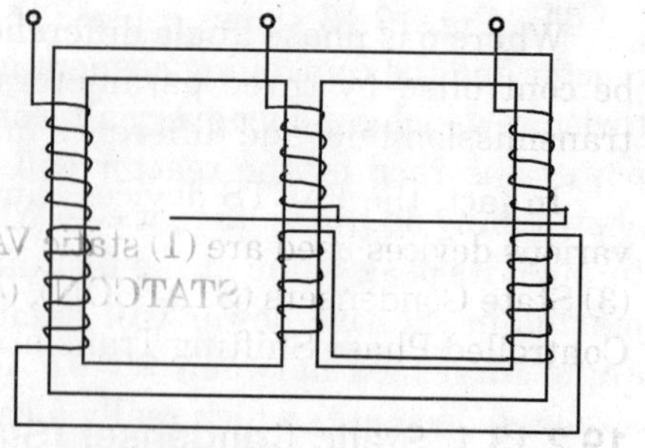

Fig. 19.15. Saturated-reactor compensator appex constant volts.

The secondary current are predominately 3rd hormonic and the circuit in thus an elementary frequency tripler. The voltage-current characteristics and hormonic performance of the above frequency tripler are not good enough to be used in power systems and such better characteristic can be obtained with higher frequency multipliers. The Fig. 19.16 shows a typical volt-amp characteristic of a saturated reactor the slope of which depends on the after-saturation inductance of the windings.

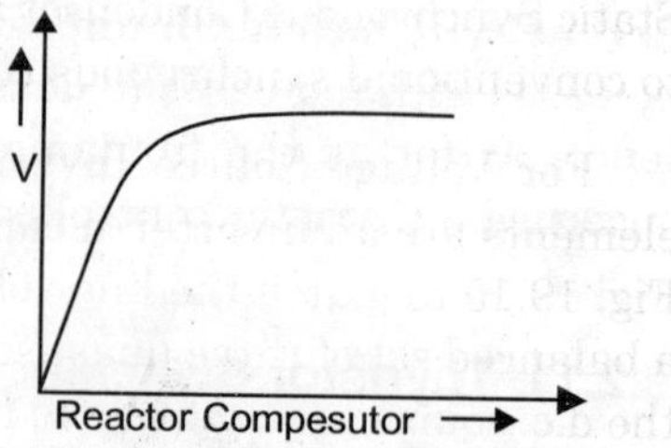

Fig. 19.16. Characteristic of a saturated Reactor compensator.

The slope of the characteristic varies between 5 and 15% this characteristic is linear above about.

19.2.13 Facts (Flexible A.C. Transmission System)

Due to ever increasing demand of electric power, the exiting transmission networks even in the developed countries are found to be weak which results in poor quality of unreliable supply. Also it is seen that in order to expand or enhance the power transfer capability of the exiting transmission network huge sum of finances are required and sometimes even difficulties are encountered in finding right-of-way for the new lines.

In fact since past few decades quite a lot of research has gone into developing new technologies but over the past few years to gain increased efficiency from the exiting power system. This newly developed programme is known as Flexible A.C. Transmission System abbreviated as FACTS. The new technologies employ high speed thyristors for switching in or out transmission line components such as capacitors reactors or phase shifting transformer for some desirable performance of the power systems.

However, the main objective of FACTS devices is to replace the exciting slow acting mechnical controls require power system operators and designes to provide generous margins to assure a stable and reliable operation of the system. As a result the exiting systems can not be made use of to their full capacity. However, with the use of fast controls, the power system margins could be reduced and power system capability could be more fully utilised while maintaining the present levels of quality and reliability. Let us explain the concept of FACTS as follows :

The most important relation between the Power Transfer, P in the Transmission (from the Sending End to the Receiving End). Receiving End Voltage V and Sending Voltage V_2 is and the Transmission line Reactance, X.

$$P = \frac{V_1 V_2}{X} \sin \delta$$

Where δ is phase angle difference between V_1 & V_2 it can be seen that the power flow can be controlled by three parameters. The voltages at the two system, the reactance of the transmission line, and difference in the voltage angles at the two ends.

In fact, the FACTS devices can be used to control one on more of these parameters. The various devices used are (1) static VAR compensators (SVC), (2) controlled series componsation, (3) State Condensers (STATCON), (4) Advanced Controlled Series Compensation, (5) Thyristor Controlled Phase Shifting Transformer. Brief outlines are given below.

19.2.13.1 Static Condenser (Statcon) and Static VAR Compensator (SV)

Fig.19.16 shows a basic circuited STATCON which is GTO (Gate Turn OFF) based compensation system. These devices are known as STATCON or Static Synchronous Condenser as these exhibit characteristics similar to conventional synchronous condensers without the moving parts.

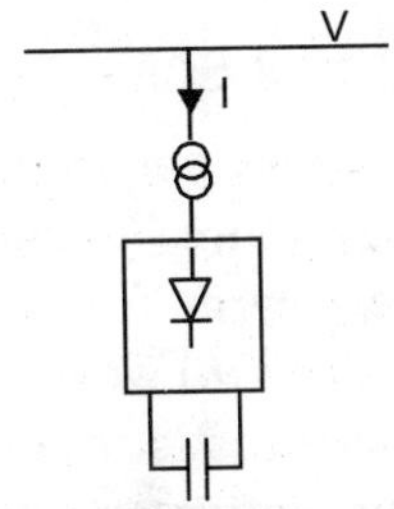

Fig. 19.17. Basic statcon

For Voltage Source Inverter (VSI) based STATCON the basic elements are an inverter, a capacitor and a transformer as shown in Fig. 19.15 to match the line voltage. A VSI inverts a d.c. voltage with a balanced set of three quasi-square voltage wave form by connecting the d.c. source sequentially to three output terminals. However, three phase a.c. generated by inverter is synchronised to the a.c. line through a small tie reactance which in fact is the leakage reactance of a transformer.

It is observed that when the inverter fundamental output voltage is higher than the system line voltage the statcon works as a capacitor and reactive VARs are generated. But when the inverter voltage is lower than the system line voltage, the STATCON acts as an inductor thereby absorbing the reactive VARs from the system. To control the reactive current, thus the magnitude of d.c. voltage is raised or lower by adjusting the phase angle of the inverter output voltage. The capacitor here in this circuit does not play an active role in the VAR generation. Infact it is only required to maintain a smooth d.c. voltage while carrying the ripple current drawn by the invertor.

However, the main difference between the SVC (STATIC VAR COMPENSATOR) and STATCON is whreas in case of SVC devices the current injected into the system depends upon the system voltage. The STATCON Current I is made perpendicular to the system voltage V with the help of a STATCON coordinator which adjusts the phase of the current source I, so that it is per pendicular to the STATCON terminal Voltage V.

19.2.13.2 SVC (Static Var Compensator)

A static VAR compensator is a shunt-connected thyristor controlled or thyristor-switched reactor (TCR)/TSR and/or thyristor-switched capacitor, or a combination of both, which inject capacitive or inductive current so as to maintain the bus voltage of the electrical power system. In the Fig. 19.18 (*a*) shows the schematic diagram of a SVC. A SVC includes a separitor-controlled or thyristor-switched reactor for absorbing reactive power and thyrsitor-switched capacitor for suppling the Reactive power. In fact the cost of a SVC is lower than STATCON or STATCON (Static Compsator). The

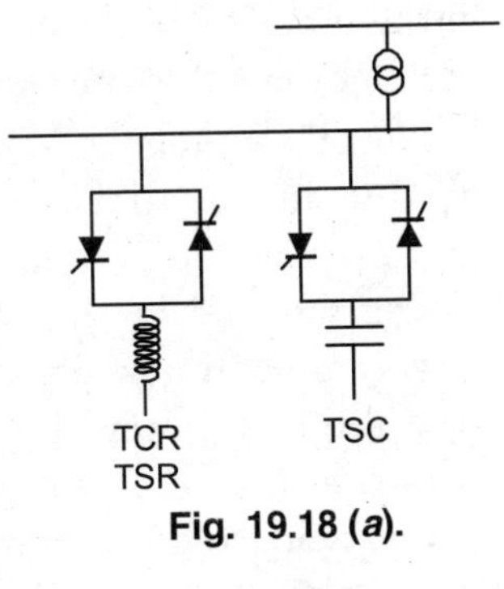

Fig. 19.18 (*a*).

application of a SVC can improve transmission capacity and the steady State stability limit and can damp the sub-synchronous oscillation. SVC may be used as a stability improvement measure under small and large disturbance conditions.

However, STATCON devices are used on the distribution system and have the following advantages. (1) The steady state loadability of the lines is improved. (2) The voltage rises due to capacitance switching is substantially reduced both in magnitude and duration. (3) Voltage variation due to customer's loading is reduced. (4) STATCON/SVC may be used for voltage conrol, MVA compensation and damping oscillation. In fact STATCON is more expesnsive the SVC on a per unit steady-state MVA basis, but it should be also noted that the performance of the STATCON outweighs the increase in cost.

19.2.13.3 Static Synchronous Generator (SSG)

Static Synchronous Generator is a combination of STATCON and an energy source to supply or absorb power. The energy source may be a battery, large d.c. storage capacitor, super conducting magnit etc. In the Fig. 19.18 (*b*) we can see a single line diagram of a STATCON connected to a storage device SSG may be used for voltage control, VAR compensation and damping oscillation under steady-state and transient conditions.

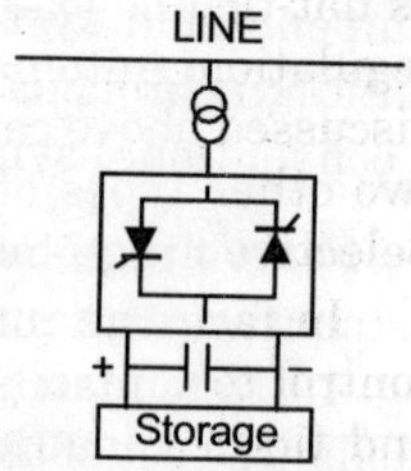

Fig. 19.18 (*b*). SSG

19.2.13.4 (Superconductor Magnetic Energy Storage) SMES

This device consists of thyristor convertor which can rapidly injected/absorb both real and/or reactive power into the system to which the SMES is connected. This device can dynamically control power in an a.c. system and thus can control voltage at the point of connection and damp out transient oscillation that are caused by large power swings in the systems. As shown in the sEREmatic diagram of Fig. 19.18 (*b*), SMES may be connected as a large device. This device may be used for voltage control, damping oscillation under large and small disturbance and automatic generation control.

19.3 LOAD FREQUENCY CONTROL

As and when the power system consists of a single machine connected to a group of loads the speed and frequency change in accordance with the governor characteristics as the load changes. If it is not important to keep frequency constant then regulation control is required.

The frequency normally would vary by about 5% between light load and full load condition. On the other hand if constant frequency is required the operator can adjust the speed of the turbine by chaging the Governor characteristic as and when desired.

Now if a change in load is taken care of two machines running in parallel as shown in Fig. 19.19 (*i*), the complexity of the system is increased. The possibility of sharing the load by the two machines may be explained the way we will now discuss the subject. Let us say that there are two power stations S_1 and S_2 interconnected through a tie-line. If the change in load is either at S_1 or S_2 interconnected through a tie-line than what happens? Now if the change in load is either at S_1 or S_2 and

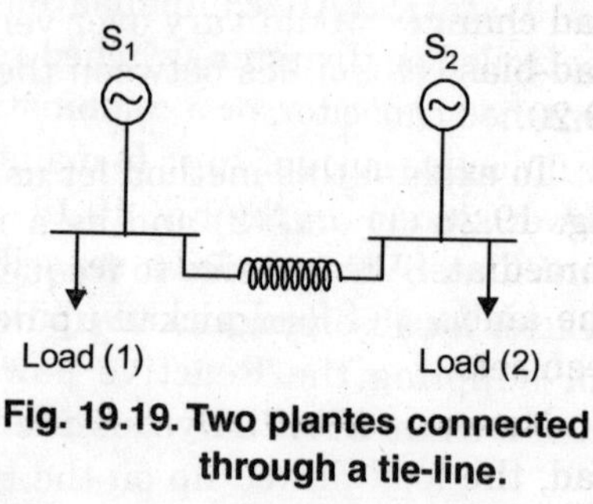

Fig. 19.19. Two plantes connected through a tie-line.

if the generation of S_1 alone is regulated to adjust this change so as to have constant frequency, the method of regulation is known as Flat Frequency Regulation under such situation the station S_2 is said to be operating on base load. The major drawback of flat frequency regulation is that S_1 must absorb all load changes for the entire load changes at station S_2 sinc the generator at S_2 would maintain its output constant. then the operation of generator S_2 on base load has the advantages when S_2 is much more efficient than the other station and it is desirable to obtain maximum output of S_2.

However, the other possibility of sharing the change in load is that both S_1 and S_2 would regulate their generations to maintain the frequency constant. This is known as parallel frequency regulation. The third possibility is that when the change in a particular area is taken care of by the generator in that area the generation for keeping constant frequency is known as flat-tie line loading control. This arrangement has the advantage that load swings on station regulation. Automatic equipment permits various types of system control. The various methods discussed above can be performed with the help of automatic control equipments. Besides these two other types of controls which are widely used in automatic arrangements. They are (*a*) Selective Frequency control and (*b*) Tie-line load-bias control.

In fact, the common method of operating a large interconnected system assign frequency control to control system, the other system are then controlled on the same system frequency and tie-line loading. The tie-line loading as the basis of automatic control is used in three different ways. One of these is known as Selective Frequency Control. In this case each system is the group takes care of the load changes on its own system and does not aid the other systems in the group of changes outside its own limits.

It has to be stated here that the most commonly used method is the tie-line load-bias control in which all power systems in the interconnection aid in regulating frequency regardless of where the frequency change originates. The equipment consists of a master load frequency controller and a tie-line recorder measuring the power input on the tie as for selective frequency control. The tie-line instrument biases the load frequency controller by changing the control until the desired relationship exists between tie-line loading and system frequency. In the Fig. 19.20, the case is illustrated.

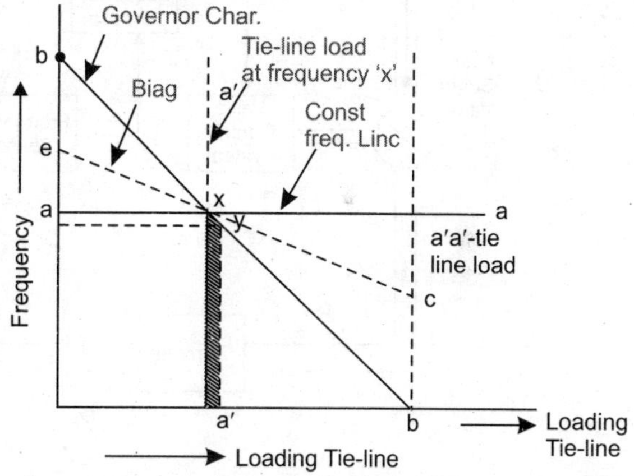

Fig. 19.20. Tie-line loading frequency characteristics.

The **solid** line *bb* represents a characteristic similar to the governor characteristic. The solid line *aa* represents a constant frequency characteristic. If the station is trying to maintain constant frequency, the tie-line load changes would vary over very wide limit where if it were working along *a′a′*, the tie-line load-bias control lies between these two extreme limits, such as *bb* or *cc* as shown in the Fig. 19.20.

To explain this method let us consider that there is increase in load on station S_1(refer to Fig. 19.20 (1) and (2) and as a result there is reduction of frequency, the station S_2 being immediately responsive to frequency, increases its generation to restore frequency to normal. The amount of load picked up depends on the bias for which the regulating equipment has been set.

Now say, if initially x is the operating frequency and if it drops to y due to increase in load, the load picked up on the tie-line will be shown by the shaded area under xy. The tie-line has definitely gone off schedule, but this power is contributed to the interconnection in a direction to improve system regulation. And the amount of bias is adjustable, depending on the extent to which station S_2 is scheduled to contribute to system regulation.

However, in a large interconnected system, as has been stressed previously that manual regulation is not feasible and therefore load frequency equipment is installed for each generator. Similarly for voltage control also voltage regulation equipment is installed on each generator. In the Fig. 19.21 we see the schematic diagram of load frequency and excitation voltage regulators of a turbo-ganerator. The controllers are set for a particular operating condition and they take care of small changes in load demand with voltage and frequency exceeding the pre-specified limits. If the operating condition change materially the controllers must be shown re-set either manually or automatically. In fact, it is known that small changes in load depend upon the change in rotor angle "δ" and is independent of the bus voltage whereas the bus voltage is dependent on machine excitation, *i.e.*, on the reactive generation Q and is independent of rotor angle δ. Therefore the two controls, *i.e.*, load frequency and excitation voltage controls are non-interactive for small changes and can be modelled and analysed independently. Besides, the load frequency controller is slow acting because of the large time constant contributed by the turbine and generator Moment of Inertia. And excitation voltage control in fault acting in the time constant of the field winding is relatively smaller, thus the transient in excitation voltage control vanish much faster and do not affect the dynamics of load frequency control.

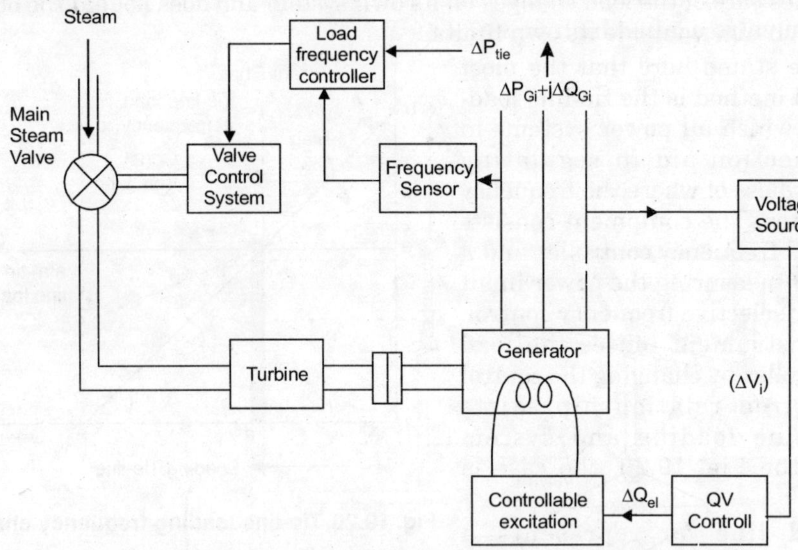

Fig. 19.21. Schematic diagram of L.F. and Q.V. regulators.

However, here we shall consider only the load frequency control aspect of regulation. It is to be noted here that the regulator designed for control should not be insensitive to fast random changes, otherwise the system will be prone to hunting, resulting in excessive wear and tear of control equipment and the rotating machines.

Infact the main objective of the load frequency controller is to exert control of frequency and at the same time of real power exchange via the outgoing lines. The change in frequency and the tie-line real power exchange via the outgoing is a measure of the change in rotor angle δ, *i.e.*, the error $\Delta\delta_i$ to be corrected. The error signals, *i.e.*, $\Delta\delta_i$ and ΔP_{tie} are amplified mixed and transformed into a real power command signal $\Delta\delta_{Gi}$ which is sent to the prime mover to call for an increment in the torque. The prime mover therefore brings change in the generator output by an amount. ΔP_{Gi} which will change the values of Δf_i and ΔP_{tie}. The process continues till the deviation Δf_i and ΔP_{tie} are well below the specified tolerances.

19.3.1 System of Speed Governing

In the Fig. 19.22 we can see the schematic diagram of a speed governing system which controls the real power flow in the power system infact, the speed governor system consists of the following parts :

1. Speed governor
2. Linkage mechanism
3. Hydraulic Amplifier
4. Speed changer

In this section we shall not discuss these parts in detail, but from the schematic diagram shown in Fig. 19.22 we can see the working system of speed governing system, where governor senses the change in speed/freq. As speed increases fly balls of Governor move upwards and point B on linkage moves downwards. It goes down when speed decreases. In the link mechanism ABC is rigid pivoted at B, but link CDE pivoted at D provides movement to the control valve as per the change in speed. It also sends feedback from steam valve movement by link DE. Speed changer gives or steady state power output setting for the turbine. When goes down it opens upper pilot valve to adovit more steam to turbine hence more power output. Opposit happens with upward movement of speed changer. Hydraulic amplifier helps to open or close the steam valve against high pressure steam.

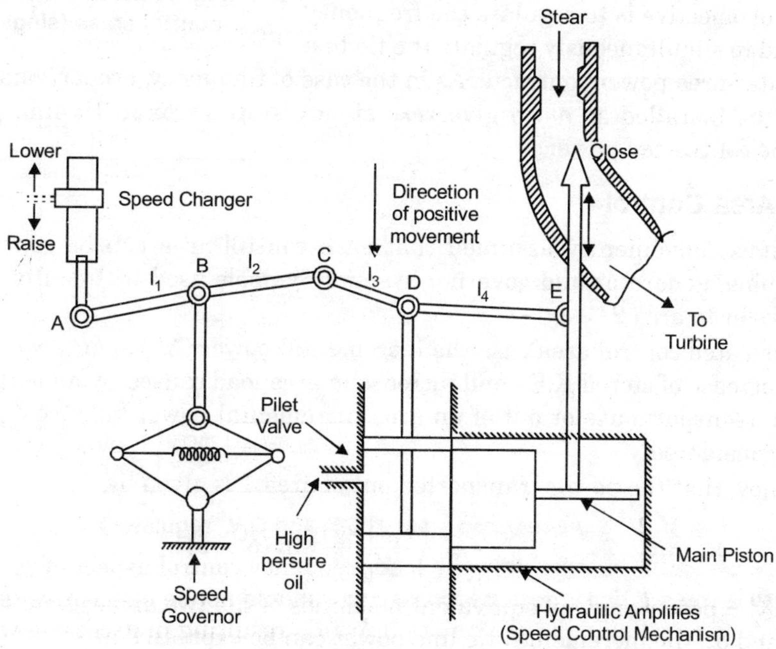

Fig. 19.22. Turbine speed governing system.

19.3.2 Load Frequency Control and Economic Despatch Control Power System

In power system load frequency control with integral controller achieves zero steady frequency error and fast dynamic response, but it exercises no control over the relative loadings of various generating station *i.e.*, economic despatch, of the control area. For example, if a sudden small increase in load, say, 1%, occurs in the control area, the load frequency control changes the speed changer settings of the governors of all generating units of the area so that, together these units match the load and the frequency return to the scheduled value (this action takes place in a few seconds. However, in the process of this change the loadings of various generating units change in a manner independent of economic loading considerations. In fact, some units in the process may even get overloaded. Some control over loading of individual units can be exercised by adjusting the gain factors included in the signal representing integral of the area control error as fed to individual units. However, this is not satisfactory.

A satisfactory solution is achieved by using independent controls for load frequency and economic despatch. While the load frequency controller is a fast acting control (in fact, a few seconds), and regulates the system around an operating point the economic despatch controller is a slow acting control, while adjusts the speed changer setting every minute (or say half a minute) in accordance with a command signal generated by the control economic despatch computer.

19.3.3 Control-Area Load Frequency Control

In the power system, an extended power system can be divided into a number of load frequency control areas interconnected by means of tie lines. Without loss of generality we shall consider a two-area case connected by a single tie-line as shown in Fig. 19.23.

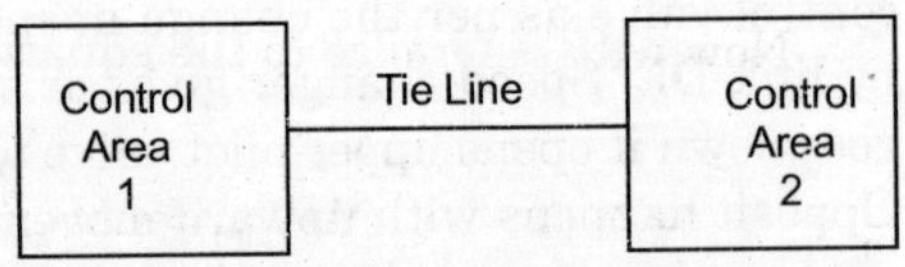

Fig. 19.23. Two interconnected control areas (single tie line).

Then control objective is to regulate the frequency of each area and to simultaneously regulate the tie line power as per inter-area power contracts. As in the case of frequency, proportional plus integral controller will be installed so as to give zero steady state error in tie-line power flow as compared to the contracted power.

19.3.3.1 Two-Area Control

However, it is conveniently assumed that each control area can be represented by an equivalent turbine, generator and governor system. Symbols used with suffix 1 refer to area 1 and suffix 2 refer to area 2.

Say in an isolated control area case the incremental power $(\Delta P_G - \Delta P_D)$ was accounted for by the rate of increase of stored K.E., and increase in area load caused by increase in frequency. Since a tie line transports use or out of an area incremental power balance equation of each area are to be considered.

Now, we know that the power transported out of area 1 is given by,

$$P_{\text{tie, 1}} = \frac{|V_1||V_2|}{X_{12}} \sin(\delta_1^0 - \delta_2^0) \qquad \text{...(19.25)}$$

where δ_1^0, δ_2^0 = power angle of equivalent machines of the two areas. Now, for incremental changes in δ_1 and δ_2, the incremental tie line power can be expressed as

$$P_{\text{tie, 1 (P.U.)}} = T_{12}(\Delta\delta_1 - \delta\Delta_2) \qquad \text{...(19.26)}$$

where $$T_{12} = \frac{|V_1||V_2|}{P_{r1} \cdot X_{12}} \cos(\delta_1^0 - \delta_2^0) = \text{Synchronous coefficient}$$

Since incremental power angles are integrals of incremental frequencies we can write equation (19.26) as

$$\Delta P_{\text{tie, 1}} = 2\pi T_{12} = (f\,\Delta f_1\,dt - f\,\Delta f_2\,dt) \qquad \text{...(19.27)}$$

where Δf_1 and Δf_2 are incremental frequency changes of areas 1 and 2 respectively.

Similarly the incremental tie line power out of area 2 is given by

$$\Delta P_{\text{tie } 2} = 2\pi T_{21}(\int \Delta f_2\, dt - \int \Delta f_1\, dt) \qquad \text{...(19.28)}$$

where
$$T_{21} = \frac{|V_2||V_1|}{P_{r2}X_{21}}\cos(\delta_2^0 - \delta_1^0) = \left(\frac{P_{r1}}{P_{r2}}\right)T_{12} = a_{12}T_{12} \qquad \text{...(19.29)}$$

Then with reference to equation $\Delta^G_{P(\text{PU})} - \Delta^P_{D(\text{PU})} = \frac{24}{f^\circ}\frac{d}{dt}(\Delta f) + B^{\Delta f}_{(\text{PU})}$ the incremental frequency changes of areas 1 and 2, respectively,

Exactly same way the incremental tie line power out of area 2 is given by

$$\Delta P_{\text{tie } 2} = 2\pi T_{2'}(\int \Delta f_2\, dt - \int \Delta f_1\, dt) \qquad \text{...(19.30)}$$

where
$$T_{21} = \frac{|V_2||V_1|}{P_{12}X_{21}}\cos(\delta_2^0 - \delta_1^0) = \left(\frac{P_{r1}}{P_{r2}}\right)T_{12} = a_{12}\, T_{12} \qquad \text{...(19.31)}$$

Now with reference to the equation,

$$\Delta P_{G(\text{PU})} - \Delta P_{D(\text{PU})} = \frac{24D\,d}{f^0\,dt}(\Delta f) + B^{\Delta f}_{(\text{PU})} \qquad \text{...(19.32)}$$

The incremental power balance equation for area 1 can be written as

$$\Delta P_{G1} - \Delta P_{D1} = \frac{24_1}{f_1^0}\cdot\frac{d}{dt}(\Delta f_1) + B_1\,\Delta f_1 + \Delta P_{\text{tie, }1} \qquad \text{...(19.33)}$$

However, it may be noted that all quantities other than frequency are in per unit in equation (19.33).

Now taking the Laplace transform of equation (19.33) and reorganising, we get,

$$\Delta F_1(S) = [\Delta P_{G1}(S) - \Delta P_{D1}(S) - \Delta P_{\text{tie}}(S)]\times\frac{K_{PS1}}{1+T_{PS1}S} \qquad \text{...(19.34)}$$

where, as defined in the equation

$$\Delta F(S) = [\Delta P_G(S) - \Delta P_D(S)]\times\left(\frac{K_{PS}}{1+T_{PS}S}\right) \qquad \text{...(19.35)}$$

$$\therefore \qquad K_{PS1} = 1/B_1$$

and
$$T_{PS1} = 2H_1/B_1 f^0 \qquad \text{...(19.36)}$$

Now compared to equation (19.35) above of the isolated control Area case, the only change is the appearance of the signal $\Delta P_{\text{tie } 1}(S)$ as can be seen in the Fig. 19.24.

Then taking the Laplace transfor of equation (19.27), the signal $\Delta P_{\text{tie, }1}(S)$ is obtained as

$$\Delta P_{\text{tie, }1}(S) = \frac{2\pi T_{12}}{S}[\delta F_{1(S)} - \Delta F_{2(S)}] \qquad \text{...(19.37)}$$

The corresponding block diagram is shown in Fig. 19.24

Then for the control area 2, $\Delta P_{tie,2\,(S)}$ is given by equation (19.38)

$$\Delta P_{tie,\,2(S)} = -2\pi\, a_{12} T_{12}\, [\Delta F_{1\,(S)} - \Delta F_{2(S)}] \quad ...(19.38)$$

This equation is also shown in Fig. 19.24

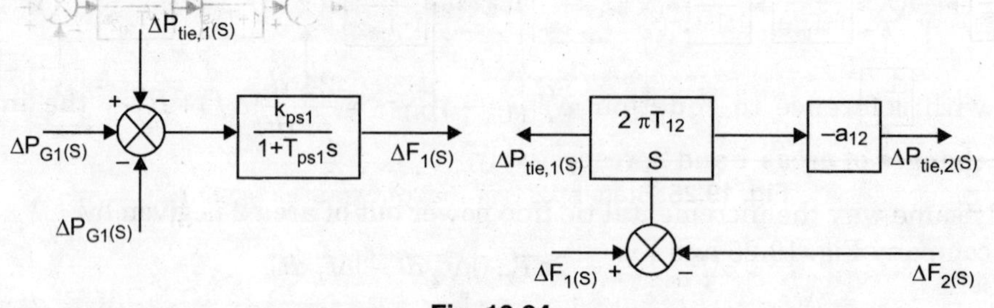

Fig. 19.24

19.3.3.2 Optimal (Two-Area) Load Frequency Control

In the power system, the modern control theory is applied. In this discussion to design an optimal load frequency controller for a two-area system. In fact, in accordance with modern control terminology ΔP_{C1} and ΔP_{C2} will be referred to as control inputs say U_1 and U_2. In the conventional approach U_1 and U_2 were provided by the integral of ACEs (Area Control Error). In modern control theory approach U_1 and U_2 will be created by a linear combination of all the system states (full state feedback). For a formulating the state variable model for this purpose the conventional feedback loops are opened and each time constant represented by a separate block as shown in Fig. 19.25. State variables are defined as the outputs of all blocks having either an integrator or a time constant. We immediately notice that the system has nine state variables.

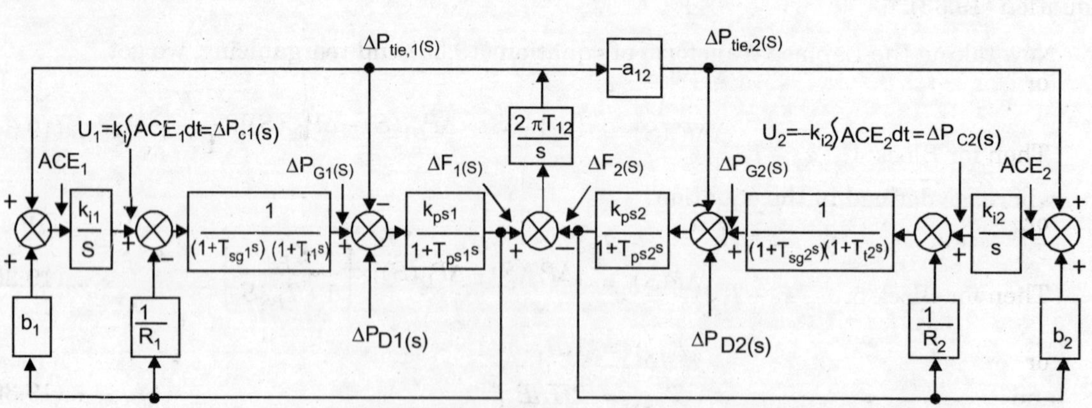

Fig. 19.25. composite block diagram of two-area load frequency control (feedback loops provided with integral of respective area control errors).

Now, before presenting the optimal design, we must formulate the state module. This is achieved below by writing the differential equation describing each individual block of Fig. 19.25 in term of state variables (note the differential equation are written by replacing S by d/dt).

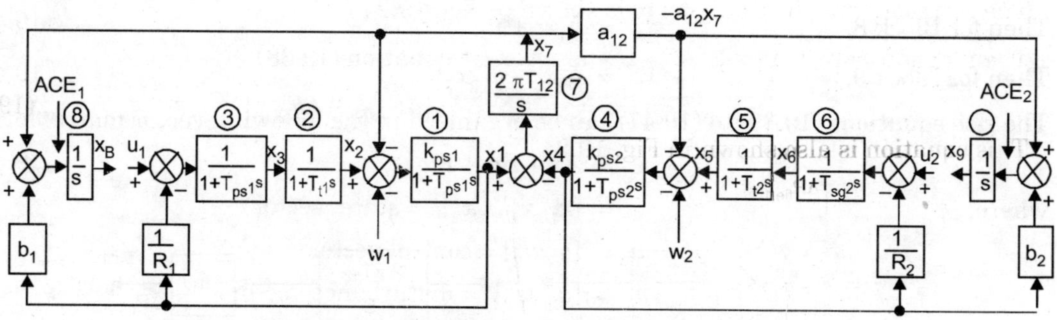

Fig. 19.26. State space model of two-area system.

Now company Fig. 19.25 and Fig. 19.26

$$x_1 = \Delta f_1 \qquad x_4 = \Delta f_2 \qquad x_8 = \int ACE, dt$$
$$x_2 = \Delta P_{G1} \qquad x_5 = \Delta P_{G2} \qquad x_9 = \int ACE_2\, dt$$
$$u_1 = \Delta P_{G1} \qquad u_2 = \Delta P_{G2}$$
$$w_1 = \Delta P_{D1} \qquad w_2 = \Delta P_{D2}$$

For Block 1, $\quad x_1 + P_{PS1}\dot{x}_1 = K_{PS1}(x_2 - x_7 - w_1)$

$$\dot{x}_1 = -\frac{1}{T_{PS1}}\cdot x_1 + \frac{K_{PS1}}{T_{PS1}}\cdot x_2 - \frac{K_{PS1}}{T_{PS1}}x_7 - \frac{K_{PS1}}{T_{PS1}}w_1 \qquad ...(19.39)$$

Then for Block 2, $\quad x_2 + T_{Z1}\dot{x}_2 = x_3$

or
$$\dot{x}_2 = -\frac{1}{T_{Z1}}x_2 + \frac{1}{T_{Z1}}x_3 \qquad ...(19.40)$$

Then for Block 3, $\quad x_3 + T_{Sg1}\dot{x}_3 = -\frac{1}{R_1}x_1 + U_1$

or
$$\dot{x}_3 = -\frac{1}{R_1 T_{Sg1}}\cdot \dot{x}_3 - \frac{1}{T_{Sg1}}x_3 + \frac{1}{T_{Sg1}} + \frac{1}{T_{sg1}}\cdot u_1 \qquad ...(19.41)$$

Then for Block 4, $\quad x_4 + T_{PS2}\dot{x}_4 = K_{PS2}(x_5 + a_{12}x_7 - w_2)$

or,
$$\dot{x}_4 = \frac{1}{T_{PS2}}\cdot x_4 + \frac{K_{PS2}}{T_{PS2}}\cdot x_5 + \frac{a_{12}K_{PS2}}{T_{PS2}}\cdot x_7 - \frac{K_{PS2}}{T_{PS2}}w_2 \qquad ...(19.42)$$

Then for Block 5, $\quad x_5 + T_{12}\dot{x}_5 = x_5$

or
$$\dot{x}_5 = -\frac{1}{T_{12}}x_5 + \frac{1}{T_{Z2}}x_6 \qquad ...(19.43)$$

Then for Block 6, $\quad x_6 + T_{Sg}\dot{x}_6 = -\frac{1}{R_2}x_4 + u_2$

or,
$$\dot{x}_6 = -\frac{1}{R_2 T_{Sg2}}\cdot x_4 - \frac{1}{T_{Sg2}} - x_6 + \frac{1}{T_{Sg2}}u_2 \qquad ...(19.44)$$

Then for Block 7,
$$\dot{x}_7 = 2\pi T_{12}x_1 - 2\pi T_{12}x_4 \qquad ...(19.45)$$

Then for Block 8, $\dot{x}_8 = b_1 x_1 + x_7$ ...(19.46)

Then for Block 9, $\dot{x}_9 = b_2 x_4 - a_{12} x_7$ ...(19.47)

The rise equations (19.39) to (19.47) can be organised in the following vector matrix form,

$$\dot{x} = Ax + Bu + Fw \quad ...19.48)$$

where,

$$x = [x_1, x_2 \ldots x_9]^T = \text{state vector}$$

$$u = [u_1\ u_2] = \text{control vector}$$

$$w = [w_1\ w_2]^2 = \text{disturbance vector}$$

In the connectional control scheme of Fig. 19.27 the control inputs u_1 and u_2 are constructed as under from the state variable x_8 and x_9 only.

Here,

$$u_1 = -k_{i1}\, x_8 - k_{i1} \int ACE_1 \cdot dt$$

$$u_2 = -k_{i2}\, x_9 - k_{i2} \int ACE_2 \cdot dt$$

In the optimal control scheme the control inputs u_1 and u_2 are generated by means of feedbacks from all the wire states with feedback constants to be determined in accordance with an optimality criterion on examination of equation (19.48) reveals that our model is not in the standard form employed in optimal theory. The standard form is

$$x = Ax + Bu$$

which does not contain the disturbances term F_w present in equation (19.48). Moreover, a constant disturbance vector w would drive some of system states and the control vector u to constant steady values, while the cost function employed in optimal control requires that the system state and control vectors wave zero steady state values for the cost function to have a minimum. For illustrations we consider two identical control areas with the following system parameters

$$T_{Sg} = 0.4s,\ T_1 = 0.5\ s,\ T_S = 20s$$

$$R = 3,\ K_{b,\,s} = 1/B = 100,\ b = 0.425,\ k = 0.09,\ a_{12} = 1,\ 2\pi\, T_{12} = 0.0\ s$$

Computer solution for the matrix k is presented below, while the system dynamic response is plotted in Figs. 19.27 and 19.28. These figures also give for comparison the dynamic response for the case of integral control action only. The improvement performance achieved by the optimal controller is evident from these figures. The values of k are :

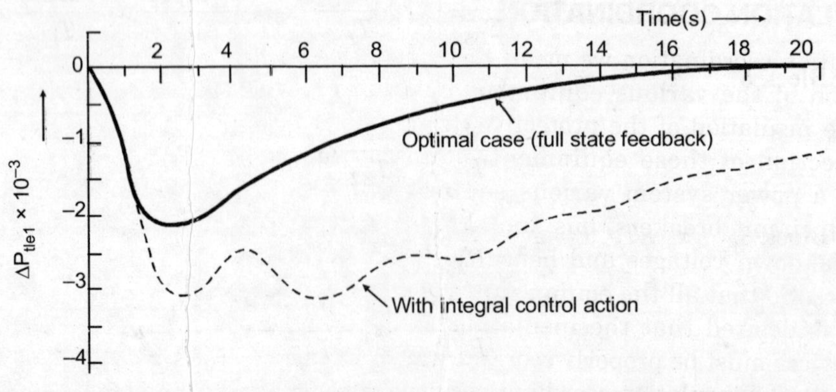

Fig. 19.27. Change in tie line power due to step load (0.01 pu) change in area 1

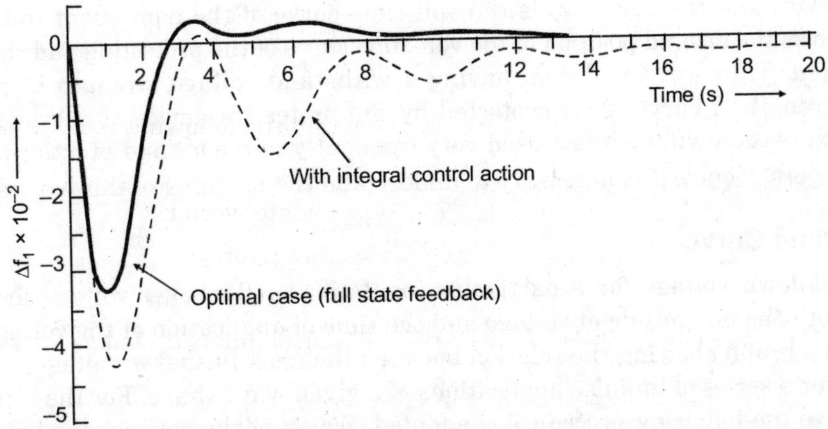

Fig. 19.28. Change in frequency of area 1 due to step load (0.01 pu) change in area

$$k = \begin{bmatrix} 0.5286\ 1.1419\ 0.6813 - 0.0046 - 0.0211 - 0.0100 - 0.7437 - 0.9999\ 0.0000 \\ -0.0046 - 0.0211 - 0.0100\ 0.5286\ 1.1419\ 0.6813\ 0.7437\ 0.0000\ 0.9999 \end{bmatrix}$$

As the control areas extend over vast geographical regions there are two way of obtaining fuel state information in each area for control purposes.

Due to the above mentioned difficulties encountered in implementation of an optimal control load frequency scheme, it is preferable to use a sub-optimal scheme employing only local states of the area. The design of such structurally constrained optimal control schemes is beyond the scope of this book. The convention control scheme is the local output feedback scheme in which feedback signal deviced from changes in local frequency and tie line power is employed.

1. Transport the state information to the distant area over communication channels. This is expensive.
2. To generate all the states locally, including the distant states by means of 'observer' by processing the local output signal. An observer being it self a high order system, renders the overall system highly complex and may, it fact, result in impairing system stability and dynamic response which is meant to be improved through optimal control.

19.4 I NSULATION COORDINATION

By insulation coordination we mean the correlation the insulation of the various equipment in a power system to the insulation of the protective devices used for the protection of these equipment against over voltages. In a power system various equipment like transformers, circuit breakers, bus supports etc. have different breakdown voltages and hence the volt time characteristics. So that all the equipments are properly protected it is desired that the insulation of various protective devices must be properly coordinated. In fact, the basic concept of insulation coordination is shown in Fig. 19.29. Curve (1) is the volt-time curve of the

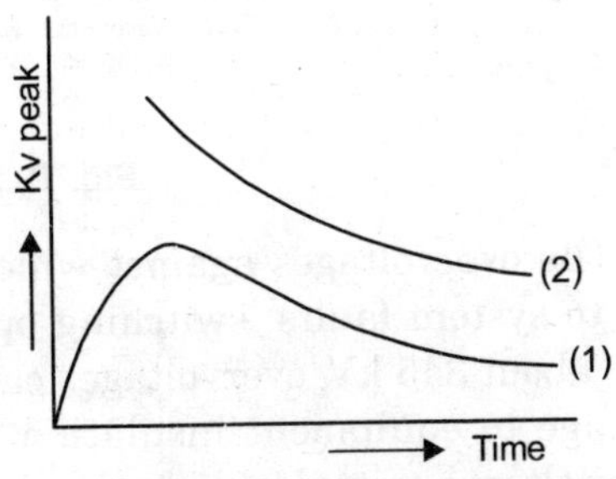

Fig. 19.29. Volt-time curve (1) (Protective device and) volt-time curve (2) (device to be protected).

protective device and the curve (2) is the volt-time curve of the equipment to be protected. Fig. 19.29 shows the desired position of the volt-time curve of the protecting and the equipment to be protected. Thus any insulation having a withstand voltage strength is excess of the insulation strength of curve (2) is protected by the protective device of curve (1). The volt-time curve expression will be infact used very frequently in theory and practice by the power system engineers. Hence, it is necessary to understand the meaning of this expression.

19.4.1 Volt-time Curve

The breakdown voltage for a particular insulation or flashover voltage for a gap is a function of both the magnitude of voltage and the time of application of the voltage. The volt-time curve is a graph showing the relation between the crest flashover voltages and the time to flashover for a series of impulse applications of a given wave shape. For the construction of volt-time curve the following procedure is adopted. Waves of the same shape but of different peak values are applied to the insulation whose volt-time curve is required. If flashover occurs on the front of the wave, the flashover point gives one point on the volt-time curve. The other possibility is that the flashover occurs just at the peak value of the wave; this gives another point on the *V-T* curve. the third possibility is that the flashover occurs on the tail side of the wave. In this case to find the point on the *V-T* curve, draw a horizontal line from the peak value of this wave and also draw a vertical line passing through the point where the flashover takes place. The intersection of the horizontal and vertical lines gives the point on the *V-T* curve. This procedure is nicely shown in Fig. 19.30.

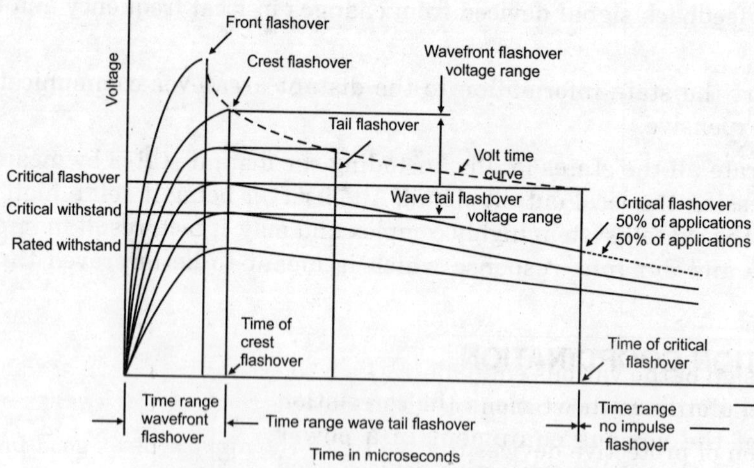

Fig. 19.30. volt-time curve (construction).

The overvoltages against which coordination is required could be caused on the system due to system faults, switching operation or lightning surges. For lower voltages, normally upto about 345 kV, overvoltages caused by system faults or switching operations do not cause damage to equipment insulation although they may be detrimental to protective devices. Overvoltages caused by lightning are of sufficient magnitude to affect the equipment insulation whereas for voltages above 345 kV it is these switching surges which are more dangerous for the equipments than the lightning surges.

The problem of coordinating the insulation of the protective equipment involves not only guarding the equipment insulation but also it is desired that the protecting equipment should not be damaged.

To assit in the process of insulation coordination, standard insulation levels have been recommended. These insulation levels are defined as follows :

Basic impulse insulation levels (BIL) are reference levels expressed in impulse crest voltage with a standard wave not longer than 1.2/50 μ sec wave. Apparatus insulation as demonstrated by suitable tests shall be equal to or greater than the basic insulation level.

The problem of insulation coordination can be studied under three steps :

1. Selection of a suitable insulation which is a function of reference class voltage (*i.e.*, 1.05 × 0 operating voltage of the system). Table 19.1 gives the BIL for various reference class voltages.

Table 19.1. Basic Impulse Insulation Levels

Reference Class kV	Standard Basic Impulse Level kV	Reduced Insulation levels
23	150	
34.5	200	
46	250	
69	350	
92	450	
115	550	450
138	650	550
161	750	650
196	900	
230	1050	900
287	1300	1050
345	1550	1300

2. The design of the various equipments such that the breakdown or flashover strength of all insulation in the station equals or exceeds the selected level as in (1).
3. Selection of protective devices that will give the apparatus as good protection as can be justified economically.

The above procedure requires that the apparatus to be protected shall have a withstand test value not less than the kV magnitude given in the second column of Table 19.2, irrespective of the polarity of the wave positive or negative and irrespective of how the system was grounded.

The third column of the table gives the reduced insulation levels which are used for selecting insulation levels of solidly grounded systems and for normally by much less than in a higher voltage station because the high voltage surges will flashover the line insulation of low voltage line and not reach the station.

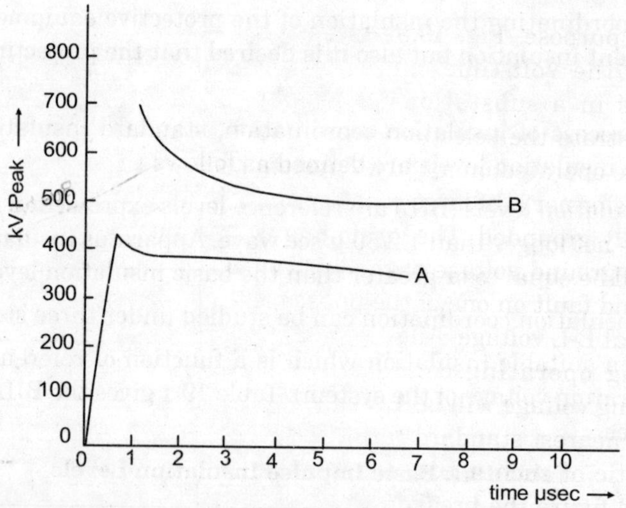

Fig. 19.31 Coordination of transformer insulation with lightning arrester : A—Lightning arrester 109 kV, B—Transformer insulation withstand characteristics.

The traditional approach to insulation coordination requires the evaluation of the highest overvoltages to which an equipment may be subjected during operation and selection of standardized value of withstand impulse voltage with suitable safety margin. However, it is realized that overvoltages are a random phenomenon and it is uneconomical to design plant with such a high degree of safety that they sustain the infrequent ones. it is also known that insulation designed on this basis does not give 100% protection and insulation failure may occur even in well designed plants and, therefore, it is desired to limit the frequency of insulation failures to the most economical value taking into account equipment cost and service continuity. Insulation coordination, therefore, should be based on evaluation and limitation of the risk oc failure than on the a priori choice of a safety margin.

The modern practice, therefore, is to make use of probabilistic concepts and statistical procedures especially for very high voltage equipments which might later on be extended to all cases where a close adjustment of insulation to system conditions proves economical. The statistical methods even though laboruous are quite useful.

19.4.2 Overvoltage Protection

The causes of overvoltages in the system have been studied extensively in earlier chapters. Basically, there are two sources: (*i*) external overvoltages due to mainly lightning, and (*ii*) internal overvoltages mainly due to switching operation. The system can be protected against external overvoltages using what are known as shielding methods which do not allow an are path to systems operating above 345 kV where switching surges are of more in portance than the lightning surges. At 345 kV, the switching voltage is considered to be 2.7 p.u., *i.e.*, 345 × 2.7 = 9.31.5 kV which corresponds to the lightning level. At 500 kV, however, 2.7 p.u. will mean 2.7 × 500 = 1350 kV switching voltage which exceeds the lightning voltage level. Therefore, the ratio of switching voltage to operating voltage is reduced by using the switching resistances between the C.B. contacts. For 500 kV, it has been possible to obtain this ratio as 2.0 and for 765 kV it is 1.7. With further increase in operating voltages it is hoped that the ratio could be brought to 1.5. So, for switching voltages the reduced levels in third column are used *i.e.*, for 345 kV, the standard BIL is 1550 kV but if the equipment can withstand even 1425 kV or 1300

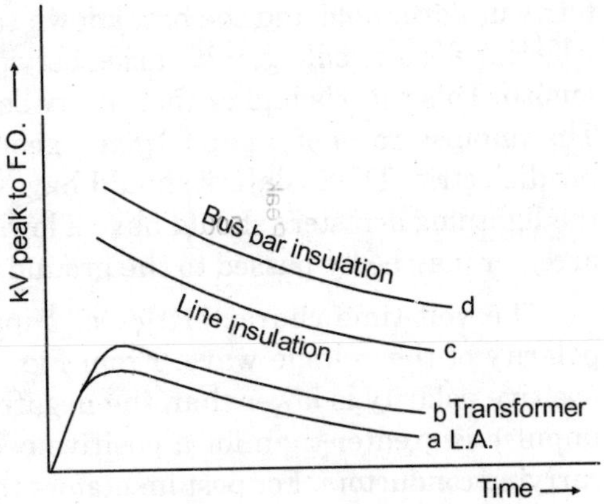

Fig. 19.32. Volt-time curves.

kV it will serve the purpose. Fig. 19.32 gives the relative position of the volt-time curves of the various equipments in a substation for proper coordination. To illustrate the selection of the BIL of a transformer to be operated on a 138 kV system assume that the transformer is of large capacity and its star point is solidly grounded. The grounding is such that the line-to-ground voltage of the healthy phase during a ground fault on one of the phases is say 74% of the normal L-L voltage. Allowing for 5% overvoltage during operating conditions, the arrester rms operating voltage will be 1.05 × 0.74 × 138 = 107.2 kV. The nearest standard rating is 109 kV. The characteristic of such a L.A. is shown in Fig. 19.32. From the figure the breakdown value of the arrester is 400 kV. Assuming a 15% margin plus 35 kV between the insulation levels of L.A. and the transformer, the insulation level of transformer should be at least equal to 400 + 0.15 × 400 + 35 = 495 kV. From is 550 kV) the insulation level of transformer is 550 kV; therefore a lightning arrester of 109 kV rating can be applied.

It is to be noted that low voltage lines are not as highly insulated as higher voltage lines so that lightning surges coming into the station would form between the line conductor and ground, thereby giving inherent protection in the line design. For protection against internal voltages normally non-shielding methods are used which allow an arc path between the ground structure and the line conductor but means are provided to quench the arc. The use of ground wire is a shielding method whereas the use of spark gaps, and lightning arresters are the non-shielding methods. We will study first the non-shielding methods and then the shielding methods. However, the non-shielding methods can also be used for external overvoltages.

The non-shielding methods are based upon the principle of insulation breakdown as the overvoltage is incident on the protective device; thereby a part of the energy content in the overvoltage is discharged to the ground through the protective device. The insulation breakdown is not only a function of voltage but it depends upon the time for which it is applied and also it depends upon the shape and size of the electrodes used. The steeper the shape of the voltage wave, the larger will be the magnitude of voltage required for breakdown; this is because an expenditure of energy is required for the rupture of any dielelctric, whether gaseous, liquid or solid, and energy involves time. The energy criterion for various insulations can be compared in terms of a common term known as Impulse Ratio which is defined as the ratio of breakdown voltage due to an impulse of specified shape to the breakdown voltatge at power frequency. The impulse ratio for sphere gap is unity because this gap has a

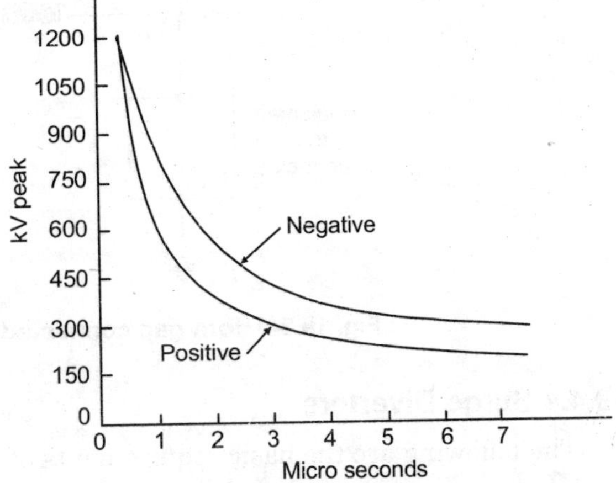

Fig. 19.33. Volt-time curves of gaps for positive and negative polarity.

fairly uniform field and the breakdown takes place on the field ionization phenomenon mainly whereas for a needle gap it varies between 1.5 to 2.3 depending upon the frequency and gap lendth. This ratio is higher than unity because of the nonuniform field between the electrodes. The impulse ratio of a gap of given geometry and dimension is greater with solid than with air dielectric. The inslators should have a high impulse ratio for an economic design whereas the lightning arresters should have a low impulse ratio so that a surge incident on the lightning arrester may be by-passed to the ground instead of passing it on to the apparatus.

The volt-time characteritics of gaps having one electrode grounded depend upon the polarity of the voltage wave. From Fig. 19.33 it is seen that the volt-time characteristic for positive polarity is lower than the negative polarity, *i.e.*, the breakdown voltage for a negative impulse is greater than for a positive because of the nearness of earthed metal or of current carrying conductors. For post insulators the negative polarity wave has a high breakdown value whereas for suspension insulators the reverse is true.

19.4.3 Horn Gap

The horn gap consists of two horn-shaped rods separated by a small distance. One end of this is connected to the line and the other to the earth as shown in Fig. 19.34, with or without a series resistance. The choke connected between the equipment to be protected and the horn gap serves two purposes: (*i*) The steepness of the wave incident on the equipment to be protected is reduced. (*ii*) It reflects the voltage surge back on to the horn.

Whenever a surge voltage exceeds the breakdown value of the gap as discharge takes place and the energy content in the rest part of the wave is by-passed to the ground. An arc is set up between the gap, which acts like a flexible conductor and rises upwards under the influence of the electromagnetic forces, thus increasing the length of the arc which eventually blows out.

There are two major drawbacks of the horn gap : (*i*) The time of operation of the gap is quite large as compared to the modern protective gear. (*ii*) If used on isolated neutral the horn gap may constitute a vicious kind of arcing ground. For these reasons, the horn gap has almost vanished from important power lines.

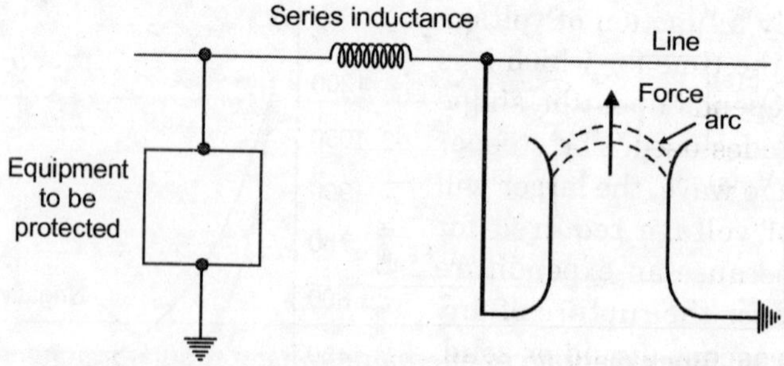

Fig. 19.34. Horn gap connected in the system for protection.

19.4.4 Surge Diverters

The following are the basic requirements of a surge diverter :

(*i*) It should not pass any current at normal or abnormal (normally 5% more than the normal voltage) power frequency voltage.

(*ii*) It should breakdown as quickly as possible after the abnormal high frequency.

(*iii*) It should not only protect the equipment but should also discharge the surge current without damaging itself.

(*iv*) It should interrupt the power frequency flow current after the surge is discharged to ground.

19.4.5 Automatic Voltage Control

Fig. 19.35 gives the schematic diagram of an automatic voltage regulator of a generator. It basically consists of a main exciter which excites the alternator field to control the output voltage. The exciter field is automatically controlled through error $e = V_{ref} - V_T$, suitably amplified through voltage and power amplifiers. It is a type-0 system which requires a constant error e for a specified voltage at generator terminals. The block diagram of the system is given in Fig. 19.32. The function of important components and their transfer function is given below:

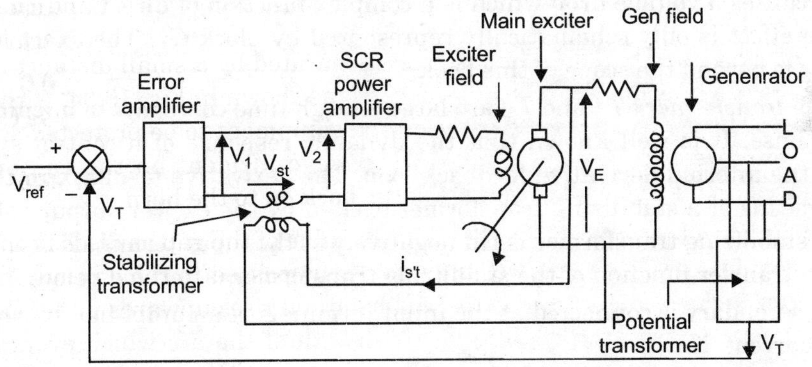

Fig. 19.35. Schematic diagram of alternator voltage regulator scheme.

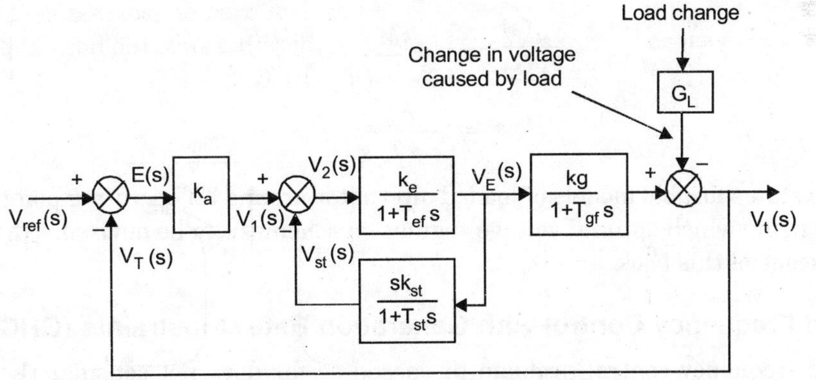

Fig. 19.36. Block diagram of alternator voltage regulator scheme.

Potential transformer : It gives a sample of terminal voltage V_T.

Differencing device : It gives the actuating error

$$e = V_{ref} - V_T$$

The error initiates the corrective action of adjusting the alternator excitation. Error wave form issuppressed carrier modulated, the carrier frequency being the system frequency of 50 Hz.

Error amplifier : It demodulates amplifies the error signal. Its gain is K_a.

SCR power amplifier and exciter field : It provides the necessary power amplification to the signal for controlling the exciter field. Assuming the amplifier time constant to be small enough to be neglected, the overall transfer function of these two is

$$\frac{K_e}{1+T_{ef}s}$$

where T_{ef} is the exciter field time constant.

Alternator : Its field is excited by the main exciter voltage v_E. Under no load it produces a voltage proportional to field current. The no load transfer function is

$$\frac{K_g}{1+T_{gf}s}$$

where T_{gf} = generator field time constant

The load causes a voltage drop which is a complex function of direct and quadrature axis currents. The effect is only schematically represented by block G_L. The exact load model of the alternator is beyond the scope of this book.

Stabilizing transformer : T_{ef} and T_{gf} are large enough time constants to impair the system's dynamic response. It is well known that the dynamic response of a control system can be improved by the internal derivative feedback loop. The derivative feedback in this system is provided by means of a stabilizing transformer excited by the exciter output voltage v_E. The output of the stabilizing transformer is fed negatively at the input terminals of the SCR power amplifier. The transfer function of the stabilizing transformer is derived below.

Since the secondary is connected at the input terminals of an amplifier, it can be assumed to draw zero current. Now

$$v_E = R_1 i_{st} + L_1 \frac{di_{st}}{dt}$$

$$\frac{V_{st}(s)}{V_E(s)} = \frac{sM}{R_1 + sL_1} = \frac{sM/R_1}{1+T_{st}S}$$

$$= \frac{sK_{st}}{1+T_{st}s}$$

Accurate state variable models of loaded alternator around an operating point are available in literature using which optimal voltage regulation schemes can be devised. This is, of course, beyond the scope of this book.

19.4.6 Load Frequency Control with Generation Rate Constraints (GRCs)

The load frequency control problem discussed so far does not consider the effect of the restrictions on the rate of change of power generation. In power systems having steam plants, power generation can change only at a specified maximum rate. The generation rate (from safety considerations of the equipment) for reheat units is quite low. Most of the reheat units have a generation rate around 3%/min. Some have a generation rate between 5 to 10% min. If these constraints are not considered, system is likely to chase large momentary disturbances. This results in undur wear and tear of the controller. Several methods have been proposed to consider the effect of GRCs for the design of automatic generation controllers. When GRC is considered, the system dynamic model becomes non-linear and linear control techniques cannot be applied for the optimization of the controller setting.

If the generation rates denoted by P_{Gi} are included in the state vector, the system order will be altered. Instead of augmenting them, while solving state equations, it may be verified at each step if the GRCs are violated. Another way of considering GRCs for both areas is to add limiters to the governors as shown in Fig. 19.37, *i.e.*, the maximum rate of valve opening or closing speed is restricted by the limiters. Here $T_{ag} g_{max}$ is the power rate limit imposed by valve or gate control. In this model

$$|\Delta Y_E| < g_{max}$$

The banded values imposed by the limiters are selected to restrict the generation rate by 10% per minute.

The GRCs result in larger deviations in ACEs as the rate at which generation can change in the area is constrained by the limits imposed. Therefore, the duration for which the power needs to be imported increases considerably as compared to the case where generation rate is in the hydro area normally remains below the safe limit and therefore GRCs for all the hydro plants can be ignored.

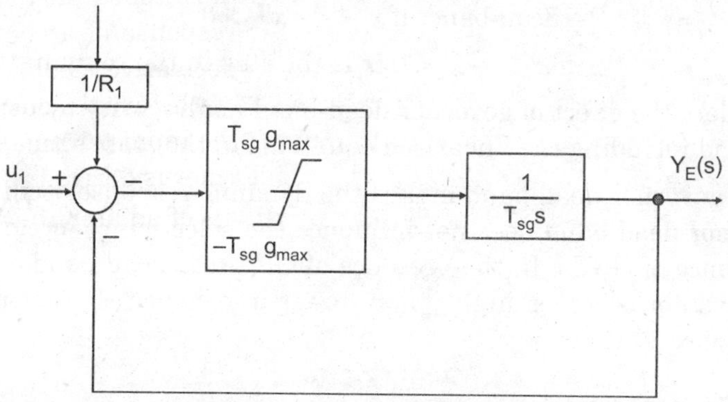

Fig. 19.37. Governor model with GRC.

19.4.7 Speed Governor Dead-Band and its Effect on AGC

The effect of the speed governor dead-band is that for a given position of the governor control valves, an increase/decrease in speed can occur before the position of the valve changes. The governor dead-band can materially affect the system response. In AGC studies, the dead-band effect indeed can be significant, since relatively small signals are under considerations.

The speed governor characteristic, though non-linear, has been approximated by linear characteristics in earlier analysis. Further, there is another non-linearity introduced by the dead-band in the governor operation. Mechanical friction and backlash and also valve overlaps in hydraulic relays cause the governor dead-band. Due to this, through the input signal increases, the speed governor may not immediately react until the input reaches a particular value. Similar action takes place when the input signal decreases. Thus the governor dead-band is defined as the total magnitude of sustained speed change within which there is no change in valve position. The limiting value of dead-band is specified as 0.06%. It was shown by Concordia *et. al.*, that one of the effects of governor dead-band is to increase the apparent steady-state speed regulation R.

The effect of the dead-band may be included in the speed governor control loop block diagram as shown in Fig. 19.38. Considering the worst case for the dead-band (*i.e.*, the system starts responding after the whole dead-band is traversed) and examining the dead-band block in Fig. 19.38.

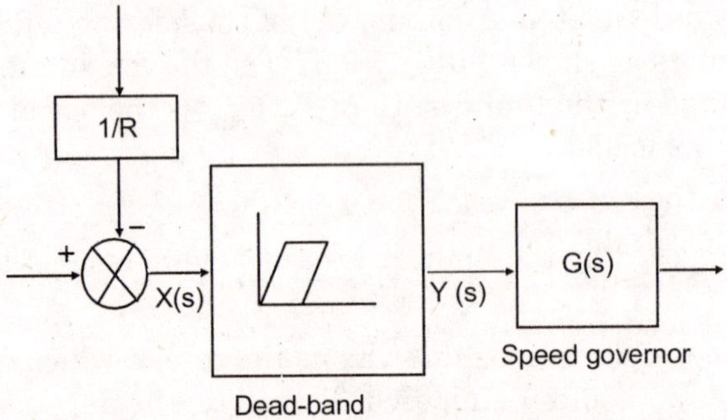

Fig. 19.38. Dead-band in speed-governor control loop.

$$\begin{aligned} y^{(r+1)} &= x^{(r)} \text{ if } x^{(r+1)} - x^{r} \le \text{dead-band} \\ &= x^{(r+1)} - \text{dead-band; if } x^{(r+1)} - x^{(r)} > 0 \\ &= x^{(r+1)}, \text{ if } x^{r+1} - x^{r} < 0 \ (r \text{ is the step in the computation}) \end{aligned}$$

Reference considers the effect of governor dead-band nonlinearity by using the describing function approach and including the linearised equations in the state space model.

The presence of governor dead-band makes the dynamic response oscillatory. It has been seen that the governor dead-band does not influence the selection of integral controller gain settings in the presence of GRCs. In the presence of GRC and dead band even for small load perturbation, the system becomes highly non-linear and hence the optimization problem becomes rather complex.

19.4.8 Digital LF Controllers

In recent years, increasingly more attention is being paid to the question of digital implementation of the automatic generation control algorithms. This is mainly due to the facts that digital control turns out to be more accurate and reliable, compact in size, less sensitive to noise and drift and more flexible. it may also be implemented in a time shared fashion by using the computer systems in load despatch centre, if so desired. The ACE, a signal which is used for AGC is available in the discrete form, *i.e.*, there occurs sampling operation between the system and the controller. Unlike thus to resort to the discrete-time analysis for optimization of the AGC strategies.

Discrete-Time Control Model

The continuous-time dynamic system is described by a set of linear differential equations

$$\dot{x} = Ax + Bu + \Gamma P$$

where x, u, p are state, control and disturbance vectors respectively and A, B and Γ are constant matrices associated with the above vectors.

The discrete-time behaviour of the continuous-time system is modelled by the system of first order linear difference equations.

$$x(k+1) = \phi x(k) + \psi u(k) + \gamma p(k)$$

where $x(k)$, $u(k)$ and $p(k)$ are the state, control and disturbance vectors and are specified at $t = kT$, $k = 0, 1, 2, \ldots$ etc. and T is the sampling period. ϕ, ψ and γ are the state, control and disturbance transition matrices and they are evaluated using the following relations.

$$\phi = e^{AT}$$

$$\psi = (e^{AT} - I)\, A^{-1} B$$

$$\gamma = (e^{AT} - I)\, A^{-1} \Gamma$$

where A, B and Γ are the constant matrices associated with x, u and p vectors in the corresponding continuous-time dynamic system. The matrix e^{AT} can be evaluated using various well-documented approaches like Sylvestor's expansion theorem, series expansion technique etc. The optimal digital load frequency controller design problem is discussed in detail in Ref. [7].

QUESTIONS

1. What is Economic Load Dispatch? Discuss with reference to problem.
2. State what are the constrains in Power System.
3. What is incremental efficiency? What is the condition for optimum operation?
4. Derive an expression for optimum Load Dispatch with transmission losses.
5. Discuss Automatic Load Dispatching with schematic a diagram.
6. What is PLCC? Explain where it is used.
7. State, what is the purposes of compensation in Power System what is meant by (*a*) Load Compensation, (*b*) Reactive Compensation, (*c*) Surge Impedance Loading.
8. Prove compensating line Reactance $X' = Z_C \sin \beta l$, where Z_C is surge impedance.
9. What are the important aspects of compensation in power system? Also state breifly discuss various types of compensators.
10. What is FACT, STATCON, SVC.
11. What is meant by Load Frequency control? Discuss.
12. Explain with a schematic diagram the system of speed Governing.
13. Explain optimal (Two-Area) Load Frequency Control.
14. What is meant by Insulation Coordination in power system? Explain with Volt-Time Curve.
15. What is Basic Impulse level of an equipment used in High Voltage system?
16. What is the purpose of Horn gap.
17. What is a Surge Diverter and what are the basic requirement of Surge Diverters?
18. What is Automatic Voltage Control? Discuss with schematic diagram of alternator Voltage regulator scheme.

Appendix–I

PROTECTION OF POWER SYSTEM APPARATUS AGAINST SURGES

DISCUSS LIGHTING ARRESTOR (SURGE DIRECTION)

Terminal equipments in a substations and other important set ups are protected against surges by surge directors or Lighting Arrestor.

A surge director or a Lighting Arrestor is connected in shunt between the line and ground. Ideally it should :

1. become conducting at voltage above directors rating.
2. restrict the voltage across the terminals to the design value.
3. become non-conducting again when the Line-To-Neural Voltage becomes lower than the design value. That means it should not permit any power follow-on current.
4. not conduct any current at neural or somewhat above normal power frequency voltages.

There are three types of Lightning Arrestors as :

1. Rod-Gap
2. Expulsion Type of lighting Arrestor
3. Valve Type (Non-Linear) Arrestor.

These are briefly discussed below :

1. Rod-Gap Type : This is constituted of a plain air gap between two square rods (of area 1 cm^2) bent at right angles and connected between the line and earth. The gap may also be in the from of horn or arcing rings. When the surge voltage reaches the design value of the gap, an arc appears in the gap providing an ionised path to ground, essentially a short-circuit. For a given gap the time to breakdown varies inversely with applied voltage. It is normally recommended that a rod gap should be so set that it breakdown at a voltage and less than 30% below the voltage withstand level of the equipment to be protected.

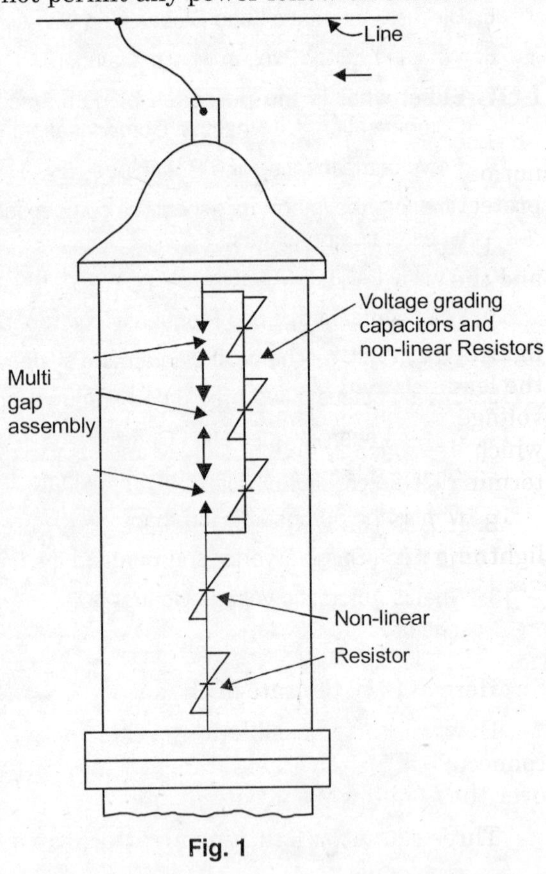

Fig. 1

2. Expulsion Type : It is the improved version of the Rod-Gap type. In this type the series gap is set to arc over at a specified voltage of the equipment to be protected. This device has the desired self clearing property. However this is not now much used.

3. Valve Type (Non Linear) LA : This type consists of non linear resistors in series with spark-gaps. The spark-gap assembly acts as a fast switch which gets ionised (conducting) at specified voltage. The nonlinear resistor elements are stacked, one over the other, in two or three section. The problem of non-uniform voltage distributor across gap is solved by capacitors and non-linear resistors connected in parallel across each gap. As a result the gap assembly has a degree of consistancy of breakdown (flash over) voltage. The entire assembly, as shown in the Fig. 3, is placed in poreclain housing properly sealed to keep out dust and moisture.

The non-linear resistors are made of loosely packed silicon carbide. The modern trend is to use Zinc oxide, which gives more desirable characteristics.

Because of loose packing of silicon carbide particles, the conduction process through it is mainly by means of short over and hence the non linear resistor characteristics. The static characteristic is especially expressed as

$$I = kV^{\propto}$$

where,
I = discharge current
V = voltage across the element
$\propto$ = an exponent more than unity (= 4)
K = constant

K and $\propto$ depends upon the material, packing and dimensions of the element.

LOCATION OF LIGHTING ARRESTORS

Location of Lighting Arrestors is a very important activity in power system. In fact the normal practice is to locate the lighting arrestor as close as possible to the equipment to the protection for the following reasons :

1. Reducing the chances of surges entering the circuit between the protective equipment and the equipment is to be protected.

2. If there is a distance between the two, a steep fronted wave after being incident on the lightning arrestor, which sparks over voltage, enters the transformar after the travelling over the lead between the two. The wave suffers reflection at the terminal and therefore, the total voltage at the terminal of the transformer is the sum of reflected and the incident voltage which approaches nearly twice the incident voltage. If the lightning arrestor is right at the terminals this could not occur.

3. If L is the inductance of the lead between the two and IR the residual voltage of the lightning arrestor the voltage incident at the transformer will be

$$V = IR + L\frac{di}{dt}$$

Here di/dt is the rate of change of the surge current.

However, it is possible to provide some separation between the two, where a capacitor is connected at the terminals of the equipment to be protected. This reduces the stress distribution over the winding of the equipment.

Three classes of lightning arrestors are available as stated below :

1. Station Type : The voltage ratings of such arrestors vary from 3kV to 312kV and are designed to discharge current not less than 100,000 Amps. They are used for the protection of substation and power transformers.

2. Line Type : The voltage ratings vary from 20kV to 73kV and can discharge current between 65000 amps and 100,000 amps. They are used for the protection of distribution transformers small power transformers and sometimes small substations.

3. Distribution Type : The voltage ratings vary for 8kV to 15kV can discharge, current upto 65,000 Amp. They are used mainly for pole mounted substations for the protection of distribution transformers upto and including the 15kv classification.

RATING OF LIGHTNING ARRESTOR

A lightning arrestor is expected of discharge surge current of very large magnitude, thousands of amperes, but since the time is very short term of micro seconds, the energy that dissipated through the lightning arrestor is small compared with what would have been if a few amperes of power frequency current had been flown for a few cycles. Therefore the main consideration in selecting the rating of L.A is the line-to-ground dynamic voltage to which the L.A may be subjected for any condition of system operation. An allowance of 5% is normally assumed, to take into account the light operating condition under no load at the far end of the line due to ferranti effect and the sudden loss of load on water wheel generators. This means an L.A. of 105% is used on a system where the line-to-ground voltage may reach line-to-line value during line-to-ground fault condition.

The overvoltages on a system depend on the neutral grounding condition which is determined by the parameters of the system. Infact, a system is said to be solidly grounded if $\frac{R_0}{R_L} \leq 1$ and $\frac{X_0}{X_L} \leq 3$.

Surge absorbers are used to absorb the surges. Infact a surge absorber is an electrical device which absorber energy contained in a travelling wave corona is a means of absorbing energy in the form of carona loss. A short length of cable between the equipment and the overhead line absorbs energy in the travelling wave because of its high capacitance and low inductance. Another method of absorbing energy is the use of Ferranti Swege absorber which consists of an air core inductor connected in series with the line and surrounded by an earthed metallic sheet called dissipator which is insulated from the inductor by the air.

Example : *A lightning arrestor is required to protect a 5000 kVA 66kV/11kV transformer. The transformer is connected to a 66KV system and the voltage fluctuations are of the order of 10%. The neutral of the transformer and that of network are directly and solidly grounded. Determine a suitable arrestor rating with reference to the following table :*

Arrestor Rating kV (rms)	*Crest kV Impulse spark over (1/50 μ sec wave)*	*Crest kV Impulse Voltage drop across Arrestor (10 × 20 μ sec current wave) at 10,000 Amp*
50	172	170
60	215	220
73	258	264

Take the following into consideration :

(a) *Bill (Basic Impulse Level) for 66kV : 350 kV.*

(b) *Dynamic over voltage = 1.3 times system operating voltage.*

(c) *Power frequency breakdown voltage = 1.5 times the arrestor rating.*

(d) *Allow 5% margin for lower limit of arrestor rating calculation. Check the margin available on the arrestor choose with reference to the BIL.*

Solution : Voltage rating of the arrestor = 66 × 1.1 × 0.8 × 1.05

[as fluctuation is 10% i.e., ± 5% = 1.05 × Rating Voltage]

= 60.98kV

In this calculation 1.05 allows 5% margin for lower limit of the arrestor rating.

From the table, hence choose arrestor with reference to the rating of 60kV. The discharge voltage of 60KV arrestor (from the table) = 220kV.

But as per the example BIL = 350kV

∴ Protective margin available = 350 – 220 = 130kV

Power frequency breakdown voltage = 60 × 1.5 = 90kV

$$\therefore \text{Dynamic over-voltage} = 1.3 \times \frac{66}{\sqrt{3}} = 50\text{kV}$$

As 90kV > 50kV, hence Arrestor rating is suitable.

The location of an arrestor in a substation depend on the probability of flashover on the line per one hundred km.

The crest of surges and arrestor characteristics are to be considered. It is not necessary to install one arrestor adjacent to each equipments in the substation. One arrestor at the entry of the incoming feeder may serve the purpose well. Lightning Arrestor requirement reduced particularly when the equipments are in an indoor substation are connected through a cable. Surges are attunnated in cables and hence for indoor substation, particularly in the medium voltage range. When fed through cable, the Arrestor duty reduces.

Example : *An O.H. line is in series with a cable of 5km length. The cable is open at for end. The surge impedence of cable and line are 40 Ω and 400 Ω respectively. A surge of peak value of 100KV travels along the O.H. line towards the junction. Find the voltage distribution at the instant 1 μ sec and 50 μ sec after the surge reaches the function. Propagation velocity in the line cable are,* $v_{line} = 3 \times 10^5 km/sec$, $v_{cable} = 1.4 \times 10^5 km/sec$.

Solution : (*a*) Reflected voltage in O.H. line

$$e_r = \left(\frac{z_2 - z_1}{z_2 + z_1}\right) \cdot e_f = \frac{40 - 40}{40 + 400} \cdot 100 = -81.8 \text{ kV}$$

Fig. 2

Transmitted (Reflected) voltage in cable

$$e_t = \frac{2z_2}{z_2 + z} \cdot e_f = \frac{2 \times 400}{400 + 40} = 18.2 \text{ kV}$$

Voltage distribution 1 μ sec after surge reaches the junction. Distance travelled by the transmitted wave in cable in 1 μ sec.

$$= v_{cable} \times \text{time} = (1.4 \times 10^5 \times 10^3) \times (1 \times 10^{-6}) = 140 \text{ m}$$

Distance travelled by the reflected wave in O.H. line in 1 μ sec

$$= v_{line} \times \text{time} = (3 \times 10^5 \times 10^3) \times (1 \times 10^{-6}) = 300 \text{ m}$$

(*b*) Voltage distribution 50 μ sec after the wave reaches the junction.

Time taken by transmitted wave to reach for end of the cable (open end) of the cable of 5 km length $= \dfrac{5 \times 10^3}{1.4 \times 10^5 \times 10^3} = 35.7$ μ sec. This voltage is reflected at open end with same sign and equal magnitude (18.2 KV), thus doubling the voltage at open end.

In remaining (50 – 35.7) = 14.3 μ sec this reflected voltage from open end of cable travels a distance of $(14.3 \times 10^{-6}) \times (1.4 \times 10^5) = 2$ km in the cable. Also the distance covered by the reflected wave from the function into O.H. line in 50 μ sec is $3 \times 10^5 \times 50 \times 10^{-6} = 15$ km.

Appendix-II

WORKED EXAMPLES ON

1. Performance of Transmission Lines.

Example 1. *A 132 kV line is fed through a 33 kV/132 kV transformer from a constant 33 kV supply as shown in Figure. Total impedance of line and transformers at 132 kV is (25 + j60) Ω. Both transformers are equipped with tap changing facilities which are so arranged that position of off nominal settings is unity.*

If the load on system is 100 MW at 0.6 p.f. lagging calculate the setting of tap changer required to maintain the voltage of load bus at 33 kV.

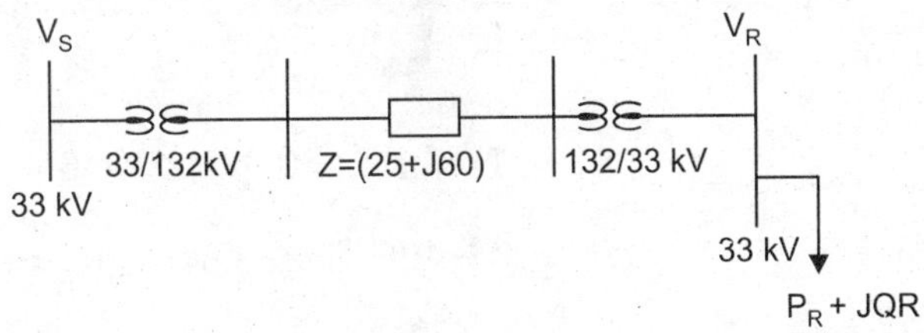

Fig. 1

Solution : Here,

$$n_1 = \frac{132}{33} n_2 = \frac{132}{33}$$

$$\frac{n_2}{n_1} = 1$$

$$ts^2\left(1-\frac{PR_R + XQ_R}{n_1 n_2 V_S V_R}\right) = \frac{V_R n_2}{V_s n_1}$$

$$ts^2\left(1-\frac{25\times100+60\times133.3}{33\times33\times\dfrac{132}{33}\times\dfrac{132}{33}}\right) = 1$$

∴ $$ts = 1.58$$

Hence with $$t_s\, t_r = 1$$

$$tr = \frac{1}{1.58}$$

Example 2. *A 50 Hz, three phase, 275 kV, 400 km line has following parameters per phase.*

$$Resistance = 0.035\ \Omega/km;$$
$$Inductance = 1\ mH/km$$
$$Capacitance = 0.01\ \mu F/km$$

If the line is supplied at 275 kV, determine the MVA rating of shunt reactor having negligible losses that would be required to main 275 kV at the receiving end when the line is delivering no load. Use nominal 'π' representation of line and circle diagram method.

Solution :

$$R = 0.035 \times 400 = 14\Omega$$
$$L = 1 \times 10^{-3} \times 400 = 0.4H$$
$$X = 314 \times 0.4 = 125\ \Omega$$
$$C = 0.01 \times 10^{-6} \times 400$$
$$= 4 \times 10^{-6}\ F$$
$$Y = j\omega c = j1.256 \times 10^{-a}$$
$$Z = R + jx = 14 + j\ 125.6$$
$$= 126.37\ \angle 83.64^{\circ}$$

For Π' representation

$$A = 1 + \frac{1}{2}YZ = 0.927\ \angle 0.546^{\circ} = \alpha$$
$$B = Z = 126.37\ \angle 83.64^{\circ} = \beta$$

Receiving end Circle Diagram

Horizontal coordinate of centre

$$= -\left|\frac{A}{B}\right| |tV_R|^2 \cos(\beta - \alpha)$$
$$= -66.7\ \text{MW}$$

Vertical coordinate of centre

$$= -\left|\frac{A}{B}\right| |V_R|^2 \sin(\beta - \alpha)$$
$$= -550.525\ \text{MVAR} = OC_R = OC_R/83 - 6.546^{\circ}$$

Radius of Circle

$$= \frac{|V_S||V_R|}{|B|} = 598.44\ \text{MVA}$$

Selecting the same scale for both axes.

(say 1 cm = 100 MW = 100 MVAR = 100 MVA)

draw the circle diagram as shown in fig. A3/1 From circle diagram we get

Mm = Capacity of compensating equipment (Q_c).

(Q_c) = 47 MVAR (lagging) (Converted to the scale chosen)

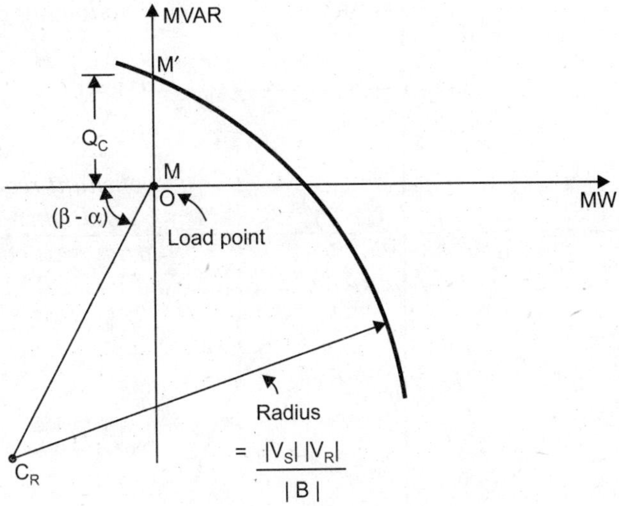

Fig. 2

Example 3. *A three phase overhead line has resistance and reactance of 5 and 20 Ω respectively. The load at the receiving end is 30 MW, 0.85 p.f. lagging at 33 kV. Find the voltage at the sending end. What will be the kVAR rating of the compensating equipment inserted at the receiving end so as to maintain a voltage of 33 kV at each end. Find also the maximum load that can be transmitted. Use circle diagram method.*

Solution :

$$R = 5\ \Omega, X = 20\ \Omega$$

$$Z = 5 + j20 = 20.6 \angle 75.96^\circ\ \Omega$$

$$A = 1 \angle 0, B = Z = 20.6 \angle 75.96\ \Omega$$

Horizontal coordinate of Centre

$$= -\left|\frac{A}{B}\right| |V_R^2| \cos(\beta - \alpha)$$

$$= 12.816 \text{ MW}$$

Vertical coordinate of Centre

$$= -\left|\frac{A}{B}\right| |V_R|^2 \sin(\beta - \alpha)$$

$$= 51.25 \text{ MVAR}$$

load point (m) located by calculating

$$P_D = 30 \text{ MW}, Q_D = P_D \tan \Phi = 18.59 \text{ MVAR}$$

Select same scale for both axes.

C_R is the centre of circle. To deliver the required load (load point M) the receiving end circle must pass through point M.

$$\text{Hence radius} = C_R M = \frac{|V_S||V_R|}{|B|}$$

Measure $C_R M$ and covert it to actual value

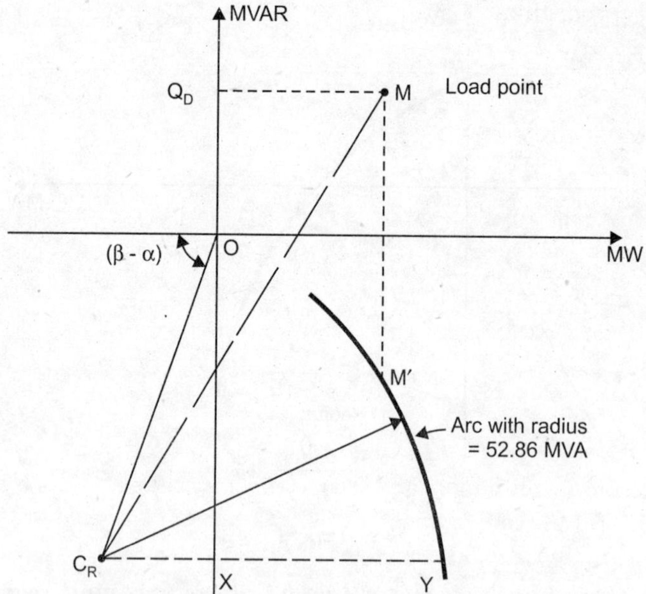

Fig. AP3.2

$$C_RM = 81.6 \text{ MVA} = \frac{|V_S| \times 33}{20.6}$$

$$\therefore \quad |V_S| = 52.86 \text{ kV}$$

$$\text{In second case } V_S = 33 \text{ kV} = V_R$$

The centre remains same at C_R

Radius of Circle

$$= \frac{V_S V_R}{B} = \frac{33 \times 33}{20.6} = 52.86 \text{ MVA}$$

Convert it to the chosen scale in cm and draw an arc of this radius with centre as C_R.

Capacity of compensating equipment

$$= Q_C = MM' = 37.8 \text{ MVAR}$$

Maximum load that can be supplied

$$= xy = 40.26 \text{ MW.}$$

Example 4. *A three phase overhead line has resistance and reactance of 5 and 25 Ω respectively. The load at the receiving end is 15 MW. 33 kV, 0.8 p.f. lagging. Find the Capacity of the compensating equipment needed to deliver this load with sending end voltage of 33 kV.*

Calculate the extra load of 0.8 lagging p.f. which can be delivered with the compensating equipment (of capacity calculated above) installed, if the receiving end voltage is permitted to drop to 28 kV. Use circle diagram method.

Solution :

$$A = \angle 0, B = 5 + j\,25$$

$$= 25.49 \angle 78.69°$$

Part I : Horizontal coordinate

$$= -\left|\frac{A}{B}\right| |V_R|^2 \cos(\beta - \alpha) = -8.378 \text{ MW}$$

Vertical coordinate of circle

$$= -\left|\frac{A}{B}\right| |V_R|^2 \sin(\beta - \alpha)$$

$$= -41.89 \text{ MVAR}$$

Radius of Circle

$$= \frac{|V_S||V_R|}{|B|} = 42.722 \text{ MVA}$$

Load $\quad P_D = 15$ MW, $Q_D = 11.25$ MVAR

Select some scale on graph paper for vertical (MVAR) axis and horizontal (MW) axis and radius (MVA). Locate the centre C_{R1} and load point M as shown in figure AP3.3.

Draw an are of radius as calculated above from the centre C_{R1}

∴ Capacity of compensating equipment

$$Q_C = MM' = 17 \text{ MVAR}$$

Part II : $|V_S| = 33$ kV, $|V_R| = 28$ kV

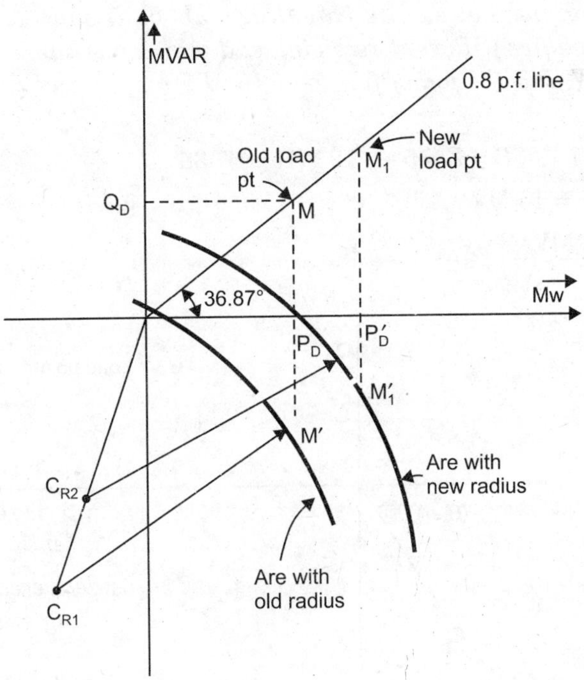

Fig. AP3.3

Locate the centre C_{R2} by calculating

Horizontal coordinate

$$= -\left|\frac{A}{B}\right| |V_R|^2 \cos(\beta - \alpha)$$

$$= 6.03 \text{ MW}$$

Vertical coordinate

$$= -\left|\frac{A}{B}| V_R |^2 \sin(\beta - \alpha)\right|$$

$$= 30.15 \text{ MVAR}$$

$$\text{New radius} = \frac{|V_S||V_R|}{|B|} = \frac{33 \times 28}{25.49}$$

$$= 36.24 \text{ MVA}$$

Draw an are of new radius from centre C_{R2}. Adjust the length M_1M_1 equal to MM' as shown in figure. [Where $Q_C = MM' = 17$ MVAR) Refer to Fig. AP3.3

The vertical line M_1M_1' intersects horizontal axis at P_D which is a new load.

$\therefore$ Extra load $= P_D' - P_D = 18.5 - 15 = 3.55$ MW

Example 5. *A 3-phase line has resistance = 5 Ω and inductive reactance = 12 Ω.*

(i) Determine V_s when the load at receiving end is 20 MVA,0.8 p.f. lagging at 33 kV.

(ii) If the sending end voltage as calculated above is maintained constant and by synchronous phase modifier (SPM) at receiving end which has same rating at zero load and for full load of 16 MW find the p.f. of full load 16 MW output and rating of Spm. Use circle diagram method.

Solution : (*i*) $A = 1 \angle 0$, $B = Z = 5 + j\,12 = 13 \angle 67.38°$

Load $P_D = 20 \times 0.8 = 16$ MW

$Q_D = 20 \times 0.6 = 12$ MVAR : $V_R = 33$ kV

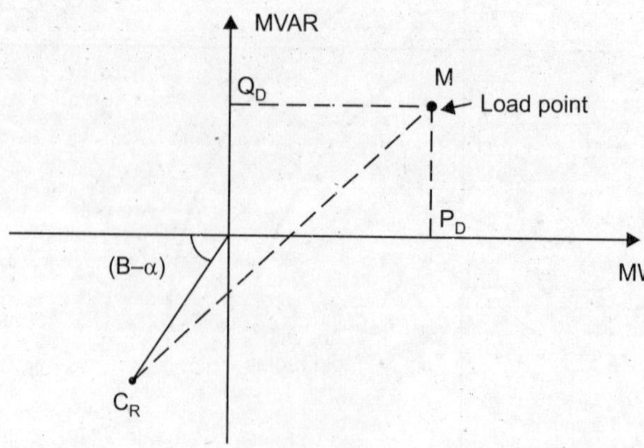

Fig. AP3.4

Horizontal coordinate of centre

$$= -\left|\frac{A}{B}\right| |V_R|^2 \cos(\beta - \alpha)$$

$$= 32.219 \text{ MW}$$

Vertical coordinate of centre

$$= \left|\frac{A}{B}\right| |V_R|^2 \sin(\beta - \alpha)$$

$$= 77.325 \text{ MVAR}$$

Locate the Centre C_R and load point M on a graph paper by selecting suitable scale, join C_RM.

Radius of circle passing through load point M is

$$C_RM = \frac{|V_S||V_R|}{|B|}$$

Hence $\qquad |V_S| = 40 \text{ kV}$

(*ii*) Here $\qquad V_S = 40 \text{ kV}, V_R = 33 \text{ kV}$

Centre of circle C_R is same as calculated in part 1 Radius of Circle

$$= \frac{|V_S||V_R|}{|B|} = \frac{40 \times 33}{13} = 101.54 \text{ MVA}$$

At no load capacity of synchronous phase modifier is $Q_C = 19$ MVAR (lagging)

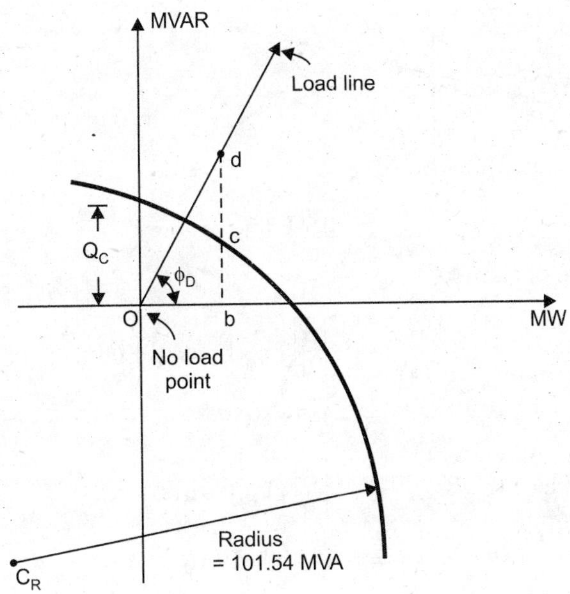

Fig. AP3.5

For full load of 16 MW, take ob = 16 MW

Draw vertical line at b cutting circle at 'C' Locate point 'd' such that $cd = Q_C = 19$ MVAR = Capacity of SPM. Join od. Hence p.f. of load = $\cos \Phi_D = 0.4586$ (lagging).

Example 6. *A 3 phase line has resistance of 5Ω/ph and Inductive reactance of 15 Ω/ph. Determine the load, the line will be supplying at 132 kV, 0.8 p.f. lagging if the sending end voltage V_S = 140 kV.*

If a s.p.m. is connected in parallel with load to improve the p.f. to 0.95 lagging, find the leading MVAR supplied by spm with V_S and V_R unchanged. Use circle diagram method.

Solution :

$$R = 5\ \Omega, X = 15\ \Omega, V_R = 132 \text{ kV},$$

$$V_S = 140 \text{ kV}$$

$$A = 1\angle 0, B = Z = 5 + j\,15 = 15.8\angle 71.6°\ \Omega.$$

$$\theta_D = \text{Load p.f. angle} = \cos^{-1} 0.8 = 36.87° \text{ p.f. at point 'b'}.$$

Refer to Fig. AP3.6

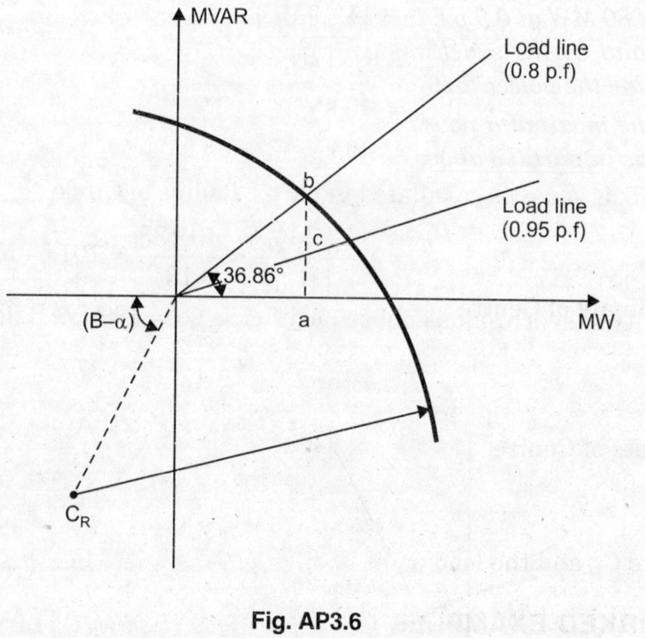

Fig. AP3.6

Horizontal coordinate of Centre

$$= -\left|\frac{A}{B}\right| |V_R|^2 \cos(\beta - \alpha)$$

$$= -347.8 \text{ MW}$$

Vertical coordinate of centre $= -\left|\frac{A}{B}\right| |V_R|^2 \sin(\beta - \alpha)$

$$= -1045.6 \text{ MVAR}$$

Radius of Circle $= \dfrac{|V_S||V_R|}{|B|} = 1168.8$ MVA

Choose suitable scale for MW. MVAR axis and for MVA (Same scale).

Locate the centre C_R on graph paper. Draw a load line at an angle (36.87°) to the horizontal axis as shown.

Draw an arc from centre C_R cutting the load line of 0.8 p.f. at point 'b'.

Draw a vertical line 'ba'.

$\therefore$ Load of 0.8 p.f. $=$ oa $= 60$ MW

Now Draw a load line of p.f. $= 0.95$

Then leading MVAR supplied by spm

$$= Q_C = bc = 20 \text{ MVAR}$$

Example 7. *The generalized circuit constants of a transmission line are as follows.*

$$A = D = 0.895 \angle 1.4°.$$

$$B = 182.5 \angle 78.6° \text{ ohm.}$$

(i) If the line supplies a load of 50 MW at 0.9 p.f. lagging at 215 kV, find the sending end voltage and hence the regulation of the line.

(ii) For a load of 80 MW at 0.9 p.f. lagging at 215 kV, determine the reactive power supplied by the line and by the synchronous capacitor, if the sending end voltage is 236 kV. Also determine the power factor of the line at the receiving end

(iii) Determine the maximum power that can be transmitted if the sending and receiving voltage are as in part (ii) above.

Solution : $V_R = 215$ kV.

(i) $P_D = 50$ MW, $Q_D = P_D = \tan \Phi = 24.216$ MVAR

Horizontal Coordinate of Centre $= -\dfrac{|A|}{|B|} |V_R|^2 \cos(\beta - \alpha)$

$= 50.223$ MW

Vertical Coordinate of Centre $= -\dfrac{|A|}{|B|} |V_R|^2 \sin(\beta - \alpha)$

$= 221.05$ MVAR

Locate the Centre C_R and the load point M on the graph paper by choosing suitable scale.

2. WORKED EXAMPLES ON POWER SYSTEM STABILITY

Example 1. *Determine the critical clearing angle for the network shown in fig. A three phase fault takes place at 'B' and the breakers A and B operate simultaneously. The generator is delivering 1 pu. Power before the fault takes place.*

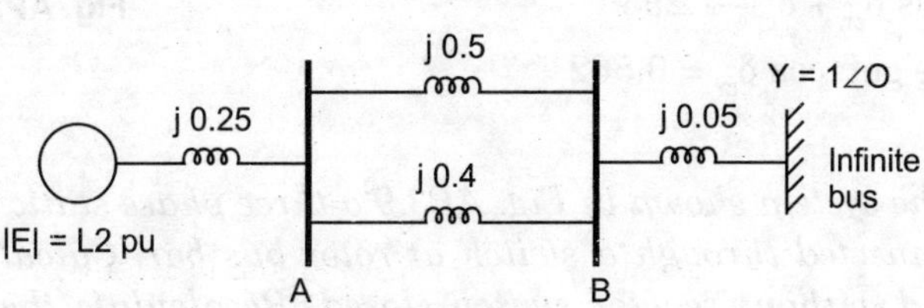

Fig. AP3.7

Solution : During prefault condition

Transfer reactance

$$X_{12} = 0.25 + \frac{0.5 \times 0.4}{0.5 + 0.4} + 0.05 = j\,0.522 \text{ pu.}$$

$$P_{el} = \frac{EV}{X_{12}}\sin\delta = \frac{12 \times 1}{0.522}\sin\delta = 2.3\sin\delta$$

$$\delta_0 = \sin^{-1}\left(\frac{1}{2.3}\right) = 25.8° = 0.45 \text{ radians}$$

During Fault

$$P_{elf} = 0$$

Post fault (with faulted line opened)

$$X_{12} = j0.25 + j0.5 + j\,0.05$$
$$= j\,0.8$$

$$P_{elf} = \frac{1.2 \times 1}{(0.8)}\sin\delta$$
$$= 1.5 \sin \delta.$$

The power angle curves and the critical case is shown in Fig. AP3.8

$$A_1 = P_m(\delta_{cr} - \delta_0)$$
$$= 1.0\,(\delta_{cr} - 0.45) = \delta_{cr} - 0.45$$

$$A_2 = \int_{\delta_{cr}}^{\delta max} (P_{elf} - Pm)d\delta$$

$$= \int_{\delta cr}^{\delta max} (1.5\sin\delta - 1)d\delta$$

where $\delta_{max} = 180 - \sin^{-1}\left(\frac{1}{1.5}\right) = 138.2°$

$= 2.41$ rad

$$\therefore \quad A_2 = -1.5[\cos 138.2 - \cos\delta_{cr}] - (2.41 - \delta_{cr})$$
$$= 1.5 \cos \delta_{cr} + \delta_{cr} - 1.293$$

With $A_1 = A_2$ we get: $\cos \delta_{cr} = 0.562$

$\therefore \quad \delta_{cr} = 55.8°$

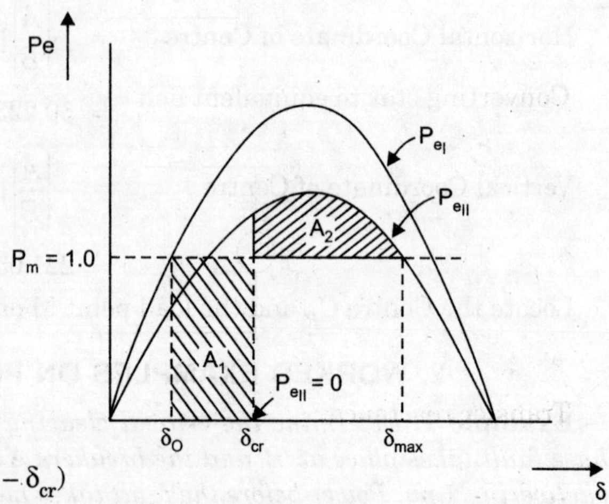

Fig. AP3.8

Example 2. *In the system shown in Fig. AP3.9 a three phase static capacitive reactance of I pu per phase is connected through a switch at rotor bus bar. Calculate the limit of steady state power with and without reactor switch closed. Recalculate the power limit with the Capacitive reactor replaced by an inductive reactor of same value. Assume the internal voltage of the generator to be 1.2 pu and that of the motor to be 1.0 pu.*

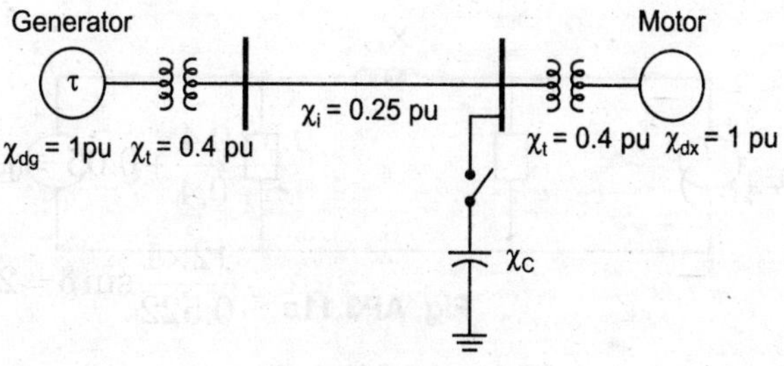

Fig. AP3.9

Solution : 1. Without reactor: Refer to eqvt. circuit in Fig. AP3.10 and Fig. AP3.10a.

Transfer reactance

$$X_{12} = 1 + 0.1 + 0.25 + 0.1 + 1 = 2.45 \text{ pu.}$$

$$P_{\max} = \frac{E_g E_m}{X_{12}} = \frac{1.2 \times 1}{2.45} = 0.489 \text{ pu.}$$

2. With capacitive reactor

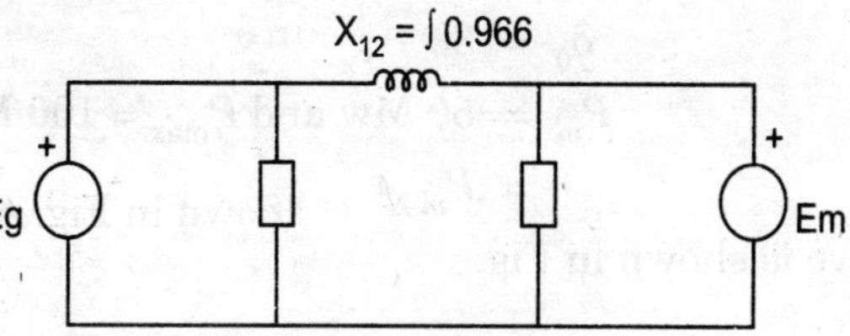

Fig. AP3.10

Converting star to equivalent delta we get

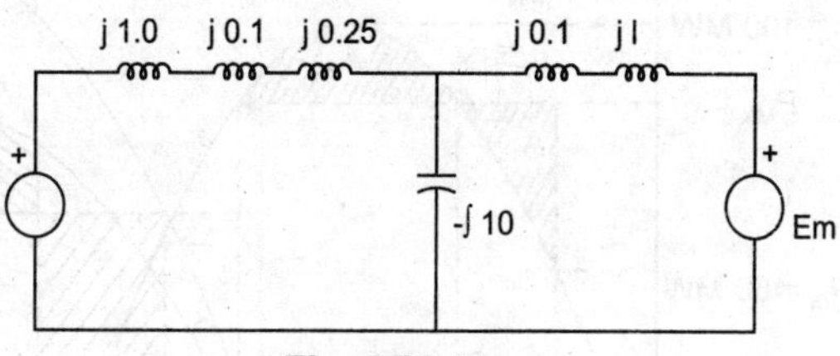

Fig. AP3.10a

Transfer reactance $\qquad X_{12} = j\,0.965$

$$\therefore \qquad P_{\max} = \frac{1.2 \times 1}{0.965} = 1.244 \text{ pu.}$$

3. With inductive reactor Refer to Fig. AP3.11 and Fig. AP3.11a

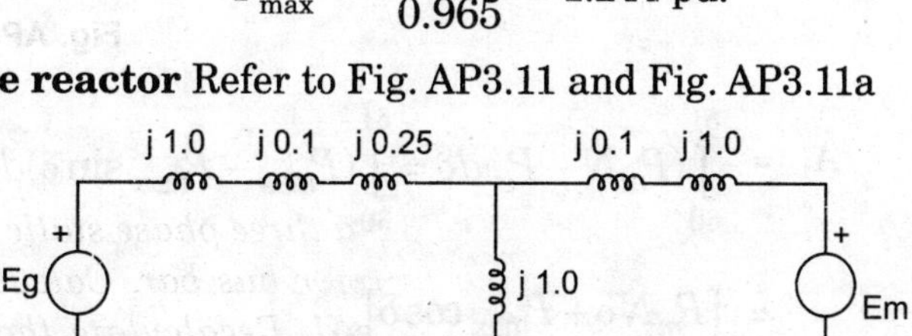

Converting star to delta

Fig. AP3.11

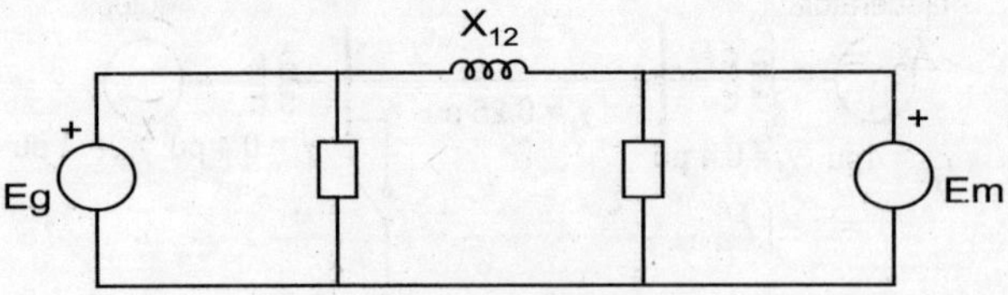

Fig. AP3.11a

Transfer reactance $X_{12} = j\,3.935$ pu.

$$P_{max} = \frac{1.2\times1}{3.935} = 0.304 \text{ pu.}$$

Example 3. *A power deficient area receives 50 Mw over a tie line from another area. The maximum steady state capacity of the line is 100 Mw. Find the allowable sudden load that can be switched on without loss stability.*

Solution : Initial Condition

$P_e = P_{max} \sin \delta_0$ or $50 = 100 \sin \delta_0$

$\therefore$ $\delta_0 = 30°$.

Initial $P_m = 50$ Mw and $P_{max} = 100$ Mw

Let the new load $= P_{mN}$

The power angle curve is shown in Fig.

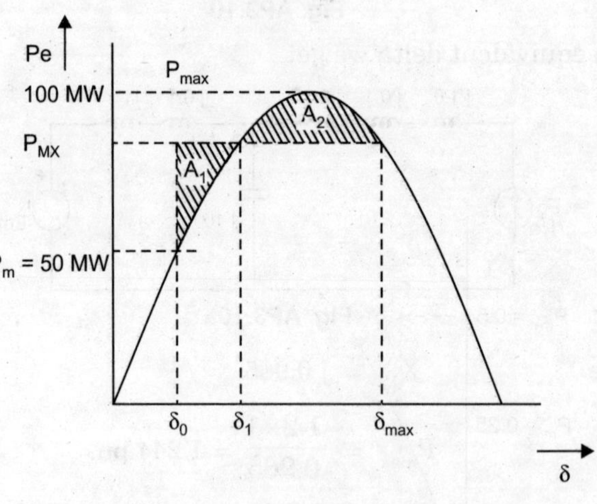

$$A_1 = \int_{\delta 0}^{\delta 1} (P_mN - P_e)d\delta = \int_{\delta 0}^{\delta 1} (P_{max} - P_{max} \sin\delta)d\delta$$

$$= [P_mN\delta + P_{max} \cos\delta]_{\delta 0}^{\delta_1}$$

$$= P_mN\,\delta_1 - 0.5235\,P_mN + 100\cos\delta_1 - 86.6$$

$$A_2 = \int_{\delta_1}^{\delta_{max}} (P_{max} \sin\delta - P_m N) d\delta$$

$$= -[P_{max} \cos\delta]_{\delta_1}^{\delta_{max}} - [P_m N \delta]_{\delta_1}^{\delta_{max}}$$

$$= P_{max} \cos\delta_1 - P_{max} \cos\delta_{max} + P_m N \delta_1 - P_m N \delta_{max}$$

Using $P_m N = P_{max} \sin\delta_1 = 100 \sin\delta_1$.

and $P_{max} = 100, \delta_{max} = \pi - \delta_1$.

By $A_1 = A_2$ we get

$(100 \sin\delta_1)\delta_1 - 0.5235 \times 100 \sin\delta_1 + 100 \cos\delta_1 - 86.6 = 100 \cos\delta_1 - 100 \cos(\pi - \delta_1) + (100 \sin\delta_1)\delta_1 - 100 \sin\delta_1(\pi - \delta_1)$

$\therefore \quad \cos\delta_1 + 0.86 = (2.618 - \delta_1) \sin\delta_1$.

By trial and error procedure it can be seen that $\delta_1 = 60°$ satisfies above equation.

$\therefore \quad P_m N = 100 \sin\delta_1 = 86.6$ Mw.

Thus the allowable sudden load that can be switched on without loss of stability, is

$$86.6 - 50 = 36.6 \text{ Mw}$$

Example 4. *A synchronous motor is receiving 25% of the power that it is capable of receiving from an infinite bus. If the load on the motor is doubled, calculated the maximum value of δ during the swinging of the rotor around its new equilibrium position.*

Solution : Initial condition

Let $P_{max} = 1. P_{e0} = 0.25$

$$\therefore \quad \delta_0 = \sin^{-1}\left(\frac{0.25}{1}\right) = 14.48°$$

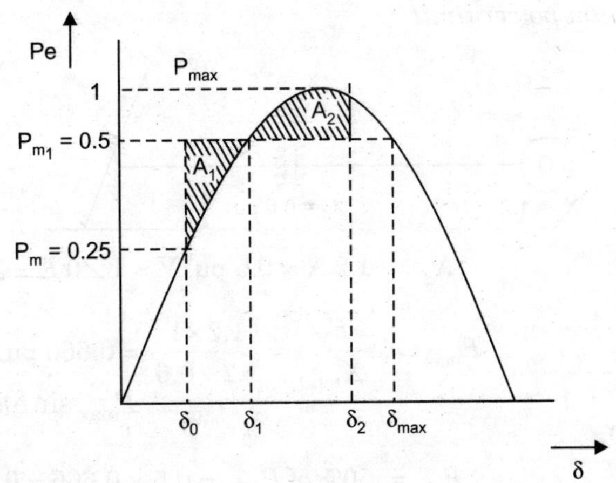

Let the rotor swings upto angle δ_2.

From Fig. $\quad \delta_1 = \sin^{-1}\left(\frac{0.5}{1}\right) = 30°$

$$A_1 = \int_{\delta_0}^{\delta_1} (0.5 - P_{max} \sin \delta) d\delta = \int_{\delta_0}^{\delta_1} (0.5 - \sin \delta) d\delta$$

$$= 0.5(\delta_1 - \delta_0) + (\cos \delta_1 - \cos \delta_0)$$

$$= 0.1686$$

$$A_2 = \int_{\delta_1}^{\delta_2} (P_{max} \sin \delta - 0.5) d\delta = (\sin \delta - 0.5) d\delta$$

$$= (\cos \delta_1 - \cos \delta_2) - 0.5\,(\delta_2 - \delta_2)$$

$$= (0.866 - \cos \delta_2) + 0.26179 - 0.5\,\delta_2.$$

With $A_1 = A_2$ we get

$$0.866 - \cos \delta_2 + 0.26179 - 0.5\,\delta_2 = 0.1686$$

$$\cos \delta_2 + 0.5\,\delta_2 = 0.959$$

By trial and error method

$$\delta_2 = 68° = \text{Angle of rotor swing.}$$

For statility the rotor can swing upto a maximum angle of

$$\delta_{max} = 180 - \delta_1 = 150°$$

Hence the machine will remain stable

Example 5. *A synchronous generator of reactance 1.2 pu is connected to an infinite bus bar through a transformer and line of total reactance 0.6 pu. The infinite bus bar voltage is 1 pu and the generator no load voltage is 1.2 pu. The inertia constant of generator is 4 mW be negligible. The system frequency is 50 Hz. Calculate the frequency of natural oscillations if the generator is loaded to*

(i) 50% and

(ii) 80% of its maximum power limit

Solution :

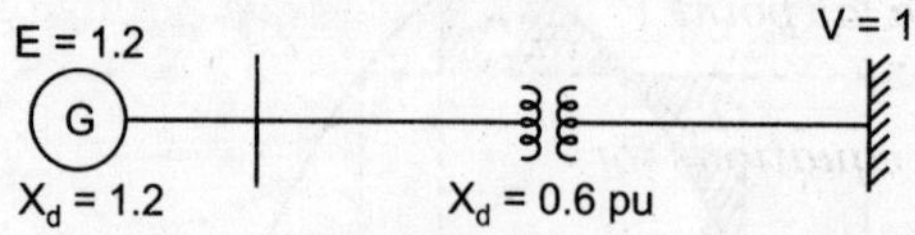

$$X_d = 1.2\,X = 0.6 \text{ pu. } V = 1\angle 0\; E = 1.2$$

$$P_{max} = \frac{E_V}{X_{total}} = \frac{1.2 \times 1}{1.2 + 0.6} = 0.666 \text{ pu.}$$

(i) With 50% loading

$$P_{e0} = 50\% \text{ of } P_{max} = 0.5 \times 0.666 = 0.333$$

Now $P_{e0} = P_{max} \sin \delta_0$

$\therefore$ $\delta_0 = \sin^{-1}(0.5) = 30°$

$$\text{Synchronising coefficient} = \left(\frac{\partial P_e}{\partial \delta}\right)_{\delta=\delta_0}$$

$$= \frac{\delta(P_{max} \sin \delta)}{\partial \delta} |_{\delta=30°} = 0.666 \cos \delta |_{\delta=30°}$$

$$= 0.666 \cos 30 = 0.577$$

$$M = \frac{GH}{\pi f}$$

With $G = 1$ pu, $M = \dfrac{H}{\pi \times 50} = 0.025$ pu sec/rad.

Natural frequency of oscillations

$$= \sqrt{\frac{1}{M}\left(\frac{\partial P_e}{\partial \delta}\right)_0} = \sqrt{\frac{1}{0.025} \times 0.577} = 4.762 \text{ rad/sec.}$$

***(ii)* For 80% loading**

$$\delta_0 = \sin^{-1}(0.8) = 53.13°.$$

Synchronous coefficient

$$= \left(\frac{\partial P_e}{\partial \delta}\right)_{\delta=5313} = 0.667 \cos 53.13 = 0.4$$

Frequency of natural oscillations

$$= \sqrt{\frac{1}{M}\left(\frac{\partial P_e}{\partial \delta}\right)_0}$$

$$= \sqrt{\frac{1}{0.025} \times 0.4} = 4 \text{ rad/sec.}$$

Example 6. *For the system shown in figure, the system is loaded to pu when a line to ground fault occurs on phase 'a' at point P. The fault is cleared by opening of the pole of C.B. in faulted phase.*

Determine the power angle equations for

(i) Prefault

(ii) During fault

(iii) Post fault conditions

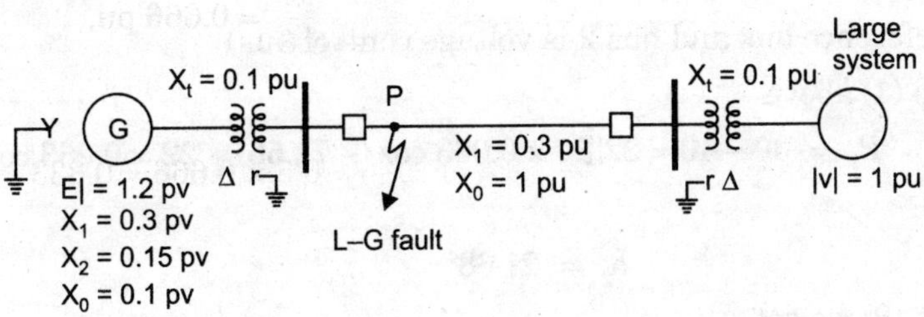

Solution : Let $V_1 < 0$ be taken as expence $\delta_1 = 0$. $S_{G1} = P_{G1} + jQ_{G1}$

Equivalent generations at each bus are

$$S_1 = P_1 + jQ_1 = (P_{G1} - P_{D1}) = j(Q_{G1} - Q_{D1})$$
$$S_2 = P_2 + jQ_2 = (P_{G2} - P_{D2}) = j(Q_{G2} - Q_{D2})$$

Transmission line series impedance

$$= Z_L = (5 + j\,15)\Omega$$

Transmission line series admittance

$$= Y_L = \frac{1}{Z_L} = 0.063 \;\angle -71.56° \text{ (mho)}$$

$$\therefore \quad Y_{BUS} = \begin{bmatrix} Y_{11} & Y_{12} \\ Y_{21} & Y_{22} \end{bmatrix}$$

$$= \begin{bmatrix} 0.063\angle -71.56 - (0.063\angle -71.56 \\ -(0.063\angle 71.56)(0.063\angle -71.56 \end{bmatrix}$$

$$= \begin{bmatrix} 0.063\angle -71.56 - 0.063\angle -108.43 \\ -(0.063\angle 108.43 0.063\angle -71.56 \end{bmatrix}$$

S.L.F.E. for two bus system are

$$P_1 = V_1[V_1 Y_{11} \cos(\theta_{11}) + V_2 Y_{12} \cos(\theta_{12} + \delta_1 - \delta_2)] \quad ...(1)$$
$$P_2 = V_2[V_2 Y_{21} \cos(\theta_{21} + \delta_1 - \delta_2) + V_2 Y_{22} \cos(\theta_{22})] \quad ...(2)$$
$$Q_1 = -V_1[V_1 Y_{11} \sin(\theta_{11}) + V_2 Y_{12} \sin(\theta_{12} + \delta_2 - \theta_1)] \quad ...(3)$$
$$Q_2 = -V_2[V_2 Y_{21} \sin(\theta_{21} + \delta_1 - \delta_2) + V_2 Y_{22} \sin(\theta_{22})] \quad ...(4)$$

The above equations are written from the general S.L.F.E. as follows

$$P_i = |V_1 \sum_{k=1}^{2} |V_k||Y_{ik}| \cos(\theta_{ik} + \delta_k - \delta_i)$$

$$Q_i = |V_i \sum_{k=1}^{2} |V_k||Y_{ik}| \sin(\theta_{ik} + \delta_k - \delta_i) \quad (i = 1, 2)$$

Now since the powers at each station are equalized

Let $$P_{G1} = \frac{P_{D1} + P_{D2}}{2} = \frac{40 + 20}{2} = 30 \text{ MW}$$

(i.e. Bus 1 is reference bus and bus 2 is voltage control bus)

Using equation (1) above

$$P_i = 30 - 40 = 22[22 \times 0.063 \cos(-71.56) + 22 \times 0.063 \cos(108.43 + \delta_2)]$$

Solving we get

$$\delta_2 = 21.68°$$

Using equation (2) we get

$$P_2 = P_{G2} - 20 = 22\,[22 \times 0.063 \cos(108.43 - 21.68) + 22 \times 0.063 \cos(-71.56)]$$

Solving we get

$$P_{G2} = 31.37 \text{ MW}$$

Using equation [3]

$$Q_1 = Q_{G1} - 39 = -22[22 \times 0.063 \sin(-71.56) + 22 \times 0.063 \sin(108.43 + 21.68)]$$

$$\therefore \quad Q_{G1} = 35.6 \text{ MVAR}$$

Using equation (4)

$$Q_2 = Q_{G2} - 15 = 22[22 \times 0.063 \sin(108.43 - 21.68) + 22 \times 0.063 \sin(-71.56)]$$

$$\therefore \quad Q_{G2} = 13.48 \text{ MVAR}$$

Thus

$$S_{G1} = P_{G1} + jQ_{G1} = (30 + j35.6) \text{ MVA}$$
$$= 46.55 \angle 49.87° \text{ MVA}$$
$$S_{G2} = P_{G2} + jQ_{G2} = (31.37 + j\,13.48) \text{ MVA}$$
$$= 34.14 \angle 23.25 \text{ MVA}$$

Hence station power factors are

$$\cos \Phi_1 = \cos(49.87) = 0.64 \text{ (lagging)}$$
$$\cos \Phi_2 = \cos(23.25) = 0.92 \text{ (lagging)}$$

EXAMPLES ON LOAD FLOW STUDY

Example 1. *Two generators are connected by a transmission line having resistance of 5 Ω and reactance of 15 Ω. The desired voltage profile is $|V_1| = |V_2| = 22$ kV. The loads at two buses are 40 MW, 0.8 p.f. lagging and 20 MW, 0.8 p.f. lagging. The station loads are equalised by the flow of power in transmission line. Estimate the torque angle and station power factors using Load Flow analysis.*

Solution : Let $V_1 \angle 0$ be taken as reference

$$\delta_1 = 0$$
$$S_{G1} = P_{G1} + jQ_{G1}$$

replace to the Fig.

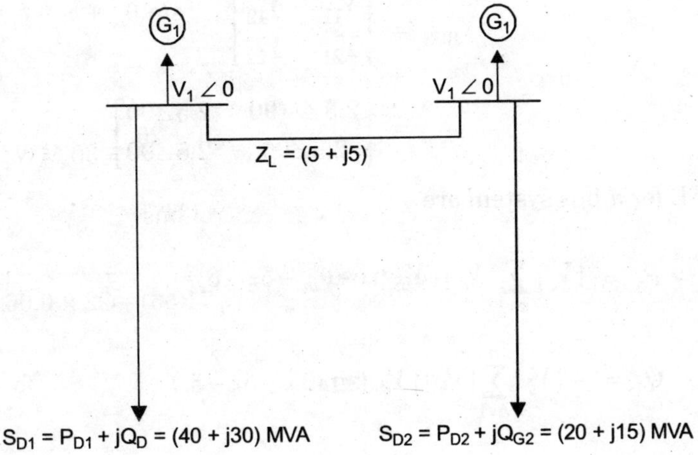

Example 2. *Sample two bus power system is shown in figure.*

$$S_{G2} = (0.3 + j\,0.15)\text{ pu.}$$
$$S_{D1} = (0.2 + j\,0.1)\text{ pu.}$$
$$S_{D}2 = (1 + j\,0.6)\text{ pu.}$$
$$V_1 = 1.02\text{ pu.}$$

Find the bus voltage V_2 and complex power S_{G1}.

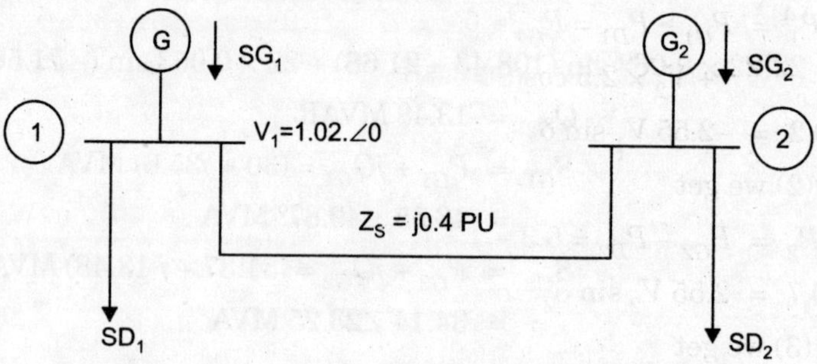

Solution :

$$S_{G1} = P_{G1} + jQ_{G1},\ S_{G2} = P_{G2} + jQ_{G2}$$
$$= 0.3 + j\,0.15$$
$$S_{D1} = P_{D1} + jQ_{D1} = 0.2 + j\,0.1$$
$$S_{D2} = P_{D2} + jQ_{D2} = 1 + j\,0.6$$

Let bus 1 voltage is reference.

Hence $\quad V_1 = 1.02 \angle 0$ pu

$$V_2 = V_2 \angle \delta_2 \text{ pu.l}$$

Line series impedance

$$= Z_s = j\,0.4 = 0.4 \angle 90° \text{ pu.}$$

Line series admittance $\quad = \dfrac{1}{Z_s} = 2.5\angle -90°$ pu.

$$Y_{BUS} = \begin{bmatrix} Y_{11} & Y_{12} \\ Y_{21} & Y_{22} \end{bmatrix}$$

$$= \begin{bmatrix} 2.5\angle -90 & 2.5\angle 90 \\ 2.5\angle 90 & 2.5\angle 90 \end{bmatrix}$$

General SLFE for n bus system are

$$P_1 = |V_i| \sum_{k=1}^{n} |V_k||Y_{ik}|\cos(\theta_{ik} + \delta_k - \delta_i)$$

$$Q_1 = -|V_i| \sum_{k=1}^{n} |V_k||Y_{ik}|\sin(\theta_{ik} - \delta_k - \delta_i) \qquad \text{for } i = 1, 2, \ldots n$$

Hence for two bus system ($n = 2$) the equations are

$$P_1 = V_1[V_1 Y_{11} \cos(\theta_{11}) + V_2 Y_{12} \cos(\theta_{12} + \delta_2 - \delta_2)] \quad ...(1)$$

$$P_2 = V_2[V_2 Y_{21} (\cos \theta_{21} + \delta_1 - \delta_2) + V_2 Y_{22} \cos(\theta_{22})] \quad ...(2)$$

$$Q_1 = -V_1[V_1 Y_{11} \sin(\theta_{11}) + V_2 Y_{12} \sin(\theta_{12} + \delta_2 - \delta_1)] \quad ...(3)$$

$$Q_2 = -V_2[V_2 Y_{21} \sin(\theta_{21} + \delta_1 - \delta_2) + V_2 Y_{22} \sin(\theta_{22})] \quad ...(4)$$

Using equation (1) we get

$$P_1 = P_{G1} - P_{D1} = P_{G1} - 0.2 = 1.02\,[1.02 \times 2.5 \cos(-90) + V_2 \times 2.5 \cos(90 + \delta_2)]$$

$$\therefore \quad P_{G1} - 0.2 = -2.55\, V_1 \sin \delta_2 \quad ...(5)$$

Using equation (2) we get

$$P_2 = P_{G2} - P_{D2} = 0.3 - 1 = 2.55\, V_2 \sin \delta_2$$

$$\therefore \quad -0.7 = 2.55\, V_2 \sin \delta_2 \quad ...(6)$$

Using equation (3) we get

$$Q_1 = Q_{G1} - Q_{D1} = Q_{G1} = 0.1 = 2.6 - 2.55\, V_2 \cos \delta_2$$

$$\therefore \quad Q_{G1} - 0.1 = 2.6 - 2.55\, V_2 \cos \delta_2 \quad ...(7)$$

Using equation (4) we get

$$Q_2 = Q_{G2} - Q_{D2} = 0.15 - 0.6 = 2.5\, V^2{}_2 - 2.55\, V_2 \cos \delta_2$$

$$\therefore \quad -0.45 = 2.5V^2{}_2 - 2.55\, V_2 \cos \delta_2 \quad ...(8)$$

From equation (5) and (6) we get

$$P_{G1} - 0.2 = 0.7$$

$$\therefore \quad P_{G1} = 0.9 \text{ pu.}$$

Equation (5) now becomes

$$0.7 = -2.55\, V_2 \sin \delta_2 \quad ...(9)$$

$$\therefore \quad \sin \delta_2 = \frac{-0.2745}{V_2}$$

From equation (8) we get

$$\cos \delta_2 = \quad ...(10)$$

Solving equation (9) and (10) by trial and error method or simultaneously we get

$$V_2 = 0.55 \text{ pu.}$$

Hence equation (7) gives

$$Q_{G1} = 1.49 \text{ pu.}$$

$$\therefore \quad S_{G1} = P_{G1} + jQ_{G1} = (0.9 + j\, 1.49) \text{ pu.}$$

Example 3. *Two 11 kV generating stations are interconnected by a link having $X = 9\ \Omega$ and $R = 0\ \Omega$. Load L_1 is 7 MW at 0.9 p.f. lagging and load L_2 is 10 MW at 0.8 p.f. lagging. Active power supplied by each station is the same. Find station power factors and reactive power loss in the link. Station bus voltages are 11 kV.*

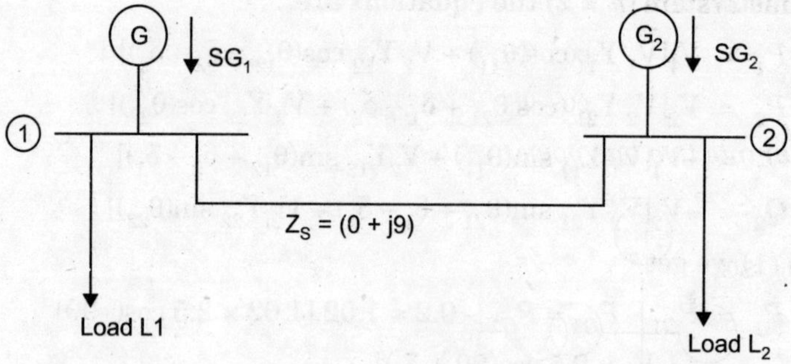

Solution:

$$S_{G1} = P_{G1} + JQ_{G1}, S_{G2} = P_{G2} + JQ_{G2}$$
$$S_{D1} = P_{D1} + JQ_{D1} = (7 + j\,3.39) \text{ MVA}$$
$$S_{D2} = P_{D2} + JQ_{D2} = (10 + j\,7.5) \text{ MVA}$$
$$|V_1| = |V_2| = 11 \text{ kV}$$
$$P_{G1} = P_{G2} = \frac{7+10}{2} = 8.5 \text{ MW}$$

Assume
$$V_1 = V_1\angle 0 = 11 < 0 \text{ kV (i.e. } \delta_1 = 0)$$
$$Y_{BUS} = \begin{bmatrix} Y_{11} & Y_{12} \\ Y_{21} & Y_{22} \end{bmatrix} = \begin{bmatrix} 0.11\angle -90 & 0.11\angle 90 \\ 0.11\angle 90 & 0.11\angle -90 \end{bmatrix}$$

Using S.L.F.E.

$$P_1 = P_{G1} - P_{D1} = V_1[V_1\,Y_{11}\,(\cos\theta_{11}) + V_2\,Y_{12}\cos(\theta_{12} + \delta_2 - \delta_2)]$$
$$8.5 - 7 = 11[11 \times 0.11\cos(-90) + 11 \times 11 \times 0.11\cos(90 + \delta_1)]$$
$$\therefore \quad \delta_2 = 6.47°$$
$$Q_1 = Q_{G1} - Q_{D1} = -V_1\,[V_1\,Y_{11}\sin(\theta_{11}) + V_2\,Y_{12}\sin(\theta_{12} + \delta_2 - \delta_1)]$$
$$Q_{G1} = 3.39 = -11[11 \times 0.11\sin(-90) + 11 \times 0.11\sin(90 - 6.47)]$$
$$Q_{G1} = 3.47 \text{ MVAR}$$
$$Q_2 = Q_{G2} - Q_{D2} = -V_2[V_2\,Y_{21}\sin(\theta_{21} + \delta_1 - \delta_2) + V_2\,Y_{22}\sin(\theta_{22})]$$
$$\therefore \quad Q_{G2} - 7.5 = -11[11 \times 0.11\sin(90 + 6.47) + 11 \times 0.11\sin(-90)]$$
$$\therefore \quad Q_{G2} = 7.58 \text{ MVAR}$$
$$Q_{G1} + Q_{G2} = Q_{D1} + Q_{D2} + Q_L$$
$$\therefore \quad Q_L = \text{Reactive power loss in line} = 0.165 \text{ MVAR}$$
$$S_{G1} = P_{G1} + jQ_{G1} = 8.5 + j3.47 = 9.18\angle 22.21° \text{ MVA}$$
$$S_{G2} = P_{G2} + jQ_{G2} = 8.5 + j7.58 = 11.39\angle 41.72° \text{ MVA}$$

Station power factors are

$$\cos \Phi_1 = \cos (22.21) = 0.93 \text{ (lagging)}$$
$$\cos \Phi_2 = \cos (41.72) = 0.75 \text{ (lagging)}$$

Example 4. *Sample two bus power system network is showin in Figure.*

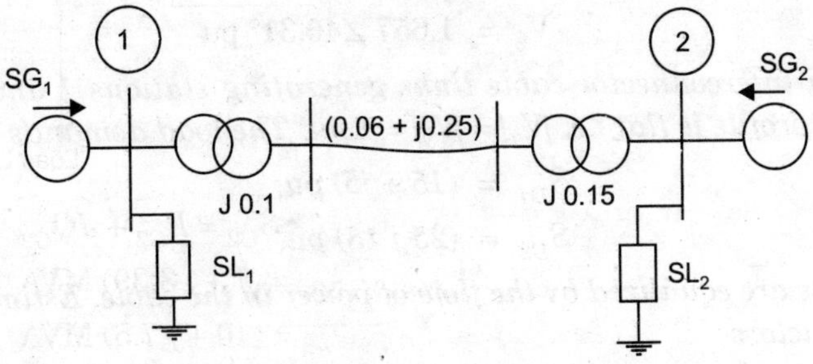

Impedance of elements are marked in the Fig.

$$S_{G2} = 0.4 + j0.2 = \text{Generation at bus 2}$$
$$S_{L2} = (0.7 + j0.3) = \text{Load at bus 2}$$
$$S_{L1} = (0.3 + j0.15) = \text{Load at bus 1}$$

Carry out three iterations of Load Flow Analysis to determine V_2. *Take bus (1) as slack bus and* $V_1 = 1.1 \angle 0$ *pu.*

Solution :

$$S_2 = (P_{G2} - P_{D2}) + j(Q_{G2} - Q_{D2})$$
$$= -0.3 - j0.1 = P_2 + jQ_2$$

$$Y_{BUS} = \begin{bmatrix} Y_{11} & Y_{12} \\ Y_{21} & Y_{22} \end{bmatrix}$$

$$= \begin{bmatrix} 1.985\angle -83.15 & 1.985\angle -83.15 \\ -1.985\angle -83.15 & 1.985\angle -83.15 \end{bmatrix}$$

Using equation (Gauss Seidal method)

$$V_P = \frac{1}{Y_{PP}}\left[\frac{P_P - jQ_P}{V_P} - \sum_{\substack{q=1 \\ q \neq p}}^{n} Y_{pq} V_q\right]$$

$(p = 1.2 \ldots n)$

Initially let $V_2 = 1 \angle 0$

Iteration 1

$$V_2 = \frac{1}{Y_{22}}\left[\frac{P_2 - jQ_2}{V_2} - Y_{21}V_1\right]$$

$$= 1.036 \angle -7.95° \text{ pu.}$$

Iteration 2

$$V_2 = \frac{1}{Y_{22}}\left[\frac{P_2 - jQ_2}{V_2} - Y_{21}V_1\right]$$

Put $V_2 = 1.036 \angle 42.16°$ pu, we get

$$V_2 = 1.657 \angle 40.31° \text{ pu}$$

Example 5. *An interconnector cable links generating stations 1 and 2 as shown in Fig. The desired voltage profile is flat i.e. $|V_1| = |V_2| = 1$ pu. The load demands at two buses are*

$$S_{D1} = (15 + j5) \text{ pu:}$$

$$S_{D2} = (25 j\ 15) \text{ pu.}$$

The station loads are equalized by the flow of power in the cable. Estimate the torque angles and station power factors

(a) For cable Z = (0. j0.05) pu.

(b) For cable Z = (0.005 + j 0.05) pu.

It is given that generator G_1 can generate maximum of 20 pu real power.

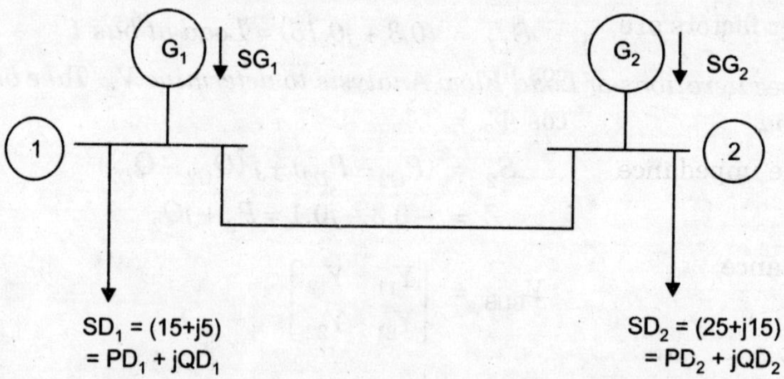

Solution : (*a*) Cable impedance $Z = j0.05$ pu

Since the cable resistance is zero, there is no real power loss in the cable.

$$\therefore \quad P_{G1} + P_{G2} = P_{D1} + P_{D2}$$

For equalization of station loads

$$P_{G1} = P_{G2} = 20 \text{ pu}$$

Voltage at bus 2 is taken as reference

i.e. $V_2 = 1 \angle 0\ (\delta_2 = 0)\ V_1 = 1 \angle \delta_1$

SLFE for two bus system are

$$P_1 = P_{G1} - P_{D1} = V_1[V_1 Y_{11} \cos(\theta_{11}) + V_2 Y_{12} \cos(\theta_{12} + \delta_2 - \delta_1)] \quad ...(1)$$

$$Q_1 = Q_{G1} - Q_{D1} = -V_1[V_1 Y_{11} \sin(\theta_{11}) + V_2 Y_{12} \sin(\theta_{12} + \delta_2 - \delta_1)] \quad ...(2)$$

$$P_2 = P_{G2} - P_{D2} = V_2[V_2 Y_{21} \cos(\theta_{21} + \delta_1 - \delta_2) + V_2 Y_{22} \sin(\theta_{22})] \quad ...(3)$$

$$Q_2 = Q_{G2} - Q_{D2} = -V_2[V_2 Y_{21} \sin(\theta_{21} + \delta_1 - \delta_2) + V_2 Y_{22} \sin(\theta_{22})] \quad ...(4)$$

$$Y_{BUS} = \begin{bmatrix} Y_{11} & Y_{12} \\ Y_{21} & Y_{22} \end{bmatrix} = \begin{bmatrix} 20\angle-90 & 20\angle 90 \\ 20\angle 90 & 20\angle-90 \end{bmatrix}$$

Using equation (1)

$$P_1 = 20 - 15 = 20\cos(-90) + 20\cos(90 - \delta_1)$$
$$\delta_1 = 14.47°$$

Using equation (2)

$$Q_1 = Q_{G1} - 5 = [20\sin(-90) + 20\sin(90 - 14.47)]$$
$$Q_{G1} = 5.634 \text{ pu}$$

Using equation (4)

$$Q_2 = Q_{G2} - 15 = -[20\sin(90 + 14.47) + 20\sin(-90)]$$
$$\therefore \quad Q_{G2} = 15.634 \text{ pu.}$$

Hence station generations are

$$S_{G1} = P_{G1} + JQ_{G1} = (20 + J5.634) = 20.77\angle 15.73° \text{ pu.}$$
$$S_{G2} = P_{G2} + JQ_{G2} = (20 + J\,5.634) = 25.38\angle 38° \text{ pu.}$$

Station power factors are

$$\cos\Phi_1 = \cos(15.73) = 0.9625 \text{ (lagging)}$$
$$\cos\Phi_2 = \cos(38) = 0.788 \text{ (lagging)}$$

(*b*) Here cable impedance

$$Z = (0.005 + j0.05) = 0.0502\angle 84.3° \text{ pu}$$

Cable admittance

$$Y = 19.92\angle -84.3° \text{ pu}$$

$$Y_{BUS} = \begin{bmatrix} Y_{11} & Y_{12} \\ Y_{21} & Y_{22} \end{bmatrix} = \begin{bmatrix} 19.92\angle-84.3 & 19.92\angle 95.7 \\ 19.92\angle 95.7 & 19.92\angle-84.3 \end{bmatrix}$$

It is given maximum real power that station 1 can generate is 20 pu.

Hence let $\quad P_{G1} = 20$ pu.

Let $\quad V_2 = 1\angle\delta_1$

Using equation (1) above we get

$$\delta_1 = 14.4° \text{ (torque angle)}$$

Using the equation (2), (3) and (4) above we get

$$Q_{G1} = 5.13 \text{ pu}, Q_{G2} = 16.12 \text{ pu},$$
$$P_{G2} = 20.10 \text{ pu.}$$

Hence station generations are

$$S_{G1} = P_{G1} + jQ_{G1} = (20 + j5.13) = 20.647\angle 14.38°$$
$$S_{G2} = P_{G2} + jQ_{G2} = (20.10 + j16.12) = 25.76\angle 38.729°$$

$\therefore$ Station power factors are

$$\cos\Phi_1 = \cos(14.38) = 0.968 \text{ (lagging)}$$

$$\cos\Phi_2 = \cos(38.729) = 0.78 \text{ (lagging)}$$

Real power loss in the cable

$$P_L = \Sigma Q_G - \Sigma P_D = 40.1 - 40 = 0.1 \text{ pu.}$$

Reactive power loss in the cable.

$$Q_L = \Sigma Q_G - \Sigma Q_D = 21.25 - 20 = 1.25 \text{ pu.}$$

Fig. below shows the flow of powers through the cable

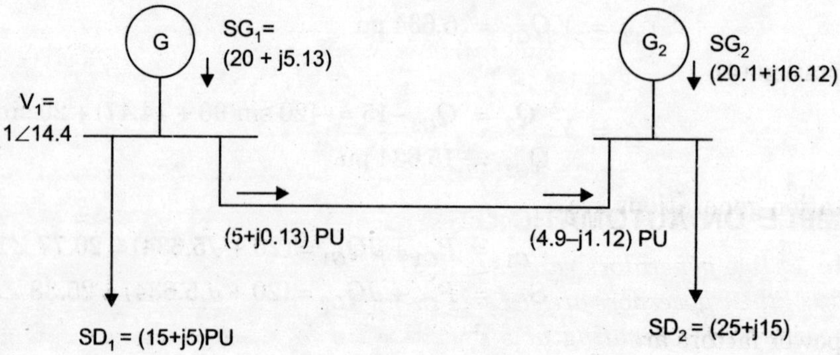

Example 6. *Find the* Y_{BUS} *for the power system shown in Figure below.*

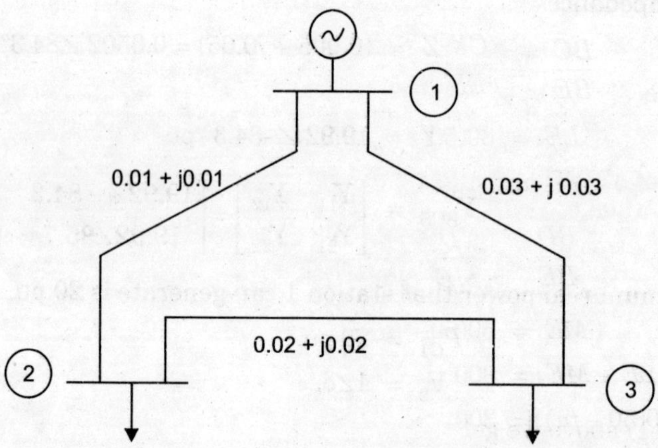

The impedance of the line are shown in fig. Use inspection method.

Solution:

$$Y_{BUS} = \begin{bmatrix} Y_{11} & Y_{12} & Y_{13} \\ Y_{21} & Y_{22} & Y_{23} \\ Y_{31} & Y_{32} & Y_{33} \end{bmatrix}$$

$$Y_{11} = \frac{1}{(0.01 + j0.01)} + \frac{1}{(0.03 + j0.03)} = (67.16 - j\,67.16) \text{ mho}$$

$$Y_{22} = \frac{1}{(0.01+j0.01)} + \frac{1}{(0.02+j0.02)} = (75.5 - j\,75.5)\text{ mho}$$

$$Y_{33} = \frac{1}{(0.02+j0.02)} + \frac{1}{(0.03+j0.03)} = (41.66 - j\,41.66)\text{ mho}$$

$$Y_{21} = Y_{12} = \frac{1}{(0.01+j0.01)} = (-50.5 + j\,50.5)\text{ mho}$$

$$Y_{13} = Y_{31} = -\frac{1}{(0.03+j0.03)} = (-\,16.66 + j\,16.66)\text{ mho}$$

$$Y_{23} = Y_{32} = -\frac{1}{(0.02+j0.02)} = (-25 + j\,25)\text{ mho}$$

EXAMPLE ON AUTOMATIC GENERATION AND VOLTAGE CONTROL

Example 1. *Two generators rated 100 MW and 200 MW are operating in parallel. Drop characteristics of their governors are 4% and 5% respectively from no load to full load. Assuming that the generators are operating at 50 Hz at no load, determine the load shared by each generator for a total load of 300 MW and the system frequency at the load.*

Solution : f_n = New operating frequency

From similarity of ΔBCX and ΔBEL

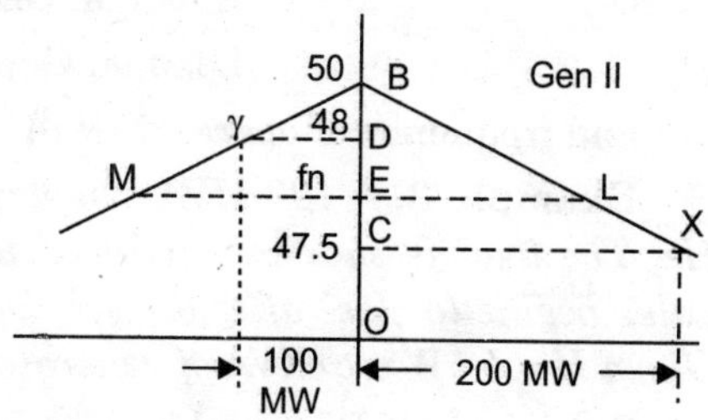

$$\frac{BC}{BE} = \frac{CX}{LE} \Rightarrow \frac{2.5}{50-f_n} = \frac{200}{LE}$$

$$\therefore \quad LE = 80(50-fn)$$

From similarity of Δ BYD and Δ BME

$$\frac{BD}{BE} = \frac{YD}{ME} \Rightarrow \frac{2}{50-f_n} = \frac{100}{ME}$$

$$ME = 50(50-fn)$$

But $\quad LE + ME = 300$

$$\therefore 80(50-fn) + 50(50-fn) = 300$$

$$\therefore \quad fn = 47.693\text{ Hz}$$

$$LE = \text{Load on GEN II} = 184.56\text{ MW}$$

$$ME = \text{Load on GEN I} = 115.35\text{ MW}$$

Example 2. *Two generators rated 200 MW and 400 MW are operating in parallel. The droop characteristics (load mV_s freq) of their governors are 4% and 5% respectively, from no load, to full load. Assuming that the generators are operating at 50 Hz at no load, how would a load of 600 MW be shared between them? What will be the system frequency at this load?*

Repeat the problem if both governors have droop of 4%.

Solution : Drop of 4% means frequency drops by 4% from 50 Hz at no load to full load.

$$\frac{BC}{AB} = \frac{PQ}{PA} \Rightarrow \frac{200}{2} = \frac{PQ}{50 - f_n}$$

$$\therefore \quad 50 - f_n = \frac{PQ}{100} \quad ...(1)$$

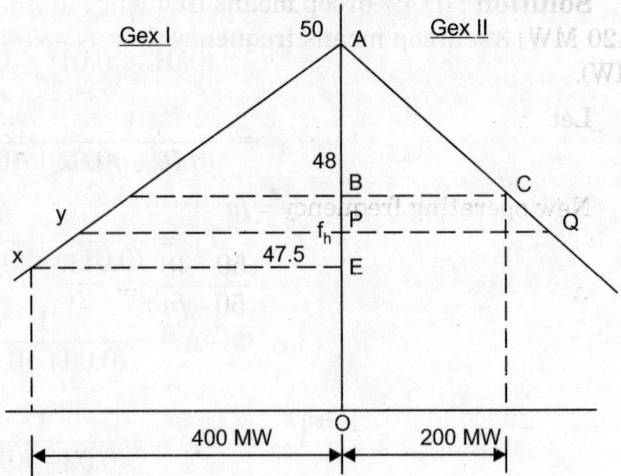

$$\frac{EX}{EA} = \frac{PY}{PA} \Rightarrow \frac{400}{2.5} = \frac{PY}{50 - f_n}$$

$$\therefore \quad 50 - f_n = (600 - PQ)\frac{2.5}{400} \quad ...(2)$$

From eq. (1) and (2)

$$\frac{PQ}{100} = (600 - PQ)\frac{2.5}{400}$$

$\therefore \quad PQ$ = Load on Gen II = .230.76 MW (overloaded)

PY = Load on Gen I = 600 – PQ = 369.23 MW

New frequency fn = 47.69 Hz

If both governors have droop of 4% and total load to be shared is 600 MW then it will be shared in proportion of their ratings.

i.e. Load on Gen I = 400 MW

Load on Gen II = 200 MW

and frequency of operation will be fn = 48 Hz

Example 3. *A 100 MVA synchronous generator operates on full load at a frequency of 50 Hz. The load is suddenly reduced to 50 MW. Due to time lag in governor system, the steam valve begins to close after 0.4 sec. Determine the change in frequency that occurs in this time. Given H = 5 kW sec/KVA of generator capacity.*

Solution : K.E. stored in rotating parts of generator and turbine

$$= 5 \times 100 \times 10^3$$

$$= 5 \times 10^5 \text{ kW sec.}$$

Excess power input to generator before steam values begins to close = 50 MW

Excess power input to rotating parts in 0.4 sec.

Stored KE $\propto$ (frequency)^2

$(5 \times 10^5 + 20 \times 10^3) \propto f^2$

Hence
$$\frac{f^2}{(50)^2} = \frac{5\times10^5 + 20\times10^3}{5\times10^5}$$

Hence f = 51 Hz = frequency at the end of 0.4 sec.

Example 4. *Two generators rated 120 MW and 250 MW are operating in parallel. Governor settings on the machines are such that they have 4% and 3% droops respectively. Determine*

(i) Load taken by each machine for a total load of 200 MW.

(ii) Percentage adjustment in no load speed to be made by the speeder motor if the machines are to share the load equally.

Solution : (*i*) 4% droop means frequency drops from 50 Hz on no load to 48 Hz at full load (120 MW) 3% droop means frequency drop from 50 Hz at no load to 48.5 Hz on full load (250 MW).

Let P_1 = ab = load on Gen I

P_2 = *bc* = load on Gen II

New operating frequency = *fn*

$$\therefore \quad \frac{50-48}{50-fn} = \frac{120}{P_1}$$

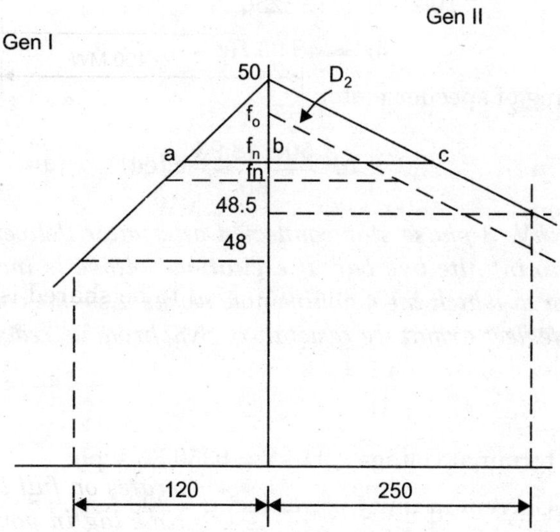

and $$\frac{50-48.5}{50-fn} = \frac{250}{P_2}$$

Let $50 - fn = \Delta f$ = drop in frequency

Hence $$\frac{2}{\Delta f} = \frac{120}{P_1}$$

and $$\frac{1.5}{\Delta f} = \frac{250}{P_2}$$

Also $P_1 + P_2$ = 200 MW

Solving above equations we get

Δf = 0.88 Hz. Hence *fn* = 49.12 Hz

P_1 = 52.94 MW, P_2 = 147.06 MW

(*ii*) If the machines are to share the load equally then either mechanical Input of Gen I has to be increased or mechanical Input of Gen 2 has to be reduced. Then only they will operate in parallel with equal frequency.

Let mechanical Input of Generator 2 is reduced. Its new droop characteristic is represented by D_2.

$$P_1 = P_2 = 100 \text{ MW at frequency} = fn$$

$$\frac{50-48}{50-f_n} = \frac{120}{100}$$

$$\therefore \quad f_n = 48.33 \text{ Hz}$$

Hence no load frequency of Gen 2 will be (say fo)

$$\frac{\text{fo} - 48.33}{100} = \frac{50-48.5}{250}$$

$$\therefore \quad \text{fo} = 48.93 \text{ Hz}$$

$\therefore$ % change in setting of speeder motor

$$= \frac{50-48.93}{50} \times 100 = 2.14\%$$

Example 5. *An 11 kV, 3 phase star connected alternator delivers 200 A at unity power factor when connected to infinite bus bar. If excitation voltage is increased by 25%, find the current and power factor at which the machine now works. Assume constant steam supply and unchanged efficiency. Neglect armature resistance. Synchronous reactance is 8 ohms. Infinite bus bar voltage is 11 kV.*

Solution :

Let V_t = terminal voltage = 11 kV = 6350.85 V/ph.

I_{a1} = Current at unity power factor = 200 A X_s = 8 Ω.

E_n = No load generated emf corresponding to I_{a1} at angle δ_1.

E_{f2} = Increased E_f due to increased excitation.

I_{a2} = corresponding armature current with ϕ_2 as p.f. angle

Figure below shows the phasor diagram at constant power output but with change in excitation voltage.

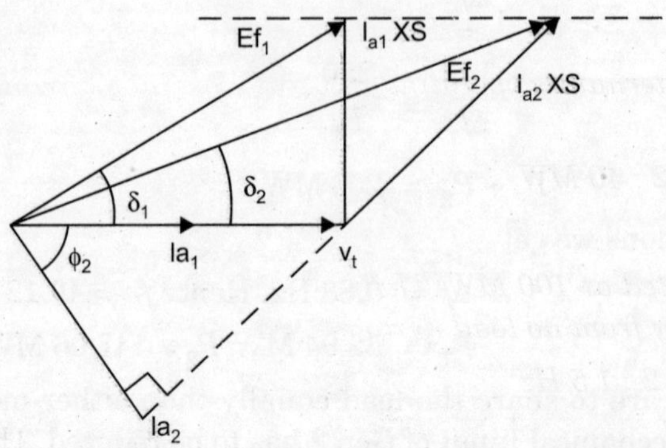

Since $$P = V_t I_a \cos \phi \text{ and is constant in both cases}$$

$$I_{a1} \cos \phi_1 = I_{a1} \cos \phi_2 \quad ...(1)$$

Also $$E_n \sin \delta_1 = E_{f2} \sin \delta_2 \quad ...(2)$$

$$E_{f1} = \sqrt{(V_t^2) + (I_{a1} X_s)^2}$$

$$= \sqrt{(6350.85)^2 + (200 \times 8)^2}$$

$$= 6549.3 \text{ V/ph.}$$

$$\delta_1 = \tan^{-1} \frac{(200 \times 8)}{(6350.85)} = 14.14°$$

$$E_{t2} = 25\% \text{ higher than } E_{f1}$$

$$= 1.25 \times 6549.3$$

$$= 8186 \text{ V/ph.}$$

From equation (2)

$$\sin \delta_2 = \frac{6549.3 \sin(14.14)}{8186}$$

$$= 0.195$$

$\therefore$ $$\delta_2 = 11.27°$$

From phasor diagram

$$E_{f2} \cos \delta_2 = V_t + I_{a2} X_s \sin \phi_2$$

$$I_{a2} \sin \phi_2 = \frac{8186 \cos(11.27) - 6350.85}{8} - 209.6 \quad ...(3)$$

$$I_{a2} \cos \phi_2 = I_{a1} \cos\phi_1 = 200 \quad ...(4)$$

$\therefore$ $$\frac{209 - 63}{200} = \frac{I_{a2} \sin\phi_2}{I_{a2} \cos\phi_2}$$

Hence $$\phi_2 = 46.35°$$

$\therefore$ $$\cos \phi_2 = 0.69 \text{ lagging}$$

and I_{a2} $$= 289.75 \text{ A}$$

Example 6. *Three alternators operating in parallel at a bus frequency of 50 Hz are loaded as follows.*

ALT 1 : 40 MW, ALT 2 : 40 MW

ALT 3 : 60 MW

Each alternator is rated at 100 MW. Their governor settings are so adjusted as to give the following fall in frequency from no load to rated load.

ALT 1 : 1.25 Hz, ALT 2 : 1.5 Hz

ALT 3 : 2 Hz.

An O.H. line 190 miles having a surge impedence of 500 Ω is short circuited a end. A steady voltage of 3 kV is suddenly applies as the others end.

Neglecting the line resistance explain with neat sketches the variation of voltage & current over the line at t = 0.001 sec, t = 0.002 sec, t = 0.003 sec and t = 0.004 sec after the voltage is applied to the line.

What is the voltage and current at sending end of line 0.004 sec after the application of sudden end of line 0.004 sec after the application of sudden voltage.

Assume the velocity of wave propagation to be 1.9 × 10^5 miles sec.

Solution : Line length $= 190$ miles

$$u = 1.9 \times 10^5 \text{ miles/sec.}$$

∴ Time taken by the wave to travel one full length of line

$$= \frac{190}{1.9 \times 15^5} = 0.001 \text{ sec.}$$

Magnitude of associated current wave

$$I = \frac{3 \times 10^3}{500} = 6 \text{ Amp}$$

The voltage and current distribution over the line for various time instants is shown below:

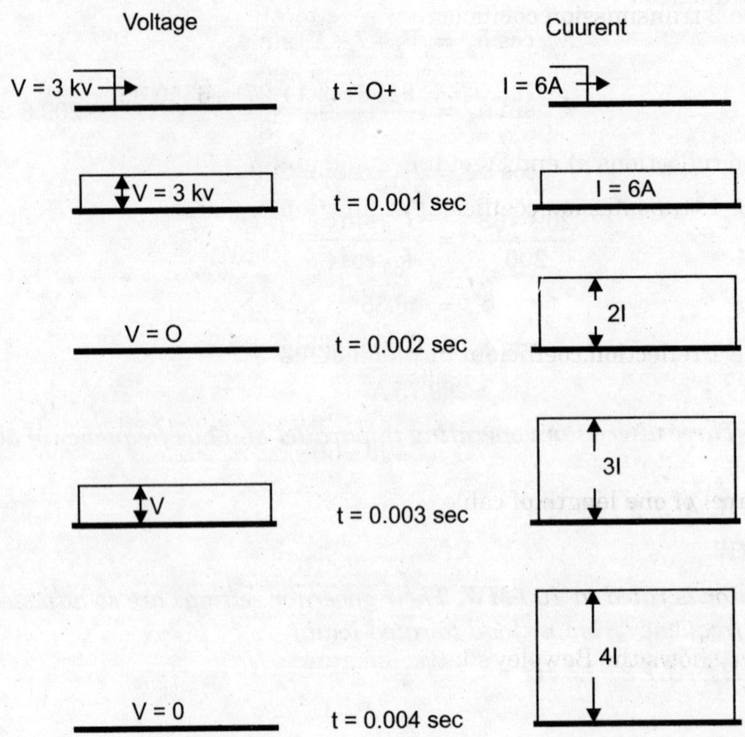

After 0.004 sec, $V = 0$

and $I = 4 \times 6 = 24$ Amp.

Example 2. *Two O.H. lines of surge impedance 400 Ω each has a 2 km length of U.G. cable of surge impedance 60 Ω between them. The velocity of wave propagation in the cable is 2×10^8 m/sec. Calculate the values of transmission and reflection coefficients of surges at each junction. Plot the voltage at first junction against time for 30 μ sec. Consider that a surge of 500 kV travels along the first line towards the first junction. Neglect the reflections from the remote end of second line.*

Solution : This is a problem to be solved using Bewley's lattice diagram. Line 1 to cable reflection coefficient at junction *A*

$$= \rho_{v1} = \frac{60-400}{60+400} = -0.739$$

Line 1 to cable transmission coefficient at junction A

$$= \tau_{v1} = \frac{2\times 60}{60+400} = 0.2608$$

Cable to line 2 reflection coefficient at junction B

$$= \rho_{v2} = \frac{400-60}{400+60} = 0.739$$

Cable to line 2 transmission coefficient at junction B

$$= \tau_{v2} = \frac{2\times 400}{400+60} = 1.739$$

There are no reflections at end 'X' of line 1 and end 'Y' of line 2.

Cable to line 1 transmission coefficient at junction A.

$$\rho_{v3} = \frac{2\times 400}{460} = 1.739$$

Cable to line 1 reflection coefficient at junction A

$$\rho_{v3} = \frac{400-60}{400+60} = 0.739$$

Time for travel of one length of cable

$$= T = \frac{2\times 10^3}{2\times 10^8} = 10\mu \text{ sec}$$

Figure above shows the Beweley's lattic diagram.

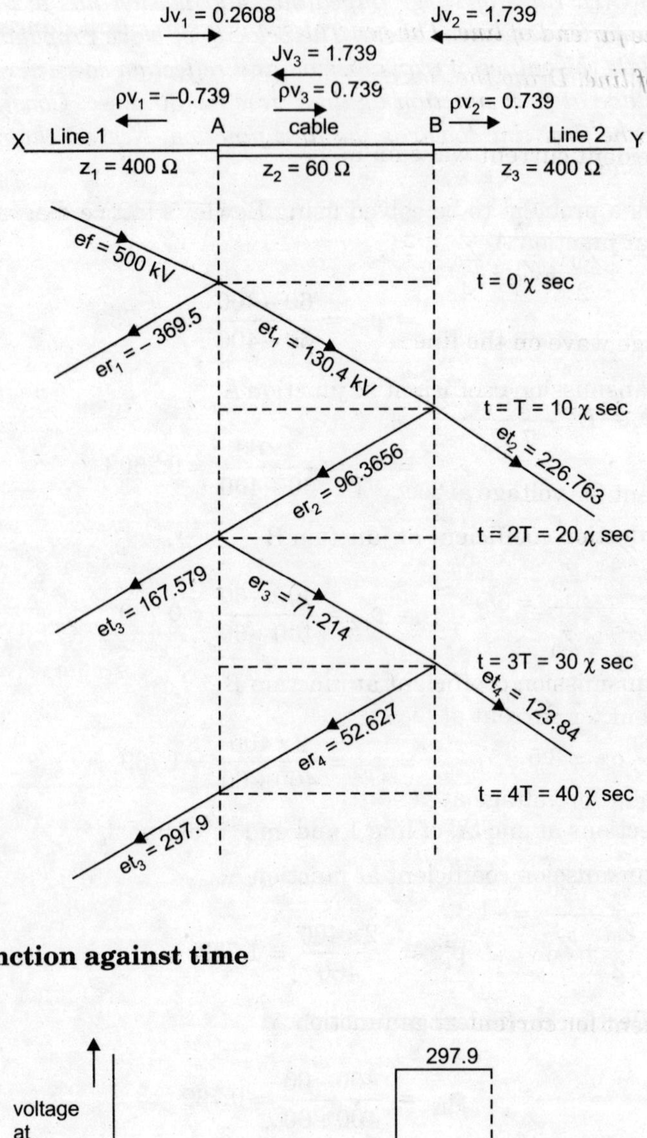

Voltage of Junction against time

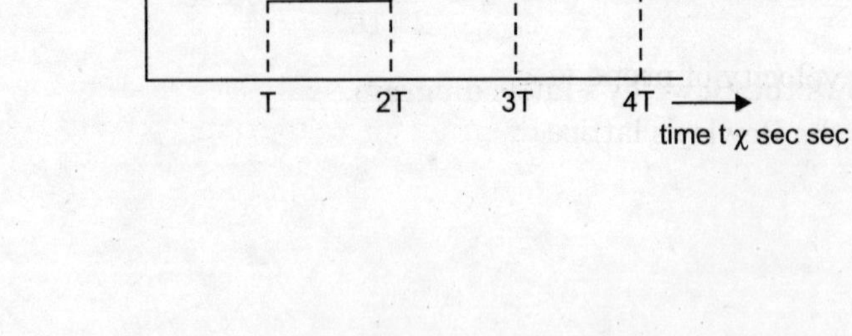

Example 3. *A generator with a generated voltage E_g is connected across the lossless transmission line of x meters. There is a load of surge impedance equal to the $\frac{1}{4}$ of surge impedance of line at the far end of line. The generator has a surge impedance equal to the $\frac{1}{2}$ of the surge impedance of line. Draw the Bewley's lattice diagram for voltage and current waves on the line*

Solution : The incident current wave on line is

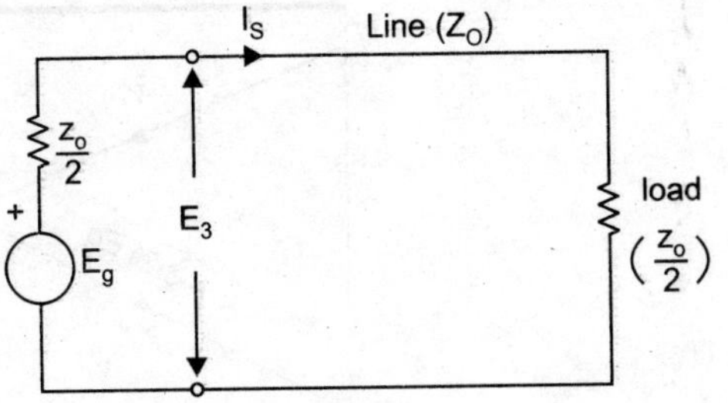

$$I_s = i_f = \frac{E_g}{\frac{Z_0}{2} + Z_0 + \frac{Z_0}{4}} = \frac{4\,E_g}{7\,Z_0}$$

The incident voltage wave on the line is

$$E_s = e_f = E_g - \frac{4\,E_g}{7\,Z_0} \times \frac{Z_0}{2} = \frac{5}{7} Eg$$

Reflection coefficient for voltage at load end

$$\rho_{v1} = \frac{\frac{Z_0}{4} - Z_0}{\frac{Z_0}{4} + Z_0} = 3/5$$

Reflection coefficient for current at load end

$$\rho_{f1} = -\rho_{v1} = 3/5$$

Reflection coefficient for voltage at generator end

$$\rho_{v2} = \frac{\frac{Z_0}{2} - Z_0}{\frac{Z_0}{2} + Z_0} = -1/3$$

Reflection coefficient for current at generator end

$$\rho_{12} = -\rho_{v2} = \frac{1}{3}$$

Time taken for a wave to travel full distance x of line

$$= \frac{x}{u} = T \text{ sec}$$

where u = Surge velocity of propagation

Fig. below shows the Bewley's lattice diagram for voltage and current waves on transmission line

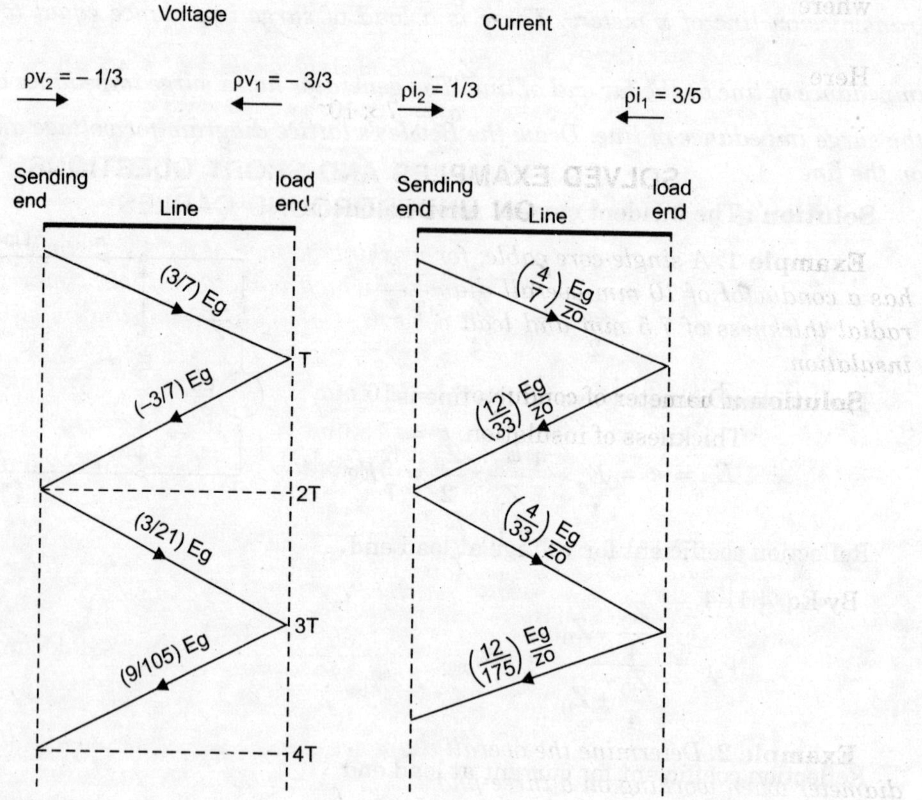

Voltage at load end Vs time

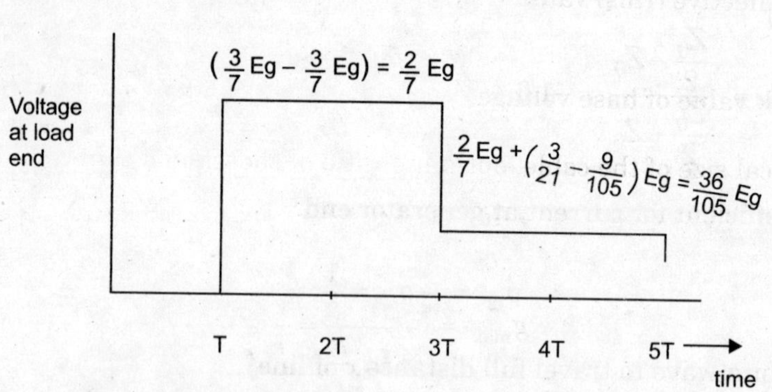

Example 9. *A travelling wave of maximum voltage of 200 kV enters an O.H. line of surge impedance of 500 Ω. The line has a conductor resistance of 7 Ω/km. What will be the value of maximum voltage of wave at 100 km? What is the power and heat loss of the wave during the time required to travel this distance. Neglect the losses in the insulators and assume the wave propagation velocity to be* 3×10^5 *km/sec.*

Solution : Since the line has resistance of 7 Ω/km the wave will get attenuated as it travels along the line. At any distance x the voltage of wave will be $V = e^{-\alpha x} V_0$

where $$a = \frac{R+Z_0^2G}{2Z_0}$$

Here $$Z_0 = 500\ \Omega.\ R = 7,\ G = 0$$

$$\therefore \quad a = 7 \times 10^{-3}$$

SOLVED EXAMPLES AND SHORT QUESTIONS ON UNDERGROUND CABLES

Example 1. *A single-core cable, for working voltage of 6.5 kV (between core and sheath), has a conductor of 10 mm overall diameter, which is insulated with impregnated paper to a radial thickness of 7.5 mm and lead covered. Calculate the maximum electric stress on the insulation.*

Solution : Diameter of conductor = 10 mm

Thickness of insulation, t = 7.5 mm

$$r = 5\text{ mm},\ R = r + t = 5 + 7.5 = 12.5\text{ mm}$$

$$\text{In}\ \frac{R}{r} = \text{In}\ \frac{12.5}{5} = 0.916$$

By Eq. 4.11.4,

$$g_{max} = \frac{V}{r\text{In}\dfrac{R}{r}} = \frac{6.5}{5 \times 0.916} = 1.419\text{ kV/mm (rms).}$$

Example 2. *Determine the overall diameter of a single-core cable and its most economical diameter when working on a three-phase, 275 kV system. The maximum permissible stress in the dielectric is not to exceed 15 kV/mm.*

Solution : Effective (rms) value of phase voltage $= \frac{275}{\sqrt{3}}$ kV

Peak value of hase voltage $= \frac{275}{\sqrt{3}} \times \sqrt{2} = 224.5$ kV

For economical size of the cable, optimum ratio of sheath and core radii is given by

$$\frac{R}{r} = e$$

and $$g_{max} = \frac{V}{r\ \text{In}\ \dfrac{R}{r}} = \frac{V}{r}$$

Now, $$g_{max} = 15\text{ kV/mm}$$

$$\therefore \quad 15 = \frac{224.5}{r};\ r = \frac{224.5}{15} = 14.96\text{ mm}$$

Economical core diameter, $D = 2r = 2 \times 14.96 = 29.92$ mm

Also, for economical size of the cable

$$R = e\,r = 2.718 \times 14.96 = 40.66\text{ mm}$$

Inner diameter of the sheath $= 2R = 2 \times 40.66 = 81.32$ mm.

Q. State What grading of Cables develop expression for capatance of Cables

Ans. We have seen that the stress is greatest at the surface of conductor or the innermost part of the dielectric. It goes on decreasing as the outermost layer of dielectric is reached. It is minimum at the sheath. If the dielectric be chosen according to the maximum stress the thickness of the cable would increase considerably. Moreover, the benefit of the chosen dielectric would not be achieved fully at the outer layers. If all the dielectric were equally stressed the thickness of the insuration could be considerably reduced. This would reduce the cost of the cable. Moreover, such a construction would enable the heat generated in the conductor to be more easily conducted to the sheath, so increasing the current rating. The method of equalizing the stress in the dielectric of the cable is called *grading of cable.* There are two methods of grading the cables : (1) Capacitance grading, and (2) Intersheath grading.

Capacitance Grading or Dielectric Grading

The dielectric stress is given by Eq. as

$$g_x = \frac{q}{2\pi\varepsilon_0\varepsilon_r x} = \frac{\text{constant}}{\varepsilon_r x} \quad \text{...(1)}$$

If it were possible to decrease ε_r as the radius x of the insulation increased in such a way that $\varepsilon_r x$ remained constant the stress would remain uniform and minimum insulation would be required. In *capacitance grading,* the homogeneous dielectric is replaced by layers of dielectrics having different values of relative permittivity such that $\varepsilon_r x$ remains constant. To get uniform stress infinite number of dielectrics will be required. To get uniform stress infinite number of dielectrics will be required. In practice two or thee layers of insulation with suitable permittivities may be used for obtaining good results. Fig. 1 shows a cable having three dielectrics of relative permittivities ε_1, ε_2, and ε_3, such that $\varepsilon_3<\varepsilon_2<\varepsilon_1$. Let r_1, r_2 and R be the outer radii of dielectrics.

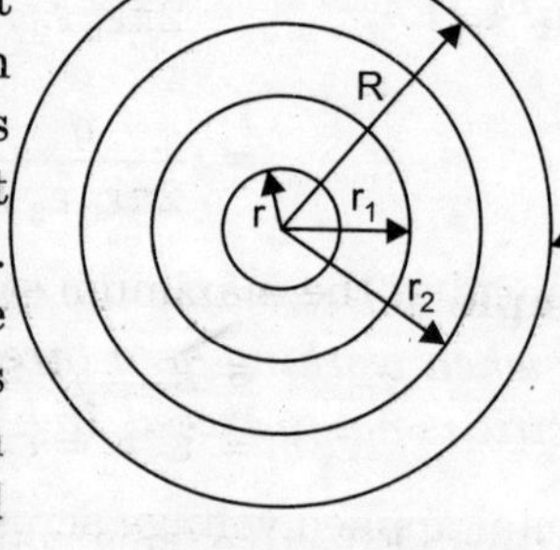

Dielectric grading.

$$V_1 = \int_r^{r_1} gx\,dx = \int_r^{r_1} \frac{q}{2\pi\varepsilon_0\,\varepsilon_1 x}\,dx$$

$$= \frac{q}{2\pi\varepsilon_0\,\varepsilon_1} \text{ In } \frac{r_1}{r} \quad \text{...(2)}$$

$$= \frac{q}{2\pi\varepsilon_0\,\varepsilon_1 r}.r \text{ In}\frac{r_1}{r} = g_{\max 1}\, r \text{ In } \frac{r_1}{r} \quad \text{...(3)}$$

Similarly, the potential difference between r_1 and r_2, i.e., across the middle layer

$$V_2 = g_{\max 2}\, r_1 \text{ In } \frac{r_2}{r_1} \quad \text{...(4)}$$

and the potential difference between r_2 and R, i.e., across the the outer layer

$$V_3 = g_{\max 3}\, r_2 \text{ In } \frac{R}{r_2} \quad \text{...(5)}$$

The total potential difference between core and earthed sheath

$$V = V_1 + V_2 + V_3$$

$$= g_{max1}\, r \ln \frac{r_1}{r} + g_{max2}\, r_1 \ln \frac{r_2}{r_1} + g_{max3}\, r_2 \ln \frac{R}{r_2} \quad ...(6)$$

The capacitance of the cable

$$C = \frac{q}{V} = q + \left[\frac{q}{2\pi\varepsilon_0}\left(\frac{1}{\varepsilon_1}\ln\frac{r_1}{r} + \frac{1}{\varepsilon_2}\ln\frac{r_2}{r_1} + \frac{1}{\varepsilon_3}\ln\frac{R}{r_2}\right)\right]$$

$$= \frac{2\pi\varepsilon_0}{\frac{1}{\varepsilon_1}\ln\frac{r_1}{r} + \ln\frac{r_2}{r_1} + \frac{1}{\varepsilon_3}\ln\frac{R}{r_2}} \quad ...(7)$$

The maximum stresses are given by

$$g_{max\,1} = \frac{q}{2\pi\varepsilon_0\,\varepsilon_1 r} \quad ...(8)$$

$$g_{max\,2} = \frac{q}{2\pi\varepsilon_0\,\varepsilon_2 r_1} \quad ...(9)$$

$$g_{max\,3} = \frac{q}{2\pi\varepsilon_0\,\varepsilon_3 R} \quad ...(10)$$

In order that the maximum stress is the same in each layer

$$g_{max\,1} = g_{max\,2} = g_{max\,3} = g_{max} \text{ (say)} \quad ...(11)$$

or, $$\varepsilon_1 r = \varepsilon_2 r_1 = \varepsilon_3 R \quad ...(12)$$

The total applied voltage across the cable

$$V = g_{max}\left(r \ln\frac{r_1}{r} + r_1 \ln\frac{r_2}{r_1} + r_2 \ln\frac{R}{r_2}\right) \quad ...(13)$$

It is to be noted that g_{max} represents the peak value of electric stress and, therefore, all the voltages in the above relations are peak values and not rms values.

Intersheath Grading

In this method of grading, the same insulating meterial is used throughout the total thickness of the cable but it is divided into two or more layers by providing intersheaths. Intersheaths are *thin* metallic cylindrical sheaths concentric with the conductor and placed between the conductor and the outside sheath. These intersheaths are maintained at suitable potentials by connecting them to tappings from the supply transformer.

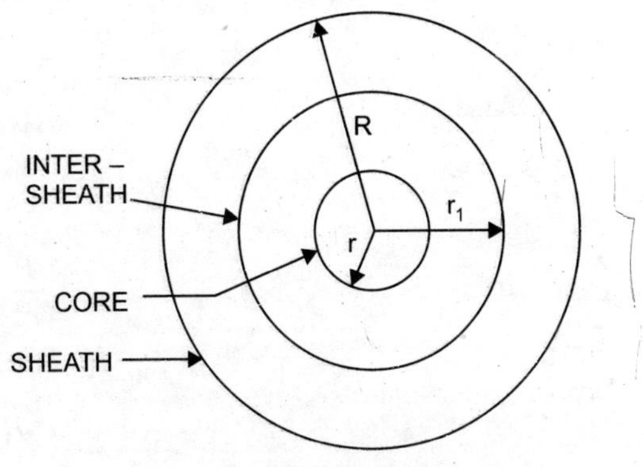

Intersheath grading.

Consider a cable with one intersheath only as shown in Fig. 2.

Let

$$r = \text{radius of the core}$$
$$r_1 = \text{radius of the intersheath}$$
$$R = \text{radius of the outer sheath}$$
$$V_1 = \text{voltage between the core and the intersheath}$$
$$V_2 = \text{voltage between the intersheath and outer sheath}$$
$$V = \text{applied voltage between the core and the sheath.}$$

The maximum potential gradient in first layer

$$g_{max\,1} = \frac{V_i}{r \text{ In } \frac{r_1}{r}} \quad ...(14)$$

The maximum potential gradient in second layer

$$g_{max\,2} = \frac{V_2}{r_1 \text{ In } \frac{R}{r_1}} \quad ...(15)$$

If the two potential gradients are equal

$$\frac{V_1}{r \text{ In } \frac{r_1}{r}} = \frac{V_2}{r_1 \text{ In } \frac{R}{r_1}} = g_{max} \text{ (say)} \quad ...(16)$$

Also,

$$V = V_1 + V_2 = g_{max}\left(r \text{ In } \frac{r_1}{r} + r1 \text{ In } \frac{R}{r_1}\right) \quad ...(17)$$

For economical size of the cable

$$\frac{r_1}{r} = e$$

and

$$g_{max} = \frac{V_1}{r} \quad ...(18)$$

$$r_1 = e\,r = \frac{eV_1}{g_{max}} \quad ...(19)$$

Also,

$$g_{max} = \frac{V_2}{r_1 \text{ In } \frac{R}{r_1}} : \frac{V_1}{r} = \frac{V_2}{r1 \text{ In } \frac{R}{r_1}}$$

In

$$\frac{R}{r1} = \frac{V_2}{V_1}.\frac{r}{r_1} = \frac{V_2}{eV_1} = \frac{V - V_1}{eV_1} = \frac{V}{eV_1} - \frac{1}{e} ; \frac{R}{r_1} = \frac{V}{e^{ev_1}} - \frac{1}{e}$$

$$R = r1e^{\frac{v}{ev_1} - \frac{1}{e}} = \frac{eV_1}{g_{max}} e^{\frac{V}{ev_1} - \frac{1}{e}}$$

$$= \frac{V_1}{g_{max}} e^{1-\frac{1}{e}} e^{\frac{V}{eV_1}} \quad \text{...(20)}$$

$$= A V_1 e^{\frac{V}{eV_1}}$$

where A is a constant equal to $\frac{1}{g_{max}} e^{1-\frac{1}{e}}$

For minimum value of R

$$\frac{dR}{dV_1} = 0;$$

$$AV_1\left(\frac{-V}{eV_1^2}\right)e^{\frac{V}{eV_1}} + Ae^{\frac{V}{eV_1}} = 0$$

$$Ae^{\frac{V}{eV_1}}\left(1 - \frac{V}{eV_1}\right) = 0$$

$$V_1 = \frac{V}{e}$$

From Eqs. (18) to (21)

$$r = \frac{V_1}{g_{max}} = \frac{V}{e\, g_{max}} \quad \text{...(22)}$$

$$r_1 = \frac{V_1 e}{g_{max}} = \frac{V}{g_{max}} \quad \text{...(23)}$$

$$R = \frac{V}{g_{max}} e^{(1-1/e)} \quad \text{...(24)}$$

It should be remembered that $e = 2.718$

$\therefore e^{(1-1/e)} = (2.718)^{(1-1/2.718)} = (2.718)^{1.718/2.718} = (2.718)^{0.632} = 1.881$

$$R = 1.881 \frac{V}{g_{max}} \quad \text{...(25)}$$

Limitations of Grading

Cable grading is of theoretical interest only. Practical difficulties are encountered in the application of grading methods. The main disadvantage of the capacitance grading is that the range of permittivity values of insulating materials available for cable insulation is limited. Moreover, the permittivities of the layers may not remain constant during the service period of the cable. Consequently, the stress distribution may change and may even cause insulation breakdown at the normal operating voltage.

The intersheaths, being very thin, are liable to be damaged during transportation or installation. High local stress may be developed at the points of damage leading to breakdown. Also, thin intersheaths are not able to carry the charging current of long cable lines and thus the current carrying capacity of the cable is reduced. Intersheath grading also presents difficulties in fixing the intersheath potentials. For these reasons the present trend is to avoid grading, wherever possible.

Example 3. *A single-core, lead sheathed cable joint has a conductor of 10 mm diameter and two layers of different insulating materials, each 10 mm thick. The relative permittivities are 3 (inner) and 2.5 (outer). Calculate the potential gradient at the surface of the conductor when the potential difference between the conductor and the lead sheathing is 60 kV.*

Solution : $r = \dfrac{10}{2} = 5$ mm, $t = 10$ mm

$$r_1 = r + t = 5 + 10 = 15 \text{ mm}$$

$$R = r + 2t = 5 + 20 = 25 \text{ mm}$$

$$\varepsilon_1 = 3, \varepsilon_2 = 2.5, V = 60 \text{ kV}$$

$$g_{max\,1} = \frac{q}{2\pi\varepsilon_0 \varepsilon_1 r}; \qquad g_{max\,2} = \frac{q}{2\pi\varepsilon_0 \varepsilon_2 r_1}$$

$$\therefore \quad \frac{r_1 g_{max2}}{r g_{max1}} = \frac{\varepsilon_1}{\varepsilon_2}$$

$$V = V_1 + V_2 = g_{max\,1}\, r \ln\frac{r_1}{r} + g_{max2}\, r_1 \ln\frac{R}{r_1} = r g_{max1}\left(\ln\frac{r_1}{r} + \frac{g_{max2} r_1}{g_{max1} r}\ln\frac{R}{r_1}\right)$$

$$= r g_{max1}\left(\ln\frac{r_1}{r} + \frac{\varepsilon_1}{\varepsilon_2}\ln\frac{R}{r_1}\right)$$

$$60 = 5 g_{max\,1}\left(\ln\frac{15}{5} + \frac{3}{2.5}\ln\frac{25}{15}\right) = 5 g_{max\,1}(1.098 + 0.613) = 8.558\, g_{max\,1}$$

$$g_{max\,1} = \frac{60}{8.558} = 7.01 \text{ kV/mm}$$

Example 4. *A single-core, lead covered cable is to be designed for 66 kV to earth. Its conductor radius is 10 mm and its insulating materials A, B and C have relative permittivities of 5, 4 and 3 respectively and corresponding maximum permissible stresses of 3.8, 2.6 and 2.0 kV/mm (rms) respectively. Find the minimum diameter of the lead sheath.*

Solution :

$$g_{max\,A} = \frac{q}{2\pi\varepsilon_0 \varepsilon_A r} \qquad g_{max\,B} = \frac{q}{2\pi\varepsilon_0 \varepsilon_B r_1} \qquad g_{max\,C} = \frac{q}{2\pi\varepsilon_0 \varepsilon_C r_2}$$

$$\varepsilon_A r g_{max\,A} = \varepsilon_B r_l g_{max\,B} = \varepsilon_C r_2 g_{max\,C}$$

$$5 \times 10 \times 3.8 = 4r_1 \times 2.6 = 3r_2 \times 2$$

$$r_1 = \frac{5\times10\times3.8}{4\times2.6} = 18.3 \text{ mm}; \; r_2 = \frac{5\times10\times3.8}{3\times2} = 31.7 \text{ mm}$$

$$V_1 = g_{\max A}\, r \ln \frac{r_1}{r} = 3.8\times10\times \ln \frac{18.3}{10} = 38\times0.604 = 22.95 \text{ KV}$$

$$V_2 = g_{\max B}\, r_1 \ln \frac{r_2}{r_1} = 2.6\times18.3 \ln\frac{31.7}{18.3} = 2.6 \times 18.3 \times 0.549 = 26.12 \text{ kV}$$

$$V_3 = g_{\max C}\, r_2 \ln \frac{R}{r_2} = 2 \times 31.7 \ln \frac{R}{r_2} = 63.4 \ln \frac{R}{r_2}$$

Also, $V = V_1 + V_2 + V_3$

$$66 = 22.95 + 26.12 + 63.4 \ln \frac{R}{r_2}$$

$$\ln\frac{R}{r_2} = \frac{66 - 22.95 - 26.12}{63.4} = \frac{16.93}{63.4} = 0.2670\,; \quad \frac{R}{r_2} = 1.306$$

$$R = 1.306\, r_2 = 1.306 \times 31.7 = 41.4 \text{ mm}$$

Diameter of sheath = $2R = 2 \times 41.4 = 82.8$ mm.

Example 5. *A 60 kV (rms) single-core metal sheathed cable is to be graded by means of a metallic intersheath. The safe electric stress of the insulating material is 4 kV/mm (rms). (a) Calculate the diameter of the intersheath and the voltage at which it must be maintained in order to obtain minimum overall diameter. Calculate also the corresponding conductor diameter. (b) Compare the conductor diameter obtained in (a) with that of an ungraded cable working under the same conditions.*

Solution : (*a*) $V = 60$ kV (rms), $g_{\max} = 4$ kV/mm (rms)

$$V_1 = \frac{V}{e} = \frac{60}{2.718} = 22.1 \text{ kV}$$

$$r = \frac{V}{e\, g_{\max}} = \frac{60}{2.718\times4} = 5.5 \text{ mm}$$

Diameter of core $2r = 2 \times 5.5 = 11$ mm

Radius of intersheath, $r_1 = \dfrac{V}{g_{\max}} = \dfrac{60}{4} = 15$ mm

Diameter of intersheath, $d_1 = 2r_1 = 2 \times 15 = 30$ mm

$$V_2 = V - V_1 = 60 - 22.1 = 37.9 \text{ kV}$$

$$R = 1.881 \frac{V}{g_{\max}} = 1.881\times\frac{60}{4} = 28.2 \text{ mm}$$

Minimum overall diameter of the cable,

$$D = 2R = 2 \times 28.2 = 56.4 \text{ mm.}$$

(b) Cable without intersheath

For economic cable size

$$\frac{R}{r} = e = 2.718;\ \text{In}\ \frac{R}{r} = 1$$

$$V = g_{max}\, r\ \text{In}\ \frac{R}{r} = g_{max}\, r$$

Diameter of conductor $= 2r = 2 \times 15 = 30$ mm

$$R = e\,r = 2.718 \times 15 = 40.77 \text{ mm}$$

$$D = 2R = 2 \times 40.77 = 81.54 \text{ mm.}$$

Q. Find expression of Cable Capacitance

Ans. The capacitance of a cable transmission line is very much larger than that of an overhead line of the same length due to the following reasons:

1. The distance between the conductors is very small.
2. The distance between the cores and the earthed sheath is also small.
3. The permittivity of the cable insulation is usually 3 to 5 times greater than that of air insulation around the conductors of overhead line,

It is easier to calculate the capacitance of an overhead system accurately if its configuration be known, but for a cable system such a calculation is only approximate. Therefore, it is a practice to perform actual tests on a cable to find its capacitance.

The approximate method of cable capacitance is based on the assumption that the cable dielectric is perfectly homogeneous. The insulation of a cable is, however, far from being homogenous or uniform. The capacitance of a single-core cable is found from the

$$C = \frac{\varepsilon_r}{18\,\text{In}\,\frac{R}{r}}\ \mu F/km$$

$$= \frac{\varepsilon_r}{18 \times 10^9\ \text{In}\ \frac{R}{r}}\ F/m$$

$$= \frac{\varepsilon_r}{18\ \text{In}\ \frac{R}{r}}\ \mu F/km$$

Q. State What is Charging Current or Capacitive Current

Ans. The capacitance of a cable determines the charging current, charging kVA, and the dielectric loss. The charging current resticts the use of cables on extra high voltage lines. For such lines additional reactive kVA is to be supplied to compensate this effect. The regulation and stability are affected by the charging current.

The current carrying capacity of an a.c. cable is also reduced by charging current. The maximum permissible load current goes on decreasing with greater lengths of an a.c. cable. With a d.c. cable there is negligible charging under normal operating conditions. The current carrying capacity of d.c. cable is, therefore, independent of its length.

QUESTIONS AND SOLVED EXAMPLES ON PER UNIT ON REPRESENTATION

Q. State what are advantages of per unit system.

Ans. Per unit system of computation has the following advantages :

1. The ordinary parameters (current, impedance, losses, etc.) vary considerably with the variation of physcial size, terminal voltage, power rating, etc., while the per unit parameters are independent of these quantities over a wide range of the same type of apparatus. In other words, the per unit impedance values for apparatus of like ratings lie within a narrow range.
2. Per unit values provide more meaningful information.
3. The chance of confusion between line and phase values in a three-phase balanced system is reduced. A per unit phase quantity has the same numerical value as the corresponding per unit line quantity regardless of the three-phase connection whether star or delta.
4. The impedance of machines are specified by the manufacturers in terms of per unit values.
5. The per unit impedance referred to either side of a single-phase transformer is the same.
6. The per unit impedance referred to either side of a three-phase transformer is the same regardless of the three-phase connections whether they are $\Delta - \Delta$, $Y - Y$ or $\Delta - Y$.
7. The computational efforts in power systems is very much reduced with the use of per unit quantities. Usually, the per unit quantities being of the order of unity or less can easily be handled with a digital computer. Manual calculations are also simplified. Per unit quantities simplify theoretical deductions and give them more generalised forms.

Q. Draw one-line diagrams of a power system with impedance diagram.

Ans. A single-line or one-line diagram is a representation of the essentials of a system in a most simplified form. As already discussed, power systems are basically composed of a set of generating plants, a transmission network, and a combination of industrial, commercial and residential loads. Now-a-days, mostly power is generated by three-phase synchronous generators, transmitted by three-phase transmission lines and distributed through three-phase system networks.

A balanced three-phase system is studied on a per phase basis. A single-phase circuit consists of one of the three-phase conductors and a neutral return conductor. The loop impedance of a single-phase circuit may be supposed to be concentrated in one conductor only with the impedance of the return conductor. The loop impedance of a single-phase circuit may be supposed to be concentrated in one conductor only with the impedance of the return conductor assumed to be zero. Thus, a three-phase balanced system is effectively replaced by a single-line diagram. Fig. 1 shows a one-line diagram of a power system. G_1, G_2 and G_3 are three synchronous generators. G_1 is grounded through a resistor, G_2 and G_3 through reactors. T_1 and T_2 are two transformers. Circuit breakers are numbered 1.8.

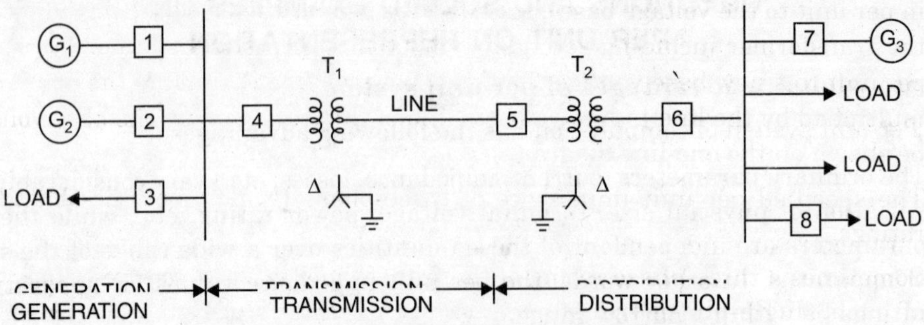

Fig. 1. Single-line diagram of a power system.

In the *impedance diagram,* the different components of the power system are replaced by their equivalent circuits. The synchronous generator is replaced by a constant voltage source behind proper impedance. The transformer is replaced by its equivalent circuit. The transmission line is replaced by nominal π-equivalent circuit. The above is only true for normal operation (i.e., for steady state analysis) when both excitation and three-phase network are balanced and they can be analysed on per phase basis. In many power system studies the synchronous generator resistance, resistance of transformer windings, The impedance diagram then becomes the *reactance diagram* Fig. 2.

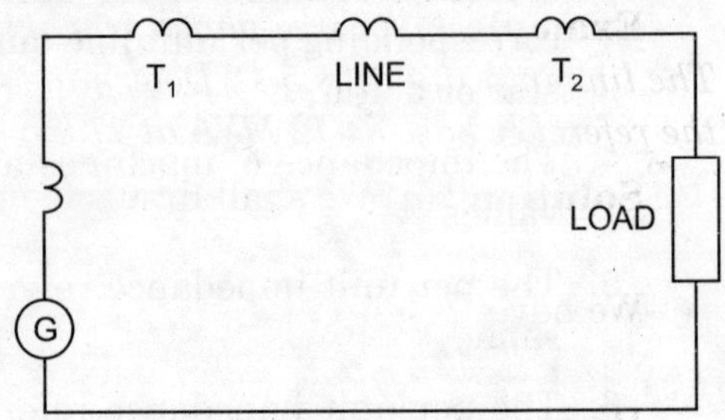

Fig. 2. Reactance diagram.

The assumptions are true in many power system studies. The ratings of line, machines and other equipment are given on the one-line diagram in either actual or per unit values.

The information supplied by a single-line diagram varies according to the requirement. For example, the one-line diagram for load flow studies may not include circuit breakers. For stability studies circuit breakers and relay positions are shown in the one-line diagram. In short-circuit studies three-separate diagrams to represent positive, negative and zero sequence networks are shown. The one-line diagram may be converted to an impedance diagram and *vice versa.*

Q. State the steps for preparation of impedance diagrams.

Ans. In order to prepare a single-line impedance diagram of and specifications of the generators, transformers lines, motors and loads are required. The following considerations are made while preparing impedance diagram:

1. Base values of voltage and V A are selected in part of the system. Unless otherwise mentioned, a given value of base voltage in a three-phase system is the line-to-line voltage and given value of base VA is the total three-phase VA.
2. The base VA is the same in all parts of the system. After selecting a convenient value of base VA for the whole system the rated value of voltage of a particular section is chosen as the base voltage. The base voltages of other sections are then assigned according to line-to-line voltage ratios of transformers. All impedance in the system before the transformer is reached, including the primary impedance, are expressed

in per unit to the voltage base chosen for the primary side. All the impedances beyond the transformer including secondary impedance of the transformer are specified in per unit to a new base voltage. This new base voltage is equal to the primary voltage multiplied by the line-to-line voltage ratio of the transformer. The base voltages may be shown on the one-line diagram.

3. The specified per unit impedance of three-phase transformers are based on their ratings.
4. Sometimes a three-phase rating of the transformer is equal to three times the rating of each single-phase transformer. The per unit impedance of the three-phase transformer is equal to the per unit impedance of the single-phase transformer.
5. The per unit impedance of all sections should be calculated on proper base with the help of the p.u. equation.

Example 1. *A three-phase synchronous generator delivers 10 MVA at a voltage of 10.5 kV. The line impedance is 5 Ω. Determine the voltage drop in the line in per unit and in volts. Use the reference base as 12 MVA at 11 kV.*

Solution : (*a*) We shall first solve this problem using actual voltage and MVA.

We have $$S_{3\varphi} = \sqrt{3} V_l I_l \text{ VA}: \ I_l = \frac{S_{3\varphi}}{\sqrt{3} V_l}$$

Here $$S_{3\varphi} = 10 \text{ MVA} = 10 \times 10^6 \text{ VA}$$
$$V_l = 10.5 \text{ kV} = 10.5 \times 10^3 \text{ V}$$

∴ $$I_l = \frac{S_{3\varphi}}{\sqrt{3} V_l} = \frac{10 \times 10^6}{\sqrt{3} \times 10.5 \times 10^3} = 549.86 \text{ A}$$

$$I_{ph} = I_l = 549.86 \text{ A}$$

Voltage drop in the line per phase = $Z_{ph} I_{ph}$ = 5 × 549-86 = 2749.3 V

(*b*) Now we shall use the per unit method to solve this problem.

We have $$S_b = 12 \text{ MVA};$$
$$V_{lb} = 11 \text{ kV}$$

$$V_{pu} = (V_l)_{pu} = \frac{V_l}{V_{lb}} = \frac{10.5}{11}$$

$$S_{pu} = \frac{S_{3\varphi}}{(S_b)_{3\varphi}} = \frac{10}{12}$$

Current in per unit, $$I_{pu} = \frac{S_{pu}}{V_{pu}} = \frac{10}{12} \times \frac{11}{10.5} = 0.8730158$$

Per unit impedance of the line, $$Z_{pu} = Z\Omega \frac{[(MVA)_b]_{3\varphi}}{(kV_l)_b^2} = Z\Omega \frac{(S_b)_{3\varphi}}{V_b^2} = 5 \times \frac{12}{(11)^2} = 0.4958677$$

Voltage drop in the line per phase in per unit

$$\Delta V_{pu} = Z_{pu} I_{pu} = 0.4958677 \times 0.8730158 = 0.4329 \text{ pu}$$

Voltage drop in the liner per phase in volts = $(\Delta V_{pu}) \times$ base voltage per phase

$$= 0.4329 \times \frac{1100}{\sqrt{3}} = 2749.36 \text{ V}$$

It is to be noted that the voltage drop in the line impedance is the voltage per phase, and hence the per unit voltage should be multiplied by the phase value of the base voltage.

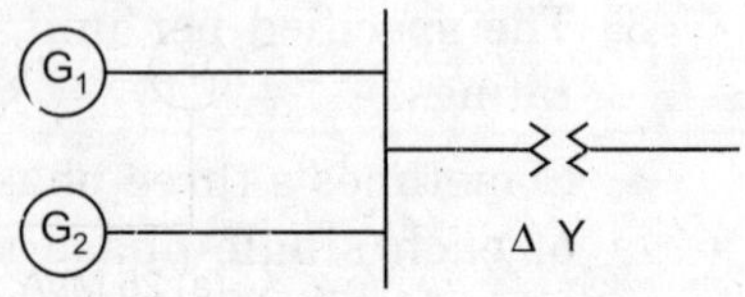

Fig. 3. Pertaining to Example 2.

Example 2. *Assuming S bases of 25 and 60 MVA calculate the through impedance in ohms between the generator and the output terminals of the transformer for the system shown in Fig. 3. The specification of the generators and transformer are given in Table below.*

Generator I	*Generator 2*	*Transformer*
30 MVA	25 MVA	60 MVA
11 kV	11 kV	11 kV (delta); 66 kV (star)
$X'' = 0.20$ pu	$X'' = 0.25$ pu	$X = 0.10$ pu

Solution : Let $V_{b2} = 11$ kV (say)

$V_{b1} = 11$ kV.

We have
$$Z_{2pu} = Z_{1pu}\left(\frac{S_{b2}}{S_{b1}}\right)\left(\frac{V_{b1}}{V_{b2}}\right)^2$$

On 25 MVA base, for generator 1, $X'' = 0.20 \times \frac{25}{30}\left(\frac{11}{11}\right)^2 = 0.1667$ pu

On 25 MVA base, for generator 2, $X'' = 0.25 \times \frac{25}{25}\left(\frac{11}{11}\right)^2 = 0.25$ pu

On 60 MVA base, for generator 1, $X'' = 0.20 \times \frac{60}{30}\left(\frac{11}{11}\right)^2 = 0.40$ pu

Calculation of per unit impedance of the tranformer

On 25 MVA base, $X = 0.10 \times \frac{25}{60}\left(\frac{11}{11}\right)^2 = 0.041667$ pu

On 60 MAV base, $X = 0.1$ pu

The reactance diagrams for S bases of 25 and 60 MVA are given in Fig. 4.

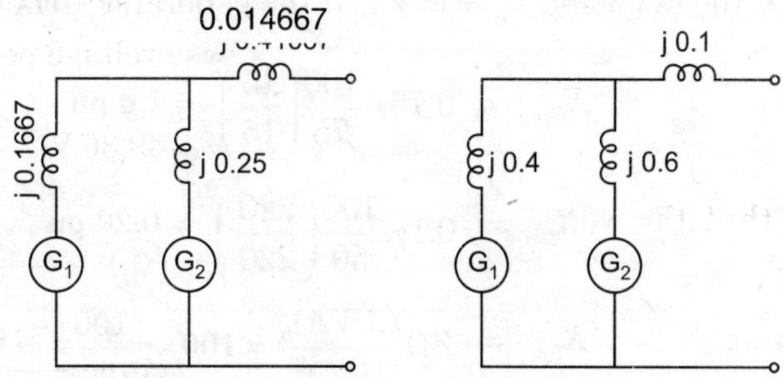

(a) 25 MVA, 11 kV BASE (b) 60 MVA, 11 kV BASE

Fig. 4. Reactance diagrams

Calculation of through impedance (reactance) per unit

(*a*) 25 MVA base

$$Z_{pu} = [j\,0.1667 \,\|\, j\,0.25] + j\,0.041667$$

$$= j\frac{0.1667 \times 0.25}{0.1667 + 0.25} + j0.041667 = j\,0.100012 + j\,0.041667 = j\,0.141679 \text{ pu}$$

(*b*) 60 MVA base

$$Z_{pu} = [j\,0.4 \,\|\, j\,0.6] + j\,0.1 = j\frac{0.4 \times 0.6}{0.4 + 0.6} + j0.1 = j\,0.34 \text{ pu}$$

Calculation of impedances (reactances) in ohms

The base impedance is given by

$$Z_b = \frac{(kV_l)_b^2}{(MVA)_b}$$

For 25 MVA base, $Z_b = \dfrac{(11)^2}{25} = 4.84\ \Omega$

For 60 MVA base, $Z_b = \dfrac{(11)^2}{60} = 2.01667\ \Omega$

Actual impedance in ohms = per unit impedance × base impedance

Actual impedance in ohms on 25 MVA base = $j\,0.141679 \times 4.84 = j\,0.6857263\ \Omega$

Actual impedance in ohms on 60 MVA base = $j\,0.34 \times 2.01667 = j\,0.6856678\ \Omega$

This problem illustrates the arbitrary selection of the *S* base, provided the value selected is used consistently throughout the network.

Example 3. *For the system shown in Fig. 5 determine the generator voltage.*

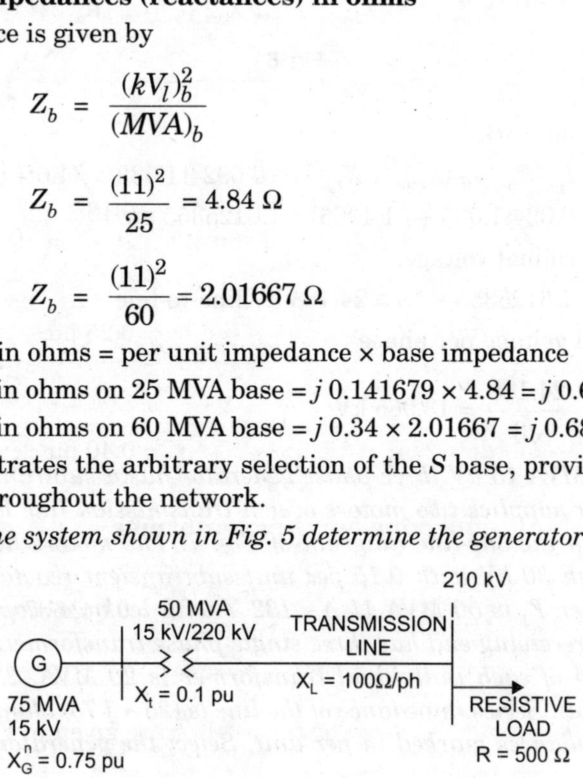

Fig. 5. Pertaining to Example 3

Solution : Let $S_b = 100$ MVA and $V_b = 15$ kV at the generator terminals.

For generator, $$X_{Gpu} = 0.75 \times \frac{100}{75}\left(\frac{15}{15}\right)^2 = 1.0 \text{ pu}$$

For transformer, $$X_{tpu} = 0.1 \times \frac{100}{50}\left(\frac{220}{220}\right)^2 = 0.20 \text{ pu}$$

For transmission line, $$X_{Lpu} = Z\Omega \frac{(MVA)_b}{(kV_l)_b^2} = 100 \times \frac{100}{(220)^2} = 0.2066 \text{ pu}$$

For the resistive load $$R_{pu} = Z\Omega \frac{(MVA)_b}{(kV_l)_b^2} = 500 \times \frac{100}{(220)^2} = 1.033 \text{ pu}$$

The system current in per unit is given by

$$I_{pu} = \frac{V_{pu}}{R_{pu}} = \frac{210/220}{1.033} = 0.924 \text{ pu}$$

Let I_{pu} be taken as the reference phasor.

$$I_{pu} = 0.924 + j0 = 0.924\angle 0°$$

The per unit impedance diagram is shown in Fig. 6

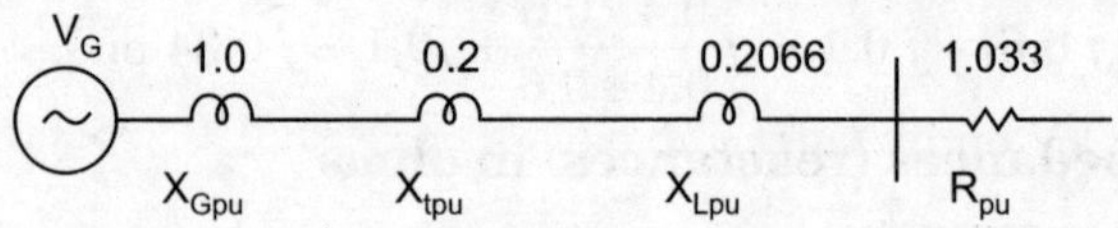

Fig. 6

By KVL,

voltage drop in the network

$$= I_{pu}[R_{pu} + j(X_{Gpu} + X_{Lpu})] = 0.0924[1.033 + j(1.0 + 0.2 + 0.2066)]$$
$$= 0.924(1.033 + j\,1.4066) = 1.6125355 \text{ pu.}$$

Actual generator terminal voltage,

$$V_G = 1.6125355 \times 15 = 24.188 \text{ kV line-to-line}$$

Generator terminal voltage per phase

$$= \frac{24.188}{\sqrt{3}} = 13.965 \text{ kV}$$

Example 7. *A 50 MVA 15 kV three-phase generator has a subtransient reactance of 0.20 per unit. The generator supplies two motors over a transmission line having transformers at both ends, as shown on the one-line diagram of Fig. 7. The motors have rated inputs of 30 MVA and 20 MVA, both 30 kV with 0.15 per unit subtransient reactance. The rating of the sending-end transformer T_1 is 50 MVA 11 Δ – 132 Y with leakage reactance of 0.10 per unit. Transformer T_2 at the receiving end has three single-phase transformers connected as a three-phase unit. The rating of each individual transformer is 20 MVA, 33/76 kV with leakage reactance of 0.12 per unit. Series impedance of the line is (25 + j 75) ohms. Draw the impedance diagram with all impedances marked in per unit. Select the generator rating as the base in the generator circuit.*

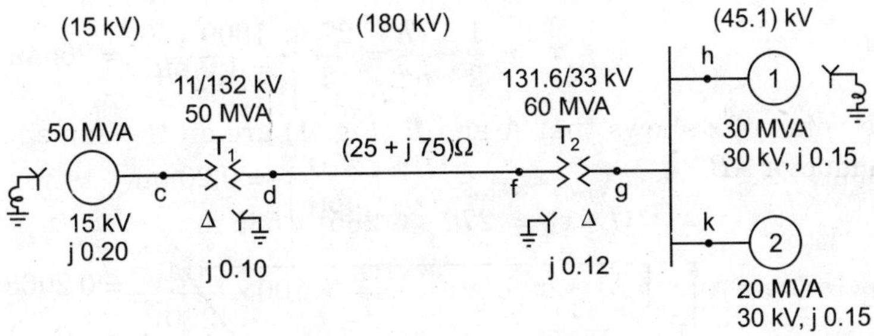

Fig. 7. One-line diagram of Example 7

Solution : *Step 1 Selection of base MVA and base voltage in kV.*

Let us take a base MVA of 50. Therefore the selected value S_b = 50 MVA shall remain the same in all parts of the system.

Let us select the base voltage in the generator circuit as V_b = 15 kV. The base voltage on the primary side of the transformer T_1 is kV. We know that for each circuit separated by transformers the base voltage changes in proportions to the turns ratio. The turns ratio is equal to the voltage ratio of the transformer. In order to determine the base kV for various components of the system we use the following formula:

$$\text{Base kV on hv side} = \text{base kV on Iv side} \times \frac{(kV)hv}{(kV)iv}$$

or

$$\text{base kV on iV side} = \text{base kV hv side} \times \frac{(kv)iv}{(kV)hv}$$

Base voltage of the transmission line = base voltage on the secondary side of T_1

$$= 15 \times \frac{132}{11} = 180 \text{ kV}$$

The base voltage in the motor circuit is determined with 180 kV applied to the primary of transformer T_2.

Base voltage in the motor circuit = base voltage on the secondary side of the transformer T_2

$$= (\text{base kV on primary of } T_2) \times (\text{turns ratio of } T_2)$$

$$= 180 \times \frac{33}{\sqrt{3} \times 76} = 45.1 \text{ kV}$$

SOLVED EXAMPLES AND SHORT QUESTION ON SAG

Example 1. *On SAG a transmission line conductor at a river crossing is supported from two towers at heights of 30 m and 90 m above water level. The horizontal distance between the towers is 270 m. If the tension in the conductor is 1800 kgf and the conductor weights 1.0 kgf per meter, find the clearance between the conductor and the water at a point midway between the towers. Assume parabolic configuration.*

Solution : $I = 270$ m; $T = 1800$ kgf; $w = 1.0$ kgf; $h = 90 - 30 = 60$ m

$$x = \frac{1}{2} - \frac{Th}{wl} = \frac{270}{2} - \frac{1800 \times 60}{1 \times 270} = 265 \text{ m}$$

The negative value of x shows that A and B (Fig. 1) are on the same side of the lowest point O of the conductor AB.

$$(I - x) = 270 - (-265) = 535 \text{ m}$$

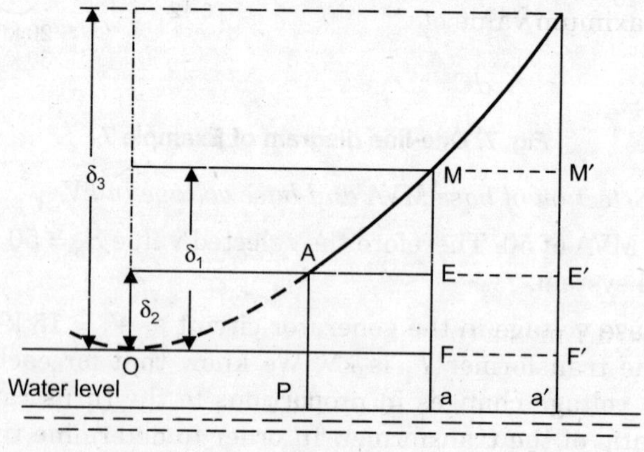

Fig. 1

Let M be the point midway between the towers. The horizontal distance of M from the lowest point is given by

$$x_1 = 265 + \frac{270}{2} = 400 \text{ m}$$

Height of M above O, $\delta_1 = \dfrac{wx^2 l}{2T} = \dfrac{1.0 \times (400)^2}{2 \times 1800} = 44.44$ m

Height of A above O, $\delta_2 = \dfrac{wx^2}{2T} = \dfrac{1 \times (265)^2}{2 \times 1800} = 19.05$

Height of B above O, $= \dfrac{w(l-x)^2}{2T} = \dfrac{1 \times (535)^2}{2 \times 1800} = 79.50$ m

Clearance between the conductor and water at point M

$$QM = QE + EM = QE + (FM + FE)$$
$$= AP + (\delta_1 - \delta_2) = 30 + (44.44 - 19.50) = 54.94 \text{ m}$$

Alternatively

$QM = Q'M' = Q'B = Q'B - (F'B - F'M') = Q'B - (\delta_3 - \delta_1) = 90 - (79.50 - 44.44) = 54.94$ m

Clearance of the lowest point from the ground

$$OG = BD - (y1 + GF) = 39 - (16.9 + 3.8) = 18.3 \text{ m}$$

Equation of the slope of the ground

$$y = mx + C = x \tan \alpha - OG = \frac{1}{20}x - 18.3$$

Clearance from ground at any point $P(x, y)$

$$C_1 = y - \left(\frac{1}{20}x - 18.3\right) = \frac{wx^2}{2H} - \frac{x}{20} + 18.3$$

For minimum or maximum value of C_1.

$$\frac{dC_1}{dx} = 0; \quad \frac{2wx}{2H} - \frac{1}{20} = 0$$

$$x = \frac{H}{20w} = \frac{2200}{20 \times 1.492} = 73.7 \text{ m}$$

Since $\frac{d^2C_1}{dx^2} > 0, x = 73.7$ m is a condition for minimum.

$$\text{Minimum clearance, } C_1 = \frac{wx^2}{2H} - \frac{x}{20} + 18.3$$

$$= \frac{1.492 \times (73.7)^2}{2 \times 2200} - \frac{73.7}{20} + 18.3 = 1.84 - 3.68 + 18.3 = 16.46 \text{ m}$$

Q. State what is the SAG Template

Ans. The use of a sag template is essential to allocate the position and height of the supports correctly on the profile. It is usually made of transparent celluloid, perspex, or some times card board. The following curves are marked on it:

1. Hot template curve or hot curve
2. Ground clearance curve
3. Support foot or tower curve
4. Cold curve or uplift curve

Hot Curve

Hot curve is obtained by plotting the sags at maximum temperature against span lengths. It shows where the supports must be located in order to maintain the prescribed ground clearance.

Ground Clearance Curve

The clearance curve is below the hot curve. It is drawn parallel to the hot curve and at a vertical distance equal to the ground clearance as prescribed by the regulations for the given line.

Support Foot Curve

This curve is drawn for locating the position of the supports for tower lines. It shows the height from the base of the standard support to the point of attachment of the lowest conductor. For wood or concrete pole lines this curve is not required to be drawn since they can be put on any convenient position.

Cold Curve or Uplift Curve

Uplift curve is obtained by plotting the sags at minimum temperature without wind or ice against span lengths. Uplift curve is drawn to check whether uplift of conductors occurs on any support. The uplift conductors may occur at low temperatures when one support is much lower than either of the adjoining ones.

Q. Write note on reparation of the SAG template.

Ans. The above mentioned curves are first drawn on a squared paper on the same scale as the line profile. Suitable scales are selected. With the help of sharp-pointed probe, the curves are then transferred to transparent celluloid or perspex. The celluloid or perspex is then cut along the line of maximum, sag, i.e., the hot curve.

Fig. 2 illustrates a typical sag template. New sag templates are to be prepared for lines with different ruling span, voltage and loading conditions.

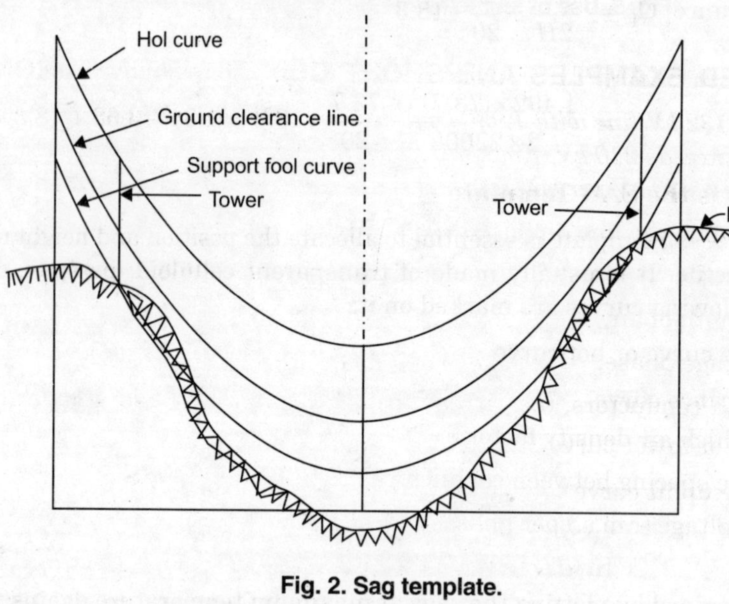

Fig. 2. Sag template.

Q. With figure explain method of using the template.

Ans. The positions of the supports of normal height are marked on the profile plan. For single supports the template is then placed over the profile plane in such a way that the maximum sag line touches the tops of the single supports and the axis is kept vertical. The clearance line marked on the template should not cut any intermediate points between the supports on the profile in which case the necessary clearance is not maintained. In such a case the height of the support or supports should be increased. For tower line the support-foot curve gives the best position of a tower. The use of sag template for single support lines is shown in Fig. 3.

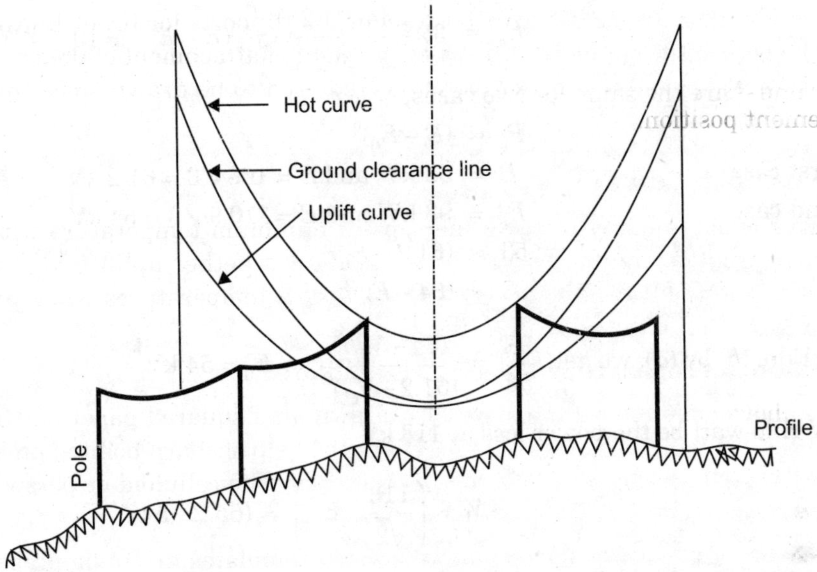

Fig. 3. Use of sag template for single support lines.

SOLVED EXAMPLES AND SHORT QUESTIONS ON CORONA

Example 1. *A 132 kV line with 1.956 cm dia. conductors is built so that corona takes place if the lines voltage exceeds 210 kV (r.m.s.). If the value of potential gradient at which ionisation occurs can be taken as 30 kV per cm, find the spacing between the conductors.*

Solution : Assume the line 3-phase

Conductor radius, $r = 1.956/2 = 0.978$ cm.

Dielectric strength of air, $g_0 = 30/\sqrt{2} = 21.2$ kV per cm.

Disruptive voltage/phase, $V_c = 210/\sqrt{3} = 121.25$ kV

Assume smooth conductors, (*i.e.*, irregularity factor $m_0 = 1$) and standard pressure and temperature for which air density factor $\delta = 1$.

Let d cm be the spacing between conductors

$\therefore$ Disruptive voltage (r.m.s.) per phase, $V_c = m_0 g_0 \delta r$

$$\text{In } (d/r)\, kV = 1 \times 21.2 \times 1 \times 0.978 \times \log_e (d/r)$$

or $$121.25 = 20.733 \text{ In } (d/r) \text{ or } \log_e \frac{d}{r} = \frac{121.25}{20.733} \text{ } 5.848$$

or $$2.3 \log_{10} d/r = 5.838 \text{ or } \log_{10} d/r = 5.848/2.3 = 2.5426$$

or $$d/r = \text{Antilog } 2.5426 = 348.8$$

$\therefore$ $$\text{Conductor spacing } d = 348.8 \times r = 348.8 \times 0.978 = 341 \text{ cm.}$$

Example 2. *A certain 3-phase equilateral transmission line has a total corona loss of 53 kW at 106 kV and a loss of 98 kW at 110.9 kV. What is the distruptive critical voltage? What is the corona loss at 113 kV?*

Solution : The power loss due to corona for 3-phase is given by:

$$P_c = 3\times\frac{244(f+25)}{\delta}\sqrt{\frac{r}{d}}(E-E_0)^2\times 10^{-5}\text{ kW/km}$$

As, f, δ, r and d are the same for two cases,

$$P_c \propto (E-E_0)^2$$

(*i*) For first case, $P = 53$ kW and $E = 106/\sqrt{3} = 61.2$ kV

For second case, $P_c = 98$ kW and $E = 110.9/\sqrt{3} = 64$ kV

$\therefore$ $53 \propto (61.2-E_0)^2$ ...(*a*)

and $98 \propto (64-E_0)^2$ ...(*b*)

Dividing Eqn. (*b*) by (*a*), we get, $\frac{98}{53} = \frac{(64-V_C)^2}{(61.2-V_C)^2}$ or $E_0 = 54$ kV

(*ii*) Let W Kilo-watt be the power loss at 113 kV

$$W \propto \left(\frac{113}{\sqrt{2}} - E_0\right)^2 \propto (65.2-54)^2 \quad ...(c)$$

Dividing Eqn. (*c*) by (*a*), we get $\frac{W}{53} = \frac{(65.2-54)^2}{(61.2-54)^2}$

$\therefore$ $W = (11.2/7.2)^2 \times 53 = 128$ kW

Example 3. *A conductor with 3.0 cm dia is passed centrally through a porcelain bushing $\epsilon_r = 4$ having internal and external diameter of 3.5 cm and 9 cm respectively. The voltage between the conductor and an earthed clamp surrounding the porcelain is 25 kV r.m.s. Determine whether cornoa will be present in the air space round the conductor.*

Solution : Assume, $g_{1\max}$ = Max^m gradient on the surface of conductor

$g_{2\max}$ = Max^m gradient on the inner side of porcelain

$$g_{1\max} = \frac{q}{2\pi\epsilon_0 r} \text{ and } g_{2\max} = \frac{r}{2\pi\epsilon_0\epsilon_r r_1}$$

$\therefore$ $g_1 \max^r = g_{2\max}\epsilon_r r_1$

or, $g_{1\max} = \frac{\epsilon_r r_1}{r} g_{2\max} = \frac{4\times 1.75}{1.5} g_{2\max}$

or, $g_{1\max} = 4.67\, g_{2\max}$

i.e., $g_{2\max} = 0.214\, g_{1\max}$

Now, $25 = g_{1\max}\, r \text{ In } \frac{1.75}{1.5} + g_{2\max}\times 1.75 \text{ In } \frac{4.5}{1.75}$

$= 0.231\, g_{1\max} + 0.354\, g_{1\max}$

$= 0.585\, g_{1\max}$

$\therefore$ $g_{1\max} = \frac{25}{.585} = 42.75$ kV/cm

since, the gradient exceeds 21.1 kV/cm, corona will be present.

Example 4. *A single phase overhead line consists of two conductors of dia 2 cm with a spacing of 1.5 m between centres. Determine line voltage for commencing of corona. Dielective strength of air = 21 kV/cm.*

Solution : Desruptive critical voltage (phase)

$$V_d = 21.1\delta r \text{ In } \frac{d}{r} = 21.1 \times 1 \times \text{In} \frac{150}{1}$$

$$= 105.72 \text{ kV}$$

Q. What are the factors affecting corona loss.

Ans. 1. Effect of frequency : Corona loss is directly proportional to system frequency.

2. Effect of system voltage : As the electric field increases with greater potential differences hence corona loss increases with higher system voltage in a steep way. Fig. 1 represents the profile of a corona loss curve against voltage.

3. Effect of conductivity of Air : The conductivity of air increases if number of ions in air increases. During rain and thunderstorms, ion content in the air enhances and thus corona loss becomes high during bad weather conditions.

4. Effect of conductor diameter : The electric field intensity reduces if the conductor radius increases. Hence with greater conductor diameters, electric field intensity reduces resulting in lower corona loss.

5. Effect of load current : Flow of load current increases the temperature of the conductor. Thus it prevents deposition of dew or snow on conductor's surface. This reduces corna loss.

6. Effect of conductor's surface. Roughness of the conductor's surface results in a field distortion and gives rise to high potential gradient causing higher corona loss. Stranding of conductors increases corona loss.

7. Effect of atmospheric condition : Rain drop and dust deposition increase the corona loss, particularly local corona discharges. If the density of air decreases corona loss increase. Fig. 2 represents the effect of rain.

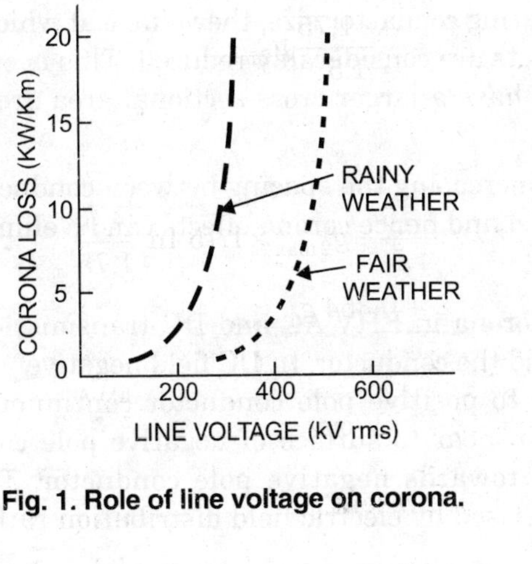

Fig. 1. Role of line voltage on corona.

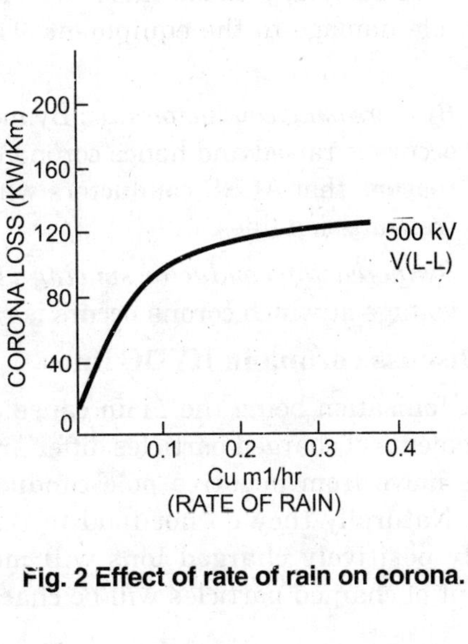

Fig. 2 Effect of rate of rain on corona.

Q. Discuss radio interference of corona.

Ans. Wireless signals are adversely affected by corona discharges. Amplitude modulated broadcasting, carrier signal, aviation, navigation etc. are affected due to interference of corona discharge on communication channel. The effects of corona discharge may create interference to communication line within even few kilometers. Above a value of particular system voltage the corona interference is extremely high. The rate of increase of radio interference is governed by conductor surface and its diameter of the line conductors. However, a transmission line with smooth surface in good weather condition may cause little radio interference problem.

The radio interference field created from transmission line conductors varies as the inverse of radio frequency. Hence, high frequency transmission like television, radar communication, microwave channel, FM broadcasting etc., are less likely to be affected. The radio interference due to corona is of importance for lines above 200 kV. Line design is thus largely affected due to corona.

Q. What are the aspects of corona on real life AC transmission.

Ans. The biggest disadvantage of corona is powerloss. Corona loss reduces the transmission efficiency. However, this reduction is not much in well designed lines. Corona loss increases sharpely during thunderstorm, fog and rain.

Third harmonic current flow has been noticed due to corona. This is due to non-sinusoidal corona current. This increases corona loss. Additionally, harmonics cause inductive interference with parallel communicating circuits. However, there is one advantage of corona. There may be loss of induced energy in form of corona due to indirect lighting on the conductor. Thus, conductor is saved from possibility of lighting.

Line designer should take into account the critical voltage and at least the critical voltage should be 10% higher than the operating voltage. Thus spacing between conductors should be more and radio interference aspects of corona should be taken care of.

Methods of Reducing Corona Effect : It has been seen that intense corona effects are observed at a working voltage of 33 kV or above. Therefore, careful design should be made to avoid corona on the sub-stations or bus bars rated for 33 kV and higher voltage otherwise highly ionised air may cause flash over in the insulators or between the phases, causing considerable damage to the equipment. The corona effects can be reduced by the following methods:

(*i*) *By increasing conductor size* : By increasing conductor size, the voltage at which corona occurs is raised and hence corona effects are considerably reduced. This is one of the reasons that ACSR conductors which have a larger cross-sectional area are used in transmission lines.

(*ii*) *By increasing conductor spacing* : By increasing the spacing between conductors, the voltage at which corona occurs is raised and hence corona effects can be eliminated.

Q. Discuss corona in HVDC lines.

Ans. Ionisation being the main cause of corona in EHV AC and DC transmission lines, the movements of charged particles differ around the conductor. In DC field negatively charged particles move from negative pole conductor to positive pole conductor continuously and steadily. Naturally they do not tend to remain near to surface of negative pole conductor. Similarly positively charged ions will move towards negative pole conductor. Thus the movement of charged particles will be characterised by electric field distribution rather than the surface gradient. In AC transmission lines, the particles will be attracted and repelled during alternate half cycles resulting in significant charged clouds around the conductors.

In HVDC lines, the corona loss is lesser due to absense of factor ✓2 in the voltage waveform. Corona over insulators and metal surface are negligible. However, the corona loss in HVDC system is an important factor and should be always taken into account when designing a line. Corona loss in HVDC lines during bad weather conditions does not increase abruptly like that in AC lines. Also, the corona loss in monopolar HVDC line is minimum.

The critical line voltage (U_c) for monopolar line at which corona will be initiated is given by the following relations.

$$U_c = E_c . r . m \log \frac{2H}{r} \text{ kV (DC)}$$

where

E_c = Critical stress kV/cm (in DC E_c = 14 kV/cm)

m = Surface irregularity factor (0.8 to 0.92)

H = Height of conductor above earth (cm)

r = Conductor radius (cm)

For bipolar line,

$$U_c = 2 E_c r m \ln \frac{D}{r}\left[\frac{1}{\sqrt{1+\left(\frac{D}{2H}\right)^2}}\right]^{kV}$$

where,

D = pole to pole spacing.

The corona loss in bipolar line is given by

$$P_{dc} = P_0 + a(E - E_c)^2 \text{ kW/km}$$

where

P = Corona loss at critical gradient = 0.3 – 0.5 kW/km

E = Actual surface stress kV/cm

a = Constant

SOLVED EXAMPLES ON POWER SYSTEM STABILITY AND PROTECTION

Example 1. *A synchronous machine is delivering 1.0 pu power to an infinite bus through a double-circuit transmission line shown in Fig. (a). The direct axis transient reactance of the generator is 0.3 pu. The per unit reactance for each line is 0.5. All reactances are gives to a base of the machine rating. One of the transmission lines experiences a solid three phase fault to ground, during which occurrence the system reactances are as shown in Fig. 1 (c). Determine the critical clearing angle before which the circuit breakers of the faulted line should operate if the stability is to be maintained.*

Solution : *Prefault condition.*

Refect to equivalence circuit in Fig. (*b*) of the transmission line of Fig. (*a*)

∴ From Fig. (*b*), the transfer reactance between *a* and *c* is given by

$$X = 0.3 + (0.5 \parallel 0.5) = 0.3 + \frac{0.5 \times 0.5}{0.5 + 0.5} = 0.55 \text{ pu}$$

The equation of the prefault power-angle curve is

$$P_{el} = \frac{EV_c}{X}\sin\delta = \frac{1.4\times 1}{0.55}\sin\delta$$

or

$$P_{el} = 2.545 \sin \delta$$

Fig. 19.25 (*h*) shows the prefault curve A. At point *a* on the prefault curve.

$$P_{el} = 1, \delta = \delta_0$$

$$\therefore \quad 1 = 2.545 \sin \delta_0$$

$$\sin \delta_0 = \frac{1}{2.545} = 0.39285, \delta_0 = 23.13° = 0.4037 \text{ rad}$$

$$\cos \delta_0 = 0.9196$$

Condition During Fault

The condition of the network during fault is shown in Fig. (*c*).

In order to determine the transfer reactance between Fig. (a) and Fig. (c) we have to use delta-star transformation and then star-delta transformation.

In delta to star transformation in Fig. (*d*)

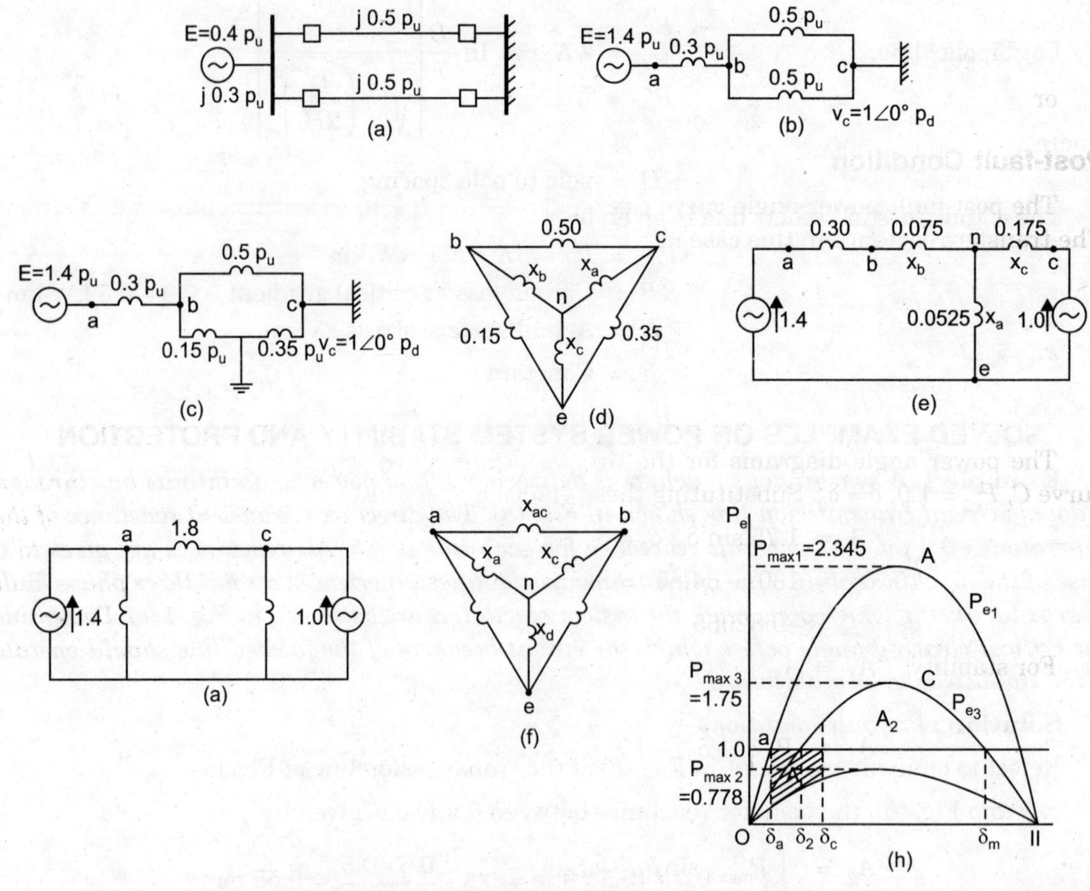

Illustrating Example

$$X_b = \frac{0.5 \times 0.15}{0.5 + 0.15 + 0.35} = 0.075 \text{ pu}$$

$$X_c = \frac{0.5 \times 0.35}{0.5 + 0.15 + 0.35} = 0.175 \text{ pu}$$

$$X_e = \frac{0.15 \times 0.35}{0.5 + 0.15 + 0.35} = 0.0525 \text{ pu}$$

The equivalent circuit of Fig. (*c*) is shown in Fig. (*e*).

Star-delta transformation is done with the help of Fig. (*f*)

$$X_{ac} = X_a + X_c + \frac{X_a X_c}{X_e} = 0.375 + 0.175 + \frac{0.375 \times 0.175}{0.0525} = 1.8 \text{ pu}$$

In the final star to delta transformation, only the reactance between the nodes *a* and *c* need be calculated as the other two shunt reactances do not contribute to active power transfer from *a* to *c*. The equivalent circuit of Fig. (*e*) is shown in Fig. (*g*)

The power-angle curve during the fault is, therefore, given by

$$P_{e2} = \frac{EV_c}{X_{ac}} \sin \delta = \frac{1.4 \times 1.0}{1.8} \sin \delta$$

or $$P_{e2} = 0.778 \sin \delta$$

Post-fault Condition

The post-fault power-angle curve can be derived from Fig. (*b*) with one line out of circuit. The transfer reactance in this case is

$$X = 0.3 + 0.5 = 0.8 \text{ pu}$$

$\therefore$ $$P_{e3} = \frac{EV_c}{X} \sin \delta = \frac{4 \times 1.0}{0.8} \sin \delta$$

$$P_{e3} = 1.75 \sin \delta$$

The power angle diagrams for the three conditions are shown in Fig. (*h*). At point *b* of curve *C*, $P_{e3} = 1.0$, $\delta = \delta_2$. Substituting these values in Eq. we get.

$$1 = 1.75 \sin \delta_2, \delta_2 = 34.85°$$

$$\delta_m = 180° - 34.85° = 145.5° = 2.5333 \text{ rad}$$

$$\cos \delta_m = -0.82065$$

For stability, $A_1 = A_2$

$$A_1 = P_s(\delta_c - \delta_0) - \int_{\delta_0}^{\delta_c} P_{\max 2} \sin \delta \, d\sigma = P_s(\delta_c - \delta_0) + P_{\max 2}(\cos \delta_c - \cos \delta_0)$$

$$A_2 = \int_{\delta_c}^{\delta_m} P_{\max 3} \sin \delta \, d\delta - P_s(\delta_m - \delta_c) = P_{\max 3}(\cos \delta_c - \cos \delta_m) - P_s(\delta_m - \delta_c)$$

Equating the two areas and simplifying we get

$$\cos\delta_c = \frac{P_s(\delta_m - \delta_0) - P_{\max 2}\cos\delta_0 + P_{\max 3}\cos\delta_m}{P_{\max 3} - P_{\max 2}}$$

$$= \frac{[1(2.5333 - 0.4037) - 0.778 \times 0.9196 + 1.75(-0.82065)]}{(1.75 - 0.778)}$$

$$= \frac{(2.1296 - 0.71545 - 1.4361)}{0.972} = 0.02258$$

$$\delta_c = 91.3^\circ$$

Example 2. *A 50 Hz. 4 pole twrbo-generator rated at 20 MVA and 13.2 kV has an inertia constant of 9 KW see / kVA :*

1. *Find the Kinetic Energy Stored by the rotor at Synchronous speed.*
2. *If the input power less rational losses is 20 MW and output power is 15 MW, find the acceleration in degree per sec² and r.p.m.*

Solution : Let us denote Machine rating in MVA by G and inertia constant by H. Than

$$H = \frac{\text{Stored energy in mega-joules at synchronous speed}}{G}$$

or $$\text{Stored energy} = GH = 20 \times 9 = 180 \text{ MJ}$$

We know that

$$\text{Sotred energy} = \frac{1}{2}I\omega^2 = \frac{1}{2}M\omega \text{ Joule}$$

where M is angular momentum in J-sec/radian and ω is in rad/sec.

[When M is expressed in M J-sec/elect degree with ω determined by synchronous speed of the machine M is called the inertia constant.

Thus with M in M J-sec/elect degree and $\omega = 360\,f$ (where f is frequency in Hz) in elect degrees/sec, Stored energy $= \frac{1}{2}I\omega^2 = \frac{1}{2}M\omega$ in MJ. But GH is also stored energy in MJ.

Therefore, $$GH = \frac{1}{2}M\omega = \frac{1}{2}M(360f) = 180\,fM$$

or $$M = \frac{GH}{180f} \text{ MJ-sec/elect degree]}$$

Here we have $$M = \frac{20 \times 9}{180 \times 50} = \frac{1}{50}$$

Input power less rotational losses $= P_s = 20$ MW

Output Power $= P_e = 15$ MW

Therefore, accelerating power $P_a = P_s - P_e = 20 - 15 = 5$ MW

But $$P_a = M\frac{d^2\delta}{dt^2}$$

So that acceleration $\alpha = \dfrac{d^2\delta}{dt^2} = \dfrac{5\times 50}{1}$ = 250 elect degree/sec^2 = 125 degree/sec^2

This corresponds to $60\times\dfrac{125}{360}$ = 20.83 rpm/sec.

Example 3. *In Example 10.21 assuming that acceleration computed is constant for a period of 10 cycles what is the change in torque angle in that period? Also find the speed in rpm at the end of 10 cycles.*

Assume that generator is synchronized to large system and has to accelerating torque before the beginning of 10 cycles period.

Solution : For a 50 Hz system 10 cycles correspond to 0.20 sec. We have found α = 250 elect degree/sec^2.

Therefore change in $\delta = \dfrac{1}{2}\times 250\times(0.20)^2$ = 5 elelct degrees

Synchronous speed $= \dfrac{120\times 5}{4}$ = 1,500 rpm

Since α was found to be 20.83 rpm/sec in the previous example, speed at the end of 10 cycles = 1,500 + 20.83 × 0.20 = 1500 + 4.17 = 1,504.17 rpm.

Example 4. *The generator of Example 10.21 is delivering rated MVA at 0.75 p.f. lagging when a fault reduces its electric power output to 50%. Neglecting losses and assuming constant power input to the shaft what is the accelerating torque at the time of occurrence of the fault?*

Solution :

Pre-fault output power = 0.75 × 20 = 15 MW

Post fault output power = 0.50 × 15 = 7.50 MW

Therefore, accelerating power P_a = 7.50 MW

But $P_a = T_a\omega$

Here $\omega = 2\pi\times\dfrac{1.500}{60}$ = 50 π rad/sec,

so that $T_a = \dfrac{7.50\times 10^6}{50\pi}$ = 47,745 N-m

Example 5. *An alternator with xd = 0.1 p.u. is delivering power of 0.8 p.u. at a terminal voltage of 1.05 p.u. The inertia constant (H) is 4 MJ/MVA. This alternator is suddenly synchronised to an infinite bus through a transformer of reactance 0.1 p.u. and a double circuit transmission line of 0.4 p.u. reactances each. Assuming the infinite bus voltage to be 1.0 p.u. determine the swing equation.*

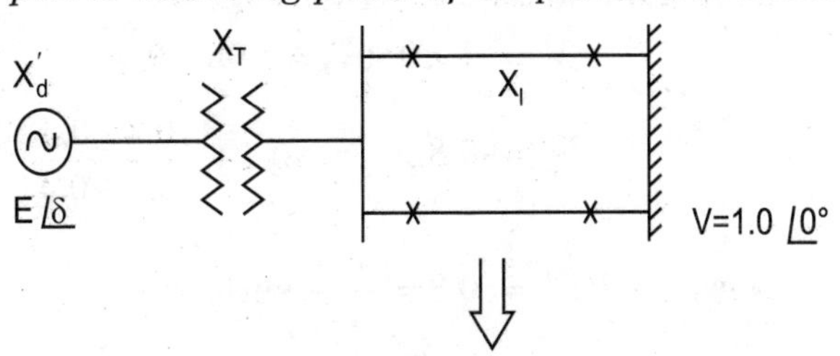

Fig. 2.

Solution : The generator on the infinite bus has been drawn in Fig. 2 and the equivalent circuit has been shown in Fig. 3.

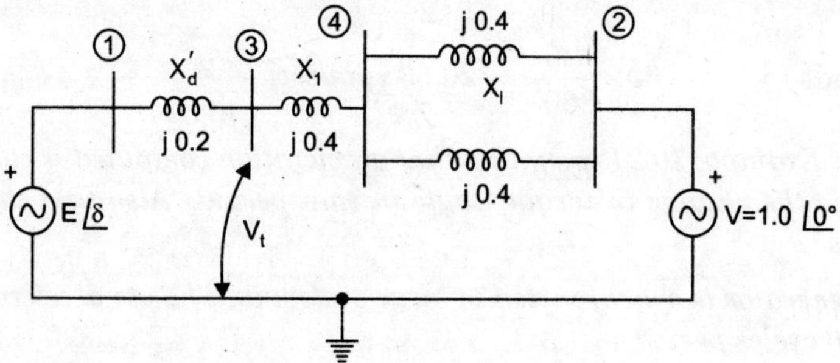

Fig. 3

From the equivalent circuit,

$$x_{12} = x'_d + x_t + \frac{x_l \times x_l}{x_l + x_l}$$

$$\begin{bmatrix} x'_d = \text{transient } d\text{-axis reactance of alternator} \\ x_t = \text{reactance of transformer} \\ x_l = \text{reactance of transmission line} \end{bmatrix}$$

$$= j\,0.2 + j\,0.1 + j\,0.2 = j\,0.5 \text{ p.u.}$$

$$\therefore \quad y_{12} = \frac{1}{j0.5} = -j\,2.0 \text{ p.u.} = y_{21}$$

$$\begin{bmatrix} y_{12} = \text{admittances connected between buses 1 \& 2} \\ y_{10}, y_{20} = \text{admittances connected with buses 1 and 2 respectively} \\ \text{Also } y_{10} = 0; y_{20} = 0 \end{bmatrix}$$

$$\therefore \quad [Y] = \begin{bmatrix} Y_{11} & Y_{12} \\ Y_{21} & Y_{22} \end{bmatrix} = \begin{bmatrix} (y_{12} + y_{10}) & -(y_{12}) \\ -(y_{12}) & (y_{21} + y_{20}) \end{bmatrix} = \begin{bmatrix} -j2 & j2 \\ j2 & -j2 \end{bmatrix}.$$

Also, observing $[Y]$, $\theta_{11} = -\frac{\pi}{2}$ and $\theta_{12} = \frac{\pi}{2}$

Now, $P_e = EVY_{12} \sin\delta = E \times 1.0 \times (2) \sin\delta = 2E \sin\delta$ ...(*i*)

During initial condition [Ref. Fig. 4]

$$V = 1\angle 0^\circ;\ V_t = 1.05\angle\theta_t \text{ p.u.}$$

$$P_e = 0.8,\ x_{32} = j0.1 + j\frac{0.4 \times 0.4}{0.4 + 0.4} = j\,0.3 \text{ p.u.}$$

Now, $P_e^{(0)} = 0.8 = \dfrac{V_t . V}{x_{32}} \sin\theta_t$

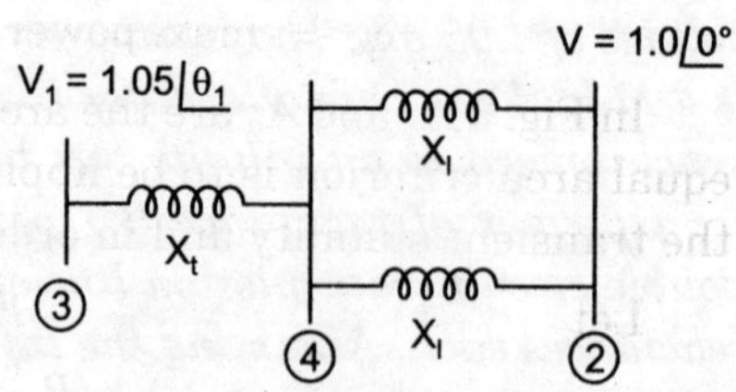

Fig. 4

When superscript (0) indicates initial condition.

or, $$0.8 = \frac{1.05\times1}{0.3}\sin\theta_t$$

$\therefore$ $$\theta_t = 13.2°$$

$\therefore$ Current $$I = \frac{V_t - V}{Z} = \frac{1.05\angle13.2° - 1\angle0°}{j0.3} = 0.804\ \angle-5.3°\ \text{p.u.}$$

However, $$E\angle\delta = V_t\angle\theta_t + Ixd'$$
$$= 1.05\ \angle13.2° + (0.804\ \angle-5.3°)\ (0.2\ \angle90°)$$
$$= 1.04 + j\ 0.4 = 1.114\ \angle\ 21°\ \text{p.u.}$$

From equation (*i*),

$$P_e = 2E\sin\delta = 2\times1.114\sin\delta = 2.228\sin\delta$$

[This is the electrical output during transient process].

$\therefore$ Swing equation

$$\frac{2H}{\omega_s}\frac{d^2\delta}{dt^2} = P_m - P_e \text{ becomes,}$$

$$\frac{2\times4}{2\pi\times50}\frac{d^2\delta}{dt^2} = P_e^{(0)} - P_e = (0.8 - 2.228\sin\delta)$$

[$\therefore$ P_m remains unchanged druing transient process and is equal to initial electrical power output $P_e^{(0)}$].

or $$\frac{d^2\delta}{dt^2} = \frac{314}{8}(0.8 - 2.228\sin\delta)$$
$$= 39.25\ (0.8 - 2.228\sin\delta)\ \text{rad/sec}^2$$

Example 6. *A synchronous generator supplies power to a targe grid through a parallel transmission line network. Suddenly a fault occurs at one of the lines and the power output drops considerably. Breakers at both the sides clear the fault and the power transfer is restored to a new value. Find the expression for critical clearing angle, initial angle and maximum power angle.*

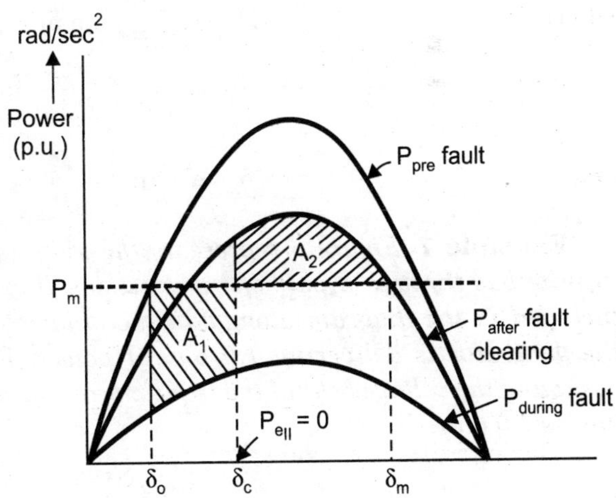

Fig. 5

Solution : δ_0 = initial power angle in steady state condition

δ_c = critical clearing angle

δ_m = max. power angle

In Fig. 5 A_1 and A_2 are the areas for which equal area criterion is to be applied to check the transient stability and in order to determine the critical clearing angle.

Let $$P_{pre\ fault} = P^{(0)} = P_{\max}\sin\delta$$
$$P_{during\ fault} = P^{(f)} = \gamma_1 P_{\max}\sin\delta$$

and $$P_{after\ fault\ clearing} = P = \gamma_2 P_{\max}\sin\delta$$

where γ_1 and γ_2 are the ratios of the maximum power that can be transferred during and after the fault, to the maximum power that can be transmitted before that fault.

Smaller values of γ_1 result in greater disturbance to the system as low γ_1 indicates low power transfer during fault (*i.e.*, higher P_a) causing higher acceleration energy.

In this problem, refering Fig. 5

$$A_1 = P_m(\delta_c - \delta_0) - \int_{\delta_0}^{\delta_c} \gamma_1 P_{max} \sin\delta \, d\delta$$

i.e.,
$$A_1 = P_m(\delta_c - \delta_0) + \gamma_1 P_{max} (\cos\delta_c - \cos\delta_0) \quad ...(a)$$

Also, from the same figure,

$$A_2 = \int_{\delta_c}^{\delta_m} \gamma_2 P_{max} \sin\delta \, d\delta - P_m(\delta_m - \delta_c)$$

$$= \gamma_2 P_{max} (\cos\delta_c - \delta_m) - P_m(\delta_m - \delta_c) \quad ...(b)$$

Transient stability will be maintained provided $A_1 = A_2$

$\therefore$ From equation (a) and (b),

$$P_m(\delta_c - \delta_0) + \gamma_1 P_{max} (\cos\delta_c - \cos\delta_0) = \gamma_2 P_{max} (\cos\delta_c - \delta_m) - P_m(\delta_m - \delta_c)$$

or,
$$P_m\delta_c - P_m\delta_0 + \gamma_1 P_{max} \cos\delta_c - \gamma_1 P_{max} \cos\delta_0 = \gamma_2 P_{max} \cos\delta_c - \gamma_2 P_{max} \cos\delta_m - P_m\delta_m + P_m\delta_c$$

or,
$$(\gamma_1 - \gamma_2) \cos\delta_c = P_m(\delta_0 - \delta_m) + \gamma_1 P_{max} \cos\delta_0 - \gamma_2 P_{max} \cos\delta_m$$

Solving for δ_c,

$$\delta_c = \cos^{-1}\left[\frac{(P_m / P_{max})(\delta_m - \delta_0) + \gamma_1 \cos\delta_0}{\gamma_2 - \gamma_1}\right]$$

where,
$$P_m = P_{max} \sin\delta_c = \gamma_2 P_{max} \sin\delta_m$$

$\therefore$
$$\delta_0 = \sin^{-1}\frac{P_m}{P_{max}} \quad \text{for } \delta_0 < 90°$$

and
$$\delta_m = \sin^{-1}\frac{P_m}{\gamma_2 P_{max}} \quad \text{for } \delta_m > 90°$$

Example 7. *Figure 6 represents the single line diagram of a generator connected to an infinite bus through a pair of transmission lines. The transient reactance of the generator is included in the diagram alongwith the other reactances. Assume a 3-phase fault at F when the generator is delivering 1.0 p.u. of power. Find the value of the critical clearing angle assuming the voltage behind the transient reactance to be 1.20 p.u. and the voltage at the infinite bus as 1.0 unit.*

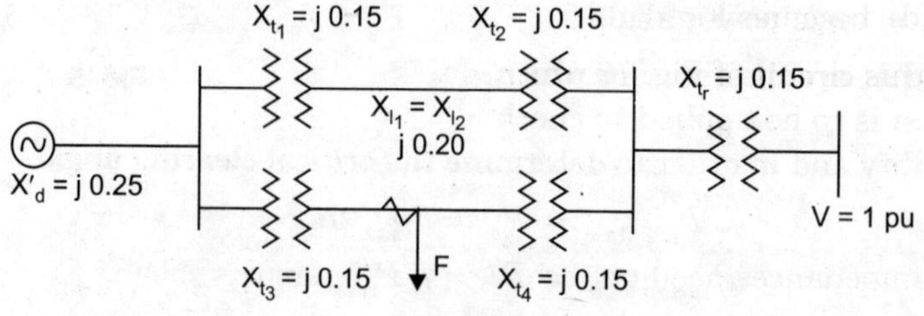

Fig. 6

Solution : The equivalent circuit has been show in Fig. 7

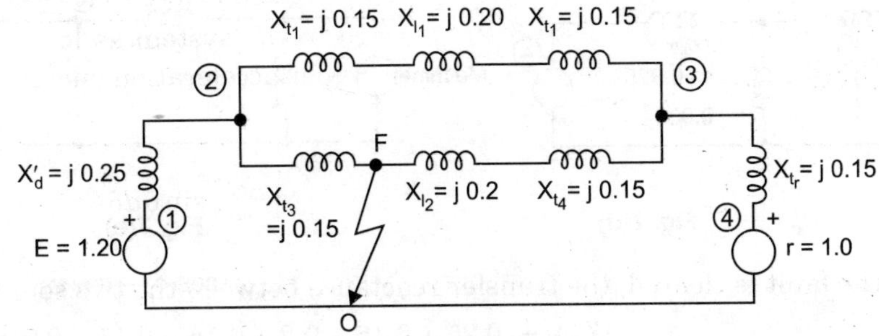

Fig. 7(a)

Before the fault F was open,

$$\therefore \qquad |X_{13}^{(0)}| = 0.25 + \frac{0.15+0.2+0.15}{2} + 0.15 = 0.65 \text{ p.u.}$$

As soon as fault occurs, the impedance diagram will look like as in Fig. 7. (*a*) Network reduction transforms this to the diagram as shown in Fig. 8.

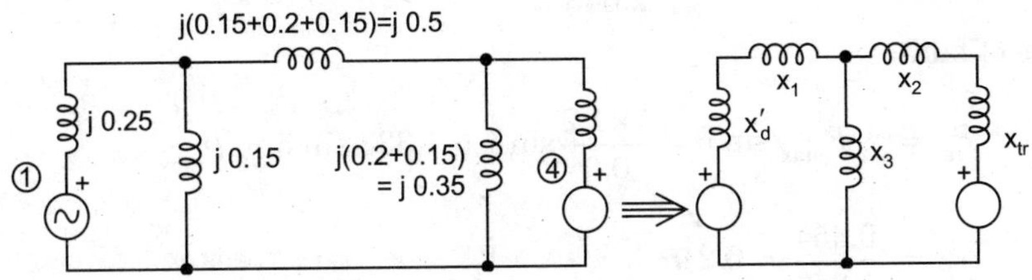

Fig. 7(b) **Fig. 7(c)**

During the fault, the circuit can be represented as shown in Fig. E 10.1 (*c*). This is possible by reducing the π network of Fig. 8 to equivalent *T* network.

Here
$$|x_1| = \frac{0.5\times0.15}{0.5+0.15+0.35} = 0.075 \text{ p.u.}$$

$$|x_2| = \frac{0.5\times0.35}{0.5+0.15+0.35} = 0.175 \text{ p.u.}$$

$$|x_3| = \frac{0.15\times0.35}{0.5+0.15+0.35} = 0.053$$

Finally, Fig. 7(d) becomes Fig. 7(e).

Transforming this circuit of Fig. 10 to an equivalent Δ, Fig. 11 results

where
$$|x_{14}^{(f)}| = 0.325 + 0.325 + \frac{(0.325)^2}{0.053} = 2.643 \text{ p.u.}$$

Shunt branch impedances need not be calculated for this problem.

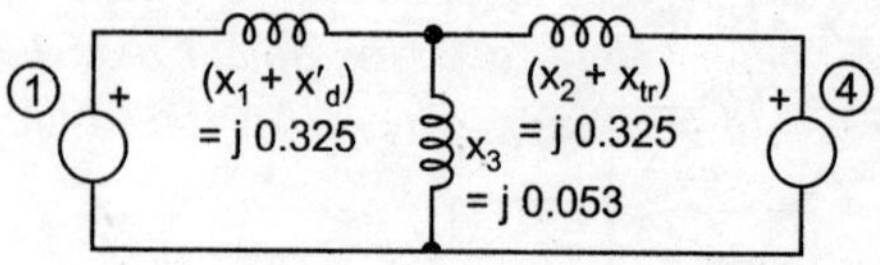

Fig. 7(d)

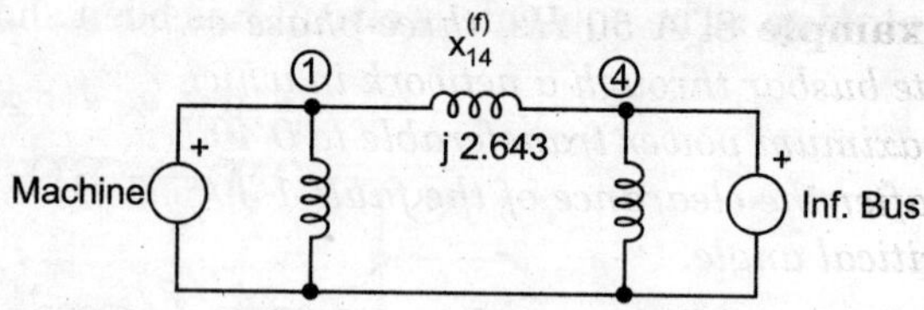

Fig. 7(e)

As soon as the fault is cleared, the transfer reactance between the two source voltage are

$$|X_{14}| = \underset{(xd)}{0.25} + \underset{(x_t)}{0.15} + \underset{(xl_1)}{0.2} + \underset{(x_{t2})}{0.15} + \underset{(x_{tr})}{0.15} = 0.9 \text{ p.u.}$$

Next we compute the steady state power equations during different conditions as

before fault $\quad P_e^{(0)} = P_{max} \sin\delta = \dfrac{1\times1.2}{0.65}\sin\delta = 1.846 \sin\delta$

during fault, $\quad P_e(f) = \gamma_1 P_{max} \sin\delta = \dfrac{1\times1.2}{2.643}\sin\delta = 0.454 \sin\delta$

after clearance of fault,

$$P_e = \gamma_2 P_{max} \sin\delta = \frac{1\times1.2}{0.9}\sin\delta = 1.333 \sin\delta$$

Hence, $\quad \gamma_1 = \dfrac{0.454}{1.846} = 0.246$

and $\quad \gamma_2 = \dfrac{1.333}{1.846} = 0.722$

Also, $\quad \delta_0 = \sin^{-1}\dfrac{P_m}{P_{max}} = \sin^{-1}\dfrac{1}{1.846} = 0.572 \text{ rad} = 32.77°$

$$\delta_m = \sin^{-1}\frac{P_m}{\gamma_2 P_{max}} = \sin^{-1}\frac{1}{0.722\times1.846} = 131°\ 38° = 2.293 \text{ rad}$$

As $\quad \delta_c = \cos^{-1}\left[\dfrac{\dfrac{P_m}{P_{max}}(\delta_m - \delta_0) + \gamma_2\cos\delta_m - \gamma_m - \gamma_1\cos\delta_0}{\gamma_2 - \gamma_1}\right]$

Hence, here, $\quad \delta_c = \cos^{-1}\left[\dfrac{\dfrac{1.0}{1.846}(2.293 - 0.572) + 0.722\cos 131.38° - 0.246\cos 32.77°}{0.722 - 0.246}\right]$

$$= \cos^{-1}\left[\frac{0.2476}{0.476}\right] = 58.66°$$

Example 8. *A 50 Hz, three-phase synchronous generator delivers 1.00 p.u. power to an infinite busbar through a network in which resistance is negligible. A fault occurs which reduces the maximum power transferable to 0.40 p.u. whereas, before the fault, this power was 1.8 p.u. and, after the clearance of the fault 1.30 p.u. By the use of the equal area criterion, determine the critical angle.*

Solution : We have

$$P_m = 1.00 \text{ p.u.}$$
$$P_{max} = 1.80 \text{ p.u}$$
$$\gamma_1 P_{max} = 0.40 \text{ p.u.}$$
$$\gamma_2 P_{max} = 1.30 \text{ p.u.}$$

Here,
$$\delta_0 = \sin^{-1}\left(\frac{P_m}{P_{max}}\right) = \sin^{-1}\left(\frac{1.0}{1.8}\right) = 33.8°$$

and
$$\delta_f = 180° - \sin^{-1}\left(\frac{P_m}{\gamma_2 P_{max}}\right) = 180° - \sin^{-1}\left(\frac{1.0}{1.30}\right) = 129.6°$$

Also
$$\gamma_1 = \frac{\gamma_1 P_{max}}{P_{max}} = \frac{0.40}{1.80} = 0.222$$

$$\gamma_2 = \frac{\gamma_2 P_{max}}{P_{max}} = \frac{1.30}{1.80} = 0.722$$

We know
$$\delta_c = \cos^{-1}\left[\frac{1}{\gamma_2 - \gamma_1}\{(\delta_f - \delta_0)\sin\delta_0 + (\gamma_2\cos\delta_f - \gamma_1\cos\delta_0\}\right]$$

$\therefore$
$$\delta_c = \cos^{-1}\left[\frac{1}{(0.722 - 0.222)}\left\{\frac{\pi(129.6° - 33.8°)}{180°}\right\}\sin 33.8° + (0.722\cos 129.6° - 0.222\cos 33.8)\}\right]$$

$\therefore$
$$\delta_c = \cos^{-1}[\{+0.93 + (-0.645)\}/0.50] = \cos^{-1} 0.57$$

$\therefore$
$$\delta_c = 55.20°$$

Example 9. *A synchronous generator is operating at an infinite bus and supplying 45% of its peak power capacity. As soon as a fault occurs, the reactance between the generator and the line becomes four times its value before the fault. The peak power that can be delivered after the fault is cleared is 70% of the original maximum value. Determine the critical clearing angle.*

Solution : As per the problem,

$$P_{max}\sin\delta_0 = 0.45\, P_{max}\ (= P_m)$$

$\therefore$
$$\delta = 26.74°$$

But
$$P_{max} = \frac{EV}{X}$$

During fault "X" becomes $4X$ (as given in the problem).

$$\therefore \quad \gamma_1 P_{max} \sin \gamma_m = \frac{EV}{4X} \sin \delta_m = \frac{1}{4} P_{max} \sin \delta_m \left[\text{Substituting } P_{max} \text{ for } \frac{EV}{X}\right]$$

$$\therefore \quad \gamma_1 = 0.25$$

After clearance of the fault, $\gamma_2 = 0.7$

or, $$\gamma_2 P_{max} \sin \delta_m = P_m = 0.45 P_{max}$$

$$\delta_m = 40°$$

i.e., $$\delta_m = 180° - 40 = 140° = 2.44 \text{ rad.}$$

$$\therefore \quad \cos \delta_c = \frac{1}{0.7 - 0.25}[0.45(2.44 - 0.467) + 0.7 \cos 140° - 0.25 \cos 26.74°]$$

$$= \cos 73°$$

$$\therefore \quad \delta_c = 73°$$

Example 10. *A synchronous motor of negligible resistance is receiving 25% of power that is capable of receiving from an infinite bus. If the load on the motor is suddenly doubled, calculate the maximum value of power angle during the swinging of the motor around its new equilibrium position.*

Solution : Input electrical power to the motor is given by

$$P_e = P_m \sin \delta_0$$

where, P_m is the maximum power which the motor is capable to receive and is equal to 1.00 p.u.

As per the problem

$$0.25 P_m = P_m \sin \delta_0$$

$$\therefore \quad \delta_0 = 14.5° = 0.25 \text{ rad.}$$

If the load on the motor is suddenly doubled, let δ' be the new power angle

$$\therefore \quad 2 \times 0.25 P_m = P_m \sin \delta_2$$

$$\therefore \quad \delta' = 30°; 0.525 \text{ rad.}$$

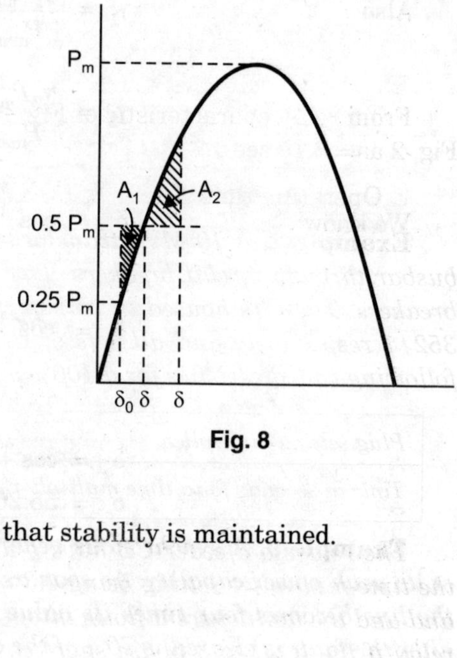

Fig. 8

Let the maximum power angle be δ [Ref. Fig. 8 such that stability is maintained.

Applying equal area criterion,

$$-A_1 = A_2 \text{ [for motor]}$$

or, $$-\int_{\delta_0}^{\delta'} (0.5P_m - P_m \sin \delta) d\delta = \int_{\delta'}^{\delta} (0.5P_m - \sin \delta) d\delta$$

or, $$-P_m(0.5\delta + \cos \delta)\Big|_{\delta_0}^{\delta'} = P_m(0.5\delta + \cos \delta)\Big|_{\delta'}^{\delta}$$

or, $$-(0.5\delta + \cos \delta)\Big|_{0.25}^{0.525} = 0.5\delta + \cos \delta\Big|_{0.525}^{\delta}$$

or, $$-0.0335 = 0.5\,\delta + \cos \delta - 0.262 - 0.866$$

$\therefore \quad \cos \delta_3 + 0.5\, \delta_3 = 1.095$

This gives (by trial and error method)

$$\delta = 0.81 \text{ elect. rad.}$$

$$= 46.5°.$$

SOLVED EXAMPLES ON PROTECTIVE RELAY IN POWER SYSTEM

Example 1. *An IDMT overcurrent relay has a current setting of 150% and has a time multiplier setting of 0.5. The relay is connected in the circuit through a C.T. (current Transformer of ratio 500% 5 amps. calculate the time of operation calculate the time of operation of the relay if the circuit carries a fault current of 6000 A. The relay characteristic is show in Fig. 2.*

Solution : Fault current in found as a

$$= 6000 \times \frac{5}{500} = 60 \text{ A.}$$

$$\text{Plug setting multiplier (P.S.M.)} = \frac{\text{Actual 10 km current in relay}}{\text{Setting current}}$$

$$= 60 \times \frac{1}{5 \times 1.5} = 8$$

From relay characteristic of Fig. 2, against the P.S.M. of 8 we get time from the graph of Fig. 2 as = 3.15 see.

$\therefore$ Operating time $= 3.15 \times 0.5 = 1.575$ see.

Example 2. *A 10 MVA transformer called upon to operate at 25% overload, feeds 11 kV busbar through circuit breakers. The 11 kV busbars feed two outgoing feeders through other breakers. The CTs housed in transformer and feeder circuit breaker has ratios of 525/1 and 352/1 respectively and all sets of C.T.S. feed induction type over curent relays having the following characteristics for a 100% plug setting as shown on Fig. 1.*

Plug setting multiplier	2	3.6	5	10	15	20
Time in seconds for a time multiplier of 1	10	6	3.9	2.8	2.2	2.1

The relays on feeder CB have plug setting of 100% for overload and 10% on earth fault, the time multiplier setting being 0.3 in all cases. A fault occurs on one of the outgoing feeders and and the fault current flows into it via transformer. Find the operating time of transformer relay if the fault is 3 phase short circuit with a fault current magnitude of 5250 A in each phase.

Solution : Fault current $= 5250$ A

$$= 5250 \times \frac{1}{525} = 10 \text{ A}$$

$$\therefore \quad \text{Plug-setting multiplier} = 10 \times \frac{1}{(1.25 \times 1)} = 8$$

Time corresponding to the multiplier from the graph drawn from the tabular data = 3.15 sec.

∴ Operating time = 3.15 × .3 = 0.945 sec.

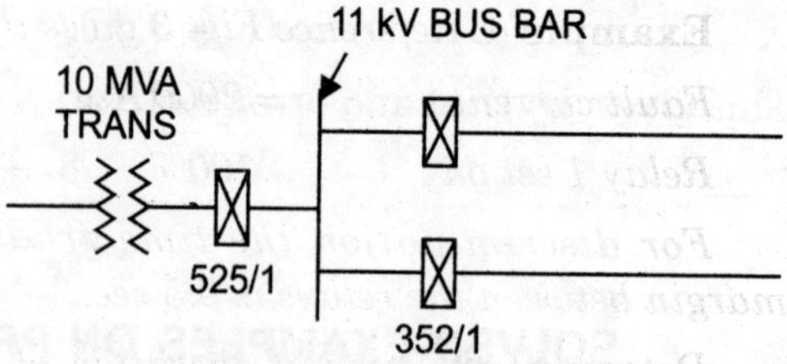

Fig. 1

Example 3. *A 20 MVA transformer which may be called upon to operate at 30% overload, feeds 11 kV busbars through a circuit breaker. Other circuit breakers supply outgoing feeders. The transformer circuit breaker is equipped with 1000/5 A CTs and the feeder circuit breakers with 400/5 A CTs and all sets of CTs feed induction type over current relays. The relays on the feeder circuit breakers have a 125% plug setting and a 0.3 time setting. If a 3 ph. fault current of 5000 A flows from the transformer to one of the feeders, find the operating time of the feeder relay, the minimum plug setting of the transformer relay and its time setting assuming a discriminative time margin of 0.5 sec.*

Solution : *Feeder* Sec. current

$$= 5000 \times \frac{5}{400} = 62.5 \text{ A}$$

Plug setting multiplier

$$= 62.5 \times \frac{1}{1.25 \times 5} = 10$$

Time from Fig 2. = 2.8 sec

Operating time = 0.3 × 2.8 = 0.84 secs.

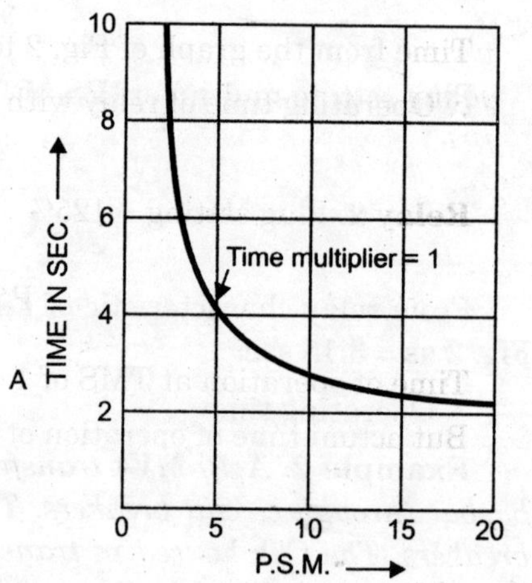

Fig. 2

Transformer. Overload current

$$= (1.3 \times 20) \times 10^3 \times \frac{1}{\sqrt{3} \times 11} = 360A$$

$$\text{Sec. current} = 1360 \times \frac{5}{1000} = 6.81 \text{ A}$$

$$\text{Plug setting multiplier} = 6.81 \times \frac{1}{P.S. \times 5}$$

where P.S. means plug setting. Since the transformer relay must not operate to overload current, plug setting (P.S.) > 6.81/5 > 1.36 > or 136%.

The plug settings are restricted to standard values (see text) in intervals of 2% so the nearest value is 15%.

Sec. Fault current $= 5000 \times \frac{5}{1000} = 25$ A

Plug setting multiplier $= 25 \times \frac{1}{5 \times 1.5} = 3.33$

Time from Fig. E 2.1 is 7 sec.

Time setting $= 0.84 + 0.5 + \frac{1.34}{7} = 0.191$

Example 4. *Reference Fig. 3 gives that*

Fault current = *2000 A* *CT Ratio* *200/1*

Relay 1 set on *100%* *Relay 2 set on* *125%*

For discrimination the time-grading margin between the relays is 0.5 sec.

Determine the time of operation of the two relays assuming that both the relays have the characteristic as shown in Fig. 3 and that the relay no. 1 has a time multiplier setting = 0.2.

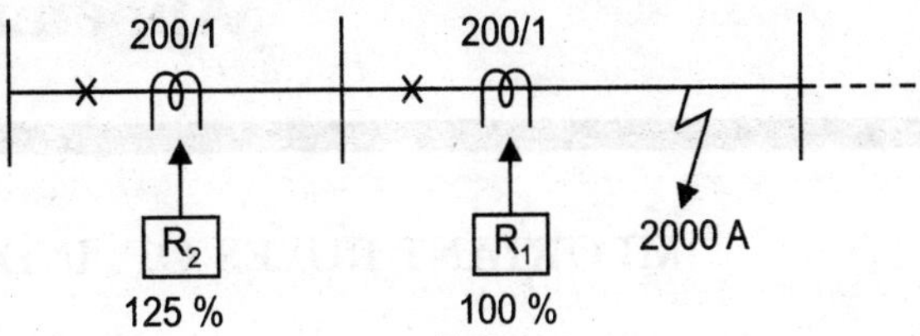

Fig. 2

Solution : Relay 1. Plug setting = 100%

$$P.S.M. = \frac{2000}{200} = 10$$

Time from the graph of Fig. 2 for a PSM of 10 = 2.8 sec

∴ Operating time of relay with T.S.M. of 0.2

$$= 2.8 \times 0.2 = 0.56 \text{ sec.}$$

Relay 2. Plug setting = 125%

$$P.S.M. = \frac{2000}{200} \times \frac{1}{1.25} = 8$$

Time of operation at TMS of 1 = 3.15 sec.

But actual time of operation of relay 2

= time of operation of relay 1 + time grading margin

= 0.56 + 0.5 = 1.06 sec.

$$\therefore \qquad T.S.M. = \frac{1.06}{3.15} = .3364$$

Appendix–III

IMPORTANT RULES OF INDIAN ELECTRICITY RULES

App.1 Indian Electricity Rules

A few important/relevant Indian Electricity Rules (required to know by students as per WBUT syllabus) are mentioned below:

Rule 33. Earthed terminal or consumer's permises

1. The supplier shall provide and maintain on the consumer's permises for the consumer's use a suitable earthed terminal in an accessible position at or near the point of commencement of supply as defined under rule 58.

Provided that in the case of medium, high or extra high voltage installation, the consumer shall, in addition to the afore-mentioned earthing arrangement, provide his own earthing system with an independent electrode,

Provided further that the supplier may not provide any earthed terminal in the case of installations already connected to his system on or before the date to be specified by the State Government in this behalf if he is satisfied that the consumer's earthing arrangement is efficient.

2. The consumer shall take all reasonable precautions to prevent mechanical damage to the earthed terminal and its lead belonging to the supplier.

3. The supplier may recover from the consumer the cost of installation of such earthed terminal on the basis laid down in the sub-rule (2) of rule 82.

Rule 46. Periodical Inspection and testing of consumer's installation

1. (*a*) Where an installation is already connected to supply system of the supplier, every such installation shall be periodically inspected and tested at invervals not exceeding five years either by the Inspector or by the supplier as may be directed by the State Government in this behalf or in the case of installations in mines, oil-fields and railways by the Central Government.

   (*b*) Where the supplier is directed by the Central or the State Government, as the case may be, to inspect and test the installation, he shall report on the condition of the submit a copy of such report to the Inspector.

2. (*a*) The fees for such inspection and test shall be determined by the Central or the State Government, as the case may be, in the case of each class of consumers and shall be payable by the consumer in advance.

   (*b*) In the event of the failure of any consumer to pay the fees on or before the dete speci disconnected under the direction of Inspector. Such disconnection however, shall not be made by the supplier without giving to the consumer even clear day's notice in writing of his in intention so to do.

3. Notwithstanding the provisions of this rule, the consumer shall at all times be solely responsible for the maintenance of his installation in such condition as to be free from danger.

Rule 47. Testing of consumer's installation

1. Upon receipt of an application for a new or additional supply of energy and before connecting the supply or reconnecting the same after a period of six months, the supplier shall inspect and test the applicant's installation.

The supplier shall maintain a record of test results obtained at each supply point to a consumer, in a form to be approved by the inspector.

2. If as a result of such inspection and test, the supplier is satisfied that the installation is likely to constitute danger, he shall serve on the applicant a notice in writing requiring him to make such modifications as are necessary to render the installation safe. The supplier may refuse to connect or reconnect the supply until the required modifications have been completed and he has been notified by the applicant.

Rule 61. Connection with earth

1. The following provisions shall apply to the connection with earth of systems at low voltage in cases where the voltage normally exceeds 125 volts and of systems at medium voltage.

(*a*) The neutral conductor of a three-phase four-wire system and the middle conductor of a two-phase three-wire system shall be earthed by not less than two separate and distinct connections with earth both at the generating station and at the sub-station. It may also be earthed at one or more points along the distribution system or service line in addition to any connection with earth which may be at the consumer's premises.

(*b*) In case of a system comprising electric supply lines having concentric cables, the external conductor of such cable shall be earthed by two separate and distinct connections with earth.

(*c*) The connection with earth may include a link by means of which the connection may be temporarily interrupted for the purpose of testing or for locating a fault.

(*d*) (*i*) In a direct current three-wire system the middle conductor shall be earthed at the generating station only, and the current from the middle conductor to earth shall be continuously recorded by means of a recording ammeter, and if at any time the current exceeds one-thousandth part of the maximum supply current, immediate steps shall be taken to improve the insulation of the system.

(*ii*) Where the middle conductor is earthed by means of a service-breaker with a resistance connected in parallel the resistance shall not exceed 10 ohms and on the opening of the circuit-breaker immediate steps shall be taken to improve the insulation of the system and the circuit-breaker shall be reclosed as soon as possible.

(*iii*) The resistance shall be used only as a protection for the ammeter in the case of earths on the system and until such earths are removed, immediate steps shall be taken to locate and remove the earth.

(*e*) In the case of an alternating system, there shall not be inserted in the connection, with earth any impedance (other than that required solely for the operation of switch-gear or instruments), cut-out or circuit-breaker, and the result of any test made to ascertain whether current (if any) passing through the connection with earth is normal, shall be duly recorded by the supplier.

(*f*) No person shall make connection with earth by the aid of, nor shall he kept it contact with, any water main not belonging to him except with the consent of the owner thereof and of the Inspector.

(*g*) Alternating current systems which are connected with earth as aforesaid may be electrically interconnected.

Provided that each connection with earth is bounded to the metal sheathing and metallic armouring (if any) of the electric supply lines concerned.

2. The frame of every generator, stationary motor, and so far as is practicable, portable motor, and the metallic parts (not intended as conductors) of all transformers and any other apparatus used for regulating or controlling energy and all medium voltage energy consuming apparatus shall be earthed by the owner by two separate and distinct connections with earth.

3. All metal casings or metallic coverings containing or protecting any electric supply line or apparatus shall be connected with earth and shall be so joined and connected across all junction boxes and other openings as to make good mechanical and electrical connection throughout their whole length:

Provided that where the supply is at low voltage, the sub-rule shall not apply to the isolated wall tubes or to breackets, switches, fans regulator covers or other fittings (other than portable hand lamps and portable and transportable apparatus) unless provided with earth terminal.

This sub-rule shall come into force immediately in the case of new installations and in case of existing installation; the provision of this sub-rule shall be complied with before the expiry of a period of two years from the commencement of those rules.

4. All earthing systems shall, before electric supply lines or apparatus are energised, be tested for electrical resistance to ensure efficient earthing.

5. All earthing systems belonging to the supplier shall, in addition, be tested for resistance on dry day during the dry season not less than once every two years.

6. A record of every earth test made and the result there of shall be kept by the supplier for a period of not less than two years after the day of testing and shall be available to the Inspector when required.

Rule 67. Connection with earth

1. The following provisions shall apply to the connection with earth of three-phase systems for use at high or extra-high voltages:

In the case of star connected systems with earthed neutrals or delta connected systems with earthed artifical neutral point:

(*a*) The neutral point shall be earthed by not less than two separate and distinct connections with earth each having its own electrode at the generating station and at the sub-station and may be earthed at any other point, provided that no interference of any description is caused by such earthing;

(*b*) In the event of an appreciable harmonic current flowing in the netural connections so as to cause interference with communication circuits, the generator or transformer neutral shall be earthed through a suitable impedance.

2. Single-phase or extra-high voltage systems shall be earthed in a manner approved by the Inspector.

3. In the case of a system comprising electric supply lines having concentric cables, the external conductor shall be the one to be connected with earth.

4. Where a supplier purposes to connect with earth an existing system for use at high or extra-high voltage which has not either been so connected with earth, he shall give not less than fourteen days notice in writing together with particulars to the telegraph authority of the proposed connection with earth.

5. Where the earthing lead and earth connection are used only in connecting earthing guards created under high or extra high voltage overhead lines where they cross a telecommunication line or a railway line, and where such lines are equipped with earth leakage relays of a type and setting approved by the Inspector, the resistance shall not exceed 25 ohms.

6. In so far as the provisions of rule 61 are consistent with the provisions of this rule, all connections with earth shall also comply with the provisions of that rule.

Appendix–IV

Q. Discuss Bewley's Lattice Diagram (Repeated Reflection of Surges).

Ans. Bewlay's Lattice Diagram provides a simple and convenient method to study the effects of these multiple or repeated Reflection of surges. In fact surges on a power system overhead line are subject to repeated reflected us at terminations, function, towers and similar discontinuities. To study the subject we can consider the case of a short cable connected to the overhead line, where say,

z_1 = surge impedance of the cable
z_2 = surge impedance of the cable
p_1 = line-to-cable reflection factor
τ_1 = line-to-cable transmission (or reflection) factor
p_2 = cable-to-line reflection factor
τ_2 = cable-to-line transmission (refrection) factor
p_c = reflection factor at the for and of the cable
τ_c = transmission (refrection) factor at the far and of cable

From the chapter ten of thin book on travelling waves, we know :

$$p_1 > \frac{z_2 - z_1}{z_2 + z_1}, \quad \tau_1 = \frac{2z_2}{z_2 + z_1}, \quad p_2 = \frac{z_1 - z_2}{z_1 + z_2}, \quad \tau_2 = \frac{2z_1}{z_1 + z_2}$$

But if the far end of the cable is open, when $p_c = 1$ and $\tau_c = 0$. Now, say a rectangular wave of amplitude e_1 originates in the overhead line on reaching the function of the line and the cable, the wave is divided into two parts. One of the waves reflected back into the overhead line and the other transmitted (or Refracted) into the cable.

The transmitted wave in the cable travels along the cable and it reaches its far end, where it is reflected back. Then after covering a distance of twice the length of the cable the wave reaches back at the junction. But at the junction once again thin wave is split into two parts one transmitted along the line and the other is reflected back into the cable. This process continues. The amplitudes of the successive waves in the cable goes on decreasing.

Bewley's Lattice diagram provides a simple and convenient method to study the effects of these multiple or expected reflection and corresponding refrections or transmissions. This diagram gives the picture of the position and direction of every incident (or original), reflected and referred (or transmitted) wave on the system at every instant.

Lattice diagram is a Distance-Time graph. The distance along the system is in the direction of original (or incident) surge is taken horizontally. The time scale is taken vertically downwards. The slopes of the line give the propogation velocities in the system. The sloping lines thus form a lattice in the distance-time plane. The amplitude of each wave is marked on its line in term of the original wave and the coefficient of reflection and refrection (or transmission).

In the Fig. AP the lattice diagram or zig-zag diagram is shown, where *AB* represents the original wave of amplitude e_1. The slope of *AB* gives the velocity of propagation along the overhead line. When the surge, reaches the junction *B* it splits up into waves *BC* and *BD*. The wave $p_1 \cdot e_1$ represented by *BC* is reflected back into the overhead line and the wave $\tau_1 \cdot e_1$ represented by *BD* in the diagram is transmitted into the cable. The slope of *BD* gives velocity in the cable. The process is repeated and continues the surge dies down.

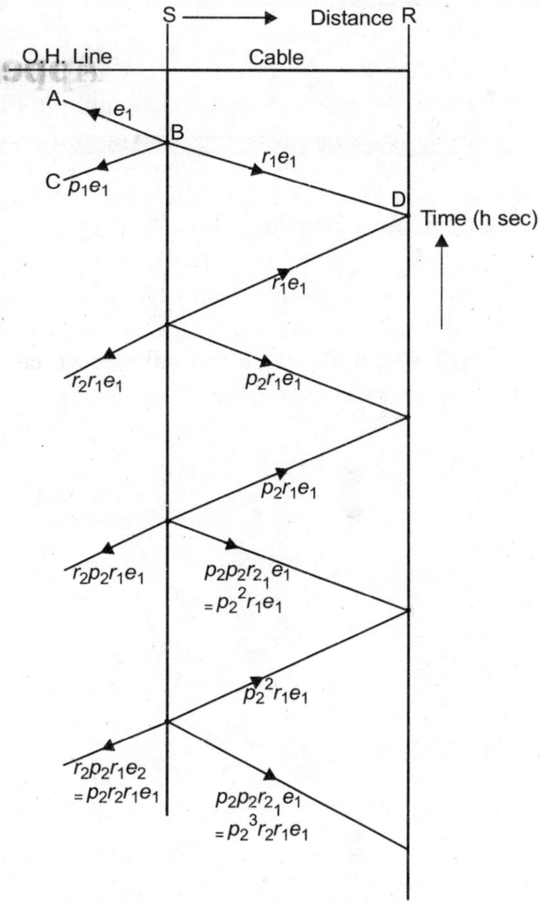

Index

A

B

C

D

E

F

G

H

I

K

L

M

Win Prizes!

ATTENTION READERS

We request you for your frank assessment regarding some of the aspects of the book given as under :

POWER SYSTEM AND PROTECTION

By

DR. N.K. DATTA

Please fill up the given space in neat capital letters.

(i) What topics of your syllabus that are important from you examination point of view are not covered in the book ?

....

....

....

....

(ii) What are the chapters wherein the treatment of the subject matter is not systematic or organised or updated ?

....

....

....

....

(iii) Have you came across misprints/mistakes/factual inaccuracies in the book ? Please specify the chapters and the page numbers.

....

....

....

....

(iv) What other books on the same subject have you found/heart better that the present book ? Please specify in terms of price and quality.

....

....

....

....

(v) Further suggestions and comments for the improvement of the book:

....

....

....

....

(PS: If need be please attach a separate sheet to write your views.)

Other Details :

(i) Who recommended you the book ? (Please tick in the box near the option relevant to you.)

☐ Teacher ☐ Friends ☐ Bookseller

....

....

....

....

(ii) Name of the recommending teacher, his designation and address.

....

....

....

....

(iii) Name and address of the bookseller you purchased the book form :

....

....

....

....

(iv) Name and address of your institution (Please mention the University or Board, as the case may be)

....

....

....

....

(v) Your name and complete postal address :

....

....

....

....

(vi) Course you are enrolled in :

....

....

....

....

The best assessment will be awarded each month. The award will be in the form of our publications as decided by the Editorial Board.

Please mail the filled up coupon at your earliest to:

S.K. KATARIA & SONS®
4760-61/23, Ansari Road, Daryaganj, Delhi-110002
Phone : 23243489, 23269324
E-mail: katson_sanjeev@yahoo.co.in
katariabooks@yahoo.com